## Group I. Amino Acids with Apolar R Groups

| | R Groups | Common Group |
|---|---|---|
| Alanine<br>Ala<br>A<br>$M_r$ 89[a] | | $CH_3$—C—COO⁻ |
| Valine<br>Val<br>V<br>$M_r$ 117 | | (CH₃)₂CH—C—COO⁻ |
| Leucine<br>Leu<br>L<br>$M_r$ 131 | | (CH₃)₂CH—CH₂—C—COO⁻ |
| Isoleucine<br>Ile<br>I<br>$M_r$ 131 | | CH₃—CH₂—CH(CH₃)—C—COO⁻ |
| Proline<br>Pro<br>P<br>$M_r$ 115 | | |
| Phenylalanine<br>Phe<br>F<br>$M_r$ 165 | | —CH₂—C—COO⁻ |
| Tryptophan<br>Trp<br>W<br>$M_r$ 204 | | —C—CH₂—C—COO⁻ |
| Methionine<br>Met<br>M<br>$M_r$ 149 | | CH₃—S—CH₂—CH₂—C—COO⁻ |

## Group II. Amino Acids with Uncharged Polar R Groups

| | R Groups | Common Group |
|---|---|---|
| Glycine<br>Gly<br>G<br>$M_r$ 75 | | H—C—COO⁻ |
| Serine<br>Ser<br>S<br>$M_r$ 105 | | HO—CH₂—C—COO⁻ |
| Threonine<br>Thr<br>T<br>$M_r$ 119 | | CH₃—CH(OH)—C—COO⁻ |
| Cysteine<br>Cys<br>C<br>$M_r$ 121 | | HS—CH₂—C—COO⁻ |
| Tyrosine<br>Tyr<br>Y<br>$M_r$ 181 | | HO——CH₂—C—COO⁻ |
| Asparagine<br>Asn<br>N<br>$M_r$ 132 | | H₂N—CO—CH₂—C—COO⁻ |
| Glutamine<br>Gln<br>Q<br>$M_r$ 146 | | H₂N—CO—CH₂—CH₂—C—COO⁻ |

## Group III. Amino Acids with Charged R Groups

| | R Groups | Common Group |
|---|---|---|
| Aspartic acid<br>Asp<br>D<br>$M_r$ 133 | | ⁻OOC—CH₂—C—COO⁻ |
| Glutamic acid<br>Glu<br>E<br>$M_r$ 147 | | ⁻OOC—CH₂—CH₂—C—COO⁻ |
| Lysine<br>Lys<br>K<br>$M_r$ 146 | | H₃N⁺—(CH₂)₄—C—COO⁻ |
| Arginine<br>Arg<br>R<br>$M_r$ 174 | | H₂N—C(NH₂⁺)—NH—(CH₂)₃—C—COO⁻ |
| Histidine<br>(at pH 6.0)<br>His<br>H<br>$M_r$ 155 | | HC=C—CH₂—C—COO⁻ |

# Principles of Biochemistry

**Geoffrey L. Zubay**

*Columbia University*

**William W. Parson**

*University of Washington*

**Dennis E. Vance**

*University of Alberta*

**WCB** **Wm. C. Brown Publishers**

Dubuque, Iowa•Melbourne, Australia•Oxford, England

**Book Team**

Editor *Elizabeth M. Sievers*
Developmental Editor *Robin P. Steffek*
Publishing Services Coordinator *Julie Avery Kennedy*
Photo Editor *Lori Hancock*
Permissions Coordinator *Karen L. Storlie*

## Wm. C. Brown Publishers
A Division of Wm. C. Brown Communications, Inc.

Vice President and General Manager *Beverly Kolz*
Vice President, Publisher *Kevin Kane*
Vice President, Director of Sales and Marketing *Virginia S. Moffat*
Vice President, Director of Production *Colleen A. Yonda*
National Sales Manager *Douglas J. DiNardo*
Marketing Manager *Patrick E. Reidy*
Advertising Manager *Janelle Keeffer*
Production Editorial Manager *Renée Menne*
Publishing Services Manager *Karen J. Slaght*
Royalty/Permissions Manager *Connie Allendorf*

## Wm. C. Brown Communications, Inc.

President and Chief Executive Officer *G. Franklin Lewis*
Senior Vice President, Operations *James H. Higby*
Corporate Senior Vice President, President of WCB Manufacturing *Roger Meyer*
Corporate Senior Vice President and Chief Financial Officer *Robert Chesterman*

Cover Images:
    Main Cover and Volume 1: Image copyright 1994 by the Scripps Research Institute/
    Molecular Graphics Images by Michael Pique using software by Yng Chen, Michael
    Connolly, Michael Carson, Alex Shah, and AVS, Inc. Endonuclease III by Kuo,
    McRee, Fisher, O'Handley, Cunningham, and Tainer

    Volume II: Courtesy of Dr. Klaus Piontek

    Volume III: Courtesy of Dr. John Kuriyan

Freelance Permissions Editor Karen Dorman
Copyediting and Production by York Production Services
Design by York Production Services
Composition by York Graphic Services

The credits section for this book begins on page C-1 and is considered
an extension of the copyright page.

**Contributors to the End-of-Chapter Problems**

    Chapters 2–24
        Hugh Akers
        Lamar University

    Chapters 25–31
        Caroline Breitenberger
        Ohio State University

A Times Mirror Company

Library of Congress Catalog Card Number: 94–70034 Complete
                                  94–70098 Volume One, Two, and Three

ISBN 0–697–24169–6—Volume One
      0–697–24170–X—Volume Two
      0–697–24171–8—Volume Three
      0–697–14275–2—Complete

Printed in the United States of America by Wm. C. Brown Communications, Inc.,
2460 Kerper Boulevard, Dubuque, IA 52001

10   9   8   7   6   5   4   3   2   1

This text is dedicated to:

Bongsoon
Polly, and
Jean

# Publisher's Note

*Principles of Biochemistry,* first edition, is available as a single, full-length casebound text or in three paperback volumes, which may be used separately or together in any combination that best suits your course needs.

| Binding Option | Description | ISBN |
|---|---|---|
| Principles of Biochemistry, 1/E (Casebound) | The full-length text, Chapters 1–31, plus four Supplements. | 0-697-14275-2 |
| Principles of Biochemistry, 1/E Volume 1: Energy, Proteins, and Catalysis (Paperback) | Chapters 1–10 | 0-697-24169-6 |
| Principles of Biochemistry, 1/E Volume 2: Metabolism (Paperback) | Chapters 11–24, plus Supplements 1 and 2. | 0-697-24170-X |
| Principles of Biochemistry, 1/E Volume 3: Molecular Genetics (Paperback) | Chapters 25–31, plus Supplements 3 and 4. | 0-697-24171-8 |

# Brief Contents

# Volume Three

## Molecular Genetics

# Extended Contents

---

**Chapter 8**

---

How Enzymes Work  154

---

**Chapter 9**

---

Regulation of Enzyme Activities  175

---

**Chapter 10**

---

Vitamins and Coenzymes  198

**Part 4**

# Metabolism of Carbohydrates  225

---

**Chapter 11**

---

Metabolic Strategies  227

**Chapter 19**

Biosynthesis of Membrane Lipids   436

**Chapter 20**

Metabolism of Cholesterol   459

---

**Chapter 24**

---

Integration of Metabolism and Hormone Action 562

---

**Supplement 1**

---

Principles of Physiology and Biochemistry: Neurotransmission 602

---

**Supplement 2**

---

Principles of Physiology and Biochemistry: Vision 614

---

**Chapter 28**

---

RNA Synthesis and Processing  700

---

**Chapter 29**

---

Protein Synthesis, Targeting, and Turnover  730

---

**Chapter 30**

---

Regulation of Gene Expression in Prokaryotes  768

---

**Chapter 31**

---

Regulation of Gene Expression in Eukaryotes  800

# List of Supplementary Text Elements

## Notes on Nutrition   598

**Methods of Biochemical Analysis***

*\* Please Note:* Most of the material on **Methods of Biochemical Analysis** and **Biochemical Experiments** presented in this textbook is **integrated** into the main part of the text. This material is **"flagged"** throughout the text by the **lab door icon. Supplemental biochemical methods material which can be identified within the text by the blue-tabbed pages is listed below.**

# Preface

As the subject of biochemistry has grown, the need for texts that are appropriate for different types of courses has also grown. There is a need for texts at three distinct levels: the "comprehensive" level that includes a great deal of detail, the "middle" level, and the "short course" level. Most up-to-date texts are written at the "comprehensive" level despite the fact that most students would benefit more from a less detailed, "middle" level text that emphasizes *principles*. Currently, there are very few up-to-date "middle" level texts available. The main goal of our text, as indicated by the title, is to serve the needs of the *principles-oriented, "middle"* level audience.

In addition to presenting material that we think is important for the "middle" level audience, we have tried to write this text with the following goals in mind:

- to give a balanced presentation that explains the biochemistry of both prokaryotes and eukaryotes;

- to indicate the biomedical significance of defects in metabolism;

- to describe metabolism with special emphasis on how pathways are organized and how they are regulated;

- to give the student a feel for the process of scientific research by describing how some major biochemical discoveries were made; and finally,

- to present the techniques used by biochemists during the course of their research.

## Unique Features of *Principles of Biochemistry*

Some of the unique features of our text can be classified as *general features* in that they apply to the entire text, while other *content-specific features* apply to an individual chapter or group of chapters.

## *General Features*

1. **The text is available as a single, full-length (31 Chapters; plus 4 Supplements) casebound text or in three paperback volumes (Volume One: Chapters 1–10, Volume Two: Chapters 11–24, plus Supplements 1 & 2; and Volume Three: Chapters 25–31, plus Supplements 3 & 4).** It is generally useful for students to bring their texts to class, but students of biochemistry are not inclined to do this because of the large size of the texts. The three volume binding format makes it convenient for students, and professors alike, to carry only one third of the textbook around at any one time. An additional advantage of the three volume binding format is that each of the volumes may be used separately or together in any combination that best suits the needs of the biochemistry course. The three volumes therefore allow greater flexibility for professors and greater manageability and affordability for students.

2. ***Eight icons**, graphic images that are intended to help the textbook reader remember and mentally cross-reference* 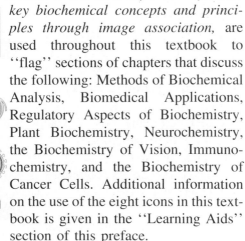 *key biochemical concepts and principles through image association,* are used throughout this textbook to "flag" sections of chapters that discuss the following: Methods of Biochemical Analysis, Biomedical Applications, Regulatory Aspects of Biochemistry, Plant Biochemistry, Neurochemistry, the Biochemistry of Vision, Immunochemistry, and the Biochemistry of Cancer Cells. Additional information on the use of the eight icons in this textbook is given in the "Learning Aids" section of this preface.

## *Content-Specific Features*

1. Chapter 6: Detailed step-by-step descriptions of the purification schemes for two proteins are presented to give the student an understanding of how different purification procedures may be effectively combined.

2. Chapters 7–9: Emphasis on basic enzymology—enzyme kinetics, mechanisms and regulation.

3. Chapter 10: An exclusive chapter devoted to vitamins and coenzymes.

4. Chapter 11: A comprehensive description of basic metabolic strategies that govern substrate and energy flow in metabolic pathways.

5. Chapter 12: A single cohesive chapter on all major aspects of glucose metabolism—glycolysis, gluconeogenesis, and regulation.

6. Chapters 14 and 18: Correct values for ATP production resulting from catabolism. Other texts do not take into account the energy required for ATP, ADP, and $P_i$ transport across the inner mitochondrial membrane. This makes a substantial difference in the calculated energy yields.

7. Chapter 20: A chapter exclusively devoted to cholesterol metabolism.

8. Chapter 22: A chapter exclusively devoted to amino acid metabolism in vertebrates.

9. Chapters 12–16, 17–23: Organization of carbohydrate metabolism, lipid metabolism, and the metabolism of nitrogen-containing compounds, respectively, into self-contained units.

10. Chapters 30 and 31: Separate chapters devoted to the regulation of gene expression in prokaryotes and eukaryotes, respectively.

## Learning Aids for the Student

In order for students to succeed in their study of biochemistry, they must be able to understand the material presented, utilize the text as a tool for learning, and enjoy reading the text. Therefore, we have included many aids to make the study of biochemistry efficient and enjoyable.

The text chapters contain the following:

***Chapter Opening Outline and Overview*** The chapter opening outline and overview were written to help students preview how the chapter is organized and what major concepts are to be covered in the chapter.

***Declarative Statement Headings*** Clear, informative headings help introduce students to each important topic.

***Underlined Terms and Key Concepts*** Underlined important terms and key concepts are easy for students to locate.

The second law of thermodynamics is that the universe inevitably proceeds from states that are more ordered to states that are more disordered. This phenomenon is measured by a thermodynamic function called entropy, which is denoted

***Numbered Equations*** Important equations within a chapter are numbered so that they can be easily located and referenced by students.

***Summary Tables*** Summary tables are designed to help students more easily understand and utilize important facts or characteristics presented for a specific topic.

***Eight Concept and Applications Icons*** Throughout the text, the student will find text sections "flagged" by eight different graphic icons or images. These icons are intended

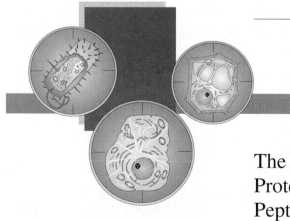

CHAPTER

3

# The Building Blocks of Proteins: Amino Acids, Peptides, and Polypeptides

### Chapter Outline

Amino Acids
   *Amino Acids Have Both Acid and Base Properties*
   *Aromatic Amino Acids Absorb Light in the Near-Ultraviolet*
   *All Amino Acids except Glycine Show Asymmetry*
Peptides and Polypeptides
Determination of Amino Acid Composition of Proteins
Determination of Amino Acid Sequence of Proteins
Chemical Synthesis of Peptides and Polypeptides

*Amino acids have common features that permit them to be linked together into polypeptide chains and uncommon features that give each polypeptide chain its unique character.*

I n the middle of the nineteenth century, the Dutch chemist Gerardus Mulder extracted a substance common to animal tissues and the juices of plants, which he believed to be "without doubt the most important of all substances of the organic kingdom, and without it life on our planet would probably not exist." At the suggestion of the famous Swedish chemist Jons Jakob Berzelius, Mulder named this substance protein (from the Greek *proteios,* meaning "of first importance") and assigned to it a specific chemical formula to help the student remember and mentally cross-reference the following concepts and applications:

 **Methods of Biochemical Analysis**

 **Biomedical Applications**

 **Regulatory Aspects of Biochemistry**

 **Plant Biochemistry**

 **Neurochemistry**

 **Biochemistry of Vision**

 **Immunochemistry**

 **Biochemistry of Cancer Cells**

***Boxed Readings***    Thought-provoking boxed readings on relevant topics are featured in various chapters—for example, Box 20A, Metabolism and Heart Disease.

| BOX | |
|---|---|
| 20A | Cholesterol Metabolism and Heart Disease |

***Molecular Graphics and Illustration Program*** Molecular graphics images of key molecules enable students to "see" and interpret three-dimensional structures.

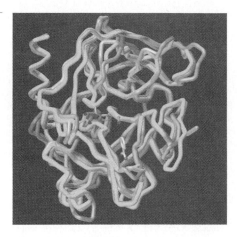

**Figure 8.6a**

Trypsin, Chymotrypsin, Elastase

***End-of-Chapter Enumerated Summary*** This concise summary of chapter concepts is designed to serve as a guide for chapter study.

***End-of-Chapter Selected Readings*** Each chapter concludes with carefully selected references that contain further information on the topics covered in that chapter.

***End-of-Chapter Problems*** These problems (over 400) are designed to help students access their understanding of the chapter's basic concepts. Brief solutions for the odd-numbered problems are found in the back of the text.

***End-of-Chapter Methods of Biochemical Analysis*** Supplemental information on experimental techniques and procedures are presented at the end of some chapters. The "Methods of Biochemical Analysis" can be easily located throughout the text as their text pages are tabbed with a blue band. The lab door icon is used to cross-reference the

## Summary

In this chapter we discussed the ways in which biochemistry parallels ordinary chemistry and those in which it is quite different. The chief points to remember are the following.

1. The basic unit of life is the cell, which is a membrane-enclosed, microscopically visible object.
2. Cells are composed of small molecules, macromolecules, and organelles. The most prominent small molecule is water, which constitutes 70% of the cell by weight. Other small molecules are present only in quite small amounts; they are precursors or breakdown products of macromolecules or coenzymes. There are four types of macromolecules: lipids, carbohydrates, proteins, and nucleic acids.
3. Noncovalent intermolecular forces largely determine the folded structure adopted by a macromolecule, particularly the relative affinity of different groupings on the macromolecule for water. In general, hydrophobic groupings are buried within the folded macromolecular structure, whereas hydrophilic groupings are located on the surface, where they can interact with water.
4. Biochemical reactions utilize a limited number of elements, most prominently carbon, hydrogen, oxygen, nitrogen, sulfur, and phosphorus. Many biochemical reactions are simple organic reactions.

"Methods of Biochemical Analysis" to pertinent text sections within a chapter.

Methods of Biochemical Analysis **3 A**

## Measurement of Ultraviolet Absorption in Solution

The general quantitative relationship that governs all absorption processes is called the Beer-Lambert law:

$$I = I_0 10^{-\epsilon cd}$$

where $I_0$ is the intensity of the incident radiation, $I$ is the intensity of the radiation transmitted through a cell of thickness $d$ (in centimeters) that contains a solution of concentration $c$ (expressed either in moles per liter or in grams per 100 ml), and $\epsilon$ is the extinction coefficient, a characteristic of the substance being investigated (see fig. 3.7).

Light absorption is measured by a spectrophotometer as shown in the illustration. The spectrophotometer usually is capable of directly recording the absorbance $A$, which is related to $I$ and $I_0$ by the equation

$$A = \log_{10}\left(\frac{I_0}{I}\right)$$

Hence $A = \epsilon cd$, and $A$ is a direct measure of concentration. We can see from figure 3.7 that the $\epsilon$ values are largest for tryptophan and smallest for phenylalanine.

Because protein absorption maxima in the near-ultraviolet (240–300 nm) are determined by the content of the

**Figure 1**

Schematic diagram of a spectrophotometer for measuring light absorption. Laboratory instruments for making measurements are much more complex than this, but they all contain the same basic components: a light source, a monochromator, a sample, and a detector. $\lambda$ is the wavelength of the light, $I_0$ and $I$ are the incident light intensity and the transmitted light intensity, respectively, and $d$ is the thickness of the absorbing solution.

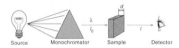

aromatic amino acids and their respective values, most proteins have absorption maxima in the 280-nm region. By contrast, absorption in the far-ultraviolet (around 190 nm) is shown by all polypeptides regardless of their aromatic amino acid content. The reason is that absorption in this region is due primarily to the peptide linkage.

***Common Abbreviations in Biochemistry***    These abbreviations can be found at the end of the text.

***Organic Chemistry for Biochemistry***    This appendix, which describes the similarities and the differences between organic chemistry and biochemistry, is also located at the end of this text.

***Appendix A: Some Major Discoveries in Biochemistry***    Summarizes major research discoveries from the past and present.

***Appendix B: Career Paths in the Biological Sciences***    A useful starting point for career information for biochemistry students.

***Appendix C: Answers to Selected Problems***    Brief solutions are provided to help the student determine if they are on the right track.

***Glossary***    Over 700 important biochemical terms are defined in this end-of-book glossary.

***Index***    An easy-to-use, comprehensive index is provided.

In addition to these chapter aids, this text also features the following:

***Four "Principles of Physiology and Biochemistry" Supplements***    These supplements cover the topics of **neurotransmission, vision, immunobiology, and carcinogenesis.** The Neurotransmission and Vision supplements, Supplements 1 and 2, are located at the end of Part 6 of the text. The Immunobiology and Carcinogenesis supplements, Supplements 3 and 4, are located at the end of Part 7 of the text. Physiological discussions within the text that relate to these topics are also "flagged" by the corresponding concept icon.

***Page-Referenced Listing of Supplementary Learning Aids***
This helpful ready reference tool is located in the frontmatter of this text.

***Illustrated Guide to the "Comparative Sizes of Biomolecules, Viruses, and Cells"***    This highly visual guide is located at the end of the text.

***Endsheets with Reference Material***    The endsheets of the text contain useful easily accessible reference information.

## Organization of the Text

This text is divided into seven parts: part 1, An Overview of Biochemical Structures and Reactions That Occur in Living Systems; part 2, Protein Structure and Function; part 3, Catalysis; part 4, Metabolism of Carbohydrates; part 5, Metabolism of Lipids; part 6, Metabolism of Nitrogen-Containing Compounds; and part 7, Storage and Utilization of Genetic Information.

The subjects presented in each of the seven parts, the level of the discussion, the organization of the parts, and the organization within each of the chapters are all designed to satisfy the teaching needs of introductory biochemistry courses.

We may think of biochemistry as being divided into three main areas: biomolecular structure, intermediary metabolism, and nucleic acid and protein metabolism. It would be possible to segregate the coverage of these topics in a biochemistry text and present them as three consecutive blocks. However, we did not take this approach in the development of this text because we felt it would not be good pedagogy for two reasons: (1) students get bored if they sit through a third of a year where they only hear about structure; and (2) students often forget what was said about the structure of a biomolecule by the time they read the section on its metabolism and therefore have to review it before going on to the metabolism material.

Our general approach in writing *Principles of Biochemistry* was to integrate the presentation of biomolecular structure and metabolism throughout the text. We were successful in applying this integrative approach to presenting biomolecular structure and metabolism in five of the seven parts of the text. Specifically, amino acid structure and protein structure are discussed in part 2, well before considering their metabolism in parts 6 and 7, respectively. This deviation from our general approach is essential because one must understand amino acid structure before one can understand the function of proteins and enzymes, which is discussed in early chapters (chapters 5, 7, 8, and 9 of parts 2 and 3). An additional advantage to discussing protein structure early in the text is that most students find it exciting. Notwithstanding the early discussion of basic protein structure in chapter 4, additional discussions pertaining to protein structure are included within the text where they are most appropriate to furthering the student's understanding protein function. Such discussions can be found in chapter 17 where membrane proteins are discussed and in chapters 30 and 31 where gene regulatory proteins are discussed.

The structures of sugars and polysaccharides are covered in the appropriate chapters within part 4 just prior to discussing their metabolism. Similarly, the structures of lipids are presented in the lipid metabolism chapters found in part 5. Nucleotide structures are addressed in chapter 23 before considering their metabolism. Finally, nucleic acid and nucleoprotein structures are examined in the first chapter (chapter 25) of part 7 prior to the discussion of the roles these molecules play in nucleic acid and protein metabolism in the six subsequent chapters.

Metabolism is regulated at the level of the gene and at the level of the enzyme. Regulation is emphasized in this text both because of its importance and because it reflects some of the most unique characteristics of biochemical reactions. Of necessity, our treatment of the subject of regulation at the gene level is confined primarily to part 7, as this is where the metabolism of nucleic acids is discussed. We cover metabolic regulation at the level of the enzyme throughout this text. Specifically, the regulation of the oxygen binding capacity of hemoglobin and the process of mus-

cle contraction is discussed in chapter 5. An entire chapter, chapter 9, is devoted to regulatory enzymes. With respect to the metabolism of carbohydrates (chapter 12) and fatty acids (chapter 18), the regulation of the catabolic and anabolic pathways of these biomolecules is presented together within the respective chapters. We feel that this approach simplifies the discussion of the regulation of these opposing pathways.

## Detailed Chapter-by-Chapter Description of the Text

Having indicated some aspects of the overall organization of this text, we now turn to a more detailed chapter-by-chapter description of the text. Part 1, An Overview of Biochemical Structures and Reactions That Occur in Living Systems, contains only two chapters. Chapter 1, Cells, Biomolecules and Water, gives an introductory presentation of the structure of cells and the molecules from which cells are composed. The central role of water in determining the structures that are formed is emphasized. A general description of the principles that govern cellular organization and determine the course of these reactions is presented. An introductory survey of the types of molecules that make up the cell is given. Finally, the overall strategy of living systems and the evolution of living systems is briefly discussed.

The main function of chapter 1 is to present an overview of the principles and facts that are presented in greater depth in various parts of the text. Most of the material in this chapter is presented in a good introductory biology course. However, there are a number of students coming from a chemistry background that may not have had the benefits of such a course. Reference should also be made at this point to the appendix "Organic Chemistry for Biochemistry" (located in the endmatter of the text), which gives a brief review of the similarities and differences between organic chemistry and biochemistry. It is assumed that all students reading this text will have had a year of general chemistry and one term of organic chemistry. Or, at the very least, that they will be taking organic chemistry concurrently with biochemistry.

In chapter 2, thermodynamics is presented in a way that is most relevant to the consideration of biochemical phenomena. We chose to present the rather difficult topic of thermodynamics in the beginning of the text because thermodynamics is an important consideration in all aspects of biochemistry. The central role of ATP as the main carrier of free energy in biochemical systems is emphasized.

The discussion of thermodynamics presented in chapter 2 gives a perspective of bioenergetic considerations, which are discussed at many points throughout the text.

Thermodynamic applications that relate to biomolecular structure and biochemical reactions are also elaborated on in specific sections of subsequent chapters.

Part 2, Protein Structure and Function, contains four chapters that relate to the structures and functions of proteins. In chapter 3, The Building Blocks of Proteins: Amino Acids, Peptides, and Polypeptides, we discuss basic structural and chemical properties of amino acids, peptides and polypeptides. In chapter 4, The Three-Dimensional Structures of Proteins, we describe how and why polypeptide chains fold into long fibrous molecules in some cases, or into compact globular molecules in other cases. In chapter 5, Functional Diversity of Proteins, we turn to the question of how protein structure relates to protein function. To explore this question, two protein systems, hemoglobin and the actin-myosin complex are examined in detail. In chapter 6, Methods for Purification and Characterization of Proteins, the primary goal is to acquaint the reader with the techniques used for protein purification. The first part of chapter 6 presents methods for protein fractionation. In the second part of this chapter, purification procedures for two proteins, UMP synthase and lactose carrier protein, are presented so that the student can see how different purification steps are combined for maximum effectiveness.

Part 3, Catalysis, is divided into four chapters. The first three chapters in part 3 deal with the properties and functions of enzymes. The last chapter explains the essential properties of coenzymes and cofactors.

Enzyme kinetics is studied for two reasons: (1) it is a practical concern to determine the activity of the enzyme under different conditions; (2) frequently the analysis of enzyme kinetics gives information about the mechanism of enzyme action. Chapter 7, Enzyme Kinetics, begins with an introductory section on the discovery of enzymes, basic enzyme terminology and a description of the six main classes of enzymes and the reactions they catalyze. The remainder of the chapter deals with basic aspects of chemical kinetics, enzyme-catalyzed reactions and various factors that affect the kinetics.

Chapter 8, How Enzymes Work, starts with a description of the basic chemical mechanisms that are exploited by enzymes. The latter half of this chapter presents a detailed description of how three enzymes—chymotrypsin, RNase, and triose phosphate isomerase—exploit these basic mechanisms of enzyme catalysis.

Most enzymes spontaneously process substrates when present, as long as inhibitory factors do not prevent this from happening. A few enzymes, known as regulatory enzymes, do not react spontaneously with their substrates unless signaled to do so by overiding metabolic conditions. Chapter 9, Regulation of Enzyme Activities, describes a wide range of mechanisms that are used to control the activity of regulatory enzymes. This chapter concludes with a detailed description of how the activity of three enzymes— phosphofructokinase, aspartate transcarbamylase and glycogen phosphorylase—are regulated.

Frequently enzymes act in concert with small molecules, coenzymes or cofactors, which are essential to the function of the amino acid side chains of the enzyme. Coenzymes or cofactors are distinguished from substrates by the fact that they function as catalysts. They are also distinguishable from inhibitors or activators in that they participate directly in the catalyzed reaction. Chapter 10, Vitamins and Coenzymes, starts with a description of the relationship of water-soluble vitamins to their coenzymes. Next, the functions and mechanisms of action of coenzymes are explained. In the concluding sections of this chapter, the roles of metal cofactors and lipid-soluble vitamins in enzymatic catalysis are briefly discussed.

Part 4, Metabolism of Carbohydrates, begins with a general overview in chapter 11, Metabolic Strategies. This chapter also relates strongly to parts 5 and 6. Chapter 11 starts with an explanation of how biochemical reactions are organized into energy-generating and energy requiring pathways and a handful of principles that explain the design of metabolic pathways. The next five chapters describe the metabolism of carbohydrates insofar as they relate to energy-generating catabolism and energy-consuming biosynthesis. Chapter 12, Glycolysis, Gluconeogenesis, and the Pentose Phosphate Pathway, deals with all aspects of glucose metabolism. After a brief section describing structures, glycolysis, the pathway for the breakdown of glucose, is examined. This is followed by a shorter section on gluconeogenesis, the synthesis of glucose from three carbon compounds. While glycolysis produces energy, gluconeogenesis consumes energy. These two processes are discussed side-by-side so that we can consider the closely related question of the regulation of these pathways. This chapter concludes with a short section on an oxidative route for glucose catabolism, the pentose phosphate pathway which serves multiple purposes.

Under anaerobic conditions the breakdown of glucose stops at the three carbon compound stage. Further catabolism requires oxygen as described in chapter 13, The Tricarboxylic Acid Cycle. The tricarboxylic acid cycle is an energy-producing process although this is not immediately obvious because the major products of the cycle outside of $CO_2$—$H_2O$ and a single ATP molecule—are the reduced forms of the coenzymes $NAD^+$ and FAD. These coenzymes contain the potential chemical energy for ATP production, but an elaborate process of membrane-associated electron transport and proton transport must precede the synthesis of ATP. This process is described in chapter 14, Electron Transport and Oxidative Phosphorylation.

Most of the energy that is used to drive biochemical processes originates from the sun. The way in which solar energy is harnessed to produce chemical energy and to fix atmospheric carbon dioxide into reduced organic compounds is described in chapter 15, Photosynthesis.

Up to this point in part 4, the focus has been on the roles that sugars and carbohydrates play in energy metabolism. Polymeric carbohydrates are major structural components in plant and bacterial cell walls and in the extracellular matrix of vertebrates. Branched-chain oligosaccharides covalently linked to proteins are used to give proteins unique signatures that guide them to their final destination within a cell and facilitate specific interactions between free proteins (ligands) and proteins attached to cells (receptors). In chapter 16, Structures and Metabolism of Oligosaccharides and Polysaccharides, the structures and functions of some of these polymeric carbohydrates are discussed.

Part 5, Metabolism of Lipids, comprises four chapters that deal with the structure and metabolism of lipids. In chapter 17, Structure and Function of Biological Membranes, we start by examining the constituents of membranes with the aim of developing a general model for membrane structure. We then turn to the question of how cells transport materials across membranes.

In chapter 18, Metabolism of Fatty Acids, we discuss the synthesis and breakdown of fatty acids. The chapter starts with a discussion of fatty acid breakdown. A second section covers the pathway for fatty acid biosynthesis. Finally, we consider the regulatory mechanisms that determine the conditions under which each of these processes occurs. As in the case of glucose metabolism, it is convenient to discuss the synthesis and breakdown in the same chapter so that the closely related topic of regulation can be considered alongside.

Thus far we have been concerned with the metabolism of fatty acids in relationship to the storage and release of energy. In chapter 19, Biosynthesis of Membrane Lipids, we focus on the metabolism of lipids that serve other roles. Many types of lipids are essential membrane components. A number of lipids also function as metabolic signals in response to hormonal signals. These lipid molecules are known as second messengers.

The final chapter in part 5, chapter 20, Metabolism of Cholesterol, deals with the synthesis of cholesterol and some of its derivatives, the steroid hormones and the bile acids. This chapter considers the structure, function and metabolism of these molecules. Also, the health-related concerns associated with cholesterol excess are addressed.

Part 6, Metabolism of Nitrogen-Containing Compounds, is concerned mostly with the metabolism of amino acids and nucleotides. Chapter 24, the last chapter in this part, deals with the integration of metabolism.

Amino acid metabolism is divided into two chapters. Amino acids are best known as the building blocks of proteins. In addition to this role, amino acids serve as precursors to many important low molecular weight compounds including nucleotides, porphyrins, parts of lipid molecules and precursors for several coenzymes. Amino acids also serve as the ''vehicles'' for converting inorganic forms of nitrogen and sulfur to organic forms. As an alternative energy source amino acid catabolism can be coupled to the regeneration of ATP from ADP of AMP. In chapter 21, Amino Acid Biosynthesis and Nitrogen Fixation in Plants and Microorganisms, we focus on the biosynthesis of amino acids and their role in bringing inorganic nitrogen and sulfur into the biological world. In addition, nonprotein amino acids are discussed briefly.

Amino acid metabolism in vertebrates contrasts sharply with amino acid metabolism in plants and microorganisms. Most striking is the fact that plants and microorganisms can synthesize all twenty amino acids required for protein synthesis whereas vertebrates can only synthesize about half this number. This leads to complex nutritional needs for vertebrates, which are discussed in chapter 22, Amino Acid Metabolism in Vertebrates. Vertebrate amino acid degradation pathways are also discussed in chapter 22 along with the existence of many pathological states that result from enzyme deficiencies in the degradative pathways.

Chapter 23, Nucleotides, deals with the biosynthesis of ribonucleotides, deoxyribonucleotides, the roles of these biomolecules in metabolic processes, and the pathways for their degradation. Medically related topics such as nucleotide metabolism deficiencies or the use of nucleotide analogs in chemotherapy are also considered.

Chapter 24, Integration of Metabolism and Hormone Action, explains the organization strategies used to integrate metabolic processes in a multicellular organism. Like the first chapter in part 4, the content of chapter 24 relates to all of the chapters on metabolism (chapters 11–24). This chapter emphasizes the fact that hormones and closely related growth factors play a dominant role in regulating metabolic activities in different tissues.

Part 6 concludes with two brief, informative supplements that integrate physiological and biochemical principles as they apply to the ''nonmetabolic'' functions of amino acids and lipids: Supplement 1—Principles of Physiology and Biochemistry: Neurotransmission; and Supplement 2—Principles of Physiology and Biochemistry: Vision.

Part 7, Storage and Utilization of Genetic Information, is composed of seven chapters, plus two supplements. In this part, we examine the means by which genetic information is replicated and expressed in the cell. This discussion includes a description of the relevant structures, their function, and their metabolism. In addition, this part includes a chapter that explains the procedures by which DNA is experimentally manipulated to build new genes or new combinations of genes. In chapter 25, Structures of Nucleic Acids and Nucleoproteins, we begin by considering the key experiments that revealed the genetic significance of DNA. At every stage in this chapter and the subsequent chapters of this section, explanations of genetic observations essential to the understanding of biochemical phenomena are given.

From the complementary duplex structure of DNA described in chapter 25, it is a short intuitive hop to a model for replication that satisfies the requirement for one round of DNA duplication for every cell division. In chapter 26, DNA Replication, Repair, and Recombination, key experiments demonstrating the semiconservative mode of replication in vivo are presented. This is followed by a detailed examination of the enzymology of replication, first for how it occurs in bacteria and then for how it occurs in animal cells. Also included in this chapter are select aspects of the metabolism of DNA repair and recombination. The novel process of DNA synthesis using RNA-directed DNA polymerases is also considered. First discovered as part of the mechanisms for the replication of nucleic acids in certain RNA viruses, this mode of DNA synthesis is now recognized as occurring in the cell for certain movable genetic segments and as the means whereby the ends of linear chromosomes in eukaryotes are synthesized.

Before going into the processes for the utilization of genetic information in the cell, DNA manipulation and some of its applications are considered. In chapter 27, DNA Manipulation and Its Applications, DNA sequencing is the first subject to be explained. Different approaches for amplifying and isolating specific genes are also explored. Following this, methods for restructuring existing DNA sequences are described. Finally, some applications of new technologies are considered. These applications focus on the mapping of the human globin gene family and the gene responsible for the genetically inherited disease, cystic fibrosis.

In chapter 28, RNA Synthesis and Processing, the DNA-directed synthesis of RNA is considered. As with the other chapters in part 7, a balanced presentation of how these processes occur in bacteria and eukaryotes is given. First, the structures of different classes of RNA are described. The RNA classes include messenger RNA, transfer RNA, and ribosomal RNA. A single enzyme is responsible for the transcription of the RNAs in bacteria. The initial transcripts for transfer RNA and ribosomal RNA undergo extensive modification after synthesis, while the messenger RNA is used without modification. In eukaryotes, even the messenger RNA undergoes extensive modifications before

it can be utilized. Also, in eukaryotes the process of transcription is much more complicated, involving several RNA polymerases and many more protein subunits which are associated with the polymerases. Every attempt is made to treat these complications without presenting an overwhelming amount of detail. RNA synthesis in certain viruses is also discussed and the fascinating subject of RNA enzymes is briefly considered. Lastly, selective inhibitors of RNA polymerases that have diagnostic value or medical value in chemotherapy are discussed.

In chapter 29, Protein Synthesis, Targeting, and Turnover, the processes of protein synthesis and transport are described. First the process whereby amino acids are ordered and polymerized into polypeptide chains is described. Next, posttranslational alterations of newly synthesized polypeptides is considered. This is followed by a discussion of the targeting processes whereby proteins migrate from their site of synthesis to their target sites of function. Finally, proteolytic reactions that result in the return of proteins to their starting materials, the amino acids, are considered.

Regulation of gene expression is on the cutting edge of research in molecular biology and biochemistry. This topic is so expansive that we chose to divide our coverage of it into two chapters: one that considers bacteria and another that considers eukaryotes. In chapter 30, Regulation of Gene Expression in Prokaryotes, the classical systems of the *lac* and *trp* operons and regulation of bacteriophage lambda are considered, with particular emphasis on the nature of gene regulatory proteins and how they interact with DNA. While most forms of regulation of gene expression that are understood involve regulation at the transcriptional level, one excellent example of translational control, the regulation of ribosomal protein synthesis, is also presented in chapter 30. Chapter 30 ends with a comprehensive summary section on the different types of regulatory proteins that are used to regulate transcription in prokaryotes.

The subject of regulation of gene expression in eukaryotes is complex and diffuse because so many different types of systems have been studied and the level of research effort in this field has reached unprecedented heights. Studies on this subject are truly at the cutting edge of modern biological investigations. In chapter 31, Regulation of Gene Expression in Eukaryotes, regulation of gene expression is first examined in yeast, a unicellular organism. We then examine regulatory mechanisms prevalent in multicellular eukaryotes. A section on the types of regulatory proteins most frequently found in eukaryotic systems is presented next. Finally, modes of regulation specific to developmental processes are considered.

Part 7 concludes with two brief, informative supplements that integrate key physiological and biochemical principles as they apply to two specialized processes that utilize genetic information: Supplement 3—Principles of Physiology and Biochemistry: Immunobiology and Supplement 4—Principles of Physiology and Biochemistry: Carcinogenesis and Oncogenes.

# Ancillary Materials
## *For the Instructor*

An *Instructor's Manual with Test Item File* contains suggestions on how to utilize the text in different course situations and detailed, worked-out solutions for the even-numbered problems found in the text chapters. These answers are **not** included in the *Student's Solutions Manual* that accompanies the text. In addition, this manual offers an average of 30 objective test questions for each chapter which can be used to generate exams. (ISBN 14276)

*Classroom Testing Software* is offered free upon request to adopters of this text. The software provides a database of questions for preparing exams. No programming experience is required. The software is available in IBM and Macintosh formats: IBM DOS 3.5 (ISBN 14278), IBM DOS 5.25 (ISBN 14279), WINDOWS 3.5 (ISBN 14280), and MAC 3.5 (ISBN 14282).

A set of *150 full-color acetate transparencies* is available free to adopters. These acetates feature key illustrations that can be used to enhance your classroom lectures. (ISBN 14277)

A set of *150 full-color projection slides* derived from the transparency illustrations is also available free to adopters. (ISBN 22871)

*Electronic acetates,* computerized image files for a majority of the text illustrations, will also be available free to adopters upon request. The electronic acetates will be available in a Mac/Windows (ISBN 26204). These electronic acetates can be clearly projected on large lecture hall screens using a LCD projection system.

A set of *175 transparency masters* is available free to adopters upon request. These black-and-white versions of in-text tables and illustrations can be used to prepare course handouts or additional transparency acetates. (ISBN 22872)

## *For the Student*

A *Student's Solutions Manual* by Hugh Akers, Lamar University, and Caroline Breitenberger, Ohio State University, provides detailed, worked-out solutions for the odd-numbered problems found in the text. This solutions manual can help students to better understand how to solve the problems in the text and prepare for exams. (ISBN 22870)

A *Student Study Art Notebook,* a lecture companion containing the illustrations from the text that correspond to the acetate transparency images, is designed to help students spend more of their lecture time listening to the professor and less time copying down art from the overhead transparencies. A copy of this student study art notebook is packaged FREE with each new text. (ARTPAK ISBN 27016)

## *Computer Software and CD-ROM for Instructors and Students*

 **NOTE: PLEASE ALSO REFER TO THE CD-SAMPLER THAT WAS PACKAGED WITH YOUR COPY OF *PRINCIPLES OF BIOCHEMISTRY***

*Gene Game Software,* by Bill Sofer of The State University of New Jersey–Rutgers, is an interactive software game that tests students' critical-thinking skills and knowledge of the scientific method as they attempt to work through "dry" lab protocols to clone a fictitious "Fountain of Youth" gene. The software provides direct feedback and hints to the student as protocols are completed. Protocols used and the results obtained are automatically recorded in the program's "lab notebook." Contact your bookstore, or call Wm. C. Brown Publishers at 1-800-338-5578 to place an order or request more information on this challenging, interactive software game. (ISBN 24893)

*Biochemical Pathways Software,* also created by Bill Sofer, is an easy-to-use tutorial review software program for Macintosh that provides quizzes/memory exercises that test the students' knowledge of Glycolysis and the TCA Cycle. Contact your bookstore, or call Wm. C. Brown Publishers at 1-800-338-5578 to place an order or request more information on this software. (ISBN 25100)

*Molecules of Life CD-ROM,* developed by a talented team of professionals from Purdue University, is a four CD-ROM set that allows the user to visualize key biomolecular structures through *interactive* animations, simulations, drills and tutorials. The set includes material on (1) Amino acids and Proteins, (2) Carbohydrates and Lipids, (3) Nucleic acids, and (4) Metabolism and Photosynthesis. The four CD-ROMs are available individually, or as a set. Contact your bookstore, or call Wm. C. Brown Publishers at 1-800-338-5578 to place an order or request more information on this software (Four CD set/MAC ISBN 27235) (Four CD set/WINDOWS ISBN 27264).

## Acknowledgments

We wish to thank our biochemistry colleagues and contributing authors to the third edition of *Biochemistry;* Dr. Raymond L. Blakley, Dr. James W. Bodley, Dr. Ann Baker Burgess, Dr. Richard Burgess, Dr. Perry Frey, Irving Geis, Dr. Lloyd L. Ingram, Dr. Gary R. Jacobson, Dr. Julius Marmur, Dr. Richard Palmiter, Dr. Milton H. Saier, Jr., Dr. Pamela Stanley and Dr. H. Edwin Umbarger, for providing us with a wealth of comprehensive information from which to draw from in the development of *Principles of Biochemistry.* Special recognition is also due to Dr. Raymond L. Blakley for his work in writing chapter 23, and to Dr. Perry A. Frey, Dr. Gary R. Jacobson, Dr. Pamela Stanley, and Dr. H. Edwin Umbarger for their invaluable feedback and proofing efforts during the many phases of writing and publishing this text. In addition, we thank Hugh Akers, Lamar University, and Caroline Breitenberger, Ohio State University, for their insightful contributions to the end-of-chapter problems in this text. We are also indebted to Michael Pique, The Scripps Research Institute, and Holly Miller, Wake Forest University Medical Center, for their special contributions to the development and generation of many of the molecular graphics images in this text.

My fellow authors and I would also like to extend a special thank you to our many colleagues across the country for reviewing the text manuscript and making many helpful suggestions. The reviewers included:

Hugh Akers
Lamar University

Richard M. Amasino
UW–Madison

Dean R. Appling
The University of Texas at Austin

John N. Aronson
University of Arizona

Paul Austin
Hanover College

Derek Baisted
Oregon State University

Terry A. Barnett, Ph.D.
Southwestern College

E. J. Behrman
The Ohio State University

Paul Arthur Berkman
Ohio State University

Frank O. Brady, Ph.D.
University of South Dakota School of Medicine

Caroline A. Breitenberger
Ohio State University

Ronald W. Brosemer
Washington State University

Oscar P. Chilson
Washington University

David P. Chitharanjan
University of Wisconsin, Stevens Point

Alan D. Cooper
Worcester State College

Rick H. Cote
University of New Hampshire

Mukul C. Datta
Tuskegee University

Lawrence C. Davis
Kansas State University

Dr. Paul H. Demchick
Barton College

Michael W. Dennis
Montana State University-Billings

Kathleen A. Donnelly, Ph.D.
Russell Sage College

Lawrence K. Duffy
University of Alaska Fairbanks

John R. Edwards
Villanova University

Alfred T. Ericson
Emporia State University

Robert J. Evans
Illinois College

David Fahrney
Colorado State University

H. Richard Fevold
University of Montana

Christopher Francklyn
University of Vermont College of Medicine

Edward A. Funkhouser
Texas A & M University

Edwin J. Geels
Dordt College

Darrel Goll
University of Arizona

Dr. Eugene Gooch
Elon College

Milton Gordon
University of Washington, Seattle

Joan M. Griffiths
Cornell University

Lonnie J. Guralnick
Western Oregon State College

James H. Hageman
New Mexico State University

B. A. Hamkalo
University of California, Irvine

Kenneth D. Hapner
Montana State University

Gerald W. Hart
University of Alabama at Birmingham

Terry L. Helser
S.U.N.Y College at Oneonta

Pui Shing Ho
Oregon State University

Joel Hockensmith
University of Virginia

Daniel Holderbaum
Case Western Reserve University

Charles F. Hosler, Jr.
University of Wisconsin-La Crosse

Larry Jackson
Montana State University

Ralph A. Jacobson
CAL POLY

John R. Jefferson
Luther College

Colleen B. Jonssen
New Mexico State
University

Floyd W. Kelly
Casper College

Mary B. Kennedy
California Institute of
Technology

R. L. Khandelwal
University of Saskatchewan

Nazir A. Khatri
Franklin College of Indiana

Ramaswamy
Krishnamoorhi
Kansas State University

James I. Lankford
St. Andrews Presbyterian
College

Daniel J. Lavoie
Saint Anselm College

Franklin R. Leach
Oklahoma State University

Carol Leslie
Union University

Michael Leung
State University of New
York/Old Westbury

Randolph V. Lewis
University of Wyoming

Albert Light
Purdue University,
West Lafayette

Donald R. Lueking
Michigan Technological
University

Dr. Celia L. Marshak
University of San Diego
1994
(Emeritus, San Diego State
University)

Lynn M. Mason
Lubbock Christian
University

Harry R. Matthews
University of California at
Davis

Martha McBride
Norwich University

William L. Meyer
University of Vermont

Holly Miller
Wake Forest University
Medical Center

Michael J. Minch
University of the Pacific

Debra M. Moriarity
University of Alabama in
Huntsville

Mary E. Morton
College of the Holy Cross

Melvyn W. Mosher
Missouri Southern State
College

Stephen H. Munroe
Marquette University

Richard M. Niles
Marshall University School
of Medicine

Jennifer K. Nyborg
Colorado State University

William R. Oliver
Northern Kentucky
University

Dr. Richard Steven Pappas
Georgia State University

Raymond Earl Poore,
Ph.D.
Jacksonville State
University

William T. Potter
The University of Tulsa

Gary L. Powell
Clemson University

Michael Eugene Pugh
Bloomsburg University of
Pennsylvania

Paul D. Ray
University of North Dakota

Philip Reyes
University of New Mexico
School of Medicine

John M. Risley
The University of North
Carolina at Charlotte

H. Alan Rowe
Norfolk State University

John E. Robbins
Montana State University

Norman G. Sansing
University of Georgia

Roy A. Scott, III
The Ohio State University

Steven E. Seifried
John A. Burns School
of Medicine
University of Hawaii

Ralph Shaw
Southeastern Louisiana
University

J. M. Shively
Clemson University

Roger D. Sloboda
Dartmouth College

Deborah Kay Smith
Meredith College

Thomas Sneider
Colorado State University

Wesley E. Stites
University of Arkansas

Eric R. Taylor
University of Southwestern
Louisiana

Martin Teintze
Montana State University

Arrel D. Toews
University of North
Carolina at Chapel Hill

H. Edwin Umbarger
Purdue University

Harry van Keulen
Cleveland State University

Robert J. Van Lanen
Saint Xavier University

Charles Vigue
University of New Haven

William H. Voige
James Madison University

Raymond E. Waldner
Palm Beach Atlantic
College

Arthur C. Washington
Tennessee State University

Daniel Weeks
University of Iowa

Steven M. Wietstock,
Ph.D.
Alma College

Steven Woeste
Scholl College of Podiatric
Medicine

Robert Zand
University of Michigan

We are grateful for the assistance of the editorial staff at Wm. C. Brown Publishers, especially Kevin Kane, publisher, Liz Sievers, our editor, Robin Steffek, developmental editor, Julie Kennedy, in-house production services coordinator, and Lori Hancock, photo editor. In taking this text from the raw manuscript stage to a production-ready stage, the authors have received tremendous assistance from Robin Steffek. Her input covers a wide range of activities from the meticulous checking of the manuscript for accuracy to numerous suggestions for modifications and special learning aids. She has pursued this project with great enthusiasm, skill, and dedication.

Once the project was ready for the production stage, we knew that a highly technical multicolor text of this magnitude would require an outstanding production team and a leader to see that every element in the final product would attain the highest quality. Laura Skinger directed the York Production Services team with confidence and determination to turn out the best possible final product. Whenever we the authors complained that something was not quite the way we wanted it, Laura would most willingly see that the concern was properly addressed without hesitation. In addition, she went way beyond our concerns to produce a text in which we could take great pride.

We hope very much that this text will be interesting and educational for students and a help to their instructors. We would appreciate any comments and suggestions from our readers. If you should find errors, please notify us so that we can make corrections in the second printing.

*Geoffrey L. Zubay    William W. Parson    Dennis E. Vance*

# An Overview of Biochemical Structures and Reactions that Occur in Living Systems

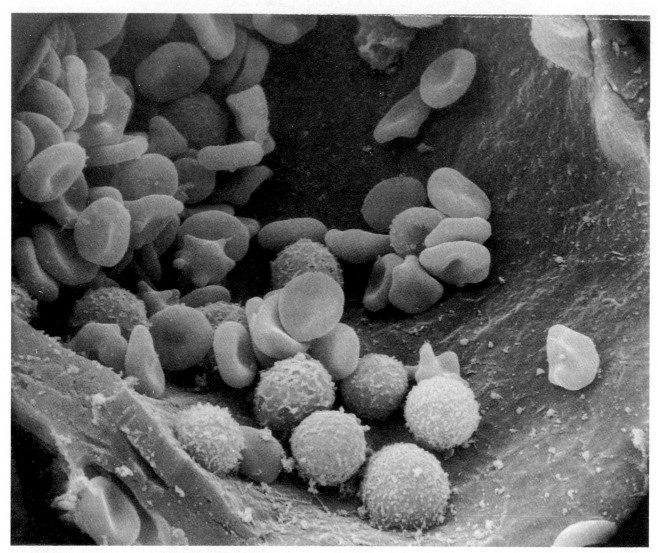

Scanning electron micrograph of erythrocytes and leucocytes in a small blood vessel. Erythrocytes are the biconcave cells with a relatively smooth outer membrane structure. Most of the protein in erythrocytes is hemoglobin, which transports oxygen from the lungs to other tissues. Leucocytes are the rounded cells that possess filamentous protrusions. They are involved in the immune response. The complex surface structures are required for interaction with other cells. Despite the dramatic difference in composition and function these cells are both derived from the same pluripotential hematopoietic stem cells. (From R. G. Kessel and R. H. Kardon, *Tissues and Organs*, W.H. Freeman, Copyright © 1979.)

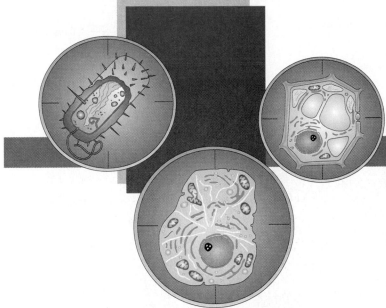

# Cells, Biomolecules, and Water

*The most unique feature of the living cell is the way in which so many reactions are organized to serve a single purpose.*

Aside from a number of striking geological features, the most prominent and unique aspect of the Earth's surface is that it is virtually covered with living organisms. In general biology we are introduced to the extraordinary diversity of living organisms, a diversity so great that it is generally acknowledged that there are more species than could ever be classified. Life seems incredibly complicated. If we limit our inspection of living things to the gross organismic level or what we can see with the naked eye, it is very easy to be overwhelmed by this complexity. Furthermore we will not find explanations for why organisms are constructed the way they are at this level of inquiry. If we probe more deeply into organismic structure, we find that all organisms are composed of much smaller units called cells. If we probe even further we find that cells are composed of a limited number of small molecules and macromolecules. We also find that the macromolecules, which make up the bulk of the solid matter of cells, are constructed from a small number of building blocks that are closely related in structure. It is at the molecular level that we find the ultimate explanations for organismic structure and behavior. We may revel in the discovery that life is not as complicated as we first thought.

Thanks to the investigations of many biochemists over the past half century we are on the verge of a thorough understanding of life at the molecular level. This is a most

## Figure 1.1

Generalized representations of the internal structures of animal and plant cells (eukaryotic cells). Cells are the fundamental units in all living systems, and they vary tremendously in size and shape. All cells are functionally separated from their environment by the plasma membrane that encloses the cytoplasm. Plant cells have two structures not found in animal cells: a cellulose cell wall, exterior to the plasma membrane, and chloroplasts. The many different types of bacteria (prokaryotes) are all smaller than most plant and animal cells. Bacteria, like plant cells, have an exterior cell wall, but it differs greatly in chemical composition and structure from the cell wall in plants. Like all other cells, bacteria have a plasma membrane that functionally separates them from their environment. Some bacteria also have a second membrane, the outer membrane, which is exterior to the cell wall.

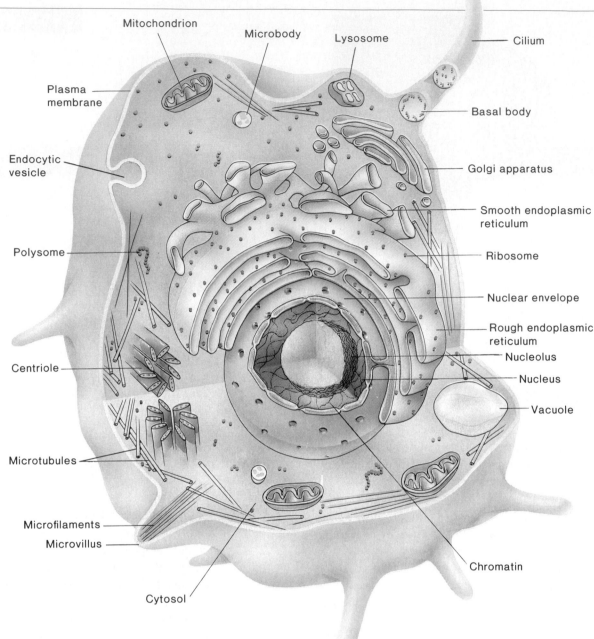

satisfying and exciting time for biochemists; their investigations and accomplishments place biochemistry in the forefront of the biological sciences. The object of this text is to acquaint the beginning student of biochemistry with the basic facts and principles of this subject.

In the introduction to this part of the text, we said that certain common principles underlie all of biochemistry. Let's look briefly now at some of these principles.

## The Cell Is the Fundamental Unit of Life

Microscopic examination of any organism reveals that it is composed of membrane-enclosed structures called cells. The enclosing membrane is called the cell membrane, or the plasma membrane. Cells vary enormously in size and shape, but even the largest cells would have to be much larger to be visible to the naked eye. Within this tiny object thousands of chemical reactions are taking place, all regulated, all designed to serve a specific function. Collectively these reactions serve the function of maintaining the cell and permitting it to replicate when the time is right. Perhaps the most amazing thing about the living cell is that so much organized activity takes place in such a small space. Figure 1.1 shows prototypical animal and plant cells, along with some common shapes and sizes of bacteria. Bacteria are single-

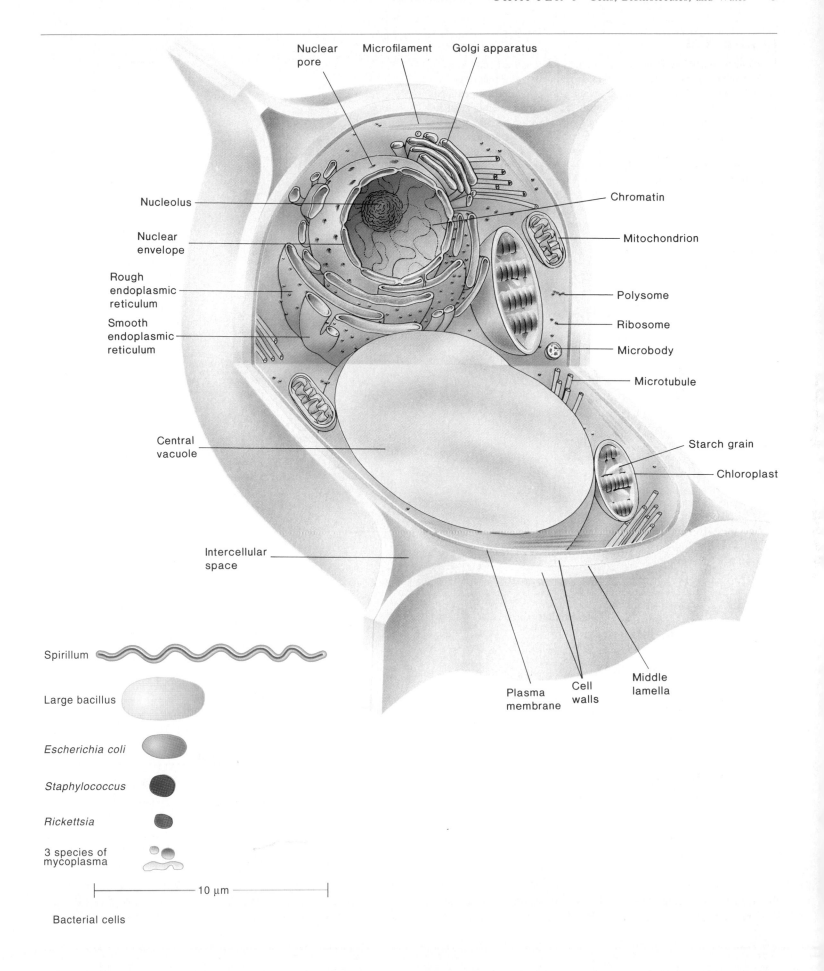

Nuclear pore

Microfilament

Golgi apparatus

Nucleolus

Nuclear envelope

Rough endoplasmic reticulum

Smooth endoplasmic reticulum

Central vacuole

Intercellular space

Chromatin

Mitochondrion

Polysome

Ribosome

Microbody

Microtubule

Starch grain

Chloroplast

Plasma membrane

Cell walls

Middle lamella

Spirillum

Large bacillus

Escherichia coli

Staphylococcus

Rickettsia

3 species of mycoplasma

10 μm

Bacterial cells

## Figure 1.2

Specialized cell types found in the human. Although all cells in a
multicellular organism have common constituents and functions,
specialized cell types have unique chemical compositions, structures,
and biochemical reactions that establish and maintain their specialized
functions. Such cells arise during embryonic development by the
complex processes of cell proliferation and cell differentiation. Except
for the sex (germ) cells, all cell types contain the same genetic
information, which is faithfully replicated and partitioned to daughter
cells. Cell differentiation is the process whereby some of this genetic
information is activated in some cells, resulting in the synthesis of
certain proteins and not other proteins. Thus, specialized cells come
to have different complements of enzymes and metabolic capacities.

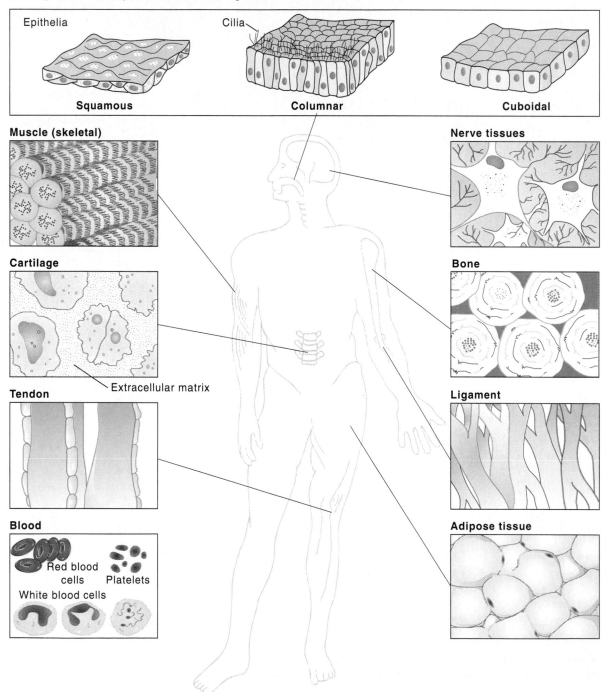

celled organisms, but sometimes the cells are connected into long chains. In multicellular organisms the cells associate to form specialized tissues.

The plasma membrane is a delicate, semipermeable, sheetlike covering for the entire cell. Forming an enclosure prevents gross loss of the intracellular contents; the semipermeable character of the membrane permits the selective absorption of nutrients and the selective removal of metabolic waste products. In many plant and bacterial (but not animal) cells, a cell wall encompasses the plasma membrane. The cell wall is a more porous structure than the plasma membrane, but it is mechanically stronger because it is constructed of a covalently cross-linked, three-dimensional network. The cell wall maintains a cell's three-dimensional form when it is under stress.

The contents enclosed by the plasma membrane constitute the cytoplasm. The purely liquid portion of the cytoplasm is called the cytosol. Within the cytoplasm are a number of macromolecules and larger structures, many of which can be seen by high-power light microscopy or by electron microscopy. Some of the structures are membranous and are called organelles. Organelles commonly found in plant and animal cells include the nucleus, the mitochondria, the endoplasmic reticulum, the Golgi apparatus, and the lysosomes (see fig. 1.1). Chloroplasts are an important class of organelles found in many plant cells but never in animal cells. Each type of organelle is a specialized biochemical factory in which certain biochemical products are synthesized. In addition to organelles, animal and plant cells contain a collection of filamentous structures termed the cytoskeleton, which is important in maintaining the three-dimensional integrity of the cell.

As we will see, the evolutionary tree is bisected into a lower prokaryotic domain and an upper eukaryotic domain. The terms prokaryote and eukaryote refer to the most basic division between cell types. The fundamental difference is that eukaryotic cells contain a membrane-bounded nucleus, whereas prokaryotes do not. The cells of prokaryotes usually lack most of the other membrane-bounded organelles as well. Plants, fungi, and animals are eukaryotes, and bacteria are prokaryotes. The biochemical functions associated with organelles are frequently present in bacteria, but they are usually located on the inner plasma membrane.

Cells are organized in a variety of ways in different living forms. Prokaryotes of a given type produce cells that are very similar in appearance. A bacterial cell replicates by a process in which two identical daughter cells arise from an identical parent cell. Simple eukaryotes can also exist as single nonassociating cells. Eukaryotes of increasing complexity can contain many cells with specialized structures and functions. For example, humans contain about $10^{14}$

### Table 1.1

The Approximate Chemical Composition of a Bacterial Cell

|  | Percent of Total Cell Weight | Number of Types of Each Molecule |
|---|---|---|
| Water | 70 | 1 |
| Inorganic ions | 1 | 20 |
| Sugars and precursors | 3 | 200 |
| Amino acids and precursors | 0.4 | 100 |
| Nucleotides and precursors | 0.4 | 200 |
| Lipids and precursors | 2 | 50 |
| Other small molecules | 0.2 | ≈200 |
| Macromolecules (proteins, nucleic acids, and polysaccharides) | 22 | ≈5,000 |

cells of more than a hundred different types. Specialized cells make up the skin, connective tissue, nerve tissue, muscles, blood, sensory functions, and reproductive organs (fig. 1.2). In such a complex organism, the capacity of different cells for replication is limited. When a skin cell or a muscle cell precursor replicates, it makes more cells of the same type. The only cells in a complex eukaryote capable of reproducing an entire organism are the germ cells, that is, the sperm and the egg.

## Cells Are Composed of Small Molecules, Macromolecules, and Organelles

Of the many different types of molecules in the various organelles and the cytosol that constitute the living cell, water is by far the most abundant, constituting about 70% by weight of most living matter (table 1.1). As a result, most other components exist essentially in an aqueous environment.

Except for water, most of the molecules found in the cell are lipids or macromolecules, which can be classified into four different categories: lipids, carbohydrates, proteins, and nucleic acids. Each type of macromolecule possesses distinct chemical properties that suit it for the functions it serves in the cell.

**Figure 1.3**

The structures of common lipids. (*a*) The structures of saturated and unsaturated fatty acids, represented here by stearic acid and oleic acid. (*b*) Three fatty acids covalently linked to glycerol by ester bonds form a triacylglycerol. (*c*) The general structure for a phospholipid consists of two fatty acids esterified to glycerol, which is linked through phosphate to a polar head group. The polar head group may be any one of several different compounds—for example, choline, serine, or ethanolamine.

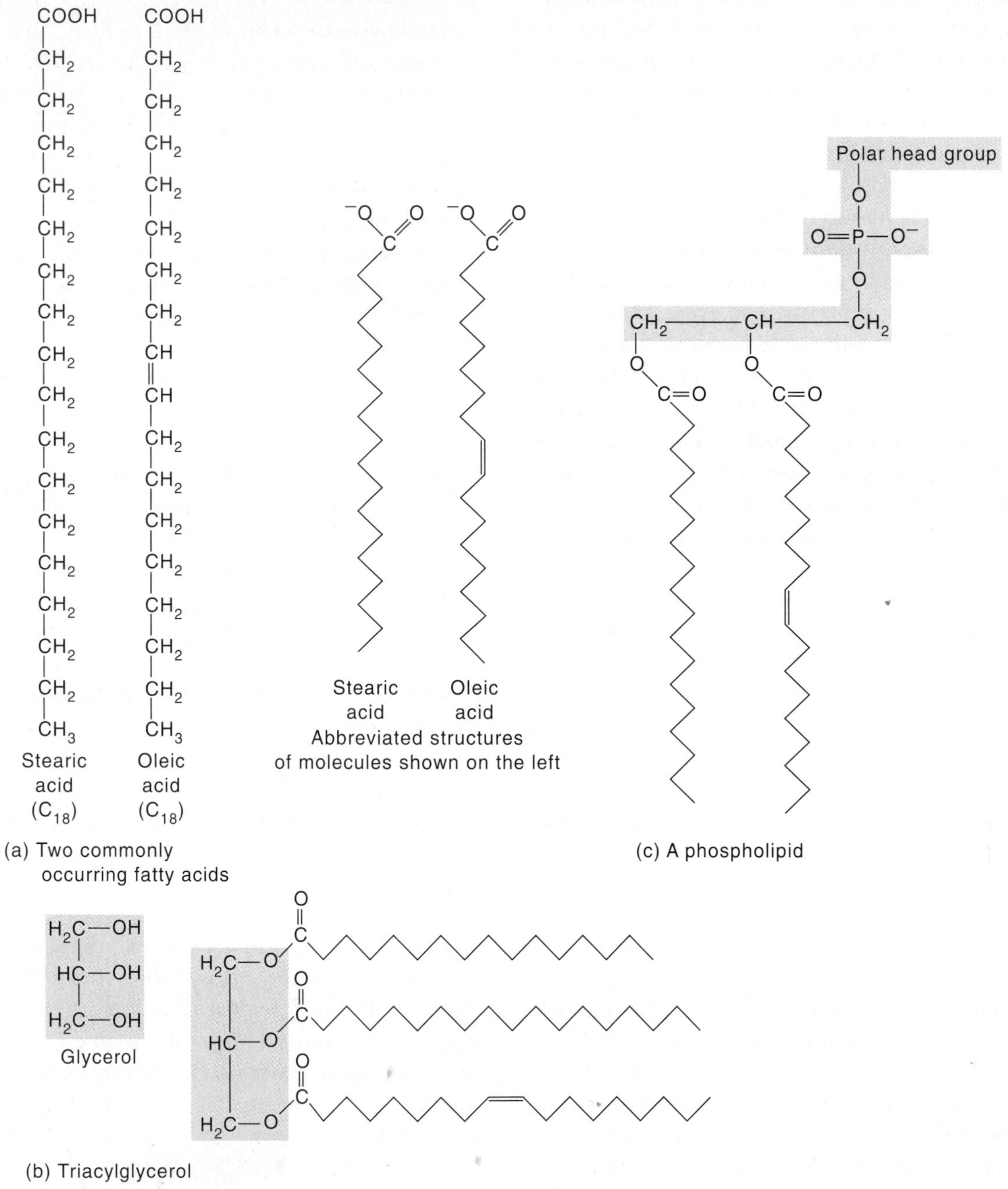

(a) Two commonly occurring fatty acids

(b) Triacylglycerol

(c) A phospholipid

Lipids are primarily hydrocarbon structures (fig. 1.3). They tend to be poorly soluble in water and are therefore particularly well suited to serve as a major component of the various membrane structures found in cells. Lipids also serve as a compact means of storing chemical energy to drive the metabolism of the cell.

Carbohydrates, like lipids, contain a carbon backbone, but they also contain many polar hydroxyl (—OH) groups and are therefore very soluble in water. Large carbohydrate molecules called polysaccharides consist of many small, ringlike sugar molecules; these sugar monomers are attached to one another by glycosidic bonds in a linear or branched array to form the sugar polymer (fig. 1.4). In the cell, such polysaccharides often form storage granules that may be readily broken down into their component sugars. With further chemical breakdown these sugars release chemical energy and may also provide the carbon skeletons for the synthesis of a variety of other molecules. Important structural functions are also served by polysaccharides. Linear polysaccharides form a major component of plant cell walls, and bacterial cell walls are composed of linear polysaccharides that are cross-linked by short polypeptide chains.

Proteins are the most complex macromolecules found in the cell. They are composed of linear polymers called polypeptides, which contain amino acids connected by peptide bonds (fig. 1.5). Each amino acid contains a central carbon atom attached to four substituents: (1) a carboxyl group, (2) an amino group, (3) a hydrogen atom, and (4) an R group. The R group gives each amino acid its unique characteristics. Twenty different amino acids occur in proteins. Some R groups are charged, some are neutral but still polar, and some are apolar.

The linear polypeptide chains of a protein fold in a highly specific way that is determined by the sequence of amino acids in the chains. Many proteins are composed of two or more polypeptides. Certain proteins function in structural roles. Some structural proteins interact with lipids in membrane structures. Others aggregate to form part of the cytoskeleton that helps to give the cell its shape. Still others are the chief components of muscle or connective tissue. Enzymes constitute yet another major class of proteins, which function as catalysts that accelerate and direct biochemical reactions.

Nucleic acids are the largest macromolecules in the cell. They are very long, linear polymers, called polynucleotides, composed of many nucleotides. A nucleotide contains (1) a five-carbon sugar molecule, (2) one or more phosphate groups, and (3) a nitrogenous base. It is the nitrogenous base that gives each nucleotide a distinct character (fig. 1.6). Five different types of nitrogenous bases are found in the two main types of nucleic acids, deoxyribonucleic acid (DNA) and ribonucleic acid (RNA). DNA contains the genetic information that is inherited when cells divide and organisms reproduce. This genetic information is used in the cell to make ribonucleic acids and proteins.

In addition to water and the macromolecules and organelles, the cytosol contains a large variety of small molecules that differ greatly in both structure and function. These never make up more than a small fraction of the total cell mass despite their great variety (see table 1.1). One class of small molecules consists of the monomer precursors of the different types of macromolecules. These monomers are derived by a series of chemical modifications from the nutrients absorbed through the cell membrane. The intermediate molecules between nutrients and monomers are present in small concentrations in the cytosol. Another class of molecules found in the cytosol includes molecules formed as side products in important synthetic reactions and as degradation products of the macromolecules. Finally, the cytosol contains small bioorganic molecules known as coenzymes, which act in concert with the enzymes in a highly specific manner to catalyze a wide variety of reactions.

## Macromolecules Fold into Complex Three-Dimensional Structures

The complex folding of biomacromolecules rarely entails making or breaking covalent linkages. Rather the folding process is dictated by the primary structure and the way in which different elements of the macromolecule interact with each other and with water. The forces that determine folding are noncovalent in character. As a rule, specific interactions amount to only a fraction of the interaction energy that occurs when a covalent bond is made or broken, but because so many of these interactions occur and their effects are additive, the energies involved can be quite large.

## *Water Is a Primary Factor in Determining the Type of Structures that Form*

Water, as we have seen, is the major component of living systems, and it interacts with many biomolecules. Some molecules are water-loving, or hydrophilic, others are water-abhorring, or hydrophobic, and still others are amphipathic, or in between. What properties of a molecule make it hydrophilic or hydrophobic? First, consider the molecular properties of water and how water interacts with itself.

An individual water molecule has a significant dipole that is due to the greater electronegativity of the oxygen

## Figure 1.4

Monomers and polymers of carbohydrates. (*a*) The most common carbohydrates are the simple six-carbon (hexose) and five-carbon (pentose) sugars. In aqueous solution, these sugar monomers form ring structures. (*b*) Polysaccharides are usually composed of hexose monosaccharides covalently linked together by glycosidic bonds to form long straight-chain or branched-chain structures.

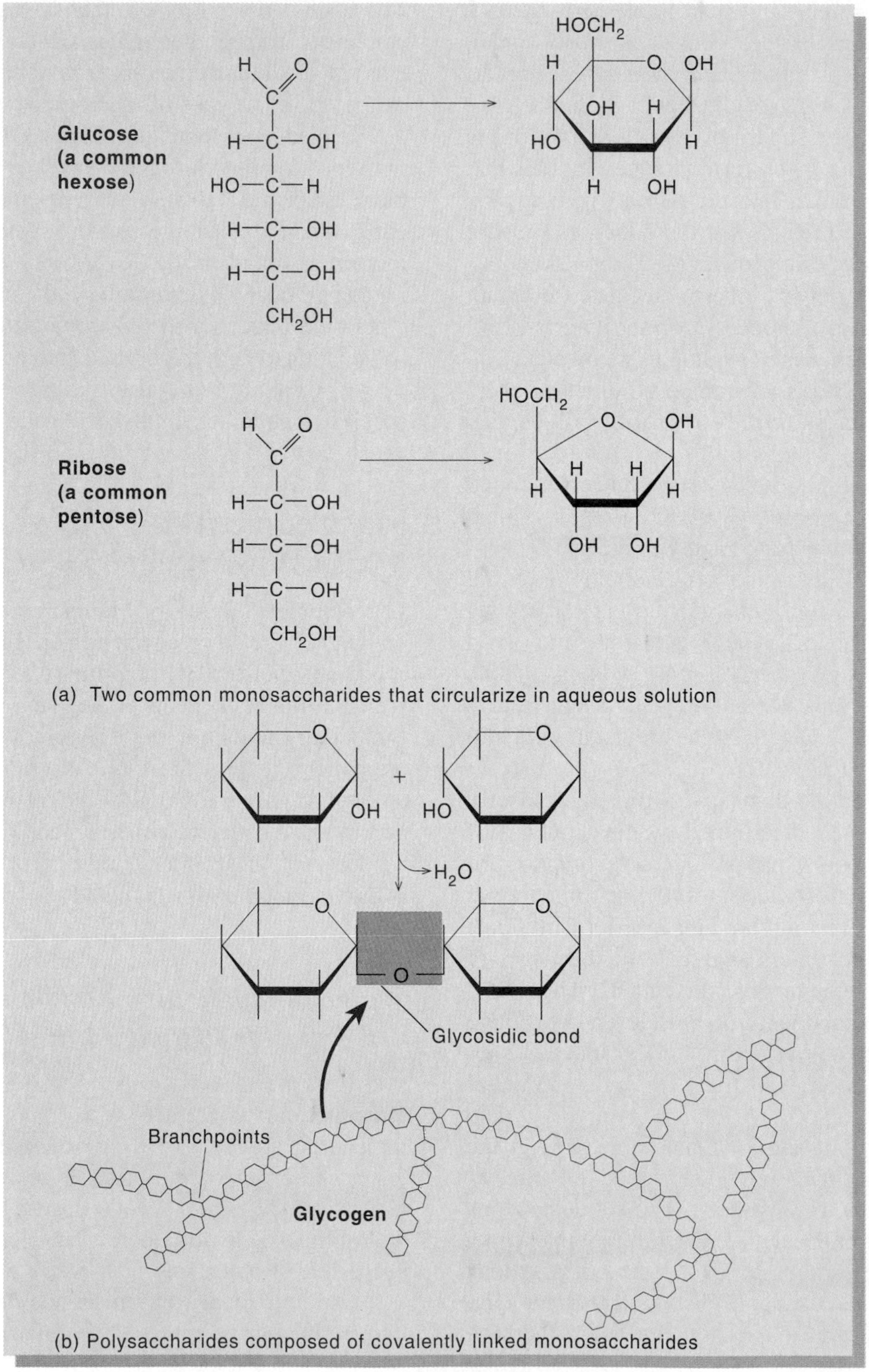

(a) Two common monosaccharides that circularize in aqueous solution

(b) Polysaccharides composed of covalently linked monosaccharides

## Figure 1.5

Amino acids and the structure of the polypeptide chain. Polypeptides are composed of L-amino acids covalently linked together in a sequential manner to form linear chains. (a) The generalized structure of the amino acid. The zwitterion form, in which the amino group and the carboxyl group are ionized, is strongly favored. (b) Structures of some of the R groups found for different amino acids. (c) Two amino acids become covalently linked by a peptide bond, and water is lost. (d) Repeated peptide bond formation generates a polypeptide chain, which is the major component of all proteins.

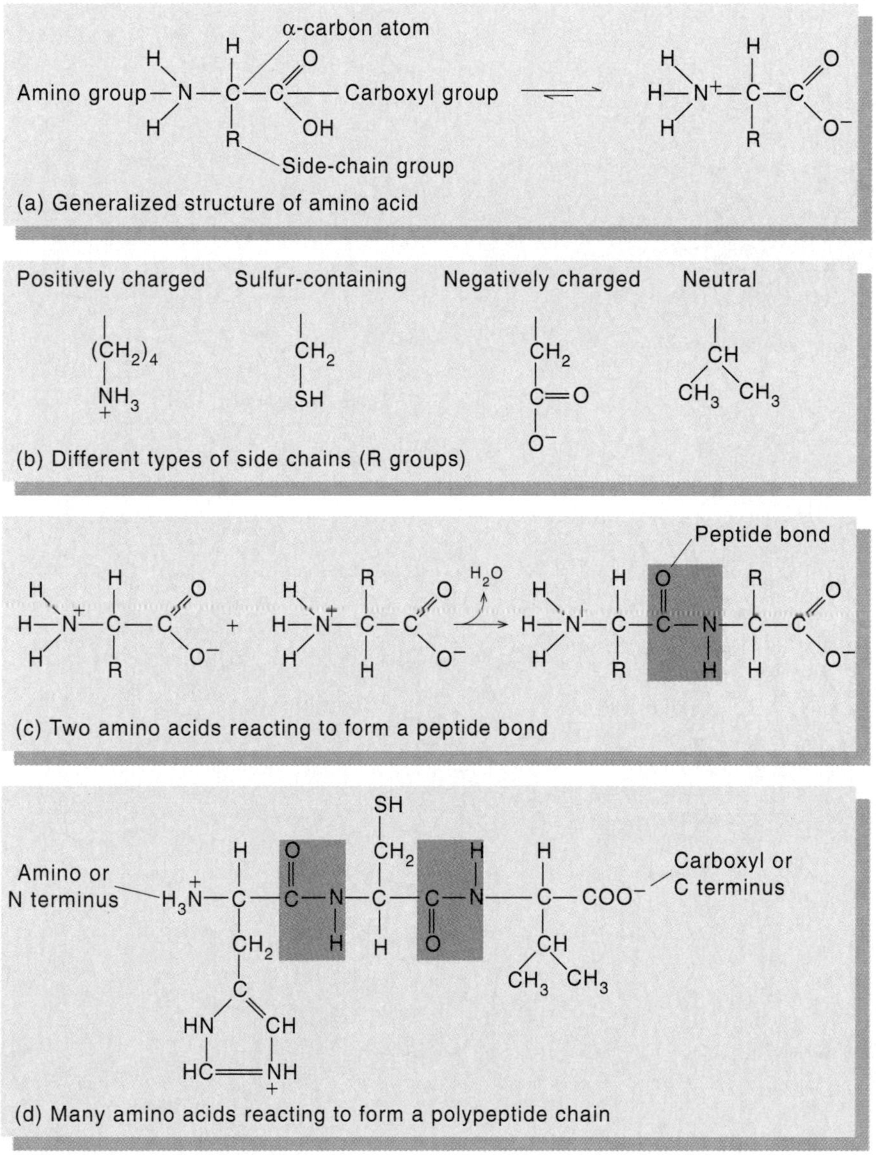

(a) Generalized structure of amino acid

(b) Different types of side chains (R groups)

(c) Two amino acids reacting to form a peptide bond

(d) Many amino acids reacting to form a polypeptide chain

## Figure 1.6

The structural components of nucleic acids. Nucleic acids are long linear polymers of nucleotides, called polynucleotides. (*a*) The nucleotide consists of a five-carbon sugar (ribose in RNA or deoxyribose in DNA) covalently linked at the 5′ carbon to a phosphate, and at the 1′ carbon to a nitrogenous base. (*b*) Nucleotides are distinguished by the types of bases they contain. These are either of the two-ring purine type or of the one-ring pyrimidine type. (*c*) When two nucleotides become linked they form a dinucleotide, which contains one phosphodiester bond. Repetition of this process produces a polynucleotide.

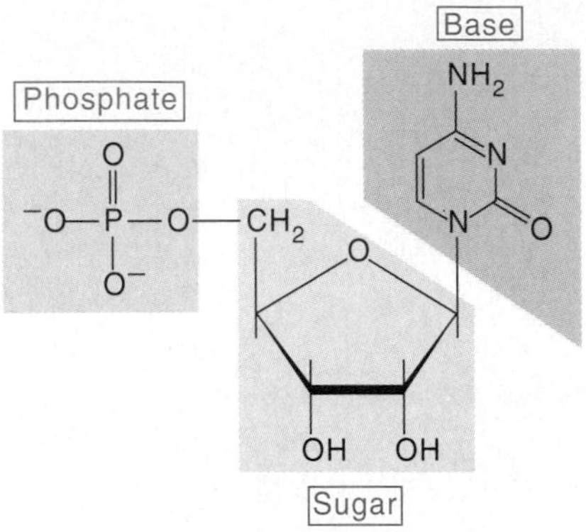

(a) Generalized structure of a nucleotide

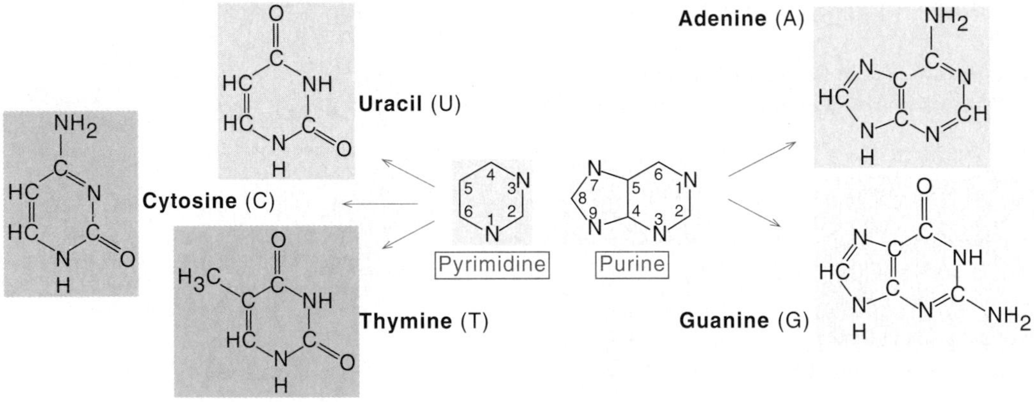

(b) Different bases found in nucleotides

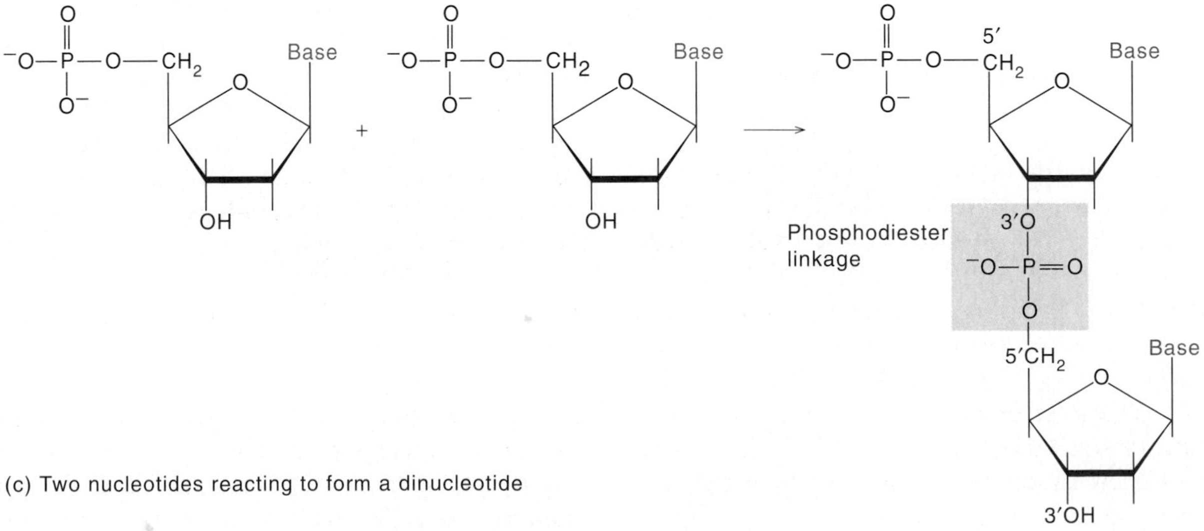

(c) Two nucleotides reacting to form a dinucleotide

**Figure 1.7**

The structure of water and the interaction of water with other water molecules.

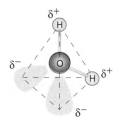

(a) Single water molecule

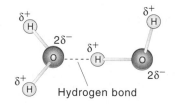

Hydrogen bond

(b) Two interacting water molecules

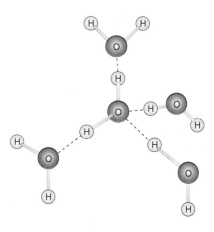

(c) Cluster of interacting water molecules

**Figure 1.8**

The arrangement of molecules in an ice crystal. Water molecules are oriented so that one proton along each oxygen–oxygen axis is closer to one or the other of the two oxygen atoms.

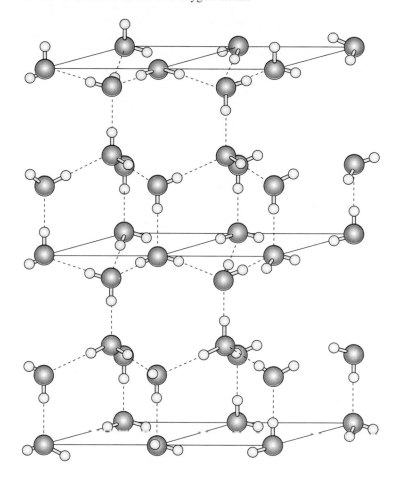

atom over the hydrogen atoms (fig. 1.7). This dipole leads to strong interactions between water molecules, in the form of hydrogen bonds. A hydrogen bond is a noncovalent interaction between polar molecules, one of which is an unshielded proton. In solid water, or ice, the polar forces hold the individual molecules together in a regular three-dimensional lattice (fig. 1.8). Most of the hydrogen bonds present in ice are also present in liquid water. Hence water is a highly hydrogen-bonded structure, not too different from ice, but with a somewhat less regular structure in which the individual molecules have greater mobility.

The dipolar properties of water molecules affect the interaction between water and other molecules that dissolve in water. For example, a favorable interaction accounts for the high solubility of sodium chloride in water (fig. 1.9). The kinds of ion–dipole interactions that take place between water and simple ions such as $Na^+$ and $Cl^-$ are also important in the interactions between the charged, or polar, groups on biomolecules and water. Thus biomolecules that contain charged residues, hydrogen-bond-forming substituents, or other kinds of polar groups are hydrophilic. In the form of small molecules such groups tend to be very soluble in water. When attached to biopolymers they determine which parts of the molecule will be oriented on the exposed surface, where they can make contact with water.

Apolar groups such as neutral hydrocarbon side chains do not contain significant dipoles or the capacity for forming hydrogen bonds. Consequently, they have nothing to gain by interacting with water, as evidenced by their poor

## Figure 1.9

The water molecule is composed of two hydrogen atoms covalently bonded to an oxygen atom with tetrahedral ($sp^3$) electron orbital hybridization. As a result, two lobes of the oxygen $sp^3$ orbital contain pairs of unshared electrons, giving rise to a dipole in the molecule as a whole. The presence of an electric dipole in the water molecule allows it to solvate charged ions because the water dipoles can orient to form energetically favorable electrostatic interactions with charged ions.

## Figure 1.10

Clathrate structures are ordered cages of water molecules around hydrocarbon chains. A portion of the cage structure of $(nC_4H_9)_3S^+F^- \cdot 23\ H_2O$ is shown. The trialkyl sulfur ion nests within the hydrogen-bonded framework of water molecules. In the intact framework, each oxygen is tetrahedrally coordinated to four others. One such oxygen atom and its associated hydrogens are shown by the arrow. (Illustration copyright by Irving Geis. Reprinted by permission.)

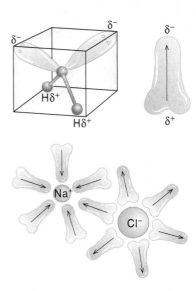

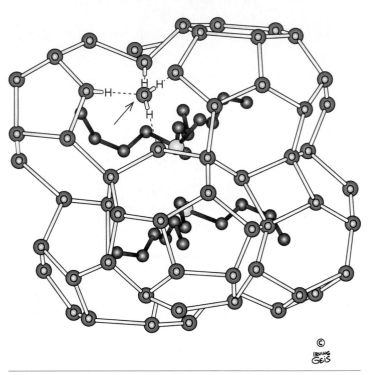

solubility in water. When such hydrophobic molecules are present in water, the water forms a rigid clathrate (cagelike) structure around them (fig. 1.10). Apolar groups in biopolymers tend to bury themselves within the structure of the biopolymer, where they are in the proximity of other apolar groups and avoid contact with water.

Some structures illustrating these principles for macromolecular interaction are shown in figures 1.11 through 1.13. Phospholipids (see fig. 1.3c), which have a hydrophilic polar group on one end and long hydrophobic side chains attached to it, produce multimolecular aggregates in an aqueous environment (fig. 1.11). These phospholipid aggregates form monomolecular layers at the air–water interface or bilayer vesicles within the water. In all of these structures, the polar head groups of the lipid are in contact with water, whereas the apolar side chains are excluded from the solvent structure.

As another example of polarity effects on macromolecular structure, consider polypeptide chains, which usually contain a mixture of amino acids with hydrophilic and hydrophobic side chains. Enzymes fold into complex three-dimensional globular structures with hydrophobic residues located on the inside of the structure and hydrophilic residues located on the surface, where they can interact with water (fig. 1.12).

## Figure 1.11

Structures formed by phospholipids in aqueous solution. Phospholipids may form a monomolecular layer at the air–water interface, or they may form spherical aggregations surrounded by water. A vesicle consists of a double molecular layer of phospholipids surrounding an internal compartment of water.

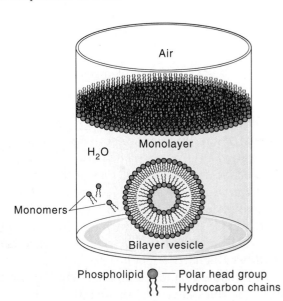

## Figure 1.12

A graphic representation of a three-dimensional model of the protein, cytochrome *c*. Amino acids with nonpolar, hydrophobic side chains (color) are found in the interior of the molecule, where they interact with one another. Polar, hydrophilic amino acid side chains (gray) are on the exterior of the molecule, where they interact with the polar aqueous solvent. (Illustration copyright by Irving Geis. Reprinted by permission.)

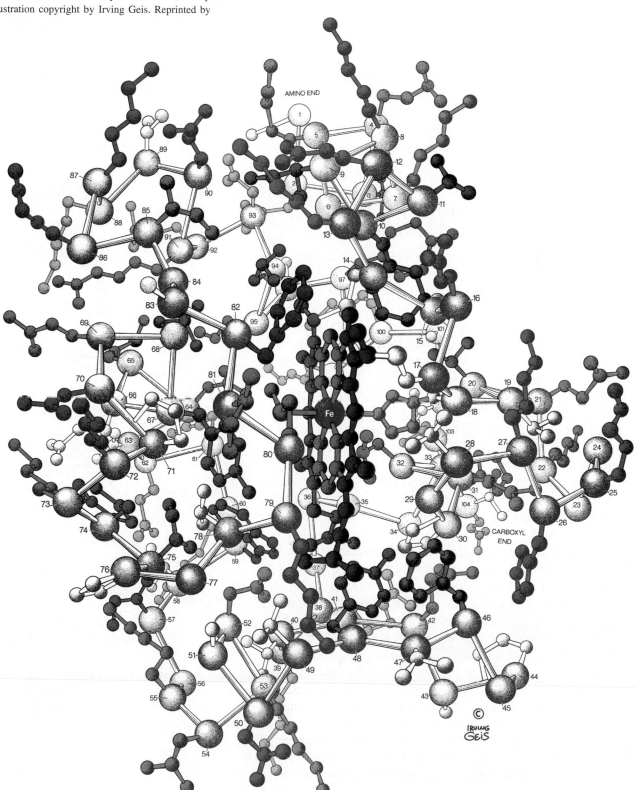

**Figure 1.13**

The right-handed helical structure of DNA. DNA normally exists as a two-chain structure held together by hydrogen bonds (dashed lines) formed between the bases in the two chains. Along the chain the planar surfaces of these bases interact and, together with the hydrogen bonds, contribute to the stability of the two-chain structure. The negatively charged phosphate groups are on the outside of the structure, where they interact with water, ions, or charged molecules. (Illustration copyright by Irving Geis. Reprinted by permission.)

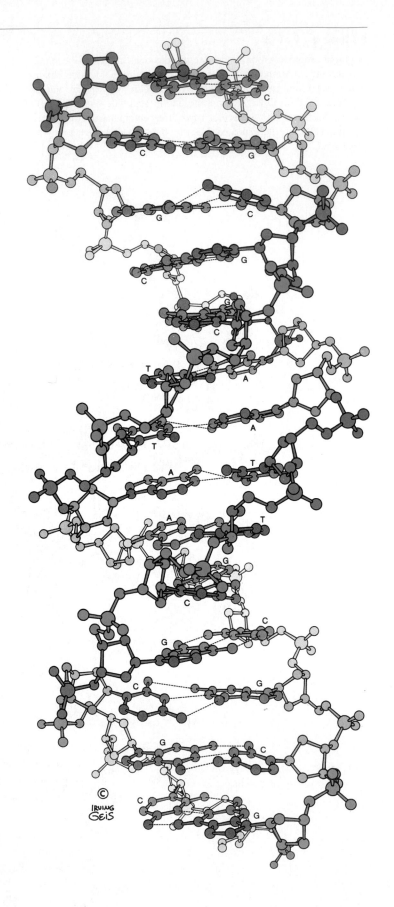

DNA forms a complementary structure of two helically oriented polynucleotide chains (fig. 1.13). The polar sugar and phosphate groups are situated on the surface, where they can interact with water; the nitrogenous bases from the two chains form intermolecular hydrogen bonds in the core of the structure.

## Biochemical Reactions Are a Subset of Ordinary Chemical Reactions

Even though the total number of biochemical reactions is very large, it is still much smaller than the potential number of reactions that occur in ordinary chemical systems. This simplification results partly from the fact that only a limited number of elements account for the vast majority of substances found in living cells. The elements of major importance, in order of decreasing numerical abundance, are hydrogen (H), carbon (C), oxygen (O), nitrogen (N), phosphorus (P), and sulfur (S). Certain metal ions are also important; these include $Na^+$, $K^+$, $Mg^{2+}$, $Ca^{2+}$, $Zn^{2+}$, and $Fe^{2+}$ or $Fe^{3+}$. Other metals and elements that are needed in very small amounts are iodine, cobalt, molybdenum, selenium, vanadium, nickel, chromium, tin, fluorine, silicon, and arsenic. In some cases we don't know the biological roles of these "trace elements" but only that they are needed by some organisms for normal growth or development.

The types of covalent linkages most commonly found in biomolecules are also quite limited (table 1.2). Only 16 different types of linkages account for more than 95% of the linkages found in biomolecules. All the elements can form single or double bonds, except for hydrogen, which only makes single bonds; all the elements exist primarily in one valence state, except for carbon and sulfur, which are frequently found in more than one valence state (table 1.3). Despite this overall simplicity, many other valence states can be found in unusual cases, and some of these are very important. For example, the biochemistry of nitrogen involves consideration of all the valence states of nitrogen from +5 to 0 to −3. A major source of nitrogen available to biosystems is gaseous nitrogen found in the atmosphere (valence state 0). Biochemical reactions convert gaseous nitro-

## Table 1.2

Types of Covalent Linkages Most Commonly Found in Biomolecules

|  | H | C | O | N | P | S |
|---|---|---|---|---|---|---|
| **H** | | | | | | |
| **C** | —C—H | —C—C—<br>C=C | | | | |
| **O** | —O—H | —C—O—<br>C=O | | | | |
| **N** | N—H | —C—N=<br>C=N— | — | | | |
| **P** | — | — | P—O<br>P=O | — | | |
| **S** | —S—H | —C—S— | S—O—<br>S=O | — | — | —S—S— |

## Table 1.3

Most Common Valences Displayed by Atoms in Covalent Linkages

| Element | Valence |
|---|---|
| H | +1 |
| C | −4 to +4 |
| O | −2 |
| P | +5 |
| N | −3 |
| S | +6, −2, −1 |

gen into other forms of nitrogen by reactions which occur uniquely in a select group of microorganisms.

Biochemical reactions involving the different classes of substances use a limited number of functional groups, some of which are illustrated in figure 1.14. All of the functional groups depicted are electrostatically neutral in organic solvents. In water or the cell cytosol, however, many of these functional groups either lose or gain protons to become charged species (as shown on the left side of fig. 1.14). Most of the reactive groups in biomolecules contain one or more of these functional groups or ones closely

## Figure 1.14

Different functional groups found in biomolecules. This figure includes the major functional groups. Other functional groups are found in minor amounts.

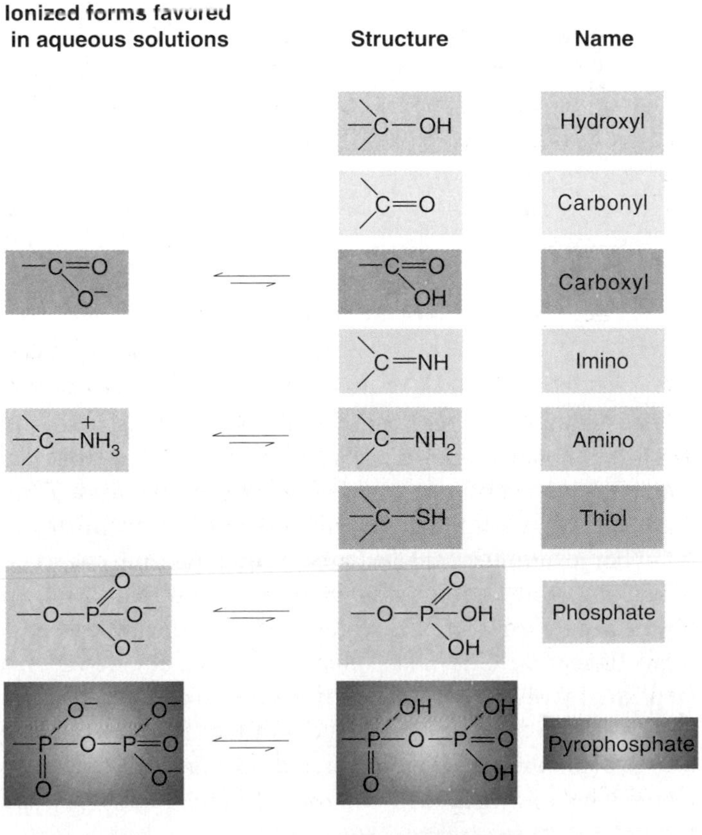

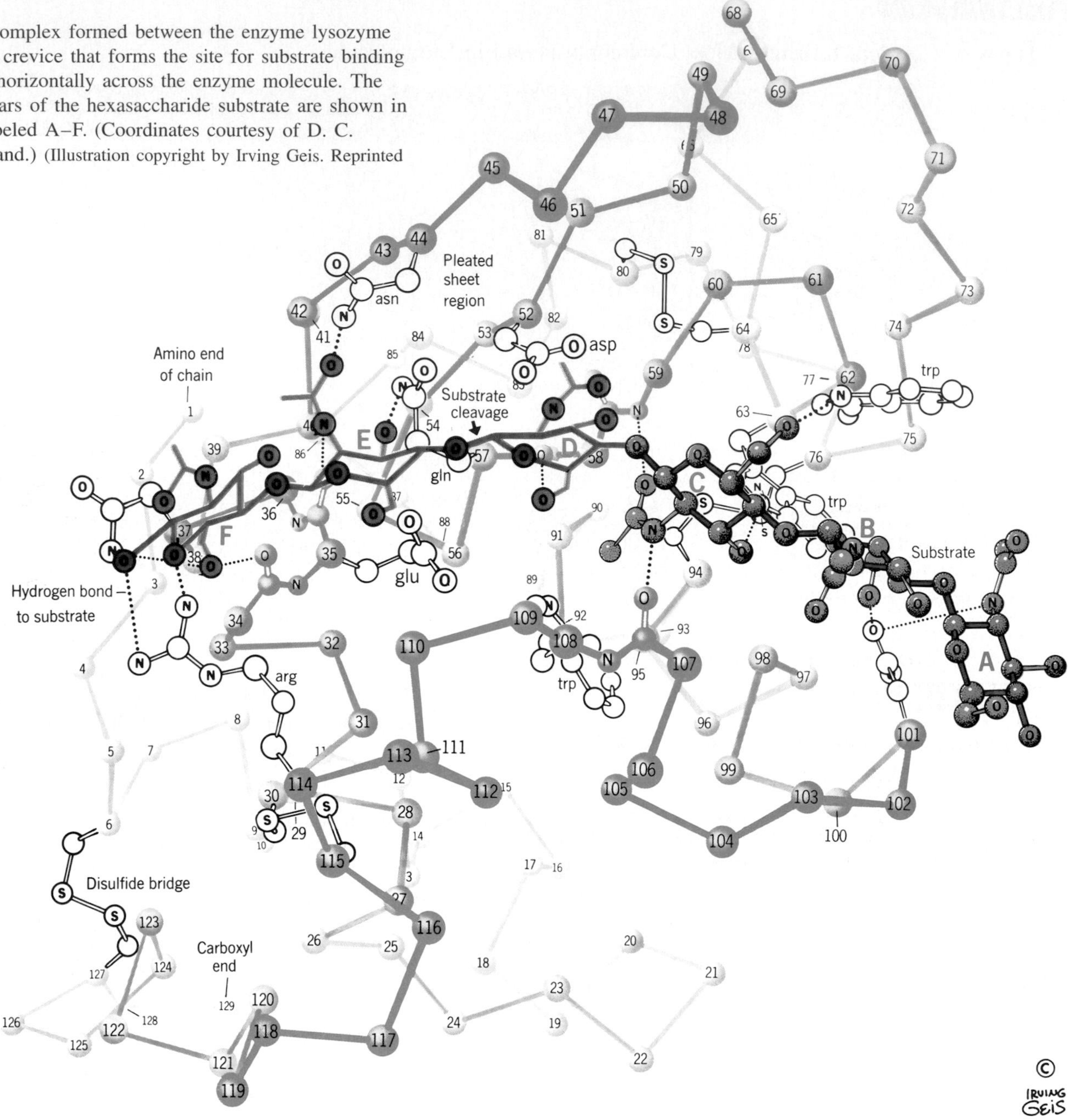

**Figure 1.15**

The structure of the complex formed between the enzyme lysozyme and its substrate. The crevice that forms the site for substrate binding (the active site) runs horizontally across the enzyme molecule. The individual hexose sugars of the hexasaccharide substrate are shown in a darker color and labeled A–F. (Coordinates courtesy of D. C. Philips, Oxford, England.) (Illustration copyright by Irving Geis. Reprinted by permission.)

related to these groups. Many cellular reactions involving these functional groups closely resemble reactions that take place in nonliving systems under different conditions. These extracellular reactions are studied in organic chemistry.

For example, peptide bond formation can occur between two amino acids by a dehydration resulting from simple heating as depicted in figure 1.5c. In the cell, peptide bond formation also takes place, but several intermediate steps are involved and the reaction takes place not by dehydrating but in the wet environment of the cytosol. Similarly the phosphodiester bond depicted in figure 1.6c is not formed by a simple dehydration reaction in the cell but

rather when one nucleotide in the triphosphate form loses a pyrophosphate group as it becomes linked to the hydroxyl group of another nucleotide.

# Biochemical Reactions Take Place on the Catalytic Surfaces of Enzymes

Although biochemical reactions resemble ordinary chemical reactions, they differ in some important ways. Chemical reactions are frequently carried out in nonaqueous solvents, using elevated temperatures and pressures, acids or bases, or

## Figure 1.16

The different fates of an amino acid. Depending on which enzymes are present and active and on the needs of the organism, an amino acid can be metabolized in different ways. Each of these conversions involves one or more steps, and usually each step requires a specific enzyme.

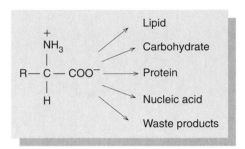

## Figure 1.17

Flow of energy in the biosphere. The sun's rays are the ultimate source of energy. These rays are absorbed and converted into chemical energy (ATP) in the chloroplasts. The chemical energy is used to make carbohydrates from carbon dioxide and water. The energy stored in the carbohydrates is then used, directly or indirectly, to drive all the energy-requiring processes in the biosphere.

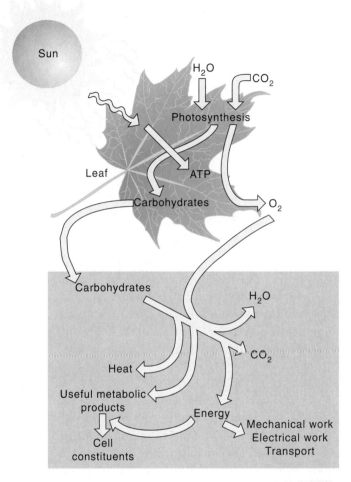

other harsh reagents—conditions that would destroy the functional organization of a living cell. Biochemical reactions usually take place under very mild conditions in aqueous solution. However, many chemical reactions do not proceed at reasonable rates under such conditions. Biochemical reactions proceed at substantially faster rates because of the very special nature of the enzyme catalysts that accelerate them.

Enzymes are structurally complex, highly specific catalysts; each enzyme usually catalyzes only one type of reaction. The enzyme surface binds the interacting molecules, or substrates, so that they are favorably disposed to react with one another (fig. 1.15). The specificity of enzyme catalysis also has a selective effect, so that only one of several potential reactions takes place. For example, a simple amino acid can be used in the synthesis of any of the four major classes of macromolecules or can simply be secreted as waste product (fig. 1.16). The fate of the amino acid is determined as much by the presence of specific enzymes as by its reactive functional groups.

## Many Biochemical Reactions Require Energy

An appreciable amount of energy is needed to build a cell. Even maintaining a cell in a steady nongrowing state requires energy input. Chemical energy is needed to drive many biochemical reactions, to do mechanical work, and for transport of substances across the plasma membrane. The ultimate source of energy that drives a cell's reactions is sunlight (fig. 1.17). Light energy is converted into chemical energy in the chloroplasts of plant cells or in the photosynthetic structures of certain microorganisms. The main form

of chemical energy produced in the chloroplast is a nucleotide containing three phosphoric acid groups attached in sequence, adenosine triphosphate, or ATP (fig 1.18). Organisms that cannot harness the light rays of the sun themselves to make ATP are able to make ATP from the breakdown of organic nutrients originating from plants or other organisms.

Most biochemical reactions fall into one of two classes: degradative or synthetic. Degradative, or catabolic, reactions result in the breakdown of organic compounds to simpler substances. Synthetic, or anabolic, reactions lead to the assembly of biomolecules from simpler molecules. Anabolic processes require energy to drive them. This energy is usually supplied by coupling the energy-requiring biosynthetic reactions to energy-releasing catabolic reactions.

**Figure 1.18**

The structures of ATP and ADP and their interconversion. The two compounds differ by a single phosphate group.

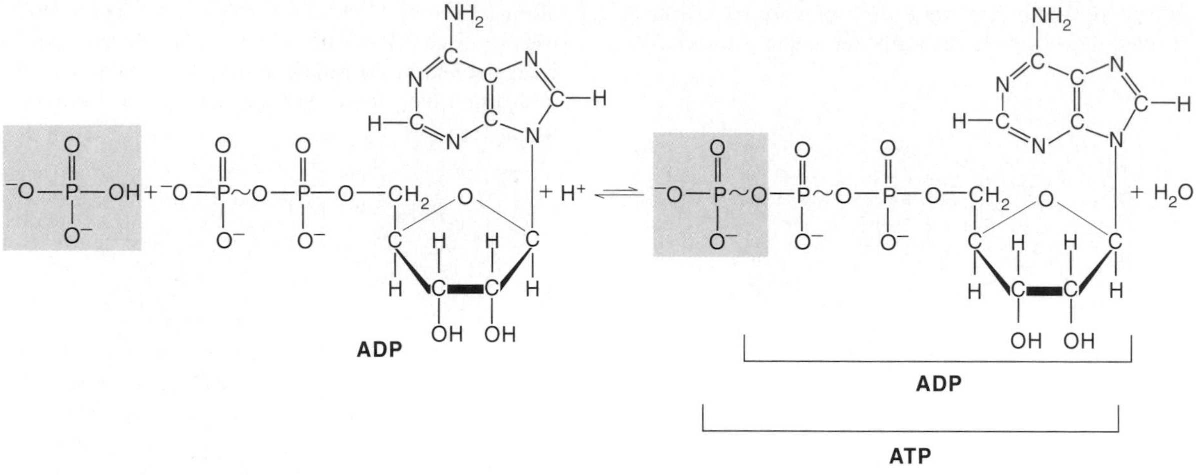

## Biochemical Reactions Are Localized in the Cell

Biochemical reactions are organized so that different reactions occur in different parts of the cell. This organization is most apparent in eukaryotes, where membrane-bounded structures are visible proof for the localization of different biochemical processes. For example, the synthesis of DNA and RNA takes place in the nucleus of a eukaryotic cell. The RNA is subsequently transported across the nuclear membrane to the cytoplasm, where it takes part in protein synthesis. Proteins made in the cytoplasm are used in all parts of the cell. A limited amount of protein synthesis also occurs in chloroplasts and mitochondria. Proteins made in these organelles are used exclusively in organelle-related functions. Most ATP synthesis occurs in chloroplasts and mitochondria. A host of reactions that transport nutrients and metabolites occur in the plasma membrane and the membranes of various organelles. The localization of functionally related reactions in different parts of the cell concentrates reactants and products at sites where they can be most efficiently utilized.

## Biochemical Reactions Are Organized into Pathways

Most biochemical reactions are integrated into multistep pathways using several enzymes. For example, the breakdown of glucose into $CO_2$ and $H_2O$ involves a series of reactions that begins in the cytosol and continues to completion in the mitochondrion. A complex series of reactions like this is referred to as a biochemical pathway (fig. 1.19).

**Figure 1.19**

Summary diagram of the breakdown of glucose to carbon dioxide and water in a eukaryotic cell. As depicted here, the process starts with the absorption of glucose at the plasma membrane and its conversion into glucose-6-phosphate. In the cytosol, this six-carbon compound is then broken down by a sequence of enzyme-catalyzed reactions into two molecules of the three-carbon compound pyruvate. After absorption by the mitochondrion, pyruvate is broken down to carbon dioxide and water by a sequence of reactions that requires molecular oxygen.

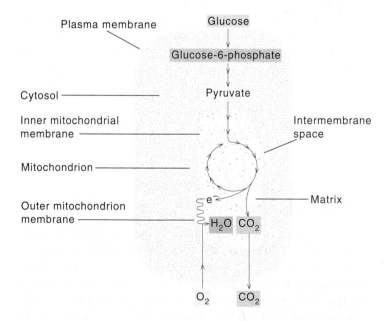

Synthetic reactions, such as the biosynthesis of amino acids in the bacterium *Escherichia coli,* are similarly organized into pathways (fig. 1.20). Frequently pathways have branchpoints. For example, the synthesis of the amino acids threo-

**Figure 1.20**

Synthesis of various amino acids from oxaloacetate. Each arrow represents a discrete biochemical step requiring a unique enzyme. Thus aspartic acid is produced in one step from oxaloacetate, whereas isoleucine is produced in five steps from threonine.

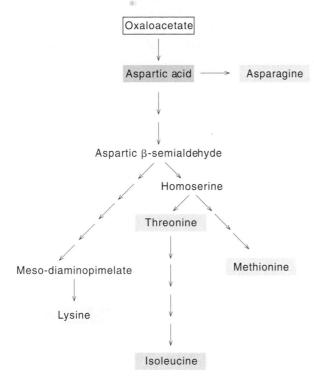

nine and lysine starts with oxaloacetate. After three steps, a branchpoint is reached with the formation of the organic compound aspartic-$\beta$-semialdehyde. One branch of this pathway leads to the synthesis of the amino acid lysine, and another branch leads to the synthesis of the amino acids methionine, threonine, and isoleucine.

To understand the role of each biochemical reaction we must identify its position in a pathway and also consider how that pathway interacts with others.

## Biochemical Reactions Are Regulated

Hundreds of biochemical reactions take place even in the cells of relatively simple microorganisms. Living systems have evolved a sophisticated hierarchy of controls that permits them to maintain a stable intracellular environment. These controls ensure that substances required for maintenance and growth are produced in amounts that are adequate without being excessive. Biochemical controls have developed in such a way that the cell can make adjustments in response to a changing external environment. Adjustments are needed because the temperature, ionic strength, acid concentration, and concentration of nutrients present in the

external environment vary over much wider limits than could be tolerated inside the cell.

The rate of intracellular reactions is a function of the availability of substrates and enzymes. Enzyme activity is controlled at two different levels. First, the rate of a catalyzed reaction is regulated by the amount of the catalyzing enzyme present in the cell. Control of enzyme amounts is usually accomplished by regulating the rate of enzyme synthesis; in some cases the rate of enzyme degradation is also regulated. We can think of controls that regulate the total amount of enzyme present as coarse controls. They define the limits of possible enzyme activity as being anywhere from 0 to 100% of the full activity of the enzyme. Fine controls that act directly on enzymes are also present. Only certain special enzymes, called regulatory enzymes, are susceptible to this second type of regulation. Regulatory enzymes usually occupy key points in biochemical pathways, and their state of activity frequently is decisive in determining the utilization of the pathway.

The underlying principle in regulation is maintaining a favorable intracellular environment in the most economical manner. The cell makes products in the amounts that are needed. Each pathway is regulated in a somewhat different way, ensuring that biochemical energy and substrates are efficiently utilized.

## Organisms Are Biochemically Dependent on One Another

Between 3 and 4 billion years ago the first self-replicating molecules appeared on earth. These entities had to have the capacity for extracting nutrients from the chemical compounds that existed in prebiotic times. We have some general notions about what types of substances were present at that time. One of the most important substances that was not present at that time in significant amounts was molecular oxygen, $O_2$. Currently this form of oxygen is required by all forms of life visible to the naked eye.

The $O_2$ used by most organisms is ultimately converted by them into $CO_2$. Oxygen is utilized at a rapid rate and it would soon disappear if it were not for special classes of photosynthetic organisms that are constantly producing more $O_2$ by the oxidation of water.

The oxygen story is an example of the dependence of one class of organisms on another for certain chemicals. A similar situation exists with the elements carbon and nitrogen, which must be converted from gaseous forms, $CO_2$ and $N_2$, to organic forms usable by most organisms. Reduced carbon compounds are constantly being lost by oxidation to gaseous $CO_2$. The supply of organic carbon compounds required by all forms of life is replenished by photosynthetic organisms; these include most plants and certain microorganisms. Similarly, nitrogen in organic molecules is constantly being lost to the atmosphere in the form of gaseous nitrogen. The reactions required for the conversion of nitrogen to a reduced form more usable to the majority of organisms occurs in only a limited number of microorganisms; yet without these nitrogen-fixing organisms life as we know it would soon vanish.

As we ascend the evolutionary tree, we find increasingly complex multicellular forms. Such organisms generally require more complex nutrients, which must ultimately be supplied to them by simpler living forms. Bacteria like *E. coli* can make all of their own amino acids from a reduced form of nitrogen, such as $NH_3$, and a reduced form of carbon, such as glucose. Humans, on the other hand, must receive most of their amino acids as nutrients. Humans and other complex organisms have gained new biochemical capacities, which permit them to synthesize the components associated with highly specialized differentiated tissues. At the same time, they have lost many of the biochemical systems required to survive on simpler nutrients.

Many biochemical reactions of great importance take place in only a limited number of organisms. This fact increases the complexity of the study of biochemistry. We must learn many reactions; we must also be aware of the biochemical potentials of different organisms. This is the only way we can understand the biochemical interdependency of organisms.

## Information for the Synthesis of Proteins Is Carried by the DNA

DNA contains the genetic information transmitted to each daughter cell when cells divide. The DNA usually exists in the form of nucleoprotein (DNA-protein) complexes called chromosomes. A prokaryotic cell contains a single chromosome. Prior to cell division this chromosome duplicates and segregates so that an identical complement of DNA goes to each of two newly formed daughter cells.

Eukaryotic cells are more complex than prokaryotic cells and usually contain more DNA, which is partitioned between several chromosomes. In both prokaryotes and eukaryotes, almost all cells of the same organism contain the same number of chromosomes. In eukaryotes most of the chromosomes are localized in the nucleus. Thus the DNA is isolated from the main body of the cytoplasm—a unique feature of eukaryotes and the primary distinction between prokaryotes and eukaryotes. Some organelles, notably the mitochondria and the chloroplasts, contain a single circular chromosome.

Eukaryotic chromosomes are detectable by light microscopy at the stage just prior to cell duplication. At this stage, called mitosis, chromosomes appear as elongated refractile structures that can be seen to segregate in equal numbers and types to each of the daughter cells before cell division (fig. 1.21). Each chromosome carries hereditary (genetic) information necessary for the synthesis of specific compounds essential for cell maintenance, growth, and replication. Each chromosome contains a single very long DNA molecule composed of $10^6$ or more nucleotides in a specific sequence. The sequence of nucleotides in the chromosomal DNA determines the sequence of amino acids in the protein polypeptide chains of the organism. The relationship between base sequences and resultant amino acids is known as the genetic code. Each grouping of three bases, called a triplet, represents a specific amino acid and is called a codon. The genetic code ensures that the organism's characteristics are reflected by the sequence of nucleotides in its DNA. When chromosomes replicate, the DNA replicates precisely, so that the same nucleotide sequence is passed along to each of the daughter cells resulting from mitosis and cell division.

The DNA does not transfer its genetic information directly to protein. Rather, this information passes through an intermediary, the messenger RNA (mRNA). The mRNA is made on a DNA template in the nucleus of a eukaryotic cell and then passes into the cytoplasm, where it serves in turn as a template for the synthesis of the polypeptide chain. The overall process of information transfer from DNA to mRNA (transcription) and from mRNA to protein (translation) is depicted in figure 1.22.

## Biochemical Systems Have Been Evolving for Almost Four Billion Years

Biochemical systems are conservative and opportunistic. They tend to evolve one step at a time, using a readily accessible route that leads to an advantage. To appreciate biochemical systems today it is useful to have some understanding of how they came to be.

The earth was formed by a process of accretion about 4.6 billion years ago. Initially it was a molten mass lacking the gravitational pull to retain its gases at the prevalent elevated temperatures. And yet, within a mere 700 million years of the planet's birth, as calculated from the isotopic record of sediments, cellular life almost certainly existed. What raw materials were available to bring about this amazing turn of events? What were the sources of energy used to drive the necessary reactions? Where did the important reactions take place? Was it in the atmosphere, in the oceans, on dry land, or all three?

Table 1.4 shows a distribution of the major elements found in the earth's crust, the ocean water, and the human body. The composition of the human body, which is reason-

**Figure 1.21**

Mitosis and cell division in eukaryotes. After DNA duplication has occurred, mitosis is the process by which quantitatively and qualitatively identical DNA is delivered to daughter cells formed by cell division. Mitosis is traditionally divided into a series of stages characterized by the appearance and movement of the DNA-bearing structures, the chromosomes. (*a*) Premitosis. (*b*) through (*h*) Successive stages of mitosis. (*i*) Postmitosis.

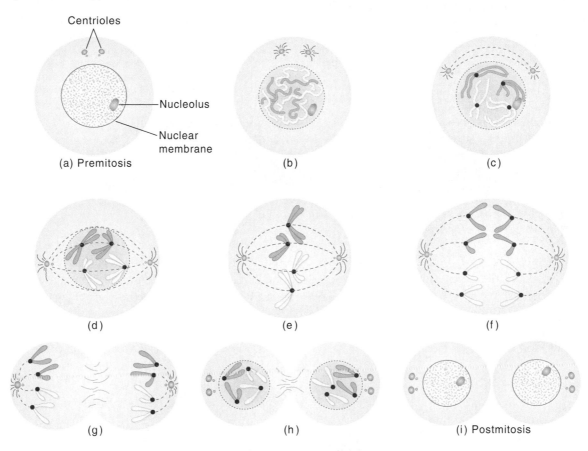

ably representative of living organisms, differs appreciably from that of the earth's crust. The four most abundant elements in the human body are hydrogen, carbon, nitrogen, and oxygen. Of these, only oxygen belongs to the class of elements in highest abundance that make up the earth's crust. Nevertheless, the elements needed to make living things were present in sufficient quantities in the earth's crust and its primitive oceans and the atmosphere.

The original water and air associated with the newly formed planet were lost because of the high temperatures. As the earth cooled, water and various gases on the surface and in the atmosphere were produced by an outgassing process. Today the earth is only about 0.5% water by weight, but because of water's low density, most of it is present on the earth's surface, where it has a major impact on the environment. Water cycles through its gaseous form in the atmosphere and its liquid form in the oceans and bodies of fresh water.

Geological evidence indicates that appreciable amounts of the total water mass have always been present as liquid water. Thus the temperature of the planet has for the most part been between 0° and 100°C, a range that is conducive to the formation of biomolecules and the origin and propagation of life. The earth has also been kind to living things in other ways. The buffering action of various clays and minerals is believed to have maintained the pH level of the oceans between 8.0 and 8.5, which is close to the pH inside living cells.

Unlike the temperature and pH of the surface water, the composition of the atmosphere has changed drastically since the origin of life. In fact, the processes taking place in living things are primarily responsible for these changes. Today's atmosphere, which is about one millionth of the mass of the earth itself, is mainly composed of nitrogen (78%) and oxygen (21%). Most of the remaining atmosphere is argon (0.9%), water (variable up to 4%), and car-

**Figure 1.22**

Transfer of information from DNA to protein. The nucleotide sequence in DNA specifies the sequence of amino acids in a polypeptide. DNA usually exists as a two-chain helical structure. The information contained in the nucleotide sequence of only one of the DNA chains is used to specify the nucleotide sequence of the messenger RNA molecule (mRNA). This sequence information is used in polypeptide synthesis. A three-nucleotide sequence in the mRNA molecule codes for a specific amino acid in the polypeptide chain. (Illustration copyright by Irving Geis. Reprinted by permission.)

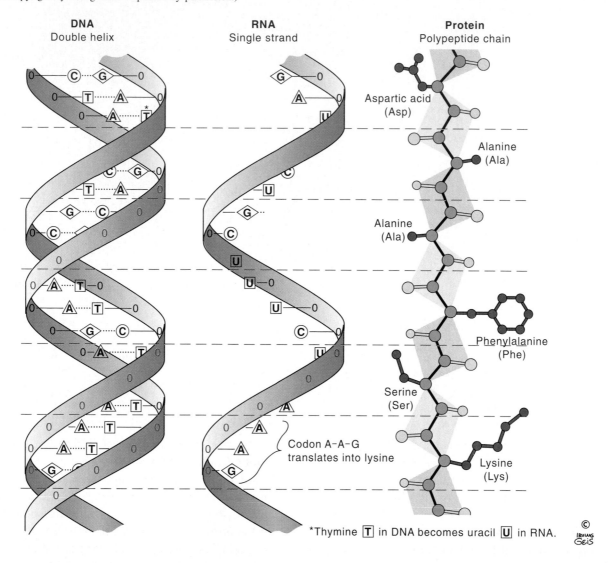

*Thymine T in DNA becomes uracil U in RNA.

bon dioxide (0.034%). The gases that made up the primitive atmosphere were quite different; especially conspicuous was the absence of gaseous oxygen. Most of the oxygen in the present atmosphere is due to the oxidation of water by photosynthetic organisms. The main forms of carbon and nitrogen in the primitive atmosphere were probably $CO_2$ and $N_2$ as they are today. In addition, and probably of great significance to the origin of life, small amounts of the more reduced forms of carbon and $H_2$ gas were present. Thus the primitive earth probably had a weakly reducing atmosphere as contrasted with today's highly oxidizing atmosphere. This situation was most favorable to the origin of life, because organic compounds that enter the biomass tend to be in a reduced state and they are readily oxidized in the presence of gaseous oxygen. It is generally believed that the first organics formed in this primitive atmosphere and then rained down to form larger bioorganic molecules in the liquid phase.

## Table 1.4

Distribution of the 24 Elements Used
in Biological Systems[a]

| Element | Atomic No. | Earth's Crust | Ocean | Human Body |
|---|---|---|---|---|
| Hydrogen (H) | 1 | 2,882 | 66,200 | 60,562 |
| Carbon (C) | 6 | 56 | 1.4 | 10,680 |
| Nitrogen (N) | 7 | 7 | <1 | 2,440 |
| Oxygen (O) | 8 | 60,425 | 33,100 | 25,670 |
| Fluorine (F) | 9 | 77 | <1 | <1 |
| Sodium (Na) | 11 | 2,554 | 290 | 75 |
| Magnesium (Mg) | 12 | 1,784 | 34 | 11 |
| Silicon (Si) | 14 | 20,475 | <1 | <1 |
| Phosphorus (P) | 15 | 79 | <1 | 130 |
| Sulfur (S) | 16 | 33 | 17 | 130 |
| Chlorine (Cl) | 17 | 11 | 340 | 33 |
| Potassium (K) | 19 | 1,374 | 6 | 37 |
| Calcium (Ca) | 20 | 1,878 | 6 | 230 |
| Vanadium (V) | 23 | 4 | <1 | <1 |
| Chromium (Cr) | 24 | 8 | <1 | <1 |
| Manganese (Mn) | 25 | 37 | <1 | <1 |
| Iron (Fe) | 26 | 1,858 | <1 | <1 |
| Cobalt (Co) | 27 | 1 | <1 | <1 |
| Nickel (Ni) | 28 | 3 | <1 | <1 |
| Copper (Cu) | 29 | 1 | <1 | <1 |
| Zinc (Zu) | 30 | 2 | <1 | <1 |
| Selenium (Se) | 34 | <1 | <1 | <1 |
| Molybdenum (Mo) | 42 | <1 | <1 | <1 |
| Iodine (I) | 53 | <1 | <1 | <1 |

[a] Amounts are given in atoms per 100,000.

The origin of life probably occurred in three phases (fig. 1.23): (1) The earliest phase was a period of chemical evolution during which the compounds needed for the nucleation of life must have been formed. These compounds include the most important class of biological macromolecules, the nucleic acids. In this phase of evolution, the synthesis of nucleic acids was "noninstructed." (2) As soon as some nucleic acids were present, physical forces between them must have led to an "instructed" synthesis, in which the already formed molecules served as templates for the synthesis of new polymers. It seems likely that "feedback loops" selected out certain nucleic acids for preferential synthesis. At some point during this period of instructed synthesis more nucleic acids and possibly protein macromolecules were formed. The products of this phase of mo-

## Figure 1.23

The origin of life probably occurred in three overlapping phases: Phase I, chemical evolution, involved the noninstructed synthesis of biological macromolecules. In phase II, biological macromolecules self-organized into systems that could reproduce. In phase III, organisms evolved from simple genetic systems to complex multicellular organisms. The arrow pointing from left to right emphasizes the unidirectional nature of the overall process.

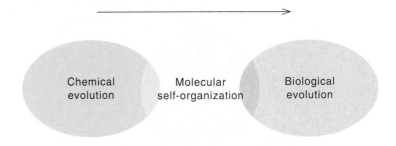

lecular self-organization must sooner or later have begun to resemble the complex organized units that we observe in the self-reproducing biosynthetic cycles of living cells. (3) In the final phase of the origin of life we find the beginnings of the divergent process of biological evolution, the development of the simplest single-celled organisms, and their differentiation into complex multicellular beings.

## All Living Systems Are Related through a Common Evolution

The classical view of evolution, based on morphological differences among organisms, can be diagrammed as a branching tree in which all existing organisms are shown at the tips of the branches (fig. 1.24). An evolutionary tree starts from the simple ancestral prokaryotic cell, which branches off in three main directions into the archaebacteria, the eubacteria, and the eukaryotes. Each of these kingdoms has continued to branch in elaborate ways; we have shown only some of the main branchpoints in figure 1.24. Prokaryotes for the most part have remained as relatively undifferentiated single-celled organisms containing a single chromosome. By contrast, eukaryotes have changed dramatically. Although the organisms on many branches of the eukaryotic part of the evolutionary tree have remained as relatively undifferentiated single-celled forms, significant numbers of eukaryotic organisms have evolved into multicellular forms in which the individual cells of the total organism have differentiated to serve different functions. This process has given rise to plants, animals, and fungi.

A somewhat different view of biological evolution has arisen from a comparison of nucleic acid sequences in different organisms. The best-known sequences for such stud-

**Figure 1.24**

Classical evolutionary tree. All living forms have a common origin, believed to be the ancestral prokaryote. Through a process of evolution some of these prokaryotes changed into other organisms with different characteristics. The evolutionary tree indicates the main pathways of evolution.

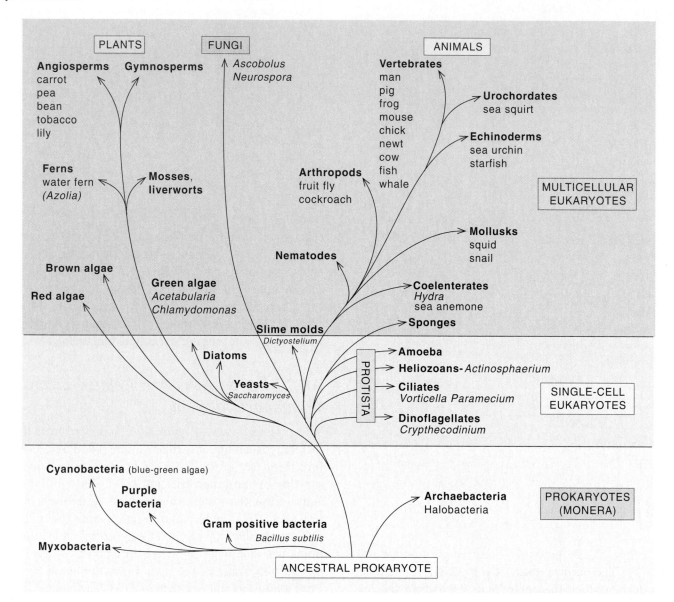

ies have come from the 16S ribosomal RNAs in different organisms. This comparative study has provided us with the evolutionary pattern shown in figure 1.25, in which distances are proportional to sequence differences. A most striking characteristic of this unrooted tree is the distinctness of the three primary kingdoms as evidenced by the large sequence distances that separate each of the kingdoms. Although we do not know the position of the root of the tree, it seems likely that the archaebacteria are closer to the common ancestor of all three kingdoms than are the eubacteria or the prokaryotes.

The organisms most familiar to us, the multicellular plants and animals, occupy a shallow domain within the eukaryotic line of descent. True, the developmental programs of the multicellular forms have generated an incredible diversity in form and function. Nevertheless, both bacteria and unicellular eukaryotes span far greater evolutionary histories. This fact is reflected in the greater biochemical diversity that we find in the unicellular microorganisms.

**Figure 1.25**

An evolutionary tree can be constructed by comparing the complete sequences of 21 different 16S and 16S-like ribosomal RNAs (rRNAs). The scale bar represents the number of accumulated nucleotide differences (mutations) per sequence position in the rRNAs of the various organisms. (Source: From N. R. Pace, G. J. Olsen and C. R. Woese, *Cell* 5:325, 1986.)

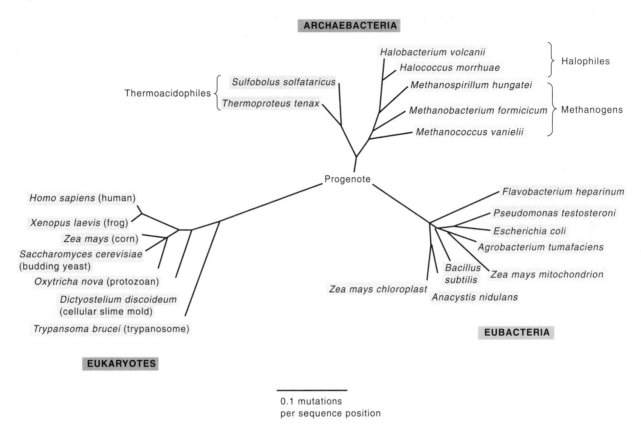

## Summary

In this chapter we discussed the ways in which biochemistry parallels ordinary chemistry and those in which it is quite different. The chief points to remember are the following.

1. The basic unit of life is the cell, which is a membrane-enclosed, microscopically visible object.
2. Cells are composed of small molecules, macromolecules, and organelles. The most prominent small molecule is water, which constitutes 70% of the cell by weight. Other small molecules are present only in quite small amounts; they are precursors or breakdown products of macromolecules or coenzymes. There are four types of macromolecules: lipids, carbohydrates, proteins, and nucleic acids.

3. Noncovalent intermolecular forces largely determine the folded structure adopted by a macromolecule, particularly the relative affinity of different groupings on the macromolecule for water. In general, hydrophobic groupings are buried within the folded macromolecular structure, whereas hydrophilic groupings are located on the surface, where they can interact with water.
4. Biochemical reactions utilize a limited number of elements, most prominently carbon, hydrogen, oxygen, nitrogen, sulfur, and phosphorus. Many biochemical reactions are simple organic reactions.

5. Biochemical reactions are carried out under very mild conditions in aqueous solvent. The reactions can proceed under these conditions because of the highly efficient nature of protein enzyme catalysts.

6. Biochemical reactions frequently require energy. The most common source of chemical energy used is adenosine triphosphate (ATP). The splitting of a phosphate from the ATP molecule can provide the energy needed to make an otherwise unfavorable reaction proceed in the desired direction.

7. Biochemical reactions of different types are localized to different parts in the cell.

8. Biochemical reactions are frequently organized into multistep pathways.

9. Biochemical reactions are regulated according to need by controlling the amount and activity of enzymes in the system.

10. Most organisms depend on other organisms for their survival. Frequently this is because a given organism cannot make all of the compounds needed for its growth and survival.

11. The specific properties of any protein are due to the specific sequence of amino acids in its polypeptide chains. This sequence is determined by the genetic information carried by the sequence of DNA nucleotides. DNA transfers the information to messenger RNA, which serves as the template for protein synthesis.

12. Shortly after the earth was formed and had cooled to a reasonable temperature, chemical processes produced compounds that would be used in the development of living cells. Nucleic acids are the most important compounds for living cells, and it is believed that they played the central role in the origin of life.

13. Evolutionary trees based on morphology or biochemical differences indicate that all living systems are related through a common evolution.

## Selected Readings

Becker, W. M., *The World of the Cell.* Menlo Park, Calif.: Benjamin/Cummings, 1986. A very readable cell biology book that could be referred to while taking biochemistry.

de Duve, C., *Blueprint for a Cell.* Burlington, N.C.: California Biological Supply Co., 1991. A short book on the origin of life that contains an excellent reference list.

Dickerson, R. E., Chemical evolution and the origin of life. *Sci. Am.* 239(3):70–86, 1978.

Doolittle, R. F., The genealogy of some recently evolved vertebrate proteins. *Trends Biochem. Sci.* 10:233–237, 1985.

Kimura, M., The neutral theory of molecular evolution. *Sci. Am.* 241(5):98–126, 1979.

Schopf, J. W., The evolution of the earliest cells. *Sci. Am.* 229(3):10–138, 1978. An authoritative account from a foremost geologist.

Stillinger, F. H., Water revisited. *Science* 209:451–457, 1980. A reminder of the central importance of water to the origin of life.

Wilson, A. C., The molecular basis of evolution. *Sci. Am.* 253(4):164–173, 1985.

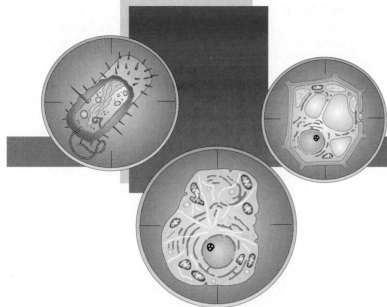

# Thermodynamics in Biochemistry

*All living systems obey the laws of thermodynamics: A pathway that requires energy usually consumes ATP while a pathway that releases energy usually produces ATP.*

The primary usefulness of thermodynamics to biochemists lies in predicting whether particular chemical reactions could occur spontaneously. A simple illustration is to predict what compounds could possibly serve as energy sources for an organism. You are aware from everyday experience that oxidation of organic molecules by molecular oxygen releases energy. For example, wood or coal burns with a large output of heat. Similarly, organisms can obtain energy by oxidizing carbohydrates, fats, or proteins. Some organisms oxidize hydrocarbons, some oxidize reduced forms of sulfur, and others oxidize iron. But no organisms live by oxidizing molecular nitrogen, and the explanation lies in thermodynamics. The reaction cannot occur spontaneously. This example illustrates the importance of thermodynamics in controlling all life. Because organisms live by extracting chemical energy from their surroundings, thermodynamics is not an esoteric subject. It is a matter of life or death.

We say that thermodynamics determines whether a process "could" occur, because thermodynamics tells us only whether the process is possible, not whether it actually does occur in a finite period of time. The rate at which a thermodynamically possible reaction occurs depends on the detailed mechanism of the process. For a biochemical process to occur rapidly, appropriate enzymes must be available. The distinction between spontaneity and speed is more critical for biochemists than it is for chemists, because if a chemical reaction does not proceed rapidly a chemist can change the pressure or temperature or increase the concentration of the reactants. A living organism is under more rigid constraints: It must function at a fixed temperature and pressure and within a limited range of concentrations of reactants.

In this chapter we discuss thermodynamic quantities. We then expand the discussion to show how the concept of free energy is used in predicting biochemical pathways, and we explore the central role of ATP in providing energy for biochemical reactions.

## Thermodynamic Quantities

The properties of a substance can be classified as either intensive or extensive. Intensive properties, which include density, pressure, temperature, and concentration, do not depend on the amount of the material. Extensive properties, such as volume and weight, do depend on the amount. Most thermodynamic properties are extensive including energy $(E)$, enthalpy $(H)$, entropy $(S)$, and free energy $(G)$.

Energy, enthalpy, entropy, and free energy are all properties of the state of a substance. This means that they do not depend on how the substance was made or how it reached a particular state. In a chemical reaction, it is the difference between the initial and final states that is important; the pathway taken to get the initial state to the final state has no bearing on whether the overall reaction releases or consumes energy (fig. 2.1).

## The First Law of Thermodynamics: Any Change in the Energy of a System Requires an Equal and Opposite Change in the Surroundings

Of the thermodynamic properties we just listed, energy is probably the most familiar. Energy can be equated with the capacity to do work. The energy of a molecule includes the internal nuclear energies and the molecular electronic, translational, rotational, and vibrational energies. Electronic energies, which reflect the interactions among the electrons and nuclei, usually are much larger than the translational, rotational, and vibrational energies.

In biochemistry, we are concerned not so much with the absolute energies of molecules as with changes in energy that occur in the course of reactions. It is easier to evaluate a change in energy $(\Delta E)$ than to calculate the absolute energies of the reactants or products, because many of

The change in energy of a system depends only on the initial and final states, not on the path by which the system gets from one state to the other. This diagram illustrates the conversion of a phosphate ester of glucose (glucose-6-phosphate) to free glucose and inorganic phosphate ion $(P_i)$ by two different pathways. Although the two routes proceed through intermediate compounds that differ in energy (A, B, C, and D), the overall energy change $(\Delta E)$ is the same. However, the work done and the amount of heat absorbed or released generally is not the same for the two paths.

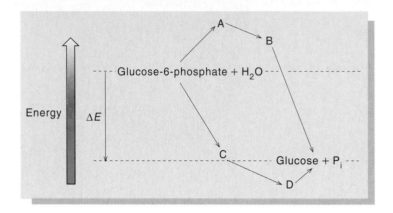

the terms that contribute to the total energy do not change much during a chemical reaction. Electronic energies usually dominate $\Delta E$, as they do the total energies. Good estimates of the energy change resulting from a chemical reaction can usually be obtained by calculating the difference between the bond energies of the reactants and those of the products (see table 4.2).

The first law of thermodynamics says that the total amount of energy in the universe is constant. Energy can undergo transformations from one form to another; for example, the chemical energy of a molecule can be transformed into thermal, electric, or mechanical energy. But any change in the total energy of one part of the universe is matched by an equal and opposite change in another part, so that the overall energy of the universe remains constant.

To apply the first law to a chemical reaction, we must take into account all the energy changes that occur. This means including any changes in the surroundings, as well as in the system of interest (fig. 2.2). The system might be a reaction occurring in a test tube or a living cell; the surroundings are all the rest of the universe. In practical terms, however, we usually need to consider only the immediate surroundings, because only these are likely to be influenced by what happens in the system. According to the first law, the overall energy remains constant even though energy in some form may flow from the system to its surroundings or from the surroundings to the system.

**Figure 2.2**

A system and its surroundings. Heat flow into the system is designated as a positive quantity ($q$), and work that the system does on the surroundings is designated as a positive quantity ($w$). The first law of thermodynamics relates $q$ and $w$ to changes in the energy of the system. Any change in the energy of the system ($\Delta E_{sys}$) is balanced by an opposite change in the energy of the surroundings ($\Delta E_{sur}$), so that the overall energy change ($\Delta E_{tot}$) is zero.

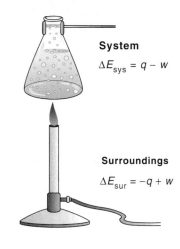

System

$\Delta E_{sys} = q - w$

Surroundings

$\Delta E_{sur} = -q + w$

$$\Delta E_{tot} = \Delta E_{sys} + \Delta E_{sur} = 0$$

If the amount of matter in a system is constant, there are only two means by which the system can gain or lose energy: transfer of heat or performance of work. The energy of a system increases if the system absorbs heat from its surroundings or if the surroundings do work on the system; it decreases if the system gives off heat to the surroundings or does work. A common way of stating the first law of thermodynamics is to equate the change in energy of the system ($\Delta E$) to the difference between the heat absorbed by the system ($q$) and the work done by the system ($w$):

$$\Delta E = q - w \qquad (1)$$

Both $q$ and $w$ depend on the path of the reaction, but their difference $\Delta E$ is independent of the path and therefore defines a state function.

The relative energy of a compound can be measured in a bomb calorimeter, a device in which the compound is thoroughly combusted and the resulting heat is measured. Energies (and enthalpies, discussed next) are usually given in units of kilocalories per mole or kilojoules per mole. One kilojoule (kJ) is the amount of energy needed to apply 1 newton (N) of force over 1 km; 1 kilocalorie (kcal) is the heat needed to raise the temperature of 1 kg of water from 14.5° to 15.5°C. One kcal is equivalent to 4.184 kJ. Physicists generally prefer to use kilojoules; chemists, including biochemists, use both units.

It is customary to record energies of molecules with reference to the energies of their constituent elements. The standard energy of formation reported for a molecule, $\Delta E^{\circ}_{f}$, is equal to the energy change associated with the formation of the molecule from the elements in their standard states. The energy change in any other reaction can be obtained by subtracting the energies of formation of the reactants from the energies of formation of the products. The complex organic molecules found in cells have relatively weak chemical bonds, and thus higher energies, than $H_2O$, $CO_2$, and the other small molecules from which they are formed.

Enthalpy is a function of state that is closely related to energy but is usually more pertinent for describing the thermodynamics of chemical or biochemical reactions. The change in enthalpy ($\Delta H$) is related to the change in energy ($\Delta E$) by the expression

$$\Delta H = \Delta E + \Delta(PV) \qquad (2)$$

where $\Delta(PV)$ is the change in the product of the pressure ($P$) and volume ($V$) of the system. $\Delta H$ is the amount of heat absorbed from the surroundings if a reaction occurs at constant pressure and no work is done other than the work of expansion or contraction of the system. (The work done when a system expands by $\Delta V$ against a constant pressure $P$ is $P\,\Delta V$. This type of work is generally not very useful in biochemical systems.) In most biochemical reactions, little change occurs in either pressure or volume, so the difference between $\Delta H$ and $\Delta E$ is relatively small.

In most chemical reactions that occur spontaneously, the enthalpy of the system decreases. If no work is done, the system gives off heat to the surroundings. But changes in enthalpy or energy do not provide a reliable way of determining whether a reaction can proceed spontaneously. For example, although LiCl and $(NH_4)_2SO_4$ both dissolve readily in water, the former process releases heat, whereas the latter absorbs heat. A mixture of solid $(NH_4)_2SO_4$ and water proceeds spontaneously to a state of higher enthalpy. This reaction is driven by an increase in the entropy of the system.

To assess the potential for a reaction to occur we must consider the entropy as well as the enthalpy.

## The Second Law of Thermodynamics: In Any Spontaneous Process the Total Entropy of the System and the Surroundings Increases

The second law of thermodynamics is that the universe inevitably proceeds from states that are more ordered to states that are more disordered. This phenomenon is measured by a thermodynamic function called entropy, which is denoted by the symbol $S$. A reaction in which entropy increases ($\Delta S$ is positive) proceeds in preference to one in which entropy decreases.

Entropy is an index of the number of different ways that a system could be arranged without changing its energy. If a system could be arranged in $\Omega$ different ways, all with the same energy, the absolute entropy per molecule would be

$$S = k \ln \Omega \qquad (3)$$

where $k$ is Boltzmann's constant ($k = 3.4 \times 10^{-24}$ cal/degree Kelvin). For a mole of substance,

$$S = Nk \ln \Omega = R \ln \Omega \qquad (4)$$

Here $N$ is the number of molecules in a mole ($6 \times 10^{23}$) and $R$ is the gas constant ($R \approx 2$ cal/(degree K · mole). Quantitative values for entropies are usually given in entropy units (1 eu = 1 cal/degree K).

The underlying idea here is that the more ways a particular state could be obtained, the greater is the probability of finding a system in that state. A system that is highly disordered could be obtained in many different arrangements that are all energetically equivalent. Thus a state in which molecules are free to move about and rotate into many different orientations or conformations is favored over a state in which motion is more restricted. The second law makes the remarkably general assertion that the total entropy change in any reaction that occurs spontaneously must be greater than zero. Note, however, that this statement specifies the total entropy, which means that we must consider the entropy change in the surroundings as well as that in the system. The entropy of the system can decrease if the entropy of the surroundings increases by a greater amount.

The absolute entropies of small molecules can be calculated by statistical mechanical methods. Table 2.1 shows the results of such calculations for liquid propane. The largest contributions to the entropy come from the translational and rotational freedom of the molecule, and much smaller contributions from vibrations; electronic terms are insignificant. Although exact calculations of this type become intractable for large biological molecules, the relative sizes of the contributions from different types of motions are similar to those in small molecules. Thus entropy is associated primarily with translation and rotation. This relationship is very different from enthalpy, in which electronic terms are dominant and translational and rotational energies are comparatively small.

Statistical mechanical calculations show that the translational entropy of a molecule depends on $\frac{3}{2}R \ln M_r$ (plus some smaller terms), where $R$ is the gas constant and $M_r$ is the molecular weight. Suppose that a molecule Y undergoes

## Table 2.1

Contributions to the Entropy of Liquid Propane at 231 K

|  | kcal/(degree K · mole) |
|---|---|
| Translational entropy | 36.04 |
| Rotational entropy | 23.38 |
| Vibrational entropy | 1.05 |
| Electronic entropy | 0.00 |
| Total | 60.47 |

a dimerization reaction so that its molecular weight doubles:

$$2\,Y \longrightarrow Y_2$$

Intuitively, we expect dimerization to decrease the entropy because the two monomeric units can no longer move independently. We can calculate the effect quantitatively as a function of the molecular weight as follows. If $M_r$ is the molecular weight of the monomer, then the change in translational entropy in going from the monomeric state to the dimer is approximately

$$\Delta S \approx \frac{3}{2}R \ln 2M_r - 2\left(\frac{3}{2}R \ln M_r\right)$$

$$= \frac{3}{2}R \ln 2 - \frac{3}{2}R \ln M_r$$

$$= -\frac{3}{2}R \ln\left(\frac{M_r}{2}\right) \qquad (5)$$

From this equation we see that the decrease of the translational entropy resulting from dimerization is a logarithmic function of the molecular weight.

Structural features that make molecules more rigid reduce rotational and vibrational contributions to entropy. Thus the formation of a double bond or ring decreases the entropy even when the molecular weight is unchanged. The formation of comparatively rigid macromolecular structures from flexible polypeptide or polynucleotide chains also requires an entropy decrease, although this can be offset by increases in the entropy of the surrounding water molecules (see chapter 4).

The entropy of a compound depends strongly on the physical state of the material. A gas has more translational and rotational freedom than a liquid, and a liquid has more freedom than a solid. As a result, entropy increases when a solid melts or a liquid vaporizes.

It can be shown that the increase in the entropy of a system that undergoes an isothermal, reversible process is

$$\Delta S = \frac{\Delta H}{T} \qquad (6)$$

where $T$ is the absolute temperature in degrees kelvin. An isothermal process is one that occurs at constant temperature. A reversible process is one that proceeds infinitely slowly through a series of intermediate states in which the system is always at equilibrium. For any real process occurring at a finite rate, the system is not strictly at equilibrium. As a consequence the $\Delta S$ is usually somewhat larger than the value given by equation (6).

From equation (6), the entropy increase on vaporization or melting can be determined simply from the heat of vaporization divided by the boiling point, or the heat of fusion divided by the melting point. The entropy increase on vaporization of water is 26 eu/mole and that on melting of ice is 5.3 eu/mole. These values are consistent with our intuition that the increase in translational and rotational freedom is much greater in going from a liquid to a gas than in going from a solid to a liquid.

The entropy of a solution is increased by the mixing of solvents, and it is decreased by interactions among the solvent molecules or interactions of solutes with the solvent. The mixing of two miscible liquids is a thermodynamically favorable process because it increases the number of positions available to the molecules. The entropy change on going from the unmixed liquids to the mixed state can be calculated from the expression

$$\Delta S = n_a R \ln \frac{1}{X_a} + n_b R \ln \frac{1}{X_b}$$
$$= -n_a R \ln X_a - n_b R \ln X_b \qquad (7)$$

where $n_a$ and $n_b$ are the number of moles of $A$ and $B$ that are mixed, and $X_a$ and $X_b$ are the corresponding mole fractions in the final solution. Because $X_a$ and $X_b$ are always less than 1, $\ln X_a$ and $\ln X_b$ are negative. This means that the dilution of each component resulting from the mixing makes a positive contribution to the entropy. Equation (7) applies to the mixing of ideal solutions, in which no interactions occur among the molecules. Any intermolecular interactions decrease the entropy by restricting the system's translational and rotational freedom.

Solvation, the interaction of a solute with the solvent, makes an important negative contribution to the entropy of a solution. Solvation can take the form of hydrogen bonding to donor or acceptor groups on the solute, or of a looser clustering of solvent molecules oriented around the solute (fig. 2.3). In general, the entropy of solvation by water be-

**Figure 2.3**

The entropy decrease resulting from solvation. When a salt is dissolved in water, the entropy of the dissociated cations and anions increases because of the increased possibilities for translation and rotation. But at the same time the movement of water molecules becomes restricted in the vicinity of the ions. The net effect is frequently a decrease in the entropy of the solution. Such a decrease in entropy can occur if the solution releases heat to the surroundings, because this increases the entropy of the surroundings.

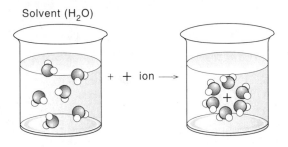

Solvent ($H_2O$)

comes more negative with an increase in the charge or polarity of the solute. Small ions are solvated more strongly than large ions with the same charge, and anions are solvated more strongly than cations.

It is noteworthy that enthalpy depends mainly on electronic interactions, while entropy depends mainly on translation and rotation; solvation affects both enthalpy and entropy. Enthalpies and entropies of solvation usually tend to oppose each other. For charged species, the more negative (favorable) the enthalpy of solvation, the more negative (unfavorable) the entropy of solvation.

From our earlier discussion, you might expect that the dissociation of a proton from a carboxylic acid, which increases the number of independent particles, would lead to an increase in entropy. However, this effect is more than counterbalanced by solvation effects. The charged anion and proton both ''freeze out'' many of the surrounding molecules of water (fig. 2.4). Thus the ionization of a weak acid decreases the number of mobile molecules and so leads to a decrease in entropy. The entropy of ionization of a typical carboxylic acid in water is about $-22$ eu/mole. The entropy of dissociation of a proton from a quaternary ammonium group ($R-NH_3^+ \longrightarrow R-NH_2 + H^+$) is usually smaller, because in this case the dissociation does not alter the number of charged species in the solution.

A different type of solvation effect occurs when an apolar molecule is added to water. The result is a decrease in entropy but not because of favorable interactions between the molecule and the solvent. The water orients on the surface of the apolar molecule to form a relatively rigid cage held together by hydrogen bonds (see fig. 1.10). This effect plays important roles in governing the folding of proteins and determining the structure of biological membranes.

**Figure 2.4**

An ionization reaction often decreases the entropy of a solution, instead of increasing it as one might at first expect, because clustering of water molecules around the ions can result in a net decrease in the number of free water molecules.

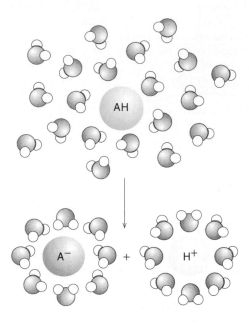

**Figure 2.5**

The entropy of binding of gas molecules to a solid catalyst is negative because of the restricted movements of the adsorbed molecules. By contrast, the entropy change on binding of a substrate to an enzyme is frequently positive. This effect arises because the restricted movement of the substrate is more than compensated for by the release of bound water from the enzyme and the substrate.

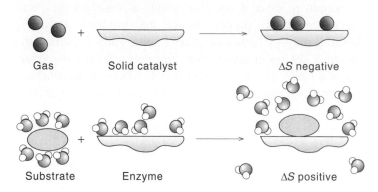

Substantial changes in entropy can occur when a small molecule binds to a protein or other macromolecule. Of particular interest is binding of a substrate or inhibitor to an enzyme. It is instructive to compare the entropy changes here with those that accompany the binding of gas molecules to the surface of a solid catalyst. When gas molecules adsorb on a solid surface, there is a large decrease in entropy (fig. 2.5). The translational entropy of the gas disappears, and the thermodynamics of the adsorbed molecules becomes more like that of a solid. This negative change in entropy is an obstacle to the industrial use of solid catalysts for reactions of gases. To make the reaction proceed, the unfavorable entropy change must be overcome by increasing the pressure or tailoring the catalyst so that the enthalpy of interaction with the gas is strongly negative. In contrast, the binding of a substrate to an enzyme frequently has a positive $\Delta S$ because water molecules are displaced when the substrate binds (fig. 2.5). Binding of an ester substrate to pepsin, for example, produces an entropy increase of 20.6 eu/mole, and binding of urea to urease produces an entropy increase of 13.3 eu/mole. Because of the favorable entropy change, the binding of the substrate may occur spontaneously even if $\Delta H$ is unfavorable.

Another important entropy effect that occurs when an enzyme and substrate combine can be called the chelation effect. This effect is best understood by discussing a relatively simple case of metal chelation. Cadmium ion tends to be quadrivalent, so if there is one amino group in a ligand molecule, as in methylamine, the cadmium can bind to four molecules. If there are two amino groups, as in ethylenediamine, the cadmium combines with two ligand molecules, as shown in table 2.2. Notice that the entropy change is much more unfavorable for the combination with four methylamines than for that with two ethylenediamines. Water molecules are released from the cadmium ion when the ligands bind, but the entropy increase from this release is about the same whether methylamine or ethylenediamine is added. The less favorable entropy change resulting from association with methylamine is due to the larger number of molecules that must attach to the cadmium in this case.

In general, the chelation effect means that a molecule with $n$ points of attachment to another molecule binds more strongly than $n$ molecules with one point of attachment, even though the enthalpy change upon binding at each point is the same. The chelation effect is important in substrate binding to enzymes, where there typically are multiple points of attachment. Several weak interactions can produce an overall tight binding because of the additive contributions of the small, favorable enthalpy changes and the lack of a proportional decrease in entropy. This effect is even greater in the binding of two proteins or the binding of a protein to a nucleic acid, because many points of interaction usually exist between these large molecules.

Table 2.2

Standard Enthalpy and Entropy Changes on Forming Complexes between Cadmium Ion and Methylamine or Ethylenediamine (en)

| Reaction[a] | $\Delta H°$ (kcal/mole) | $\Delta S°$ (eu/mole) | $T \Delta S°$ (kcal/mole) | $\Delta G°$ (kcal/mole) |
|---|---|---|---|---|
| $Cd^{2+} + 4\ CH_3NH_2 \longrightarrow Cd(CH_3NH_2)_4{}^{2+}$ | −13.7 | −16.0 | −4.77 | −8.94 |
| $Cd^{2+} + 2\ en \longrightarrow Cd(en)_2{}^{2+}$ | −13.5 | −3.3 | −0.98 | −12.50 |

[a] en = ethylenediamine; temperature = 25°C.

*Source:* Data from Spike and Parry, Thermodynamics of chelation I. The statistical factor in chelate ring formation, *J. Amer. Chem. Soc.* 75:2726, 1953.

The final column in table 2.2 indicates the changes in free energy accompanying the reactions of cadmium ion with the two amine compounds. Free energy is a function of both enthalpy and entropy; it provides the most useful indication of whether a reaction can proceed spontaneously, as explained in the next section.

## Free Energy Provides the Most Useful Criterion for Spontaneity

We have seen that a system tends toward the lowest enthalpy and the highest entropy. The tendency for the enthalpy of a system to decrease can be explained simply by the second law of thermodynamics. If no work is done, an enthalpy decrease means a transfer of heat from the system to the surroundings, and if the surroundings are at a lower temperature, such a flow of heat will be driven by an increase in the entropy of the surroundings. But enthalpy changes do not afford a reliable rule for determining whether a reaction can proceed spontaneously, because the enthalpy of a system can increase if the entropy of the system also increases. The second law does provide a reliable rule, but its application is often difficult because it requires that we consider the entropy changes in the surroundings as well as in the system.

A more convenient function for predicting the direction of a reaction was discovered by Josiah Gibbs. He was the first to appreciate that in reactions occurring at equilibrium and constant temperature, the change in entropy of the system is numerically equal to the change in enthalpy divided by the absolute temperature. This relationship is the one already presented in equation (6). The equation can be transposed to

$$\Delta H - T\Delta S = 0 \qquad (8)$$

In search of a criterion for spontaneity, Gibbs proposed a new function called the free energy, defined by the equation

$$\Delta G = \Delta H - T\Delta S \qquad (9)$$

Here $\Delta H$ and $\Delta S$ are the changes in enthalpy and entropy in the system alone, not including the surroundings. For a reaction occurring at equilibrium, such as the melting of ice at 0°C, the change in free energy is zero. For the same reaction occurring at a higher temperature, say 10°C, the term $T\Delta S$ is larger, making $\Delta G$ negative. Ice at 10°C melts spontaneously. In the reverse reaction, conversion of water to ice at 10°C, there is a positive change in free energy. This process does not occur spontaneously. Gibbs proposed that a reaction can occur spontaneously if, and only if, $\Delta G$ is negative. If $\Delta G$ is zero, the system is in equilibrium and no net reaction occurs in either direction.

Free energy, like energy, enthalpy, and entropy, is a state function and an extensive property of a system. If the free energy change is favorable (negative), and a good pathway exists, a reaction does occur. If no pathway exists for the conversion, a catalyst may be added that provides an acceptable pathway. However, if the free energy change is unfavorable (positive), no catalyst can ever make the reaction proceed.

## Applications of the Free Energy Function

The free energy function dominates most discussions of thermodynamics in biochemistry. Not only does the sign of $\Delta G$ determine the direction in which a reaction proceeds, but the magnitude of $\Delta G$ indicates just how far the reaction must proceed before the system comes to equilibrium. This is because the standard free energy change $\Delta G°$ has a simple relationship to the equilibrium constant. We elaborate on these points in the following sections. Despite its usefulness, however, many people find the free energy function difficult to grasp intuitively. The reason is that $\Delta G$ is a composite of enthalpic and entropic terms, which often make opposite contributions.

## Values of Free Energy Are Known for Many Compounds

The standard free energy of formation of a compound, $\Delta G°_f$, is the difference between the free energy of the compound in its standard state and the total free energies of the elements of which the compound is composed, again when the elements are in their standard states. The standard states usually are chosen to be the states in which the elements or molecules are stable at 25°C and 1 atmosphere pressure. For oxygen and nitrogen, these are the gases $O_2$ and $N_2$; for solid elements such as carbon, they are the pure solids. For most solutes, the standard states are taken to be 1 M solutions. However, in biochemistry the standard state for hydrogen ion in solution is usually defined as a $10^{-7}$ M solution because this is close to the concentration in most systems of interest to biochemists.

Standard free energies of formation are known for thousands of compounds. They usually are given in units of kcal/mole or kJ/mole. The values for a few compounds of biological interest are collected in table 2.3. By subtracting the sum of the free energies of formation of the reactants from the sum of the free energies of formation of the products, it is possible to calculate the standard free energy change in any reaction for which all the free energies of formation are known.

From the values listed in table 2.3, we can calculate the standard free energy change for the reaction

$$\text{Oxaloacetate}^{2-} + \text{H}^+ (10^{-7} \text{ M}) \longrightarrow$$
$$\text{CO}_2(g) + \text{pyruvate}^-$$

as

$$\Delta G° = -113.44 - 94.45 - (9.87 - 190.62)$$
$$= -7.4 \text{ kcal/mole}$$

The free energy change, when all the reactants and products are in their standard states (1 M oxaloacetate dianion and pyruvate anion, $10^{-7}$ M hydrogen ion, and 1 atm $CO_2$), is $-7.4$ kcal/mole. The negative value of $\Delta G°$ means that the reaction proceeds spontaneously under these conditions. However, some of the concentrations are not very realistic. At pH 7, carbon dioxide is present partly in the form of the bicarbonate anion, rather than as gaseous $CO_2$. To take this into account, we can add the standard free energy change

### Table 2.3

Standard Free Energies of Formation of Some Compounds of Biological Interest

| Substance | $\Delta G°_f$ (kcal/mole) | $\Delta G°_f$ (kJ/mole) |
|---|---|---|
| Lactate ions (1 M) | −123.76 | −516 |
| Pyruvate ions (1 M) | −113.44 | −474 |
| Succinate dianions (1 M) | −164.97 | −690 |
| Glycerol (1 M) | −116.76 | −488 |
| Water | −56.69 | −280 |
| Acetate anions (1 M) | −88.99 | −369 |
| Oxaloacetate dianions (1 M) | −190.62 | −797 |
| Hydrogen ions ($10^{-7}$ M) | −9.87[a] | −41[a] |
| Carbon dioxide (gas) | −94.45 | −394 |
| Bicarbonate ions (1 M) | −140.49 | −587 |

[a] This is the value for hydrogen ions at a concentration of $10^{-7}$ M. The free energy of formation at unit activity (1 M) is 0.

for the reaction of $CO_2$ with water to give the bicarbonate anion plus a proton:

$$\text{CO}_2(g) + \text{H}_2\text{O} \longrightarrow \text{HCO}_3^- + \text{H}^+$$

This calculation yields a correction of $-140.49 - 9.87 - (-56.69 - 94.45) = 0.8$ kcal/mole. The free energy change for the reaction of oxaloacetate to form pyruvate and 1 M bicarbonate ions instead of $CO_2$ is $-7.4 + 0.8 = -6.6$ kcal/mole.

The preceding calculation illustrates the point that the standard free energy change for a reaction can be found by adding or subtracting the free energies of any other reactions that combine to give the desired reaction. Another example is the calculation of the standard free energy of hydrolysis of ATP to adenosine diphosphate (ADP) and $P_i$ at pH 7. This calculation can be done by combining the free energy change for the hydrolysis of glucose-6-phosphate with the free energy change for forming glucose-6-phosphate from glucose and ATP, as shown in table 2.4. We return to these reactions in a later section.

## The Standard Free Energy Change in a Reaction Is Related Logarithmically to the Equilibrium Constant

In biochemistry we are most concerned with reactions occurring in aqueous solution. Suppose we have a chemical reaction with the stoichiometry

**Table 2.4**

Calculating the Standard Free Energy of ATP Hydrolysis ($\Delta G^{\circ\prime}$) by Adding the Free Energies of Two Other Reactions

| Reaction[a] | $\Delta G^{\circ\prime}$ (kcal/mole) |
|---|---|
| Glucose + $ATP^{4-}$ $\longrightarrow$ glucose-6-phosphate$^{2-}$ + $ADP^{3-}$ + $H^+$ | $-5.4$ |
| Glucose-6-phosphate$^{2-}$ + $H_2O$ $\longrightarrow$ glucose + $HPO_4^{2-}$ | $-3.0$ |
| $ATP^{4-}$ + $H_2O$ $\longrightarrow$ $ADP^{3-}$ + $HPO_4^{2-}$ + $H^+$ | $-8.4$ |

[a] The values of $\Delta G^{\circ\prime}$ are for reactions at pH 7 in the absence of $Mg^{2+}$. In the presence of 10 mM $Mg^{2+}$, $\Delta G^{\circ\prime}$ for ATP hydrolysis is about $-7.5$ kcal/mole.

$$aA + bB \longrightarrow cC + dD$$

where $a$, $b$, $c$, and $d$ refer to the moles of A, B, C, and D, respectively. The free energy change in the reaction is

$$\Delta G = G_{\text{final state}} - G_{\text{initial state}} \quad (10)$$

If the reaction occurs at constant temperature and pressure, equation (10) can be expressed as the difference $\Delta G^{\circ}$ between the standard free energies of the products and reactants plus a correction for the concentrations:

$$\Delta G = \Delta G^{\circ} + RT \ln \frac{[C]^c[D]^d}{[A]^a[B]^b} \quad (11)$$

The last term in equation (11) is the correction for concentration and as such is an entropic contribution to $\Delta G$. It is derived by using equation (7) to find the entropy changes associated with diluting the reactants and products from their standard states (1 M) to the actual concentrations in the solution. (In a rigorous treatment, we should use activities instead of concentrations in this formula, but for simplicity we ignore the difference, keeping in mind that it can be substantial in some cases.) If the concentrations of the reactants exceed those of the products, so that the ratio $[C]^c[D]^d/[A]^a[B]^b$ is less than 1, the logarithm is negative, making $\Delta G$ more negative than $\Delta G^{\circ}$ and favoring the reaction in the forward direction. A concentration ratio greater than 1 favors the reverse reaction.

When the reaction comes to equilibrium,

$$\frac{[C]^c[D]^d}{[A]^a[B]^b} = K_{\text{eq}} \quad (12)$$

where $K_{\text{eq}}$ is the equilibrium constant for the reaction. We also know that at equilibrium $\Delta G = 0$. Therefore, from equation (11),

$$\Delta G^{\circ} = -RT \ln K_{\text{eq}} \quad (13)$$

**Table 2.5**

Relationship between $\Delta G^{\circ}$ and $K_{\text{eq}}$ (at 25°C)

| $\Delta G^{\circ}$ (kcal/mole)[a] | $K_{\text{eq}}$ |
|---|---|
| $-6.82$ | $10^5$ |
| $-5.46$ | $10^4$ |
| $-4.09$ | $10^3$ |
| $-2.73$ | $10^2$ |
| $-1.36$ | $10$ |
| $0$ | $1$ |
| $1.36$ | $10^{-1}$ |
| $2.73$ | $10^{-2}$ |
| $4.09$ | $10^{-3}$ |
| $5.46$ | $10^{-4}$ |
| $6.82$ | $10^{-5}$ |

[a] $\Delta G^{\circ}$ values at 25°C are calculated from the equation

$$\begin{aligned} \Delta G^{\circ} &= -RT \ln K_{\text{eq}} \\ &= -1.98 \times 298 \times 2.3 \log K_{\text{eq}} \\ &= -1364 \log K_{\text{eq}} \end{aligned}$$

Thus the standard free energy change for a reaction can be used to obtain the equilibrium constant. Conversely, if we know $K_{\text{eq}}$, we can find $\Delta G^{\circ}$. Because of the logarithmic relationship $K_{\text{eq}}$ has a very steep dependence on $\Delta G^{\circ}$ (table 2.5). A reaction that proceeds to 99% completion is, for most practical purposes, a quantitative reaction. It requires an equilibrium constant of 100 but a standard free energy change of only $-2.7$ kcal/mole, which is little more than half the standard free energy change for the formation of a hydrogen bond.

Equations (11) and (13) are two of the most important thermodynamic relationships for biochemists to remember. If the concentrations of reactants and products are at their equilibrium values, there is no change in free energy for the reactions going in either direction. Living cells, however, maintain some compounds at concentrations far from the equilibrium values, so that their reactions are associated with large changes in free energy. We expand on this point in chapter 11.

We mentioned that biochemists usually define the standard state of protons as $10^{-7}$ M and report values of free energy and equilibrium constants for solutions at pH 7. These values are designated by a prime and written as $\Delta G^{\circ\prime}$, $\Delta G'$ and $K'_{eq}$. Unprimed symbols are used to designate values based on a standard state of 1 M for protons (pH 0). For a reaction that releases one proton, the relationship between $K'_{eq}$ and $K_{eq}$ is $K'_{eq} = 10^7 K_{eq}$. In evaluating the standard free energies $\Delta G^{\circ\prime}$ and $\Delta G^{\circ}$, it is critical to use the equilibrium constants $K'_{eq}$ and $K_{eq}$, respectively, because these can be very different quantities.

## Free Energy Is the Maximum Energy Available for Useful Work

The free energy change gives a quantitative measure of the maximum amount of useful work that could be obtained from a reaction that occurs at constant temperature and pressure. By "useful" work, we mean work other than the unavoidable work of expansion or contraction against the fixed pressure of the surroundings. If $\Delta G$ is zero, the system is at equilibrium, which means that we could not obtain any useful work from the process. If $\Delta G$ is less than zero, the process could yield useful work as the system proceeds spontaneously toward equilibrium. If $\Delta G$ is greater than zero, the process is headed away from equilibrium, and we have to perform work on the system to drive it in this direction. The farther the reactants are from equilibrium, the larger the value of $-\Delta G$, and the larger the amount of work we might obtain from the reaction. However, $-\Delta G$ gives only the maximum amount of useful work. Remember that work and heat, unlike $\Delta G$, $\Delta H$, and $\Delta S$, are not functions of state. The amount of work actually obtained depends on the path the process takes, and it can be zero even if $-\Delta G$ is large.

## Biological Systems Perform Various Kinds of Work

To sustain and propagate life requires that cells do various types of work. This work takes three major forms, related to three broad categories of cellular activities:

1. Mechanical work: changes in location or orientation. Mechanical work is done whenever an organism, cell, or subcellular structure moves against the force of gravity or friction. As examples, consider the contracting muscles that propel a runner up a hill, the swimming of a flagellated protozoan in a pond, the migration of chromosomes toward the opposite poles of the mitotic spindle, and the movement of a ribosome along a strand of messenger RNA.

2. Concentration and electrical work: movements of molecules and ions across membranes. Concentration work, the movement of a molecule or ion across a membrane against a prevailing concentration gradient, establishes the localized concentrations of specific materials on which most essential life processes depend. Concentration work is sometimes referred to as osmotic work. Examples include the uptake of amino acids from the blood by muscle cells, pumping of sodium ions out of a marine microorganism, and movement of nitrate from the soil into the cells of a plant root. Electrical work is required to move a charged species across a membrane against an electric potential gradient. Although the most dramatic example of this is the generation of large potential differences in the electric organ of the electric eel, electrical work is done by almost all types of cells. It underlies the mechanisms of excitation of nerve and muscle cells and the conduction of impulses along axons.

3. Synthetic work: changes in chemical bonds. Synthetic work is necessary for the formation of the complex organic molecules of which cells are composed. As we have seen, these are in general molecules of higher enthalpy and lower entropy than the simple molecules available to organisms from their environment, so free energy must be expended in their synthesis. Synthetic work is most obvious during periods of growth of an organism, but it also occurs in nongrowing, mature organisms, which must continuously repair and replace existing structures. The continuous expenditure of energy to elaborate and maintain ordered structures that were created out of less-ordered raw materials is one of the most characteristic properties of living cells.

## Favorable Reactions Can Drive Unfavorable Reactions

In this section we consider the value of the free energy function in understanding how energetically unfavorable reactions are coupled with energetically favorable ones to do synthetic work. In table 2.4, we made use of the principle that the free energies of all the components of a solution are additive. In general, if the free energy changes associated with two reactions A $\longrightarrow$ B and C $\longrightarrow$ D are $\Delta G_{AB}$ and $\Delta G_{CD}$, the free energy change accompanying the combined process A + C $\longrightarrow$ B + D is simply $\Delta G_{AB} + \Delta G_{CD}$. It is this principle that allows living organisms to synthesize complex molecules with high enthalpies and low entropies. Thermodynamically unfavorable reactions can be driven by coupling them to favorable processes.

From equation (13) you can see that whereas free energy changes combine additively, equilibrium constants combine multiplicatively. If the equilibrium constants for the reactions A $\longrightarrow$ B and C $\longrightarrow$ D are $K_{AB}$ and $K_{CD}$, the equilibrium constant for A + C $\longrightarrow$ B + D is $K_{AB}K_{CD}$.

There are numerous ways to achieve a coupling of favorable and unfavorable reactions. As an example, let's consider the formation of glucose-6-phosphate and water from glucose and inorganic phosphate ion ($P_i$):

$$\text{Glucose} + P_i \longrightarrow \text{glucose-6-phosphate} + H_2O \quad (14)$$

This reaction has an unfavorable $\Delta G^{\circ\prime}$ of +3.0 kcal/mole at 298 K ($K_{eq}$ is 0.0062); the reaction does not occur spontaneously. On the other hand, the reaction

$$\text{ATP} + H_2O \longrightarrow \text{ADP} + P_i + H^+ \quad (15)$$

has a highly favorable $\Delta G^{\circ\prime}$ of about $-8.4$ kcal/mole ($K_{eq} \approx 1.35 \times 10^6$). If reactions (14) and (15) are combined to give the reaction

$$\text{Glucose} + \text{ATP} \longrightarrow \text{G-6-P} + \text{ADP} + H^+ \quad (16)$$

the overall $\Delta G^{\circ\prime}$ is 3.0 − 8.4, or −5.4 kcal/mole, and the overall equilibrium constant is $8.7 \times 10^3$. The combined reaction is thermodynamically favorable (fig. 2.6). A cell can use this reaction to synthesize glucose-6-phosphate, provided that it has a source of ATP.

To take advantage of such a thermodynamic combination of favorable and unfavorable processes, a cell must have a catalytic mechanism for actually linking the two reactions. The breakdown of ATP and ADP and $P_i$, reaction (15), would be fruitless if it occurred independently of reaction (14). In many cells, the combined reaction (16) is catalyzed by an enzyme that facilitates the transfer of phosphate from ATP directly to glucose (hexokinase). This is a common motif in biosynthetic processes. But the coupling mechanism does not have to be so direct.

Another common mechanism for coupling an unfavorable reaction to a favorable one is simply to arrange for one of the reactions to precede or follow the other.

**Figure 2.6**

The formation of glucose-6-phosphate (G-6-P) has a positive $\Delta G^{\circ\prime}$; the hydrolysis of ATP and ADP has a negative $\Delta G^{\circ\prime}$. If the two reactions are combined, the overall $\Delta G^{\circ\prime}$ is negative.

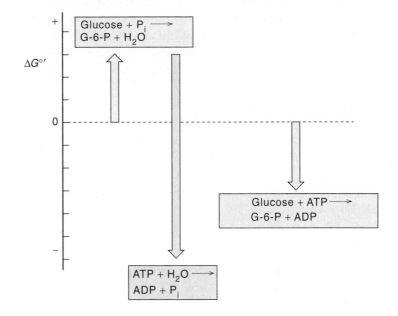

As an example, consider the following sequence of reactions:

$$\text{Acetyl-CoA} + \text{oxaloacetate} \xrightarrow{1} \text{citryl-CoA} \xrightarrow{2}$$
$$\text{citrate} + \text{coenzyme A} \quad (17)$$

The first step has a $\Delta G^{\circ\prime}$ of −0.05 kcal/mole, which is close to zero; it does not occur to any great extent unless the concentrations of acetyl-coenzyme A (acetyl-CoA) and oxaloacetate are greater than the concentration of citryl-CoA. The second step, however, has a highly favorable $\Delta G^{\circ\prime}$ of −8.4 kcal/mole. When the two steps are combined, $\Delta G^{\circ\prime}$ for the overall reaction is about −8.3 kcal/mole, and the equilibrium constant lies far in the forward direction. These two reactions are catalyzed by the enzyme citrate synthase, by a mechanism that ensures that they always occur together.

In the case of reactions that occur sequentially, with one step pulling or pushing the other, it is not necessary for the two steps to be catalyzed by the same enzyme, although this can help to speed up the overall sequence. For example, in many organisms the reactions shown in (17) are preceded by a reaction in which malate is converted to oxaloacetate:

$$\text{Malate}^{2-} + \text{NAD}^+ \longrightarrow$$
$$\text{oxaloacetate}^{2-} + \text{NADH} + H^+ \quad (18)$$

This step is catalyzed by a separate enzyme, malate dehydrogenase. It has a very unfavorable $\Delta G^{\circ\prime}$ of about $+7$ kcal/mole. When this reaction is followed by reaction (17), the combined $\Delta G^{\circ\prime}$ is about $-1.3$ kcal/mole, so the equilibrium constant for the overall sequence is favorable. You can view this simply as an illustration of the principle of mass action: (17) pulls (18) along by removing one of the products.

This last point deserves additional emphasis. It is important to keep in mind that what determines whether a process occurs spontaneously is $\Delta G$, not $\Delta G^{\circ}$. As we saw in equation (11), the actual free energy change depends on the concentrations of the reactants and products. The hydrolysis of ATP (reaction 15), for example, has a $\Delta G^{\circ\prime}$ of about $-8.4$ kcal/mole, but the actual $\Delta G$ in the cytosol of living cells is typically more negative than this by between 5 and 6 kcal/mole because the concentration ratio $[ADP][P_i]/[ATP]$ is much less than 1.

## ATP as the Main Carrier of Free Energy in Biochemical Systems

Virtually all living organisms use ATP for transferring free energy between energy-producing and energy-consuming systems. Processes that proceed with large negative changes in free energy, such as the oxidative degradation of carbohydrates or fatty acids, are used to drive the formation of ATP, and the hydrolysis of ATP is used to drive biosynthetic reactions and other processes that require increases in free energy. In the human body, about 2.3 kg of ATP is formed and consumed every day in the course of these reactions.

### The Hydrolysis of ATP Yields a Large Amount of Free Energy

ATP can be hydrolyzed at two different sites, as shown in figure 2.7. Hydrolysis of the linkage between the $\beta$ and $\gamma$ phosphate groups yields ADP and $P_i$. Hydrolysis between the $\alpha$ and $\beta$ phosphates gives adenosine monophosphate (AMP) and pyrophosphate ion ($HP_2O_7^{3-}$). The standard free energy change ($\Delta G^{\circ\prime}$) is about $-8.4$ kcal/mole in either case. The formation of AMP and pyrophosphate, however, can be pulled forward by hydrolysis of the pyrophosphate to give two equivalents of $P_i$. This secondary reaction has a $\Delta G^{\circ\prime}$ of about $-7.9$ kcal/mole and is catalyzed by pyrophosphatase enzymes present in most types of cells. The overall $\Delta G^{\circ\prime}$ of about $-16.3$ kcal/mole for breakdown of ATP to AMP and 2 $P_i$ makes this process effectively irreversible ($K_{eq} \approx 1 \times 10^{12}$). Cells commonly use this series of reactions in biosynthetic processes in which a reversal is intolerable, such as in the synthesis of nucleic acids. The simpler hydrolysis to produce ADP and $P_i$ allows some re-

versibility but has the advantage that less free energy is needed to resynthesize ATP from ADP than from AMP.

The standard free energies for the hydrolysis of ATP, ADP, or pyrophosphate depend on the pH, on the ionic strength, and also on the concentration of $Mg^{2+}$, which binds to both the reactants and the products and is required as a cosubstrate by most enzymes that use ATP. At pH 7, the principal ionic forms of ATP, ADP, and $P_i$ have net charges of $-4$, $-3$, and $-2$, respectively (see fig. 2.7). Because the hydrolysis of $ATP^{4-}$ to give $ADP^{3-} + HPO_3^{2-}$ is accompanied by release of a proton, raising the pH makes the hydrolysis more favorable. Increasing the $Mg^{2+}$ concentration makes the hydrolysis less favorable. $\Delta G^{\circ\prime}$ decreases in magnitude from $-8.4$ kcal/mole in the absence of $Mg^{2+}$ to about $-7.7$ kcal/mole in the presence of 1 mM $Mg^{2+}$, and to about $-7.5$ kcal/mole at 10 mM. The value $-7.5$ kcal/mole is probably close to the $\Delta G^{\circ\prime}$ under typical physiological conditions, and is used elsewhere in this text. But remember that the low ratio of $[ADP][P_i]$ to $[ATP]$ can make $\Delta G$ for hydrolysis of ATP in living cells substantially more negative than the $\Delta G^{\circ\prime}$. In resting cells, the enzymatic reactions that synthesize ATP usually are more than adequate to keep up with the processes that consume it.

Why is $\Delta G^{\circ\prime}$ for the hydrolysis of ATP so negative? The answer involves several different factors. At pH 7 in the presence of 10 mM $Mg^{2+}$, $\Delta H^{\circ\prime}$ and $-T \Delta S^{\circ\prime}$ for ATP hydrolysis are both negative, and the two terms make similar contributions to the overall value of $\Delta G^{\circ\prime}$. The favorable entropy change results partly from the fact that the proton released in the reaction is diluted into a solution with a very low proton concentration. In addition, solvent water molecules probably are less highly ordered around ADP and $P_i$ than they are around ATP. The negative $\Delta H^{\circ\prime}$ results partly from the fact that the negatively charged oxygen atoms in ATP tend to repel each other. The phosphoric anhydride bond in ATP also is weakened by competition between the phosphorus atoms, which both tend to pull electrons away from the bridging oxygen. Another consideration is that the major products of ATP hydrolysis at pH 7 ($ADP^{3-}$ and $HOPO_3^{2-}$) have a larger total number of resonance forms than the reactant $ATP^{4-}$ does (fig. 2.8). The increased resonance lowers the energy of $ADP^{3-}$ and $HOPO_3^{2-}$ relative to $ATP^{4-}$. The magnitudes of all of these effects vary with the pH, because protonation of the oxygen atoms in ATP, ADP, or $P_i$ relieves some of the repulsive electrostatic interactions, decreases the contributions that some of the resonance forms make to the structure, and decreases the ordering of nearby water molecules. At very low pH, when ATP, ADP, and $P_i$ are all fully protonated, $\Delta G^{\circ}$ probably becomes slightly positive, favoring ATP formation rather than hydrolysis.

**Figure 2.7**

Alternative sites of ATP hydrolysis. The charged species shown are the main ones present at physiological pH and ionic strength. The phosphate groups of ATP are referred to as $\alpha$, $\beta$, and $\gamma$ as indicated. Under physiological conditions, ATP and ADP also bind $Mg^{2+}$ (not shown).

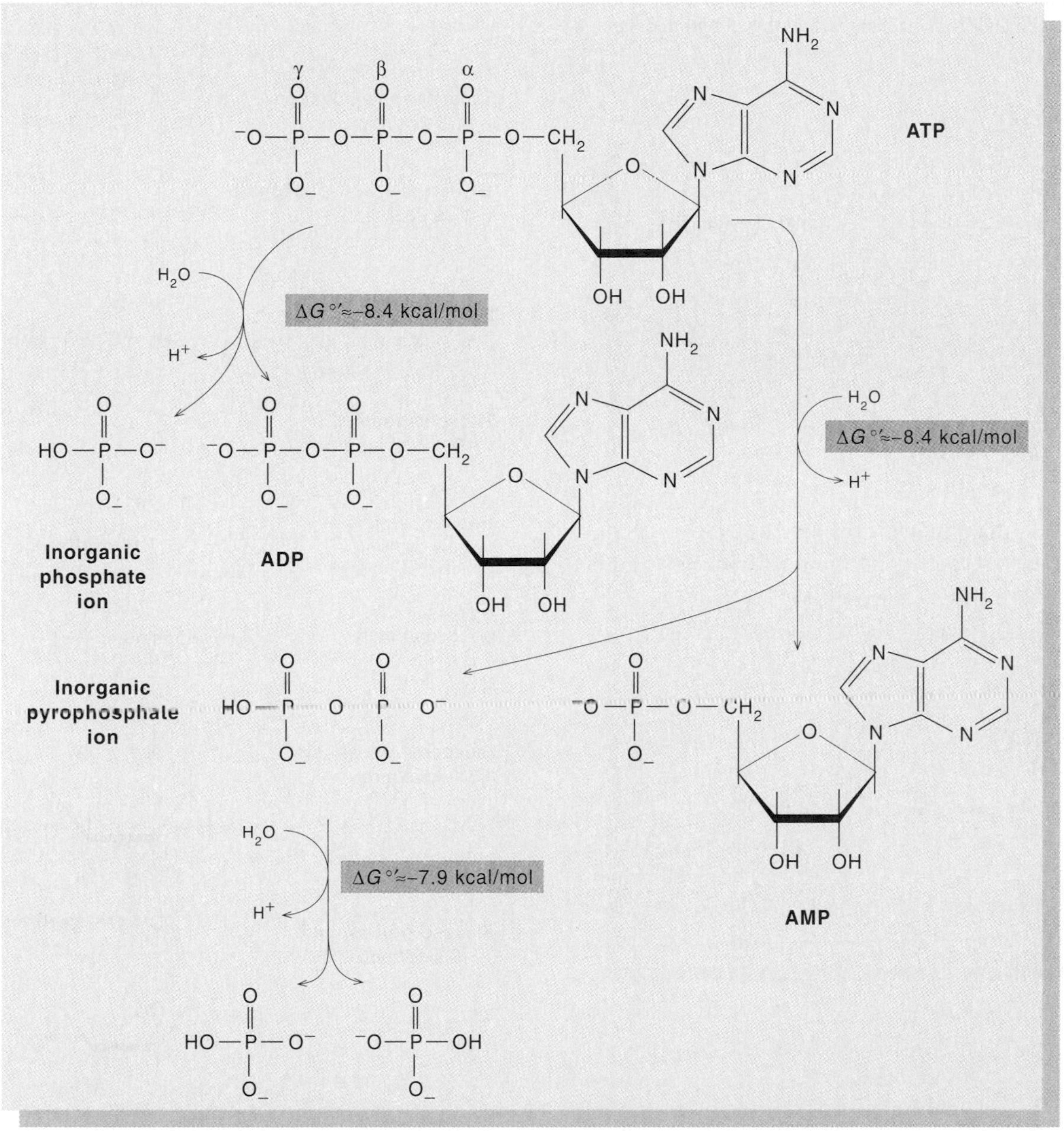

In addition to ATP and other nucleoside triphosphates, cells use a variety of other organic phosphate compounds in energy metabolism. These include acetyl phosphate, glycerate-1,3-bisphosphate, phosphoenolpyruvate, phosphocreatine, and phosphoarginine. Figure 2.9 shows the structures of these compounds and some simpler phosphate esters. The compounds are ranked in the figure in order of their standard free energies of hydrolysis, with the materials having the most negative values of $\Delta G^{\circ\prime}$ at the top. Given equal concentrations of reactants and products, any compound in the figure could be synthesized, in principle, at the expense of any of the compounds above it. Thus ADP could be phos-

## Figure 2.8

The phosphate groups of ATP, ADP, and $P_i$ can be written in a variety of resonance forms. This figure shows the major resonance forms of the $\beta$ phosphate group in ATP and of the same group in ADP. The hydrolysis of ATP to ADP results in an increase in the number of resonance forms available to this group. This increase contributes to the negative $\Delta H^{\circ\prime}$ of hydrolysis of ATP. At physiological pH there are also favorable contributions to $\Delta G^{\circ\prime}$ from the release of a proton and from the disordering of water around the polyphosphate chain.

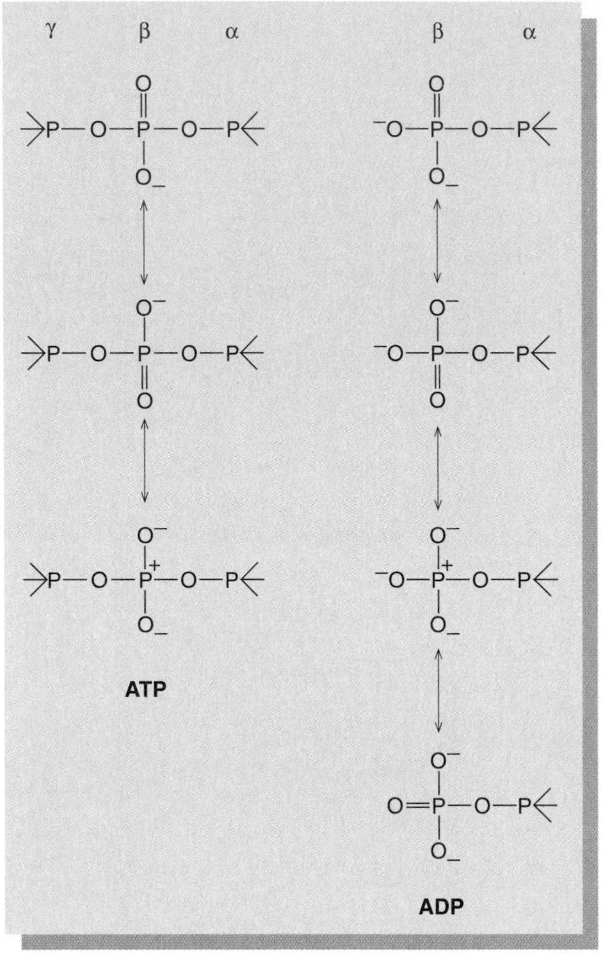

## Figure 2.9

Standard free energies of hydrolysis for some common phosphorylated compounds. The $\Delta G^{\circ\prime}$ value refers to hydrolysis of the phosphate group indicated by the symbol $\textcircled{P}$. Note that ATP occupies an intermediate position among these compounds. Given equal concentrations of the reactants and products, ADP can accept a phosphate group from any of the compounds above it, and ATP can donate a phosphate to the unphosphorylated forms of the compounds below it.

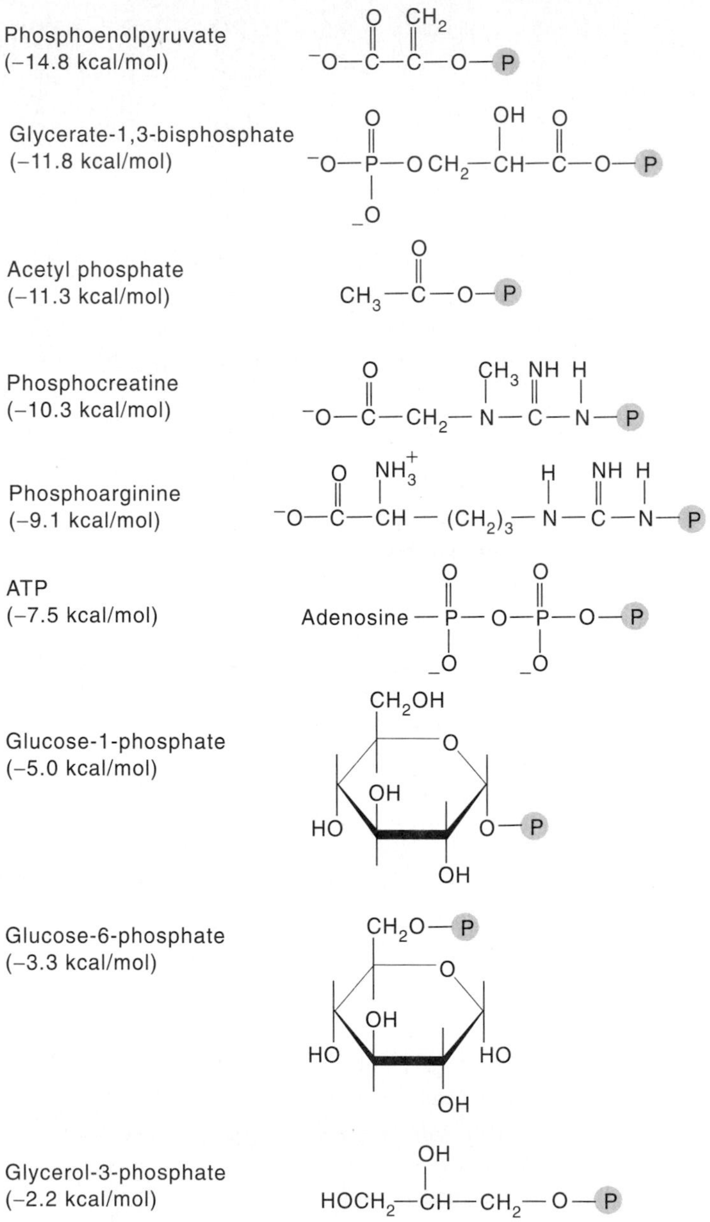

phorylated to ATP by phosphocreatine, glycerate-1,3-bisphosphate, or phosphoenolpyruvate, and ATP can be used to convert glucose to glucose-6-phosphate, or glycerol to glycerol phosphate. The reverse processes occur only if the ratio of reactants to products is sufficiently high, but this is not necessarily an unusual circumstance. In cells of some tissues, the reaction

$$\text{Creatine} + \text{ATP} \longrightarrow \text{creatine phosphate} + \text{ADP} + \text{H}^+ \quad (19)$$

occurs readily in either direction in response to changes in the concentrations of the reactants and products. The same is true of the reaction

$$\text{Glycerate-3-phosphate} + \text{ATP} \longrightarrow \text{glycerate-1,3-bisphosphate} + \text{ADP} + \text{H}^+ \quad (20)$$

Most of the phosphorylated materials shown in figure 2.9 participate in only a few biochemical reactions, and many of them are formed only in certain types of cells or organisms. ATP, in contrast, is formed by all living things, and it participates in literally hundreds of different enzymatic reactions. What is it about ATP that has made this molecule such a universal currency of free energy in biology? Several considerations are relevant here. First, $\Delta G^{\circ\prime}$ for hydrolysis of ATP to ADP is large enough so that this reaction releases a substantial amount of free energy, enough to drive many of the reactions that are important for biosynthetic pathways. At the same time, the $\Delta G^{\circ\prime}$ is small enough so that ATP itself can be synthesized readily at the expense of available nutrients. We touched on this point when discussing the relative merits of hydrolysis at the $\alpha$-$\beta$ or $\beta$-$\gamma$ positions in ATP.

The second consideration returns us to the distinction between spontaneity and speed. Although the hydrolysis of ATP has a large negative $\Delta G^{\circ\prime}$, it also is critical that ATP is a relatively stable compound in aqueous solution. It does not hydrolyze rapidly under physiological conditions of pH and temperature. The hydrolysis, though far downhill thermodynamically, is slowed by a substantial activation barrier. But the barrier must be of such a nature that it can easily be overcome enzymatically. This feature allows the free energy of hydrolysis to be channeled quickly and selectively into reactions in which it is needed but to be conserved when energy is not in demand.

Third, the products of the hydrolysis of ATP provide opportunities for coupling to a wide variety of chemical reactions. We saw how the phosphate group is incorporated into glucose-6-phosphate, and later chapters provide many illustrations of similar enzymatic processes. We also will see reactions in which the pyrophosphate group or the adenylyl group (AMP) is incorporated into the products. Such a broad array of reactions could not be driven by a molecule that decomposed to release an inert material such as $N_2$.

Finally, the adenine and ribosyl groups of ATP, ADP, and AMP provide additional structural features that allow these molecules to bind to enzymes and thus to participate in regulating enzymatic activities. This may be part of the reason that no known organisms base their energy-transfer reactions entirely on inorganic pyrophosphate or other polyphosphate compounds without a nucleoside moiety.

## Summary

In this chapter we discussed some principles of thermodynamics as they relate to biochemical reactions. The following points are of greatest importance.

1. Thermodynamics is useful in biochemistry for predicting whether a given reaction can occur and, if so, how much work a cell can obtain from the process.

2. The thermodynamic quantities energy, enthalpy, entropy, and free energy are properties of the state of a system. Changes in these quantities depend only on the difference between the initial and final states, not on the mechanism whereby the system goes from one state to the other.

3. Energy is the capacity to do work.

4. The first law of thermodynamics says that energy cannot be created or destroyed in a chemical reaction. If the energy of a system increases, the surroundings must lose an equivalent amount of energy, either by the transfer of heat or by the performance of work.

5. The energy of a molecule includes translational, rotational, and vibrational energy, as well as electronic and nuclear energy. Electronic terms usually account for most of the change in energy $\Delta E$ in a chemical reaction.

6. The change in enthalpy $\Delta H$ is given by the expression $\Delta H = \Delta E + \Delta(PV)$. For most biochemical reactions, $\Delta E$ and $\Delta H$ are nearly equal. The organic molecules found in cells generally have much higher enthalpies than the simpler molecules from which they are built.

7. In most reactions that proceed spontaneously, the enthalpy of the system decreases. If no work is done, the system gives off heat to the surroundings. But in some spontaneous reactions, heat is absorbed in the absence of work and the enthalpy of the system increases. Such reactions invariably show an increase in the entropy of the system.

8. Entropy is a measure of the order in a system: Systems that are highly ordered have low entropies. The entropy of a molecule depends mainly on translational and rotational freedom. Biological macromolecules generally have much lower entropies than their building blocks.

9. The second law of thermodynamics states that an overall increase must take place in the entropy of the system and its surroundings in any process that occurs spontaneously. An isolated system proceeds spontaneously to states of increasingly greater entropy (greater disorder).

10. The change in the free energy of a system is defined as $\Delta G = \Delta H - T \Delta S$, where $T$ is the absolute temperature. A reaction at constant pressure and temperature can occur spontaneously if, and only if, $\Delta G$ is negative. The maximal amount of useful work that can be obtained from a reaction is equal to $-\Delta G$.

11. For a reaction in solution, $\Delta G$ depends on the standard free energy change ($\Delta G°$) and on the concentrations of the reactants and products. The standard free energy change is related to the equilibrium constant by the expression $\Delta G° = -RT \ln K_{eq}$. Increasing the concentration of the reactants relative to the concentration of the products makes $\Delta G$ more negative.

12. Reactions that are thermodynamically unfavorable can be coupled to favorable reactions. The coupling of the reactions may be direct or sequential, as in a biochemical pathway.

13. ATP is the main coupling agent for free energy in living cells. The free energy provided by the hydrolysis of ATP is used to drive many reactions that would not occur spontaneously by themselves.

14. Several features make ATP particularly well suited for its role. First, hydrolysis of ATP to ADP and $P_i$ or to AMP and $PP_i$ releases a considerable amount of free energy. Second, ATP does not hydrolyze rapidly by itself, but it can be hydrolyzed readily in enzymatically catalyzed reactions. This difference allows the free energy of hydrolysis to be channeled into reactions in which it is needed but to be conserved when energy is not in demand. Third, the products of the hydrolysis of ATP provide opportunities for coupling to a wide variety of chemical reactions. Finally, the adenine and ribosyl groups of ATP, ADP, and AMP provide additional structural features that allow these molecules to bind to a large number of enzymes and thus to participate in regulating enzymatic activities.

## Selected Readings

Alberty, R. A., and F. Daniels, *Physical Chemistry,* 5th ed. New York: Wiley, 1975.

Cantor, C. R., and P. R. Schimmel, *Biophysical Chemistry.* San Francisco: Freeman, 1980.

Ingraham, L. L., and A. B. Pardee, Free Energy and Entropy in Metabolism. In D. M. Greenberg (ed.), *Metabolic Pathways,* vol. 1. New York: Academic Press, 1967.

Tinoco, I., Jr., K. Sauer, and J. C. Wang, *Physical Chemistry, Principles and Applications in Biological Sciences,* 2d ed. Englewood Cliffs, N.J.: Prentice-Hall, 1985.

Van Holde, K. E., *Physical Biochemistry,* 2d ed. Englewood Cliffs, N.J.: Prentice-Hall, 1985.

## Problems

1. In some respects, thermodynamic considerations are more important to a biochemist than to a chemist. Why is this so?

2. Name three extensive and three intensive properties that relate to thermodynamic quantities. What is the basic difference between the two types of properties?

3. What is meant by a state function? Why is enthalpy a state function?

4. Why can we equate internal energy and enthalpy for most biochemical reactions?

5. In thermodynamics it is important to distinguish between the total system and the system being studied. Why?

6. Why do enthalpies and entropies of solvation tend to negate one another?

7. Why does the entropy of a weak acid decrease on ionization?

8. Transfer of a hydrophobic molecule (e.g., a hydrophobic amino acid and side chain) from an aqueous to a nonaqueous environment is entropically favorable. Explain.

9. As we will see in chapter 13, oxaloacetate is formed by the oxidation of malate.

$$\text{L-Malate} + NAD^+ \longrightarrow$$
$$\text{Oxaloacetate} + NADH + H^+$$

The reaction has a $\Delta G^{\circ\prime}$ of +7.0 kcal/mole. Suggest reasons that the reaction proceeds in the direction of oxaloacetate production in the cell.

10. Cite three factors that make ATP ideally suited to transfer energy in biochemical systems.

11. You wish to measure the $\Delta G^{\circ\prime}$ for hydrolysis of ATP,

$$ATP + H_2O \longrightarrow ADP + P_i + H^+$$

but the equilibrium for the hydrolysis lies so far toward products that analysis of the ATP concentration at equilibrium is neither practical nor accurate. However, you have the following data that allow calculation of the value indirectly.

$$\text{Creatine phosphate} + ADP + H^+ \longrightarrow$$
$$ATP + \text{Creatine} \quad K_{eq}' = 59.5 \quad \textbf{(P1)}$$

$$\text{Creatine} + P_i \longrightarrow$$
$$\text{Creatine phosphate} + H_2O$$
$$\Delta G^{\circ\prime} = +10.5 \text{ kcal/mole} \quad \textbf{(P2)}$$

Assume that $2.3RT = 1.36$ kcal/mole

(a) Calculate the value of $\Delta G^{\circ\prime}$ for reaction (P1).

(b) Calculate the $\Delta G^{\circ\prime}$ for the hydrolysis of ATP.

12. The hydrolysis of lactose (D-galactosyl-$\beta$-(1,4) D-glucose) to D-galactose and D-glucose occurs with $\Delta G^{\circ\prime}$ of −4.0 kcal/mole.

(a) Calculate $K_{eq}'$ for the hydrolytic reaction.

(b) What are the $\Delta G^{\circ\prime}$ and $K_{eq}'$ for the synthesis of lactose from D-galactose and D-glucose?

(c) Lactose is synthesized in the cell from UDP-galactose plus D-glucose and is catalyzed by lactose synthase. Given that $\Delta G^{\circ\prime}$ of hydrolysis of UDP-galactose is −7.3 kcal/mole, calculate $\Delta G^{\circ\prime}$ and $K_{eq}'$ for the reaction

$$\text{UDP-galactose} + \text{D-glucose} \longrightarrow \text{Lactose} + UDP$$

13. For each of the following reactions, calculate $\Delta G^{\circ\prime}$ and indicate whether the reaction is thermodynamically favorable as written.

(a) Glycerate-1,3-bisphosphate + Creatine $\longrightarrow$
     Phosphocreatine + 3-Phosphoglycerate

(b) Glucose-6-phosphate $\longrightarrow$ Glucose-1-phosphate

(c) Phosphoenolpyruvate + ADP $\longrightarrow$ Pyruvate + ATP

14. Assume that an individual needs 2,500 Calories per day (1 Calorie = 1,000 calories or 1 kcal) to meet energy requirements. For simplicity, consider that the energy needs of this individual are met with glucose (not that unrealistic, considering that most of the world meets these needs mainly with starches). Glucose, and carbohydrates in general contain about 4 Calories per gram. The ATP yield during catabolism is 30 moles of ATP per mole of glucose. What mass (in pounds) of $K_4ATP$ are synthesized per day by this individual?

15. Proteins that serve as gene repressors frequently have two identical binding sites that interact with complementary sites on the DNA. If a repressor protein is cut in half without damaging either of its DNA binding sites, how is the binding to DNA affected? How is the binding affected if one of the two binding sites on the DNA is eliminated by changing the nucleotide sequence? Discuss the enthalpy, entropy, and free energy effects.

# Protein Structure and Function

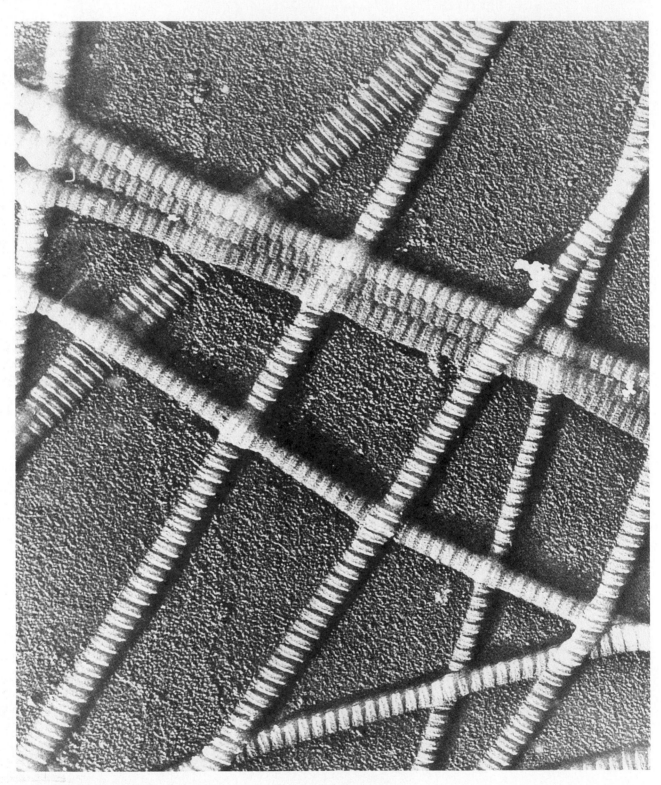

An electron micrograph of collagen fibrils from skin. The banded appearance of the fibrils arises from a staggered arrangement of collagen molecules. Each collagen molecule in turn is composed of three polypeptide chains that interact by hydrogen bonds. (Courtesy of Jerome Gross, Massachusetts General Hospital.)

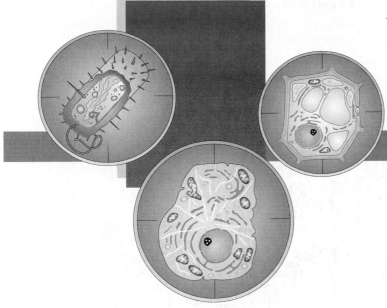

# The Building Blocks of Proteins: Amino Acids, Peptides, and Polypeptides

*Amino acids have common features that permit them to be linked together into polypeptide chains and uncommon features that give each polypeptide chain its unique character.*

In the middle of the nineteenth century, the Dutch chemist Gerardus Mulder extracted a substance common to animal tissues and the juices of plants, which he believed to be "without doubt the most important of all substances of the organic kingdom, and without it life on our planet would probably not exist." At the suggestion of the famous Swedish chemist Jons Jakob Berzelius, Mulder named this substance protein (from the Greek *proteios,* meaning "of first importance") and assigned to it a specific chemical formula ($C_{40}H_{62}N_{10}O_{12}$). Although he was wrong about the chemistry of proteins, he was right about their being indispensable to living organisms. The term "protein" endures.

Proteins are the most abundant of cellular components. They include enzymes, antibodies, hormones, transport molecules, and even components for the cytoskeleton of the cell itself. Proteins are also informational macromolecules, the ultimate heirs of the genetic information encoded in the sequence of nucleotide bases within the chromosomes. Structurally and functionally, they are the most diverse and dynamic of molecules and play key roles in nearly every biological process. Proteins are complex macromolecules with exquisite specificity; each is a specialized player in the orchestrated activity of the cell. Together they tear down and build up molecules, extract energy, repel invaders, act as delivery systems, and even synthesize the genetic apparatus itself.

In the first three chapters of part 2 we discuss the basic structural and chemical properties of proteins. In this chapter we concentrate on the structural and chemical properties of amino acids, peptides, and polypeptides—the building blocks of proteins. From our presentation you will learn the following:

1. Certain acidic and basic properties are common to all amino acids found in proteins except for the amino acid proline.

2. Side chains give amino acids their individuality. These side chains serve a variety of structural and functional roles.

3. The $\alpha$-carboxyl group of one amino acid can react with the $\alpha$-amino group of another amino acid to form a dipeptide.

4. Many amino acids, reacting in a similar way, can become linked to form a linear polypeptide chain.

5. The amino acid sequence in a polypeptide can be determined by a process of partial breakdown into manageable fragments, followed by stepwise analysis proceeding from one end of the chain to the other.

6. Polypeptide chains with a prespecified sequence can be synthesized by well-established chemical methods.

## Amino Acids

Every protein molecule can be viewed as a polymer of amino acids. There are 20 common amino acids. Figure 3.1a shows the structure of a single amino acid. At the center is a tetrahedral carbon atom called the alpha ($\alpha$) carbon ($C_\alpha$). It is covalently bonded on one side to an amino group ($NH_2$) and on the other side to a carboxyl group (COOH). A third bond is always to a hydrogen, and the fourth bond is to a variable side chain (R). In neutral solution (pH 7), the carboxyl group loses a proton and the amino group gains one. Thus an amino acid in solution, while neutral overall, is a doubly charged species called a zwitterion (fig. 3.1b).

The structures of the 20 amino acids commonly found in proteins are listed in table 3.1. All of these amino acids except proline have a protonated $\alpha$-amino group ($-NH_3^+$) attached to the $\alpha$ carbon. In proline one of the N—H linkages is replaced by an N—C linkage forming part of a cyclic

### Figure 3.1

Amino acid anatomy. (a) Uncharged amino acid. (b) Doubly charged zwitterion. (Illustration copyright by Irving Geis. Reprinted by permission.)

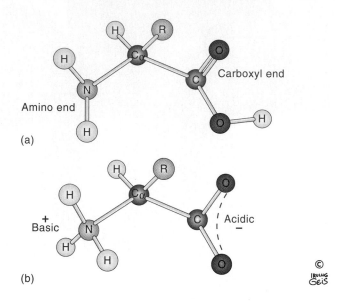

structure. Various ways of classifying amino acids according to their R groups have been proposed. In table 3.1 we divide the amino acids into three categories. The first category contains eight amino acids with relatively apolar R groups; the second category contains seven amino acids with uncharged polar R groups; and the third category contains five amino acids with R groups that normally exist in the charged state.

Amino acids are often abbreviated by three-letter symbols; when this proves to be too cumbersome (as in certain kinds of charts and figures), one-letter symbols are used. Both designations are given in table 3.1, together with the molecular weight ($M_r$) of each amino acid.

In addition to the 20 commonly occurring $\alpha$-amino acids, a variety of other amino acids are found in minor amounts in proteins and in nonprotein compounds. The unusual amino acids found in proteins result from modification of the common amino acids. In a few cases these amino acids are incorporated directly into the polypeptide chains during synthesis. Most frequently the amino acid is modified after incorporation. The unusual amino acids found in nonprotein compounds are extremely varied in type and are formed by a number of different metabolic pathways (see chapter 21).

## Table 3.1

Structure of the Twenty Amino Acids Found in Proteins

| Group I. Amino Acids with Apolar R Groups | | | Group II. Amino Acids with Uncharged Polar R Groups | | | Group III. Amino Acids with Charged R Groups | | |
|---|---|---|---|---|---|---|---|---|
| | *R Groups* | *Common Group* | | *R Groups* | *Common Group* | | *R Groups* | *Common Group* |
| Alanine Ala A $M_r$ 89[a] | $CH_3$ | H, C, COO⁻, NH₃⁺ | Glycine Gly G $M_r$ 75 | H | H, C, COO⁻, NH₃⁺ | Aspartic acid Asp D $M_r$ 133 | ⁻O—C(=O)— | CH₂, C, H, COO⁻, NH₃⁺ |
| Valine Val V $M_r$ 117 | $CH_3$, $CH_3$, CH | H, C, COO⁻, NH₃⁺ | Serine Ser S $M_r$ 105 | HO—CH₂ | H, C, COO⁻, NH₃⁺ | Glutamic acid Glu E $M_r$ 147 | ⁻O—C(=O)— | CH₂—CH₂, C, H, COO⁻, NH₃⁺ |
| Leucine Leu L $M_r$ 131 | CH₃, CH₃, CH—CH₂ | H, C, COO⁻, NH₃⁺ | Threonine Thr T $M_r$ 119 | CH₃—CH—, OH | H, C, COO⁻, NH₃⁺ | Lysine Lys K $M_r$ 146 | H₃N⁺—(CH₂)₄ | H, C, COO⁻, NH₃⁺ |
| Isoleucine Ile I $M_r$ 131 | CH₃—CH₂—CH—, CH₃ | H, C, COO⁻, NH₃⁺ | Cysteine Cys C $M_r$ 121 | HS—CH₂ | H, C, COO⁻, NH₃⁺ | Arginine Arg R $M_r$ 174 | H₂N—C(—NH₂⁺)—NH—(CH₂)₃ | H, C, COO⁻, NH₃⁺ |
| Proline Pro P $M_r$ 115 | H₂C, H₂C, H₂C—N⁺(H)(H), C—COO⁻, H | | Tyrosine Tyr Y $M_r$ 181 | HO—⟨C₆H₄⟩—CH₂ | H, C, COO⁻, NH₃⁺ | Histidine (at pH 6.0) His H $M_r$ 155 | HC=C—CH₂, HN⁺, NH, CH | H, C, COO⁻, NH₃⁺ |
| Phenylalanine Phe F $M_r$ 165 | ⟨C₆H₅⟩—CH₂ | H, C, COO⁻, NH₃⁺ | Asparagine Asn N $M_r$ 132 | NH₂—C(=O)—CH₂ | H, C, COO⁻, NH₃⁺ | | | |
| Tryptophan Trp W $M_r$ 204 | ⟨indole⟩—C—CH₂, CH, N—H | H, C, COO⁻, NH₃⁺ | Glutamine Gln Q $M_r$ 146 | NH₂—C(=O)—CH₂—CH₂ | H, C, COO⁻, NH₃⁺ | | | |
| Methionine Met M $M_r$ 149 | CH₃—S—CH₂—CH₂ | H, C, COO⁻, NH₃⁺ | | | | | | |

[a] Molecular weights ($M_r$) in this text are expressed in units of grams per mole.

## *Amino Acids Have Both Acid and Base Properties*

The charge properties of amino acids are very important in determining the reactivity of certain amino acid side chains and in the properties they confer on proteins. The charge properties of amino acids in aqueous solution may best be considered under the general treatment of acid–base ionization theory. We find this treatment useful at other points in the text as well.

Recall that water can be considered a weak acid (or a weak base) because it dissociates into a proton and a hydroxide ion, according to the equilibrium

$$H_2O \rightleftharpoons H^+ + OH^- \tag{1}$$

The equilibrium expression for this reaction is

$$K_{eq} = \frac{[H^+][OH^-]}{[H_2O]} \tag{2}$$

Because water dissociates to such a small extent, the concentration of undissociated water is high and does not vary significantly for chemical reactions in aqueous solution. Therefore, the denominator in this equation is effectively constant, with a value of 55.5. The constant $K_w$ for the dissociation of water is redefined by the expression

$$K_w = [H^+][OH^-] = 10^{-14} \text{ (mole/l)}^2 \tag{3}$$

at 25°C.

In pure water we expect equal amounts of $H^+$ ("hydrogen ion") and $OH^-$ ("hydroxide ion"). From equation (3) we can calculate the concentration of $H^+$ or $OH^-$ in pure water to be $10^{-7}$ M. Therefore, a solution with an $H^+$ concentration of $10^{-7}$ M is defined as neutral. An $H^+$ concentration greater than $10^{-7}$ M indicates an acidic solution; an $H^+$ concentration less than $10^{-7}$ M indicates a basic solution. Rather than deal with exponentials, it is convenient to express the $H^+$ concentration on a pH scale, the term "pH" being defined by the equation

$$pH = \log(1/[H^+]) = -\log[H^+] \tag{4}$$

According to this definition a neutral solution has a pH of 7. Other values of pH and corresponding $H^+$ and $OH^-$ concentrations are given in table 3.2.

The most common equilibria that biochemists encounter are those of acids and bases. The dissociation of an acid may be written as

$$HA \rightleftharpoons H^+ + A^- \tag{5}$$

The equilibrium constant for this reaction is called the acid dissociation constant $K_a$ written as

$$K_a = \frac{[H^+][A^-]}{[HA]} \tag{6}$$

### Table 3.2

The pH Scale

| pH | [H$^+$] | [OH$^-$] |
|----|---------|----------|
| 0  | $10^0$    | $10^{-14}$ |
| 1  | $10^{-1}$ | $10^{-13}$ |
| 2  | $10^{-2}$ | $10^{-12}$ |
| 3  | $10^{-3}$ | $10^{-11}$ |
| 4  | $10^{-4}$ | $10^{-10}$ |
| 5  | $10^{-5}$ | $10^{-9}$  |
| 6  | $10^{-6}$ | $10^{-8}$  |
| 7  | $10^{-7}$ | $10^{-7}$  |
| 8  | $10^{-8}$ | $10^{-6}$  |
| 9  | $10^{-9}$ | $10^{-5}$  |
| 10 | $10^{-10}$ | $10^{-4}$  |
| 11 | $10^{-11}$ | $10^{-3}$  |
| 12 | $10^{-12}$ | $10^{-2}$  |
| 13 | $10^{-13}$ | $10^{-1}$  |
| 14 | $10^{-14}$ | $10^0$     |

Strong acids in aqueous solution dissociate completely into anions and protons. The concentration of hydrogen ion $[H^+]$ is therefore equal to the total concentration $C_{HA}$ of the acid HA that is added to the solution. Thus the pH of the solution of a strong acid is simply $-\log C_{HA}$.

The pH of the solution of a weak acid is a function of both the $C_{HA}$ and the acid dissociation constant. The dissociation constant of a weak acid may be written in terms of the species present in the equation for the acid dissociation constant.

First solving equation (6) for $[H^+]$ gives

$$[H^+] = \frac{K_a[HA]}{[A^-]} \tag{7}$$

Taking the logarithm of both sides and changing signs gives us

$$-\log[H^+] = -\log K_a + \log\frac{[A^-]}{[HA]} \tag{8}$$

Substituting pH for $-\log[H^+]$ and $pK_a$ for $-\log K_a$ in equation (8), we obtain the Henderson-Hasselbach equation:

$$pH = pK_a + \log\left[\frac{A^-}{HA}\right] = pK_a + \log\left[\frac{base}{acid}\right] \tag{9}$$

**Figure 3.2**

The dependence of pH on the equivalents of base added to a typical weak acid. Note that at the $pK_a$, $[A^-] = [HA]$.

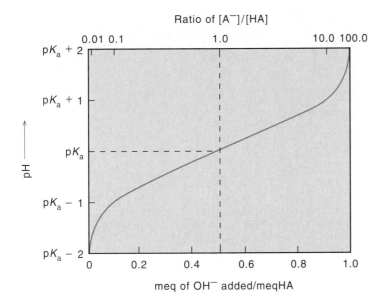

**Figure 3.3**

Titration curve of alanine. The predominant ionic species at each cardinal point in the titration is indicated.

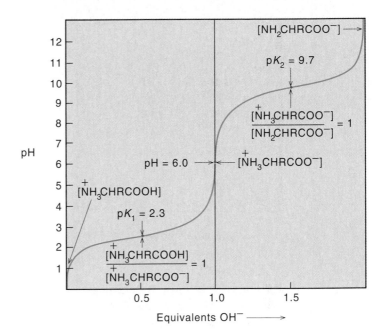

The Henderson-Hasselbach equation is useful for calculating the molar ratio of base (proton acceptor) to acid (proton donor) for a given pH and p$K$ or for calculating the p$K$, given the ratio of base (proton acceptor) to acid (proton donor). It can be seen that when the concentration of anion or base is equal to the concentration of undissociated acid (i.e., when the acid is half neutralized), the pH of the solution is equal to the p$K$ of the acid.

The values of p$K$ for a particular molecule are determined by titration. A typical pH dependence curve for the titration of a weak acid by a strong base is shown in figure 3.2. The concentration of the anion equals the concentration of the acid when the acid is exactly half neutralized. Note that at this point on the curve, the pH is least sensitive to the quantity of added base (or acid). Under these conditions, the solution is said to be buffered. Biochemical reactions are typically highly dependent on the pH of the solution. Therefore, it is frequently advantageous to study reactions in buffered solutions. The ideal buffer is one that has a p$K$ numerically equivalent to the working pH.

A simple amino acid with a nonionizable R group gives a complex titration curve with two inflection points. For an example, see the titration of alanine, shown in figure 3.3. At very low pH, alanine carries a single positive charge on the $\alpha$-amino group. The first inflection point occurs at a pH of 2.3. This is the p$K$ for titration of the carboxyl group, $pK_1$ ($-COOH \longrightarrow -COO^-$). At a pH of 6.0, alanine has an equal amount of positive and negative charge. This value is referred to as the isoelectric point (pI), or the isoe-

lectric pH. As the titration continues, a second inflection point is reached at a pH of 9.7. The p$K$ at this point, $pK_2$, the $[-NH_3^+]$ and $[-NH_2]$ are equal.

Amino acids with an ionizable R group show even more complex titration curves, indicative of three ionizable groups (fig. 3.4). The p$K$ for the ionizable side chain, $pK_R$, is usually readily distinguishable from the p$K$ values for the ionizable $\alpha$-carboxyl and $\alpha$-amino groups, $pK_1$ and $pK_2$, respectively, because the $\alpha$-amino groups have numerical values close to the comparable p$K$ values of alanine (see fig. 3.3 and table 3.3). Note that the only ionizable R group with a $pK_R$ in the vicinity of 7, where most biological systems function, is that for histidine. This means that although other ionizable groups are usually fully charged under biological conditions, the side chain of histidine can be fully charged, uncharged, or partially charged, depending on the precise situation. This variability has major implications for the way the histidine side chain functions in enzyme catalysis. The side chain can serve as either a proton donor or a proton acceptor (see discussion in chapter 8).

An additional point should be noted from table 3.3. Whereas the amino acid side chains (R groups) that are normally charged at physiological pH are restricted to five amino acids (aspartic acid, glutamic acid, lysine, arginine, and sometimes histidine), a number of potentially ionizable R groups are part of other amino acids. These include cysteine, serine, threonine, and tyrosine. The ionization reac-

## Figure 3.4

Titration curves of glutamic acid, lysine, and histidine. In each case, the pK of the R group is designated $pK_R$.

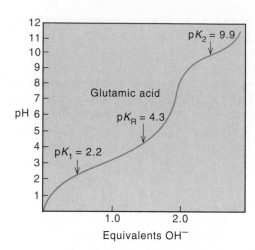

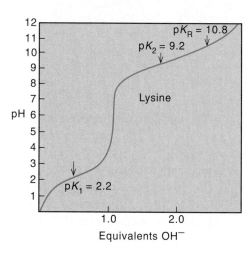

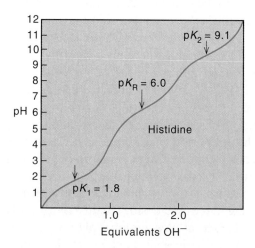

## Table 3.3

Values of pK for the Ionizable Groups of the Twenty Amino Acids Commonly Found in Proteins

| Amino Acid | $pK_1$ ($\alpha$—COOH) | $pK_2$ ($\alpha$—NH$_3^+$) | $pK_R$ (R Group) |
|---|---|---|---|
| Alanine | 2.35 | 9.87 | — |
| Arginine | 1.82 | 8.99 | 12.48 |
| Asparagine | 2.1 | 8.84 | — |
| Aspartic acid | 1.99 | 9.90 | 3.90 |
| Cysteine | 1.92 | 10.78 | 8.33 |
| Glutamic acid | 2.10 | 9.47 | 4.07 |
| Glutamine | 2.17 | 9.13 | — |
| Glycine | 2.35 | 9.78 | — |
| Histidine | 1.80 | 9.33 | 6.04 |
| Isoleucine | 2.32 | 9.76 | — |
| Leucine | 2.33 | 9.74 | — |
| Lysine | 2.16 | 9.18 | 10.79 |
| Methionine | 2.13 | 9.28 | — |
| Phenylalanine | 2.16 | 9.18 | — |
| Proline | 1.95 | 10.65 | — |
| Serine | 2.19 | 9.21 | ≈13 |
| Threonine | 2.09 | 9.10 | ≈13 |
| Tryptophan | 2.43 | 9.44 | — |
| Tyrosine | 2.20 | 9.11 | 10.13 |
| Valine | 2.29 | 9.74 | — |

tions for all of the potentially ionizable side chains are indicated in figure 3.5.

The acidic and basic groups within a protein can be titrated just like free amino acids to determine their number and their $pK_a$ values. A titration curve for $\beta$-lactoglobulin is shown in figure 3.6. This protein contains 94 potentially ionizable groups. The protein is positively charged at low pH and negatively charged at high pH. At intermediate pH values a point is found where the sum of the positive side-chain charges exactly equals the sum of the negative charges, so that the net charge on the protein is zero. This value, as we have noted, is the isoelectric point (pI) of the protein; for $\beta$-lactoglobulin the pI is about 5.2. The isoelectric point is not an invariant quantity. The binding of charged species present in the solution could raise or lower the pI, depending on their charge.

**Figure 3.5**

Equilibrium between charged and uncharged forms of amino acid side chains.

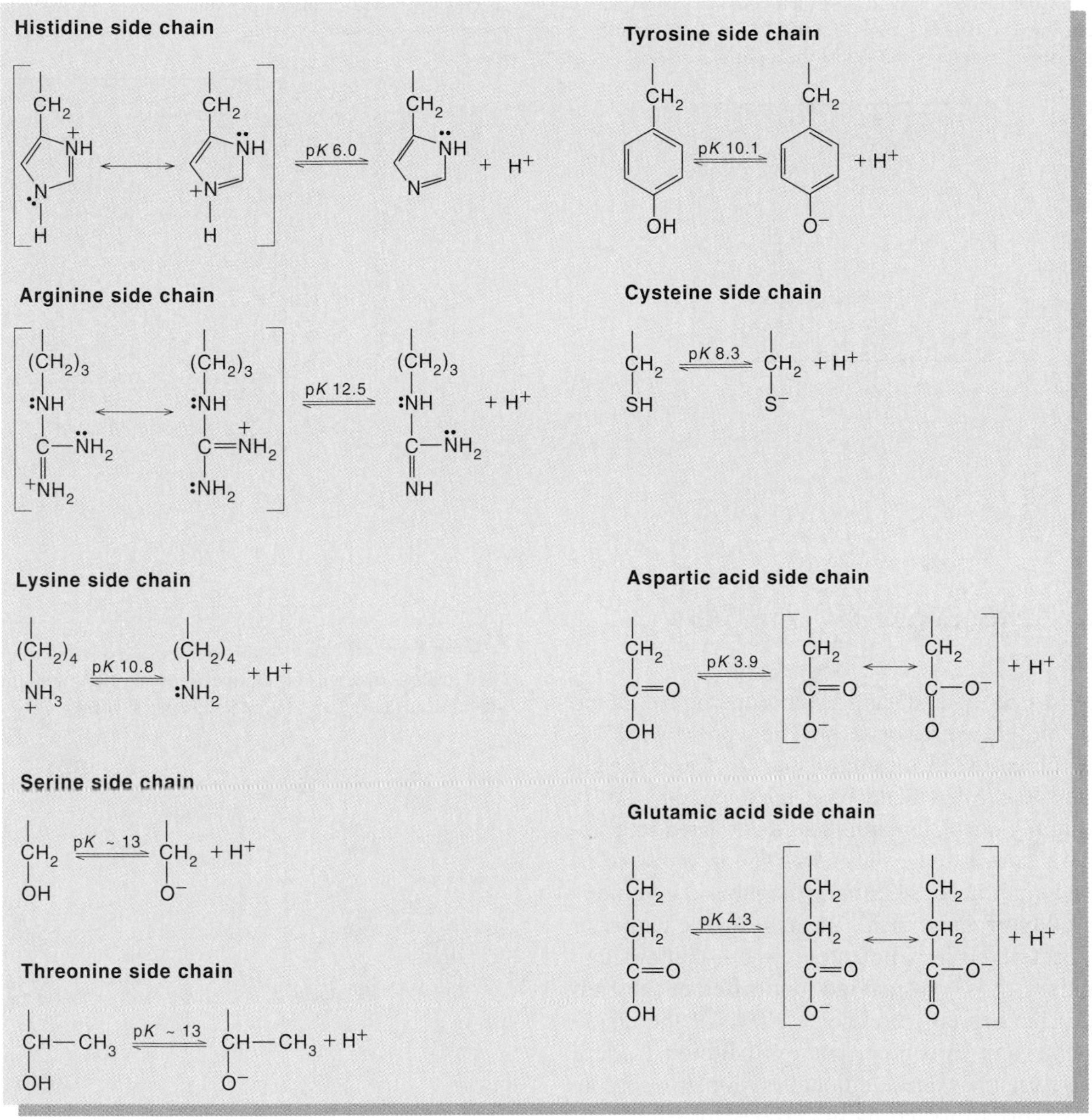

## Aromatic Amino Acids Absorb Light in the Near-Ultraviolet

The aromatic amino acids phenylalanine, tyrosine, and tryptophan all possess absorption maxima in the near-ultraviolet (fig. 3.7). These absorption bands arise from the interaction of radiation with electrons in the aromatic rings. The near-ultraviolet absorption properties of proteins are determined solely by their content of these three aromatic amino acids. In solution, UV absorption can be quantified with the help of a conventional spectrophotometer and used as a measure of the concentration of proteins (see Methods of Biochemical Analysis 3A).

**Figure 3.6**

Titration curve of β-lactoglobulin. At very low values of pH (<2) all ionizable groups are protonated. At a pH of about 7.2 (indicated by horizontal bar) 51 groups (mostly the glutamic and aspartic amino acids and some of the histidines) have lost their protons. At pH 12 most of the remaining ionizable groups (mostly lysine and arginine amino acids and some histidines) have lost their protons as well.

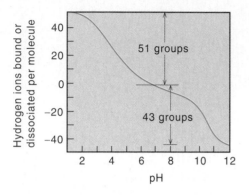

**Figure 3.7**

Ultraviolet absorption spectra of tryptophan (Trp), tyrosine (Tyr), and phenylalanine (Phe) at pH 6. The molar absorptivity is reflected in the extinction coefficient, with the concentration of the absorbing species expressed in moles per liter. (Source: From D. B. Wetlaufer, *Adv. Protein Chem.* 17:303–390, 1962.)

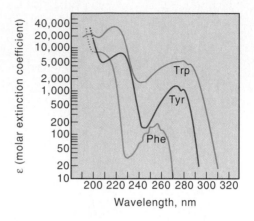

## All Amino Acids except Glycine Show Asymmetry

One of the most striking and significant properties of amino acids is their chirality, or handedness. The word "chiral" is related to the Greek word meaning hand. Just as the right hand is related to the left hand by a mirror image, so, in general, a naturally occurring amino acid is related to a stereoisomer by its mirror image. The observation is true of 19 out of the 20 amino acids; the one exception is glycine.

The chirality of amino acids stems from the chiral, or asymmetric, center, the α-carbon atom. The α-carbon atom is a chiral center if it is connected to four different substituents. Thus glycine has no chiral center. Two of the amino acids, isoleucine and threonine, possess additional chiral centers because each has one additional asymmetric carbon. You should be able to locate these carbons by simple inspection.

Two structures that constitute a stereoisomeric pair are referred to as enantiomers. The two enantiomers for alanine are illustrated in figure 3.8. These two isomers are called L-alanine and D-alanine, according to the way in which the substituents are arranged about the asymmetric carbon atom. The naming by D and L (for "dextrorotatory" and "levorotatory"; see chapter 6) refers to a convention established by Emil Fischer many years ago. According to this convention all amino acids found in proteins are of the L form. Some D-amino acids are found in bacterial cell walls and certain antibiotics.

**Figure 3.8**

The covalent structure of alanine, showing the three-dimensional structure of the Ⓛ and Ⓓ stereoisomeric forms.

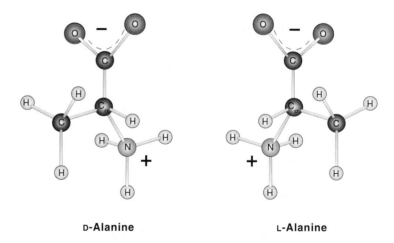

D-Alanine                    L-Alanine

## Peptides and Polypeptides

Amino acids can link together by a covalent peptide bond between the α-carboxyl end of one amino acid and the α-amino end of another. Formally, this bond is formed by the loss of a water molecule, as shown in figure 3.9. The peptide bond has partial double-bond character owing to resonance effects; as a result, the C—N peptide linkage and all of the atoms directly connected to C and N lie in a planar configuration called the amide plane. In the following chap-

## Figure 3.9

Formation of a dipepetide from two amino acids. (*a*) Two amino acids. (*b*) A peptide bond (CO—NH) links amino acids by joining the $\alpha$-carboxyl group of one with the $\alpha$-amino group of another. A water molecule is lost in the reaction. It is conventional to draw dipeptides and polypeptides so that their free amino terminus is to the left and their free carboxyl terminus is to the right. The amide plane refers to six atoms that lie in the same plane. (Illustration copyright by Irving Geis. Reprinted by permission.)

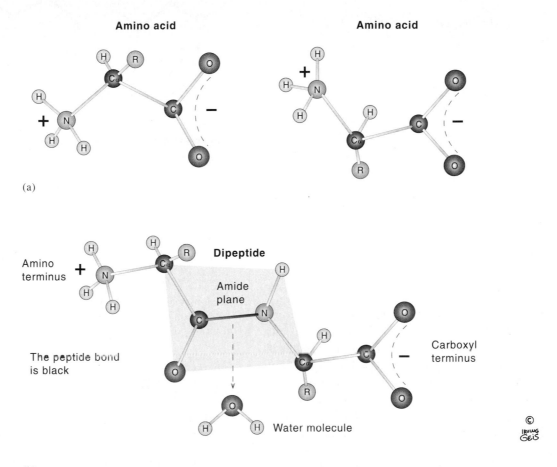

ter we see that this amide plane, by limiting the number of orientations available to the polypeptide chain, plays a major role in determining the three-dimensional structures of proteins.

Any number of amino acids can be joined by successive peptide linkages, forming a polypeptide chain. The polypeptide chain, like the dipeptide, has a directional sense. One end, called the N-terminal, or amino-terminal, end, has a free $\alpha$-amino group, whereas the other end, the C-terminal, or carboxyl-terminal, end, has a free $\alpha$-carboxyl group. The sequence of main-chain atoms from the N-terminal end to the C-terminal end is N—$C_\alpha$—C—N and so on, and in the opposite direction it is C—$C_\alpha$—N—C and so on. Short polypeptide chains, up to a length of about 20 amino acids, are called peptides or oligopeptides if they are fragments of whole polypeptide chains. A small protein mole-

cule may contain a polypeptide chain of only 50 amino acids; a large protein may contain chains of 3,000 amino acids or more. One of the larger single polypeptide chains is that of the muscle protein myosin, which consists of approximately 1,750 amino acid residues. Figure 3.10 shows a section of a polypeptide chain as a linear array with $\alpha$ carbons and planar amides alternating as repeating units of the main chain. Different side chains are attached to each $\alpha$ carbon.

In addition to the covalent peptide bonds formed between adjacent amino acids within a polypeptide chain, covalent disulfide bonds can be formed within the same polypeptide chain or between different polypeptide chains (fig. 3.11). Such disulfide linkages have an important stabilizing influence on the structures formed by many proteins (see chapter 4).

## Figure 3.10

A polypeptide chain, with the backbone shown in color and the amino acid side chains in outline. The polypeptide chain is oriented so that the C-terminal end (not shown) is to the left. (Illustration copyright by Irving Geis. Reprinted by permission.)

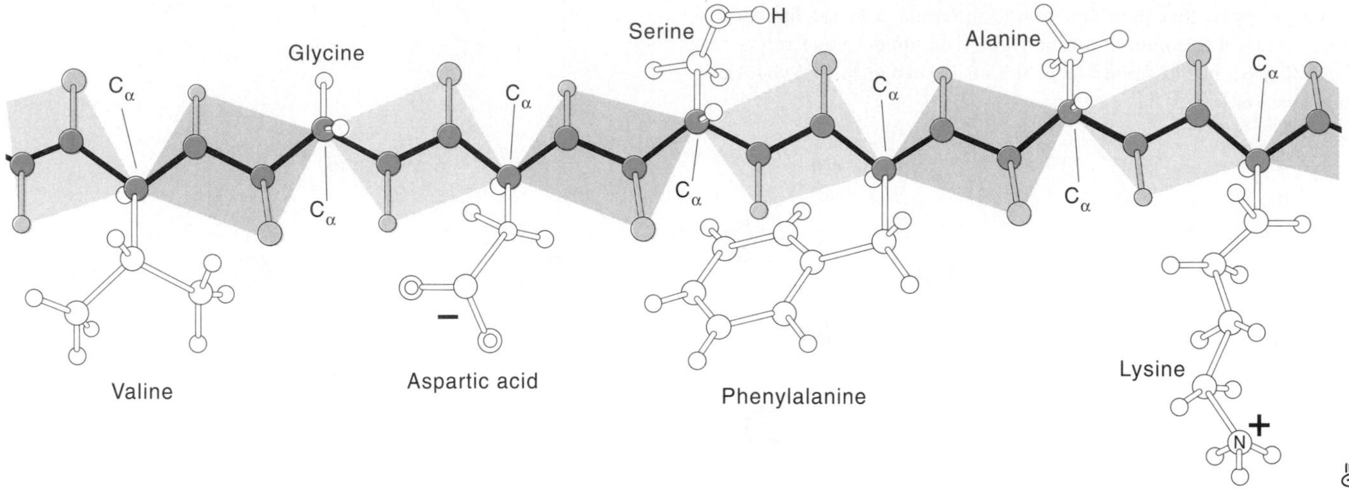

# Determination of Amino Acid Composition of Proteins

Each protein is uniquely characterized by its amino acid composition and sequence. A protein's amino acid composition is defined simply as the number of each type of amino acid composing the polypeptide chain. To discover a protein's amino acid composition it is necessary to (1) break down the polypeptide chain into its constituent amino acids, (2) separate the resulting free amino acids according to type, and (3) measure the quantities of each amino acid.

Cleavage of the peptide bonds is usually achieved by boiling the protein in 6-$N$ HCl; this treatment causes hydrolysis of the peptide bonds and the consequent release of free amino acids (fig. 3.12). Although acid hydrolysis is the most frequently used means of breaking a protein into its constituent amino acids, it results in the partial destruction of the indole ring of tryptophan. Consequently, the amount of tryptophan in the protein must be estimated by an alternative method (e.g., spectroscopic absorption) when using acid hydrolysis. In addition, acid hydrolysis results in the loss of ammonia from the side-chain amide groups of glutamine and asparagine, with the consequent production of

## Figure 3.11

Disulfide bonds can form between two cysteines. The cysteines can exist in the cytosol as free amino acids (as shown), in which case they give rise to cystine, or they can be on polypeptide chains. In the latter instance, they can be on the same polypeptide chains or different polypeptide chains. In either case the formation of covalent disulfide bonds stabilizes structural relationships.

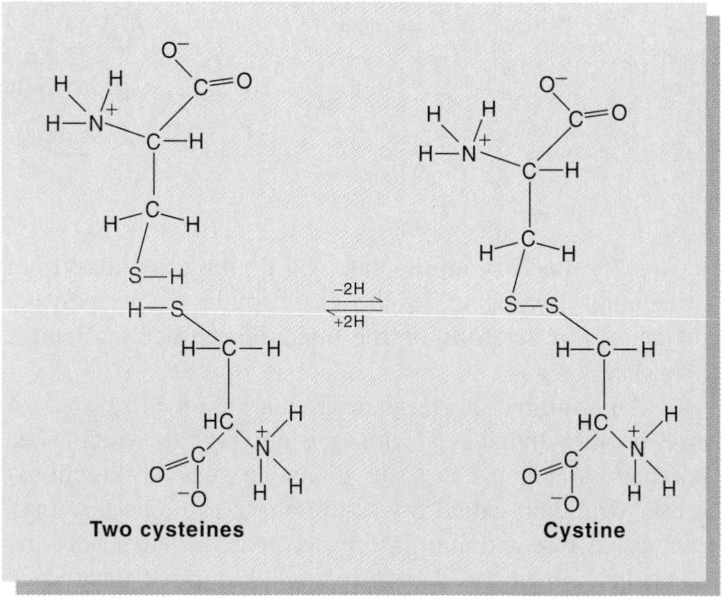

**Figure 3.12**

Acid hydrolysis of a protein or polypeptide to yield amino acids.

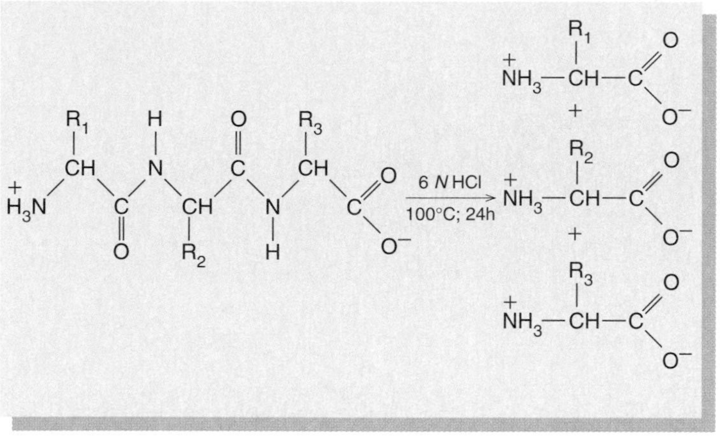

**Figure 3.13**

Migration of aspartic acid $\ominus$ and lysine $\oplus$ through a column with a higher affinity for aspartic acid. Views show the column at successively increasing time intervals after starting the elution.

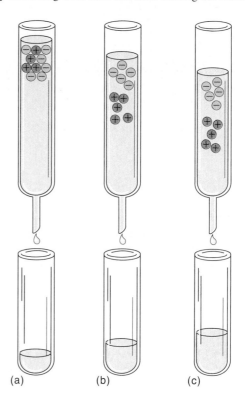

(a)  (b)  (c)

glutamic and aspartic acids. Therefore, estimates of amino acid composition based on acid hydrolysis show glutamine and glutamic acid combined and measured as glutamic acid. Similarly, asparagine and aspartic acid are combined and measured as aspartic acid.

Separation of amino acids for quantitative analytical purposes is usually achieved by ion-exchange chromatography. The general efficacy of chromatographic techniques is based on a difference in affinity between each compound to be separated and an immobile phase or resin. The resin consists of some relatively chemically inert polymer, which has weakly basic side-chain constituents that are positively charged at pH 7. If we were to add some of this resin to a solution containing free aspartic acid and lysine at pH 7, the negatively charged aspartic acid would have a higher affinity for the resin than would the positively charged lysine. If we then pump a solution of these two amino acids through a column containing such a positively charged resin, the progress of the aspartic acid through the column would be retarded relative to the lysine, owing to the greater affinity of the aspartic acid for the resin (fig. 3.13).

Ion-exchange resins exist that have differential binding affinities for all the naturally occurring amino acids. Such resins are effective in separating a solution of amino acids into its components. We must emphasize that the details of the forces responsible for the differential binding of amino acids to an ion-exchange resin are quite complicated and depend additionally on side-chain polarity, on subtle

differences in the p$K$ values of $\alpha$-amino and $\alpha$-carboxyl groups, on solvation effects, and on other factors. To enhance the separation properties of the column, such separation techniques frequently exploit changes in the pH of the solution buffer (eluting buffer) used to remove the compounds of interest. For example, a column might initially be run with the eluting buffer at a pH that results in some amino acids being so strongly bound to the resin that they are essentially immobile. However, after the separation and elution of the less strongly bound amino acids, the pH of the eluting buffer can be appropriately shifted to lessen the charge difference between the resin and the strongly bound amino acids. These amino acids can then be eluted and separated according to the newly established pattern of resin-binding affinities.

Quantitative determination of the separated amino acids is achieved by their reaction with ninhydrin to produce

## Figure 3.14

Reaction of ninhydrin with an amino acid yields a colored complex.
The ninhydrin reaction permits qualitative location of amino acids in
chromatography and quantitative assay of separated amino acids.

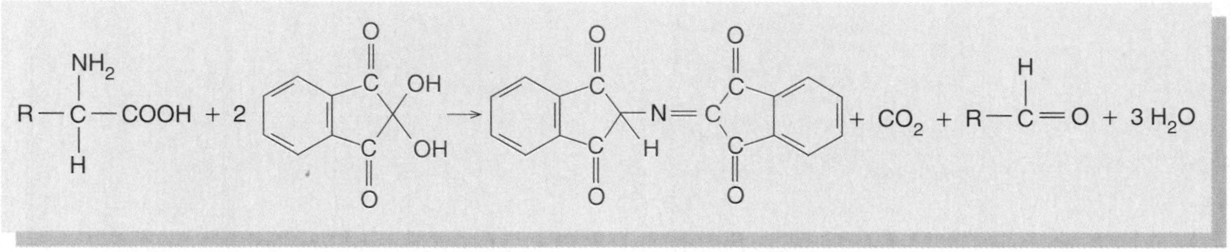

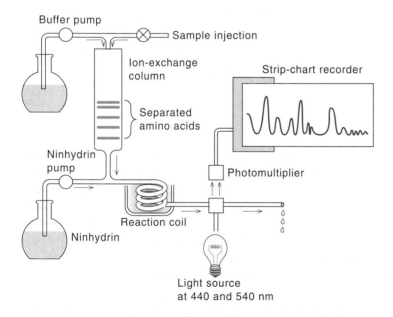

## Figure 3.15

Schematic diagram of an amino acid analyzer. The amino acids are
passed through an ion-exchange column and thereby separated. Eluted
fractions are mixed and reacted with ninhydrin. The intensity of the
resulting colored product is measured in a spectrophotometer, and the
results are displayed on a recording chart.

a colored reaction product. This product is measured spectrophotometrically. As shown in figure 3.14, the ninhydrin
reaction abstracts an amino group from each amino acid, so
that the amount of colored product formed is proportional to
the amount of amino acid initially present.

Currently, measurements of amino acid composition
are usually carried out on an amino acid analyzer, a device
that automates the previously described operations. As illus-

trated in figure 3.15, the amino acid analyzer consists of an
ion-exchange column through which the appropriate eluting
buffer is pumped after the amino acids are introduced at the
top of the column. As the separated amino acids emerge,
they are mixed with ninhydrin solution and passed through a
heated coil of tubing to allow the formation of the colored
ninhydrin reaction product. The separated ninhydrin reaction products then pass through a cell that measures their
optical absorbance at 540 and 440 nm and plots the results
on a strip-chart recorder. The absorbance is measured at two
wavelengths because proline, which is substituted at its
amino group, forms a different ninhydrin reaction product,
with an absorption maximum that is correspondingly different from that of the remaining amino acids.

Usually the amino acid analyzer is first standardized
by running through it a sample containing known quantities
of amino acids to account for any differences in their ninhydrin reaction properties. In this way it is possible to relate
directly the amount of amino acid present to the amount of
colored product formed, as measured by the area under the
''peak'' produced on the strip-chart recorder (see fig. 3.15).
Similarly, the amino acid hydrolysate of a protein of unknown composition can be run through the analyzer, and the
relative peak areas can be used to estimate the ratios of the
different amino acids present.

Conversion of the relative ratios of amino acids into an
estimate of actual composition requires some additional information concerning the protein's molecular weight; for
example, an analysis giving relative ratios of Ala (1.0), Gly
(0.5), and Lys (2.0) could correspond to composition $Ala_2$-
$Gly$-$Lys_4$ or any multiple thereof. The required information
is usually available, and in any case, an estimation of composition based on a minimum molecular weight of the protein is always possible.

# Determination of Amino Acid Sequence of Proteins

The most important properties of a protein are determined by the sequence of amino acids in the polypeptide chain. This sequence is called the primary structure of the protein. We know the sequences for thousands of peptides and proteins, largely through the use of methods developed in Fred Sanger's laboratory and first used to determine the sequence of the peptide hormone insulin in 1953. Knowledge of the amino acid sequence is extremely useful in a number of ways: (1) it permits comparisons between normal and mutant proteins (see chapter 5); (2) it permits comparisons between comparable proteins in different species and thereby has been instrumental in positioning different organisms on the evolutionary tree (see fig. 1.24); (3) finally and most important, it is a vital piece of information for determining the three-dimensional structure of the protein.

Determining the order of amino acids involves the sequential removal and identification of successive amino acid residues from one or the other free terminal of the polypeptide chain. However, in practice it is extremely difficult to get the required specific cleavage reaction of the desired products to proceed with 100% yield. This obstacle becomes significant when sequencing long polypeptides, because the fraction of the total material of minimum polypeptide chain length becomes constantly smaller as the successive removal of terminal residues continues. Conversely, the amino acid released from the polypeptide chain becomes increasingly contaminated with amino acids released from previously unreacted chains.

Because of this fundamental chemical limitation, the polypeptide chain must be broken down into sequences short enough for the chemistry to produce reliable results. The short sequences are then reassembled to obtain the overall sequence. The steps actually involved in protein sequencing (fig. 3.16) are

1. purification of the protein
2. cleavage of all disulfide bonds
3. determination of the terminal amino acid residues
4. specific cleavage of the polypeptide chain into small fragments in at least two different ways
5. independent separation and sequence determination of peptides produced by the different cleavage methods
6. reassembly of the individual peptides with appropriate overlaps to determine the overall sequence

The first step, protein purification, is discussed in chapter 5. Once the protein is pure, sequence analysis can begin, with cleavage of the disulfide bonds. Cleavage is achieved by oxidizing the disulfide linkages with performic acid (fig. 3.17). Sometimes this step results in the production of two or more polypeptide chains, in which case the individual chains must be separated.

The third step is to determine the polypeptide chain end groups. If the polypeptide chains are pure, then only one N-terminal and one C-terminal group should be detected. The amino-terminal amino acid can be identified by reaction with fluorodinitrobenzene (FDNB) (fig. 3.18). Subsequent acid hydrolysis releases a colored dinitrophenol (DNP)-labeled amino-terminal amino acid, which can be identified by its characteristic migration rate on thin-layer chromatography or paper electrophoresis. A more sensitive method of end-group determination involves the use of dansyl chloride (see Methods of Biochemical Analysis 3B).

Chemical methods for carboxyl end-group determination are considerably less satisfactory. Treatment of the peptide with anhydrous hydrazine at 100°C results in conversion of all the amino acid residues to amino acid hydrazides except for the carboxyl-terminal residue, which remains as the free amino acid and can be isolated and identified chromatographically. Alternatively, the polypeptide can be subjected to limited breakdown (proteolysis) with the enzyme carboxypeptidase. This results in release of the carboxyl-terminal amino acid as the major free amino acid reaction product. The amino acid type can then be identified chromatographically.

Step 4 involves breaking down the polypeptide chain into shorter, well-defined fragments for subsequent sequence analysis. Fragmentation can be achieved by the use of endopeptidases, which are enzymes that catalyze polypeptide chain cleavage at specific sites in the protein. Figure 3.19 shows the specificity of four endopeptidases commonly used for this purpose. Another specific chemical method for polypeptide chain cleavage involves reaction with cyanogen bromide. This reaction cleaves specifically at the methionine residues, with the accompanying conversion of free carboxyl-terminal methionine to homoserine lactone (fig. 3.20). Although this methionine reaction product differs from the 20 naturally occurring amino acids, it is nevertheless readily identified by subsequent conversion to homoserine.

Peptides resulting from cleavage of the intact protein are generally separated by column chromatography. The isolated peptides may then be analyzed (step 5) to determine both their amino acid composition and their sequence. Se-

# Figure 3.16

Steps involved in the sequence determination of the B chain of
insulin. Amino acids are represented here by their single-letter codes
(see table 3.1).

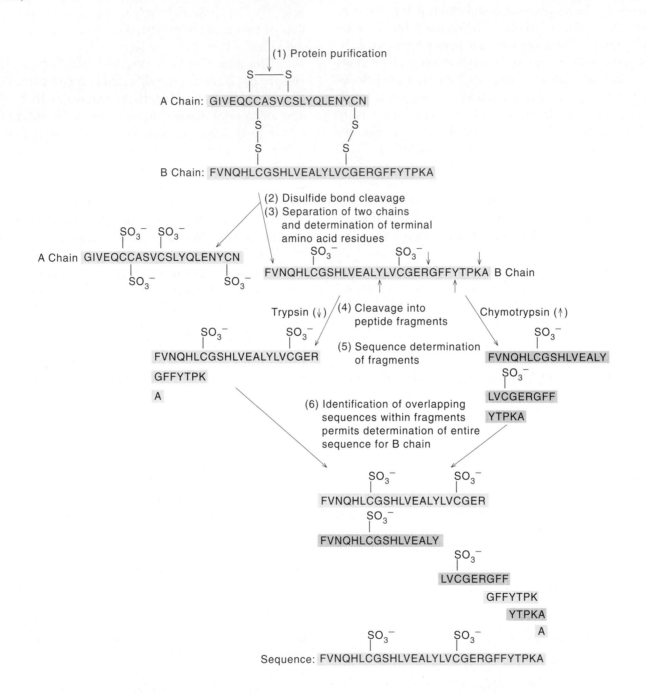

**Figure 3.17**

Disulfide cleavage reactions. Prior to sequence analysis inter- and intrachain disulfide linkages are irreversibly cleaved by one of the two procedures shown.

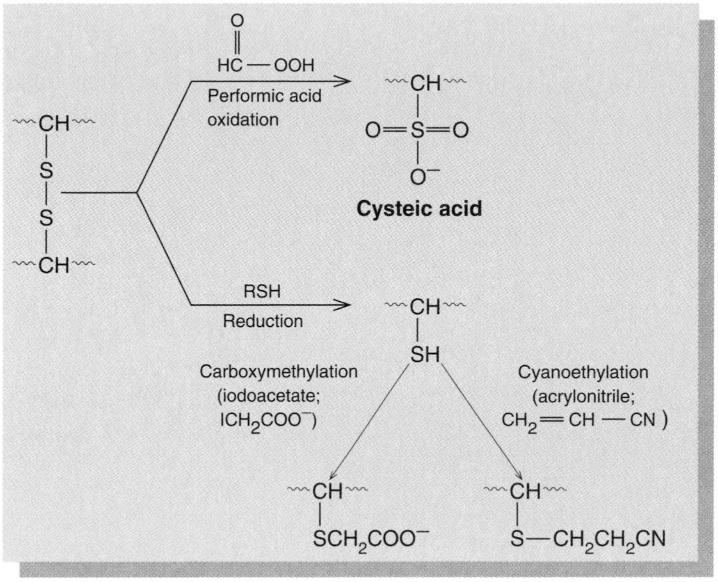

**Figure 3.18**

Polypeptide chain end-group analysis. (*a*) Amino-terminal group identification. A more sensitive method, the dansyl chloride method, is described in Methods of Biochemical Analysis 3B. (*b*) Carboxyl-terminal group identification. Identification of this amino acid is considerably more difficult.

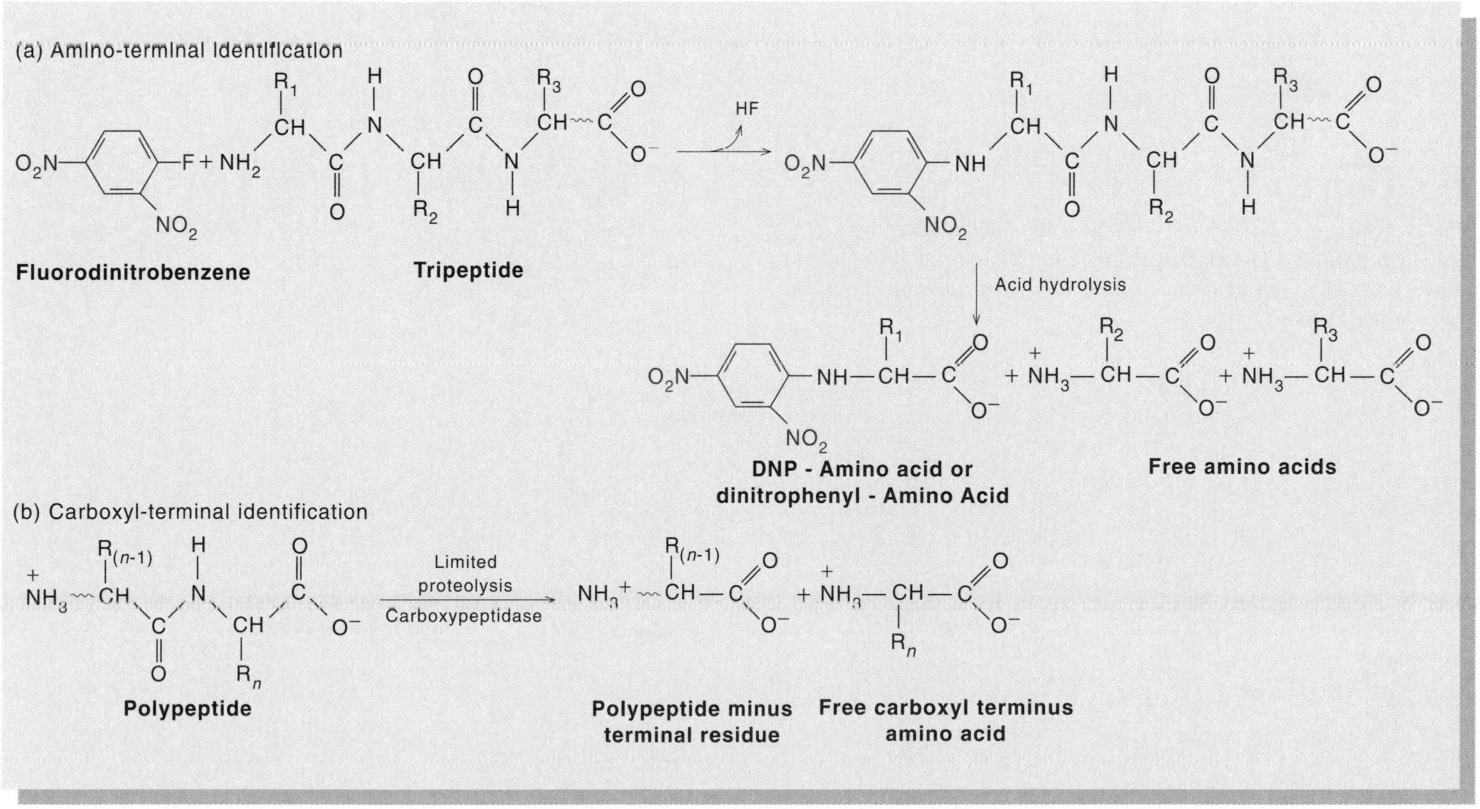

## Figure 3.19

Site of action of some endopeptidases used for polypeptide chain cleavage prior to sequence analysis. Of the four different enzymes used, trypsin is used most frequently because of its high specificity.

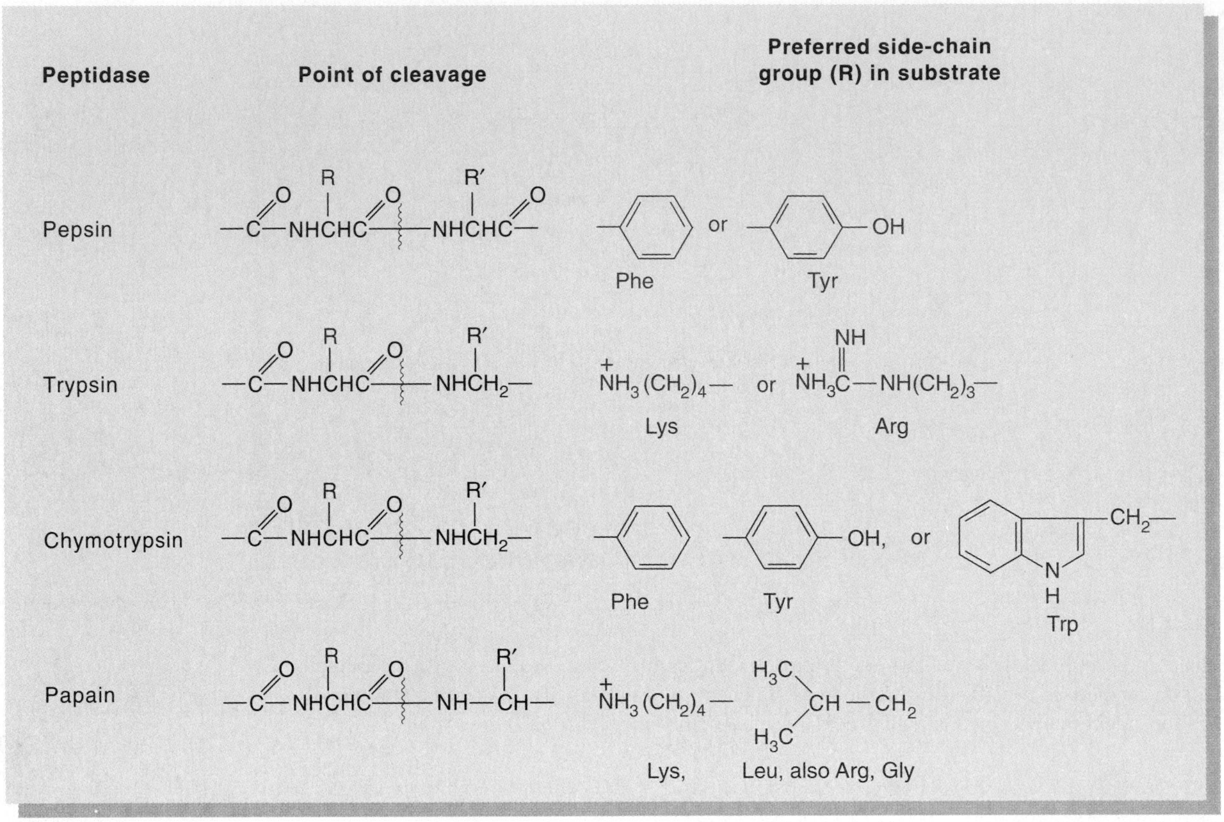

| Peptidase | Point of cleavage | Preferred side-chain group (R) in substrate |
|---|---|---|

## Figure 3.20

The cleavage of polypeptide chains at methionine residues by cyanogen bromide. The cleavage reaction is accompanied by the conversion of the newly formed free carboxyl-terminal methionine to homoserine lactone.

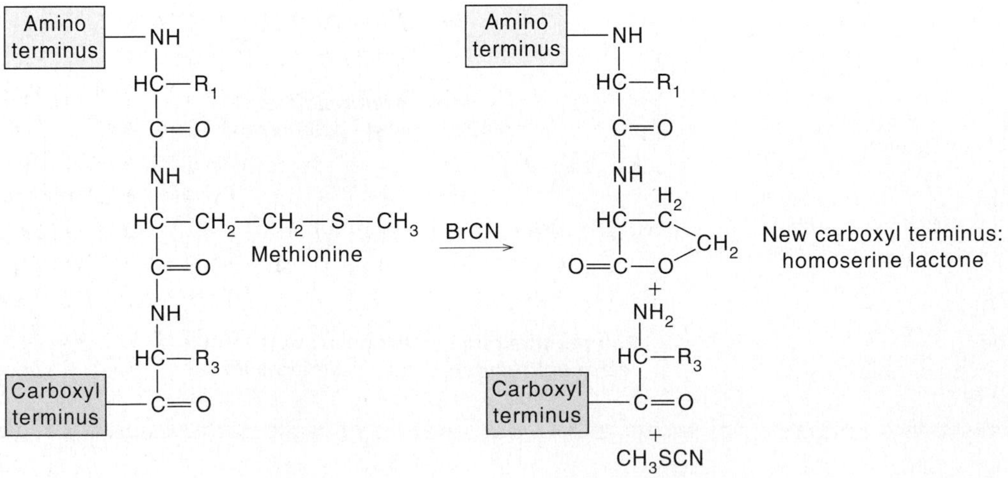

**Figure 3.21**

The Edman degradation method for polypeptide sequence determination. The sequence is determined one amino acid at a time, starting from the amino-terminal end of the polypeptide. First the polypeptide is reacted with phenylisothiocyanate to form a polypeptidyl phenylthiocarbamyl derivative. Gentle hydrolysis releases the amino-terminal amino acid as a phenylthiohydantoin (PTH), which can be separated and detected spectrophoto-metrically. The remaining intact polypeptide, shortened by one amino acid, is then ready for further cycles of this procedure. A more sensitive reagent, dimethylaminoazobenzene isothiocyanate, can be used in place of phenylisothiocyanate. The chemistry is the same.

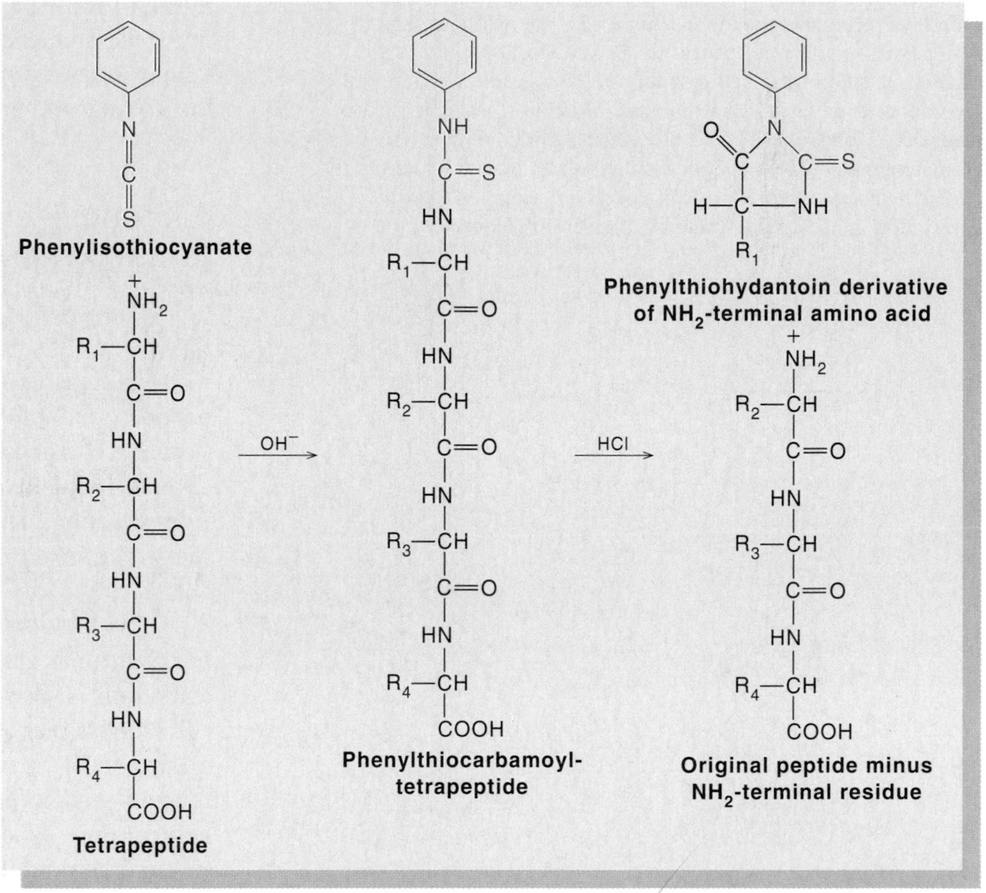

quence determination involves the stepwise removal and identification of successive amino acids from the polypeptide amino terminal by means of the Edman degradation (fig. 3.21). This process is carried out by reacting the free amino-terminal group with phenylisothiocyanate to form a peptidyl phenylthiocarbamyl derivative. Gentle hydrolysis with hydrochloric acid releases the amino-terminal amino acid as a phenylthiohydantoin (PTH) derivative. The remaining intact peptide, shortened by one amino acid, is then ready for further cycles of this procedure. The PTH-amino acid can be identified by its properties on thin-layer chromatography (fig. 3.22). High pressure liquid chromatography (HPLC) is the method of choice where quantitative results on small amounts of material are required. The general use of HPLC is described in chapter 6.

Devices called sequenators are available that automate the Edman degradation procedure. The success of these devices depends in large part on the technical innovation of covalently linking the peptide to be sequenced to glass beads. Attachment of the peptide through its carboxyl-terminal group to this immobile phase facilitates the complete removal of potentially contaminating reaction products during successive stages of the degradation.

Finally, having established the sequences of the individual peptides, it is necessary only to establish how they are connected together in the intact protein (step 6). It is at this stage that we see why the preceding sequence analysis was performed on peptides obtained by two different specific cleavage methods. This approach makes it possible to piece together the overall sequence, because the two sets of results produce overlapping sequences. That is, the free amino and carboxyl residues of peptides originally interconnected in the intact protein and liberated by one specific cleavage method recur in the internal sequences of the peptides liberated by a second specific method.

Once the protein's primary sequence has been determined, the location of disulfide bonds in the intact protein can be established by repeating a specific enzymatic cleavage on another sample of the same protein in which the disulfide bonds have not previously been cleaved. Separation of the resulting peptides shows the appearance of one new peptide and the disappearance of two other peptides, when compared with the enzymatic digestion product of the material whose disulfide bonds have first been chemically cleaved. In fact, these difference techniques are generally useful in the detection of sites of mutations in protein mole-

**Figure  3.22**

Thin-layer chromatography of amino acid–phenylthiohydantoin derivatives on silica gel plates. (*a*) Separation is done in a 98:2 mixture of chloroform and ethanol. (*b*) This is followed by further separation using an 88:2:10 mixture of chloroform, ethanol, and methanol. More sophisticated procedures, using column chromatography, give superior resolution and improved sensitivity. Automated sequencers always use such procedures. A general description of the use of columns is given in chapter 6.

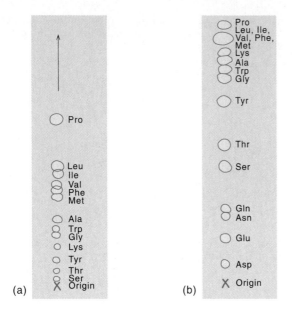

cules of previously known sequence, because a single substitution generally affects the chromatographic properties of only a single peptide released during proteolytic digestion.

Great progress has been made in recent years in devising procedures for sequencing the DNA that encodes for proteins (see chapter 27). Knowing the sequence of coding triplets in DNA allows us to read off the amino acid sequence of the corresponding protein. Nevertheless, such studies have produced the remarkable observation that some eukaryotic DNA sequences coding for proteins are not continuous but instead contain untranslated intervening DNA sequences. Although these results have profound implications for protein evolution, they obviously confound the general applicability of DNA-sequencing methods for the purposes of protein primary structure determination. In cases like this, the usual solution has been to isolate the mRNA for the protein and use this to make a DNA carrying the same sequence. This procedure circumvents the intervening sequence problem because the mRNA carries only the coding sequences (see chapter 28).

## Chemical Synthesis of Peptides and Polypeptides

Knowledge about the structure–function interrelationships in proteins and peptides has encouraged biochemists to develop techniques for synthesizing peptides and proteins with predetermined sequences. To synthesize a peptide in the laboratory, we must overcome several problems related to preventing undesired groups from reacting. The amino and carboxyl groups that are to remain unlinked must be blocked; so must all reactive side chains.

After blocking those groups to be protected, the carboxyl group is activated. It is of interest that carboxyl-group activation is also employed in natural biosynthesis in the cell (see chapter 29). After peptide synthesis, the protecting groups must be removed by a mild method. The overall process—comprising protection, activation, coupling, and unblocking—is shown in figure 3.23.

An important variation of the usual methods of peptide synthesis involves attaching a protected (*t*-butoxycarbonyl group) amino acid to a solid polystyrene resin, removal of the amino protecting group, condensation with a second protected amino acid, and so on. In the last step, the finished peptide is cleaved from the resin. This method (outlined in figure 3.24) has the advantage that cumbersome purification between steps, often resulting in serious losses, is replaced by mere washing of the insoluble resin. Because each reaction is essentially quantitative, very long peptides, and even proteins, can be synthesized by this method. Indeed, Li synthesized a 39-amino-acid protein hormone, adrenocorticotropic hormone, by this method, and Robert Merrifield synthesized bovine pancreatic ribonuclease, which contains 129 amino acids in a single polypeptide chain. A number of variants of ribonuclease that contain one or more changes in amino acid sequence also have been made by this method. The importance of the Merrifield process was underscored by the awarding of a Nobel Prize to Merrifield in 1984.

**Figure 3.23**

Schematic diagram illustrating the chemical method for peptide synthesis. First the amino acids to be linked are selected. The carboxyl group and the amino group that are to be excluded from peptide synthesis are protected (steps 1 and 1′). Next the amino acid containing the unprotected carboxyl group is carboxyl-activated (step 2). This amino acid is mixed and reacted with the other amino acid (step 3). Protecting groups are then removed from the product (step 4).

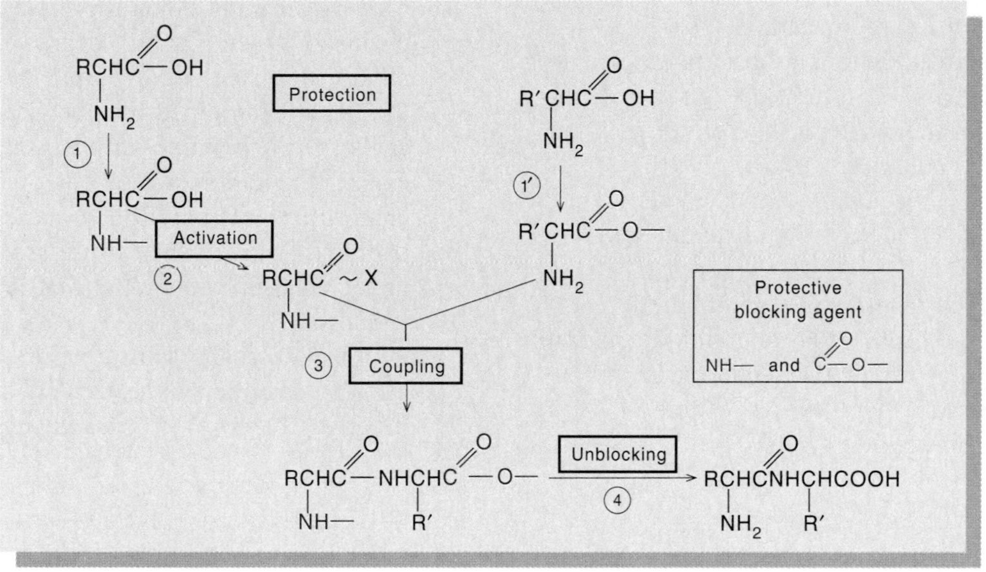

**Figure 3.24**

Merrifield procedure for solid-state dipeptide synthesis. (1) Polymer is activated. (2) Amino acid containing a *t*-butoxycarbonyl (BOC)-protecting group is carboxyl-linked to the polymer. This amino acid will be the carboxyl-terminal amino acid in the final peptide. (3) The BOC protecting group is removed from the polymer-linked amino acid. (4) A second amino acid, containing a BOC on its α-amino group and a dicyclohexylcarbodiimide (DCC)-activated group, is reacted with the column-bound amino acid to form a dipeptide. (5) The dipeptide is released from the polymer and the BOC-protecting group by adding hydrogen bromide (HBr) in trifluoroacetic acid.

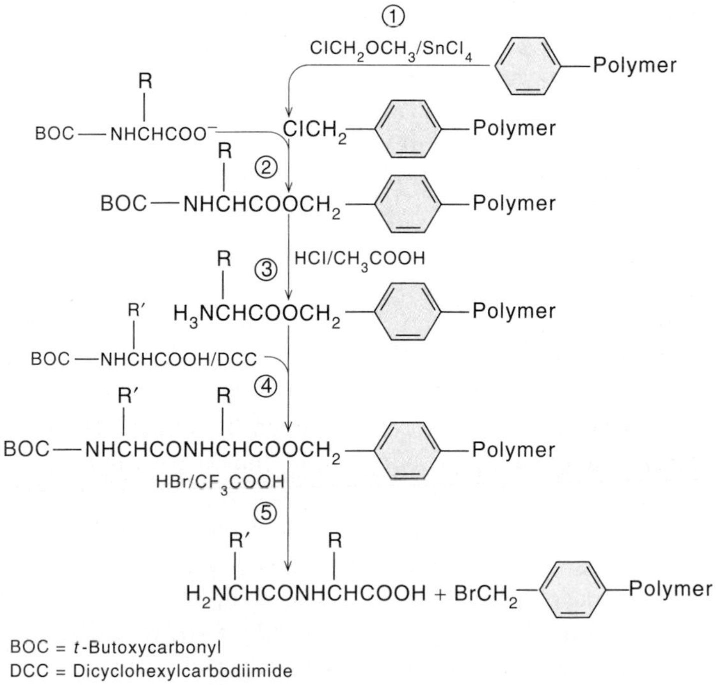

BOC = *t*-Butoxycarbonyl
DCC = Dicyclohexylcarbodiimide

## Summary

In this chapter we dealt with some of the fundamental properties of amino acids and polypeptide chains. The following points are especially important.

1. Nineteen of the 20 amino acids commonly found in proteins have a carboxyl group and an amino group attached to an $\alpha$-carbon atom; they differ in the side chain attached to the same $\alpha$ carbon.
2. All amino acids have acidic and basic properties. The ratio of base to acid form at any given pH can be calculated from the p$K$ with the help of the Henderson-Hasselbach equation.
3. All amino acids except glycine are asymmetric and therefore can exist in at least two different stereoisomeric forms.
4. Peptides are formed from amino acids by the reaction of the $\alpha$-amino group from one amino acid with the $\alpha$-carboxyl group of another amino acid.
5. Polypeptide formation involves a repetition of the process involved in peptide synthesis.

6. The amino acid composition of proteins can be discovered by first breaking down the protein into its component amino acids and then separating the amino acids in the mixture for quantitative estimation.
7. The amino acid sequences of proteins can be discovered by breaking down the protein into polypeptide chains and then partially degrading the polypeptide chains. For each polypeptide chain fragment, the sequence is determined by stepwise removal of amino acids from the amino-terminal end of the polypeptide chain. Two different methods of forming polypeptide chain fragments are used so as to produce a map of overlapping fragments, from which the sequence of undegraded polypeptide chains in the proteins can be deduced.
8. Polypeptide chains with a predetermined amino acid sequence can be synthesized by chemical methods involving carboxyl-group activation.

## Selected Readings

Barrett, G. C. (ed.), *Chemistry and Biochemistry of Amino Acids.* New York: Chapman and Hall, 1985. A recent and authoritative volume on this classical subject.

Gray, W. R., End group analysis using dansyl chloride. *Methods. Enzymol.* 25:121–138, 1972. This volume of *Methods in Enzymology* contains several chapters on end-group analysis.

Hunkapiller, M. W., J. E. Strickler, and K. J. Wilson, Contemporary methodology for protein structure determination. *Science* 226:304–311, 1984.

Kent, S. B. H., Chemical synthesis of peptides and proteins. *Ann. Rev. Biochem.* 57:957–989, 1988. Comprehensive and up-to-date.

Merrifield, B., Solid phase synthesis. *Science* 232:341–347, 1986.

Sanger, R., Sequences, sequences and sequences. *Ann. Rev. Biochem.* 57:1–28, 1988.

## Problems

1. A typical protein is 16% nitrogen by weight. How well does this percentage value match with the formula for proteins proposed by Berzelius?
2. (a) A 10mM solution of a weak monocarboxylic acid has a pH of 3.00. Calculate the values for $K_a$ and p$K_a$ for this carboxylic acid.
   (b) You add 0.06 g of NaOH ($M_r = 40$) to 1,000 ml of the acid solution in part (a). Calculate the final pH assuming no volume change.
3. A buffer was prepared by dissolving 3.71 g of citric acid and 2.91 g of KOH in water and diluting to a final volume of 250 ml. What is the pH of this buffer? What is the [H$^+$]? Use 3.14, 4.77, and 6.39 for the p$K_a$'s of citric acid.
4. Given the p$K_a$ values in the text, predict how the titration curves for glutamic acid and glutamine differ.
5. Calculate the isoelectric point for histidine, aspartic acid, and arginine. Calculate the fractional charge for

each ionizable group on aspartate at pH equal to the pI. Do these calculations verify the isoelectric point of aspartic acid?

6. Which of the naturally occurring amino acid side chains are charged at pH 2? pH 7? pH 12? (Consider only those amino acids whose side chains have >10% charge at the pH examined.)

7. Amino acids are sometimes used as buffers. Indicate the appropriate pH value(s) of buffers containing aspartic acid, histidine, and serine.

8. Polyhistidine is insoluble in water at pH 7.8 but is soluble at pH 5.5. Explain this observation. Would you expect the polymer to be soluble at pH 10?

9. As indicated in the text, 19 of the 20 amino acids have a chiral $\alpha$ carbon with only glycine (R = H) lacking chirality. Two of the protein amino acids have a second chiral carbon. Can you identify them?

10. A mixture of alanine, glutamic acid, and arginine was chromatographed on a weakly basic ion-exchange column (positively charged) at pH 6.1. Predict the order of elution of the amino acids from the ion-exchange column. Are the amino acids separated from each other? Explain.

    Suppose you have a weakly acidic ion-exchange column (negatively charged), also at pH 6.1. Predict the order of elution of the amino acids from this column. Propose a strategy for separating the amino acids using one or both columns. Explain your rationale. (Assume only ionic interactions between the amino acids and the ion-exchange resin.)

11. You have a peptide that is a potent inhibitor of nerve conduction and you wish to obtain its primary sequence. Amino acid analysis reveals the composition to be Ala(5); Lys; Phe. Reaction of the intact peptide with FDNB releases free DNP-alanine on acid hydrolysis. $\epsilon$-DNP-lysine (but not $\alpha$-DNP-lysine) is also found. Tryptic digestion gives a tripeptide (composition Lys, Ala(2)) and a tetrapeptide (composition Ala(3), Phe). Chymotryptic digestion of the intact peptide releases a hexapeptide and free alanine. Derive the peptide sequence.

12. From a rare fungus you have isolated an octapeptide that prevents baldness, and you wish to determine the peptide sequence. The amino acid composition is Lys(2), Asp, Tyr, Phe, Gly, Ser, Ala. Reaction of the intact peptide with FDNB yields DNP-alanine plus 2 moles of $\epsilon$-DNP-lysine on acid hydrolysis. Cleavage with trypsin yields peptides the compositions of which are (Lys, Ala, Ser) and (Gly, Phe, Lys), plus a dipeptide. Reaction with chymotrypsin releases free aspartic acid, a tetrapeptide with the composition (Lys, Ser, Phe, Ala), and a tripeptide the composition of which, following acid hydrolysis, is (Gly, Lys, Tyr). What is the sequence?

# Measurement of Ultraviolet Absorption in Solution

The general quantitative relationship that governs all absorption processes is called the Beer-Lambert law:

$$I = I_0 10^{-\epsilon cd}$$

where $I_0$ is the intensity of the incident radiation, $I$ is the intensity of the radiation transmitted through a cell of thickness $d$ (in centimeters) that contains a solution of concentration $c$ (expressed either in moles per liter or in grams per 100 ml), and $\epsilon$ is the extinction coefficient, a characteristic of the substance being investigated (see fig. 3.7).

Light absorption is measured by a spectrophotometer as shown in the illustration. The spectrophotometer usually is capable of directly recording the absorbance $A$, which is related to $I$ and $I_0$ by the equation

$$A = \log_{10}\left(\frac{I_0}{I}\right)$$

Hence $A = \epsilon cd$, and $A$ is a direct measure of concentration. We can see from figure 3.7 that the $\epsilon$ values are largest for tryptophan and smallest for phenylalanine.

Because protein absorption maxima in the near-ultraviolet (240–300 nm) are determined by the content of the

**Figure 1**

Schematic diagram of a spectrophotometer for measuring light absorption. Laboratory instruments for making measurements are much more complex than this, but they all contain the same basic components: a light source, a monochromator, a sample, and a detector. $\lambda$ is the wavelength of the light, $I_0$ and $I$ are the incident light intensity and the transmitted light intensity, respectively, and $d$ is the thickness of the absorbing solution.

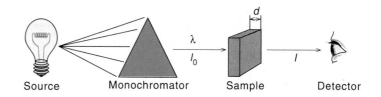

Source      Monochromator      Sample      Detector

aromatic amino acids and their respective values, most proteins have absorption maxima in the 280-nm region. By contrast, absorption in the far-ultraviolet (around 190 nm) is shown by all polypeptides regardless of their aromatic amino acid content. The reason is that absorption in this region is due primarily to the peptide linkage.

# The Dansyl Chloride Method for N-Terminal Amino Acid Determination

The dansyl chloride method provides an alternative to the Sanger method for N-terminal amino acid determination. Because it is considerably more sensitive than the Sanger method, it has become the method of choice. The reaction is diagrammed in the illustration. A polypeptide is treated with dansyl chloride to give an *N*-dansyl peptide derivative. This derivative is hydrolyzed to yield a highly fluorescent *N*-dansyl-amino acid, which is detected chromatographically.

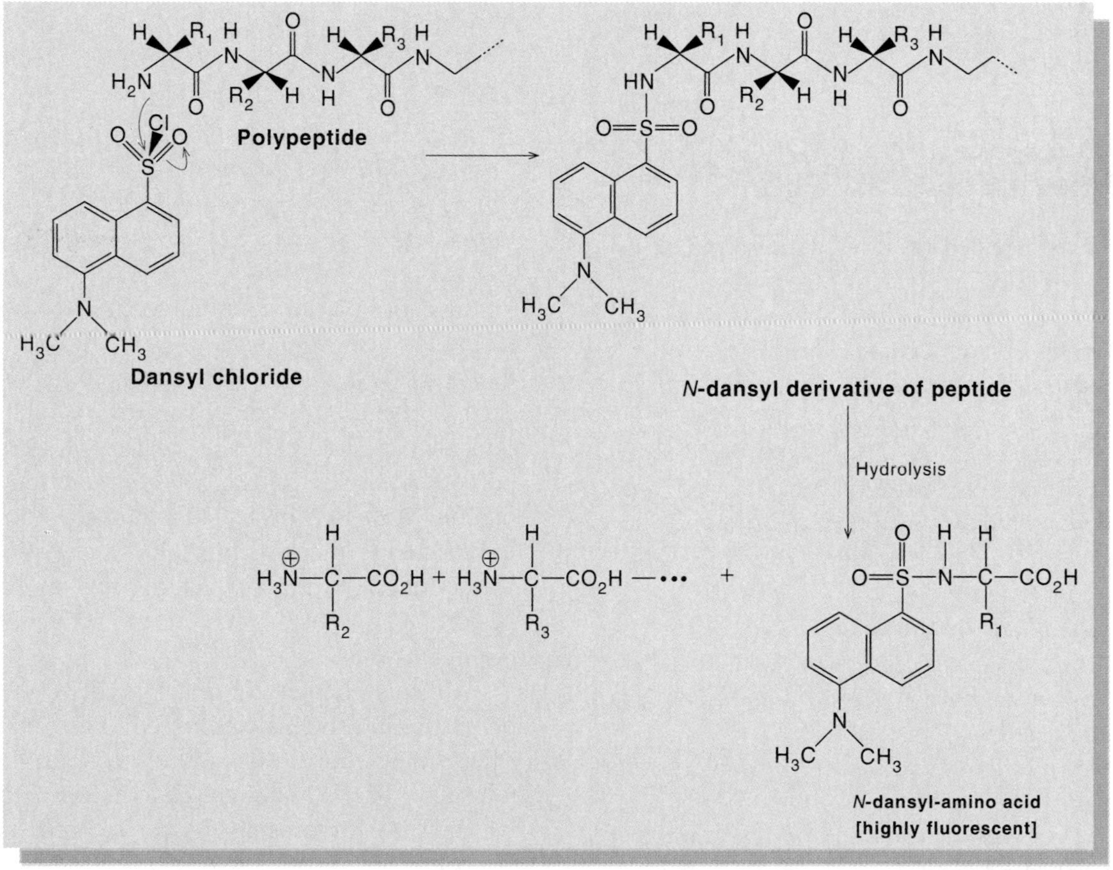

**Polypeptide**

**Dansyl chloride**

**N-dansyl derivative of peptide**

Hydrolysis

*N*-dansyl-amino acid
[highly fluorescent]

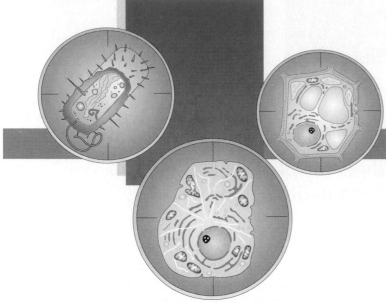

# The Three-Dimensional Structures of Proteins

*Proteins adopt the most stable folded structures;
this is a function of the way in which the individual
amino acid residues interact with each other and
with water.*

T he enormous structural diversity of proteins begins with the amino acid sequences of polypeptide chains. Each protein consists of one or more unique polypeptide chains, and each of these polypeptide chains is folded into a three-dimensional structure. The final folded arrangement of the polypeptide chain in the protein is referred to as its conformation. Most proteins exist in unique conformations exquisitely suited to their function. It is the availability of a wide variety of conformations that permits proteins as a group to perform a broader range of functions than any other class of biomolecules.

In this chapter we deal primarily with the structural properties of proteins, and in the following chapter we consider the functional diversity of proteins. Traditionally proteins have been divided into two groups: fibrous and globular. Fibrous proteins aggregate to form highly elongated structures having the shape of fibers or sheets. Each protein unit that makes up these aggregated structures is built from a repeating structural motif, giving the molecules a simple structure that is relatively easy to analyze. By contrast, globular proteins are more complex, containing one or more polypeptide chains folded back on themselves many times to give an approximately spherical shape. We consider the relatively simple fibrous proteins first.

## Figure 4.1

Resonance and the planar structure of the peptide bond. (*a*) Two major hybrids contribute to the structure of the peptide bond. In structure 1 the C—N bond is a single bond with no overlap between the nitrogen lone electron pair and the carbonyl carbon. The carboxyl carbon is $sp^2$-hybridized and is therefore planar, whereas the nitrogen is $sp^3$-hybridized and pyramidal. By contrast, in structure 2 there is a double bond between the amide nitrogen and the carbonyl carbon; also, the nitrogen atom bears a charge of $+1$ and the carbonyl oxygen bears a charge of $-1$. Both the carboxyl carbon and the amide nitrogen are $sp^2$-hybridized, both are planar, and all six atoms lie in the same plane. (*b*) The structure of the peptide bond is a compromise between the two resonating hybrids, structures 1 and 2. (*c*) Dimensions of the peptide bond and surrounding linkages. The C—N bond length of 1.325 Å is significantly less than the length of a single C—N bond, 1.47 Å. (Illustration for part *c* copyright by Irving Geis. Reprinted by permission.)

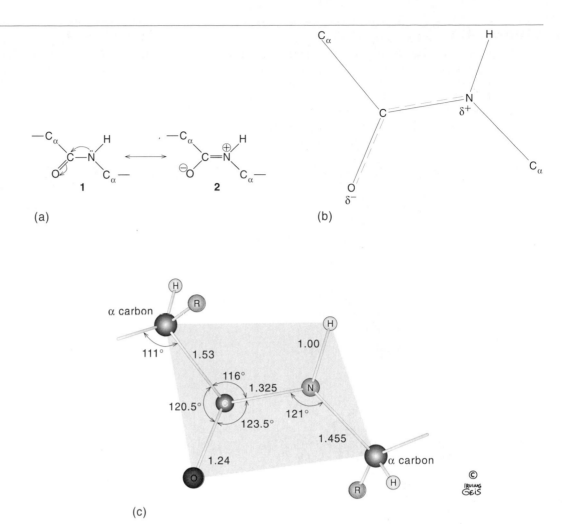

# Pauling and Corey Provided the Foundation for Our Understanding of Fibrous Protein Structures

Linus Pauling and Robert Corey examined the structures of crystals formed by amino acids and short peptides before they ventured into the world of proteins. From their crystallographic investigations of amino acids and peptides, they formulated two rules that describe the ways in which amino acids and peptides interact with one another to form noncovalently bonded crystalline structures. These rules laid the foundations for our understanding of how amino acids in protein polypeptide chains interact with one another.

Rule number one was that the peptidyl C—N linkage and the four atoms to which the C and the N atoms are directly linked always form a planar structure as though the C—N linkage is a double bond rather than a single bond as normally written. Pauling reasoned that the C—N linkage is a resonating structure with partial double-bond character

(fig. 4.1), so that it locks the peptide grouping into a planar conformation. This property is extremely important because it greatly reduces the flexibility in the polypeptide chain. With this grouping in a rigid conformation, the only flexibility remaining in the polypeptide backbone results from rotation about the carbon that joins adjacent peptide planar groups (fig. 4.2).

The second rule that Pauling and Corey formulated was that peptide carbonyl and amino groups always form the maximum number of hydrogen bonds. Recall (see chapter 1) that in water hydrogen bonds are formed between a partially unshielded proton from one molecule and an oxygen atom that originates from another molecule (see fig. 1.7). Nitrogen atoms can also serve as H bond acceptors. The attraction between the H bond donor and the H bond acceptor is strongest along the lone pair orbital axis of the acceptor atom. As a rule the angle between an N or O acceptor and an N—H or O—H donor is close to 180° (fig. 4.3). Thus a hydrogen bond brings two interacting groups close

## Figure 4.2

Basic dimensions of a dipeptide. The conformational degrees of freedom of a polypeptide chain are restricted to rotations about the single-bond connections between the adjacent planar transpeptide groups to $C_\alpha$, that is, the $C_\alpha$—$C_2$ and $C_\alpha$—$N_1$ single bonds. The corresponding rotations are represented by $\psi$ and $\phi$, respectively, which have values of 180° for the fully extended configuration shown. (Illustration copyright by Irving Geis. Reprinted by permission.)

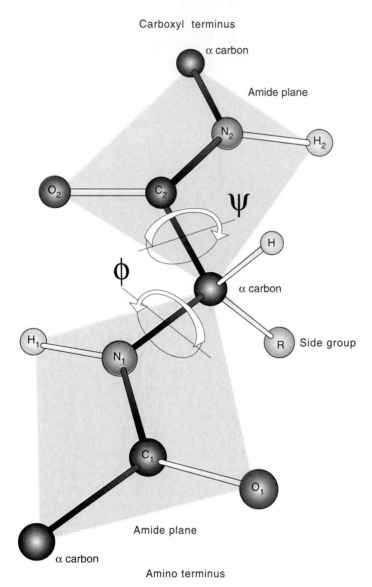

## Figure 4.3

Major hydrogen-bond donor and acceptor groups found in proteins. Note that the angle between the O acceptor and the N—H donor is 180° in the hydrogen-bond complex.

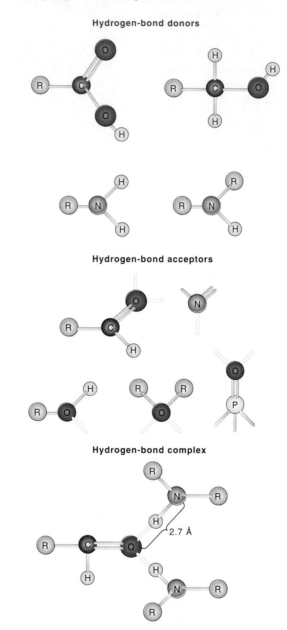

together and orients them in a certain way. This second rule further limits the number of conformations available to polypeptide chains.

Following their investigations on amino acid and peptide crystals Pauling and Corey turned their attention to the x-ray diffraction patterns of a number of fibrous proteins. A vast number of fibrous proteins exist in nature, but the majority of them give diffraction patterns that fall into one of three types: the $\alpha$ pattern (see Methods of Biochemical Analysis 4A), the $\beta$ pattern, and the collagen pattern. Fiber diffraction data only give information about the repeating units of a protein structure because of the lack of three-dimensional order in the fibers; but because fibrous protein polypeptide chains are arranged in simple repetitious units the overall structure could be deduced. The first type of

## Table 4.1

Radii for Covalently Bonded and Nonbonded Atoms

| Covalent Bond Radii (in Å) | | | | Van der Waals Radii (in Å) | | | |
| --- | --- | --- | --- | --- | --- | --- | --- |
| *Element* | *Single Bond* | *Double Bond* | *Triple Bond* | *Element* | | | |
| Hydrogen | 0.30 | | | Hydrogen | 1.2 | | |
| Carbon | 0.77 | 0.67 | 0.60 | Carbon | 2.0 | | |
| Nitrogen | 0.70 | | | Nitrogen | 1.5 | | |
| Oxygen | 0.66 | | | Oxygen | 1.4 | | |
| Phosphorus | 1.10 | | | Phosphorus | 1.9 | | |
| Sulfur | 1.04 | | | Sulfur | 1.8 | | |

diffraction pattern, which was observed for a subgroup of the keratins, was consistent with a helical arrangement of the polypeptide chains; this pattern indicated that there are two regularly repeating units along the helix axis: one with a repeat of 5.4 Å and one with a repeat of 1.5 Å.

## The Structure of the α-Keratins Was Determined with the Help of Molecular Models

Space-filling molecular models were used in an attempt to build structures compatible with the x-ray data. These models were designed so that covalently linked atoms were accurately spaced according to known dimensions (table 4.1). Moreover, the individual atoms made as hard spheres were of a size so that nonbonded atoms could get no closer to one another than their van der Waals radii would normally allow (see table 4.1). It should be recalled from chemistry that the average separation between two nonbonded atoms is known as their van der Waals separation. If nonbonded atoms get closer than this they repel each other greatly, and favorable interaction between them falls less rapidly if the distance is larger than this. Pauling and Corey tried to arrange the polypeptide chains so as to maximize the number of peptide hydrogen bonds in a way that was consistent with the x-ray diffraction data. By trial and error they came to the conclusion that the most acceptable structure was a right-handed helical coil (fig. 4.4). This structure, known as the alpha (α) helix, has a rigid, regularly repeating backbone structure with an advance of 1.5 Å per amino acid residue along the helix axis explaining the corresponding spot in the diffraction (see Methods of Biochemical Analysis 4A). The diffraction spots resulting from the helix could be explained by

a spacing of 5.4 Å between adjacent turns of the polypeptide backbone along the helix axis. In fig. 4.4 we see three ways of representing this structure. In (*a*) we see a ball-and-stick model in which the planar groups are highlighted and the hydrogen bonds are represented as dashed lines. In (*b*) we see a space-filling model similar to the type that Pauling and Corey used. And in (*c*) we see a wire model. The latter two models are shown in both side and top views.

The space-filling model (*b*) shows that the α helix is a tightly packed structure with no unfilled cavities whether one looks at a profile or down the helix. The wire model (*c*) illustrates the helical structure best. But for purposes of discussion, the ball-and-stick model (*a*) is the most suitable. Careful inspection shows that the polypeptide backbone follows the path of a right-handed helical spring in which each residue's carbonyl group forms a hydrogen bond with the amide NH group of the residue four amino acids further along the polypeptide chain. All residues in the α helix have nearly identical conformations so they lead to a regular structure in which each 360° of helical turn incorporates approximately 3.6 amino acid residues and rises 5.4 Å along the helix axis direction. This rise accounts for the observed advance per amino acid residue along the helix axis of 5.4/3.6 = 1.5 Å.

Although alternative helical arrangements having different hydrogen-bonding patterns and different geometries are possible, the α helix is by far the most commonly observed. The unique stability of the α helix is related to the formation of regularly arranged hydrogen bonds between all the carbonyl and amino groups and to the tight packing achieved in folding the chain to form the structure.

Fibrous proteins in which the α helix is a major structural component are found in hair, scales, horns, hooves, wool, beaks, nails, and claws—proteins referred to as α-

## Figure 4.4

Three ways of projecting the $\alpha$-helix. (a) This simple ball-and-stick model highlights the planar peptides. The interpeptide hydrogen bonds are shown by dashed lines, and the amino acid side chains are indicated by R groups. Approximately two turnings of the helix are shown. There are about 3.6 residues per turn. In (b) and (c) we see identical projections of a side view using space-filling and wire models, respectively. The space-filling models use van der Waals radii for the atoms. (Illustration for part a copyright by Irving Geis. Reprinted by permission.)

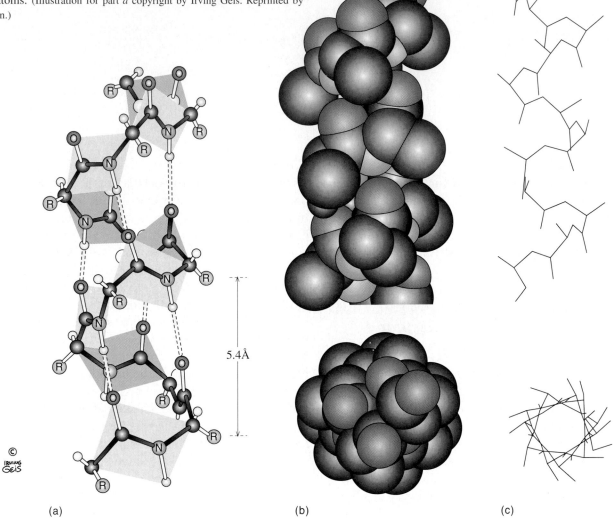

5.4Å

(a)                    (b)                    (c)

keratins. When individual $\alpha$ helices aggregate in this side-by-side fashion, they usually form long cables in which the individual helices are spirally twisted so that the resulting cable has an overall left-handed twist (fig. 4.5). The formation of a cable with twisted $\alpha$ helices results from optimization of packing among the amino acid side-chain residues between helices. In figure 4.6 we see how the side-chain residues of an $\alpha$ helix are arranged in a spiral fashion so that residues falling on the same side of a helix do not lie along a line parallel to the helix axis. As a consequence the packing of helices is optimized when helices interact at an angle of about 18°. Obviously, if $\alpha$ helices involved in such a packing interaction were straight, they would soon separate. However, their packing interaction can be preserved if the helices are slightly twisted around each other, with the resultant formation of the left-twisted cable structure characteristic of the keratins. The coiled-coil character of such fibers consequently represents a trade-off between some local deformations that coil the $\alpha$ helix and the optimization of extended side-chain packing interactions in the cable as a whole.

**Figure 4.5**

The assembly of hair α-keratin from one α helix to a protofibril, to a microfibril, and finally, to a single hair. (Illustration copyright by Irving Geis. Reprinted by permission.)

α helix

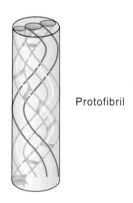

Protofibril

Microfibril

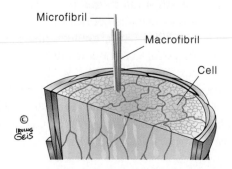

Microfibril

Macrofibril

Cell

**Figure 4.6**

Coiling of α helices in α-keratins. Residues on the same side of an α helix form rows that are tilted relative to the helix axis. Packing helices together in fibers is optimized when the individual helices wrap around each other so that rows of residues pack together along the fiber axis. Helices in coiled coil (c) are oriented in parallel.

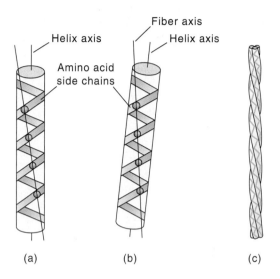

The springiness of hair and wool fibers results from the tendency of the α-helical cables to untwist when stretched and spring back when the external force is removed. In many forms of keratin, the individual α helices, or fibers, are covalently linked by disulfide bonds formed between cysteine residues of adjacent polypeptide chains. In addition to giving added strength to the fibers, the pattern of these covalent interactions serves to influence and fix the extent of curliness in the hair fiber as a whole. Chemical reactions and mechanical processes involving reductive cleavage, reorganization, and reoxidation of these interhelix disulfide bonds form the basis of the "permanent wave."

## The β-Keratins Form Sheetlike Structures with Extended Polypeptide Chains

Pauling and Corey noticed that certain fibrous proteins give radically different diffraction patterns and behave macroscopically as sheets rather than fibers. Diffraction patterns of fibrous proteins from different sources reveal a repeat along the direction of the extended polypeptide chain of either 13.0 or 14.0 Å, suggesting two closely related structures. Pauling and Corey interpreted these patterns as resulting from extended polypeptide chains lying side by side either in a parallel or an antiparallel fashion (fig. 4.7). Although both of these structures exist in nature, the antiparallel sheet is more common in fibrous proteins. This structure

## Figure 4.7

Two forms of the $\beta$-sheet structure: (*a*) the antiparallel and (*b*) the parallel $\beta$ sheet. The advance per two amino acid residues is indicated for each structure.

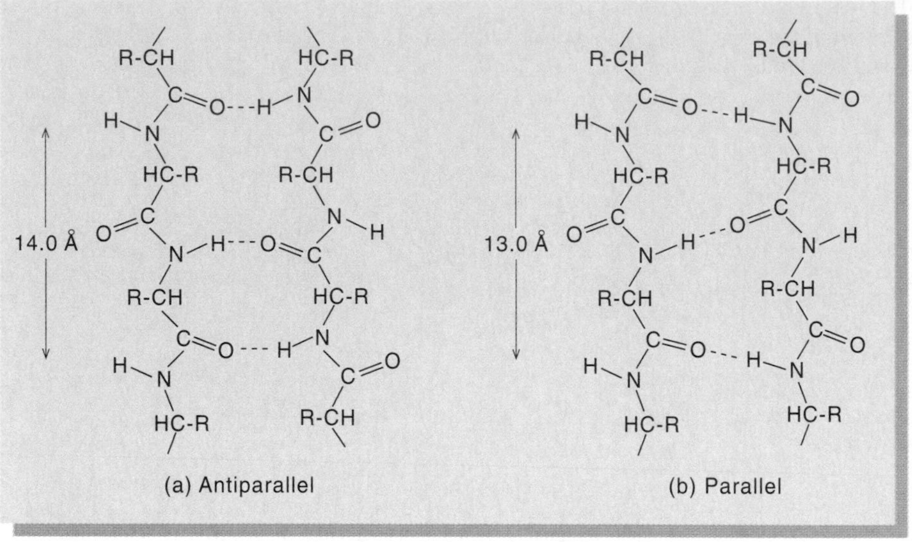

14.0 Å

(a) Antiparallel

13.0 Å

(b) Parallel

## Figure 4.8

The antiparallel $\beta$ sheet. This structure is composed of two or more polypeptide chains in the fully extended form, with hydrogen bonds formed between the chains. Hydrogen bonds are shown as dashed lines. (Illustration copyright by Irving Geis. Reprinted by permission.)

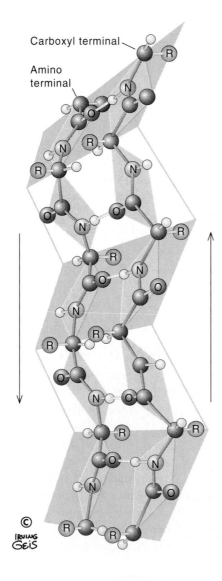

Carboxyl terminal

Amino terminal

is known as the antiparallel $\beta$-pleated sheet (fig. 4.8). Regular hydrogen bonds form between the peptide backbone amide NH and carbonyl oxygen groups of adjacent chains. The $\beta$ sheet can be extended into a multistranded structure simply by adding successive chains in the appropriate directions to the sheet. Parallel and antiparallel $\beta$-pleated sheets are both composed of polypeptide chains that have conformations pointing alternate R groups to opposite sides of the sheet but that have their peptide planes nearly in the sheet plane to allow for good interchain hydrogen bonding. Nevertheless, the chain conformation that produces the best interchain hydrogen bonding in parallel sheets is slightly less extended than that for the antiparallel arrangement. As a result, the parallel sheet has both a shorter repeat period 6.5 Å (versus 7.0 for the antiparallel structure) and a more pronounced pleat.

The best known $\beta$-keratin structure in nature is that found in certain silks. Silks are composed of stacked antiparallel $\beta$ sheets (fig. 4.9). Sequence analysis of silk proteins shows them to be composed largely of glycine, serine, and alanine, in which every alternate residue is glycine. Since the side-chain groups of a flat antiparallel sheet point alternately upward and downward from the plane of the sheet, all the glycine residues are arranged on one surface of each sheet and all the substituted amino acids are on the other. Two or more such sheets can consequently be packed intimately together to form an arrangement of stacked sheets in which two adjacent glycine-substituted or alanine-substituted sheet surfaces interlock with each other (fig. 4.9*b*). Owing to the extended conformations of the polypeptide chains in the $\beta$ sheets and the interlocking of the side chains between sheets, silk is a mechanically rigid material that resists stretching.

## Figure 4.9

The three-dimensional architecture of silk (*a*). The side chains of one sheet nestle quite efficiently between those of neighboring sheets (*b*). (Illustration copyright by Irving Geis. Reprinted by permission.)

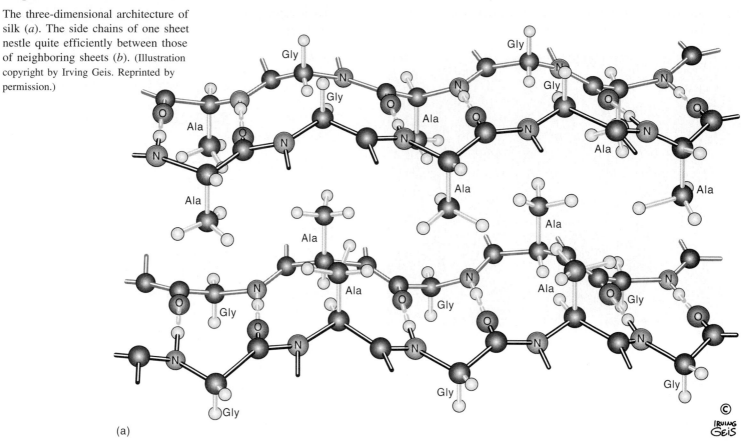

(a)

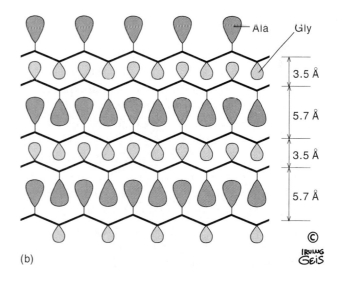

(b)

## Collagen Forms a Unique Triple-Stranded Structure

Collagen is a particularly rigid and inextensible protein that serves as a major constituent of tendons and connective tissues. In the electron microscope it can be seen that collagen fibrils have a distinctive banded pattern with a periodicity of 680 Å (fig. 4.10). These fibrils are of varying thicknesses depending on their source and the mode of preparation. Individual fibrils are composed of collagen molecules 3,000 Å long that aggregate in a staggered side-by-side fashion (fig. 4.11).

Collagen has a most unusual amino acid composition in which glycine, proline, and hydroxyproline are the dominant amino acids. Further characterization of the polypeptide chains shows that these amino acids are arranged in a repetitious tripeptide sequence, Gly-X-Y, in which X is frequently a proline and Y is frequently a hydroxyproline. This unusual amino acid sequence and the unique diffraction pattern of collagen were strong indications for a totally different type of fibrous protein. Pauling attempted to determine the structure of collagen by his molecular model approach but failed. At a meeting in Cambridge, England, in 1958 where the correct structure was presented, Pauling jested

**Figure 4.10**

An electron micrograph of collagen fibrils from skin. (Courtesy of Jerome Gross, Massachusetts General Hospital.)

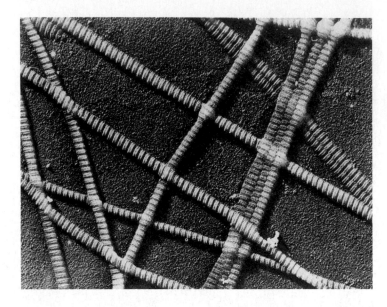

**Figure 4.11**

The banded appearance of collagen fibrils in the electron microscope arises from the schematically represented staggered arrangement of collagen molecules (*above*) that results in a periodically indented surface. $D$, the distance between cross striations is $\approx 680$ Å so that the length of a 3,000-Å-long collagen molecule is 4.4$D$. (Line art courtesy of Karl A. Piez. Photomicrograph © Michael C. Webb/Visuals Unlimited.)

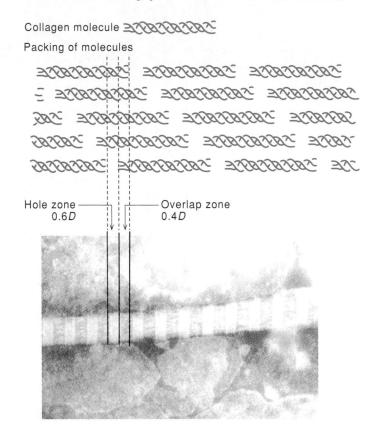

that the structure he had proposed might be more stable than the one found in nature. The natural structure was found by Ramachandran. The repeating proline residue excluded the possibility that the polypeptide chains in collagen could adopt either an $\alpha$-helical or a $\beta$-sheet conformation. Instead, individual collagen polypeptide chains assume a left-handed helical conformation and aggregate into three-stranded cables with a right-handed twist (fig. 4.12). When viewed down the polypeptide chain axis (fig. 4.13*b*), the successive side-chain groups point toward the corners of an equilateral triangle. The glycine at every third residue is required because there is no room for any other amino acid inside the triple helix where the glycine R groups are located. The three collagen chains do not form hydrogen bonds among residues of the same chain. Instead, the collagen chains within each three-stranded cable form interchain hydrogen bonds. This produces a highly interlocked fibrous structure that is admirably suited to its biological role, which is to provide rigid connections between muscles and bones as well as structural reinforcement for skin and connective tissues.

Although living organisms contain additional types of fibrous proteins, as well as polysaccharide-based structural motifs, we focused here on the three arrangements that are the most widely distributed. Two of these, the $\alpha$-keratins and the $\beta$-keratins incorporate polypeptide secondary structures that also commonly occur in globular proteins. Colla-

gen, in contrast, is a protein that evolution has developed to play more specialized roles.

## Globular Protein Structures Are Extremely Varied and Require a More Sophisticated Form of Analysis

It is not surprising that repeating structures with long-range order were the first protein structures to be understood. The demands on the available technology were minimal. Much more sophisticated technology was required to interpret the diffraction patterns of proteins that have shorter range repetition in their structures (see Methods of Biochemical Analysis 4B). Among the crystallographers who struggled over

## Figure 4.12

The triple helix of collagen. (Illustration copyright by Irving Geis. Reprinted by permission.)

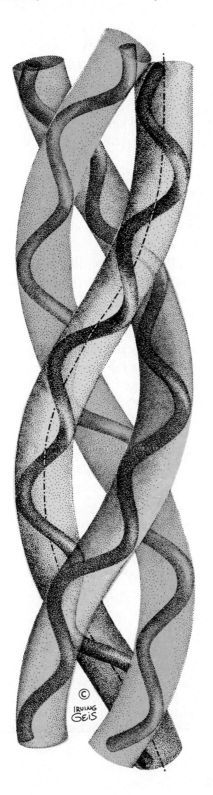

## Figure 4.13

The basic coiled-coil structure of collagen. Three left-handed single-chain helices wrap around one another with a right-handed twist. (*a*) Ball-and-stick single-collagen chain. (*b*) View from the top of the helix axis. Note that glycines are all on the inside. In this structure the C=O and N—H groups of glycine protrude approximately perpendicularly to the helix axis so as to form interchain hydrogen bonds. (Illustration for part *a* copyright by Irving Geis. Reprinted by permission.)

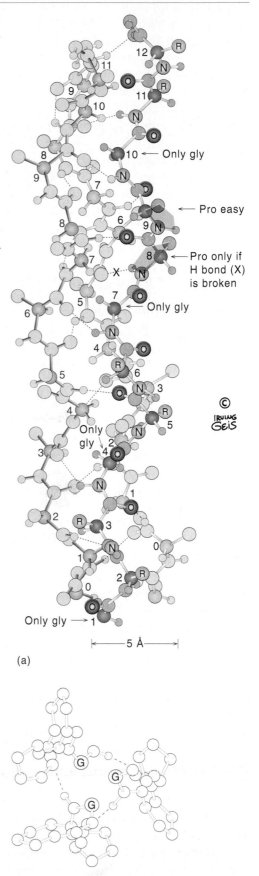

(a)

(b)

these structures were two who emerged as the major pioneers: John Kendrew and Max Perutz. In the 1950s they led their research teams to determine the first of such structures—myoglobin and hemoglobin. Since that time enormous advances in protein chemistry and computer technology have systematized the necessary research and greatly reduced the amount of work and time required to determine a protein structure. Accurate structural determinations have now been made for more than 500 different proteins. From this wealth of data, patterns of structure are becoming apparent that suggest, among other things, that the overall folding arrangements of proteins may some day be predictable from the amino acid sequences of the polypeptide chains.

# Folding of Globular Proteins Reveals a Hierarchy of Structural Organization

In the analysis of fibrous protein structures little mention was made of the importance of their interaction with water in determining the final folded structures of the proteins. This is because most of the side chains in fibrous proteins are exposed to water except when they interact with each other to form multimolecular aggregates. Then the relative affinity between other similar side chains and water becomes a major issue. In the case of globular proteins the interaction of amino acid side chains with water is a major issue from the start because globular proteins have many of their amino acid side chains buried in the interior of their folded structures. Hence, in our analysis of the structure of globular proteins we must be aware of the structural considerations that are important in the determination of fibrous proteins but also of additional considerations, first raised in chapter 1, that relate to the interaction of the amino acid side chains with water.

We may think of the structure of globular proteins at four levels (fig. 4.14). Ultimately, the entire three-dimensional structure is determined by a protein's amino acid sequence, or primary structure. At a higher level of organization, the regularly repeating geometry of the hydrogen-bonding groups of the polypeptide backbone leads to the formation of regular hydrogen-bonded secondary structures. Secondary structural elements include regions of $\alpha$ helix, $\beta$ sheets, and bends. Association between elements of secondary structure in turn results in the formation of structural domains, the properties of which are determined both by chiral characteristics of the polypeptide chain and by the nature of the amino acid side chains. The side chains of the residues in each domain pack together in a way that tends to optimize favorable interactions between side chains and between side chains and the surrounding water.

Further association of domains results in the formation of the protein's tertiary structure—the overall folding of the polypeptide chain in three dimensions. Finally fully folded protein subunits can pack together to form quaternary structures.

## Visualizing Folded Protein Structures

Because globular proteins have nonrepeating structures, it is essential to have a means for displaying the entire three-dimensional structure in sufficient detail, and yet not too much detail, so that the overall structural design can be appreciated.

The primary data of protein crystallography yield a three-dimensional electron-density map, which must be interpreted in terms of a three-dimensional model of all atom positions in the protein. Such modeling is usually done by computer graphics.

In figure 4.15 we see four different presentations that show selected parts of the protein ribonuclease and that highlight features of special interest. Each method of presentation has its advantages. The space-filling model (see fig. 4.15a) is excellent for displaying the volume occupied by molecular constituents and the shape of the outer surface. Figure 4.15b shows a stereo pair of a space-filling model. When the pair is viewed with stereo glasses, the three-dimensional illusion is striking. Figure 4.15c shows the polypeptide chain, with N and O atoms labeled and with dotted lines representing hydrogen bonds. In addition, all $\alpha$-carbon positions are numbered. Figure 4.15d shows an abstraction of the polypeptide chain in which the $\beta$ strands are characterized as flat arrows and the $\alpha$ helices as spiral ribbons. This simplified style has proved useful in classifying and comparing proteins according to the folding patterns of their secondary and tertiary structures.

## Primary Structure Determines Tertiary Structure

Throughout our discussion of protein structures we assumed that structures form because they represent the most stable way of arranging the polypeptide chains. The first direct support for this notion for a globular protein came from the studies of Christian Anfinsen. Anfinsen unfolded pancreatic ribonuclease, an enzyme containing 124 amino acid residues with four disulfide bridges, and then found conditions under which it could be refolded into its original native structure. First, the enzyme was denatured in a solution containing a hydrogen-bond-breaking reagent (urea) and 2-mercaptoethanol (a thiol reagent that reduces disulfides to sulfhydryls), thus cleaving the cova-

## Figure 4.14

Hierarchies of protein structures. (Illustration for part *c* copyright by
Irving Geis. Reprinted by permission.)

(a) Primary structure (amino acid sequence in the protein chain)

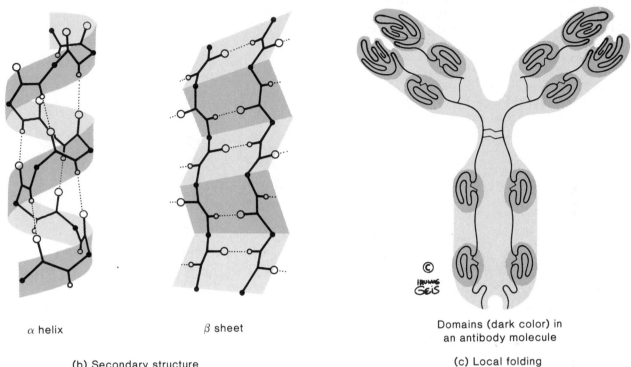

$\alpha$ helix

$\beta$ sheet

Domains (dark color) in
an antibody molecule

(b) Secondary structure

(c) Local folding

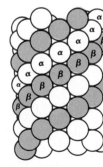

One complete protein chain
($\beta$ chain of hemoglobin)

The four separate chains
of hemoglobin assembled
into an oligomeric protein

$\alpha$ (white) and $\beta$ (color)
tubulin molecules in a
microtubule

(d) Tertiary structure

(e) Quaternary structure

(f) Quaternary structure

## Figure 4.15

Four ways of representing the three-dimensional structure of the
protein pancreatic ribonuclease. (*a*) Space-filling model. (*b*) The
stereo pair of space-filling models illusion can be seen without stereo
glasses by the following method. With your eyes about 10 inches
from the page, stare at the drawings below as if you were looking
straight ahead at a far-away object. A double image forms and the
central pair drifts together and fuses. Then the illusion becomes
apparent. Adjust the page, if necessary, so that a horizontal line is
perfectly parallel with the eyes. As an aid to seeing one image with
each eye, use two cardboard tubes from toilet-paper rolls. Close the
right eye and focus the left eye on the left image. Do the same for
the other eye. Now, with both eyes together, the three-dimensional
illusion should appear. (*c*) Ball and stick model. Dotted lines
represent hydrogen bonds, and numbers indicate α-carbon positions.
(*d*) Ribbon model. The β-strands are represented by the flat red
arrows. The α helices are shown as spiral ribbons. (Illustrations for
parts *a–d* copyright by Irving Geis. Reprinted by permission.)

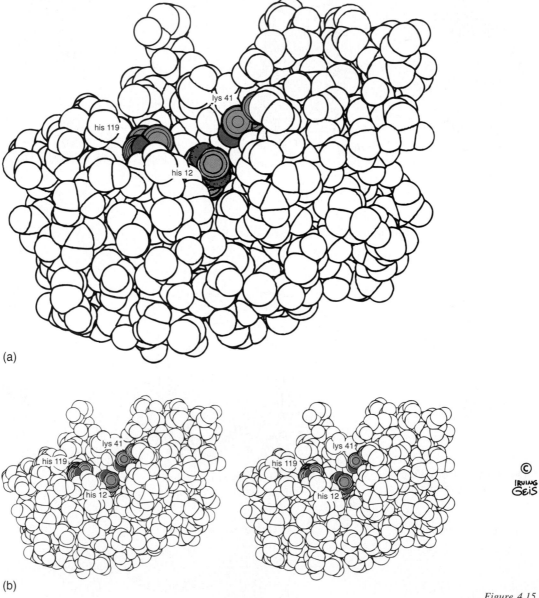

(a)

(b)

*Figure 4.15 continued on
facing page.*

*Continued*

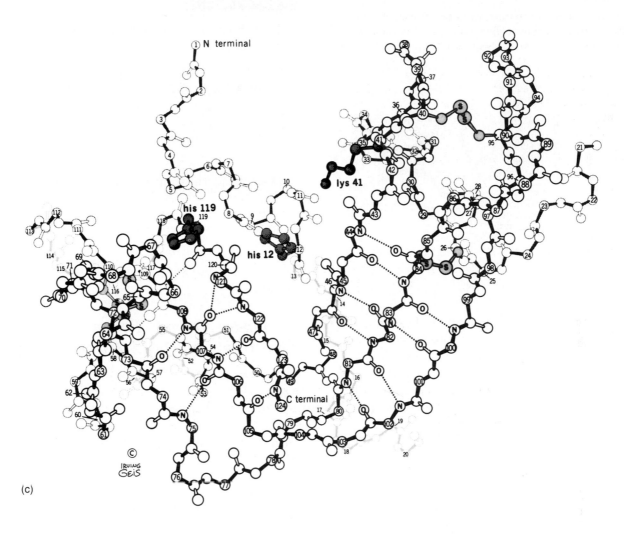

(c)

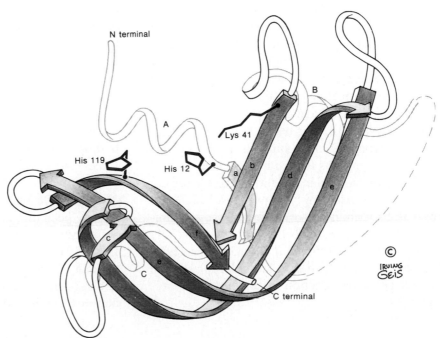

(d)

**Figure 4.16**

Schematic representation of an experiment to demonstrate that the information for folding into a biologically active conformation is contained in the protein's amino acid sequence.

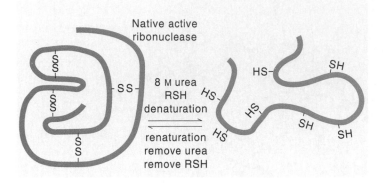

**Table 4.2**

Bond Energies between Some Atoms of Biological Interest

| Energy Values for Single Bonds (kcal/mole) | | | | | |
|---|---|---|---|---|---|
| C—C | 82 | C—H | 99 | S—H | 81 |
| O—O | 34 | N—H | 94 | C—N | 70 |
| S—S | 51 | O—H | 110 | C—O | 84 |

| Energy Values for Multiple Bonds (kcal/mole) | | | | | |
|---|---|---|---|---|---|
| C=C | 147 | C=N | 147 | C=S | 108 |
| O=O | 96 | C=O | 164 | N≡N | 226 |

lent cross-links (fig. 4.16). These conditions have been used since as a general means of denaturing proteins in a manner that completely disrupts the conformation without causing precipitation. The reduced, denatured ribonuclease was enzymatically inactive. Renaturation was carried out by removing the urea, which resulted in refolding of the protein. Finally the mercaptoethanol was removed to permit the air oxidation of the reduced disulfides back to disulfide cross-links. The result of this series of manipulations was an almost complete recovery of the enzymatic activity of the original ribonuclease molecule.

Thus it appears that the information for folding a protein to the native conformation is embodied in the amino acid sequence, because of the many possible disulfide-paired ribonuclease isomers, only one was formed in major yield. Further studies have given similar results with many other proteins.

Despite the elegance and simplicity of Anfinsen's experiment, the current indications are that special proteins accelerate the folding process and may return unfolded or incorrectly folded proteins to their native states. For example, two types of polypeptide-chain-binding (PCB) proteins have been discovered: proteins related to the ''heat shock'' protein with a subunit molecular weight of 70,000 (designated hsp70) and the GroEL family of proteins. The hsp70 proteins appear to be ubiquitous, occurring in bacteria, mitochondria, and the eukaryotic cytosol. They are especially abundant in the endoplasmic reticulum. The GroEL proteins are limited to bacteria, mitochondria, and chloroplasts. We have more to say about the function of some of these proteins in chapter 29.

## Secondary Valence Forces Are the Glue That Holds Polypeptide Chains Together

The studies of Anfinsen not only showed that folding can be a spontaneous process predetermined by the primary amino acid sequence, but they suggested that folded structures are thermodynamically more stable. Thus the folding of globular proteins should be understandable in terms of the types of forces that exist between polypeptide chains and water.

When forces involved in determining protein conformation are being considered, we can usually ignore covalent bond energies, despite their large magnitude (table 4.2). This is because covalent bonds (except for disulfide bonds) are not made or broken when polypeptide chains fold into their native three-dimensional conformations. The bonds that are affected (made or broken) on folding are, by and large, of the noncovalent type involving secondary valence forces. Typically the intermolecular bond energies between noncovalently linked atoms range in value from 0.1 to 6 kcal/mol. The intermolecular forces between noncovalently linked atoms may be grouped into four categories: hydrophobic effects, electrostatic forces, van der Waals forces, and hydrogen bonds (H bonds).

*Hydrogen Bonds.* Many characteristics of H bonds have already been discussed because of the major role they play in stabilizing fibrous proteins. Additional aspects of H bonds to be considered relate more closely to the structures of globular proteins. Evidence for the existence of an H bond comes from the observation of a decreased distance between donor and acceptor groups forming the H bond.

Thus from the van der Waals radii given in table 4.1, we can calculate the distances between nonbonded H and O atoms (2.6 Å) and between nonbonded H and N atoms (2.7 Å). When an H bond is present, this distance is usually reduced by about 0.8 Å in both cases. Some important H-bond donors and acceptors are shown in figure 4.3.

Polypeptides carry a number of H-bond donor and acceptor groups, both in their backbone structure and in their side chains. Water also contains a hydroxyl H-donor group and an oxygen H-acceptor group for making H bonds (see fig. 1.7). Formation of the maximum number of H bonds between a polypeptide chain and water requires the complete unfolding of the polypeptide chain. However, it is not obvious that such an unfolding would result in a net energy gain. The reason is that water is a highly H-bonded structure, and for every H bond formed between water and protein, an H bond within the water structure itself must be broken. The compromise followed by most proteins is to maximize the number of intramolecular H bonds between the backbone peptide groups but to keep most of the potential H-bond-forming side chains of the protein near the protein–water interface where they can interact directly with water.

***Van der Waals Forces.*** Van der Waals interactions are of two types: one attractive and one repulsive. Attractive van der Waals forces involve interactions among induced dipoles that arise from fluctuations in the electron charge densities of neighboring nonbonded atoms. Such interactions amount to 0.1–0.2 kcal/mol; despite their small size, the large number of such interactions that occur when molecules come close together makes such interactions quite significant. Van der Waals forces favor close packing in folded protein structures.

Repulsive van der Waals interactions occur when noncovalently bonded atoms or molecules come very close together. An electron–electron repulsion arises when the charge clouds between two molecules begin to overlap. If two molecules are held together exclusively by van der Waals forces, their average separation is governed by a balance between the van der Waals attractive and repulsive forces. This distance is known as the van der Waals separation. Some van der Waals radii for biologically important atoms are given in table 4.1. The van der Waals separation between two nonbonded atoms is given by the sum of their respective van der Waals radii.

***Hydrophobic Effects.*** Hydrogen bonds and van der Waals forces are of major importance in determining the secondary structures formed by fibrous proteins. To understand the complex folded structures found in globular proteins additional types of interactions between amino acid side chains and water must also be considered. The so-called hydrophobic effects, that lead to the interaction of hydrophobic groups in proteins, are the hardest type of noncovalent interactions to appreciate. Whereas H bonds and van der Waals forces relate primarily to enthalpic factors, hydrophobic effects relate primarily to entropic factors. Furthermore, the entropic factors mainly concern the solvent not the solute.

As a first step to understanding the nature of these effects it is useful to consider the interaction between a small hydrophobic molecule, like hexane and water. It is tempting to ascribe the low water solubility of hexane to van der Waals attractive forces between these small hydrophobic molecules. However, thermodynamic measurements indicate that this is not the case. The hydrophobic molecule hexane has a small favorable enthalpy for solution in water. In spite of this, hexane is poorly soluble in water due to an unfavorable entropic factor. What is this mysterious factor? Owing to the weak enthalpic interactions between hexane and water, the water withdraws slightly in the region of the apolar hydrophobic molecule and forms a relatively rigid hydrogen-bonded network with itself (for example, see the clathrate structure illustrated in fig. 1.10). The network effectively restricts the number of possible orientations of water molecules directly surrounding the dissolved hexane molecules. This ordering of water constitutes an energetically unfavorable entropic effect.

The same type of entropic effect plays a major role in directing the folding of globular proteins. About half of the amino acid side chains in proteins are hydrophobic (e.g., alanine, valine, isoleucine, leucine, and phenylalanine). Entropic effects strongly favor internal locations for these side chains where they are free from contacts with water (e.g., see fig. 1.12, which shows the location of polar and apolar side chains in cytochrome $c$).

The native folded state of a globular protein reflects a delicate balance between opposing energetic contributions of large magnitude. Whereas entropic factors favor the packing of hydrophobic side chains into the interior regions of globular proteins, enthalpic factors favor placing hydrophilic side chains on the surface of the protein where they can interact with water. In the limited number of cases where data are available, the overall entropy of folding appears to be slightly negative (unfavorable) and the overall enthalpy is also slightly negative (favorable). On balance, folding is opposed by the entropy change but favored by the enthalpy change and occurs in most cases because the latter factor outweighs the former.

***Electrostatic Forces.*** Electrostatic forces are of three main types: charge–charge interactions, charge–dipole interactions, and dipole–dipole interactions. The energy of interaction between two charges $Q_1$ and $Q_2$ is proportional to the

## Table 4.3

Energy Dependence of Interaction on the Distance of Separation of the Interacting Species

| Range of Interaction | Type of Interaction |
|---|---|
| $1/R$ | Charge–charge |
| $1/R^2$ | Charge–dipole |
| $1/R^3$ | Dipole–dipole |
| $1/R^6$ | Van der Waals (dipole-induced dipole) attractive forces |
| $1/R^{12}$ | Van der Waals repulsive forces |

product of the charges and inversely proportional to the distance $R$ between them (table 4.3):

$$\text{energy of interaction} \propto \frac{Q_1 Q_2}{R}$$

In solution this interaction is reduced by the dielectric constant of the surrounding medium:

$$\text{energy of interaction} \propto \frac{Q_1 Q_2}{\epsilon R}$$

If the two charges in question are buried within a protein, their interaction energy can be substantially increased because the dielectric constant in the regions inaccessible to water is much lower than the dielectric constant of water. The long-range nature of charge–charge interactions has led to the speculation that such forces can be important in accelerating interaction between proteins, between proteins and nucleic acids, and between proteins and small molecules, such as coenzymes and substrates (see chapter 10).

Favorable charge–charge interactions between oppositely charged amino acids are less significant in determining protein folding than are the ion–dipole interactions between the charged groups of amino acid side chains and water. With very few exceptions side chains containing charged groups as well as polar side chains are located on the protein surface at the protein–water interface.

## Domains Are Functional Units of Tertiary Structure

Within a single folded chain or subunit, contiguous portions of the polypeptide chain often fold into compact local units called domains, each of which might consist, for example of a four-helix cluster or a "barrel" or an antiparallel $\beta$ sheet (fig. 4.17). Sometimes the domains within a protein are very

## Figure 4.17

Examples of three types of structural domains found in many proteins: (*a*) the four-helix cluster; (*b*) the $\beta$ barrel; and (*c*) the antiparallel $\beta$ sheet. (Reprinted by permission of Jane S. Richardson.)

**FOUR HELIX CLUSTER**

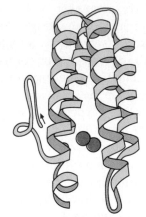

(a)          Myohemerythrin

**$\beta$-BARREL SHAPE**

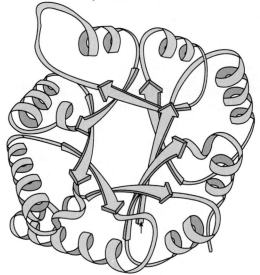

(b)          Triose phosphate isomerase

**ANTIPARALLEL $\beta$ SHEET**

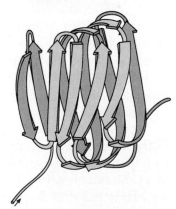

(c)          Tomato bushy stunt virus domain 3

**Figure 4.18**

Papain, a protein in which the domains are very different from each other. (Reprinted by permission of Jane S. Richardson.)

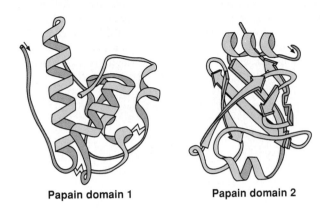

Papain domain 1          Papain domain 2

**Figure 4.20**

Schematic backbone drawing of the elastase molecule, showing the similar β-barrel structures of the two domains. (Reprinted by permission of Jane S. Richardson.)

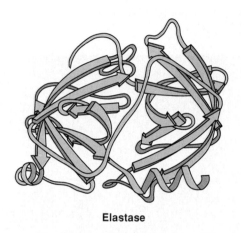

Elastase

**Figure 4.19**

Rhodanese domains 1 and 2 as an example of a protein with two domains that resemble each other extremely closely. Rhodanese is a liver enzyme that detoxifies cyanide by catalyzing the formation of thiocyanate from thiosulfate and cyanide. (Reprinted by permission of Jane S. Richardson.)

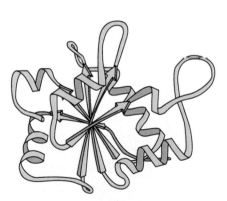

Rhodanese domain 1

Rhodanese domain 2

different from each other, as within the protease papain (fig. 4.18), but often they resemble each other very closely, as in rhodanese (fig. 4.19).

The separateness of two domains within a subunit varies all the way from independent globular domains joined only by a flexible length of polypeptide chain, to domains with tight and extensive contact and a smooth globular surface for the outside of the entire subunit, as in the proteolytic enzyme elastase (fig. 4.20). An intermediate level of domain separateness, characterized by a definite neck or cleft between the domains, is found in phosphoglycerate kinase (fig. 4.21).

Domains as well as subunits can serve as modular bricks to aid in efficient assembly of the native conformation. Undoubtedly, the existence of separate domains is important in simplifying the protein-folding process into separable, smaller steps, especially for very large proteins. There is no strict upper limit on folding size. Indeed, known domains vary in size all the way from about 40 residues to more than 400. Furthermore, it has been estimated that there may be more than a thousand basically different types of domains.

Another important function of domains is to allow for movement. Completely flexible hinges would be impossible between subunits because they would simply fall apart. However, flexible hinges can exist between covalently linked domains. Limited flexibility between domains is often crucial to substrate binding, allosteric control (discussed in chapter 9), or assembly of large structures. In hexokinase, the two domains within the individual subunits hinge toward each other on binding of the substrate glucose, enclosing it almost completely (fig. 4.22). In this manner glucose can be bound in an environment that excludes water

## Figure 4.21

The dumbbell domain organization of phosphoglycerate kinase, with a relatively narrow neck between two well-separated domains. (Copyright 1994 by the Scripps Research Institute/Molecular Graphics Images by Michael Pique using software by Yng Chen, Michael Connolly, Michael Carson, Alex Shah, and AVS, Inc. Visualization advice by Holly Miller, Wake Forest University Medical Center.)

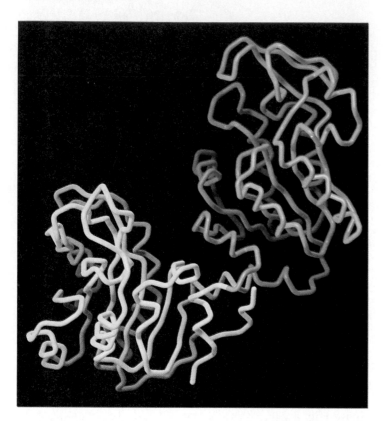

## Figure 4.22

Schematic representation of the change in conformation of the hexokinase enzyme on binding substrate. E and E′ are the inactive and active conformations of the enzyme, respectively. G is the sugar substrate. Regions of protein or substrate surface excluded from contact with solvent are indicated by a crinkled line. Figure 8.3 presents a more detailed view of the hexokinase molecule. (Source: From W. S. Bennett and T. A. Steitz, Glucose-induced conformational changes in yeast hexokinase, *Proc. Natl. Acad. Sci. USA* 75:4848, 1978.)

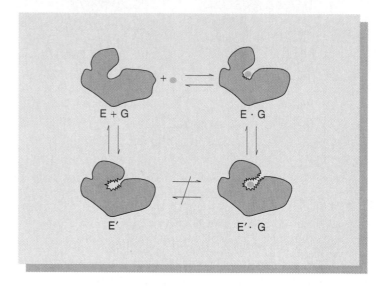

## Figure 4.23

Relative probabilities that any given amino acid occurs in the $\alpha$-helical, $\beta$-sheet, or $\beta$-hairpin-bend secondary structural conformations.

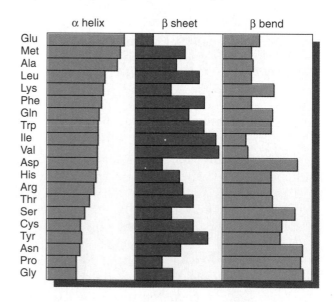

as a competing substrate (see chapter 12 for further details on the hexokinase reaction).

## *Predicting Protein Tertiary Structure*

We began the discussion of globular protein tertiary structure by pointing out that the secondary and tertiary structure is determined by the primary structure and that this is probably a reflection of the fact that the native folded conformation is the most stable structure that can be formed. If this is so, then it should be possible to predict a protein's structure from its primary sequence. At this juncture, such predictions remain an elusive goal. However, most proteins are made of a limited number of domains, which tend to reappear in many different proteins. Since this is the case, it may be possible to predict the structures of many proteins in the future by using the information accumulated from x-ray diffraction studies of related proteins.

It is clear that certain amino acids tend to form particular secondary structures. As shown in figure 4.23 glutamic acid, methionine, and alanine appear to be the strongest $\alpha$-helix formers, whereas valine, isoleucine, and tyrosine are the most probable $\beta$-sheet formers. Proline, glycine, asparagine, aspartic acid, and serine occur most frequently in so-called $\beta$-bend conformations—an unfolded segment that permits a sharp change in direction.

This type of information is of value in the prediction of secondary structural regions of proteins from their amino acid sequences. The observed frequencies of occurrence of an amino acid in a given conformation provide estimates of the probabilities that the same amino acid behaves similarly in a sequence the actual secondary structure of which is unknown. To predict the secondary structure from the sequence, it is consequently necessary only to plot the probabilities for the individual amino acids sequentially, or better, to plot a local average over a few adjacent residues. Plotting such an average accounts for the cooperative nature of secondary-structure formation. The sequences Gly-Pro-Ser and Ala-His-Ala-Glu-Ala, for example, give high joint probabilities for being, respectively, in $\beta$-bend and $\alpha$-helical conformations. However, comparisons of predicted versus directly observed polypeptide conformations give mixed results. This situation is a consequence of two facts: that several amino acids are somewhat ambiguous in their secondary-structure-forming tendencies, and that strong $\beta$-bend formers occasionally turn up in the middle of $\alpha$ helices.

## Quaternary Structure Involves the Interaction of Two or More Proteins

Although many globular proteins function as monomers, biological systems abound with examples of more complex protein assemblies (table 4.4). The higher order organization of globular subunits to form a functional aggregate is referred to as quaternary structure. Protein quaternary structures can be classified into two fundamentally different types. The first involves the assembly of proteins (sometimes referred to as subunits because they constitute a part of the final structure) that have very different structures. Examples range from the hormone insulin, which has two different subunits, to complex assemblies such as ribosomes, which contain 20 or more nonidentical protein subunits in addition to one or more RNA components. The organization of quaternary structures depends on the specific nature of the interactions between each molecular subunit and its neighbors. Each intermolecular interaction generally occurs only once within a given aggregate arrangement, so that the overall complex structure has a highly irregular geometry.

A second, commonly observed pattern of quaternary structure is for a molecular aggregate to have multiple cop-

**Table 4.4**

Molecular Weight and Subunit Composition of Selected Proteins

| Protein | Molecular Weight | Number of Subunits | Function |
|---|---|---|---|
| Glucagon | 3,300 | 1 | Hormone |
| Insulin | 11,466 | 2 | Hormone |
| Cytochrome $c$ | 13,000 | 1 | Electron transport |
| Ribonuclease A (pancreas) | 13,700 | 1 | Enzyme |
| Lysozyme (egg white) | 13,900 | 1 | Enzyme |
| Myoglobin | 16,900 | 1 | Oxygen storage |
| Chymotrypsin | 21,600 | 1 | Enzyme |
| Carbonic anhydrase | 30,000 | 1 | Enzyme |
| Rhodanese | 33,000 | 1 | Enzyme |
| Peroxidase (horseradish) | 40,000 | 1 | Enzyme |
| Hemoglobin | 64,500 | 4 | Oxygen transport |
| Concanavalin A | 102,000 | 4 | Unknown |
| Hexokinase (yeast) | 102,000 | 2 | Enzyme |
| Lactate dehydrogenase | 140,000 | 4 | Enzyme |
| Bacteriochlorophyll protein | 150,000 | 3 | Enzyme |
| Ceruloplasmin | 151,000 | 8 | Copper transport |
| Glycogen phosphorylase | 194,000 | 2 | Enzyme |
| Pyruvate dehydrogenase (E. coli) | 260,000 | 4 | Enzyme |
| Aspartate carbamoyltransferase | 310,000 | 12 | Enzyme |
| Phosphofructokinase (muscle) | 340,000 | 4 | Enzyme |
| Ferritin | 440,000 | 24 | Iron storage |
| Glutamine synthase (E. coli) | 600,000 | 12 | Enzyme |
| Satellite tobacco necrosis virus | 1,300,000 | 60 | Virus coat |
| Tobacco mosaic virus | 40,000,000 | 2,130 | Virus coat |

## Figure 4.24

Tobacco mosaic virus structure (TMV). (*a*) Diagram of TMV structure, an example of a helical virus. The nucleocapsid (protein shell) is composed of a helical assembly of 2,130 identical protein subunits (protomers) with the RNA of the virus spiraling on the inside. (*b*) An electron micrograph of the negatively stained helical capsid (400,000×). In negative staining the virus is immersed in a pool of a heavy-metal salt that is much more electron-dense than the virus. The result is that the darker portions of the figure that surround the virus appear denser than those parts of the figure where the less dense nucleoprotein is located. (© Dennis Kunkel/Phototake.)

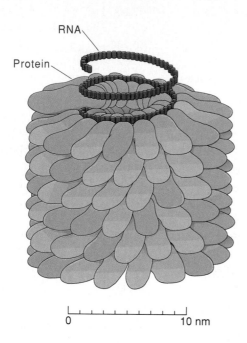

(a)

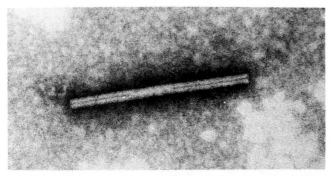

(b)

ies of the same kind of subunits. Owing to the recurrence of specific structural interactions between the subunits, such aggregates typically form regular geometric arrangements. Some outstanding examples of such aggregates are the rodlike viruses and the polyhedral viruses.

## Figure 4.25

The structure of the capsid (protein shell) for an icosahedral virus such as tomato bushy stunt virus. Pentons (P) are located at the 12 vertices of the icosahedron. Hexons (H), of which there are 20, form the edges and faces of the icosahedron. Each penton is composed of five protein subunits and each hexon is composed of six protein subunits. In all, the structure contains 180 protein subunits.

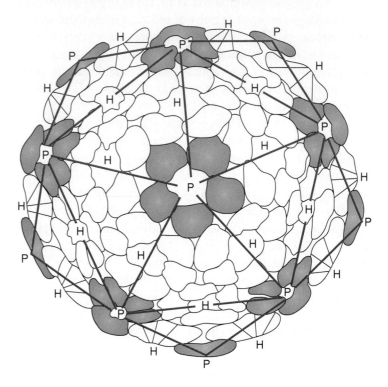

In a typical rodlike virus, the helical aggregation of the protein-coat subunits form a cylindrical container for the virus's nucleic acid (fig. 4.24). The assembly of symmetrical aggregates in a polyhedral virus reflects the structural stabilization that occurs when all the subunits interact in geometrically similar ways, that is, essentially like the atoms in a salt crystal. However, one of the surprising results from x-ray crystallography is that such assemblies often contain distinctly different types of quaternary interactions, even between chemically identical subunits. In the 180-subunit polyhedral protein coat of tomato bushy stunt virus, for example, groups of 5 subunits are in contact around a fivefold symmetry axis, whereas other groups of 6 subunits have a distinct but similar contact around a threefold axis (fig. 4.25). The versatility that permits such nonequivalent associations allows assembly of larger and more complex structures and may be even more common in biological structures too large to have been examined crystallographically.

# Summary

In this chapter we introduced some of the basic principles that govern protein structure. The discussion of protein structures begun in this chapter is continued in many other chapters in this text in which we consider structures designed for specific purposes. In chapter 5 we examine the protein structures for two systems: the protein that transports oxygen in the blood and the proteins that constitute muscle tissue. In chapters 8 and 9 we discuss structures of specific enzymes. In chapters 17 and 24 we consider proteins that interact with membranes. In chapters 30 and 31 we study regulatory proteins that interact with specific sites on the DNA. And finally, in supplement 3 we examine the structures of immunoglobin molecules.

In this chapter we also introduced the subject of the three-dimensional structures of proteins. This is an important subject to keep in mind throughout the text. Our discussion focused on the following points.

1. Most proteins may be divided into two groups: fibrous and globular. Fibrous proteins usually serve structural roles. Globular proteins function as enzymes and in many other capacities.
2. The three most prominent groups of fibrous proteins are the $\alpha$-keratins, the $\beta$-keratins, and collagen.
3. The $\alpha$-keratins are composed of right-handed helical polypeptide chains in which all the peptide NH and carbonyl groups form intramolecular hydrogen bonds. When these helical coils interact, they form left-handed coiled coils.
4. The $\beta$-keratins consist of extended polypeptide chains in which adjacent polypeptides are oriented in either a parallel or an antiparallel fashion. Sheets formed from such extended polypeptide chains may be stacked on top of one another.
5. Collagen fibrils are composed of extended polypeptide chains that are coiled in a left-handed manner. Three of these chains interact by hydrogen bonding and coil together into a right-handed cable. Collagen fibrils are composed of a staggered array of many such cables interacting in a side-by-side manner.
6. The structures of fibrous proteins are determined by the amino acid sequence, by the principle of forming the maximum number of hydrogen bonds, and by the steric limitations of the polypeptide chain, in which the peptide grouping is in a planar conformation.
7. X-ray diffraction provides data from which we can deduce the dimensions of the polypeptide chains in proteins. The use of x-ray techniques is, however, limited to molecules that can be oriented to achieve two- or three-dimensional order.
8. Fibrous proteins may achieve two-dimensional order, but they usually do not achieve three-dimensional order. Therefore, the diffraction pattern of fibrous proteins gives information about the regularly repeating elements along the long axis of the fibers but tells us very little about the orientation of amino acid side chains.
9. Many globular proteins can be crystallized to achieve three-dimensional order. Study of the crystals of a globular protein can lead to a complete determination of its three-dimensional structure.
10. The forces that hold globular proteins together are the same as those that hold fibrous proteins together, but there is less emphasis on regularity and more emphasis on burying the hydrophobic regions in the interior of the protein.
11. The secondary structures found in the keratins recur in smaller patches in globular proteins. Such regions of secondary structure are folded into a seemingly endless array of tertiary structures.
12. Tertiary structures can be understood in terms of a limited number of domains.
13. Quaternary structures are formed between nonidentical subunits to give irregular macromolecular complexes or between identical subunits to give geometrically regular structures.

## Selected Readings

Anfinsen, B. C., Principles that govern the folding of protein chains. *Science* 181:223–230, 1973. Nobel Prize recounting by the man who showed that proteins fold spontaneously into their native structures.

Baron, M., D. G. Norman, and I. D. Campbell, Protein modules. *Trends Biochem. Sci.* 16:13–17, 1991.

Branden, Carl, and John Tooze, *Introduction to Protein Structure.* New York and London: Garland Publishing, 1991.

Cantor, C. R., and P. R. Schimmel, *Biophysical Chemistry,* vols. 1, 2, and 3. New York: Freeman, 1980. Includes several chapters (2, 5, 13, 17, 20, and 21) on the principles of protein folding and conformation.

Chothia, C., Principles that determine the structures of proteins. *Ann. Rev. Biochem.* 53:537–572, 1984.

Chothia, C., and A. V. Finkelstein, The classification and origins of protein folding patterns. *Ann. Rev. Biochem.* 59:1007–1039, 1990.

Chothia, C., and A. Leak, Helix movements in proteins. *Trends Biochem. Sci.* 10:116–118, 1985.

Cohen, C., and D. A. D. Parry, $\alpha$-Helical coiled coils—a widespread motif in protein. *Trends Biochem. Sci.* 11:245–248, 1986.

Creighton, T. E., *Proteins, Structures and Molecular Principles.* New York: Freeman, 1984. Very readable and reasonably comprehensive.

Dorit, R. L., L. Schoenbach, and W. Gilbert, How big is the universe of exons? *Science* 250:1377–1381, 1990. Predicts that there are between 1,000 and 7,000 different kinds of domains in all proteins found in nature.

Farber, G. K., and G. A. Petsko, The evolution of $\alpha/\beta$ barrel enzymes. *Trends Biochem. Sci.* 15:228–234, 1990.

Fasman, G. D., Protein conformation prediction. *Trends Biochem. Sci.* 14:295–299, 1989.

Fersht, A. R., The hydrogen bond in molecular recognition. *Trends Biochem. Sci.* 12:301–304, 1987.

Hogle, J. M., M. Chow, and D. J. Filman, The structure of polio virus. *Sci. Am.* 256(3):42–49, 1987.

Karplus, M., and J. A. McCannon, The dynamics of proteins. *Sci. Am.* 254(4):42–51, 1986. A reminder that proteins are not rigid inflexible structures.

Pauling, L., *The Nature of the Chemical Bond,* 3d ed. Ithaca: Cornell University Press, 1960. A classic on molecular structure.

Pauling, L., and R. B. Corey, Configurations of polypeptide chains with favored orientations around single bonds: two new pleated sheets. *Proc. Natl. Acad. Sci. USA* 37:729–740, 1953. Classic paper.

Pauling, L., R. B. Corey, and H. R. Branson, The structure of proteins: two hydrogen-bonded helical configurations of the polypeptide chain. *Proc. Natl. Acad. Sci. USA* 27:205–211, 1951. Another classic paper.

Richardson, J. S., and D. C. Richardson, The *de novo* design of protein structures. *Trends Biochem. Sci.* 14:304–309, 1989.

Rose, C. D., A. R. Geselowizt, G. J. Lesser, R. H. Lee, and M. H. Zehfus, Hydrophobicity of amino acid residues in globular proteins. *Science* 229:834–838, 1985.

Rossman, M. G., and P. Argos, Protein folding. *Ann. Rev. Biochem.* 50:497–532, 1981.

Rossman, M. G., and J. E. Johnson, Icosahedral RNA virus structure. *Ann. Rev. Biochem.* 58:533–573, 1989.

Sali, A., J. P. Overington, M. S. Johnson, and T. L. Bundell, From comparisons of protein sequences and structures to protein modelling and design. *Trends Biochem. Sci.* 15:235–240, 1990.

Tonegawa, S., The molecules of the immune system. *Sci. Am.* 253(4):122–130, 1985.

Valegard, K., L. Liljas, K. Fridborg, and T. Unge, The three-dimensional structure of the bacterial virus MS2. *Nature* 345:36–41, 1990.

Wuthrich, K., Protein structure determination in solution by nuclear magnetic resonance spectroscopy. *Science* 243:45–50, 1989. The most effective technique for determining protein fine structure in cases where x-ray diffraction cannot be used.

Wright, P. E., What can two-dimensional NMR tell us about proteins? *Trends Biochem. Sci.* 14:255–259, 1989.

Yang, J. T., Protein secondary structure and circular dichroism: a practical guide. *Chemtracts, Biochem. Mol. Biol.* 1:484–490, 1990.

## Problems

1. The principal force driving the folding of some proteins is the movement of hydrophobic amino acid side chains out of an aqueous environment. Explain.

2. Outline the hierarchy of structural organization in proteins.

3. What is the role of loops or short segments of "random" structure in a protein whose structure is primarily $\alpha$-helical?

4. What are some consequences of changing a hydrophilic residue to a hydrophobic residue on the surface of a globular protein? What are the consequences of changing an interior hydrophobic residue to a hydrophilic residue in the protein?

5. Some proteins are anchored to membranes by insertion of a segment of the N terminus into the hydrophobic interior of the membrane. Predict (guess) the probable structure of the sequence (Met-Ala-(Leu-Phe-Ala)$_3$-(Leu-Met-Phe)$_3$-Pro-Asn-Gly-Met-Leu-Phe). Why would this sequence be likely to insert into a membrane?

6. Suppose that every other Leu residue in the peptide shown in problem 5 was changed to Asp. Would this necessarily alter the secondary structure? Explain whether insertion into the membrane would be altered.

7. If you had several helical springs, how could you determine whether each spring was right- or left-handed?

8. "Left- and right-handed $\alpha$ helices of polyglycine are equally stable." Defend or refute this statement.

9. Urea

$$H_2NCNH_2$$
(with O double-bonded to C)

and guanadinium chloride ($[(H_2N)_2C{=}NH_2]^+Cl^-$) are commonly used denaturants (cause loss of protein conformation). Provide explanations for how these substances might disrupt protein structures.

10. Many proteins (e.g., important metabolic enzymes) are insoluble in water and are found "attached" to membranes within cells. What amino acid residues do you expect to find on the "side" of the protein that "attaches" to the membrane?

11. Often the enzymes mentioned in problem 10 are purified with the aid of detergents such as sodium dodecylsulfate

$$CH_3(CH_2)_{11}-O-S-O^-\ ^+Na$$
(with O double-bonded above and below S)

What is the function of the detergent?

12. It might be argued that in protein structure, as in everyday life, it is a "right-handed world." Use examples of protein structure discussed in this chapter to support this contention.

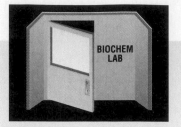

# X-Ray Diffraction: Applications to Fibrous Proteins

X-ray diffraction played a major role in the discovery of the structure of fibrous proteins. In most cases the fibers under study are oriented in two dimensions by stretching. In this analysis we illustrate how the technique is used to study the $\alpha$ form of the synthetic polypeptide poly-L-alanine.

A stretched fiber containing many poly-L-alanine molecules is suspended vertically and exposed to a collimated monochromatic beam of $CuK_\alpha$ x-rays, as shown in figure 1(a). Only a small percentage of the x-ray beam is diffracted; most of the beam travels through the specimen with no change in direction. A photographic film is held in back of the specimen. A hole in the center of the film allows the incident undiffracted beam to pass through.

Coherent diffraction occurs only in certain directions specified by Bragg's law: $2d \sin \theta = n\lambda$, where $d$ is the distance between identical repeating structural elements, $\theta$ is the angle between the incident beam and the regularly spaced diffracting planes, $\lambda$ is the wavelength of x-rays used, and $n$ is the order of diffraction, which may equal any integer but is usually strongest for $n = 1$. For small $\theta$, $\sin \theta \approx \theta$ and $d \approx 1/\theta$, so that a spot far out on the photographic film is indicative of a repeating element of small dimension.

Figure 1(b) shows the diffraction pattern obtained when the fiber axis is normal to the beam. Note the strong off-vertical reflection at 5.4 Å (arrow). A different diffraction pattern (c) is obtained when the fiber axis is inclined to the beam at 31°. Note the strong reflection at 1.5 Å in the upper part of the diagram (arrow).

## Figure 1

(a) Experimental arrangement for obtaining x-ray diffraction pattern shown in (b). (b) Diffraction pattern of the $\alpha$ form of a cluster of poly-L-alanine molecules oriented vertically. (c) Diffraction pattern of the same fiber bundle with the fiber axis inclined to the beam at 31°. (b and c Brown & Trotter, 1956. Source: C. H. Bamford et al., *Synthetic Polypeptides*. Copyright © 1956, Academic Press, Orlando, Fla.)

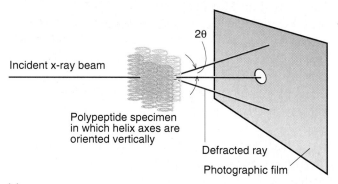

Incident x-ray beam

$2\theta$

Polypeptide specimen
in which helix axes are
oriented vertically

Defracted ray

Photographic film

(a)

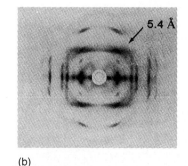

5.4 Å

(b)

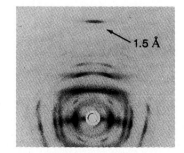

1.5 Å

(c)

# Three-Dimensional Protein Structure Can Be Analyzed by X-Ray Diffraction of Protein Crystals

The quality of the information potentially available through x-ray diffraction depends on the degree or extent of order of the protein molecules in a given sample. A sample under investigation usually consists of a hydrated purified protein. If the individual protein molecules are packed in a random order relative to one another, the distances between atoms or groups of atoms can be derived, but we cannot tell how these spacings are ordered in three dimensions.

If the protein is fibrous, it is often possible to learn its two-dimensional order. For example, when a fibrous sample consisting of long $\alpha$ helices is stretched, the helix axes become oriented in the direction of stretching. The resulting fibrous bundle gives a characteristic x-ray diffraction pattern indicating an ordered molecular arrangement along the helix axis. As we have seen, the pitch of the $\alpha$ helix (5.4 Å) and the advance per residue along the helix axis (1.5 Å) were detected by this technique. But the orientation of specific atoms in the helix cannot be determined from diffraction patterns of stretched fibers because of the lack of three-dimensional order. To deduce the correct three-dimensional structure for the $\alpha$ helix (and the $\beta$ sheet), it was necessary to work with molecular models. (Currently, computer programs are available for such purposes.) By trial and error, model structures were built that were consistent with the steric limitations of the polypeptide chain that had the helix pitch and the advance per residue indicated by the diffraction pattern of stretched fibers.

The most information about a protein's structure is obtained from ordered three-dimensional protein crystals; this is the main interest of x-ray crystallographers. The goal in x-ray crystallography is to obtain a three-dimensional image of a protein molecule in its native state at a sufficient level of detail to locate its individual constituent atoms. The way this is done can most easily be appreciated by considering the more familiar problem of how we obtain a magnified image of an object in a conventional light microscope. In a light microscope, light from a point source is projected on the object we wish to examine. When the light waves hit the object, they are scattered so that each small part of the ob-

**Figure 1**

Schematic diagram of the procedures followed for image reconstruction in light microscopy (top) and x-ray crystallography (bottom).

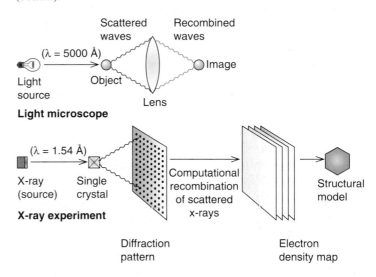

ject essentially serves as a new source of light waves. The important point is that the light waves scattered from the object contain information about its structure. The scattered waves are collected and recombined by a lens to produce a magnified image of the object (fig. 1).

Given this picture, we might ask what prevents us from simply putting a protein molecule in place of our object and viewing its magnified image. The basic problem here is one of resolution. The resolution, or extent of detail, that can be recovered from any imaging system depends on the wavelength of light incident on the object. Specifically, the best resolution obtainable equals $\lambda/2$, or one-half the wavelength of the incident light. Because $\lambda$ lies in the range of 4,000–7,000 Å for visible light, a visible-light microscope clearly does not have the resolving power to distinguish the atomic structural detail of molecules. What we need is a form of incident radiation with a wavelength comparable to interatomic distances. X-rays emitted from excited metal atoms, with wavelengths in the range of one to a few angstroms, would be most suitable.

However, simply replacing a visible-light source with an x-ray source does not solve all the problems. For example to get a three-dimensional view of a protein, some provision must be made for looking at it from all possible angles, an obvious impossibility when dealing with a single molecule. Furthermore, when x-rays interact with proteins, very few of the rays are scattered. Most x-rays pass through the protein, but a relatively large number of them interact destructively with the protein, so that a single molecule would be destroyed before scattering enough x-rays to form a useful image. Both these problems are overcome by replacing a single protein molecule with an ordered three-dimensional array of many molecules that scatters x-rays essentially as if it were one molecule. The ordered array of protein molecules forms a single crystal, so the general technique is called protein x-ray crystallography.

The problems do not end here, because although the protein crystal readily scatters incident x-rays, no lens materials are available that can recombine the scattered x-rays to produce an image. Instead, the best that can be done is to directly collect the scattered x-rays in the form of a diffraction pattern. Although recording the diffraction pattern results in loss of some important information, experimental techniques have been developed for recovering the lost information. Eventually the scattered waves can be mathematically recombined in a computational analog of a lens. By collecting the diffraction pattern of the crystal in many orientations, it is possible to construct a three-dimensional image of the protein molecule.

Crystals suitable for protein x-ray studies may be grown by a variety of techniques, which generally depend on solvent perturbation methods for rendering proteins insoluble in a structurally intact state. The trick is to induce the molecules to associate with each other in a specific fashion to produce a three-dimensionally ordered array. A typical protein crystal useful for diffraction work is about 0.5 mm on a side and contains about $10^{12}$ protein molecules (an array $10^4$ molecules long along each crystal edge). Note especially that, because protein crystals are from 20 to 70% solvent by volume, crystalline protein is in an environment that is not substantially different from free solution.

The x-ray radiation usually employed for protein crystallographic studies is derived from the bombardment of a copper target with high-voltage (50 kV) electrons, producing characteristic copper x-rays with $\lambda = 1.54$ Å. Figure 2 shows, in schematic fashion, the x-ray diffraction pattern from a protein crystal. Several features about this pattern bear explanation. First, as you can see, the diffraction pattern consists of a regular lattice of spots of different intensities. The spots are due to destructive interference of waves

## Figure 2

Schematic view of an x-ray diffraction pattern. The spacing of the spots is reciprocally related to the dimensions of the repeating unit cell of the crystal. The symmetry of the spots (e.g., the mirror planes in the sample shown) and the pattern of missing spots (alternating spots along the mirror axes) give information on how molecules are arranged in the unit cell. Information concerning the structure of the molecule is contained in the intensities of the spots. Spots closest to the center of the film arise from large-scale or low-resolution structural features of the molecule, whereas those farther out correspond to progressively more detailed features. Circles show 5-Å and 3-Å regions of resolution. Mirror axes are labeled $m$. Spacing of vertically oriented spots $b*$ and horizontally oriented spots $a*$ are reciprocally related to $b$ and $a$, the dimensions of the unit cell.

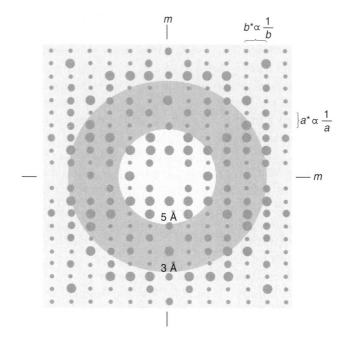

scattered from the repeating unit of the crystal. For the crystal with the diffraction pattern shown, the repeating unit (or crystal unit cell) contains four symmetrically arranged protein molecules. Corresponding symmetrical features appear in the spot intensity pattern. Further, the lattice spacing of the diffraction spots is inversely proportional to the actual dimensions of the crystal's repeating unit or unit cell. Consequently, both the crystal's unit-cell dimensions and general molecule packing arrangement can be derived from inspection of the crystal's diffraction pattern.

Information concerning the detailed structural features of the protein is contained in the intensities of the diffraction spots. All the atoms in the protein structure make individual contributions to the intensity of each diffraction spot. Therefore, to deduce the three-dimensional structure, all the spots must be measured, either by scanning the x-ray films with a densitometer or by measuring the diffraction spots individually with a scintillation counter.

Initial studies of a protein's tertiary structure are generally carried out at low resolution, that is, using intensity data near the origin (center) of the diffraction pattern. Diffraction data near the origin reflect large-scale structural features of the molecule, whereas those nearer the edge correspond to progressively more detailed features. Figure 3 provides examples of electron-density maps calculated at different resolutions to show how various levels of structural detail appear at different degrees of resolution.

A powerful aspect of protein crystallography is that once the native structure is known, various cofactors or enzyme substrate analogs can be bound to the molecule in the crystal. By simply measuring the diffraction intensities, we can compute a new map that allows direct and explicit examination of the structural interactions between the native protein and its substrate or cofactor molecules. Detailed analysis of these interactions has provided much of the foundation for our current understanding of many protein catalytic and functional properties.

**Figure 3**

Views of crystallographic electron-density maps, showing how the structural detail revealed depends on the resolution of the data used to compute the maps. The actual molecular structure is inserted in its true position in the electron-density maps.

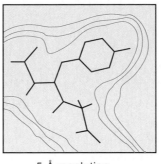

5-Å resolution

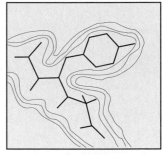

3-Å resolution

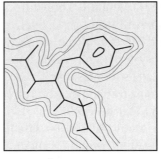

2-Å resolution

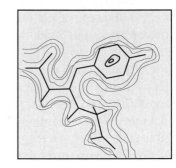

1.5-Å resolution

# Radiation Techniques for Examining Protein Structure

The Anfinsen experiment raises the question: Must we always use x-ray diffraction to determine a protein's structure? In the case of ribonuclease the structure is known from x-ray diffraction, but at the time of Anfinsen's investigation it was not. He relied heavily on the fact that the reconstituted enzyme had the same activity as the native enzyme. This remains an acceptable approach as long as one is working with an enzyme that has a demonstrable catalytic activity, but what of the many proteins that do not?

Because of such limitations, x-ray diffraction is still the most powerful tool for determining protein structure. However, apart from the enormous amount of work it requires to use, x-ray diffraction suffers from two disadvantages: (1) it requires that a protein be available in the crystalline form, which is not always the case; and (2) there is no absolute assurance that a protein's structure in the crystalline form is the same as its structure in solution, which may be more like the environment in the cell.

Fortunately, there are at least a dozen other techniques, less fussy in their demands, that may be used to investigate protein structure either in the solid state or the solution state (table 4C). The information obtained by these procedures is very extensive. As you can see from the table, it covers virtually every aspect of protein structure, with considerable overlap between what is yielded by different techniques. Excellent references explaining the use of other radiation techniques are given at the end of this chapter.

**Table 4C**

Radiation Techniques for Examining Protein Structure

| Technique | Information Obtained |
|---|---|
| (a) X-ray diffraction | Detailed atomic structure |
| (b) Nuclear magnetic resonance spectrometry (NMR) | Structure of specific sites; ionization state of individual residues |
| (c) Electron paramagnetic resonance (EPR) | Structure of specific sites; this includes structures of small molecules whether they be carbohydrates, lipids, small proteins or complexes between segments of DNA and proteins |
| (d) Spectrometry | |
| (e) Optical absorption spectroscopy | Measurement of concentrations or rate of reactions |
| (f) Infrared absorption spectroscopy | Type and extent of secondary structure |
| (g) Light scattering | Molecular weight and size |
| (h) Ultraviolet absorption spectroscopy | Concentration and conformation |
| (i) Fluorescence | Proximity between specific sites |
| (j) Raman scattering | Structure of specific sites |
| (k) Optical rotatory dispersion (ORD) | Type and extent of secondary structure |
| (l) Circular dichroism (CD) | Type and extent of secondary structure |
| (m) Fluorescence polarization | Molecular weight, shape, flexibility, and orientation of secondary-structure units |

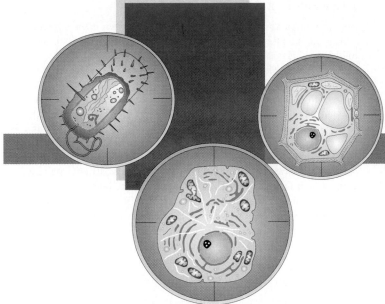

# Functional Diversity of Proteins

*Each protein is exquisitely suited to carry out a specific function.*

Now that some of the major types of protein structures have been described it is appropriate to turn to the question of how protein structure relates to protein function. To explore this question, two protein systems, hemoglobin and the actin-myosin complex, are examined in detail.

## Hemoglobin Is an Allosteric Oxygen-Binding Protein

Hemoglobin is the principal soluble protein in the red blood cells (erythrocytes). It is an oxygen-binding protein, the main task of which is to pick up oxygen as the blood passes through the lungs and to release it in the capillaries of other tissues. A closely related protein, myoglobin, serves to store oxygen temporarily in the muscles and other peripheral tissues, where the oxygen is consumed by cellular metabolism. Myoglobin consists of a single polypeptide subunit with a molecular weight of 16,900 and a bound cofactor, heme. Hemoglobin has an $\alpha_2\beta_2$ subunit structure, with a heme bound to each of the four subunits. The $\alpha$ and $\beta$ subunits are structurally very similar to each other and to the single subunit of myoglobin. (The $\alpha$ subunits have 141 amino acids; the $\beta$ subunits, 146.) The heme

## Figure 5.1

The heme pocket. The helices of hemoglobin (and myoglobin) form a hydrophobic pocket for the heme and provide an environment where the iron atom reversibly binds oxygen. The chemical structure of heme is shown in figure 5.10 and is described in atomic detail in chapters 10 and 14. (Illustration copyright by Irving Geis. Reprinted by permission.)

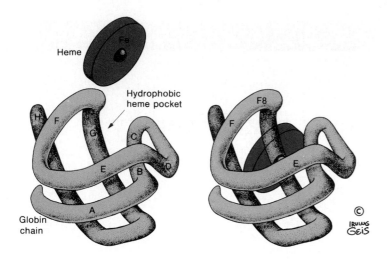

## Figure 5.2

Equilibrium curves measure the affinity for oxygen of hemoglobin and of the simpler myoglobin molecule. Myoglobin, a protein of muscle, has just one polypeptide chain and resembles a single subunit of hemoglobin. The vertical axis gives the amount of oxygen bound to one of these proteins, expressed as a percentage of the total amount that can be bound. The horizontal axis measures the partial pressure of oxygen in a mixture of gases with which the solution is allowed to reach equilibrium. For myoglobin, the equilibrium curve is hyperbolic. Myoglobin absorbs oxygen readily but becomes saturated at a low pressure. The hemoglobin curve is sigmoidal. Initially, hemoglobin is reluctant to take up oxygen, but its affinity increases with oxygen uptake. At arterial oxygen pressure, both molecules are nearly saturated, but at venous pressure, myoglobin gives up only about 10% of its oxygen, whereas hemoglobin releases roughly half. At any partial pressure, myoglobin has a higher affinity than hemoglobin, which allows oxygen to be transferred from blood to muscle.

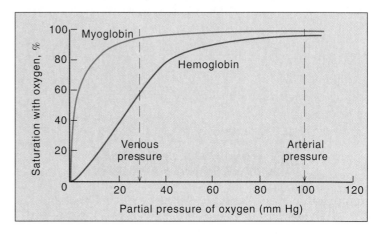

cofactor in each subunit consists of a porphyrin (protoporphyrin IX) with a central iron atom. (The structure of heme is shown in fig. 5.1 and is discussed in more detail in chapters 10 and 14.)

Figure 5.1 shows the overall folding plan of the polypeptide chains of hemoglobin and myoglobin. Each chain has eight stretches of $\alpha$ helix, which are labeled A–H. These helical segments fold together to form a pocket that surrounds the heme. The amino acid side chains lining the heme-binding pocket are predominantly apolar, and thus are well suited to binding the hydrophobic porphyrin ring. This apolar, water-free environment allows the $Fe^{2+}$ iron atom to bind $O_2$ reversibly while minimizing the opportunity for oxidation of the iron to the +3 state.

Although the components of myoglobin and hemoglobin are remarkably similar, their physiological responses are very different. On the basis of weight, each molecule binds about the same amount of oxygen at high oxygen tensions (pressures). At low oxygen tensions, however, hemoglobin releases bound oxygen much more readily. These differences are reflected in the oxygen-binding curves of the purified proteins in aqueous solution (fig. 5.2). The oxygen-binding curve for myoglobin (Mb) is hyperbolic in shape, as is expected for a simple one-to-one association of myoglobin and oxygen:

$$Mb + O_2 \rightleftharpoons MbO_2$$

$$K_f = \frac{[MbO_2]}{[Mb][O_2]} = \text{equilibrium formation constant}$$

If $y$ is the fraction of myoglobin molecules saturated, and if the oxygen concentration is expressed in terms of the partial pressure of oxygen $[O_2]$,[a] then

$$K_f = \frac{y}{[1 - y][O_2]} \qquad \text{and} \qquad y = \frac{K_f[O_2]}{1 + K_f[O_2]} \quad \textbf{(1)}$$

This is the equation of a hyperbola, as shown in figure 5.2.

Hemoglobin (Hb) behaves differently. Its sigmoidal binding curve can be fitted by an association-constant expression with a greater-than-first-power dependence on the oxygen concentration:

$$K_f = \frac{[HbO_2]}{[Hb][O_2]^n} \qquad \text{and} \qquad y = \frac{K_f O_2{}^n}{1 + K_f O_2{}^n} \quad \textbf{(2)}$$

Under physiological conditions the value of $n$ is around 2.8. This indicates that the binding of oxygen molecules to the

---

[a] Partial pressure is usually indicated by a lower-case p to the left. For simplicity the p has been deleted.

four hemes in hemoglobin does not occur independently: Binding to any one heme is affected by the state of the other three. The first oxygen attaches itself with the lowest affinity, and successive oxygens are bound with a somewhat higher affinity, and so on. The exact value of $n$ in equation 2 is a function of other factors discussed later on. In general, a value of $n > 1$ indicates cooperative binding of $O_2$ (or positive cooperativity), a value of $n < 1$ indicates anticooperative binding (or negative cooperativity), and a value of $n = 1$ indicates no cooperativity.

The cooperative binding of oxygen by hemoglobin is ideally suited to the conditions involved in oxygen transport. In the lungs, where the oxygen tension is relatively high, hemoglobin can become nearly saturated with oxygen. In the tissues, however, where the oxygen tension is relatively low, hemoglobin can release about half its oxygen (see fig. 5.2). If myoglobin were used as the oxygen transporter, less than 10% of the oxygen would be released under similar conditions. The positive cooperativity associated with oxygen binding to hemoglobin is a special case of allostery in which the binding of ''substrate'' to one site stimulates or inhibits the binding of ''substrate'' to another site on the same multisubunit protein. Allostery is also commonly observed in regulatory proteins (chapter 9).

## The Binding of Certain Factors to Hemoglobin Influences Oxygen Binding in a Negative Way

The combination of Hb with $O_2$ is not only a function of the oxygen tension but also depends on the pH, $CO_2$ concentration, and glycerate-2,3-bisphosphate (GBP or BPG). GBP (fig. 5.3) binds preferentially to the deoxygenated form of hemoglobin with a dissociation constant of about $10^{-5}M^{-1}$. Its dissociation constant with $HbO_2$ is about $10^{-3}M^{-1}$. Since the concentration of GBP and hemoglobin are both about 5 mM in the erythrocyte, we expect most of the deoxy form to be complexed with GBP and much of the oxyhemoglobin to be free of GBP. The net effect of the GBP is to shift the oxygen-binding curve to higher oxygen tensions (fig. 5.4). This shift is not sufficient to lower the binding of oxygen at the high oxygen tensions in the capillary tissues of the lungs, but it is sufficient to cause a substantially greater release of oxygen at the lower oxygen tensions that exist in the peripheral tissues.

Carbon dioxide ($CO_2$) and hydrogen ions ($H^+$) both have negative effects on oxygen binding also. These effects are closely related and together are known as the Bohr effect. A substantial amount of $CO_2$ generated by decarboxylation reactions of intermediary metabolism diffuses from cells through interstitial fluid into the blood plasma, with

**Figure 5.3**

The structure of glycerate-2,3-bisphosphate, an allosteric effector for hemoglobin oxygen release.

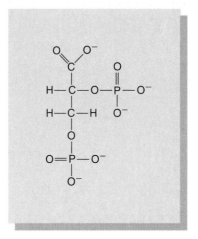

**Figure 5.4**

Oxygen-binding curve for hemoglobin as a function of the partial pressure of oxygen. Two curves are shown: one in the absence and one in the presence of glycerate-2,3-bisphosphate (GBP). GBP decreases the affinity between oxygen and hemoglobin, as shown by the displacement of the binding curve to high oxygen concentrations in its presence.

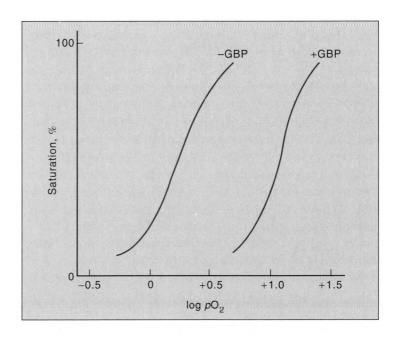

only a small fraction becoming hydrated as carbonic acid. This is because the nonenzymatic hydration of $CO_2$ to form $H_2CO_3$ is a slow reaction, with a half-time of about 10 s. On

## Figure 5.5

Transport of oxygen ($O_2$) and carbon dioxide ($CO_2$) in the circulatory system. In most tissues $O_2$ is released and $CO_2$ is withdrawn by the red blood cells; in the lungs these processes are reversed.

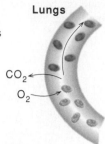

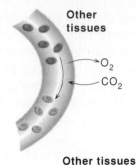

**Lungs**

$$HCO_3^- + H^+ \rightleftharpoons H_2CO_3 \xrightarrow{\text{Carbonic anhydrase}} CO_2 + H_2O$$

$$O_2 + HHb^+ \rightleftharpoons H^+ + HbO_2$$

$$HHb^+ + O_2 + HCO_3^- \rightleftharpoons HbO_2 + CO_2 + H_2O$$

**Other tissues**

$$H^+ + HbO_2 \rightleftharpoons HHb^+ + O_2$$

$$CO_2 + H_2O \xrightarrow{\text{Carbonic anhydrase}} H_2CO_3 \rightleftharpoons H^+ + HCO_3^-$$

$$HbO_2 + CO_2 + H_2O \rightleftharpoons HHb^+ + O_2 + HCO_3^-$$

entry into erythrocytes, hydration of $CO_2$ occurs in a reaction that is catalyzed rapidly by carbonic anhydrase, an enzyme that is localized in the erythrocytes. At the pH of blood (about 7.4) the carbonic acid largely dissociates into $H^+$ and $HCO_3^-$:

$$CO_2 + H_2O \xrightarrow{\text{Carbonic anhydrase}} H_2CO_3 \rightleftharpoons H^+ + HCO_3^-$$

The release of protons through these two consecutive reactions explains how increasing concentrations of $CO_2$ cause a lowering of pH. Lowering the pH decreases the affinity of hemoglobin for oxygen, because protons, like GBP, bind preferentially to deoxyhemoglobin. For example, at pH 7.6 and an oxygen tension of 40 mm Hg, hemoglobin retains more than 80% of its oxygen; if the pH is lowered to 6.8, only 45% of the oxygen is retained. The negative effect of $CO_2$ on oxygen binding is mainly due to the tendency of $CO_2$ to lower the pH (i.e., raise the $H^+$ concentration); however, $HCO_3^-$ also binds preferentially to deoxyhemoglobin, and this binding also promotes the release of $O_2$.

The Bohr effect is closely related to the major roles that hemoglobin plays in disposing of the $CO_2$ produced in tissues, and in controlling the blood pH. While oxygen is being delivered to the tissues in the venous blood, the $CO_2$ is being removed from the tissues (fig. 5.5). The $CO_2$ that diffuses into the erythrocytes is rapidly converted into carbonic acid, which in turn dissociates into $H^+$ and $HCO_3^-$. The protons produced by this dissociation would lower the pH and reverse the dissociation if it were not for the buffer-

ing action of the hemoglobin. Loss of oxygen decreases the acid dissociation constant of the hemoglobin so that it picks up the excess protons. This serves two purposes: it helps to keep the pH constant and it enables the blood to absorb more $CO_2$. This $CO_2$ is ultimately released when the blood reaches the lungs and the hemoglobin again becomes oxygenated. Another important point is that the majority of the $HCO_3^-$ produced in the erythrocyte diffuses into the venous blood. Indeed, the pH of the blood system is controlled within narrow limits by the buffering action of the bicarbonate and the hemoglobin with minor assistance from other proteins in the blood.

One must marvel at the way various factors work in concert so that hemoglobin can be useful in multiple roles: oxygen deliverer, carbon dioxide remover, and pH stabilizer. From the explanation we have given it should be clear why the hemoglobin is confined to cells in the blood rather than being present as a free plasma protein. The intracellular carbonic anhydrase and GBP of the erythrocytes are essential for efficient hemoglobin function.

## X-Ray Diffraction Studies Reveal Two Conformations for Hemoglobin

X-ray diffraction studies on fully oxygenated hemoglobin and deoxygenated hemoglobin have shown that the molecule is capable of existing in at least two states, with significant differences in tertiary and quaternary structures (fig. 5.6). Further studies on partially oxygenated hemoglobin may indicate additional intermediate structures between

### Figure 5.6

Three-dimensional structure of oxy- and deoxyhemoglobin as determined by x-ray crystallography. This is a view down the twofold symmetry axis, with the $\beta$ chains on top. In the oxy–deoxy transformation (quaternary motion) $\alpha_1\beta_1$ and $\alpha_2\beta_2$ dimers move as units relative to each other. This allows glycerate-2,3-bisphosphate to bind to the larger central cavity in the deoxy conformation. A close-up of the binding site is shown in figure 5.9. (Illustration copyright by Irving Geis. Reprinted by permission.)

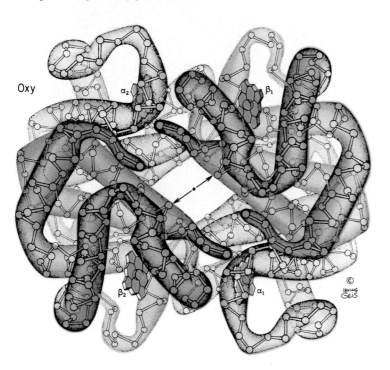

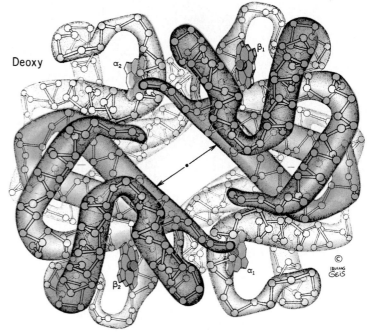

these two extremes. Presently the two-state model serves as a useful conceptual framework for explaining the cooperative nature of oxygen binding.

The hemoglobin tetramer is composed of two identical halves (dimers), with the $\alpha_1\beta_1$ subunits in one dimer and the $\alpha_2\beta_2$ subunits in the other. The subunits within the dimers are tightly held together while the dimers themselves are capable of motion with respect to each other (fig. 5.7). The interface between the movable dimers contains a network of salt bridges and hydrogen bonds when hemoglobin is in the deoxy conformation (fig. 5.8). The quaternary transformation that takes place on binding of oxygen causes the breakage of these bonds.

The effects of $H^+$, $HCO_3^-$, and glycerate-2,3-bisphosphate on oxygen binding can be understood in terms of their stabilizing effect on the deoxy conformation. The decreased oxygen binding as the pH is lowered from 7.6 to 6.8 suggests the involvement of histidine side chains, because these are the only side-chain groups in proteins that commonly have a $pK_a$ in this range. Certain histidines in the charged form make salt linkages that contribute to the stability of the deoxy form (see fig. 5.8). As the pH is lowered, these histidines tend to become charged, which increases the stability of the deoxy form. Such a change should inhibit a structural transition to the oxy form and thereby lower the affinity of the protein for oxygen. Similarly, glycerate-2,3-bisphosphate binds most strongly to the deoxy form (fig. 5.9) and thereby discourages the transition to the oxy form, which lowers the affinity for oxygen. Bicarbonate ions bind to the amino terminal $\alpha$-amino groups in hemoglobin; this binding also favors the deoxy conformation. Binding of $HCO_3^-$ is favored by the high $HCO_3^-$ concentration created by the carbonic anhydrase reaction in the red cells. Hemoglobin thus serves as a carrier of $HCO_3^-$ from tissues to the lungs, where $CO_2$ is discharged.

## Figure 5.7

The deoxy–oxy shift on binding oxygen in one hemoglobin molecule. The projection shown in this figure is approximately perpendicular to the one shown in figure 5.6. The $\alpha_1\beta_1$ dimer moves as a unit relative to the $\alpha_2\beta_2$ dimer. The interface between the two dimers is crucial to the cooperativity effect in hemoglobin. The interface is not visible in this figure (see fig. 5.8). (Illustration copyright by Irving Geis. Reprinted by permission.)

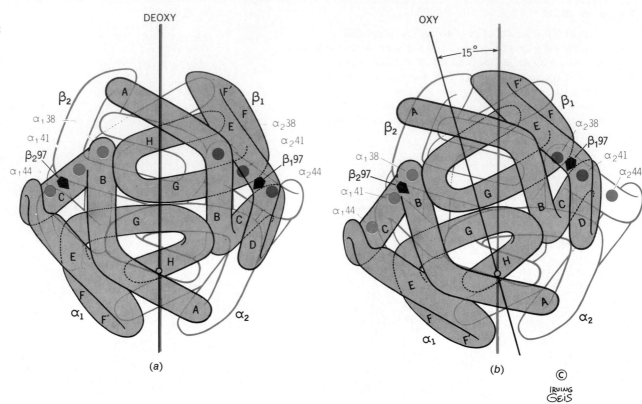

(a)

(b)

## Figure 5.8

(a) The $\alpha_1\beta_2$ (and $\alpha_2\beta_1$) interface is shown schematically at the lower left and in detail (b). This is the regulatory zone of the hemoglobin molecule, which contains crucial hydrogen bonds and salt bridges. (b) All the hydrogen bonds and salt bridges shown here (dotted lines) exist only in the deoxy state, with the exception of $\alpha_1 41$–$\beta_2 40$ and $\alpha_1 94$–$\beta_2 102$, which exist only in the oxy state. Only the $\alpha$ carbons are shown in the backbone structure of the hemoglobin, except in the region of the interface. (Illustration for part b copyright by Irving Geis. Reprinted by permission.)

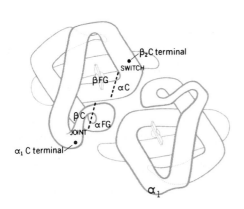

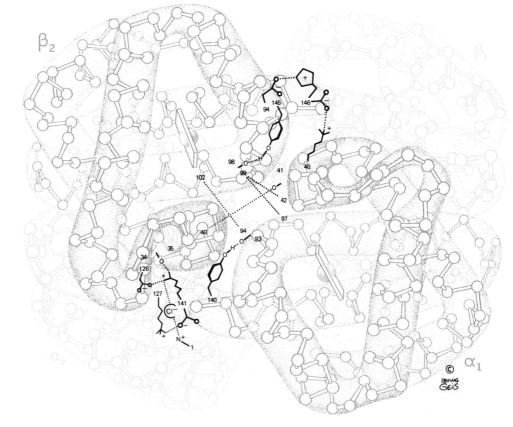

**Figure 5.9**

The binding of glycerate-2,3-bisphosphate in the central cavity of deoxyhemoglobin between $\beta$ chains. The surrounding positively charged residues are the amino terminal, His 2, Lys 82, and His 143. (Illustration copyright by Irving Geis. Reprinted by permission.)

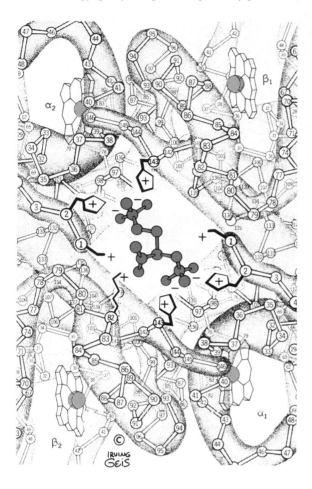

**Figure 5.10**

A close-up view of the iron–porphyrin complex with the F helix in deoxyhemoglobin. Note that the iron atom is displaced slightly above the plane of the porphyrin. (Illustration copyright by Irving Geis. Reprinted by permission.)

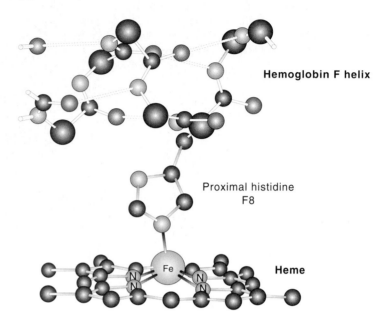

**Figure 5.11**

Downward movement of the iron atom and the complexed polypeptide chain on binding oxygen. The structure is shown before (black) and after (blue) binding oxygen. Movement of His F8 is transmitted to valine FG5, straining and breaking the hydrogen bond to the penultimate tyrosine. Only the $\alpha$ chain is shown here.

## Changes in Conformation Are Initiated by Oxygen Binding

The oxygen binding at the heme group itself initiates the tertiary- and quaternary-structure changes that are responsible for the cooperative effects seen in hemoglobin. The heme group contains an $Fe^{2+}$ ion located near the center of a porphyrin. This $Fe^{2+}$ makes four single bonds to the nitrogens in the porphyrin ring, and a fifth bond to a histidine side chain of the F helix, histidine F8 (fig. 5.10). When oxygen is present it binds at the sixth coordination position of the iron, on the other side of the heme. Movement of the iron on oxygen binding (or release) pulls the F8 histidine and the F helix to which it is covalently attached (fig. 5.11). The tertiary-structure change in the F helix of one subunit causes strains elsewhere in this subunit and in the other

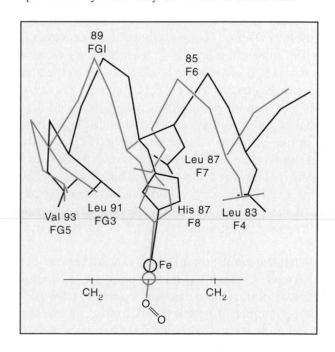

## Figure 5.12

Invariant residues in the $\alpha$ and $\beta$ chains of mammalian hemoglobin. The blue dots, indicating the positions of invariant residues, line the heme pockets as well as the crucial $\alpha_1\beta_2$ interface. The invariant residues have been found in about 60 species. There are 43 invariant positions in the hemoglobin molecule. (Illustration copyright by Irving Geis. Reprinted by permission.)

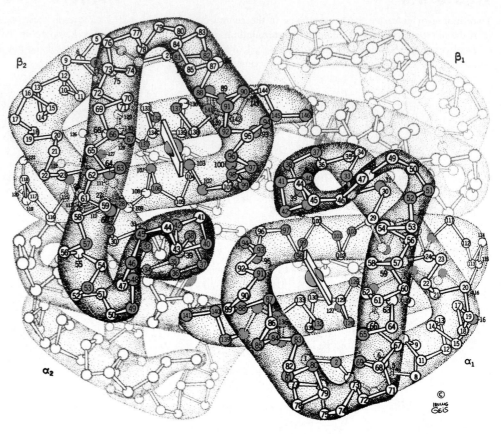

three subunits as well. The resulting structural reorganization favors the binding of additional oxygen at other unoccupied sites in the tetramer.

The $Fe^{2+}$ ion is well suited to its job in hemoglobin, not only because it has a natural affinity for $O_2$, but also because it changes its electron structure in a significant way when it binds $O_2$. $Fe^{2+}$ is normally paramagnetic as a result of the four unpaired electrons in its outer $(3d)$ electron orbitals. As such, it is too large to sit precisely in the plane of a porphyrin, as studies with model compounds have shown. $Fe^{2+}$ is also paramagnetic when it is pentacoordinated in deoxyhemoglobin, and as expected, it is displaced from the plane of the porphyrin by a few tenths of an angstrom. When $O_2$ binds, the $Fe^{2+}$ becomes hexacoordinated and diamagnetic (no unpaired electrons). This change reflects a major reorganization of its $3d$ orbitals, which decreases the radius of the $Fe^{2+}$ so that it can move to an energetically more favorable position in the center of the porphyrin (see fig. 5.11).

The structural arguments advanced here to explain oxygen binding by hemoglobin are supported by comparisons of the amino acid sequences of $\alpha$ and $\beta$ chains for a large number of hemoglobins from different species. Data from 60 species of $\alpha$ chains and 66 species of $\beta$ chains

reveal 43 invariant positions in the hemoglobin molecule. These invariants are plotted on a map of the hemoglobin structure in figure 5.12. Here the invariant positions are shown by blue dots. In a sense, the positions of these dots provide a diagram of the working machinery of hemoglobin, because they line the heme pockets, where oxygen is bound, and the crucial $\alpha_1\beta_2$ interface, which changes its orientation when oxygen binds. Electrostatic forces and hydrogen bonds stitch this inferface together in the deoxy conformation when the molecule gives up its oxygen to the tissues. If changes occur, resulting from mutation at any of these positions, then trouble is likely to develop. Figure 5.13 shows the positions of pathological mutations in hemoglobin that result from changes in the heme pockets or in the $\alpha_1\beta_2$ interface.

As a rule, the invariant amino acids are the only critical loci for hemoglobin function. One striking exception occurs at position 6 of the $\beta$ chain. If a hydrophobic valine residue is substituted for glutamic acid (a charged side chain), disaster results. The specific consequence of this alteration is to cause hemoglobin tetramers to aggregate when they are in the deoxy state. The aggregates form long fibers that stiffen the normally flexible red blood cell. The resulting distortion of the red cells leads to capillary occlu-

## Figure 5.13

Positions of mutations in hemoglobin that produce a pathological condition. Comparison with figure 5.12 shows that these mutations, in general, show the same pattern as the distribution of the invariant positions. Dark circles around a numbered position indicate positions of abnormal residues, solid black dot indicates the valine $\beta_6$ mutation in sickle cell anemia. Heavy black circles indicate a hemoglobin that (is easily) oxidized to the $Fe^{3+}$ form (methemoglobin), and a jagged black perimeter indicates unstable hemoglobin. Dark blue indicates increased oxygen affinity; light blue indicates decreased oxygen affinity. (Illustration copyright by Irving Geis. Reprinted by permission.)

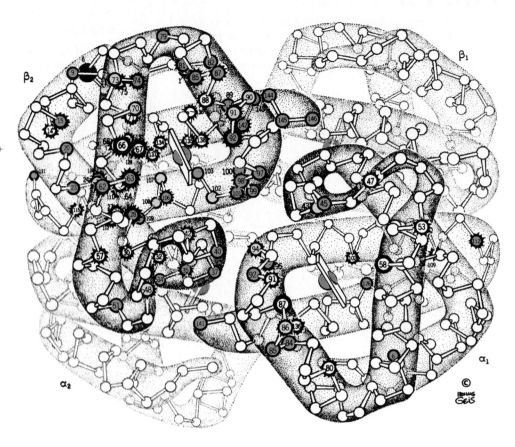

sion, which prevents proper delivery of oxygen to the tissues. This pathological condition is known as sickle cell anemia.

## *Alternative Theories on How Hemoglobins and Other Allosteric Proteins Work*

We have spent a great deal of time describing the function of hemoglobin because it is the best understood regulatory protein and provides us with a model system for understanding in general terms how other allosteric proteins work. The first indication that hemoglobin was an allosteric protein came from the sigmoidal shape of its oxygen-binding curve (see fig. 5.2). Allosteric proteins are usually composed of two or more subunits. Different ligands may bind to quite different sites or to quite similar sites as in the case of hemoglobin.

Two quite different models were proposed about 25 years ago to explain the unusual nature of the hemoglobin oxygen-binding curve (fig. 5.14). These models can also be used as a starting point for discussing other allosteric proteins. The first model, introduced by Jacque Monod, Jeffery Wyman, and Pierre Changeux in 1965, is called the symme-

try model. In this model hemoglobin can exist in only two conformations, one with all four of the subunits within a given tetramer in the low-affinity form and one with all four subunits in the high-affinity form (see fig. 5.14a). In this model, the hemoglobin molecule is always symmetrical, meaning that all the subunits are either in one state or the other and that all of the binding sites within a given tetramer have identical affinities. The binding of oxygen to one of the subunits favors the transition to the high-affinity form. The greater the number of oxygens binding to the tetramer, the more likely is the transition from the low-affinity form to the high-affinity form.

The second model, proposed by Koshland, Nemethy, and Filmer in 1966, is referred to as the sequential model (see fig. 5.14b). In this model the binding of an oxygen molecule to a given subunit causes that subunit to change its conformation to the high-affinity form. Because of its molecular contacts with its neighbors, the change increases the probability that another subunit in the same molecule will switch to the high-affinity form and bind a second oxygen more readily. The binding of the second oxygen has the same type of enhancing effect on the remaining unoccupied oxygen binding sites.

## Figure 5.14

Alternative models for hemoglobin allostery. (*a*) In the symmetry model hemoglobin can exist in only two states. (*b*) In the sequential model hemoglobin can exist in a number of different states. Only the subunit binding oxygen must be in the high-affinity form.

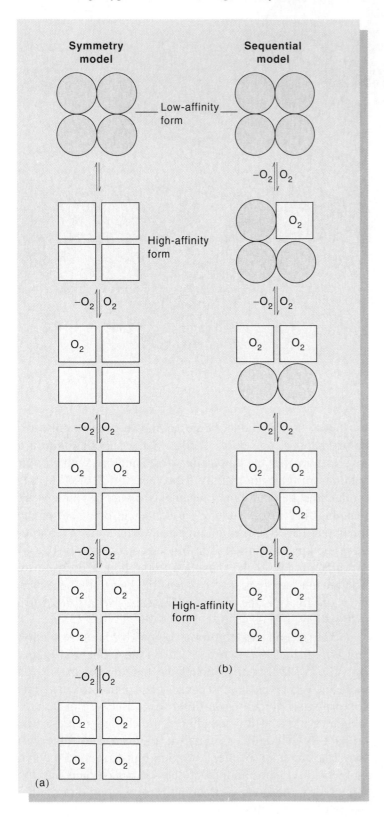

(a)

(b)

Either of these models (or something in between) could account for the sigmoidal oxygen-binding curve of hemoglobin, and either is consistent with the fact that deoxygenated and fully oxygenated hemoglobin have different conformations. The only way to discriminate rigorously between the two models is to obtain structural information on partially oxygenated hemoglobin. This information is still lacking, so no final judgment can yet be made. Even when the situation is fully resolved for hemoglobin, there is no assurance that other allosteric proteins work in the same way.

## Muscle Is an Aggregate of Proteins Involved in Contraction

Vertebrate skeletal muscle represents a remarkable example of a supermolecular aggregate capable of undergoing a reversible reorganization. Voluntary muscle tissue is arranged into fibers that are surrounded by an electrically excitable membrane called the sarcolemma (fig. 5.15). Each fiber is composed of many myofibrils, which when viewed in the light microscope present a striated and banded appearance. As shown in figure 5.16*a* myofibril exhibits a longitudinally repeating structure called the sarcomere. This 23,000-Å long repeating unit is characterized by the appearance of several distinct bands: the less optically dense band being referred to as the I band, and the denser one as the A band. Furthermore, a dense line appears in the center of the I band, called the Z line; and a dense narrow band somewhat similar in appearance also occurs in the center of the A band, called the M line. Adjacent to the M line are regions of the A band that appear less dense than the remainder and are termed the H zone.

Transverse sections of the sarcomere reveal that these patterns result from the interdigitation of two sets of filaments (see fig. 5.16). For example, when a sarcomere is sectioned in the I band, a somewhat disordered arrangement of thin filaments (each about 70 Å in diameter) is seen. In contrast, when sectioned in the H zone, a hexagonal array of thick filaments (each about 150 Å in diameter) is apparent. The substantive observation is that a transverse section in the dense region of the A band shows a regularly packed array of interdigitating thick and thin filaments. It was this observation that led Hugh Huxley (1990) to propose that the process of muscle contraction involved sliding of the thick and thin filaments past each other (fig. 5.17).

## Figure 5.15

The hierarchy of muscle organization. A voluntary muscle such as the bicep is a composite of many fibers connected to tendons at both ends. Each muscle fiber is composed of several myofibrils that are surrounded by an electrically excitable membrane (sarcolemma).

Myofibrils exhibit longitudinally repeating structures called sarcomeres. The fine structure of the sarcomere is described in figure 5.16.

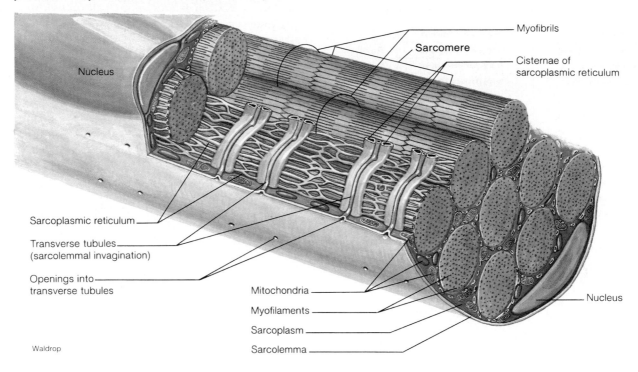

Waldrop

## Figure 5.16

Electron micrograph of a striated muscle sarcomere showing the appearance of filamentous structures when cross-sectioned at the locations illustrated below. (Electron micrograph courtesy of Dr. Hugh Huxley, Brandeis University.)

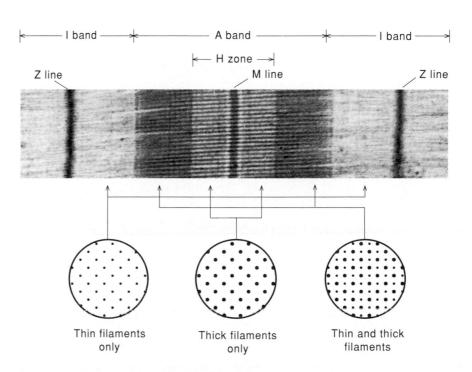

## Figure 5.17

The sliding-filament model of muscle contraction. During contraction, the thick and thin filaments slide past each other so that the overall length of the sarcomere becomes shorter.

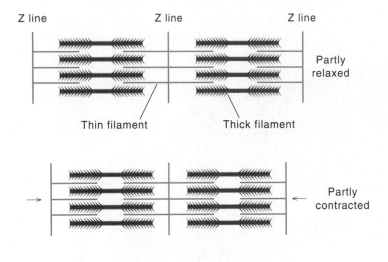

Partly relaxed

Thin filament          Thick filament

Partly contracted

## Table 5.1

### Principal Proteins of Vertebrate Skeletal Muscle

| Protein | $M_r$ | Subunits | Function |
|---------|-------|----------|----------|
| Myosin | 510,000 | 2 × 223,000 (heavy chains) 22,000–18,000 (light chains) | Major component of thick filaments |
| Actin | 42,000 | One type | Major component of thin filaments |
| Tropomyosin | 64,000 | 2 × 32,000 | Rodlike protein that binds along the length of actin filaments |
| Troponin | 69,100 | 30,500 (TN-T) 20,800 (TN-I) 17,800 (TN-C) | Complex of three protein subunits involved in the regulation of muscle contraction |

## Figure 5.18

A molecular view of muscle structure. (*a*) Segment of actin-tropomyosin-troponin. (*b*) Segment of myosin. (*c*) Integration of thin filaments (actin) and thick filaments (myosin) in a muscle fiber.

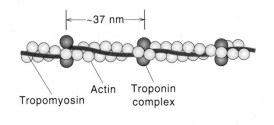

|←  ~37 nm  →|

Tropomyosin      Actin    Troponin complex

(a)

Two identical heavy chains

Light chains (nonidentical)

|←  150 nm  →|

(b)

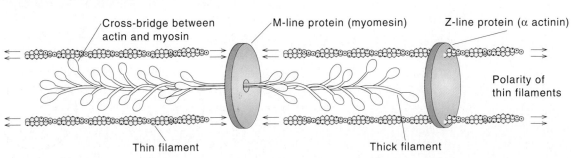

Cross-bridge between actin and myosin

M-line protein (myomesin)

Z-line protein (α actinin)

Polarity of thin filaments

Thin filament          Thick filament

(c)

Subsequent analyses have shown that the thin filaments are composed of three proteins (fig. 5.18 and table 5.1). The main filamentous structure consists of an aggregate of globular actin molecules that takes on the form of a right-handed double helix. The individual actin molecules have a molecular weight of 42,000. Every turn of the actin helix incorporates 14 actin molecules and two molecules of a 360-Å long filamentous molecule of tropomyosin (TM) that fits into the two grooves of the actin double helix. The TM molecule is a dimer of two identical $\alpha$-helical chains, which wind around each other in a coiled coil. Each TM dimer spans seven actin monomers, and a succession of TM dimers extends the full length of the thin filament. Two molecules of troponin (TN) bind to the actin filament at each helical repeat. Troponin is a complex of three nonidentical subunits: TN–C (a calcium-binding subunit) TN–T (a TM-binding subunit), and TN–A (an "inhibitory" subunit). The TN–TM proteins form a regulatory complex the properties of which are discussed below.

Thick filaments are composed of myosin, a large molecule containing two identical heavy chains (223 kD) and two different light chains (22 and 18 kD). The structural organization of myosin is illustrated in figure 5.18. The molecule has two identical globular head regions that incorporate the light chains and a significant fraction of the heavy chains. The tails of the heavy chains form very long $\alpha$ helices that wrap around each other to form left-handed coiled coils. The long $\alpha$-helical structure is favored by the absence of proline over a region of more than a thousand residues and by the abundance of helix formers, such as leucine, alanine, and glutamate (see fig. 4.23). The coiled coil is favored by repeating seven-residue units in which every first and fourth residue has hydrophobic side chains that interact best in the coiled-coil conformation (see fig. 4.6). The individual myosin molecules can be cleaved into fragments by partial degradation with various proteases. Separation of such fragments has made it possible to demonstrate that globular head regions contain binding sites for actin and that the head pieces catalyze the hydrolysis of adenosine triphosphate (ATP). The $\alpha$-helical coiled coils form the backbone of the thick filament. They also form an arm that can provide a flexible extension or hinge connecting the globular head to the body of the thick filament. The thick filament contains many myosin molecules oriented in a staggered fashion (see fig. 5.18).

Given this marvelous piece of molecular architecture, the question remains as to how it works. The answer lies in the observation that actin cyclically binds the globular myosin head group to form cross-bridges in a reaction that depends on the myosin-catalyzed hydrolysis of ATP. The cyclic binding and release of actin from myosin is driven by the energy-releasing hydrolysis of ATP in a manner that causes rearrangement of the actin–myosin cross-bridges. When muscle is completely relaxed, there is a minimum number of cross-bridges and the muscle is fully stretched. When the muscle is activated and under tension, it contracts and more cross-bridges are formed as the region of overlap between actin and myosin increases. At each stage of the contraction process it is essential to break the existing bridges with the help of ATP hydrolysis before new ones can be formed. It is important to realize that although ATP allows new bridges to be formed, the ATP is required to break the existing bridges, not to form the new ones. Cross-bridge formation is energetically favored in the absence of ATP.

A likely scenario for the contraction process is shown in figure 5.19. The ATP that binds to myosin causes the bridges to break or weaken. This bound ATP is rapidly hydrolyzed to ADP and $P_i$, but the hydrolysis products are not immediately released by the myosin. The $P_i$ is released first, but its release requires effective contact with the actin—the regulated step in muscular contraction. Once the $P_i$ has been released, a strong bridge forms between the myosin and the actin. This step is followed by a structural change in the myosin that leads to the translocation of the myosin relative to the actin filament and finally to ADP dissociation. The translocation step is referred to as the power stroke in muscular contraction because it is at this point that the energy ultimately donated by the ATP is expended in the form of a complex structural change. Further contraction requires fresh ATP followed by bridge dissociation and ATP hydrolysis. If conditions are right, the $P_i$ dissociates and the bridges reform; this time they reform at points further along the actin. Cyclic repetition of this process results in a net increase in the number of actin–myosin bridges and further contraction of the sarcomere.

As indicated above, the process of contraction is regulated or triggered by the TN–TM system. Because voluntary muscles are under the conscious control of the animal, one expects a signal from the central nervous system to initiate the process of contraction. A nerve impulse communicated to the muscle causes a depolarization of the sarcolemma membrane that surrounds the muscle fibers, which in turn causes a release of $Ca^{2+}$ from the endoplasmic reticulum in cytoplasm of the muscle cell (fig. 5.20). The $Ca^{2+}$ ions form a complex with the TN–C component of the troponin molecule (see table 5.1). This process induces changes within the TM complex that overcome the inhibitory effect of the TN–I subunit. Then, through TN–T, a signal is sent to TM that triggers the contraction event. The precise nature of this signal from troponin to tropomyosin is unclear; it appears to involve a movement of the tropomyosin on the actin surface. This movement evidently leads to an allosteric transition of the actin that encourages

# Figure 5.19

Steps in the contraction process. Because contraction is a cyclical process, the choice of a starting point is somewhat arbitrary. Five frames are shown; the first two and the last two frames are identical to make the cyclical nature of the process clear. In the first frame (*a*), the myosin head groups contain the hydrolysis products of a single ATP molecule, ADP and $P_i$. A structural transition in the actin leads to contact between the actin and the myosin and the release of $P_i$. The release of the $P_i$ is the rate-limiting step in muscular contraction.

In the second frame (*b*), strong bridges form between actin and myosin. This is followed by a structural alteration in the myosin molecules and an effective translocation of the thick filament relative to the thin filament in (*c*). During this process the ADP is released. After the translocation step, the bridge structure is broken by the binding of ATP, which is rapidly hydrolyzed to ADP and $P_i$. Each thick filament has about 500 myosin heads, and each head cycles about five times per second in the course of a rapid contraction.

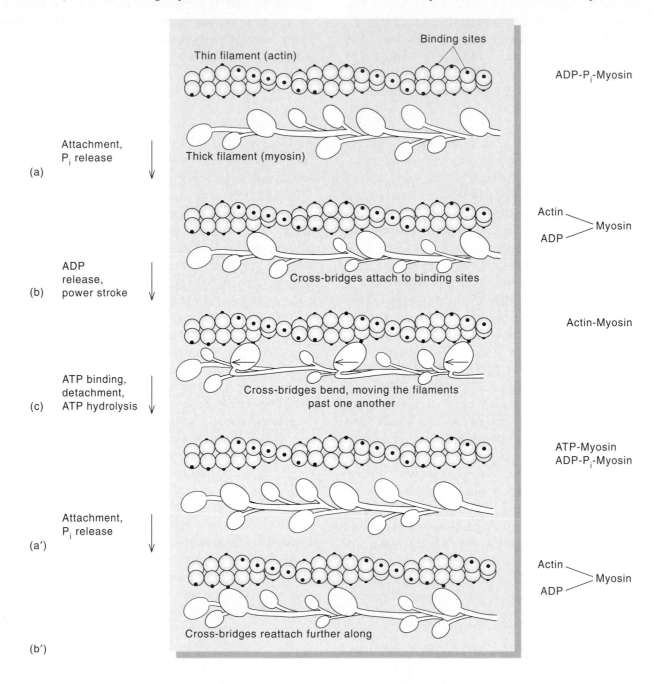

**Figure 5.20**

The effect of calcium on muscle contraction. Binding of calcium to the TN—TM–actin complex produces a shift in the location of TM, which produces an allosteric transition in actin. The allosteric transition in actin facilitates the release of $P_i$ from myosin, which strengthens the interaction between actin and myosin.

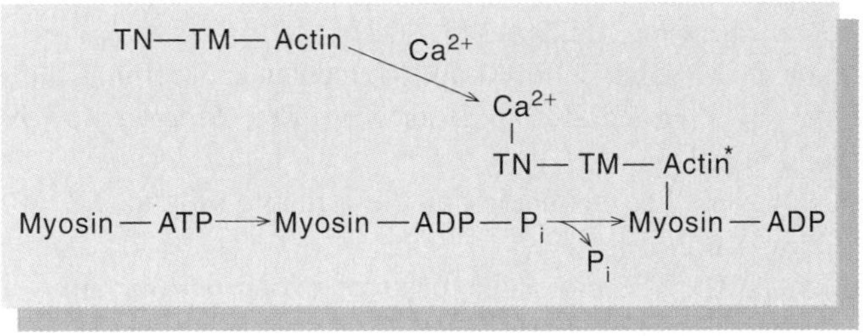

more favorable contact between the complementary binding sites on actin and myosin. As a result, a strong bridge is formed and the $P_i$ is released from the myosin.

It is noteworthy that for some time after death, muscles enter a state of rigor in which they are fully contracted and the maximum number of bridges are formed. Rigor is probably due to the depletion of ATP and a considerable discharge of calcium from the sarcoplasmic reticulum. In living tissue, the cytosolic $Ca^{2+}$ concentration is restored to resting levels within 30 ms of receiving a signal, and the myofibrils relax.

## Summary

In this chapter we considered the relationship between structural and functional properties for two protein systems.

1. Hemoglobin is a tetramer made of two almost identical subunits. The function of hemoglobin is threefold: to transport $O_2$ from the lungs to the tissues where it is consumed, to transport $CO_2$ from the tissues where it is produced to the lungs where it is expelled, and to maintain the blood pH over a narrow range. Cooperative interactions among the subunits allow hemoglobin to pick up the maximum amount of oxygen at high oxygen tensions in the lung tissue and to deliver the maximum amount of oxygen to the oxygen-consuming tissues.

2. Muscle is an aggregate of several different proteins. Its main protein components are organized as overlapping filaments of two types: thin filaments, composed mainly of actin molecules, and thick filaments, composed of myosin molecules. The process of muscular contraction entails a sliding of the two types of filaments past each other. In a fully contracted myofibril the actin and myosin filaments show a maximum overlap with each other. The contraction process involves the breakage and reformation of bridges between the actin and myosin molecules in a reaction that requires the expenditure of ATP.

## Selected Readings

Akers, G. K., M. C. Doyle, D. Myers, and M. A. Daugherty, Molecular code for cooperativity in hemoglobin. *Science* 255:54–63, 1992. A new model for hemoglobin allostery.

Allen, R. D., The microtubule as an intracellular engine. *Sci. Am.* 238(2):42–49, 1987.

Baldwin, J., Structure and cooperativity of haemoglobin. *Trends Biochem. Sci.* 5:224–228, 1980.

Berg, H. C., How bacteria swim. *Sci. Am.* 233(2):36–44, 1975.

Cantor, C. R., and P. Schimmel, *Biophysical Chemistry.* New York: Freeman, 1980. Especially see volume 2, entitled *Techniques for Study of Biophysical Structure and Function.*

Caplan, A. I., Cartilage. *Sci. Am.* 251(4):84–94, 1984.

Dickerson, R. E., and I. Geis, *Hemoglobin.* Menlo Park, Calif.: Benjamin/Cummings, 1983. A magnificent presentation of every facet of hemoglobin biochemistry and genetics.

Doolittle, R., Proteins. *Sci. Am.* 253(4):88–96, 1985. Overview emphasizing evolutionary considerations.

Eisenberg, D., and D. Crothers, *Physical Chemistry and Its Applications to the Life Sciences.* Menlo Park, Calif.: Benjamin/Cummings, 1979.

Eyre, D. R., M. A. Pdaz, and P. M. Gallop, Cross-linking in collagen and elastin. *Ann. Rev. Biochem.* 53:717–748, 1984.

Gething, M. J., and J. Sambrook, Protein folding in the cell. *Nature* 355:33–45, 1992.

Huxley, H. E., Sliding filaments and molecular motile systems. *J. Biol. Chem.* 265:8347–8352, 1990.

Hynes, R. O., Fibronectins. *Sci. Am.* 254(6):42–51, 1986.

Ingram, V. M., Gene mutation in human haemoglobin: the chemical difference between normal and sickle-cell haemoglobin. *Nature* 180:326–328, 1957. A classic paper.

Karplus, M., and J. A. McCammon, The dynamics of proteins. *Sci. Am.* 254(4):42–51, 1986. A reminder that proteins are dynamic structures.

Lawn, R. M., and G. A. Vehar, The molecular genetics of hemoglobin. *Sci. Am.* 254(3):48–65, 1986.

Martin, G. R., R. Timpl, P. K. Muller, and K. Kuhn, The genetically distinct collagens. *Trends Biochem. Sci.* 10:285–287, 1985. A brief, authoritative account of the most abundant protein found in vertebrates.

Pauling, L., H. A. Itano, S. J. Singer, and I. C. Wells, Sickle-cell anemia: a molecular disease. *Science* 110:543–548, 1949. A classic paper.

Pollard, T. D., and J. A. Cooper, Actin and actin-binding proteins. *Ann. Rev. Biochem.* 55:987–1036, 1986. Discusses structure and function.

Rayment, I., H. M. Holden, M. Whittaker, C. B. Yohn, M. Lorenz, K. C. Holmes, and R. A. Milligan, Structure of the Actin-Myosin Complex and Its Implications for Muscle Contraction. *Science* 261:58–65, 1993.

Salemme, R., Structure and function of cytochromes *c*. *Ann. Rev. Biochem.* 46:299–329, 1977.

Steinert, P. M., and D. R. Roop, Molecular and cellular biology of intermediate filaments. *Ann. Rev. Biochem.* 57:593–626, 1988.

Tonegawa, S., The molecules of the immune system. *Sci. Am.* 253(4):122–130, 1985.

Wilson, A. C., The molecular basis of evolution. *Sci. Am.* 253(4):164–170, 1985.

## Problems

1. In the latter half of the 1800s one of the first suggestions that proteins were large molecules came from ashing experiments in which hemoglobin was converted to $Fe_2O_3$. These procedures suggested a molecular mass of hemoglobin greater than 15,900, a number unheard of at that time. Why did the researchers of the past century, who were excellent analytical chemists, deviate so significantly from the molecular mass of 64,500 now known for hemoglobin? What quantity of $Fe_2O_3$ results from the ashing of 1.00 g of hemoglobin?

2. Carbonic acid in the blood readily dissociates into hydrogen and bicarbonate ions. If the serum pH of 7.4 equals that inside the erythrocytes, what percentage of the carbonic acid is ionized? Use a value of 6.4 for the first $pK_a$ of carbonic acid.

3. In addition to oxygen, hemoglobin subunits can also carry carbon dioxide. This is performed by covalent addition of $CO_2$ to the N termini of the hemoglobin chains to produce a carbonate structure. Propose reactions for this process utilizing (a) $CO_2$ and (b) $HCO_3^-$.

4. Figure 5.13 indicates locations of mutations that have been shown to produce pathological conditions. The majority of the types of mutations that have been discovered in human hemoglobins have been mutations in which either amino acid residues bearing charges are replaced with ones with no charge or in which uncharged amino acids are replaced with charged amino acids. Do you think this represents a basic biological principle or is it an artifact of the detection process? Explain.

5. Sickle-cell anemia becomes most apparent during a sickle-cell crisis when the soft tissues are often acutely painful. Using information provided in the text on the molecular-cellular effects of the sickling of red blood cells, can you provide an explanation for the origin of the pain?

6. The hemoglobin present in a fetus is analogous to the $\alpha_2\beta_2$ tetramer of the adult, but the two $\beta$ chains have been replaced with comparable $\gamma$ chains. Considering the relevant biology, which hemoglobin type (adult or fetal) do you expect to have the greater affinity for oxygen?

7. Use the information presented in problem 6 above to propose a possible "future genetic engineering solution" to sickle-cell anemia. Are there any deleterious ramifications of your proposal?

8. Carbon 2 in glycerate-2,3-bisphosphate is in the D configuration. Would you expect the L configuration of glycerate-2,3-bisphosphate to have the same effect on the biochemistry of hemoglobin? Why or why not?

9. Figures 5.10 and 5.11 show a histidyl F8 residue interacting with the heme iron on deoxy- and oxyhemoglobin. Would you expect the imidazole nitrogen of the histidyl group to be protonated or unprotonated during this interaction? Why?

10. Would you expect the blood–hemoglobin system to transport more moles of $O_2$ or $CO_2$? Why?

11. The protein tropomyosin (TM) is composed of two identical chains of $\alpha$ helix (see table 5.1) that are in turn twisted around each other in a helical structure. Consider the average amino acid residue weight to be 105 daltons and use typical $\alpha$-helix dimensions (fig. 4.4) to calculate the length of each chain in TM. Explain any discrepancy observed between the calculated length and the observed length of 360 Å.

12. Bryan Allen made aviation history by pedaling the *Gossamer Albatross* from near Folkestone, England, to Cap Gris Nez, France, from 4:51 AM to 7:40 AM on June 12, 1979. During this flight he continually produced about a third of a horsepower (*Nat. Geographic* 156:5,640, 1979). Considering that the energy available from the hydrolysis of ATP is 7.5 kcal/mole (see fig. 2.9), determine the number of moles of ATP required for this flight. For simplicity, assume that the muscles are 50% efficient in the conversion of chemical energy into mechanical energy and that one horsepower equals the energy expenditure of 178 cal/s.

13. What is the cause of rigor after death?

14. Explain how muscular contraction is regulated.

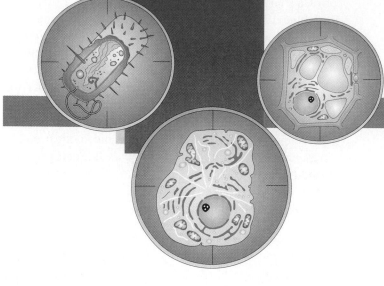

# Methods for Characterization and Purification of Proteins

*Proteins can be isolated and characterized according to their size, shape, and charge.*

In the previous three chapters we described the structures of amino acids and proteins, and in two cases we examined how these structures relate to their function. Some of the methods for structure determination were also discussed (e.g., sequence analysis in chapter 3 and x-ray diffraction in chapter 4). To analyze the structure of a protein we must isolate it from the complex mixture in which it exists in whole cells. The primary object of this chapter is to acquaint you with techniques used for protein purification. Because these procedures are often used for protein characterization as well, they will add to your repertoire of methods for protein characterization.

In the first part of this chapter methods for protein fractionation are discussed in isolation. Success in protein purification depends on picking a number of procedures and combining them in an effective order. Two examples are given in which trains of procedures are combined to purify specific proteins from crude whole-cell extracts.

# Methods of Protein Fractionation and Characterization

Many more techniques for protein analysis exist than can be covered in a single chapter. The emphasis here is on some of the more popular procedures and the principles involved in their use.

## Differential Centrifugation Divides a Sample into Two Fractions

A typical crude broken-cell preparation contains disrupted cell membranes, cellular organelles, and a large number of soluble proteins, all dispersed in an aqueous buffered solution. The membranes and the organelles can usually be separated from one another and from the soluble proteins by differential centrifugation. Differential centrifugation divides a sample into two fractions: the pelleted fraction, or sediment, and the supernatant fraction, that is, the fraction that is not sedimented. The two fractions may then be separated by decantation.

According to its purpose, differential centrifugation involves the use of different speeds and different times of centrifugation (table 6.1). For example, if the protein of interest is in the mitochondrial fraction, the crude cell lysate is centrifuged first at $4,000 \times g$ for 10 min to remove cell membranes, nuclei, and (in the case of plant material) chloroplasts. The supernatant from this step contains, among other elements, the mitochondria, and is decanted and recentrifuged at $15,000 \times g$ for 20 min to obtain a sediment primarily containing mitochondria. If ribosomes instead of mitochondria are the goal, then the crude lysate is centrifuged at $30,000 \times g$ for 30 min, and the resulting supernatant is decanted and centrifuged at $100,000 \times g$ for 180 min to obtain a ribosomal sediment. If obtaining the soluble protein fraction is the goal, then the entire lysate is centrifuged at $100,000 \times g$ for 180 min, and the resulting supernatant, containing the soluble protein, is carefully decanted for further processing.

## Differential Precipitation Is Based on Solubility Differences

Once a crude extract of protein has been made, it is common to separate this mixture into different fractions by a precipitation step. The solubility of a protein reflects a delicate balance between different energetic interactions—both internally within the protein and between the protein and the surrounding solvent. Consequently, the choice of solvent can affect both the solubility and the structure of a protein.

## Table 6.1

Sedimentation Conditions for Different Cellular Fractions

| Fraction Sedimented | Centrifugal Force ($\times g$) | Time (min) |
| --- | --- | --- |
| Cells (eukaryotic) | 1,000 | 5 |
| Chloroplasts; cell membranes; nuclei | 4,000 | 10 |
| Mitochondria; bacteria cells | 15,000 | 20 |
| Lysosomes; bacterial membranes | 30,000 | 30 |
| Ribosomes | 100,000 | 180 |

*A Minimum of Solubility Occurs at the Isoelectric Point* Proteins typically have charged amino acid side chains on their surfaces that undergo energetically favorable polar interactions with the surrounding water. The total charge on the protein is the sum of the side-chain charges. However, the actual charge on the weakly acidic and basic side-chain groups also depends on the solution pH. In fact, the acidic and basic groups within the protein can be titrated just like free amino acids (see fig. 3.6) to determine their number and their p$K$ values.

Proteins tend to show a minimum solubility at their isoelectric pH a fact that is apparent for $\beta$ lactoglobulin (fig. 6.1). The decrease in solubility at the isoelectric pH reflects the fact that the individual protein molecules, which all have similar charges at pH values away from their isoelectric points, cease to repel each other. Instead, they coalesce into insoluble aggregates.

*Salting In and Salting Out* Proteins also show a variation in solubility that depends on the concentration of salts in the solution. These frequently complex effects may involve specific interactions between charged side chains and solution ions or, particularly at high salt concentrations, may reflect more comprehensive changes in the solvent properties. For the effect of salt concentration on the solubility of $\beta$-lactoglobulin, again see figure 6.1. Most globulins are sparingly soluble in pure water. The increase in solubility that occurs after adding salts such as sodium chloride is often referred to as salting in.

The effects of four different salts on the solubility of hemoglobin at pH 7 can be seen in figure 6.2. All four salts produce the salting-in effect with this protein; two of them, sodium sulfate and ammonium sulfate, also produce a greatly decreased solubility of the protein at high salt con-

**Figure 6.1**

Solubility of β-lactoglobulin as a function of pH and ionic strength. The isoelectric pH (pI) for this protein is about 5.2, which corresponds to the point of minimum solubility.

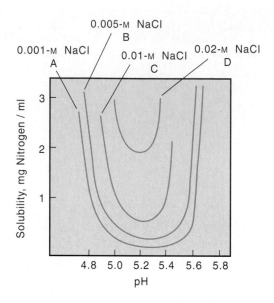

**Figure 6.2**

Solubility of horse carbon monoxide hemoglobin in different salt solutions. The addition of a moderate amount of salt (salting in) is required to solubilize this protein. At high concentrations, certain salts compete more favorably for solvent, decreasing the solubility of the protein and thus leading to its precipitation (salting out). (Source: E. J. Cohn and J. T. Edsall, *Proteins, Amino Acids, and Peptides as Ions and Dipolar Ions.* Copyright © 1942, Reinhold, New York, N.Y.)

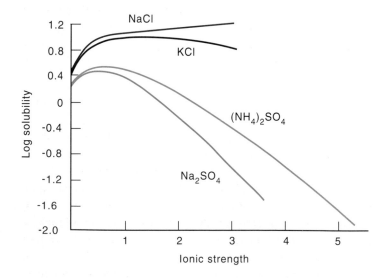

centrations. This result is called salting out and occurs with salts that effectively compete with the protein for available water molecules. At high salt concentrations, the protein molecules tend to associate with each other because protein–protein interactions become energetically more favorable than protein–solvent interactions.

Each protein has a characteristic salting-out point, a fact we can exploit to make protein separations in crude extracts. For this purpose $(NH_4)_2SO_4$ is the most commonly used salt because it is very soluble and is generally effective at lower concentrations than many other salts.

Typically, the desired protein precipitates over a range of salt concentrations. If it precipitates in the range of 20–30% by weight, we first add ammonium sulfate to a concentration slightly below 20% and then centrifuge to remove by sedimentation any proteins that precipitate in the 0–20% range. To the supernatant from this centrifugation we then add more ammonium sulfate, to 30%. Centrifugation at this point brings down the desired protein, as well as other proteins that precipitate in this range of salt concentrations. The supernatant is discarded and the sediment saved for further purification.

## Column Procedures Are the Most Versatile and Productive Purification Methods

Column procedures are the most effective and most varied of purification methods. Common to all column procedures is the use of a glass cylinder with an opening at the top and bottom. The cylinder is filled with a column of hydrated material, and the protein sample is applied to the top of the column. Then further buffer is passed through the column. In most column procedures, proteins bind differentially to the column material, and a change in the elution buffer causes them to be eluted differentially according to their degree of affinity for the column material. The eluant exiting from the bottom of the column is collected in equal-sized fractions with the help of an automatic fraction collector (fig. 6.3). Each fraction is analyzed, and fractions containing an appreciable amount of the desired protein are pooled for further purification.

Many variations on this basic procedure are in common use:

1. In gel-exclusion chromatography a cross-linked dextran without any special attached functional groups is used for the column substrate (fig. 6.4). Large molecules flow more rapidly through this type of column than small ones. The dextrans have different degrees of cross-linking, making them effective over different size ranges.

2. Ion-exchange chromatography makes use of the fact that proteins differ enormously in their affinity for positively or negatively charged columns. However, the cross-linked resins used in amino acid fractionation (see chap-

## Figure 6.3

Collecting fractions during column chromatography. Column material and elution procedure are chosen to effect optimal separation of the desired protein.

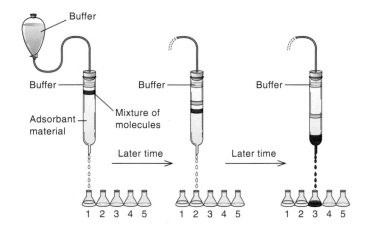

## Figure 6.4

Polydextran column showing separation of small and large molecules. The column material is immersed in solvent, which penetrates the gel particles. A separation is initiated by layering a small sample containing different-sized proteins on the top of the column. This sample is pushed through the column by opening the stopcock at the bottom and adding further solvent at the top to keep up with the flow. As shown, the small protein molecules can penetrate the gel particles but the big ones cannot. Therefore the big proteins move through the column much more rapidly, and a separation of the two proteins results.

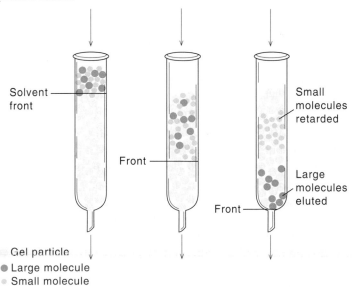

**Table 6.2**

## Some Column Materials for Ion-Exchange Chromatography of Proteins

| Matrix | Functional Groups on Column |
| --- | --- |
| Phosphocellulose (PC) | $-PO_3^-$ |
| Carboxymethyl cellulose (CMC) | $-CH_2-COO^-$ |
| Diethylaminoethyl cellulose (DEAE) | $-(CH_2)_2-\overset{+}{N}H\begin{smallmatrix}CH_2-CH_3\\CH_2-CH_3\end{smallmatrix}$ |

ter 3) are rarely used for protein separations because proteins are too large to penetrate the resin beads. Instead, finely divided celluloses containing either positively or negatively charged groups are most commonly used (table 6.2). The affinity of a protein for the column material is proportional to the salt concentration required to release the protein from the material. Typically, a column is loaded with protein solution at a low ionic strength so that most of the proteins bind to the column. After loading, elution is initiated by gradually increasing the salt concentration of the elution buffer. Proteins are eluted in the order of increasing affinity.

3. Finely divided celluloses may also be used in the column in conjunction with attached hydrophobic groups such as octyl alcohol. In this case, it is proteins with exposed hydrophobic centers that bind to the column with varying affinities. These proteins may be eluted in order of decreasing affinity for the column by increasing the level of free octyl alcohol in the eluting buffer.

4. Affinity chromatography makes use of chemical groups that have a special binding affinity for the proteins that are being sought. For example, many enzymes bind preferentially to cofactors, such as adenosine triphosphate or pyridine nucleotides. Often, the separation of such enzymes from other proteins can be achieved on a column that has one of these cofactors attached. As the mixture passes through the column, proteins that have specific binding affinities for the cofactor bind to the column, and other proteins pass through. The cofactor-binding protein can then be eluted with a solution containing the same soluble cofactor.

5. High-performance liquid chromatography (HPLC) is not so much a new type of chromatography as a new way of applying old chromatographic techniques, using extremely high pressures. The same principles are in-

## Figure 6.5

Gel electrophoresis for analyzing and sizing proteins. (*a*) Apparatus for slab-gel electrophoresis. Samples are layered in the little slots cut in the top of the gel slab. Buffer is carefully layered over the samples, and a voltage is applied to the gel for a period of usually 1–4 h. (*b*) After this time the proteins have moved into the gel at a distance proportional to their electrophoretic mobility. The pattern shown indicates that different samples were layered in each slot. (*c*) Results obtained when a mixture of proteins was layered at the top of the gel in phosphate buffer, pH 7.2, containing 0.2% SDS. After electrophoresis the gel was removed from the apparatus and stained with Coomassie Blue. The protein and its molecular weight are indicated next to each of the stained bands. (*d*) The logarithm of the molecular weight against the mobility (distance traveled) shows an approximately linear relationship. (Source: Data of K. Weber and M. Osborn.)

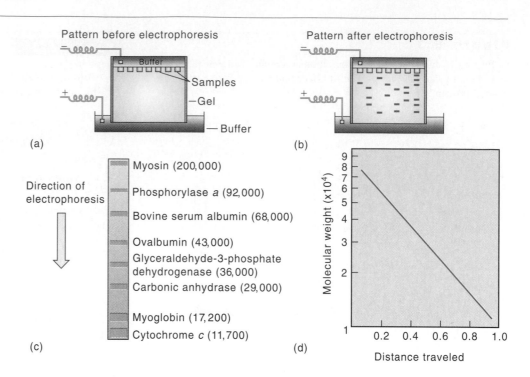

## Electrophoresis Is Used for Resolving Mixtures

Gel electrophoresis is the best way to analyze mixtures and assess purity. Gel electrophoresis separates proteins according to their size and their charge. It is almost always performed in aqueous solution supported by a gel system. The gel is a loosely cross-linked network that functions to stabilize the protein boundaries between the protein and the solvent, both during and after electrophoresis, so that they may be stained or otherwise manipulated.

The most popular method of electrophoretic separation by gels employs sodium dodecyl sulfate (SDS). This method not only gives an index of protein purity but yields an estimate of the protein subunit molecular weights. The mixture of proteins to be characterized is first completely denatured by adding SDS (a detergent) and mercaptoethanol and by briefly heating the mixture. Denaturation is caused by the association of the apolar tails of the SDS molecules with protein hydrophobic groups. Any cystine disulfide

volved, but column materials usually consist of more finely divided particles made of physically stronger materials, which can withstand pressures of 5,000–10,000 psi without changing their structure. The column apparatus itself must also be designed to withstand high pressures. The main advantage of HPLC is that superior resolution of eluted substances can be achieved.

bonds are cleaved by a disulfide interchange reaction with mercaptoethanol.

The resulting unfolded polypeptide chains have relatively large numbers of SDS molecules bound to them. The success of the technique for molecular weight estimation depends on two facts: (1) Each bound SDS molecule contributes one negative charge to the denatured protein complex, so that the charge of the protein in its native state is effectively masked by the more numerous charged groups of the associated detergent molecules. (2) The total number of detergent molecules bound is proportional to the polypeptide chain length or, equivalently, the protein's molecular weight. As a result, the SDS-denatured protein molecules acquire net negative charges that are approximately proportional to their molecular weights (fig. 6.5).

Another electrophoretic method, called isoelectric focusing, is frequently used for characterizing proteins based on differences in their isoelectric points. The apparatus usually consists of a narrow tube containing a gel and a mixture of ampholytes, which are small molecules with positive and negative charges. The ampholytes have a wide range of isoelectric points, and are allowed to distribute in the column under the influence of an electric field. This step creates a pH gradient from one end of the gel to the other, as each particular ampholyte comes to rest at a position coincident with its isoelectric point. At this stage, a solution of proteins is introduced into the gel. The proteins migrate in the electric field until each reaches a point at which the pH resulting from the ampholyte gradient exactly equals its own isoelectric point. Isoelectric focusing provides a way of both accurately determining a protein's isoelectric point and effecting separations among proteins, the isoelectric points of which may differ by as little as a few hundredths of a pH unit.

An elegant extension of the electrophoretic method of separation involves combining isoelectric focusing with SDS gel electrophoresis to produce a two-dimensional electrophoretogram. This technique is most valuable for the analysis of very complex mixtures. First, the sample is run in a one-dimensional pH gradient gel (isoelectric focusing). The resulting narrow strip of gel, containing the partially separated mixture of proteins, is placed alongside a square slab of SDS gel. An electrical field is imposed so that the sample moves at right angles to its motion in the first gel. Figure 6.6 shows the separation of total *E. coli* protein into more than 1,000 different components.

Although electrophoresis is the method of choice for assessing protein purity, it is not frequently used as a purification step, at least not yet. This is because of the small amounts that can usually be analyzed and because of the denaturing conditions that are often used in electrophoretic analysis.

## Sedimentation and Diffusion Are Used for Size and Shape Determination

We have seen that sedimentation by centrifugation is used for protein purification. The methods of centrifugation can be used quantitatively for the assessment of protein size and shape.

In a centrifugal force field, protein molecules slowly migrate toward the bottom of a centrifuge tube at a rate that is proportional to their molecular weight (fig. 6.7). The rate of sedimentation may be recorded by optical methods that do not interfere with the operation of the centrifuge. From this rate, we can obtain the sedimentation constant *s*. This constant equals the rate at which a molecule sediments, divided by the gravitational field (angular acceleration in a spinning rotor), as defined by the equation

$$s = \frac{dx/dt}{\omega^2 x} \quad \textbf{(1)}$$

where $dx/dt$ is the rate at which the particle travels at distance $x$ from the center of rotation, $\omega$ is the angular velocity of the rotor in radians per second (hence $\omega^2 x$ is the angular acceleration), and $t$ is the time of centrifugation in seconds. The sedimentation constant is usually given in Svedberg units (S); one $S = 10^{-13} s$.

From the sedimentation constant we can obtain the molecular weight, provided we have certain other information. This additional information includes the frictional co-

### Figure 6.6

Two-dimensional SDS isoelectric-focusing gel electrophoresis. First the sample is run in a one-dimensional pH gradient, partially separating the sample along a strip of gel. Then the strip of gel containing the sample is placed alongside an SDS gel, and the proteins are permitted to further separate by moving in the second dimension, at right angles to the first separation. Sample shown is total *E. coli* protein; individual proteins are detected by autogradiography. (Source: Photograph provided by Patrick O'Farrell. See O'Farrell, in *J. Biol. Chem.* 250:4007, 1975.)

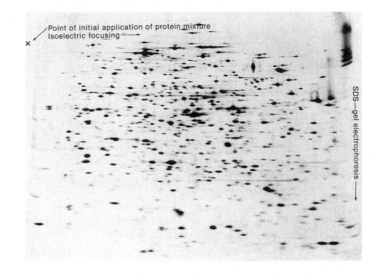

### Figure 6.7

Apparatus for analytical ultracentrifugation. (*a*) The centrifuge rotor and method of making optical measurements. (*b*) The optical recordings as a function of centrifugation time. As the light-absorbing molecule sediments, the solution becomes transparent.

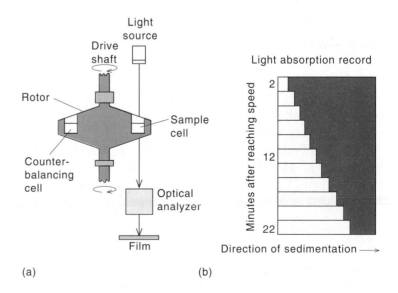

## Table 6.3

Physical Constants of Some Proteins

| Protein | Molecular Weight | Diffusion Constant $(D \times 10^7)$ | Sedimentation Constant (S) | pI[a] (Isoelectric) |
|---|---|---|---|---|
| Cytochrome $c$ (bovine heart) | 13,370 | 11.4 | 1.17 | 10.6 |
| Myoglobin (horse heart) | 16,900 | 11.3 | 2.04 | 7.0 |
| Chymotrypsinogen (bovine pancreas) | 23,240 | 9.5 | 2.54 | 9.5 |
| β-Lactoglobulin (goat milk) | 37,100 | 7.5 | 2.9 | 5.2 |
| Serum albumin (human) | 68,500 | 6.1 | 4.6 | 4.9 |
| Hemoglobin (human) | 64,500 | 6.9 | 4.5 | 6.9 |
| Catalase (horse liver) | 247,500 | 4.1 | 11.3 | 5.6 |
| Urease (jack bean) | 482,700 | 3.46 | 18.6 | 5.1 |
| Fibrinogen (human) | 339,700 | 1.98 | 7.6 | 5.5 |
| Myosin (cod) | 524,800 | 1.10 | 6.4 | — |
| Tobacco mosaic virus | 40,590,000 | 0.46 | 198 | — |

[a] pI = −log I = the isoelectric point, that is, the pH at which the protein carries no net charge (see fig. 6.1).

efficient ($f$) of the protein and the density of the protein. The coefficient $f$ is bigger for larger proteins and, for proteins of the same molecular weight, it is larger for elongated, rodlike molecules. The density of the protein is important because of the buoyancy factor, $1 - \bar{v}_p\rho_s$, which takes into account the density difference between solvent ($\rho_s$) and the volume of water displaced per gram of protein ($\bar{v}_p$). The equation that relates $s$, $M$, and $f$ is

$$s = \frac{M(1 - \bar{v}_p\rho_s)}{Nf} \qquad (2)$$

where $N$ is Avogadro's number. Thus, in order to estimate the molecular weight from the sedimentation constant we must have a means of determining $\bar{v}_p$ and $f$.

For most proteins $\bar{v}_p$ is about 0.75 ml/g, so its value does not present much of a problem. The frictional coefficient, however, is a sensitive function of the shape, varying over a wide range, and we must usually know its value if we need a serious estimate of the molecular weight. The value of $f$ is usually found by working with the diffusion constant $D$, which is related to the frictional constant by

$$D = \frac{RT}{Nf} \quad \text{or} \quad f = \frac{RT}{ND} \qquad (3)$$

where $R$ is the gas constant and $T$ is the absolute temperature. Substituting this expression for $f$ in equation (2) and transposing leads to

$$M_r = \frac{RTs}{D(1 - \bar{v}_p\rho_s)} \qquad (4)$$

The diffusion constant $D$ is a function of both molecular weight and shape. It can be measured by observing the spread of an initially sharp boundary between the protein solution and a solvent as the protein diffuses into the solvent layer. Once we know the value of the diffusion constant, we can combine the information with the sedimentation data and calculate the molecular weight of the protein.

Representative sedimentation and diffusion data, together with the calculated molecular weights, are presented in table 6.3. The sedimentation constants are usually larger the greater the molecular weight. Diffusion constants are usually inversely proportional to the molecular weights. Exceptions arise because of proteins with unusual shapes. For example, the globular protein urease and the rodlike protein myosin have similar molecular weights. Yet their sedimentation and diffusion constants both differ by about a factor of 3. Rapid diffusion can be a highly desirable property for an enzyme that must pass rapidly from one point to another. Such a protein benefits from a globular shape, which is quite common for enzymes. By contrast, a rodlike protein can be advantageous for creating a cytoskeletal boundary within the cytoplasm. Myosin is used for just such purposes in many cell types.

## Protein Purification Procedures

To characterize proteins, we must usually first isolate the protein under study from the complex mixture of proteins found in the organism. Once we decide to purify a particular protein, we must weigh several factors. For example, how much material is needed? What level of purity is required? The starting material should be readily available and should contain the desired protein in relative abundance. If the protein is part of a larger structure, such as the nucleus, the mitochondria, or the ribosome, then it is advisable to isolate the large structure first from a crude cell extract.

Purification must usually be performed in a series of steps, using different techniques at each step. Some purification techniques are more useful when handling large amounts of material, whereas others work best on small amounts. A purification procedure is arranged so that the techniques that are best for working with large amounts are used during early steps in the overall purification. The suitability of each purification step is evaluated in terms of the amount of purification achieved by that step and the percent recovery of the desired protein.

Combining techniques introduces new considerations and new problems. If two purification techniques each give 10-fold enrichment for the desired protein when executed independently on a crude extract, it does not mean that they give 100-fold enrichment when combined. In general, they give somewhat less. As a rule, purification techniques that combine most effectively usually are based on different properties of the protein. For example, a technique based on size fractionation is more effectively combined with a technique based on charge than with another technique based on size fractionation.

Throughout the purification process we must have a convenient means of assaying for the desired protein, so we can know the extent to which it is being enriched relative to the other proteins in the starting material. In addition, a major concern in protein purification is stability. Once the protein is removed from its normal habitat, it becomes susceptible to a variety of denaturation and degradation reactions. Specific inhibitors are sometimes added to minimize attack by proteases on the desired protein. During purification it is usual to carry out all operations at 5°C or below. This temperature control minimizes protease degradation problems and decreases the chances of denaturation.

In their natural habitat, proteins are usually surrounded by other proteins and organic factors. When these are removed or diluted as during purification, the protein becomes surrounded by water on all sides. Proteins react differently to a pure aqueous environment; many are destabilized and rapidly denatured. A common remedial measure is to add 5%–20% glycerol to the purification buffer. The organic surface of the glycerol is believed to simulate the environment of the protein in the intact cell. Two other ingredients that are most frequently added to purification buffers are mercaptoethanol and ethylenediaminetetraacetate (EDTA). The mercaptoethanol inhibits the oxidation of protein —SH groups, and the EDTA chelates

divalent cations. Divalent cations, even in trace amounts, can lead to aggregation problems or activate degradative enzymes.

The following two examples of purification show how various techniques can be effectively combined to produce purified proteins with minimum effort and loss of activity.

## Purification of an Enzyme with Two Catalytic Activities

The last two steps in the biosynthesis of the mononucleotide uridine 5′-monophosphate (UMP) are catalyzed by (1) orotate phosphoribosyltransferase (OPRTase) and (2) orotate 5′-monophosphate decarboxylase (OMPDase).

Mary Ellen Jones and her colleagues set out to purify the enzyme or enzymes involved in these two reactions. Their main goal was to determine whether the two reactions are carried out by one protein or more than one. Their findings indicated that the two reactions were both catalyzed by the same enzyme, consisting of a single polypeptide chain. To demonstrate this fact, it was necessary to monitor both enzyme activities at each step in the purification and show that both activities copurified. For this purpose, Jones used specific enzyme assays for both enzyme activities. All fractions were assayed for both enzymatic activities at each stage of the purification.

The main data associated with the purification are summarized in table 6.4. This table indicates the total protein obtained in each step, the number of enzyme units[a] for each enzyme, and the ratio of enzyme units to total protein, called the specific activity. In the absence of enzyme inactivation, the specific activity should be directly proportional to the enrichment. The percent recovery refers to the amount of enzyme activity in the indicated fraction, as compared with the amount present in fraction 1. This number is usually less than 100%. The apparent losses may reflect actual losses of enzyme during purification, or they may reflect inactivation (usually due to unknown causes) of the enzyme during purification.

The nine steps involved in the purification of UMP synthase from starting tissue are outlined in figure 6.8. All steps were carried out at 0–5°C. About 200 g of Ehrlich ascites cells, a mammalian tumor rich in the desired enzymes, was suspended in buffer and processed in a tissue homogenizer, which mechanically breaks down the tissue

---

[a]Enzyme units are proportional to the amount of enzyme activity. The relationship between enzyme units and absolute amount of enzyme need not concern us here.

## Table 6.4

Outline of Purification of UMP Synthase from Ehrlich's Ascites Carcinoma

| Fraction | Volume (ml) | Protein (mg) | OMPDase[a] | | | OPRTase[a] | | | Ratio of OMPDase to OPRTase |
| --- | --- | --- | --- | --- | --- | --- | --- | --- | --- |
| | | | Units[b] | Sp. Act.[c] | Percent Recovery | Units[b] | Sp. Act.[c] | Percent Recovery | |
| 1. Streptomycin fraction | 1040 | 11,700 | 40.4 | 0.0034 | | 20.5 | 0.0018 | | 2.0 |
| 2. Dialyzed $(NH_4)_2SO_4$ fraction | 144 | 311 | 24.3 | 0.0078 | 60 | 8.7 | 0.0028 | 42 | 2.8 |
| 3. Affinity column eluate (concentrated) | 0.475 | 0.51 | 4.0 | 7.8[d] | 10 | 0.35 | 0.69 | 3.3 | 11.4 |

[a] OMPDase = orotate 5′-monophosphate decarboxylase; OPRTase = orotate phosphoribosyltransferase.

[b] Units refer to total amount of enzyme activity.

[c] Specific activity refers to the units of enzyme activity divided by the total protein.

[d] This value represents a 2,300-fold enrichment from fraction 1.

and the cell membranes (step 1). Then EDTA and an —SH reagent were added to this total cell lysate. Solid streptomycin sulfate was also added with stirring (step 2). Streptomycin sulfate aggregates nucleic acids so that they may be more easily removed by centrifugation. The resulting slurry was subjected to high-speed centrifugation, and the resulting supernatant (see table 6.4, fraction 1) was carefully decanted for further processing (step 3).

Preliminary experiments had shown that the desired enzymes were in the 18.5%–28% $(NH_4)_2SO_4$ fraction. This knowledge served as the basis for the next three steps. First 239 g of solid $(NH_4)_2SO_4$ was added to 1040 ml of supernatant (step 4). The resulting precipitate was removed by centrifugation (step 5). Then an additional 120 g of $(NH_4)_2SO_4$ was added to the supernatant (step 6). The resulting slurry was centrifuged. This time the supernatant was discarded after centrifugation, leaving the sediment containing the enzyme activity for further processing (step 7). The sediment was resuspended in a dilute buffer for column chromatography (see table 6.4, fraction 2). Inspection of table 6.4 indicates only about a twofold increase in specific activity between fractions 1 and 2.

The main purification was achieved by two column steps, carried out in series. The first column was an affinity column containing an analog of UMP, 6-azauridine 5′-monophosphate (azaUMP), covalently attached to an agarose column support system. In dilute buffer, greater than 99% of the protein in fraction 2 is retained on this column. After thorough rinsing of the column with dilute buffer, $5 \times 10^{-5}$ M azaUMP was added to the buffer. This addition resulted in the elution of the UMP synthase (step 8). Then the column eluant carrying the two enzyme activities associated with UMP synthase was resuspended in pure buffer and passed over a phosphocellulose column (step 9). Phosphocellulose was chosen because the negatively charged phosphate groups result in the retention of proteins by electrostatic attraction alone; they also resemble the phosphate groups in the naturally occurring enzyme substrate, OMP. Thus, the phosphocellulose column may be thought of as an ion-exchange column and an affinity column combined. In ordinary ion-exchange chromatography, the protein, after column loading, is eluted by increasing the ionic strength with a simple inorganic salt. In this example, Jones used a more specific method to elute the enzyme, which involved adding $10^{-5}$ M azaUMP and $2 \times 10^{-5}$ M OMP to the original loading buffer. The addition does not substantially increase the ionic strength of the buffer. Therefore the only phosphocellulose-bound proteins that are likely to be eluted by this treatment are those with an especially high affinity for either of these nucleotides (azaUMP or OMP). This fact should greatly favor selective elution of those enzymes that carry specific sites for binding these nucleotides. The two column steps together resulted in an enzyme preparation (see table 6.4, fraction 3) that was approximately 2,300-fold purified over the starting material, as measured by the increase in specific activity.

The final product (see table 6.4, fraction 3) was examined by SDS gel electrophoresis and found to contain a single band with an estimated molecular weight of 51,000. The denaturing conditions of an SDS gel are expected to dissociate a multisubunit protein. Hence the SDS gel result indicated that the enzyme contains one type of polypeptide chain, but it does not tell us whether the enzyme contains one or more of these chains. Sedimentation analysis on a

## Figure 6.8

Outline of purification scheme for UMP synthase from Ehrlich's ascites tumor cells of mice.

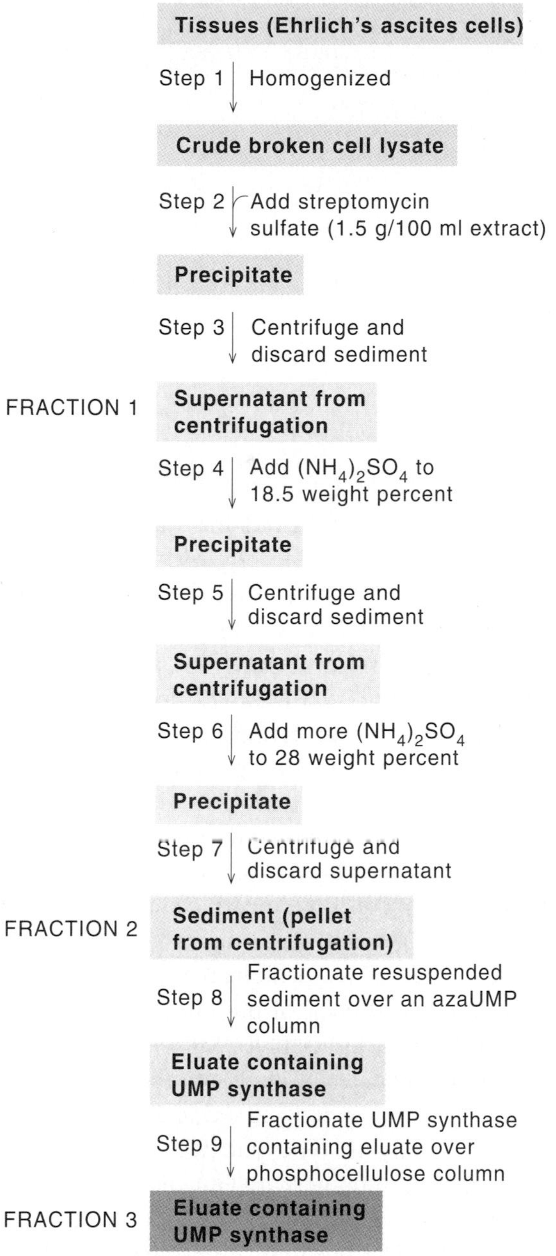

```
              Tissues (Ehrlich's ascites cells)

        Step 1 │ Homogenized
               ↓
              Crude broken cell lysate

        Step 2 ┌ Add streptomycin
               ↓ sulfate (1.5 g/100 ml extract)

              Precipitate

        Step 3 │ Centrifuge and
               ↓ discard sediment

FRACTION 1    Supernatant from
              centrifugation

        Step 4 │ Add (NH₄)₂SO₄ to
               ↓ 18.5 weight percent

              Precipitate

        Step 5 │ Centrifuge and
               ↓ discard sediment

              Supernatant from
              centrifugation

        Step 6 │ Add more (NH₄)₂SO₄
               ↓ to 28 weight percent

              Precipitate

        Step 7 │ Centrifuge and
               ↓ discard supernatant

FRACTION 2    Sediment (pellet
              from centrifugation)

               ┌ Fractionate resuspended
        Step 8 │ sediment over an azaUMP
               ↓ column

              Eluate containing
              UMP synthase

               ┌ Fractionate UMP synthase
        Step 9 │ containing eluate over
               ↓ phosphocellulose column

FRACTION 3    Eluate containing
              UMP synthase
```

sucrose density gradient in a nondenaturing buffer indicated a single band with an estimated molecular weight of about 50,000. These two results taken together demonstrate that the enzyme in its native state contains a single polypeptide chain. The purity of the enzyme was also confirmed by isoelectric focusing and two-dimensional electrophoresis, with isoelectric focusing in the first direction followed by SDS gel electrophoresis in the second direction.

The conclusion drawn from these results was that both of the enzyme activities associated with UMP synthase—OMPDase and OPRTase—are contained in a single protein. The basis for the conclusion was that both enzyme activities are always present in the same fractions throughout the multistep purification. However, inspection of table 6.4 indicates a possible objection to this interpretation. In the columns showing percent recovery, it can be seen that substantial amounts of enzyme activity are lost for both enzymes during purification, but that considerably more activity is lost for the OPRTase. These losses could be due to actual loss of enzymes during purification or to some sort of inactivation of the enzyme sites. The preferential loss of OPRTase activity is emphasized by the last column in table 6.4, which gives the ratio of the two enzyme activities in the different fractions. Considered alone, these data could indicate that a separate catalytic unit of OPRTase is lost during purification. Jones thinks this is unlikely for two reasons: (1) the activity appears in no fractions other than with OMPDase during purification, and (2) the OPRTase activity is notably unstable. It appears, then, that both enzyme activities exist at distinct sites on a single protein. The greater loss in activity of one enzyme activity over the other is attributed to a greater sensitivity of one reaction site over the other on the enzyme surface.

## Purification of a Membrane-Bound Protein

The second purification procedure we examine illustrates an unusual approach to the purification of a membrane-bound protein. The lactose carrier protein of *E. coli* is normally tightly bound to the plasma membrane. This protein is involved in the active transport of the dissaccharide lactose across the cytoplasmic membrane. When lactose carrier protein is present, the intracellular concentration of lactose can achieve levels 1,000-fold higher than those found in the external medium. Ron Kaback devised a simple yet elegant procedure for the purification of this protein.

Purification of the membrane-bound lactose carrier protein is a very different problem from the purification of the soluble OMP synthase. Both the approach to purification and the assays for the protein during purification are quite novel. The assay involves reconstituting a transport system with membranes that are free of lactose carrier protein, then adding the partially purified carrier protein and radioactively labeled lactose. The activity in this assay system is proportional to the transport of radioactive lactose across the membrane in the cell-free reconstituted system.

The results of the purification steps are tabulated in table 6.5, and the purification procedure is outlined in

**Table 6.5**

Purification of the Lactose Carrier Protein

| Fraction | Protein (mg) | Percent Recovery (Total Protein) | Percent Recovery (Carrier Protein) | Purification Factor |
|---|---|---|---|---|
| 1. Membrane fraction | 12.5 | 100 | 100 | 1.0 |
| 2. Urea-extracted membrane | 5.6 | 45 | 76 | 1.7 |
| 3. Urea/cholate-extracted membrane | 2.6 | 21 | 61 | 2.9 |
| 4. Octylglucoside extract | 0.4 | 3.2 | 38 | 12 |
| 5. DEAE[a] column peak | 0.056 | 0.4 | 14 | 35 |

[a] DEAE = diethylaminoethyl cellulose

figure 6.9. In this procedure advantage was taken of the fact that the carrier protein in its native state is firmly bound to the cytoplasmic membrane. Thus, the first step consisted of isolating these membranes from the rest of the cell constituents. Starting from the membrane fraction only 35-fold purification was required to achieve pure carrier protein. This rapid result was possible because a special strain of *E. coli*, containing about 100 times the normal carrier protein, was used as starting material for the purification. The high initial content was engineered by putting the carrier protein gene on a multicopy plasmid, which was then inserted into the cell—a procedure described in chapter 27.

Bacterial cells are much tougher than mammalian cells, requiring a more stringent procedure for cell disruption. In this case, the cells were placed in a so-called French pressure cell and bled through an orifice from very high pressures ($\approx$ 10,000 psi) to atmospheric pressure (step 1). Under these conditions the cells literally explode, fragmenting their membranes and releasing the cytoplasmic contents. The membranes were pelleted by a brief centrifugation, leaving a supernatant containing DNA, ribosomes, and cytoplasmic protein, which was removed by decantation (step 2). The pelleted membrane fraction was next resuspended and extracted with 5 M urea and then reextracted with 6% sodium cholate. The pellet obtained by high-speed centrifugation from these two extractions still contained most (61%) of the carrier protein in the more rapidly sedimenting membrane fraction (fraction 3), although 79% of the total membrane-bound protein was released by these treatments (steps 3 and 4).

At this point, the carrier protein was released from a suspension of the membranes by addition of the hydrophobic reagent octylglucoside in the presence of *E. coli* phospholipid (step 5). It is believed that the octyl part of octylglucoside competes effectively with the membrane for binding to hydrophobic centers on the carrier protein. The *E. coli* phospholipid facilitates dissociation of the carrier protein from the membrane fraction. The solubilized octylglucoside-containing extract was fourfold enriched in carrier protein after a high-speed centrifugation to remove residual membrane and membrane-bound proteins.

Finally, the octylglucoside-containing extract was passed over a positively charged diethylaminoethyl (DEAE) sepharose column (sepharose is a form of cross-linked polydextran) in a buffer containing 10 mM potassium phosphate, 20 mM lactose, and 0.25 mg of washed *E. coli* lipid per milliliter. The carrier protein passed through this column as a symmetrical peak of protein (step 6). Most of the remaining protein in the extract adsorbed to the positively charged column. The fractions containing the bulk of the carrier protein activity were judged to be pure by SDS gel electrophoretic analysis. The purified protein contained a single polypeptide chain with an estimated molecular weight of 33,000.

The two purifications described here are as different as the two proteins involved. No two purifications are exactly alike, but the principles of purification, as stated at the outset, are quite similar. Fortunately, an almost endless variety of purification techniques exists. This variety is both helpful and challenging because a great deal of knowledge and creativity are required to exploit it. In addition to professional expertise, the two most important things required to make a purification possible are an unambiguous assay for the protein in question and a means of stabilizing the protein during purification.

**Figure 6.9**

Outline of purification procedure for lactose carrier protein from *E. coli.*

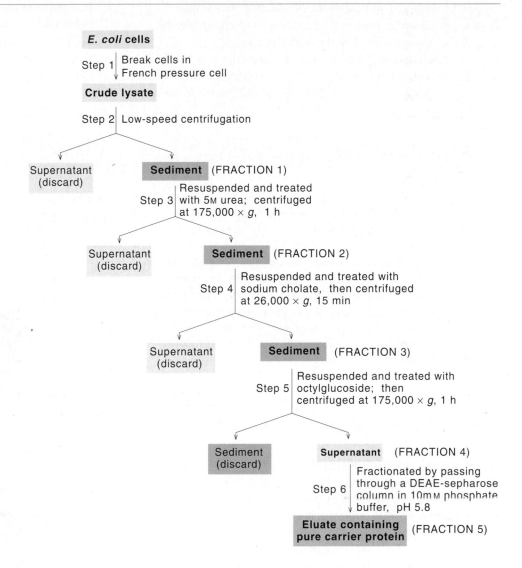

## Summary

In this chapter we considered the different means of purification that enable us to study individual proteins in isolation. The following points are the most important.

1. Whereas proteins must be studied *in vivo* in their normal habitat, to characterize them in great detail they must also be isolated in pure form. Protein purification is a complex art, and a great variety of purification methods are usually applied in sequence for the purification of any protein.

2. Frequently the first step in protein purification is differential sedimentation of broken cell parts. In this way soluble proteins may be separated from organelle-sequestered proteins.

3. Following differential sedimentation, proteins may be separated into crude fractions by the addition of increasing amounts of $(NH_4)_2SO_4$. Specific proteins characteristically precipitate in a limited range of salt concentrations.

4. Column procedures are useful in fractionating proteins with different affinity properties and sizes. By the use of different column materials in conjunction with specific eluting solutions, highly purified protein preparations can be obtained.

5. Gel electrophoresis, which separates proteins according to their size and charge, can be used in purification or more frequently as a means of assaying the

purity during a purification. Isoelectric focusing, which separates proteins according to their isoelectric points, can be used for the same purpose.
6. The molecular weights of soluble proteins can be roughly estimated by SDS gel electrophoresis or can be rigorously determined by sedimentation techniques.

7. The purification of two proteins, UMP synthase from mammalian tumor cells, and lactose carrier protein from *E. coli* bacteria is described in detail to illustrate how different fractionation methods can be combined most effectively.

## Selected Readings

Cantor, C. R., and P. Schimmel, *Biophysical Chemistry,* New York: W. H. Freeman, 1980. See especially volume 2, entitled *Study of Biophysical Structure and Function.*

Chaif, B. T., and S. B. H. Kent, Weighing naked proteins: Practical, high-accuracy mass measurement of peptides and proteins. *Science* 257:1885–1893, 1992. Proteins with molecular masses of as much as 100 kd or more can be analyzed at picomole sensitivities to give simple mass spectra corresponding to the intact molecule. This development has allowed unprecedented accuracy in the determination of protein molecular weights.

Eisenberg, D., and D. Crothers, *Physical Chemistry and Its Applications to the Life Sciences,* Menlo Park, Calif.: Benjamin/Cummings, 1979.

*Methods in Enzymology,* New York: Academic Press. A continuing series of over 250 volumes that discusses most methods at the professional level but is still understandable for students.

Scopes, R., *Protein Purification: Principles and Practice,* 2d ed. New York: Springer-Verlag, 1987. A recent general treatment of this subject.

## Problems

1. Why are "salting out" procedures often used as an initial purification step following the production of a crude extract by centrifugation?
2. Rarely is DEAE-cellulose used above a pH of about 8.5. Can you provide a reason(s) why?
3. Given that the only structural difference between phosphorylase a and phosphorylase b is that phosphorylase a has a covalently bound phosphate on serine 14, do you expect phosphorylase a and phosphorylase b to elute as a single peak on DEAE-cellulose chromatography? What if gel filtration was utilized?

4. A method for the purification of 6-phosphogluconate dehydrogenase from *E. coli* is summarized in the table. For each step, calculate the specific activity, percentage yield, and degree of purification (*n*-fold). Indicate which step results in the greatest purification. Assume that the protein is pure after gel (Bio-Gel A) exclusion chromatography. What percentage of the initial crude cell extract protein was 6-phosphogluconate dehydrogenase?

### Table for Problem 4

| Step | Volume (ml) | Total Protein (mg) | Total Units | Specific Activity (U/mg) | Yield (%) | Purification (*n*-fold) |
|---|---|---|---|---|---|---|
| Cell extract | 2,800 | 70,000 | 2,700 | | | |
| ((NH$_4$)$_2$SO$_4$) fractionation | 3,000 | 25,400 | 2,300 | | | |
| Heat treatment | 3,000 | 16,500 | 1,980 | | | |
| DEAE chromatography | 80 | 390 | 1,680 | | | |
| CM-cellulose | 50 | 47 | 1,350 | | | |
| Bio-Gel A | 7 | 35 | 1,120 | | | |

5. Although used effectively in the 6-phosphogluconate dehydrogenase isolation procedure, heat treatment cannot be used in the isolation of all enzymes. Explain.

6. Assume that the isoelectric point (pI) of 6-phosphogluconate dehydrogenase is 6. Explain why the buffer used in the DEAE-cellulose chromatography must have a pH greater than 6 but less than 9 for the enzyme to bind to the DEAE resin.

7. Will the 6-phosphogluconate dehydrogenase bind to the CM-cellulose in the same buffer pH range used with the DEAE-cellulose? Explain. In what pH range would you expect the dehydrogenase to bind to CM-cellulose? Explain.

8. Examine the isolation procedure shown in problem 4 and explain why gel exclusion chromatography is used as the final step rather than as the step following the heat treatment.

9. A student isolated an enzyme from anaerobic bacteria and subjected a sample of the protein to SDS polyacrylamide gel electrophoresis. A single band was observed on staining the gel for protein. His adviser was excited about the result, but suggested that the protein be subjected to electrophoresis under nondenaturing (native) conditions. Electrophoresis under nondenaturing conditions revealed two bands after the gel was stained for protein. Assuming the sample had not been mishandled, offer an explanation for the observations.

10. A salt-precipitated fraction of ribonuclease contained two contaminating protein bands in addition to the ribonuclease. Further studies showed that one contaminant had a molecular weight of about 13,000 (similar to ribonuclease) but an isoelectric point 4 pH units more acidic than the pI of ribonuclease. The second contaminant had an isoelectric point similar to ribonuclease but had a molecular weight of 75,000. Suggest an efficient protocol for the separation of the ribonuclease from the contaminating proteins.

11. You have a mixture of proteins with the following properties:

Protein 1: $M_r$ 12,000,    pI = 10
Protein 2: $M_r$ 62,000,    pI = 4
Protein 3: $M_r$ 28,000,    pI = 8
Protein 4: $M_r$ 9,000,    pI = 5

Predict the order of emergence of these proteins when a mixture of the four is chromatographed in the following systems:
(a) DEAE-cellulose at pH 7, with a linear salt gradient elution.
(b) CM-cellulose at pH 7, with a linear salt gradient elution.
(c) A gel exclusion column with a fractionation range of 1,000–30,000 $M_r$, at pH 7.

12. You wish to purify an ATP-binding enzyme from a crude extract that contains several contaminating proteins. To purify the enzyme rapidly and to the highest purity, you must consider some sophisticated strategies, among them affinity chromatography. Explain how affinity chromatography can be applied to this separation, and explain the physical basis of the separation.

# Catalysis

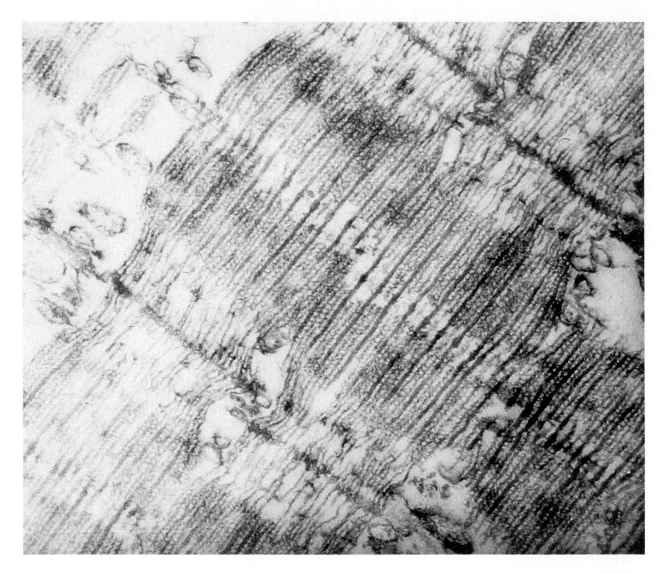

Electron micrograph of a longitudinal section of a skeletal muscle fiber showing a number of myofibrils. Muscle protein is an unusual example of a structural protein with enzymatic activity. (Courtesy of Dr. Hugh Huxley)

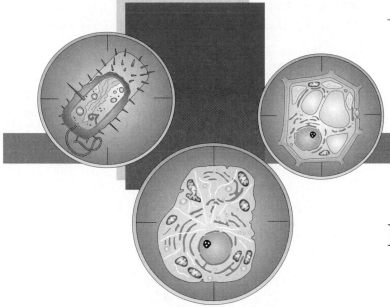

# Enzyme Kinetics

*An enzyme catalyzed reaction proceeds rapidly
under mild conditions because it lowers the
activation energy for a reaction; enzymes are
usually highly specific for the reactions they
catalyze.*

Most of this book is concerned with the reactions that occur in living cells. These reactions are catalyzed by enzymes. In this and the following three chapters we focus on the ways in which enzymes function. The present chapter deals with the kinetics of enzyme-catalyzed reactions; the next two explore the mechanisms of enzymatic catalysis. Finally in chapter 10 we examine the small cofactors that work together with many enzymes.

A catalyst is a substance that accelerates a chemical reaction without itself undergoing any net change. The rate enhancements achieved by many enzymes are extraordinarily high. For example, carbonic anhydrase, an enzyme found in red blood cells, catalyzes the reaction

$$CO_2 + H_2O \longrightarrow H_2CO_3 \qquad (1)$$

In the presence of the enzyme, this reaction occurs about $10^7$ times as rapidly as it does in the absence of the enzyme. One molecule of carbonic anhydrase can hydrate about $10^6$ molecules of $CO_2$ a second.

Kinetic analysis is one of the most basic topics of enzymology. Such studies reveal not only how fast an enzyme can function, but also its preferences for various reactants (or substrates as they usually are called), the effect of substrate concentration on the reaction rate, and the sensitivity of the enzyme to specific inhibitors or activators. By studying the alterations in rate under different conditions we often gain clues to the mechanism of the reaction.

## The Discovery of Enzymes

The existence of catalysts in biological materials was recognized as early as 1835 by Jöns Jakob Berzelius, the Swedish chemist who discovered several elements, introduced the way of writing chemical symbols, and coined the term "catalysis." Berzelius noted that potatoes contained something that catalyzed the breakdown of starch, and he suggested that all natural products are formed under the influence of such catalysts. But the chemical nature of biological catalysts was unknown, and it remained a mystery for many years. In the period between 1850 and 1860, Louis Pasteur demonstrated that fermentation, the anaerobic breakdown of sugar to $CO_2$ and ethanol, occurred in the presence of living cells and did not occur in a flask that was capped after any cells that it contained had been killed by heat. Then in 1897, Eduard Buchner discovered by accident that fermentation was catalyzed by a clear juice that he had prepared by grinding yeast with sand and filtering out the unbroken cells. Looking for a way to preserve the juice, Buchner had added sugar. It probably was a disappointment to him that the sugar was broken down rapidly and the mixture frothed with $CO_2$. But Buchner's discovery made it possible to explore metabolic processes such as fermentation in a greatly simplified system, without having to deal with the complexities of cell growth and multiplication, and without the barriers imposed by cell walls or membranes. Arthur Harden and William Young soon showed that yeast extracts contained two different types of molecules, both of which were necessary for fermentation to occur. Some were small, dialyzable, heat-stable molecules such as inorganic phosphate; others were much larger, nondialyzable molecules—the enzymes—that were destroyed easily by heat.

Although early investigators surmised that enzymes might be proteins, this remained in dispute until 1927, when James Sumner succeeded in purifying and crystalizing the enzyme urease from beans. In the 1930s, John Northrop isolated and characterized a series of digestive enzymes, generalizing Sumner's conclusion that enzymes are proteins. Since then thousands of different enzymes have been purified, and the structures of many of them have been solved to atomic resolution; almost all of these molecules have proved to be proteins. Surprisingly, however, recent work has shown that some RNA molecules also have enzymatic activity.

**Figure 7.1**

The six main classes of enzymes and the reactions they catalyze.

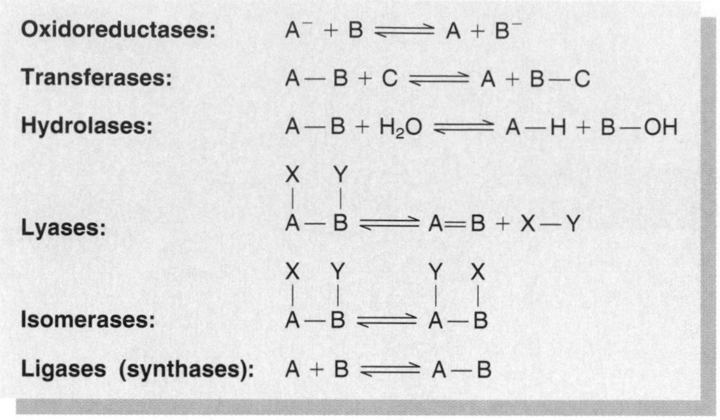

| | |
|---|---|
| **Oxidoreductases:** | $A^- + B \rightleftharpoons A + B^-$ |
| **Transferases:** | $A{-}B + C \rightleftharpoons A + B{-}C$ |
| **Hydrolases:** | $A{-}B + H_2O \rightleftharpoons A{-}H + B{-}OH$ |
| **Lyases:** | $\begin{matrix} X & Y \\ \| & \| \\ A{-}B \end{matrix} \rightleftharpoons A{=}B + X{-}Y$ |
| **Isomerases:** | $\begin{matrix} X & Y \\ \| & \| \\ A{-}B \end{matrix} \rightleftharpoons \begin{matrix} Y & X \\ \| & \| \\ A{-}B \end{matrix}$ |
| **Ligases (synthases):** | $A + B \rightleftharpoons A{-}B$ |

## Enzyme Terminology

Enzymes often are known by common names obtained by adding the suffix "-ase" to the name of the substrate or to the reaction that they catalyze. Thus, glucose oxidase is an enzyme that catalyzes the oxidation of glucose; glucose-6-phosphatase catalyzes the hydrolysis of phosphate from glucose-6-phosphate; and urease catalyzes the hydrolysis of urea. Common names also are used for some groups of enzymes. For example, an enzyme that transfers a phosphate group from ATP to another molecule is usually called a "kinase," instead of the more formal "phosphotransferase."

A systematic scheme for classifying enzymes was adopted in 1972 by the International Union of Biochemistry. In this scheme, each enzyme is designated by four numbers that indicate the main class, subclass, subsubclass, and the serial number of the enzyme in its subsubclass. The six main classes (fig. 7.1) are (1) oxidoreductases, (2) transferases, (3) hydrolases, (4) lyases, (5) isomerases, and (6) ligases, or synthases. Oxidoreductases catalyze oxidation–reduction reactions. Transferases catalyze the transfer of a functional group from one molecule to another. Hydrolases catalyze bond cleavage by the introduction of water. Lyases catalyze the removal of a group to form a double bond or the addition of a group to a double bond. Isomerases catalyze intramolecular rearrangements, and ligases catalyze reactions that join two molecules.

Many enzymes require additional small molecules called cofactors for their activity. Cofactors can be simple inorganic ions such as $Mg^{2+}$ or complex organic molecules known as coenzymes. The cofactor usually binds tightly to a special site on the enzyme. An enzyme lacking an essential cofactor is called an apoenzyme, and the intact enzyme with the bound cofactor is called the holoenzyme.

## Basic Aspects of Chemical Kinetics

Before we delve into enzyme kinetics, we need to discuss some of the basic principles that apply to the kinetics of both enzymatic and nonenzymatic reactions. Let's first consider a nonenzymatic reaction that converts a single reactant (R) into a product (P):

$$R \longrightarrow P \qquad (2)$$

To measure the velocity of the reaction ($v$), we plot the concentration of R as a function of time (fig. 7.2). The rate at any particular time ($t$) is

$$v = -\frac{d[R]}{dt} \qquad (3)$$

where $d[R]/dt$ is the slope of the plot at that time. The minus sign is needed because $v$ is defined by convention to be a positive number, whereas $d[R]/dt$ is always negative ([R] decreases with time). However, we might equally well choose to measure the increase in the concentration of the product as a function of time, in which case we express the rate as $v = d[P]/dt$.

For a simple reaction of this type, the rate at any given time usually is found to be proportional to the remaining concentration of the reactant:

$$v = k[R] \qquad (4)$$

The proportionality constant $k$ is the rate constant. The rate constant is independent of the concentration of the reactant, but it can depend on other parameters, such as temperature or pH, and, as we shall see, it may be altered by a catalyst. It has dimensions of reciprocal seconds ($s^{-1}$). Combining equations (3) and (4), we have

$$\frac{d[R]}{dt} = -k[R] \qquad (5)$$

A reaction of this type is said to follow first-order kinetics because the rate is proportional to the concentration of a single species raised to the first power (fig. 7.2). An example is the decay of a radioactive isotope such as $^{14}C$. The rate of decay at any time (the number of radioactive disintegrations per second) is simply proportional to the amount of $^{14}C$ present. The rate constant for this extremely slow nuclear reaction is $8 \times 10^{-12} \, s^{-1}$. Another example is the initial electron-transfer reaction that occurs when photosyn-

**Figure 7.2**

The kinetics of a first-order reaction in which a single reactant (R) is converted irreversibly to a product (P). The concentrations of R and P are plotted as functions of time. The rate ($v$) at any given time can be obtained from the slope of either curve:

$$v = -\frac{d[R]}{dt} = \frac{d[P]}{dt}$$

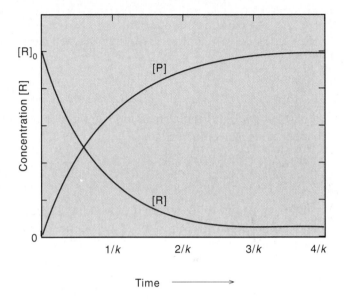

thetic organisms are excited with light (see chapter 15). In this case there actually are two reactants, the electron donor and the acceptor, but they are held close together on a protein so that they react as a unit; the excited complex simply decays spontaneously to a more stable state. This process occurs with a rate constant of $3 \times 10^{11} \, s^{-1}$, which makes it one of the fastest reactions known.

A slightly more complicated situation arises if the reaction is reversible:

$$R \underset{k_{-1}}{\overset{k_1}{\rightleftharpoons}} P \qquad (6)$$

Here $k_1$ is the rate constant for the forward reaction, and $k_{-1}$ is that for the back reaction. In a reversible reaction, the rate equation becomes

$$\frac{d[R]}{dt} = -k_1[R] + k_{-1}[P] \qquad (7)$$

The term $-k_1[R]$ is the same as in equation (5); the term $k_{-1}[P]$ describes the formation of R from P. In this case, [R] and [P] proceed from their initial values to their final, equilibrium values ([R]$_{eq}$ and [P]$_{eq}$), which generally are not zero. At equilibrium, $d[R]/dt$ must go to zero, which requires that

$$k_1[R]_{eq} = k_{-1}[P]_{eq} \tag{8}$$

Rearranging this expression gives

$$\frac{[P]_{eq}}{[R]_{eq}} = \frac{k_1}{k_{-1}} = K_{eq} \tag{9}$$

where $K_{eq}$ is the equilibrium constant.

Now consider a reaction involving two reactants, A and B:

$$A + B \xrightarrow{\ k\ } P \tag{10}$$

The rate of such a bimolecular reaction usually depends on the concentrations of both reactants:

$$\frac{d[P]}{dt} = k[A][B] \tag{11}$$

The reaction is said to follow second-order kinetics because its rate is proportional to a product of two concentrations. The kinetics are first order in either [A] or [B] alone, but second order overall. The rate constant for a second-order reaction has the dimensions of $M^{-1}s^{-1}$.

## *A Critical Amount of Energy Is Needed for the Reactants to Reach the Transition State*

The equilibrium constant $K_{eq}$ for a reaction is strictly a function of thermodynamic factors. The most pertinent thermodynamic quantity here is the change in free energy that occurs in the reaction $\Delta G$. As we saw in chapter 2, $\Delta G$ and $K_{eq}$ are related by the expression $\Delta G° = -RT \ln K_{eq}$. If the free energy of the products is less than the free energy of the reactants ($\Delta G < 0$), then the equilibrium constant favors the formation of products; if the free energy of the products exceeds that of the reactants ($\Delta G > 0$), the reverse reaction is favored. We noted above that the equilibrium constant for a one-step reaction is equal to the ratio of the rate constants in the forward and reverse directions [equation (9)]. However, the value of $K_{eq}$ does not tell us how long it takes the reaction to reach equilibrium because it says nothing about the magnitudes of the individual rate constants. These may be very large or very small. The overall $\Delta G$ for a reaction therefore indicates only whether the reaction is possible, not how rapidly the reaction occurs. To understand the kinetics, we must look into the mechanism of the reaction.

The rate at which two molecules react depends partly on how frequently the molecules collide. Collisions occur as a result of random diffusion of the reactants in the solution. The number of collisions per second is proportional to the product of the two concentrations, and the second-order rate constant $k$ for the reaction includes a proportionality factor for this relationship. The rate constant also includes a factor that gives the fraction of the collisions that are effective. If every collision were to result in a reaction, this second factor would be 1 and the rate constant for the reaction of two small molecules in aqueous solution would be about $10^{11}$ $M^{-1}s^{-1}$. Because of their large masses, proteins diffuse relatively slowly, so the frequency at which a protein and a small molecule collide is lower than the collision frequency for two small molecules. As a consequence, a reaction between a protein and a small molecule has a maximum rate constant on the order of $10^8$ to $10^9$ $M^{-1}s^{-1}$.

But not every collision results in a reaction. What determines the fraction of the collisions that are effective? A partial answer is that the colliding species must have a certain critical energy in order to surmount a barrier that separates the reactants from the products. This is illustrated schematically in figure 7.3. The surface in figure 7.3a represents the energy of a system in which a proton can be bound to either of two molecules. Suppose the proton is initially on molecule A, and we are interested in how rapidly it moves to molecule B. As the proton moves from one place to the other, its electrostatic interactions with molecule A become less favorable, and the interactions with B improve. The energy of the system goes through a maximum when the proton is at an intermediate position. At this point, the system is said to be in the transition state. The probability that a collision leads to a reaction depends, in part, on the probability that the molecules collide with enough energy to reach this state.

The amount of free energy required to reach the transition state is called the activation free energy, $\Delta G^{\ddagger}$. From equation (13) of chapter 2, we can equate $\Delta G^{\ddagger}$ to $-RT \ln K^{\ddagger}$, where $K^{\ddagger}$ is an equilibrium constant for the formation of the transition state from the reactants. The fraction of the reactants that are in the transition state at any given moment is given approximately by $K^{\ddagger}$, or $e^{-\Delta G^{\ddagger}/RT}$. We therefore can write the overall rate constant for a reaction as

$$k = k_0 e^{-\Delta G^{\ddagger}/RT} \tag{12}$$

Here $k_0$ is an intrinsic rate constant for conversion of the transition state into the products and typically depends on the nuclear vibration frequency of a bond that is being formed or broken.

## Figure 7.3

Schematic energy diagrams for a reaction in which a proton moves from one molecule to another. In drawing (a), coordinates in the plane at the bottom represent the location of the proton. The free energy of the system for a particular set of coordinates is represented by the distance of the cuplike surface above the plane. The two minima in the energy surface indicate the positions of the proton when it is bound optimally to one molecule or the other. The best route along the surface from one of these minima to the other goes through a pass, or saddle point. Drawing (b) shows a plot of the free energy as a function of distance along the optimal route over this pass. The activation free energy of the reaction ($\Delta G^{\ddagger}_a$) is the difference between the energies at the pass and at the starting point.

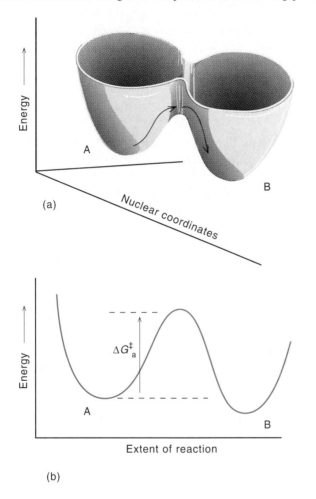

(a)

(b)

## *Catalysts Speed up Reactions by Lowering the Free Energy of Activation*

Equation (12) indicates that, if $\Delta G^{\ddagger}$ is greater than zero as is most often the case, a reaction can be sped up either by decreasing $\Delta G^{\ddagger}$ or by raising the temperature. Little can usually be done to change $k_0$. For living organisms, it gener-

## Figure 7.4

An enzyme speeds up a reaction by decreasing $\Delta G^{\ddagger}$. The enzyme does not change the free energy of the substrate (S) or product (P); it lowers the free energy of the transition state. The two vertical arrows indicate the activation free energies ($\Delta G^{\ddagger}$) of the catalyzed and uncatalyzed reactions.

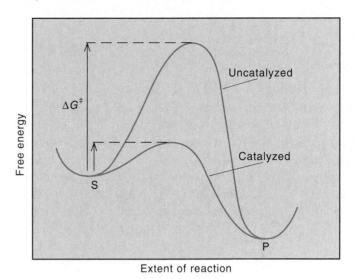

ally is impractical to raise the temperature because many organisms have little or no control over the ambient temperature and can survive only within a narrow range of temperatures. Also, changing the temperature is not a very selective way to control reaction rates because most reactions speed up when the temperature is raised. The remaining alternative is to decrease $\Delta G^{\ddagger}$.

Because rate constants depend exponentially on $-\Delta G^{\ddagger}/RT$ [equation (12)], small changes in $\Delta G^{\ddagger}$ will have a large effect on the reaction rate. At physiological temperatures, it takes a decrease of only 1.36 kcal/mole to speed up a reaction by a factor of 10, and a decrease by 8.16 kcal/mole increases the rate by a factor of $10^6$. These are relatively modest free energy changes because forming a single H bond can release anywhere from 4 to 10 kcal/mole. In addition, because $\Delta G^{\ddagger}$ depends on the structures of the reactants and products and on the detailed nature of the reaction, it affords an opportunity to control reaction rates with a great deal of specificity. Enzymes, then, must work by decreasing the activation free energies for the specific reactions that they catalyze (fig. 7.4). They do this in a variety of ways that we discuss in detail in the next chapter. Our main concern here is to describe in more general terms how the kinetic properties of enzymatic reactions are measured and characterized.

It should be kept in mind that although an enzyme can increase the rate at which a reaction occurs, it (or any other catalyst) cannot alter the overall equilibrium constant. Since $K_{eq}$ is equal to the ratio of the rate constants for the forward and reverse processes, the catalyst increases *both* of these rate constants without changing their ratio.

## Kinetics of Enzyme-Catalyzed Reactions

Kinetic analysis was used to characterize enzyme-catalyzed reactions even before enzymes had been isolated in pure form. As a rule, kinetic measurements are made on purified enzymes *in vitro*. But the properties so determined must be referred back to the situation *in vivo* to ensure they are physiologically relevant. This is important because the rate of an enzymatic reaction can depend strongly on the concentrations of the substrates and products, and also on temperature, pH, and the concentrations of other molecules that activate or inhibit the enzyme. Kinetic analysis of such effects is indispensable to a comprehensive picture of an enzyme.

### Kinetic Parameters Are Determined by Measuring the Initial Reaction Velocity as a Function of the Substrate Concentration

The usual procedure for measuring the rate of an enzymatic reaction is to mix enzyme with substrate and observe the formation of product or disappearance of substrate as soon as possible after mixing, when the substrate concentration is still close to its initial value and the product concentration is small. The measurements usually are repeated over a range of substrate concentrations to map out how the initial rate depends on concentration. Spectrophotometric techniques are used commonly in such experiments because in many cases they allow the concentration of a substrate or product in the mixture to be measured continuously as a function of time.

Measurements of reactions that occur in less than a few seconds require special techniques to speed up the mixing of the enzyme and substrate. One way to achieve this is to place solutions containing the enzyme and the substrate in two separate syringes. A pneumatic device then is used to inject the contents of both syringes rapidly into a common chamber that resides in a spectrophotometer for measuring the course of the reaction (fig. 7.5). Such an apparatus is referred to as a "stopped-flow" device because the flow stops abruptly when the movement of the pneumatic driver is arrested. In this type of apparatus it is possible to make kinetic measurements within about 1 ms after mixing of enzyme and substrate.

**Figure 7.5**

The stopped-flow apparatus for measuring enzyme-catalyzed reactions very soon after mixing enzyme and substrate.

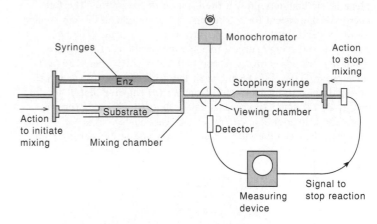

Equations (5) and (10) imply that the velocity of an uncatalyzed reaction increases indefinitely with an increase in the concentration of the reactants. With enzyme-catalyzed reactions, something very different is observed. The rate usually increases linearly with substrate concentration at low concentrations, but then levels off and becomes independent of the concentration at high concentrations (fig. 7.6). The explanation for this hyperbolic dependence on substrate concentration is straightforward. For an enzyme to affect $\Delta G^{\ddagger}$, the substrate must bind to a special site on the protein, the active site (fig. 7.7). At very low concentrations of substrate, the active sites of most of the enzyme molecules in the solution are unoccupied. Increasing the substrate concentration brings more enzyme molecules into play, and the reaction speeds up. At high concentrations, on the other hand, most of the enzyme molecules have their active sites occupied, and the observed rate depends only on the rate at which the bound reactants are converted into products. Further increases in the substrate concentration then have little effect.

### The Henri-Michaelis-Menten Treatment Assumes That the Enzyme–Substrate Complex Is in Equilibrium with Free Enzyme and Substrate

The hyperbolic saturation curve that is commonly seen with enzymatic reactions led Leonor Michaelis and Maude Menten in 1913 to develop a general treatment for kinetic analysis of these reactions. Following earlier work by Victor Henri, Michaelis and Menten assumed that an enzyme–substrate complex (ES) is in equilibrium with free enzyme

**Figure 7.6**

The reaction velocity $v$ as a function of the substrate concentration [S] for an enzyme-catalyzed reaction. At high substrate concentrations the reaction velocity reaches a limiting value, $V_{max}$. $K_m$ is the substrate concentration at which the rate is half maximal.

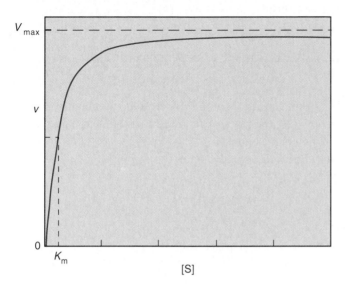

**Figure 7.7**

Thermolysin is an enzyme that hydrolyzes peptide bonds. Close-up view of the active site of thermolysin (beige) illustrating the enzyme-substrate interaction between a tight binding substrate analog, phosphoramidon (yellow), and the enzyme. (Based on the crystal structure described by D. E. Tronrud, A. F. Monzingo, and B. W. Matthews. Copyright 1994 by the Scripps Research Institute/Molecular Graphics Images by Michael Pique using software by Yng Chen, Michael Connolly, Michael Carson, Alex Shah, and AVS, Inc. Visualization advice by Holly Miller, Wake Forest University Medical Center.)

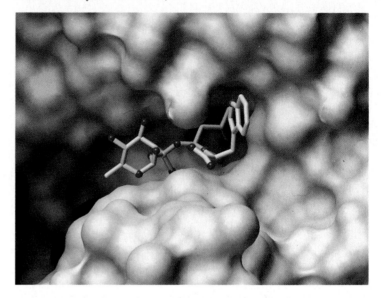

(E) and substrate (S), and that the formation of products (P) proceeds only through the ES complex:

$$E + S \underset{k_{-1}}{\overset{k_1}{\rightleftharpoons}} ES \xrightarrow{k_2} E + P \qquad (13)$$

Their objective was to relate the reaction rate to observable quantities and interpretable molecular parameters.

Because the slow step in the reaction described by equation (13) is assumed to be formation of E and P from ES, the velocity of the reaction should be

$$v = \frac{d[P]}{dt} = k_2[ES] \qquad (14)$$

A maximum velocity ($V_{max}$) is obtained when all of the enzyme is in the form of the enzyme–substrate complex. Since the total concentration of enzyme $[E_t]$ is equal to $[E] + [ES]$,

$$V_{max} = k_2[E_t] = k_2([E] + [ES]) \qquad (15)$$

Dividing equation (14) by equation (15) gives

$$\frac{v}{V_{max}} = \frac{[ES]}{[E] + [ES]} \qquad (16)$$

The fraction on the right-hand side can be evaluated by making use of the dissociation constant of the ES complex in equation (13), $K_s$:

$$K_s = \frac{[E][S]}{[ES]} \quad \text{and} \quad [ES] = \frac{[E][S]}{K_s} \qquad (17)$$

Substituting this value of [ES] into equation (16) and rearranging, gives

$$v = \frac{V_{max}[S]}{K_s + [S]} \qquad (18)$$

Equation (18) is the Henri-Michaelis-Menten equation, which relates the reaction velocity to the maximum velocity, the substrate concentration, and the dissociation constant for the enzyme–substrate complex. Usually substrate is present in much higher molar concentration than enzyme, and the initial period of the reaction is examined so that the free substrate concentration [S] is approximately equal to the total substrate added to the reaction mixture.

## Steady-State Kinetic Analysis Assumes That the Concentration of the Enzyme–Substrate Complex Remains Nearly Constant

Rather than discussing the implications of equation (18) at this point, it is useful to develop a more general expression that avoids the assumption that the enzyme–substrate complex is in equilibrium with free enzyme and substrate. To develop this expression, we introduce the concept of the steady state, which was first pro-

**Figure 7.8**

Concentrations of free enzyme (E), substrate (S), enzyme–substrate complex (ES), and product (P) over the time course of a reaction. The shaded portion of the top graph is shown in expanded form in the bottom graph. After a brief initial period (usually less than a few seconds) the concentration of ES remains approximately constant for an extended period. The steady-state approximation is applicable during this second period. Most measurements of enzyme kinetics are made in the steady state.

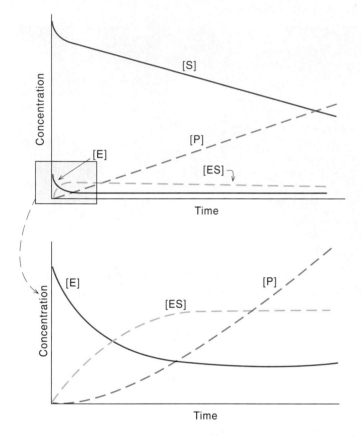

posed by G. E. Briggs and J. B. S. Haldane in 1925. The steady state constitutes the time interval when the rate of the reaction is approximately constant. The system usually reaches a steady state soon after enzyme and substrate are mixed, following a brief period when the concentration of the enzyme–substrate complex (ES) builds up (fig. 7.8). The ES concentration then remains almost constant for the duration of the steady state, while the concentrations of the substrate and product continue to change substantially.[a]

Let's now write out the reaction of equation (13) in more detail.

$$E + S \underset{k_{-1}}{\overset{k_1}{\rightleftharpoons}} ES \underset{k_{-2}}{\overset{k_2}{\rightleftharpoons}} E + P \qquad (19)$$

In this more complete scheme, the ES complex forms from E and S with rate constant $k_1$. ES can either dissociate again with rate constant $k_{-1}$ or go on to P with $k_2$. If we confine

ourselves to measuring the initial rate of the reaction in the steady state, we can continue to neglect regeneration of ES from E and P (the step involving $k_{-2}$) because the concentration of P will be too small for this back-reaction to occur at a significant rate. For the rate of formation of ES, $v_f$, we then can write

$$v_f = k_1[E][S] \qquad (20)$$

Similarly, the rate of disappearance of ES, $v_d$, is

$$v_d = k_{-1}[ES] + k_2[ES] \qquad (21)$$

If the concentration of ES is virtually constant during the steady state, the rates of formation and disappearance of ES must be nearly equal, $v_f = v_d$. We therefore can describe the situation in the steady state by combining equations (20) and (21).

$$k_1[E][S] = (k_{-1} + k_2)[ES] \qquad (22)$$

or

$$\frac{[E][S]}{[ES]} = \frac{k_{-1} + k_2}{k_1} = K_m \qquad (23)$$

The constant $K_m$ defined in equation (23) is called the Michaelis constant and is one of the key parameters in enzyme kinetics. It is a simple matter to proceed from this point to an expression comparable to the Henri-Michaelis-Menten equation (18), but with $K_m$ in place of $K_s$. First, rearranging equation (23) gives

$$[ES] = \frac{[E][S]}{K_m} \qquad (24)$$

With this expression for [ES] we can follow the same procedure that led to equation (18), this time arriving at the Briggs-Haldane equation for the reaction velocity.

$$v = \frac{V_{max}[S]}{[S] + K_m} \qquad (25)$$

Equation (25) is identical to (18) except that the more complex constant $K_m$ has replaced $K_s$. The term "Michaelis-Menten equation" is often used for either expression.

For the purpose of graphical representation of experimental data, it is convenient to rearrange equation (25). Taking the reciprocals of both sides of equation (25) gives

$$\frac{1}{v} = \frac{1}{V_{max}} + \frac{K_m}{V_{max}}\frac{1}{[S]} \qquad (26)$$

[a] The steady state is no stranger to the living cell. Most reactions in a living cell are in a steady state most of the time. Thus, the steady state used originally as a convenience by kineticists is also an appropriate way to analyze enzymatic reactions *in vivo*.

**Figure 7.9**

A plot of the reciprocal of the rate (1/v) as a function of the reciprocal of the substrate concentration (1/[S]) fits a straight line. Extrapolating the line to its intercept on the ordinate (infinite substrate concentration) gives $1/V_{max}$. Extrapolating to the intercept on the abscissa gives $-1/K_m$.

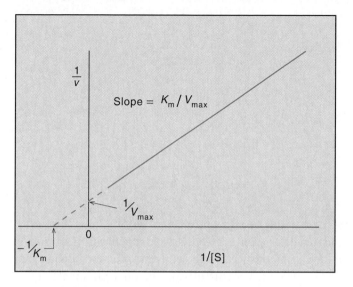

**Table 7.1**

The Michaelis Constants for Some Enzymes

| Enzyme and Substrate | $K_m$ (M) |
|---|---|
| Catalase | |
|   $H_2O_2$ | 1.1 |
| Hexokinase | |
|   Glucose | $1.5 \times 10^{-4}$ |
|   Fructose | $1.5 \times 10^{-3}$ |
| Chymotrypsin | |
|   $N$-Benzoyltyrosinamide | $2.5 \times 10^{-3}$ |
|   $N$-Formyltyrosinamide | $1.2 \times 10^{-2}$ |
|   $N$-Acetyltyrosinamide | $3.2 \times 10^{-2}$ |
|   Glycyltyrosinamide | $1.2 \times 10^{-1}$ |
| Aspartate aminotransferase | |
|   Aspartate | $9.0 \times 10^{-4}$ |
|   $\alpha$-Ketoglutarate | $1.0 \times 10^{-4}$ |
| Fumarase | |
|   Fumarate | $5.0 \times 10^{-6}$ |
|   Malate | $2.5 \times 10^{-5}$ |

This expression indicates that a plot of $1/v$ versus $1/[S]$ fits a straight line with a slope of $K_m/V_{max}$ (fig. 7.9). Such a plot is known as a Lineweaver-Burk, or double-reciprocal, plot. The intercept of the line on the ordinate occurs at $1/v = 1/V_{max}$, and the intercept on the abscissa occurs at $1/[S] = -1/K_m$. $V_{max}$ and $K_m$ thus can be determined readily from the graph.

***Significance of the Michaelis Constant, $K_m$.*** The Michaelis constant $K_m$ has the dimensions of a concentration (molarity), because $k_{-1}$ and $k_2$, the two rate constants in the numerator of equation (23), are first-order rate constants with units expressed per second ($s^{-1}$), whereas the denominator $k_1$ is a second-order rate constant with units of $M^{-1}s^{-1}$. To appreciate the meaning of $K_m$, suppose that $[S] = K_m$. The denominator in equation (25) then is equal to $2[S]$, which makes the velocity $v = V_{max}/2$. Thus, the $K_m$ is the substrate concentration at which the velocity is half maximal (fig. 7.6).

$K_m$ values for several enzyme–substrate pairs are given in table 7.1. The values vary over a wide range but typically lie between $10^{-6}$ and $10^{-1}$ M. With enzymes that can act on several different substrates, $K_m$ can vary substantially from substrate to substrate. With the enzyme chymotrypsin, for example, the $K_m$ for the substrate glycyltyrosinamide is about 50 times that for the substrate $N$-benzoyltyrosinamide.

Does the $K_m$ indicate how tightly a particular substrate binds to the active site of an enzyme? Not necessarily. According to equation (23),

$$K_m = \frac{k_{-1} + k_2}{k_1} \qquad (23')$$

whereas the dissociation constant of the ES complex is

$$K_s = \frac{k_{-1}}{k_1} \qquad (27)$$

Comparing these two expressions, we see that $K_m$ must always be larger than $K_s$. If $k_2 \gg k_{-1}$, then $K_m \approx k_2/k_1$, which is much greater than $K_s$. On the other hand, if $k_{-1} \gg k_2$, then $K_m$ is approximately equal to $K_s$. In the latter circumstance, a smaller $K_m$ means a smaller dissociation constant, which implies tighter binding to the enzyme. (This is the limiting case assumed in the Henri-Michaelis-Menten treatment, for when $k_{-1} \gg k_2$, the enzyme–substrate complex is in equilibrium with the free enzyme and substrate.) But whether or not this is a valid approximation depends on the particular enzyme and the substrate. For a multistep reaction the relationship between $K_m$ and the rate constants can be considerably more complex.

***Significance of the Turnover Number, $k_{cat}$.*** The turnover number of an enzyme, $k_{cat}$, is the maximum number of molecules of substrate that could be converted to product each second per active site. Because the maximum rate is obtained at high substrate concentrations, when all the active sites are occupied with substrate, the turnover number is a measure of how rapidly an enzyme can operate once the active site is filled. This is given simply by

$$k_{cat} = \frac{V_{max}}{[E_t]} \qquad (28)$$

Turnover numbers for some representative enzymes are listed in table 7.2. The enormous value of $4 \times 10^7$ molecules/s achieved by catalase is among the highest known; the low value for lysozyme is at the other end of the spectrum. As is the case with $K_m$, the relationship of $k_{cat}$ to individual rate constants, such as $k_2$ and $k_3$, depends on the details of the reaction mechanism.

***Significance of the Specificity Constant, $k_{cat}/K_m$.*** Under physiological conditions, enzymes usually do not operate at saturating substrate concentrations. More typically, the ratio of the substrate concentration to the $K_m$ is in the range of 0.01–1.0. If [S] is much smaller than $K_m$, the denominator of the Briggs-Haldane equation [equation (25)] is approximately equal to $K_m$, so that the velocity of the reaction becomes

$$v \approx \frac{V_{max}}{K_m}[S] \qquad (\text{when } [S] \ll K_m)$$

$$= \frac{k_{cat}}{K_m}[E_t][S] \qquad (29)$$

The ratio $k_{cat}/K_m$ is referred to as the specificity constant. Equation (29) indicates that the specificity constant provides a measure of how rapidly an enzyme can work at low [S]. Table 7.3 gives the values of the specificity constants for some particularly active enzymes.

The specificity constant $k_{cat}/K_m$ is useful for comparing the relative abilities of different compounds to serve as a substrate for the same enzyme. If the concentrations of two substrates are the same, and are small relative to the $K_m$ values, the ratio of the rates when the two substrates are present is equal to the ratio of the specificity constants.

Another use of the specificity constant is for comparing the rate of an enzyme-catalyzed reaction with the rate at which random diffusion brings the enzyme and substrate into contact. We mentioned previously that if every collision between a protein and a small molecule results in a reaction, the maximum value of the second-order rate constant is on the order of $10^8$ to $10^9$ $M^{-1}s^{-1}$. Some of the values of $k_{cat}/K_m$ in table 7.3 are in this range. The reactions

**Table 7.2**

Values of $k_{cat}$ for Some Enzymes

| Enzyme | $k_{cat}$ $(s^{-1})$ |
|---|---|
| Catalase | 40,000,000 |
| Carbonic anhydrase | 1,000,000 |
| Acetylcholinesterase | 14,000 |
| Penicillinase | 2,000 |
| Lactate dehydrogenase | 1,000 |
| Chymotrypsin | 100 |
| DNA polymerase I | 15 |
| Lysozyme | 0.5 |

these enzymes catalyze proceed at nearly the maximum possible speed, given a fixed, low concentration of substrate and given the restriction that the enzyme and substrate have to find each other by diffusion. The only practical way to go much faster is to have the substrate generated right on the enzyme or in its immediate vicinity, so that little diffusional motion is necessary.

## Kinetics of Enzymatic Reactions Involving Two Substrates

Enzymes that catalyze reactions with two or more substrates work in a variety of ways. In some cases, the intermolecular reaction occurs when all the substrates are bound in a common enzyme–substrate complex; in others, the substrates bind and react one at a time. A frequent application of kinetic measurements is to distinguish between such alternatives.

Consider a reaction in which two substrates, $S_1$ and $S_2$, are converted to products $P_1$ and $P_2$. One way for the reaction to occur is for $S_1$ to bind to the enzyme first, forming the binary complex $ES_1$. Binding of $S_2$ can then form the ternary complex $ES_1S_2$, which gives rise to the products

$$E \xrightarrow{\quad S_1 \quad} ES_1 \xrightarrow{\quad S_2 \quad} ES_1S_2 \xrightarrow{\quad P_1 + P_2 \quad} E \qquad (30)$$

This process is referred to as an ordered pathway. An alternative is a random-order pathway, in which the two substrates can bind to the enzyme in either order. Still another scheme is for $S_1$ to bind to the enzyme and be converted to

**Table 7.3**

Enzymes for Which $k_{cat}/K_m$ Is Close to the Diffusion-Controlled Association Rate

| Enzyme | Substrate | $k_{cat}$ (s$^{-1}$) | $K_m$ (M) | $k_{cat}/K_m$ (M$^{-1}$s$^{-1}$) |
|--------|-----------|----------------------|-----------|-----------------------------------|
| Acetylcholinesterase | Acetylcholine | $1.4 \times 10^4$ | $9 \times 10^{-5}$ | $1.6 \times 10^8$ |
| Carbonic | $CO_2$ | $1 \times 10^6$ | 0.012 | $8.3 \times 10^7$ |
| anhydrase | $HCO_3^-$ | $4 \times 10^5$ | 0.026 | $1.5 \times 10^7$ |
| Catalase | $H_2O_2$ | $4 \times 10^7$ | 1.1 | $4 \times 10^7$ |
| Crotonase | Crotonyl-CoA | $5.7 \times 10^3$ | $2 \times 10^{-5}$ | $2.8 \times 10^8$ |
| Fumarase | Fumarate | 800 | $5 \times 10^{-6}$ | $1.6 \times 10^8$ |
| | Malate | 900 | $2.5 \times 10^{-5}$ | $3.6 \times 10^7$ |
| Triosephosphate isomerase | Glyceraldehyde 3-phosphate | $4.3 \times 10^3$ | $4.7 \times 10^{-4}$ | $2.4 \times 10^8$ |
| $\beta$-Lactamase | Benzylpenicillin | $2.0 \times 10^3$ | $2 \times 10^{-5}$ | $1 \times 10^8$ |

Source: From Alan Fersht, *Enzyme Structure and Mechanism*, 3d ed. Copyright © 1985 by W. H. Freeman & Company, New York. Reprinted with permission.

$P_1$, leaving the enzyme in an altered form, $E'$. $S_2$ then binds to $E'$ and is converted to $P_2$, returning the enzyme to its original form.

$$E \xrightarrow{\ \ S_1\ \ } ES_1 \xrightarrow{\ \ P_1\ \ } E' \xrightarrow{\ \ S_2\ \ } E'S_2 \xrightarrow{\ \ P_2\ \ } E \quad (31)$$

This process is called the Ping-Pong mechanism to emphasize the bouncing of the enzyme between two states, E and $E'$. Ping-Pong pathways are commonly observed with enzymes that contain bound coenzymes. Interconversion of the enzyme between the two forms usually involves modification of the coenzyme.

Kinetic equations for these and other mechanisms can be worked out just as we have done for reactions involving only one substrate. Techniques for doing this are described in the references at the end of the chapter; here we simply illustrate a few of the results. For the Ping-Pong mechanism [equation (31)], the double-reciprocal form of the final expression is

$$\frac{1}{v} = \frac{1}{V_{max}}\left(1 + \frac{K_{m2}}{[S_2]}\right) + \frac{K_{m1}}{V_{max}}\frac{1}{[S_1]} \quad (32)$$

where $K_{m1}$ is the Michaelis constant for $S_1$, and $K_{m2}$ is that for $S_2$. This expression is similar to that for a one-substrate reaction (equation 26), except that the first term on the right is multiplied by the factor $(1 + K_{m2}/[S_2])$. If we measure the rate as a function of $[S_1]$, keeping $[S_2]$ constant, a plot of $1/v$ versus $1/[S_1]$ is linear, but the intercept on the ordinate (the apparent $V_{max}$) depends on $[S_2]$. Increasing $[S_2]$ in-

**Figure 7.10**

Double-reciprocal plots ($1/v$ versus $1/[S_1]$) for the Ping-Pong mechanism. Measurements made at different values of $[S_2]$ give a set of parallel straight lines. The intercepts on the ordinate depend on $[S_2]$. $K_{m1}$, $K_{m2}$, and $V_{max}$ can be obtained by replotting the intercepts as a function of $1/[S_2]$.

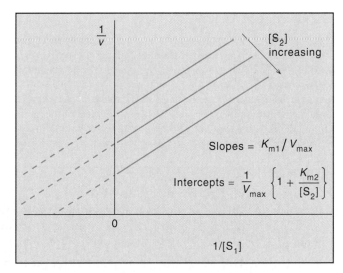

creases the apparent $V_{max}$ (fig. 7.10). From a series of such plots, measured at different values of $[S_2]$, we can find the true $V_{max}$ in addition to $K_{m1}$ and $K_{m2}$.

An ordered pathway [equation (30)] results in the expression

$$\frac{1}{v} = \frac{1}{V_{max}}\left(1 + \frac{K_{m2}}{[S_2]} + \frac{K_{m1}}{[S_1]} + \frac{K_{m2}}{[S_2]}\frac{K_{s1}}{[S_1]}\right) \quad (33)$$

## Figure 7.11

Double-reciprocal plots for an ordered pathway. Measurements made at different fixed values of $[S_2]$ give a set of lines that intersect to the left of the ordinate. The two values of $K_m$, $V_{max}$ and $K_{s1}$ can be obtained by replotting the slopes and intercepts of these lines as functions of $1/[S_2]$. A random pathway gives similar results, but can be distinguished by making such measurements for the reverse reaction $(P_1 + P_2 \rightarrow S_1 + S_2)$ in addition to the forward reaction.

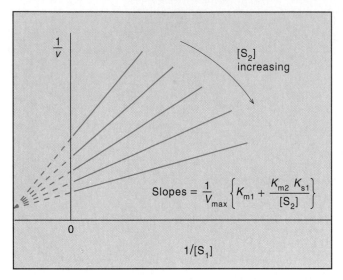

## Figure 7.12

The activity of a typical enzyme as a function of temperature. The turnover number ($k_{cat}$) increases with temperature until a point is reached at which the enzyme is no longer stable. The temperature at which $k_{cat}$ is greatest should not be interpreted as the ''optimum temperature'' for the enzyme. Because denaturation of the enzyme occurs continuously during the measurement, the position of the maximum depends on how quickly the experimenter is able to assay the enzyme activity.

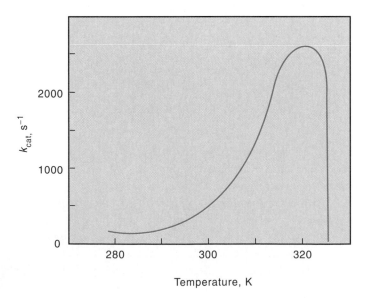

where $K_{s1}$ is the dissociation constant for $ES_1$. A plot of $1/v$ versus $1/[S_1]$ at constant $[S_2]$ is still linear, but now both the slope and the intercept depend on $[S_2]$ (fig. 7.11). Again, all of the kinetic parameters can be obtained from a series of plots measured at different $S_2$ concentrations. However, additional measurements must be made to determine which substrate binds to the enzyme first. In some cases, the substrate that binds first can be shown to form a stable enzyme–substrate complex in the absence of the other substrate.

## Effects of Temperature and pH on Enzymatic Activity

For most enzymes, the turnover number increases with temperature until a temperature is reached at which the enzyme is no longer stable (fig. 7.12). Above this point there is a precipitous, and usually irreversible, drop in activity. At lower temperatures, the temperature dependence of $k_{cat}$ can be related to the activation energy of the slowest (rate-limiting) step in the catalytic pathway [see equation (12)]. With many enzymes a 10°C rise in temperature increases $k_{cat}$ by about a factor of 2, which translates into an activation energy of about 12 kcal/mole.

Enzymes, like other proteins, are stable over only a limited range of pH. Outside this range, changes in the charges on ionizable amino acid residues result in modifications of the tertiary structure of the protein and eventually lead to denaturation. But within the range at which an enzyme is stable, both $k_{cat}$ and $K_m$ often depend on pH. The effects of pH can reflect the $pK_a$ of ionizing groups on either the enzyme or the substrate. A substrate that has an amine group, for example, may bind to the enzyme best when this group is protonated. In many cases, however, the pH dependence reflects ionizable residues that constitute the active site on the enzyme or are essential for maintaining the structure of the active site, and the optimum pH is a characteristic more of the enzyme than of the substrate. Thus, the maximum activity of chymotrypsin always occurs around pH 8, the activity of pepsin peaks around pH 2, and acetylcholinesterase works best at pH 7 or higher (fig. 7.13). The activity of papain, on the other hand, is essentially independent of pH between 4 and 8.

## Enzyme Inhibition

Most enzymes are sensitive to inhibition by specific agents that interfere with the binding of a substrate at the active site or with conversion of the enzyme–substrate complex into products. Two of the major applications of kinetic measurements are in distinguishing between

**Figure 7.13**

Enzyme activity ($k_{cat}/K_m$) as a function of pH for three different enzymes. The optimum pH usually is a characteristic of the enzyme and not the particular substrate. Often the pH sensitivity is an indication of an ionizable group at the active site.

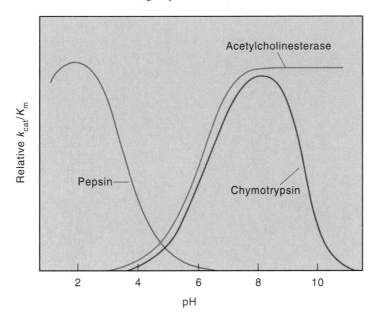

**Figure 7.14**

Types of enzyme inhibition. (a) A competitive inhibitor competes with the substrate for binding at the same site on the enzyme. (b) A noncompetitive inhibitor binds to a different site but blocks the conversion of the substrate to products. (c) An uncompetitive inhibitor binds only to the enzyme–substrate complex. (E = enzyme; S = substrate.)

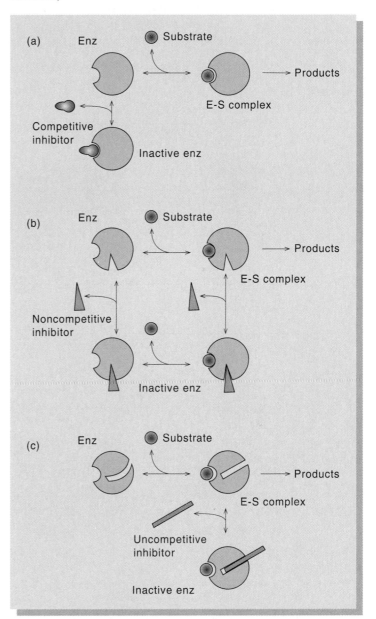

different types of inhibition and in providing quantitative information on the effectiveness of various inhibitors. Such information is essential for an understanding of how cells regulate their enzymatic activities. Comparisons of a series of inhibitors also can help to map the structure of an enzyme's active site and are a key step in the design of therapeutic drugs.

## Competitive Inhibitors Bind at the Active Site

In many cases, an inhibitor resembles the substrate structurally and binds reversibly at the same site on the enzyme. This activity is called competitive inhibition because the inhibitor and the substrate compete for binding (fig. 7.14a). The inhibitor is prevented from binding if the active site is already occupied by the substrate. As an example consider the proteolytic enzyme trypsin, which cleaves polypeptide chains at peptide linkages adjacent to basic amino acid residues. Trypsin is inhibited competitively by benzamidine. The substrate-binding site on the enzyme consists of a pocket where a positively charged lysine or arginine side chain fits snugly and interacts with a negatively charged carboxylate group (fig. 7.15). When protonated, benzamidine is positively charged and has a flat, delocalized electronic structure resembling that of an arginine side chain. The binding pocket accepts and binds a benzamidine ion reversibly in place of its regular substrate.

An expression describing enzyme kinetics in the presence of a competitive inhibitor can be derived straightforwardly. Consider the reaction that we treated previously.

$$\text{E} + \text{S} \underset{k_{-1}}{\overset{k_1}{\rightleftharpoons}} \text{ES} \underset{k_{-2}}{\overset{k_2}{\rightleftharpoons}} \text{E} + \text{P} \qquad (19')$$

## Figure 7.15

The specificity pocket of trypsin can accommodate an arginine side chain of a polypeptide substrate or a benzamidine ion, which acts as a competitive inhibitor. Asp = aspartic acid.

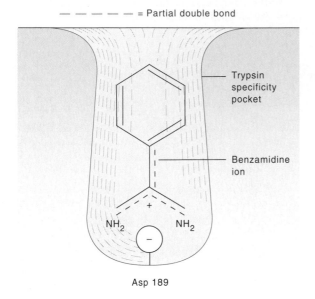

— — — — — = Partial double bond

Trypsin specificity pocket

Benzamidine ion

NH$_2$    NH$_2$

Asp 189

Polypeptide chain

C$_a$

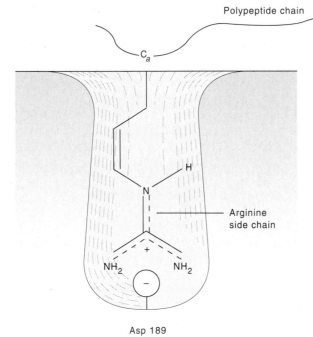

Arginine side chain

NH$_2$    NH$_2$

Asp 189

Arginine side chain

## Figure 7.16

Competitive inhibition. A series of double-reciprocal plots ($1/v$ versus $1/[S]$) measured at different concentrations of the inhibitor (I) all intersect at the same point ($1/V_{max}$) on the ordinate. The slopes of the plots and the intercepts on the abscissa are simple, linear functions of $[I]/K_i$, where $K_i$ is the dissociation constant of the inhibitor–enzyme complex.

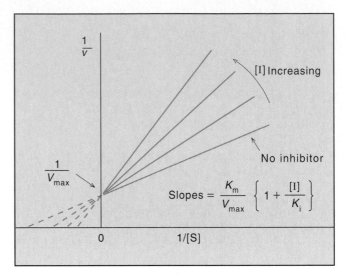

$\dfrac{1}{v}$

[I] Increasing

$\dfrac{1}{V_{max}}$

No inhibitor

$\text{Slopes} = \dfrac{K_m}{V_{max}}\left\{1 + \dfrac{[I]}{K_i}\right\}$

0          1/[S]

We now have the additional feature that the enzyme also reacts reversibly with the inhibitor (I) to give an inactive complex (EI).

$$E + I \rightleftharpoons EI \tag{34}$$

The derivation proceeds just as the derivation that led to equations (25) and (26), except that the total enzyme concentration $[E_t]$ is now $[E] + [ES] + [EI]$, instead of just $[E] + [ES]$. As a result, in place of equation 26 we end up with

$$\frac{1}{v} = \frac{1}{V_{max}} + \frac{K_m}{V_{max}}\frac{1}{[S]}\left(1 + \frac{[I]}{K_i}\right) \tag{35}$$

where $K_i$ is the dissociation constant of the enzyme–inhibitor complex.

$$K_i = \frac{[E][I]}{[EI]} \tag{36}$$

According to equation (35), a plot of $1/v$ versus $1/[S]$ is linear and passes through the same intercept on the ordinate as the plot obtained in the absence of the inhibitor ($1/V_{max}$). This is because the effect of the inhibitor disap-

## Figure 7.17

Noncompetitive inhibition. The double-reciprocal plots pass through different points on the ordinate, but intersect at the same point ($-1/K_m$) on the abscissa. The slopes and the intercepts on the ordinate are linear functions of [I]/$K_i$.

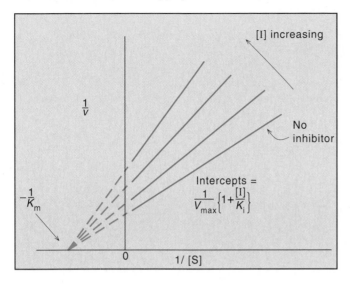

Table 7.4

### Some Inhibitors of Enzymes that Form Covalent Linkages with Functional Groups on the Enzyme

| Inhibitor | Enzyme Group that Combines with Inhibitor |
|---|---|
| Cyanide | Fe, Cu, Zn, other transition metals |
| p-Mercuribenzoate | Sulfhydryl |
| Diisopropylfluorophosphate | Serine hydroxyl |
| Iodoacetate | Sulfhydryl, imidazole, carboxyl, thioether |

pears at infinite substrate concentration for a competitive inhibitor. The slope of the double-reciprocal plot, however, depends on the product

$$\frac{K_m}{V_{max}}\left(1 + \frac{[I]}{K_i}\right)$$

instead of simply $K_m/V_{max}$ (fig. 7.16). As a result the apparent $K_m$ is altered to $K'_m$, where

$$K'_m = K_m\left(1 + \frac{[I]}{K_i}\right)$$

By measuring the slope of the plot as a function of [I] we can determine $K_i$.

## Noncompetitive and Uncompetitive Inhibitors Do Not Compete Directly with Substrate Binding

Some inhibitors bind at sites other than the enzyme's active site and do not compete directly with binding of the substrate. Instead, they act by interfering with the reaction of the enzyme–substrate complex. An inhibitor that binds to

an enzyme whether or not the active site is occupied by the substrate is termed a noncompetitive inhibitor (fig. 7.14b). A noncompetitive inhibitor decreases the maximum velocity of an enzymatic reaction without affecting the $K_m$. The inhibitor removes a certain fraction of the enzyme from operation, no matter the concentration of the substrate. Plots of $1/v$ versus $1/[S]$ in the presence of different concentrations of a noncompetitive inhibitor intersect at the same point on the abscissa ($-1/K_m$) but pass through the ordinate at different points (fig. 7.17).

Still another possibility is that the inhibitor binds only to the enzyme–substrate complex and not to the free enzyme (fig. 7.14c). This reaction is called uncompetitive inhibition. Uncompetitive inhibition is rare in reactions that involve a single substrate but more common in reactions with multiple substrates. Plots of $1/v$ versus $1/[S]$ at different concentrations of an uncompetitive inhibitor give a series of parallel lines.

## Irreversible Inhibitors Permanently Alter the Enzyme Structure

Competitive, noncompetitive, and uncompetitive inhibition are all reversible. If the inhibited enzyme is dialyzed to remove the inhibitor, the enzymatic activity recovers. There are, however, inhibitors that react essentially irreversibly, usually by forming a covalent bond to an amino acid side chain or a bound coenzyme. Examples of such inhibitors are given in table 7.4. Like reversible inhibitors, irreversible inhibitors can change either $V_{max}$ or $K_m$ or both.

Irreversible inhibitors often provide clues to the nature of the active site. Enzymes that are inhibited by iodoacetamide, for example, frequently have a cysteine in the active site, and the cysteinyl sulfhydryl group often plays an essential role in the catalytic mechanism (fig. 7.18). An example is glyceraldehyde 3-phosphate dehydrogenase, in which the catalytic mechanism begins with a reaction of the cysteine with the aldehyde substrate (see fig. 12.21). As we discuss in chapter 8, trypsin and many related proteolytic enzymes are inhibited irreversibly by diisopropyl-fluorophosphate (fig. 7.18), which reacts with a critical serine residue in the active site.

**Figure 7.18**

Iodoacetamide is an irreversible inhibitor of many enzymes that contain a cysteine residue in the active site. Diisopropylfluoro-phosphate is an irreversible inhibitor of trypsin, chymotrypsin, and several related enzymes. It reacts with a serine residue at the active site.

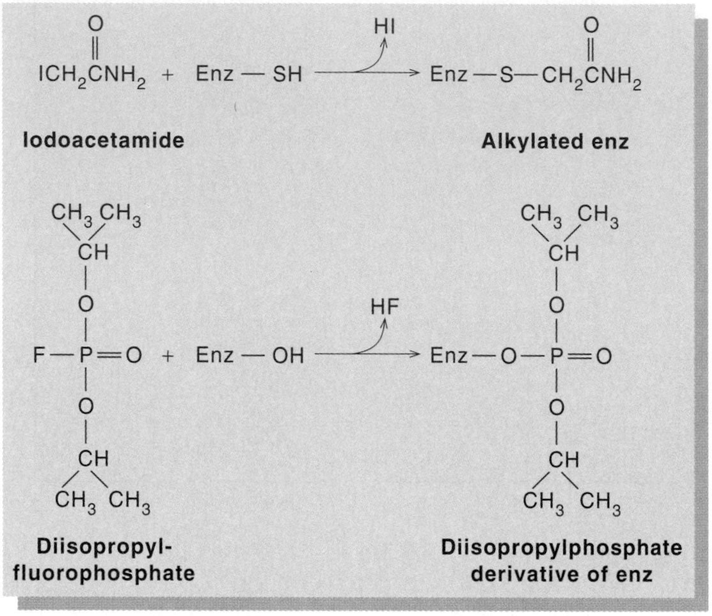

## Summary

Enzymes are biological catalysts. Kinetic analysis is one of the most broadly used tools for characterizing enzymatic reactions.

1. The rate of a reaction depends on the frequency of collisions between the reacting species and on the fraction of the collisions that produce products. The former depends on the concentrations of the reactants; the latter depends on temperature and activation free energy $\Delta G^{\ddagger}$. $\Delta G^{\ddagger}$ can be interpreted as the free energy needed to convert the reactants to a transition state. A catalyst increases the reaction rate by lowering $\Delta G^{\ddagger}$.

2. Enzyme kinetics usually are studied by mixing the enzyme and substrates and measuring the initial rate of formation of product or the disappearance of a reactant. Special techniques are necessary to measure very fast reactions. It is common to measure the rate as a function of substrate concentration, pH, and temperature.

3. Enzymes have localized catalytic sites. The substrate (S) binds at the active site to form an enzyme–substrate complex (ES). Subsequent steps transform the bound substrate into product and regenerate the free enzyme. The overall speed of the reaction depends on the concentration of ES. Shortly after the enzyme and substrate are mixed, [ES] becomes approximately constant and remains so for a period of time termed the steady state. The rate ($v$) of the reaction in the steady state usually has a hyperbolic dependence on the substrate concentration. It is proportional to [S] at low concentrations but approaches a maximum ($V_{max}$) when the enzyme is fully charged with substrate. The Michaelis constant $K_m$ is the substrate concentration at which the rate is half maximal. $K_m$ and $V_{max}$ often can be obtained from a plot of $1/v$ versus $1/[S]$. If ES is in equilibrium with the free enzyme and substrate, $K_m$ is equal to the dissociation constant for the complex ($K_s$). More generally, $K_m$ depends on at least three rate constants and is larger than $K_s$.

4. The turnover number $k_{cat}$ is the maximum number of molecules of substrate converted to product per unit

time per active site and is $V_{max}$ divided by the total enzyme concentration. The specificity constant $k_{cat}/K_m$ is a measure of how rapidly an enzyme can work at low substrate concentrations. This is usually the best index of the effectiveness of an enzyme.

5. Enzymes that catalyze reactions of two or more substrates work in a variety of ways that can be distinguished by kinetic analysis. Some enzymes bind their substrates in a fixed order; others bind in random order. In some cases binding of one substrate gives a partial reaction before the second substrate binds.

6. Enzymes can be inhibited by agents that interfere with the binding of substrate or with conversion of the ES complex into products. Reversible inhibitors are clas-

sified as competitive, noncompetitive or uncompetitive. A competitive inhibitor competes with substrate for binding at the active site. Consequently, a sufficiently high concentration of substrate can eliminate the effect of a competitive inhibitor. Noncompetitive inhibitors bind at a separate site and block the reaction regardless of whether the active site is occupied by substrate. An uncompetitive inhibitor binds to the ES complex but not to the free enzyme. These three forms of inhibition are distinguishable by measuring the rate as a function of the concentrations of the substrate and inhibitor. Irreversible inhibitors often provide information on the active site by forming covalently linked complexes that can be characterized.

## Selected Readings

*Advances in Enzymology.* New York: Academic Press. An annually published volume containing monographs on selected topics.

Boyer, P. D. (ed.), *The Enzymes.* New York: Academic Press. A continuing series of monographs on selected enzymes. See particularly the chapter entitled "Steady State Kinetics" by W. W. Cleland in vol. 2.

Fersht, A., *Enzyme Structure and Mechanism,* 2d ed. New York: W. H. Freeman & Co., 1985.

Frost, A. A., and R. G. Pearson, *Kinetics and Mechanism,* 2d ed. New York: Wiley, 1961. An excellent introduction to general chemical kinetics.

Purich, D. L., *Contemporary Enzyme Kinetics and Mechanism.* New York: Academic Press, 1983. Selected chapters from *Methods in Enzymology.* Detailed information on how to analyze kinetic data and on effects of temperature, pH, and inhibitors.

Segal, I. H., *Enzyme Kinetics.* New York: Wiley, 1975.

## Problems

1. Explain what is meant by the order of a reaction, using the reaction below as an example. What is the reaction order for each reactant? For the overall reaction? (Consider the forward and reverse reaction.)

$$A + B \rightleftharpoons 2C$$

2. In a first-order reaction a substrate is converted to product so that 87% of the substrate is converted in 7 min. Calculate the first-order rate constant. In what time is 50% of the substrate converted to product?

3. Prove that the $K_m$ equals the substrate concentration at one-half maximal velocity.

4. The Michaelis constant $K_m$ is frequently equated with $K_s$, the [ES] dissociation constant. However, there is usually a disparity between those values. Why? Under what conditions are $K_m$ and $K_s$ equivalent?

5. When quantifying the activity of an enzyme, does it matter if you measure the appearance of a product or the disappearance of a reactant?

6. An enzyme was assayed with substrate concentration of twice the $K_m$ value. The progress curve of the enzyme (product produced per minute) is shown here. Give two possible reasons why the progress curve becomes nonlinear.

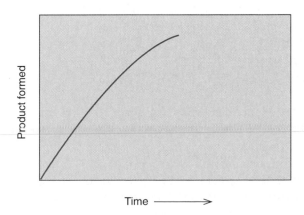

7. What is the steady-state approximation and under what conditions is it valid?

8. Assume that an enzyme-catalyzed reaction follows Michaelis-Menten kinetics with a $K_m$ of 1 $\mu$M. The initial velocity is 0.1 $\mu$M/min at 10 mM substrate. Calculate the initial velocity at 1 mM, 10 $\mu$M, and 1 $\mu$M substrate. If the substrate concentration increased to 20 mM, would the initial velocity double? Why or why not?

9. If the $K_m$ for an enzyme is $1.0 \times 10^{-5}$ M and the $K_i$ of a competitive inhibitor of the enzyme is $1.0 \times 10^{-6}$ M, what concentration of inhibitor would be necessary to lower the reaction rate by a factor of 10 when the substrate concentration is $1.0 \times 10^{-3}$ M? $1.0 \times 10^{-5}$ M? $1.0 \times 10^{-6}$ M?

10. Assume that an enzyme-catalyzed reaction follows the scheme shown:

$$E + S \underset{k_2}{\overset{k_1}{\rightleftharpoons}} ES \underset{k_4}{\overset{k_3}{\rightleftharpoons}} E + P$$

Where $k_1 = 10^9$ M$^{-1}$ s$^{-1}$, $k_2 = 10^5$ s$^{-1}$, $k_3 = 10^2$ s$^{-1}$, $k_4 = 10^7$ M$^{-1}$ s$^{-1}$, and $[E_t]$ is 0.1 nM. Determine the value of each of the following.

(a) $K_m$
(b) $V_{max}$
(c) Turnover number
(d) Initial velocity when $[S]_o$ is 20 $\mu$M.

11. A colleague has measured the enzymatic activity as a function of reaction temperature and obtained the data shown in this graph. He insists on labeling point A as the "temperature optimum" for the enzyme. Try, tactfully, to point out the fallacy of that interpretation.

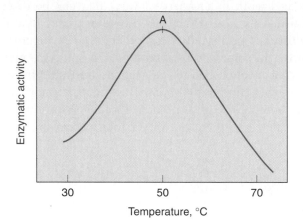

12. You have isolated a tetrameric NAD$^+$-dependent dehydrogenase. You incubate this enzyme with iodoacetamide in the absence or presence of NADH (at 10 times the $K_m$ concentration), and you periodically remove aliquots of the enzyme for activity measurements and amino acid composition analysis. The results of the analyses are shown in the table.

### Table for Problem 12

| Time (min) | (No NADH Present) | | | | (NADH Present) | | | |
| --- | --- | --- | --- | --- | --- | --- | --- | --- |
| | Activity (U/mg) | His | (Residues/mole) | Cys | Activity (U/mg) | His | (Residues/mole) | Cys |
| 0 | 1,000 | 20 | | 12 | 1,000 | 20 | | 12 |
| 15 | 560 | 18.2 | | 11.4 | 975 | 20 | | 11.4 |
| 30 | 320 | 17.3 | | 10.8 | 950 | 20 | | 10.8 |
| 45 | 180 | 16.7 | | 10.4 | 925 | 19.8 | | 10.4 |
| 60 | 100 | 16.4 | | 10.0 | 900 | 19.6 | | 10.0 |

(a) What can you conclude about the reactivities of the cysteinyl and histidyl residues of the protein?

(b) Which residue can you implicate in the active site? On what do you base the choice? Are the data conclusive concerning the assignment of a residue to the active site? Why or why not?

(c) After 1 h you dilute the enzyme incubated with iodoacetamide but no NADH. Do you expect the enzyme activity to be restored? Explain.

13. The initial velocity data shown in the table were obtained for an enzyme.

| [S] (mM) | Velocity $(\text{Ms}^{-1}) \times 10^7$ |
|---|---|
| 0.10 | 0.96 |
| 0.125 | 1.12 |
| 0.167 | 1.35 |
| 0.250 | 1.66 |
| 0.50 | 2.22 |
| 1.0 | 2.63 |

Each assay at the indicated substrate concentration was initiated by adding enzyme to a final concentration of 0.01 nM. Derive $K_m$, $V_{max}$, $k_{cat}$, and the specificity constant.

14. You measured the initial velocity of an enzyme in the absence of inhibitor and with inhibitor A or inhibitor B. In each case, the inhibitor is present at 10 $\mu$M. The data are shown in the table.

| [S] (mM) | Velocity $(\text{Ms}^{-1}) \times 10^7$ Uninhibited | Velocity $(\text{Ms}^{-1}) \times 10^7$ Inhibitor A | Velocity $(\text{Ms}^{-1}) \times 10^7$ Inhibitor B |
|---|---|---|---|
| 0.333 | 1.65 | 1.05 | 0.794 |
| 0.40 | 1.86 | 1.21 | 0.893 |
| 0.50 | 2.13 | 1.43 | 1.02 |
| 0.666 | 2.49 | 1.74 | 1.19 |
| 1.0 | 2.99 | 2.22 | 1.43 |
| 2.0 | 3.72 | 3.08 | 1.79 |

(a) Determine $K_m$ and $V_{max}$ of the enzyme.
(b) Determine the type of inhibition imposed by inhibitor A and calculate $K_i$(s).
(c) Determine the type of inhibition imposed by inhibitor B and calculate $K_i$(s).

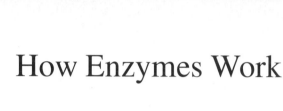

# How Enzymes Work

*The effectiveness of enzymes is due to the fact that they bring the reactants (substrates) together in a location where they are ideally oriented with respect to each other and the catalytic groups of the enzyme.*

In chapter 7 we saw that enzymes can increase the rates of reactions by many orders of magnitude. We noted that enzymes are highly specific in the reactions they catalyze and in the particular substrates they accept. In this chapter we explore the mechanisms of several enzyme-catalyzed reactions in greater detail. Our goal is to relate the activity of each of these enzymes to the structure of the active site, where the functional groups of amino acid side chains, the polypeptide backbone, or bound cofactors must interact with the substrates in such a way as to favor the formation of the transition state. We explore enzyme catalytic mechanisms in many subsequent chapters as well but usually in less detail than here.

## General Themes in Enzymatic Mechanisms

Several broad themes recur frequently in enzymatic reaction mechanisms. Among the most important of these are (1) proximity effects, (2) general-acid and general-base catalysis, (3) electrostatic effects, (4) nucleophilic or electrophilic catalysis by enzymatic functional groups, and

(5) structural flexibility. All known enzymes use at least one of these themes, and most use more than one. We start by discussing these five themes in general terms and then see how they apply to three enzymes for which the mechanisms have been studied in depth.

## The Proximity Effect: Enzymes Bring Reacting Species Close Together

The idea of the proximity effect is that an enzyme can accelerate a reaction between two species simply by holding the two reactants close together in an appropriate orientation. It has long been known that intramolecular reactions involving two groups that are tied together in a single molecule are usually much faster than the corresponding intermolecular reactions between two independent molecules. The cyclization of succinic acid to form succinyl anhydride [equation (1)], for example, occurs much more rapidly than the formation of acetic anhydride from two molecules of acetic acid [equation (2)].

$$\begin{matrix} \text{CO}_2\text{H} \\ | \\ \text{CH}_2 \\ | \\ \text{CH}_2 \\ | \\ \text{CO}_2\text{H} \end{matrix} \longrightarrow \begin{matrix} \text{O} \\ \| \\ \text{C} \\ \diagup \quad \diagdown \\ \text{CH}_2 \\ | \qquad \text{O} + \text{H}_2\text{O} \\ \text{CH}_2 \\ \diagdown \quad \diagup \\ \text{C} \\ \| \\ \text{O} \end{matrix} \qquad \textbf{(1)}$$

$$\begin{matrix} \text{CH}_3\text{CO}_2\text{H} \\ \\ \\ \text{CH}_3\text{CO}_2\text{H} \end{matrix} \longrightarrow \begin{matrix} \text{O} \\ \| \\ \text{CH}_3\text{C} \\ \diagdown \\ \quad \text{O} + \text{H}_2\text{O} \\ \diagup \\ \text{CH}_3\text{C} \\ \| \\ \text{O} \end{matrix} \qquad \textbf{(2)}$$

We cannot compare the rate constants for these two reactions directly because they are expressed in different units. The intramolecular reaction [equation (1)] is kinetically first order, whereas the intermolecular reaction [equation (2)] is second order. But suppose the two molecules of acetic acid that enter into reaction (2) are labeled isotopically to make them distinguishable and that one type of molecule is present in great excess over the other. The process is then kinetically first order in the concentration of the limiting reactant. To make the rate constant the same as for reaction (1), the more abundant species has to be present at a concentration of $3 \times 10^5$ M! This is far above any concentration that can actually be obtained.

This example from organic chemistry shows that tying two reactants together can have an enormous effect on the rate of a reaction. The effect is due largely to differences between the entropy changes that accompany the inter- and intramolecular reactions. The formation of the transition state requires a larger decrease of translational and rotational entropy in the intermolecular reaction than it does in the intramolecular reaction. In the intramolecular reaction much of this entropy decrease has already occurred during the preparation of the reactant. Enzymes that catalyze intermolecular reactions take advantage of the proximity effect by binding the reactants close together at the active site so that the reactive groups are oriented appropriately for the reaction.

## General-Base and General-Acid Catalysis Avoids the Need for Extremely High or Low pH

Chemical bonds are formed by electrons, and formation or breakage of bonds requires the migration of electrons. In broad terms, reactive chemical groups function either as electrophiles or as nucleophiles. Electrophiles are electron-deficient substances that react with electron-rich substances; nucleophiles are electron-rich substances that react with electron-deficient substances. The task of a catalyst often is to make a potentially reactive group more reactive by increasing its electrophilic or nucleophilic character. In many cases the simplest way to do this is to add or remove a proton.

As an example, consider the hydrolysis of an ester (fig. 8.1). Because the electronegativity of the oxygen atom in the C=O group is greater than that of the carbon, the oxygen has a fractional negative charge $\delta^-$, and the carbon has a fractional positive charge $\delta^+$. Hydrolysis of an ester in neutral aqueous solution can occur if the oxygen atom of $H_2O$, acting as a nucleophile, reacts with the carbonyl carbon. The initial product is an intermediate in which the carbon atom has four substituents in a tetrahedral arrangement. The reaction is completed by the breakdown of the tetrahedral intermediate to release the alcohol. However, water is a comparatively weak nucleophile, and its reaction with esters in the absence of a catalyst is very slow. Hydrolysis of esters occurs much more rapidly at high pH, when the negatively charged hydroxide ion replaces water as the reactive nucleophile (fig. 8.1a). The nucleophilic character of water itself also can be increased by interaction with a basic group other than $OH^-$ (fig. 8.1b). By offering electrons to one of the protons of the water, the base increases the electron density on the oxygen.

The term ''general base'' is used to describe any substance that is capable of binding a proton in aqueous solution. Enzymes use a variety of functional groups in this role.

## Figure 8.1

Several ways that the hydrolysis of an ester can occur. A red, curved arrow represents the movement of an electron pair from an electron donor to an acceptor. (*a*) Catalysis by free hydroxide ion. (*b*) General-base catalysis. (*c*) General-acid catalysis.

(a) Hydroxide ion catalysis

(b) General-base catalysis

(c) General-acid catalysis

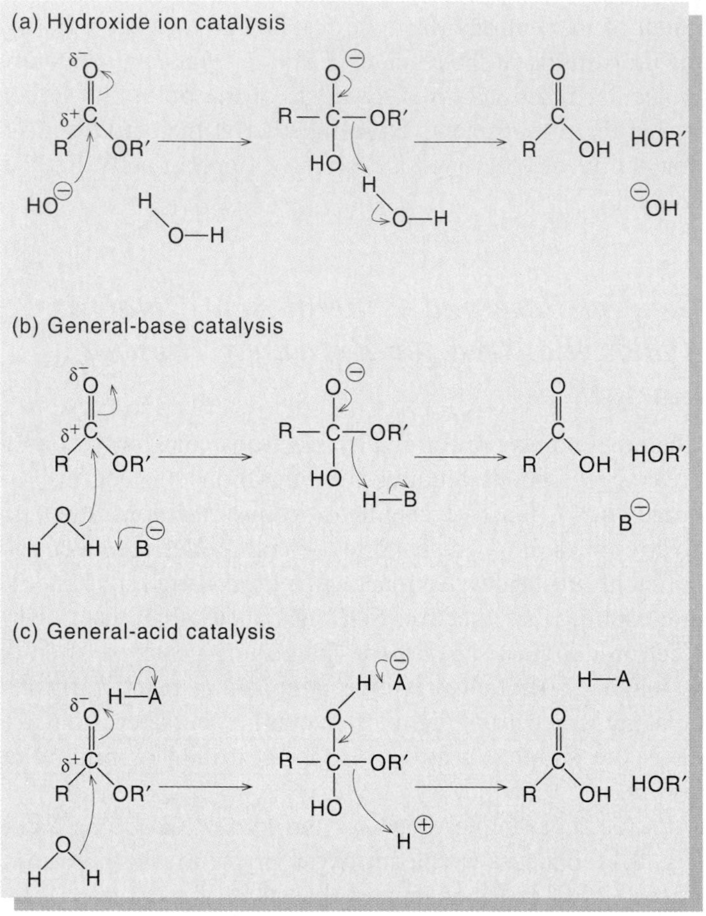

Two factors make free $OH^-$ ions themselves unsuitable for enzymatic catalysis and thus favor the use of a general base. First, the low concentration of $OH^-$ limits its availability at physiological pH. In contrast, proteins contain numerous functional groups that can serve as general bases at moderate pH or even under mildly acidic conditions. The only requirement is that the base can exist at least partly in its unprotonated form at the ambient pH. This condition can easily be met by selecting a basic group from among the ionizable or polar amino acid side chains from an amino-terminal $—NH_2$ group or a carboxy-terminal carboxylate (see table 3.3). The second advantage of using a general base instead of $OH^-$ is that a basic group provided by the protein can be positioned precisely with respect to the substrate in the active site, allowing the proximity effect to come into play. Free $OH^-$ ions are much more mobile. In exceptional cases in which $OH^-$ does act as a nucleophile in

an enzymatic reaction, it usually is tightly bound to a metal cation.

The hydrolysis of an ester also can be catalyzed by an acid (fig. 8.1*c*). The acid donates a proton to the carbonyl oxygen, increasing the positive charge on the carbon. The term "general acid" is used to refer to any substance capable of releasing a proton in solution, and again enzymes almost always use such proton donors in preference to free $H^+$ or $H_3O^+$ ions, presumably because a general acid can operate at moderate pH and is easy to fix in position.

An important point to note in figure 8.1 is that the same general acid or base that catalyzes the formation of the tetrahedral intermediate also can participate in the decomposition of the intermediate. When a general acid (HA) donates a proton to the ester oxygen it becomes a base ($A^-$), which can retrieve the proton as the intermediate breaks down. When a general base ($B^-$) removes a proton from water it becomes an acid (BH), which can provide a proton to the alcohol. Note also that general-acid and general-base catalysis are not mutually exclusive; they can both occur in a concerted manner in the same step of a reaction.

## Electrostatic Interactions Can Promote the Formation of the Transition State

The frequent use of general acids and general bases in enzymes illustrates the underlying principle that enzymes act by stabilizing the distribution of electrical charge in transition states. In the enzymatic hydrolysis of an ester, the key transition state probably resembles the tetrahedral intermediates shown in figure 8.1. To form such an intermediate, electrons must move from the nucleophile through the carbon atom of the C=O group to the oxygen. There is thus a net movement of negative charge from the nucleophile to the substrate. In the absence of a general acid or base, a charge approaching $+1$ appears on the nucleophile, and a charge approaching $-1$ appears on the C=O oxygen. A general base can stabilize this new distribution of charge by offering electrons to the nucleophile so that some of the positive charge moves to the base. By providing a proton to the C=O oxygen, a general acid can delocalize the negative charge here. But enzymes have other ways to achieve a similar stabilization. Suppose that the active site included a positively charged amino acid side chain, such as that of lysine or arginine, located near the oxygen atom of the C=O group. A positive charge in this region favors the formation of the tetrahedral intermediate. A negative charge in the region of the nucleophile has a similar effect. The interactions of such charges are termed electrostatic effects.

Electrostatic interactions can be significant even between groups whose net formal charge is zero. This is because charge distributions within molecular groups are not uniform but rather vary from atom to atom. We alluded to this previously when discussing the partial charges on the oxygen and carbon atoms of an ester (fig. 8.1). Similar considerations apply to other functional groups: The electron distributions around the nuclei leave each atom with a small net positive or negative charge. In an alcoholic —OH group, for example, the oxygen atom has a negative charge of approximately $-0.4$ atomic charge units, and the hydrogen has a charge of about $+0.4$.

As a reacting substrate is transformed into a transition state, the changing charges on its atoms interact with the charges on atoms of the surrounding protein and any nearby water molecules. The energy difference between the initial state and the transition state thus depends critically on the details of the protein structure. We see illustrations of this in the three enzymes discussed later on.

## Enzymatic Functional Groups Provide Nucleophilic and Electrophilic Catalysts

Another strategy for catalyzing the hydrolysis of an ester or an amide is to replace water by a stronger nucleophilic group that is part of the enzyme's active site. The $HOCH_2$— group of a serine residue is often used in this way. In such cases, the reaction of the serine with the substrate splits the overall reaction into a two-step process. Instead of immediately yielding the free carboxylic acid, the breakdown of the initial tetrahedral intermediate yields an intermediate ester that is attached covalently to the enzyme.

$$R-\overset{\overset{\displaystyle O}{\|}}{C}-NHR' + HOCH_2-\text{Enzyme} \longrightarrow$$

$$R-\overset{\overset{\displaystyle O}{\|}}{C}-OCH_2-\text{Enzyme} + R'NH_2 \quad (3)$$

The *acyl-enzyme* ester intermediate must be hydrolyzed by a second reaction, in which water becomes the nucleophile.

$$R-\overset{\overset{\displaystyle O}{\|}}{C}-OCH_2-\text{Enzyme} + H_2O \longrightarrow$$

$$R-\overset{\overset{\displaystyle O}{\|}}{C}-OH + HOCH_2-\text{Enzyme} \quad (4)$$

The proteolytic enzymes trypsin, chymotrypsin, and elastase, discussed later on, all work in this way.

Nucleophilic groups on enzymes participate in a variety of other types of reactions in addition to hydrolytic reactions. An example is acetoacetic acid decarboxylase, which catalyzes the reaction

$$CH_3-\overset{\overset{\displaystyle O}{\|}}{C}-CH_2-CO_2H \longrightarrow CH_3-\overset{\overset{\displaystyle O}{\|}}{C}-CH_3 + CO_2 \quad (5)$$

**Figure 8.2**

In acetoacetic acid decarboxylase, the positive charge of a protonated Schiff base intermediate pulls electrons from a nearby carbon–carbon bond, thereby releasing $CO_2$.

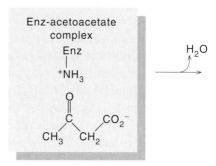

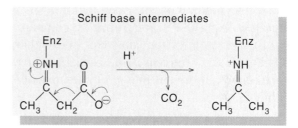

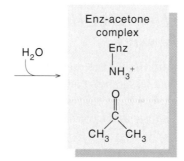

The reaction proceeds through an intermediate called a Schiff base, in which the substrate is covalently attached to the $\epsilon$-amino group of a lysine residue at the enzyme's active site.

$$CH_3-\overset{\overset{\displaystyle O}{\|}}{C}-CH_2-CO_2H + \text{Enzyme}-NH_2 \longrightarrow$$

$$CH_3-\overset{\overset{\displaystyle N-\text{Enzyme}}{\|}}{C}-CH_2CO_2H + H_2O \quad (6)$$

Protonation of the nitrogen atom of the Schiff base introduces a positive charge that pulls electrons from the nearby carbon–carbon bond, causing decarboxylation (fig. 8.2).

The mechanisms outlined in equations 3–6 illustrate a basic feature: Nucleophilic catalysis by enzymes involves the formation of an intermediate state in which the substrate is covalently attached to a nucleophilic group of the enzyme. In addition to the —CH₂OH group of serine and the ε-amino of lysine, the —CH₂SH of cysteine is often used as a nucleophile. The carboxylate of aspartate or glutamate and the imidazole group of histidine can play a similar role. Some enzymes take advantage of bound coenzymes, such as thiamine, biotin, pyridoxamine, or tetrahydrofolate, to obtain additional nucleophilic reagents (see chapter 10).

There also are numerous enzymes that use bound metal ions to form complexes with substrates. In these enzymes, the metal ion usually serves as an electrophilic functional group rather than as a nucleophile. Carbonic anhydrase, for example, contains a $Zn^{+2}$ ion that binds one of the substrates, hydroxide ion, as a ligand. The bound $OH^-$ reacts with the other substrate, $CO_2$. In alcohol dehydrogenase, and in the proteolytic enzymes thermolysin and carboxypeptidase A, a $Zn^{+2}$ ion forms a complex with a carbonyl oxygen atom of the substrate. The withdrawal of electrons by the $Zn^{+2}$ increases the partial positive charge on the carbonyl carbon and thus promotes reaction with a nucleophile.

## Structural Flexibility Can Increase the Specificity of Enzymes

Although precise positioning of the reactants is a fundamental aspect of enzyme catalysis, most enzymes undergo some change in their structure when they bind substrates. A particularly dramatic example is hexokinase, which catalyzes the transfer of a phosphate group from adenosine triphosphate (ATP) to glucose.

$$\text{ATP} + \text{glucose} \longrightarrow$$
$$\text{ADP} + \text{glucose—6—phosphate} \quad (7)$$

When hexokinase binds glucose, its structure changes in a way that brings together the elements of the active site (fig. 8.3). The enzyme literally closes like a set of jaws around the substrate! Such a structural change is often referred to as an induced fit.

Enfolding a substrate in this way can serve to maximize the favorable entropy change associated with removing a hydrophobic substrate molecule from water. It also allows the enzyme to control the electrostatic effects that promote formation of the transition state. The substrate is forced to respond to the directed electrostatic fields from the enzyme's functional groups, instead of the disordered fields from the solvent.

Structural changes also contribute to the high specificity of some enzymatic reactions. In hexokinase, the structural change induced by glucose promotes the binding of the other substrate, ATP. ATP does not bind to the enzyme

## Figure 8.3

Models of the crystallographic structure of hexokinase in the "open" (a) and "closed" (b) conformations. The enzyme (shown in blue) adopts the open conformation in the absence of substrates, but switches to the closed conformation when it binds glucose (red). Hexokinase also has been crystallized with a bound analog of ATP. In the absence of glucose, the enzyme with the bound ATP analog remains in the open conformation. The structural change caused by glucose results in the formation of additional contacts between the enzyme and ATP. This can explain why the binding of glucose enhances the binding of ATP. (Courtesy of Dr. Thomas A. Steitz.)

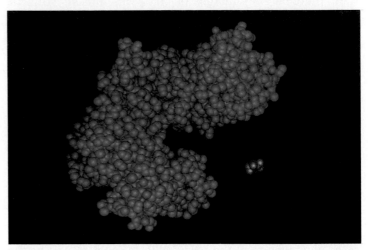

(a)

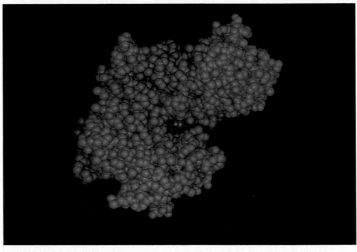

(b)

properly unless glucose is already present in the catalytic site. If ATP were to bind in the absence of glucose, the enzyme might have a tendency to catalyze the transfer of phosphate from ATP to water, resulting in a wasteful loss of ATP.

$$\text{ATP} + \text{H}_2\text{O} \longrightarrow \text{ADP} + \text{HPO}_4^{2-} + \text{H}^+ \quad (8)$$

Hexokinase does not catalyze this side reaction; it waits for glucose to bind first.

The structural changes that occur in hexokinase bring home the point that enzyme crystal structures give static snapshots of molecules that actually are highly flexible. In solution, the structure of an enzyme undergoes fluctuations that vary widely in amplitude and frequency from place to place in the protein. Vibrations and rotations involving only a few atoms occur on time scales of $10^{-13}$–$10^{-11}$ s. Somewhat larger motions, such as the flipping of the aromatic ring of a tyrosine, typically occur on scales of $10^{-9}$–$10^{-8}$ s. Major reorganizations may take $10^{-6}$–$10^{-3}$ s. All of these types of motions can be important in catalysis.

## Detailed Mechanisms of Enzyme Catalysis

In the preceding sections, we discussed five themes that occur frequently in enzyme reaction mechanisms. We now examine several representative enzymes in finer detail. We focus on three enzymes for which crystal structures have been obtained, because the most decisive advances in our understanding of enzyme reaction mechanisms have come by inspecting such structures.

### Serine Proteases: Enzymes that Use a Serine Residue for Nucleophilic Catalysis

The serine proteases are a large family of proteolytic enzymes that use the reaction mechanism for nucleophilic catalysis outlined in equations (3) and (4), with a serine residue as the reactive nucleophile. The best known members of the family are three closely related digestive enzymes: trypsin, chymotrypsin, and elastase. These enzymes are synthesized in the mammalian pancreas as inactive precursors termed zymogens. They are secreted into the small intestine, where they are activated by proteolytic cleavage in a manner discussed in chapter 9.

In the digestive system trypsin, chymotrypsin, and elastase work as a team. They are all endopeptidases, which means that they cleave protein chains at internal peptide bonds, but each preferentially hydrolyses bonds adjacent to a particular type of amino acid residue (fig. 8.4). Trypsin cuts just next to basic residues (lysine or arginine); chymotrypsin cuts next to aromatic residues (phenylalanine, tyrosine, or tryptophan); elastase is less discriminating but prefers small, hydrophobic residues such as alanine.

About half of the amino acid residues of trypsin are identical to the corresponding residues in chymotrypsin, and about a quarter of the residues are conserved in all three of the pancreatic endopeptidases (fig. 8.5). The structural simi-

**Figure 8.4**

Trypsin, chymotrypsin, and elastase—three members of the serine protease family—catalyze the hydrolysis of proteins at internal peptide bonds adjacent to different types of amino acids. Trypsin prefers lysine or arginine residues; chymotrypsin, aromatic side chains; and elastase, small, nonpolar residues. Carboxypeptidases A and B, which are not serine proteases, cut the peptide bond at the carboxyl-terminal end of the chain. Carboxypeptidase A preferentially removes aromatic residues; carboxypeptidase B, basic residues. (Illustration copyright by Irving Geis. Reprinted by permission.)

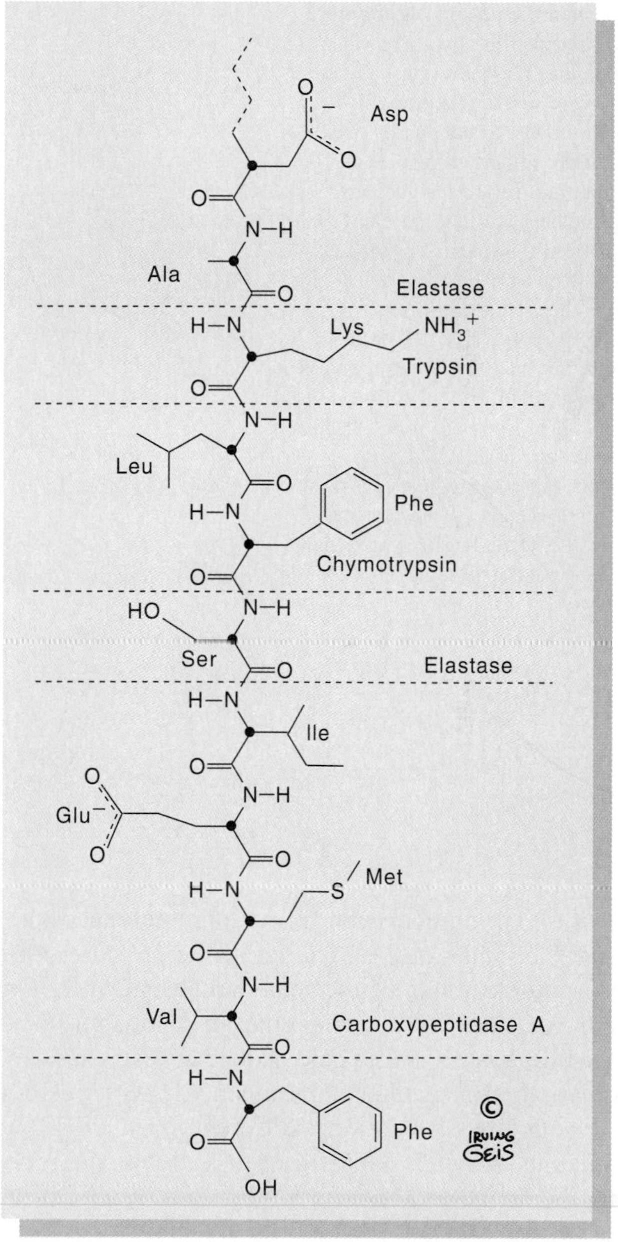

larities of trypsin, chymotrypsin, and elastase are even more evident in the crystal structures. As is shown in figure 8.6a the folding of the polypeptide chain is essentially the same in all three enzymes. These enzymes are classical illustra-

# Figure 8.5

Schematic diagrams of the amino acid sequences of chymotrypsin, trypsin, and elastase. Each circle represents one amino acid. Amino acid residues that are identical in all three proteins are in solid color. The three proteins are of different lengths but have been aligned to maximize the correspondence of the amino acid sequences. All of the sequences are numbered according to the sequence in chymotrypsin. Long connections between nonadjacent residues represent disulfide bonds. Locations of the catalytically important histidine, aspartate, and serine residues are marked. The links that are cleaved to transform the inactive zymogens to the active enzymes are indicated by parenthesis marks. After chymotrypsinogen is cut between residues 15 and 16 by trypsin and is thus transformed into an active protease, it proceeds to digest itself at the additional sites that are indicated; these secondary cuts have only minor effects on the enzymes's catalytic activity. (Illustration copyright by Irving Geis. Reprinted by permission.)

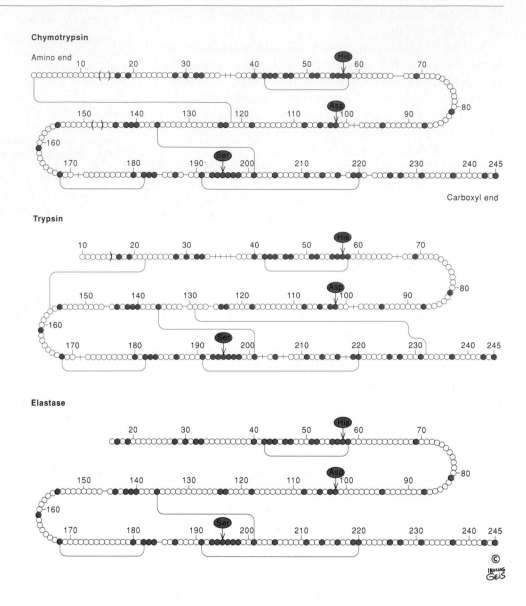

tions of diverging evolution from a common ancestor. The structures of some of the bacterial serine proteases also are homologous to those of the mammalian enzymes. On the other hand, subtilisin, a serine protease obtained from *Bacillus subtilis,* has an amino acid sequence that seems totally unrelated to the mammalian sequences. Its three-dimensional structure also is very different from those of the mammalian enzymes (see fig. 8.6b). Subtilisin probably joined the serine protease family by convergent evolution. Remarkably, a small set of critical amino acid residues comes together in the folded structure to form the essential elements of the active site in all of these proteases. Using the chymotrypsin numbering system, these are His 57, Asp 102, and Ser 195 (fig. 8.7).

That Ser 195 played an important role in the catalytic mechanism was known from early studies on the enzyme

inhibitor diisopropylfluorophosphate (see chapter 7). This inhibitor reacts irreversibly with chymotrypsin or trypsin to form an inactive derivative in which the diisopropylphosphate group is covalently attached to the serine residue (see fig. 7.18). A variety of other inhibitors react in a parallel manner with this same serine residue. As a rule, these inhibitors do not react with other serines in the enzyme or with serines in enzymes that are not part of the serine protease family. The reactivity of Ser 195 thus is not a property of serine residues in general but depends on the special surroundings of this residue in the protein. We will see shortly that it is the juxtaposition with His 57 and Asp 102 that makes the serine especially reactive.

Although His 57 is far removed from Ser 195 in the primary sequence, studies with affinity labels showed that it must be near the serine residue in the active site. A deriva-

**Figure 8.6**

(*a*) Superimposed computer-generated *c*-alpha tube structures for trypsin (green), chymotrypsin (yellow-gold) and elastase (pale pink). The side chains for the three key catalytic residues, Asp 102 (red), His 57 (lt. blue), and Ser 195 (orange) are shown for trypsin. The structure for trypsin also includes a bound inhibitor, benzamidine, in yellow. (*b*) Superimposed *c*-alpha tube structures for subtilisin (lavender) and trypsin (green), both with catalytic side chains showing. The three key catalytic residues, Asp 102 (red), His 57 (lt. blue), and Ser 195 (orange) for trypsin and Asp 32 (red), His 64 (lt. blue), and Ser 221 (orange) for subtilisin are shown. (Images (*a*) and (*b*): Copyright 1994 by the Scripps Research Institute/Molecular Graphics Images by Michael Pique using software by Yng Chen, Michael Connolly, Michael Carson, Alex Shah, and AVS, Inc. Visualization advice by Holly Miller, Wake Forest University Medical Center.)

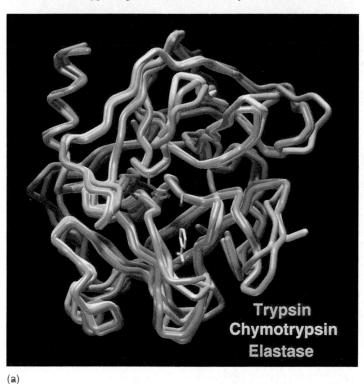

Trypsin
Chymotrypsin
Elastase

(a)

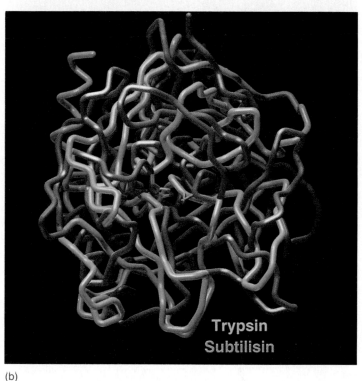

Trypsin
Subtilisin

(b)

tive of phenylalanine containing a reactive chloromethylketone group was found to inhibit chymotrypsin irreversibly by reacting with the histidine. The inhibition can be prevented by the presence of other aromatic molecules that bind competitively at the active site.

The initial evidence for the formation of an acyl-enzyme ester intermediate came from studies of the kinetics with which chymotrypsin hydrolyzed analogs of its normal polypeptide substrates. The enzyme turned out to hydrolyze esters as well as peptides and simpler amides. Of particular interest was the reaction with the ester *p*-nitrophenyl acetate. This substrate is well suited for kinetic studies because one of the products of its hydrolysis, *p*-nitrophenol, has a yellow color in aqueous solution, whereas *p*-nitrophenyl acetate itself is colorless. The change in the absorption spectrum makes it easy to follow the progress of the reaction. When rapid-mixing techniques are used to add the substrate to the enzyme, an initial burst of *p*-nitrophenol is detected within the first few seconds, before the reaction settles down to a constant rate (fig. 8.8). The amount of *p*-nitrophe-

nol released in the burst is approximately equal to the amount of enzyme present in the solution. These observations suggest that the overall enzymatic reaction occurs in two distinct steps, as shown in figure 8.9. In the first step, *p*-nitrophenol is released and the acetyl group is transferred to the enzyme, forming an acyl-enzyme intermediate. In the second step, the intermediate is hydrolyzed, and acetate is released. The finding that diisopropylfluorophosphate prevents the initial burst of *p*-nitrophenol suggested that the enzymatic group that forms the ester intermediate is Ser 195.

When the crystal structures of trypsin, chymotrypsin, and elastase with bound substrate analogs were determined, the substrate analogs were indeed located close to Ser 195 (see figs. 8.6 and 8.7). Histidine 57 sits nearby, in an orientation suggesting that the OH group of Ser 195 forms a hydrogen bond to the imidazole side chain of the histidine. Aspartic acid 102 sits on the opposite edge of the imidazole ring, where its negatively charged carboxylate group can interact with the proton on the other nitrogen of the ring.

## Figure 8.7

Computer-generated c-alpha tube model of the crystal structure of trypsin, seen from the same perspective as in figure 8.6(a). The coloring of the amino acid side chains and the inhibitor (benzamidine) is as in figure 8.6(a). (b) A close-up view of the active site of trypsin with bound benzamidine. (Images (a) and (b): Copyright 1994 by the Scripps Research Institute/Molecular Graphics Images by Michael Pique using software by Yng Chen, Michael Connolly, Michael Carson, Alex Shah, and AVS, Inc. Visualization advice by Holly Miller, Wake Forest University Medical Center.)

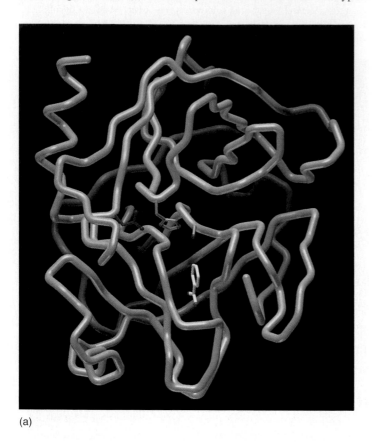

(a)

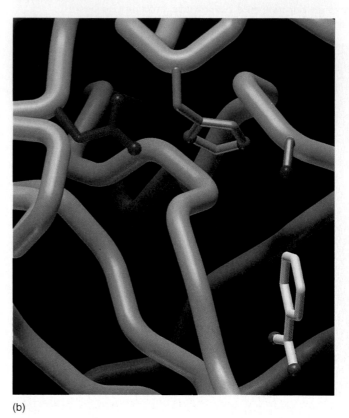

(b)

## Figure 8.8

p-Nitrophenol formation as a function of time during the hydrolysis of p-nitrophenyl acetate by chymotrypsin. A rapid initial burst of p-nitrophenol is followed by a slower, steady-state reaction. The amount of p-nitrophenol released in the burst is approximately equal to the amount of enzyme present.

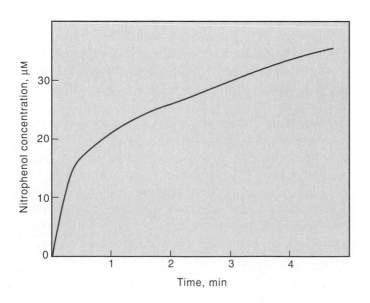

The side chains of the aspartate and histidine residues thus appear to be oriented so as to facilitate removal of the proton from the serine's OH group.

$$-\overset{\overset{\displaystyle O}{\|}}{C}-O^- \text{---} HN \overset{C=C}{\underset{C}{\diagup\diagdown}} N\text{---}HOCH_2-$$

Because withdrawing the proton would increase the nucleophilic character of the oxygen, this arrangement explains why Ser 195 is exceptionally reactive. The dependence of the enzyme kinetics on pH agrees with this picture. Enzymatic activity depends on the presence of a basic group with a $pK_a$ of about 6.8, which is in the range consistent with a histidine side chain. The $k_{cat}$ decreases abruptly if this group is protonated (fig. 8.10). Protonating His 57 prevents the histidine from forming a hydrogen bond to Ser 195, thereby decreasing the nucleophilic reactivity of the serine.

The crystal structures also provided an explanation for the different substrate specificities of trypsin, chymotrypsin,

## Figure 8.9

Steps in the hydrolysis of *p*-nitrophenyl acetate by chymotrypsin. In the hydrolysis of this and most other esters, the breakdown of the acyl-enzyme intermediate is the rate-determining step. In the hydrolysis of peptides and amides, the rate-determining step usually is the formation of the acyl-enzyme intermediate. This makes the transient formation of the intermediate more difficult to study because the intermediate breaks down as rapidly as it forms.

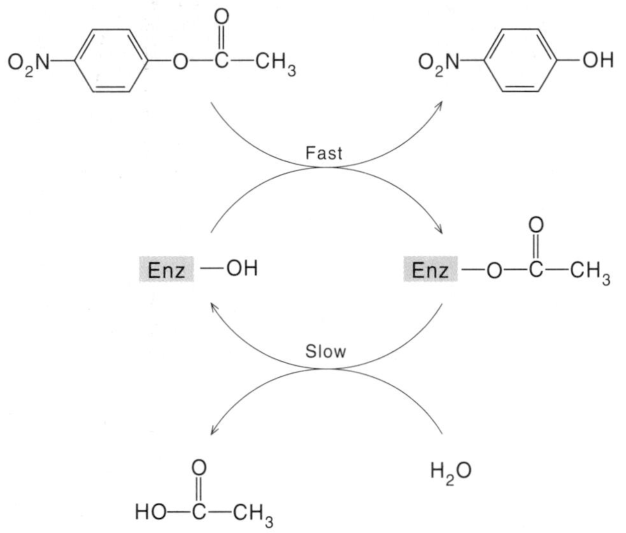

## Figure 8.10

The turnover number ($k_{cat}$) and the Michaelis constant ($K_m$) as a function of pH for the hydrolysis of *N*-acetyl-L-tryptophanamide by chymotrypsin at 25°C. The decrease in $k_{cat}$ as the pH is lowered between 8 and 6 probably reflects the protonation of His 57. The increase in $K_m$ above pH 9 probably reflects the deprotonation of Ile 16, which results in the rotation of Gly 193 out of the substrate-binding site.

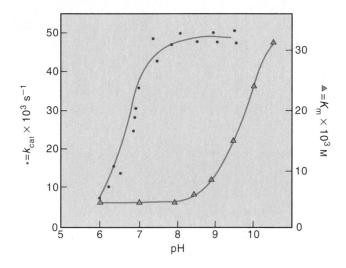

and elastase. In both trypsin and chymotrypsin, the side chain of the substrate fits snugly into a pocket (see figs. 8.6 and 8.7). At the far end of the pocket in trypsin is an aspartic acid residue. The negative charge of the aspartate carboxylate favors the binding of the positively charged side chain of lysine or arginine (see fig. 7.15). In chymotrypsin the aspartic acid residue is replaced by serine, creating a less polar environment that suits the side chain of tyrosine, phenylalanine, or tryptophan. In elastase, the binding pocket is padded with the bulky side chains of valine and threonine residues, so that it can accommodate only small substrates such as alanine.

Another important feature of the substrate-binding site in serine proteases is that the carbonyl oxygen atom of the peptide bond that is cleaved is hydrogen-bonded to one or more NH groups. In trypsin, chymotrypsin, and elastase, these are the amide NH groups of Ser 195 and Gly 193. The interaction with the two protons favors an increase in the negative charge on the oxygen, facilitating formation of a tetrahedral intermediate state, as we discussed previously in connection with general-acid catalysis and electrostatic effects. Figure 8.7 shows this part of the active site in trypsin.

The overall reaction mechanism of chymotrypsin is sketched in figure 8.11. Part *a* shows the enzyme–substrate complex, with an aromatic side chain of the substrate seated in the binding pocket and the carbonyl oxygen atom hy-

drogen-bonded to the amide NH hydrogens of Gly 193 and Ser 195. Aspartate 102 and His 57 are aligned as described above, with Ser 195 forming a hydrogen bond with the histidine. Part *b* shows a tetrahedral intermediate state. Serine 195 has released its proton to His 57 and launched a nucleophilic attack on the carbonyl carbon of the substrate. The histidine residue thus acts as a general-base catalyst in the formation of the intermediate. The movement of negative charge to the carbonyl oxygen creates an "oxyanion," which is stabilized largely by electrostatic interactions with the amide protons of Ser 195 and Gly 193. The tetrahedral oxyanion intermediate is not stable enough to be isolated, and the evidence for its existence is inferential. In part *c*, this intermediate has decomposed to form the more stable acyl-enzyme intermediate, in which the serine is linked as an ester to the carboxylic part of the substrate and the amine product has been released. Histidine 57 probably facilitates this decomposition by acting as a general acid.

To complete the reaction, the breakdown of the acyl-enzyme probably occurs as shown in parts *d*, *e*, and *f* of figure 8.11. The steps here are essentially a reversal of the steps through parts *a*, *b*, and *c*, except that water replaces the amino part of the substrate. The reaction probably proceeds by way of a tetrahedral intermediate similar to one shown in part *b* except for this substitution (part *e*).

In figure 8.11, note that Asp 102 does not actually remove a proton from His 57 in the formation of either of the tetrahedral intermediates. The effect of the aspartate's negatively charged carboxyl group probably is best viewed

## Figure 8.11

The probable mechanism of action of chymotrypsin. The six panels show the initial enzyme-substrate complex (*a*), the first tetrahedral (oxyanion) intermediate (*b*), the acyl-enzyme (ester) intermediate with the amine product departing (*c*), the same acyl-enzyme intermediate with water entering (*d*), the second tetrahedral (oxyanion) intermediate (*e*), and the final enzyme-product complex (*f*). In the transition states between these intermediates, there probably is a more even distribution of negative charge between the different oxygen atoms attached to the substrate's central carbon atom.

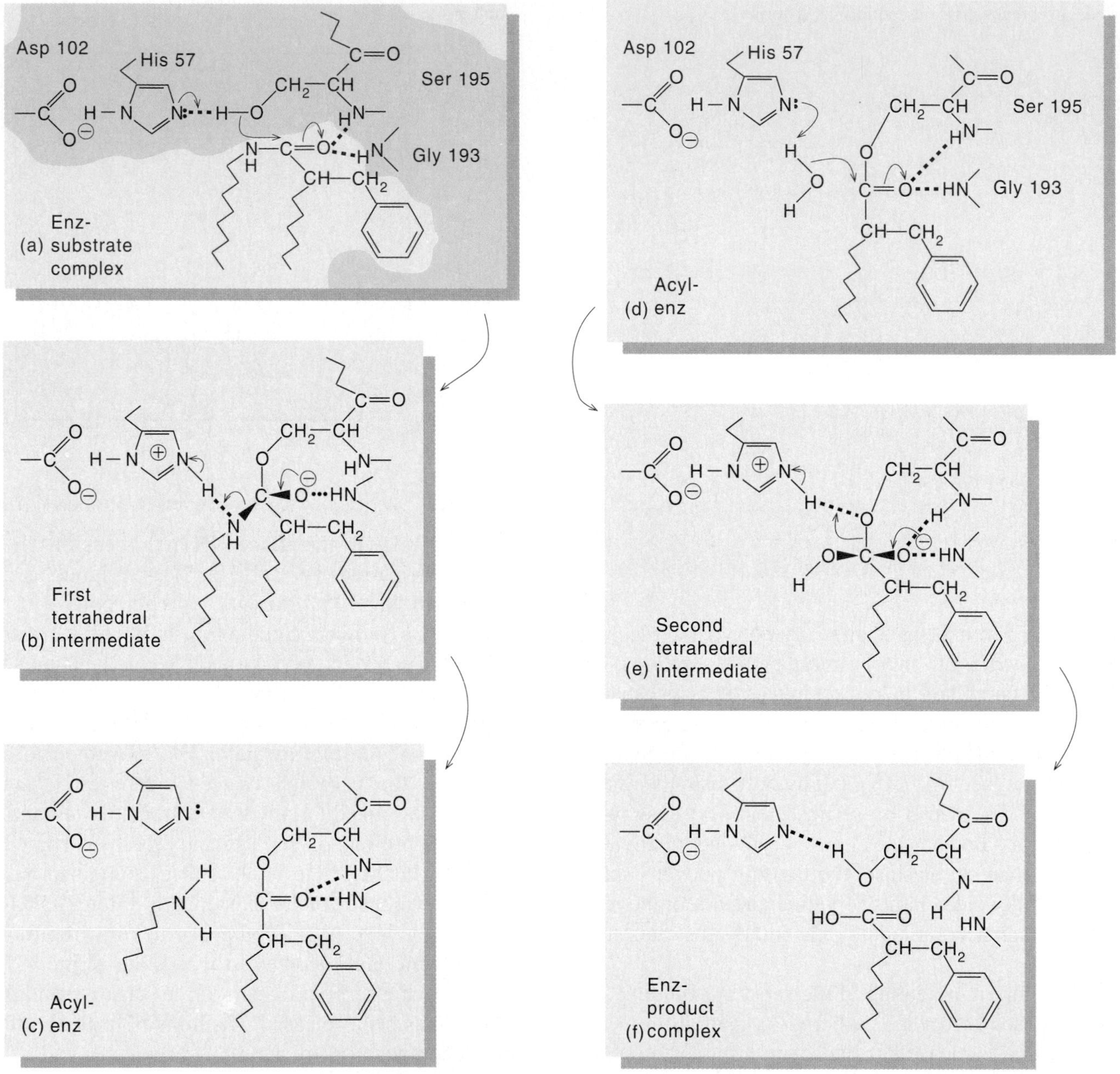

as an electrostatic effect that favors movement of positive charge from Ser 195 to His 57. The stabilization of the oxyanion intermediate by the amide protons of Gly 193 and Ser 195 can be described similarly as a favorable electrostatic interaction of the negatively charged oxygen with nearby atoms that have positive charges, rather than as an actual transfer of a proton to the oxygen. The carboxyl group of Asp 102 also helps to align the imidazole ring of

His 57 in the proper orientation for removing the proton from Ser 195.

One way to test a scheme like that shown in figure 8.11 is to investigate the effects of modifying the amino acid residues that play important roles. The inhibitory effect of the reaction of Ser 195 with diisopropylfluorophosphate was mentioned earlier (also see fig. 7.18). To examine the importance of His 57, this residue can be modified by meth-

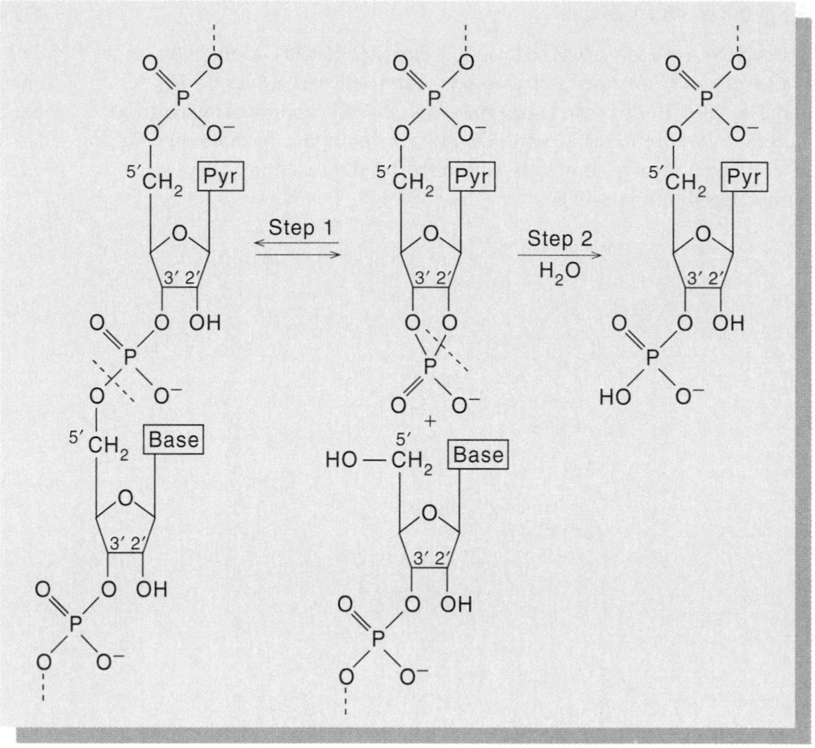

**Figure 8.12**

A portion of an RNA chain, indicating points of cleavage by pancreatic ribonuclease. ''Pyr'' refers to a pyrimidine; ''Base'' can be either a purine or a pyrimidine. The 2', 3', and 5' carbon atoms are labeled. The enzymatic reaction proceeds in two steps, with a cyclic 2',3'-phosphate diester as an intermediate.

ylation, which disrupts the interaction with Asp 102. This process decreases the activity of chymotrypsin by a factor of more than $10^3$. The importance of Asp 102 was tested by site-directed mutagenesis of trypsin and subtilisin. Replacing the aspartate by asparagine reduces $k_{cat}$ by a factor of about $10^4$.

## Ribonuclease A: An Example of Concerted Acid–Base Catalysis

Ribonucleases are a widely distributed family of enzymes that hydrolyze RNA by cutting the P—O ester bond attached to a ribose 5' carbon (fig. 8.12). A good representative of the family is the pancreatic enzyme ribonuclease A (RNase A), which is specific for a pyrimidine base (uracil or cytosine) on the 3' side of the phosphate bond that is cleaved. When the amino acid sequence of bovine RNase A was determined in 1960 by Stanford Moore and William Stein, it was the first enzyme and only the second protein to be sequenced. RNase A thus played an important role in the development of ideas about enzymatic catalysis. It was one of the first enzymes to have its three-dimensional structure elucidated by x-ray diffraction and was also the first to be synthesized completely from its amino acids. The synthetic protein proved to be enzymatically indistinguishable from the native enzyme.

RNase A is a relatively small protein, with a single polypeptide chain of 124 amino acid residues (see fig.

4.15). An early discovery by Frederick Richards that turned out to be useful was that the protein could be cleaved between residues 20 and 21 by the bacterial serine protease, subtilisin. The resulting two polypeptides were separated and purified. They were enzymatically inactive individually, but regained the activity of the native enzyme when they were recombined. This work shows that strong, noncovalent interactions occur that can hold protein chains together even when one of the peptide links is cut. It also makes it possible to modify specific amino acid residues of the two polypeptide chains independently and to explore how each residue contributes to the reassembly of the protein and the recovery of enzymatic activity.

The hydrolysis of RNA catalyzed by RNase A occurs in two distinct steps, with a 2',3'-phosphate cyclic diester intermediate (see fig. 8.12). The intermediate can be identified relatively easily, because its breakdown is much slower than its formation. Ribonucleases do not hydrolyze DNA, which lacks the 2'-hydroxyl group needed for the formation of the cyclic intermediate.

RNase A is completely inhibited if either of two histidine residues (His 12 or His 119) is modified by carboxymethylation with iodoacetate (fig. 8.13) suggesting that these histidines play important roles in the active site. In support of this conclusion, the reaction of iodoacetate with His 12 or His 119 is inhibited by cytidine-3'-phosphate and other small molecules that bind at the active site. Lysine 41 has been implicated similarly in the active site by the observation that enzymatic activity is destroyed by the reaction of

## Figure 8.13

When RNase A is treated with iodoacetate ($ICH_2CO_2^-$), the two major products obtained are carboxymethylated derivatives of His 12 and His 119. Both of these enzymes are severely inhibited, which suggests that both His 12 and His 119 are important in the active site. The enzyme also is completely inhibited by the reaction of Lys 41 with fluorodinitrobenzene.

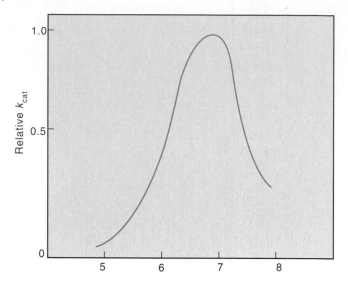

**1-carboxymethyl-His 119**

**3-carboxymethyl-His 12**

**Dinitrobenzyl-Lys 41**

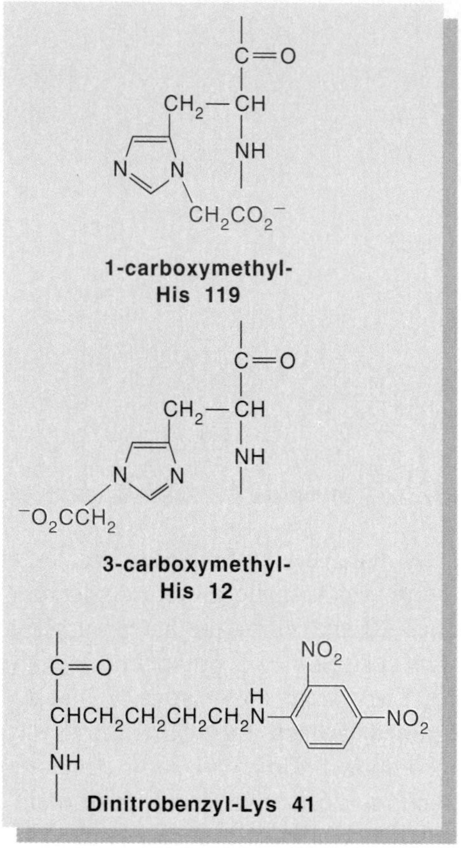

## Figure 8.14

The dependence of $k_{cat}$ of RNase A on pH. The bell-shaped curve suggests that one histidine residue must be in the protonated state and another must be unprotonated. Similar pH dependences are found for the hydrolysis of either RNA or pyrimidine nucleoside-2′,3′-phosphate cyclic diesters.

the ε-amino group of this residue with fluorodinitrobenzene (see fig. 8.13). A second lysine (Lys 7) also appears to be important for substrate binding, although the enzyme retains some activity when this residue is modified.

The enzymatic activity of RNase A shows a bell-shaped dependence on pH, with an optimum near pH 7 (fig. 8.14). The operation of the enzyme appears to require that a dissociable group with a $pK_a$ of about 6.3 be in the protonated form, and that a group with a $pK_a$ of about 8 be unprotonated. These groups have been identified as His 12 and His 119 by measuring the NMR spectrum of the enzyme as a function of pH. Spectral peaks due to histidine residues undergo characteristic shifts as the imidazole ring is protonated.

The pH dependence of the kinetics, taken with the information that histidines 12 and 119 are essential for enzymatic activity, suggests that the two histidines participate in the reaction by concerted general-base and general-acid catalysis. This supposition is supported by the crystal structure

of the enzyme. Figure 8.15 shows a model of the crystal structure of RNase S (the active enzyme obtained by cutting ribonuclease with subtilisin and recombining the two polypeptides) with a bound competitive inhibitor, "UpCH₂A." UpCH₂A resembles the dinucleotide substrate UpA but cannot be hydrolyzed because it has a methylene carbon in place of the oxygen between the P and the 5′—CH₂ of the ribose of adenosine (fig. 8.16). Similar crystal structures have been obtained with other substrate analogs. In all of the structures, His 12 and His 119 sit on opposite sides of the substrate's phosphate group, with His 12 positioned close to the 2′ OH group. Lysines 7, 14, and 66 also are located nearby, where their positively charged amino groups have favorable electrostatic interactions with the negatively charged phosphate. Near His 119 is the carboxyl group of an acidic residue, Asp 121. The electrostatic effect of Asp 121 favors the transfer of a proton from a molecule of water to His 119, much as Asp 102 of chymotrypsin induces a proton to move from Ser 195 to His 57.

The pyrimidine ring on the 3′ side of the substrate fits snugly into a groove on RNase A. Valine 43 is located on one side of the pyrimidine ring, and Phe 120 on the other. The specificity of the enzyme for pyrimidine nucleotides appears to result largely from hydrogen bonding to Thr 45. The threonine side chain can form a pair of complementary hydrogen bonds with either uracil or cytosine (fig. 8.17). Crystallographic studies have shown that if the pyrimidine is replaced by a purine, the compound can still bind to the enzyme, but the distance between His 12 and the 2′ OH of

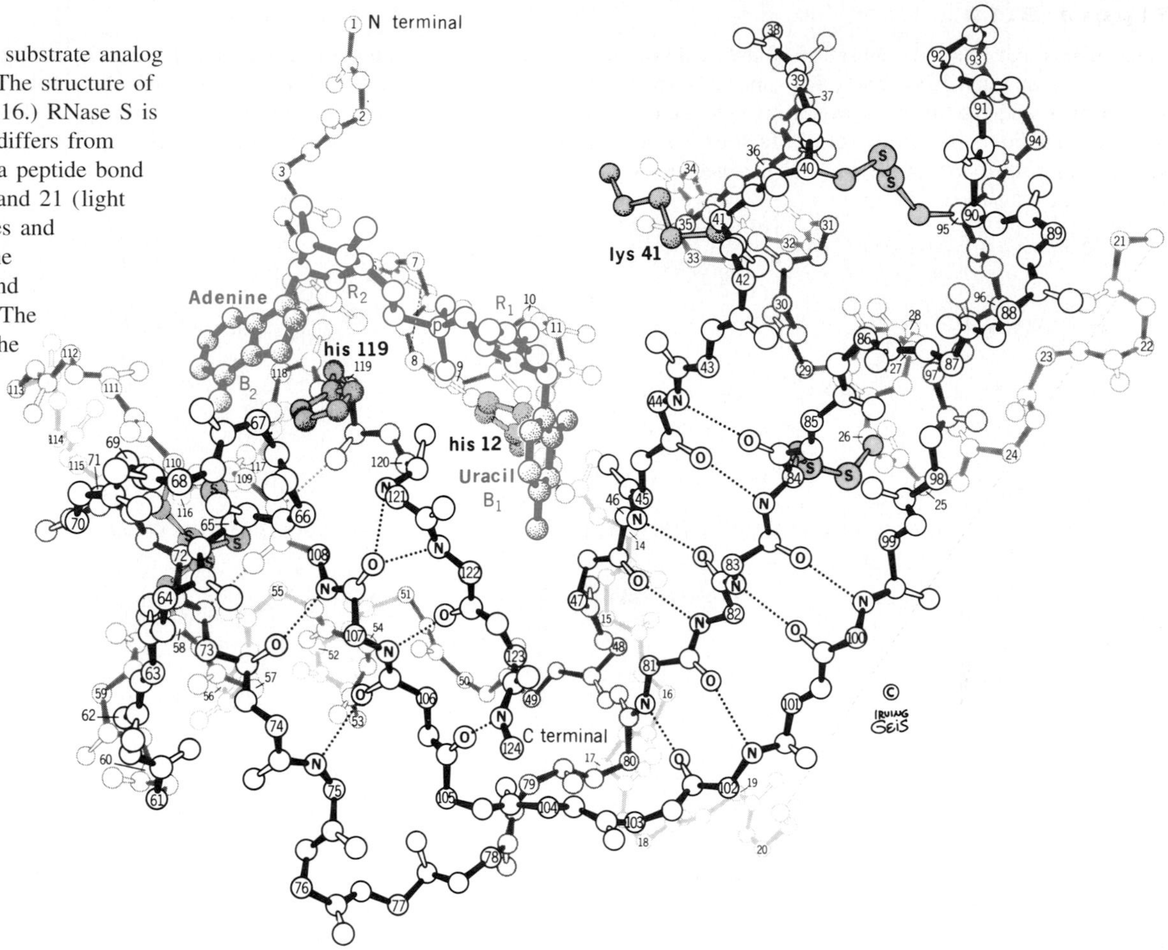

## Figure 8.15

A model of the binding of the substrate analog UpCH$_2$A (blue) to RNase S. (The structure of UpCH$_2$A is shown in figure 8.16.) RNase S is an active enzyme, although it differs from the native enzyme because of a peptide bond cleavage between residues 20 and 21 (light dashed line). The two histidines and the lysine that are crucial to the active site (His 12, His 119, and Lys 41) are indicated in gray. The dotted lines indicate some of the hydrogen bonds that maintain the protein structure.
(Illustration copyright by Irving Geis. Reprinted by permission.)

## Figure 8.16

Structure of UpCH$_2$A, a competitive inhibitor of RNase A. UpCH$_2$A differs from the dinucleotide substrate UpA in having a methylene group instead of oxygen between the phosphorus and C-5′ of the adenosine.

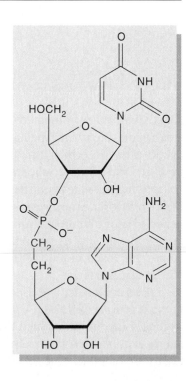

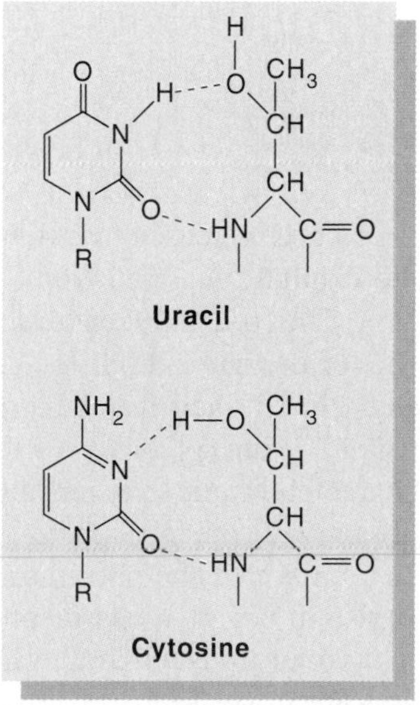

## Figure 8.17

The side chain and amide NH group of Thr 45 in RNase A can form a pair of complementary hydrogen bonds with either uracil or cytosine. These probably account for the specificity of RNase A for pyrimidine nucleotides.

Uracil

Cytosine

## Figure 8.18

The probable catalytic mechanism of RNase A. In the formation of the 2′,3′-cyclic ester, His 12 acts as a general base, and His 119 acts as a general acid. The increase in negative charge on the phosphate's oxygen atoms in the transition state is favored by electrostatic interactions with Lys 7 and Lys 41. In the breakdown of the cyclic ester, His 119 acts as a general base and His 12 acts as a general acid. The reaction proceeds through two short-lived intermediates in which the phosphorus atom is pentavalent, in addition to the more stable cyclic ester intermediate. Py = pyrimidine.

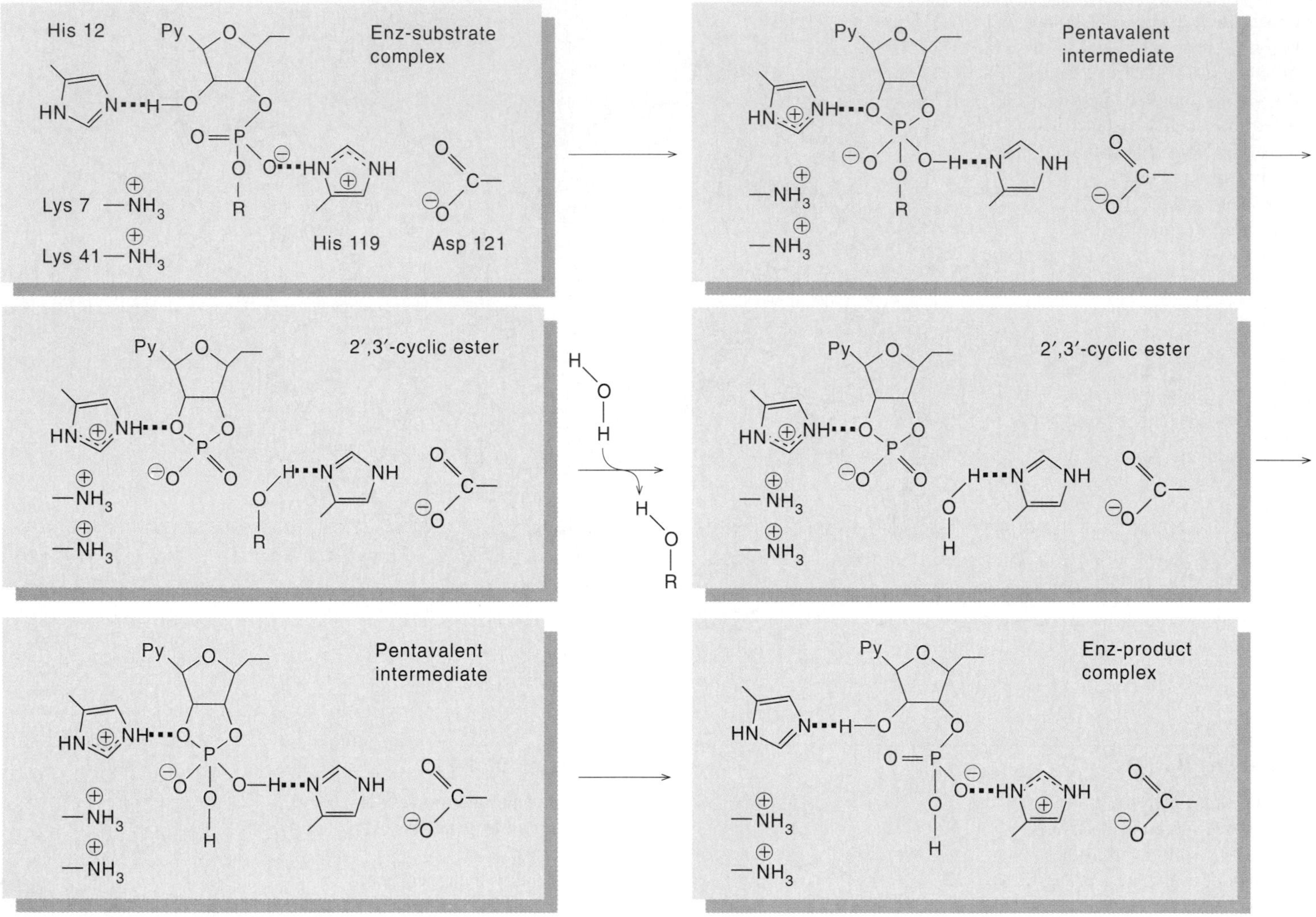

the ribose increases by about 1.5 Å. This evidently prevents the catalytic reaction from occurring.

The probable catalytic mechanism of RNase A is shown in figure 8.18. In the formation of the 2′,3′-cyclic ester, His 12 acts as a general base to remove the proton of the 2′ hydroxyl group, which increases the nucleophilic character of the oxygen atom. Reaction of the hydroxyl group with the phosphate creates a transient intermediate state in which the phosphorus atom is pentavalent. The increase in negative charge on the phosphate's oxygen atoms is favored by electrostatic interactions with Lys 7 and Lys 41. The formation of the pentavalent intermediate also is promoted by His 119, which acts as a general acid to protonate one of the phosphate oxygens. His 119 then probably participates in transferring a proton to the 5′ oxygen atom.

Expulsion of the protonated 5′ OH group converts the transient pentavalent intermediate into the more stable 2′,3′-cyclic ester, in which the substituents of the phosphorus have returned to a tetrahedral geometry.

In the breakdown of the 2′,3′-cyclic ester, the roles of the two histidines are reversed. Histidine 119 now acts as general base to remove a proton from $H_2O$, and His 12 acts as a general acid to protonate the substrate's 2′ oxygen. Again, the reaction probably proceeds by way of a pentavalent intermediate.

The pentavalent phosphoryl intermediates that figure in the reactions of RNase A probably have the geometry of a trigonal bipyramid in which three of the phosphorus atom's substituents lie in a plane. The entering and leaving oxygen atoms are on either side of the plane, forming a straight line

## Figure 8.19

(*a*) Ribbon diagram representation of the crystal structure of the complex between RNase A and uridine vanadate. The catalytic histidine side chains, histidines 12 and 119, are shown in light blue. Uridine vanadate, which contains a vanadium atom (pink) surrounded by five oxygen atoms (red) is thought to mimic the pentavalent transition state formed by a phosphoryl substrate. Uridine vanadate binds to RNase A about $10^4$ times more tightly than the corresponding uridine 2′,3′-cyclic phosphate ester. (*b*) Close-up view of the RNase A and uridine vanadate complex showing the proximity and position of histidines 12 and 119. (Copyright 1994 by the Scripps Research Institute/Molecular Graphics Images by Michael Pique using software by Yng Chen, Michael Connolly, Michael Carson, Alex Shah, and AVS, Inc. Visualization advice by Holly Miller, Wake Forest University Medical Center.)

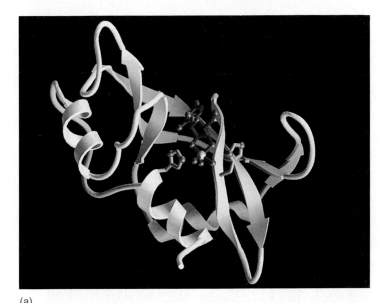

(a)

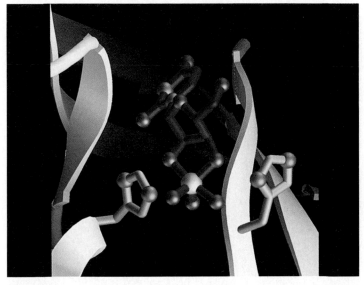

(b)

with the phosphorus. One observation that supports the formation of such pentavalent intermediates is that RNase A is inhibited strongly by uridine vanadate, in which the vanadium atom is surrounded by five oxygens with a similar geometry. Figure 8.19 shows the structure of the complex of the enzyme with the inhibitor. The disposition of histidines 12 and 119 on either side of the vanadate group in the crystal structure is consistent with the linear arrangements of the entering and leaving oxygen atoms on either side of the phosphorus atom in figure 8.18.

## *Triosephosphate Isomerase Has Approached Evolutionary Perfection*

Triosephosphate isomerase catalyzes the interconversion of dihydroxyacetone phosphate and 3-phosphoglyceraldehyde.

$$
\begin{array}{c}
CH_2OH \\
| \\
C{=}O \\
| \\
CH_2OPO_3^{2-}
\end{array}
\quad \rightleftharpoons \quad
\begin{array}{c}
H{-}C{=}O \\
| \\
H{-}C{-}OH \\
| \\
CH_2OPO_3^{2-}
\end{array}
\qquad (10)
$$

Dihydroxyacetone phosphate        Glyceraldehyde-3-phosphate

The interconversion of the two triosephosphates is an essential step in the catabolism of carbohydrates (see chapter 12). Examining the catalytic mechanism of triosephosphate isomerase is instructive. It is among the enzymes that approach evolutionary perfection in the sense that its specificity constant $k_{cat}/K_m$ is greater than $10^8 \, s^{-1} M^{-1}$, which is near the limit set by the rates of diffusion of the substrate and protein (see table 7.3). The enzyme achieves this high value for the specificity constant by having both a relatively high $k_{cat}$ and a relatively low $K_m$.

Crystal structures have been obtained for triosephosphate isomerase purified both from yeast and from chicken muscle. The proteins from the two organisms have identical residues at about half the positions in their amino acid sequences, and their three-dimensional structures are very similar. The overall structure is a $\beta$ barrel composed of eight parallel $\beta$ sheets linked by eight $\alpha$ helices (see fig. 4.17*b*). The active site is near the center of the molecule and contains a glutamic acid (Glu 165) with nearby histidine and lysine residues (fig. 8.20). Glu 165 was first implicated in the active site by the observation that it reacts irreversibly with chloroacetone phosphate, an inhibitor that is a close structural analog of the substrate dihydroxyacetone phosphate. (In chloroacetone phosphate, a chlorine atom replaces the hydroxyl group on C-1.) In crystal structures of the enzyme with bound dihydroxyacetone phosphate, the carboxyl group of Glu 165 is located close to C-1 of the substrate. Changing Glu 165 to aspartate by site-directed mutagenesis decreases the rate of the isomerization reaction catalyzed by the enzyme approximately 1,000-fold. This is a dramatic effect, considering that the substitution of Asp for Glu only entails shortening the amino acid side chain by one $-CH_2$-group.

## Figure 8.20

(a) Ribbon diagram of the crystal structure of triosephosphate isomerase from chicken muscle. This figure shows one of the two subunits of the enzyme. The active site amino acid side chains Glu 165 (red), His 95 (light blue) and Lys 13 (dark blue) are shown. (b) Close-up view of the active site of triosephosphate isomerase.

(Copyright 1994 by the Scripps Research Institute/Molecular Graphics Images by Michael Pique using software by Yng Chen, Michael Connolly, Michael Carson, Alex Shah, and AVS, Inc. Visualization advice by Holly Miller, Wake Forest University Medical Center.)

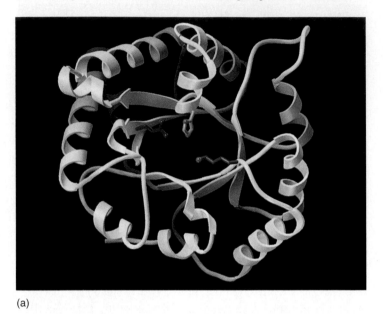

(a)

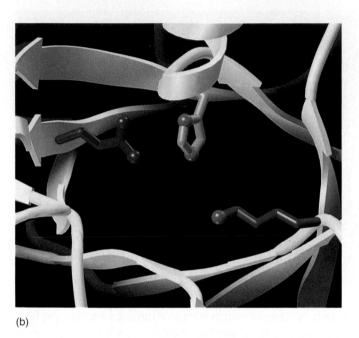

(b)

## Figure 8.21

The probable reaction mechanism of triosephosphate isomerase. The γ-carboxylate group of Glu 165 acts as a general base to remove a proton from C-1 of the substrate, dihydroxyacetone phosphate (DHAP). This generates a planar ene-diolate intermediate that has two tautomeric forms. After an interconversion of ene-diolate tautomers, the protonated glutamic acid residue returns a proton to C-2. The electrostatic effects of His 95 and Lys 12 facilitate the formation of the negatively charged intermediate. GAP = glyceraldehyde-3-P.

The rate of the isomerase reaction decreases if a basic group is protonated by lowering the pH below 6.5. It seems clear that this is the carboxyl group of Glu 165 because lowering the pH also prevents the reaction of this residue with chloroacetone phosphate. This suggests that, in the enzymatic reaction, Glu 165 acts as a general base to remove a proton from C-1 of dihydroxyacetone phosphate, converting the substrate into an ene-diolate intermediate as shown in figure 8.21. To complete the reaction, the protonated glutamic acid residue can return a proton to C-2 of the ene-diolate instead of to C-1.

Additional support for this mechanism comes from experiments in which the enzymatic reaction is run in $D_2O$. Deuterium is incorporated stereospecifically onto carbon 2 of the product.

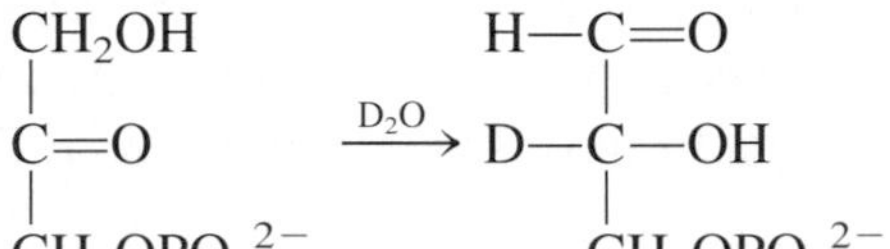

$$CH_2OH \qquad\qquad H{-}C{=}O$$
$$\begin{array}{ccc} CH_2OH & & H-C=O \\ | & \xrightarrow{D_2O} & | \\ C=O & & D-C-OH \qquad\qquad (11) \\ | & & | \\ CH_2OPO_3{}^{2-} & & CH_2OPO_3{}^{2-} \end{array}$$

In the absence of the enzyme, the carbon-bound hydrogen atom on 3-phosphoglyceraldehyde does not exchange with deuterium atoms of the solvent at any significant rate. (We are not concerned here with an exchange of the proton bound to the alcohol oxygen in the reactant or product. This proton does exchange rapidly with the solvent in either the presence or the absence of an enzyme, but the deuteron that is incorporated here can be removed immediately by putting the product back in $H_2O$.) The incorporation of deuterium can be explained as follows: After a proton is transferred from C-1 of the substrate to the carboxyl group of Glu 165, it has an opportunity to escape into the solution and to be replaced by a deuteron. A proton attached to a carboxylic acid oxygen usually exchanges rapidly with the solvent. The deuteron on Glu 165 then can be transferred to C-2 of the product.

The hydrogen-bonding pattern in the crystal and studies of the pH dependence of the reaction suggest that the histidine and lysine residues in the active site do not act as general acids in the catalytic pathway but rather serve to stabilize the negative charge on an ene-diolate intermediate by electrostatic effects (see fig. 8.21).

Assuming that the mechanism shown in figure 8.21 is generally correct, how can we determine whether the formation or the breakdown of the ene-diolate intermediate is the rate-limiting step in the overall reaction? One approach is to study the kinetics using substrates that are specifically labeled with deuterium. With [1(R)-$^2$H]-dihydroxyacetone phosphate (fig. 8.22a), $k_{cat}$ is about three times smaller than the $k_{cat}$ obtained with the $^1$H analog. An isotope effect of this magnitude is consistent with the view that the bond holding the H or D atom to C-1 must be broken in the

**Figure 8.22**

The substrates of triosephosphate isomerase can be labeled specifically with deuterium. The rate of the isomerase reaction with [1(R)-$^2$H]-dihydroxyacetone phosphate (a) as substrate is slowed by about a factor of 3 relative to the rate with ordinary dihydroxyacetone phosphate. This isotope effect indicates that breaking the H—C bond to form the ene-diolate intermediate is the rate-limiting step in conversion of dihydroxyacetone phosphate to 3-phosphoglyceraldehyde. Little or no isotope effect is obtained when the reaction is run in the opposite direction starting with [2-$^2$H]-3-phosphoglyceraldehyde (b), indicating that the breakdown of the ene-diolate must be rate-limiting in this direction. Note that although the two hydrogen atoms on C-1 of dihydroxyacetone phosphate are chemically identical, they are stereochemically distinguishable. The enzyme always removes the one that has been replaced by deuterium in (a).

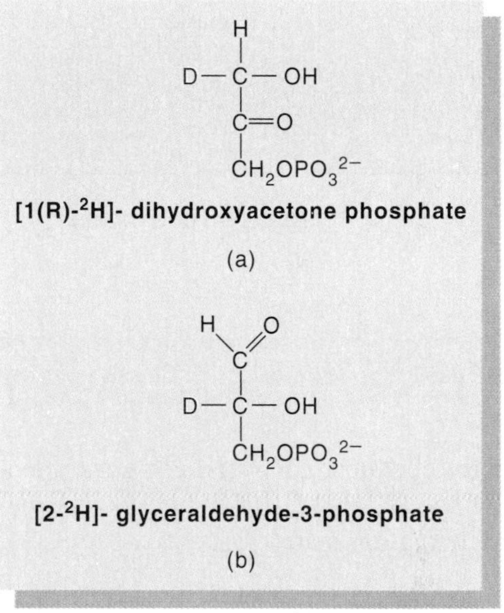

[1(R)-$^2$H]- dihydroxyacetone phosphate

(a)

[2-$^2$H]- glyceraldehyde-3-phosphate

(b)

rate-limiting step. This is consistent with the view that the formation of the ene-diolate is rate-limiting. Similar measurements with [2-$^2$H]-3-phosphoglyceraldehyde (fig. 8.22b) show that there is little or no isotope effect on $k_{cat}$ for the reverse direction, from 3-phosphoglyceraldehyde to dihydroxyacetone phosphate. In this direction, the rate-limiting step must be the breakdown of the ene-diolate intermediate.

By measuring the kinetics of the isotopic exchange reactions catalyzed by the enzyme, along with the overall kinetics of the isomerization reactions in both directions, Jeremy Knowles and his colleagues obtained rate constants for the individual steps of the catalytic mechanism. Figure 8.23 shows the pathway in the form of a free energy profile. The free energy changes for binding of the reactant or product to the enzyme are calculated on the basis of typical concentrations of dihydroxyacetone phosphate and 3-phosphoglyceraldehyde found in muscle cells (40 $\mu$M). At this low concentration of substrate, when the overall rate of the

## Figure 8.23

The calculated free energy profile of the reaction catalyzed by triosephosphate isomerase. (Enz = enzyme; DHAP = dihydroxyacetone phosphate; GAP = glyceraldehyde-3-phosphate.) The free energy changes associated with binding of DHAP and GAP to the enzyme are calculated on the assumption that DHAP and GAP are present at concentrations of 40 $\mu$M.

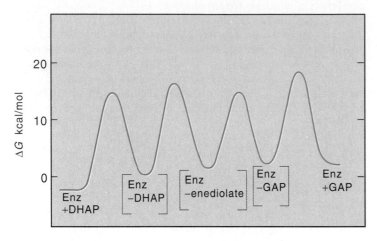

Reaction coordinate

reaction is described by the specificity constant $k_{cat}/K_m$, the rate-limiting step appears to be the formation of the enzyme–substrate complex. This step occurs with a rate constant of about $10^8$ $M^{-1}s^{-1}$ and evidently is set by the rate of diffusion of the enzyme and substrate. It can be slowed down by increasing the viscosity of the solution. Because the activation free energies for the additional steps in the interconversion of the enzyme–substrate and enzyme–product complexes are low enough so that these steps do not limit the overall rate of the reaction, there is no evolutionary pressure to increase the rate constants for these steps any further. Evolutionary changes that led to tighter binding of the substrate might actually be harmful, because lowering the free energy of the enzyme–substrate complex can increase the activation free energy for one or more of the steps in the conversion of this complex into the product. Given the fixed, low concentration of its substrate, triosephosphate isomerase appears to have reached perfection!

## Summary

Like ordinary chemical catalysts, enzymes interact with the reacting species in a manner that lowers the free energy of the transition state.

1. All known enzymatic reaction mechanisms depend on one or more of the following five themes:
   a. proximity effects (enzymes hold the reactants close together in an appropriate orientation)
   b. general-acid or general-base catalysis (acidic or basic groups of the enzyme donate or remove protons and often do first one and then the other)
   c. electrostatic effects (charged or polar groups of the enzyme favor the redistribution of electric charges that must occur to convert the substrate into the transition state)
   d. nucleophilic or electrophilic catalysis (nucleophilic or electrophilic functional groups of the enzyme or a cofactor react with complementary groups of the substrate to form covalently linked intermediates
   e. structural flexibility (changes in the protein structure can increase the specificity of enzymatic reactions by ensuring that substrates bind or react in an obligatory order and by sequestering bound substrates in pockets that are protected from the solvent).

2. The serine proteases trypsin, chymotrypsin, and elastase are very similar in structure but have substrate-binding pockets that are tailored for different amino acid side chains. The active site of each enzyme contains three critical residues: serine, histidine, and aspartate, which are positioned so that the serine hydroxyl group becomes a strong nucleophilic reagent for reaction with the substrate's peptide carbon atom. This reaction generates an acyl-enzyme intermediate. Formation and hydrolysis of the acyl-enzyme intermediate probably are promoted by electrostatic interactions that stabilize tetrahedral transition states and by general-acid and general-base catalysis.

3. Ribonuclease A hydrolyzes RNA adjacent to pyrimidine bases. The reaction proceeds through a 2′,3′-phosphate cyclic diester intermediate. Formation and breakdown of the cyclic diester appear to be promoted by concerted general-base and general-acid catalysis by two histidine residues, and by electrostatic interactions with two lysines. These reactions proceed through pentavalent phosphoryl intermediates. The geometry of these intermediates resembles the geometry of vanadate compounds that act as inhibitors of the enzyme.

4. Triosephosphate isomerase interconverts dihydroxyacetone phosphate and 3-phosphoglyceraldehyde. A glutamic acid residue probably acts as a general base to remove a proton from the substrate, forming an enediolate intermediate. Histidine and lysine side chains stabilize this intermediate electrostatically. Triosephosphate isomerase appears to have reached evolutionary perfection in the sense that it catalyzes its reaction at the maximum possible rate given the concentration of the substrate in the cell.

## Selected Readings

Albery, W. J., and J. R. Knowles, Free-energy profile of the reaction catalyzed by triosephosphate isomerase. *Biochem.* 15:5588, 5627, 1976.

Blackburn, P., and S. Moore, Pancreatic ribonuclease. *The Enzymes* 15:317, 1982.

Blow, D., Structure and mechanism of chymotrypsin. *Acc. Chem. Res.* 9:145, 1976.

Branden, C. -I., H. Jornvall, H. Eklund, and B. Furugren, Alcohol dehydrogenases. *The Enzymes* 11:104, 1975.

Breslow, R., How do imidazole groups catalyze the cleavage of RNA in enzyme models and enzymes? Evidence from ''negative catalysis.'' *Acc. Chem. Res.* 24:317, 1991.

Eklund, H., B. V. Plapp, J. -P. Samama, and C. -I. Branden, Binding of substrate in a ternary complex of horse liver alcohol dehydrogenase. *J. Biol. Chem.* 257:14349, 1982.

Fersht, A., *Enzyme Structure and Mechanism,* 2d ed. New York: Freeman, 1985.

Fersht, A. R., D. M. Blow, and J. Fastrez, Leaving group specificity in chymotrypsin-catalyzed hydrolysis of peptides: A stereochemical interpretation. *Biochem.* 12:2035, 1973.

Findlay, D., D. G. Herries, A. P. Mathias, B. R. Rabin, and C. A. Ross, The active site and mechanism of pancreatic ribonuclease. *Nature* 190:781, 1961.

Holbrook, J. J., A. Liljas, S. J. Steindel, and M. G. Rossmann, Lactate dehydrogenase. *The Enzymes* 11:191, 1975.

Holmes, M. A., D. E. Tronrud, and B. W. Matthews, Structural analysis of the inhibition of thermolysin by an active-site-directed irreversible inhibitor. *Biochem.* 22:236, 1983.

Imoto, T., L. N. Johnson, A. C. T. North, D. C. Phillips, and J. A. Rupley, Vertebrate lysozymes. *The Enzymes* 7:665, 1972.

Kelly, J. A., A. R. Sielecki, B. D. Sykes, M. N. G. James, and D. C. Phillips, X-ray crystallography of the binding of the bacterial cell wall trisaccharide NAM-NAG-NAM to lysozyme. *Nature* 282:875, 1979.

Kraut, J., How do enzymes work? *Science* 242:533, 1988.

Lolis, E., T. Alber, R. C. Davenport, D. Rose, F. C. Hartman, and G. A. Petsko, Structure of yeast triosephosphate isomerase at 1.9-Å resolution. *Biochem.* 29:6609, 1990.

Markley, J. L., Correlation proton magnetic resonance studies at 250 MHz of bovine pancreatic ribonuclease. I. Reinvestigation of histidine peak assignments. *Biochem.* 14:3546, 1975.

Page, M. I. (ed.), *Enzyme Mechanisms.* London: Royal Society of Chemistry, 1987. See particularly the chapters by M. I. Page (Theories of Enzyme Catalysis), A. L. Fink (Acyl Group Transfer—The Serine Proteinases), D. S. Auld (Acyl Group Transfer—Metalloproteinases), P. M. Cullis (Acyl Group Transfer—Phosphoryl Transfer), M. L. Sinnott (Glycosyl Group Transfer), and J. P. Richard (Isomerization Mechanisms through Hydrogen and Carbon Transfer).

Reeke, G. N., J. A. Hartsuck, M. L. Ludwig, F. A. Quiocho, T. A. Steitz, and W. N. Lipscomb, The structure of carboxypeptidase A: Some results at 2.0-Å resolution, and the complex with glycyl tyrosine at 2.8-Å resolution. *Proc. Natl. Acad. Sci. USA* 58:2220, 1967.

Rose, I. A., Mechanism of the aldose-ketose isomerase reactions. *Adv. Enzymol.* 43:491, 1975.

Shoham, M., and T. Steitz, Crystallographic studies and model building of ATP at the active site of hexokinase. *J. Mol. Biol.* 140:1, 1980.

Warshel, A., G. Naray-Szabo, F. Sussman, and J. -K. Hwang, How do serine proteases really work? *Biochem.* 28:3629, 1989.

## Problems

1. Hexokinase promotes the following reaction:

   Glucose + ATP $\longrightarrow$ ADP + Glucose-6-phosphate

   Hexokinase initially binds glucose, then the hexokinase–glucose complex binds ATP. How do you explain the purpose of this substrate-binding pattern to a fellow student without using the induced-fit concept?

2. (a) In what ways are the mechanistic features of chymotrypsin, trypsin, and elastase similar?
   (b) If the mechanisms of these enzymes are similar, what features of the enzyme active site dictate substrate specificity?

3. If a lysine were substituted for the aspartate in the trypsin side-chain-binding crevice, would you expect the enzyme to be functional? If it were functional, what effect would you predict the substitution to have on substrate specificity?

4. Plant proteolytic enzymes are cysteine proteases, that is, they have a cysteine that is critical for the catalytic activity. Plant proteases are thought to function by a mechanism reminiscent of that shown for chymotrypsin (fig. 8.11). Propose a structure for the acyl-enzyme intermediate that would exist for plant proteases.

5. How could the inhibitors presented in table 8.4 be used to indicate that an unknown protease was of plant versus nonplant origin?

6. Notice how the active site of chymotrypsin contains Asp 102, His 57, and Ser 195 (fig. 8.11), whereas the active site of RNase A contains Asp 121, Lys 41, Lys 7, His 12, and His 119 (fig. 8.18). What fundamental principle concerning protein structure is emphasized when the residue number in the active site of enzymes is considered?

7. RNase A utilizes a $2',3'$-cyclic phosphate ester as an intermediate (fig. 8.18) but yields only a $3'$-phosphate product. Within the mechanism, what controls the final position of the phosphate (i.e., why isn't the $2'$-phosphate a product)?

8. For many enzymes, $V_{max}$ is dependent on pH. At what pH do you expect $V_{max}$ of RNase to be optimal? Why?

9. RNase can be completely denatured by boiling or by treatment with chaotropic agents (e.g., urea), yet can refold to its fully active form on cooling or removal of the denaturant. By contrast, when enzymes of the trypsin family and carboxypeptidase A are denatured, they do not regain full activity on renaturation. What aspects of trypsin and carboxypeptidase A structure preclude their renaturation to the fully active form?

10. In figures 8.18 and 8.21 lysyl residues are shown in the lower right parts of the figures. An initial examination of the mechanisms indicates these lysyl groups are only observers. Why are they shown in the "mechanism," and what do they do?

11. Why do structural analogs of the transition-state intermediate of an enzyme inhibit the enzyme competitively and with low $K_i$ values?

12. Transition-state analogs of a specific chemical reaction have been used to elicit antibodies with catalytic activity. These catalytic antibodies have great promise as experimental tools as well as having commercial value. Why is it reasonable to assume that the binding site for the transition-state analog on the antibody mimics the enzyme active site? What difficulties might be encountered if a catalytic antibody is sought for a reaction requiring a cofactor (coenzyme)?

13. Using site-directed mutagenesis techniques, you isolate a series of recombinant enzymes in which specific lysine residues are replaced with aspartate residues. The enzymatic assay results are shown in the table.

| Enzyme Form | Activity (U/mg) |
|---|---|
| Native enzyme | 1,000 |
| Recombinant Lys 21 $\longrightarrow$ Asp 21 | 970 |
| Recombinant Lys 86 $\longrightarrow$ Asp 86 | 100 |
| Recombinant Lys 101 $\longrightarrow$ Asp 101 | 970 |

(a) What do you infer about the role(s) of Lys 21, 86, and 101 in the catalytic mechanism of the native enzyme?

(b) Speculate on the location of Lys 21 and Lys 101. Would you expect these residues to be conserved in an evolutionary sense?

(c) Would you expect Lys 86 to be evolutionarily conserved? Why or why not?

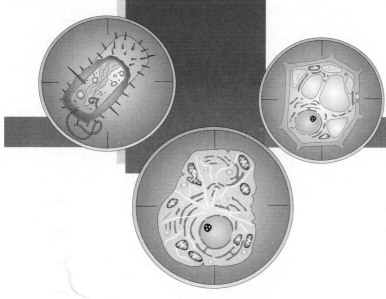

# Regulation of Enzyme Activities

*A select group of enzymes are regulated by one of two strategies: modification of the covalent structure of the enzyme or modification of their structure by the reversible binding of effector molecules.*

The living cell resembles a complex factory with many different assembly lines of worker enzymes performing specific tasks. The activities of these workers have to be regulated precisely so that resources are used efficiently and a steady flow of parts keeps all the assembly lines moving smoothly. Cells use two basic strategies for regulating their enzyme activities. The first strategy is to adjust the amount and location of key enzymes. This requires mechanisms for the control of synthesis, degradation, and transport of proteins, which we discuss in Chapters 29, 30 and 31. The second strategy is to regulate the activities of the enzymes that are on hand. We deal exclusively with the second strategy in this chapter.

Enzymes subject to regulation are a select few of the total enzymes in a cell and are so subject for two reasons. First, it is not necessary to regulate the activity of every enzyme to achieve the desired level of control. In many cases, an entire metabolic pathway can be controlled by regulating only the enzyme that catalyzes the first step in the pathway (see chapter 11). Second, elaborate structural properties are required to create an enzyme that can be regulated.

In the first part of this chapter we survey the different methods that cells use to regulate enzyme activities. Then we consider in detail how these methods apply to three of the most thoroughly characterized regulatory enzymes.

In principle, the activities of many enzymes can be altered by changes in pH. Cells do take advantage of this possibility in a few cases. Lysozyme, for example, is most active in the pH 5 region, which is characteristic of some extracellular secretions, and is much less active in the intracellular pH region near 7. The activity of lysozyme thus remains low until the enzyme is secreted. But this is not a very practical solution to the problem of regulating the activities of intracellular enzymes, because most cells must hold their pH within narrow limits. Two strategies are much more widely applicable. The first is to modify the covalent structure of the enzyme in such a way as to alter either $K_m$ or $k_{cat}$. The second is to use an inhibitor or activator, an *effector,* that binds reversibly to the enzyme and, again, alters either the $K_m$ or $k_{cat}$. Such an effector may bind either at the active site itself or at some more distant site on the enzyme. In the latter case, it is termed an allosteric effector (from the Greek words *allos,* meaning ''other,'' and *stereos,* meaning ''space,'' and enzymes that are regulated by such effectors are called allosteric enzymes. Although allosteric effectors usually are small molecules such as ATP, some proteins are inhibited or activated when they bind to other proteins.

## Partial Proteolysis Results in Irreversible Covalent Modifications

In chapter 8, we mentioned that the pancreas secretes trypsin, chymotrypsin, and elastase as inactive zymogens, which are activated by extracellular proteases. Trypsin is activated when the intestinal enzyme enteropeptidase cuts off an N-terminal hexapeptide. Trypsin in turn activates chymotrypsin by cutting it at the N-terminal end between Arg 15 and Ile 16. This type of change in the covalent structure of an enzyme is termed partial proteolysis. Delaying the activation prevents the digestive enzymes from destroying the pancreatic cells in which they are synthesized.

The crystal structures of chymotrypsin and its zymogen precursor *chymotrypsinogen* suggest an explanation for the increase in catalytic activity that results from trimming off the N-terminal end of the zymogen. Although the catalytic triad of Asp 102, His 57, and Ser 195 has a similar structure in chymotrypsinogen and chymotrypsin, the substrate-binding pocket is not properly formed in the zymogen (fig. 9.1*a*). The NH group of Gly 193 is not in position to form a hydrogen bond with the carbonyl oxygen of the sub-

**Figure 9.1**

A schematic drawing of the structural changes that occur when chymotrypsinogen (*a*) is converted to chymotrypsin (*b*). In chymotrypsinogen, the carboxylate group of Asp 194 forms a salt bridge to His 40; in chymotrypsin, the bridge goes to the new N terminal, the —NH$_3^+$ group of Ile 16. This change evidently allows Gly 193 to swing around so that its amide NH comes closer to the NH of Ser 195. The two amide groups form essential hydrogen bonds to the substrate in the enzyme–substrate complex (see figs. 8.7 and 8.11).

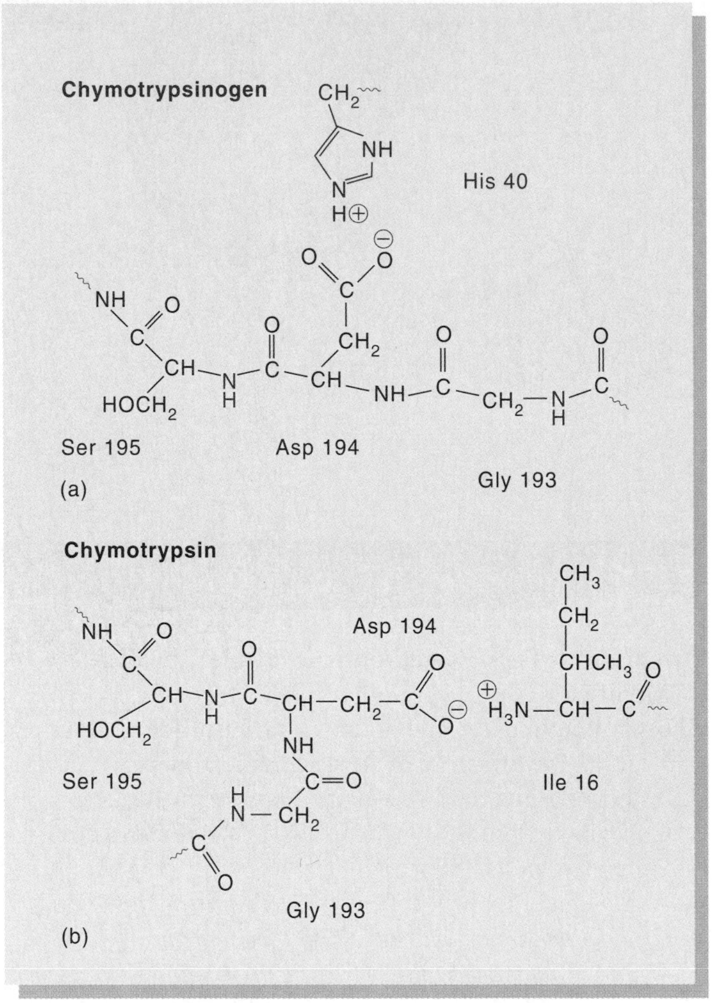

strate. The importance of this bond for the activity of the enzyme was discussed in chapter 8 (see figs. 8.7 and 8.11). The major constraint preventing the completion of the binding pocket appears to be a hydrogen bond between the carboxylate group of the neighboring residue, Asp 194, and His 40. When the zymogen is cleaved, the new N-terminal —NH$_3^+$ group of Ile 16 forms a salt bridge with Asp 194, thus allowing Gly 193 to rotate into the correct orientation (fig. 9.1*b*). A similar bridge between the N-terminal group and Asp 194 occurs in trypsin.

## Figure 9.2

The blood coagulation cascade. Each of the curved red arrows represents a proteolytic reaction, in which a protein is cleaved at one or more specific sites. With the exception of fibrinogen, the substrate in each reaction is an inactive zymogen; except for fibrin, each product is an active protease that proceeds to cleave another member in the series. Many of the steps also depend on interactions of the proteins with $Ca^{2+}$ ions and phospholipids. The cascade starts when factor XII and prekallikrein come into contact with materials that are released or exposed in injured tissue. (The exact nature of these materials is still not fully clear.) When thrombin cleaves fibrinogen at several points, the trimmed protein (fibrin) polymerizes to form a clot.

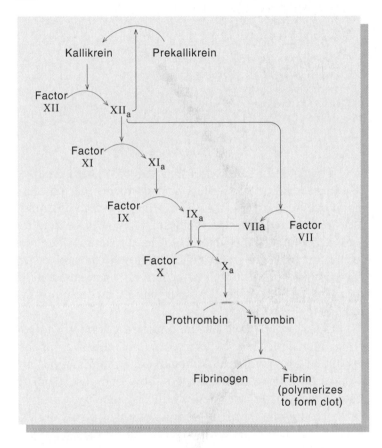

Table 9.1 lists some of the other enzymes activated by partial proteolysis. A common pattern is for proteolysis to occur in a loop that connects two different domains of the protein, relieving a constraint that interferes with the formation of the active site.

The activation of an enzyme by partial proteolysis is an irreversible process. Once the enzyme is cut, it remains active until it is degraded or inhibited by some other means. This is fine for the digestive enzymes, but it clearly raises a problem in blood coagulation: A mechanism must exist to prevent the clot from spreading away from the site of the injury and taking over the entire blood supply. Several mechanisms work to this end. The activated enzymes are diluted by the flow of the blood, degraded in the liver, and inhibited by other blood proteins that bind tightly to the active enzymes and occlude their active sites. Blood coagulation thus depends on a delicate balance between a rapid, irreversible mechanism of enzyme activation, and rapid, irreversible mechanisms for disposing of the active enzymes.

The enzymes that participate in blood clotting also are activated by partial proteolysis, which again serves to keep them in check until they are needed. The blood coagulation system involves a cascade of at least seven serine proteases, each of which activates the subsequent enzyme in the series (fig. 9.2). Because each molecule of activated enzyme can, in turn, activate many molecules of the next enzyme, initiation of the process by factors that are exposed in damaged tissue leads explosively to the conversion of prothrombin to thrombin, the final serine protease in the series. Thrombin then cuts another protein, fibrin, into peptides that stick together to form a clot.

## Phosphorylation, Adenylylation, and Disulfide Reduction Lead to Reversible Covalent Modifications

A type of covalent modification used more widely than partial proteolysis is phosphorylation of the side chains of serine, threonine, or tyrosine residues. Phosphorylation differs from partial proteolysis in being reversible. The introduction and removal of the phosphate group are catalyzed by separate enzymes (phosphorylation by a protein kinase, and dephosphorylation by a phosphatase), which are themselves usually under metabolic regulation (fig. 9.3).

**Figure 9.3**

Phosphorylations of serine side chains in enzymes are catalyzed by kinases, and dephosphorylation by phosphatases. Threonine and tyrosine side chains undergo similar reactions, but these are less common than the phosphorylation of serine.

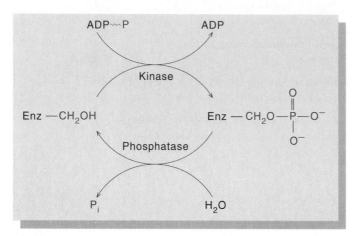

**Table 9.2**

### Some Enzymes Regulated by Phosphorylation

***Enzymes of Carbohydrate Metabolism***
Glycogen phosphorylase
Phosphorylase kinase
Glycogen synthase
Phosphofructokinase-2
Pyruvate kinase
Pyruvate dehydrogenase

***Enzymes of Lipid Metabolism***
Hydroxymethylglutaryl-CoA reductase
Acetyl-CoA carboxylase
Triacylglycerol lipase

***Enzymes of Amino Acid Metabolism***
Branched-chain ketoacid dehydrogenase
Phenylalanine hydroxylase
Tyrosine hydroxylase

In eukaryotic organisms, phosphorylation is used to control the activities of literally hundreds of enzymes, a few of which are listed in Table 9.2. These enzymes are phosphorylated or dephosphorylated in response to extracellular signals, such as hormones or growth factors. The adrenal hormone epinephrine, for example, is transmitted through the blood to muscle and adipose tissue when there is a need for muscular exertion. On reaching the target tissue, epinephrine initiates a chain of events that leads to the activation of a protein kinase. The kinase [cyclic adenosine monophosphate (cAMP)-dependent protein kinase] catalyzes the phosphorylation of one or more specific enzymes, depending on the tissue. Some enzymes are activated when they are phosphorylated and inactivated when the phosphate is removed; others are inactivated by phosphorylation. In adipose tissue, phosphorylation activates triacylglycerol lipase, an enzyme that breaks down esters of fatty acids. In muscle, phosphorylation activates glycogen phosphorylase, an enzyme that breaks down glycogen, and it stops the synthesis of glycogen by switching off glycogen synthase.

Phosphorylation also can modify an enzyme's sensitivity to allosteric effectors. Phosphorylation of glycogen phosphorylase reduces its sensitivity to the allosteric activator adenosine monophosphate (AMP). Thus, a covalent modification triggered by an extracellular signal can override the influence of intracellular allosteric regulators. In other cases, variations in the concentrations of intracellular effectors can modify the response to the covalent modification, depending on the metabolic state of affairs in the cell.

The covalent addition of an adenylyl (AMP) group to a tyrosine residue is another form of reversible, covalent modification (fig. 9.4). In *Escherichia coli,* adenylylation is used to regulate glutamine synthase, a key enzyme in nitrogen metabolism (see chapter 21). The tyrosine residue that accepts the adenylyl group is located close to the active site. The addition of the bulky and negatively charged adenylyl group inhibits the enzyme, perhaps simply by occluding the active site.

Other groups that can be attached covalently to enzymes include fatty acids, isoprenoid alcohols such as farnesol, and carbohydrates. Although such modifications are widespread, our understanding of how cells use them to regulate enzymatic activities is still fragmentary.

A reversible covalent modification that plants use extensively is the reduction of cystine disulfide bridges to sulfhydryls. Many of the enzymes of photosynthetic carbohydrate synthesis are activated in this way (table 9.3). Some of the enzymes of carbohydrate breakdown are inactivated by the same mechanism. The reductant is a small protein called "thioredoxin," which undergoes a complementary oxidation of cysteine residues to cystine (fig. 9.5). Thioredoxin itself is reduced by electron-transfer reactions driven by sunlight, which serves as a signal to switch carbohydrate metabolism from carbohydrate breakdown to synthesis. In one of the regulated enzymes, phosphoribulokinase, one of the freed cysteines probably forms part of the catalytic active site. In nicotinamide-adenine dinucleotide phosphate (NADP)-malate dehydrogenase and fructose-1,6-bis-

## Figure 9.4

The transfer of the adenylyl group from ATP to a tyrosine residue is used to regulate some enzymes. The other product of the adenylylation reaction *(top)* is inorganic pyrophosphate. Hydrolysis of the adenylyl-tyrosine ester bond releases AMP *(bottom).*

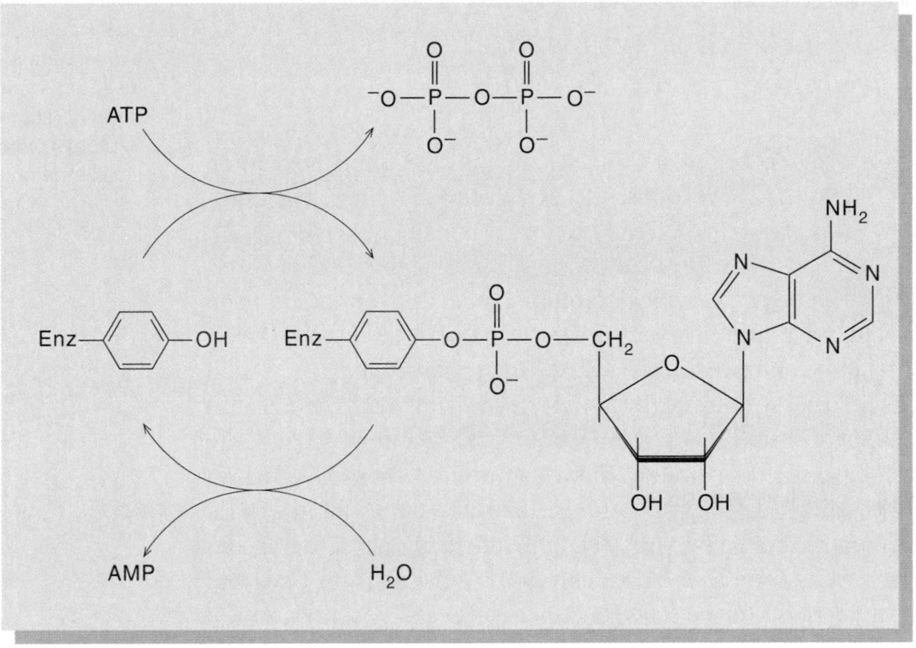

## Table 9.3

### Some Enzymes Regulated by Disulfide Reduction in Plants

**Activated by Reduction**
Fructose-1,6-bisphosphatase
Sedoheptulose-1,7-bisphosphatase
Glyceraldehyde-3-phosphate dehydrogenase
NADP-malate dehydrogenase
Phosphoribulokinase
Thylakoid ATP-synthase

**Inhibited by Reduction**
Phosphofructokinase

## Figure 9.5

Reduction of the disulfide bond of cystine is used to activate enzymes of photosynthetic carbohydrate biosynthesis in plants. The reductant is a small protein called thioredoxin. Thioredoxin also serves as a reductant for the biosynthesis of deoxynucleotides in animals and microorganisms as well as in plants.

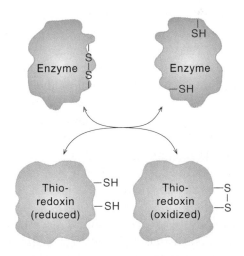

phosphatase, the cysteines both appear to be some distance from the catalytic site. How the structural changes that result from the reduction are transmitted to the active sites of these enzymes is not yet known.

Covalent modifications of enzymes allow a cell to regulate its metabolic activities more rapidly and in much more intricate ways than is possible by changing the absolute concentrations of the same enzymes. They still do not provide truly instantaneous responses to changes in conditions, however, because each modification requires the action of another enzyme, which must itself be regulated. A lag also occurs in responding to the removal or inhibition of the enzyme that causes the modification, because reversing the modification requires still another enzyme.

## Allosteric Regulation Allows an Enzyme to Be Controlled Rapidly by Materials that Are Structurally Unrelated to the Substrate

Whereas eukaryotic organisms use phosphorylation to handle responses to extracellular signals, both prokaryotic and eukaryotic organisms commonly use allosteric regulation in responding to changes in conditions within a cell. A typical circumstance that might demand such a response is a surplus or deficit of ATP or some other metabolic intermediate. Allosteric regulation enables a cell to adjust an enzymatic activity almost instantaneously in response to changes in the concentration of a metabolite because, unlike covalent modification, it does not require an intermediate enzyme. If the metabolite acts as an allosteric effector, the activity of the enzyme can increase (or decrease) as soon as the concentration of the effector rises and can decrease (or increase) again as soon as the concentration falls.

Regulation of enzymes by allosteric effectors is considerably more common than regulation by compounds that bind at the active site for two reasons: First, whereas an agent that binds at the active site usually acts as an inhibitor, a compound that binds at an allosteric site can serve as either an inhibitor or an activator, depending on the structure of the enzyme. Second, a substance that binds to an allosteric site does not need to have any structural relationship to the substrate. Consider, for example, the metabolic pathway of histidine biosynthesis in plants and bacteria. The pathway requires nine enzymes that work one after another. If histidine is already present in abundance, it is advantageous for a cell to cut off the entire pathway at the first step to avoid wasting energy or accumulating the products of the intermediate steps. The first step in the pathway is the reaction between phosphoribosylpyrophosphate and adenosine triphosphate (ATP), neither of which even vaguely resembles histidine, and yet the enzyme that catalyzes this step is strongly inhibited by histidine. Binding of histidine to an allosteric site causes a structural change that is transmitted to the active site.

Inhibition of the initial step of a biosynthetic pathway by an end product of the pathway is a recurrent theme in metabolic regulation. In addition, many key enzymes are regulated by ATP, adenosine diphosphate (ADP), AMP, or inorganic phosphate ion ($P_i$). The concentrations of these materials provide a cell with an index of whether energy is abundant or in short supply. Because ATP, ADP, AMP, or $P_i$ often are chemically unrelated to the substrate of the enzyme that must be regulated, they usually bind to an allosteric site rather than to the active site.

### Figure 9.6

Kinetics of the reaction catalyzed by phosphofructokinase. In the presence of 1.5 mM ATP, the rate has a sigmoidal dependence on the concentration of the substrate fructose-6-phosphate. Although ATP also is a substrate for the reaction, the sigmoidal kinetics seen under these conditions are associated with the binding of ATP to an inhibitory allosteric site. The kinetics become hyperbolic if a low concentration of AMP is added.

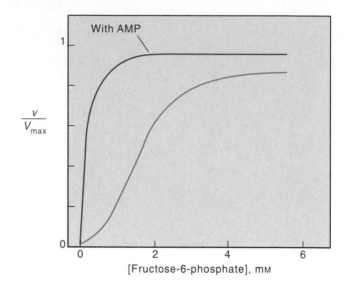

## Allosteric Enzymes Typically Exhibit a Sigmoidal Dependence on Substrate Concentration

In chapter 5 we saw that the binding of $O_2$ to any one of the four subunits of hemoglobin increases the affinity of the other subunits for $O_2$. This effect reflects cooperative changes in the tertiary and quaternary structure of the protein. Binding of $O_2$ alters the interactions among the subunits in such a way that the entire protein tends to flip into a state with increased $O_2$ affinity; binding of glycerate-2,3-bisphosphate favors a transition in the opposite direction. When Jacques Monod, Jeffreys Wyman, and Jean-Pierre Changeux first advanced the idea of allosteric enzymes in 1963, they suggested that these enzymes might contain multiple subunits and that the changes in catalytic activity caused by allosteric effectors can reflect alterations in quaternary structure. This suggestion turned out to be remarkably accurate. Although there is no reason why an enzyme consisting of a single subunit cannot be sensitive to allosteric effectors, most of the enzymes regulated in this way do have multiple subunits, and the changes in activity often can be related to interactions among the subunits.

One indication of the importance of intersubunit interactions in allosteric enzymes is that many such enzymes do not obey the classical Michaelis-Menten kinetic equation. A plot of the rate of reaction as a function of substrate concentration is not hyperbolic, as described by the Michaelis-Menten equation, but rather sigmoidal, resembling the curve for the binding of $O_2$ to hemoglobin (see fig. 5.2). Furthermore, allosteric effectors often cause the kinetics to change from one of these forms to the other, much as glycerate-2,3-bisphosphate affects the degree of cooperativity in the binding of $O_2$ to hemoglobin. Figure 9.6 shows an illustration of these effects for phosphofructokinase, which catalyzes the formation of fructose-1,6-bisphosphate from fructose-6-phosphate and ATP.

$$\text{Fructose-6-phosphate + ATP} \xrightarrow{\text{phosphofructokinase}}$$
$$\text{Fructose-1,6-bisphosphate + ADP} \quad \textbf{(1)}$$

In the presence of 1.5 mM ATP, the kinetics have a sigmoidal dependence on the concentration of fructose-6-phosphate; at very low concentrations of ATP, or in the presence of AMP, the kinetics become hyperbolic.

Phosphofructokinase has four identical subunits. To explore how sigmoidal kinetics can arise in such an enzyme, consider an enzyme that has just two such subunits, each with its own catalytically active site. We can schematize the binding of substrate to the enzyme as follows:

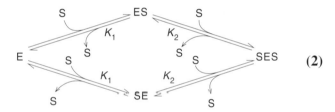

$$\textbf{(2)}$$

Here ES and SE represent complexes of the enzyme (E) with substrate (S) on the two different subunits, and SES represents the complex in which both binding sites are occupied. The two dissociation constants $K_1$ and $K_2$ are not necessarily identical because the binding of substrate to one subunit can affect the dissociation constant for the other subunit. In fact, this is just the point we want to explore.

Let's assume that the rate constant $k_{cat}$ for the formation of products on either subunit is the same, whether only that site or both catalytic sites are occupied. Suppose also that ES, SE, and SES are in equilibrium with the free enzyme and substrate. By following the same procedure that led to the Henri-Michaelis-Menten equation in chapter 7, we can derive an expression for the rate of the enzymatic reaction in terms of [S], $K_1$, and $K_2$. Here we just give the result.

$$v = \frac{V_{max}\dfrac{[S]}{K_1}\left(1 + \dfrac{[S]}{K_2}\right)}{\left(1 + \dfrac{[S]}{K_1}\right) + \dfrac{[S]}{K_1}\left(1 + \dfrac{[S]}{K_2}\right)} \quad \textbf{(3)}$$

**Figure 9.7**

Kinetics of an enzyme with two identical subunits. The curves were calculated with equation (3). The abscissa is the ratio of the substrate concentration [S] to $K_d$, where $K_d = \sqrt{K_1 K_2}$ and $K_1$ and $K_2$ are the dissociation constants for the first and second molecule of substrate. $K_d$ is taken to be the same for all three curves. For the *solid curve*, $K_1$ and $K_2$ are assumed to be identical ($K_1 = K_2 = K_d$); equation (3) then reduces to equation (4). For the *dashed curve*, $K_2 = K_1/25$ ($K_1 = 5K_d$ and $K_2 = K_d/5$). For the *dotted curve*, $K_2 = 10^{-4}K_1$ ($K_1 = 100K_d$ and $K_2 = K_d/100$); equation (3) then reduces to equation (5).

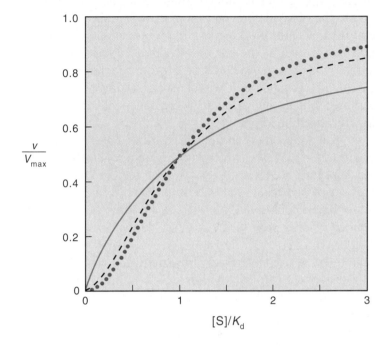

If the binding of S to one subunit does *not* affect binding to the other, then $K_2 = K_1$, and equation (3) reduces to

$$v = \frac{V_{max}[S]}{K_s + [S]} \quad \textbf{(4)}$$

where $K_s = K_1 = K_2$. This is simply the Henri-Michaelis-Menten equation [equation (18) in chapter 7]. A plot of $v$ versus (S) according to equation (4) is hyperbolic. Such a plot is shown as the solid curve in figure 9.7. The dashed curve in figure 9.7 is a plot of equation (3) when $K_2 = K_1/25$. To facilitate comparison with the solid curve, $K_1$ is taken to be $5K_d$ and $K_2$ to be $K_d/5$, where $K_d$ is the same for the two curves. This means that the overall $\Delta G°$ for the formation of SES from E + 2S is the same. (The overall equilibrium constant is $(1/K_1)(1/K_2)$, or $1/K_d^2$.) Note that the dashed curve is sigmoidal in shape. It starts out rising more slowly than the solid curve, rises steeply in the region of $[S] \approx K_d$, where $v/V_{max} \approx 0.5$, and then continues rising more slowly.

A ratio of 25 between $K_1$ and $K_2$ means that the $\Delta G°$ for binding the second molecule of substrate is only

1.9 kcal/mol more favorable than the $\Delta G°$ for binding the first molecule. This is the order of magnitude of the $\Delta G°$ for forming a single hydrogen bond. Evidently, sigmoidal kinetics might be obtained if binding the substrate causes even a relatively minor change in the conformation of the enzyme. In chapter 8, we noted that binding of glucose to hexokinase results in a pronounced conformational change that can be seen in the crystal structure. But to yield sigmoidal kinetics, it is essential that events occurring at two different binding sites be linked: Binding at one site must decrease the dissociation constant at the other site. If no such coupling occurs, the kinetics follow equation (4) and are hyperbolic even if the enzyme has multiple subunits. This situation is observed with many enzymes, including one we discussed in chapter 8. Triosephosphate isomerase exists as a dimer, but its two subunits work independently, and its kinetics are hyperbolic.

The dotted curve in figure 9.7 shows a plot of equation 6 with $K_2 = 10^{-4} K_1$, which is equivalent to a difference of 5.5 kcal/mol between the $\Delta G°$ values for binding the first and second molecules of substrate. When the ratio of the dissociation constants is this large, the kinetics can be described equally well by the simpler expression

$$v = \frac{V_{max}[S]^2}{K_1 K_2 + [S]^2} \tag{5}$$

This is the limiting form of equation (3) when $K_2 \ll [S] \ll K_1$.

It is not uncommon for enzymes that are regulated allosterically to have four or even more subunits. The active form of acetyl-coenzyme A (acetyl-CoA) carboxylase consists of linear strings of 10 or more identical subunits (see fig. 18.11). In general, the larger the number of subunits that interact such that the binding of substrate to one subunit promotes binding to others, the more steeply the enzyme's kinetics depend on [S] in the region where $v/V_{max} \approx 0.5$. To derive a more general form of equation (5), consider a protein E with $n$ identical subunits, each of which has a binding site for a substrate S. Suppose that binding of the first molecule of S strongly favors binding to all $n$ subunits, giving the overall reaction

$$E + nS \rightleftharpoons ES_n \tag{6}$$

The concentration of $ES_n$ then is

$$[ES_n] \approx \frac{[E][S]^n}{K_h} \tag{7}$$

where $K_h$ is the product of the individual dissociation constants for all $n$ steps leading to $ES_n$. The fraction of the protein that has taken up the substrate is

$$y \approx \frac{[ES_n]}{[E] + [ES_n]} = \frac{[S]^n}{K_h + [S]^n} \tag{8}$$

This is called the Hill equation, and the exponent $n$ is the Hill coefficient.

Equation (8) is an approximation because it ignores intermediate species that have some, but not all, of the binding sites occupied. Even so, the Hill coefficient provides a useful measure of cooperativity. The binding of $O_2$ to hemoglobin is described well by the Hill equation with $n \approx 2.8$. In the case of phosphofructokinase, which has four subunits, the dependence of the rate on the fructose-6-phosphate concentration at a fixed, relatively high concentration of ATP is described well with $n \approx 3.8$.

## The Symmetry Model Provides a Useful Framework for Relating Conformational Transitions to Allosteric Activation or Inhibition

Several theoretical models have been developed for relating changes in dissociation constants to conformational changes in oligomeric proteins. We discussed two such models in connection with the oxygenation of hemoglobin (see fig. 5.14), and one of them is shown again in a slightly different form in figure 9.8. The basic idea is that the protein's subunits can exist in either of two conformations: R ("relaxed") and T ("tight," or "tense"). The substrate is assumed to bind more tightly to the R form than to the T form, which implies that binding of the substrate favors the transition from T to R. Conformational transitions of the individual subunits are assumed to be tightly linked, so that if one subunit flips from T to R, the others must do the same. Binding of the first molecule of substrate thus promotes the binding of a second. Because the concerted transition of all of the subunits from T to R or back preserves the overall symmetry of the protein, this model is called the "symmetry" model. The symmetry model can account for the behavior of allosteric activators if these compounds also bind preferentially to the R state and thus stabilize the conformation that is more effective at binding the substrate. Allosteric inhibitors act by stabilizing the T state.

Using the symmetry model, the fraction of the binding sites occupied at any given substrate concentration can be described with an expression that includes the substrate dissociation constants for the two conformations ($K_R$ and $K_T$) and the equilibrium constant between the T and R conformations in the absence of substrate, $L = [T]/[R]$. Thus, the symmetry model attempts to explain the difference between $K_1$ and $K_2$ in equation (3) by introducing a third independent parameter. Considering that equation (3) can fit the experimental data for a dimeric enzyme with only two pa-

## Figure 9.8

Symmetry model for allosteric transitions of a dimeric enzyme. The model assumes that the enzyme can exist in either of two different conformations (T and R), which have different dissociation constants for the substrate ($K_T$ and $K_R$). Structural transitions of the two subunits are assumed to be tightly coupled, so that both subunits must be in the same state. $L$ is the equilibrium constant (T)/(R) in the absence of substrate. If the substrate binds much more tightly to R than to T ($K_R \ll K_T$), the binding of a molecule of substrate to either subunit pulls the equilibrium between T and R in the direction of R. Because both subunits must go to the R state, the binding of the second molecule of substrate is promoted.

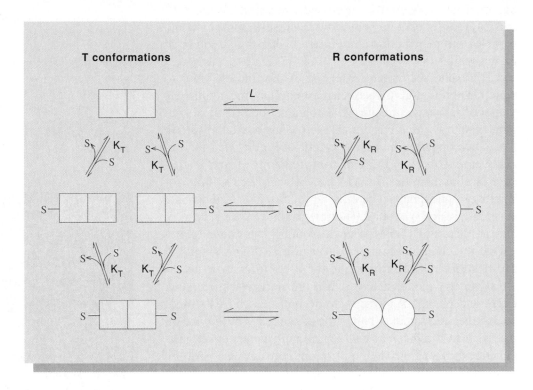

rameters, what do we gain by using a more complicated equation? The usefulness of the symmetry model is that it can be generalized to enzymes that have a larger number of subunits. The general expression still contains only four independent parameters: $K_R$, $K_T$, $L$, and the total number of subunits ($n$). But this simplicity comes with a cost: We have to assume that the entire oligomeric protein has just two different conformational states. Is it not more realistic to assume that the protein also can adopt a variety of intermediate conformations? The answer depends in part on the particular protein and in part on our goals in using any type of model. Models that allow intermediate conformations have been developed, but they of course require more independent parameters. The ''sequential'' model shown in figure 5.14 is one of these more elaborate models. Although the sequential model is more general than the symmetry model and may be more realistic, the available experimental data in most cases do not justify a distinction between the two.

The symmetry model is useful even if it does oversimplify the situation, because it provides a conceptual framework for discussing the relationships between conformational transitions and the effects of allosteric activators and inhibitors. In the following sections we consider three oligomeric enzymes that are under metabolic control and see that substrates and allosteric effectors do tend to stabilize each of these enzymes in one or the other of two distinctly different conformations.

## Phosphofructokinase: Allosteric Control of Glycolysis Is Consistent with the Symmetry Model

Phosphofructokinase catalyzes the transfer of a phosphate group from ATP to fructose-6-phosphate (equation 1). This is a major site of regulation of glycolysis, the metabolic pathway by which glucose breaks down to pyruvate (see chapter 12). As we saw in figure 9.6, the kinetics of phosphofructokinase are strongly cooperative with respect to fructose-6-phosphate. The kinetics are noncooperative with respect to the other substrate, ATP, at low concentrations, but at concentrations above 0.5 mM ATP acts as an inhibitor. The inhibitory effect results from binding to an allosteric site that is distinct from the substrate-binding site for ATP. Phosphofructokinase also is inhibited by phosphoenolpyruvate and by citrate, an intermediate in two other metabolic pathways that embark from pyruvate. On the other hand, it is stimulated by ADP, AMP, guanosine diphosphate (GDP), cAMP, fructose-2,6-bisphosphate, and a variety of other compounds. Most if not all of the activators bind to the same allosteric site as ATP and may work simply by preventing the inhibitory effect of ATP. The effects of ATP, ADP, and AMP are such that phosphofructokinase is restrained when the cell's needs for energy have been satisfied, as reflected in a high ratio of [ATP] to [ADP] and [AMP], and is unleashed when the cell needs additional energy. The effects of cAMP and fructose-2,6-bisphosphate relate to hormonal control of carbohydrate metabolism in higher organisms, and is discussed further in chapters 12 and 24. Our focus here is on how fructose-6-phosphate and some of the major allosteric effectors alter the structure of the enzyme.

Phosphofructokinase was one of the first enzymes to which Monod and his colleagues applied the symmetry model of allosteric transitions. It contains four identical subunits, each of which has both an active site and an allosteric site. The cooperativity of the kinetics suggests that the enzyme can adopt two different conformations (T and R) that have similar affinities for ATP but differ in their affinity for fructose-6-phosphate. The binding for fructose-6-phosphate is calculated to be about 2,000 times tighter in the R conformation than in T. When fructose-6-phosphate binds to any one of the subunits, it appears to cause all four subunits to flip from the T conformation to the R conformation, just as the symmetry model specifies. The allosteric effectors ADP, GDP, and phosphoenolpyruvate do not alter the maximum rate of the reaction but change the dependence of the rate on the fructose-6-phosphate concentration in a manner suggesting that they change the equilibrium constant ($L$) between the T and R conformations.

Philip Evans and his coworkers have determined the crystal structures of phosphofructokinase from two species of bacteria, *E. coli* and *Bacillus stearothermophilus*. By crystallizing the enzyme in the presence and absence of the substrate and several allosteric effectors, they obtained detailed views of both the T and R conformations. This work has led to an explanation of why phosphofructokinase appears to be constrained largely to all-or-nothing transitions between these two states, rather than adopting a series of intermediate conformations.

Figure 9.9 shows the crystal structure of two of the subunits of phosphofructokinase from *B. stearothermophilus*. In the complete enzyme, the subunits are disposed symmetrically about three mutually perpendicular axes. Each of these axes is a twofold symmetry axis, which means that rotating the entire structure by half of a full circle (180°) around the symmetry axis results in an identical structure. This rotation is shown diagramatically in figure 9.10. Because ADP is a product of the enzymatic reaction as well as an allosteric activator, it binds at both the catalytic and allosteric sites (see figs. 9.9 and 9.10). The catalytic site for fructose-6-phosphate in each subunit is at the interface of the subunit with one of its neighbors, and the allosteric site is at the interface with a different neighbor.

In the transition between the T and R conformations, the four subunits rotate by about 7° with respect to each other (see fig. 9.10). This rotation is associated with coupled rearrangements of the structures at the interfaces between adjacent subunits. Figure 9.11 shows how these rearrange-

## Figure 9.9

Computer-generated structure of two of the four subunits of phosphofructokinase from *Bacillus stearothermophilus*. The enzyme, shown as yellow and light blue tubes, was crystallized in the R conformation in the presence of the substrate fructose-6-phosphate (dark blue) and the allosteric activator ADP (pink). The magnesium ions (white/silver spheres), $Mg^{2+}$, bound to the ADP molecules are also shown. (Copyright 1994 by the Scripps Research Institute/Molecular Graphics Images by Michael Pique using software by Yng Chen, Michael Connolly, Michael Carson, Alex Shah, and AVS, Inc. Visualization advice by Holly Miller, Wake Forest University Medical Center.)

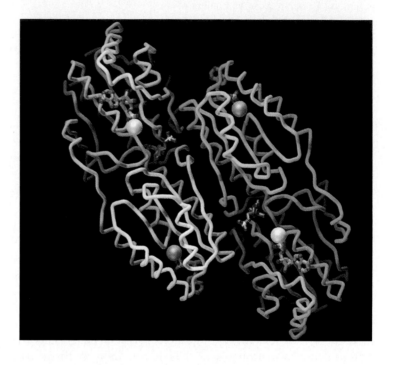

ments affect the binding site for fructose-6-phosphate. The most significant structural change in this region is an inversion of the orientation of the side chains of Glu 161 and Arg 162. In the R conformation, Arg 162 forms a hydrogen bond to the phosphate group of fructose-6-phosphate, whereas Glu 161 points in the opposite direction. In the T conformation, Arg 162 points away from the binding site, and Glu 161 inserts a negative charge into the site, where it forms a hydrogen bond with Arg 243. The change in the orientations of the negatively charged Glu 161 and the positively charged Arg 162 probably accounts for most of the difference between the dissociation constants for fructose-6-phosphate in the two states. Note that although figure 9.11*b* shows the molecule of fructose-6-phosphate bound on subunit A, Glu 161 and Arg 162 are residues of subunit D. Arginines 252 and 243 also contribute to the binding site from opposite sides of the boundary. Structural changes that occur on one of the subunits thus are intricately linked to changes on the other. The substrate-binding site for ATP, on

## Figure 9.10

Outlines of phosphofructokinase in the T (solid lines) and R (dashed lines) conformations. The enzyme contains four identical subunits (A, B, C, and D). The locations of the catalytic and allosteric sites are indicated in the two subunits closest to the viewer (A and D). The binding sites for fructose-6-phosphate (F6P) are at the interface of these subunits; the allosteric sites are at the interfaces of A with B, and of D with C. Two of the three perpendicular symmetry axes are labeled p and q. A 180° rotation about axis q interchanges the positions of subunits A and C and also interchanges B and D. A similar rotation about p interchanges A with B, and C with D. (Source: From T. Schirmer and P. R. Evans, Structural basis of the allosteric behaviour of phospho-fructokinase, *Nature* 343:140, 1990.)

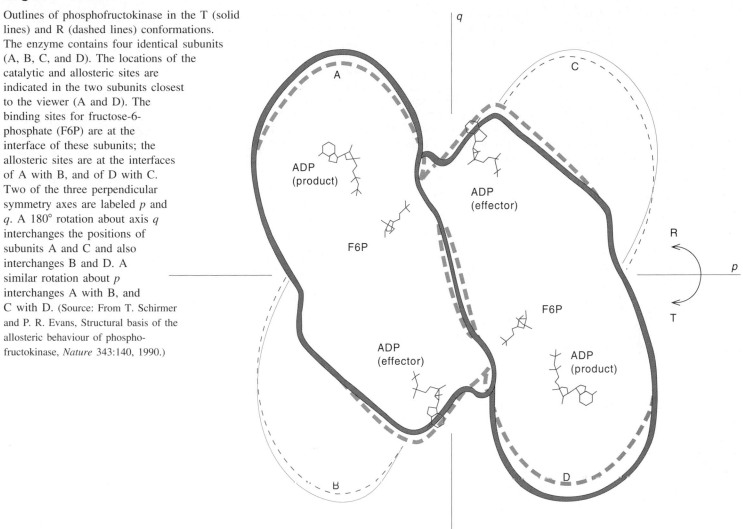

the other hand, is made up of residues from only one subunit (subunit A in figure 9.11). This probably explains why the binding of fructose-6-phosphate to the enzyme is strongly cooperative, whereas the binding of ATP as a substrate is not cooperative.

In the T conformation, Glu 161 and Arg 162 are located at the end of a stretch of polypeptide that winds up into a helical turn in the transition to the R structure (see fig. 9.11). This coiling is linked to a major structural change in an adjacent region of the interface between subunits A and D. The interface here includes a pair of antiparallel β strands, each of which is hydrogen-bonded to a parallel strand in its own subunit, as shown in figure 9.12. In the T conformation (fig. 9.12a), the β strands from the different subunits are hydrogen-bonded together directly across the interface. In the R conformation (fig. 9.12b), the strands have moved apart, and the region between them is filled by a row of hydrogen-bonded water molecules.

The insertion of water at the interface between subunits A and D is an essential component of the rotation of the subunits with respect to each other, and it appears to be an all-or-nothing effect. Intermediate conformations in which only some of the water molecules are present would have a less extensive network of hydrogen bonds and thus are probably less stable than either the R or the T conformation. The same might be said of the winding of the helical turn between residues 155 and 161; intermediates in which the helical turn is partially unwound probably would be destabilized by steric crowding or force the structure to expand in a way that leaves empty spaces in other regions. These considerations, taken with the close coupling between the individual subunits, seem to explain why all of the subunits undergo concerted transitions from the R to the T state or back, without giving appreciable concentrations of intermediate states.

Although the crystal structures offer a very plausible

## Figure 9.11

Interface between subunits A and D of phosphofructokinase near the catalytic site in (a) the T and (b) the R structures. Crystals of the enzyme in the R state were obtained in the presence of fructose-6-phosphate and ADP (see fig. 9.9); crystals in the T state were obtained in the presence of a nonphysiological allosteric inhibitor, 2-phosphoglycolate. The wavy green line represents part of the boundary between subunits A and D. The heavy green line indicates the polypeptide backbone. The side chains of Glu 161 and Arg 162 are shown in red. Note the inversion of the positions of these side chains in the two structures. (Source: From T. Schirmer and P. R. Evans, Structural basis of the allosteric behaviour of phosphofructokinase, *Nature* 343:140, 1990.)

(a) T state

(b) R state

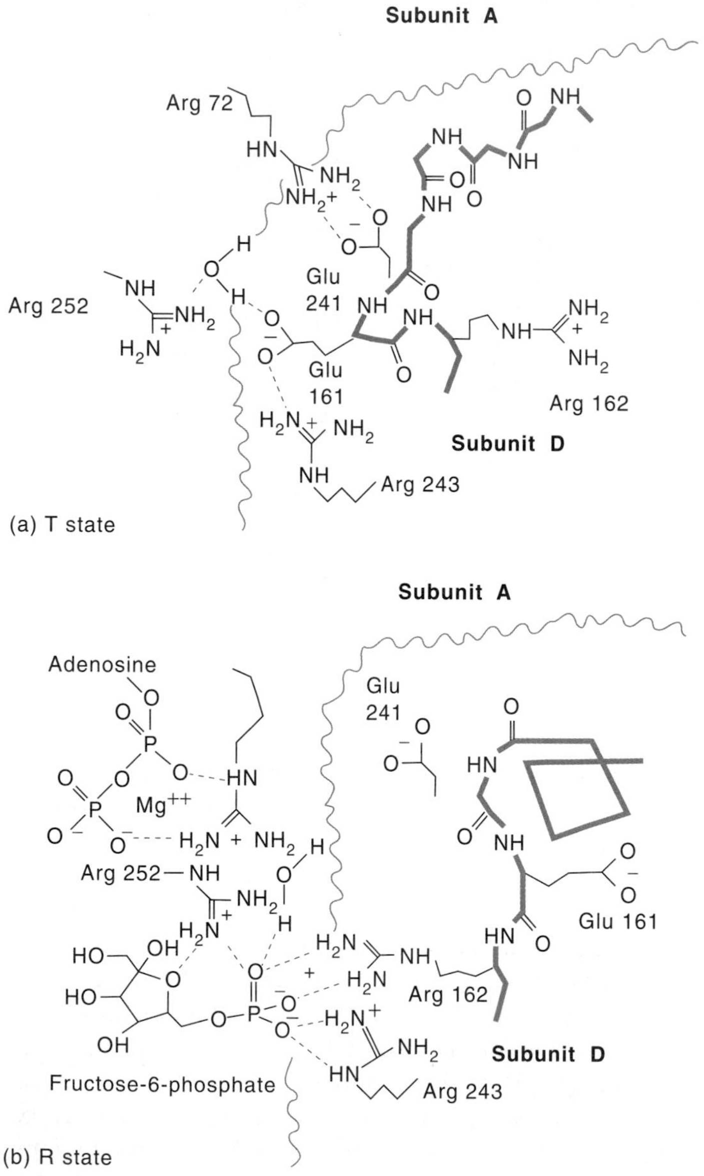

## Figure 9.12

Hydrogen bonds of the peptide backbone and the side chain of threonine 245 at the interface between subunits A and D of phosphofructokinase, in (a) the T and (b) the R structures. Note the additional molecules of water (red) between the two subunits in the R structure. (Source: From T. Schirmer and P. R. Evans, Structural basis of the allosteric behaviour of phosphofructokinase, *Nature* 343:140, 1990.)

(a) T state

(b) R state

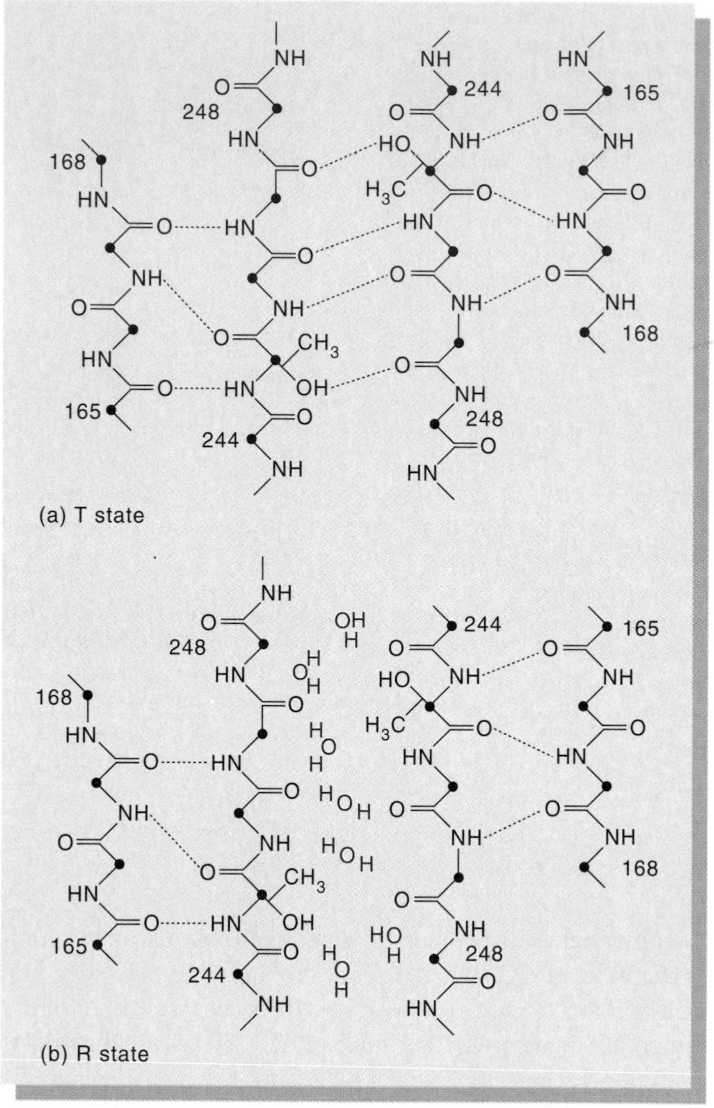

explanation for the cooperative binding of fructose-6-phosphate, it is still not clear why various allosteric effectors stabilize the protein in different conformational states. However, it is significant that the allosteric binding site lies at the interface of different subunits. The equilibrium between the R and T conformations appears to be sensitive to subtle structural changes in this region.

## Figure 9.13

The reaction catalyzed by aspartate carbamoyl transferase, and the feedback inhibition of this enzyme in *E. coli* by the end product of the pathway, CTP. The series of small arrows represents additional reaction steps in the pathway from carbamoyl aspartate to CTP. These steps are discussed in chapter 1. The upward arrow with the negative sign indicates the feedback inhibition.

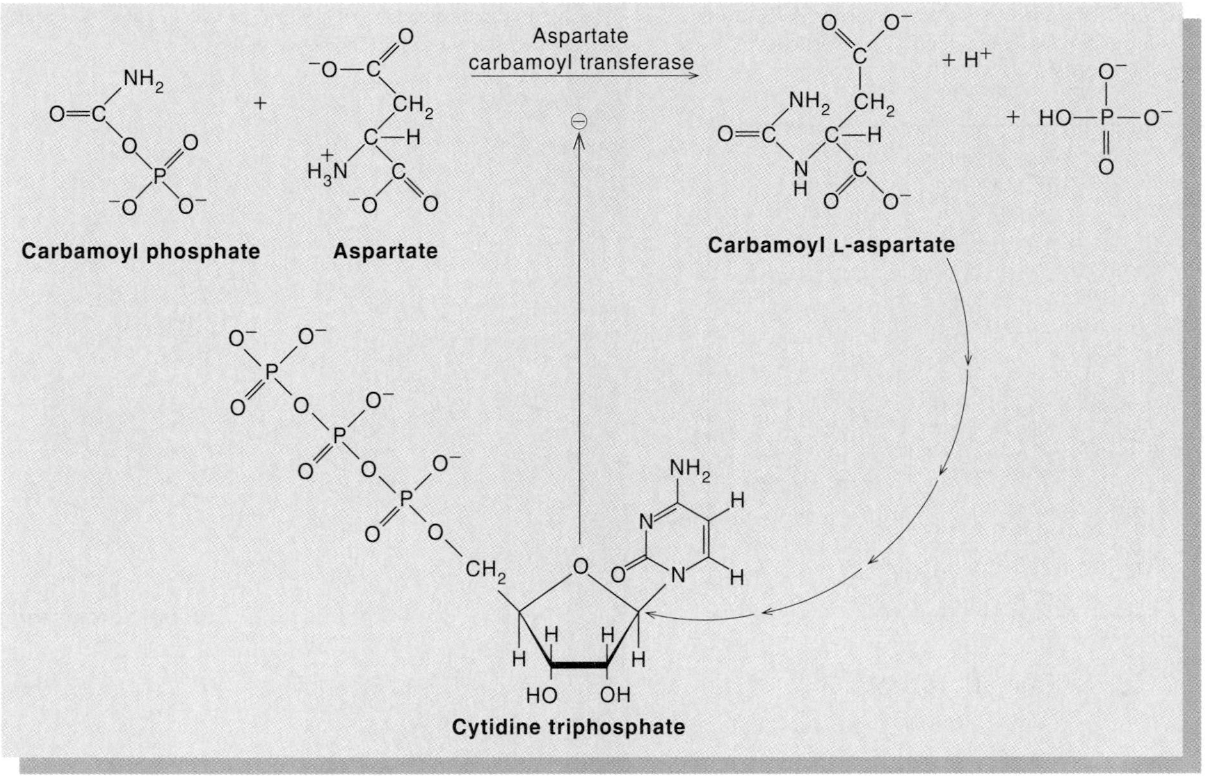

## Aspartate Carbamoyl Transferase: Allosteric Control of Pyrimidine Biosynthesis

Aspartate carbamoyl transferase, or aspartate transcarbamylase, catalyzes the transfer of a carbamoyl group

$$(H_2N-\overset{\overset{\displaystyle O}{\|}}{C}-)$$

from carbamoyl phosphate to aspartic acid to form carbamoyl aspartate (fig. 9.13). This step commits aspartate to the biosynthetic pathway for pyrimidines (see figure 23.13). Aspartate carbamoyl transferase is inhibited by cytidine triphosphate (CTP) and uridine triphosphate (UTP), the end products of the pathway, and it is stimulated by a purine, ATP. The opposing effects of CTP, UTP, and ATP serve to keep the biosynthesis of pyrimidines in balance with that of purines. This balance is important because cells need purines and pyrimidines in approximately equal amounts for the synthesis of nucleic acids.

The kinetics and physical properties of aspartate carbamoyl transferase from *E. coli* were studied in considerable detail by Howard Schachman and his colleagues. The kinetics have a sigmoidal dependence on the concentration of aspartate, as shown in figure 9.14. CTP shifts the kinetic curve to the right, and thus inhibits the enzyme strongly at low concentrations of aspartate but not at high concentrations. ATP reverses the effect of CTP, or in the absence of CTP eliminates the cooperativity altogether, making the kinetics hyperbolic instead of sigmoidal.

The behavior of aspartate carbamoylase changed dramatically when the enzyme was treated with the organic mercurial compound *p*-hydroxymercuribenzoate (see fig. 9.14). The binding of aspartate no longer showed positive cooperativity, and ATP or CTP were without effect. Exposure to mercurials was found to cause the enzyme to dissociate into two types of fragments, one of which retained the enzymatic activity but was no longer affected by CTP or

**Figure 9.14**

Effects of CTP, ATP, and mercurials on the rate of the reaction catalyzed by aspartate carbamoyl transferase. In the absence of CTP and ATP, the sigmoidal kinetics show positive cooperativity with respect to aspartate. CTP augments the positive cooperativity; ATP reverses the effect of CTP. An organic mercurial, or ATP in the absence of CTP, eliminates the cooperativity, converting the curve from sigmoidal to hyperbolic.

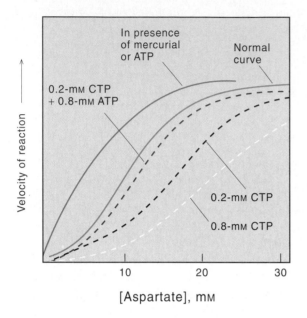

**Figure 9.15**

Subunit structure of aspartate carbamoyl transferase and the fragments produced by treating the enzyme with mercurials. In the complete enzyme *(top)*, the three sets of regulator dimers are sandwiched between two trimers of catalytic subunits (see fig. 9.17). The approximate location of the active site in each *c* subunit of the trimer facing the viewer is indicated with a *c*.

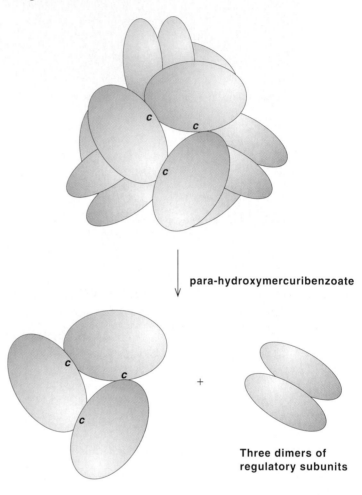

ATP. The other fragment bound CTP and ATP but had no enzymatic activity. In the native enzyme, each molecule is a complex of six catalytic (*c*) subunits and six regulatory (*r*) subunits (fig. 9.15). The fragments resulting from treatment with mercurials consist of trimers of the *c* subunit, and dimers of *r*. When the $c_6r_6$ complex was reconstituted from the fragments, the enzyme regained its sigmoidal kinetics and its sensitivity to CTP and ATP. These observations provided a convincing demonstration that the regulatory agents bind at an allosteric site and not at the active site. The two sites are on totally different subunits!

A substrate analog that proved particularly useful for studying aspartate carbamoyl transferase is *N*-phosphonacetyl-L-aspartate (PALA). PALA is structurally similar to a covalently linked adduct of carbamoyl phosphate and aspartate and thus resembles a likely intermediate in the enzymatic reaction (fig. 9.16). It binds to the enzyme in a highly cooperative manner, and its binding is promoted by ATP and opposed by CTP. The binding of PALA decreases the sedimentation and diffusion coefficients of the enzyme, indicating that the protein expands or changes shape. It also decreases the chemical reactivity of a cysteine residue in

each *c* subunit and increases the reactivity of several cysteines in each *r* subunit. CTP opposes these effects.

The changes in sedimentation coefficient and chemical reactivity caused by PALA fit the model that the enzyme exists in two distinct conformations (T and R) and that the binding of PALA to only one or two of the *c* subunits causes the entire $c_6r_6$ complex to flip from the T to the R state. The dissociation constant for PALA is higher in the T state than in the R state. The equilibrium constant $L = [T]/[R]$ has been calculated to be 250 in the absence of substrates and

**Figure 9.16**

*N*-phosphonacetyl-L-aspartate (PALA) is structurally similar to a likely intermediate in the reaction catalyzed by aspartate carbamoyl transferase. The binding of PALA to the enzyme is prevented competitively by carbamoyl phosphate, and PALA prevents the binding of aspartate. These observations support the view that PALA binds at the catalytic site.

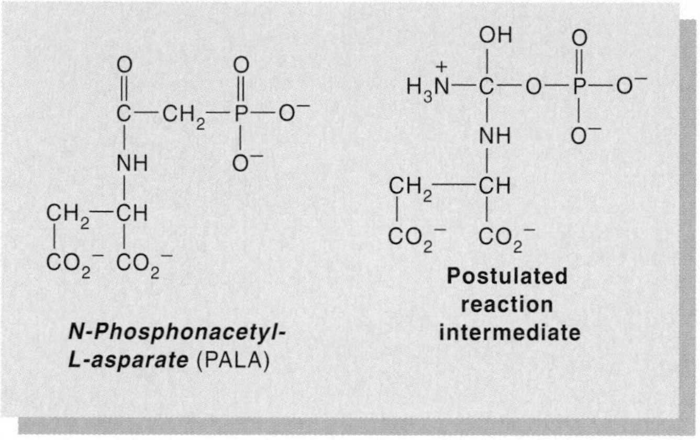

**N-Phosphonacetyl-L-asparate (PALA)**

**Postulated reaction intermediate**

allosteric effectors, 70 in the presence of ATP alone, and 1,250 in the presence of CTP alone. ATP thus shifts the equilibrium toward the conformational state that favors binding of the substrate, and CTP shifts the equilibrium in the direction of weaker binding.

Crystallization of aspartate carbamoyl transferase with and without bound PALA or CTP, by William Lipscomb and his colleagues, led to detailed pictures of the R and T conformations. Figures 9.17 and 9.18 show the crystal structures from two different perspectives. In both the R and the T conformations, the six *c* subunits are arranged in two equilateral trimers, one of which is inverted and stacked on top of the other (see figs. 9.15 and 9.17). The three *c* units in each trimer are related to each other by a threefold axis of symmetry. (Rotating the structure by one-third of a circle about this axis results in an identical structure.) The substrate-binding site on each *c* subunit is located in a pocket between two domains of the polypeptide. At the end of one of these domains the *c* subunit interacts with another *c* subunit in the same trimer, close to *its* substrate-binding site. In the other domain, it interacts more extensively with a *c* subunit in the other $c_3$ trimer. The six *r* subunits are arranged in three sets of dimers that form another equilateral triangle about the same axis of symmetry. Like the *c* subunits, each *r* subunit is folded into two domains: a peripheral domain where it interacts with its companion *r* subunit in the dimer, and a smaller domain that interacts with two adjacent *c* sub-

units. In the latter region, each *r* subunit binds an atom of $Zn^{2+}$ that evidently plays a purely structural role. The binding site for the allosteric effectors CTP and ATP is located in the peripheral domain of the *r* subunit, at a considerable distance from the active sites.

In the transition from the T to the R conformation, the two *c* trimers rotate slightly with respect to each other about the threefold symmetry axis, so that they come into a more eclipsed alignment (see fig. 9.17). The *r* dimers rotate with respect to each other about a perpendicular axis and appear to act as a lever that moves the two *c* trimers apart by about 12 Å and opens up a cavity at the center of the entire structure (see fig. 9.18).

Figure 9.19 shows some of the details of the substrate-binding site in the R structure. As we mentioned above, the binding site consists of a pocket between two domains of a *c* subunit. Arginines 167 and 229 from one domain interact with the two carboxylate groups of PALA. Arginine 105 of the other domain interacts with the phosphonate group, and presumably does the same with the phosphate of carbamoyl phosphate. Histidine 134 appears to provide a hydrogen bond to the peptide oxygen atom of PALA. If it does the same to the corresponding oxygen of carbamoyl phosphate, it can serve as a general acid in the catalytic mechanism. In addition to these residues, the active site also includes two residues from a different *c* subunit, Ser 80 and Lys 84 (see fig. 9.19). As with phosphofructokinase, the location of the active site at the interface between two subunits provides a clue to how binding of substrate to one subunit can affect the binding at another.

The structural transition to the T state disrupts the active site in two major ways. First, the domain of the *c* subunit that includes Arg 105 and His 134, which interact with carbamoyl phosphate, is pulled away from the domain that interacts with aspartate, because some of the residues in both domains are tied up in an alternative set of hydrogen bonds. Arginine 105 is hydrogen-bonded to Glu 50 in the same domain, instead of to the substrate; His 134 interacts with a residue in the other *c* trimer. In addition, the loop of the *c* subunit that contains Ser 80 and Lys 84 is pulled out of the active site by hydrogen bonds to still another *c* subunit. A baroque net of interrelationships thus links the catalytic sites of all the *c* subunits in the complex.

The transition between the R and T states also involves large changes in the conformation of the *r* subunits. These changes include both the peripheral domain where CTP or ATP binds and the domain that interfaces with the *c* subunits (see figs. 9.17 and 9.18). However, it still is not clear how the binding of CTP to the peripheral domain tips the conformational equilibrium in favor of T, whereas ATP,

# Figure 9.17

Structures of aspartate carbamoyl transferase in the T conformation (a) and the R conformation (b) viewed along the threefold symmetry axis. The enzyme contains two $c_3$ clusters and three $r_3$ clusters. The $\alpha$-carbon chains of one of the $c_3$ groups are shown in aqua; those of the other $c_3$ group are in blue. One of the $r$ subunits in each of the $r_2$ groups is shown in orange; the other, in red. The enzyme was crystallized in the T form in the absence of substrate or allosteric effectors; the R structure was obtained with bound PALA. The PALA molecules are seen in yellow in (b). Zinc ions bound to the $r$ subunits are shown in white.

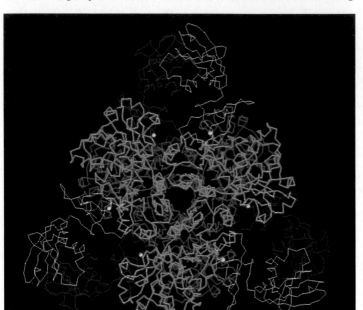

(a)

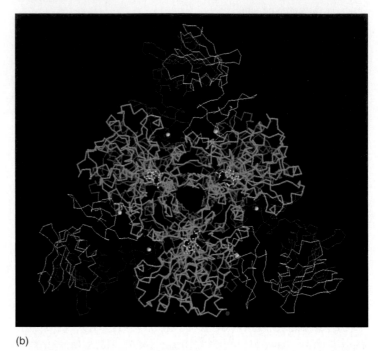

(b)

# Figure 9.18

Structures of aspartate carbamoyl transferase in the T conformation (a) and the R conformation (b) viewed along an axis perpendicular to the threefold symmetry axis. The structures and the color coding are the same as in figure 9.17. Note the expansion of the cavity between the upper and lower $c_3$ groups in the R structure.

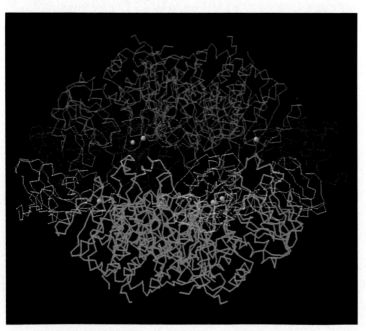

(a)

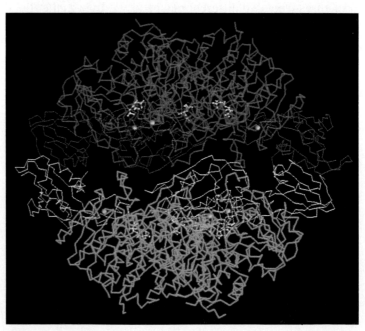

(b)

## Figure 9.19

The binding of PALA to the R conformation of aspartate carbamoyl transferase. Arg 105 and His 134 are provided by one domain of a *c* subunit, and Arg 167 and Arg 229 by the other domain. Ser 80 and Lys 84 are part of a loop of protein from a different *c* subunit. The PALA is indicated by red. The wavy green lines indicate polypeptide backbone structure.

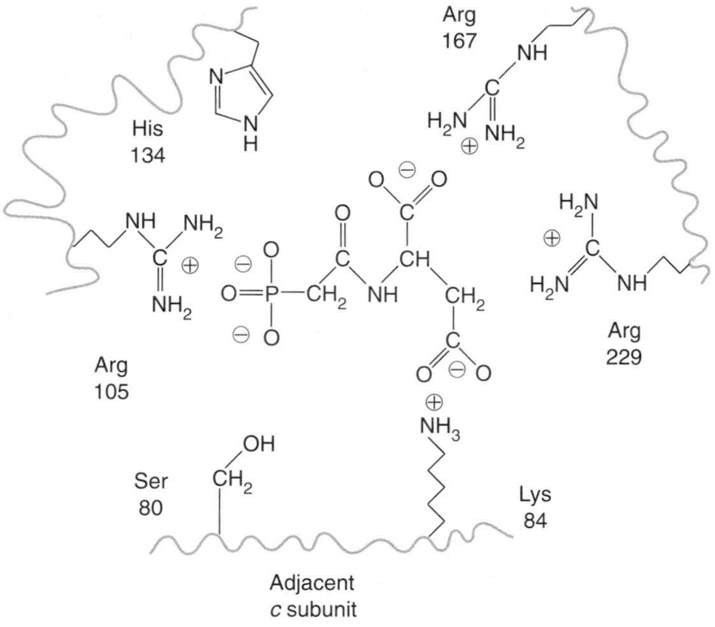

## Figure 9.20

The rate of the reaction catalyzed by glycogen phosphorylase, as a function of the concentration of its main allosteric activator, AMP. The curves shown in color were obtained in the presence of ATP. Phosphorylase *b* (lower two curves) is almost completely inactive in the absence of AMP. Its activity is half maximal at an AMP concentration of about 40 $\mu$M. ATP greatly increases the concentration of AMP required for activity. Phosphorylase *a* (upper two curves) has about 80% of its maximal activity in the absence of AMP and reaches full activity at very low AMP concentrations; it also is relatively insensitive to inhibition by ATP.

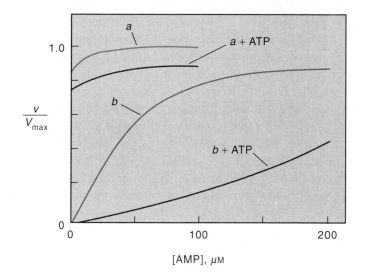

which binds to the same site as CTP, favors the formation of R. The crystal structures of the enzyme with bound ATP or CTP are very similar, both at the catalytic site and at the allosteric site.

# Glycogen Phosphorylase: Combined Control by Allosteric Effectors and Phosphorylation

Glycogen phosphorylase catalyzes the removal of a terminal glucose residue from glycogen. (The structure of glycogen is shown in figs. 12.9 and 12.11.) The glycosidic bond is cleaved by a reaction with inorganic phosphate, so that the product is glucose-1-phosphate instead of free glucose.

$$(\text{glucose})_n + P_i \longrightarrow$$

$$(\text{glucose})_{n-1} + \text{glucose-1-phosphate} \quad (9)$$

This is the first step in the metabolic breakdown of glycogen to pyruvate.

In the early 1940s, Carl Cori and Gerty Radnitz Cori discovered that phosphorylase exists in two forms, *a* and *b*,

which differ greatly in their catalytic activities. As shown in figure 9.20, phosphorylase *b* has virtually no activity in the absence of AMP. It is activated by low concentrations of AMP, but the activation is inhibited competitively by ATP. At the concentrations of AMP and ATP that prevail in resting muscle tissue, phosphorylase *b* is essentially inactive. Phosphorylase *a*, on the other hand, has about 80% of its maximal activity in the absence of AMP and becomes fully active at very low concentrations of AMP. It also is relatively insensitive to inhibition by ATP (see fig. 9.20).

The Coris found that the interconversion of phosphorylases *a* and *b* is catalyzed by another enzyme, and subsequent work by Earl Sutherland showed that this process is under hormonal control. In muscle, conversion of phosphorylase *b* to *a* is stimulated by epinephrine; in liver, it is stimulated by both epinephrine and the pancreatic hormone glucagon. The structural basis for the difference between the two forms of phosphorylase remained unknown until the late 1950s, when Edwin Krebs and Edmund Fischer showed that phosphorylase *a* has a phosphate on serine 14. This phosphate is absent in the *b* form of the enzyme. Krebs and Fischer also showed that the kinase that catalyzes the addition of the phosphate is itself regulated by a phosphorylation catalyzed by another enzyme, the cAMP-dependent protein kinase!

The main effect of AMP on either phosphorylase $b$ or phosphorylase $a$ is to decrease the $K_m$ for $P_i$. This change can be interpreted as we have interpreted the actions of allosteric effectors on phosphofructokinase and aspartate carbamoyl transferase, on the model that the enzyme can exist in two conformational states (R and T) with different affinities for the substrate. However, phosphorylase presents the additional complexity that the equilibrium constant ($L$) between the two conformational states can be altered by a covalent modification of the enzyme. In the absence of substrates, [T]/[R] appears to be greater than 3,000 in phosphorylase $b$ but to decrease to about 10 in phosphorylase $a$.

Both forms of phosphorylase are inhibited by glucose or glucose-6-phosphate. Glucose inhibits by binding at the catalytic site while glucose-6-phosphate binds at the same allosteric site as AMP and ATP. A separate inhibitory allosteric site binds adenine, adenosine, or (much more weakly) AMP.

Louise Johnson and her coworkers have determined the crystal structures of T and the R forms of muscle phosphorylase $b$ and the R form of phosphorylase $a$. In parallel with this work, Robert Fletterick and coworkers determined the structure of the T form of muscle phosphorylase $a$. The crystal structures provide an incisive look at the structural changes that accompany the transitions from the T to the R conformation and from the nonphosphorylated form of the enzyme to the phosphorylated form.

In keeping with the complexity of its allosteric and covalent regulation, phosphorylase is a large enzyme. It consists of a dimer of two identical subunits, each with a molecular weight of about 97,400 (fig. 9.21). The catalytic site is buried near the center of each subunit, at the end of a tunnel that opens to a concave surface. A binding site for glycogen on the surface can be recognized by the location of a small oligosaccharide in the crystal structure. Because the glycogen attachment site is about 30 Å from the catalytic site, the enzyme evidently clings to one branch of a glycogen particle while it chews on another branch. Near the catalytic site is a covalently bound molecule of the coenzyme pyridoxal phosphate (fig. 9.22), which probably participates as a general acid in the catalytic mechanism. The binding site for the allosteric effectors AMP, ATP, and glucose-6-phosphate is about 30 Å from the catalytic site, at one of the interfaces of the two subunits (see fig. 9.21). Serine 14, the locus of the covalent modification that converts phosphorylase $b$ to $a$, is in the same region.

When phosphorylase $b$ undergoes the transition from the T to the R conformation, major structural changes occur in a loop between residues 282 to 286, which connects two $\alpha$-helical chains (fig. 9.23). In the T form, this loop obstructs the substrate-binding site for $P_i$; in the R form, the loop is pulled out of the $P_i$ site. Some of the structural

## Figure 9.21

($a$) Ribbon diagram of the crystal structure of phosphorylase $a$ in the R state. The view is along the twofold rotational symmetry axis of the dimer, with the allosteric sites and the phosphoserine (Ser-P) on each subunit facing forward. Regions where the positions of the $C_d$ carbons differ by more than 1 Å between the R and T states are shown in orange for subunit 1 *(bottom)* and in pink for subunit 2 *(top)*. Less mobile regions of the polypeptide chain are in green for subunit 1 and in blue for subunit 2. The N-terminal residues (10–23) and the C-terminal residues (837–842) are in white (residues 1–9 are disordered and cannot be seen in the crystal structure). The bound pyridoxal phosphate (PLP) indicates the location of the catalytic site. The allosteric effector site is occupied by AMP. The first two helices at the N-terminal end (labeled $\alpha_1$ and $\alpha_2$ in subunit 1) are connected by a loop (Cap) that forms one of the interfaces between the two subunits. ($b$) Ribbon diagram of phosphorylase $b$ in the T state. The orientation of the dimer is the same as in ($a$). Regions where the $C_d$ positions differ by more than 1 Å between the R and T states are represented in red for subunit 1 *(bottom)* and in yellow for subunit 2 *(top)*; more fixed regions are in cyan and purple. The N-terminal residues and the C-terminal residues are in white. PLP is at the catalytic site and AMP at the allosteric effector site, as in ($a$). Maltopentaose is bound at the glycogen storage site. (From D. Barford, S. -H. Hu, and L. N. Johnson, *J. Mol. Biol.* 218:233, 1991. © 1991 Academic Press LTD., London England.)

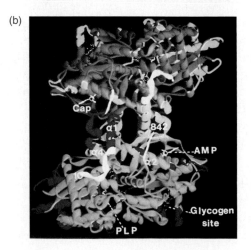

(a)

(b)

**Figure 9.22**

Pyridoxal phosphate, a prosthetic group in glycogen phosphorylase, is covalently attached to a lysine side chain of the enzyme. The phosphate group of the pyridoxal phosphate probably acts as a general acid to transfer a proton to inorganic phosphate in the enzymatic mechanism.

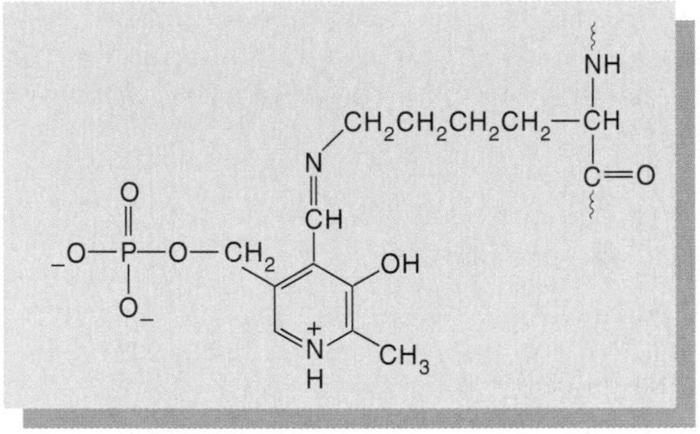

**Figure 9.23**

(a) In the T state of phosphorylase b, a loop of the polypeptide chain between residues 282 and 286 obstructs the active site. Asp 283, near the middle of this loop, inserts its negatively charged side chain into the binding site for phosphate. The locations of the aspartate and the pyridoxal phosphate prosthetic group (PLP) are indicated in red.
(b) In the R state, the subunits rotate with respect to one another. The loop from residues 282 to 286 is disordered, and Asp 283 leaves the phosphate-binding site. The green lines indicate polypeptide backbone structures. (Source: From D. Barford and L. N. Johnson, The allosteric transition of glycogen phosphorylase, *Nature* 340:609, 1989.)

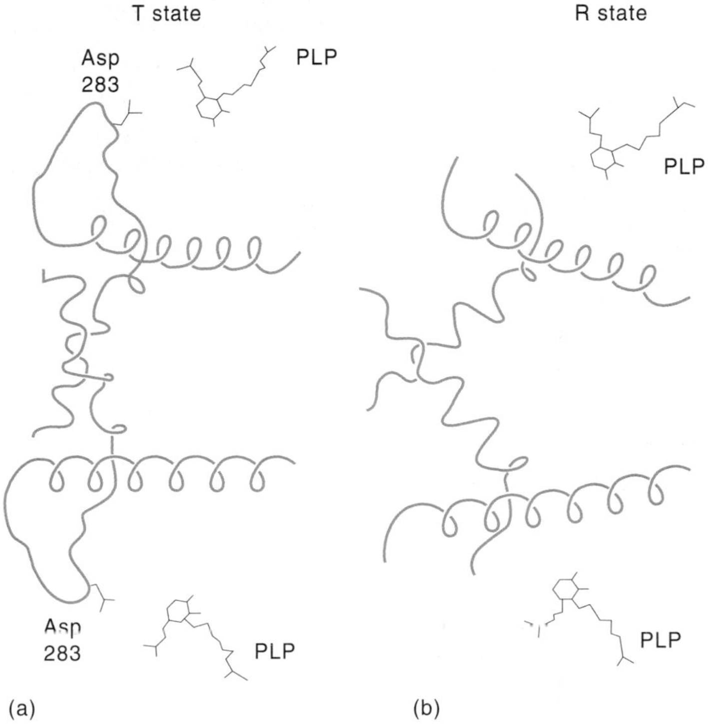

changes are shown in figure 9.24. In the R form, the substrate $P_i$ is bound to Arg 569, Gly 135, Lys 574 (not shown in the figure), and the pyridoxal phosphate. In the T form, the side chain of Asp 283 sits in this region, and Arg 569 is pulled away by hydrogen bonding to Pro 281. The replacement of the positively charged arginine side chain by the negatively charged aspartate explains why the affinity for $P_i$ is much lower in the T conformation.

The transformation from phosphorylase b to phosphorylase a includes structural changes that tighten the interactions between the two subunits. Figures 9.21 and 9.25 show some of these changes. In phosphorylase a, the phosphate group attached to each serine 14 is hydrogen-bonded to Arg 69 of its own subunit but also to Arg 43 of the other subunit. Hydrogen bonds also exist between Arg 10 and Leu 115 of different subunits and between Gln 72 and Asp 42. All of these intersubunit hydrogen bonds are missing in phosphorylase b. Instead, a bond exists between Arg 43 and a leucine in the same subunit, and one intersubunit hydrogen bond exists between His 36 and Asp 838. The N-terminal portion of the chain is ejected from this region in phosphorylase b and is replaced by residues from the C-terminal end, including Asp 838. At the nearby allosteric effector site, the pulling together of the two subunits in phosphorylase a enhances the binding of AMP but disfavors the binding of the inhibitor glucose-6-phosphate.

## Figure 9.24

Residues in the region of the substrate-binding site for phosphate in phosphorylase *b* in the T state (*a*) and the R state (*b*). The side chain of Asp 283 leaves the binding site in the R structure, and the side chain of Arg 569 becomes available to interact with the phosphate. Key side chains and bound phosphate anion are shown in red. Portions of the polypeptide backbone are drawn in heavy black. (Source: From D. Barford and L. N. Johnson, The allosteric transition of glycogen phosphorylase, *Nature* 340:609, 1989.)

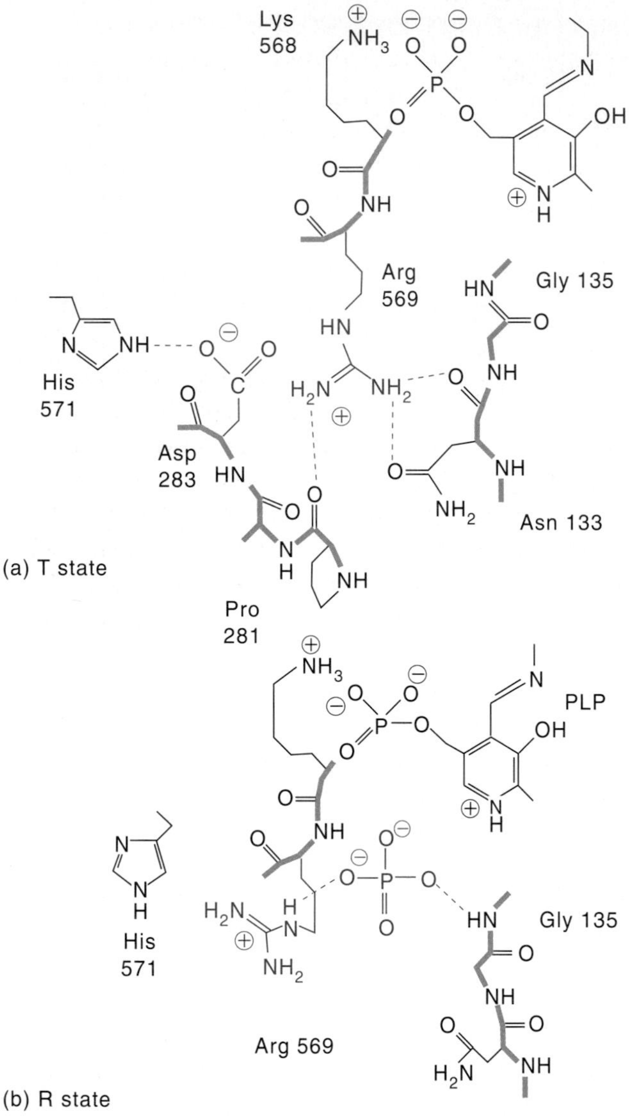

(a) T state

(b) R state

## Figure 9.25

Structural changes that accompany the conversion of phosphorylase *b* to phosphorylase *a*. This figure shows the interface between the two subunits in the region of Ser 14, the residue that gains a phosphate group in phosphorylase *a*. Portions of the polypeptide backbone and the side chains of some residues from one of the subunits are drawn in red; the other subunit is drawn in black. A larger number of hydrogen bonds link the two subunits in phosphorylase *a (top)* than in phosphorylase *b (bottom)*. (Source: From S. R. Sprang et al., Structural changes in glycogen phosphorylase induced by phosphorylation, *Nature* 336:215, 1988.)

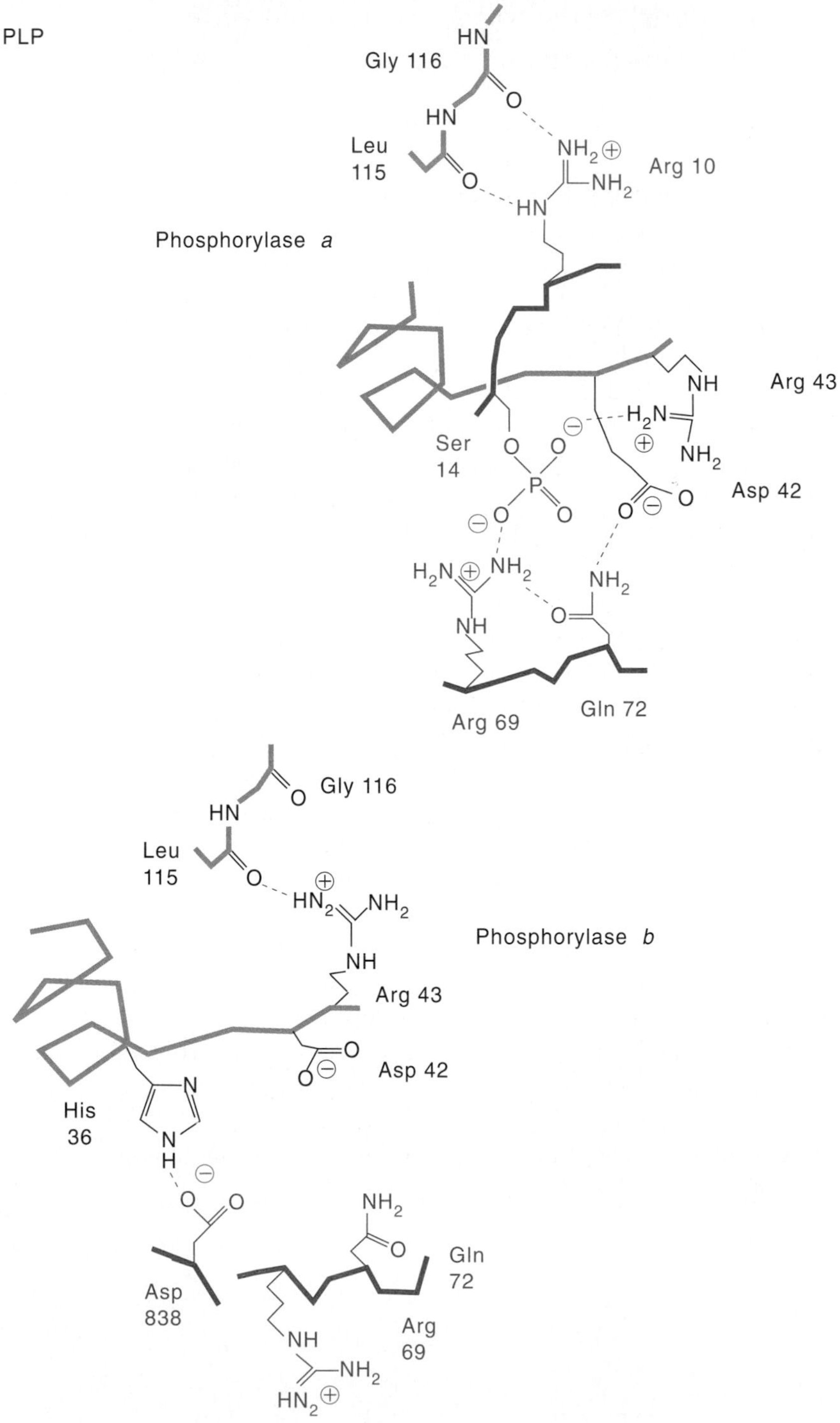

Phosphorylase *a*

Phosphorylase *b*

## Summary

Cells regulate their metabolic activities by controlling rates of enzyme synthesis and degradation and by adjusting the activities of specific enzymes. Enzyme activities vary in response to changes in pH, temperature, and the concentrations of substrates or products, but also can be controlled by covalent modifications of the protein or by interactions with activators or inhibitors.

1. Partial proteolysis, an irreversible process, is used to activate proteases and other digestive enzymes after their secretion and to switch on enzymes that cause blood coagulation. Common types of reversible covalent modification include phosphorylation, adenylylation, and disulfide reduction.

2. Allosteric effectors are inhibitors or activators that bind to enzymes at sites distinct from the active sites. Allosteric regulation allows cells to adjust enzyme activities rapidly and reversibly in response to changes in the concentrations of substances that are structurally unrelated to the substrates or products. The initial steps in a biosynthetic pathway commonly are inhibited by the end products of the pathway, and numerous enzymes are regulated by ATP, ADP, or AMP.

3. The kinetics of allosteric enzymes typically show a sigmoidal dependence on substrate concentration rather than the more common hyperbolic saturation curves. These enzymes usually have multiple subunits, and the sigmoidal kinetics can be ascribed to cooperative interactions of the subunits. Binding of substrate to one subunit changes the dissociation constant for substrate on another subunit. The extent of the cooperativity can be described by the Hill equation or by the "symmetry" model or the more general "sequential" model. The symmetry model postulates that the enzyme can exist in two conformations, T and R. It is assumed that the substrate binds more tightly to the R conformation than to the T conformation, that binding of the substrate or an allosteric effector changes the equilibrium between these conformations, and that cooperative interactions make all of the subunits switch from one conformation to the other in a concerted manner. The symmetry model provides a useful conceptual framework that is consistent with the behavior of many allosteric enzymes.

4. Phosphofructokinase, the key regulatory enzyme of glycolysis, has four identical subunits. It exhibits sigmoidal kinetics with respect to fructose-6-phosphate and is inhibited by ATP and stimulated by ADP. In the transition between the R and T conformations, the subunits rotate with respect to each other and there is a rearrangement of the binding site for fructose-6-phosphate, which is located at an interface between subunits. At another interface, antiparallel $\beta$ strands of two subunits are hydrogen-bonded together in the T structure but are separated by a row of water molecules in the R structure. This structural feature explains the cooperative nature of the conformational transition in all of the subunits: Intermediate structures probably would be less stable than either the R or the T structure.

5. Aspartate carbamoyl transferase, the first enzyme in the biosynthesis of pyrimidines, is inhibited allosterically by CTP, an end product of the pathway, and is stimulated by ATP. It has six identical catalytic ($c$) subunits and six regulatory ($r$) subunits. The $c$ and $r$ subunits can be separated by treating the enzyme with mercurials. Binding of a substrate analog to one of the $c$ subunits causes the entire $c_6 r_6$ complex to flip to the R conformation; binding of CTP to an $r$ subunit favors the T conformation. Again, the substrate-binding sites are located at interfaces between subunits, and the T to R transition results in a rotation of the subunits and brings together components of the active site.

6. Glycogen phosphorylase breaks down glycogen to glucose-1-phosphate. It exists in two forms that differ by a covalent modification. Phosphorylase $a$ is phosphorylated on a serine residue and is the more active form under cellular conditions. Phosphorylase $b$, which lacks the phosphate, is stimulated allosterically by AMP but inhibited by ATP. The enzyme has two identical subunits, with the binding site for allosteric effectors at the interface. In the T to R transition there is a repositioning of arginine and aspartate residues at the binding site for $P_i$. Conversion of phosphorylase $b$ to phosphorylase $a$ tightens the interactions between the subunits and favors the transition to the R form.

## Suggested Reading

Barford, D., S. -H. Hu, and L. N. Johnson, Structural mechanism for glycogen phosphorylase control by phosphorylation and AMP. *J. Mol. Biol.* 218:233, 1991.

Barford, D., and L. N. Johnson, The allosteric transition of glycogen phosphorylase. *Nature* 340:609, 1989.

Cséke, C., and B. B. Buchanan, Regulation of the formation and utilization of photosynthate in leaves. *Biochim. Biophys. Acta.* 853:43, 1986.

Furie, B., and B. C. Furie, The molecular basis of blood coagulation. *Cell.* 53:505, 1988.

Lipscomb, W. N., Structure and function of allosteric enzymes. *Chemtracts-Biochem. Mol. Biol.* 2:1, 1991.

Schachman, H. R., Can a simple model account for the allosteric transition of aspartate transcarbamylase? *J. Biol. Chem.* 263, 18583, 1988.

Schirmer, T., and P. R. Evans, Structural basis of the allosteric behaviour of phosphofructokinase. *Nature* 343:140, 1990.

Sprang, S. R., et al., Structural changes in glycogen phosphorylase induced by phosphorylation. *Nature* 336:215, 1988.

Stevens, R. C., J. E. Gouaux, and W. N. Lipscomb, Structural consequences of effector binding to the T state of aspartate carbamoyltransferase: Crystal structures of the unligated and ATP- and CTP-complexed enzymes at 2.6 Å resolution. *Biochem.* 29:7691, 1990.

## Problems

1. Notice the salt bond in figure 9.1b between the beta carboxyl group of Asp 194 and the N terminus (Ile 16) in the structure of chymotrypsin. Why is the N-terminus amino acid number 16?

2. Many hemophiliacs are sustained by regular injections of one of the cascade components, for example, factor VIII (fig. 9.2). Why doesn't the addition of factor VIII cause an uncontrolled clotting of the blood in these individuals?

3. While reading this chapter you should have noticed that protein regulation mechanisms fall into two major categories: Covalent modifications and allosteric regulation. Does one approach seem to be inherently less complicated? Why?

4. What similarities and differences can you find in the covalent modification of proteins shown in figures 9.3 and 9.4?

5. The cAMP-dependent protein kinases phosphorylate specific Ser (Thr) residues on target proteins. Given the availability of serine and threonine residues on the surface of globular proteins, how might a protein kinase select the "correct" residues to phosphorylate?

6. Is the interaction between an allosteric effector and an allosteric enzyme always an equilibrium?

7. The substrate concentration yielding half-maximal velocity is equal to $K_m$ for hyperbolically responding enzymes. Is this relationship true for allosterically or sigmoidally responding enzymes? How do positive or negative allosteric effectors change the substrate concentration required for half-maximal velocity?

8. ATP is both a substrate and an inhibitor of the enzyme phosphofructokinase (PFK). Although the substrate fructose-6-phosphate binds cooperatively to the active site, ATP does not bind cooperatively. Explain how ATP may be both a substrate and an inhibitor of PFK.

9. Aspartate carbamoyltransferase is an allosteric enzyme in which the active sites and the allosteric effector binding sites are on different subunits. Explain how it might be possible for an allosteric enzyme to have both kinds of sites on the same subunit.

10. Examine the relationship of aspartate car-bamoyltransferase (ACTase) activity to aspartate concentration shown in the figure. Estimate the $(S_{0.9}/S_{0.1})$ ratio for the reaction under the following conditions: (a) normal curve, (b) plus 0.2 mM CTP, (c) plus 0.8 mM ATP. Do these ratios differ significantly? Explain.

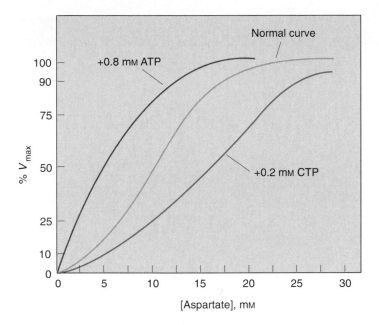

11. Why do you think the researchers who initially prepared PALA (fig. 9.16) utilized a methylene between the carbonyl and phosphoryl group? Why didn't they "just" use a "regular" phosphate group. *Hint:* Compare the postulated reaction intermediate (fig. 9.16) with the reaction shown in figure 9.13.

12. In some instances, protein kinases are activated or inhibited by low-molecular-weight modifiers. Explain how a metabolite might more effectively regulate an enzyme by modifying a protein kinase rather than directly inhibiting the target enzyme.

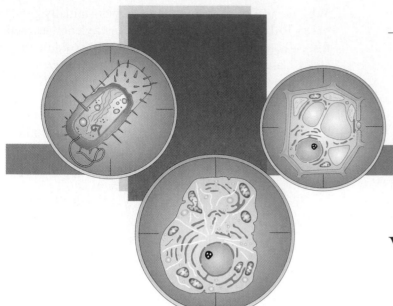

# Vitamins and Coenzymes

*Coenzymes are small organic molecules that act in concert with the enzyme to catalyze biochemical reactions.*

The types of chemical reactions that can be catalyzed by proteins alone are limited by the chemical properties of the functional groups found in the side chains of nine amino acids: the imidazole ring of histidine; the carboxyl groups of glutamate and aspartate; the hydroxyl groups of serine, threonine, and tyrosine; the amino group of lysine; the guanidinium group of arginine; and the sulfhydryl group of cysteine. These groups can act as general acids and bases in catalyzing proton transfers and as nucleophilic catalysts in group transfer reactions.

Many metabolic reactions involve chemical changes that cannot be brought about by the structures of the amino acid side chain functional groups in enzymes acting by themselves. In catalyzing these reactions, enzymes act in cooperation with other smaller organic molecules or metallic cations, which possess special chemical reactivities or structural properties that are useful for catalyzing reactions. In this chapter we introduce these small molecules and survey the range of reactions they catalyze.

It has been known since the middle of the nineteenth century that small amounts of certain substances are very important to healthy nutrition. Vitamins are organic molecules essential in small quantities for healthy nutrition in rats or humans. The list of such molecules grew as they were purified from foodstuffs and shown to cure various disorders in animals maintained on deficient diets. The name "vitamine" was given in 1911 to the first vitamin to be isolated, thiamine. When it became clear that a number of essential organic micronutrients were not amines, the -*e* was dropped.

Vitamins are divided into water-soluble and lipid-soluble groups. In addition to vitamins, vitaminlike nutrients are required in small amounts by the organism and frequently function in similar capacities to vitamins. These vitaminlike compounds are not classified as vitamins because rats and humans have a limited capacity to synthesize them, provided that the diet contains the essential precursors. Table 10.1 lists both vitamins and vitaminlike nutrients.

In this chapter we are concerned, not primarily with vitamins *per se,* but with coenzymes. Many coenzymes are modified forms of vitamins. The modifications take place in the organism after ingestion of the vitamins. Coenzymes act in concert with enzymes to catalyze biochemical reactions. Tightly bound coenzymes are sometimes referred to as prosthetic groups. A coenzyme usually functions as a major component of the active site on the enzyme, which means that understanding the mechanism of coenzyme action usually requires a complete understanding of the catalytic process.

## Water-Soluble Vitamins and Their Coenzymes

As we have noted, some vitamins are soluble in water, and others are soluble in lipids. In the following sections we survey the range of biochemical reactions in which water-soluble coenzymes participate.

### Thiamine Pyrophosphate Is Involved in C—C and C—X Bond Cleavage

The structure of thiamine pyrophosphate (TPP) is given in figure 10.1. The vitamin, thiamine, or vitamin $B_1$, lacks the pyrophosphoryl group. Thiamine pyrophosphate is the essential coenzyme involved in the actions of enzymes that catalyze cleavages of the bonds indicated in red in figure 10.1. The bond scission in figure 10.1*b* is representative of those in many $\alpha$-keto acid decarboxylations, nearly all of which require the action of TPP. The phosphoketolase reaction involves both cleavages shown in figure 10.1*c*, whereas the transketolase reaction (see fig. 12.33) involves the cleavage of the carbon–carbon bond but not the elimination of —OH. Acetolactate and acetoin arise by the formation of

**Table 10.1**

Vitamins and Vitaminlike Nutrients

| Vitamin | Function |
|---------|----------|
| **Water-Soluble Vitamins** | |
| Thiamine ($B_1$) | Precursor of the coenzyme thiamine pyrophosphate. Deficiency can cause beriberi. |
| Riboflavin ($B_2$) | Precursor of the coenzymes flavin mononucleotide and flavin adenine dinucleotide. Deficiency leads to growth retardation. |
| Pyridoxine ($B_6$) | Precursor of the coenzyme pyridoxal phosphate. Deficiency causes dermatitis in rats. |
| Nicotinic acid (niacin) | Precursor of the coenzymes nicotinamide adenine dinucleotide and nicotinamide adenine dinucleotide phosphate. Deficiency leads to pellagra. |
| Pantothenic acid | Precursor of coenzyme A (CoA). Deficiency leads to dermatitis in chickens. |
| Biotin | Precursor of the coenzyme biocytin. Deficiency leads to dermatitis in humans. |
| Folic acid | Precursor of the coenzyme tetrahydrofolic acid. Deficiency causes anemias. |
| Vitamin $B_{12}$ | Precursor of the coenzyme deoxyadenosyl cobalamin. Deficiency leads to pernicious anemia. |
| Vitamin C | Cosubstrate in the hydroxylation of proline in collagen. Deficiency leads to scurvy. |
| **Lipid-Soluble Vitamins** | |
| Vitamin A | Vision, growth, and reproduction (see supplement 2). |
| Vitamin D | Regulation of calcium and phosphate metabolism (see chapter 24). |
| Vitamin E | Antisterility factor in rats. |
| Vitamin K | Important for blood coagulation. |
| **Vitaminlike Nutrients** | |
| Inositol | Mediator of hormone action (see chapters 13, 19, and 24). |
| Choline | Important for integrity of cell membranes and lipid transport (see chapter 19). |
| Carnitine | Essential for transfer of fatty acids to mitochondria (see chapter 18). |
| $\alpha$-Lipoic acid | Coenzyme in the oxidative decarboxylation of keto acids (this chapter). |
| *p*-Aminobenzoate (PABA) | Component of folic acid (this chapter). |
| Coenzyme Q (ubiquinones) | Important for electron transport in mitochondria (see chapter 14). |

## Figure 10.1

(a) Structure of thiamine pyrophosphate and (b, c) the bonds it cleaves or forms. Reactive part of the coenzyme and the bonds subject to cleavage in (b) and (c) are indicated in red.

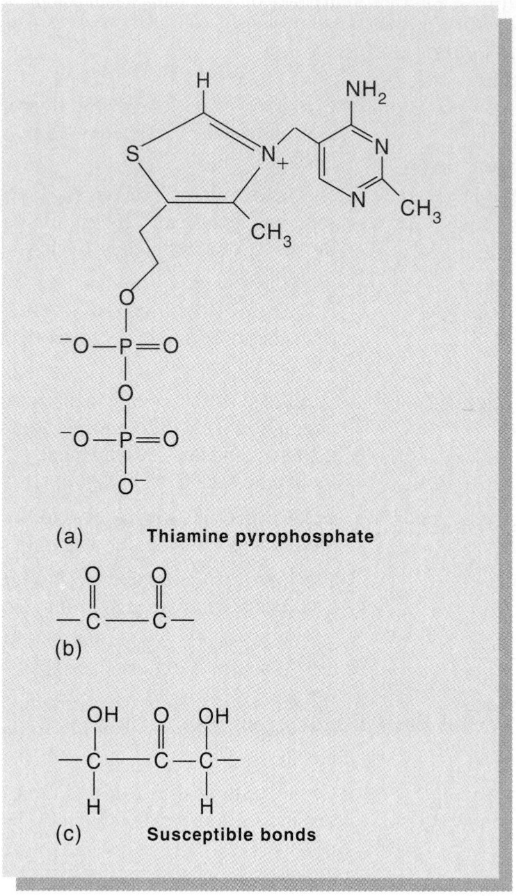

(a)   **Thiamine pyrophosphate**

(b)

(c)   **Susceptible bonds**

The active intermediate shown in equation (1) undergoes nucleophilic addition to the bond of polar carbonyl groups in substrates to produce intermediates, such as

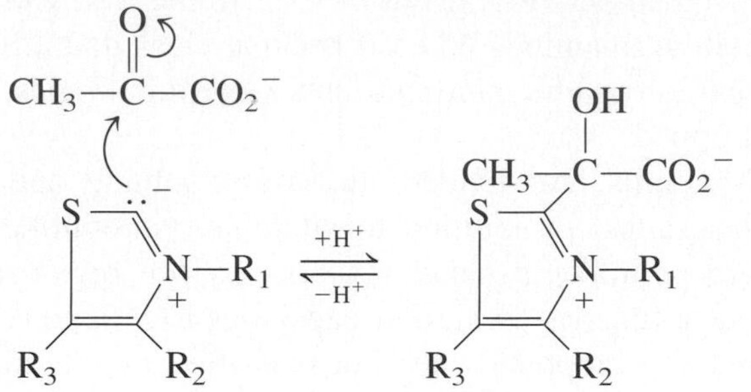

$$(2)$$

Intermediates of this type have the necessary chemical reactivity for cleaving the bonds indicated in figure 10.1b and c. The decarboxylated product of the pyruvate adduct shown in equation (2) is resonance-stabilized by the thiazolium ring (fig. 10.2a). This intermediate may be protonated to $\alpha$-hydroxyethyl thiamine pyrophosphate (fig. 10.2d); alternatively, it may react with other electrophiles, such as the carbonyl groups of acetaldehyde or pyruvate, to form the species in figure 10.2b and c; or it may be oxidized to acetyl-thiamine pyrophosphate (fig. 10.2e). The fate of the intermediate depends on the reaction specificity of the enzyme with which the coenzyme is associated.

## *Pyridoxal-5′-Phosphate Is Required for a Variety of Reactions with α-Amino Acids*

Pyridoxal-5′-phosphate is the coenzyme form of vitamin $B_6$, and has the structure shown in figure 10.3. The name vitamin $B_6$ is applied to any of a group of related compounds lacking the phosphoryl group, including pyridoxal, pyridoxamine, and pyridoxine.

Pyridoxal-5′-phosphate participates in many reactions with $\alpha$-amino acids, including transaminations, $\alpha$ decarboxylations, racemizations, $\alpha,\beta$ eliminations, $\beta,\gamma$ eliminations, aldolizations, and the $\beta$ decarboxylation of aspartic acid.

the carbon–carbon bond in figure 10.1c (the structures of these two compounds are shown in figure 10.2).

The mechanism for the bond cleavages indicated in figure 10.1b was clarified by Ronald Breslow. In one of the earliest applications of nuclear magnetic resonance to biochemical mechanisms, he demonstrated that the proton bonded to C-2 in the thiazolium ring is readily exchangeable with the protons of $H_2O$ and deuterons of $D_2O$ in a base-catalyzed reaction

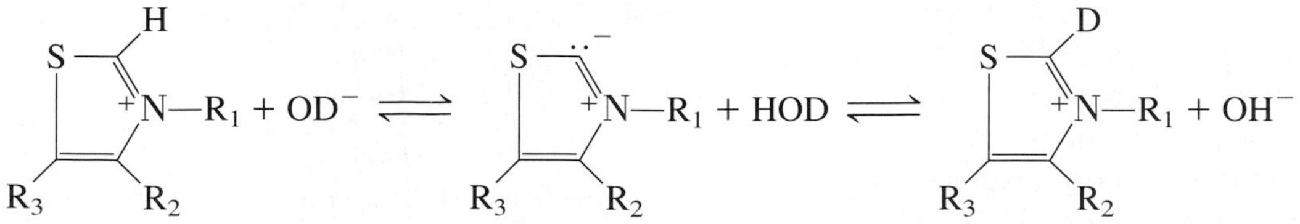

$$(1)$$

# Figure 10.2

Mechanism of thiamine pyrophosphate action. Intermediate (*a*) is represented as a resonance-stabilized species. It arises from the decarboxylation of the pyruvate-thiamine pyrophosphate addition compound shown at the left of (*a*) and in equation (2). It can react as a carbanion with acetaldehyde, pyruvate, or H⁺ to form (*b*), (*c*), or (*d*), depending on the specificity of the enzyme. It can also be oxidized to acetyl-thiamine pyrophosphate (TPP) (*e*) by other enzymes, such as pyruvate oxidase. The intermediates (*b*) through (*e*) are further transformed to the products shown by the actions of specific enzymes.

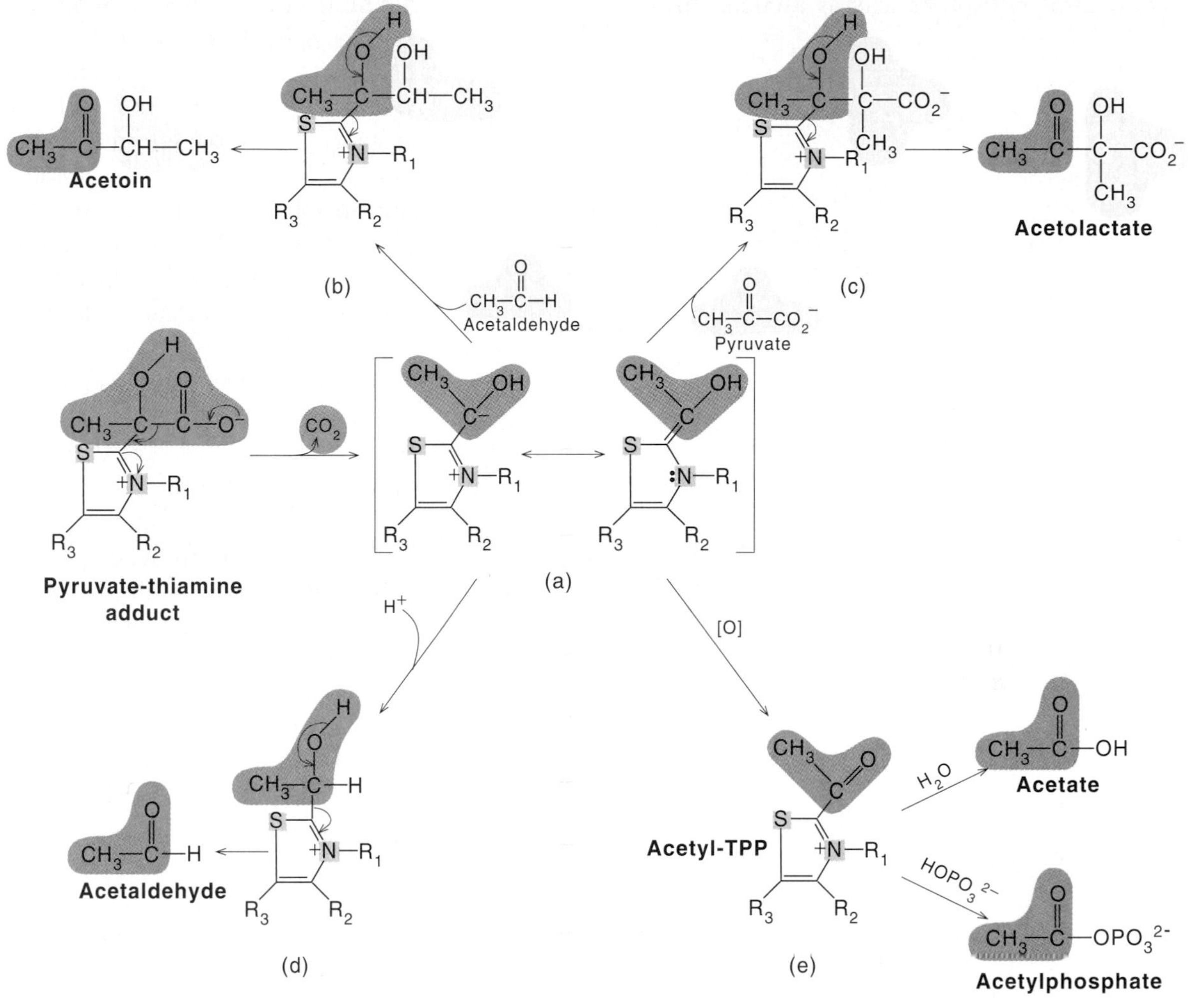

# Figure 10.3

Structures of vitamin $B_6$ derivatives and the bonds cleaved or formed by the action of pyridoxal phosphate (*a*). The reactive part of the coenzyme is shown in red in (*a*). The bonds shown in red in (*d*) are the types of bonds in substrates that are subject to cleavage.

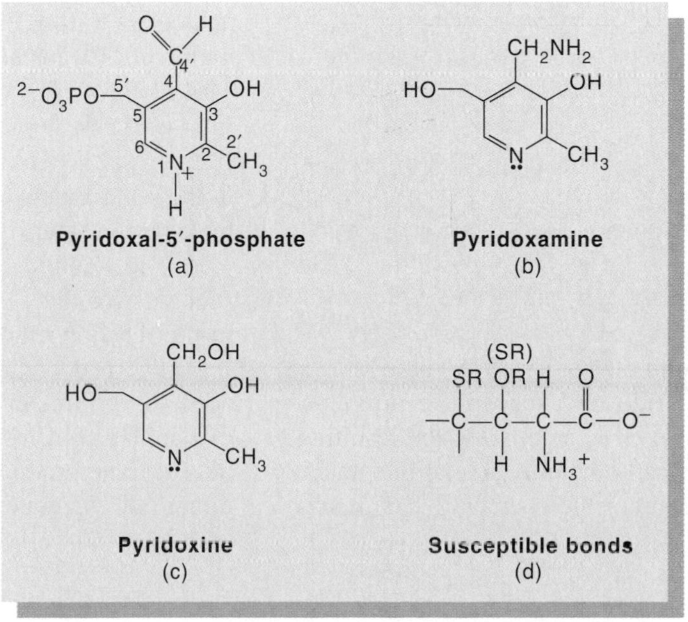

**Pyridoxal-5′-phosphate**
(a)

**Pyridoxamine**
(b)

**Pyridoxine**
(c)

**Susceptible bonds**
(d)

The following equations illustrate several reactions in which pyridoxal-5′-phosphate acts as a coenzyme.

$$R_1\text{—}\overset{\overset{\displaystyle H}{|}}{\underset{\underset{\displaystyle NH_3^+}{|}}{C}}\text{—}CO_2^- + R_2\text{—}\overset{\overset{\displaystyle O}{\|}}{C}\text{—}CO_2^- \xrightleftharpoons{\text{Transaminase}} \tag{3}$$

$$R_1\text{—}\overset{\overset{\displaystyle O}{\|}}{C}\text{—}CO_2^- + R_2\text{—}\overset{\overset{\displaystyle H}{|}}{\underset{\underset{\displaystyle NH_3^+}{|}}{C}}\text{—}CO_2^-$$

$$R\text{—}\overset{\overset{\displaystyle H}{|}}{\underset{\underset{\displaystyle NH_3^+}{|}}{C}}\text{—}CO_2^- + H^+ \xrightleftharpoons{\text{Decarboxylase}}$$

$$CO_2 + R\text{—}CH_2\text{—}NH_3^+ \tag{4}$$

$$R\text{—}\overset{\overset{\displaystyle H}{|}}{\underset{\underset{\displaystyle NH_3^+}{|}}{C}}\text{—}CO_2^- \xrightleftharpoons{\text{Racemase}} R\text{—}\overset{\overset{\displaystyle NH_3^+}{|}}{\underset{\underset{\displaystyle H}{|}}{C}}\text{—}CO_2^- \tag{5}$$

$$R\text{—}\overset{\overset{\displaystyle OH}{|}}{CH}\text{—}\overset{\overset{\displaystyle H}{|}}{\underset{\underset{\displaystyle NH_3^+}{|}}{C}}\text{—}CO_2^- \xrightleftharpoons{\text{Aldolase}}$$

$$R\text{—}CHO + \overset{}{\underset{\underset{\displaystyle NH_3^+}{|}}{CH_2}}\text{—}CO_2^- \tag{6}$$

$$^-O_2C\text{—}CH_2\text{—}\overset{\overset{\displaystyle H}{|}}{\underset{\underset{\displaystyle NH_3^+}{|}}{C}}\text{—}CO_2^- + H^+ \xrightleftharpoons{\text{Aspartate-}\beta\text{-decarboxylase}}$$

$$CO_2 + CH_3\text{—}\overset{\overset{\displaystyle H}{|}}{\underset{\underset{\displaystyle NH_3^+}{|}}{C}}\text{—}CO_2^- \tag{7}$$

These equations involve bond cleavages of the type shown in color in figure 10.3$d$. Pyridoxal-5′-phosphate promotes these heterolytic bond cleavages by stabilizing the resulting electron pairs at the $\alpha$- or $\beta$-carbon atoms of $\alpha$-amino acids. To do this, the aldehyde group of the coenzyme first reacts

with the $\alpha$-amino group of an amino acid to produce an aldimine (fig. 10.4$a$) or Schiff's base, which is internally stabilized by H bonding. Loss of the $\alpha$ hydrogen as $H^+$ produces a resonance-stabilized species (fig. 10.4$b$) in which the electron pair is delocalized into the pyridinium system. This active intermediate may undergo further reactions at the carbon to form products determined by the reaction specificity of the enzyme. If, for example, the enzyme is a racemase, the species resulting from the loss of the proton from the $\alpha$ carbon may accept a proton from the opposite side to produce, ultimately, the enantiomer of the amino acid.

When the substrate is substituted at the $\beta$ carbon with a potential leaving group, such as —OH, —SH, $—OPO_3^{3-}$ (see fig. 10.3$d$), the corresponding $\alpha$-carbanion intermediate (see fig. 10.4$b$) can eliminate the group. This is an essential step in $\alpha,\beta$ eliminations. Upon hydrolysis, the elimination intermediate produces pyridoxal-5′-phosphate and the substrate-derived enamine, which spontaneously hydrolyzes to ammonia and an $\alpha$-keto acid.

The full series of intermediates in a transamination is shown in figure 10.5$a$. After protonation at the aldimine carbon of pyridoxal-5′-phosphate (step 3), hydrolysis (step 4) forms an $\alpha$-keto acid and pyridoxamine-5′-phosphate. The reverse of this sequence with a second $\alpha$-keto acid (steps 5 through 8) completes the transamination reaction.

An intermediate analogous to that in figure 10.4$b$ but generated from glycine and so lacking the $\beta$ and $\gamma$ carbons, can react as a carbanion with an aldehyde to produce a $\beta$-hydroxy-$\alpha$-amino acid. These reactions are catalyzed by aldolases, such as threonine aldolase or serine hydroxymethyl transferase.

$\beta$-Decarboxylases (fig. 10.5$b$) generate intermediates analogous to those in figure 10.4$b$ by catalyzing the elimination of $CO_2$ instead of $H^+$ from the intermediate in figure 10.4$a$ (step 4, fig. 10.5$b$). Protonation of the $\alpha$-carbanionic intermediates by protons from $H_2O$, followed by hydrolysis of the resulting imines, produces the amines corresponding to the replacement of the carboxylate group in the substrate by a proton (steps 5 through 8, fig. 10.5$b$).

The stability of the resonance hybrid (see fig. 10.4$b$) accounts for the catalytic action of pyridoxal-5′-phosphate in the reactions shown in equations (3) through (6).

The $\beta$ decarboxylation of aspartate (equation 7) proceeds by elimination of a $\beta$-carbanionic intermediate like that in figure 10.4$d$ from the ketimine, analogous to the intermediate produced by loss of the $\alpha$ proton from the aldimine of aspartate with pyridoxal-5′-phosphate.

## Figure 10.4

Structures of catalytic intermediates in pyridoxal-phosphate–dependent reactions. The initial aldimine intermediate resulting from Schiff's base formation between the coenzyme and the $\alpha$-amino group of an amino acid (a). This aldimine is converted to the resonance-stabilized intermediate (b) by loss of a proton at the $\alpha$ carbon. Further enzyme-catalyzed proton transfers to intermediates (c) and (d) may occur, depending on the specificity of a given enzyme. The enzymes use their general acids and bases to catalyze these proton transfers.

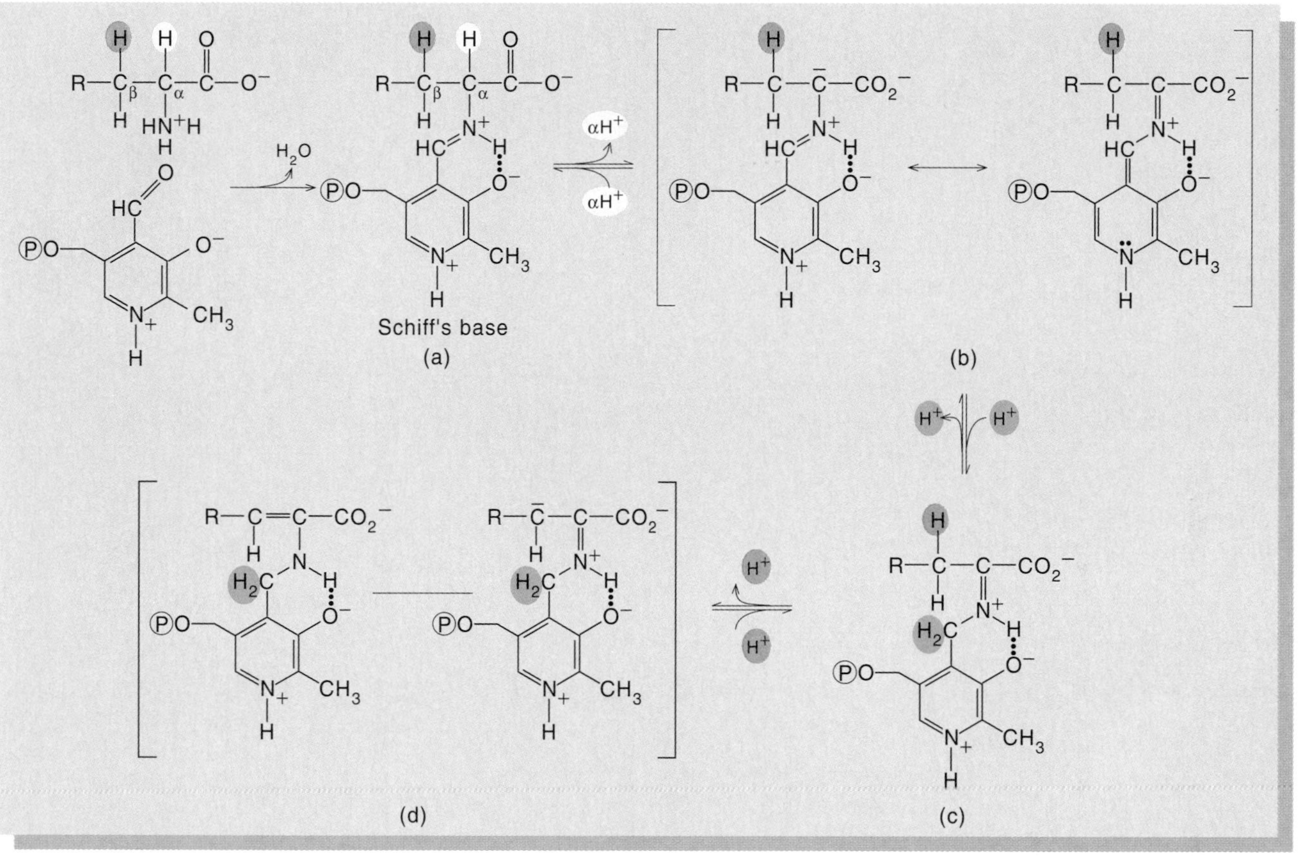

The fundamental biochemical function of pyridoxal-5'-phosphate is the formation of aldimines with $\alpha$-amino acids that stabilize the development of carbanionic character at the $\alpha$ and $\beta$ carbons of $\alpha$-amino acids in intermediates, such as those in figures 10.4b and c. Enzymes acting alone cannot stabilize these carbanions and so cannot, by themselves, catalyze reactions requiring their formation as intermediates.

## Nicotinamide Coenzymes Are Used in Reactions Involving Hydride Transfers

Nicotinamide adenine dinucleotide ($NAD^+$) is one of the two coenzymatic forms of nicotinamide (fig. 10.6). The other is nicotinamide adenine dinucleotide phosphate ($NADP^+$), which differs from $NAD^+$ by the presence of a phosphate group at C-2' of the adenosyl moiety.

The nicotinamide coenzymes are biological carriers of reducing equivalents (electrons). The most common function of $NAD^+$ is to accept two electrons and a proton ($H^-$ equivalent) from a substrate undergoing metabolic oxidation to produce NADH, the reduced form of the coenzyme. This then diffuses or is transported to the terminal-electron transfer sites of the cell and reoxidized by terminal-electron acceptors, $O_2$ in aerobic organisms, with the concomitant formation of ATP (chapter 14). Equations (8), (9), and (10) are typical reactions in which $NAD^+$ acts as such an acceptor.

# Figure 10.5

Mechanisms of action of pyridoxal phosphate: (*a*) in glutamate-oxaloacetate transaminase, and (*b*) in aspartate β-decarboxylase.

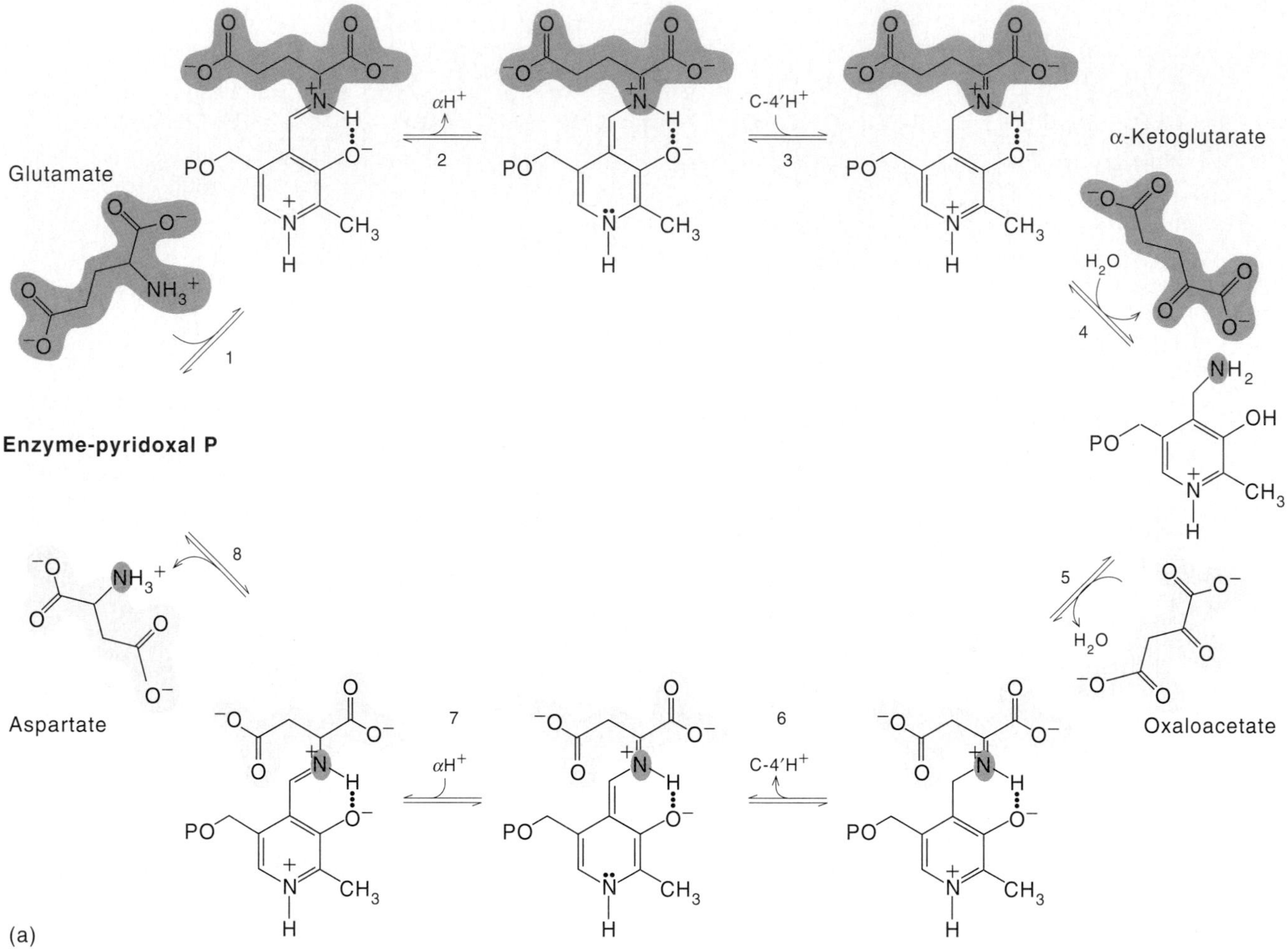

(a)

$$\text{NAD}^+ + \text{CH}_3\text{CH}_2\text{OH} \overset{\substack{\text{Alcohol} \\ \text{dehydrogenase}}}{\rightleftharpoons} \quad \text{CH}_3\!-\!\overset{\displaystyle O}{\overset{\|}{\text{CH}}} + \text{NADH} + \text{H}^+ \qquad \textbf{(8)}$$

$$^-\text{O}_2\text{C(CH}_2)_2\overset{\displaystyle \overset{+}{\text{NH}_3}}{\text{CHCO}_2{}^-} + \text{NAD}^+ + \text{H}_2\text{O} \overset{\substack{\text{Glutamate} \\ \text{dehydrogenase}}}{\rightleftharpoons} \quad ^-\text{O}_2\text{C(CH}_2)_2\overset{\displaystyle O}{\overset{\|}{\text{CCO}_2}}{}^- + \text{NADH} + \text{NH}_4{}^+ + \text{H}^+ \qquad \textbf{(9)}$$

$$\text{HPO}_4{}^{2-} + {}^{2-}\text{O}_3\text{POCH}_2\overset{\displaystyle \text{OH}}{\overset{|}{\text{CH}}}\!-\!\overset{\displaystyle O}{\overset{\|}{\text{CH}}} + \text{NAD}^+ \overset{\substack{\text{Glyceraldehyde-3P} \\ \text{dehydrogenase}}}{\rightleftharpoons} {}^{2-}\text{O}_3\text{POCH}_2\overset{\displaystyle \text{OH}}{\overset{|}{\text{CH}}}\overset{\displaystyle O}{\overset{\|}{\text{C}}}\text{OPO}_3{}^{2-} + \text{NADH} + \text{H}^+ \qquad \textbf{(10)}$$

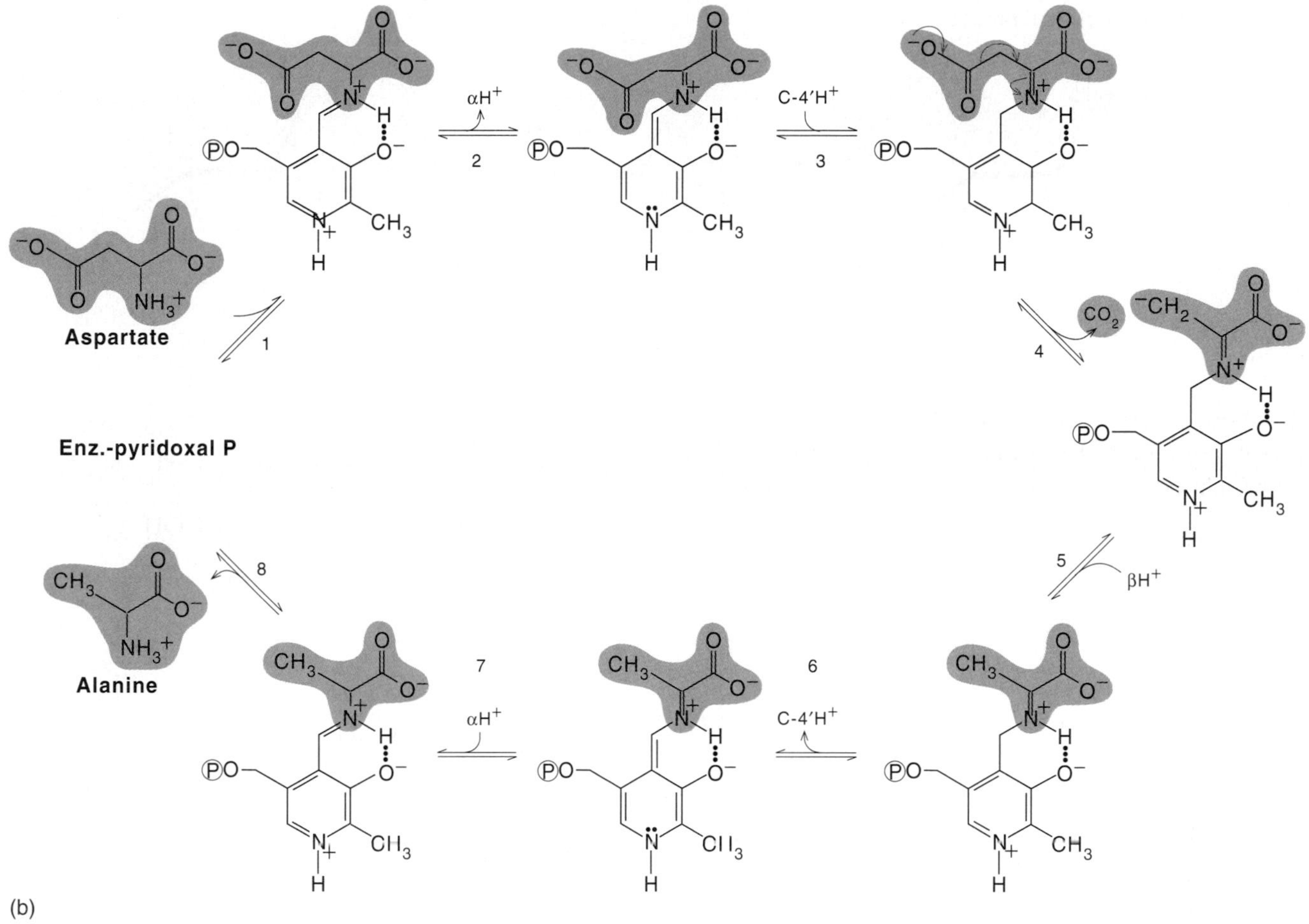

(b)

## Figure 10.6

Structures of nicotinamide and nicotinamide coenzymes. The reactive
sites of the coenzymes are shown in red.

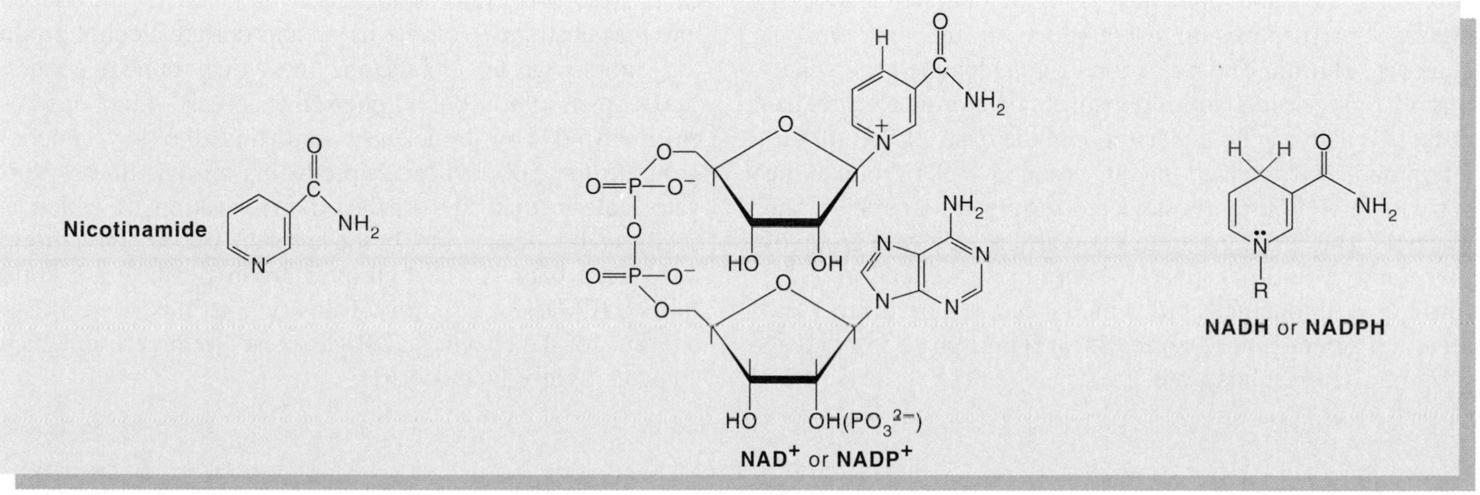

## Figure 10.7

Mechanism of $NAD^+$ action in UDPgalactose-4-epimerase. No net oxidation or reduction occurs. Only the intermediate is oxidized.

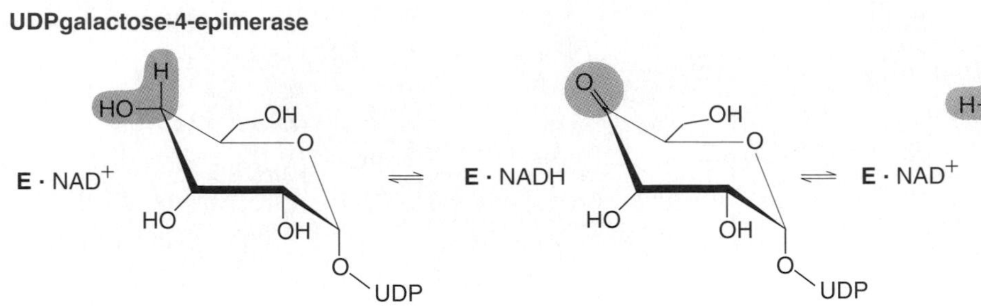

**UDPgalactose-4-epimerase**

The chemical mechanisms by which $NAD^+$ is reduced to NADH in equations (8) through (10) are probably similar, as represented in generalized forms in equation (11).

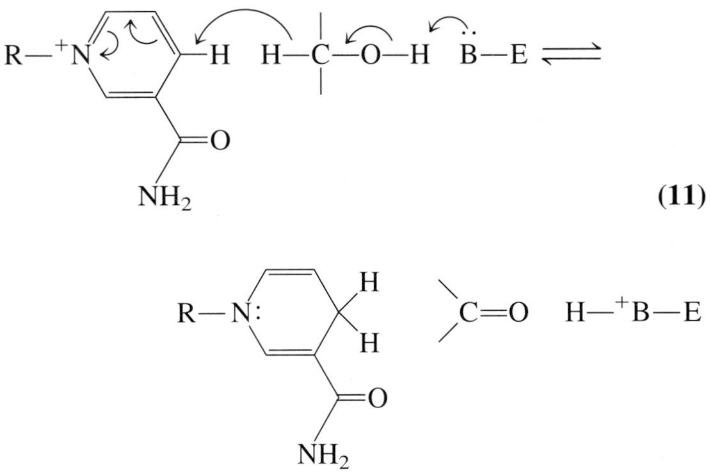

$$(11)$$

According to this formulation, the immediate oxidation product in equation (9), where $-NH_2$ replaces $-OH$ in equation (11), is the imine of $\alpha$-ketoglutarate, which quickly undergoes hydrolysis to $\alpha$-ketoglutarate and ammonia in aqueous solution. The oxidation of an aldehyde group catalyzed by glyceraldehyde-3-phosphate dehydrogenase [equation (10)] also can be understood on the basis of this formulation once it is realized that an essential $-SH$ group at the active site is transiently acylated during the course of the reaction. The $-SH$ group reacts with the aldehyde group of glyceraldehyde-3-phosphate according to equation (12), forming a thiohemiacetal, which becomes oxidized. The resulting acylenzyme then reacts with phosphate to produce glycerate-1,3-bisphosphate.

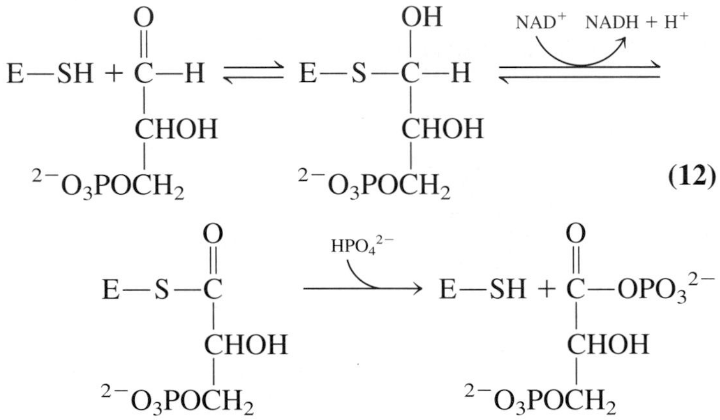

$$(12)$$

In addition to acting as a cellular electron carrier, $NAD^+$ also acts as a true coenzyme with certain enzymes. Enzymes are sometimes confronted with the problem of catalyzing such reactions as epimerizations, aldolizations, and eliminations on substrates lacking the intrinsic chemical reactivities required for these reactions to occur at significant rates. Sometimes such reactivities can be introduced into the substrate by oxidizing an appropriate alcohol group to a carbonyl group, and the enzyme is then found to contain $NAD^+$ as a tightly bound coenzyme. $NAD^+$ functions coenzymatically by transiently oxidizing the key alcohol group to the carbonyl level, producing an oxidatively activated intermediate the further transformation of which is catalyzed by the enzyme. In the last step, the carbonyl group is reduced back to the hydroxyl group by the transiently formed NADH. A reaction of this type is illustrated in figure 10.7 for the enzyme UDP-galactose-4-epimerase, which contains tightly bound $NAD^+$.

**Figure 10.8**

Structures of the vitamin riboflavin (*a*) and the derived flavin coenzymes (*b*). Like NAD$^+$ and NADP$^+$, the coenzyme pair FMN and FAD are functionally equivalent coenzymes, and the coenzyme involved with a given enzyme appears to be a matter of enzymatic binding specificity. The catalytically functional portion of the coenzymes is shown in red.

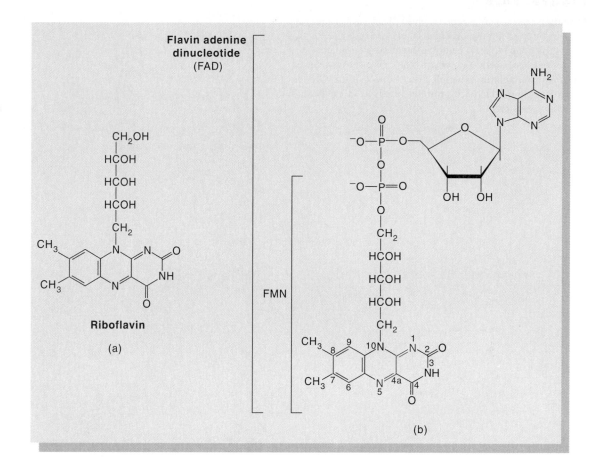

## *Flavins Are Used in Reactions Involving One or Two Electron Transfers*

Flavin adenine dinucleotide (FAD) (fig. 10.8) and flavin mononucleotide (FMN) are the coenzymatically active forms of vitamin B$_2$, riboflavin. Riboflavin is the N$^{10}$-ribityl isoalloxazine portion of FAD, which is enzymatically converted into its coenzymatic forms first by phosphorylation of the ribityl C-5' hydroxy group to FMN and then by adenylylation to FAD. FMN and FAD are functionally equivalent coenzymes, and the one that is involved with a given enzyme appears to be a matter of enzymatic binding specificity.

The catalytically functional portion of the coenzymes is the isoalloxazine ring, specifically N-5 and C-4a (see fig. 10.8*b*), which is thought to be the immediate locus of catalytic function, although the entire chromophoric system extending over N-5, C-4a, C-10a, N-1, and C-2 should be regarded as an indivisible catalytic entity, as are the nicotinamide, pyridinium, and thioazolium rings of NAD$^+$, pyridoxal phosphate, and thiamine pyrophosphate, respectively.

Flavin-containing enzymes are known as flavoproteins and, when purified, normally contain their full complements of FAD or FMN. The bright yellow color of flavoproteins is due to the isoalloxazine chromophore in its oxidized form. In a few flavoproteins, the coenzyme is known to be covalently bonded to the protein by means of a sulfhydryl or imidazole group at the C-8 methyl group and in at least one case at C-6. In most flavoproteins, the coenzymes are tightly but noncovalently bound, and many can be resolved into apoenzymes that can be reconstituted to holoenzymes by readdition of FAD or FMN.

Flavin coenzymes exist in three spectrally distinguishable oxidation states that account in part for their catalytic functions; the yellow oxidized form, the red or blue one-electron reduced form, and the colorless two electron re-

## Figure 10.9

Oxidation states of flavin coenzymes. The flavin coenzymes exist in three spectrally distinguishable oxidation states that account in part for their catalytic functions. They are the yellow oxidized form, the red or blue one-electron reduced form, and the colorless two-electron reduced form. Groups in red are those which are centrally involved in oxidation–reduction reactions.

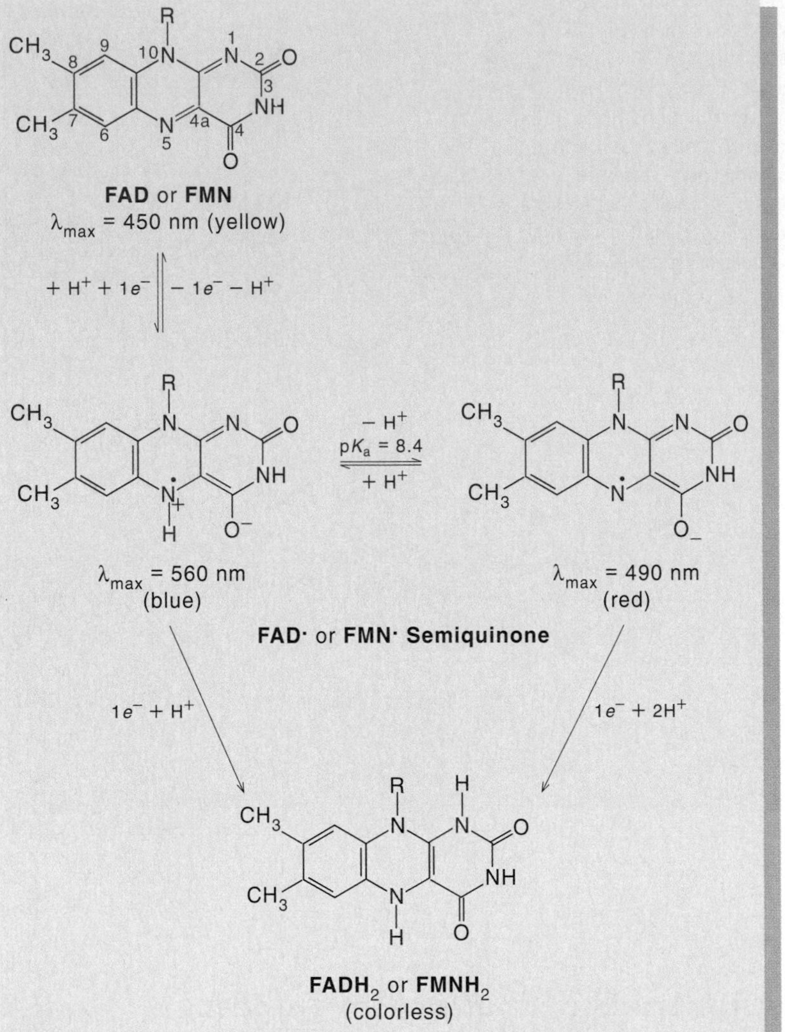

**FAD** or **FMN**
$\lambda_{max}$ = 450 nm (yellow)

$+ H^+ + 1e^-$  ⇌  $- 1e^- - H^+$

$\lambda_{max}$ = 560 nm (blue)

$- H^+$
$pK_a$ = 8.4
$+ H^+$

$\lambda_{max}$ = 490 nm (red)

**FAD·** or **FMN·** Semiquinone

$1e^- + H^+$          $1e^- + 2H^+$

**FADH$_2$** or **FMNH$_2$**
(colorless)

duced form. Their structures are depicted in figure 10.9. These and other less well defined forms often have been detected spectrally as intermediates in flavoprotein catalysis.

Flavins are very versatile redox coenzymes. Flavoproteins are dehydrogenases, oxidases, and oxygenases that catalyze a variety of reactions on an equal variety of substrate types. Since these classes of enzymes do not consist exclusively of flavoproteins, it is difficult to define catalytic specificity for flavins. Biological electron acceptors and donors in flavin-mediated reactions can be two-electron acceptors, such as $NAD^+$ or $NADP^+$, or a variety of one-electron acceptor systems, such as cytochromes ($Fe^{2+}$/ $Fe^{3+}$) and quinones, and molecular oxygen is an electron acceptor for flavoprotein oxidases as well as the source of oxygen for oxygenases. The only obviously common aspect of flavin-dependent reactions is that all are redox reactions.

Typical reactions catalyzed by flavoproteins are listed in table 10.2, which groups flavoproteins into those that do not utilize molecular oxygen as a substrate and those that do. You can best appreciate the significance of this difference when you realize that the reduced form of FAD (FADH$_2$), a likely intermediate in many flavoprotein reactions, spontaneously reacts with $O_2$ to produce $H_2O_2$. In the case of the dehydrogenases, therefore, either FADH$_2$ is not an intermediate or is somehow prevented from reacting with $O_2$. Among the dehydrogenases are two that utilize the two-electron acceptor substrates $NAD^+$ or $NADP^+$, and it is reasonable to suppose that the two-electron reduction of $NAD^+$ by an intermediate E · FADH$_2$ is involved. Also listed in table 10.2 are other dehydrogenases for which the electron acceptors from E · FADH$_2$ are not given. These enzymes are membrane-bound and transfer electrons directly to membrane-bound acceptors, mainly one-electron

## Table 10.2

Reactions Catalyzed by Flavoproteins

| Flavoprotein | Reaction |
|---|---|
| **Dehydrogenases** | |
| Glutathione reductase | $H^+ + GSSG + NADPH \rightleftharpoons 2\,GSH + NADP^+$ |
| Acyl-CoA dehydrogenases | $RCH_2CH_2COSCoA + NAD^+ \rightleftharpoons RCH{=}CHCOSCoA + NADH + H^+$ |
| Succinate dehydrogenase | $^-O_2CCH_2CH_2CO_2^- + E \cdot FAD \rightleftharpoons {}^-O_2CCH{=}CHCO_2^- + E \cdot FADH_2$ |
| D-Lactate dehydrogenase | $CH_3{-}CHOH{-}CO_2^- + E \cdot FAD \rightleftharpoons CH_3{-}CO{-}CO_2^- + E \cdot FADH_2$ |
| **Oxidases** | |
| Amino acid oxidases | $R{-}\overset{\overset{\displaystyle NH_3^+}{\mid}}{C}H{-}CO_2^- + O_2 + H_2O \longrightarrow R{-}CO{-}CO_2^- + H_2O_2 + NH_4^+$ |
| Monoamine oxidase | $R{-}CH_2\overset{+}{N}H_3 + O_2 + H_2O \longrightarrow R{-}CHO + H_2O_2 + \overset{+}{N}H_4$ |
| **Monooxygenases** | |
| Lactate oxidase | $CH_3{-}CHOH{-}CO_2^- + O_2 \longrightarrow CH_3{-}CO_2^- + CO_2 + H_2O$ |
| Salicylate hydroxylase | $2H^+ + \underset{}{\text{(salicylate)}}{-}CO_2^- + O_2 + NADH \longrightarrow \underset{}{\text{(catechol)}}{-}OH + CO_2 + NAD^+ + H_2O$ |

acceptors, such as quinones and cytochromes ($Fe^{2+}/Fe^{3+}$). The stability of the flavin semiquinone, FAD · and FMN · in figure 10.9, gives flavins the capability of interacting with one-electron acceptors in electron-transport systems.

The other classes of flavoproteins in table 10.2 interact with molecular oxygen either as the electron-acceptor substrates in redox reactions catalyzed by oxidases or as the substrate sources of oxygen atoms for oxygenases. Molecular oxygen also serves as an electron acceptor and source of oxygen for metalloflavoproteins and dioxygenases, which are not listed in the table. These enzymes catalyze more complex reactions, involving catalytic redox components, such as metal ions and metal–sulfur clusters in addition to flavin coenzymes.

A recurrent theme in many flavoprotein reactions is the probable involvement of $FADH_2$ or the reduced form of FMN ($FMNH_2$) as transient intermediates. Figure 10.10 illustrates a reasonable catalytic pathway for the first enzyme listed in table 10.2; this reaction shows the likely involvement of E · $FADH_2$ in each case. The mechanisms by which E · FAD is reduced to E · $FADH_2$ by NADPH in the for-

ward direction and by glutathione in the reverse direction are undoubtedly different.

The biochemical importance of flavin coenzymes appears to be their versatility in mediating a variety of redox processes, including electron transfer and the activation of molecular oxygen for oxygenation reactions. An especially important manifestation of their redox versatility is their ability to serve as the switch point from the two-electron processes, which predominate in cytosolic carbon metabolism, to the one-electron transfer processes, which predominate in membrane-associated terminal electron-transfer pathways. In mammalian cells, for example, the end products of the aerobic metabolism of glucose are $CO_2$ and NADH (see chapter 13). The terminal electron-transfer pathway is a membrane-bound system of cytochromes, nonheme iron proteins, and copper-heme proteins—all one-electron acceptors that transfer electrons ultimately to $O_2$ to produce $H_2O$ and $NAD^+$ with the concomitant production of ATP from ADP and $P_i$. The interaction of NADH with this pathway is mediated by NADH dehydrogenase, a flavoprotein that couples the two-electron oxidation of NADH with the one-electron reductive processes of the membrane.

**Figure 10.10**

Mechanism of the flavin-dependent glutathione reductase reaction. The first steps, not shown, involve the reduction of FAD to FADH$_2$ by NADPH and the binding of glutathione (glutathione is a sulfhydryl compound, see figure 22.15). The mechanism by which oxidized glutathione is reduced by the E · FADH$_2$ is shown.

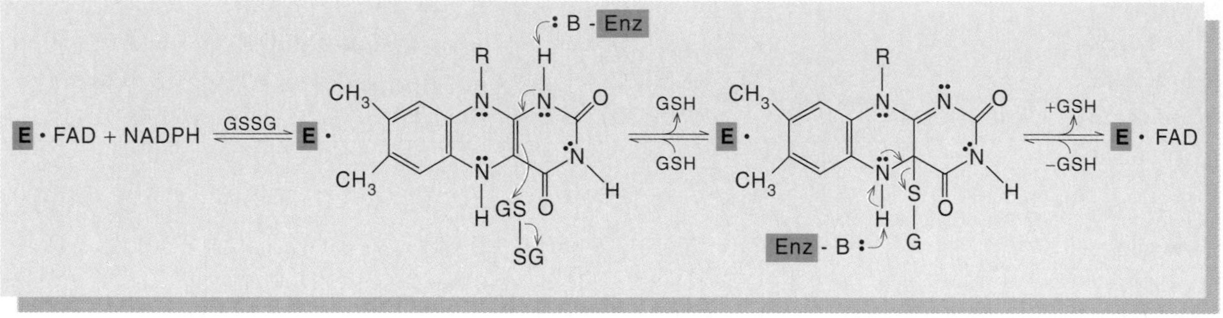

## Reactions Requiring Acyl Activation Frequently Use Phosphopantetheine Coenzymes

4′-Phosphopantetheine coenzymes are the biochemically active forms of the vitamin pantothenic acid. In figure 10.11, 4′-phosphopantetheine is shown as covalently linked to an adenylyl group in coenzyme A; or it can also be linked to a protein such as a serine hydroxyl group in acyl carrier protein (ACP). It is also found bonded to proteins that catalyze the activation and polymerization of amino acids to polypeptide antibiotics. Coenzyme A was discovered, purified, and structurally characterized by Fritz Lipmann and colleagues in work for which Lipmann was awarded the Nobel Prize in 1953.

The sulfhydryl group of the β-mercaptoethylamine (or cysteamine) moiety of phosphopantetheine coenzymes is the functional group directly involved in the enzymatic reactions for which they serve as coenzymes. From the standpoint of the chemical mechanism of catalysis, it is the essential functional group, although it is now recognized that phosphopantetheine coenzymes have other functions as well. Many reactions in metabolism involve acyl-group transfer or enolization of carboxylic acids that exist as unactivated carboxylate anions at physiological pH. The predominant means by which these acids are activated for acyl transfer and enolization is esterification with the sulfhydryl group of pantetheine coenzymes.

The mechanistic importance of activation is exemplified by the condensation of two molecules of acetyl-coenzyme A to acetoacetyl-coenzyme A catalyzed by β-ketothiolase:

$$CH_3-\overset{\displaystyle O}{\overset{\|}{C}}-SCoA + CH_3-\overset{\displaystyle O}{\overset{\|}{C}}-SCoA \rightleftharpoons$$

$$CH_3-\overset{\displaystyle O}{\overset{\|}{C}}-CH_2-\overset{\displaystyle O}{\overset{\|}{C}}-SCoA + CoASH \quad (13)$$

The two important steps of the reaction depend on both acetyl groups being activated, one for enolization and the other for acyl-group transfer. In the first step, one of the molecules must be enolized by the intervention of a base to remove an α proton, forming an enolate:

$$B:H-CH_2-\overset{\displaystyle O}{\overset{\|}{C}}-SCoA \rightleftharpoons$$

$$\overset{+}{B}-H + CH_2\overset{\displaystyle O^-}{=}\overset{\|}{C}-SCoA \quad (14)$$

The enolate is stabilized by delocalization of its negative charge between the α carbon and the acyl oxygen atom and by interactions with enzymatic groups, making it thermodynamically accessible as an intermediate. Moreover, this developing charge is also stabilized in the transition state preceding the enolate, so it is also kinetically accessible; that is, it is rapidly formed. If, by contrast, the same enolization reaction were carried out by the acetate anion, it would result in the generation of a second negative charge in the enolate, an energetically and kinetically unfavorable process.

**Figure 10.11**

Structures of the vitamin pantothenic acid (in red) and coenzyme A. The terminal —SH (in blue) is the reactive group in coenzyme A (CoASH).

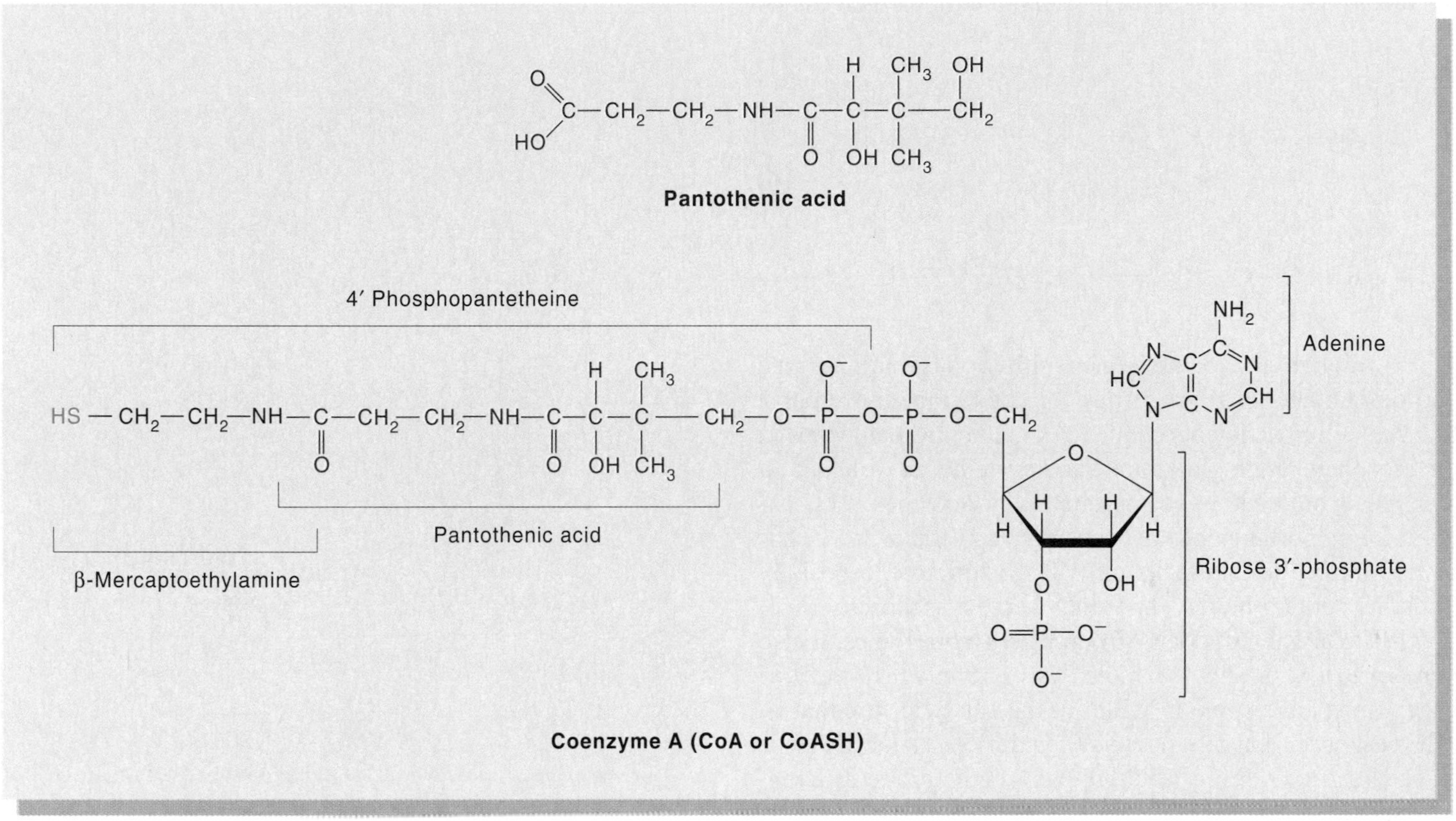

**Pantothenic acid**

**Coenzyme A (CoA or CoASH)**

The second stage of the condensation is the reaction of the enolate anion with the acyl group of a second molecule of acetyl-CoA:

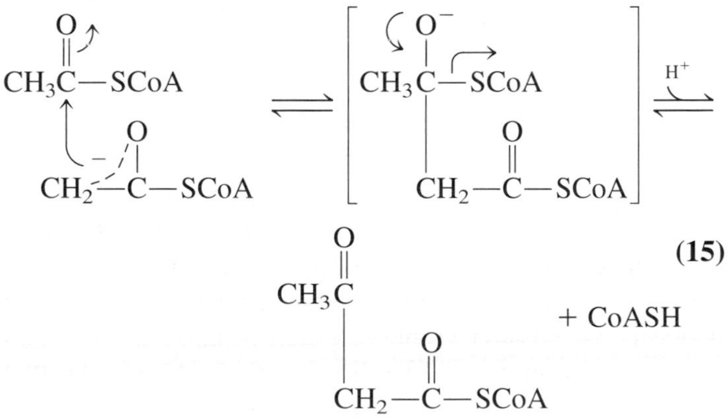

(15)

Nucleophilic addition to the neutral activated acyl group is a favored process, and coenzyme A is a good leaving group from the tetrahedral intermediate. The occurrence of this process with the acetate anion, that is, acetate reacting with an enolate anion, again provides a sharp contrast with the process of equation (15), for it would entail the nucleophilic addition of an anion to an anionic center, generating a dianionic transition state—an unfavorable process from both thermodynamic and kinetic standpoints. Moreover, the resulting intermediate would not have a very good leaving group other than the enolate anion itself, so the transition-state energy for acetoacetate formation would be high. Finally, the $K_{eq}$ for the condensation of 2 mol of acetate to 1 mol of acetoacetate is not favorable in aqueous media, whereas the condensation of 2 mol of acetyl-CoA to produce acetoacetyl-CoA and coenzyme A is thermodynamically spontaneous. The maintenance of metabolic carboxylic acids involved in enolization and acyl-group transfer reactions as coenzyme A esters provides the ideal lift over the kinetic and thermodynamic barriers to these reactions.

The foregoing discussion, in emphasizing the purely electrostatic energy barriers, does not address the question of whether there is an activation advantage in thiol esters relative to oxygen esters. Why thiol esters in preference to oxygen esters? Thiol esters are more readily enolized than oxygen esters. They are more "ketonelike" because of their electronic structures, in which the degree of resonance-electron delocalization from the sulfur atom to the acyl group is less than that of oxygen esters. As a result the charge-separated resonance form is a smaller contributor to the electronic structure in thiol esters than in oxygen esters.

$$\left[ R_1-\overset{\overset{\text{O}}{\|}}{\text{C}}-\overset{..}{\underset{..}{\text{S}}}-R_2 \longleftrightarrow R_1-\overset{\overset{\text{O}^-}{|}}{\text{C}}=\overset{+}{\text{S}}-R_2 \right] \tag{16}$$

$$\left[ R_1-\overset{\overset{\text{O}}{\|}}{\text{C}}-\overset{..}{\underset{..}{\text{O}}}-R_2 \longleftrightarrow R_1-\overset{\overset{\text{O}^-}{|}}{\text{C}}=\overset{+}{\text{O}}-R \right]$$

Although the pantetheine sulfhydryl group has the appropriate chemical properties for activating acyl groups, this characteristic is not unique to pantetheine coenzymes in the biosphere. Both glutathione and cysteine, as well as cysteamine, would serve, so the chemistry does not itself explain the importance of these coenzymes. Coenzyme A has many binding determinants in its large structure, especially in the nucleotide moiety, so it may serve a specificity function in the binding of coenzyme A esters by enzymes. It also may serve as a binding "handle" in cases in which the acyl group must have some mobility in the catalytic site, that is, if it must enolize at one site and then diffuse a short distance to undergo an addition reaction to a ketonic group of a second substrate.

One system in which pantetheine almost certainly performs such a carrier role is the fatty acid synthase from *E. coli,* in which 4'-phosphopantetheine is a component of the acyl carrier protein (see chapter 18).

## α-Lipoic Acid Is the Coenzyme of Choice for Reactions Requiring Acyl-Group Transfers Linked to Oxidation–Reduction

α-Lipoic acid is the internal disulfide of 6,8-dithiooctanoic acid, the structural formula of which is given in figure 10.12. It is the coupler of electron and group transfers catalyzed by α-keto acid dehydrogenase multienzyme complexes. The pyruvate and α-ketoglutarate dehydrogenase complexes are centrally involved in the metabolism of carbohydrates by the glycolytic pathway (chapter 12) and the tricarboxylic acid cycle (chapter 13). They catalyze two of the three decarboxylation steps in the complete oxidation of glucose, and they produce NADH and activated acyl compounds from the oxidation of the resulting ketoacids:

**Figure 10.12**

(*a*) Lipoic acid. (*b*) Reduced lipoid acid. (*c*) Lipoic acid bound to the ε-amino group of a lysine residue. Structures shown in red are those centrally involved in coenzyme reactions.

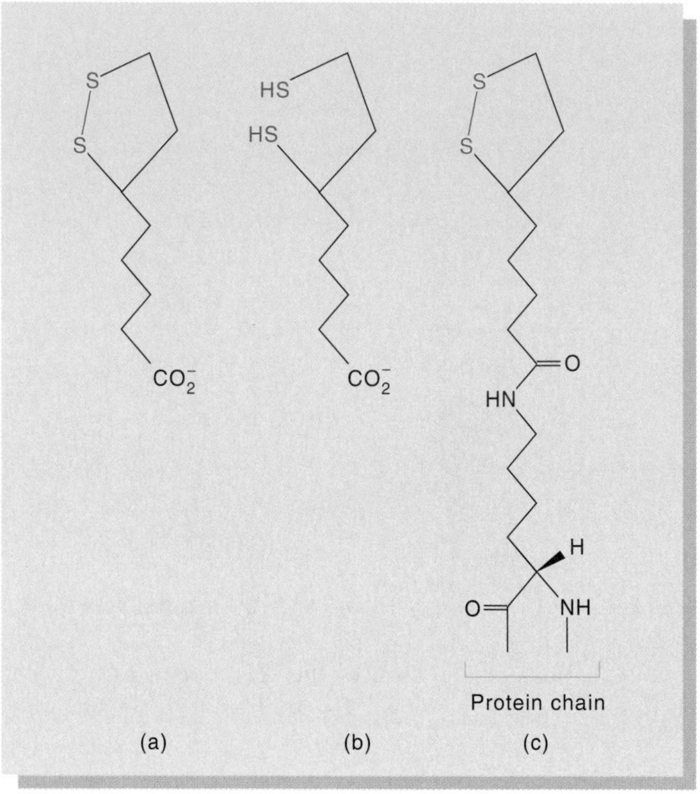

(a)    (b)    (c)

$$R-\overset{\overset{\text{O}}{\|}}{\text{C}}-CO_2^- + NAD^+ + CoASH \longrightarrow$$

$$CO_2 + NADH + R-\overset{\overset{\text{O}}{\|}}{\text{C}}-SCoA \tag{17}$$

The chemical aspect of the coenzymatic action of α-lipoic acid is to mediate the transfer of electrons and activated acyl groups resulting from the decarboxylation and oxidation of α-keto acids within the complexes. In this process, lipoic acid is itself transiently reduced to dihydrolipoic acid (see fig. 10.12), and this reduced form is the acceptor of the activated acyl groups. Its dual role of electron and acyl-group acceptor enables lipoic acid to couple the two processes.

The coenzymatic capabilities of α-lipoyl groups result from a fusion of its chemical and physical properties, the ability to act simultaneously as both electron and acyl-group acceptor, the ability to span long distances to interact with sites separated by up to 2.8 nm, and the ability to act cooperatively with other $\alpha$-lipoyl groups by disulfide interchange to relay electrons and acyl groups through distances that exceed its reach. This reaction is discussed in chapter 13.

## Biotin Mediates Carboxylations

The biotin structure shown in figure 10.13 is an imidazolone ring cis-fused to a tetrahydrothiophene ring substituted at position 2 by valeric acid. In carboxylase enzymes, biotin is covalently bonded to the proteins by an amide linkage between its carboxyl group and a lysyl-$\epsilon$-NH$_2$ group in the polypeptide chain. This arrangement places the imidazolone ring at the end of a long flexible chain of atoms extending a maximum of about 1.4 nm from the $\alpha$ carbon of lysine.

Biotin is the essential coenzyme for carboxylation reactions involving bicarbonate as the carboxylating agent. Several reactions have been described in which ATP-dependent carboxylation occurs at carbon atoms activated for enolization by ketonic or activated acyl groups. One reaction is known in which a nitrogen atom of urea is carboxylated.

A general formulation of the ATP-dependent carboxylation of an $\alpha$ carbon by $^{18}$O-enriched bicarbonate is

$$RCH_2-\overset{\overset{O}{\|}}{C}-SCoA + ATP + HC^{18}O_3^- \xrightarrow{\text{Biotinyl carboxylase}}$$

$$H^+ + R-\underset{\underset{C^{18}O_2^-}{|}}{CH}-\overset{\overset{O}{\|}}{C}-SCoA + ADP + H^{18}OPO_3^{2-} \quad (18)$$

The appearance of $^{18}$O in inorganic phosphate verifies that the function of ATP in the reaction is essentially the "dehydration" of bicarbonate.

The ATP-dependent carboxylation of biotin by bicarbonate is believed to control the transient formation of carbonic-phosphoric anhydride, or "carboxyphosphate," as an active carboxylation intermediate:

$$HO-\overset{\overset{O}{\|}}{C}-O^- \xrightarrow[\text{ADP}]{\text{ATP}} {}^-O-\underset{\underset{OH}{|}}{\overset{\overset{O}{\|}}{P}}-O-\overset{\overset{O}{\|}}{C}-O^- \xrightarrow[P_i]{\text{Biotinyl-E}} \quad (19)$$

$$N^1\text{-carboxybiotinyl-E}$$

Figure 10.13

Structures of biotinyl enzyme and $N^1$-carboxybiotin. The reactive portions of the coenzyme and the active intermediate are shown in red. In carboxylase enzymes, biotin is covalently bonded to the proteins by an amide linkage between its carboxyl group and a lysyl-$\epsilon$-NH$_2$ group in the polypeptide chain.

D-Biotinyl-protein

$N^{1'}$-Carboxybiotin

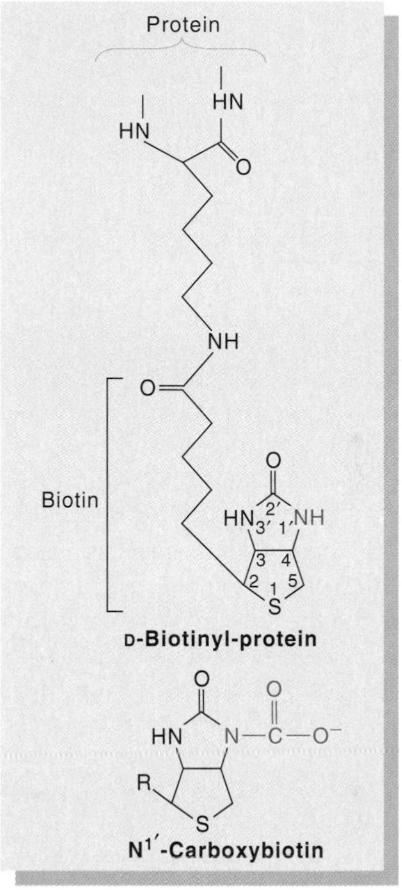

The coenzymatic function of biotin appears to be to mediate the carboxylation of substrates by accepting the ATP-activated carboxyl group and transferring it to the carboxyl acceptor substrate. There is good reason to believe that the enzymatic sites of ATP-dependent carboxylation of biotin are physically separated from the sites at which $N^1$-carboxybiotin transfers the carboxyl group to acceptor substrates, that is, the transcarboxylase sites. In fact, in the case of the acetyl-CoA carboxylase from E. coli (see chapter 18), these two sites reside on two different subunits, while the biotinyl group is bonded to a third, a small subunit designated biotin carboxyl carrier protein.

**Figure 10.14**

Structures and enzymatic interconversions of folate coenzymes. The reactive centers of the coenzymes are shown in red. The most active forms of the coenzyme contain oligo- or polyglutamyl groups.

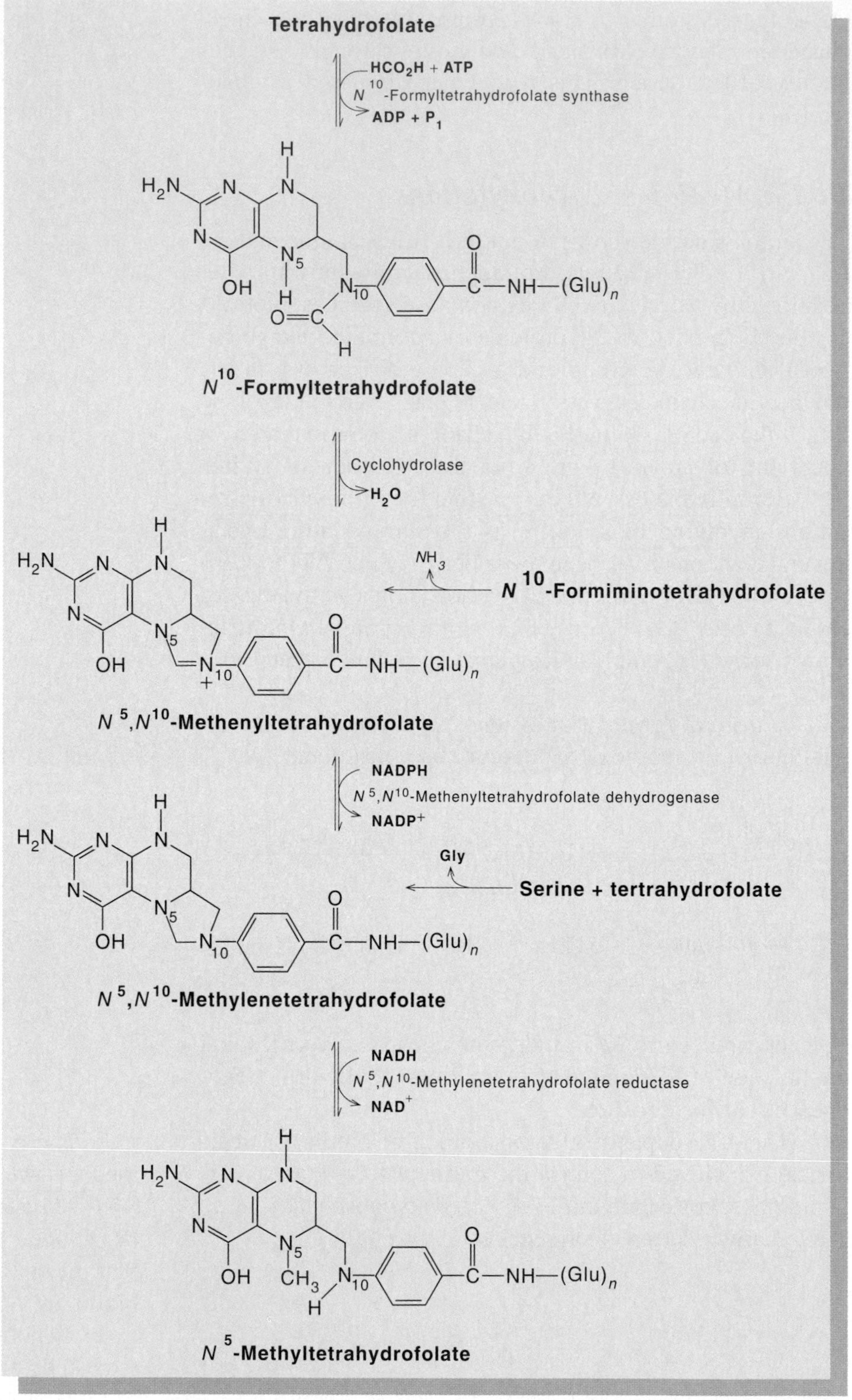

**Tetrahydrofolate**

$N^{10}$-Formyltetrahydrofolate synthase
HCO$_2$H + ATP
ADP + P$_1$

$N^{10}$-**Formyltetrahydrofolate**

Cyclohydrolase
H$_2$O

$N^{10}$-**Formiminotetrahydrofolate**
NH$_3$

$N^5,N^{10}$-**Methenyltetrahydrofolate**

$N^5,N^{10}$-Methenyltetrahydrofolate dehydrogenase
NADPH
NADP$^+$

Serine + tertrahydrofolate
Gly

$N^5,N^{10}$-**Methylenetetrahydrofolate**

$N^5,N^{10}$-Methylenetetrahydrofolate reductase
NADH
NAD$^+$

$N^5$-**Methyltetrahydrofolate**

Biotin appears to have just the right chemical and structural properties to mediate carboxylation. It readily accepts activated carboxyl groups at N$^1$ and maintains them in an acceptably stable yet reactive form for transfer to acceptor substrates. Since biotin is bonded to a lysyl group, the N$^1$-carboxyl group is at the end of a 1.6-nm chain with bond rotational freedom about nine single bonds, giving it the capability to transport activated carboxyl groups through space from the carboxyl activation sites to the carboxylation sites.

**Figure 10.15**

Involvement of folate coenzymes in one-carbon metabolism. Shown in red are the one-carbon units of the end products that originate with the reactive one-carbon units of the folate coenzymes.

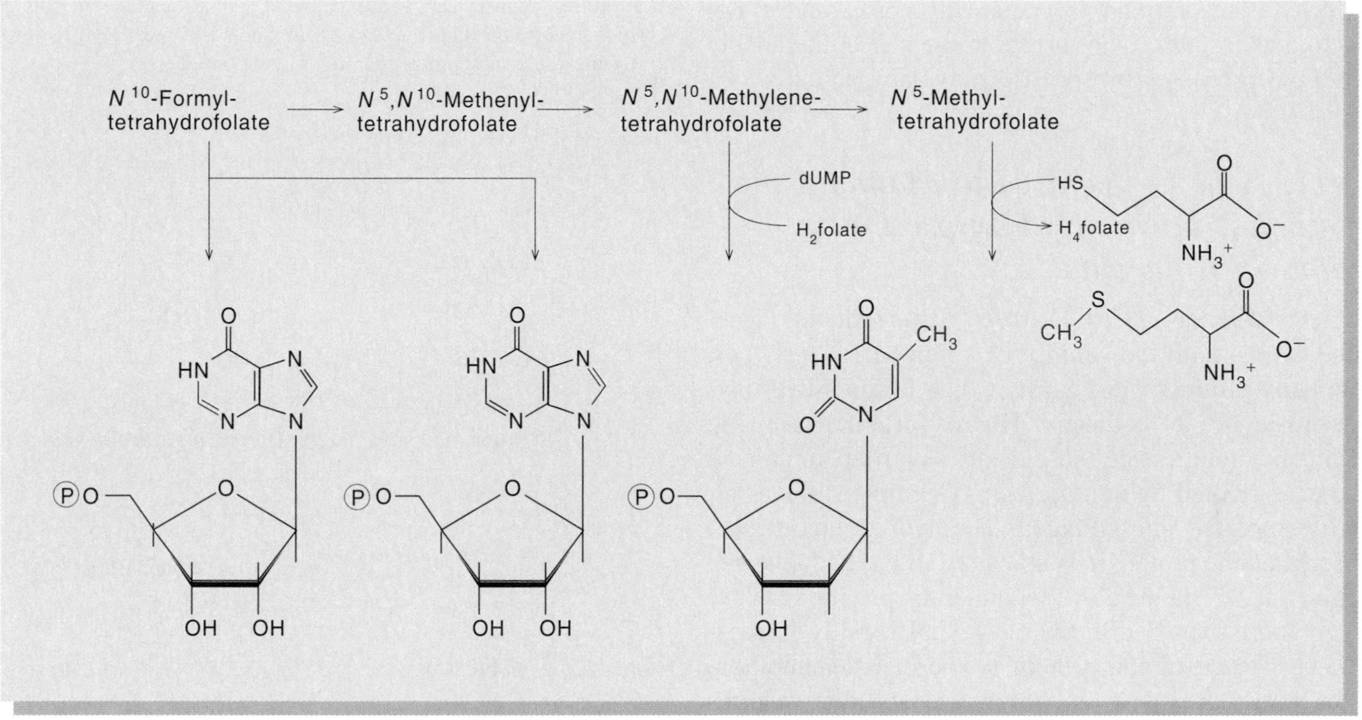

## Folate Coenzymes Are Used in Reactions for One-Carbon Transfers

Tetrahydrofolate and its derivatives $N^5,N^{10}$-methylene-tetrahydrofolate, $N^5,N^{10}$-methenyltetrahydrofolate, $N^{10}$-formyltetrahydrofolate, and $N^5$-methyltetrahydrofolate are the biologically active forms of folic acid, a four-electron oxidized form of tetrahydrofolate. The structural formulas are given in figure 10.14, which also shows how they arise from tetrahydrofolate. The structures are shown glutamylated on the carboxyl group of the $p$-aminobenzoyl group; the most active forms contain oligo- or polyglutamyl groups, linked through the $\gamma$-carboxyl groups.

The tetrahydrofolates do not function as tightly enzyme-bound coenzymes. Rather, they function as cosubstrates for a variety of enzymes associated with one-carbon metabolism. $N^{10}$-formyltetrahydrofolate is produced enzymatically from tetrahydrofolate and formate in an ATP-linked process in which formate is activated by phosphorylation to formyl phosphate: the formyl group of formyl phosphate is then transferred to $N^{10}$ of tetrahydrofolate. $N^{10}$-Formyltetrahydrofolate is a formyl donor substrate for some enzymes and is interconvertible with $N^5,N^{10}$-methenyltetrahydrofolate by the action of cyclohydrolase.

$N^5,N^{10}$-Methylenetetrahydrofolate is a hydroxymethyl-group donor substrate for several enzymes and a methyl-group donor substrate for thymidylate synthase (fig. 10.15). It arises in living cells from the reduction of $N^5,N^{10}$-methenyltetrahydrofolate by NADPH and also by the serine hydroxymethyltransferase-catalyzed reaction of serine with tetrahydrofolate.

$N^5$-Methyltetrahydrofolate is the methyl-group donor substrate for methionine synthase, which catalyzes the transfer of the five-methyl group to the sulfhydryl group of homocysteine. This and selected reactions of the other folate derivatives are outlined in figure 10.15, which emphasizes the important role tetrahydrofolate plays in nucleic acid biosynthesis by serving as the immediate source of one-carbon units in purine and pyrimidine biosynthesis.

Formaldehyde is a toxic substance that reacts spontaneously with amino groups of proteins and nucleic acids, hydroxymethylating them and forming methylene-bridge crosslinks between them. Free formaldehyde therefore wreaks havoc in living cells and could not serve as a useful hydroxymethylating agent. In the form of $N^5,N^{10}$-methylenetetrahydrofolate, however, its chemical reactivity is attenuated but retained in a potentially available form where needed for specific enzymatic action. Formate, however, is quite unreactive under physiological conditions and must be activated to serve as an efficient formylating agent. As $N^{10}$-formyltetrahydrofolate it is in a reactive state suitable for transfer to appropriate substrates. The fundamental biochemical importance of tetrahydrofolate is to maintain formaldehyde and formate in chemically poised states, not so reactive as to pose toxic threats to the cell but available for essential processes by specific enzymatic action.

## Ascorbic Acid Is Required to Maintain the Enzyme that Forms Hydroxyproline Residues in Collagen

Ascorbic acid (vitamin C; fig. 10.16) is the reducing agent required to maintain the activity of a number of enzymes, most notably proline hydroxylase, which forms 4-hydroxyproline residues in collagen. Hydroxyproline (see fig. 10.16c) is not synthesized biologically as a free amino acid but rather is created by modification of proline residues already incorporated into collagen. The hydroxylation reaction occurs as the protein is synthesized in the endoplasmic reticulum. At least a third of the numerous proline residues in collagen are modified in this way, substantially increasing the resistance of the protein to thermal denaturation.

Proline hydroxylase is a diooxygenase that requires ferrous iron as a cofactor and uses $\alpha$-ketoglutarate as its second substrate. One oxygen atom from $O_2$ is incorporated into hydroxyproline; the other goes to the $\alpha$-ketoglutarate, which decomposes to succinate and $CO_2$:

$$prolyl\text{-}peptide + {}^-O_2CCH_2CH_2\overset{\overset{O}{\|}}{C}-CO_2^- + O_2 \longrightarrow$$

$$hydroxyprolyl\text{-}peptide +$$

$$^-O_2CCH_2CH_2CO_2^- + CO_2 \quad (20)$$

Ascorbic acid is synthesized by plants and many animals but not by primates or guinea pigs. In scurvy, the disease associated with a severe deficiency of ascorbic acid, connective tissues throughout the body deteriorate. Weakening of the capillary walls results in hemorrhages, wounds heal poorly, and lesions occur in the bones.

## Vitamin $B_{12}$ Coenzymes Are Associated with Rearrangements on Adjacent Carbon Atoms

The principal coenzymatic form of vitamin $B_{12}$ is 5'-deoxyadenosylcobalamin, the structural formula of which is given in figure 10.17. The structure includes a cobalt–

**Figure 10.16**

(a) Ascorbic acid is a reductant that appears to be required to keep some enzymes in their active states. Oxidation converts ascorbic acid to (b) dehydroascorbic acid, which decomposes irreversibly by hydrolysis of the lactone ring. One enzyme that requires ascorbic acid is proline hydroxylase, which converts proline residues in collagen to (c) 4-hydroxyproline (hydroxy group shown in red). Hydroxyproline residues help to stabilize the structure of collagen.

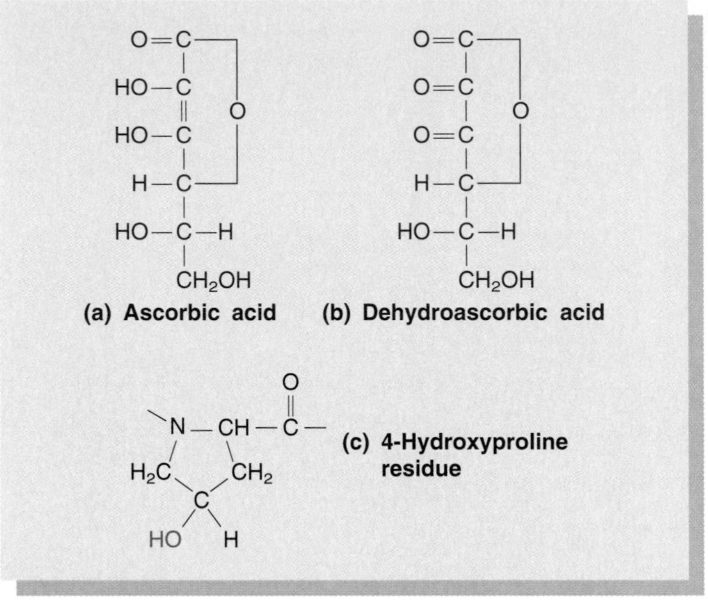

carbon bond between the 5' carbon of the 5'-deoxyadenosyl moiety and the cobalt (III) ion of cobalamin. [Note: The metal oxidation state may be denoted either as $Co^{3+}$ or as $Co(III)$. The former stresses the free-ion character, while the latter stresses the bound character.] Vitamin $B_{12}$ itself is cyanocobalamin, in which the cyano group is bonded to cobalt in place of the 5'-deoxyadenosyl moiety. Other forms of the vitamin have water (aquocobalamin) or the hydroxyl group (hydroxycobalamin) bonded to cobalt.

The vitamin was discovered in liver as the antipernicious anemia factor in 1926, but discovery of its complete structure had to await its purification, chemical characterization, and crystallization, which required more than 20 years. Even then the determination of such a complex structure proved to be an elusive goal by conventional approaches of that day and had to await the elegant x-ray crystallographic study of Lenhert and Hodgkin in 1961, for which Dorothy Hodgkin was awarded the Nobel Prize in 1964.

Most 5'-deoxyadenosylcobalamin-dependent enzymatic reactions are rearrangements that follow the pattern of equation (21), in which a hydrogen atom and another group

## Figure 10.17

Structure of 5'-deoxyadenosylcobalamin coenzyme (vitamin $B_{12}$). The reactive groups are shown in red.

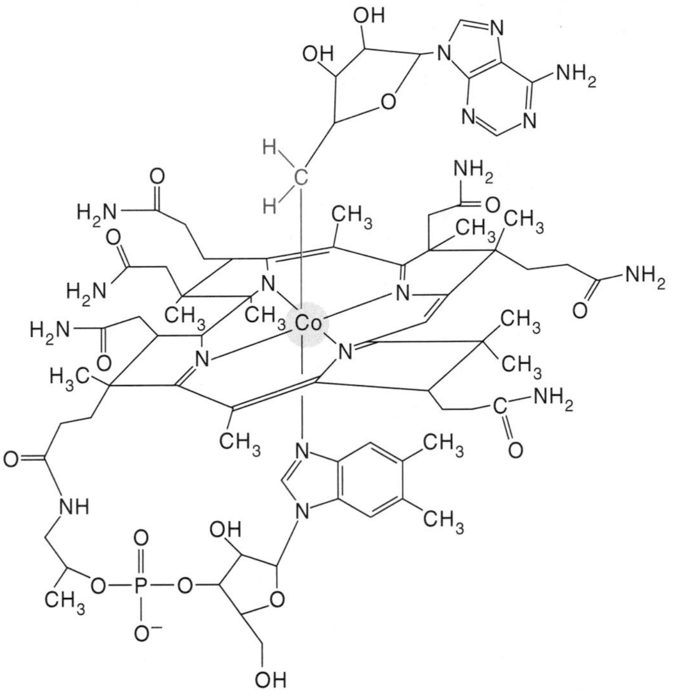

(designated X) bonded to an adjacent carbon atom exchange positions, with the group X migrating from $C_\alpha$ to $C_\beta$:

$$a-\underset{X}{\overset{b}{C_\alpha}}-\underset{H}{\overset{c}{C_\beta}}-d \rightleftharpoons a-\underset{H}{\overset{b}{C_\alpha}}-\underset{X}{\overset{c}{C_\beta}}-d \qquad (21)$$

Three specific examples of rearrangement reactions are given in equations (22) through (24). It is interesting and significant that the migrating groups —OH, —COSCoA, and —CH $(NH_3{}^+)CO_2{}^-$ have little in common and that the hydrogen atoms migrating in the opposite direction are often chemically unreactive.

$$^-O_2C-CH_2CH_2-\underset{NH_3{}^+}{CH}-CO_2{}^- \underset{\text{mutase}}{\overset{\text{Glutamate}}{\rightleftharpoons}}$$

$$^-O_2C-\underset{NH_3{}^+}{CH}-\overset{CH_3}{CH}-CO_2{}^- \qquad (22)$$

$$^-O_2C-CH_2CH_2-\overset{O}{C}-SCoA \underset{\text{mutase}}{\overset{\text{Methylmalonyl-CoA}}{\rightleftharpoons}}$$

$$^-O_2C-\overset{CH_3}{\underset{}{CH}}-\overset{O}{C}-SCoA \qquad (23)$$

$$CH_3\underset{OH}{CH}CH_2OH \xrightarrow{\text{Dioldehydrase}} CH_3CH_2\underset{OH}{CH}-OH \xrightarrow{\text{Dioldehydrase}}$$

$$CH_3CH_2CHO + H_2O \qquad (24)$$

Indeed, hydrogen migrations in all the $B_{12}$ coenzyme-dependent rearrangements proceed without exchange with the protons of water; that is, isotopic hydrogen in substrates is conserved in the products. This fact plus spectroscopic evidence implicating Co(II) and organic radicals as catalytic intermediates in the reaction have led to the proposal of the mechanism illustrated in figure 10.18.

The reaction begins by homolytic cleavage of the Co—C bond (fig. 10.18), generating Co(II) and 5'-deoxyadenosyl free radical (step 1). The radical abstracts a hydrogen atom from the substrate, the migrating hydrogen in equation (21), generating 5'-deoxyadenosine and a substrate-derived free radical as intermediates (step 2). The substrate-radical undergoes rearrangement to a product-derived free radical, which abstracts a hydrogen atom to form the final product and regenerate the coenzyme (steps 3–5).

The most fundamental property of 5'-deoxyadenosyl-cobalamin leading to its unique action as a coenzyme is the weakness of the Co—C bond. This bond has a low dissociation energy, less than 30 kcal/mole, strong enough to be essentially stable in free solution but weak enough to be broken as a result of strain induced by multiple binding interactions between the enzyme binding sites and the adenosyl and cobalamin portions of the coenzyme. The radicals resulting from cleavage of this bond and abstraction of hydrogen from substrates undergo the rearrangements characteristic of $B_{12}$-dependent reactions.

## Iron-Containing Coenzymes Are Frequently Involved in Redox Reactions

Iron as a cofactor in catalysis is receiving increasing attention. The most common oxidation states of iron are $Fe^{2+}$ and $Fe^{3+}$. Iron complexes are nearly all octahedral, and practically all are paramagnetic (as a result of unpaired electrons in the $3d$ orbital). The most common form of iron in biological systems is heme. Heme groups ($Fe^{2+}$) and hematin ($Fe^{3+}$) most frequently involve a complex with protoporphyrin IX (fig. 10.19). They are the coenzymes (prosthetic

## Figure 10.18

Hypothetical partial mechanism of vitamin $B_{12}$-dependent rearrangements. The designations Co(III) and Co(II) refer to species that are spectrally and magnetically similar to $Co^{3+}$ and $Co^{2+}$, respectively. Co(III) is diamagnetic and red, and Co(II) is paramagnetic (unpaired electron) and yellow. The metal does not undergo a change in electrostatic charge when the cobalt–carbon bond breaks homolytically (i.e., without charge separation), because one electron remains with the metal and the other with 5'-deoxyadenosine.

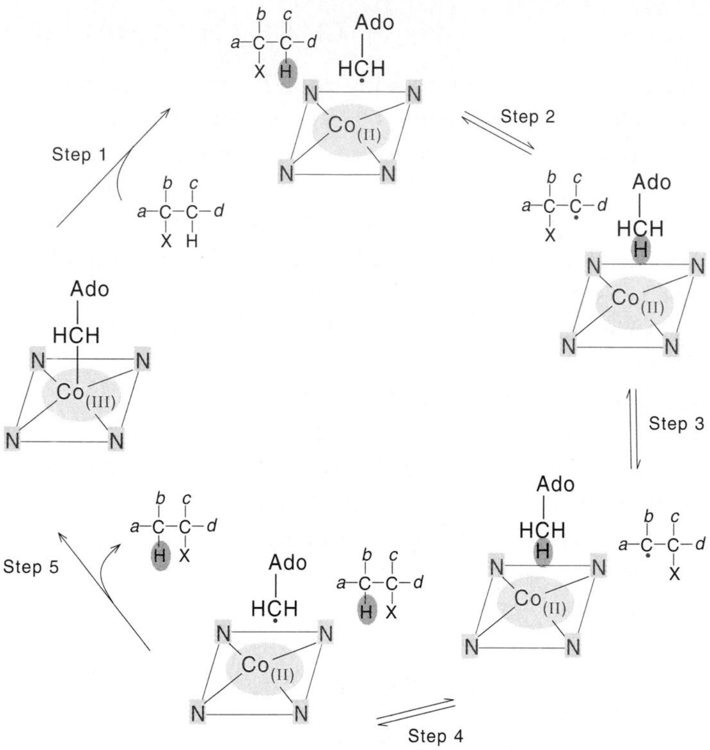

## Figure 10.19

Structure of protoporphyrin IX. This coenzyme acts in conjunction with a number of different enzymes involved in oxidation and reduction reactions.

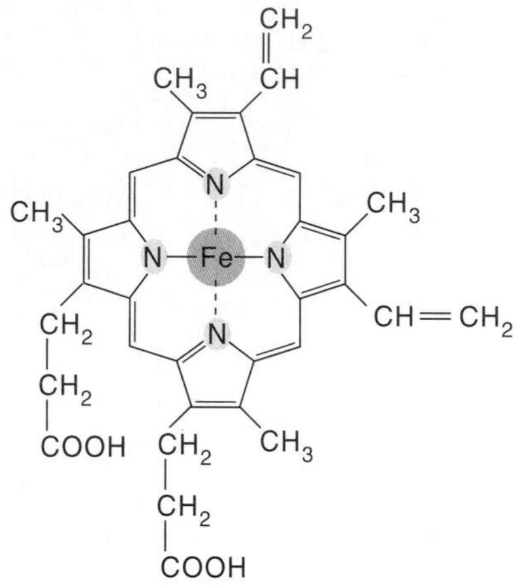

the identities of the upper axial ligands donated by the substrates. Spectral data show clearly that the heme coenzymes participate directly in catalysis; however, the mechanisms of action of hemes are not as well understood as those of other coenzymes.

Many redox enzymes contain iron–sulfur clusters that mediate one-electron transfer reactions. The clusters consist of two or four irons and an equal number of inorganic sulfide ions clustered together with the iron, which is also liganded to cysteinyl-sulfhydryl groups of the protein (fig. 10.20). The enzyme nitrogenase, which catalyzes the reduction of $N_2$ to $2 NH_3$ contains such clusters in which some of the iron has been replaced by molybdenum (see chapter 21). Electron-transferring proteins involved in one-electron transfer processes often contain iron–sulfur clusters. These proteins include the mitochondrial membrane enzymes NADH dehydrogenase and succinate dehydrogenase (chapter 13), which are flavoproteins, and the small-molecular-weight proteins ferredoxin, rubredoxin, adrenodoxin, and putidaredoxin (chapters 14, 15, and 20).

Heme coenzymes, iron–sulfur clusters, flavin coenzymes, and nicotinamide coenzymes cooperate in multienzyme systems to catalyze the chemically remarkable hydroxylations of hydrocarbons such as steroids (chapter 20). In these hydroxylation systems, the heme proteins constitute a family of proteins known as cytochrome P450, named for the wavelength corresponding to the most intense absorption band of the carbon monoxide-liganded heme, an inhib-

groups) for a number of redox enzymes, including catalase, which catalyzes dismutation of hydrogen peroxide (equation (25)), and peroxidases, which catalyze the reduction of alkyl hydroperoxides by such reducing agents as phenols, hydroquinones, and dihydroascorbate (represented as $AH_2$ in equation (26)).

$$2 H_2O_2 \rightleftharpoons 2 H_2O + O_2 \qquad (25)$$

$$R—O—O—H + AH_2 \longrightarrow A + R—O—H + H_2O \qquad (26)$$

Heme proteins exhibit characteristic visible absorption spectra as a result of protoporphyrin IX; their spectra differ depending on the identities of the lower axial ligand donated by the protein and the oxidation state of the iron as well as

## Figure 10.20

Structures of iron–sulfur clusters. Many redox enzymes contain iron–sulfur clusters that mediate one-electron transfer reactions.

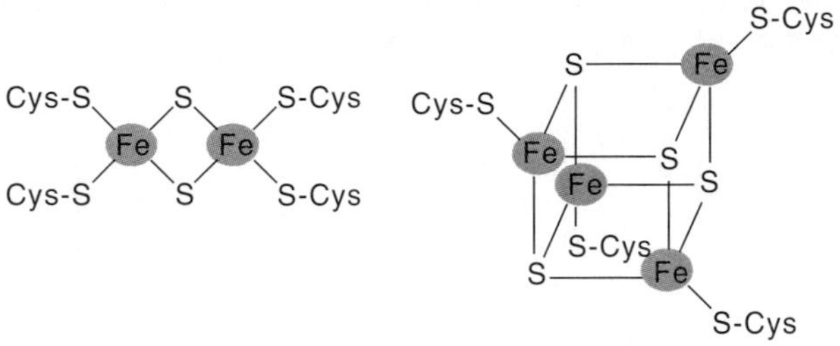

## Figure 10.21

Steps in the formation of the oxygenating species of cytochrome P450. The oxygenating species of cytochrome P450 is generated by the transfer of two electrons in discrete steps from NADPH via the flavoprotein reductase and iron–sulfur clusters to the iron porphyrin complex, together with the reaction with oxygen. The Fe(III)-peroxide then undergoes a further expulsion of water to form the oxygenating species shown in brackets. The oxygenating species is not directly observed, since it reacts quickly, but it is thought to contain Fe(IV) and a delocalized radical-cation in the porphyrin ring.

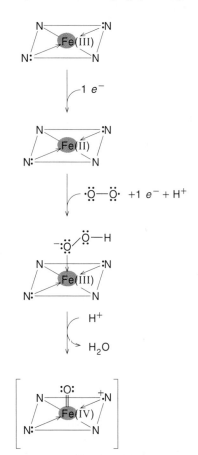

Oxygenating intermediate

ited form. The reactions catalyzed by these systems are represented in generalized form by equation (27), which also shows the fate of the two oxygens from $^{18}O_2$.

$$H^+ + NADPH + {}^{18}O_2 + R\!-\!\underset{\underset{H}{|}}{\overset{\overset{H}{|}}{C}}\!-\!R \longrightarrow$$

$$R\!-\!\underset{\underset{H}{|}}{\overset{\overset{{}^{18}OH}{|}}{C}}\!-\!R + NADP^+ + H_2{}^{18}O \quad (27)$$

One oxygen of $O_2$ is incorporated into the hydroxyl group of the product, whereas the other is incorporated into water. The enzymes usually include a cytochrome P450, an iron–sulfur cluster-containing protein, such as adrenodoxin or putidaredoxin, and a flavoprotein reductase.

In the mechanism of oxygenation by these enzymes, the flavoproteins and iron–sulfur proteins supply reducing equivalents in one-electron units from NADPH to cytochrome P450, and reduced cytochrome P450 reacts directly with $O_2$. These reactions generate a Fe(III)-peroxide complex, shown in figure 10.21, which undergoes a further dehydration to an oxygenating species of cytochrome P450. The oxygenating species is thought to be an oxo-complex of Fe(IV), in which the porphyrin ring has been oxidized to a radical-cation by loss of an electron. The positive charge and unpaired electron in figure 10.21 are stabilized by delocalization through the conjugated $\pi$-electron system of the porphyrin ring (see fig. 10.19).

The most widely accepted mechanism for cytochrome P450-catalyzed oxygenation of substrates is illustrated for a hydrocarbon substrate in figure 10.22. The oxygenating species, the oxo-Fe(IV) radical-cation, abstracts a hydrogen

atom from the hydrocarbon to form a hydrocarbon radical and a hydroxy-Fe complex. Hydrogen abstraction proceeds by the pairing of one electron from the oxo-Fe(IV) species and one electron from the carbon–hydrogen bond of the hydrocarbon. In the second step of oxygenation the hydrocarbon radical abstracts the hydroxy group from the hydroxy-Fe(III) complex by a pairing of the unpaired radical electron with one electron of the hydroxyl group, leaving cytochrome P450 in the +3 oxidation state, where it began in figure 10.21. The mechanism shown in figure 10.21 is known as the "rebound" mechanism of oxygenation. The

## Figure 10.22

The rebound mechanism for oxygenation of hydrocarbons by cytochrome P450. The oxygenating species of cytochrome P450 in figure 10.21 is a highly reactive paramagnetic species that can abstract an unactivated and unreactive hydrogen from a hydrocarbon in the first step to form a hydrocarbon radical. In the rebound step the hydrocarbon radical abstracts a hydroxyl radical from the hydroxy-porphyrin intermediate.

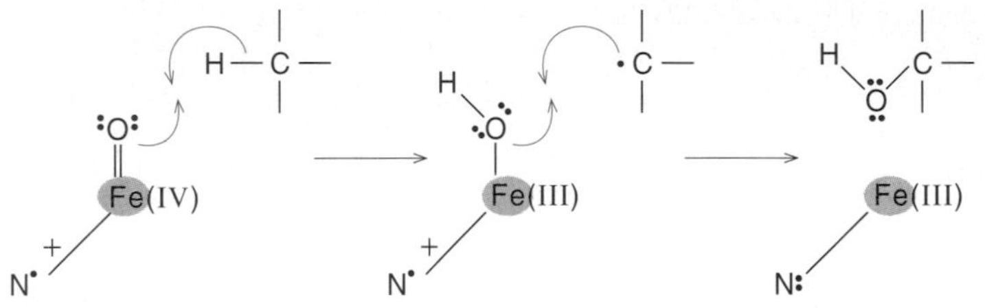

## Figure 10.23

Proposed role of $Zn^{2+}$ in carbonic anhydrase. Carbonic anhydrase catalyzes the reaction $H_2O + CO_2 \rightleftharpoons HCO_3^- + H^+$. In the enzyme, $Zn^{2+}$ forms a complex with $H_2O$. Proton displacement generates an $OH^-$ ion still bound to the $Zn^{2+}$. Nucleophilic attack of $CO_2$ by the $OH^-$ ion generates bicarbonate.

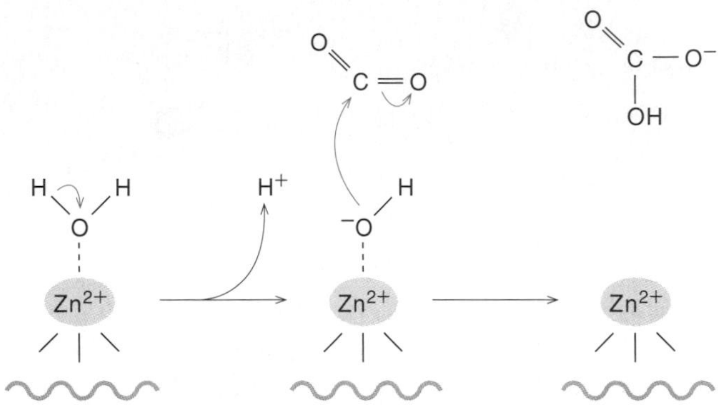

second step of the mechanism is the rebound step, and the rate constant for this step in a microsomal cytochrome P450 has been estimated to be greater than $10^{10}$ s$^{-1}$. The magnitude of this rate constant explains why the radical and hydroxy-Fe intermediates have not been detected by presently available spectroscopic methods.

## Metal Cofactors

In addition to cobalt and iron (discussed above), other metals frequently function as cofactors in enzyme-catalyzed reactions. Like coenzymes, they are useful because they offer something not available in amino acid side chains. The most important of such features of metals are their high concentration of positive charge, their directed valences for interacting with two or more ligands, and their ability to exist in two or more valence states.

The alkali metal and alkaline-earth metal ions are spherically symmetrical with respect to charge distribution, so they do not usually show directed valences. There are notable exceptions, such as the case of $Mg^{2+}$ in chlorophyll (see chapter 15), which adopts an approximately square planar distribution of orbitals. Alkali metals, $Na^+$ and $K^+$, with their single positive charges and lack of $d$-electronic orbitals for sharing, almost never make tight complexes with proteins. On rare occasions the alkaline-earth divalent cations, $Mg^{2+}$ and $Ca^{2+}$, can make strong complexes. The alkali metal and alkaline-earth cations are asymmetrically distributed in the organism, with $K^+$ and $Mg^{2+}$ being concentrated in the cytosol and $Na^+$ and $Ca^{2+}$ being concentrated in the organelles and the blood.

Most of the remaining metals found in biological systems are from the first transition series, in which the $3d$

orbitals are only partially filled. These metals all prefer structures with multiple coordination numbers, and, except for $Zn^{2+}$, which has a filled $3d$ orbital, they can exist in more than one oxidation state, a feature making them potentially useful in oxidation–reduction reactions. The stability of zinc's electronic state combined with its preference for a coordination number of 4–6 makes zinc singularly useful in enzyme reactions, in which it acts as a Lewis acid (e.g., see fig. 10.23), and in structural situations, in which it chelates nitrogen (usually in histidine side chains) and/or sulfur-containing amino acid side chains (usually cysteines) to rigidify the structural domain of a protein (see fig. 31.19).

The involvement of transition metals in biochemical reactions is discussed in later chapters.

## Lipid-Soluble Vitamins

We have seen that most water-soluble vitamins are converted by single or multiple steps to coenzymes. Our understanding of the lipid-soluble vitamins (see table 10.1) and of how they are utilized by the organism is much less extensive. We briefly discuss the structure and functions of some lipid-soluble vitamins.

Vitamin $D_3$ (cholecalciferol) can be made in the skin from 7-dehydrocholesterol in the presence of ultraviolet light (see fig. 24.13). Vitamin $D_3$ is formed by the cleavage of ring 3 of 7-dehydrocholesterol. Vitamin $D_3$ made in skin or absorbed from the small intestine is transported to the liver and hydroxylated at C-25 by a microsomal mixed-

**Figure 10.24**

Structures of vitamins $K_1$ and $K_2$. $K_1$ is found in plants, $K_2$ in animals and bacteria.

**Vitamin $K_1$**

(phylloqinone)

**Vitamin $K_2$**

(menaqinone)

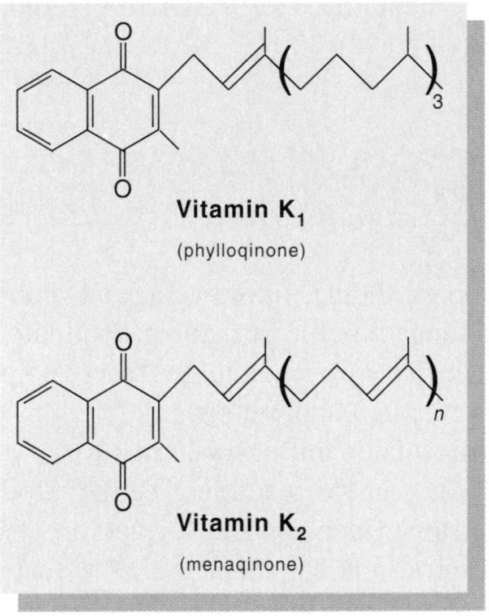

**Figure 10.25**

Vitamin-K-dependent carboxylation of a glutamic acid residue in a protein. This reaction is essential to blood clotting.

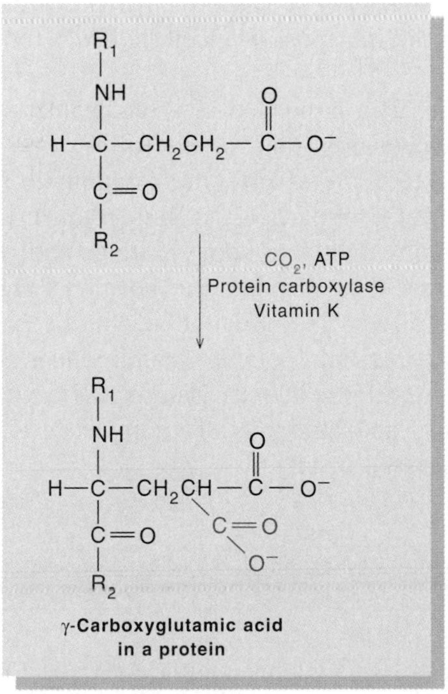

CO₂, ATP
Protein carboxylase
Vitamin K

**γ-Carboxyglutamic acid in a protein**

**Figure 10.26**

Structure of vitamin E (α-tocopherol).

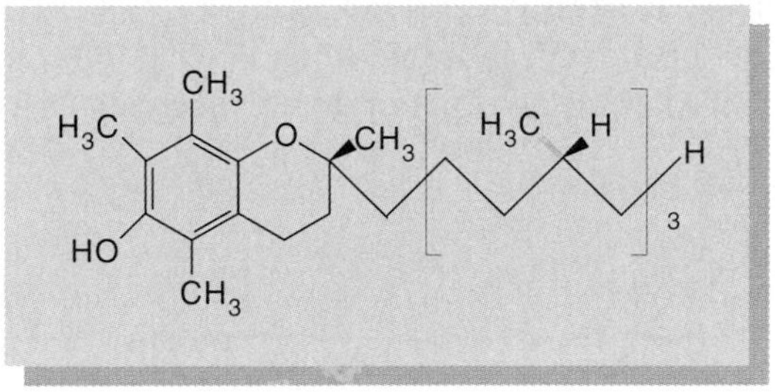

function oxidase. 25-Hydroxyvitamin $D_3$ appears to be biologically inactive until it is hydroxylated at C-1 by a mixed-function oxidase in kidney mitochondria. The 1,25-dihydroxyvitamin $D_3$ is delivered to target tissues for the regulation of calcium and phosphate metabolism. The structure and mode of action of 1,25-dihydroxyvitamin $D_3$ is analogous to that of the steroid hormones (see chapter 24).

Vitamin K was discovered by Henrik Dam in Denmark in the 1920s as a fat-soluble factor important in blood coagulation (K is for "koagulation"). The structures of vitamins $K_1$ and $K_2$ (fig. 10.24) were elucidated by Edward Doisy. Vitamin $K_1$ is found in plants, vitamin $K_2$ in animals and bacteria. How this vitamin functions in blood coagulation eluded scientists until 1974, when vitamin K was shown to be needed for the formation of γ-carboxyglutamic acid (fig. 10.25) in certain proteins. γ-Carboxyglutamic acid specifically binds calcium, which is important for blood coagulation. Such modified glutamic acid residues appear to be important in many other processes involving calcium transport and calcium-regulated metabolic sequences.

Vitamin E (α-tocopherol) (fig. 10.26) was recognized in 1926 as an organic-soluble compound that prevented sterility in rats. The function of this vitamin still has not been clearly established. A favorite theory is that it is an antioxidant that prevents peroxidation of polyunsaturated fatty acids. Tocopherol certainly prevents peroxidation *in vitro*, and it can be replaced by other antioxidants. However, other antioxidants do not relieve all the symptoms of vitamin E deficiency.

Vitamin A (*trans*-retinol) is called an isoprenoid alcohol because it consists, in part, of units of a single five-carbon compound called isoprene:

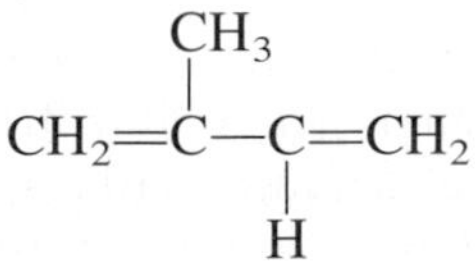

Isoprene is also a precursor of steroids and terpenes. This relationship becomes clear when we examine the biosynthesis of these compounds (see chapter 20). Vitamin A is either biosynthesized from $\beta$-carotene (see fig. S2.4) or absorbed in the diet. Vitamin A is stored in the liver predominantly as an ester of palmitic acid. For many decades, it has been known to be important for vision and for animal growth and reproduction. The form of vitamin A active in the visual process is 11-*cis*-retinal, which combines with the protein opsin to form rhodopsin. Rhodopsin is the primary light-gathering pigment in the vertebrate retina (see supplement S2).

## Summary

Coenzymes are molecules that act in cooperation with enzymes to catalyze biochemical processes, performing functions that enzymes are otherwise chemically not equipped to carry out. Most coenzymes are derivatives of the water-soluble vitamins, but a few, such as hemes, lipoic acid, and iron−sulfur clusters, are biosynthesized in the body. Each coenzyme plays a unique chemical role in the enzymatic processes of living cells.

1. Thiamine pyrophosphate promotes the decarboxylation of $\alpha$-keto acids and the cleavage of $\alpha$-hydroxy ketones.
2. Pyridoxal-5′-phosphate promotes decarboxylations, racemizations, transaminations, aldol cleavages, and elimination reactions of amino acid substrates.
3. Nicotinamide coenzymes act as intracellular electron carriers to transport reducing equivalents between metabolic intermediates. They are cosubstrates in most of the biological redox reactions of alcohols and carbonyl compounds and also act as cocatalysts with some enzymes.
4. Flavin coenzymes act as cocatalysts with enzymes in a large number of redox reactions, many of which involve $O_2$.
5. Phosphopantetheine coenzymes form thioester linkages with acyl groups, which they activate for group transfer reactions.
6. Lipoic acid mediates electron transfer and acyl-group transfer in $\alpha$-keto acid dehydrogenase complexes.
7. Biotin mediates carboxylation of activated methyl groups.
8. Phosphopantetheine, lipoic acid, and biotin, by virtue of their long, flexible structures, facilitate the physical translocation of chemically reactive species among separate catalytic sites.
9. Tetrahydrofolates are cosubstrates for a variety of one-carbon transfer reactions. Tetrahydrofolates maintain formaldehyde and formate in chemically poised states, making them available for essential processes by specific enzymatic action.
10. Heme coenzymes participate in a variety of electron-transfer reactions, including reactions of peroxides and $O_2$. Iron-sulfur clusters, composed of Fe and S in equal numbers with cysteinyl side chains of proteins, mediate other electron-transfer processes, including the reduction of $N_2$ to $2\,NH_3$. Nicotinamide, flavin, and heme coenzymes act cooperatively with iron−sulfur proteins in multienzyme systems that catalyze hydroxylations of hydrocarbons and also in the transport of electrons from foodstuffs to $O_2$.
11. Ascorbic acid is needed as a reductant to maintain some enzymes in their active forms.
12. Metal ions serve as catalytic elements in some enzymes; $Zn^{2+}$ is particularly important in this regard. $Ca^{2+}$ binds tightly to some proteins and acts to trigger intracellular responses to hormonal signals.
13. In general, less is known about the mechanisms of action of the lipid-soluble vitamins than about the coenzymes derived from water-soluble vitamins. The structures and functions of vitamins D, K, E, and A are discussed briefly.

## Selected Readings

Bruice, T. C., and S. J. Benkovic, *Bioorganic Chemistry,* vols. 1 and 2. Menlo Park, Calif.: Benjamin, 1966. A detailed discussion of the mechanisms of bioorganic reactions, including those involving coenzymes.

DiMarco, A. A., T. A. Bobik, and R. S. Wolfe, Unusual coenzymes of methanogenesis. *Ann. Rev. Biochem.* 59:355 (1990). A review of the recently characterized coenzymes required for the biosynthesis of methane in methanogenic bacteria.

Dolphin, D. (ed.), $B_{12}$, vols. 1 and 2. New York: Wiley-Interscience, 1982. Chemistry and mechanism of action of vitamin $B_{12}$.

Dolphin, D., R. Poulson, and O. Avamovic (eds.), *Pyridoxal Phosphate*, part A & part B. New York: Wiley-Interscience, 1987. Chemistry and mechanism of action of pyridoxal phosphate.

Dolphin, D., R. Poulson, and O. Avamovic (eds.), *Pyridine Nucleotide Coenzymes*, part A & part B. New York: Wiley-Interscience, 1987. Chemical, biochemical, and medical aspects of pyridine nucleotide coenzymes.

Frausto de Silva, J. J. K., and R. J. P. Williams, *The Biological Chemistry of the Elements*. Oxford: Clarendon Press, 1991.

Frey, P. A., The importance of organic radicals in enzymatic cleavage of unactivated C—H bonds. *Chem. Rev.* 90:1343, 1990. A brief review of coenzymes required to cleave unreactive C—H bonds.

Jencks, W. P., *Catalysis in Chemistry and Enzymology*. New York: McGraw-Hill, 1969. A detailed analysis of mechanisms of enzymatic and nonenzymatic reactions, including those involving coenzymes.

Knowles, J. R., The mechanism of biotin-dependent enzymes. *Ann. Rev. Biochem.* 58:195, 1989. Review of the chemical mechanism of biotin-dependent carboxylation reactions.

Ortiz de Montellano, P. R. (ed.), *Cytochrome P-450: Structure, Mechanism and Biochemistry*. New York: Plenum, 1986. A treatise on the structure and function of cytochrome P450 monooxygenases.

Phipps, D. A., *Metals and Metabolism*. Oxford Chemistry Series. Oxford: Clarendon Press, 1976. Examines the importance of metal ions in metabolic processes.

Popják, G., Stereospecificity of enzymic reactions. In P. D. Boyer (ed.), *The Enzymes*, vol. 2. New York: Academic Press, 1970. p. 115.

Walsh, C. T., *Enzymatic Reaction Mechanisms*. San Francisco: Freeman, 1977. Provides discussion of the mechanisms of enzymatic reactions. An indepth treatment of coenzymes.

## Problems

1. Which of the coenzymes listed in table 10.1 can be considered to be derivatives of AMP?

2. What structural features of biotin and lipoic acid allow these cofactors to be covalently bound to a specific protein in a multienzyme complex yet participate in reactions at active sites on other enzymes of the complex?

3. The following reactions are catalyzed by pyridoxal-5′-phosphate-dependent enzymes. Write a reaction mechanism for each, showing how pyridoxal-5′-phosphate is involved in catalysis.

   (a)

   $$CH_3—\overset{O}{\overset{\|}{C}}—COO^- + R—\underset{\underset{NH_3^+}{|}}{CH}—COO^- \longrightarrow$$

   $$CH_3—\underset{\underset{NH_3^+}{|}}{CH}—COO^- + R—\overset{O}{\overset{\|}{C}}—COO^-$$

   (b) $H_3^+N—(CH_2)_4—\underset{\underset{NH_3^+}{|}}{CH}—CO_2^- \longrightarrow$

   $$CO_2 + H_3^+N—(CH_2)_5—NH_3^+$$

   (c) $^-O_2C—CH_2—\underset{\underset{NH_3^+}{|}}{CH}—CO_2^- \longrightarrow$

   $$CO_2 + CH_3—\underset{\underset{NH_3^+}{|}}{CH}—CO_2^-$$

4. $NADP^+$ differs from $NAD^+$ only by phosphorylation of the C-2′ OH group on the adenosyl moiety. The redox potentials differ only by about 5 mV. Why do you suppose it is necessary for the cell to employ two such similar redox cofactors?

5. Thiamine-pyrophosphate-dependent enzymes catalyze the reactions shown below. Write a chemical mechanism that shows the catalytic role of the coenzyme.

   (a)

   $$CH_3—\overset{O}{\overset{\|}{C}}—CO_2^- \longrightarrow CO_2 + CH_3—\overset{O}{\overset{\|}{C}}—H$$

   (b)

   $$Ⓟ OCH_2—(CHOH)_2—\underset{\underset{OH}{|}}{CH}—\overset{O}{\overset{\|}{C}}—CH_2OH + HOPO_3^{2-} \longrightarrow$$

   $$Ⓟ O—CH_2—(CHOH)_2—CHO + CH_3—\overset{O}{\overset{\|}{C}}—OPO_3^{2-} + H_2O$$

6. What chemical features allow flavins (FAD, FMN) to mediate electron transfer from NAD(P)H to cytochromes or iron–sulfur proteins?

7. What coenzymes would you expect to participate in the following reactions (specifically indicate which are enzyme-bound coenzymes or "substrate/product-like" coenzymes).

(a) 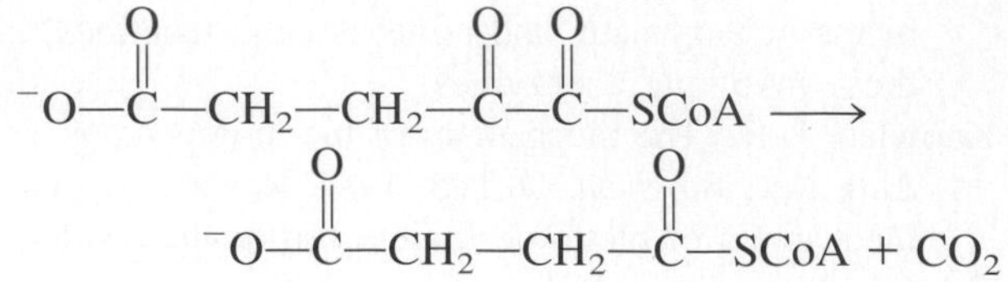 
$$\overset{O}{\underset{\|}{}}\,\overset{O}{\underset{\|}{}}\,\overset{O}{\underset{\|}{}}$$
$$^-O-C-CH_2-CH_2-C-C-SCoA \longrightarrow$$

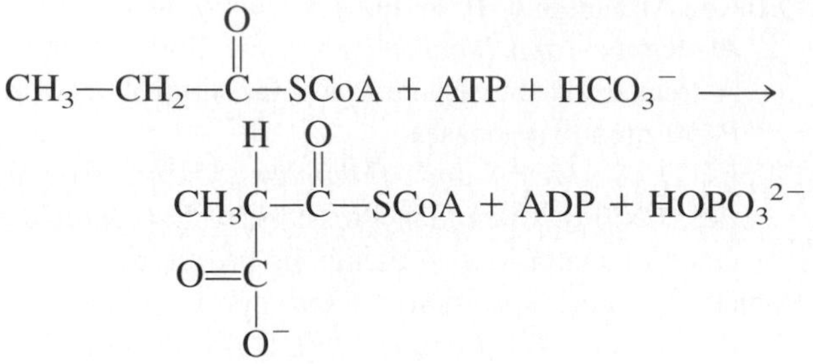

$$^-O-\overset{O}{\underset{\|}{C}}-CH_2-CH_2-\overset{O}{\underset{\|}{C}}-SCoA + CO_2$$

(b) 
$$CH_3-CH_2-\overset{O}{\underset{\|}{C}}-SCoA + ATP + HCO_3^- \longrightarrow$$

$$CH_3\overset{H}{\underset{}{C}}-\overset{O}{\underset{\|}{C}}-SCoA + ADP + HOPO_3^{2-}$$
$$\overset{O=C}{\underset{\overset{|}{O^-}}{}}$$

(c) 
$$CH_3-CH_2-CH_2-\overset{O}{\underset{\|}{C}}-SCoA \longrightarrow$$

$$CH_3-CH=CH-\overset{O}{\underset{\|}{C}}-SCoA$$

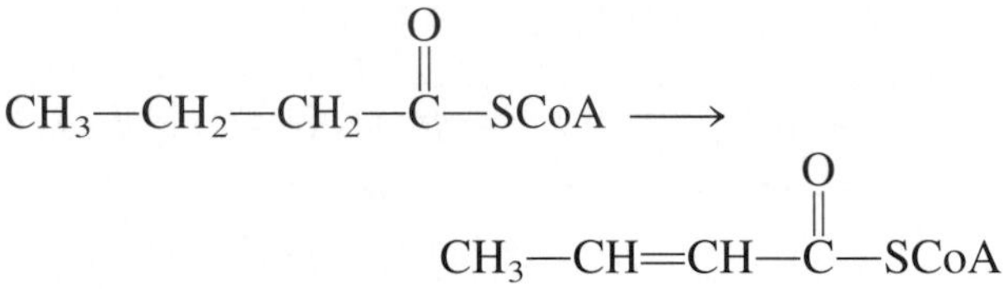

8. Write mechanisms that indicate the involvement of biotin in the following reactions:

(a)

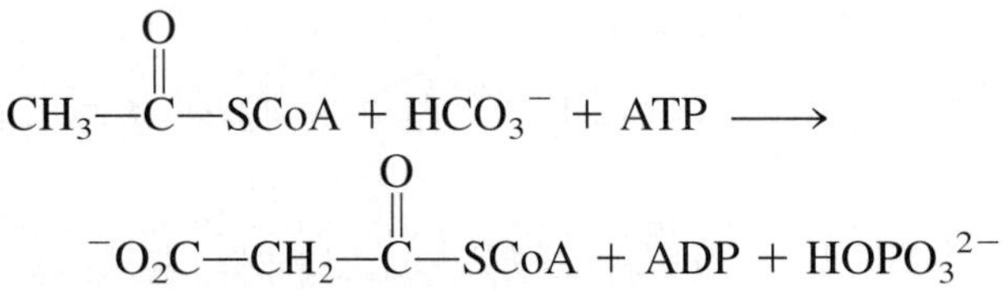

$$CH_3-\overset{O}{\underset{\|}{C}}-SCoA + HCO_3^- + ATP \longrightarrow$$

$$^-O_2C-CH_2-\overset{O}{\underset{\|}{C}}-SCoA + ADP + HOPO_3^{2-}$$

(b)
$$CH_3-CH_2-\overset{O}{\underset{\|}{C}}-SCoA$$

$$+ \,^-O_2C-CH_2-\overset{O}{\underset{\|}{C}}-CO_2^- \rightleftharpoons$$

$$^-O_2C-\underset{\overset{|}{CH_3}}{CH}-\overset{O}{\underset{\|}{C}}-SCoA + CH_3-\overset{O}{\underset{\|}{C}}-CO_2^-$$

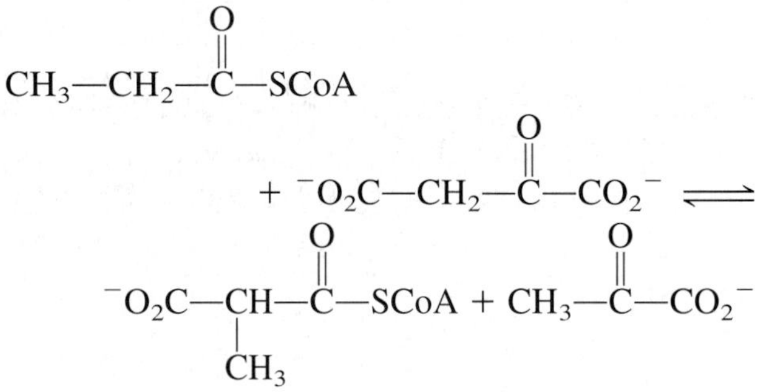

9. (a) What metabolic advantage is gained by having flavin cofactors covalently or tightly bound to the enzyme?
   (b) Would covalently bound $NAD^+$ ($NADP^+$) be a metabolic advantage or disadvantage?

10. Given the amino acid of the general structure

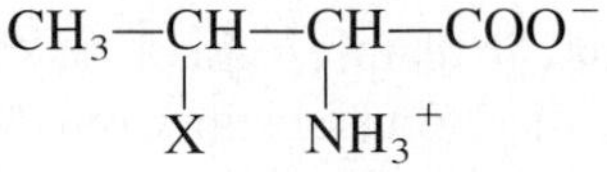

$$CH_3-\underset{\overset{|}{X}}{CH}-\underset{\overset{|}{NH_3^+}}{CH}-COO^-$$

we could use pyridoxal-5′-phosphate to eliminate X, decarboxylate the amino acid, or oxidize the α car-

bon to a carbonyl with formation of pyridoxamine-5′-phosphate. The metabolic diversity afforded by PLP, unchanneled, could wreak havoc in the cell. What other components are required to channel the PLP-dependent reaction along specific reaction pathways?

11. What coenzymes covered in this chapter are (a) biological redox agents, (b) acyl carriers, (c) both redox agents and acyl carriers.

12. For each of the following enzymatically catalyzed reactions, identify the coenzyme involved:
   (a) 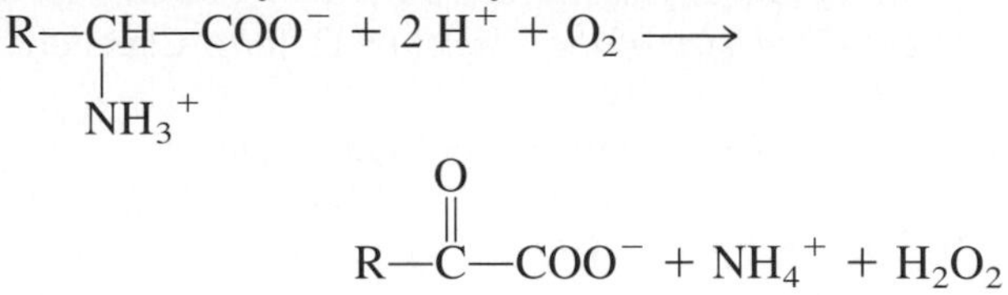 $R-\underset{\overset{|}{NH_3^+}}{CH}-COO^- + 2\,H^+ + O_2 \longrightarrow$

$$R-\overset{O}{\underset{\|}{C}}-COO^- + NH_4^+ + H_2O_2$$

   (b) $HO-CH_2-\underset{\overset{|}{NH_3^+}}{CH}-CO_2^- \longrightarrow$

$$CH_3-\overset{O}{\underset{\|}{C}}-CO_2^- + NH_4^+$$

   (c) $CH_3-CH_2-\overset{O}{\underset{\|}{C}}-SCoA + HCO_3^- + ATP \longrightarrow$

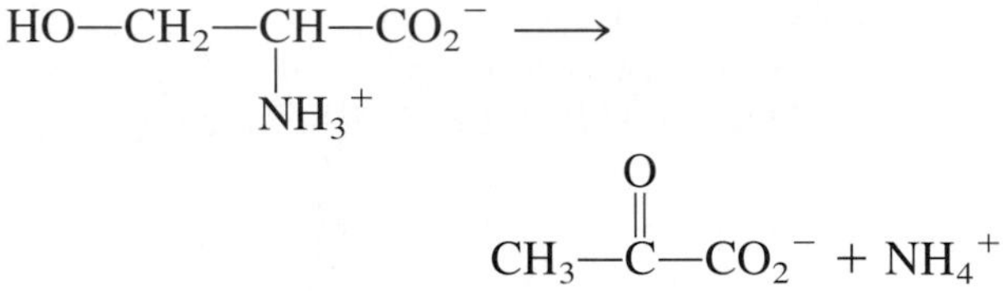

$$^-O_2C-\underset{\overset{|}{CH_3}}{CH}-\overset{O}{\underset{\|}{C}}-SCoA + ADP + P_i$$

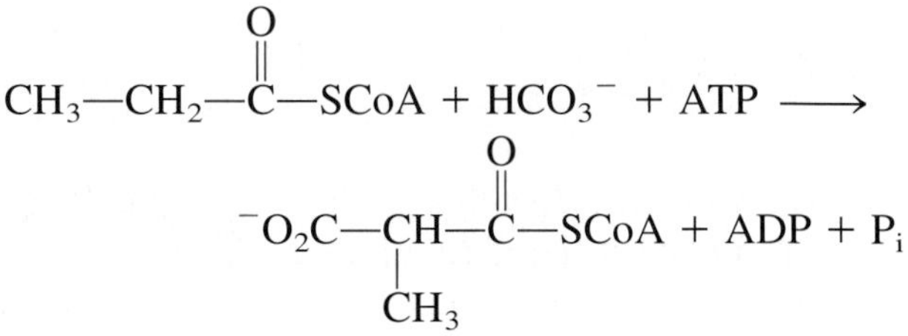

   (d) 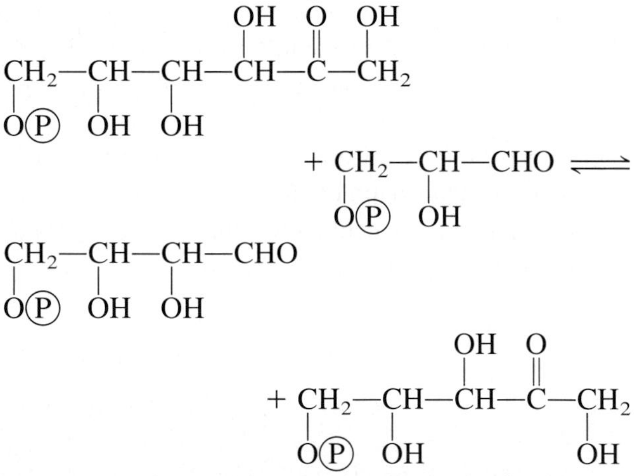 
$$\underset{\overset{|}{O\!\!P}}{CH_2}-\underset{\overset{|}{OH}}{CH}-\underset{\overset{|}{OH}}{CH}-\overset{OH}{\underset{}{CH}}-\overset{O}{\underset{\|}{C}}-\overset{OH}{\underset{}{CH_2}}$$

$$+ \underset{\overset{|}{O\!\!P}}{CH_2}-\underset{\overset{|}{OH}}{CH}-CHO \rightleftharpoons$$

$$\underset{\overset{|}{O\!\!P}}{CH_2}-\underset{\overset{|}{OH}}{CH}-\underset{\overset{|}{OH}}{CH}-CHO$$

$$+ \underset{\overset{|}{O\!\!P}}{CH_2}-\underset{\overset{|}{OH}}{CH}-\overset{OH}{\underset{}{CH}}-\overset{O}{\underset{\|}{C}}-\underset{\overset{|}{OH}}{CH_2}$$

13. The structure of ascorbic acid is shown in figure 10.16. Which hydrogen is the most acidic? Why?

14. Some bacterial toxins use $NAD^+$ as a true substrate rather than as a coenzyme. The toxins catalyze the transfer of ADP-ribose to an acceptor protein. Examine the structure of $NAD^+$ and indicate which portion of the molecule is transferred to the protein. What is the other product of the reaction?

# Metabolism of Carbohydrates

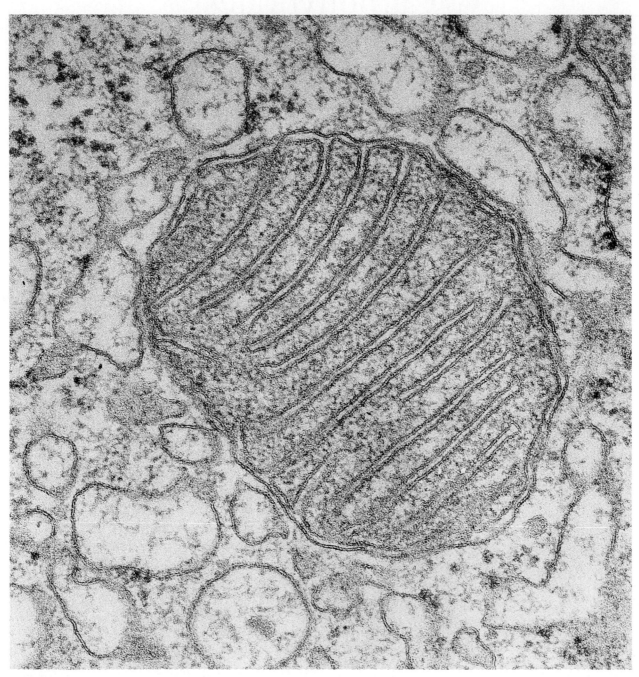

Electron micrograph of a thin section of a
liver mitochondrion. Courtesy of Dr. Daniel S.
Friend

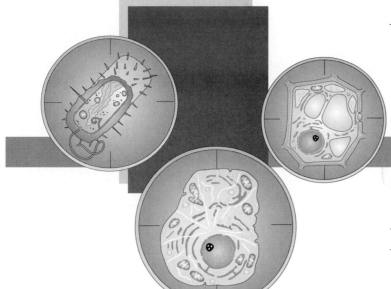

# Metabolic Strategies

*Biochemical reactions are organized into catabolic
pathways that produce energy and reducing power,
and anabolic pathways that consume these products
in the process of biosynthesis.*

In this and the next 14 chapters we consider the synthesis
and degradation of small molecules in the living cell. These
aspects of biochemistry are collectively referred to as intermediary metabolism because they focus on the small-
molecule intermediates in metabolic pathways. In this chapter the principles governing the intermediary metabolism
are discussed.

## Living Cells Require a Steady Supply of Starting Materials and Energy

Intermediary metabolism, the synthesis and degradation of
small molecules, serves two functions: It supplies the energy needed for the synthesis of macromolecules and other
energy-requiring processes, and it furnishes these processes

## Table 11.1

Metabolic Classification of Organisms

| Type of Organism | Source of ATP[a] | Source of NADPH[b] | Source of Carbon | Examples |
|---|---|---|---|---|
| Chemoautotroph | Oxidation of inorganic compounds | Oxidation of inorganic compounds | $CO_2$ | Hydrogen, sulfur, iron, and denitrifying bacteria |
| Photoautotroph | Sunlight | $H_2O$ | $CO_2$ | Cells of higher plants, blue-green algae, photosynthetic bacteria |
| Photoheterotroph | Sunlight | Oxidation of organic compounds | Organic compounds | Nonsulfur purple bacteria |
| Heterotroph | Oxidation of organic compounds | Oxidation of organic compounds | Organic compounds | All higher animals, most microorganisms, nonphotosynthetic plant cells |

[a] ATP = adenosine triphosphate.

[b] NADPH = The reduced form of nicotinamide adenine dinucleotide phosphate is the main source of reducing power.

with the necessary starting materials—amino acids for protein synthesis, fatty acids for lipid synthesis, nucleoside triphosphates for nucleic acid synthesis, and sugars for polysaccharide synthesis.

The demand for energy and starting materials varies widely in different biological processes. To meet these fluctuating needs, the rates of the reaction sequences in intermediary metabolism must be adjustable over broad ranges. Living organisms adjust these rates so that the concentrations of key metabolites are remarkably stable. This constancy is achieved by maintaining a rate of synthesis for each intermediate that balances its rate of utilization. The term steady state is used to describe this nonequilibrium situation.

## Organisms Differ in Sources of Energy, Reducing Power, and Starting Materials for Biosynthesis

To maintain a steady state and to permit growth and reproduction as well, all living cells require energy (ATP) and starting materials for biosynthesis. Reducing power (NADPH) is also required since most biosynthesis involves converting compounds to a more reduced state. Whereas the needs of different organisms are similar, the ways in which organisms satisfy their needs can be quite different. Indeed the most fundamental metabolic distinction between organisms relates to the ways in which they satisfy their basic metabolic needs (table 11.1).

Autotrophs exploit the inorganic environment without recourse to compounds produced by other organisms. Thus, all of their carbon compounds must be synthesized from inorganic compounds of carbon, carbonates, or carbon dioxide. Chemoautotrophs obtain reducing power by oxidation of inorganic materials, such as hydrogen, or reduced compounds of sulfur or nitrogen. Each species is usually specific in the electron donors it can utilize. The same reduced compounds supply electrons for regeneration of ATP by electron-transfer phosphorylation. Photoautotrophs obtain ATP by electron-transfer phosphorylation during the cycling of photochemically excited electrons. Some photoautotrophs, including plants, are also able to use the energy of sunlight to extract electrons from water; thereby photoautotrophs have achieved independence of all sources of energy except for the sun, and of all sources of electrons and carbon except for water and carbon dioxide ($CO_2$). For this reason, photoautotrophic fixation of carbon dioxide is the predominant base of the food chain in the biosphere.

Common heterotrophs depend on preformed organic compounds for all three primary needs. Although some carbon dioxide is fixed in heterotrophic metabolism, the heterotrophic cell thrives at the expense of compounds formed by other cells, and is not capable of the net conversion (fixation) of carbon dioxide into organic compounds. Some bacteria referred to as photoheterotrophs are able to regenerate ATP photochemically but cannot use photochemical reactions to supply electrons to $NADP^+$. Such organisms are like other heterotrophs in their dependence on preformed organic compounds. But because of their photochemical

**Figure 11.1**

The biosynthesis of tryptophan. Tryptophan is synthesized in five steps from chorismate. Each step requires a specific enzyme activity. The four intermediates between chorismate and tryptophan serve no function other than as precursors of tryptophan. The detailed steps of each reaction in this sequence are described in chapter 21.

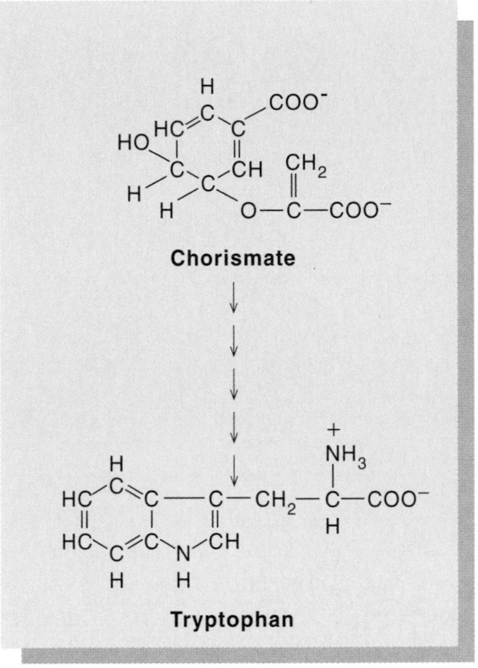

Chorismate

Tryptophan

ways. In a pathway it is the overall sequence that serves a function, not the individual reactions. For example, the conversion of the organic compound chorismate to tryptophan occurs in five discrete steps, each requiring a specific enzymatic activity (fig. 11.1). The intermediates between chorismate and tryptophan serve no function except as precursors of tryptophan. The end product, tryptophan, has many uses, including its role as a building block in protein synthesis.

A central role of breakdown of the six-carbon sugar, glucose, to the three-carbon compound, pyruvate, is to supply ATP to the cell. But, as in the case of the tryptophan pathway, the function of each reaction in glucose catabolism can only be appreciated by considering the overall sequence. This pathway involves nine enzymatically catalyzed steps in going from glucose to pyruvate (fig. 11.2). In the first and third steps ATP is actually consumed. It is only in the seventh and tenth steps that the starting amount of ATP is recovered and further ATP is synthesized. Hence, the function of this sequence is not fully served until each substrate has passed through the entire sequence.

Glucose breakdown and tryptophan biosynthesis illustrate pathways consisting of a linear series of reactions. In chapter 12 we will see that the glycolytic pathway also contains branches so that common intermediates can proceed along diverging routes after the branchpoint. We will also see that the glycolytic pathway is organized so that some of the reactions within the pathway can go in either the forward or the reverse direction.

apparatus, the photoheterotrophs are able to use available organic food more efficiently than common heterotrophs.

Although all heterotrophs depend on preformed organic compounds, they differ markedly in the numbers and types of compounds they require. Some species can make all required compounds when supplied with a single carbon source; others have lost some or many biosynthetic capabilities. Mammals must obtain about half of their amino acids from external sources. They also are unable to make several metabolic cofactors, as evidenced by their well-known nutritional requirements for vitamins. Some bacteria and some parasites have even more extensive nutritional requirements than mammals.

## Reactions Are Organized into Sequences or Pathways

To appreciate how biochemical reactions serve common functions we must examine the organization of biochemical reactions in some detail. Most of the enzyme-catalyzed reactions in living cells are organized into sequences or path-

## Sequentially Related Enzymes Are Frequently Clustered

A fundamental aspect of biochemical organizations is that the enzymes that catalyze a sequence are often clustered together in the cell. As a result, an intermediate produced in the first reaction in a pathway is passed directly to the second enzyme in the pathway, and so on. Such an arrangement might be expected to accelerate synthesis of the end product, minimize the loss of intermediates, and facilitate regulation of the pathway.

Three types of enzyme clustering are found (fig. 11.3). In the simplest situation, all of the catalytic activities for a particular pathway are found in proteins that exist as independent soluble proteins in the same cellular compartment. In such cases the intermediates must get from one enzyme to the next in the sequence by free diffusion through the cytosol or by transfer after contact between two sequentially related enzymes—one carrying the reactive intermediate and the other ready to receive it. Such is the situation for the enzymes involved in the breakdown of glucose to pyruvate.

## Figure 11.2

The breakdown of glucose to pyruvate. The conversion of glucose to pyruvate requires ten steps and one enzyme for each step. In steps 1 and 3, ATP is consumed. In steps 7 and 10, ATP is produced. The net production of ATP is 2 moles for each mole of glucose consumed. Unlike the intermediates in the tryptophan pathway, many of the intermediates between glucose and pyruvate serve as useful substances for other purposes. A full description of all the steps in this sequence is given in chapter 12.

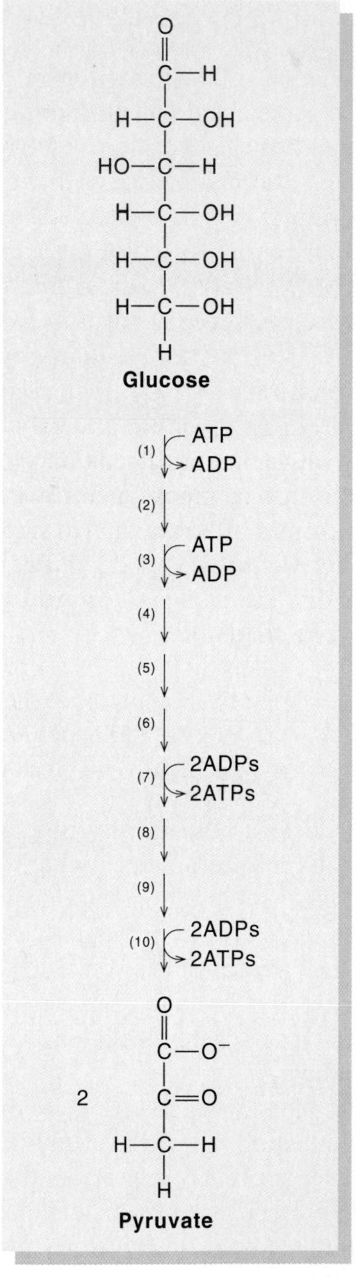

## Figure 11.3

The organization of functionally related enzymes. Functionally related enzymes are organized in three ways: (*a*) as unlinked proteins soluble in the aqueous milieu of the same cellular compartment, (*b*) as components in a multiprotein complex, or (*c*) as components on a membrane.

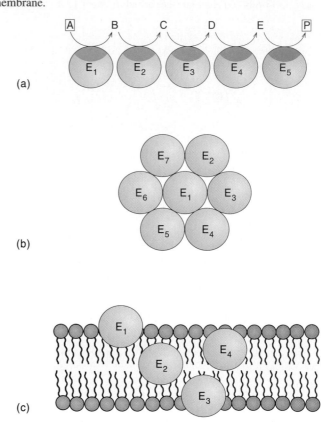

localization lead to faster rates of reactions. Second, these higher concentrations facilitate regulation of a pathway by limiting the starting materials or key intermediates to an organelle and allowing direct interaction of one enzyme with another. Fatty acid metabolism in eukaryotes superbly illustrates this type of arrangement (see chapter 18). The enzymes involved in fatty acid catabolism are all located inside the mitochondrion, whereas all enzymes involved in fatty acid synthesis are located outside the mitochondrion in the cytosol.

The second type of arrangement observed for sequentially related enzymes is exemplified by the *Escherichia coli* fatty acid synthase (see chapter 18). This synthase is a complex of most of the enzymes involved in fatty acid synthesis. The intermediates in this case are bound to the enzyme complex until synthesis is complete.

In the third type of arrangement, functionally related enzymes are membrane-bound. This is the case for the enzymes of electron transport and oxidative phosphorylation

Enzymes with related functions often are compartmentalized into specific organelles, a process that enhances metabolic efficiency in various ways. First, the higher concentrations of enzymes and intermediates resulting from

## Figure 11.4

A schematic block diagram of the metabolism of a typical aerobic heterotroph. The block labeled ''Catabolism'' represents pathways by which nutrients are converted to small-molecule starting materials for biosynthetic processes. Catabolism also supplies the energy (ATP) and reducing power (NADPH) needed for activities that occur in the second block; these compounds shuttle between the two boxes. The block labeled ''Biosynthesis'' represents the synthesis of low- to medium-molecular-weight components of the cell as well as the synthesis of proteins, nucleic acids, lipids, and carbohydrates and the assembly of membranes, organelles, and the other structures of the cell.

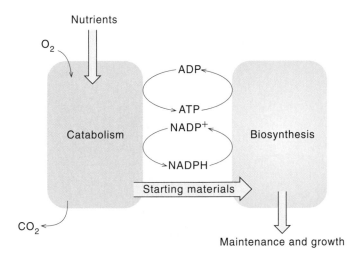

(see chapter 14). These enzymes are all bound to the inner mitochondrial membrane.

## Pathways Show Functional Coupling

Large metabolic charts have been designed to display all the major biochemical pathways. Such charts present a bewildering array of interconnected pathways, making it difficult to appreciate relationships between different pathways. The overall operational aspects of metabolism may be clarified by simpler block diagrams that omit details and focus on functional relationships. Such a functional block diagram for a typical heterotrophic aerobic cell is shown in figure 11.4. The metabolism of such a system is symbolized by two functional blocks:

1. Catabolism or degradative metabolism. Foods are oxidized to carbon dioxide. Most of the electrons liberated in this oxidation are transferred to oxygen, with concomitant production of ATP (electron-transfer phosphorylation). Other electrons are used in the regeneration of NADPH, the most frequently used reducing agent for biosynthesis. The major pathways in this block are the glycolytic sequence and the tricarboxylic acid (TCA)

cycle. These same sequences, with the pentose phosphate pathway, also supply carbon skeletons for the cell's biosynthetic processes.

2. Biosynthesis, or anabolism. This block includes much greater chemical complexity than the first. The starting materials produced by glycolysis and the TCA cycle are converted into hundreds of cell components, with NADPH serving as a reducing agent when necessary, and with ATP serving as the universal coupling agent or energy-transducing compound. Synthesis of starting materials is followed by the synthesis of macromolecules and growth.

The starting materials consumed in biosynthetic sequences must be continually replaced by catabolic processes. The energy sources and reducing sources, ATP and NADPH, are used in very different ways. When they contribute to biosynthesis, ATP is converted to ADP or AMP, and NADPH is converted to $NADP^+$. They (ADP, AMP, or $NADP^+$) must be regenerated, reduced, or rephosphorylated at the expense of substrate oxidation in the catabolic block of reactions. As figure 11.4 indicates, ATP and NADPH are the only compounds that have primary roles in coupling major functional blocks. Other coupling agents are essential but narrower in scope. For example, NADH and $FADH_2$ participate within the catabolic block in the transfer of electrons from substrate to oxygen, but they are not significantly involved in coupling between major functional blocks.

The block diagram of figure 11.4 represents metabolism in aerobic heterotrophs. The photochemical production of ATP in photoheterotrophs requires the addition of a photochemical block, with ATP production in the catabolic block being deleted. For a photoautotroph, two blocks are needed in addition to those shown in figure 11.4: (1) a photochemical block that regenerates ATP from ADP and reduces $NADP^+$ to NADPH while consuming water and producing oxygen, and (2) a block that uses the ATP and NADPH for the reduction of carbon dioxide to various products. These photosynthetic products provide the input to the catabolic block, which in this case serves only one function—the production of the starting materials for synthesis.

Another way of looking at metabolism, which emphasizes the precursor–product relationship, is depicted in figure 11.5. Here metabolism is pictured as three concentric boxes. The inner box shows the central metabolic pathways and the interconversions of various small molecules. These substances serve as the starting materials for all other metabolism. Some of the compounds formed from these intermediates are shown in the middle concentric box. These

## Figure 11.5

The overall metabolism of an aerobic heterotroph represented as three concentric boxes. The central metabolic pathways are represented in the innermost box. It is here that carbohydrate compounds are degraded to produce energy, reducing power and starting materials for biosynthesis. These pathways will be discussed in chapters 12 through 16. Only some key intermediates are shown. In general in moving out from the central box to the outer two boxes reactions require energy and reducing power, while moving in the opposite direction leads to the production of energy and reducing power. The middle box shows some of the building blocks for the larger molecules found in the outer box. The reactions involved in fatty acid metabolism, mevalonate synthesis, amino acid metabolism and nucleotide synthesis are discussed in chapters 18, 20, 21 and 22, and 23, respectively. The reactions involved in lipid synthesis, cholesterol synthesis, nucleic acid synthesis and protein synthesis are discussed in chapters 18 and 19, 20, 26 and 28, and 29, respectively.

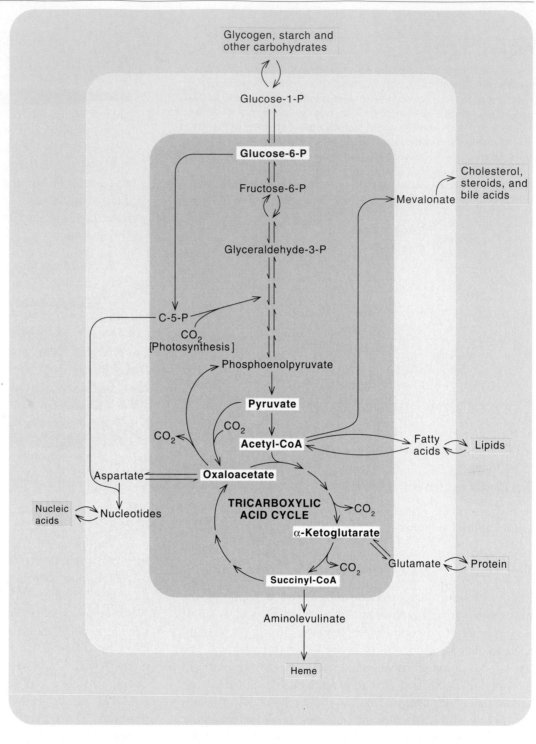

compounds serve as building blocks for the final products indicated in the outer box. Each of the next 13 chapters deals with an expanded segment of the metabolism depicted in this figure, and we repeatedly refer to this figure in the introductory passages of subsequent chapters.

## The ATP–ADP System Mediates Conversions in Both Directions

Figure 11.4 emphasizes the coupling of catabolic sequences and biosynthetic (anabolic) sequences to the ATP–ADP

cycles. In fact, the involvement of the ATP–ADP system is much greater than indicated. Most metabolic sequences within the two major blocks are tightly linked by the ATP–ADP system. To appreciate the way in which this is done we must look at individual sequences in greater detail.

In analyzing metabolism we must distinguish between a sequence and a conversion. A conversion might be the transformation of starting material A to end product Z; a sequence is the specific set of reactions by which such a conversion is carried out. The conversion of glucose to pyruvate involves a ten-step sequence (see figures 11.2 and

## Figure 11.6

The interconversion of fructose-6-phosphate and fructose-1,6-bisphosphate.

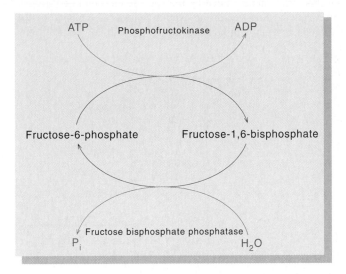

## Figure 11.7

Schematic illustration of the organization of metabolic sequences into oppositely directed pairs. The two sequences result in opposite conversions. The values for the overall equilibrium constant are a function of the conversion and the number of ATP-to-ADP conversions to which each sequence is coupled. Note the similarity to figure 11.6.

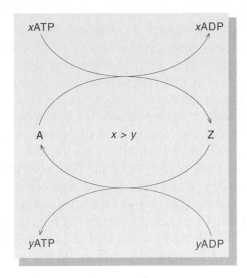

12.13). As a rule, a conversion in one direction is catabolic and produces energy in the form of ATP, whereas a conversion in the reverse direction is anabolic and requires the input of energy. Any conversion can be carried out by many different hypothetical sequences of reactions, which might be coupled to nearly any number of ATP-to-ADP or ADP-to-ATP interconversions, depending on the specificities of the enzymes that catalyze the component reactions. The value of the overall free energy change and the overall equilibrium constant depend strongly on the stoichiometry of this coupling, that is, the number of molecules of ATP converted to ADP (or of ADP converted to ATP) per mole of Z that is produced.

In fact any conversion can be made thermodynamically favorable by coupling it to a sufficient number of ATP-to-ADP conversions. This strategy is exploited in the design of metabolically paired sequences. A simple example is the interconversion of fructose-1,6-bisphosphate (F-1,6-bisP) into fructose-6-phosphate (F-6-P). This is the simplest possible type of interconversion since it involves only one reaction in either direction (fig. 11.6).

$$F\text{-}6\text{-}P + ATP \longrightarrow F\text{-}1,6\text{-}bisP + ADP; \quad K_{eq} \approx 10^4 \ (1)$$
$$\underline{F\text{-}1,6\text{-}bisP + H_2O \longrightarrow F\text{-}6\text{-}P + P_i; \qquad K_{eq} \approx 10^4 \ (2)}$$
$$ATP + H_2O \longrightarrow ADP + P_i$$

The conversion of fructose-1,6-bisphosphate to fructose-6-phosphate is an energetically favorable reaction, so it does not require any ATP-to-ADP conversions. The oppositely directed conversion of fructose-6-phosphate to fructose-1,6-bisphosphate requires the conversion of a single ATP to

ADP to make it thermodynamically favorable. Since the interconversions have dissimilar starting materials and products, it is not surprising that the two conversions use different enzymes.

From this simple example we can see that paired oppositely directed conversions can be made thermodynamically feasible in either direction by coupling to an appropriate number of ATP-to-ADP conversions. Metabolic sequences commonly occur in pairs. The two sequences of a pair connect the same compounds, say A and Z, but run in opposite directions (fig. 11.7). Invariably the reactions of two oppositely directed sequences are coupled to different numbers of ATP-to-ADP conversions, as in the previous examples, so that they are thermodynamically favorable in either direction. Usually $x > y$, where $x$ is the number of ATPs consumed in the anabolic direction, and $y$ is the number of ATPs produced in the catabolic direction.

## Conversions Are Kinetically Regulated

Pairs of oppositely directed sequences appear to be cycles (see figs. 11.6 and 11.7). Indeed, if both sequences were simultaneously active they would form a true cycle. Because the end products of one sequence are the starting materials for the other, such a cycle has no metabolic consequences except that ATP is expended. This is because more ATP is used in one sequence than is regenerated in the other. Because a cyclic operation can waste energy, such oppositely directed paired sequences are often called futile cycles. Fortunately for the metabolism, kinetic constraints usually prevent futile cycles from operating. For this reason futile cycles are commonly referred to as pseudocycles. These constraints result from the regulation of one or more enzymes for each sequence of a pair so that conversions occur only in one direction at any given time.

The interconversion of fructose-6-phosphate and fructose-1,6-bisphosphate, provides an instructive example of the regulation of an interconversion (see fig. 11.6). Each "sequence" of this pseudocycle consists of only one reaction, and the "sequences" differ by one ATP-to-ADP conversion. Clearly if both reactions in this conversion were allowed to function simultaneously a metabolic mishap would occur leading to nothing except the loss of ATP. This is prevented from occurring by the design of the enzymes that carry out these two conversions. Both of these enzymes are regulatory enzymes whose activities are modulated by small molecules that serve as signals of the metabolic state of the cell. The system is designed so that only one of the enzymes is active at any given time. Which enzyme is active depends on the metabolic state of the cell. The precise mechanisms for the regulation of these particular enzymes are discussed in chapters 9 and 12.

## Pathways Are Regulated by Controlling Amounts and Activities of Enzymes

The two most important principles governing the regulation of biochemical pathways are economy and flexibility. Organisms regulate their metabolic activities to avoid deficiencies or excesses of metabolic products. If a significant change occurs in the environment, such as a shift in the concentration or kinds of nutrients, the organism must quickly adjust its metabolism if it is to survive.

Pathways are regulated at two levels: (1) they are regulated by controlling the amounts of enzyme (see chapters 30 and 31); and (2) they are regulated by controlling the activity of existing enzymes. In this chapter we focus on the latter. Important aspects of this subject were already dealt with in chapter 9.

## *Enzyme Activity Is Regulated by Interaction with Regulatory Factors*

The most common way of regulating metabolic activity is by direct control of enzyme activity. Enzyme activities are usually regulated by noncovalent interaction with small-molecule regulatory factors (see chapter 9) or by a reversible covalent modification, such as phosphorylation or

**Figure 11.8**

Three patterns for end-product inhibition. In end-product inhibition the first reaction in the pathway (*a*) or the first reactions after a branchpoint (*b*) are inhibited by the specific end products (*c*). Sometimes the first step in the overall pathway is inhibited by the end products from both branchpoints.

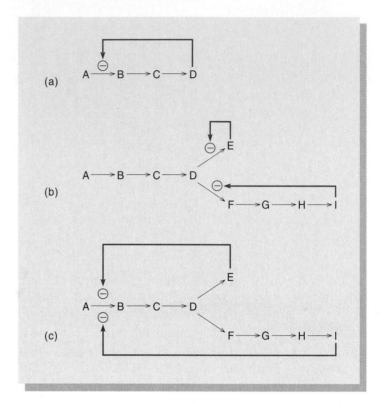

adenylation (also see chapters 9, 12, and 24) of an amino acid side chain. The effect of the regulatory factor in specific instances may be to increase or decrease the activity of the enzyme. What sets these regulatory factors in motion is a separate question. Sometimes the metabolism inside the cell is the major player in determining which regulatory factors are produced; at other times external sources control the action as in the case of hormones that are made elsewhere and influence special target cells. One instance of the effectiveness of hormonal signals is discussed in the next chapter, and the subject is discussed in general terms in chapter 24.

## *Regulatory Enzymes Occupy Key Positions in Pathways*

Enzymes that are susceptible to direct regulation occupy key positions in metabolic pathways. In a multistep pathway the first enzyme in the pathway is usually regulated, and the others are not (fig. 11.8*a*). In the case of CTP synthesis (see chapter 9), we saw that the enzyme aspartate car-

## Figure 11.9

Branchpoints in metabolic pathways sometimes occur at locations where an intermediary, S, can follow a catabolic sequence or an anabolic sequence. The first reactions after the branchpoints are catalyzed by enzymes B and A, respectively. Once the first step after the branchpoint has been taken, the metabolic intermediate is irreversibly committed to follow that pathway.

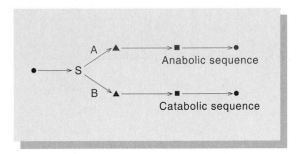

bamoyltransferase is negatively regulated by CTP. That is, excess CTP binds to the regulatory sites on this multiprotein enzyme, thereby inhibiting its activity. Aspartate carbamoyltransferase catalyzes the first step in a multistep pathway leading to CTP. Because CTP is the end product of the pathway, this type of regulation is referred to as end-product inhibition. End-product inhibition is very common in anabolic pathways.

The mechanism is clearly flexible and economical. If there is sufficient end product, there is no point in processing substrates down the pathway. To do so is a waste of both energy and materials. Furthermore, it is most effective to block the first enzyme in the pathway: this by itself rapidly reduces activity of the entire pathway. It is redundant to block any other enzymes in the pathway, because no intermediates are available to these enzymes if the first step is blocked. Blocking a middle enzyme in the pathway, instead of the first enzyme, also is quite wasteful. The result is accumulation of an intermediate that serves no useful function. Indeed, such intermediates in excess frequently have harmful effects.

In a branched-chain anabolic pathway, end-product inhibition usually regulates the first enzyme after the branchpoint (fig. 11.8b). This arrangement leads to control of either pathway after the branchpoint (fig. 11.8b). If the supply of the branchpoint substrate (D in fig. 11.8b) is limiting, the inhibition of one pathway after the branchpoint can increase the metabolic flow of the other pathway. This effect is often highly significant. Apropos of this point, anabolic sequences frequently arise as branchpoints from catabolic pathways (fig. 11.9). In such instances the control factors affect whether a catabolic intermediate is further degraded or serves as a substrate for biosynthesis.

## Regulatory Enzymes Often Show Cooperative Behavior

It is clear from what we have just said in the previous subsection that control of the rates of metabolic sequences is most effective when it is exerted at branchpoints. As you might expect, the enzymes that catalyze reactions at branchpoints have evolved features to make the partitioning highly sensitive to metabolic signals. A striking feature of a typical branchpoint enzyme (the first enzyme in a biosynthetic sequence, for example) is that the reaction that it catalyzes is of high kinetic order with respect to its substrate (see chapter 9). Whereas other enzymes along the pathway typically exhibit normal Michaelis kinetic responses to the concentrations of their substrates (first order or hyperbolic kinetics), branchpoint enzymes usually catalyze reactions of higher order (cooperative or sigmoidal kinetics). A fourth-order response is quite common. The two types of behavior are shown in figure 11.10. Any enzyme-catalyzed reaction must level off (to a kinetic order of 0) at high concentration because of saturation of the catalytic sites, but for the "normal Michaelis" enzyme the kinetic order at very low concentrations of substrate is 1, whereas for the cooperative enzyme it has a higher value, frequently 4.

A major advantage of this higher order is that it makes the system very sensitive to changes in substrate concentration. If the first reaction in the sequence responds to the fourth power of the concentration of its substrate (the starting material for the sequence), the flow of material into the sequence is much more sensitive to small changes in the concentration of the starting material than if the reaction were first order. This will help to maintain the steady state of the system.

Perhaps an even more important advantage of high kinetic order is that the regulatory responses of the enzyme are greatly strengthened. We discussed kinetic regulation of enzymes at metabolic branchpoints by negative feedback, in which the end product of the sequence binds to a regulatory site and decreases the affinity of the catalytic site for the substrate (the branchpoint metabolite). When the enzyme is constructed so as to have cooperative interactions between sites, the sensitivity to negative feedback control is greatly sharpened.

## Both Anabolic and Catabolic Pathways Are Regulated by the Energy Status of the Cell

It is not always an advantage for the concentration of an amino acid or other end product to be the only factor that determines the rate at which that end product is synthesized. If a cell is low on energy or catabolic intermediates, it may not be able to afford to synthesize amino acids, even if their

**Figure 11.10**

Reaction velocity as a function of substrate concentration for a first-order enzymic reaction (*left*) and for a cooperative enzyme with fourth-order kinetics (*right*). Substrate concentration is expressed as $[S]/[S_{0.5}]$ and velocity as a fraction of maximum velocity. At a $[S]/[S_{0.5}]$ value of 1 the substrate concentration is equal to $[S_{0.5}]$ and the reaction velocity is half of the maximum velocity. Note the much greater rate of increase over the same interval for the cooperative enzyme.

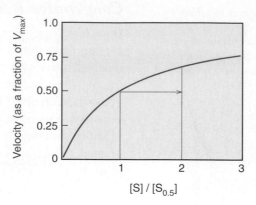
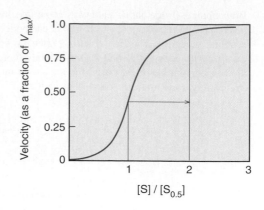

concentrations are quite low. Continued biosynthesis under such circumstances might deplete the already scarce supply of energy to the point where essential functions, such as maintenance of concentration gradients across membranes, is impaired. It is clearly advantageous for the rate of biosynthesis to be regulated by the general energy status of the cell as well as the need for a specific end product.

Because the adenine nucleotide system ATP–ADP (and occasionally ATP–AMP) couples energy into biosynthetic sequences, we might expect that this system also supplies the necessary signal indicating the energy status of the cell, to be sensed by the kinetic control mechanisms of biosynthesis. It is helpful to have a term that quantitatively expresses the energy status of the cell. The term used is the energy charge, which is defined as the effective mole fraction of ATP in the ATP–ADP–AMP pool.

$$\text{Energy charge} = \frac{(\text{ATP} + 0.5\ \text{ADP})}{(\text{ATP} + \text{ADP} + \text{AMP})} \qquad (3)$$

In this equation the 0.5 in the numerator takes into account the fact that ADP is about half as effective as ATP at carrying chemical energy. Thus, two ADPs can be converted into one ATP and one AMP by a reaction with an equilibrium constant of about 1. The highly active enzyme that catalyzes this reaction (adenylate kinase) is universally distributed

$$\text{ATP} + \text{AMP} \rightleftharpoons 2\ \text{ADP}, \qquad K_{eq} \approx 1 \qquad (4)$$

The values for the energy charge could conceivably vary from 0 to 1, as illustrated in figure 11.11. In fact, however, the values for real cells are usually held within very narrow limits. The result is a general stabilizing effect on the cellular metabolism. In much the same way, voltage regulation is necessary for the effective operation of many electronic devices. Now let us see how the limits are maintained.

Several reactions in anabolic and catabolic pathways have been found to respond to variation in the value of the

**Figure 11.11**

Relative concentrations of ATP, ADP, and AMP as a function of the adenylate energy charge. The adenylate kinase reaction was assumed to be at equilibrium, and a value of 1.2 was used for its effective equilibrium constant in the direction shown in the equation in the text.

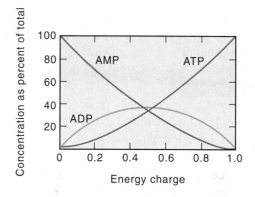

energy charge. As might be expected, the enzymes in catabolic pathways respond in a direction opposite to that of enzymes in anabolic pathways. The two responses are compared in figure 11.12. It is clear that if catabolic sequences, which lead to the generation of ATP, respond as shown by the upper curve, and biosynthetic sequences, which use ATP, respond as shown by the lower curve, then the charge is strongly stabilized at a value at which the two curves intersect. A tendency for the charge to fall is resisted by the resulting increases in the rates of catabolic sequences and the decreases in the rates of biosynthetic sequences. A tendency for the charge to rise is resisted by opposing effects.

## Strategies for Pathway Analysis

Biochemical precursor–product relationships vary in complexity. Some pathways such as the conversion of phenylalanine to tyrosine involve only one en-

**Figure 11.12**

Variation in reaction rates as a function of the energy charge. As the energy charge increases, the rate of catabolic reactions decreases. Meanwhile, the rate of anabolic reactions increases. The combined effect is to stabilize the energy charge at a value around 0.9.

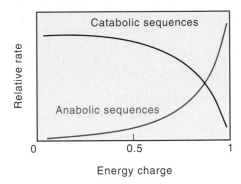

zyme-catalyzed reaction: one precursor and one product. Other pathways such as the utilization of $CO_2$ in autotrophic organisms illustrate the most complex type of relationship between precursor ($CO_2$) and products (carbon-containing compounds of the cell), a multistep pathway with many branchpoints; the complete description of all of the reactions involved in $CO_2$ utilization would include most of the synthetic reactions of the organism. Different procedures are needed to analyze simple and complex pathways.

## Analysis of Single-Step Pathways

The investigation of a single-step pathway usually begins with a crude cell-free extract from a source abundant in the enzyme that catalyzes the conversion. Two of the most popular sources of cells for many biochemical studies are rat liver and *E. coli.* To investigate a particular reaction, the precursor (substrate) is added to the extract, and the amount of product formed as a function of time is determined. Once an assay for disappearance of precursor and appearance of product has been developed, the crude extract can be processed into fractions that can be tested to see which are active in the conversion. Through further fractionation and assays it should ultimately be possible to purify the enzyme of interest. Some procedures followed in enzyme purification were discussed in chapter 6, and many procedures used to determine the mechanisms of action of the purified enzymes were considered in chapters 8 and 9.

The process of assaying for a particular reaction during purification may be complicated in cases requiring co-substrates, coenzymes, or cofactors. Usually these additional requirements are discovered by the trial-and-error procedure of adding test substances to the reaction mixture and observing whether they accelerate the reaction or lead

to the formation of more product(s). If more than one step and several enzymes are involved, then intermediates between precursor and product must also be considered and the whole process of analysis increases in complexity.

## Analysis of Multistep Pathways

Most pathways comprise a series of enzyme-catalyzed steps exemplified by the five-step pathway leading from chorismate to tryptophan (fig. 11.13). Because tryptophan synthesis does not occur in vertebrates, this reaction cannot be studied in rat liver extracts, but it can be studied in extracts made from *E. coli* bacteria. Bacteria and many other microorganisms have the additional advantage that powerful genetic methods are applicable. In general genetic methods are most useful at the beginning of a study when the goal is to determine the number of genes and related enzyme-catalyzed reactions of a pathway. Genetic methods are also useful to test conclusions drawn from *in vitro* studies by making parallel measurements *in vivo*. In the case of the tryptophan biosynthetic pathway, genetic methods were used to determine the number of gene-encoded proteins required to synthesize tryptophan. This was done by a procedure known as complementation analysis. First, a large number of bacterial mutants were isolated, which could not grow unless tryptophan was added to the growth medium. Such mutants usually have defects in one or more of the genes required for tryptophan biosynthesis. Pairwise matings are done between mutants to create partial diploid cells that carry complete sets of the tryptophan pathway genes from two different mutants. Such diploid cells no longer require added tryptophan in cases in which the defects in the two genomes are in different genes. With some mated pairs the need for tryptophan is not eliminated. This occurs in cases where the paired genomes carry defects in the same gene (fig. 11.14). Such mutants are said to belong to the same complementation group. An analysis of this sort led to a classification of most of the tryptophan-requiring mutants into a limited number of complementation groups. In an exhaustive analysis the total number of complementation groups should correspond to the number of genes required to make a functional tryptophan pathway. Since most genes encode a single polypeptide chain, the number of groups should reflect the number of polypeptide chains involved in the biosynthesis of tryptophan.

The next phase in pathway analysis involves biochemical procedures. As many metabolic intermediates as possible are isolated, their structures are determined, the order of reactions is determined, and the enzymes that catalyze the different reaction steps are isolated and characterized. The mutants isolated for the complementation analysis are also

## Figure 11.13

The tryptophan biosynthetic pathway in *E. coli*. There are five enzymatically catalyzed reactions involved in tryptophan biosynthesis and five different polypeptides associated with these reactions. Polypeptides E and D normally make a tetrameric complex, which catalyzes the first two reactions in the pathway. The next two reactions are catalyzed by a single polypeptide, C. The final reaction is catalyzed by a tetrameric complex composed of the B and A polypeptides in equal numbers.

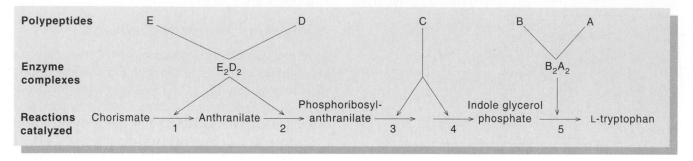

## Figure 11.14

Complementation occurs only between genomes that are defective in different genes. On the left are shown two different types of gene pairs that might arise in a mating to test for complementation. The genes *E*, *D*, *C*, *B*, and *A* are intended to represent the five genes of the tryptophan pathway, which are clustered in *E. coli*. A minus sign indicates a defective gene. On the right are shown the results of plating 10 cells on a minimal medium with and without tryptophan added. Growth in the absence of added tryptophan indicates complementation. Growth is observed after overnight incubation. Under favorable growth conditions each cell divides many times, forming a clone of identical cells.

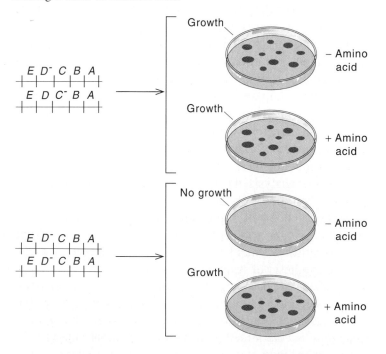

valuable aids for these biochemical experiments. The usefulness of the mutants at this stage in the investigation stems from the fact that each mutation introduces a specific block at some point in the pathway. A bacterium with one nonfunctioning enzyme in a pathway accumulates the intermediate just before the defect and makes little or no intermediates (or final product) for the remaining steps in the pathway (fig. 11.15). Intermediates often are much easier to isolate from mutants than from normal cells because of their greater abundance and stability. Ideally the structures of all intermediates isolated from different mutant cells are determined. The information resulting from such observations should give a step-by-step picture of the chemical reactions that occur in the pathway.

From this point on there are numerous ways of proceeding with the analysis. Crude extracts can be made from different mutant cells. The extract from one mutant can be used to complement the extract from another mutant in tryptophan synthesis, which can lead to an assay for a particular enzyme carried by one mutant that is missing in the other mutant. The goal at this juncture is to purify each enzyme of the pathway so that the properties of the enzymes and the reactions they catalyze can be individually scrutinized. All of the tryptophan enzymes have been isolated from *E. coli*. Studies of other systems indicate a remarkable similarity for the operation of this pathway in different microorganisms and plants.

## *Radiolabeled Compounds Facilitate Pathway Analysis*

Before radioactively labeled substrates became available, detection of intermediates and products was often quite dif-

## Figure 11.15

Comparison of the relative amounts of pathway intermediates and final products in normal and mutant organisms containing a defective enzyme in the pathway. The precursor, intermediates, and final product of the pathway are labeled A, B–D, and E, respectively. The enzymes 1, 2, 3, and 4 are indicated by the numbers over the reaction arrows. A cross through the reaction arrow indicates a defective enzyme. In the wild-type organism with no defective enzymes (*top*), intermediates are present at low concentrations compared with final product. In the mutant organism with a defective enzyme in the pathway (*bottom*), the intermediate just before the defective enzyme accumulates to an abnormally high concentration. Intermediates and products after the block are practically nonexistent. If the final product of the pathway is required for viability, it has to be supplied directly from external sources.

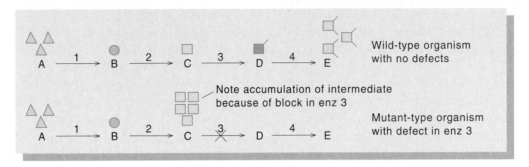

ficult. Radiolabeled compounds greatly increase the sensitivity of detection of products and especially the intermediates in a pathway. Moreover, they also facilitate determination of the order of reactions in a pathway. Thus, a radiolabeled precursor administered to a reaction mixture for a short time is expected to label the early intermediates preferentially. The same compound administered for a long time, however, is expected to accumulate in the final product of the pathway. This is the situation for an extract prepared from a wild-type cell. If the extract were prepared from a mutant cell we would expect a difference in the labeling pattern as suggested in figure 11.15.

Specific inhibitors can serve the same role as genetic blocks in the analysis of a pathway. Genetic methods were rarely used for pathway analysis until the 1940s. Radiolabeling was not used until the 1950s. In spite of this many important pathways were elucidated before this. For example, Hans Krebs discovered the complex multistep pathway known as the tricarboxylic acid cycle in the 1930s. Krebs used cell-free extracts from pigeon flight muscle which are especially rich in the enzymes of the cycle (see chapter 13). Although mutants were not available for imposing specific blocks on the passage of intermediates in this pathway, a strong inhibitor of one of the reactions was discovered. Frequently, even today, inhibitors are useful for pathway analysis, especially in cases in which mutants are not available. Inhibitors sometimes have an advantage over the use of mutant extracts for *in vitro* analysis because they can be added at various times after starting the reaction.

## Pathways Are Usually Studied Both in Vitro *and* in Vivo

Although biochemists spend most of their time studying reactions *in vitro*, the ultimate proof of the significance of a reaction or a series of reactions is that it is used *in vivo*. For the purpose of making parallel measurements, *in vivo* and *in vitro*, isotopes and genetic blocks are invaluable. Isotopes permit intermediates and products to be detected *in vivo* as well as *in vitro*. Genetic mutants permit parallel observations to be made of the effect of a specific enzyme deficiency on the specified pathway. In cases in which *in vitro* and *in vivo* analyses lead to different conclusions, the significance of the *in vitro* analysis may not have been correctly assessed, and further studies are necessary. A classic example of this occurred in early investigations of *E. coli* DNA polymerase (chapter 26). An enzyme was isolated that catalyzes the replication of DNA *in vitro*. However, subsequent mutant studies showed that this activity was not required for replication of DNA *in vivo*.

The brief description given here has emphasized the power of genetic approaches for elucidating metabolic pathways. In the last 10 years, the fields of genetics and biochemistry have been revolutionized by the techniques of DNA recombinant technology. DNA recombinant technology has made it possible to isolate virtually any gene from any cell type, even those for which genetic methods have not been accessible in the past. This means that in the future biochemists will place even greater emphasis on the genetic approach in their work. DNA recombinant technology methods are discussed in chapter 27.

## Summary

1. All cells need energy and starting materials for synthesis. Ultimately these are supplied by autotrophic organisms, especially green plants; in plants the starting materials are made from $CO_2$ and the supply of chemical energy and reducing power is dependent on the absorption of light energy. In a heterotrophic organism, the role of catabolism is to supply those basic needs from conversions (usually oxidation) of foodstuffs.

2. Metabolic chemistry is characterized by functionality. Each reaction is important because of its participation in a sequence of reactions, and each sequence interacts functionally with other sequences.

3. Sequences may be broadly classified into two main types: biosynthetic (anabolic) and degradative (catabolic). Anabolic sequences usually require energy, and catabolic sequences usually produce energy.

4. Metabolic regulatory mechanisms have evolved so as to stabilize concentrations of key metabolites under a broad range of conditions.

5. Energy is coupled from catabolic sequences of energy-requiring activities of a cell by the ATP–ADP system. In a similar manner, reducing power is coupled by the NADPH–NADP$^+$ system.

6. The stoichiometry of coupling to ATP-to-ADP conversions contributes to the overall equilibrium constant of a sequence and therefore can determine the direction of conversion that is thermodynamically favorable. Any conversion can be made favorable by coupling to an appropriate number of ATP-to-ADP conversions.

7. Metabolic sequences occur in oppositely directed pairs, which are controlled by regulatory enzymes.

8. Regulatory enzymes respond to signals in such a way that rates of biosynthesis are controlled by the need for product.

9. Metabolism is regulated primarily by adjustment of the ratios by which intermediates are partitioned at metabolic branchpoints.

10. Different procedures are used for analysis of simple and complex pathways. Analysis of single-step pathways often begins with the isolation and characterization of the enzyme involved. During enzyme isolation each purification step is monitored by a specific assay that measures the conversion of substrate to product. Multistep pathway analysis ideally begins with complementation analysis, a genetic technique that entails the isolation of mutants with genetic blocks in each step of the pathways. Once the numbers of enzymes and intermediates are established, each enzyme can be isolated with the help of a specific assay.

11. Radiolabeled compounds are most useful for pathway analysis in two respects: first, they permit detection of pathway intermediates with great sensitivity, and second, they can be used as tracers for determining the order of intermediates in a pathway.

12. Inhibitors that block specific steps in a pathway play the same role as genetic blocks in the *in vitro* analysis of a pathway.

13. A complete understanding of a pathway requires parallel investigations *in vitro* and *in vivo*. *In vitro* analyses permit a detailed study of isolated components. *In vivo* observations substantiate the biologic significance of *in vitro* observations.

## Selected Readings

Atkinson, D. E., *Cellular Energy Metabolism and Its Regulation.* New York: Academic Press, 1977. General discussion covers some topics in this and the two following chapters in somewhat greater depth than treated in our book.

Cohen, P., *Control of Enzyme Activity,* 2d ed. London and New York: Chapman and Hall, 1983. Brief discussion of some types of regulation of activity of metabolic enzymes, emphasizing regulation by covalent modification of the enzymes.

Herman, R. H., R. M. Cohn, and P. D. McNamara, *Principles of Metabolic Control in Mammalian Systems.* New York and London: Plenum Press, 1980. Discusses various aspects of metabolic control in mammals, mainly at the intracellular level. Many references.

Hochachka, P. W., and G. N. Somera, *Biochemical Adaptation.* Princeton: Princeton University Press, 1984. An excellent and extensive discussion of how biochemical processes, including many discussed in this book, are adapted by various types of organisms in fitting themselves for survival under specific and often difficult conditions.

Jencks, W. P., How does ATP make work? *Chemtracts-Biochem. Mol. Biol* 1:1–13, 1990.

Westheimer, F. H., Why nature chose phosphates. *Science* 235:1173–1178, 1987.

## Problems

1. Consider the following relationships among the four major classes of biological molecules. What similarities and differences can you see in the chemical relationships between the right and left columns?

| Small Molecule | Large Molecule |
| --- | --- |
| Nucleotide | Nucleic acid |
| Amino acid | Protein |
| Monosaccharide | Polysaccharide |
| Fatty acid | Lipid |

2. What is a metabolic pathway?
3. What is the relationship between catabolism and anabolism?
4. Besides a difference in the number of phosphate moieties, what are the major biochemical differences between $NAD^+$ and $NADP^+$?
5. What is the primary advantage of subcellular compartments to intermediary metabolism?
6. Using only figure 11.2 determine if the conversion of glucose to pyruvic acid is an oxidation or a reduction or neither.
7. Explain why each of the following statements is false in terms of efficient metabolic regulation:
   (a) Most enzymes operate *in vivo* near $V_{max}$.
   (b) End-product inhibition usually occurs at the last or next-to-last enzyme in a metabolic pathway.
   (c) Catabolic pathways tend to diverge from a single metabolite.
   (d) The enzymes regulated in a metabolic pathway usually exhibit simple Michaelis-Menten kinetics.
   (e) Energy charge is unimportant in the regulation of anabolic sequences but is of primary importance in the regulation of catabolic sequences.
   (f) Enzymes that catalyze a sequence of reactions are rarely grouped in multienzyme complexes.
   (g) Compartmentalization of metabolic pathways is seldom a regulatory strategy the cell uses.
8. How is it possible that both the glycolytic degradation of glucose to lactate and the reverse process, formation of glucose from lactate (gluconeogenesis), are energetically favorable?
9. What is the metabolic advantage of having the "committed step" of a pathway under strict regulation?
10. Theoretically, the reactions shown below constitute a futile cycle. Explain.

$$\text{Glucose} + \text{ATP} \xrightarrow{\text{Glucokinase}} \text{Glc-6-P} + \text{ADP}$$

$$\text{Glc-6-P} + \text{HOH} \xrightarrow{\text{Glucose-6-phosphatase}} \text{Glucose} + \text{phosphate}$$

In the liver cell, the enzymes are spatially separated: Glucokinase in the cytosol and glucose-6-phosphatase in the endoplasmic reticulum. Does this separation influence the futile cycle?

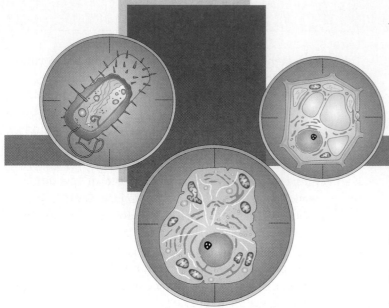

# Glycolysis, Gluconeogenesis, and the Pentose Phosphate Pathway

**Figure 12.1**

Loss of two hydrogens by glycerol leads to the formation of glyceraldehyde or dihydroxyacetone, depending on whether the two hydrogens are lost from the end or middle position, respectively.

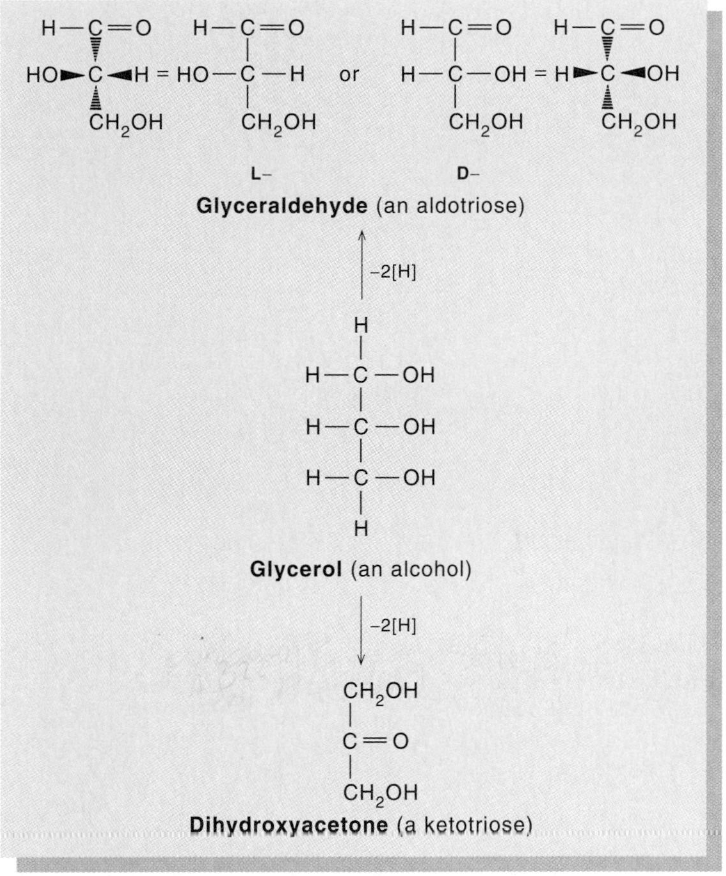

*Glycolysis entails the breakdown of hexoses to 3 carbon compounds while producing ATP. The reverse process, gluconeogenesis entails the synthesis of hexose and related energy storage polysaccharides in a process that consumes ATP.*

C arbohydrates are made from carbon skeletons richly laced with hydroxyl groups (see fig. 1.4). The hydrophilic hydroxyl groups give carbohydrates the potential for strong interactions with water; they also give them functional groups to which various substituents can be added. Finally, the hydroxyl groups provide the possibility of strong intra- or interchain interactions via hydrogen bonds. Carbohydrates play major roles in metabolism and in many of the structures found in the cell. In this chapter we focus on the bioenergetic properties of carbohydrates. First we discuss the relevant structures and then we consider their metabolism.

## Carbohydrates Important in Energy Metabolism

Small carbohydrates such as glucose, fructose, and pyruvate occupy key roles in energy metabolism and supply carbon skeletons for the synthesis of other compounds. Polymers in which glucoses are linked in chains are important as short-term energy storage compounds. Before discussing their metabolism it is essential that we consider some of the structural properties of these ubiquitous molecules.

## Monosaccharides and Related Compounds

The simplest carbohydrates, sometimes referred to as monosaccharides, or sugars, are either polyhydroxyaldehydes (aldoses) or polyhydroxyketones (ketoses). They can be derived from polyalcohols (polyols) by oxidation of one carbinol group to a carbonyl group. For example, the simple three-carbon triol, glycerol, can be converted either to the aldotriose, glyceraldehyde, or to the ketotriose, dihydroxyacetone, by loss of two hydrogens (fig. 12.1).

Since the middle carbon of glyceraldehyde is connected to four different substituents, it is a chiral center leading to two possible forms of glyceraldehyde. D-Glyceraldehyde is illustrated in figure 12.2 in the Fischer projec-

## Figure 12.2

Configurational relationships of the D-aldoses. The most important sugars are starred. Note that in the two D series the configuration about the chiral center farthest from the carbonyl is the same.

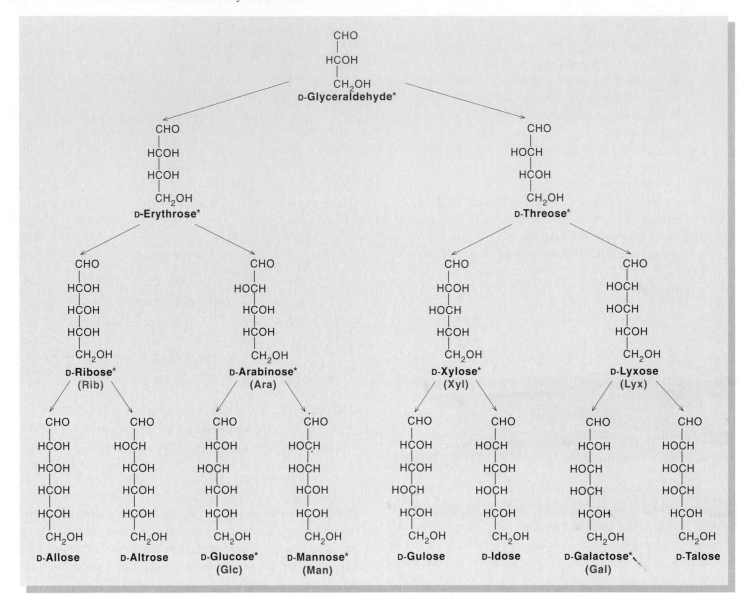

tion formula, in which the —OH group attached to the central carbon atom points to the right. If the central carbon were in the plane of the paper with tetrahedrally arranged substituents, the H and OH connected to it would project above the plane of the paper and the other two substituents would project below the plane of the paper. A molecule with such a chiral center is optically active and the two configurations of the molecule are designated d (for dextrorotatory) or l (for levorotatory) according to the way they rotate plane-polarized light (see Methods of Biochemical Analysis 12A).

For common sugars, the prefixes D or L refer to that center of symmetry most remote from the aldehyde or ketone end of the molecule. By convention, all optically active centers are related to the asymmetrical carbon of glyceraldehyde. Isomers stereochemically related to D-glyceraldehyde are designated D, and those related to L-glyceraldehyde are designated L. We may visualize the four-, five-, and six-carbon sugars (tetroses, pentoses, and hexoses) as arising from the trioses through the stepwise condensation of formaldehyde to either glyceraldehyde or dihydroxyacetone. Indeed this may be the way in which sugars arose in prebi-

**Figure 12.3**

Configurational relationships of the D-ketoses. The most important sugars are starred.

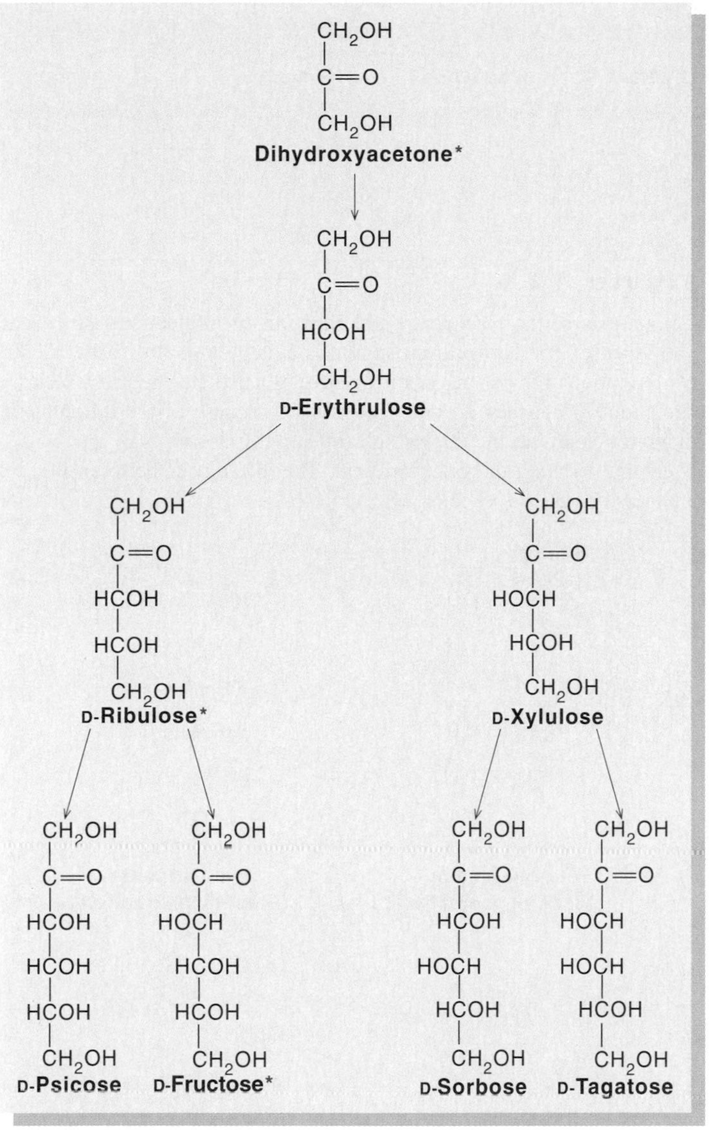

Hemiacetal formation was first observed in optical studies on D-glucose. The optical rotation ($[\alpha]_D$) of a freshly dissolved sample of D-glucose changes with time because it possesses two stereoisomeric hemiacetals (anomers) that are interconvertible in solution (fig. 12.5). One of these anomers, $\alpha$-D-glucose, has $[\alpha]_D = 113$; the other, $\alpha$-L-glucose, has $[\alpha]_D = 19$. The optical rotation of a freshly prepared solution of either of these compounds eventually approaches an intermediate value that depends on the equilibrium between the two anomers.

The $\alpha$ designation for the D series indicates that the aldehyde of the C-1 hydroxyl group is on the same side of the structure as the ring oxygen in the Fischer projection; in the $\beta$ configuration the aldehyde is on the opposite side. When the sugar is dissolved in water, the two hemiacetals are in equilibrium with the straight-chain hydrated form. Conversion of one hemiacetal form into the other is called mutarotation. Equilibrium is reached without an added catalyst in a few hours at room temperature. The open-chain form usually represents only a small fraction of the total (see fig. 12.5 for actual percentages).

Hemiacetals with five-member rings are called furanoses, and hemiacetals with six-member rings are called pyranoses. In cases in which either five- or six-member rings are possible, the six-member ring usually predominates. For example, for glucose, less than 0.5% of the furanose forms exist at equilibrium (see fig. 12.5, bottom). Both furanoses and pyranoses are more realistically represented by pentagons or hexagons in the Haworth convention shown in figure 12.6b.

Haworth structures are unambiguous in depicting configurations, but even they do not show the true spatial relationship of groups attached to rings. The normal angle between the bonds formed by the saturated carbon atoms (109°) causes the pyranose molecule to pucker into either a chairlike or boatlike conformation. For glucose and most other pyranoses the chair form (fig. 12.6c) predominates. However, we usually display pyranoses by the Haworth projection because it is easier to draw.

otic times but the biosynthesis of sugars occurs by other means. The tetroses, pentoses, and hexoses related to D-glyceraldehyde are shown in figure 12.2. The ketoses (e.g., fructose) are similarly related to dihydroxyacetone (fig. 12.3).

***Monosaccharides Cyclize to Form Hemiacetals*** Aldehydes can add hydroxyl compounds to the carbonyl group. If a molecule of water is added, the product is an aldehyde hydrate, as shown in figure 12.4. If a molecule of alcohol is added, the product is a hemiacetal; the addition of a second alcohol results in an acctal. Sugars readily form intramolecular hemiacetals in cases in which the resulting compound has a five- or six-member ring.

## In Disaccharides the Monosaccharides Are Linked by Glycosidic Bonds

Warming glucose in methanol and acid produces a mixture of two new substances: $\alpha$- and $\beta$-methylglucoside (fig. 12.7); the comparable derivatives of galactose are referred to as methylgalactosides, and the generic name is methylglycoside. The bond formed between the sugar and the alcohol is called a glycosidic bond. Note that the formation of a glycoside from a sugar and an alcohol is formally equivalent to the formation of an acetal from a hemiacetal

## Figure 12.4

Aldehydes can add $H_2O$ to form hydrates or can add alcohols to form hemiacetals and acetals.

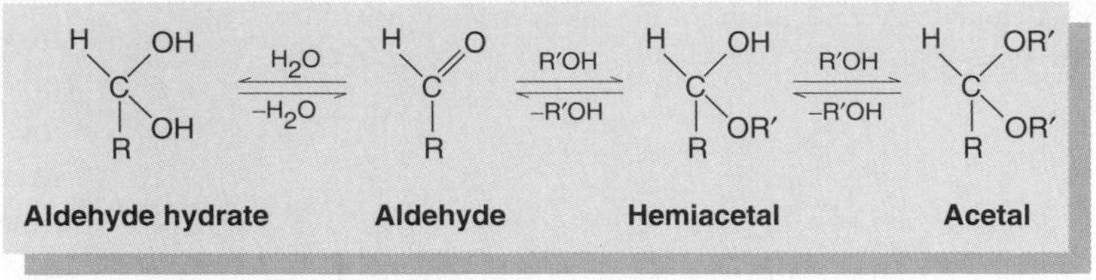

Aldehyde hydrate        Aldehyde        Hemiacetal        Acetal

## Figure 12.5

Different forms of glucose that result from dissolving glucose in water. At 25°C in water, glucose reaches an equilibrium containing about 0.02% free aldehyde, 38% $\alpha$-pyranose form ($[\alpha]_D=113$), 62% $\beta$-pyranose form ($[\alpha]_D=9$), and less than 0.5% of the furanose forms. The anomeric carbon is shown in color.

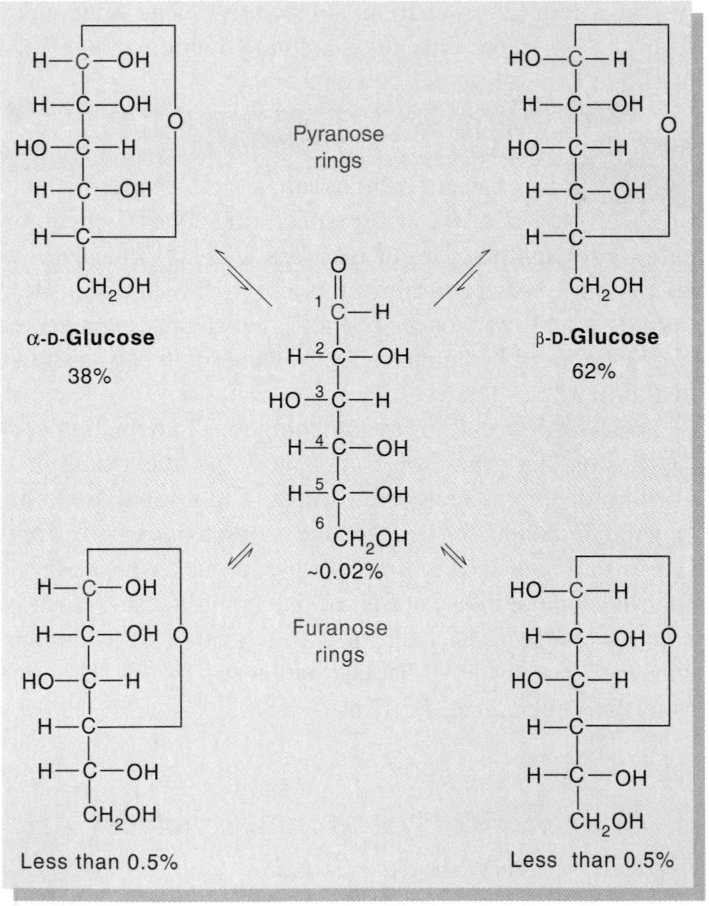

## Figure 12.6

Comparison of the Fischer (*a*) and Haworth (*b*) projections for $\alpha$- and $\beta$-D-glucose. The Haworth projection is a step closer to reality. Chair configurations for the two anomers of D-glucose are the most accurate depiction (*c*) but they are not always used because of the difficulty in drawing. Note that the largest substituent, —$CH_2OH$, is in an equatorial location in both structures. The differences between the two anomers are shown in color.

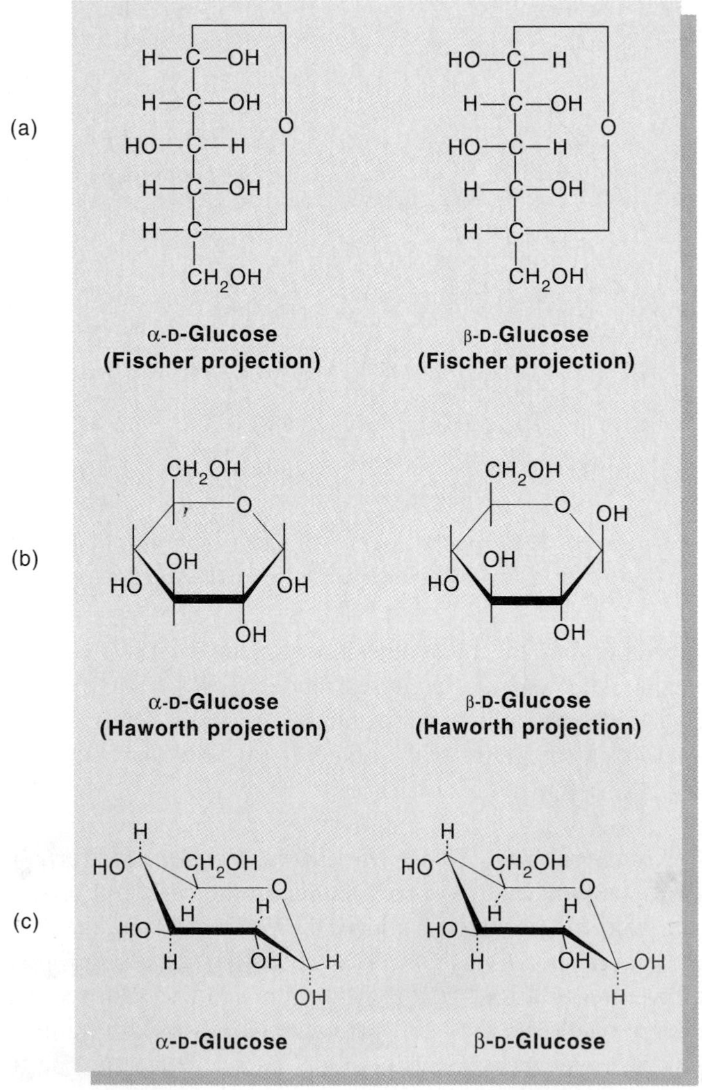

and an alcohol (see fig. 12.4). Whereas the two hemiacetal forms of glucose are in equilibrium through mutarotation, the corresponding glycosides are locked into one configuration. This form occurs because mutarotation requires an intermediate in which the reactive carbon can adopt a car-

**Figure 12.7**

Formation of methyl glucosides. Glucosides (or glycosides) are quite stable in alkali, but they hydrolyze readily in dilute acid.

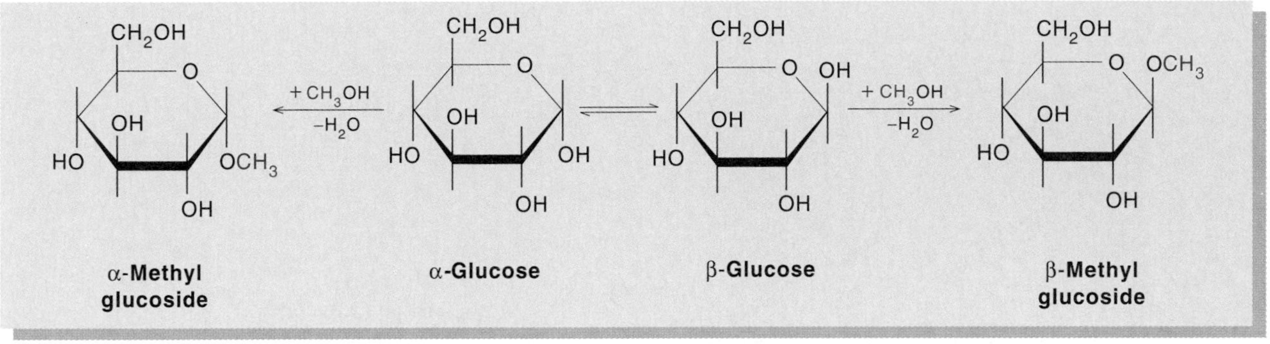

$\alpha$-Methyl glucoside      $\alpha$-Glucose      $\beta$-Glucose      $\beta$-Methyl glucoside

**Figure 12.8**

Four commonly occurring disaccharides. The configuration about the hemiacetal group has not been specified for lactose, maltose, or cellobiose because both anomers exist in equilibrium.

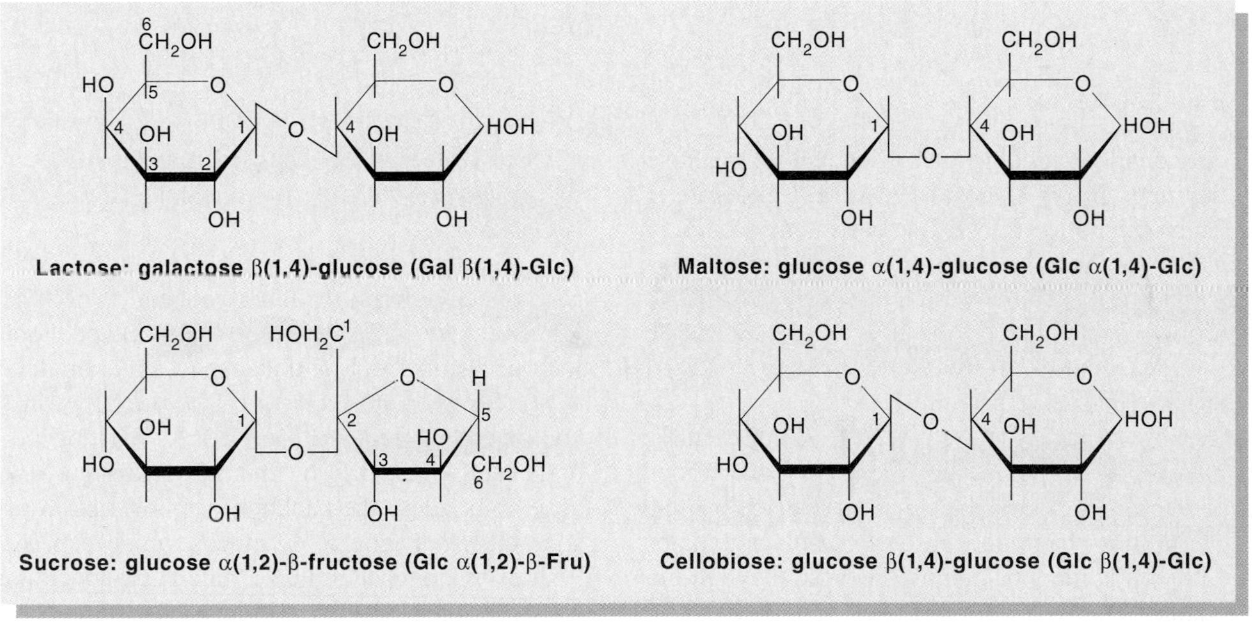

Lactose: galactose $\beta$(1,4)-glucose (Gal $\beta$(1,4)-Glc)

Maltose: glucose $\alpha$(1,4)-glucose (Glc $\alpha$(1,4)-Glc)

Sucrose: glucose $\alpha$(1,2)-$\beta$-fructose (Glc $\alpha$(1,2)-$\beta$-Fru)

Cellobiose: glucose $\beta$(1,4)-glucose (Glc $\beta$(1,4)-Glc)

bonyl structure, a configuration that is not possible once the glycoside is formed.

A glycoside can be formed with aliphatic alcohols, phenols, and hydroxy carboxylic acids, but the most important glycosides are those formed with other sugars. Monosaccharides are linked by glycosidic bonds to form disaccharides (fig. 12.8). For instance, the disaccharide maltose contains a glycosidic bond between the C-1 of one glucose molecule and the C-4 of another. The compound is said to have an $\alpha$(1,4)-glycosidic linkage because the reactive C-1 carbon of one sugar is connected to the C-4 of another, and the configuration about the reactive carbon is $\alpha$. Maltose possesses one potentially free aldehyde group and is referred to as a reducing sugar because of the potential reducing power of the aldehyde moiety. The configuration about the hemiacetal hydroxyl group is not specified in figure 12.8 because it can undergo mutarotation. Maltose is most familiar as a degradation product of starch.

**Figure 12.9**

Structure of the storage polysaccharides glycogen and starch. The main chain is $\alpha(1,4)$-linked. Side chains are connected to the main chain by $\alpha(1,6)$ linkages.

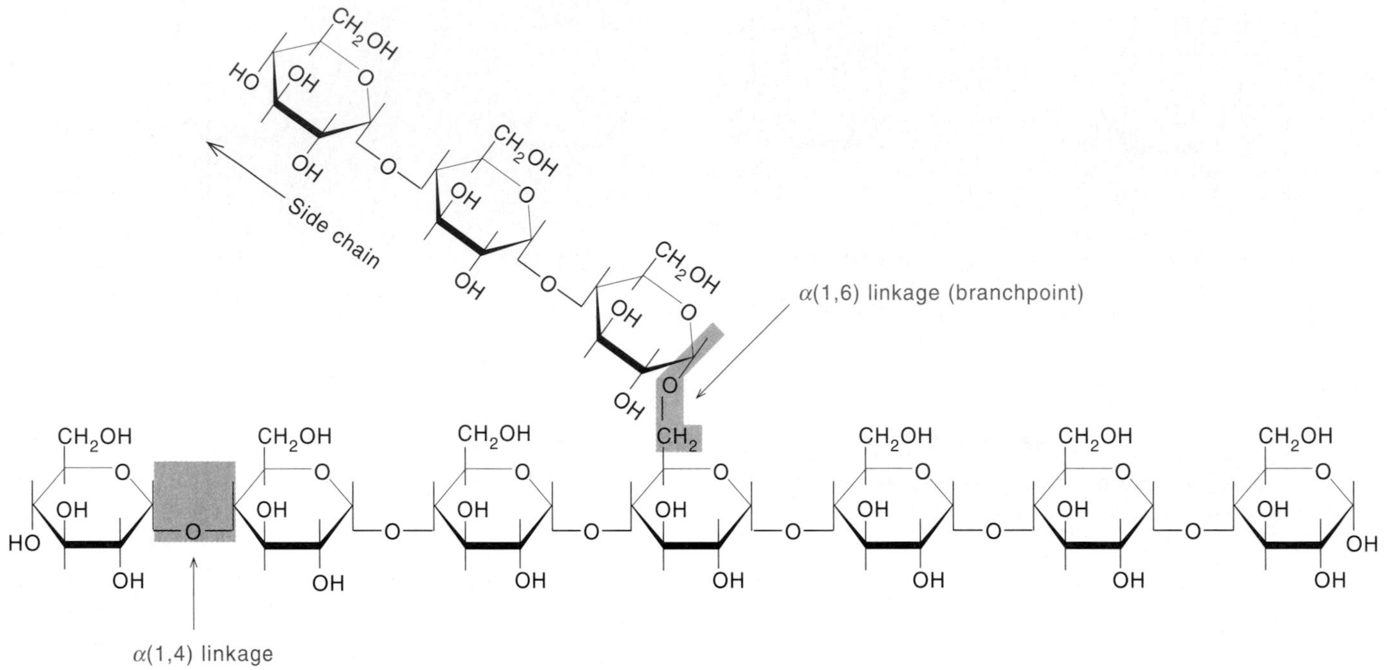

The disaccharide cellobiose is identical to maltose except for having a $\beta(1,4)$-glycosidic linkage. Cellobiose is a degradation product of cellulose. Lactose is a disaccharide found exclusively in the milk of mammals. Lactose contains a $\beta(1,4)$-glycosidic linkage between galactose and glucose. Sucrose is found in abundance in sugar beets and sugar cane; it contains an $\alpha(1,2)$-glycosidic linkage between D-glucose and D-fructose.

Most carbohydrates in nature exist as high-molecular-weight polymers called polysaccharides. Polysaccharides are composed of simple or derived sugars connected by glycosidic bonds. In this chapter we restrict ourselves to a description of homopolymers of glucose because they are the most important in energy metabolism.

## Starch and Glycogen Are Major Energy-Storage Polysaccharides

Although glucose is the most important sugar in energy metabolism in most cells, it is not present in the cell to any great extent as the free monosaccharide. Cells store glucose for future use in the form of homopolymers, thereby reducing the osmotic pressure of the stored sugar. A polysaccharide consisting of 1,000 glucose units exerts an osmotic pressure that is only 1/1,000 of the pressure that would result if the glucose units were all present as separate molecules. In the polymeric form, glucose can be stored compactly until needed. Another advantage of the poly-

saccharide over the monomer is that it contains more potential energy as will become apparent when we discuss its catabolism.

The two major polysaccharides used for energy storage are starch in plants and glycogen in animals and bacteria. Both are $\alpha(1,4)$-glycosidically linked homopolymers with occasional $\alpha(1,6)$ linkages to make branchpoints (fig. 12.9). Starch and glycogen differ primarily in their chain lengths and branching patterns. Glycogen is highly branched, with an $\alpha(1,6)$ linkage occurring every 8–10 glucose units along the backbone, giving rise in each case to short chains of about 8–12 glucose units each. Starch occurs both as unbranched amylose and as branched amylopectin. Like glycogen, amylopectin has $\alpha(1,6)$ branches, but these occur less frequently along the molecule (once every 20 or so residues) and give rise to longer side chains (lengths of 20–25 glucose units are common). Starch deposits are usually about 10%–30% amylose and 70%–90% amylopectin.

## Cellulose Serves a Structural Function Despite Its Similarity in Composition to the Energy-Storage Polysaccharides

Cellulose, the major component of plant cell walls, is also a homopolymer of glucose. However, the glycosidic linkages in cellulose are in the $\beta$ configuration instead of the $\alpha$ configuration, as in starch and glycogen.

## Figure 12.10

Fibril arrangements in the cell wall of *Valonia* (12,000×). (Electron micrograph from A. Frey-Wyssling and K. Mühlethaler, *Ultrastructural Plant Cytology*, Elsevier Science Publishers, Amsterdam, 1965, p. 298. Reprinted with permission from Elsevier Science Publishers.)

Cellulose is insoluble in water because of the high affinity of the polymer chains for one another. Its individual polymeric chains have molecular weights of 50,000 or greater. The molecular chains of cellulose interact in parallel bundles of about 2,000 chains. Each bundle constitutes a single microfibril. Many microfibrils arranged in parallel constitute a macrofibril, which can be seen under the light microscope. Figure 12.10 shows the inner cell walls of the plant *Valonia*; the fibrils in the wall are almost pure cellulose.

## *The Configurations of Glycogen and Cellulose Dictate Their Roles*

It is a remarkable fact that the main energy-storage polysaccharides and the main structural polysaccharides found in nature both have a primary structure of (1,4)-linked polyglucose. Why should two such closely related compounds be used in totally different roles? A closer look at the stereochemistry of the $\alpha$ and $\beta$ glycosidic linkage for polyglucose indicates why this is so.

Recall that D-glucose exists in the chair form of a pyranose ring (see fig. 12.6c). The ring has a rigid character to it. We can think of it as a structural building block in a polysaccharide chain, just as we think of the rigid planar peptide grouping as a structural building block in a polypeptide chain. It is also possible to specify two torsional angles $\phi$ and $\psi$ for rotation about the glycosidic C—O linkage (fig. 12.11). These angles are used extensively to discuss polypeptide configurations in proteins (see fig 4.1). They have

limited value in discussing polysaccharide structures, because much less information is available about such structures. Nevertheless, it is clear that only the $\beta(1,4)$-linked polyglucose has the capacity to form straight chains (see fig. 12.11a). A straight chain can be created by flipping each glucose unit by 180° relative to the previous one. This process results in an almost fully extended polysaccharide chain, which is characteristic of the structure of cellulose. Evidently this conformation is energetically favored. By contrast, $\alpha(1,4)$-linked units in a polyglucose cause a natural turning of the chain (see fig. 12.11b). Consistent with this fact is the observation that amylose adopts a coiled helical configuration. Indeed, one of the first helical structures to be discovered (1943) was the left-handed helix of amylose wound around molecules of iodine (fig. 12.12). This structure is responsible for the characteristic blue color of the amylose–iodine complex.

The extended-chain form of polyglucose has been exploited in nature for structural purposes, leaving the coiled form for use as an energy-storage macromolecule. Correlated with this functional difference is the omnipresence of degrading enzymes for glycogen and starch and the very limited phylogenetic distribution of comparable enzymes for cellulose. Cellulose is degraded in the gastrointestinal tract of herbivores, such as the cow, or in insects, such as termites, by a protozoan that synthesizes the enzyme cellulase. Humans do not possess this enzyme and hence cannot degrade cellulose.

## The Synthesis and Breakdown of Sugars

Having covered the basic structural features of carbohydrates, we can now discuss sugar metabolism. First we describe how sugars are degraded to produce energy and carbon compounds that can be used in biosynthesis. Then we consider the reverse process of gluconeogenesis—synthesis of sugars from smaller carbon compounds. Next we consider how these two opposing pathways are regulated with a particular focus on how simultaneous operation of the two pathways is prevented. Finally we consider an alternative mode of sugar breakdown that is most important in supplying reducing power for biosynthesis and ribose for nucleic acids.

## Overview of Glycolysis

The process of glycolysis is without a doubt the single most ubiquitous pathway in all energy metabolism. Glycolysis can be characterized as a nearly universal process because it occurs in almost every living cell. It can occur in the ab-

## Figure 12.11

Energetically favored conformations of $\beta(1,4)$-linked D-glucose (*a*) and $\alpha(1,4)$-linked D-glucose (*b*). Note that in the $\beta(1,4)$ configuration in (*a*), alternating residues are flipped 180° relative to one another so that long straight chains result. In the $\alpha(1,4)$ configuration (*b*), the chain has a natural curvature. (Illustration copyright by Irving Geis. Reprinted by permission.)

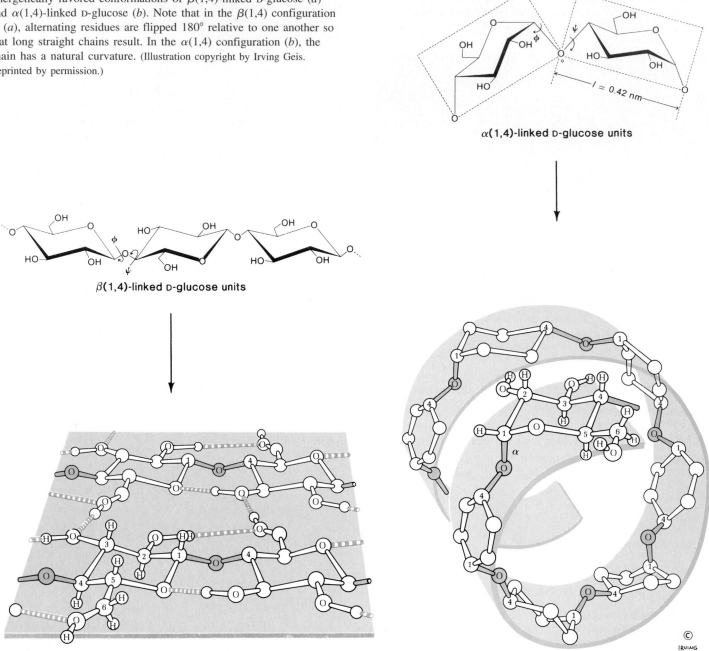

(a)

(b)

sence of oxygen. It is generally regarded as a primitive process, both in the sense that it occurs in the cytosol rather than being compartmentalized into specific organelles within eukaryotic cells, and also because it probably arose early in biological history, before the advent of eukaryotic organelles.

The glycolytic pathway was the first major metabolic sequence to be elucidated. Much of the definitive work was done in the 1930s by the German biochemists, Gustav Embden, Otto Meyerhof, and Otto Warburg. Because of their contributions the alternative name, Embden-Meyerhof pathway, is sometimes used for the glycolytic pathway.

**Figure 12.12**

Structure of the helical complex of amylose with iodine ($I_2$). The amylose forms a left-handed helix with six glucosyl residues per turn and a pitch of 0.8 nm. The iodine molecules ($I_2$) fit inside the helix parallel to its long axis.

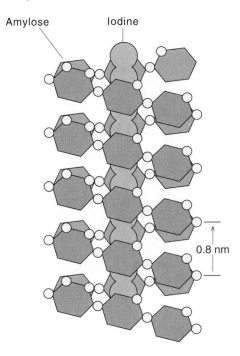

Amylose            Iodine

0.8 nm

The glycolytic pathway appears in detail in figure 12.13. The essence of the process is suggested by the name, since glycolysis comes from the Greek roots *glykos,* meaning "sweet," and *lysis,* meaning "loosing." Literally, glycolysis is the loosing, or splitting, of something sweet, that is, the starting sugar. From figure 12.13 it is clear that the actual splitting occurs at step 4 (starting from glucose). At this point, a six-carbon sugar is cleaved to yield two three-carbon compounds, one of which, glyceraldehyde-3-phosphate, is the only oxidizable molecule in the entire pathway. Subsequent to the cleavage of step 4, two successive ATP-generating steps occur: one at step 7 and the other at step 10. These steps represent the energy payoff of the process, since these are the only ATP-yielding reactions of the pathway under anaerobic conditions.

Except for glycerate-1,3-bisphosphate all of the intermediates in the pathway are pictured as belonging to one of three metabolic pools. Within each metabolic pool the intermediates are readily interconvertible and usually present in relative concentrations close to their equilibrium values. Between the pools the concentrations of the intermediates can be very different because of the lack of rapid interconversion and also because the equilibrium values are often very large or very small.

## Three Hexose Phosphates Constitute the First Metabolic Pool

The three hexose phosphates—glucose-1-phosphate, glucose-6-phosphate and fructose-6-phosphate—constitute a single metabolic pool (fig. 12.14). This pool can be replenished by generation of any of its components (also see fig. 16.2 for conversions of other hexoses to members of this pool). Glucose-1-phosphate is the first product in the utilization of storage polysaccharides; glucose-6-phosphate is the first hexose phosphate formed when free glucose is metabolized; and fructose-6-phosphate is the first hexose phosphate formed when carbohydrate is made *de novo* by gluconeogenesis or photosynthesis.

## Phosphorylase Converts Storage Carbohydrates to Glucose Phosphate

The first step in the cell's utilization of stored starch or glycogen involves the removal of a terminal glucose residue, a reaction catalyzed by the enzyme glycogen phosphorylase. Overall, the reaction involves same-side displacement at C-1 of the terminal residue, with an incoming phosphate group replacing the remainder of the polysaccharide molecule. The product generated is therefore glucose-1-phosphate with retention of configuration about the C-1 residue (fig. 12.15).

Since glycogen is a branched structure (see fig. 12.9) there must be a mechanism for removing the branches prior to the action of phosphorylase. In the process of glycogen breakdown a debranching enzyme transfers $\alpha(1,4)$-linked residues from the branchpoints to the nonreducing end of another branchpoint thereby forming a new $\alpha(1,4)$ linkage and making more glucose units available for phosphorylase-catalyzed phosphorolysis. The remaining $\alpha(1,6)$ bond linking the ultimate glycosyl residue at the branch to the main chain is hydrolyzed by the same debranching enzyme.

Digestion of dietary starch or glycogen in the intestine follows a different course. Interior bonds of the polysaccharide are hydrolyzed by enzymes secreted into the intestine by the pancreas and the intestinal mucosal cells; the macromolecule is split into progressively smaller fragments. Individual sugar residues are not removed from the polysaccharide or its major fragments. The degradation proceeds to the disaccharide maltose. The final step entails hydrolysis of maltose to free glucose.

The difference between the modes of hydrolysis inside cells and in the intestine reflects biological needs and func-

## Figure 12.13

The glycolytic pathway from glucose to pyruvate, indicating two anaerobic options (ethanol or lactate) and one aerobic option (TCA cycle). The red arrow indicates how NADH formed in glycolysis could be reoxidized to $NAD^+$ when pyruvate is converted to an end product of ethanol or lactate under anaerobic conditions. The three metabolic pools are screened.

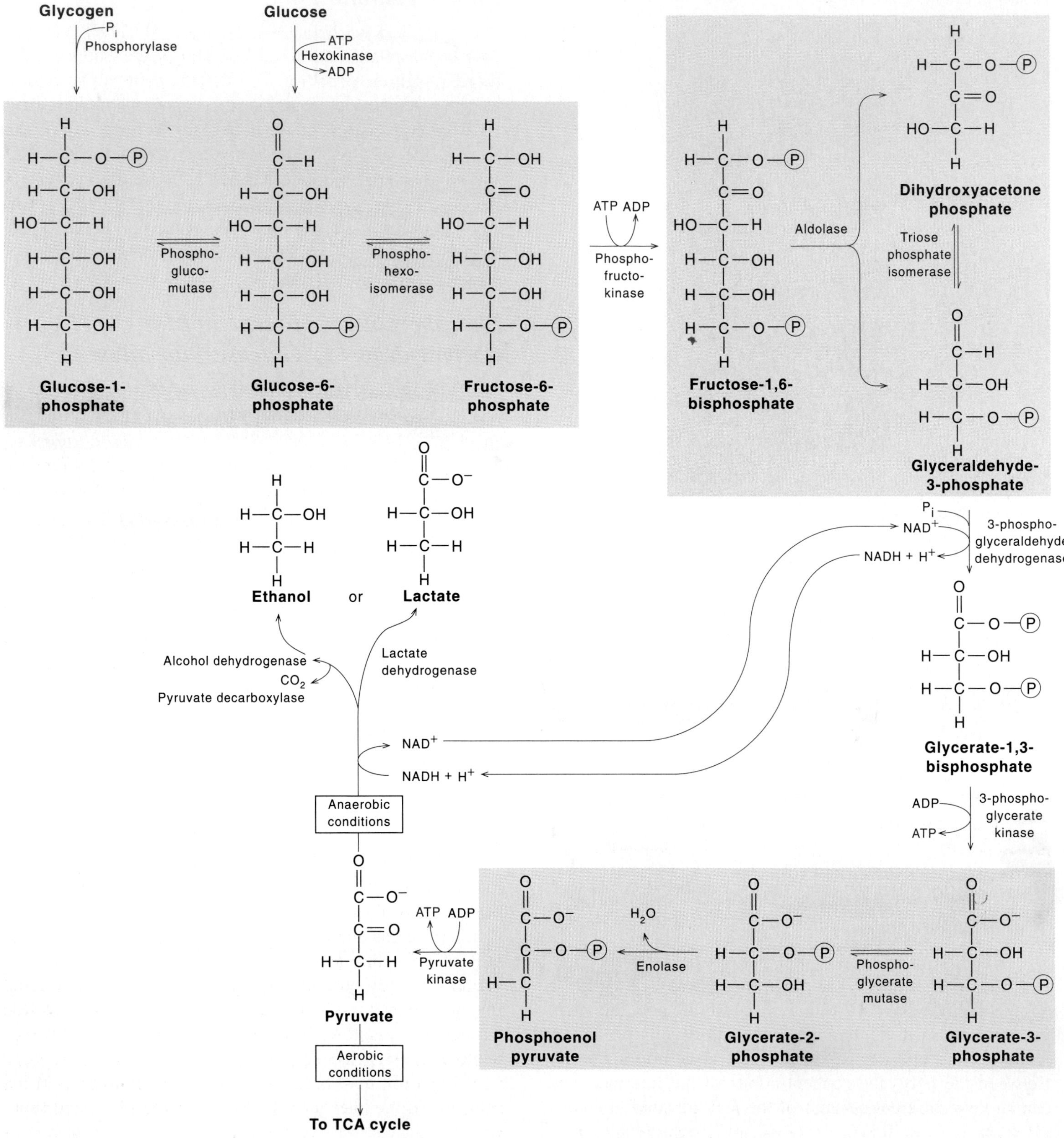

## Figure 12.14

The hexose monophosphate pool. The equilibrium percentages of the three hexose monophosphates are indicated. Horizontal flow arrows indicate major routes of replenishment and utilization of the three hexose monophosphates.

Phosphorolysis of storage polysaccharides $\longrightarrow$ Glucose-1-phosphate (3%) $\longrightarrow$ Polysaccharide synthesis

Phosphorylation of glucose by ATP $\longrightarrow$ Glucose-6-phosphate (65%) $\longrightarrow$ Pentose phosphate pathway

Gluconeogenesis or photosynthesis $\longrightarrow$ Fructose-6-phosphate (32%) $\longrightarrow$ Glycolysis

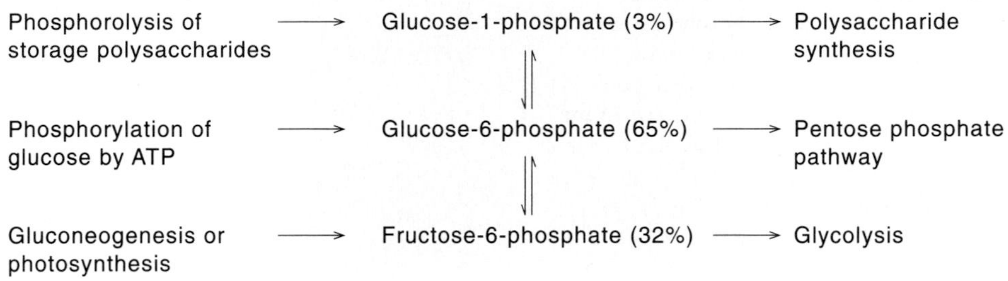

## Figure 12.15

The reaction catalyzed by phosphorylase. An oxygen of a phosphate ion attacks C-1 of the terminal glucosyl unit of starch or glycogen, displacing the macromolecule and generating glucose-1-phosphate. The reaction proceeds with retention of configuration, a fact suggesting that it may involve an oxonium ion intermediate. Phosphorylase is a complex regulatory enzyme, the allosteric properties of which were discussed in chapter 9.

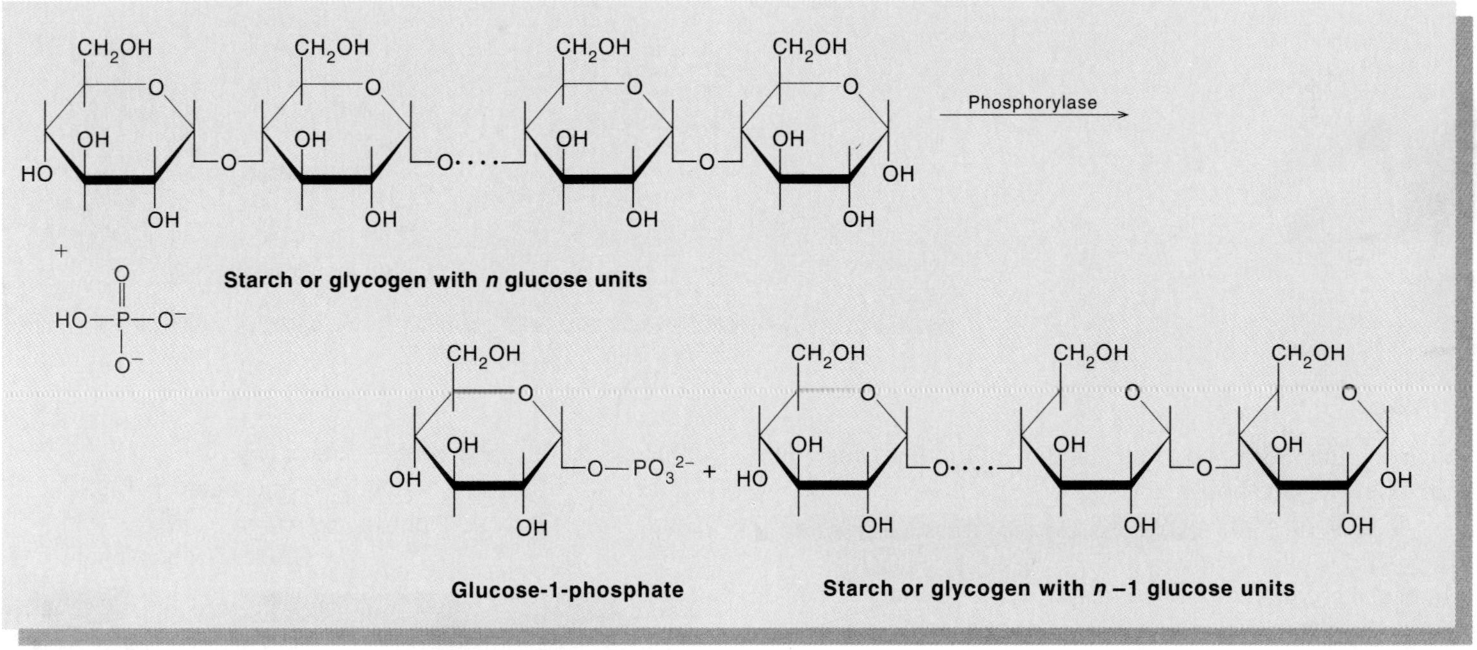

**Starch or glycogen with *n* glucose units**

**Glucose-1-phosphate**     **Starch or glycogen with *n* −1 glucose units**

tions. Phosphorylated sugars do not cross biological membranes readily. In the cell, phosphorylated hexoses are desirable because they are not lost by diffusion out of the cell. In contrast, most cells are able to take up nonphosphorylated sugars from their surroundings. Because the very reason for hydrolysis of polysaccharides in the intestine is to make their component residues available for absorption into the body, unphosphorylated glucose is preferable.

### Hexokinase and Glucokinase Convert Free Sugars to Hexose Phosphates

Glucose obtained by mammals from the diet through intestinal hydrolysis of lactose, sucrose, glycogen, or starch is brought into the hexose phosphate pool through the action of hexokinase. This enzyme catalyzes phosphorylation at the oxygen attached to C-6 of glucose (fig. 12.16). As usual in metabolism, the source of the phosphate group is ATP.

Glucose + ATP $\longrightarrow$
$$\text{Glucose-6-phosphate} + \text{ADP} + \text{H}^+ \quad (1)$$

The standard free energy change is about $-5$ kcal/mol, and the equilibrium constant is about 5,700. Thus, equilibrium considerations indicate a potential for this reaction to occur. The potential can be converted to reality only by an enzyme with appropriate kinetic properties. Hexokinases purified from various tissues typically have a Michaelis constant for glucose between 10 and 20 $\mu$M. Thus, by the expenditure of ATP, hexokinase can convert glucose in the micromolar

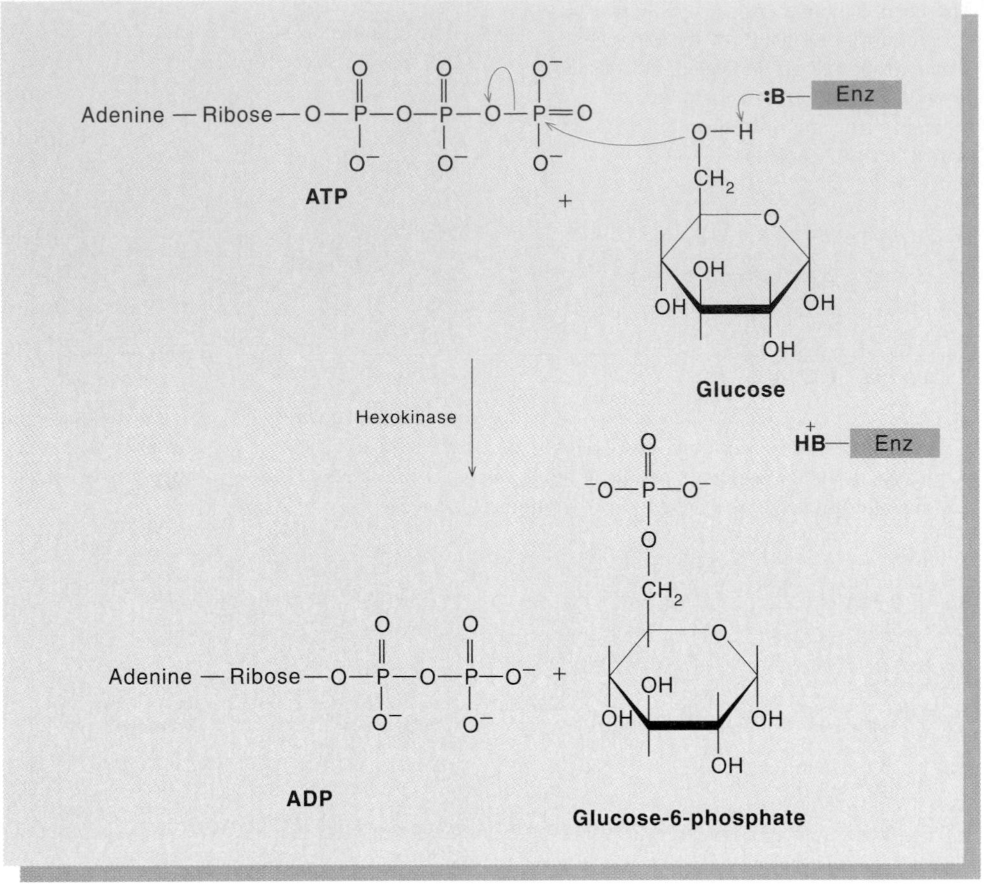

**Figure 12.16**

The reaction catalyzed by hexokinase. Attack on the terminal phosphorus atom of ATP is probably facilitated by proton removal by a negatively charged group in the catalytic site of the enzyme (:B in the figure). It is also facilitated by the fact that the terminal phosphate of ATP is an excellent leaving group.

range to glucose-6-phosphate in the millimolar range for utilization in metabolism.

The liver contains another enzyme, namely glucokinase, that catalyzes the same reaction as hexokinase. Its Michaelis constant (about 10 mM) is 1,000 times larger than hexokinase; thus glucokinase can function only when the concentration of glucose is much higher. This enzyme probably is active only when blood glucose is high, as after a meal, and the liver is taking up glucose for conversion to glycogen.

## *Phosphoglucomutase Interconverts Glucose-1-phosphate and Glucose-6-phosphate*

The enzyme phosphoglucomutase catalyzes the interconversion of glucose-1-phosphate and glucose-6-phosphate (fig. 12.17). The interconversion occurs in two discrete steps by way of a phosphorylated enzyme and glucose-1,6-bisphosphate. The reactions are

$$\text{Glucose-1-phosphate} + \text{Enz-P} \rightleftharpoons$$
$$\text{Glucose-1,6-bisphosphate} + \text{Enz} \quad \textbf{(2)}$$

$$\text{Glucose-1,6-bisphosphate} + \text{Enz} \rightleftharpoons$$
$$\text{Glucose-6-phosphate} + \text{Enz-P} \quad \textbf{(3)}$$

Sum:      $$\text{Glucose-1-phosphate} \rightleftharpoons$$
$$\text{Glucose-6-phosphate} \quad \textbf{(4)}$$

where Enz represents phosphoglucomutase, and Enz-P represents phosphoglucomutase phosphorylated at a specific serine residue in the active site. The enzyme cycles between the free and the phosphorylated states at each conversion of glucose-1-phosphate to glucose-6-phosphate (or of glucose-6-phosphate to glucose-1-phosphate).

## *Phosphohexoseisomerase Interconverts Glucose-6-phosphate and Fructose-6-phosphate*

Glucose-6-phosphate and fructose-6-phosphate interconvert readily in weak alkaline solution. Presumably this is because the intermediate enediol is stabilized by ionization in an alkaline medium (fig. 12.18). In the cytosol glucose-6-phosphate and fructose-6-phosphate are interconverted by the action of the enzyme phosphohexoseisomerase. The re-

## Figure 12.17

The reaction catalyzed by phosphoglucomutase. The enzyme apparently can bind glucose phosphates in two ways, allowing it to transfer phosphoryl groups to and from either the oxygen atom at C-1 or the oxygen at C-6. The direction of reaction is driven by mass action, depending on the relative concentrations of glucose-1-phosphate and glucose-6-phosphate.

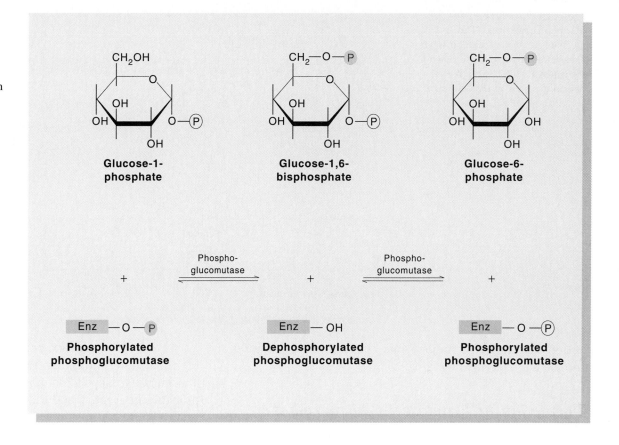

## Figure 12.18

Mechanism of the interconversion of glucose-6-phosphate and fructose-6-phosphate. Loss of a proton from the oxygen attached to C-2 of the intermediate enediol leads to fructose-6-phosphate. **A** and **B** represent catalytic groups on the enzyme. It is not always known what specific groups are involved in a catalysis. In this case the **HA** group originates from a glutamate on the enzyme.

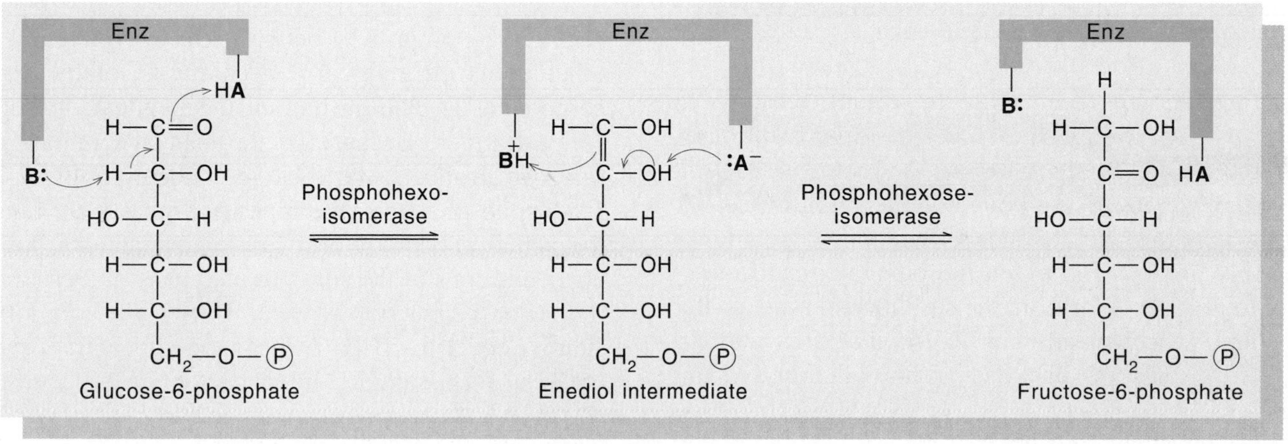

**Figure 12.19**

The reaction catalyzed by phosphofructokinase. The mechanism of this reaction is very similar to the hexokinase reaction shown in figure 12.16. **:B** is a proton acceptor at the active site.

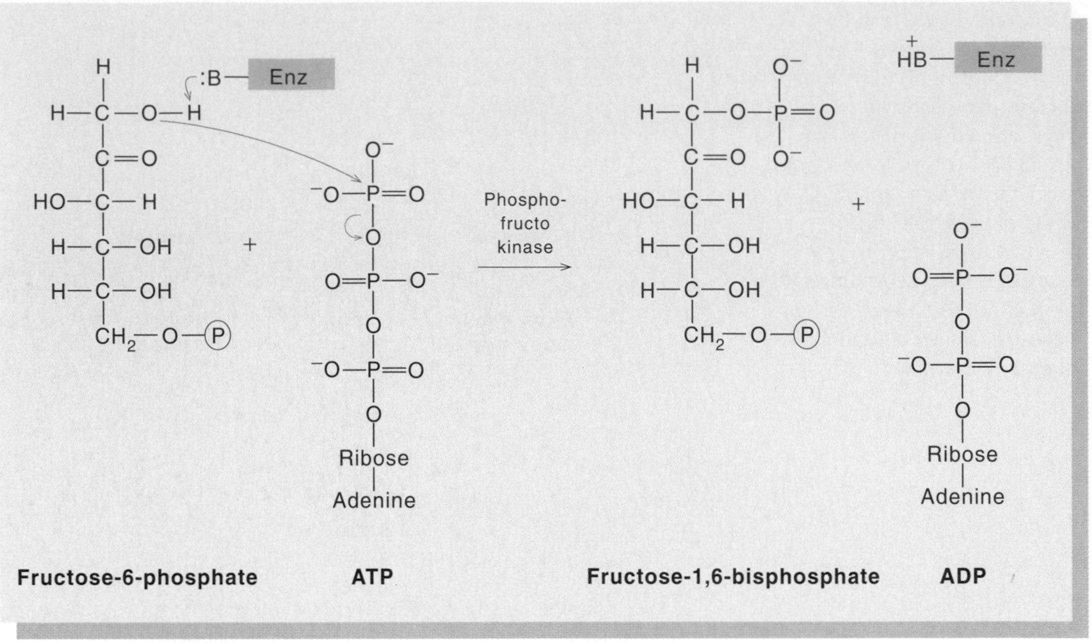

| Fructose-6-phosphate | ATP | | Fructose-1,6-bisphosphate | ADP |

action resembles the nonenzymic conversion by going via the enediol intermediate.

## Formation of Fructose-1,6-bisphosphate Signals a Commitment to Glycolysis

Fructose-6-phosphate is converted to fructose-1,6-bisphosphate by transfer of a phosphoryl group from ATP in a reaction catalyzed by phosphofructokinase (fig. 12.19). Since net regeneration of ATP is a major function of the catabolism of carbohydrates, it may seem strange that an early step of glucose breakdown consumes ATP. However, there are good reasons for this. If fructose-6-phosphate were the substrate for aldolase, which cleaves the six-carbon sugar to two trioses, only one of the products would be phosphorylated. The unphosphorylated product might need to be phosphorylated immediately for protection against loss by diffusion. Phosphorylation before cleavage is even more effective.

Another, perhaps even more important, reason why phosporylation occurs before cleavage has to do with regulation. The reaction by which material is removed from the hexose phosphate pool is the point at which control must be exerted for maximal effectiveness. Metabolic regulation is possible only at steps for which the physiological ratio of precursor to product is far from the equilibrium ratio, so that kinetic control mechanisms can cause increases and decreases in the rates of reactions without thermodynamic constraints.

Under physiological conditions the concentration ratio of fructose-bisphosphate to fructose-6-phosphate varies considerably, depending on metabolic conditions, but is probably between 5 and 0.2. At equilibrium for the phosphofructokinase reaction, the ratio would be about $10^4$. Since the concentration ratio is far from equilibrium, no thermodynamic limitations exist for the reaction.

## Fructose-1,6-bisphosphate and the Two Triose Phosphates Constitute the Second Metabolic Pool in Glycolysis

The metabolic pool that consists of fructose-1,6-bisphosphate and the two triose phosphates—glyceraldehyde-3-phosphate and dihydroxyacetone phosphate (DHAP)—is somewhat different from the other two pools of intermediates in glycolysis because of the nature of the chemical relationships between these compounds. In the other pools the relative concentrations of the component compounds at equilibrium are independent of the absolute concentrations. Because of the cleavage of one substrate into two products, the relative concentrations of fructose-1,6-bisphosphate and the triose phosphates are functions of the actual concentrations. For such reactions, the relative concentrations of the split products must increase with dilution. (For the reaction A $\rightleftharpoons$ B + C, the equilibrium constant is equal to [B][C]/[A]. If the concentration of A decreases, for example, by a factor of 4, equilibrium is

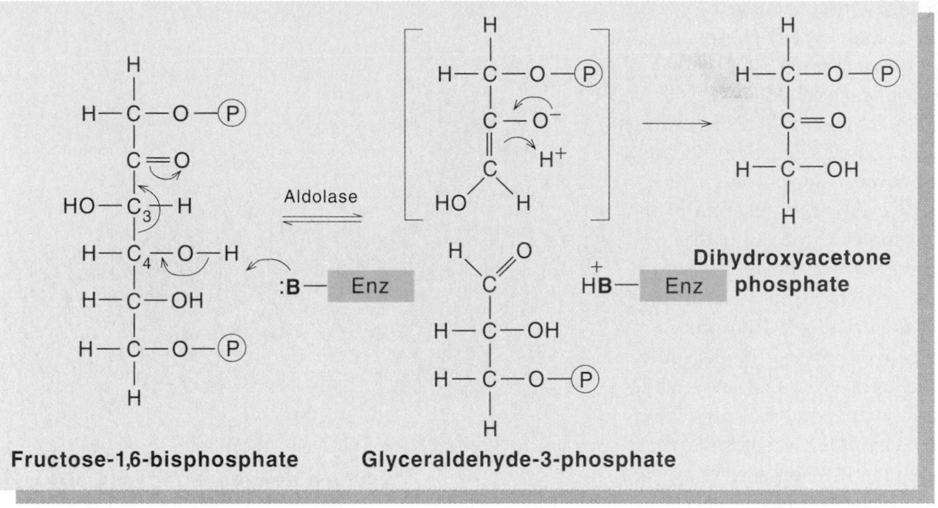

**Figure 12.20**

Cleavage of fructose-1,6-bisphosphate, an aldolase-catalyzed reaction. The aldolase reaction entails a reversal of the familiar aldol condensation. The first step involves abstraction of the hydrogen of the C-4 hydroxyl group, followed by elimination of an enolate anion.

Fructose-1,6-bisphosphate      Glyceraldehyde-3-phosphate

restored when the concentrations of B and C decrease by a factor of only 2). Except for that difference, interconversions of the compounds in this pool are similar to those in the other pools because reactions within the pool are close to equilibrium and can go rapidly in either direction.

## Aldolase Cleaves Fructose-1,6-bisphosphate

Fructose-1,6-bisphosphate is cleaved by aldolase into two molecules of triose phosphate. This reaction represents the reversal of an aldol condensation (fig. 12.20). Most aldolases are highly specific for the ''upper'' end of the substrate molecule, requiring a phosphate group at C-1, a carbonyl at C-2, and specific steric configurations at C-3 and C-4. The nature of the remainder of the molecule is unimportant as far as the enzyme action is concerned.

## Triose Phosphate Isomerase Interconverts the Two Trioses

The isomeric triose phosphates, glyceraldehyde-3-phosphate and dihydroxyacetone phosphate, bear the same relationship to each other as do glucose-6-phosphate and fructose-6-phosphate. Their interconversion, catalyzed by triose phosphate isomerase, is equally facile (see fig. 12.13). Dihydroxyacetone phosphate is a starting material for the synthesis of the glycerol moiety of fats (chapter 19), but only glyceraldehyde-3-phosphate is used in glycolysis. Thus, under ordinary circumstances nearly all of the dihydroxyacetone phosphate that is formed in the cleavage of

fructose bisphosphate is converted to glyceraldehyde-3-phosphate by triose phosphate isomerase. All six carbons of the hexoses are thereby made available for the later steps of carbohydrate catabolism. The mechanism of action of triosephosphate isomerase is described in chapter 8.

## The Conversion of Triose Phosphates to Phosphoglycerates Occurs in Two Steps

The oxidation of glyceraldehyde-3-phosphate to glycerate-3-phosphate is the first energy-yielding (ATP-producing) reaction in the glycolytic pathway. The mechanism by which an inorganic phosphate ion is taken up and transferred to adenosine diphosphate (ADP) during the conversion of glyceraldehyde-3-phosphate to glycerate-3-phosphate requires two separate enzyme-catalyzed reactions.

The first reaction is catalyzed by the enzyme 3-phosphoglyceraldehyde dehydrogenase.

$$\text{Glyceraldehyde-3-phosphate} + \text{NAD}^+ + \text{P}_i \longrightarrow$$
$$\text{Glycerate-1,3-bisphosphate} + \text{NADH} + \text{H}^+ \quad \textbf{(5)}$$

This reaction is initiated by the condensation of an —SH group of a specific cysteine residue at the catalytic site of the enzyme with the aldehyde carbonyl forming a sulfhydryl adduct, or thiohemiacetal (fig. 12.21, step 1). A pair of electrons, along with a proton (thus in effect a hydride ion) are then donated to a NAD$^+$ molecule that is tightly bound nearby, converting the tetrahedral hemiacetal into a thioester (step 2). Thioesters provide considerably more free energy on hydrolysis than comparable oxygen esters, facili-

## Figure 12.21

The reaction catalyzed by glyceraldehyde-3-phosphate dehydrogenase (3-phosphoglyceraldehyde dehydrogenase). This interesting and complex reaction consists of several steps. The enzyme first catalyzes a reaction of the substrate with a sulfhydryl group of a cysteine residue of the enzyme itself. The substrate is then oxidized from the aldehyde level of oxidation to the carboxylic acid level while still attached covalently to the enzyme. Displacement of the enzyme by inorganic phosphate ion liberates the product, glycerate-1,3-bisphosphate. The bound NADH of the enzyme, which became reduced when the substrate was oxidized, then transfers a pair of electrons to an unbound $NAD^+$, and the enzyme is ready for another catalytic cycle.

Glyceraldehyde-3-phosphate

3-Phosphoglyceraldehyde dehydrogenase

Glycerate-1,3-bisphosphate

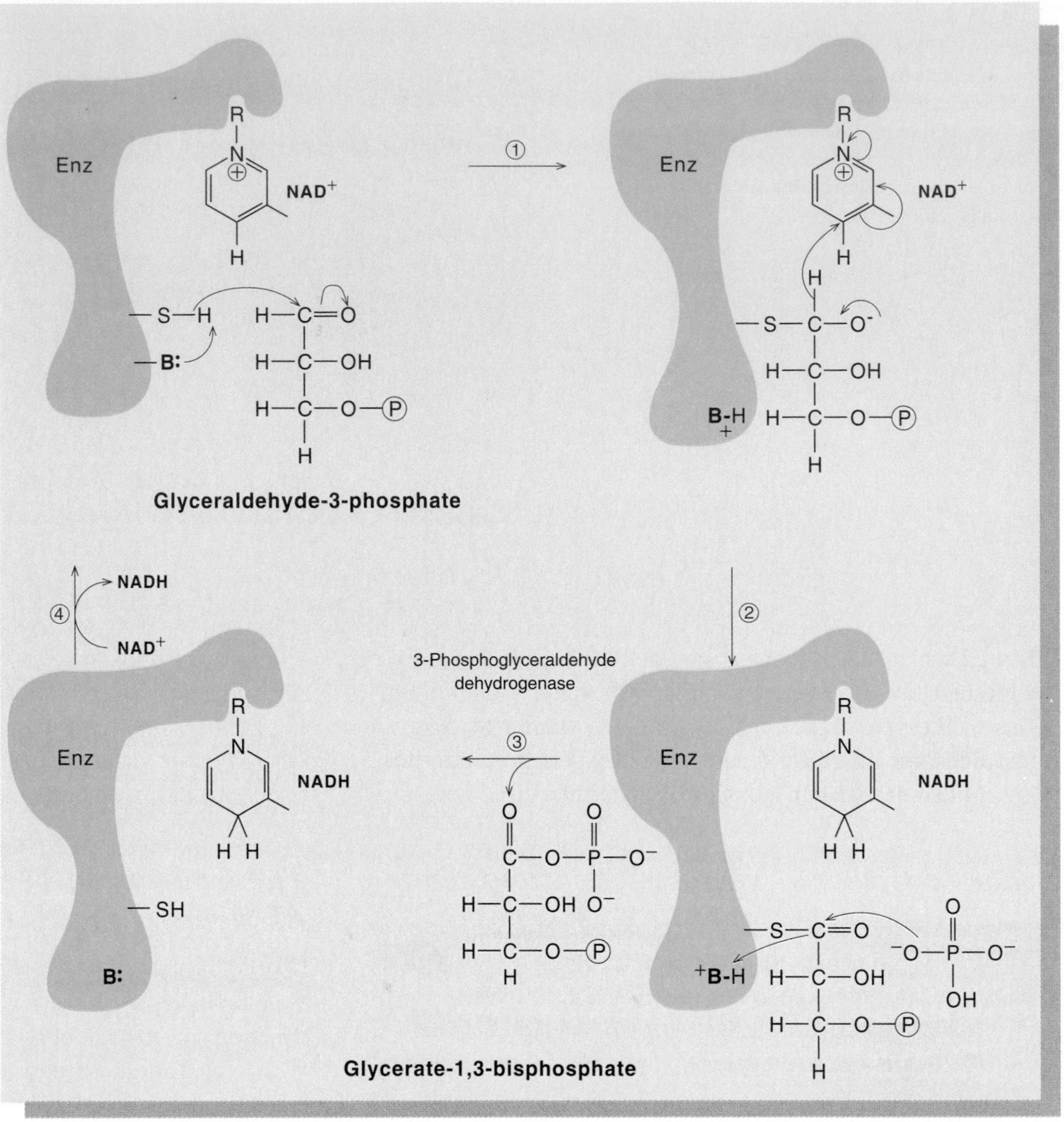

tating the next step (step 3), an attack by an oxygen of a phosphate ion on the carbonyl carbon of the thioester. The sulfur atom, and thus the enzyme molecule, is displaced from covalent linkage to the reaction intermediate, and the product, glycerate-1,3-bisphosphate, is released. The hydride ion is passed on to an $NAD^+$ molecule in solution that binds to the enzyme and dissociates after reduction. Thus an NADH molecule is generated in solution, and the tightly bound $NAD^+$ is again in the oxidized form ready to participate in another catalytic cycle.

In the second step leading to glycerate-3-phosphate, a phosphate group is transferred from glycerate-1,3-bisphosphate to ADP. This reaction is catalyzed by 3-phosphoglycerate kinase.

$$H^+ + \text{Glycerate-1,3-bisphosphate} + ADP \rightleftharpoons$$
$$\text{Glycerate-3-phosphate} + ATP \quad (6)$$

In the two steps catalyzed by 3-phosphoglyceraldehyde dehydrogenase and 3-phosphoglycerate kinase, the oxidation of glyceraldehyde-3-phosphate to glycerate-3-phosphate is coupled to the regeneration of ATP.

Since two molecules of glyceraldehyde-3-phosphate are produced from each molecule of hexose, the oxidation of the aldehyde to glycerate-3-phosphate leads to the production of two molecules of ATP per molecule of glucose or other hexose consumed. At this point the energy (ATP) gained is equal to the energy invested if the starting material was a free hexose. The two phosphate groups supplied by ATP in the hexokinase and phosphofructokinase reactions are still contained in glycerate-3-phosphate. If storage glycogen or starch was the starting material, only one phosphate group is supplied by ATP, and one of the two molecules of ATP regenerated in the phosphoglycerate kinase step represents a net gain.

**Figure 12.22**

Interconversion of glycerate-3-phosphate and glycerate-2-phosphate, catalyzed by phosphoglyceromutase. The reaction closely resembles that catalyzed by phosphoglucomutase (see fig. 12.17) except that the phosphate binds to a histidine side chain instead of a serine side chain. The enzyme can transfer a covalently bound phosphoryl group either to the oxygen on C-2 of glycerate-3-phosphate or to the oxygen on C-3 of glycerate-2-phosphate. The resulting glycerate-2,3-bisphosphate, in turn, can donate either of its phosphoryl groups to the enzyme. These catalytic capabilities provide for interconversion of the two monophosphoglycerates in either direction.

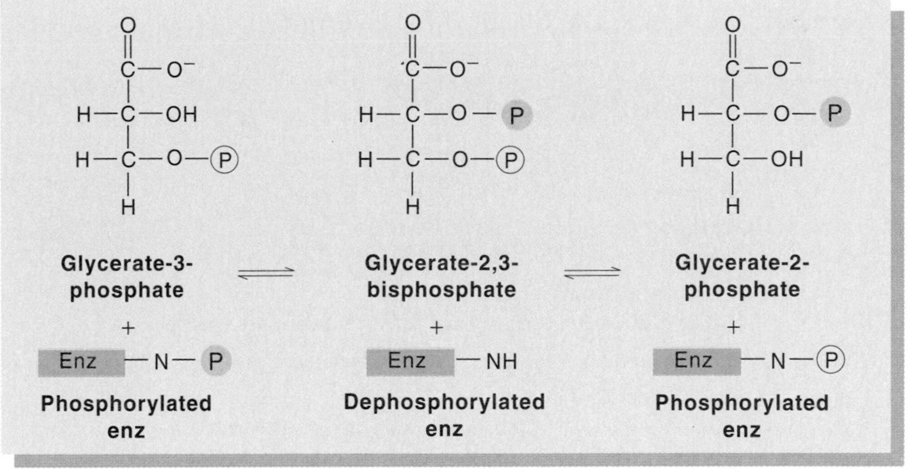

## The Three-Carbon Phosphorylated Acids Constitute a Third Metabolic Pool

Glycerate-3-phosphate, glycerate-2-phosphate, and phosphoenolpyruvate (PEP) make up another equilibrium group of metabolites that, like the hexose phosphates or the triose phosphates, function as a single metabolic pool (see fig. 12.13).

The first interconversion between glycerate-3-phosphate and glycerate-2-phosphate is similar to the reaction catalyzed by phosphoglucomutase (fig. 12.22). The enzyme is transiently phosphorylated in the course of the reaction. The second interconversion in the three-carbon pool is between glycerate-2-phosphate and phosphoenolpyruvate. This reaction is catalyzed by enolase and entails a dehydration (see fig. 12.13).

## Conversion of Phosphoenolpyruvate to Pyruvate Generates ATP

The final energy payoff in the glycolytic pathway occurs in the hydrolysis of phosphoenolpyruvate to pyruvate and the concomitant phosphorylation of ADP to ATP. Two molecules of ATP are produced for each molecule of hexose phosphate consumed, bringing the net yield of ATP to two molecules for each molecule of glucose (two molecules of ATP are regenerated in the phosphoglycerate kinase step and two in this step, and two are consumed in the hexokinase and phosphofructokinase steps).

Glucose + 2 NAD$^+$ + 2 ADP + 2 P$_i$ $\longrightarrow$

2 Pyruvate + 2 NADH + 2 H$^+$ + 2 ATP + 2 H$_2$O    (7)

The net yield is three ATPs for each glucose catabolized when the substrate is storage glycogen or starch. These are

the total yields of ATP from the anaerobic catabolism of sugars. The 10 steps from glucose to pyruvate are reviewed in table 12.1. It can be seen that overall glycolysis goes with a significant drop in free energy, about −25 kcal/mol of glucose. In calculating the free energy drop at each step, we do not get a realistic picture from the standard free energies. This is because the standard free energies do not take into account the concentrations of the intermediates. If we take the intermediates into account, we see that a sizeable drop in free energy occurs at only three steps (see table 12.1 and fig. 12.23). Most of the other steps proceed with little or no change in free energy. Our discussion of regulation will show that the process of glycolysis is regulated at precisely these three steps.

## The NAD$^+$ Reduced in Glycolysis Must Be Regenerated

The pyruvate produced in glycolysis must be metabolized further if glycolysis is to continue. This is essential because the NAD$^+$ reduced for each molecule of phosphoglyceraldehyde that was oxidized needs to be regenerated (see fig. 12.13). Because equimolar amounts of NADH and pyruvate are produced in glycolysis, a simple way to reoxidize the NADH is by transfer of electrons to pyruvate, forming lactate (fig. 12.24). This is the solution employed by many kinds of organisms. Indeed, even skeletal muscle tissue uses this strategy in times of exertion when the oxygen supply cannot keep up with the energy needs of the tissue. The reduction of pyruvate is catalyzed by the enzyme lactate dehydrogenase.

Pyruvate + NADH + H$^+$ $\longrightarrow$ Lactate + NAD$^+$    (8)

**Table 12.1**

Reactions, Enzymes, and Standard Free Energies for Steps in the Glycolytic Pathway

| Step | Reaction | Enzyme | $\Delta G^{\circ\prime}$ | $\Delta G^{\prime a}$ |
|------|----------|--------|---------|---------|
| 1 | Glucose + ATP $\longrightarrow$ Glucose-6-phosphate + ADP + H$^+$ | Hexokinase | −4.0 | −8.0 |
| 2 | Glucose-6-phosphate $\rightleftharpoons$ Fructose-6-phosphate | Phosphohexose isomerase | +0.4 | −.60 |
| 3 | Fructose-6-phosphate + ATP $\longrightarrow$ <br> Fructose-1,6-bisphosphate + ADP + H$^+$ | Phosphofructokinase | −3.4 | −5.3 |
| 4 | Fructose-1,6-bisphosphate $\rightleftharpoons$ <br> Dihydroxyacetone phosphate + Glyceraldehyde-3-phosphate | Aldolase | +5.7 | −.31 |
| 5 | Dihydroxyacetone phosphate $\rightleftharpoons$ Glyceraldehyde-3-phosphate | Triose phosphate isomerase | +1.8 | +.60 |
| 6 | Glyceraldehyde-3-phosphate + P$_i$ + NAD$^+$ $\rightleftharpoons$ <br> Glycerate-1,3-bisphosphate + NADH + H$^+$ | Phosphoglyceraldehyde dehydrogenase | +1.5 | −.41 |
| 7 | H$^+$ + Glycerate-1,3-bisphosphate + ADP $\rightleftharpoons$ <br> Glycerate-3-phosphate + ATP | 3-Phosphoglycerate kinase | −4.5 | +.31 |
| 8 | Glycerate-3-phosphate $\rightleftharpoons$ Glycerate-2-phosphate | Phosphoglyceromutase | +1.1 | +.19 |
| 9 | Glycerate-2-phosphate $\rightleftharpoons$ Phosphoenolpyruvate + H$_2$O + H$^+$ | Enolase | +0.4 | −.79 |
| 10 | Phosphoenolpyruvate + ADP + H$^+$ $\longrightarrow$ Pyruvate + ATP | Pyruvate kinase | −7.5 | −4.0 |

Net reaction:
$$C_6H_{12}O_6 + 2\ NAD^+ + 2\ ADP + 2\ P_i \longrightarrow 2\ C_3H_4O_3 + 2\ NADH + 2\ H^+ + 2\ ATP + 2\ H_2O$$

| | $-27.8^b$ | $-23.0$ |

[a] $\Delta G$ values are for the human erythrocyte using approximate values for concentrations.
[b] To obtain this sum for the standard free energy of the overall reaction per mole of glucose, the indicated free energies for reactions 6–10 must all be doubled. This is because each of these reactions occurs twice for every glucose consumed.

**Figure 12.23**

Energy profile in going from glucose to pyruvate by the glycolytic pathway. The three large energy drops are between glucose and glucose-6-phosphate, fructose-6-phosphate and fructose-1,6-bisphosphate, and phosphoenolpyruvate (PEP) and pyruvate. The actual reactions are described in table 12.1.

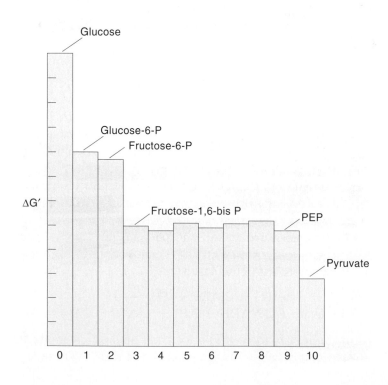

**Figure 12.24**

Regeneration of NAD$^+$ by reduction of pyruvate to lactate. Because NAD$^+$ is a necessary participant in the oxidation of glyceraldehyde-3-phosphate to glycerate-1,3-bisphosphate, glycolysis is possible only if there is a way by which NADH can be reoxidized.

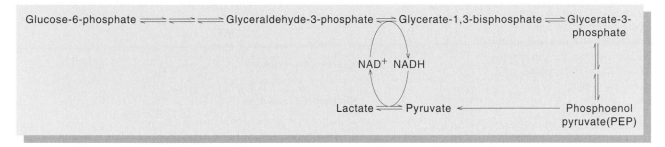

Yeast and some other kinds of organisms use a different strategy. They produce ethanol and $CO_2$, rather than lactate, from the anaerobic catabolism of sugars. Unlike true facultative anaerobes such as *E. coli*, which can live anaerobically for an indefinite period, yeasts can live only for a few generations in the total absence of oxygen because they require molecular oxygen for synthesis of membrane components. Rather than a true alternative life style, yeasts use alcohol fermentation as part of a very effective competitive strategy. When fruits ripen and fall from the tree or bush, yeasts are likely to be among the first invaders. In the anaerobic interior of the fruit, yeasts rapidly convert sugars to ethanol, which they excrete. As ethanol accumulates, the growth of most other microorganisms is discouraged, but yeasts can continue to grow until the concentration of ethanol reaches about 12%. When the softened fruit breaks open, a thorough-going change in the energy metabolism of the yeast cells occurs. They begin to take up the ethanol they had previously excreted, and oxidize it to $CO_2$, with a much larger production of ATP than in the anaerobic stage (about 14 moles of ATP per mole of ethanol, compared with 2 moles of ATP per mole of glucose in the anaerobic stage). By rapidly converting the available sugars into a compound that cannot be metabolized for energy by most other organisms and that is in fact toxic to them, yeasts gain an advantage over competing microorganisms and also gain energy in the process. They then can exploit the waste product of the first stage of growth as the main nutrient for the later aerobic stages of growth. This example illustrates that the organization of even the most central metabolic pathways can, like other properties of an organism, be shaped by competition.

## Summary of Glycolysis

Glycolysis consists of a chain of 10 reactions that starts from the 6-carbon compound glucose and ends with the 3-carbon compound pyruvate (see fig. 12.13 and table 12.1). All of the intermediates in this pathway are phosphorylated, and in the process of degradation two ATP molecules are made from two ADP molecules. This is the energy payoff of the pathway. In effect, the energy of the carbon–carbon bond is expended to create the immediately useful form of chemical energy, ATP. This pathway does not require oxygen, and it is found in nearly all cells, whether they are aerobic or anaerobic. Glucose is frequently not the starting material for glycolysis, and so differences between organisms and differences in the same organism in different environments show up at the starting point and the termination point of the pathway. One oxidation step in the pathway results in the conversion of NAD$^+$ to NADH. This NAD$^+$ must be regenerated if glycolysis is to continue. Under anaerobic conditions some reaction must occur involving the reduction of pyruvate to regenerate the NAD$^+$.

The glycolytic pathway resembles a series of "lakes" (metabolic pools) connected by short "rivers" (the reactions between the pools). This pattern is reflected in the ways that functional metabolic relationships have evolved. Reactions involving ATP and ADP occur in the interconnecting reactions, or rivers. Clearly this is where they are expected, because an ATP-linked reaction within a metabolic pool makes no more sense than a hydroelectric power plant in the middle of a lake.

**Figure 12.25**

Relationships in glycolysis and gluconeogenesis. Points at which ATP is produced or consumed are indicated. Compounds in the same metabolic pools are indicated by purple boxes. Three small pseudocycles (Ia, II, III) in the paired sequences occur between glycogen and pyruvate, or between glycogen and glucose (Ib, II, III). Only enzymes that are unique to either glycolysis or gluconeogenesis are indicated (screened in blue).

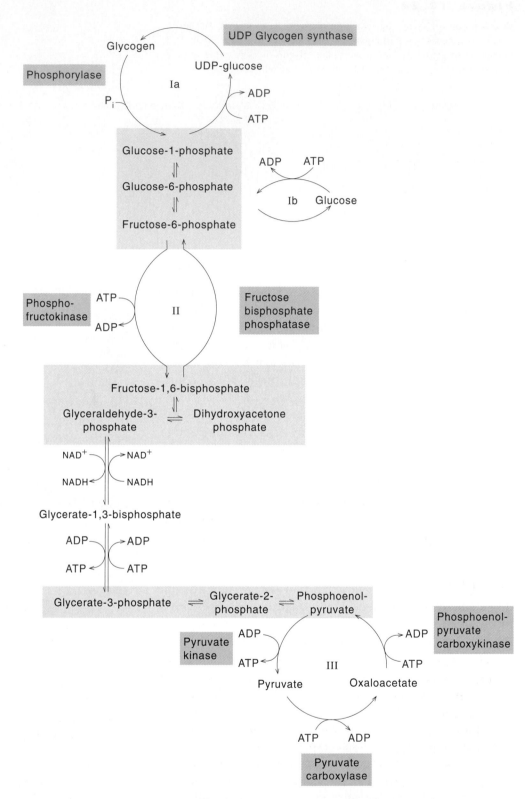

## Gluconeogenesis

Gluconeogenesis embraces the pathways from C-3 carbon sources, such as lactate, pyruvate, or amino acids, to hexoses or storage polysaccharides.

Glycolysis and gluconeogenesis constitute a set of oppositely directed conversions with different ATP-to-ADP stoichiometries. In most oppositely directed conversions the two sequences and the associated enzymes are entirely separate. Not so in these two sequences. Only at three points, all

**Figure 12.26**

The pyruvate–phosphoenolpyruvate pseudocycle. One molecule of ATP is regenerated in the conversion of phosphoenolpyruvate to pyruvate, catalyzed by pyruvate kinase, which occurs in glycolysis. Conversion of pyruvate to phosphoenolpyruvate, which is necessary in gluconeogenesis, must be coupled to two ATP-to-ADP conversions if it is to be thermodynamically feasible. The first ATP is used indirectly, in the production of the active carboxyl transfer agent, biotin-COO⁻ (chapter 10). Transfer of a carboxyl group to pyruvate produces oxaloacetate. Decarboxylation of oxaloacetate aids in the attack by the carbonyl oxygen on the terminal phosphorus atom of ATP or GTP, producing phosphoenol-pyruvate and ADP or GDP. This figure expands on the reactions shown in the lower part of figure 12.25 (III).

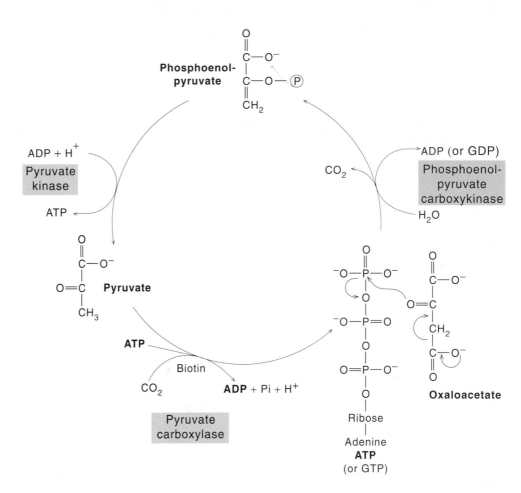

## Gluconeogenesis Consumes ATP

The organization of glycolysis and gluconeogenesis as a series of connected metabolic pools makes it possible for most of the same enzymes to function in both directions. Only the reactions connecting the metabolic pools require different enzymes and a coupling to the ATP–ADP system to make them thermodynamically feasible in the direction of gluconeogenesis.

Flux within a pool is determined by simple mass-action considerations. For example, when glycerate-3-phosphate is being produced in the phosphoglycerate kinase reaction and phosphoenolpyruvate is being removed by the action of pyruvate kinase, the flux in the 3-carbon carboxylate pool is in the glycolytic direction, from glycerate-3-phosphate to phosphoenolpyruvate. When the pattern of activation of the regulatory enzymes is different, so that phosphoenolpyruvate is being produced and glycerate-3-phosphate is being consumed, the flux in the

outside of the three pools, are the oppositely directed reactions of glycolysis and gluconeogenesis catalyzed by different enzymes (fig. 12.25).

same pool, catalyzed by the same enzymes, proceeds in the gluconeogenic direction, from phosphoenolpyruvate to glycerate-3-phosphate. Similarly for the other pools, the direction of net flux within the pool depends on which components are flowing into the pool and which are leaving it.

In three of the reactions connecting the pools we find a big drop in free energy in the glycolytic direction (see table 12.1). Clearly if cells are to conduct these reactions in the reverse direction, ATP must be pumped into the system and different enzymes will be required.

## Conversion of Pyruvate to Phosphoenolpyruvate Requires Two High Energy Phosphates

The first step in the gluconeogenic direction involves the formation of phosphoenolpyruvate from pyruvate. Reversal of the pyruvate kinase reaction requires at least two ATP-to-ADP conversions. The means by which this is done is shown in figure 12.26.

The first step in the conversion of pyruvate to PEP entails the carboxylation of pyruvate to form oxaloacetate. This reaction is catalyzed by pyruvate carboxylase. As in many other enzymic carboxylation reactions, the immediate $CO_2$ donor is a carboxylated derivative of the coenzyme biotin (see chapter 10). Equation (9) depicts the reaction for the production of oxaloacetate:

$$\text{Pyruvate}^- + \text{ATP}^{4-} + \text{HOCO}_2^- \xrightarrow{\text{Enz-biotin}}$$
$$\text{Oxaloacetate}^{2-} + \text{ADP}^{3-} + \text{P}_i^{2-} + \text{H}^+ \quad (9)$$

The carboxylation of pyruvate supplies a significant portion of the thermodynamic push for the next step in the sequence. This is because the free energy change for decarboxylation of $\beta$-keto carboxylic acids such as oxaloacetate is large and negative. The oxaloacetate formed from pyruvate by carboxylation is converted to phosphoenolpyruvate in a reaction catalyzed by phosphoenolpyruvate carboxykinase. In many species, including mammals, this reaction involves a GTP-to-GDP conversion.

$$\text{Oxaloacetate} + \text{GTP} \xrightarrow[\text{carboxykinase}]{\text{phosphoenolpyruvate}}$$
$$\text{Phosphoenolpyruvate} + \text{GDP} + \text{CO}_2 \quad (10)$$

The use of GTP, UTP or CTP in a metabolic reaction or sequence is energetically equivalent to the use of ATP, because the nucleoside diphosphate that is produced is rephosphorylated at the expense of ATP in the reaction catalyzed by nucleoside diphosphate (NDP) kinase.

$$\text{ATP} + \text{NDP} \rightleftharpoons \text{ADP} + \text{NTP} \quad (11)$$

NDP may be CDP, GDP, or UDP, and NTP the corresponding triphosphate. The enzyme is unusual in its lack of specificity for the acceptor nucleoside diphosphate.

If we add the equations for the reactions catalyzed by pyruvate carboxylase, phosphoenolpyruvate carboxykinase, and nucleoside diphosphate kinase, we obtain the overall reaction for conversion of pyruvate to phosphoenolpyruvate.

$$\text{H}_2\text{O} + \text{Pyruvate} + 2\,\text{ATP} \longrightarrow$$
$$\text{Phosphoenolpyruvate} + 2\,\text{ADP} + \text{P}_i + 2\,\text{H}^+ \quad (12)$$

Thus in effect two molecules of ATP are used to reverse the conversion that yields one molecule of ATP in the glycolytic direction.

## Conversion of Phosphoenolpyruvate to Fructose-1,6-bisphosphate Uses the Same Enzymes as Glycolysis

The interconversions of the phosphorylated three-carbon acids—phosphenolpyruvate, glycerate-2-phosphate, and glycerate-3-phosphate—have already been discussed in this chapter. This pool is linked to the fructose-1,6-bisphos-

phate/triose phosphate pool by the two reactions that lead to the conversion of glycerate-3-phosphate to glyceraldehyde-3-phosphate. The linkage is organized differently from the reactions that connect most other metabolic pools. In this case the same enzymes function in both directions.

A glance at the energy chart (see table 12.1) shows that this is energetically feasible. Thus, free energy changes very little in any of the steps between phosphoenolpyruvate and fructose-1,6-bisphosphate. Furthermore, in the glycolytic direction, one of these reactions results in the reduction of an $NAD^+$ to an NADH, and the other results in the conversion of an ADP to an ATP. In the gluconeogenic direction, these conversions are reversed. Low ratios of ATP to ADP and NADH to $NAD^+$ favor glycolysis, while high ratios favor gluconeogenesis.

## Fructose-bisphosphate Phosphatase Converts Fructose-1,6-bisphosphate to Fructose-6-phosphate

Fructose-1,6-bisphosphate is converted to fructose-6-phosphate by hydrolysis of the phosphoryl ester bond at C-1 in a reaction catalyzed by fructose bisphosphate phosphatase. The standard free energy change for this reaction is about $-4$ kcal/mol, corresponding to an equilibrium constant of about $10^3$. Thus, the two conversions (the phosphorylation of fructose-6-phosphate to form fructose 1,6-bisphosphate with ATP as the phosphate donor, and the hydrolysis of fructose-bisphosphate to form fructose-6-phosphate) are both thermodynamically favored under any conditions that are likely to exist in a living cell. These two reactions constitute a pseudocycle and, consistent with the principles enunciated in the previous chapter, the pathways have evolved so the number of ATP-to-ADP conversions is greater in one direction than in the other.

## Hexose Phosphates Can Be Converted to Storage Polysaccharides

When fructose-6-phosphate is generated by gluconeogenesis or photosynthesis (see chapter 15), an equimolar amount of glucose-1-phosphate is usually removed from the hexose monophosphate pool by conversion to storage polysaccharide (glycogen in animals and many kinds of microorganisms; starch in green plants).

The first step in the conversion of glucose-1-phosphate to starch or glycogen is activation at the expense of a nucleoside triphosphate. In plants the activated form is ADP-glucose, formed by transfer of an AMP moiety from ATP to the phosphate of glucose-1-phosphate by the en-

**Figure 12.27**

The elongation step in starch synthesis. (*a*) The activated form of glucose in starch synthesis is ADP-glucose. This is formed from glucose and ATP as shown. (*b*) The ADP-glucose does not react directly with the elongating starch molecule. Rather, the starch synthase produces an oxonium ion intermediate, which is attacked by the terminal C-4 hydroxyl on the growing polymer.

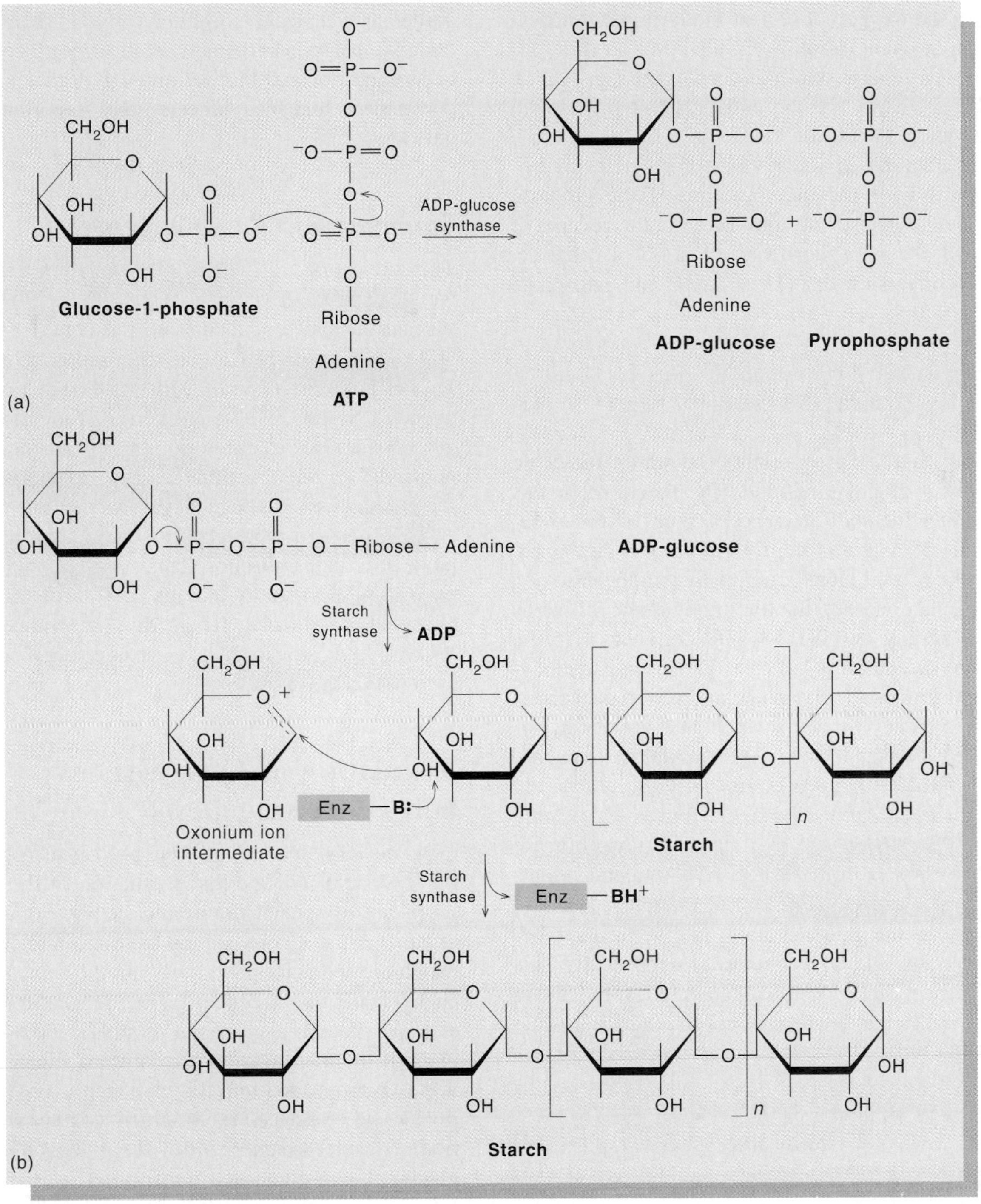

zyme ADP-glucose synthase. An oxygen atom of the phosphoryl group of glucose-1-phosphate attacks the α phosphorus atom of ATP, displacing pyrophosphate (fig. 12.27). Since the bond broken and the bond formed are both pyrophosphate bonds, the change in standard free energy for this reaction is close to 0, and the equilibrium constant is near 1. However, if the pyrophosphate is hydrolyzed by the enzyme pyrophosphatase, as probably occurs under most conditions in living cells, the standard free energy change for the two reactions combined is large and negative, about $-7.5$ kcal/mol.

In the reaction catalyzed by starch synthase, ADP-glucose decomposes to generate an oxonium ion. This species reacts with the oxygen of C-4 of the terminal glucosyl residue of a starch chain, resulting in the addition of a glucosyl unit to the polymeric starch molecule (see fig. 12.27). The standard free energy for the elongation step in starch synthesis is about $-3$ kcal/mol.

Assuming that the pyrophosphate formed is hydrolyzed, the equation for the incorporation of the glucosyl residue of glucose-1-phosphate into starch is the reverse of the equation for the phosphorolysis of a polysaccharide, except for the conversion of ATP to ADP and phosphate ion.

$$\text{Glucose-1-phosphate} + \text{ATP} + \text{Starch}^n \longrightarrow$$
$$\text{Starch}^{n+1} + \text{ADP} + 2\,P_i + H^+ \quad \textbf{(13)}$$

In this equation, $\text{Starch}^{n+1}$ represents the starch molecule after addition of a glucosyl residue. The reactions in this conversion, which include cleavage of both of the pyrophosphate bonds of ATP and the formation of a new pyrophosphate bond, are a bit more complex than in the case of a simple kinase reaction, but the thermodynamic effect is merely that of adding an ATP-to-ADP conversion in the direction of polysaccharide synthesis. Thus, the pseudocycle that connects glucose-1-phosphate and starch is energetically equivalent to any other in which two oppositely directed conversions differ by one ATP-to-ADP conversion.

In most animals, many bacterial species, yeasts and other fungi, glycogen serves the same function as starch does in plants. Glycogen resembles starch in consisting of glucose residues linked primarily by $\alpha(1,4)$-acetal bonds, but it contains a larger number of $\alpha(1,6)$ branches. Bacteria use ADP-glucose as the glucosyl donor in glycogen synthesis, while vertebrates use UDP-glucose. Energetically, this makes no difference. Since UTP is regenerated from UDP at the expense of ATP, the net reaction is the same as when ADP-glucose is used.

$$H^+ + \text{Glucose-1-phosphate} + \text{UTP} \rightleftharpoons$$
$$\text{UDP-glucose} + PP_i + H_2O \quad \textbf{(14)}$$
$$H_2O + PP_i \longrightarrow 2\,P_i + H^+ \quad \textbf{(15)}$$
$$\text{UDP-glucose} + \text{Glycogen}^n \longrightarrow$$
$$\text{Glycogen}^{n+1} + \text{UDP} + H^+ \quad \textbf{(16)}$$
$$\text{UDP} + \text{ATP} \rightleftharpoons \text{UTP} + \text{ADP} \quad \textbf{(17)}$$

Sum: $\text{Glucose-1-phosphate} + \text{ATP} + \text{Glycogen}^n \longrightarrow$
$$\text{Glycogen}^{n+1} + \text{ADP} + 2\,P_i + H^+ \quad \textbf{(18)}$$

In these equations $\text{glycogen}^{n+1}$ represents the glycogen molecule after addition of a glucosyl residue.

Because of its roles in the synthesis of glycogen, in isomerization of hexose phosphates, and as a precursor for numerous biosynthetic intermediates, UDP-glucose is regarded as a central hexose derivative in mammalian metabolism. In bacteria and plants, both ADP-glucose (production of storage polysaccharide) and UDP-glucose (sugar interconversions and biosynthesis) play important roles as precursors.

## Summary of Gluconeogenesis

The conversions in gluconeogenesis are the reverse of what we observed in glycolysis. Thus, glucose or storage polysaccharides are produced from pyruvate. Overall, the conversion of pyruvate to storage polysaccharide "costs" 7 ATPs per 6-carbon unit. Most of the reactions in gluconeogenesis simply involve a reversal of identical reactions in glycolysis. Only at three points, all outside the metabolic pools, do we find reactions in gluconeogenesis that use different enzymes: (1) the conversion of pyruvate to phosphoenolpyruvate, (2) the conversion of fructose-1,6-bisphosphate to fructose-6-phosphate, and (3) the conversion of hexose phosphate to storage polysaccharide (or hexose phosphate to glucose). These three reactions have different ATP-to-ADP stoichiometries in the two directions, which ensures that the equilibrium is favorable in either direction.

## Regulation of Glycolysis and Gluconeogenesis

Several features of the glycolysis–gluconeogenesis system indicate that regulation of these conversions are of special importance to the organism. In many organisms these conversions have the highest fluxes of all metabolic sequences. Not only are the fluxes high, but they change direction frequently. In the mammalian liver, for example, massive glycogen synthesis takes place after a meal. If the meal was high in protein, much of the glucose that is incorporated into the glycogen is produced from pyruvate and oxaloacetate via gluconeogenesis. During a period of fasting much or all of the glycogen is converted to glucose-1-phosphate and metabolized to pyruvate.

When focusing on the energetics of these pathways, it is appropriate to consider the pathway as a whole. But when considering problems of regulation, it is more useful to look at the small pseudocycles where the regulatory enzymes are located (see fig. 12.25). At any one of these crucial points it would be a metabolic disaster for the conversions to operate simultaneously in both directions because this would merely result in the degradation of ATP. Furthermore these control points must respond promptly to changing metabolic needs to preserve the metabolic harmony of the steady state.

We will focus our discussion on the well-understood regulatory enzymes that modulate the flux between glycogen and the hexose monophosphate pool (pseudocycle Ia), and between fructose-6-phosphate and fructose-1,6-phosphate (pseudocycle II). Some aspects of the regulation between the 3-carbon pool and pyruvate (pseudocycle III) are discussed in the next chapter.

Sites of metabolic control in intact cells can be identified by examining changes in metabolite concentrations under conditions when metabolic fluxes change abruptly. Experiments of this nature were carried out by Oliver Lowry in the 1960s. Lowry showed that depriving mouse brain tissue of $O_2$ resulted in a dramatic increase in the rate of glycolysis. Since the glycolytic pathway is a linear series of reactions, the flux through each enzymatic step in the pathway must have increased. Measurements of the metabolites showed that the concentration of fructose-6-phosphate, the substrate of the phosphofructokinase reaction, had decreased, whereas the concentration of fructose-1,6-phosphate, the product of this reaction, had increased. By themselves, these changes in the concentrations of its substrate and product would tend to decrease the rate of phosphofructokinase activity. Because the rate of the reaction increased, it is clear that phosphofructokinase must have been influenced by factors other than its substrate and product concentrations.

As an analogy, consider a river with a dam somewhere between its source and its terminus. If the flood gates of the dam are opened, the flow of water transiently increases everywhere in the river. At sites near the end of the river, the increase in the flow can be explained simply by the increased pressure of water coming from upstream; at sites near the source, it can be attributed to a reduced pressure downstream. Only at the site of the dam is the situation reversed: the flow of water increases here in spite of the fact that the pressure downstream has risen and the pressure upstream has fallen.

Phosphofructokinase and the other enzymes that regulate pseudocycles I, II, and III of glycolysis are influenced by both intracellular and extracellular signals. We consider some of the intracellular signals first.

## How Do Intracellular Signals Regulate Energy Metabolism?

The intracellular signals for the glycolysis–gluconeogenesis pathways consist of small-molecule allosteric effectors, the concentrations of which reflect the energy charge: ATP and citrate are two of the effectors that reflect a high energy charge (we will hear more about citrate in the next chapter); AMP is one of the effectors that indicates a low energy charge. The way in which these signals operate is exemplified by a consideration of the well-studied case of phosphofructokinase (see chapter 9). This enzyme has a site(s), separate from the catalytic site, for binding small-molecule allosteric effectors. Binding of allosteric effectors at the regulatory site changes the conformation of the enzyme. This conformational change affects the catalytic site. There are two conformations: One favors catalytic activity, the other does not. ATP and citrate, which indicate a high energy charge, favor the less active form of the enzyme, whereas AMP, an indication of a low energy charge, favors the more active form of the enzyme.

Fructose bisphosphate phosphatase, the enzyme paired with phosphofructokinase at pseudocycle II, is affected in just the opposite way by some of the same allosteric effectors. The overall effect of these allosteric effectors is to encourage glycolysis when the energy charge is low and to encourage gluconeogenesis when the energy charge is high.

The paired enzymes that regulate flux at pseudocycle I (Ia) are affected in a parallel way by the same small-molecule allosteric effectors. A summary of the results for the four regulatory enzymes associated with pseudocycles Ia and II is presented in table 12.2.

## Hormonal Controls Can Override Intracellular Controls

Hormonal signals coming from other cells are more difficult to understand than allosteric effectors because they sometimes seem to go against the interests of the cell they influence. This influence is possible because hormonal signals respond to systemic needs.

In liver cells the response to energy charge, or to signals that reflect the energy charge, can be quite different from the responses of other cells because liver cells are involved in the regulation of the energy needs for the entire organism. In particular, liver cells secrete glucose so as to maintain a reasonably constant level of blood glucose. A low blood glucose level indicates a systemic need for glu-

**Table 12.2**

Activities of Some of the Enzymes that Control the Fluxes in the Glycolysis–Gluconeogenesis Pathways

| Intracellular Controls | Pseudocycle I | | Pseudocycle II | |
|---|---|---|---|---|
| | Glycogen phosphorylase | Glycogen synthase | Phosphofructokinase | Fructose bisphosphate phosphatase |
| High energy charge | ↓ | ↑ | ↓ | ↑ |
| Low energy charge | ↑ | ↓ | ↑ | ↓ |
| **Hormonal Controls** | | | | |
| Low blood glucose | ↑ | ↓ | ↓ | ↑ |

↑ = increase.  ↓ = decrease.

cose or energy. This need triggers an extracellular signal (discussed later on), which elevates liver cell activities of both glycogen phosphorylase and fructose bisphosphate phosphatase and simultaneously lowers the activities of glycogen synthase and phosphofructokinase. The net effect of this stimulation is to encourage glucose production both by breakdown of glycogen and by gluconeogenesis (see Table 12.2).

Since the liver's metabolic functions are not dictated by local needs so much as by the general need for controlling the blood glucose level, we might ask what additional regulatory inputs are required to respond to these systemic needs? As in many other cases in which the activities of different tissues are correlated for the good of the whole organism, hormones are involved at this additional level of control. A low blood glucose level is first sensed by the pancreas, whose $\alpha$ cells respond by releasing the peptide hormone glucagon into the blood. Glucagon binds to specific receptors in the plasma membranes of liver cells. This binding initiates a chain of reactions that leads to the conversion of glycogen phosphorylase from its less active form (phosphorylase $b$) to its hormone-activated form (phosphorylase $a$). Phosphorylase $a$ is not only more active than phosphorylase $b$, it is largely insensitive to AMP (see chapter 9). The net result is that glucagon directs the liver cells to respond to the systemic (organism-side) need for raising the blood glucose level by breaking down glycogen to hexose phosphate intermediates.

## Hormonal Effects of Glucagon Are Mediated by Cyclic AMP

Glucagon itself never enters the liver cells, so it must somehow trigger an event to occur on the cytosolic side of the plasma membrane. A breakthrough in our understanding of

this phenomenon came in the 1950s when Earl Sutherland found that the membrane-containing fraction of broken liver cells produces a soluble factor in the presence of ATP and glucagon, which stimulates the activity of glycogen phosphorylase. Further analyses showed that the factor was $3',5'$-cAMP synthesized by a membrane-bound enzyme (fig. 12.28). The detailed mechanism for how cAMP synthesis is triggered will be discussed in chapter 24. Here we focus on the effects of the cAMP so produced.

Cyclic AMP triggers a cascade of reactions that ultimately lead to glycogen breakdown. The immediate action of cAMP is to activate a protein kinase that phosphorylates a number of proteins, including phosphorylase kinase. Phosphorylation of phosphorylase kinase converts it from an inactive to an active form, which catalyzes the conversion of phosphorylase $b$ to phosphorylase $a$ (see chapter 9). The cascade of effects triggered by glucagon is shown in figure 12.29.

The same cAMP-dependent protein kinase that is responsible for phosphorylating phosphorylase kinase also catalyzes the phosphorylation of glycogen synthase. Whereas phosphorylation of glycogen phosphorylase leads to increased activity, the phosphorylation of glycogen synthase decreases its activity. As a result when glycogen breakdown is stimulated in response to glucagon, glycogen synthesis is inhibited. In this way the simultaneous operation of both enzymes associated with pseudocycle Ia is prevented.

When the systemic need has been satisfied there should be a way of reversing the hormone triggered process. All of the effects we have seen are promptly reversed when the need for them has ended. Cyclic AMP is hydrolyzed to $5'$-AMP by an enzyme called cyclic AMP phosphodiesterase, and protein phosphatases catalyze the hydrolytic re-

**Figure 12.28**

Synthesis of cyclic AMP. A catalytic site on adenylate cyclase (:B) removes a proton from the C-3 oxygen, which then attacks the $\alpha$-phosphate and displaces the pyrophosphate group. This reaction occurs on the inner plasma membrane (see fig. 12.29).

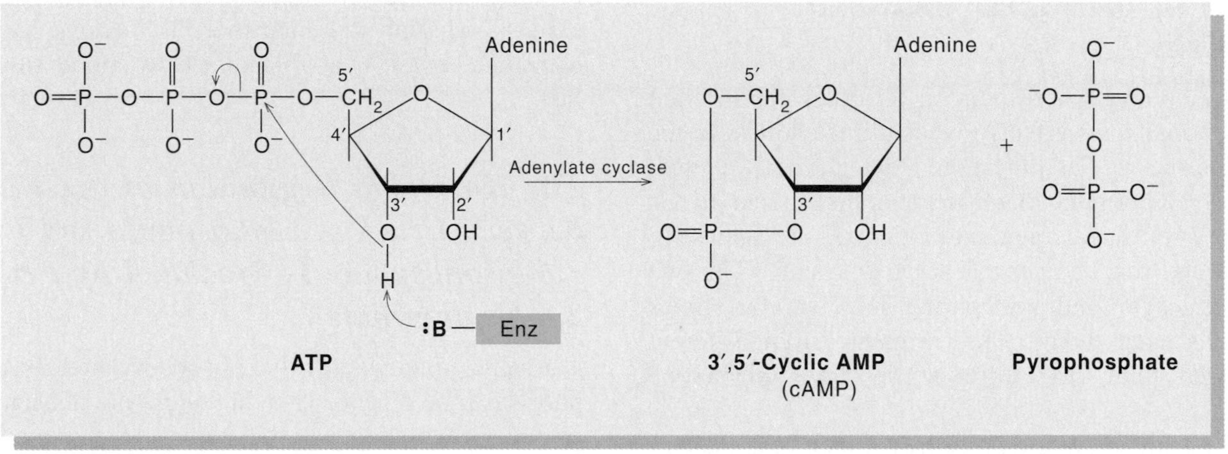

**Figure 12.29**

Molecular basis for the stimulation of glycogen breakdown by the hormones epinephrine and glucagon. The epinephrine or the glucagon activates adenylate cyclase, which activates protein kinase, which activates phosphorylase kinase. Phosphorylase kinase in turn activates phosphorylase, which cleaves glucose from glycogen. This cascade of regulatory interactions results in a very large amplification, so that a small amount of hormone can have a large effect on the rate of breakdown of glycogen.

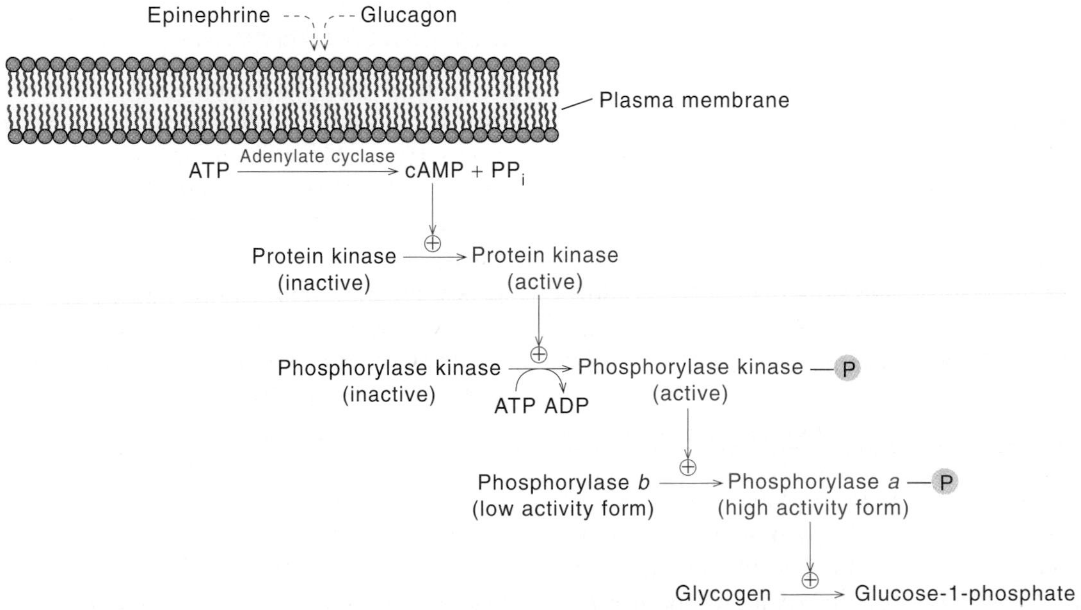

moval of the phosphoryl groups that were attached to proteins by protein kinase and phosphorylase kinase.

## The Hormone Epinephrine Stimulates Glycolysis in Both Liver Cells and Muscle Cells

The hormone epinephrine responding to somewhat different needs of the organisms affects glycogen metabolism in the liver in the same way as glucagon. Epinephrine is a small tyrosine-derived hormone that is made in the adrenal glands (see chapter 24). It is liberated into the blood in response to distress signals from the central nervous system. The two hormones, glucagon and epinephrine, have similar effects on liver cells even though they originate from different sources and bind to different plasma membrane receptors.

Despite some similarities in their effects, the physiological significance of these two hormones is quite different. Glucagon is part of a system designed to stabilize the blood glucose level and is, therefore, involved in maintaining the status quo. By contrast epinephrine is an alarm signal that leads to deviations from the status quo. Its release, triggered by the central nervous system in response to fear or anger, prepares the body for a sudden high energy output. Epinephrine is often termed the ''fight or flight'' hormone. Its function in activating phosphorylase is to build up glycolytic intermediates, thereby eliminating any time lag in energy mobilization if and when the organism begins to fight or to flee.

Whereas epinephrine and glucagon have very similar effects on glycogen metabolism in liver cells, only epinephrine affects glycogen metabolism in skeletal muscle cells. This difference occurs because muscle cells carry plasma receptors for epinephrine but not for glucagon. Muscle is strictly a consumer tissue and does not contribute to the stabilization of the blood glucose level; thus it should not, and does not, break down its glycogen reserves in response to low blood glucose. By contrast it is appropriate that muscle tissue should respond to epinephrine because a buildup of glycolytic intermediates in the muscle tissue itself prepares the organism for a sudden burst of muscle activity.

Another indication of the difference between the metabolic roles of muscle, a consumer tissue and liver, a contributer tissue, is the enzyme glucose-6-phosphatase. This enzyme is required for the production of neutral glucose. It is present in liver and kidney but not in strict consumer tissues, such as muscle and brain. Without glucose-6-phosphatase the glucose-6-phosphate cannot be converted to dephosphorylated glucose, which is necessary for

secretion into the blood stream. This enzyme and hexokinase constitute a pseudocycle for the interconversion of glucose and glucose-6-phosphate (see fig. 12.25). It is clear that glucose-6-phosphatase must be regulated so that it is only active when the blood glucose level is low. A deficiency of this enzyme is associated with von Gierke's disease, a condition in which there is massive liver enlargement and severe hypoglycemia (low blood sugar) after a fast.

## The Hormonal Regulation of the Flux between Fructose-6-phosphate and Fructose-1,6-bisphosphate Is Mediated by Fructose-2,6-bisphosphate

The same protein kinase that phosphorylates glycogen phosphorylase and glycogen synthase does not phosphorylate the enzymes of pseudocycle II. Rather an enzyme gets phosphorylated that catalyzes the synthesis of a potent allosteric effector of the two relevant enzymes, phosphofructokinase and fructose bisphosphate phosphatase. In the unphosphorylated form this enzyme synthesizes fructose-2,6-bisphosphate. Phosphorylation converts it into a degradative enzyme for the same compound. Fructose-2,6-bisphosphate is an activator of phosphofructokinase and an inhibitor of fructose bisphosphate phosphatase. As a result the net effect of glucagon on pseudocycle II is to stimulate fructose bisphosphate phosphatase while inhibiting phosphofructokinase (see table 12.2 and fig. 12.30).

## Summary of the Regulation of Glycolysis and Gluconeogenesis

Control of glycolysis and gluconeogenesis is confined to three points, all located outside the pools. These control points are organized into three small pseudocycles (see fig. 12.25). The ATP stoichiometries are different for going in opposite directions in these pseudocycles. This difference ensures that the reactions are thermodynamically feasible in either direction, but it does not dictate which reactions are active and which are inactive under any particular metabolic conditions. The activity status is determined by small-molecule allosteric factors, which interact with the regulatory enzymes that catalyze the reactions within each of the pseudocycles. The enzymes are designed in such a way that only one arm of a pseudocycle is active at any given time.

In all unicellular organisms and in most cells of multicellular organisms, these regulatory enzymes are sensitive to inhibition and activation by small molecules that reflect the metabolic state of the cell. When an effector binds to the regulatory site of an enzyme, it encourages the enzyme to

**Figure 12.30**

Some major points of regulation in the glycolytic–gluconeogenic pathways. Normal cellular controls favor glycogen breakdown and glycolysis when the energy charge is low. When the blood glucose level is low, hormonal controls on the liver cells favor glycogen breakdown and gluconeogenesis. The target sites for control factors are indicated by red arrows, with a plus or a minus at the arrowhead to indicate activation or inhibition, respectively.

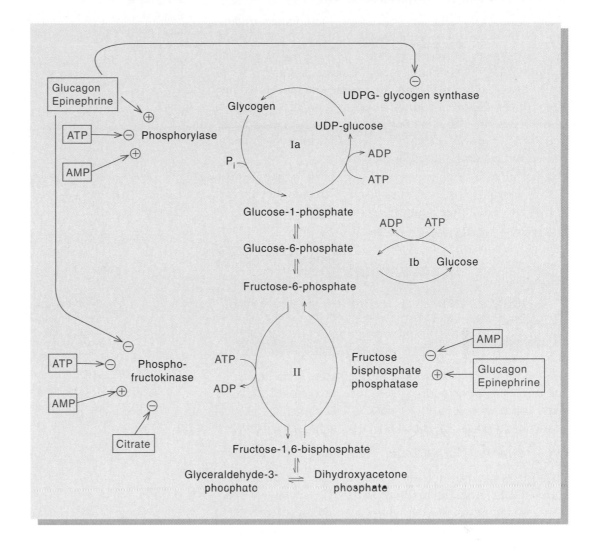

adopt one of two possible conformations. The conformation adopted may be the active or the inactive conformation, depending on the enzyme and the allosteric effector. For example, phosphofructokinase, which converts fructose-6-phosphate to fructose-1,6-bisphosphate, is active when binding to AMP. This is likely to happen when the AMP level is relatively high and energy is in short supply. The oppositely directed enzyme in the same pseudocycle, fructose bisphosphate phosphatase, adopts an inactive conformation when binding to the same allosteric effector. In this way the pseudocycle is controlled so that it is only active in the direction favoring the metabolic state of the cell.

In some tissues of multicellular organisms, cells have a dual responsibility: maintaining themselves and serving other cells in the same organism. For example, liver cells must maintain themselves, but they also have an organismwide responsibility to maintain the blood glucose level. If the blood glucose level falls below normal, a signal is sent to the liver from the pancreas in the form of the

hormone glucagon. Glucagon binds to specific receptors on the outside surface of the plasma membrane of the liver cell, which stimulates the formation of cAMP by an enzyme binding to the inner side of the plasma membrane. The cAMP triggers a series of phosphorylations inside the cell, which ultimately results in the phosphorylation of both glycogen phosphorylase and glycogen synthase. While the phosphorylation activates phosphorylase it inhibits the synthase, thereby promoting the formation of hexose phosphates. In general the metabolic signal delivered by a hormonal signal overrides any metabolic signals that may have originated from inside the cell.

Glucagon and epinephrine also regulate pseudocycle II so as to stimulate gluconeogenesis while inhibiting glycolysis. They do this through a chain of reactions that results in a lowering of the concentration of the allosteric effector fructose-2,6-bisphosphate. This effector stimulates phosphofructokinase while it inhibits fructose bisphosphate phosphatase.

# The Pentose Phosphate Pathway

Many kinds of organisms and some mammalian organs, notably liver, possess an alternative pathway for the oxidation of hexoses which results in a pentose phosphate and carbon dioxide. This pentose can be used as a precursor of the ribose found in nucleic acids or other sugars containing from three to seven carbon atoms which are needed in smaller amounts. The first and third reactions in the pentose phosphate pathway generate NADPH which is a major source of reducing power in many cells.

Cells differ considerably in their use of the pentose phosphate pathway. In muscle, a tissue in which carbohydrates are utilized almost exclusively for generation of mechanical energy, the enzymes of the pentose phosphate pathway are lacking. By contrast, red blood cells are totally dependent on the pentose phosphate pathway as a source of NADPH for which they need to keep the iron of hemoglobin in its normal +2 valence state. A deficiency in glucose-6-phosphate dehydrogenase, the first enzyme in the pentose phosphate pathway, can lead to the wholesale destruction of red blood cells and a condition known as hemolytic anemia.

## Two NADPH Molecules Are Generated by the Pentose Phosphate Pathway

NADPH is required for many biosynthetic sequences. It is generated in different kinds of cells by a variety of reactions, including an NADP$^+$-linked oxidation of malate to pyruvate and $CO_2$ and transfer of hydride ion from NADH to NADP$^+$ in a mitochondrial reaction that is driven by metabolic energy. However, in many cases, including in the mammalian liver, a major part of the NADPH requirement is met by oxidation of glucose-6-phosphate to ribulose-5-phosphate and $CO_2$. The four electrons that are released by the oxidation are transferred to two molecules of NADP$^+$.

The four reactions involved in this conversion are shown in figure 12.31. The first oxidation, catalyzed by glucose-6-phosphate dehydrogenase at C-1, converts the hemiacetal derivative of the aldehyde group to the lactone of the corresponding acid, 6-phosphogluconic acid. After hydrolysis of the lactone, the second oxidation at C-3, converts the secondary alcohol to a ketone. The expected product, 3-keto-6-phosphogluconic acid, is decarboxylated yielding ribulose-5-phosphate.

At this point the metabolic function of the sequence, when it is serving to supply electrons for biosynthesis, is fulfilled. Two molecules of NADPH are generated for each molecule of glucose-6-phosphate oxidized. It is only necessary to convert ribulose-5-phosphate, the end product of the

**Figure  12.31**

Stage 1 of the pentose phosphate pathway. Net Reaction: Glucose-6-phosphate + 2 NADP$^+$ → Ribulose-5-phosphate + $CO_2$ + 2 NADPH + 2 H$^+$.

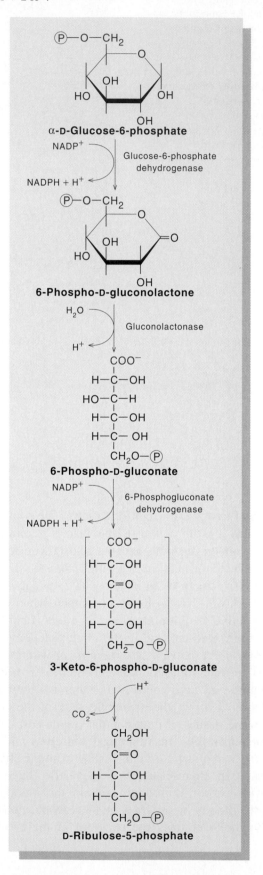

**Figure 12.32**

The transaldolase-catalyzed conversion of fructose-6-phosphate and erythrose-4-phosphate to glyceraldehyde-3-phosphate and sedoheptulose-7-phosphate. This is a two-step conversion. The first step is similar to the aldolase reaction except that the dihydroxyacetone produced is held at the catalytic site while the aldose product diffuses away and is replaced by another aldose molecule. The second step involves an aldol condensation.

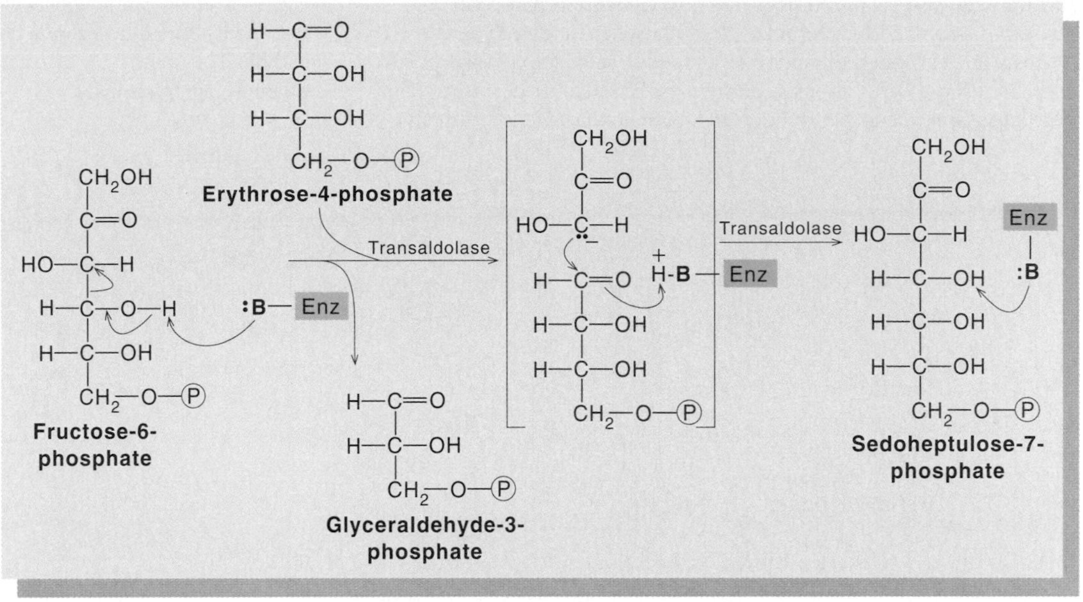

oxidative sequence, to compounds in the mainstream of metabolism.

## Transaldolase and Transketolase Catalyze the Interconversion of Many Phosphorylated Sugars

The key enzymes involved in these conversions are transaldolase and transketolase. The two enzymes are similar in their substrate specificities. Both require a ketose as a donor and an aldose as an acceptor. The steric requirements at positions C-1 through C-4 are the same as the requirements of aldolase in the glycolytic pathway, except that aldolase requires phosphorylation at C-1, and both transaldolase and transketolase require a free hydroxyl group at C-1.

Transaldolase catalyzes a two-step conversion. The first step, an aldol cleavage of the bond between C-3 and C-4 of a ketose, is essentially identical to the reaction catalyzed by aldolase. However, the dihydroxyacetone that is produced in the transaldolase reaction from carbons 1, 2, and 3 is not released. Rather, it is held at the catalytic site while the glyceraldehyde-3-phosphate produced diffuses away and is replaced by erythrose-4-phosphate. An aldol condensation then generates the second product of the reaction, a ketose that contains the first three carbon atoms of the original ketose attached to C-1 of the acceptor aldose (fig. 12.32).

The reaction catalyzed by transketolase superficially resembles the transaldolase reaction in that the substrate

specificity is identical, but in this case the cleavage is between carbons 2 and 3 of the ketose. A two-carbon moiety is retained on the enzyme following cleavage and is subsequently transferred to an acceptor aldose. This reaction is chemically quite different from the transaldolase reaction. Aldol condensations and cleavages, which occur readily in mildly alkaline solution, are feasible because protons on a carbon atom adjacent to a carbonyl carbon are moderately acidic. The carbanion formed on dissociation of such a proton can participate in a nucleophilic addition to the carbonyl carbon of another molecule of aldehyde or ketone, as in the transaldolase reaction. A ketol condensation, in contrast, would involve the carbonyl carbon as the nucleophile. That is not energetically feasible because the polarity of the carbonyl bond precludes a negative charge on carbon.

Transketolase is one of several enzymes that catalyze reactions of intermediates with a negative charge on what was initially a carbonyl carbon atom. All such enzymes require thiamine pyrophosphate (TPP) as a cofactor (chapter 10). The transketolase reaction is initiated by addition of the thiamine pyrophosphate anion to the carbonyl of a ketose phosphate, for example xylulose-5-phosphate (fig. 12.33). The adduct next undergoes an aldol-like cleavage. Carbons 1 and 2 are retained on the enzyme in the form of the glycolaldehyde derivative of TPP. This intermediate condenses with the carbonyl of another aldolase. If the reactants are xylulose-5-phosphate and ribose-5-phosphate, the products are glyceraldehyde-3-phosphate and the seven-carbon ketose, sedoheptulose-7-phosphate (see fig. 12.33).

**Figure 12.33**

The transketolase-catalyzed conversion of xylulose-5-phosphate and ribose-5-phosphate to glyceraldehyde-3-phosphate and sedoheptulose-7-phosphate. Although the aldolase and ketolase reactions superficially resemble each other, they proceed by very different mechanisms. This is because in the aldolase reaction the carbon adjacent to a carbonyl acts as a nucleophilic agent, whereas the ketolase reaction involves an intermediate with a negative charge on what was originally a carbonyl carbon. The latter type of reaction is more complex and requires thiamine pyrophosphate.

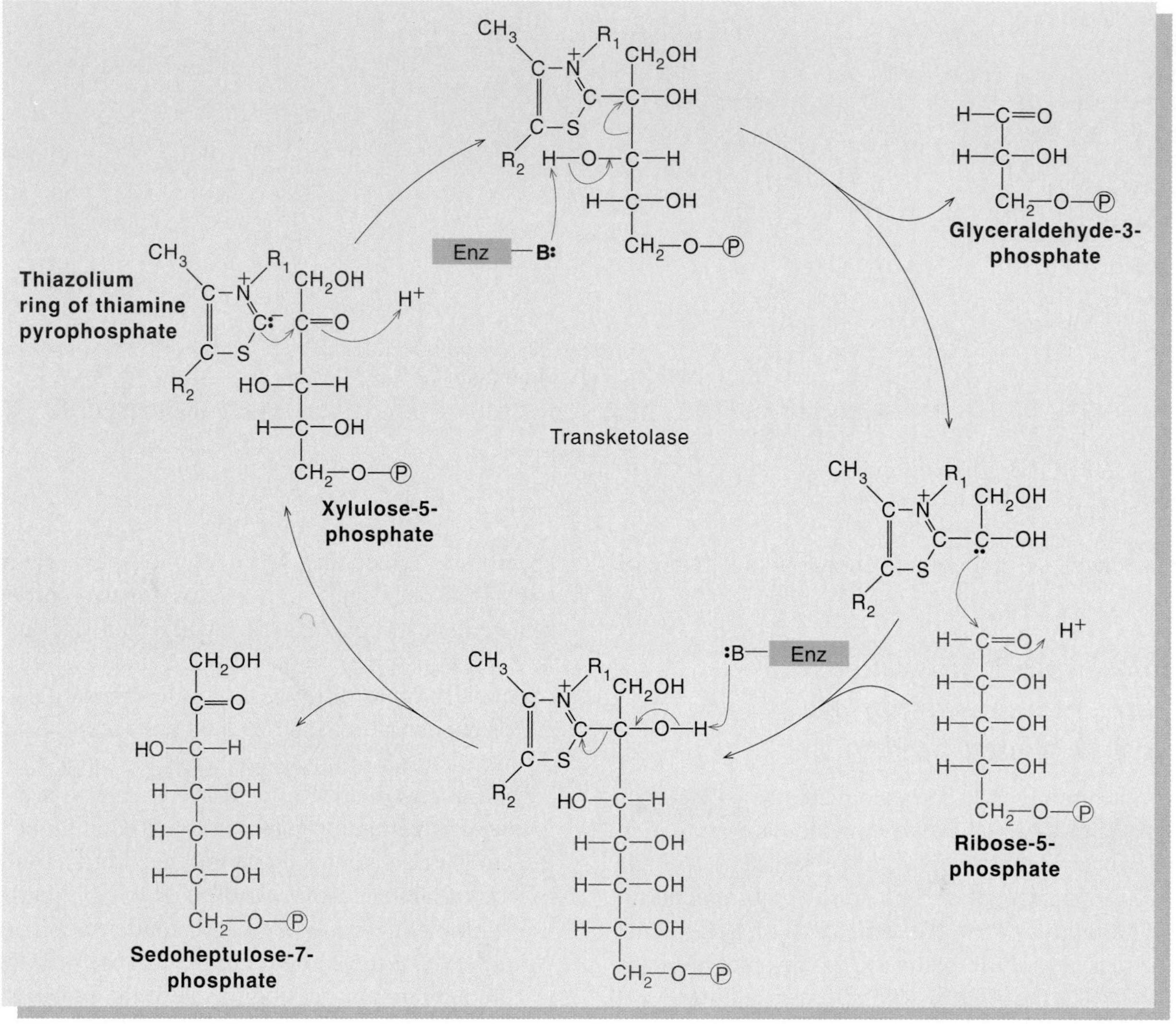

## Production of Ribose-5-phosphate and Xylulose-5-phosphate

Both transaldolase and transketolase require a ketose phosphate as the donor molecule and an aldose phosphate as the acceptor. Furthermore, both enzymes require the same steric configuration at carbons 3 and 4 as is found in glucose and fructose. Ribulose-5-phosphate, the first pentose phosphate to be formed in the pentose phosphate pathway, does not have the correct configuration to serve as a substrate for either transaldolase or transketolase. However, both a suitable donor ketose and an acceptor aldose can be made by isomerizations of ribulose-5-phosphate, and enzymes that catalyze those isomerizations are found in cells that possess the pentose phosphate pathway (fig. 12.34).

Because of the possible alternative pathways, which probably occur simultaneously, no single set of reactions can uniquely describe the pentose phosphate pathway. One possible set of pathways is shown in figure 12.34. If the triose phosphate formed is converted to hexose phosphate, the overall pathway can be seen as regenerating five molecules of hexose phosphate for each six used initially.

$$6 \text{ Glucose-6-phosphate} + 12 \text{ NADP}^+ \longrightarrow$$
$$6 \text{ CO}_2 + 5 \text{ Fructose-6-phosphate}$$
$$+ 12 \text{ NADPH} + 10 \text{ H}^+ \quad \textbf{(19)}$$

## Figure 12.34

Stage 2 of the pentose phosphate pathway. The groups in red are those transferred in transketolase-catalyzed reactions. The groups in bold type are transferred in the transaldose-catalyzed reactions. All of the reaction arrows in this figure are double-headed to indicate that the reactions can go in either direction with little change in free energy.

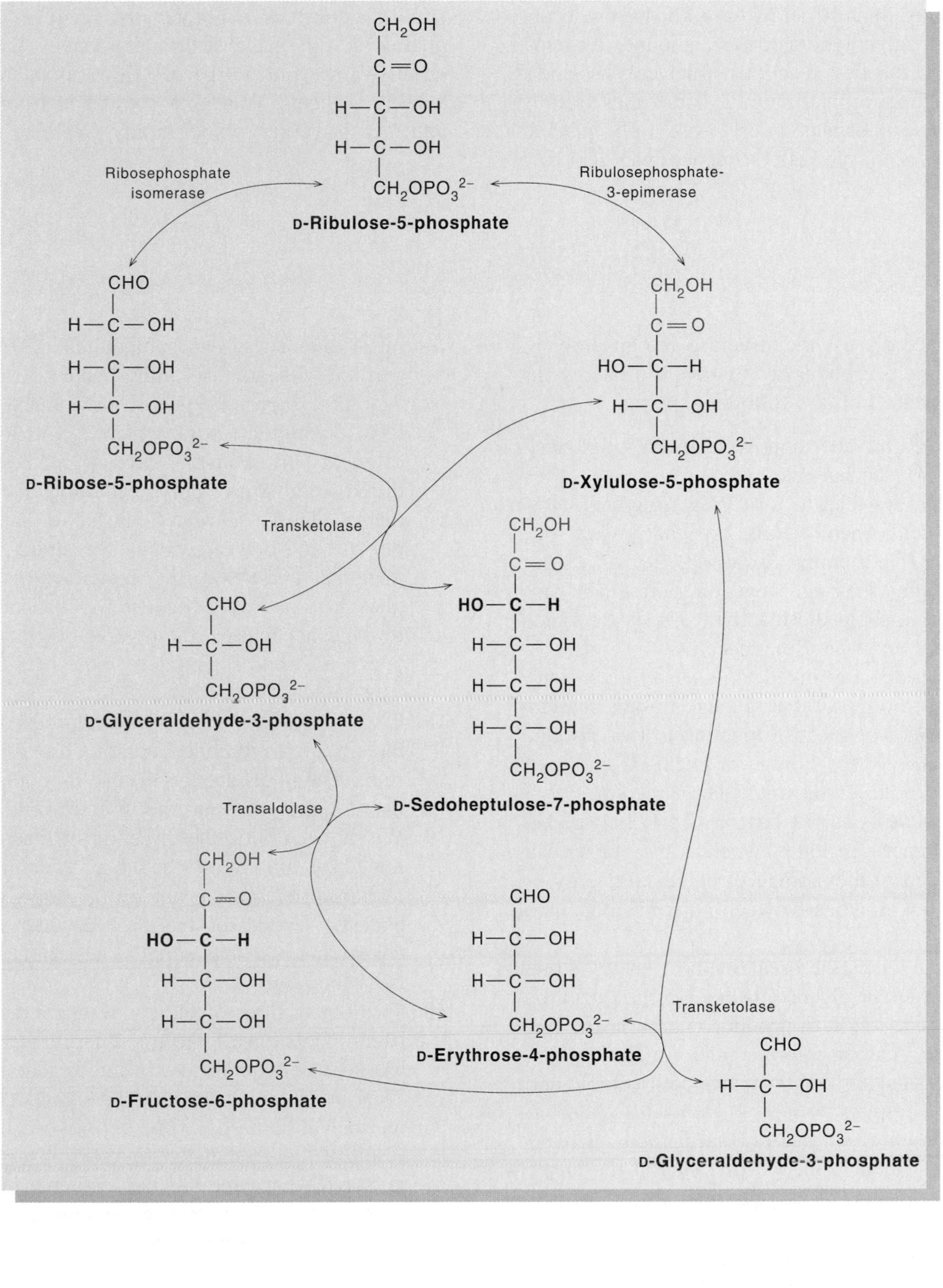

The pathway was first formulated in this form.

Alternatively, it is possible to write a sequence of reactions, including the action of phosphofructokinase and aldolase on seven-carbon intermediates, in which the carbon of ribulose-5-phosphate is converted mainly to glyceraldehyde-3-phosphate. Such a pathway, with the triose phosphate entering the glycolytic sequence, amounts to a bypass, or shunt, around the first reactions of glycolysis, and the name hexose monophosphate shunt is sometimes used. Any amount of ribose-5-phosphate or erythrose-4-phosphate that may be needed for biosynthetic sequences can also be obtained from this oxidative pentose phosphate pathway.

Because the transaldolase and transketolase reactions are symmetrical with respect to types of bonds cleaved and formed, their equilibrium constants are near 1. The pool of sugar phosphates is thus near equilibrium in cells that contain these enzymes. In a water-flow analogy, the sugar phosphates and the reactions that interconvert them resemble a large swamp, with ill-defined flows along many interconnecting channels. Water may be fed in from any direction and may leave the swamp in any direction.

## Summary

In this chapter we discuss the structure and anaerobic metabolism of the carbohydrates involved in energy metabolism. We focused on the following points.

1. Carbohydrates include monosaccharides, oligosaccharides, and polysaccharides.
2. Monosaccharides tend to form ring structures known as intramolecular hemiacetals, especially when the product is a five-member (furanose) ring or a six-member (pyranose) ring. Depending on which way the ring forms about the reactive carbon, the structure is called an $\alpha$- or $\beta$-hemiacetal.
3. Monosaccharides are linked together by glycosidic bonds to form oligosaccharides and polysaccharides.
4. Polysaccharides of glucose function in two distinct roles. Some serve for storage of chemical energy, and others fill structural roles. The structural polysaccharides like cellulose have a $\beta(1,4)$ linkage between monomeric residues, whereas the polysaccharides involved in the storage of chemical energy have an $\alpha(1,4)$ linkage between adjacent monomeric residues.
5. Glycolysis involves the breakdown of energy-storage polysaccharides or glucose to the 3-carbon acid, pyruvate. Its function is to produce energy in the form of ATP and 3-carbon intermediates for further metabolism. Glycolysis occurs under both aerobic and anaerobic conditions.
6. The breakdown of storage polysaccharides yields glucose-1-phosphate, but the first step in the metabolism of free glucose is phosphorylation to glucose-6-phosphate. These two compounds are freely interconvertible with fructose-6-phosphate.
7. The three sugar phosphates just named make up the hexose phosphate pool, which is a major crossroad of energy metabolism. Glucose-1-phosphate is produced from storage polysaccharides and is used in their synthesis; glucose-6-phosphate is produced from glucose and is the precursor for blood glucose in mammals and the starting material for the pentose phosphate pathway; fructose-6-phosphate is produced by gluconeogenesis, the reverse of glycolysis, in which 6-carbon sugars are synthesized from pyruvate.
8. Glycolysis and gluconeogenesis proceed through three pools of intermediates that are close to equilibrium. The pools are the hexose monophosphate pool, the aldolase pool, and the three-carbon pool.
9. Within each pool, the direction of reaction is determined by simple mass-action considerations. The direction of overall conversion (glycolysis or gluconeogenesis) depends on the rates at which the connecting reactions supply and remove materials from the pool.
10. Two out of three of the connecting reactions between pools have large free energies of reaction and are far from equilibrium in the metabolizing cell. They are subject to regulatory signals that indicate the metabolic needs of the cell or organism. As a result of those regulatory interactions, polysaccharide is stored when nutrients are available in excess of current needs, and storage compounds are broken down to supply energy when necessary.

11. In mammals, the liver is a special organ. It is a major site of storage of glycogen, and it regulates the level of glucose in the blood. When the blood glucose level is low, the liver replenishes it by hydrolysis of glucose-6-phosphate, which is made simultaneously by breakdown of glycogen and by gluconeogenesis. When the liver's supply of glycogen is depleted, blood glucose is replenished exclusively by gluconeogenesis. Hormonal signals trigger the liver to respond to the need for maintaining the blood glucose level.

12. The pentose phosphate pathway serves a variety of functions: (1) the production of NADPH for biosynthesis; (2) the production of ribose, required mainly for nucleic acid synthesis, and (3) the interconversion of a variety of phosphorylated sugars.

## Selected Readings

Atkinson, D. E., *Cellular Energy Metabolism and Its Regulation.* New York: Academic Press, 1977. A general discussion, covering some topics in this and the preceding and following chapters in somewhat greater depth than the treatment in our book.

Beitner, R., *Regulation of Carbohydrate Metabolism.* Boca Raton, Fla.: CRC Press, 1985. Two-volume discussion, at a rather advanced level, of many aspects of the regulation of mammalian carbohydrate metabolism with frequent discussions of clinical conditions. Many references.

Dawes, E. A., *Microbial Energetics.* Glasgow and London: Blackie [New York: Chapman and Hall], 1986. Discussion of bacterial metabolism, with emphasis on the central pathways and the generation and use of ATP. A good source for learning something of the diversity of metabolic adaptations among different groups of bacteria.

Hers, H. G., and L. Hue, Gluconeogenesis and related aspects of glycolysis. *Ann. Rev. Biochem.* 52:617–653, 1983.

Hochachka, P. W., and G. N. Somera, *Biochemical Adaptation.* Princeton, N.J.: Princeton University Press, 1984. An excellent and extensive discussion of how biochemical processes, including many discussed in this book, are adapted by various types of organisms in fitting themselves for survival under specific and often difficult conditions.

Hoffman, E., Phosphofructokinase—a favorite of enzymologists and students of metabolic regulation. *Trends Biochem. Sci.* 3:145–147, 1978.

Katz, J., and R. Rognstad, Futile cycling in glucose metabolism. *Trends Biochem. Sci.* 3:171–174, 1978.

Pilkis, S. J., M. R. El-Maghrabi, and T. H. Claus, Hormonal regulation of hepatic gluconeogenesis and glycolysis. *Ann. Rev. Biochem.* 57:755–784, 1988.

Roehrig, K. L., *Carbohydrate Biochemistry and Metabolism.* Westport, Ct.: Avi Publishing Co., 1984. Carbohydrate metabolism discussed at a rather elementary level. Covers several topics that are not included in this book, such as disorders of carbohydrate metabolism with brief discussions of many types of human genetic diseases in which carbohydrate metabolism is impaired.

Srivastava, D. K., and S. A. Bernhard, Metabolite transfer via enzyme–enzyme complexes. *Science* 234:1081–1087, 1986. Unusual mode of transfer observed for intermediates in the glycolytic pathway.

Van Schaftingen, E., Fructose-2,6-bisphosphate. *Adv. Enzymol.* 59:315–395, 1987. An allosteric effector that regulates the flow between fructose-6-phosphate and fructose-1,6-bisphosphate.

## Problems

1. Why are there more D-aldohexoses (fig. 12.2) than D-ketohexoses (fig. 12.3)?

2. Draw the structures of $\alpha$-sophorose, $\alpha$-melibiose, and lactulose. Sophorose is two glucoses linked $\beta$ (1-2). Melibiose is a galactose linked to a glucose by an $\alpha$ (1-6) bond. Lactulose is galactose linked $\beta$ (1-4) to fructose.

3. A tool that has often been used to deduce ring sizes and linkages between sugars is methylation with methyl iodide. The methylated sugar is then hydrolyzed and the location of the methyl groups determined. A disaccharide called trehalose has been obtained from a number of vegetable sources. Trehalose is composed only of glucose, and when

methylated and hydrolyzed, only 2,3,4,6-tetramethyl-glucose results. Using the same procedure, maltose (fig. 12.8) gives 2,3,4,6-tetramethylglucose and 2,3,6-trimethylglucose in equal molar ratios. Propose a structure(s) for trehalose.

4. The substance glucuronic acid can be considered to be glucose with carbon 6 oxidized to the carboxylic acid level. Glucuronic acid $\beta$-glycosides are the metabolic fate of a number of obnoxious compounds found in plants. Draw the structure of glucuronic acid. From its name can you guess where it was first found?

5. Aldoses and ketoses that have the potential to form open chains (i.e., have hemiacetal or hemiketal groups) are referred to as reducing sugars. This name is derived from the fact they can readily reduce $Ag^+$ to Ag or $Cu^{2+}$ to $Cu^{1+}$. Examine the disaccharides in figure 12.8 and determine which substances are reducing sugars. Is trehalose (problem 3) a reducing sugar?

6. The polysaccharide inulin is an energy-storage substance found in the tubers of a number of plants. The major part of inulin is composed of fructoses linked $\beta$ (2-1). Draw the structure of inulin.

7. Which glucose carbons (use glucose carbon numbers) are lost as $CO_2$ in the conversion of pyruvate into ethanol (fig. 12.13).

8. In the conversion of pyruvate to phosphoenolpyruvate in gluconeogenesis (fig. 12.26) an addition of $CO_2$ is followed by a decarboxylation. Why would nature add an item only to remove it in the next step? Is the carbon added the same as the one removed?

9. Glucose can be purchased with essentially any specific carbon labeled with $^{14}C$. If your objective is to assess the relative importance of glycolysis versus the pentose phosphate pathway in a particular tissue, how might you use radioactive glucose to answer this question?

10. After reading about glycolysis and gluconeogenesis do you find anything unusual about the name pyruvate kinase?

11. An intermediate in the interconversion of glycerate-3-phosphate to glycerate-2-phosphate is glycerate-2,3-bisphosphate. Where have we encountered this bisphosphate before?

12. The common yeast genus *Rhodotorula* is thought to be missing the enzyme phosphofructokinase yet seems to survive nicely. Can you propose a pathway around phosphofructokinase?

13. Why are the mechanisms of the enzymes that interconvert glucose-6-phosphate and fructose-6-phosphate and the enzyme that interconverts dihydroxyacetone phosphate and glyceraldehyde-3-phosphate virtually identical? Compare the mechanisms of phosphoglycerate mutase and phosphoglucomutase. Why are they so similar?

14. Suppose that you have isolated a facultative microorganism that you are growing anaerobically in a medium containing a carbohydrate. Explain, on the basis of your knowledge of metabolism, why each of the following statements about the fermentation is false:
    (a) The culture must be growing on glucose because bacteria ferment few other compounds.
    (b) The products of the fermentation must be more highly oxidized than the substrates, otherwise no energy is conserved.
    (c) The culture cannot be producing any $CO_2$.

15. A yeast culture is fermenting glucose to ethanol. To ensure that the $CO_2$ released during fermentation is radiolabeled, what carbon(s) of glucose must be labeled with $^{14}C$?

16. Assume that a mutant form of glyceraldehyde-3-phosphate dehydrogenase was found to hydrolyze the oxidized enzyme-bound intermediate with water rather than phosphate.
    (a) Write a chemical reaction that describes the hydrolysis, showing the structures of the products.
    (b) What would be the effect, if any, on the ATP yield from glycolysis of glucose to lactate?
    (c) What would be the effect of the mutation on an obligate aerobic microorganism?

17. Suppose that you are seeking bacterial mutants with altered triose phosphate isomerase (TPI). The organism of interest is known to use the glycolytic pathway with the production of lactate.
    (a) Explain why the absence of TPI would be lethal to an organism fermenting glucose exclusively through the glycolytic pathway.
    (b) Suppose that you have an organism that uses glycolysis and an oxidative pathway as energy

sources. Mutants of that organism having only 10% of the TPI activity present in the wild-type cells grew slowly on glucose under anaerobiosis but grew faster aerobically. Explain the metabolic basis of the observation.

(c) You constructed a plasmid that would direct the synthesis of dihydroxyacetone phosphate phosphatase and introduced the plasmid into the mutant organism described in part (b). Predict whether the plasmid-bearing organism with an active DHAP phosphatase could grow on glucose or glycerol either anaerobically or aerobically. What are the metabolic considerations you used to make your predictions?

18. There are two sites of ADP phosphorylation in glycolysis. These processes are called substrate level phosphorylations. Arsenate ($AsO_4^{3-}$), an analog of phosphate, uncouples ATP formation resulting from glyceraldehyde-3-phosphate oxidation but not that resulting from dehydration of glycerate-2-phosphate. Explain.

19. The disaccharide sucrose can be cleaved by either of two methods:

$$\text{Sucrose} + H_2O \xrightarrow{\text{Invertase}} \text{glucose} + \text{fructose}$$

$$\text{Sucrose} + P_i \xrightarrow{\underset{\text{phosphorylase}}{\text{Sucrose}}} \text{glucose-1-phosphate} + \text{fructose}$$

(a) Given that the $\Delta G^{\circ\prime}$ value for the invertase reaction is $-7.0$ kcal/mole, calculate the $\Delta G^{\circ\prime}$ value for the sucrose phosphorylase catalyzed reaction. Assume that the $\Delta G^{\circ\prime}$ for hydrolysis of glucose-1-phosphate is $-5$ kcal/mole. Based on the calculated value of $\Delta G^{\circ\prime}$, calculate the equilibrium constant for sucrose phosphorylase at 25°C.

(b) Explain the metabolic advantage to the cell of cleaving sucrose with phosphorylase rather than with invertase.

20. What concentration of glucose would be in equilibrium with 1.0 mM glucose-6-phosphate, assuming that hexokinase is present and the concentration ratio of ATP to ADP is 5? Is it reasonable to expect an actively metabolizing cell to maintain the concentration of glucose necessary to sustain the concentration of glucose-6-phosphate at 1 mM?

21. 2-Phosphoglycerate and phosphoenolpyruvate differ only by dehydration between C-2 and C-3, yet the difference in the $\Delta G^{\circ\prime}$ of hydrolysis is about $-12$ kcal/mole. How does dehydration "trap" so much chemical energy?

22. Elevated pyruvate concentration inhibits the heart muscle lactate dehydrogenase (LDH) isoenzyme but not the skeletal muscle LDH isoenzyme.
(a) What would be the consequence of having only the heart isoenzyme form in skeletal muscle?
(b) Would there necessarily be a negative consequence if the skeletal muscle isoenzyme were the only LDH in heart muscle?

23. We stated that the equilibrium constant for the pyruvate kinase reaction is $10^6$. Assume that the steady-state concentration of ATP is 2 mM and of ADP is 0.2 mM. Calculate the concentration ratio of pyruvate to PEP under these conditions. Does your calculation support or refute the assertion that the pyruvate kinase reaction is metabolically irreversible?

24. In gluconeogenesis, the thermodynamic barrier imposed by pyruvate kinase is overcome by coupling two separate reactions for the synthesis of PEP from pyruvate.
(a) Write the two chemical reactions used to bypass the pyruvate kinase reaction.
(b) Calculate the overall $\Delta G^{\circ\prime}$ of the two reactions you wrote in part (a). (Assume that GTP is the thermodynamic equivalent of ATP). What can you now surmise about the feasibility of PEP formation from pyruvate by this route?

25. Write a chemical reaction for the $NADP^+$-dependent oxidation of 6-phosphogluconate to ribulose-5-phosphate.

26. Fructose-2,6-bisphosphate is a potent activator of the liver phosphofructokinase (PFK-1) and a potent inhibitor of liver fructose-1,6-bisphosphate phosphatase (FBPase-1). Fructose-2,6-bisphosphate is the product of a second phosphofructokinase (PFK-2) and is hydrolyzed to fructose-6-phosphate by FBPase-2. The activities of PKF-2 and FBPase-2 reside on a single, bifunctional protein in liver. The bifunctional protein is under glucagon control imposed via cAMP. (H. -G. Hers, *Arch. Biol Med. Exp.* 18:243–251, 1985.)

(a) Under what metabolic conditions is PKF-2 active? FBPase-2?

(b) Gluconeogenesis in liver is stimulated by the hormone glucagon. The activity of PFK-2/FBPase-2 bifunctional enzyme under glucagon regulation shifts from an active PFK-2 to an inactive PFK-2. Inactivation of the PFK-2 alone would still not be adequate to stimulate gluconeogenesis sufficiently for the organism. Explain.

(c) Cyclic AMP-dependent phosphorylation of PFK-2/FBPase-2 not only inhibits PFK-2 but stimulates FBPase-2. Under these conditions, gluconeogenesis is sufficiently rapid to meet cellular demand. Explain.

(d) What would you predict as the relative activities of the following enzymes in the liver of a rat made diabetic through chemical means (administration of alloxan or streptozotocin): PFK-2, FBPase-2, PFK-1, FBPase-1, pyruvate carboxylase, PEP carboxykinase?

27. Muscle pyruvate kinase (PK) responds hyperbolically to its substrate, PEP, but the liver form of the enzyme responds sigmoidally. Fructose-1,6-bisphosphate is an allosteric activator of liver pyruvate kinase, but it apparently has no effect on the muscle enzyme.

(a) If liver PK responded hyperbolically to PEP and were otherwise unregulated, how might gluconeogenesis be affected?

(b) What is the metabolic advantage of having the liver PFK activated by fructose-1,6-bisphosphate?

# Methods for Structural Analysis:
## Polarized Light and Polarimetry

Light is a form of electromagnetic radiation that oscillates sinusoidally in space and time. The oscillating electric and magnetic fields of light are perpendicular to each other and are both in a plane perpendicular to the direction of the light ray. In an unpolarized beam, the fields are equally strong with all different orientations in the plane (figure 1). A light beam is said to be polarized if the orientations of the fields in the plane are fixed. In linearly polarized light the orientation of the fields does not vary along the direction of the beam. In circularly polarized light the orientation of the fields gradually changes in either a right-handed or left-handed manner along the direction of the beam.

A polarimeter is an instrument for studying the inter-action of polarized light with optically active substances (figure 2). A cylindrical tube is filled with a solution containing the substance of interest. A monochromatic beam of polarized light is passed through the solution, and the effects on the polarized beam after passing through the solution are measured.

The specific rotation is defined as $[\alpha] = (100 \times A)/(c \times l)$, where $A$ is the observed rotation in degrees, $c$ is the concentration of the optically active substance in grams per 100 ml of solution, and $l$ is the path length in decimeters of the solution through which the rotation is observed. Light from a sodium lamp (596 nm) is commonly used for the measurement.

**Figure 1**

Unpolarized and polarized light.

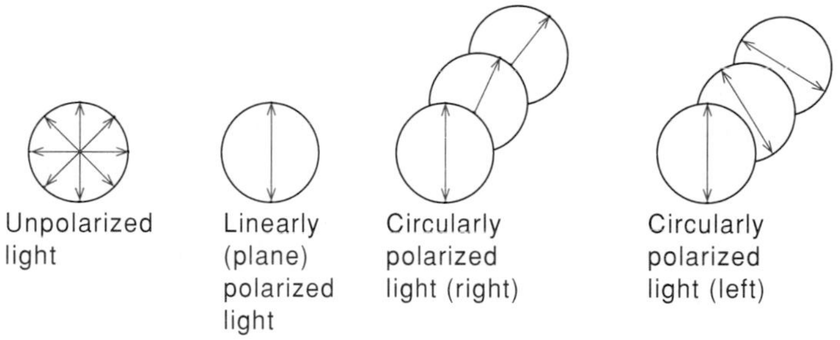

Unpolarized
light

Linearly
(plane)
polarized
light

Circularly
polarized
light (right)

Circularly
polarized
light (left)

**Figure 2**

Simple polarimeter for measuring
rotation of linearly polarized light.

Plane polarized
light

Rotated plane polarized
light

Monochromatic
light

Nicol
prism

Sample
tube

Nicol prism mounted on
circular dial and rotated to give
maximum of transmitted light

$\alpha$

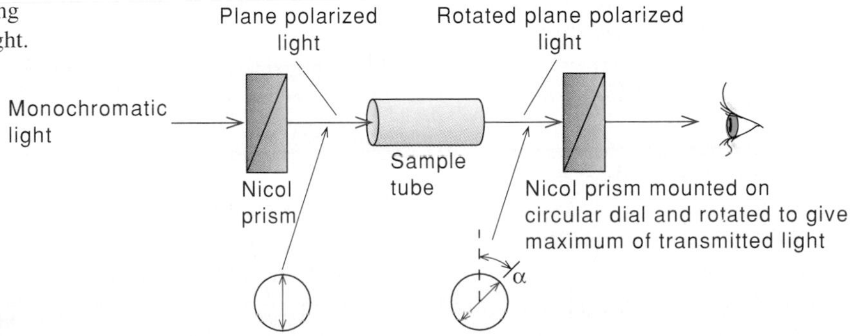

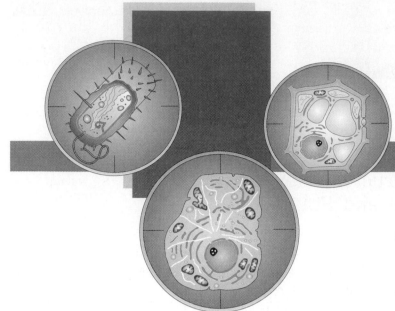

# The Tricarboxylic Acid Cycle

*The tricarboxylic acid (TCA) cycle catabolizes pyruvate to $CO_2$ and $H_2O$ in an oxygen-requiring process.*

Only a small fraction of the total free energy content of glucose is released under anaerobic conditions. This is because no net oxidation of organic substrates can occur in the absence of oxygen. Catabolism under anaerobic conditions means that every oxidative event in which electrons are removed from an organic compound must be accompanied

## Figure 13.1

Outline of the main reactions of carbohydrate metabolism considered in this chapter. Most of these reactions (in black) belong to the tricarboxylic acid cycle, which provides a means for catabolizing three carbon units all the way to $CO_2$. This process, which requires oxygen, yields considerably more energy and reducing power than simple glycolysis. Two of these reactions are part of the glyoxylate bypass, a means of synthesizing four-carbon and six-carbon units from the two-carbon level of acetyl-CoA. The reactions of the glyoxylate bypass are found in plants, fungi, and microorganisms but not in vertebrates. Citrate formed in the mitochondrion can be transferred to the cytosol, where it leads to the formation of acetyl-CoA and NADPH (not shown) used in biosynthesis.

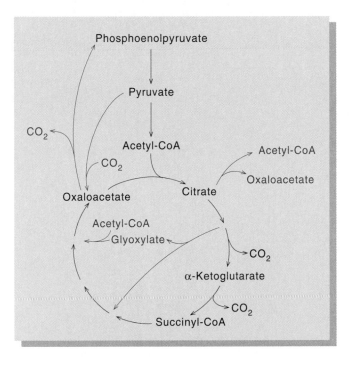

by a reductive event in which electrons are returned to another organic compound, often closely related to the first compound. The cell operating under anaerobic conditions must content itself with the generation of only two ATP molecules per molecule of glucose fermented. Most of the energy of the glucose molecule remains untapped. Given access to oxygen, however, the cell can do much more with the oxidizable organic molecules available to it, and the energy yield increases dramatically. In this chapter the lower half of the central metabolic pathways requiring oxygen is explored (fig. 13.1). With oxygen available as the electron acceptor, the carbon atoms of glucose (or another substrate) can be oxidized fully to $CO_2$, and all the electrons that are removed during the multiple oxidation events are transferred ultimately to oxygen. In the process, the ATP yield per glucose is close to 15 times greater than that possible under anaerobic conditions. Therein lies the advantage of the aerobic way of life.

It is not surprising then, that aerobic processes capable of extracting further energy from pyruvate have come to play so prominent a role in energy metabolism. The overall process of aerobic respiratory metabolism and its distinguishing characteristics are (1) the use of oxygen as the ultimate electron acceptor, (2) the complete oxidation of organic substrates to $CO_2$ and water, and (3) the conservation of much of the free energy as ATP.

The oxygen required for aerobic metabolism actually serves as the terminal electron acceptor only, providing for the continuous reoxidation of reduced coenzyme molecules (most prominent of which are NADH and $FADH_2$). It is these coenzyme molecules (in their oxidized form) that serve as electron carriers for the stepwise oxidation of organic intermediates derived from pyruvate. Aerobic respiratory metabolism can therefore be thought of in terms of two separate but intimately linked processes: the actual oxidative metabolism, in which electrons are removed from organic substrates and transferred to coenzyme carriers, and the concomitant reoxidation of the reduced coenzymes by transfer of electrons to oxygen, accompanied indirectly by the generation of ATP.

Under aerobic conditions, the glycolytic pathway becomes the initial phase of glucose catabolism (fig. 13.2). The other three components of respiratory metabolism are the tricarboxylic acid (TCA) cycle, which is responsible for further oxidation of pyruvate, the electron-transport chain, which is required for the reoxidation of coenzyme molecules at the expense of molecular oxygen, and the oxidative phosphorylation of ADP to ATP, which is driven by a proton gradient generated in the process of electron transport. Overall, this leads to the potential formation of approximately 30 molecules of ATP per molecule of glucose in the typical eukaryotic cell.

In this chapter, discussion focuses on the TCA cycle and its central role in the aerobic catabolism of carbohydrates. Chapter 14 explains how the free energy present in the reduced coenzymes that are generated by glycolysis and the TCA cycle is conserved as ATP during the companion process of electron transport and oxidative phosphorylation.

## Discovery of the TCA Cycle

The discovery of the TCA cycle began with a series of biochemical experiments performed in the early 1900s on anaerobic suspensions of minced animal tissues. The experiments established that the suspensions contained enzymes that could transfer hydrogen atoms from various low-molecular-weight organic acids to other reducible compounds, such as the dye, methylene blue (methylene blue was a convenient indicator in these experiments because it is converted from a blue to a colorless form by

**Figure 13.2**

The components of respiratory metabolism include glycolysis, the tricarboxylic acid (TCA) cycle, the electron-transport chain, and the oxidative phosphorylation of ADP to ATP. Glycolysis converts glucose to pyruvate; the TCA cycle fully oxidizes the pyruvate (by means of acetyl-CoA) to $CO_2$ by transferring electrons stepwise to coenzymes; the electron-transport chain reoxidizes the coenzymes at the expense of molecular oxygen; and the energy of coenzyme oxidation is conserved in the form of a transmembrane proton gradient that is then used to generate ATP.

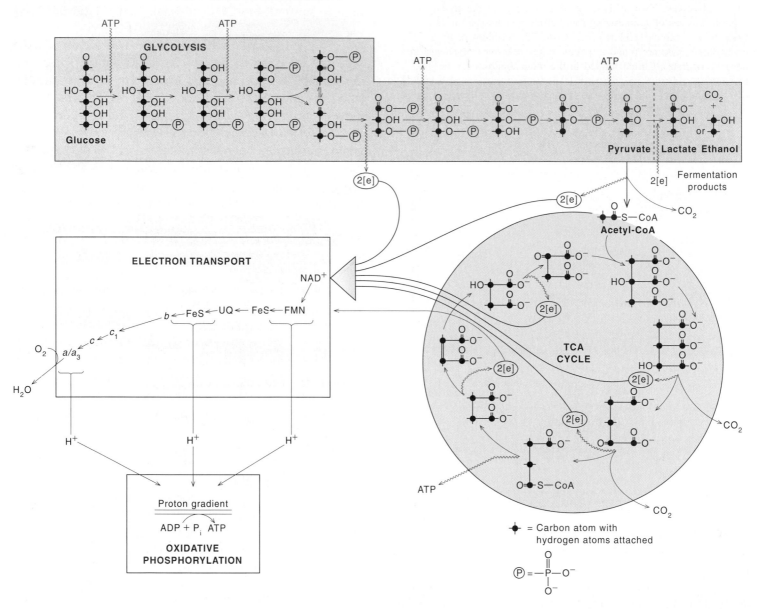

reduction). Only a few organic acids were active in the reduction: succinate, fumarate, malate, and citrate. It was later observed that in the presence of oxygen the same suspensions oxidized these acids to $CO_2$ and water.

Albert Szent-Györgyi found that when small amounts of these organic acids were added to a tissue suspension in the presence of glucose, considerably more oxygen was consumed than was required to oxidize the added acid. He concluded that in the presence of oxygen the organic acids had a catalytic effect on the oxidation of glucose or other carbohydrates in the tissue slices. Szent-Györgyi also ob-

served that a specific inhibitor of succinate dehydrogenase, malonate, blocked the utilization of oxygen (respiration) by muscle suspensions. This finding suggested that oxidation of succinate is an indispensable reaction in muscle oxidative metabolism. When the same response was found in many other tissues, the phenomenon seemed to have fundamental and perhaps universal significance.

Next, Hans Krebs studied the interrelationships between the oxidative metabolism of different organic acids. For his experiments he used slices of pigeon flight muscle, which are particularly active in oxidative metabolism. He found a select group of organic acids that was oxidized very rapidly by extracts from the muscle tissue. His list included the organic acids that Szent-Györgyi had found plus oxaloacetate, $\alpha$-ketoglutarate, isocitrate, and *cis*-aconitate. Like Szent-Györgyi, Krebs found that catalytic amounts of the same organic acids stimulated the oxidation of pyruvate or carbohydrate, and that their catalytic effect was blocked by malonate. Because malonate specifically inhibited the conversion of succinate to fumarate by succinate dehydrogenase, he concluded that this reaction was one of a series of steps essential for the complete oxidation of pyruvate. Additional steps in the process were presumed to involve the other organic acids that had been found to catalyze the process. The organic acids with catalytic potential could be arranged in a chain related by biochemical conversions:

$$\text{Citrate} \to \textit{cis-}\text{aconitate} \to \text{isocitrate} \to \alpha\text{-ketoglutarate} \to$$
$$\text{succinate} \to \text{fumarate} \to \text{malate} \to \text{oxaloacetate} \quad \textbf{(1)}$$

But if this chain of reactions were to act catalytically, then some way must exist for regenerating all of the intermediates of the chain from a single intermediate. A cyclical process would be the simplest way of doing this. For the chain to operate as a cycle, one or more additional reactions needed to be discovered. Krebs found that in the absence of oxygen, small amounts of citrate could be formed from added oxaloacetate and pyruvate. Could this reaction be the missing link in the cyclical process? Krebs thought so, and proposed that the first step in pyruvate oxidation involved its condensation with oxaloacetate to form citrate and $CO_2$. The oxidation of more complex carbohydrates then could be explained by their prior conversion to pyruvate through the glycolytic pathway.

Further support for the cyclical nature of the chain came from observations of malonate-inhibited muscle slices. In such preparations the addition of any of the intermediates of the chain led to accumulation of succinate. Even the addition of fumarate, the immediate product of succinate oxidation, led to accumulation of succinate.

When the oxidation of substantial amounts of pyruvate was blocked by malonate, stoichiometric amounts of pyruvate would react if either oxaloacetate or its precursors, malate or fumarate, was added. None of the precursors to succinate in the chain was effective in this regard. This strongly supported Krebs' hypothesis that oxidation of pyruvate involved its condensation with oxaloacetate and the subsequent series of conversions in the chain described by Krebs. The Krebs cycle in its originally proposed form is shown in figure 13.3. It is also known as the tricarboxylic acid (TCA) cycle or the citrate cycle.

**Figure 13.3**

Original tricarboxylic acid (TCA) cycle proposed by Krebs. This cycle is also called the citrate cycle, or the Krebs cycle. To start the cycle in operation, pyruvate loses one of its carbons and condenses with a four-carbon dicarboxylic acid, oxaloacetic acid, to form a six-carbon tricarboxylic acid, citrate. In one turning of the cycle, two carbons are lost as $CO_2$, thus returning the citrate to oxaloacetate. The conversion blocked by malonate is indicated by a red bar.

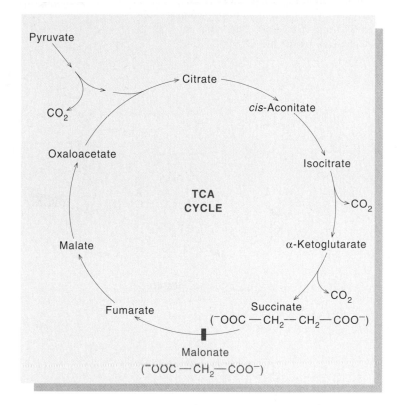

## Steps in the TCA Cycle

We now know that the tricarboxylic acid cycle (fig. 13.4) begins with acetyl-coenzyme A which results from the oxidative decarboxylation of pyruvate (acetyl-CoA) or the oxidative cleavage of fatty acids, (see chapter 18). Regardless of its source, acetyl-CoA transfers its acetyl group to oxaloacetate, thereby generating citrate. In a cyclic series of reactions, the citrate is subjected to two successive decarboxylations and several oxidative events, leaving a four-carbon compound from which the starting oxaloacetate is eventually regenerated. Each turn of the cycle involves the entry of two carbons from acetyl-CoA and the release of two carbons as $CO_2$. As a result, the cycle is balanced with respect to carbon flow and functions without net consumption or buildup of oxaloacetate (or any other intermediate), unless side reactions occur that either feed carbon into the cycle or drain it off into alternative pathways.

## Figure 13.4

The TCA cycle. In each turn of the cycle, acetyl-CoA from the glycolytic pathway or from $\beta$ oxidation of fatty acids enters and two fully oxidized carbon atoms leave (as $CO_2$). ATP is generated at one point in the cycle, and coenzyme molecules are reduced. The two $CO_2$ molecules lost in each cycle originate from the oxaloacetate of the previous cycle rather than from incoming acetyl from acetyl-CoA. This point is emphasized by the use of color.

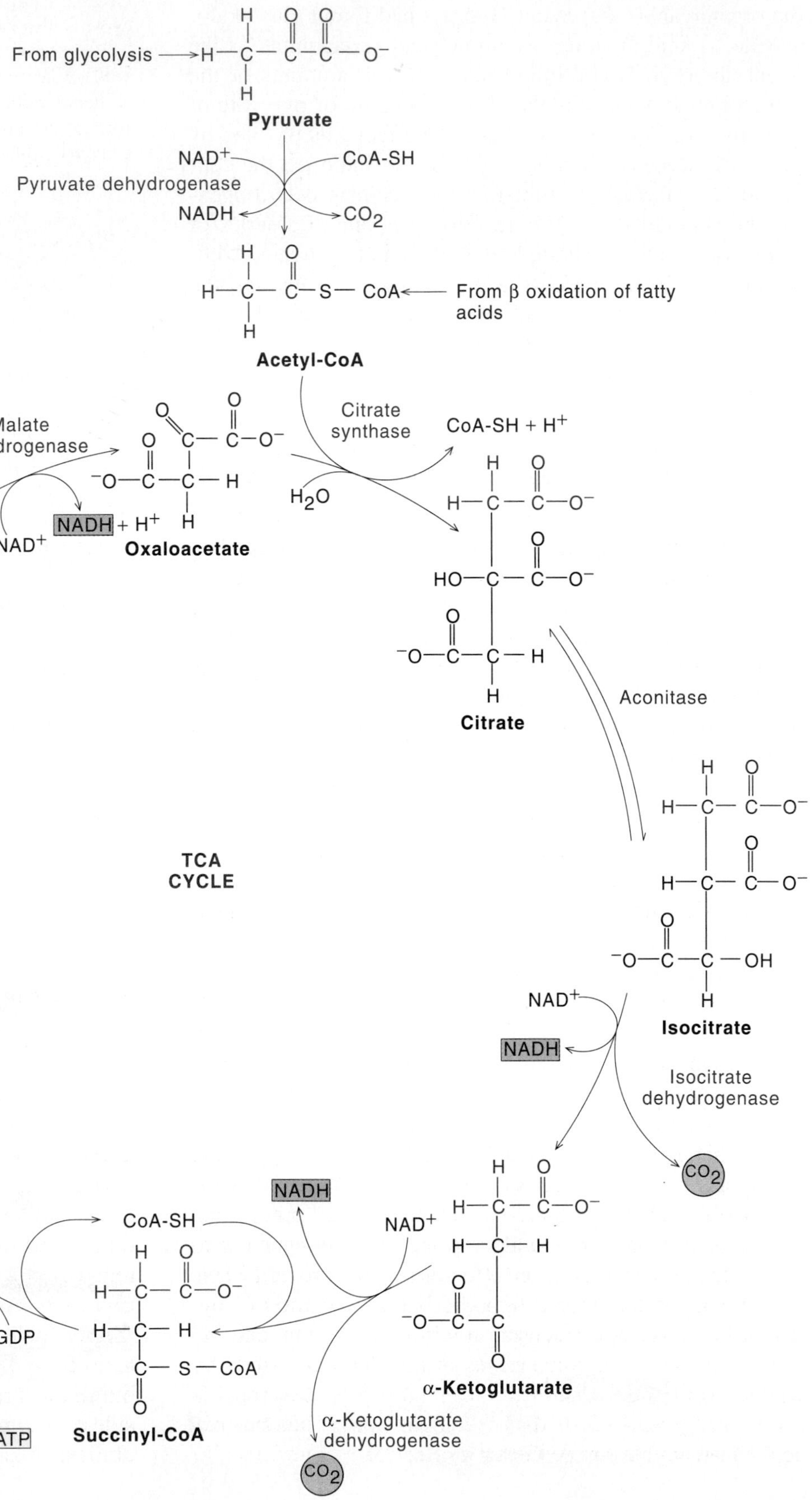

## The Oxidative Decarboxylation of Pyruvate Leads to Acetyl-CoA

Although the acetyl-CoA with which the cycle begins may be derived catabolically from fatty acids or amino acids, the major source of acetyl-CoA in most cells is the pyruvate available from the glycolytic breakdown of carbohydrate. The gap from the glycolytic pathway to the TCA cycle is bridged by the oxidative decarboxylation of pyruvate (with which glycolysis ends) to yield acetate in the form of acetyl-CoA (with which the TCA cycle commences). The decarboxylation of pyruvate is catalyzed by a cluster of three enzymes called the pyruvate dehydrogenase complex. In eukaryotic cells, this enzyme complex is located in the mitochondria, as are all the other reactions of aerobic energy metabolism beyond pyruvate. Since the glycolytic pathway occurs in the cytosol, it is as pyruvate that the carbon derived from glucose (or other carbohydrate substrates) enters the mitochondria.

Decarboxylation of an $\alpha$-keto acid like pyruvate is a difficult reaction for the same reason as are the ketol condensations (see fig. 12.33): Both kinds of reactions require the participation of an intermediate in which the carbonyl carbon carries a negative charge. In all such reactions that occur in metabolism, the intermediate is stabilized by prior condensation of the carbonyl group with thiamine pyrophosphate. In figure 13.5 thiamine pyrophosphate and its hydroxyethyl derivative are written in the doubly ionized ylid form rather than the neutral form because this is the form that actually participates in the reaction even though it is present in much smaller amounts.

In the first step of the conversion catalyzed by pyruvate decarboxylase, a carbon atom from thiamine pyrophosphate adds to the carbonyl carbon of pyruvate. Decarboxylation produces the key reactive intermediate, hydroxyethyl thiamine pyrophosphate (HETPP). As shown in figure 13.5, the ionized ylid form of HETPP is resonance-stabilized by the existence of a form without charge separation. The next enzyme, dihydrolipoyltransacetylase, catalyzes the transfer of the two-carbon moiety to lipoic acid. A nucleophilic attack by HETPP on the sulfur atom attached to carbon 8 of oxidized lipoic acid displaces the electrons of the disulfide bond to the sulfur atom attached to carbon 6. The sulfur then picks up a proton from the environment as shown in figure 13.5. This simple displacement reaction is also an oxidation–reduction reaction, in which the attacking carbon atom is oxidized from the aldehyde level in HETPP to the carboxyl level in the lipoic acid derivative. The oxidized (disulfide) form of lipoic acid is converted to the reduced (mercapto) form. The fact that the two-carbon moiety has become an acyl group is shown more clearly after dissocia-

tion of thiamine pyrophosphate (TPP), which generates acetyl lipoic acid.

Further transfer of the acyl group to coenzyme A is catalyzed by the same enzyme. This displacement reaction produces reduced lipoic acid. A third enzyme, dihydrolipoyl dehydrogenase, catalyzes oxidation of this product back to the disulfide form. The electrons lost in that oxidation are transferred first to an enzyme-bound flavin (not shown in the figure) and then to $NAD^+$.

The overall equation for the conversion catalyzed by the pyruvate dehydrogenase complex is

$$Pyruvate + NAD^+ + CoA \longrightarrow$$
$$Acetyl\text{-}CoA + NADH + CO_2 \quad (2)$$

The standard free energy change for this conversion is about $-8$ kcal/mol.

### The Nature of the Pyruvate Dehydrogenase Complex.
Many molecules of each of the three enzymes that participate in the conversion of pyruvate to acetyl-coenzyme A are organized into a giant enzyme complex. In mammals the complex has a molecular weight of about $9 \times 10^6$; it contains 60 molecules of the transacetylase and perhaps 20–30 each of the other two enzymes. The complex in *Escherichia coli* is smaller and contains fewer molecules of each enzyme (fig. 13.6). Each molecule of transacetylase within the multienzyme complex contains two molecules of lipoic acid covalently bound to the enzyme through an amide bond to the $\epsilon$-amino group of a lysine residue. The disulfide of lipoic acid is thus at the end of a long chain and can sweep over a considerable area on the surface of the subunit to which it is attached and also reach the catalytic sites of neighboring molecules of pyruvate decarboxylase and dihydrolipoyl dehydrogenase. Thus, it appears that the catalytic activities of the complex depend on the ability of the sulfur atoms of the tethered lipoic acid to visit successively the three types of catalytic sites contained in the complex. Pyruvate binds at the decarboxylase site and is converted to HETPP as discussed previously; then the disulfide group of oxidized lipoic acid picks up the two-carbon moiety, oxidizing it to an acetyl group, and carries it to a site on the transacetylase subunit, where the acetyl group is transferred to coenzyme A. In the form of acetyl-CoA, the two-carbon unit is free from the enzymic tether and can diffuse away as the primary product of the sequence of reactions that are catalyzed by the complex. The reduced lipoic acid is oxidized at a site on the dehydrogenase and is ready to pick up another acetyl group from the decarboxylase site. The NADH that is formed in the oxidation of dihydrolipoic acid dissociates from the enzyme and is available to the electron-transfer machinery of the mitochondrion.

## Figure 13.5

The conversion of pyruvate to acetyl-CoA. The reactions are catalyzed by the enzymes of the pyruvate dehydrogenase complex. This complex has three enzymes: pyruvate decarboxylase, dihydrolipoyl transacetylase, and dihydrolipoyl dehydrogenase. In addition, five coenzymes are required: thiamine pyrophosphate, lipoic acid, CoASH, FAD, and NAD$^+$. Lipoic acid is covalently attached to the transacetylase component of the complex by an amide bond between the carboxyl group of lipoic acid and the terminal amino group of a lysine residue of the enzyme. In fact one or more lipoic acids are involved in each transfer reaction for reasons indicated in chapter 10 (see fig. 10.12). Reactants and products shown in red.

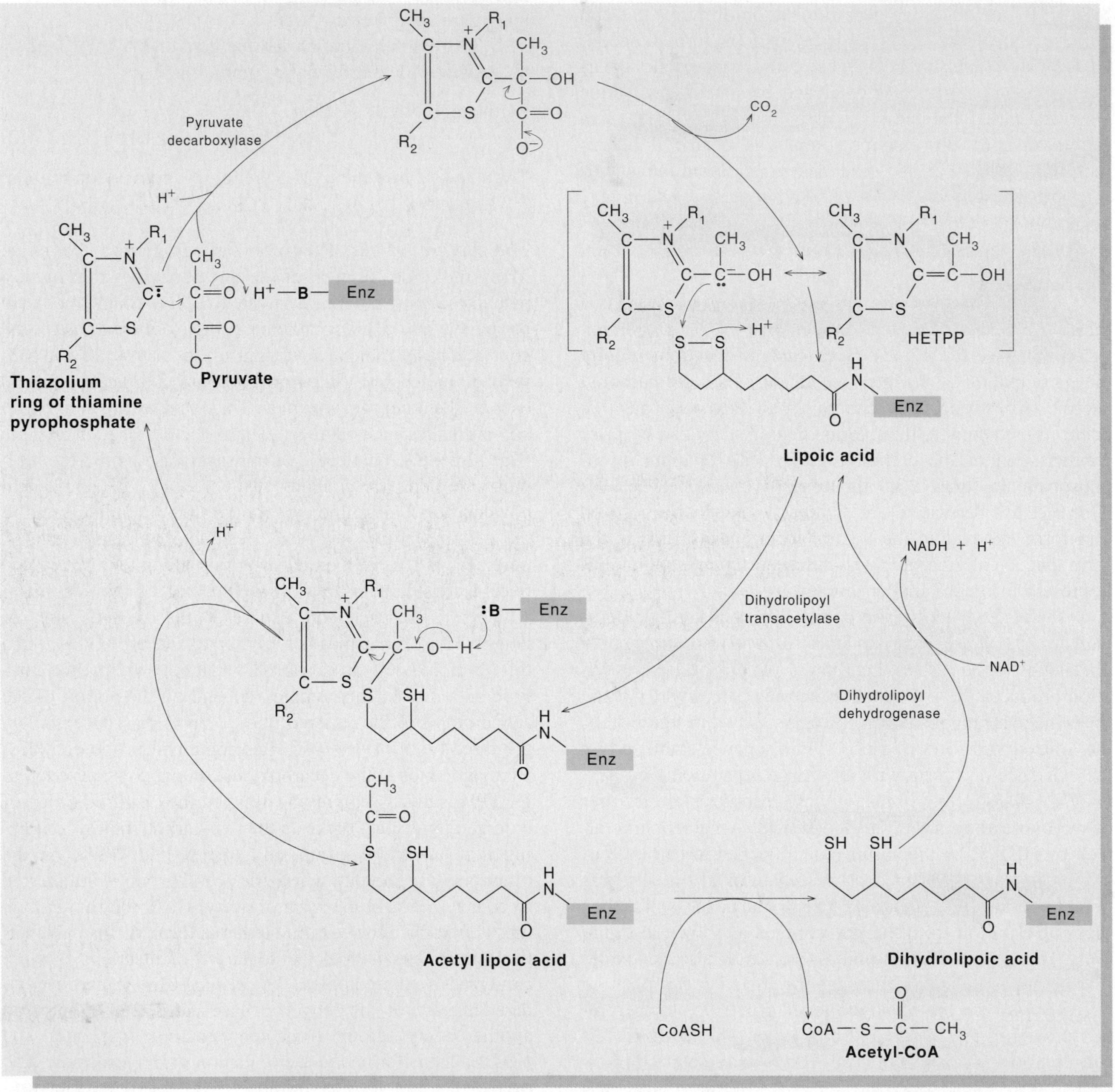

**Figure 13.6**

Interactions of α-lipoyl groups in the pyruvate dehydrogenase complex. The cubic structure represents the 24 subunits of dihydrolipoyl transacetylase, which constitutes the core of the complex. Two of the 48 lipoyl groups in the core are shown interacting with one of the 24 pyruvate decarboxylases (E₁ · TPP) and one of the 12 dihydrolipoyl dehydrogenase (E₂ · FAD) subunits. Note the interaction of the lipoyl groups in relaying electrons over the long distance between TPP and FAD. The lipoic acid must interact at active sites over distances too long to be covered by a single fully extended coenzyme. This problem is solved by use of a shuttle system involving two α-lipoyl groups for each transfer.

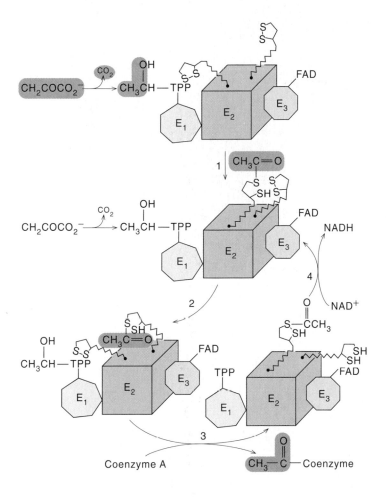

At this point in the oxidation of glucose, four electrons per glucose molecule have been lost in the oxidation of glyceraldehyde-3-phosphate and four more in the conversion of pyruvate to acetyl-CoA. Thus, of the total of 24 electrons lost in the oxidation of glucose to $CO_2$, 16 remain to be transferred to oxidizing agents in the course of the oxidation of two molecules of acetyl-CoA. A major func-

tion of the TCA cycle is to mobilize these electrons for use in electron-transfer phosphorylation (chapter 14).

## Citrate Synthase Is the Gateway to the TCA Cycle

The first step of the TCA cycle is the reaction catalyzed by citrate synthase, in which acetyl-CoA enters the cycle and citrate is formed.

$$\text{Acetyl-CoA} + \text{oxaloacetate} \longrightarrow \text{Citrate} + \text{CoA} \quad (3)$$

This reaction is an aldol condensation, in which a carbanion generated at C-2 of the acetyl group (by loss of a proton to water or to an acceptor group on the enzyme) adds to the carbonyl group of oxaloacetate (fig. 13.7). Coenzyme A is released from the product while it is still bound to the enzyme, so that the products of the reaction are free coenzyme A and citrate.

## Aconitase Catalyzes the Isomerization of Citrate to Isocitrate

In the first reaction within the cycle, citrate is converted to its isomer isocitrate (fig. 13.8).

$$\text{Citrate} \rightleftharpoons \textit{cis}\text{-aconitate} \rightleftharpoons \text{isocitrate} \quad (4)$$

Aconitase, the enzyme that catalyzes this isomerization, is named for the fact that the unsaturated compound formed by removing $H_2O$ from either citrate or isocitrate, *cis*-aconitate, can also serve as substrate or product. Aconitase catalyzes the attainment of equilibrium between citrate, isocitrate, and *cis*-aconitate. The three compounds may be considered as belonging to the same metabolic pool.

## Isocitrate Dehydrogenase Catalyzes the First Oxidation in the TCA Cycle

The first oxidative conversion of the TCA cycle is catalyzed by isocitrate dehydrogenase. This conversion takes place in two steps: oxidation of the secondary alcohol to a ketone (oxalosuccinate), followed by a β decarboxylation to produce α-ketoglutarate (fig. 13.9).

$$\text{Isocitrate} + NAD^+ \longrightarrow$$
$$\alpha\text{-ketoglutarate} + NADH + CO_2 \quad (5)$$

$NAD^+$ is the electron acceptor for the oxidative step, and $Mg^{2+}$ or $Mn^{2+}$ is required for the decarboxylation. The oxalosuccinate that is presumably an intermediate does not dissociate from the enzyme.

## Figure 13.7

Formation of citrate, catalyzed by citrate synthase. A carbanion of acetyl-CoA, generated by loss of a proton to water (or to an acceptor group on the enzyme), adds to the carbonyl group of oxaloacetate. The immediate product of the condensation is probably citryl-coenzyme A, but the products that dissociate from the catalytic site of the enzyme are citrate and free coenzyme A.

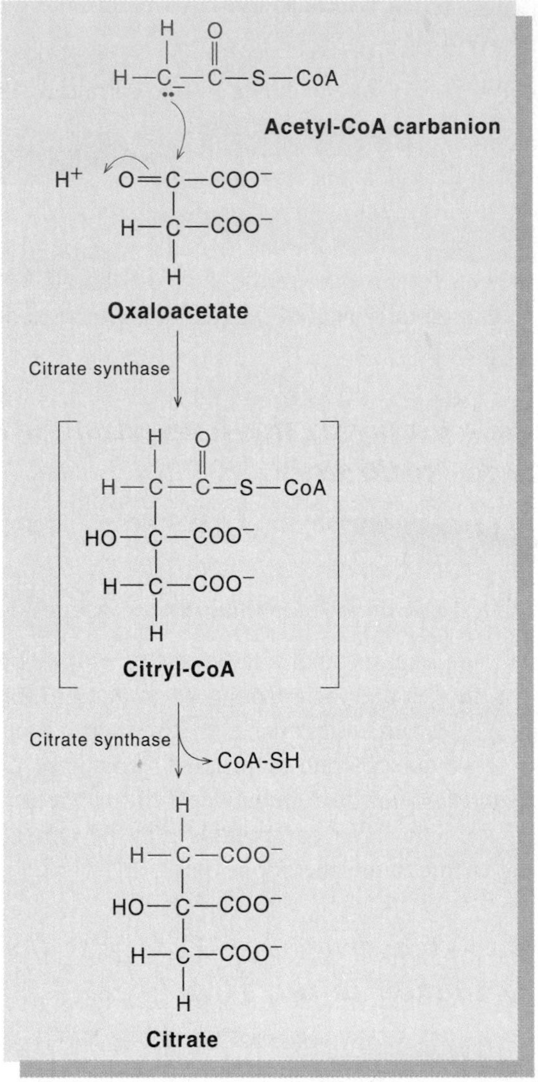

## Figure 13.8

Interconversion of citrate, *cis*-aconitate, and isocitrate, catalyzed by aconitase. At equilibrium the relative concentrations of these compounds are about 90, 6, and 4, respectively.

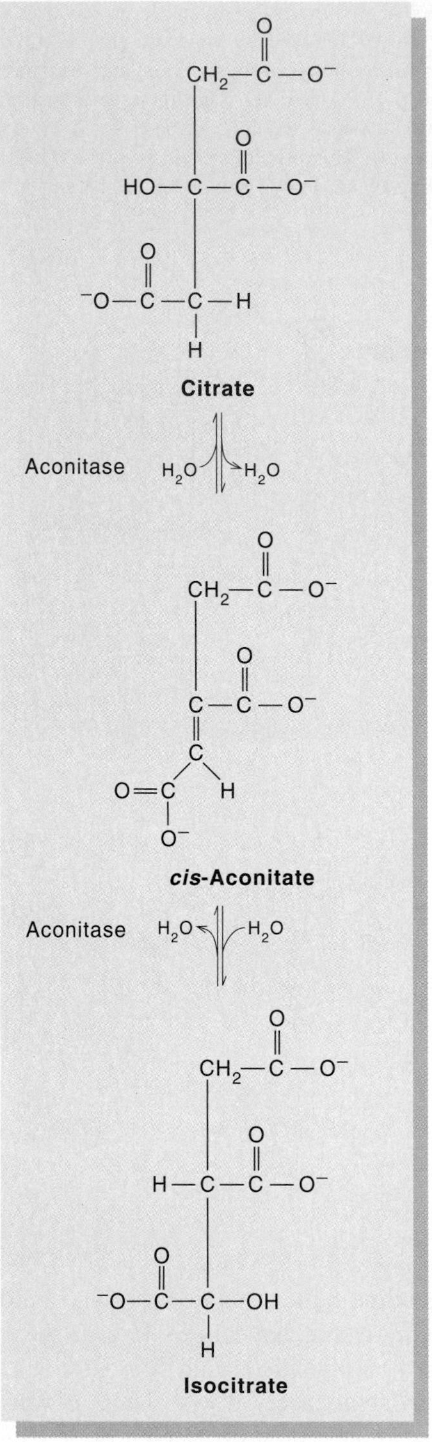

## α-Ketoglutarate Dehydrogenase Catalyzes the Decarboxylation of α-Ketoglutarate to Succinyl-CoA

The second of two oxidative decarboxylation reactions of the TCA cycle is catalyzed by α-ketoglutarate dehydrogen-ase. The reaction sequence is analogous to that of pyruvate dehydrogenase, complete with the conversion of some of the energy of the oxidation in a coenzyme-containing deriv-ative, succinyl-CoA.

$$\alpha\text{-Ketoglutarate} + NAD^+ + CoA \longrightarrow$$
$$Succinyl\text{-}CoA + NADH + CO_2 \quad \textbf{(6)}$$

The enzyme complex involved in this reaction also is very similar to the pyruvate dehydrogenase complex. Indeed, the same dihydrolipoyl dehydrogenase subunit is used in both complexes. The product in both cases is the coenzyme A

## Figure 13.9

The oxidative decarboxylation of isocitrate to $\alpha$-ketoglutarate, catalyzed by mitochondrial isocitrate dehydrogenase. The intermediate, oxalosuccinate, is not released from the enzyme. :B represents a catalytic side chain from the enzyme.

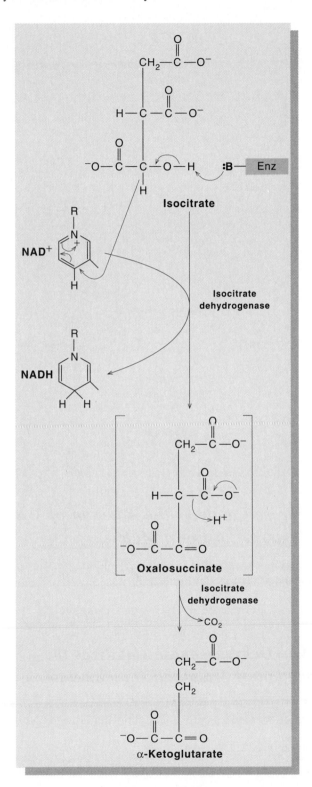

**Isocitrate**

**Isocitrate dehydrogenase**

**Oxalosuccinate**

**Isocitrate dehydrogenase**

$\alpha$-**Ketoglutarate**

ester of the acid containing one less carbon atom than the substrate, in this case succinyl-CoA (see fig. 13.4).

## Succinate Thiokinase Couples the Conversion of Succinyl-CoA to Succinate with the Synthesis of GTP

The thioester bond makes succinyl-CoA an activated intermediate. Although some succinyl-CoA is used in the synthesis of heme in animals, most of it is retained in the TCA cycle where it leads to the regeneration of the oxaloacetate needed to keep the cycle operating. The thioester is converted to succinate in a coupled reaction that results in the formation of GTP.

$$P_i + \text{succinyl-CoA} + \text{GDP} \longrightarrow \text{Succinate} + \text{GTP} \quad (7)$$

The reaction is complex and involves an intermediate in which a phosphate is attached to a histidine residue of the succinate thiokinase enzyme. Probably CoA is first displaced by inorganic phosphate, forming succinyl phosphate. A nitrogen atom of a specific histidine residue then attacks phosphorus, displacing succinate and forming an $N$-phosphoryl derivative. In the final step GDP attacks the phosphorus atom of that derivative forming GTP. The role of GTP in this reaction is played by ATP in some organisms.

## Succinate Dehydrogenase Catalyzes the Oxidation of Succinate to Fumarate

In terms of carbon atoms, we might say that because two carbons entered the cycle as the acetyl group and two left as $CO_2$, the functional part of the cycle is finished at this stage, and thus, the four carbons of succinate must merely be converted to oxaloacetate to serve again as an acetyl acceptor. However, that assumption would be erroneous. When the TCA cycle serves as a means of oxidizing acetyl groups, that is, when it is not producing biosynthetic starting materials at a significant rate, its function is to supply electrons to the electron-transfer phosphorylation system. When we reach succinate, four electrons have been transferred to $NAD^+$. Thus, of the eight electrons that are removed in the oxidation of acetate to $CO_2$, four remain in succinate. In terms of its oxidative function, the cycle is only half finished at this point.

The next step in the TCA cycle, the oxidation of succinate to fumarate, involves insertion of a double bond into a saturated hydrocarbon chain.

$$\text{Succinate} + \text{FAD} \longrightarrow \text{Fumarate} + \text{FADH}_2 \quad (8)$$

This is not an easy reaction in organic chemistry. It is, however, a very important type of reaction in metabolic chemis-

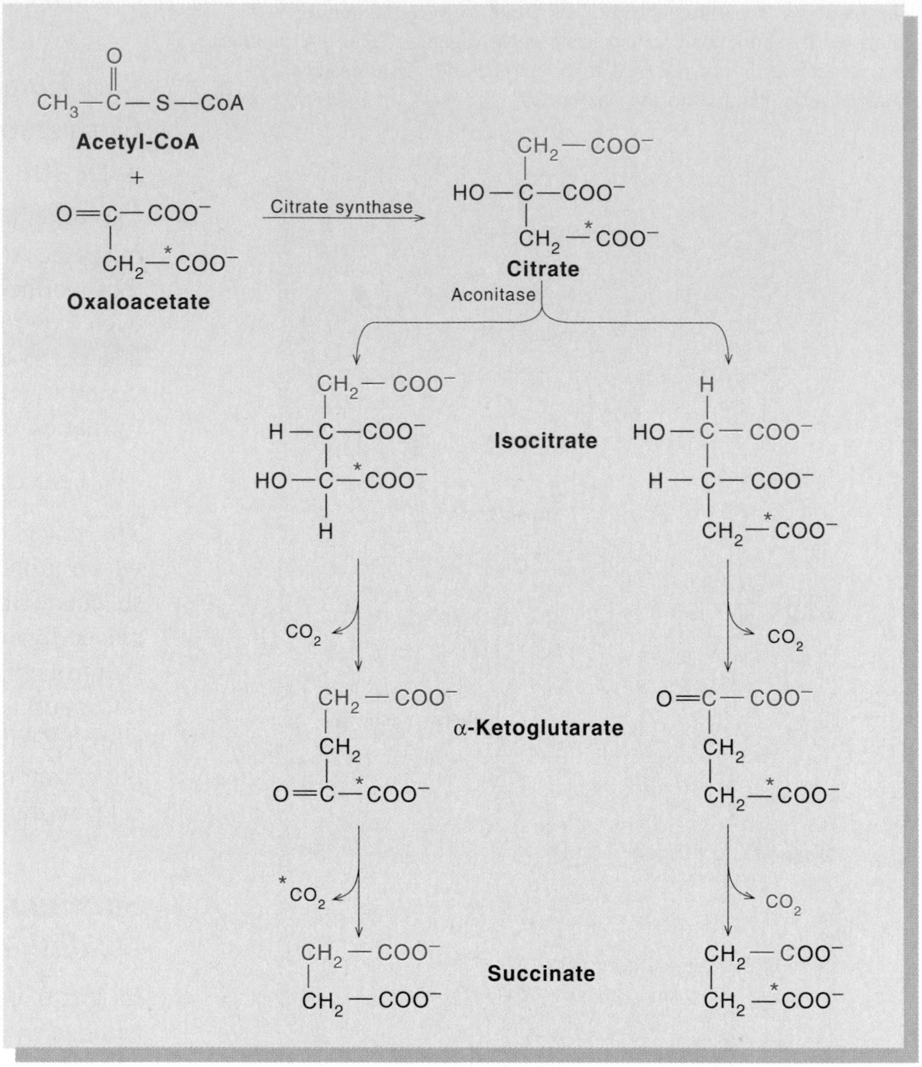

**Figure 13.10**

Stereochemical relationships in the synthesis and metabolism of citrate. When oxaloacetate labeled with $^{13}C$ in the carbonyl group $\beta$ to the keto group (*) was used as substrate, the researchers expected that half of the label would be found in succinate and half in $CO_2$. That prediction was based on the assumption that the two —$CH_2$—$COO^-$ arms of citrate must be equivalent in every way. In fact, all of the label was found in $CO_2$. Thus, only the intermediates shown on the left were produced. This result shows that both the condensation of acetyl-CoA with oxaloacetate and the isomerization of citrate are stereospecific reactions. The carbon atoms supplied by acetyl-CoA are shown in red. Neither of those atoms is lost in the first turn of the cycle after their entry.

try and is an integral step in the oxidation of carbohydrates, fats, and several amino acids.

Because of the nature of the reaction, a strong oxidizing agent is required. The electron acceptor that is used in most oxidative steps of catabolism, $NAD^+$, is not a strong enough oxidizing agent to give a reasonable equilibrium constant. Flavoproteins are stronger oxidizing agents than $NAD^+$, and succinate dehydrogenase, which catalyzes the oxidation of succinate to fumarate, is a flavoprotein enzyme. The oxidation of succinate by the electron acceptor of succinate dehydrogenase is about 16 kcal (65 kJ) more favorable than it would be if the electron acceptor were $NAD^+$. This makes the equilibrium constant more favorable by a factor of about $10^{11}$. In this example we see how important the choice of cofactors can be. In general, FAD is a better oxidizing agent than $NAD^+$, and NADH is a better reducing agent than $FADH_2$.

## Fumarase Catalyzes the Addition of Water to Fumarate to Form Malate

Fumarate is converted to L-malate by stereospecific addition of water across the double bond.

$$\text{Fumarate} + H_2O \longrightarrow \text{L-malate} \qquad (9)$$

## Malate Dehydrogenase Catalyzes the Oxidation of Malate to Oxaloacetate

The final oxidation step of the cycle involves the conversion of malate to oxaloacetate by malate dehydrogenase.

$$\text{L-Malate} + NAD^+ \longrightarrow$$
$$\text{Oxaloacetate} + NADH + H^+ \qquad (10)$$

This enzyme uses $NAD^+$ as the oxidizing agent.

## Stereochemical Aspects of TCA Cycle Reactions

Some early applications of isotopic tracer techniques to the TCA cycle led to a new generalization concerning the stereochemistry of interaction between enzymes and certain types of substrate.

In the early 1940s, before the discovery of $^{14}C$ and when acetyl-coenzyme A was unknown, two research groups used the stable isotope $^{13}C$ and mass spectrometers to study carbon flow in the TCA cycle. Pyruvate and $^{13}CO_2$ were added to pigeon liver preparations to form carboxy-labeled oxaloacetate. Malonate was added to stop the TCA cycle at succinate. The expected result was that half of the $^{13}C$ would be found in succinate and the other half in $CO_2$ (fig. 13.10).

To the surprise of the biochemical community, it was found that although pyruvate was consumed and succinate was produced, there was no $^{13}C$ in the succinate. Further experiments showed that the isotope had been incorporated but was all lost in the oxidative decarboxylation of $\alpha$-ketoglutarate to succinate. Since the citrate molecule has a plane of symmetry, it was taken for granted that the two $-CH_2-COO^-$ arms of the molecule must be chemically equivalent and that the $-OH$ group would have an equal chance of migrating into either arm in the aconitase reaction. In that case, the label also would have an equal chance of being lost or retained in the decarboxylation of $\alpha$-ketoglutarate. For a few years this experimental result was thought to prove that citrate could not be an intermediate of the cycle and that some derivative of citrate must be the immediate product of the condensation of oxaloacetate with an unknown active two-carbon intermediate.

Then, in 1948, Ogston suggested that citrate was not necessarily excluded by the isotopic evidence, because the two $-CH_2-COO^-$ arms might actually not be equivalent when citrate was the substrate for an enzymic reaction. He pointed out that if the substrate were attached to the enzyme at three points, its orientation would be fixed by those attachments, and it would be impossible for the two identical arms to exchange positions. Thus, only one of them occupy the position that allowed it to participate in the reaction (fig. 13.11).

Ogston's concept is a valid and important generalization. Three constraints are necessary to fix an object in three-dimensional space, but they need not all be points of attachment. In principle the two $a$ groups of a molecule of the type $Ca_2bd$ could be distinguished by an enzyme if the three constraints were one point of attachment, one pocket, into which $b$ could fit but $d$ could not, and the position of the reactive groups of the catalytic site.

### Figure 13.11

Citrate is shown with three of the substituents from the central carbon atom making contact with the enzyme surface. Binding in this way makes the two $-CH_2-COO^-$ groups nonequivalent, as they must be in the TCA cycle.

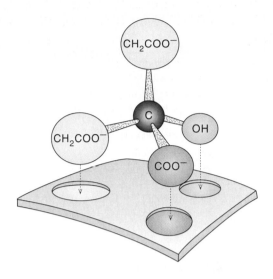

Ogston's contribution led to an interesting extension of concepts concerning stereochemistry of enzyme action. Compounds of the type $Ca_2bd$ are termed prochiral, and it is recognized that an enzyme that either synthesizes such a compound or uses it as a substrate nearly always does so stereospecifically. In the case of citrate synthase, for example, it is inherently likely that the planar carbonyl carbon of oxaloacetate lies flat on an enzyme surface and that only one side of the atom is available for attack by acetyl-coenzyme A.

## ATP Stoichiometry of the TCA Cycle

By summing the component reactions of the TCA cycle, we arrive at the following overall reaction:

$$\text{Acetyl-CoA} + 2\,H_2O + 3\,NAD^+ + FAD + ADP + P_i \longrightarrow$$
$$2\,CO_2 + 3\,NADH + 2\,H^+ + FADH_2 + CoA + ATP \quad (11)$$

Looking at this summary reaction, you may wonder why it doesn't reflect the substantially greater ATP yield that is supposed to be characteristic of aerobic metabolism. The answer is that energy is stored in the reduced coenzyme molecules on the right-hand side of the reaction. Reoxidation of these compounds liberates a large amount of free energy. It is only as the electrons are transferred stepwise from the coenzymes to molecular oxygen that the coupled

**Table 13.1**

Reactions of the TCA Cycle

| Reaction | Enzyme | $\Delta G^{\circ\prime}$ (kcal/mol) |
|---|---|---|
| 1. Acetyl-CoA + oxaloacetate + $H_2O$ $\longrightarrow$ citrate + CoA | Citrate synthase | −7.7 |
| 2. Citrate $\rightleftharpoons$ $cis$-aconitase $\rightleftharpoons$ isocitrate | Aconitase | +1.5 |
| 3. Isocitrate + $NAD^+$ $\longrightarrow$ $\alpha$-ketoglutarate + NADH + $CO_2$ | Isocitrate dehydrogenase | −5.0 |
| 4. $\alpha$-Ketoglutarate + $NAD^+$ + CoA $\longrightarrow$ Succinyl-CoA + NADH + $CO_2$ | $\alpha$-Ketoglutarate dehydrogenase | −8.0 |
| 5. Succinyl-CoA + GDP + $P_i$ $\longrightarrow$ succinate + GTP + CoA | Succinate thiokinase | −0.7 |
| 6. Succinate + FAD $\longrightarrow$ fumarate + $FADH_2$ | Succinate dehydrogenase | 0.0 |
| 7. Fumarate + $H_2O$ $\longrightarrow$ L-malate | Fumarase | −0.9 |
| 8. L-malate + $NAD^+$ $\longrightarrow$ oxaloacetate + NADH + $H^+$ | Malate dehydrogenase | +7.1 |
| Acetyl-CoA + 2 $H_2O$ + 3 $NAD^+$ + FAD + ADP + $P_i$ $\longrightarrow$ 2 $CO_2$ + 3 NADH + 2 $H^+$ + $FADH_2$ + CoA + ATP | | −13.7 |

generation of ATP occurs. We will see that the oxidative phosphorylation system regenerates approximately 2.5 molecules of ATP for each pair of electrons from NADH, and approximately 1.5 molecules of ATP for each pair of electrons from $FADH_2$ (see chapter 14). Thus, in aerobic mitochondrial metabolism, oxidation of 1 mol of acetyl groups leads to regeneration of 10 mol of ATP.

If we take pyruvate rather than acetyl-CoA as the starting point, the equation for the TCA cycle is

$$\text{Pyruvate} + 4\,NAD^+ + FAD$$
$$+ \text{ADP (or GDP)} + P_i + 2\,H_2O \longrightarrow$$
$$3\,CO_2 + 4\,NADH + 2\,H^+$$
$$+ FADH_2 + \text{ATP (or GTP)} \quad \textbf{(12)}$$

and we see that the mitochondrial part of carbohydrate metabolism yields about 12.5 mol of ATP per mol of pyruvate or 25 mol per mol of glucose.

## Thermodynamics of the TCA Cycle

In the metabolic pathway for the oxidation of pyruvate four reactions occur (those catalyzed by pyruvate dehydrogenase, citrate synthase, isocitrate dehydrogenase, and $\alpha$-ketoglutarate dehydrogenase) for which the equilibrium constants are large. For four others (those catalyzed by aconitase, succinate thiokinase, succinate dehydrogenase, and fumarase), the equilibrium constant is fairly close to 1, and for one reaction (catalyzed by malate dehydrogenase), it is much less—about $10^{-5}$ (table 13.1). It seems strange that such an unfavorable reaction should occur in a sequence

(the TCA cycle) that has one of the largest fluxes in metabolism. This is possible only because the following reaction, catalyzed by citrate synthase, is so thermodynamically favorable. The equilibrium constant for this reaction is about $5 \times 10^5$.

The highly unfavorable free energy for the conversion of malate to oxaloacetate may actually serve an important function in facultative aerobes. In facultative aerobes operating in the absence of $O_2$ the Krebs cycle cannot operate in the normal way. In spite of this the organism still has a need for the intermediates for biosynthesis normally supplied by the Krebs cycle. This situation is remedied by operating the first part of the cycle to $\alpha$-ketoglutarate or succinyl-CoA in the forward direction while operating the last part of the cycle from oxaloacetate to succinate or succinyl-CoA in the reverse direction. In this way the organism is able to synthesize the intermediates for biosynthesis while balancing its needs for oxidation and reduction. The highly unfavorable free energy for the conversion of malate to oxaloacetate becomes a highly favorable free energy for the conversion of oxaloacetate to malate so that the reactions between oxaloacetate and succinyl-CoA have a favorable overall free energy for operating in the reverse direction (see table 13.1).

Before leaving the subject of thermodynamics it should be pointed out that most of the arguments in this section are based on standard free energy data rather than the actual free energies which would require knowing the concentrations of reactants and products in the mitochondria. This is a serious limitation we must live with until there is a reliable means for estimating these concentrations.

## The Amphibolic Nature of the TCA Cycle

A major function of the TCA cycle is the oxidation of acetate to $CO_2$ with concomitant conservation of the energy of oxidation as reduced coenzymes and eventually as ATP. Strictly speaking, then, the TCA cycle has but a single substrate, acetyl-CoA. In most cells, however, there is considerable flux of four-, five-, and six-carbon intermediates into and out of the cycle, which occurs in addition to the catabolic function of the cycle. Such branchpoint reactions serve two main purposes: (1) to provide for the synthesis of compounds derived from any of several intermediates of the cycle, and (2) to replenish and augment the supply of intermediates in the cycle as needed. Because the TCA cycle can function both in a catabolic mode and as a source of precursors for anabolic pathways, it is often called an amphibolic pathway (from the Greek, *amphi,* meaning ''both''). Some of the main biosynthetic pathways that begin with intermediates in the TCA cycle, as well as the ways in which the supply of intermediates in the cycle are replenished, are indicated in figure 13.12. We discuss these pathways briefly here; they are discussed in greater detail in subsequent chapters (for fats, see chapters 18, 19, and 20, for amino acids, see chapters 21 and 22).

Four intermediates in the TCA cycle serve as starting points for compounds outside of the cycle. Oxaloacetate and $\alpha$-ketoglutarate are used in the synthesis of several amino acids, succinyl-CoA is used in heme synthesis; and citrate is the source of the acetyl-CoA in the cytosol, which is used for the synthesis of fats and other lipids and some amino acids. These are the major drains on the TCA cycle.

Reactions that replenish the intermediates in the TCA cycle are termed anaplerotic, from a Greek root that means ''filling up.'' It is not necessary to replenish the intermediate that is used in a biosynthetic pathway directly, because any intermediate can be replenished by a feeding-in process from any point in the cycle.

When the carbohydrates are being metabolized, TCA cycle intermediates are replenished by production of oxaloacetate from pyruvate. In mammals, this reaction is catalyzed by pyruvate carboxylase, and one ATP-to-ADP conversion is associated with the carboxylation. Other properties of this reaction are discussed later in this chapter in connection with regulation of the TCA cycle and related metabolic sequences.

In prokaryotic organisms and some eukaryotes, oxaloacetate is fed into the cycle by carboxylation of phosphoenolpyruvate. Energetically, the carboxylation of phosphoenolpyruvate, which is catalyzed by phosphoenolpyruvate carboxylase, is equivalent to the sum of the pyruvate kinase and pyruvate carboxylase reactions, which are used by mammals.

$$H^+ + \text{Phosphoenolpyruvate} + \text{ADP} \longrightarrow \text{Pyruvate} + \text{ATP} + H_2O \quad (13)$$

$$H_2O + \text{Pyruvate} + \text{ATP} + CO_2 \longrightarrow \text{Oxaloacetate} + \text{ADP} + P_i + H^+ \quad (14)$$

Sum: $\text{Phosphoenolpyruvate} + CO_2 \longrightarrow \text{Oxaloacetate} + P_i \quad (15)$

## The Glyoxylate Cycle Permits Growth on a Two-Carbon Source

Usually, condensation of acetyl-CoA with oxaloacetate to form citrate seals the metabolic fate of the acetyl carbons—the inevitable result of the TCA cycle is their oxidation and eventual release as $CO_2$. However, in certain species of bacteria, protozoa, fungi, and plants, the formation of citrate is also a starting point for formation of sugars and other cellular components from two-carbon compounds. The pathway that permits this to happen is known as the glyoxylate cycle. The glyoxylate cycle bypasses the two steps of the TCA cycle in which $CO_2$ is released. Furthermore, two molecules of acetyl-CoA are taken in per turn of the cycle, rather than just one as in the TCA cycle. The net result is the conversion of two molecules of two carbons each (i.e., the acetate of acetyl-CoA) into one four-carbon compound, succinate.

Comparison of the glyoxylate cycle with the TCA cycle reveals that two of the five reactions of the glyoxylate cycle are unique to this cycle, whereas the other three reactions are common to both cycles (fig. 13.13). In plant seedlings and many other eukaryotic organisms that possess this capability, the enzymes of the glyoxylate cycle are compartmentalized in specialized organelles called glyoxysomes.

## Oxidation of Other Substrates by the TCA Cycle

The TCA cycle, strictly speaking, has only one input fuel—acetyl-CoA. Catabolism of carbohydrates and fats leads to the production of acetyl-CoA, so that the TCA cycle is ideally suited to serve as the major oxidative sequence in the catabolism of those types of compounds. However, degradation of the amino acids that result from the hydrolysis of protein produces a number of intermediates, among which are $\alpha$-ketoglutarate, succinyl-CoA, and oxaloacetate (chapter 22). $\alpha$-Ketoglutarate and succinyl-CoA, can be oxidized

# Figure 13.12

The TCA cycle, showing some of the branchpoint pathways (in red) that either drain or replenish the TCA intermediates.

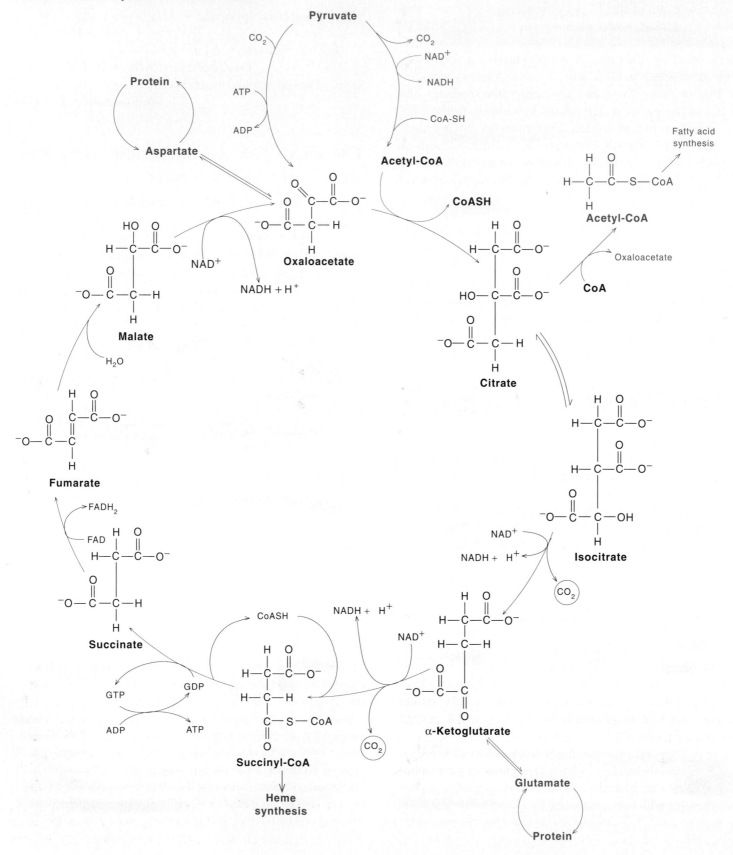

## Figure 13.13

Comparison of the TCA cycle and the glyoxylate cycle. In the TCA cycle one molecule of acetyl-CoA is oxidized to two molecules of $CO_2$. In the glyoxylate cycle (red) two molecules of acetyl-CoA are converted to one molecule of oxaloacetate. As indicated, the glyoxylate cycle uses some of the enzymes of the TCA cycle. Only enzymes operative in the glyoxylate cycle are shown. In plant cells, enzymes of the glyoxylate cycle are located in specialized organelles called glyoxysomes. In yeasts and other eukaryotic organisms, these enzymes are located in the cytosol and are structurally distinct molecules called isozymes.

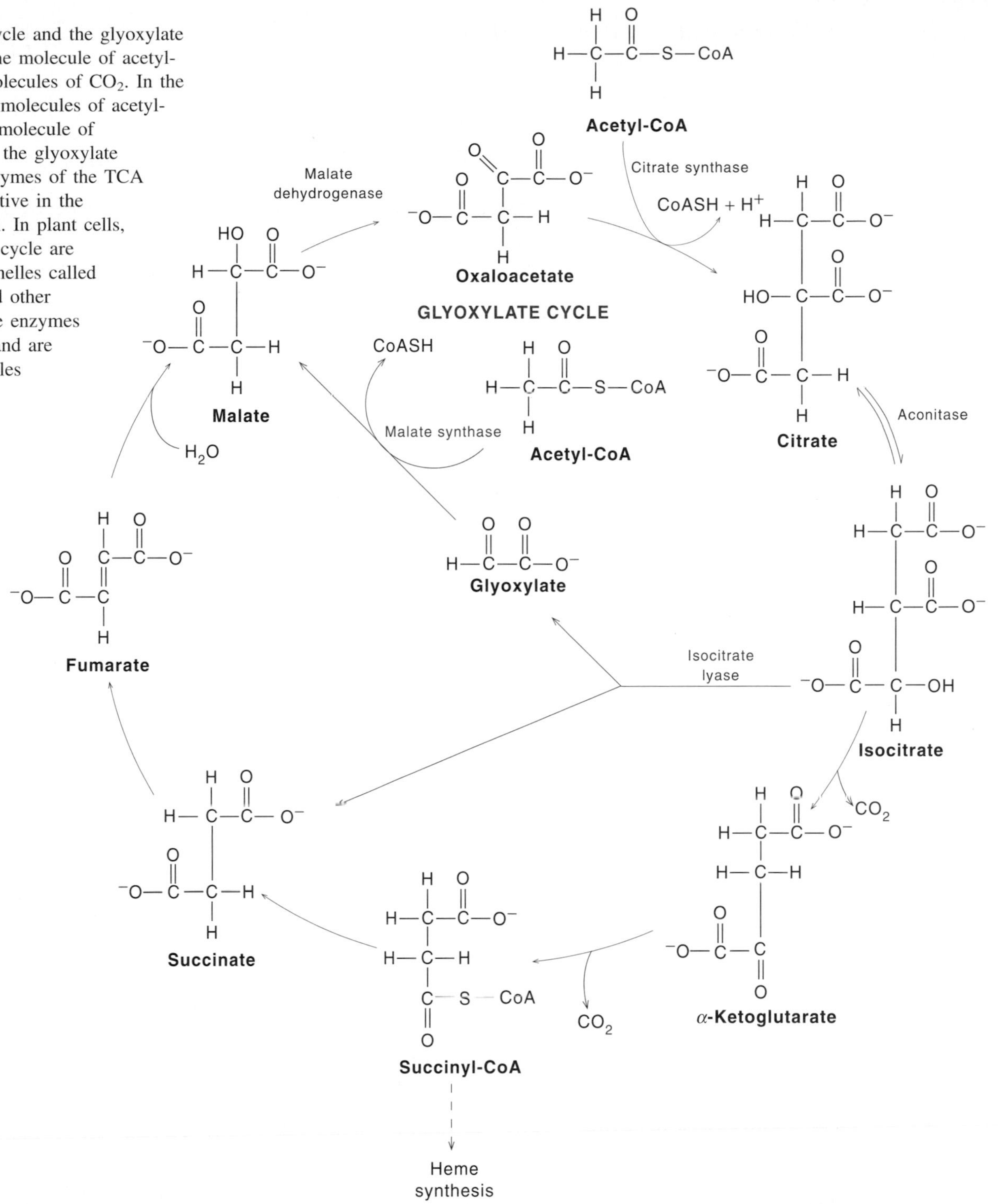

to oxaloacetate, but the cycle as such cannot oxidize oxaloacetate further. In the presence of acetyl-CoA, each molecule of oxaloacetate used in the synthesis of citrate is regenerated in the cycle; thus, no net oxidation of oxaloacetate occurs in that case either.

The problem of how to oxidize oxaloacetate is solved by the action of phosphoenolpyruvate carboxykinase, which we discussed in connection with gluconeogenesis (see equa-

tion 10 in chapter 12). This enzyme catalyzes the conversion of oxaloacetate to phosphoenolpyruvate with the help of ATP or GTP, and thus permits the total oxidation of oxaloacetate to $CO_2$ by the enzymes of the TCA cycle.

In addition to the amino acids that are converted to intermediates of the TCA cycle, others are converted to pyruvate or acetyl-CoA and thus enter the cycle in the usual way. In fact, all 20 of the protein amino acids are metabo-

# Figure 13.14

Major regulatory sites of the TCA cycle, with activators (⊕) and inhibitors (⊖) of specific reactions shown in red.

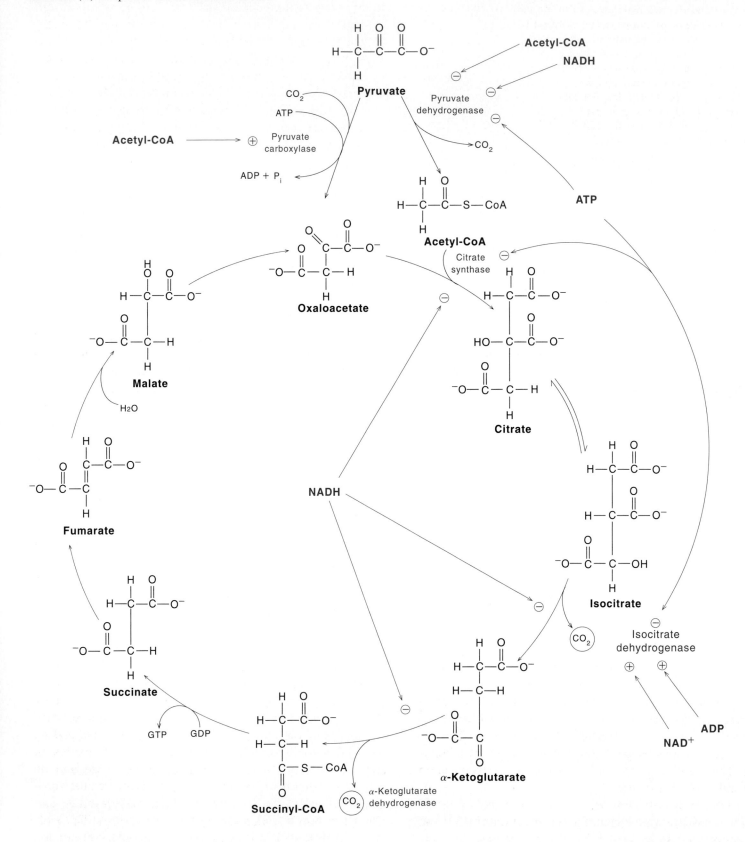

lized by way of the TCA cycle (see fig. 22.1). Thus, although it is often thought of as part of the pathway for carbohydrate metabolism, the TCA cycle is actually the central oxidative sequence for all three of the major types of carbon and energy sources: carbohydrates, fats, and proteins.

## The TCA Cycle Activity Is Regulated at Metabolic Branchpoints

The TCA cycle is carefully regulated to ensure that its level of activity relates to cellular needs. The cycle serves two functions: (1) furnishing reducing equivalents (as NADH and to a lesser extent as $FADH_2$) to the electron-transport chain, and (2) by means of side reactions, providing substrates for biosynthesis reactions. Both of these functions are reflected in the regulation of the cycle.

In its primary role as a means of oxidizing acetyl groups to carbon dioxide and water, the TCA cycle is sensitive both to the availability of its substrate, acetyl-CoA, and to the accumulated levels of its principal end products, NADH and ATP. Actually the ratio of NADH to $NAD^+$ and the energy charge or the ATP-to-ADP ratio are more important than the individual concentrations. Other regulatory parameters to which the TCA cycle is sensitive include the ratios of acetyl-CoA to free CoA, acetyl-CoA to succinyl-CoA, and citrate to oxaloacetate. The major known sites for regulation are shown in figure 13.14. These include two enzymes outside the TCA cycle (pyruvate dehydrogenase and pyruvate carboxylase) and three enzymes inside the TCA cycle (citrate synthase, isocitrate dehydrogenase, and $\alpha$-ketoglutarate dehydrogenase). As might be suspected, each of these sites of regulation represents an important metabolic branchpoint.

### The Pyruvate Branchpoint Partitions Pyruvate between Acetyl-CoA and Oxaloacetate

Acetyl-CoA is the only compound that can enter the TCA cycle when the cycle is operating purely oxidatively, but one molecule of oxaloacetate must enter for each molecule of citrate, $\alpha$-ketoglutarate, or succinyl-CoA that is removed for use in biosynthesis. It follows that pyruvate is a major metabolic branchpoint in a cell that is living on carbohydrate. The partitioning of pyruvate between decarboxylation to acetyl-CoA and carboxylation to oxaloacetate is, in effect, partitioning between the two major metabolic uses of pyruvate: oxidation of carbon for regeneration of ATP and conversion to starting materials for biosynthesis.

Some of the regulatory interactions that affect partitioning at pyruvate are shown in figure 13.14. We focus first on the effects of acetyl-CoA, which is a negative modifier for pyruvate dehydrogenase and a very strong positive modifier for pyruvate carboxylase. To illustrate the regulatory roles of those effects, consider a mitochondrion in which the TCA cycle is functioning only oxidatively, so that no input of oxaloacetate is required. If conditions change and $\alpha$-ketoglutarate begins to be removed for biosynthesis, the rate at which oxaloacetate is regenerated by the cycle is reduced by the rate of $\alpha$-ketoglutarate removal. The citrate synthase reaction can proceed no more rapidly than the rate at which oxaloacetate, one of its substrates, is supplied, so the rate of citrate synthesis, too, decreases. Consequently, acetyl-CoA is used more slowly, causing its concentration to increase slightly. The increase in acetyl-CoA concentration leads to a decrease in the activity of the pyruvate dehydrogenase complex and an increase in the activity of pyruvate carboxylase. As a result of those effects, the partitioning of pyruvate is changed so as to produce more oxaloacetate, thus replenishing the cycle and allowing the cycle to continue to function in spite of the removal of $\alpha$-ketoglutarate.

Since the system responds to the concentration of acetyl-CoA and thus indirectly to the concentration of oxaloacetate, it adjusts automatically to maintain a functional concentration of oxaloacetate regardless of whether the intermediate removed from the cycle is citrate, $\alpha$-ketoglutarate, succinyl-CoA, oxaloacetate itself, or any combination of these. The same effects cause the rate of conversion of pyruvate to oxaloacetate to decrease when the rate of removal of biosynthetic precursors decreases.

The negative effects of acetyl-CoA on pyruvate dehydrogenase activity are supplemented by ATP and NADH. These effects are in the correct direction to cause the rate of acetyl-CoA synthesis to vary with the need for electrons and for regeneration of NADH and ATP.

In addition to regulation by these small-molecule modifiers, the pyruvate dehydrogenase complex, at least in mammals, is subject to regulation by covalent modification. Each of the giant complexes contains several molecules of a protein kinase and a protein phosphorylase. The kinase catalyzes phosphorylation of specific serine hydroxyl groups of the pyruvate decarboxylase portion of the complex, and the phosphorylase catalyzes hydrolytic removal of these phosphoryl groups. The phosphorylated enzyme is relatively inactive. Thus, the phosphorylation–dephosphorylation system also contributes to regulation of the rate of conversion of pyruvate to acetyl-CoA. The action of the kinase, and the resultant decrease in the activity of the pyruvate dehydrogenase complex, is favored by high ATP-to-ADP and NADH-to-$NAD^+$ ratios and by high acetyl-CoA. These effects act in the same direction as the direct effects of the

**Figure 13.15**

Cycling between mitochondria and cytosol is the supply of acetyl-CoA for use in biosynthetic sequences in the cytosol. Citrate moves from the interior of the mitochondria to the cytosol. In the cytosol it is cleaved to acetyl-CoA and oxaloacetate by citrate lyase. The equilibrium constant for this reaction is favorable because an ATP-to-ADP conversion is involved. Most of the oxaloacetate is reduced to malate. The malate may be taken up by mitochondria or oxidized to pyruvate and $CO_2$, generating NADPH for use in biosynthetic sequences in the cytosol. The pyruvate enters the mitochondria, where it may be converted to oxaloacetate or acetyl-CoA by the usual routes (see fig. 13.14).

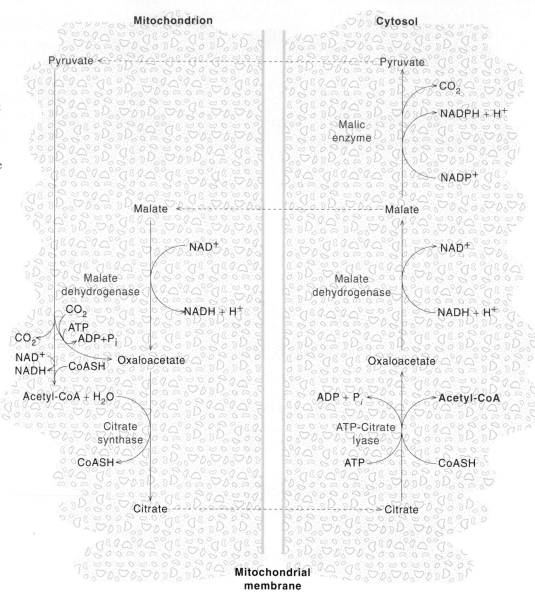

modifier metabolites on the enzymes of the complex, so the two types of regulation reinforce one another.

## Citrate Synthase Is Negatively Regulated by NADH and the Energy Charge

Thus far we have discussed the two most important enzymes regulating the supplies of acetyl-CoA and oxaloacetate for the TCA cycle. This leaves us with the three enzymes, all within the cycle, that regulate the activity of the cycle. The first of these, citrate synthase, catalyzes the formation of citrate from acetyl-CoA and oxaloacetate. Regulation, by energy charge and other parameters, of the rate of glycolysis and of the pyruvate dehydrogenase reaction play

important roles in controlling the rate of citrate synthesis. For example, yeast citrate synthase has been shown to respond sensitively to variation in the value of the energy charge. Synthesis of citrate is also favored by a low NADH-to-$NAD^+$ ratio. This is clearly metabolically desirable, since the operation of the cycle leads to an increase in this ratio.

## Isocitrate Dehydrogenase Is Regulated by the NADH-to-$NAD^+$ Ratio and the Energy Charge

The equilibrium constant for the conversion of citrate to isocitrate is small, and the interconversion is rapid, so these two intermediates make up a metabolic pool. Since regulation is not expected within a metabolic pool for reasons discussed in chapters 10 and 11, the next possible regulatory site as we go around the TCA cycle is the conversion of isocitrate to $\alpha$-ketoglutarate. This reaction is catalyzed by isocitrate dehydrogenase, which is a highly regulated enzyme. In yeast the activity of isocitrate dehydrogenase is strongly cooperative (sigmoidal) with respect to the isocitrate concentration. The enzyme is stimulated by $NAD^+$ and AMP and inhibited by NADH, making it very sensitive to the NADH-to-$NAD^+$ ratio. The properties of the mammalian enzyme are similar to those of the yeast enzyme, with the exception that ADP replaces AMP as a positive modifier.

To understand why isocitrate dehydrogenase is so intensely regulated we must consider reactions beyond the TCA cycle, and indeed beyond the mitochondrion (fig. 13.15). Of the two compounds citrate and isocitrate, only citrate is transported across the barrier imposed by the mitochondrial membrane. Citrate that passes from the mitochondrion to the cytosol plays a major role in biosynthesis, both because of its immediate regulatory properties and because of the chain of covalent reactions it initiates. In the cytosol citrate undergoes a cleavage reaction in which acetyl-CoA is produced. The other cleavage product, oxaloacetate, can be utilized directly in various biosynthetic reactions or it can be converted to malate. The malate so formed can be returned to the mitochondrion, or it can be converted in the cytosol to pyruvate, which also results in the reduction of $NADP^+$ to NADPH. The pyruvate is either utilized directly in biosynthetic processes, or like malate, can return to the mitochondrion.

The acetyl-CoA produced in the cytosol from citrate breakdown is used in biosynthetic reactions, including the synthesis of lipids, some amino acids, cofactors, and pigments. This is the only source of acetyl-CoA in the cytosol because the acetyl-CoA produced in the mitochondrion cannot diffuse across the mitochondrial membrane. The NADPH produced indirectly in the cytosol from the citrate is also of great importance in biosynthetic reactions.

In addition to its importance in providing cytosolic acetyl-CoA and NADPH, citrate also serves as a major regulator of the rate of fatty acid synthesis. As we shall see (chapter 18) citrate is a strong positive modifier of the first reaction in fatty acid synthesis. It should be remembered (see chapter 12) that citrate also is a negative modifier of phosphofructokinase and thereby exerts a negative effect on glycolysis, which also occurs in the cytosol.

The effect of small-molecule modifiers on isocitrate dehydrogenase is appropriate in that a high-energy charge favors inhibition of isocitrate dehydrogenase and thus favors an accumulation of mitochondrial citrate. This leads to an increased flow of citrate from the mitochondrion to the cytosol where the citrate can exert its multiple positive effects on biosynthesis and its negative effects on glycolysis.

## $\alpha$-Ketoglutarate Dehydrogenase Is Negatively Regulated by NADH

$\alpha$-Ketoglutarate is also a branchpoint metabolite, since it can be transaminated to form glutamate (see fig. 13.12 and chapter 21). Glutamate is needed in protein synthesis directly and also is a precursor of a number of other amino acids and peptides. Thus, it is consumed in the cytosol in large amounts in a cell that is synthesizing protein rapidly. We expect that the reaction catalyzed by an $\alpha$-ketoglutarate dehydrogenase is regulated so as to retain carbon in the TCA cycle when energy is in short supply. Conversely, when the energy supply is high it is advantageous to allow the concentration of $\alpha$-ketoglutarate to rise, facilitating the formation of glutamate and exit from the cycle. The balance is achieved by negative regulation of $\alpha$-ketoglutarate dehydrogenase in response to NADH. When NADH is low, the enzyme competes more strongly for $\alpha$-ketoglutarate and thus maximizes the regeneration of ATP while tending to decrease biosynthetic activities that use ATP. No adequate test of the effect of adenine nucleotides on this enzyme has been reported.

## Summary

This chapter is mainly concerned with the contribution of the tricarboxylic acid cycle to carbohydrate metabolism. The TCA cycle is the main source of electrons for oxidative phosphorylation, and thereby the major energetic sequence in the metabolism of aerobic cells or organisms. It serves as the main distribution center of metabolism, receiving carbon from the degradation of carbohydrates, fats, and proteins and, when it is appropriate, supplying carbon compounds for the synthesis of carbohydrates, fats, or proteins. Every aspect of the metabolism of an aerobic organism is directly dependent on the TCA cycle.

1. The TCA cycle begins with acetyl-CoA, which is obtained either by oxidative decarboxylation of pyruvate available from glycolysis or by oxidative cleavage of fatty acids.

2. The acetyl-CoA transfers its acetyl group to oxaloacetate, thereby generating citrate. In a cyclic series of reactions, the citrate is subjected to two successive decarboxylations and four oxidative events, leaving a four-carbon compound malate from which the starting oxaloacetate is regenerated.

3. Only a single ATP is directly generated by a turn of the TCA cycle. Most of the energy produced by the cycle is stored in the form of reduced coenzyme molecules, NADH and $FADH_2$. Reoxidation of these compounds (see chapter 14) liberates a large amount of free energy, which is captured in the form of ATP.

4. Some of the main biosynthetic pathways begin with intermediates in the TCA cycle. When intermediates in the cycle are used as starting materials for biosynthesis, they must be replenished to keep the cycle operating. When carbohydrates are being metabolized, TCA cycle intermediates are replenished by production of oxaloacetate from pyruvate.

5. The glyoxylate cycle permits growth on a two-carbon source. The glyoxylate cycle bypasses the two steps of the TCA cycle in which $CO_2$ is released. Furthermore, two molecules of acetyl-CoA are taken in per turn of the cycle rather than just one, as in the TCA cycle. The net result is the conversion of two molecules of two carbons each into one four-carbon compound, succinate. Two additional enzymes are needed for operation of the glyoxylate cycle: isocitrate lyase and malate synthase.

6. Degradation of amino acids produces a number of intermediates, among which are $\alpha$-ketoglutarate, succinyl-CoA, and oxaloacetate. $\alpha$-Ketoglutarate and succinyl-CoA can be oxidized to oxaloacetate, but the cycle as such cannot oxidize oxaloacetate further. Oxaloacetate is oxidized further by first converting it to phosphoenolpyruvate. This permits the total oxidation of oxaloacetate to $CO_2$ by the enzymes of the TCA cycle.

7. The TCA cycle is regulated to ensure that its level of activity corresponds closely to cellular needs. In its primary role as a means of oxidizing acetyl groups to $CO_2$ and water, the TCA cycle is sensitive both to the availability of its substrate, acetyl-CoA, and to the accumulated levels of its principal end products, NADH and ATP. Other regulatory parameters to which the TCA cycle is sensitive include $NAD^+$, ADP, acetyl-CoA, succinyl-CoA, and citrate. The major known sites for regulation of the cycle include two enzymes outside the cycle (pyruvate dehydrogenase and pyruvate carboxylase) and three enzymes inside the cycle (citrate synthase, isocitrate dehydrogenase, and $\alpha$-ketoglutarate dehydrogenase). All of these sites of regulation represent important metabolic branchpoints.

## Selected Readings

Atkinson, D. E., *Cellular Energy Metabolism and Its Regulation.* New York: Academic Press, 1977. A solid, detailed discussion of energy metabolism by an author who is prominent in this area.

Broda, E., *The Evolution of Bioenergetic Processes.* New York: Pergamon Press, 1975. An excellent discussion of cellular energetics from an evolutionary perspective.

Gest, H., Evolutionary roots of the citric acid cycle in prokaryotes. *Biochem. Soc. Symp.* 54:3–16, 1987. A fascinating recount of how the evolution of the cycle is traced by studying the way in which enzymes of the cycle are used in present day microorganisms.

Hers, H. G., and Hue, L., Gluconeogenesis and related aspects of glycolysis. *Ann. Rev. Biochem.* 52:617–653, 1983.

Krebs, H. A., The history of the tricarboxylic acid cycle. *Perspect. Biol. Med.* 14:154, 1970. An engaging historical account by the man who masterminded much of it.

Mehlman, M., and Hanson, R. W., *Energy Metabolism and the Regulation of Metabolic Processes in Mitochondria.* New York: Academic Press, 1972. The title speaks for itself—a good reference.

Williamson, J. R., and Cooper, R. V., Regulation of the citric acid cycle in mammalian systems. *FEBS Lett.* 117(Suppl.):K73, 1980. A well-rounded review of TCA cycle regulation as a contemporary research theme from a symposium dedicated to Hans Krebs.

## Problems

1. In the conversion of isocitrate to $\alpha$-ketoglutarate, oxidation and decarboxylation steps occur. Figure 13.9 shows the oxidation step first. Can you suggest a reason why the oxidation step is first?

2. In the late 1930s the tricarboxylic acid cycle (fig. 13.3) contained cis-aconitate, but more modern versions of the cycle (since the late 1950s) do not. The involvement of cis-aconitate is still shown in many nonbiochemistry texts. Can you explain why cis-aconitate is deleted from modern versions of the tricarboxylic acid cycle?

3. Notice in the initial steps of the tricarboxylic acid cycle, citrate is converted to isocitrate, which is then oxidized. Why didn't nature "just" oxidize citrate and save an enzymatic step?

4. Ethylene glycol (ethane-1,2-diol), a major component of antifreeze, is readily metabolized by many organisms that have the glyoxylate cycle. Can you propose a sequence of catabolic steps that can explain the catabolism of ethylene glycol?

5. Many organisms can live on glutamate as their sole carbon and nitrogen source. Assuming glutamate is converted into $\alpha$-ketoglutarate, produce a scheme that completely oxidizes glutamate.

6. Notice the intermediate in the reaction of citrate synthase (fig. 13.7). Do you think at some time in the future evolution will produce a variety of citrate synthase that recovers the energy in the thioester, analogous to the production of GTP (ATP) by succinate thiokinase (page 291)? Would this energy recovery have any effect on the thermodynamics of the tricarboxylic acid cycle?

7. Summarize in the simplest words the portion of the tricarboxylic acid cycle that is bypassed by the glyoxylate cycle?

8. The reactions catalyzed by isocitrate dehydrogenase and $\alpha$-ketoglutarate dehydrogenase are both oxidative decarboxylation reactions. How similar are the reactions?

9. What effect would the following have on the control of the citric acid cycle: (a) a sudden influx of acetyl-CoA from the degradation of fatty acids, and (b) a sudden need for heme biosynthesis?

10. Do all of the metabolites associated with the tricarboxylic acid cycle have free access across the mitochondrial membrane?

11. Using your knowledge of metabolism, determine whether the following statements are true or false and explain the reasoning behind your decision.
    (a) Dihydrolipoamide dehydrogenase catalyzes the only oxidation–reduction reaction in the pyruvate dehydrogenase complex.
    (b) Hydrolysis of the thioester bond of acetyl-CoA yields insufficient energy to drive phosphorylation of ADP.
    (c) The methyl group of each acetyl-CoA molecule entering the TCA cycle is derived from the methyl group of pyruvate.
    (d) Even if aconitase were unable to discriminate between the two ends of the citrate molecule, the $CO_2$ released would still come from the oxaloacetate rather than the acetyl-CoA substrate of the citrate synthase reaction.
    (e) Malate cannot be converted to fumarate because the TCA cycle is unidirectional.

12. Assume that you have a buffered solution containing pyruvate dehydrogenase and all the enzymes of the TCA cycle but none of the cycle intermediates.
    (a) If you add 3 $\mu$moles each of pyruvate, CoASH, $NAD^+$, GDP, and $P_i$, how much $CO_2$ evolves? What other products form?
    (b) In addition to the reagents in (a), you add 3 $\mu$moles each of the TCA cycle intermediates. How much $CO_2$ evolves? Explain.
    (c) If you were to add an electron acceptor that reoxidized NADH to the system described in (a), would there be increased $CO_2$ evolution? Why or why not?
    (d) Explain the effect on $CO_2$ evolution of adding the NADH-reoxidizing system to the system described in (b), assuming that you also added excess GDP and $P_i$?

13. What would you expect to be the metabolic consequences of the following mutations in yeast?
    (a) Inability to synthesize malate synthase.
    (b) Pyruvate carboxylase that is not activated by acetyl-CoA.
    (c) Pyruvate dehydrogenase that is inhibited by acetyl-CoA more strongly than is the wild-type enzyme.

14. The substrate hydroxypyruvate

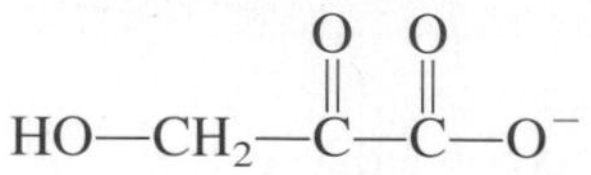

$$HO-CH_2-\overset{\overset{\displaystyle O}{\|}}{C}-\overset{\overset{\displaystyle O}{\|}}{C}-O^-$$

Hydroxypyruvate

is metabolized to pyruvate in a five-step process requiring the four intermediates whose structures are shown here.

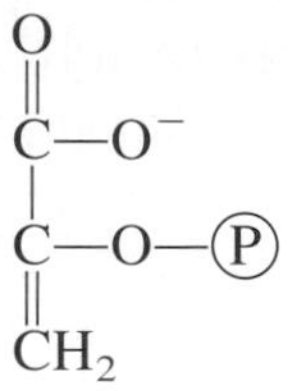

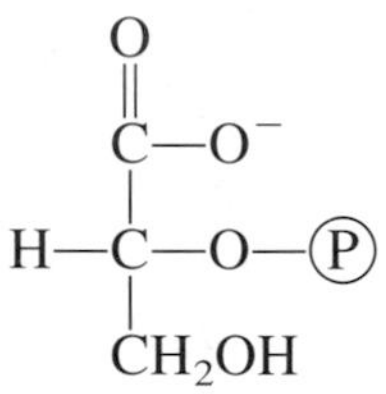

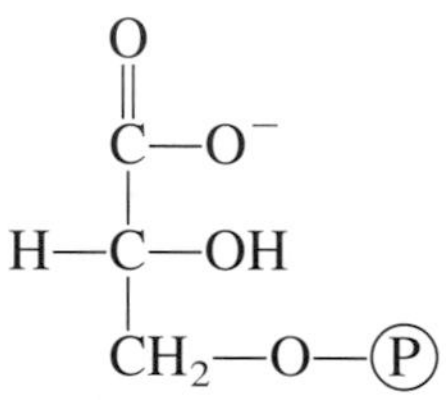

$$\begin{array}{c} \overset{\overset{\displaystyle O}{\|}}{C}-O^- \\ \overset{\overset{}{|}}{C}-O-\text{P} \\ \overset{\overset{}{\|}}{CH_2} \\ A \end{array} \qquad \begin{array}{c} \overset{\overset{\displaystyle O}{\|}}{C}-O^- \\ H-\overset{\overset{}{|}}{C}-OH \\ \overset{\overset{}{|}}{CH_2OH} \\ B \end{array}$$

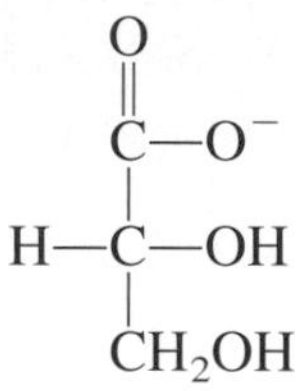

$$\begin{array}{c} \overset{\overset{\displaystyle O}{\|}}{C}-O^- \\ H-\overset{\overset{}{|}}{C}-O-\text{P} \\ \overset{\overset{}{|}}{CH_2OH} \\ C \end{array} \qquad \begin{array}{c} \overset{\overset{\displaystyle O}{\|}}{C}-O^- \\ H-\overset{\overset{}{|}}{C}-OH \\ \overset{\overset{}{|}}{CH_2-O-\text{P}} \\ D \end{array}$$

The letter designating the intermediate does not necessarily reflect the order in which it is used. The metabolic conversion requires NADH and catalytic quantities of both ATP and ADP. Assume that the pathway begins with an NADH-mediated reaction.

(a) Designate the order in which the intermediates are used in the metabolism of hydroxypyruvate to pyruvate.

(b) Write an overall equation for the metabolism of hydroxypyruvate to pyruvate.

(c) Name each of the intermediates (A–D) and indicate which are intermediates in the glycolytic pathway.

(d) Explain why only catalytic rather than stoichiometric amounts of ADP and ATP are required in the pathway.

15. Under anaerobic conditions, *E. coli* synthesizes an NADH-dependent fumarate reductase rather than succinate dehydrogenase, the flavoprotein that oxidizes succinate to fumarate.

(a) Write an equation for the reaction catalyzed by fumarate reductase.

(b) NADH produced by the glyceraldehyde-3-phosphate dehydrogenase reaction is reoxidized by reducing an organic intermediate. Rather than reduce pyruvate to lactate, anaerobic *E. coli* utilize the fumarate reductase. However, under anaerobiosis, the activity of $\alpha$-ketoglutarate dehydrogenase is virtually nonexistent. Show how fumarate is formed, using reactions beginning with PEP and including the necessary TCA cycle enzymes. (Spiro, S., and J. R. Guest, *TIBS*. 16:310–314 (1991).)

(c) What is the metabolic advantage to anaerobic *E. coli* in using the fumarate reductase pathway rather than lactate dehydrogenase to reoxidize NADH?

16. Consider the glyceraldehyde-3-phosphate dehydrogenase-phosphoglycerokinase enzymes of glycolysis and the succinate thiokinase of the TCA cycle. Compare the mechanisms of incorporation of inorganic phosphate into the respective nucleoside diphosphates.

17. The pyruvate dehydrogenase complex may have been regulated by phosphorylation of any one of the three different enzymes in the complex, yet regulation occurs on the first enzyme of the complex. How is regulation of the complex consistent with the regulation observed in metabolic pathways whose enzymes are not physically associated?

18. Although there is no net synthesis of glucose from acetyl-CoA in mammals, acetyl-CoA has two major functions in gluconeogenesis. Explain the functions of acetyl-CoA in the synthesis of glucose from lactate in mammalian liver.

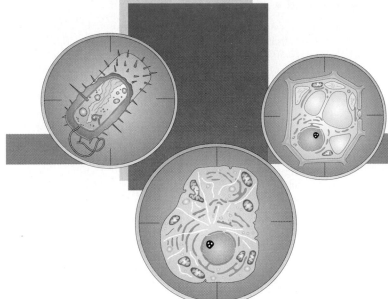

# Electron Transport and Oxidative Phosphorylation

*Potential energy stored in the reduced coenzymes, NADH and FADH₂, produced in the reactions of the TCA cycle is used to drive the synthesis of large amounts of ATP.*

I n the previous two chapters, we described the catabolism of glucose to pyruvate by glycolysis and the further breakdown of pyruvate to $CO_2$ and $H_2O$ in the TCA cycle. These are oxidative processes. As the carbohydrate chain is chopped into progressively smaller fragments, electrons are transferred to the coenzymes $NAD^+$ and FAD. Glucose catabolism involves six such oxidative reactions: one in glycolysis, one in the conversion of pyruvate to acetyl-CoA, and the remaining four in the TCA cycle. In five of these reactions, $NAD^+$ is reduced to NADH; in the sixth, FAD bound to succinate dehydrogenase is reduced to $FADH_2$. In later chapters we see that fatty acids and amino acids also are catabolized by oxidative pathways that use these same coenzymes as electron acceptors.

The continued availability of $NAD^+$ and FAD for use as electron acceptors depends on reoxidation of the reduced coenzymes. It is in this process of coenzyme reoxidation that molecular oxygen finally enters the picture, because the terminal electron acceptor for NADH and $FADH_2$ oxidation is $O_2$.

Reoxidation of NADH and $FADH_2$ at the expense of molecular oxygen can be summarized by the following overall reactions:

$$NADH + H^+ + \frac{1}{2}O_2 \longrightarrow NAD^+ + H_2O$$

$$\Delta G^{\circ\prime} = -52.6 \text{ kcal/mol} \quad \textbf{(1)}$$

$$FADH_2 + \frac{1}{2}O_2 \longrightarrow FAD + H_2O$$

$$\Delta G^{\circ\prime} = -43.4 \text{ kcal/mol} \quad \textbf{(2)}$$

The most important aspect of these reactions is the large amount of free energy they release. In fact, most of the free energy available from the oxidation of glucose is still present in the reduced coenzymes generated during glycolysis and the TCA cycle. Complete oxidation of 1 mol of glucose to $CO_2$ and $H_2O$ results in the formation of 10 mol of NADH and 2 mol of $FADH_2$, which release $10 \times 52.6 + 2 \times 43.4$, or a total of 613 kcal when they are reoxidized. Since the total $\Delta G^{\circ\prime}$ for glucose oxidation is about $-686$ kcal/mol, it is clear that about 90% of this free energy is tapped by the cell only when the reduced coenzymes are reoxidized. Cells that live aerobically use the transfer of electrons from NADH or $FADH_2$ to $O_2$ to drive the formation of ATP, and this process accounts for most of the ATP yield in aerobic metabolism.

Our task in this chapter is twofold: to understand how NADH and $FADH_2$ are reoxidized by the transfer of electrons to $O_2$ and to consider how cells couple these reactions to the synthesis of ATP. The first of these processes, electron transport, does not occur by a direct reaction of the coenzymes with $O_2$, but rather by a stepwise flow of electrons through a chain of intermediate electron carriers in the mitochondrial inner membrane (fig. 14.1). In the course of

## Figure 14.1

Overview of electron transport and oxidative phosphorylation. Multiprotein electron-transfer complexes in the mitochondrial inner membrane reoxidize the NADH and $FADH_2$ that are generated by the TCA cycle and other mitochondrial reactions. These complexes pass electrons stepwise to molecular oxygen, and this flow is coupled indirectly but tightly to the synthesis of ATP. As electrons move through some of the electron-transfer complexes, protons are taken up from the solution in the mitochondrial matrix space and released at the outer surface of the inner membrane, setting up a gradient of pH and electric potential across the membrane. Protons flow back into the mitochondrion through an ATP-synthase enzyme, which uses the free energy liberated from the proton influx to drive ATP synthesis.

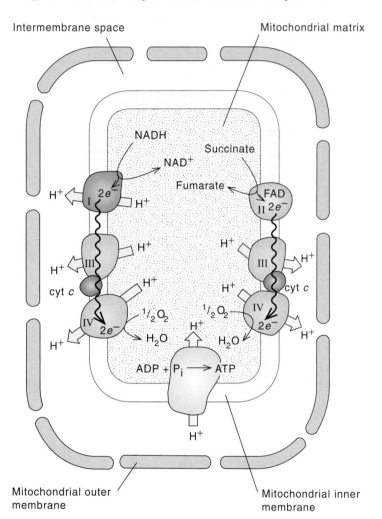

these reactions, protons are taken up from the solution on one side of the membrane and released on the opposite side. ATP synthesis is driven by the flow of protons back across the membrane. ATP synthesis that is linked to electron-transfer reactions in this way is called oxidative phosphorylation.

**Figure 14.2**

(*a*) Thin-section electron micrograph of a mitochondrion in a frog kidney cell. (Courtesy of Dr. J. Luft, University of Washington.) (*b*) Schematic cross-sectional view of a mitochondrion. The cristae are extensions of the inner membrane, which separates the matrix from the intermembrane space. The intermembrane space communicates with the cytosol through pores in the outer membrane.

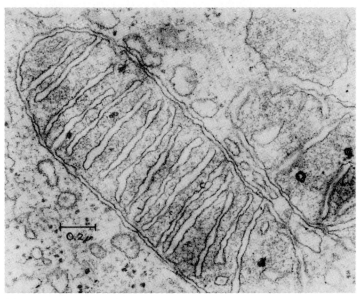

(a)

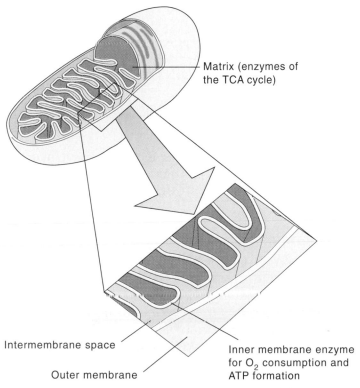

Matrix (enzymes of the TCA cycle)

Intermembrane space

Outer membrane

Inner membrane enzyme for $O_2$ consumption and ATP formation

(b)

# The Mitochondrial Electron-Transport Chain

In eukaryotic cells, electron transport and oxidative phosphorylation occur in mitochondria. Mitochondria have both an outer membrane and an inner membrane with extensive infoldings called cristae (fig. 14.2). The inner membrane separates the internal matrix space from the intermembrane space between the inner and outer membranes. The outer membrane has only a few known enzymatic activities and is permeable to molecules with molecular weights up to about 5,000. By contrast, the inner membrane is impermeable to most ions and polar molecules, and its proteins include the enzymes that catalyze oxygen consumption and formation of ATP. The role of mitochondria in $O_2$ uptake, or respiration, was demonstrated in 1913 by Otto Warburg but was not fully confirmed until 1948, when Eugene Kennedy and Albert Lehninger showed that mitochondria carry out the reactions of the TCA cycle, the transport of electrons to $O_2$, and the formation of ATP.

## *A Bucket Brigade of Molecules Carries Electrons from the TCA Cycle to $O_2$*

The electron carriers that participate in the flow of electrons to $O_2$ are a structurally diverse group. Occupying a central position are a series of heme-containing proteins, the cytochromes. Hemes are porphyrins with iron at the center (see chapter 10). Three main types of cytochromes, *a*, *b*, and *c*, exist and are distinguished by different substituents on the periphery of the porphyrin ring and different modes of attachment of the porphyrin to the protein (fig. 14.3). Cytochrome *c* is a small, water-soluble protein associated loosely with the inner mitochondrial membrane. Another *c* cytochrome ($c_1$), two *b* cytochromes ($b_L$ and $b_H$), and two *a* cytochromes (*a* and $a_3$) are embedded in the membrane as parts of large complexes described below.

The Fe atoms of the cytochromes undergo oxidation and reduction during respiration, cycling between the ferrous ($Fe^{2+}$) and ferric ($Fe^{3+}$) oxidation states. The absorption spectra of the oxidized and reduced forms differ (fig. 14.4). In the 1930s, David Keilin used this property to measure the oxidation–reduction states of cytochromes in living cells. Under anaerobic conditions, the cytochromes rapidly became reduced; in the presence of $O_2$, they became oxidized. Certain molecules that inhibited respiration (CO, $N_3^-$, or $CN^-$) blocked the oxidation; other inhibitors (amytal, rotenone, and malonate) blocked the reduction. Keilin found that the transfer of electrons from cytochrome *c* to $O_2$

## Figure 14.3

Iron protoporphyrin IX (heme) is found in the *b*-type cytochromes and in hemoglobin and myoglobin. In heme *c*, cysteine residues of the protein (R) are attached covalently by thioether links to the two vinyl (—CH=CH₂) groups of protoporphyrin IX. Heme *c* is found in the *c* cytochromes. In heme A, which is found in the *a* cytochromes, a 15-carbon isoprenoid side chain is attached to one of the vinyls, and a formyl group replaces one of the methyls.

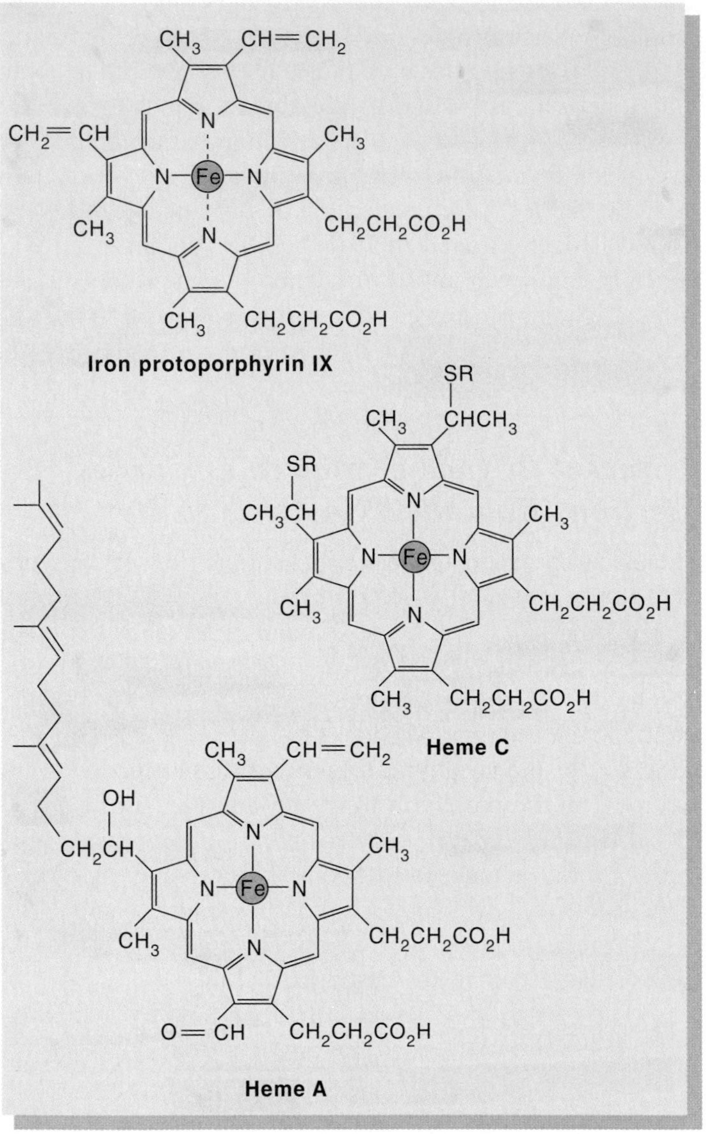

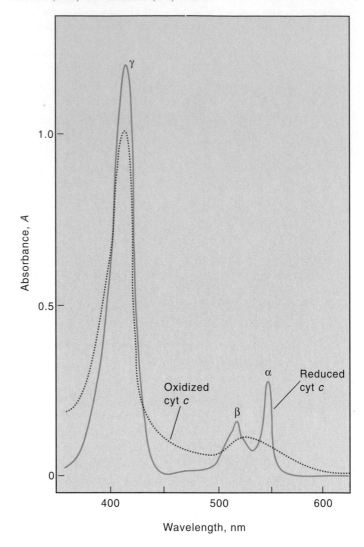

**Iron protoporphyrin IX**

**Heme C**

**Heme A**

## Figure 14.4

Optical absorption spectra of a 10 $\mu$M solution of cytochrome *c* in the reduced (blue) and oxidized (red) states.

The *b* cytochromes and cytochrome $c_1$ fit into this scheme between reducing substrates and cytochrome *c*. The idea thus developed that the respiratory apparatus includes a chain of cytochromes that operate in a defined sequence. The next question was whether the cytochromes are all bound together in a giant complex, or whether they diffuse independently in the membrane. Before we address this point, we need to consider three other types of electron carriers that participate in the electron-transport chain: flavoproteins, iron–sulfur proteins, and ubiquinone.

The dehydrogenases that remove electrons from succinate or NADH contain flavins as prosthetic groups. NADH

required another component, which he called "cytochrome oxidase." By 1940, it was clear that cytochrome oxidase was identical to an enzyme that Warburg had discovered and that it involved cytochromes *a* and $a_3$. The two *a* cytochromes appeared to work in series in passing electrons from cytochrome *c* to $O_2$.

$$\text{cytochrome } c \text{ (Fe}^{2+}) \quad \text{cytochrome } a \text{ (Fe}^{3+}) \quad \text{cytochrome } a_3\text{(Fe}^{2+}) \quad \frac{1}{4}O_2 + H^+$$

$$\text{cytochrome } c \text{ (Fe}^{3+}) \quad \text{cytochrome } a \text{ (Fe}^{2+}) \quad \text{cytochrome } a_3\text{(Fe}^{3+}) \quad \frac{1}{2}H_2O \qquad \textbf{(3)}$$

dehydrogenase contains flavin mononucleotide (FMN); succinate dehydrogenase contains covalently bound flavin adenine dinucleotide (FAD). Flavins can undergo one-electron reduction to semiquinone forms or two-electron reduction to the dihydroflavins, $FMNH_2$ and $FADH_2$ (see figs. 10.8 and 10.9).

The mitochondrial inner membrane also has a flavoprotein, glycerol-3-phosphate dehydrogenase, that oxidizes glycerol-3-phosphate to dihydroxyacetone phosphate, reducing a bound FAD. In the matrix space are at least eight other flavoprotein dehydrogenases, including enzymes that participate in the oxidation of fatty acids (see chapter 18). These dehydrogenases feed electrons to still another flavoprotein in the membrane.

NADH dehydrogenase and succinate dehydrogenase also contain Fe atoms that are bound by the S atoms of cysteine residues of the protein, in association with additional, inorganic sulfide atoms. Structures of these complexes are shown in figure 10.19. Succinate dehydrogenase has three iron–sulfur centers, one with a [2Fe–2S] cluster, one with [4Fe–4S], and one with a cluster containing 3 Fe atoms and 3 (or possibly 4) sulfides. Iron–sulfur centers undergo one-electron oxidation–reduction reactions.

The final electron carrier to be considered is ubiquinone (UQ), a benzoquinone with a long, hydrophobic side chain (fig. 14.5). The concentration of UQ in the mitochondrial inner membrane far exceeds that of the cytochromes. In heart mitochondria, for example, the concentration of UQ is about seven times that of cytochrome $a_3$. Because of its hydrophobic character, the UQ is able to move freely in the phospholipid bilayer of the membrane.

Ubiquinone undergoes a two-electron reduction to the dihydroquinone or quinol, $UQH_2$ (see fig. 14.5). It also can accept a single electron and stop at the semiquinone, which can be either anionic (UQ·⁻) or neutral (UQH·), depending on the pH and on the nature of the binding site when the semiquinone is bound to a protein. The · in UQ·⁻ or UQH· indicates that the semiquinone contains an unpaired electron.

The reduction of UQ can be measured by the disappearance of an absorption band at 275 nm. Using this technique, it was shown that adding a substrate such as succinate caused a rapid reduction of essentially all the UQ present in the inner membrane; the resulting $UQH_2$ could be reoxidized by the cytochrome system in the presence of $O_2$. To determine whether UQ is a *necessary* participant in electron transport from succinate to $O_2$, the quinone was removed from mitochondria by selective extraction with an organic solvent. The depleted mitochondria were incapable of respiration but recovered this activity when UQ was added back.

### Figure 14.5

Structures of ubiquinone in its oxidized state (UQ) and in the fully reduced dihydroquinone or quinol state ($UQH_2$). The UQ found in eukaryotes has 50 carbons (10 prenyl units) in the side chain; shorter side chains are found in some bacteria.

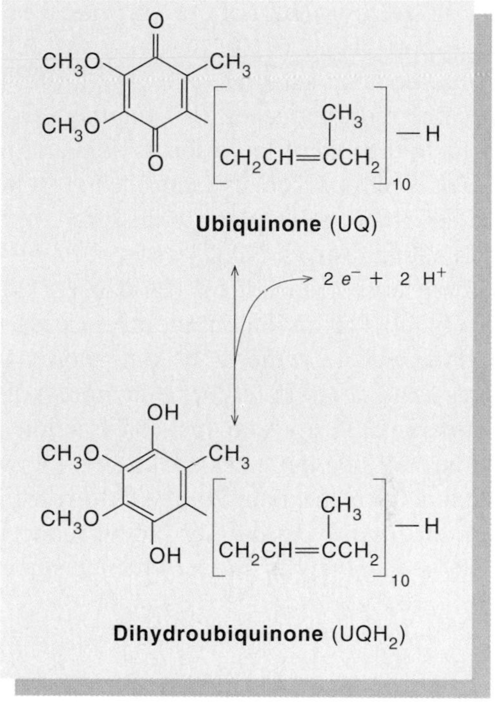

## The Sequence of Electron Carriers Was Deduced from Kinetic Measurements

With the help of spectrophotometric techniques, Britton Chance and others showed that any of the normal respiratory substrates can reduce all of the cytochromes in the membrane. Either succinate or NADH-linked substrates also can reduce all of the UQ. The flavins, however, are reduced only partially by succinate but almost completely by a combination of succinate and malate. These observations fit the idea that all of the cytochromes and UQ are part of a common network to which various flavoprotein dehydrogenases can feed electrons.

Chance and others also measured the rates at which the different electron carriers became oxidized, following addition of $O_2$ to anaerobic mitochondria, or became reduced after addition of a suitable substrate. When $O_2$ was added, cytochrome $a_3$ became oxidized first, followed by cytochrome $a$, the $c$ cytochromes, the $b$ cytochromes, and flavins, in that order. Later work showed that UQ became oxidized at about the same rate as the $b$ cytochromes. Observations of this sort led to the proposal that the electron carriers are arranged in the following sequence:

$$\text{succinate} \rightarrow \text{FAD}$$

$$\text{NADH} \rightarrow \text{FMN} \rightarrow [\text{UQ, cyt } b] \rightarrow [\text{cyt } c_1, \text{cyt } c] \rightarrow \text{cyt } a \rightarrow \text{cyt } a_3 \rightarrow O_2 \qquad \textbf{(4)}$$

This scheme was supported and refined by examining the effects of specific inhibitors of individual steps in the electron-transport chain. If CO or $CN^-$ was added in the presence of a reducing substrate and $O_2$, all of the electron carriers became more reduced. This fits the idea that these inhibitors act at the end of the respiratory chain, preventing the transfer of electrons from cytochrome to $O_2$. If amytal (a barbiturate) or rotenone (a plant toxin long used as a fish poison) was added instead, $NAD^+$ and the flavin in NADH dehydrogenase were reduced, but the carriers downstream became oxidized. The antibiotic antimycin caused $NAD^+$, flavins, and the $b$ cytochromes to become more reduced, but cytochromes $c$, $c_1$, $a$, and $a_3$ all became more oxidized. The situation here is analogous to the construction of a dam across a stream: When the gates are closed, the water level rises upstream from the dam, and falls downstream. The observation that antimycin did not inhibit reduction of UQ showed that the quinone fits into the chain upstream of cytochromes $c$, $c_1$, $a$, and $a_3$.

## Redox Potentials Give a Measure of Oxidizing and Reducing Strengths

We now see that mitochondria contain a variety of molecules—cytochromes, flavins, ubiquinone, and iron–sulfur proteins—all of which can act as electron carriers. To discuss how these carriers cooperate to transport electrons from reduced substrates to $O_2$, it is useful to have a measure of each molecule's tendency to release or accept electrons. The standard redox potential, $E°$, provides such a measure. Redox potentials are thermodynamic properties that depend on the differences in free energy between the oxidized and reduced forms of a molecule. Like the electric potentials that govern electron flow from one pole of a battery to another, $E°$ values are specified in volts. Because electron-transfer reactions frequently involve protons also, an additional symbol is used to indicate that an $E°$ value applies to a particular pH; thus, $E°'$ refers to an $E°$ at pH 7.

Consider the transfer of electrons from a reduced molecule, $D_{red}$, to an oxidized acceptor, $A_{ox}$. $\Delta G°$ for the reaction is proportional to the difference between the $E°$ values of the two molecules ($\Delta E°$).

$$D_{red} + A_{ox} \longrightarrow D_{ox} + A_{red} \qquad \textbf{(5)}$$

$$\Delta G° = -n\mathcal{F} \Delta E° = -n\mathcal{F}(E_A° - E_D°) \qquad \textbf{(6)}$$

Here $n$ is the number of electrons transferred from the donor to the acceptor, and $\mathcal{F}$ is the Faraday constant (23,060 cal $volt^{-1}$ $mol^{-1}$ or 96.5 kJ $volt^{-1}$ $mol^{-1}$). Note that the overall reaction is spontaneous ($\Delta G°$ negative) if $\Delta E°$ is positive, that is, if electrons move from the molecule with the more negative $E°$ value to that with the more positive value. In other words, negative $E°$ values are associated with strong reductants, and positive values with strong oxidants. Electrons flow spontaneously in the direction of more positive potential. With $n = 2$, a $\Delta E°$ of 0.1 V corresponds to a $\Delta G°$ of $-4.61$ kcal/mol.

$\Delta E°$ values can be measured by separating the two redox couples into different solutions connected by a salt bridge and a voltmeter (fig. 14.6). If both solutions are under standard conditions, the meter senses a voltage difference equal to the $\Delta E°$. Standard conditions are chosen to be 1 M concentrations of all reactants and products except for $H^+$, $OH^-$, and $H_2O$. By the law of mass action, a solution containing a redox couple $D_{ox}/D_{red}$ can be made more oxidizing by increasing the concentration ratio $[D_{ox}]/[D_{red}]$, or more reducing by decreasing this ratio.

Equation (6) defines the *difference* between two $E°$ values. To set the individual values, we need to choose a particular redox couple as a reference. The reference used most commonly by biochemists is the standard hydrogen half-cell, in which protons at pH 0 are reduced to $H_2$ at a pressure of 1 atm ($2 H^+ + 2 e^- \longrightarrow H_2$). By convention, this is assigned an $E°$ value of zero. Relative to the standard hydrogen half-cell, the $NAD^+/NADH$ couple at pH 7.0 has an $E°'$ of $-0.32$ V, and the pyruvate/lactate couple has an $E°'$ of $-0.19$ V. Table 14.1 gives the $E°'$ values of additional redox couples, including components of the respiratory chain. These values are listed in order of increasing $E°'$, which means that under standard conditions a given couple reduces any of the couples below it in the table.

The sequence of carriers in the respiratory chain should be generally consistent with the relative $E°$ values of the carriers because, given equal concentrations of reactants and products, electrons flow from a carrier with a more negative $E°$ value to one with a more positive value. The sequence of flavoproteins, UQ, and cytochromes that we presented in equation (4) agrees with this expectation. This proposition is demonstrated by plotting the $E°'$ values as a function of the carriers' suggested positions in the chain (fig. 14.7). However, the $E°$ values do not dictate the de-

## Table 14.1

$E^{\circ\prime}$ Values of Biochemical Redox Couples

| Redox Couple | $E^{\circ\prime}$ (V) | $n^a$ |
|---|---|---|
| Succinate + $CO_2$ + 2 $H^+$ + 2 $e^-$ $\rightleftharpoons$ $\alpha$-ketoglutarate + $H_2O$ | −0.67 | 2 |
| Glycerate-3-phosphate + 2 $H^+$ + 2 $e^-$ $\rightleftharpoons$ glyceraldehyde-3-P + $H_2O$ | −0.55 | 2 |
| $\alpha$-Ketoglutarate + $CO_2$ + 2 $H^+$ + 2 $e^-$ $\rightleftharpoons$ isocitrate | −0.38 | 2 |
| $NAD^+$ + $H^+$ + 2 $e^-$ $\rightleftharpoons$ NADH | −0.32 | 2 |
| FMN + 2 $H^+$ + 2 $e^-$ $\rightleftharpoons$ $FMNH_2$ | −0.22[b] | 2 |
| FAD + 2 $H^+$ + 2 $e^-$ $\rightleftharpoons$ $FADH_2$ | −0.22[b] | 2 |
| Pyruvate + 2 $H^+$ + 2 $e^-$ $\rightleftharpoons$ lactate | −0.19 | 2 |
| Oxaloacetate + 2 $H^+$ + 2 $e^-$ $\rightleftharpoons$ malate | −0.17 | 2 |
| Fumarate + 2 $H^+$ + 2 $e^-$ $\rightleftharpoons$ succinate | −0.03 | 2 |
| Cytochrome $b_L(Fe^{+3})$ + $e^-$ $\rightleftharpoons$ cytochrome $b_L(Fe^{+2})$ | −0.03 | 1 |
| UQ + $H^+$ + 2 $e^-$ $\rightleftharpoons$ UQH $\cdot$ | +0.03[c] | 1 |
| Cytochrome $b_H(Fe^{+3})$ + $e^-$ $\rightleftharpoons$ cytochrome $b_H(Fe^{+2})$ | +0.05 | 1 |
| UQ + 2 $H^+$ + 2 $e^-$ $\rightleftharpoons$ $UQH_2$ | +0.11[c] | 2 |
| UQH $\cdot$ + $H^+$ + $e^-$ $\rightleftharpoons$ $UQH_2$ | +0.19[c] | 1 |
| Rieske Fe-S $(Fe^{+3})$ + $e^-$ $\rightleftharpoons$ Fe-S $(Fe^{+2})$ | +0.28 | 1 |
| Cytochrome $c_1(Fe^{+3})$ + $e^-$ $\rightleftharpoons$ cytochrome $c_1(Fe^{+2})$ | +0.23 | 1 |
| Cytochrome $c(Fe^{+3})$ + $e^-$ $\rightleftharpoons$ cytochrome $c$ $(Fe^{+2})$ | +0.24 | 1 |
| Cytochrome $a(Fe^{+3})$ + $e^-$ $\rightleftharpoons$ cytochrome $a(Fe^{+2})$ | +0.28 | 1 |
| Cytochrome $a_3(Fe^{+3})$ + $e^-$ $\rightleftharpoons$ cytochrome $a_3(Fe^{+2})$ | +0.35 | 1 |
| $O_2$ + 4 $H^+$ + 4 $e^-$ $\rightleftharpoons$ 2 $H_2O$ | +0.82 | 4 |

[a] $n$ is the number of electrons transferred.

[b] This value is for the free coenzyme. $E^{\circ\prime}$ values for flavoproteins range from −0.3 to 0 V.

[c] For UQ in aqueous ethanol.

## Figure 14.6

Apparatus for measuring the difference between the $E^\circ$ values of two redox couples. The cell on the left contains equimolar concentrations of NADH and $NAD^+$; that on the right, equimolar concentrations of $FMNH_2$ and FMN. If both solutions are at pH 7, the voltmeter senses the difference between the two $E^{\circ\prime}$ values (0.10 V, negative on the left, in this example). To determine the $E^{\circ\prime}$ of one of the redox couples, the other couple is replaced by a standard redox couple.

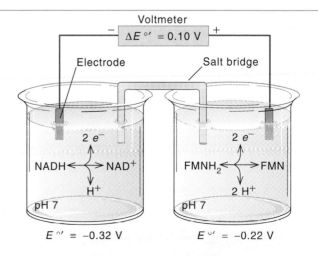

## Figure 14.7

$E^{\circ\prime}$ values of some of the electron carriers in the respiratory chain, plotted as a function of the positions of the carriers in the chain. The diagram includes components of the cytochrome $bc_1$ complex that are discussed later in the chapter. For simplicity, it shows a single $E^{\circ\prime}$ value for ubiquinone, although the $UQH_2/UQH\bullet$ and $UQH\bullet/UQ$ redox couples have different values. The $E^{\circ\prime}$ for the FMN associated with NADH dehydrogenase is not known accurately; the value shown is for free FMN. The $E^{\circ\prime}$ values and positions for the iron–sulfur centers in NADH dehydrogenase also are uncertain; only two of the six to eight iron–sulfur centers in this region of the respiratory chain are shown. The right-hand scale gives the standard free energy change ($\Delta G^{\circ\prime}$) for transferring two equivalents of electrons from a carrier with the corresponding $E^{\circ\prime}$ value to $O_2$. Sites of action of some inhibitors (amytal, rotenone, CO and $CN^-$) are shown.

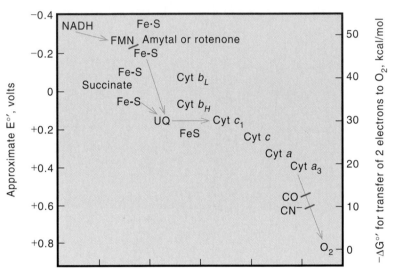

Approximate position in chain

## Figure 14.8

The multisubunit complexes of the respiratory chain. Complexes I (NADH dehydrogenase) and II (succinate dehydrogenase) transfer electrons from NADH and succinate to UQ. Complex III (the cytochrome $bc_1$ complex) transfers electrons from $UQH_2$ to cytochrome $c$, and complex IV (cytochrome oxidase), from cytochrome $c$ to $O_2$. The arrows represent paths of electron flow. NADH and succinate provide electrons from the matrix side of the inner membrane, and $O_2$ removes electrons on this side. Cytochrome $c$ is reduced and oxidized on the opposite side of the membrane, in the lumen of a crista or in the intermembrane space.

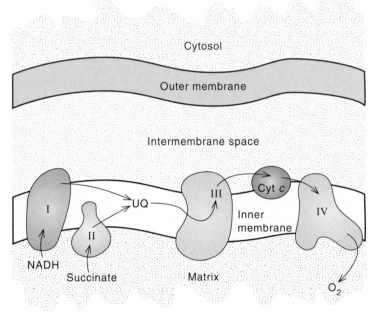

catalyzes the reduction of UQ by NADH; the succinate dehydrogenase complex (complex II), catalyzes reduction of UQ by succinate. The cytochrome $bc_1$ complex (complex III) transfers electrons from reduced ubiquinone ($UQH_2$) to cytochrome $c$. And finally, cytochrome oxidase (complex IV) transfers electrons from cytochrome $c$ to $O_2$. Each of these complexes contains numerous polypeptide subunits (table 14.2).

***Complex I: The NADH Dehydrogenase Complex.*** The NADH dehydrogenase complex is the largest complex in the mitochondrial inner membrane. It has some 26 different polypeptides, including approximately seven iron–sulfur centers and a flavoprotein with bound FMN. NADH probably reacts with the FMN, reducing it to $FMNH_2$, and the reduced flavin then transfers electrons to an iron–sulfur center (fig. 14.9). Electrons then move from one iron–sulfur center to another, and eventually to UQ.

NADH dehydrogenase is oriented in the inner membrane so that its binding site for NADH faces inwardly toward the matrix space (see fig. 14.8). This orientation is

tailed organization of the respiratory chain because electrons can move in the direction of more negative $E^\circ$ if the reactants are present at sufficiently high concentrations relative to the products. Furthermore, even if electron transfer from one particular carrier to another is thermodynamically favorable, it does not occur unless the two carriers are able to make contact. The contacts between carriers in the respiratory chain depend on how the hemes, flavins, quinones, and iron–sulfur centers are positioned in the proteins that bind them.

## *Most of the Electron Carriers Occur in Large Complexes*

By disrupting the mitochondrial inner membrane with detergents, Yousef Hatefi, David Green, and others found that all of the major electron carriers except for cytochrome $c$ and UQ occur in the form of four large complexes (fig. 14.8). The isolated NADH dehydrogenase complex (complex I)

**Table 14.2**

Components of the Mitochondrial Electron-Transport Complexes

| Complex | Approximate Molecular Weight[a] | Polypeptides | Prosthetic Groups |
|---|---|---|---|
| I NADH dehydrogenase | $1 \times 10^6$ | at least 26 | FMN, Fe-S centers |
| II Succinate dehydrogenase | $1 \times 10^5$ | 2 Fe-S proteins, 2 others | FAD, Fe-S centers |
| III Cytochrome $bc_1$ complex | $2 \times 10^5$ | 2 cytochromes, 1 Fe-S protein, 6–8 others | $b$ and $c$ type hemes (cyt $b_L$, $b_H$, $c_1$), Fe-S center |
| IV Cytochrome oxidase | $2 \times 10^5$ | up to 13 | $a$ type hemes (cyt $a$, $a_3$), 2 Cu |

[a] Complex III occurs as a dimer in the membrane; the molecular weight listed is for a monomer. The same is true of complex IV.

**Figure 14.9**

Electron transfer through complex I. The reduction of FMN to $FMNH_2$ by NADH requires the uptake of one proton from the matrix. $FMNH_2$ subsequently transfers electrons to a series of iron–sulfur centers and releases protons to the solution in the intermembrane space. When the non–sulfur centers reduce UQ to $UQH_2$, two more protons are taken up from the matrix.

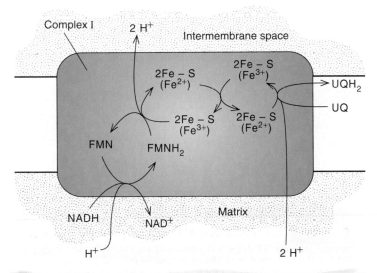

**Figure 14.10**

Complex II: The succinate dehydrogenase complex.

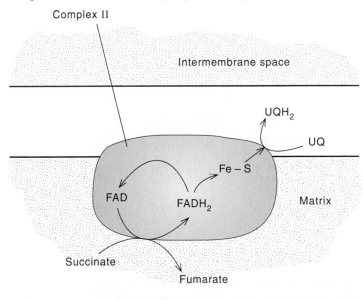

appropriate for oxidizing the NADH generated in the matrix by the TCA cycle. The $UQH_2$ formed by complex I diffuses to complex III in the membrane's phospholipid bilayer (see fig. 14.8), but the protons that are taken up when the quinone undergoes reduction come from the solution on the matrix side of the membrane. In the course of the electron-transfer reactions, protons also are released to the solution in the intermembrane space. These proton movements take on special significance when we discuss the formation of ATP

***Complex II: The Succinate Dehydrogenase Complex.***
Succinate dehydrogenase is the only enzyme of the TCA cycle that is embedded in the inner membrane. Its four subunits include two iron–sulfur proteins, one of which also has a covalently attached FAD. As in NADH dehydrogenase, the substrate-oxidation site is on the matrix side of the membrane (fig. 14.10).

## Figure 14.11

Electron transport through complex III: the Q cycle. In this figure, the thin black arrows represent the path of electron flow; the thick green arrows represent the paths of protons and of UQ in its various oxidation states. UQH$_2$ is oxidized to UQ in two steps at an enzyme site on the side of the inner membrane facing the intermembrane space (top part of the figure), transferring the first electron to the Fe-S protein and the second electron to cytochrome $b_L$. The stoichiometries indicated are for two molecules of UQH$_2$ undergoing these reactions. One of the two molecules of UQ that are produced diffuses to a site on the matrix side of the membrane (bottom), where it is reduced back to UQH$_2$. The two electrons required for the reduction come through the $b$ cytochromes. Antimycin inhibits this reaction. The UQH$_2$ generated at the reduction site diffuses back to the oxidation site, where it joins the pool of UQH$_2$ coming from complexes I and II. Electrons leaving the UQH$_2$ oxidation site reduce cytochrome $c$ (top), which dissociates from the complex to go to cytochrome oxidase. Protons also are released to the solution in the intermembrane space. Protons are taken up from the matrix at the UQ reduction site.

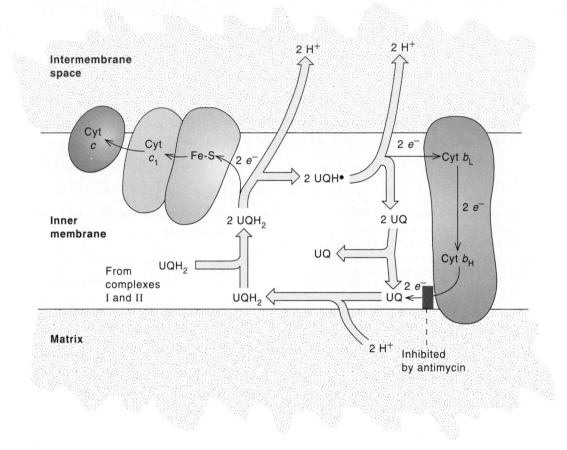

***Complex III: The Cytochrome bc$_1$ Complex.*** The cytochrome $bc_1$ complex contains two $b$-type cytochromes: $b_L$ and $b_H$. Both of the $b$ hemes are on a single 30-kd polypeptide. The complex also contains cytochrome $c_1$, an iron–sulfur protein, and between four and six additional subunits.

Electrons moving through complex III from UQ to cytochrome $c$ follow a circuitous path (fig. 14.11). UQH$_2$ must transfer one of its electrons to the iron–sulfur protein before it can transfer an electron to cytochrome $b_L$.

Oxidation of UQH$_2$ by the iron–sulfur protein generates the semiquinone UQH•, which then serves as the reductant for cytochrome $b_L$. The reduced iron–sulfur protein transfers an electron to cytochrome $c_1$ and on to cytochrome $c$. Meanwhile, the reduced cytochrome $b_L$ passes an electron to cytochrome $b_H$, which then contributes the electron for reduction of another molecule of UQ at a second site in the complex.

$$\text{UQH}_2 \xrightarrow[\text{Fe-S (Fe}^{+3})\ \text{Fe-S (Fe}^{+2})]{\text{H}^+} \text{UQH•} \xrightarrow[\text{cyt } b_L(\text{Fe}^{+3})\ \text{cyt } b_L(\text{Fe}^{+2})]{\text{H}^+} \text{UQ} \quad (7)$$

$$
\begin{array}{ccc}
\text{cyt } b_L\,(\text{Fe}^{+2}) & \text{cyt } b_H\,(\text{Fe}^{+3}) & \frac{1}{2}\text{UQH}_2 \\
& & \\
\text{cyt } b_L\,(\text{Fe}^{+3}) & \text{cyt } b_H\,(\text{Fe}^{+2}) & \frac{1}{2}\text{UQ} + \text{H}^+ \quad (8)
\end{array}
$$

## Figure 14.12

A schematic drawing of the hemes and Cu atoms in complex IV. $Cu_A$ and heme $a$ are shown at the top, and $Cu_B$ and heme $a_3$ are shown at the bottom. Molecules of cytochrome $c$ bring one electron at a time to the complex, reducing $Cu_A$ and heme $a$. Electrons move from these components to heme $a_3$ and $Cu_B$. When heme $a_3$ and $Cu_B$ have both been reduced, two electrons can be transferred to an $O_2$ molecule, one from heme $a_3$ and one from $Cu_B$. (b) Transfer of two more electrons causes the bridging $^-O$—$O^-$ bond to split, with one of the $O^{2-}$ atoms remaining on heme $a_3$ and the other on $Cu_B$. (c) Two of the incoming protons from the matrix convert the $O^{2-}$ atoms to $OH^-$ ions, and two more protons result in the conversion of the $OH^-$ ions into water molecules.

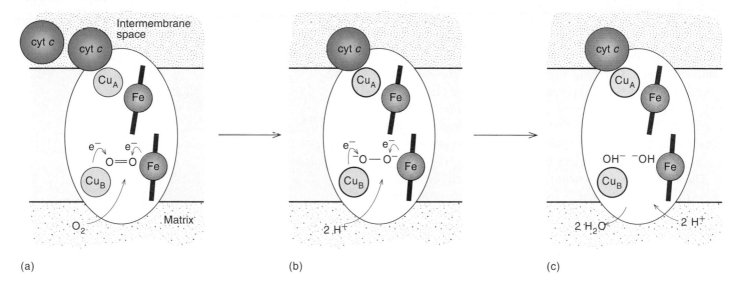

(a)                          (b)                          (c)

If all of these steps occur twice, so that two electrons pass through the $b$ cytochromes, a molecule of UQ is reduced fully to $UQH_2$. At this point, we may seem to be back to where we started. But note that for each $UQH_2$ that is regenerated, two molecules of $UQH_2$ are oxidized. There is a net oxidation of one $UQH_2$ to UQ, and two electrons proceed to cytochrome $c$ and on their way to $O_2$ (see fig. 14.11).

This scheme for electron transport through the cytochrome $bc_1$ complex is known as the Q cycle. The role of the iron–sulfur protein was confirmed by extracting this protein from the cytochrome $bc_1$ complex. If the iron–sulfur protein is removed, electron transfer from $UQH_2$ to cytochrome $c_1$ is prevented. This shows that the iron–sulfur protein operates upstream of cytochrome $c_1$, even though its $E°$ is more positive than that of the cytochrome (see fig. 14.7 and table 14.1).

As shown in figure 14.11, the sites at which $UQH_2$ and $UQH•$ undergo oxidation face the intermembrane space, whereas the UQ reduction site is on the matrix side. UQ and $UQH_2$ evidently diffuse through the membrane from one site to the other. Antimycin blocks electron transfer from cytochrome $b_H$ to UQ at the reduction site; this inhibitor was particularly helpful in clarifying the steps of the Q cycle.

Oxidation and reduction of UQ by the cytochrome $bc_1$ complex results in the uptake of protons from the solution on the matrix side of the inner membrane and the release of protons to the intermembrane space. Two protons are taken up on the matrix side, and four protons are released on the opposite side for each pair of electrons that proceed to cytochrome $c$ (see fig. 14.11). We return to this point later in the discussion.

***Complex IV: Cytochrome Oxidase.*** Cytochrome oxidase contains two atoms of Cu in addition to the hemes of cytochromes $a$ and $a_3$. The Cu atoms undergo one-electron oxidation–reduction reactions between the cuprous ($Cu^+$) and cupric ($Cu^{2+}$) states. One of the Cu atoms ($Cu_B$) is close to the Fe of cytochrome $a_3$ (fig. 14.12). The other ($Cu_A$) is associated with cytochrome $a$, but not so intimately. Oxidation of cytochrome $c$ takes place on the side of the membrane facing the intermembrane space, whereas the reduction of $O_2$ by cytochrome $a_3$ and $Cu_B$ occurs on the matrix side.

Reduction of one molecule of $O_2$ to 2 $H_2O$ requires the addition of four electrons. Adding one electron to $O_2$ to produce the superoxide radical ($O_2^{-•}$) is thermodynamically difficult and has to be pulled along by a further reduction to the level of peroxide ($H_2O_2$) or $H_2O$. The enzyme facilitates this process by binding $O_2$ to $Cu_B$ as well as to cytochrome $a_3$ and then transferring two electrons at a time (see fig. 14.12).

## Reconstitution Experiments Demonstrate the Need for Carriers to Mediate Electron Transfer between Complexes

Having described the four complexes that make up the respiratory chain, we now turn to the question of how electrons move between the complexes. To address this question, purified complexes were recombined so that they reconstituted longer stretches of the respiratory chain. First, the complexes are mixed in the presence of phospholipids and a detergent. When the detergent is removed by dilution, the phospholipids assemble into vesicles and the proteins are incorporated into the phospholipid bilayer. If UQ is added, vesicles that contain complex I and complex III catalyze electron transfer from NADH to cytochrome c. In the presence of cytochrome c, vesicles containing complex III and complex IV transfer electrons from $UQH_2$ to $O_2$.

These reconstitution experiments supported the model for electron transfer shown in figure 14.8. In this model the complexes do not bind to each other directly. Instead, movement of electrons from complexes I and II to complex III is mediated by diffusion of $UQH_2$ from one complex to the other within the phospholipid bilayer. Similarly, electrons move from complex III to complex IV by the diffusion of reduced cytochrome c along the surface of the membrane. Remember that cytochrome c differs from the other cytochromes in being a water-soluble protein. It is attached loosely to the membrane surface by electrostatic interactions.

There is additional evidence that the electron-transfer complexes are not connected in fixed chains. If most of the cytochrome oxidase complexes in the membrane are inhibited with CO, the few molecules that remain uninhibited are still able to catalyze oxidation of all the cytochrome c by $O_2$. This suggests that cytochrome c can diffuse from one cytochrome oxidase complex to another, rather than remaining bound to an individual complex. Also, cytochrome c, UQ, and the complexes themselves move about at different rates, which means that they cannot all stay stuck together.

## Oxidative Phosphorylation

Evidence that oxidative reactions can drive the formation of ATP was obtained about 1940 by Herman Kalckar, who showed that aerobic cells make ATP from ADP and $P_i$ by a process that depends on respiration. At that time it was unclear that this oxidative phosphorylation occurred in mitochondria or that it involved NADH. Resolution of these questions had to await development of methods for preparing mitochondria free of other cellular constituents, and for presenting NADH on the matrix side of the inner membrane. When these methods were devised in

### Figure 14.13

Approximately 2.5 molecules of ADP can be phosphorylated to ATP for each pair of electrons that traverse the electron-transport chain from NADH to $O_2$. About 1.5 molecules of ATP are formed for a pair of electrons that enter the chain via succinate dehydrogenase or other flavoproteins such as glycerol-3-phosphate dehydrogenase. Approximately one molecule of ATP is formed for each pair of electrons that enters via cytochrome c. Electron flow through each of complexes I, III, and IV thus is coupled to phosphorylation.

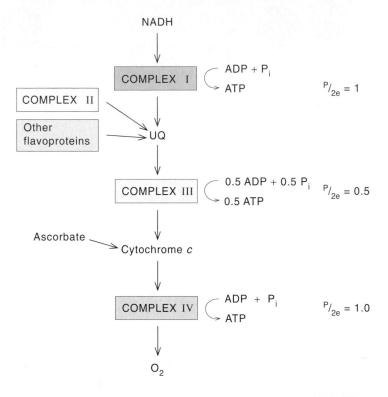

the late 1940s, Eugene Kennedy and Albert Lehninger found that mitochondria do oxidize endogenous NADH and that at least two ATPs are generated for each NADH oxidized.

How do mitochondria couple such different types of reactions as electron transfer and the formation of a phosphate anhydride bond? We explore this question in the following sections.

## Electron Transfer Is Coupled to ATP Formation at Three Sites

The number of molecules of ATP formed per pair of electrons transferred down the respiratory chain to $O_2$ is termed the P-to-O ratio. When NADH is used as the reducing substrate, the measured P-to-O ratio is about 2.5; with succinate it is about 1.5. Reductants that feed electrons directly to cytochrome c give a P to O of about 1.0. This last observation indicates that complex IV, which conducts electrons

## Figure 14.14

The rate of respiration by a suspension of mitochondria can be increased dramatically by the addition of a small amount of ADP. An oxidizable substrate (succinate) and $P_i$ first are added at the times indicated by the vertical arrows, but the respiratory rate remains low until ADP is added. A second period of rapid respiration is obtained when more ADP is added, indicating that essentially all of the added ADP gets used up. The P-to-O ratio can be determined by dividing the amount of ADP added by the amount of oxygen taken up during the period of rapid respiration.

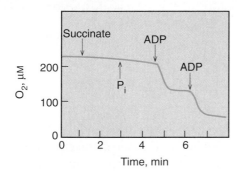

## Figure 14.15

Structures of two uncouplers of oxidative phosphorylation, 2,4-dinitrophenol (DNP) and carbonylcyanide-*p*-trifluoromethoxyphenylhydrazone (FCCP). DNP has a weakly dissociable proton on the oxygen atom; FCCP has a similar proton on one of the nitrogens.

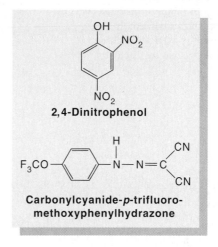

**2,4-Dinitrophenol**

**Carbonylcyanide-*p*-trifluoro-methoxyphenylhydrazone**

from cytochrome *c* to $O_2$, has a site where the energy of the electron-transfer reactions is coupled to ATP synthesis. Because the electron-transport chains for NADH and succinate converge at UQ, the higher P-to-O ratio obtained with NADH suggests that another such coupling site exists in complex I, between NADH and UQ (see figs. 14.8 and 14.13). Finally, the intermediate P-to-O ratio obtained with succinate indicates that a coupling site occurs in complex III, but probably not in complex II. Thus, the respiratory chain appears to have three distinct coupling sites for ATP synthesis.

The contribution of the individual coupling sites to the overall P-to-O ratios may be estimated from the differences in the P-to-O ratios observed when different reducing substrates are used. The difference in P-to-O ratios observed when NADH and succinate are used (2.5 − 1.5) suggests that 1.0 ATP is synthesized per electron pair at the first coupling site (complex I), while the difference observed when succinate and ascorbate are used (1.5 − 1.0) suggests that 0.5 ATPs are synthesized per electron pair at the second coupling site (complex III). Finally the P-to-O ratio obtained when ascorbate is used suggests that one ATP is synthesized per electron pair at the third coupling site (complex IV).

By incorporating the purified electron-transport complexes into phospholipid vesicles along with the mitochondrial ATP-synthase enzyme that is described below, Efraim Racker and his coworkers verified the capacity of the individual complexes I, III, and IV to support the formation of ATP. In figure 14.7, you can see that the flow

of two electrons through each of these complexes involves a sufficiently negative $\Delta G^{\circ\prime}$ to drive the phosphorylation of ADP to ATP. ($\Delta G^{\circ\prime}$ for the reaction ADP + $P_i \longrightarrow$ ATP + $H_2O$ in the presence of 10 mM $Mg^{2+}$ is about 7.5 kcal/mol, which is equivalent to a $\Delta E^{\circ\prime}$ of about 0.16 V for $n = 2$ electrons/molecule.)

Respiration and phosphorylation usually are tightly coupled processes; if phosphorylation stops, so, as a rule, does respiration and *vice versa*. As an example, respiration slows down greatly if mitochondria run out of ADP (fig. 14.14). Addition of a small amount of ADP in the presence of an oxidizable substrate and excess $P_i$ causes a brief period of brisk respiration that can be repeated by adding more ADP. Experiments of this type provide one means for estimating P-to-O ratios. The P-to-O ratio is proportional to the amount of ADP consumed to reduce one oxygen of $O_2$ to one oxygen of water.

## Uncouplers Release Electron Transport from Phosphorylation

The tight coupling of respiration and phosphorylation can be disrupted by molecules known as uncouplers. Uncouplers include a diverse group of molecules structurally, but they are all lipophilic weak acids (fig. 14.15). If an uncoupler is added to mitochondria in the presence of an oxidizable substrate, $O_2$ uptake commences immediately and continues until essentially all of the $O_2$ in the solution is used up (fig. 14.16*a*). This happens even in the absence of added ADP or $P_i$. The free energy released in

**Figure 14.16**

(*a*) Addition of an uncoupler to a
suspension of mitochondria causes brisk $O_2$
consumption, which continues until all the
$O_2$ in the solution is used up. An oxidizable
substrate (succinate) is added before the
uncoupler, but $P_i$ is not required.
(*b*) Oligomycin, an inhibitor of the ATP-
synthase, blocks the stimulation of
respiration caused by ADP. It does not
block the stimulation caused by an
uncoupler.

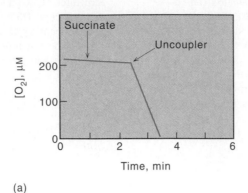

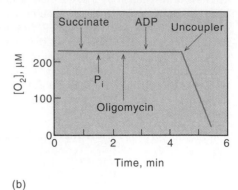

(a)

(b)

the electron-transfer reactions is lost as heat rather than
being captured in ATP.

If an uncoupler somehow caused the breakdown of an
intermediate form of an electron carrier, the electron carrier
would be set free and electron transport to $O_2$ could continue. However, something evidently is different about oxidative phosphorylation compared with the substrate-level
phosphorylation catalyzed by 3-phosphoglyceraldehyde
dehydrogenase (see chapter 12), because uncouplers have
no effect on the latter reaction. Nor do they affect other
soluble enzymes that make or use ATP. On the other hand, a
molecule that acts as an uncoupler at any one of the three
coupling sites of oxidative phosphorylation invariably has a
similar effect at the other two sites. This suggests that uncouplers cause the breakdown of something that is generated at all three sites.

Another class of phosphorylation inhibitors functions
by blocking the ATP-synthase enzyme directly. Inhibitors
of this type include the antibiotic oligomycin. If mitochondria are treated with oligomycin, ADP no longer is able to
increase the rate of respiration (fig. 14.16*b*). Oligomycin
does not block the stimulation of respiration caused by an
uncoupler, demonstrating that the two types of inhibitors of
phosphorylation act by different mechanisms.

## The Chemiosmotic Theory Proposes
## That Phosphorylation Is Driven by
## Proton Movements

In 1961, Peter Mitchell suggested a radically new theory to
explain the coupling mechanism of oxidative phosphorylation. Mitchell proposed that the component generated at all
three coupling sites is not a high-energy chemical species

but rather an electrochemical potential gradient for protons
across the mitochondrial inner membrane. Figure 14.17 illustrates the basic idea. Electron transport down the respiratory chain results in the movement of protons across the
membrane, from the mitochondrial matrix to the intermembrane space. The removal of protons from the matrix causes
the pH in this region to rise. Because protons are positively
charged, the matrix also becomes negatively charged with
respect to the intermembrane space and the cytosol. The
differences in pH and electrical potential across the membrane provide a driving force that tends to pull protons back
into the matrix. Part of the free energy decrease associated
with the electron-transfer reactions thus is exchanged for an
electrochemical potential difference between the solutions
on the two sides of the inner membrane. Mitchell suggested
that protons move back across the membrane through special channels in an ATP-synthase, which uses this inward
flow of protons to drive the formation of ATP. This is called
the chemiosmotic theory because it emphasizes that chemical reactions can drive, or be driven by, movements of molecules or ions between osmotically distinct spaces separated
by membranes. Let's examine some of the evidence supporting the theory.

## Electron Transport Creates an
## Electrochemical Potential Gradient for
## Protons across the Inner Membrane

Peter Mitchell and Jennifer Moyle showed that protons are pumped out of mitochondria during respiration. When $O_2$ is added to a suspension of mitochondria, the pH decreases in the solution surrounding the

## Figure 14.17

According to the chemiosmotic theory, flow of electrons through the electron-transport complexes pumps protons across the inner membrane from the matrix to the intermembrane space. This raises the pH in the matrix and leaves the matrix negatively charged with respect to the intermembrane space and the cytosol. Protons flow passively back into the matrix through a channel in the ATP-synthase, and this flow drives the formation of ATP.

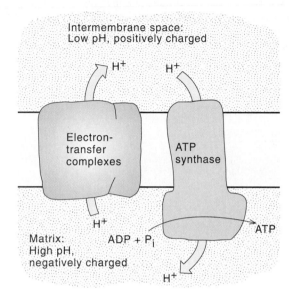

mitochondria (fig. 14.18). If the integrity of the inner membrane is broken by the addition of a detergent, this pH change is not seen, indicating that the protons that appear outside the mitochondria probably emerge from the matrix space inside. A detergent makes the membrane leaky, allowing any protons that are pumped out to go right back in.

Mitchell and Moyle calculated that mitochondria depleted of ADP build up a pH gradient of about 0.05 pH units across the inner membrane. More recent measurements indicate that approximately 10 protons are pumped out for each pair of electrons moving down the respiratory chain from NADH to $O_2$, and approximately 6 protons for a pair of electrons coming from succinate.

The pH gradient created by respiration collapses abruptly if an uncoupler is added (see fig. 14.18). In the absence of an uncoupler, the pH change caused by a brief period of respiration decays slowly, in agreement with the view that the mitochondrial inner membrane blocks the free diffusion of protons between the matrix and the intermembrane space. Mitchell and Moyle reinforced this conclusion by experiments in which they added a small amount of HCl to a suspension of mitochondria and measured the rate at which protons leaked through the membrane. The leakage occurs on a time scale of minutes, which is much slower than the milliseconds-to-seconds time scale of oxidative phosphorylation. The idea that the membrane is largely impermeable to protons fits well with what we know about the structure of biological membranes (see chapter 17). An ion such as $H_3O^+$ cannot pass readily through the hydrocarbon region of the phospholipid bilayer. In the absence of an uncoupler, the conductance of the phospholipid bilayer to protons is about $10^6$ times lower than that of the aqueous phases on either side of the membrane.

The fact that uncouplers are lipophilic weak acids (see above) explains their ability to collapse transmembrane pH gradients. Their lipophilic character allows uncouplers to diffuse relatively freely through the phospholipid bilayer. Because they are weak acids, uncouplers can release a proton to the solution on one side of the membrane and then diffuse across the membrane to fetch another proton. The chemiosmotic theory thus provides a simple explanation of the effects of uncouplers on oxidative phosphorylation.

The pumping of protons across the membrane by the respiratory chain creates a transmembrane electrical potential gradient, in addition to a pH gradient. This was shown by examining the movements of lipophilic anions and cations. If the matrix space becomes negatively charged with respect to the region outside of the mitochondrion, the electrical difference tends to pull positively charged ions into the matrix and to push negatively charged ions out. Whether or not a particular ion actually moves across the membrane depends on whether the ion can pass through the phospholipid bilayer. In some lipophilic ions such as the triphenylmethylphosphonium ion $(C_6H_5)_3CH_3P^+$ the charge is sufficiently buried by apolar groups that the ion can move across the membrane relatively freely. Another lipophilic cation that passes rapidly through phospholipid bilayers is the complex of the ionophore valinomycin with $K^+$ (fig. 14.19). Respiring mitochondria take up lipophilic cations, and they extrude lipophilic anions. These observations support the idea that the efflux of protons driven by electron transport makes the interior of the mitochondrion negatively charged relative to the solution outside.

One way to estimate the electrical potential difference across the mitochondrial membrane is to measure the concentrations of a lipophilic ion after the ion has come to equilibrium. The larger the electrical potential difference ($\Delta\psi$), the larger the ratio of the concentrations on the two sides. Such measurements indicate respiring mitochondria generate a $\Delta\psi$ on the order of $-0.15$ V, inside negative.

## Figure 14.18

Respiring mitochondria extrude protons. A weakly buffered suspension of mitochondria is provided with an oxidizable substrate and allowed to use up all the $O_2$ in the solution. The traces show measurements of the pH of the suspension, with pH decreases plotted upward. When a small amount of $O_2$ is added (upward-pointing arrow), respiration can occur for a few seconds. The pH of the solution decreases suddenly at this point (Exp. A). The pH change persists for a period of several minutes, long after the $O_2$ has been used up. If an uncoupler is added, the pH returns abruptly to nearly its original level. Experiment B was done in the presence of a detergent that disrupted the inner membrane, making the membrane permeable to protons. Respiration still occurs under these conditions, but the pH change is not seen. In Exp. A, the number of protons that move across the membrane can be estimated from the pH change, and can be related to the number of O atoms consumed. In this experiment, about six protons appear to be translocated per O atom. However, this is an underestimate because part of the pH gradient is dissipated by the phosphate/$OH^-$ transport system discussed later in the chapter. Larger $\Delta H^+$ to O ratios are measured if the phosphate transporter is inhibited. For more details see P. Mitchell and J. Moyle, *Nature,* 208:147–151, 1965; and *Biochem. J.* 105:1147–1162, 1967; also A. Alexandre et al., *J. Biol. Chem.* 255:10721–10730, 1980.

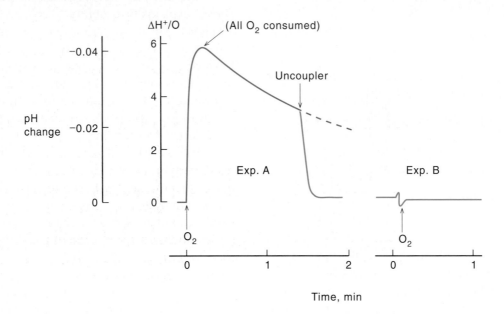

## Figure 14.19

Valinomycin is one of a group of compounds termed ''ionophores,'' which facilitate movement of ions across phospholipid membranes. Valinomycin has the cyclic structure [-D-valine-lactate-L-valine-hydroxyvalerate-]₃. This figure shows the structure of the complex formed between valinomycin and $K^+$. The complex is soluble in organic solvents and can pass through the phospholipid bilayer of the mitochondrial inner membrane. $K^+$ dissociates reversibly from the complex in the aqueous solution on either side of the membrane.

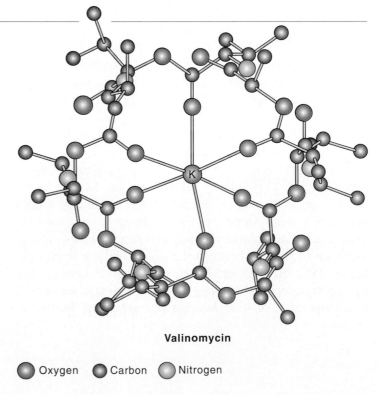

**Valinomycin**

🔴 Oxygen  ⚫ Carbon  ⚪ Nitrogen

**Figure 14.20**

A general scheme showing how electron transport can result in proton translocation. A, B, D, and E are electron carriers that bind protons when they are reduced. A and D undergo reduction to $AH_2$ and $DH_2$ on the matrix side of the membrane, picking up protons from the solution on this side. They then transfer protons along with electrons to B and E. $BH_2$ and $EH_2$ are reoxidized on the opposite side of the membrane, releasing protons here. Electrons move without protons from $BH_2$ to D.

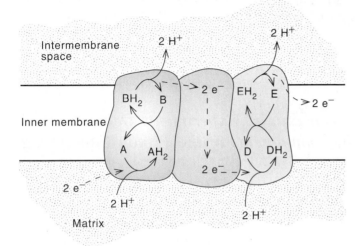

## How Do the Electron-Transfer Reactions Pump Protons across the Membrane?

NADH, flavins, and quinones all bind protons in addition to electrons when they undergo reduction. Suppose that an electron-transfer reaction in which protons are taken up from the solution occurs on the matrix side of the inner membrane, and a reaction that releases protons to the solution occurs facing the intermembrane space. If protons are transferred along with electrons between the two sites, the flow of electrons ferries protons across the membrane (fig. 14.20). The key point here is that, unlike reactions in free solution, enzymatic reactions in organized structures such as membranes can have a directional, or vectorial, character.

This basic idea accounts well for the proton translocation that occurs in complex III. In the Q cycle (see fig. 14.11), $UQH_2$ is oxidized to UQ on the outside of the inner membrane, and UQ is reduced to $UQH_2$ on the matrix side. Hydrogen atoms move across the membrane by the diffusion of $UQH_2$ from one of these catalytic sites to the other. The net transfer of two electrons from $UQH_2$ to cytochrome c results in the uptake of two protons from the matrix and release of four protons to the intermembrane space.

The mechanism of proton translocation in complexes I and IV is not yet understood. Here, the electron-transfer reactions may cause protein conformational changes that open gates for proton movement first on one side of the membrane and then on the other.

## Flow of Protons Back into the Matrix Drives the Formation of ATP

The chemiosmotic theory postulates that protons moving back into the matrix via an ATP-synthase drive the formation of ATP. Evidence for this is that an electrochemical potential gradient for protons can support the formation of ATP in the absence of electron-transfer reactions. A transient pH gradient that pulls protons into the matrix can be set up by first incubating mitochondria at pH 9, so that the inside becomes alkaline, and then quickly lowering the pH of the suspension medium to 7 (fig. 14.21).

Current estimates are that three protons move into the matrix through the ATP-synthase for each ATP that is synthesized. We see below that one additional proton enters the mitochondrion in connection with the uptake of ADP and $P_i$ and export of ATP, giving a total of four protons per ATP. How does this stoichiometry relate to the P-to-O ratio? When mitochondria respire and form ATP at a constant rate, protons must return to the matrix at a rate that just balances the proton efflux driven by the electron-transport reactions. Suppose that 10 protons are pumped out for each pair of electrons that traverse the respiratory chain from NADH to $O_2$, and 4 protons move back in for each ATP molecule that is synthesized. Because the rates of proton efflux and influx must balance, 2.5 molecules of ATP (10/4) should be formed for each pair of electrons that go to $O_2$. The P-to-O ratio thus is given by the ratio of the proton stoichiometries. If oxidation of succinate extrudes six protons per pair of electrons, the P-to-O ratio for this substrate is 6/4, or 1.5. These ratios agree with the measured P-to-O ratios for the two substrates.

Let's now consider how much free energy is released by moving protons into the mitochondrion. Is it really enough to drive the synthesis of ATP? The free energy change depends both on the ratio of the proton concentrations on the two sides of the membrane and on the difference between the electric potentials on the two sides.

$$\Delta G_{H^+} = G_{H^+(in)} - G_{H^+(out)} = RT \ln\left\{\frac{[H^+]_{in}}{[H^+]_{out}}\right\} + \mathscr{F}\,\Delta\psi$$

$$= -2.3RT\,\Delta pH + \mathscr{F}\,\Delta\psi \quad (9)$$

## Figure 14.21

An electrochemical potential gradient for protons can be set up across the mitochondrial inner membrane in the absence of electron transfer. In step 1, mitochondria are incubated at high pH in the presence of KCl to reduce the proton concentration inside and to load the inside with $K^+$ and $Cl^-$. In step 2, the pH and the KCl concentration outside the mitochondria are decreased. Valinomycin (Val) is added to carry $K^+$ ions out. Because the counterion $Cl^-$ cannot move rapidly across the membrane, the efflux of $K^+$ down its concentration gradient leaves the inside of the mitochondrion with a net negative charge relative to the outside. This difference in electric potential together with the higher pH inside favors the influx of protons. Protons moving inward through the ATP-synthase can drive the formation of ATP.

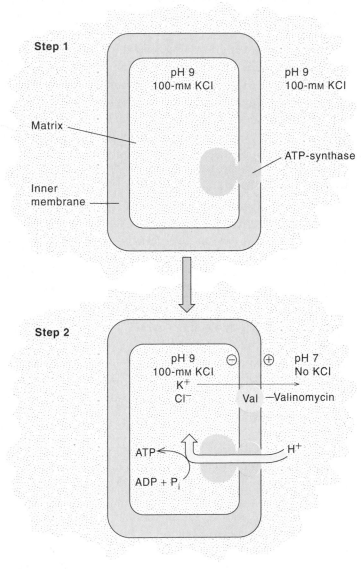

Here $\mathscr{F}$ is the Faraday constant, $\Delta pH = pH_{in} - pH_{out}$, and $\Delta\psi$ is the electric potential difference (the electric potential inside minus that outside). If the pH in the matrix space is 0.05 pH units above that in the cytosol, the term $-2.3RT\,\Delta pH$ amounts to $-0.07$ kcal/mol. If $\Delta\psi$ is $-0.15$ V, $\mathscr{F}\,\Delta\psi$ is $-3.46$ kcal/mol. This gives a total $\Delta G_{H^+}$ of $-3.53$ kcal/mol, with the term $\mathscr{F}\,\Delta\psi$ making the dominant contribution. If four protons move across the membrane for each ATP molecule synthesized, the $\Delta G$ associated with proton translocation is $4 \times (-3.53)$, or $-14.1$ kcal/mol of ATP ($-58.4$ kJ/mol), which is about twice as large as the $\Delta G^{\circ\prime}$ of ATP hydrolysis ($-7.5$ kcal/mol, or $-31.4$ kJ/mol). The free energy decrease thus is large enough to account for the high [ATP]/[ADP][$P_i$] ratio that mitochondria generate under physiological conditions.

## Reconstitution Experiments Demonstrate the Components of the Proton-Conducting ATP-Synthase

Studies by Efraim Racker and his colleagues showed that the catalytic sites of the ATP-synthase reside in spherical complexes that line the matrix side of the mitochondrial inner membrane (fig. 14.22). This part of the ATP-synthase is referred to as $F_1$ and it consists of five different polypeptide subunits with a total molecular weight of about 360,000. $F_1$ is attached loosely to a base-piece of three to five polypeptides ($F_0$), which is embedded in the inner membrane (fig. 14.23).

If the purified $F_1 - F_0$ complex is incorporated into a phospholipid vesicle and an electrochemical potential gradient for protons is set up across the membrane, the complex can synthesize ATP. To show this, Efriam Racker and Walter Stoeckenius generated a proton gradient by incorporating a light-driven proton pump, bacteriorhodopsin from *Halobacterium halobium* (see fig. 17.12), into membrane vesicles along with the $F_0 - F_1$ complex. When the vesicles were illuminated, ATP was formed (fig. 14.24). This experiment demonstrated that the ATP-synthase does not have to be attached directly to the electron-transport complexes of the respiratory chain, because no electron carriers were included in the liposomes. It provided strong evidence that the coupling of phosphorylation to electron transport is mediated by an electrochemical potential gradient that can be generated at one site on the membrane and used at relatively distant sites.

$F_0$ appears to contain the proton-conducting channel of the ATP-synthase. If $F_0$ alone is incorporated into liposomes, it makes the membranes leaky to protons. The multiple copies of one of the small subunits of $F_0$ may assemble to form a tubular channel across the membrane (see fig. 14.23).

**Figure 14.22**

(*a*) Negatively stained electron micrograph of part of a mitochondrion, showing head-pieces of the $F_1$ ATP-synthase lining the inner membrane. The wormlike white area is a crista; the large dark areas are the matrix. The $F_1$ head-pieces show up as white spots projecting from the membrane of the crista into the matrix. (From B. Chance and D. Parsons, Cytochrome function in relation to inner membrane structure of mitochondria, *Science* 142:1176, 1963, © 1963 by the AAAS.) (*b*) Negatively stained electron micrograph of mitochondrial membrane vesicles. These vesicles are capable of oxidative phosphorylation. $F_1$ head-pieces line the outer surfaces of the membranes. (*c*) Electron micrograph of purified $F_1$. (*d*) Mitochondrial membrane vesicles that were depleted of $F_1$ by treatment with urea and trypsin. These membranes contain a functional electron-transfer chain but do not form or hydrolyze ATP. Note that the surfaces of the membranes appear smooth. (Micrographs, *b*, *c*, and *d* courtesy of Dr. E. Racker.)

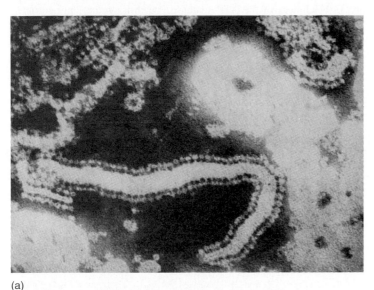
(a)

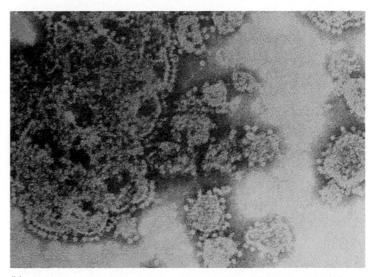
(b)

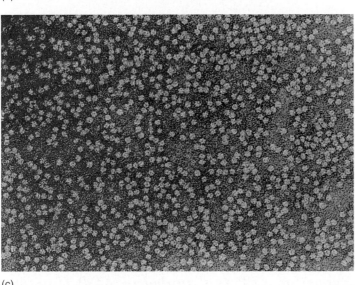
(c)

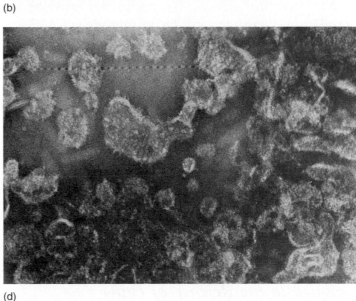
(d)

How the movement of protons through $F_0$ drives the formation of ATP in $F_1$ is not known. However, studies by Paul Boyer, Harvey Penefsky, and others have shown that the enzyme binds ADP and ATP very tightly. The reaction

$$\text{ADP} + P_i \rightleftharpoons \text{ATP} + H_2O \qquad (10)$$

occurs between the bound nucleotides on the enzyme, even in the absence of proton flow. $P_i$ evidently reacts directly with ADP to give ATP in a single step, without the formation of a phosphorylated enzyme intermediate. Remarkably, the equilibrium constant for the formation of ATP is close to 1 when the nucleotides are bound to the enzyme. When the nucleotides are free in solution, the equilibrium constant is about $10^{-5}$, which is highly unfavorable for the formation of ATP.

**Figure 14.23**

Components of the proton-conducting ATP-synthase. The $F_1$ head-piece includes three $\alpha$ and three $\beta$ subunits and one copy each of three other subunits ($\gamma$, $\delta$, and $\epsilon$). $F_o$ includes a cluster of 9–12 copies of a small peptide, which appears to form a transmembrane channel for protons.

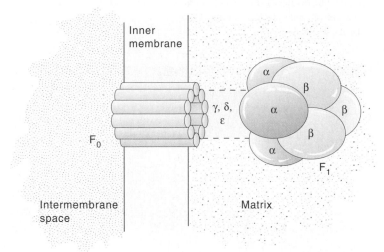

Proton movements do not have much effect on the equilibrium constant for the formation of ATP from bound ADP and $P_i$ on the synthase. Instead, they affect the release of ATP from the enzyme. The nucleotides evidently are bound in an environment that favors formation of ATP, and proton movements change the enzyme's conformation in such a way that the ATP is released.

## Transport of Substrates, $P_i$, ADP, and ATP into and out of Mitochondria

One of the principles underlying the chemiosmotic theory is that the mitochondrial inner membrane is relatively impermeable to ions. This raises the question of how $P_i$, adenine nucleotides, and substrates, such as pyruvate and citrate, move into and out of mitochondria.

### Uptake of $P_i$ and Oxidizable Substrates Is Coupled to the Release of Other Compounds

Mitochondria can take up $P_i$ from the cytosol even when the concentration of $P_i$ inside exceeds that outside. This process is catalyzed by a protein that links the uptake of $P_i$ to an outward movement of $OH^-$. Phosphate probably enters in the form of $H_2PO_4^-$, carrying one negative charge. The exchange of $P_i$ for $OH^-$ is therefore electrically neutral. It is, however, sensitive to the pH gradient. Because $OH^-$ and

**Figure 14.24**

ATP synthesis can be obtained in an artificial system in which the mitochondrial ATP-synthase is incorporated into membrane vesicles together with bacteriorhodopsin, a membrane protein obtained from *Halobacterium halobium*. When bacteriorhodopsin is excited with light, it pumps protons across the membrane (see chapter 17). The solution inside the vesicle thus becomes acidic and positively charged relative to the external solution. Protons can move back out through the ATP-synthase, thereby driving the formation of ATP. The formation of ATP is blocked by uncouplers or oligomycin. Note that the $F_1$ head-piece of the ATP-synthase, which is on the inner surface of the mitochondrial inner membrane, is on the outside of the membrane in this artificial system. The ATP-synthase also faces outward in submitochondrial particles made by breaking the inner membrane with sonic oscillations. In all cases, ATP synthesis requires protons to move through $F_0$ in the direction of $F_1$.

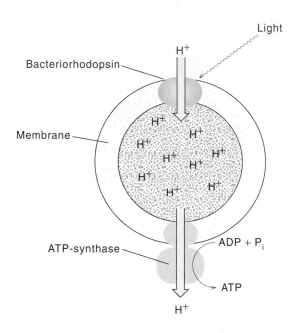

$H^+$ are in equilibrium with $H_2O$, an outward movement of $OH^-$ is effectively equivalent to an inward movement of $H^+$. It is thermodynamically downhill in respiring mitochondria because the pH is higher in the matrix than in the cytosol (fig. 14.25a).

### Export of ATP Is Coupled to ADP Uptake

ATP, which is produced in the mitochondria largely for use elsewhere in the cell, is exported by a transport system that exchanges ATP for ADP. This exchange is not electrically neutral, since the net charge on ATP is approximately $-4$, whereas that on ADP is about $-3$ at pH 7. The outward movement of a molecule of ATP in exchange for uptake of an ADP removes one negative charge from the matrix and is driven by the electrical gradient, $\Delta\psi$ (fig. 14.25b). The

## Figure 14.25

(*a*) The phosphate/hydroxide exchange protein carries a phosphate ion ($H_2PO_4^-$) in one direction across the mitochondrial inner membrane in exchange for an $OH^-$ ion moving in the opposite direction. These movements are reversible, but a net efflux of $OH^-$ is thermodynamically downhill because the respiratory chain pumps protons out of the matrix and raises the pH there. Because $OH^-$ efflux is linked to $H_2PO_4^-$ uptake, phosphate is concentrated in the matrix. (*b*) The ADP/ATP exchange protein exchanges $ADP^{-3}$ for $ATP^{-4}$. An outward movement of ATP removes one negative charge from the matrix and is favored because proton pumping by the respiratory chain gives the matrix a negative charge relative to the cytosol.

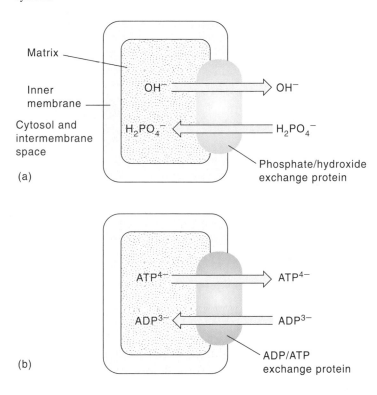

(a)

(b)

ATP/ADP exchange protein, or adenine nucleotide translocator, is actually the most abundant protein in the mitochondrial inner membrane.

The combined effect of exchanging extramitochondrial $ADP^{-3}$ and $H_2PO_4^-$ for mitochondrial $ATP^{-4}$ and $OH^-$ is to move one proton into the mitochondrial matrix for every molecule of ATP that the mitochondrion releases into the cytosol. This proton translocation must be considered, along with the movement of protons through the ATP synthase, to account for the P-to-O ratio of oxidative phosphorylation. If three protons pass through the ATP synthase, and the adenine nucleotide and $P_i$ transport systems move one additional proton, then four protons in total move into the matrix for each ATP molecule provided to the cytosol.

## Electrons from Cytosolic NADH Are Imported by Shuttle Systems

The mitochondrial inner membrane has no transport system for $NAD^+$ or NADH. In animal cells, most of the NADH that must be oxidized by the respiratory chain is generated in the mitochondrial matrix by the TCA cycle or the oxidation of fatty acids. However, NADH also is generated by glycolysis in the cytosol. If $O_2$ is available, it clearly is advantageous to reoxidize this NADH by the respiratory chain, rather than by the formation of lactate or ethanol as described in chapter 12. This is evident from the findings that approximately 2.5 molecules of ATP can be formed for each NADH oxidized in the mitochondria, whereas no ATP is made when NADH is oxidized by the cytosolic lactate dehydrogenase or alcohol dehydrogenase.

Animal cells use several types of "shuttle" system to transfer electrons from cytosolic NADH to the respiratory chain. These shuttle systems consume energy so that instead of 2.5 only about 1.5 molecules of ATP are produced for every cytosolic NADH.

## Complete Oxidation of Glucose Yields about 30 Molecules of ATP

The shuttles that operate between the cytosol and mitochondria must be taken into account when we calculate the total yield of ATP from a molecule of glucose that is oxidized to $CO_2$ and $H_2O$. To make such a calculation, recall first that two molecules of ATP are made from each molecule of glucose in glycolysis, and two more are made via GTP in the TCA cycle (fig. 14.26 and table 14.3). Two molecules of NADH are produced in the cytosol by glycolysis, and eight molecules of NADH are generated in the mitochondrial matrix by the pyruvate dehydrogenase complex and the TCA cycle. Reoxidation of the eight molecules of mitochondrial NADH by the respiratory chain drives the export of approximately 20 molecules of ATP to the cytosol (2.5 for each molecule of NADH). The two molecules of succinate proceeding through the succinate dehydrogenase reaction also transfer electrons to the respiratory chain, generating about three molecules of ATP (1.5 per molecule of succinate). The two molecules of NADH formed in the cytosol could contribute about three ATPs ($2 \times 1.5$). Exporting the two molecules of ATP formed in the TCA cycle requires the uptake of two protons by the mitochondria, which reduces the total yield of ATP by about 0.5. (Remember that the adenine nucleotide transporter and the ATP-synthase together must import four protons for each ATP they provide to the cytosol.) This leaves a grand total of 29.5 or 31 molecules of ATP. Some texts quote a value of 38

Table 14.3

Balance Sheet for Oxidation of One Molecule of Glucose

| | NADH | FADH$_2$ | ATP |
|---|---|---|---|
| **CYTOPLASMIC REACTIONS** | | | |
| *Glycolysis* | | | |
| 1. Glucose ⟶ glucose-6-P | | | −1 |
| 2. Fructose-6-P ⟶ fructose-1,6-P | | | −1 |
| 3. (2) Glyceraldehyde-3-P ⟶ glycerate-1,3-bisphosphate | +2 | | |
| 4. (2) Glycerate-1,3-bisphosphate ⟶ glycerate-3-P | | | +2 |
| 5. (2) Phosphoenolpyruvate ⟶ pyruvate | | | +2 |
| **MITOCHONDRIAL REACTIONS** | | | |
| *Pyruvate Dehydrogenase* | | | |
| (2) Pyruvate ⟶ acetyl-CoA | +2 | | |
| *TCA Cycle* | | | |
| 1. (2) Isocitrate ⟶ α-ketoglutarate | +2 | | |
| 2. (2) α-Ketoglutarate ⟶ succinyl-CoA | +2 | | |
| 3. (2) Succinyl-CoA ⟶ succinate | | | +1.5[a] |
| 4. (2) Succinate ⟶ fumarate | | +2 | |
| 5. (2) Malate ⟶ oxaloacetate | +2 | | |
| *Electron Transport–Oxidative Phosphorylation* | | | |
| 1. Reoxidation of NADH produced in glycolysis | −2 | | +3[b] |
| 2. Reoxidation of NADH produced by pyruvate dehydrogenase | −2 | | +5 |
| 3. Reoxidation of NADH produced in TCA cycle | −6 | | +15 |
| 4. Reoxidation of FADH$_2$ produced in TCA cycle | | −2 | +3 |
| Totals | 0 | 0 | 29.5 |

[a] This number is 1.5 instead of 2 to take into account the energy required to export ATP to the cytosol.

[b] This number is 3 instead of 5 because the glycerol phosphate shuttle bypasses complex I.

molecules of ATP per glucose, but this is based on unrealistically high P-to-O ratios.

The actual yields of ATP under physiological conditions are probably less than these limiting values. Some of the free energy that is stored in the form of ΔpH and Δψ may be lost by leakage of protons or other ions through the membrane or may be used to drive reactions other than the formation of ATP. In addition, NADH that is formed in the cytosol may be used there for reductive biosynthetic reactions, rather than contributing electrons to the respiratory chain. Even so, it is clear that respiration allows cells to make ATP in amounts that are far greater than the amounts provided by glycolysis.

## Figure 14.26

Complete oxidation of a molecule of glucose to $CO_2$ and $H_2O$ generates approximately 30 molecules of ATP. Reoxidation of eight molecules of NADH in the mitochondrial matrix yields about 20 molecules of ATP (2.5 ATP per molecule of NADH). Reoxidation of two molecules of $FADH_2$ bound to succinate dehydrogenase yields 3 molecules of ATP (1.5 per molecule of $FADH_2$). Two molecules of ATP are produced in glycolysis, and the TCA cycle produces two more ATP via GTP. The latter two molecules of ATP are equivalent to about 1.5 molecules of ATP in the cytosol because exporting them from the mitochondria requires proton uptake. (This proton uptake uses some of the energy stored in the electrochemical gradient for protons and thus decreases the total amount of ATP that can be made by the ATP synthase.) Oxidizing the two molecules of NADH formed by glycolysis generates three molecules of ATP if a glycerol–phosphate shuttle is used. Slightly more ATP can be formed by using a malate shuttle.

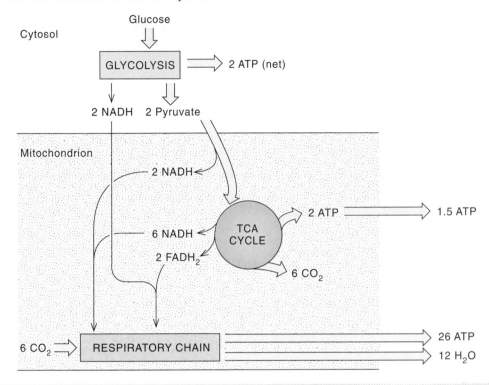

## Summary

In eukaryotes, most of the reactions of aerobic energy metabolism occur in mitochondria. An inner membrane separates the mitochondrion into two spaces: the internal matrix space and the intermembrane space. An electron-transport system in the inner membrane oxidizes NADH and succinate at the expense of $O_2$, generating ATP in the process. The operation of the respiratory chain and its coupling to ATP synthesis can be summarized as follows:

1. Electron transfer to $O_2$ occurs stepwise, through a series of flavoproteins, cytochromes (heme proteins), iron–sulfur proteins, and a quinone. Most of the electron carriers are collected in four large complexes, which communicate via two mobile carriers—ubiquinone (UQ) and cytochrome $c$. Complex I transfers electrons from NADH to UQ, and complex II transfers electrons from succinate to UQ. Both of these complexes contain flavins and numerous iron–sulfur centers. Complex III, which contains three cytochromes (cytochromes $b_L$, $b_H$, and $c_1$) and one iron–sulfur protein, passes electrons from reduced ubiquinone ($UQH_2$) to cytochrome $c$. Complex IV contain two cytochromes ($a$ and $a_3$) and two Cu atoms, and transfers electrons from cytochrome $c$ to $O_2$. The transfer of electrons through complex III occurs by a cyclic series of reactions (the Q cycle), in which $UQH_2$ and UQ undergo oxidation and reduction at two distinct sites.

2. As electrons move through complexes I, III, and IV, protons are taken up from the matrix and released on the cytosolic side of the membrane. This raises the pH of the matrix and leaves the matrix negatively charged relative to the cytosol, creating an electrochemical potential difference that tends to pull protons from the cytosol back into the matrix.

3. Proton influx through the $F_0$ base-piece of the ATP-synthase in the inner membrane drives the formation of ATP by causing the release of bound ATP from the catalytic site on the $F_1$ head-piece of the enzyme.

Approximately 2.5 molecules of ATP are synthesized for each pair of electrons that pass down the electron-transport chain from NADH to $O_2$.
4. Uncouplers dissipate the electrochemical potential gradient by carrying protons across the membrane; respiration then occurs rapidly even in the absence of phosphorylation.
5. The electrochemical potential gradient also drives the uptake of $P_i$ and ADP into the mitochondrial matrix and the export of ATP to the cytosol.

## Selected Readings

Boyer, P., A perspective of the binding change mechanism for ATP synthesis. *FASEB J.* 3:2164, 1989. A review of work on the ATP-synthase.

Chan, S. I., and P. M. Li, Cytochrome *c* oxidase: Understanding nature's design of a proton pump. *Biochem.* 29:1, 1990. A review of the operation of complex IV.

Chance, B., and G. R. Williams, Respiratory enzymes in oxidative phosphorylation II. Difference spectra. *J. Biol. Chem.* 217:395, 1955. One of a series of papers developing kinetic techniques for elucidating the sequence of electron carriers in the respiratory chain.

Ernster, L. (ed.), *Bioenergetics.* Amsterdam: Elsevier, 1984. A collection of reviews covering electron transport, the ATP-synthase, translocation of ions across the mitochondrial inner membrane, thermogenesis in brown fat, and other topics in bioenergetics.

Hinkle, P. C., A. Kumar, A. Resetar, and D. L. Harris, Mechanistic stoichiometry of mitochondrial oxidative phosphorylation. *Biochem.* 30:3576, 1991. Measurements of P-to-O ratios and a discussion of the amount of ATP synthesized during the oxidation of glucose.

Mitchell, P., and J. Moyle, Stoichiometry of proton translocation through the respiratory chain and adenosine triphosphatase systems of rat liver mitochondria. *Nature* 208:147, 1965. The initial observations that electron transport moves protons outward across the mitochondrial inner membrane and that ATP hydrolysis does the same.

Reynafarje, B., and A. L. Lehninger, The $K^+$/site and $H^+$/site stoichiometry of mitochondrial electron transport. *J. Biol. Chem.* 254:6331, 1978. Valinomycin is used to allow $K^+$ to move inward across the inner membrane in response to the electric potential difference created by $H^+$ efflux. The number of protons pumped out is found to be larger than previously estimated.

Trumpower, B., Function of the iron–sulfur proteins of the cytochrome $bc_1$ segment in electron transfer and energy-conserving reactions of the mitochondrial respiratory chain. *Biochim. Biophys. Acta* 639:129, 1981. A review of the structure and operation of complex III and the Q cycle.

## Problems

1. Compare and contrast the role of heme in hemoglobin–myoglobin versus the cytochromes.

2. In the reaction catalyzed by dihydrolipoyl dehydrogenase, one of three enzymes in the pyruvate dehydrogenase complex (fig. 13.5), electrons flow from oxidized lipoic acid to enzyme-bound FAD to $NAD^+$. Compare the flow of electrons in the latter part of this scheme (FAD to $NAD^+$) to the flow of electrons in the electron transport scheme (Complex I). Is there a distinct difference in the flow of electrons in the two schemes? If so, can you provide a possible explanation for this difference?

3. Calculate the ATP yield (mole/mole) of the complete oxidation of each of the following: (a) pyruvate, (b) glyceraldehyde-3-phosphate, (c) the acetyl portion of acetyl-CoA.

4. Why have the classical methods of enzymology (purification and characterization of proteins) been of limited help in deducing how the electron-transport system functions?

5. Why did nature pick iron, and to a much lesser degree copper, rather than calcium, magnesium, potassium, and so forth for roles in the electron-transport scheme?

6. Compare iron–sulfur proteins, flavoproteins, and quinones with respect to the following:
   (a) Chemical nature of the functional group that undergoes oxidation–reduction.
   (b) Number of reducing equivalents per redox center involved in electron donor/acceptor reactions of physiological importance. Indicate if semiquinones are formed and include them in the reduction scheme.
   (c) Stoichiometry of protons taken up per electron.

7. (a) Describe how heme is bound to the protein portion of the $a/a_3$-, *b*-, and *c*-type cytochromes.

(b) Although three types of cytochromes occur in rat liver mitochondria, CO and $CN^-$ inhibit electron transfer only at the cytochrome $a/a_3$ complex. Why do these inhibitors interact with cytochrome $a/a_3$ but not with cytochrome $b$ or cytochrome $c$?

8. Calculate the standard redox potential change ($\Delta E^{\circ\prime}$) and the standard free energy change ($\Delta G^{\circ\prime}$) for the following reactions at pH 7.0. Write a balanced equation for each reaction.

    (a) Cyt $c(Fe^{2+})$ + cyt $a_3(Fe^{3+}) \rightarrow$
    $$\text{cyt } c(Fe^{3+}) + \text{cyt } a_3(Fe^{2+})$$

    (b) 4 cyt $c(Fe^{2+})$ + $O_2$ + 4 $H^+ \rightarrow$
    $$4 \text{ cyt } c(Fe^{3+}) + 2 \text{ HOH}$$

    (c) Oxidation of succinate by succinate: Cytochrome $c$ reductase.

9. (a) In biological oxidation–reduction reactions, does the stoichiometry of electron transfer (reducing equivalents per mole) differ among the 1Fe, 2Fe-2S, and 4Fe-4S centers?

    (b) 4Fe-4S centers function in electron transport over a wide range of reduction potentials. Nothing inherent in the iron–sulfur cluster suggests this range of reduction potentials. Therefore, what other component(s) must dictate reduction potential?

10. What percentage of cytochrome $c$ will be in the oxidized form in a solution held at +0.30 V and pH 7.0?

11. Given the standard reduction potentials for cytochrome $c$ and ubiquinone at pH 7.0 (see text), calculate the corresponding values at pH 6.0 and 8.0.

12. Rotenone, which blocks the transfer of electrons from $FMNH_2$ of the NADH dehydrogenase to ubiquinone, is a potent insecticide and fish poison.

    (a) Explain why rotenone is lethal to insects and fish.

    (b) Would you expect the use of rotenone as an insecticide to be potentially hazardous to other animals (e.g., humans)? Why or why not?

    (c) If isolated mitochondria are respiring with succinate as substrate, do you expect a change in $O_2$ consumption on addition of rotenone? If $\beta$-hydroxybutyrate is the respiratory substrate?

13. (a) Explain the necessity of having a ubiquinone concentration in excess of other mitochondrial electron-transfer components.

    (b) Suppose that you are examining muscle tissue mitochondria in which the UQ content is well below normal. You find that concentrations of all other electron-transfer components are within the normal range. Predict the effect of UQ deficiency on oxidation of: (i) NADH-producing substrates, (ii) succinate, (iii) ascorbate plus a redox mediator.

    (c) Explain why UQ-deficient muscle tissue has greater than normal concentrations of lactate. (Ogasahara, Engel, Frens, et al., *Proc. Natl. Acad. Sci. USA* 86:2379–2382, 1989.)

14. (a) Explain what is meant by "tightly coupled" mitochondria. How can we determine whether mitochondria are tightly coupled?

    (b) What is the importance of "respiratory control" in oxidation of metabolites?

    (c) In what metabolic circumstance is it advantageous for the organism to have mitochondria uncoupled?

15. The uncoupling reagent, 2,4-dinitrophenol (2,4-DNP), is highly toxic to humans, causing marked increase in metabolism, body temperature, profuse sweating, and in many instances, collapse and death. For a brief period in the 1940s, however, doses of 2,4-DNP presumed to be sublethal were prescribed as a means of weight reduction in humans.

    (a) Explain why administration of 2,4-DNP results in increased metabolic rate as evidenced by increased $O_2$ consumption.

    (b) How are the metabolic events resulting from administration of 2,4-DNP pertinent to regulation of glycolysis and the TCA cycle?

    (c) Why does consumption of 2,4-DNP lead to hyperthermia and profuse sweating?

    (d) Explain how 2,4-DNP uncouples oxidative phosphorylation.

16. A suspension of mitochondria is incubated with pyruvate, malate, and $^{14}C$-labeled triphenylmethylphosphonium [TPP] chloride under aerobic conditions. The mitochondria are rapidly collected by centrifugation, and the amount of $^{14}C$ that they contain is measured. In a separate experiment, the volume of the mitochondrial matrix space was determined so that the concentration of TPP cation in the matrix can be calculated. The internal concentration is found to be 1,000 times greater than that in the external solution.

    (a) What is the apparent electric potential difference ($\Delta\Psi$) across the inner membrane? Express your answer in the appropriate units, and indicate which side of the membrane is positive.

    (b) Qualitatively, how might $\Delta\Psi$ be affected by the addition of an uncoupler?

17. Differentiate between electrogenic and neutral transport systems in mitochondria. How is electrogenic transport influenced by the membrane potential? What is the effect of neutral transport on the pH gradient?

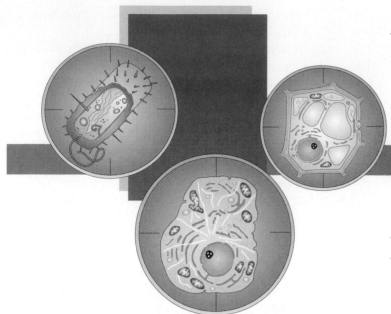

# Photosynthesis

*In photosynthesis light energy is converted into chemical energy.*

Heterotrophic organisms, including animals, fungi, and most types of bacteria, live by degrading complex molecules provided by other organisms. Life on earth obviously could not continue indefinitely in this manner without a mechanism for synthesizing complex molecules from simple ones. The energy that sustains this synthesis comes almost entirely from the sun, and is captured in the process of photosynthesis. Plants and photosynthetic bacteria annually convert, or "fix," about $10^{11}$ tons of carbon from $CO_2$ into organic compounds. In spite of the impressive magnitude of this conversion, the total amount of fixed carbon on earth is decreasing as a result of consumption. As our reserves of energy diminish, it becomes increasingly important that we understand how photosynthesis works and how our activities affect it.

An overall equation for $CO_2$ fixation as it occurs in plants is

$$6\,CO_2 + 6\,H_2O + \text{light} \rightarrow C_6H_{12}O_6 + 6\,O_2 \qquad (1)$$

**Figure 15.1**

Photosynthesis harnesses the energy of sunlight for biosynthesis. In plants, the reactions of photosynthesis occur in chloroplasts. Pigment-protein antenna complexes in the thylakoid membrane absorb light and pass energy to the reaction centers of two distinct photosystems, where excited chlorophyll molecules (P680 in photosystem II, P700 in photosystem I) transfer electrons to a series of electron acceptors. Photosystem I generates strong reductants that ultimately reduce $NADP^+$; photosystem II generates a strong oxidant that splits $H_2O$, releasing $O_2$. Electrons flow from photosystem II to photosystem I through the cytochrome $b_6f$ complex. The electron-transfer reactions release protons in the thylakoid lumen and take up protons from the stroma, setting up a transmembrane pH gradient. Flow of protons back across the membrane through an ATP-synthase drives the formation of ATP. NADPH and ATP provided by the electron-transfer reactions are used to convert $CO_2$ to carbohydrates. $CO_2$ combines with ribulose-1,5-bisphosphate to form two molecules of glycerate-3-phosphate. ATP and NADPH are used to reduce the glycerate-3-phosphate to glyceraldehyde-3-phosphate and to recycle some of the glyceraldehyde-3-phosphate to ribulose-1,5-bisphosphate. The remaining glyceraldehyde-3-phosphate is exported from the chloroplast or converted to other carbohydrates. Many of the photosynthetic reactions resemble reactions seen in earlier chapters: The formation of hexoses and pentoses from trioses resembles reactions of gluconeogenesis and the pentose phosphate pathway (see chapter 12); the cytochrome $b_6f$ complex is similar to the mitochondrial complex III, and the proton-conducting ATP-synthase is similar to the mitochondrial ATP-synthase (see chapter 14). The major difference between the operation of mitochondria and chloroplasts is the direction of electron flow. In mitochondria, electrons flow from reduced organic compounds to $O_2$; photosynthetic organisms use the energy of light to push electrons in the opposite direction.

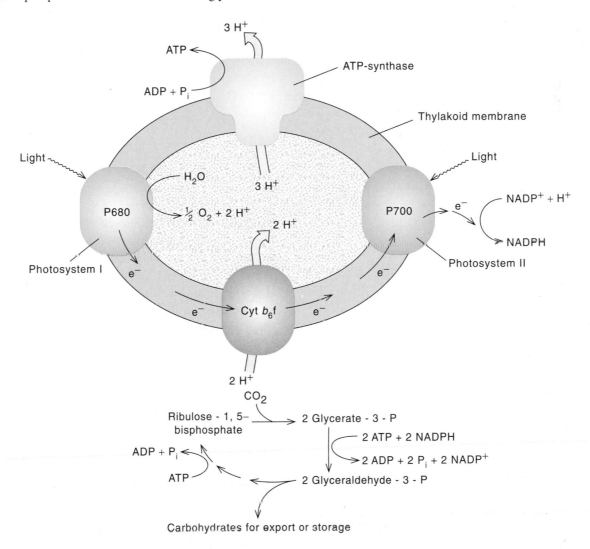

Electrons and protons are removed from $H_2O$, $O_2$ is evolved, and $CO_2$ is reduced to the level of a carbohydrate (fig. 15.1). One group of photosynthetic bacteria, the cyanobacteria, carry out identical reactions. Other types of photosynthetic bacteria carry out similar overall processes but do not evolve $O_2$; these species use materials other than $H_2O$ as a source of electrons.

If we leave out the light, the equilibrium for the synthesis of glucose from $CO_2$ and $H_2O$ lies vanishingly far to the left. The equilibrium constant at 27°C is $10^{-496}$! In this

## Figure 15.2

(a) Electron micrograph of a chloroplast in a lettuce leaf. The organelle is shaped like a flattened sausage with a width of about 10 $\mu$M and a thickness of about 3 $\mu$M. It is surrounded by a double outer membrane or envelope (E). The stroma (S) contains DNA, ribosomes, and the soluble enzymes of $CO_2$ fixation. Extending throughout the stroma is the thylakoid membrane, which is differentiated into stacked regions or grana (G) and unstacked stromal lamellae (SL). (Courtesy of Dr. Charles Arntzen.) (b) A schematic drawing of the chloroplast membrane systems. The highly folded thylakoid membrane separates the thylakoid lumen from the stroma.

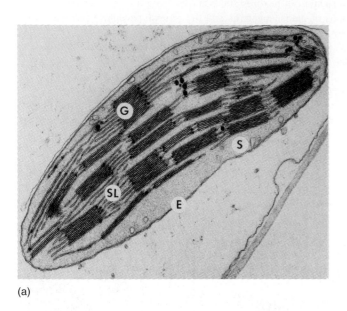

(a)

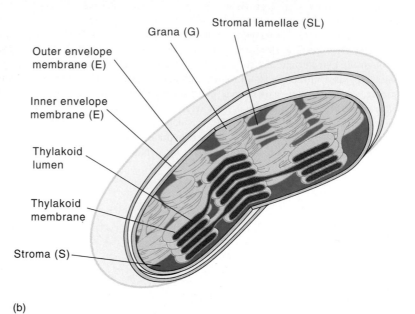

(b)

chapter we explore how photosynthetic organisms use light to drive the reaction in the direction of carbohydrates against this enormous thermodynamic gradient.

## The Photochemical Reactions of Photosynthesis Take Place in Membranes

Although important differences exist between photosynthesis in bacteria and plants, the photochemical reactions that capture the energy of light are basically the same. In plants, the reactions of photosynthesis take place in specialized organelles, the chloroplasts. Figure 15.2a is an electron micrograph of a chloroplast from a lettuce leaf. Chloroplasts are bounded by an envelope of two membranes, and they have a third internal membrane called the thylakoid membrane. In electron micrographs of thin-sectioned chloroplasts, the thylakoid membrane has the appearance of a large number of separate sheets of flattened vesicles. It actually is a single highly folded membrane that encloses a distinct compartment, the thylakoid lumen (fig. 15.2b). In places, the folded membrane stacks into disklike structures called grana. The chlorophyll found in chloroplasts is bound to proteins in the thylakoid membrane, and it is here that the initial conversion of light into chemical en-

ergy occurs. The thylakoid membrane also contains a collection of electron carriers and an ATP-synthase that are similar to those in the mitochondria. The enzymes responsible for the actual fixation of $CO_2$ and the synthesis of carbohydrates reside in the stroma that surrounds the thylakoid membrane (see fig. 15.2).

Algae are members of the plant kingdom and contain chloroplasts similar to those of higher plants, but prokaryotic photosynthetic organisms do not have chloroplasts. In prokaryotes, the photochemical reactions occur in the plasma membrane, which has extensive invaginations resembling the cristae of the mitochondrial inner membrane (fig. 15.3). Table 15.1 summarizes the main distinctions between the various types of photosynthetic organisms.

## Photosynthesis Depends on the Photochemical Reactivity of Chlorophyll

With the exception of certain archaebacteria, all known photosynthetic organisms take advantage of the photochemical reactivity of one or another type of chlorophyll (fig. 15.4). Chlorophylls resemble hemes but differ from them in four major respects:

1. The central metal atom is magnesium rather than iron.

**Table 15.1**

Properties of Photosynthetic Organisms

| Group | Evolve $O_2$ | Contain Chloroplasts | Type of Chlorophyll | Number of Photosystems |
|---|---|---|---|---|
| Algae and higher plants | yes | yes | chlorophyll *a* and *b* | 2 |
| Cyanobacteria | yes | no | chlorophyll *a* and *b* | 2 |
| Purple bacteria | no | no | bacteriochlorophyll *a* or *b* | 1 |

**Figure 15.3**

Cross-sectional view of a cell of *Rhodospirillum rubrum,* a purple photosynthetic bacterium. A double-membrane system surrounds the cell. The inner membrane is extensively invaginated into tubules (arrows). These look circular when they are cut in cross section. (Courtesy of Dr. Gerald Peters.)

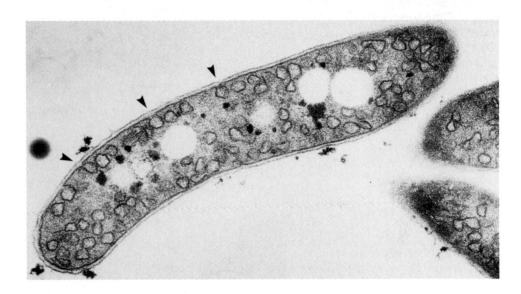

2. Chlorophylls have an additional ring (ring V).

3. In the case of chlorophyll *a* and chlorophyll *b*, the two major chlorophylls in plants and cyanobacteria, one of the pyrrole rings (ring IV) is reduced by the addition of two hydrogens. In bacteriochlorophylls *a* and *b*, which occur in the purple and green bacteria, two of the rings are reduced (rings II and IV).

4. The propionyl side chain of ring IV is esterified with a long-chain isoprenoid alcohol.

Chlorophylls *a* and *b* contain the alcohol phytol; bacteriochlorophylls *a* and *b* have either phytol or geranylgeraniol, depending on the species of bacteria. Photosynthetic organisms also contain small amounts of pheophytins or bacteriopheophytins, which are the same as the corresponding chlorophylls or bacteriochlorophylls except that two hydrogens replace the $Mg^{2+}$ (see fig. 15.4). We will see that pheophytins and bacteriopheophytins play special roles as electron carriers in photosynthesis.

The reduction of ring IV in chlorophylls *a* or *b* changes the optical absorption spectrum of the molecule dramatically. Whereas the long-wavelength absorption band of a cytochrome is relatively weak (see fig. 14.4), chlorophyll *a* has an intense absorption band at 676 nm (fig. 14.5). Chlorophyll *b* has a similar band at 642 nm. Bacteriochlorophylls *a* and *b* have strong absorption bands in the region of 770 nm (see fig. 15.5). The chlorophylls thus absorb red or near infrared light very well.

## Light Is Composed of Photons

Before we discuss how chlorophylls can transform light into chemical energy, let's review some of the basic properties of light. Light is an oscillating electromagnetic field (fig.

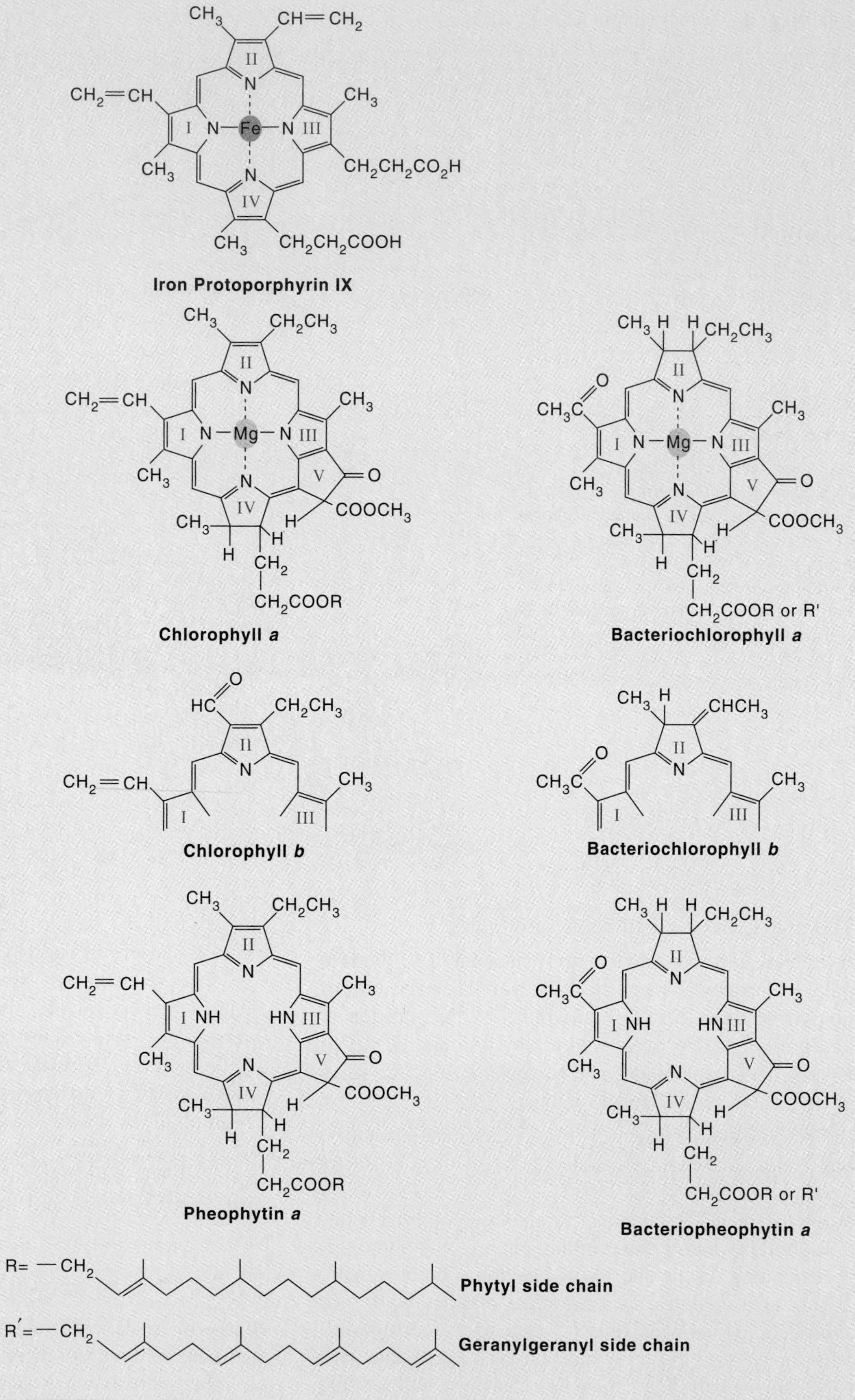

# Figure 15.4

Structures of protoporphyrin IX (the prosthetic group of hemoglobin, myoglobin, and the *c*-type cytochromes), and several types of chlorophyll and bacteriochlorophyll. Chlorophyll *a* and chlorophyll *b* are the main types of chlorophyll in plants and the cyanobacteria. Purple photosynthetic bacteria contain either bacteriochlorophyll *a* or *b*, depending on the bacterial species. (The green photosynthetic bacteria contain still another form of bacteriochlorophyll.) Chlorophyll *b* and bacteriochlorophyll *b* are the same as chlorophyll *a* and bacteriochlorophyll *a* except for the substituents on ring II. Pheophytins and bacteriopheophytins are the same as the corresponding chlorophylls or bacteriochlorophylls, except that two hydrogen atoms replace Mg.

**Iron Protoporphyrin IX**

**Chlorophyll *a***

**Bacteriochlorophyll *a***

**Chlorophyll *b***

**Bacteriochlorophyll *b***

**Pheophytin *a***

**Bacteriopheophytin *a***

**Phytyl side chain**

**Geranylgeranyl side chain**

## Figure 15.5

Absorption spectra of chlorophyll *a* and bacteriochlorophyll *a* in ether. Note that the long-wavelength absorption bands are much stronger than the α band of reduced cytochrome *c* (see fig. 14.4). In the chlorophyll–protein complexes found in photosynthetic organisms, the long-wavelength absorption band generally is shifted to even longer wavelengths. This probably reflects interactions between neighboring chlorophylls, which are bound to the proteins as dimers or larger groups.

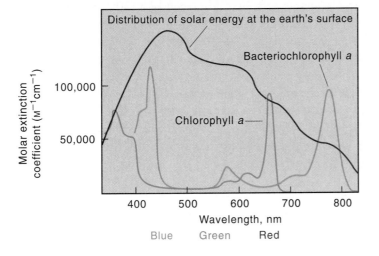

## Figure 15.6

Light is an oscillating electromagnetic field. The lengths of the arrows in this diagram represent the strength of the electric field at a particular time, as a function of position in a ray of light proceeding along the *y* axis.

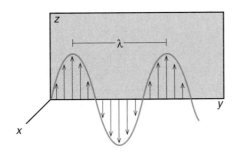

15.6). It interacts with matter in packets, or quanta, called *photons*, each of which contains a definite amount of energy. The relationship between the energy ε of a photon and the frequency ν of the oscillating electromagnetic field is given by the equation

$$\epsilon = h\nu$$

where *h* is Planck's constant ($4.12 \times 10^{-15}$ eV · s). A mole of photons is called an einstein, and the energy per einstein is

$$\mathscr{E} = N\epsilon = Nh\nu \tag{2}$$

where *N* is Avogadro's number ($6.02 \times 10^{23}$ photons/einstein). The frequency ν is the number of oscillations per second at a given point in space. The wavelength λ of the oscillations (the distance between successive peaks in the amplitude of the field) depends on both ν and the velocity *c* at which the peaks move through space

$$\lambda = \frac{c}{\nu} \tag{3}$$

In a vacuum, light travels with a velocity $c = 3 \times 10^{10}$ cm/s.

Photon energies often are stated in units of electron volts (eV). One eV is the energy required to move an electron against a potential difference of 1 V and is equivalent to 23,060 kcal/mol (96,480 kJ/mol). Blue light with a wavelength of 450 nm ($\nu = 6.7 \times 10^{14}$ s$^{-1}$) has an energy of 2.75 eV, or 64 kcal/einstein, and far red light (700 nm) has an energy of 1.77 eV (41 kcal/einstein).

## Photons of Light Interact with Electrons in Molecules

The electrons in a molecule reside in a set of molecular orbitals, each of which is characterized by a particular energy. In an isolated molecule, the orbital energies depend mainly on the interactions of the electrons with each other and with the nuclei. A molecule can have a variety of energies, depending on how its electrons are distributed among the available orbitals. For an organic molecule with $2n$ electrons, the lowest overall energy usually is obtained when two electrons with antiparallel spins occur in each of the first *n* orbitals, leaving all the orbitals with higher energies empty (fig. 15.7). This is the ground state of the molecule.

When light interacts with a molecule, and a photon is absorbed, an electron moves from one of the occupied molecular orbitals to an unoccupied orbital with a higher energy (see fig. 15.7). Two requirements must be met for this to occur. First, the difference between the energies of the two orbitals must be the same as the photon's energy, $h\nu$. This is why a given type of molecule absorbs light of some wavelengths and not of others. Second, the two orbitals must have different geometrical symmetries and must be oriented appropriately with respect to the oscillating electrical field of the light. This requirement explains why some absorption bands are stronger than others.

If chlorophyll (or any other molecule) absorbs a photon, it is excited to a state that lies above the ground state in energy. An excited molecule can return to the ground state by releasing energy in several different ways (see fig. 15.7). One possibility is to emit a photon; this is fluorescence. Another decay mechanism is to transfer the energy to a

**Figure  15.7**

When a molecule absorbs light, an electron is excited to a molecular orbital with higher energy. The horizontal bars in this diagram represent molecular orbitals for electrons. Each orbital can hold two electrons with antiparallel spins (arrows pointing upward or downward). Only the top few of the filled orbitals are shown here. Absorption of light raises an electron from one of these orbitals to an orbital that is normally unoccupied. For this to occur, the energy of the photon must match the difference between the energies of the two orbitals. The choice of an upward or downward arrow is arbitrary, but no change of spin occurs during the excitation. As long as the spins of the two unpaired electrons remain antiparallel, so that the molecule has no net electronic spin, the molecule is said to be in an excited "singlet" state. An excited molecule can return directly to the ground state by giving off energy as fluorescence or heat, or by transferring energy to another nearby molecule, or it can transfer an electron to another molecule (see fig. 15.8).

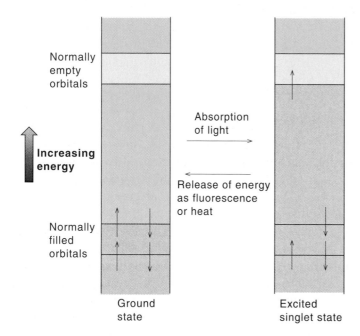

neighboring molecule. This phenomenon, resonance energy transfer, plays an important role in photosynthesis, as we will see. An excited molecule also can transfer an electron to a neighboring molecule.

Electron transfer can be a favorable path for the decay of an excited molecule, because an electron in the upper, normally unoccupied orbital is bound less tightly than one in a lower, normally filled orbital. If the absorption of a photon increases the energy of the molecule by $\epsilon$ electron volts, where $\epsilon \approx h\nu$, it makes the standard redox potential for removing an electron from the excited molecule ($E°$) more negative by approximately $\epsilon$ volts, compared with the $E°$ for the molecule in the ground state. In the case of chlorophyll $a$, $E°$ for oxidation in the ground state is about $+0.5$ V, and $h\nu$ for the long-wavelength absorption band is 1.7 eV. $E°$ of the excited molecule is therefore about $-1.2$ V. This means that in the excited state chlorophyll $a$ is

an extremely strong reductant. For comparison, recall that $NAD^+/NADH$ has an $E°'$ of only $-0.32$ V (see table 14.1). The basic principle underlying the photochemical process is that excitation causes chlorophyll to release an electron. The chlorophyll is oxidized by the electron loss while the molecule receiving the electron becomes reduced. The oxidized chlorophyll returns to its ground state by extracting an electron from a third molecule (fig. 15.8).

The idea that light drives the formation of oxidants and reductants was first advanced by C. B. van Niel in the 1920s. It was strengthened through experiments done by Robin Hill in 1939. Hill discovered that isolated chloroplasts evolved $O_2$ if illuminated in the presence of an added electron acceptor such as ferricyanide, $Fe(CN)_6^{3-}$. The electron acceptor became reduced in the process. Because no fixation of $CO_2$ occurred under these conditions, this experiment demonstrated that the photochemical reactions of photosynthesis can be separated from the reactions that involve $CO_2$ fixation.

## Photooxidation of Chlorophyll Generates a Cationic Free Radical

The photooxidation of chlorophyll can be detected by changes in the optical absorption spectrum of the molecule. Oxidation results in the loss of the chlorophyll's characteristic absorption bands. Measurements of this sort were made first in the 1950s by Bessel Kok and Louis Duysens. Kok found that illumination of chloroplasts caused an absorbance decrease at 700 nm (fig. 15.9). He suggested that this reflected the photooxidation of a reactive chlorophyll complex, which he called P700. (P stood for "pigment," and 700 for the wavelength at which the unoxidized complex had its main absorption band.) Duysens made similar observations on purple photosynthetic bacteria. Here the reactive bacteriochlorophyll complex absorbed at 870 nm, and Duysens called it P870. A second type of reactive complex in chloroplasts, P680, was discovered by H. Witt and his colleagues. Subsequent investigations showed that P700 and P680 were parts of two distinct photochemical systems: photosystem I and photosystem II.

Photooxidation of P870, P700, or P680 generates a cationic radical ($P870^{+\bullet}$, $P700^{+\bullet}$, or $P680^{+\bullet}$) in which the positive charge and the unpaired electron are delocalized over the conjugated ring system of the chlorophyll. In fact, the electron spin resonance (ESR) and electron nuclear double resonance (ENDOR) spectra of $P870^{+\bullet}$ indicate that the unpaired electron is delocalized over two bacteriochlorophyll $a$ molecules, which form a closely interacting pair. The strong interaction between the two bacteriochlorophylls explains why the absorption band of the complex is at

## Figure 15.8

The photochemical process that initiates photosynthesis is an electron-transfer reaction. The horizontal bars and vertical arrows represent molecular orbitals and electrons, as in figure 15.7. Absorption of light increases the free energy of a chlorophyll complex (Chl) by $h\nu$, making the transfer of an electron to an acceptor (A) thermodynamically favorable. The oxidized chlorophyll complex ($Chl^+$) extracts an electron from a donor (D).

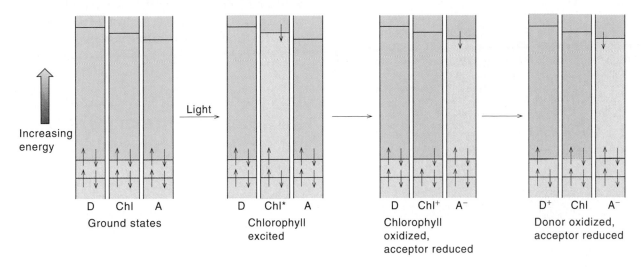

Increasing energy

Light

| D | Chl | A |
| Ground states | | |

| D | Chl* | A |
| Chlorophyll excited | | |

| D | Chl⁺ | A⁻ |
| Chlorophyll oxidized, acceptor reduced | | |

| D⁺ | Chl | A⁻ |
| Donor oxidized, acceptor reduced | | |

## Figure 15.9

When a suspension of chloroplasts is illuminated, its optical absorbance decreases in the regions around 430 and 700 nm. The absorbance changes reflect the oxidation of a special chlorophyll complex (P700). Addition of a chemical oxidant such as potassium ferricyanide causes similar absorbance changes. (Source: From B. Kok, Partial purification and determination of oxidation reduction potential of the photosynthetic chlorophyll complex absorbing at 700 mμ, *Biochim. et Biophys. Acta* 48:527, 1961.)

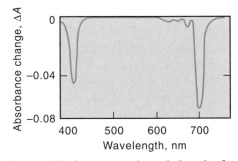

870 nm, whereas the long-wavelength band of monomeric bacteriochlorophyll *a* in solution is at 770 nm (see fig. 15.5). P700 and P680 probably consist of similar dimers of chlorophyll *a*, although the arrangements of the chlorophylls in these dimers may be somewhat different.

## The Reactive Chlorophyll Is Bound to Proteins in Reaction Centers

The chlorophyll or bacteriochlorophyll that undergoes photooxidation is bound to a protein in a complex called a reaction center. Reaction centers have been purified by disrupting chloroplasts or bacterial membranes with detergents. This was first achieved with purple photosynthetic bacteria. When they are excited with light, purified bacterial reaction centers can carry out the initial photochemical transfer of an electron from P870 to a series of electron acceptors.

Reaction centers of purple bacteria typically contain three polypeptides, four molecules of bacteriochlorophyll, two bacteriopheophytins, two quinones, and one nonheme iron atom. In some bacterial species, both quinones are ubiquinone. In others, one of the quinones is menaquinone (vitamin $K_2$), a naphthoquinone that resembles ubiquinone in having a long side chain (fig. 15.10). Reaction centers of some species, such as *Rhodopseudomonas viridis*, also have a cytochrome subunit with four *c*-type hemes.

The crystal structure of reaction centers from *R. viridis* was determined by Hartmut Michel, Johann Deisenhofer, Robert Huber, and their colleagues in 1984. This was the first high-resolution crystal structure to be obtained for an integral membrane protein. Reaction centers from another species, *Rhodobacter sphaeroides*, subsequently proved to have a similar structure. In both species, the bacteriochlorophyll and bacteriopheophytin, the iron atom and the quinones are all on two of the polypeptides, which are folded into a series of α helices that pass back and forth across the cell membrane (fig. 15.11*a*). The third polypeptide resides largely on the cytoplasmic side of the membrane, but it also has one transmembrane α helix. The cytochrome subunit of the reaction center in *R. viridis* sits on the external (periplasmic) surface of the membrane.

Figure 15.11*b* shows the arrangement of the pigments in greater detail. Two of the four bacteriochlorophylls are

**Figure 15.10**

The structures of ubiquinone, menaquinone (vitamin $K_2$), plastoquinone, and phylloquinone (vitamin $K_1$). Purple photosynthetic bacteria contain ubiquinone, menaquinone, or both, depending on the bacterial species; chloroplasts contain plastoquinone and phylloquinone.

Ubiquinone

Menaquinone

Plastoquinone

Phylloquinone

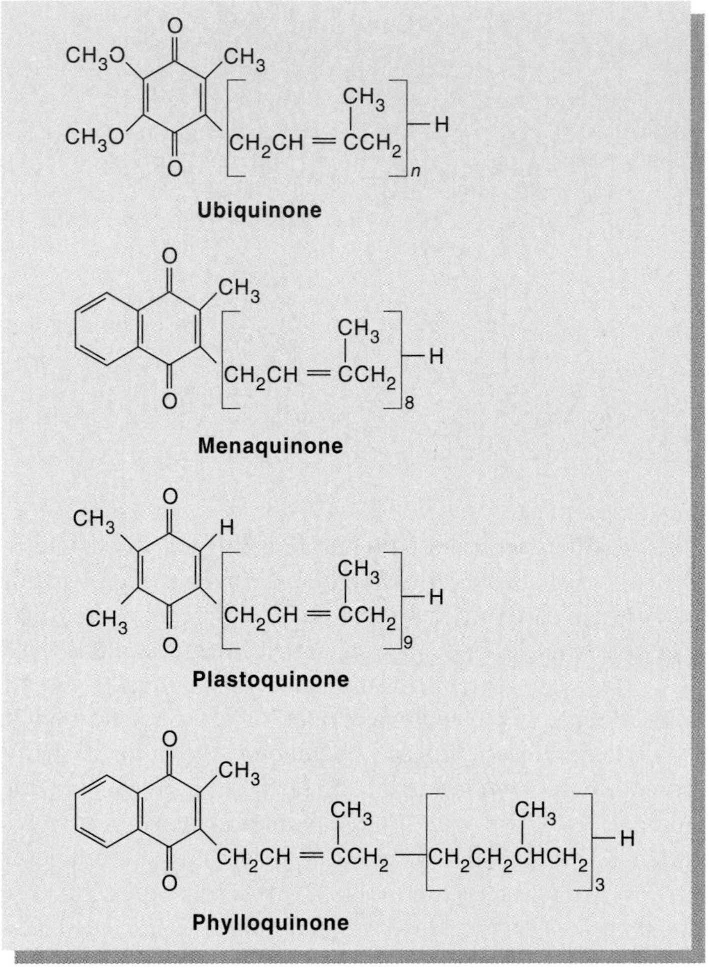

packed closely together. Studies of the reaction center's optical absorption spectrum indicate that this is the bacteriochlorophyll dimer that releases an electron when the reaction center is excited with light.

Although the structures of the plant reaction centers are not yet known in detail, photosystem II reaction centers resemble reaction centers of purple bacteria in several ways. The amino acid sequences of their two major polypeptides are homologous to those of the two polypeptides that hold the pigments in the bacterial reaction center. Also, the reaction centers of photosystem II contain a nonheme iron atom and two molecules of plastoquinone, a quinone that is closely related to ubiquinone (see fig. 15.10), and they contain one or more molecules of pheophytin $a$ and several

molecules of chlorophyll $a$ in addition to the two that probably make up P680.

The reaction center of photosystem I is larger and more complex. It contains two large polypeptides and at least seven other smaller subunits. The reactive chlorophyll $a$ dimer P700 resides on the two main polypeptides, along with about 60 additional molecules of chlorophyll $a$, two quinones, and an iron–sulfur center.

## In Purple Bacterial Reaction Centers, Electrons Move from P870 to Bacteriopheophytin and Then to Quinones

Let's now consider the acceptors that extract an electron from P870 in the purple bacterial reaction center. The first acceptor the reduction of which can be well resolved kinetically is one of the two bacteriopheophytins. When reaction centers are excited with a short flash of light, an excited singlet state of the reactive bacteriochlorophylls (P870*) is formed essentially instantaneously. This state decays in about $3 \times 10^{-12}$ s, and as it does, a P870$^+_\bullet$BPh$^-_\bullet$ radical pair is created. Here P870$^+_\bullet$ is the radical formed by removing an electron from P870, and BPh$^-_\bullet$ is the anionic radical formed by adding an electron to a bacteriopheophytin (fig. 15.12). The creation of the radical-pair state can be detected spectrophotometrically by the loss of absorption bands of P870 and the bacteriopheophytin and the formation of new absorption bands attributable to the two radicals. The bacteriopheophytin that undergoes reduction is the one on the right in figures 15.11$a$ and $b$. Differences in their amino acid environments make the two bacteriopheophytins spectroscopically distinguishable.

The P870$^+_\bullet$BPh$^-_\bullet$ radical pair disappears in about $2 \times 10^{-10}$ s when an electron jumps from BPh$^-_\bullet$ to one of the quinones ($Q_A$). This leaves the reaction center in the state P870$^+_\bullet$Q$_A^-_\bullet$, where $Q_A^-_\bullet$ is the anionic semiquinone (see fig. 15.12).

The electron carriers that participate in these first few steps are all fixed in position in the reaction center so that the initial steps in electron transfer involve electron movements only. One indication of this is that the electron-transfer reactions, in addition to being phenomenally fast, are almost independent of temperature. The reactions also are amazingly efficient. This can be expressed in terms of the quantum yield of P870$^+_\bullet$Q$_A^-_\bullet$, which is the number of moles of P870$^+_\bullet$Q$_A^-_\bullet$ formed per einstein of light absorbed. The measured quantum yield in purified reaction centers is $1.02 \pm 0.04$. Essentially every time the reaction center is excited, an electron moves from P870 to $Q_A$.

## Figure 15.11

(a) Structure of the reaction center of *Rhodopseudomonas viridis*. The α-carbon chains of the two main polypeptides are shown in yellow and gold. Each of these subunits has five α-helical regions that pass back and forth across the phospholipid bilayer of the cell membrane. A third subunit, shown in light yellow, sits on the cytosolic side of the membrane but also has one transmembrane α helix. The cytochrome subunit, which resides on the outer (periplasmic) surface of the membrane, is shown in purple at the top, with its four heme groups in red. The four bacteriochlorophylls are represented in shades of blue and green, the two bacteriopheophytins in white, the iron atom in red, the two quinones in purple, and a carotenoid molecule in orange. The phytyl side chains of the bacteriochlorophylls and bacteriopheophytins have been truncated for clarity; the isoprenoid tail of one of the quinones ($Q_B$) is disordered and not visible in the crystal structure. (Based on the crystal structure described by J. Deisenhofer, O. Epp, K. Miki, R. Huber, and H. Michel.) (b) An expanded view of some of the components of the *R. viridis* reaction center. The color coding is as in (a). The bacteriochlorophyll dimer that undergoes photooxidation is at the top. These two molecules are about 3 Å apart where they overlap in ring I. In *R. viridis*, one of the two quinones ($Q_A$, on the right in the figure) is menaquinone and the other ($Q_B$, left) is ubiquinone.

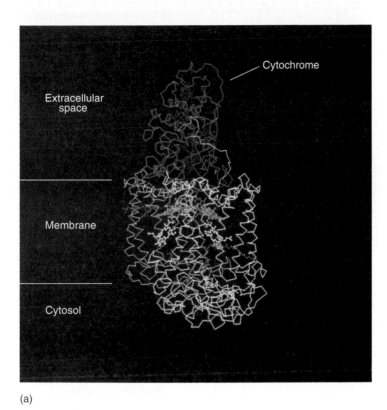

(a)

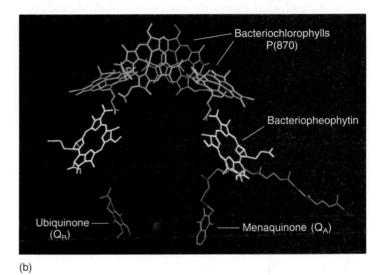

(b)

## A Cyclic Electron-Transport Chain Moves Protons Outward across the Membrane That Drive the Formation of ATP

From $Q_A^{-\bullet}$, an electron moves to the second quinone bound to the reaction center ($Q_B$ in fig. 15.13). In the meantime, a cytochrome replaces the electron that was removed from P870, preparing the reaction center to operate again. In *R. viridis*, the electron donor is the bound cytochrome with four hemes shown at the top of figure 15.11a. In other species, it often is a soluble *c*-type cytochrome with a single heme, resembling mitochondrial cytochrome *c*.

When the reaction center is excited a second time, a second electron is pumped from P870 to the bacteriopheophytin and on to $Q_A$ and $Q_B$. This places $Q_B$ in the fully reduced form, $Q_B^{2-}$. Uptake of two protons transforms the reduced quinone to the uncharged quinol, $QH_2$, which dissociates from the reaction center into the phospholipid bilayer of the cell membrane. The protons taken up in the formation of $QH_2$ come from the cytosol (see fig. 15.13).

$QH_2$ is reoxidized by a cytochrome $bc_1$ complex that is very similar to complex III of the mitochondrial respiratory chain. Electrons move through the cytochrome $bc_1$ complex to a *c*-type cytochrome, which then diffuses to the reaction center and provides an electron for the reduction of P870$^{+\bullet}$ (see fig. 15.13). As it does in mitochondria, movement of electrons through the cytochrome $bc_1$ complex results in the release of protons on the outer surface of the membrane. The proton extrusion can be explained by the Q cycle that we discussed in connection with the mitochondrial complex (see fig. 14.11). Flow of electrons from $QH_2$ back to P870 thus pumps protons outward across the cell membrane, generating a transmembrane electrochemical potential gradient. The inside of the cell becomes negatively charged relative

**Figure 15.12**

The initial electron-transfer steps in reaction centers of purple photosynthetic bacteria. (P870* = the first excited singlet state of P870; BPh = bacteriopheophytin; $Q_A$ = menaquinone or ubiquinone, depending on the bacterial species.) In this scheme, various states of the photosynthetic apparatus are positioned vertically according to their free energies, with the states that have the highest free energies at the top.

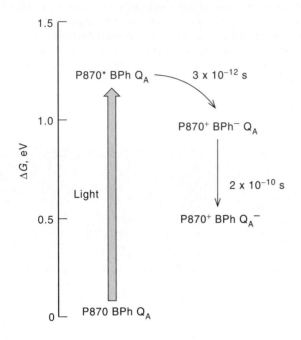

**Figure 15.13**

In purple photosynthetic bacteria, electrons return to $P870^+$ from the quinones $Q_A$ and $Q_B$ via a cyclic pathway. When $Q_B$ is reduced with two electrons, it picks up protons from the cytosol and diffuses to the cytochrome $bc_1$ complex. Here it transfers one electron to an iron–sulfur protein and the other to a $b$-type cytochrome and releases protons to the extracellular medium. The electron-transfer steps catalyzed by the cytochrome $bc_1$ complex probably include a Q cycle similar to that catalyzed by complex III of the mitochondrial respiratory chain (see fig. 14.11). The $c$-type cytochrome that is reduced by the iron–sulfur protein in the cytochrome $bc_1$ complex diffuses to the reaction center, where it either reduces $P870^+$ directly or provides an electron to a bound cytochrome that reacts with $P870^+$. In the Q cycle, four protons probably are pumped out of the cell for every two electrons that return to P870. This proton translocation creates an electrochemical potential gradient across the membrane. Protons move back into the cell through an ATP-synthase, driving the formation of ATP.

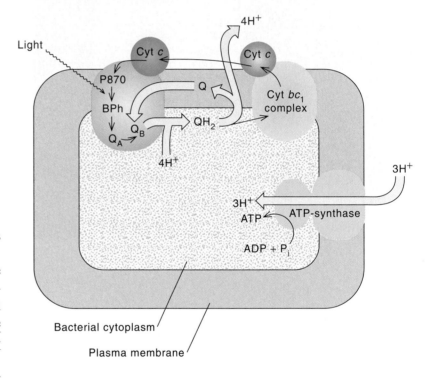

to the external medium, and the cytosolic pH becomes higher than the external pH.

Flow of protons back into the bacterial cell, down the electrochemical potential gradient, is mediated by an ATP-synthase resembling the ATP-synthase of the mitochondrial inner membrane (see chapter 14). As in mitochondria, the movement of protons through the $F_0$ base-piece of the enzyme drives the formation of ATP (see fig. 15.13).

Note that the electron-transport system of purple photosynthetic bacteria is cyclic. Excitation of the reaction center with light creates a strong reductant (BPh$\overset{-}{\bullet}$) and a strong oxidant (P870$\overset{+}{\bullet}$), and electrons return from BPh$\overset{-}{\bullet}$ to P870$\overset{+}{\bullet}$ via quinones, the cytochrome $bc_1$ complex, and $c$-type cytochromes. This cyclic flow of electrons results in the formation of ATP but no net oxidation or reduction. However, biosynthesis of carbohydrates from $CO_2$ requires a reductant, such as NADPH, in addition to ATP. Photosynthetic bacteria obtain the necessary reductant by using ATP to energize the transfer of electrons from succinate or other substrates to $NADP^+$. These reactions are carried out by dehydrogenase complexes in the cell membrane.

## An Antenna System Transfers Energy to the Reaction Centers

The reactive chlorophyll or bacteriochlorophyll molecules of P700, P680, or P870 account for only a small fraction of the total pigment in photosynthetic membranes. Chloroplasts contain about 300 chlorophylls per molecule of P700 and P680, and the cell membranes of purple photosynthetic bacteria have from 25 to several hundred bacteriochlorophylls per P870 molecule, depending on the species. Most of the chlorophyll or bacteriochlorophyll is not photochemi-

## Figure 15.14

(a) A molecule in an excited state (Chl$^*_D$) can transfer its energy to another molecule (Chl$_A$). The donor molecule returns to its ground state (Chl$_D$), and the acceptor is elevated to an excited state (Chl$^*_A$). This requires that the difference in energy between Chl$^*_D$ and Chl$_D$ (the excitation energy of Chl$_D$) be the same as the difference between Chl$^*_A$ and Chl$_A$ (the excitation energy of Chl$_A$). When the energies match in this way, a resonance can occur between the two states {Chl$^*_D$ Chl$_A$} and {Chl$_D$ Chl$^*_A$}. (b) The energy of light absorbed by pigment-protein complexes in the antenna system hops rapidly from complex to complex by resonance energy transfer until it is trapped in an electron-transfer reaction in a reaction center.

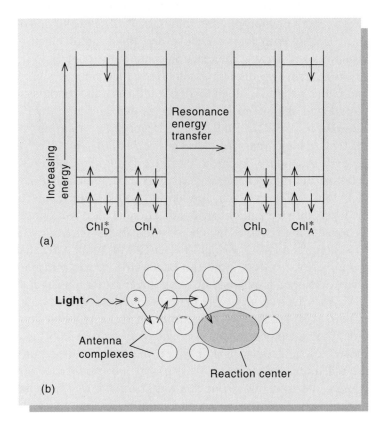

(a)

(b)

## Figure 15.15

The amount of $O_2$ released when a suspension of algae (*Chlorella pyrenoidosa*) was excited with a train of short flashes of light, as a function of (a) the time between the flashes and (b) the intensity of the flashes. Each flash lasted about 10 $\mu$s. The total amount of $O_2$ released was measured after several thousand flashes and was divided by the number of flashes and by the amount of chlorophyll in the suspension. (What happens on the first few flashes is discussed later.) The measurements in (a) were made with flashes comparable to the strongest flashes used in (b). For the measurements shown in (b), the flashes were spaced 20 ms apart. Note that the maximum amount of $O_2$ released per flash was only one molecule of $O_2$ per several thousand molecules of chlorophyll. With saturating flashes, the amount of $O_2$ evolution is limited by the concentration of reaction centers, whereas most of the chlorophyll in the algae is part of the antenna system.

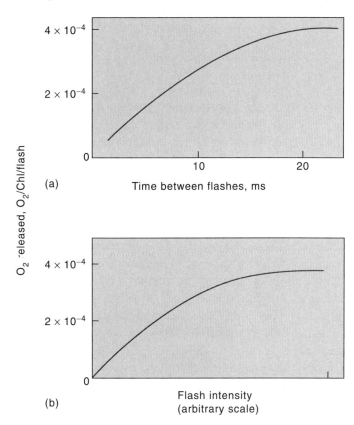

(a)

(b)

cally active. Instead, it serves as an antenna. When one of the molecules in the antenna is excited with light, it can transfer its energy to a neighboring molecule by resonance energy transfer. Energy absorbed anywhere in the antenna migrates rapidly from molecule to molecule until it is trapped by an electron-transfer reaction in a reaction center (fig. 15.14). Measurements of the lifetime of fluorescence from the antenna pigments indicate that the energy is trapped in a reaction center within about $10^{-10}$ s in purple bacteria and within about $5 \times 10^{-10}$ s in chloroplasts.

The distinction between the antenna and the reaction centers grew out of experiments done by Robert Emerson and William Arnold in the 1930s. They measured the amount of $O_2$ that evolved when they excited suspensions of green algae with a train of short flashes of light. To obtain the maximum $O_2$ per flash, they had to allow a period of about $2 \times 10^{-2}$ s of darkness between successive flashes. The amount of $O_2$ evolved per flash decreased if the flashes were spaced more closely together (fig. 15.15a). Each flash evidently generated a product that had to be consumed before the photosynthetic apparatus was ready to work again. If the next flash arrived too soon, its energy was wasted,

**Figure 15.16**

Structures of β-carotene, a major carotenoid in many types of plants, and spheroidene, a common carotenoid of photosynthetic bacteria. About 350 different natural carotenoids exist. These vary widely in color, depending principally on the number of conjugated double bonds.

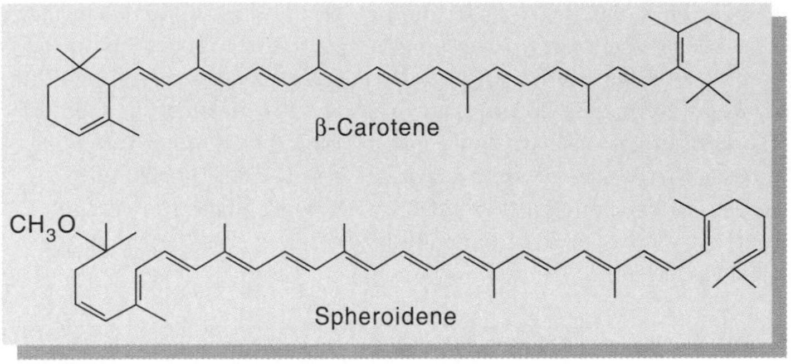

β-Carotene

Spheroidene

mainly as heat and fluorescence. We know now that the chlorophyll complexes that undergo oxidation in the reaction centers (P700 and P680 in algae) have to return to their reduced states before they can react again. The acceptors that remove electrons from the chlorophyll complexes also have to be reoxidized.

Emerson and Arnold found that the amount of $O_2$ evolved per flash increased as they excited the algae with a larger and larger number of photons on each flash but reached a plateau when they made the flashes sufficiently strong (see fig. 15.15b). Even with flashes of saturating intensity and optimal timing, the amount of $O_2$ was very small relative to the chlorophyll content of the algae. The cells contained about 2500 molecules of chlorophyll for each molecule of $O_2$ they evolved. This was very puzzling at the time because it was generally accepted that chlorophyll was intimately involved in the photochemistry of $CO_2$ fixation and $O_2$ evolution. Now it is clear that most of the chlorophyll is part of the antenna system. When the flash intensity is high, the amount of $O_2$ evolution that can occur on each flash is limited by the concentration of reaction centers, which is much smaller than the total concentration of chlorophyll.

The chlorophyll or bacteriochlorophyll molecules that comprise the antenna are bound noncovalently to integral membrane proteins. Each polypeptide typically carries only a few pigment molecules. These small complexes aggregate into larger arrays that allow excitations to hop rapidly from complex to complex. Antenna systems also contain a variety of carotenoids and other pigments. These accessory pigments fill in the antenna's absorption spectrum in regions where chlorophylls do not absorb well. As shown in figure 15.5, chlorophyll a absorbs red or blue light well but not

green. Carotenoids, which are long, linear polyenes (fig. 15.16), have absorption bands in the green region. The energy they absorb is transferred to the chlorophyll molecules of the antenna and from there to the reaction centers. This transfer is thermodynamically downhill.

## Chloroplasts Have Two Photosystems Linked in Series

We mentioned earlier that chloroplasts contain two reactive chlorophyll complexes: P700 and P680. Together with their antennae and their initial electron acceptors and donors, the reaction centers that contain P700 or P680 form two distinct assemblies called photosystem I and photosystem II. Much evidence indicates that photosystems I and II operate in series, as shown in figure 15.17. Excitation of P700 (photosystem I) generates a strong reductant, which transfers electrons to $NADP^+$ by way of several secondary electron carriers. Excitation of P680 (photosystem II) generates a strong oxidant, which oxidizes $H_2O$ to $O_2$. The reductant formed in photosystem II injects electrons into a chain of carriers that connect the two photosystems. This scheme, which is called the Z scheme, was first suggested by R. Hill and F. Bendall in 1960. Note that the Z scheme differs from the bacterial electron-transport chain discussed previously in being linear rather than cyclic.

The electron acceptors on the reducing side of photosystem II resemble those of purple bacterial reaction centers. The acceptor that removes an electron from P680 is a molecule of pheophytin a. The second and third acceptors are plastoquinones (see fig. 15.10). As in bacterial reaction centers, electrons move one at a time from the first quinone to the second. When the second quinone becomes doubly reduced, it picks up protons from the stromal side of the thylakoid membrane and dissociates from the reaction center.

The chain of carriers between the two photosystems includes the cytochrome $b_6f$ complex and a copper protein, plastocyanin. Like the mitochondrial and bacterial cytochrome $bc_1$ complexes, the cytochrome $b_6f$ complex contains a cytochrome with two b-type hemes (cytochrome $b_6$), an iron–sulfur protein, and a c-type cytochrome (cytochrome f). As electrons move through the complex from reduced plastoquinone to cytochrome f, plastoquinone probably executes a Q cycle similar to the cycle we presented for UQ in mitochondria and photosynthetic bacteria (see figs. 14.11 and 15.13). The cytochrome $b_6f$ complex provides electrons to plastocyanin, which transfers them to P700⁺˙ in the reaction center of photosystem I. The electron carriers between P700 and $NADP^+$ and between $H_2O$ and P680 are

## Figure 15.17

The Z scheme. [$(Mn)_4$ = a complex of four Mn atoms bound to the reaction center of photosystem II; $Y_Z$ = tyrosine side chain; Phe $a$ = pheophytin $a$; $Q_A$ and $Q_B$ = two molecules of plastoquinone; Cyt $b/f$ = cytochrome $b_6f$ complex; PC = plastocyanin; Chl $a$ = chlorophyll $a$; Q = phylloquinone (vitamin $K_1$); Fe-$S_x$, Fe-$S_A$, and Fe-$S_B$ = iron–sulfur centers in the reaction center of photosystem I; FD = ferredoxin; FP = flavoprotein (ferredoxin-NADP oxidoreductase).] The sequence of electron transfer through Fe-$S_A$ and Fe-$S_B$ is not yet clear.

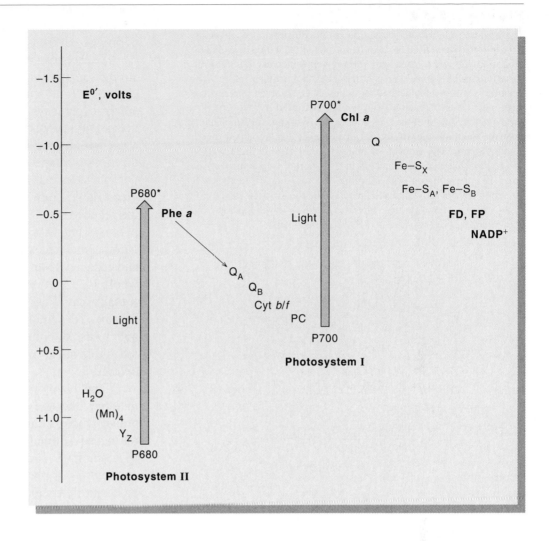

discussed later, after we discuss some of the evidence supporting the Z scheme.

The initial observations that led to the proposed Z scheme came from measurements of how the quantum yield of $O_2$ evolution depends on the excitation wavelength. The quantum yield of $O_2$ evolution is the molar ratio of $O_2$ evolved to photons absorbed. In green algae, the quantum yield is nearly independent of wavelength between 400 and 675 nm but falls off drastically in the far-red region near 700 nm (fig. 15.18). This seems odd, because most of the absorbance in the far-red region is due to chlorophyll $a$. How can light absorbed by chlorophyll $a$ be used less efficiently than light absorbed by carotenoids or other accessory pigments? A clue came from Emerson's finding in 1956 that 700-nm light is used more efficiently if it is superimposed on a background of weak blue-green light (see fig. 15.18). It was found subsequently that the blue-green light actually does not have to be presented simultaneously with the red light. Illumination with blue-green light improves the utilization of far-red light even if the blue-green light is turned off several seconds before the red light is turned on.

These observations could be explained by the Z scheme if the absorption spectra of the antennas associated with photosystems I and II are different. Because the two photosystems must operate in series, light is used most efficiently when the flux of electrons through photosystem II is equal to that through photosystem I. If light of a particular wavelength excites one photosystem more frequently than the other, some of the light is wasted.

It turns out that photosystem I can absorb and use virtually all wavelengths of light up to about 740 nm. Photosystem II, however, is not excited well by far-red light. The pool of electron carriers between the two photosystems is, therefore, drained of electrons if algae are illuminated with far-red light in the absence of added reductants. P700 remains oxidized and is unable to respond to the light. Photosystem II does absorb blue-green light well, so illumination with green light replenishes the pool. Because electrons

## Figure 15.18

The quantum yield of $O_2$ evolution from a suspension of algae (*Chlorella pyrenoidosa*) as a function of the excitation wavelength. Measurements were made without any supplementary light (lower curve) and with supplementary blue-green light (upper curve). The quantum yield was calculated as moles of $O_2$ evolved per einstein of light incident on the sample; $O_2$ evolution caused by the supplementary light alone was subtracted. Without the supplementary light the quantum yield falls off precipitously in the far-red region above 680 nm. The antenna of photosystem I absorbs light well in this region, but that of photosystem II does not. Supplementary blue-green light, which is absorbed well by protosystem II, increases the quantum yield with which the far-red light can be used. (Source: From R. Emerson et al., Some factors influencing the long-wavelength limit of photosynthesis, *Proc. Natl. Acad. Sci. USA* 43:133, 1957.)

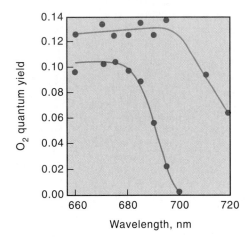

## Figure 15.19

Illumination of a suspension of red algae (*Porphyridium cruentum*) with far-red light (680 nm) causes an oxidation of cytochrome *f*, as measured by an absorbance change at 422 nm. (The box below the trace indicates the period when the far-red light was on.) Green (562-nm) light superimposed on top of the far-red light causes the cytochrome to become more reduced. Additional measurements showed that green light also could cause cytochrome oxidation if it was turned on in the absence of the far-red light when the cytochrome was initially reduced. This result indicates that the green light can drive either photosystem I or photosystem II. Far-red light can cause only cytochrome oxidation, because it is absorbed only by photosystem I.

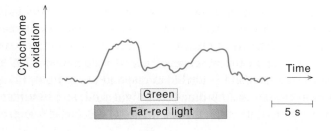

remain in the pool when the blue-green light is turned off, far-red light can be used effectively for a time even after a period of darkness.

Subsequent experiments by Louis Duysens provided strong support for the idea that photosystem I withdraws electrons from carriers between the two photosystems, whereas photosystem II feeds electrons to these components. Duysens and his colleagues showed that illuminating algae with far-red light causes cytochrome *f* to become oxidized, as we expect if far-red light drives photosystem I but not photosystem II (fig. 15.19). When a supplementary green light is turned on, some of the cytochrome returns to the reduced state. This makes sense if green light can drive photosystem II. During the illumination, individual cytochrome molecules cycle repeatedly between the oxidized and reduced states, so that the cytochrome population as a whole is in a steady state of partial oxidation. Green light actually can cause either oxidation or reduction of cytochrome *f*, depending on the initial conditions, so it must be able to excite either photosystem II *or* photosystem I. Far-red light, however, can only cause oxidation of the cytochrome.

The electron carriers in photosystem II are located mainly in the stacked, granal regions of the thylakoid membrane, whereas those of photosystem I are mainly in the unstacked stromal lamellae (fig. 15.20). The cytochrome $b_6 f$ complex is found in both regions of the membrane. The distribution of photons between the two photosystems depends partly on the relative numbers of antenna complexes in the two regions of the membrane and can change as a result of movements of antenna complexes from one region to another. Movements of one type of antenna complex appear to be regulated by phosphorylation of the protein. If chloroplasts are illuminated with light that is absorbed better by photosystem I than by photosystem II, the mobile antenna complex moves into the stacked regions, and a larger fraction of the excitations then is passed to photosystem II.

If the reaction centers of photosystem I and photosystem II are segregated into separate regions of the thylakoid membrane, how can electrons move from photosystem I to photosystem II? Evidently the plastoquinone that is reduced in photosystem II can diffuse rapidly in the membrane, just as ubiquinone does in the mitochondrial inner membrane. Plastoquinone thus carries electrons from photosystem II to the cytochrome $b_6 f$ complex. Plastocyanin acts similarly as a mobile electron carrier from the cytochrome $b_6 f$ complex to the reaction center of photosystem I, just as cytochrome *c* carries electrons from the mitochondrial cytochrome $bc_1$ complex to cytochrome oxidase and as a *c*-type cytochrome provides electrons to the reaction centers of purple bacteria (see fig. 15.13).

**Figure 15.20**

The photosystem II complexes are located mainly in the stacked (granal) regions of the thylakoid membrane, whereas photosystem I complexes are most abundant in the unstacked stromal lamellae. The cytochrome $b_6 f$ complex is located in both regions of the membrane. Some types of antenna complexes also reside in both regions and can shift from one region to the other in response to a change in conditions. The ATP-synthase occurs mainly in the unstacked regions. These conclusions came from studies in which thylakoid membranes were fragmented gently by ultrasound and then separated by density-gradient centrifugation into fractions representing stacked and unstacked membranes.

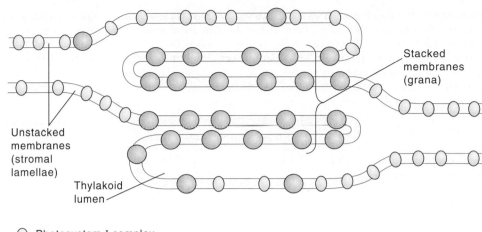

Stacked membranes (grana)

Unstacked membranes (stromal lamellae)

Thylakoid lumen

○ Photosystem I complex

● Photosystem II complex

## Photosystem I Reduces NADP$^+$ by Way of Iron–Sulfur Proteins

On the reducing site of photosystem I, the initial electron acceptor appears to be a molecule of chlorophyll $a$ (see fig. 15.17). The second acceptor probably is a quinone, phylloquinone (vitamin $K_1$, fig. 15.10). In these respects, photosystem I resembles photosystem II and purple photosynthetic bacteria, which use pheophytin $a$ or bacteriopheophytin $a$ followed by a quinone. From this point on, photosystem I is different: its next electron carriers consist of iron–sulfur proteins instead of additional quinones.

Photosystem I contains three iron–sulfur clusters firmly associated with the reaction center. These are designated Fe-$S_X$, Fe-$S_A$, and Fe-$S_B$ in figure 15.17. The cysteines of Fe-$S_X$ are provided by the two main polypeptides of the reaction center, which also bind P700 and its initial electron acceptors; Fe-$S_A$ and Fe-$S_B$ are on a separate polypeptide. The quinone that is reduced in photosystem I probably transfers an electron to Fe-$S_X$, which in turn reduces Fe-$S_A$ and Fe-$S_B$. From here, electrons move to ferredoxin, a soluble iron–sulfur protein found in the chloroplast stroma, then to a flavoprotein (ferredoxin-NADP oxidoreductase), and finally to NADP$^+$.

## O$_2$ Evolution Requires the Accumulation of Four Oxidizing Equivalents in the Reaction Center of Photosystem II

Oxidation of two molecules of $H_2O$ to $O_2$ requires removing four electrons for each $O_2$ produced. According to the Z scheme each electron must traverse the photochemical reac-

tions of both photosystem I and photosystem II, so at least eight photons ($2 \times 4$) must be absorbed for each $O_2$ released. Experimental measurements of the number of photons that are needed, the quantum requirement, are needed on the order of 8–12. (The quantum requirement is the reciprocal of the quantum yield of $O_2$ evolution.) The quantum requirement exceeds eight if some of the photons that are absorbed are lost as heat or fluorescence from the antenna or if some of the electrons pumped through the photosystems are not removed to NADP$^+$ but instead cycle back into the electron-transport chain between the two photosystems.

If the photosystem II reaction center transfers only one electron at a time, how does it assemble the four oxidizing equivalents needed for oxidation of $H_2O$ to $O_2$? One possibility is that several different photosystem II reaction centers cooperate, but this seems not to happen. Instead, each reaction center progresses independently through a series of oxidation states, advancing to the next state each time it absorbs a photon. In this event $O_2$ evolution would occur only when a reaction center has accumulated four oxidizing equivalents. Support for this conclusion comes from measurements made by Pierre Joliot of the amount of $O_2$ evolved on each flash when chloroplasts are excited with a series of short flashes after a period of darkness. No $O_2$ is released on the first or second flashes (fig. 15.21), but on the third flash, there is a burst of $O_2$. After this, the amount of $O_2$ released on each flash oscillates, going through a maximum every fourth flash.

Bessel Kok pointed out that this pattern can be explained if the photosystem II reaction center cycles through five different oxidation states, $S_0$ through $S_4$, as shown in figure 15.22. When the system reaches state $S_4$, $O_2$ is given

**Figure 15.21**

$O_2$ evolution from a suspension of chloroplasts that was excited with a series of 24 flashes after having been kept in the dark for several minutes. Little or no $O_2$ is evolved on the first two flashes. $O_2$ evolution peaks on the third flash and on every fourth flash thereafter. The oscillations in $O_2$ evolution are damped, and after many flashes the yields converge on the level indicated by the horizontal line.

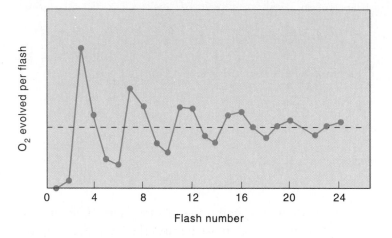

**Figure 15.22**

The five oxidation states of the $O_2$-evolving apparatus. One electron ($e^-$) is removed photochemically in each of the transitions between states $S_0$ and $S_4$. $S_4$ decays spontaneously, releasing $O_2$. Protons are released at several steps of the cycle.

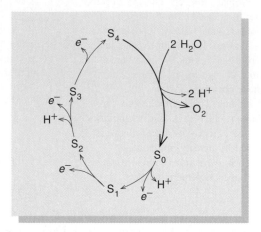

off, and the reaction center returns to state $S_0$. The fact that the first burst of $O_2$ comes on the third flash instead of the fourth can be explained if the reaction centers relax mainly into $S_1$ rather than into $S_0$ during the dark period before the flashes.

The component that undergoes oxidation as photosystem II progresses from one S state to the next is a complex of four atoms of manganese bound to the two central polypeptides of the reaction center. P680$^{+}$ draws electrons from the Mn complex by way of a tyrosine in one of these polypeptides (see fig. 15.17). In the course of this reaction, the phenolic side chain of the tyrosine is oxidized transiently to a free radical. Although the structure of the Mn complex is not yet known, most of the transitions of the complex probably represent sequential oxidations of Mn atoms from the Mn(III) level to Mn(IV).

The $E°$ of the P680$^{+}$/P680 redox couple must be greater than $+1$ V, because the oxidized Mn complex that P680$^{+}$ generates is powerful enough to oxidize water. At pH 5, which is approximately the pH of the thylakoid lumen, the $O_2/H_2O$ couple has an $E°$ of $+0.94$ V. P700$^{+}$/P700, by contrast, has an $E°'$ of only about $+0.5$ V. It is surprising that P680$^{+}$ is a much stronger oxidant than P700$^{+}$, since they both consist of chlorophyll $a$ dimers. Apparently, the redox potential of the chlorophyll radical could depend strongly on the surroundings in the protein.

## Flow of Electrons from $H_2O$ to NADP$^+$ Drives Proton Transport into the Thylakoid Lumen; Protons Return to the Stroma through an ATP-Synthase

As in mitochondria and purple photosynthetic bacteria, the flow of electrons through the chloroplast's cytochrome $b_6f$ complex pumps protons across the thylakoid membrane. When plastoquinone undergoes reduction in photosystem II and in the cytochrome $b_6f$ complex, protons are taken up from the stromal side of the membrane. When the plastoquinone is reoxidized, protons are released into the thylakoid lumen. This lowers the pH in the lumen, and makes the inside positively charged with respect to the stroma. Four protons probably move across the membrane for each pair of electrons that proceed to photosystem I. Two more protons are released in the lumen for each molecule of $H_2O$ that is oxidized to $\frac{1}{2} O_2$, and one proton is taken up from the stroma for each molecule of NADP$^+$ that is reduced to NADPH (fig. 15.23).

Proton movement back out through the thylakoid membrane is conducted by an ATP-synthase, and this movement drives the formation of ATP. The chloroplast ATP-synthase is structurally very similar to the mitochondrial ATP-synthase (see fig. 14.24). Its head-piece (CF$_1$)

## Figure 15.23

Transport of two electrons from photosystem II through the cytochrome $b_6f$ complex to photosystem I results in the movement of four protons from the chloroplast stroma to the thylakoid lumen. The proton translocation probably occurs in a Q cycle resembling that illustrated in figure 14.11. Two more protons are released in the lumen for each molecule of $H_2O$ that is oxidized to $O_2$, and one additional proton is removed from the stroma for each molecule of

$NADP^+$ reduced to NADPH. (Two protons are taken up when the flavoprotein is reduced by photosystem I; one ends up on NADPH, and the other is returned to the solution.) The flow of protons from the thylakoid lumen back to the stroma through an ATP-synthase ($CF_0$-$CF_1$) drives the formation of ATP. The abbreviations used in this figure are the same as in figure 15.17.

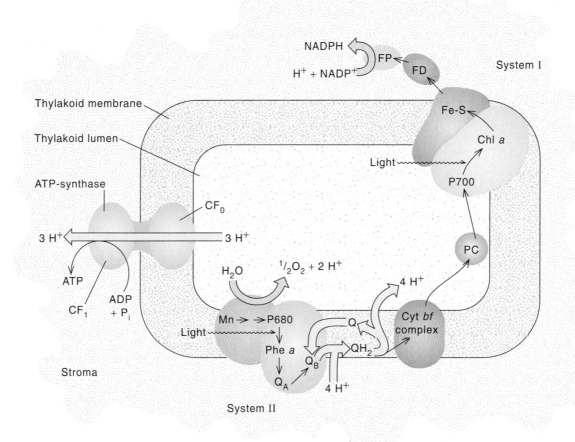

contains the catalytic sites; its hydrophobic base-piece ($CF_0$) includes the proton-conducting channel.

Observations in chloroplasts played a key role in the development of the chemiosmotic theory of oxidative phosphorylation, which we discussed in chapter 14. André Jagendorf and his colleagues discovered that if chloroplasts are illuminated in the absence of ADP, they developed the capacity to form ATP when ADP was added later, after the light was turned off. The amount of ATP synthesized was much greater than the number of electron-transport assemblies in the thylakoid membranes, so the energy to drive the phosphorylation could not have been stored in an energized

form of one of the electron carriers. Protons were taken up from the solution outside the thylakoid membranes during the illumination, and the ability to form ATP correlated with the magnitude of the pH gradient across the membrane. If a pH gradient in the right direction was set up across the thylakoid membrane by raising the pH on the outside, chloroplasts could make ATP without any illumination at all (fig. 15.24).

Formation of ATP by photosynthetic systems is often called photophosphorylation. Although photophosphorylation and oxidative phosphorylation are very similar, they do differ in a few details. The respiratory chain pumps protons

**Figure 15.24**

Chloroplasts can use an electrochemical potential gradient for protons to form ATP in the dark. In a two-step experiment, spinach chloroplasts first were equilibrated at an acidic pH (pH 4.0). The pH of the solution then was raised quickly to a value between 6.7 and 8.8, as indicated on the abscissa of the graph, and ADP and $P_i$ were added. Because the pH in the thylakoid lumen was buffered at pH 4, the sudden increase in the external pH created a pH gradient of between 2.7 and 4.8 pH units across the thylakoid membrane. Proton efflux through the ATP-synthase can drive the formation of ATP. Under optimal conditions, one molecule of ATP was formed for approximately every five chlorophylls in the sample (see the ordinate scale on the graph). For comparison, chloroplasts contain only about one molecule of P700 and one of P680 per 400 chlorophylls. (Source: From A. Jagendorf and E. Uribe, ATP formation caused by acid-base transition of spinach chloroplasts, *Proc. Natl. Acad. Sci. USA* 55:197, 1966.)

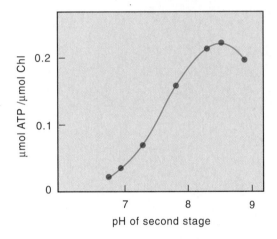

outward across the inner membrane, from the mitochondrial matrix to the intermembrane space (see chapter 14); the chloroplast electron-transport chain pumps protons into the thylakoid lumen. The electrochemical potential gradient for protons across the mitochondrial membrane consists mainly of an electric potential ($\Delta\psi$); in chloroplasts it consists mainly of a pH gradient ($\Delta$pH), which can be greater than 3 pH units during strong illumination. In both cases, however, the catalytic head-piece of the ATP-synthase is on the side of the membrane that becomes more alkaline. The relative importance of $\Delta$pH and $\Delta\psi$ probably depends on the permeability of the membrane to other ions such as $Cl^-$. Movement of $Cl^-$ into the thylakoid lumen reduces $\Delta\psi$, allowing for the buildup of a larger pH gradient.

In the Z scheme, photosystem II, the cytochrome $b_6f$ complex and photosystem I operate in series to move electrons from $H_2O$ to $NADP^+$ and to create an electrochemical potential gradient for protons across the thylakoid membrane. In addition to this linear pathway, chloroplasts in some plant species may use a cyclic electron-transfer scheme that includes photosystem I and the cytochrome $b_6f$

**Figure 15.25**

The reductive pentose cycle, or Calvin cycle. The number of arrows drawn at each step in the diagram indicates the number of molecules proceeding through that step for every three molecules of $CO_2$ that enter the cycle. The entry of three molecules of $CO_2$ results in the formation of one molecule of glyceraldehyde-3-phosphate (box on right), and requires the oxidation of six molecules of NADPH to $NADP^+$ and the breakdown of nine molecules of ATP to ADP.

complex but not photosystem II. If the concentration of $NADP^+$ is low, electrons ejected by photosystem I can be passed from one of the iron–sulfur proteins to a $b$-type cytochrome, and from there to the plastoquinone pool between the two photosystems (see fig. 15.17). From plastoquinone, electrons return to P700 via the cytochrome $b_6f$ complex and plastocyanin. The Q cycle that is included within this cyclic electron pathway results in proton translocation across the thylakoid membrane and thus contributes to the electrochemical potential gradient that drives the ATP-synthase.

## Carbon Fixation Utilizes the Reductive Pentose Cycle

The ATP and NADPH generated by the photosynthetic electron-transfer reactions are used for fixation of $CO_2$. Reactions in which $CO_2$ is incorporated into carbohydrates were discovered in the 1950s by Melvin Calvin and his coworkers in some of the first biochemical studies employing radioactive tracers. A key experiment was to expose algae to a brief period of illumination in the presence of $^{14}CO_2$ and then to disrupt the cells and search for organic molecules that had become labeled with $^{14}C$. The material that turned out to be labeled most rapidly was glycerate-3-phosphate, and almost all of its $^{14}C$ was in the carboxyl group. Further experiments showed that the glycerate-3-phosphate is formed from ribulose-1,5-bisphosphate. A molecule of $CO_2$ is incorporated in the process, so that one molecule of the five-carbon sugar ribulose-1,5-bisphosphate yields two molecules of glycerate-3-phosphate (fig. 15.25). The reaction is catalyzed by ribulose bisphosphate carboxylase in the chloroplast stroma.

Other enzymes in the stroma can convert glycerate-3-phosphate back to ribulose-1,5-bisphosphate. The reactions of this reductive pentose cycle, or Calvin cycle, are shown in figure 15.25. 3-Phosphoglycerate is first phosphorylated to glycerate-1,3-bisphosphate at the expense of ATP and then reduced to glyceraldehyde-3-phosphate by NADPH. (Note that the nucleotide specificity in the reductive step differs from that of the cytosolic glyceraldehyde-3-phos-

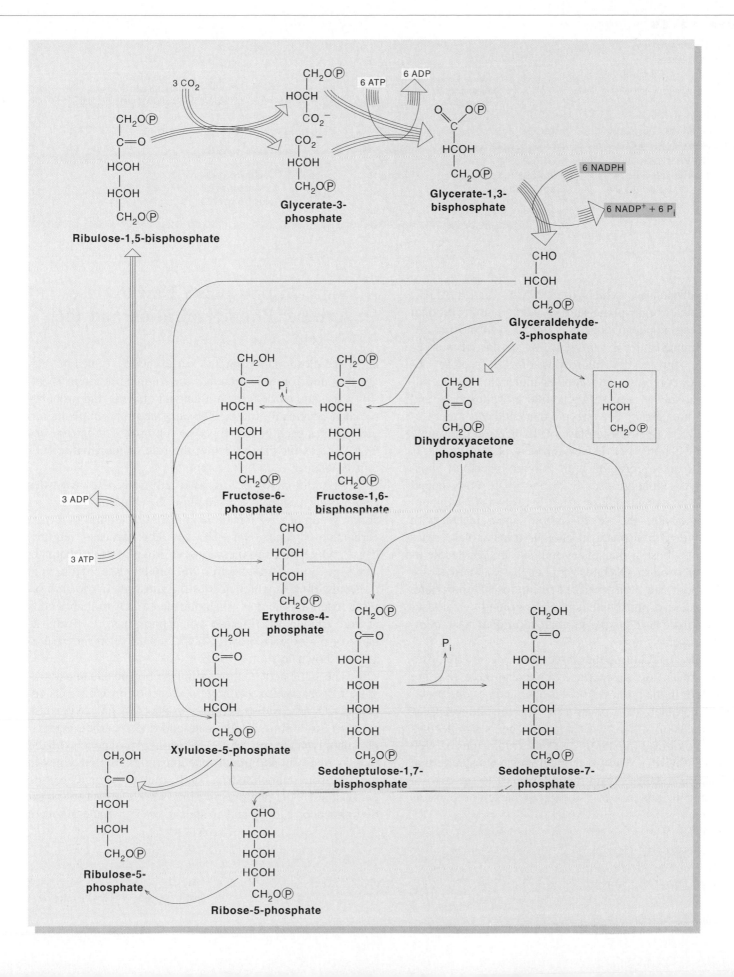

**Figure 15.26**

The reaction catalyzed by ribulose bisphosphate carboxylase involves 2-carboxy-3-ketoarabinitol-1,5-bisphosphate as an enzyme-bound intermediate. The intermediate probably forms by the addition of $CO_2$ to the enolate of ribulose-1,5-bisphosphate. The substrate is known to be $CO_2$ rather than bicarbonate.

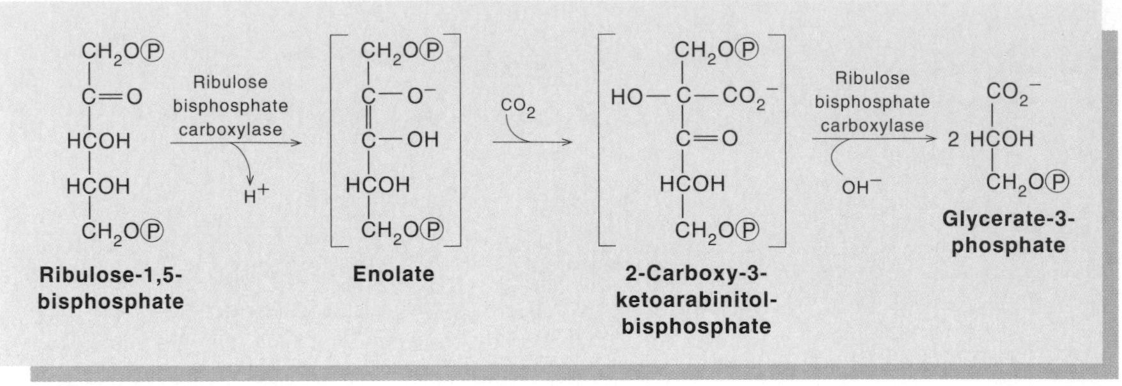

phate dehydrogenase, which uses $NAD^+$ and NADH.) Glyceraldehyde-3-phosphate and its isomerization product dihydroxyacetone phosphate can combine to form fructose-1,6-bisphosphate under the influence of aldolase. Fructose-6-phosphate, formed by hydrolysis of the fructose-1,6-bisphosphate, combines with another molecule of glyceraldehyde-3-phosphate, generating xylulose-5-phosphate and erythrose-4-phosphate. The enzyme that catalyzes this reaction is similar to the transketolase of the pentose phosphate pathway (see chapter 12). The erythrose-4-phosphate that is formed goes on to combine with dihydroxyacetone phosphate to give sedoheptulose-1,7-bisphosphate in a second reaction catalyzed by aldolase. After hydrolysis to sedoheptulose-7-phosphate, the seven-carbon sugar reacts with glyceraldehyde-3-phosphate in another transketolase reaction, forming ribose-5-phosphate and a second molecule of xylulose-5-phosphate. Xylulose-5-phosphate and ribose-5-phosphate both can be isomerized to ribulose-5-phosphate. Finally, ribulose-5-phosphate is phosphorylated to ribulose-1,5-bisphosphate at the expense of an additional ATP, completing the cycle.

In figure 15.25, note that three molecules of ribulose-1,5-bisphosphate are regenerated for every three that are carboxylated. In the process, there is a net gain of one molecule of glyceraldehyde-3-phosphate. The metabolic cost is nine molecules of ATP converted to ADP and six molecules of NADPH oxidized to $NADP^+$, or three molecules of ATP and two of NADPH. The glyceraldehyde-3-phosphate that is produced is exported from the chloroplast to the cytosol or converted to hexoses for storage in the chloroplast as starch.

# Ribulose-Bisphosphate Carboxylase-Oxygenase: Photorespiration and the C-4 Cycle

Ribulose-bisphosphate carboxylase typically accounts for more than half the soluble protein in a leaf. It is the world's most abundant enzyme and probably the most abundant protein. The enzyme found in plants has eight copies each of two types of subunits. The larger subunit contains the catalytic site; the role of the smaller subunit is unclear.

In addition to serving as a substrate, $CO_2$ activates ribulose-bisphosphate carboxylase by binding to the $\epsilon$-amino group of a lysyl residue in the large subunit to form a carbamate (lysine—NH—$CO_2^-$). This process requires $Mg^{2+}$, which binds to the carboxyl group of the carbamate. The $Mg^{2+}$ in turn forms part of the binding site for a second molecule of $CO_2$, which acts as the substrate in the carboxylase reaction. The reactive molecule of $CO_2$ probably adds to the enolate of ribulose-1,5-bisphosphate to form 2-carboxy-3-ketoarabinitol-1,5-bisphosphate as an intermediate, as shown in figure 15.26.

The same active site of ribulose bisphosphate carboxylase also catalyzes a competing reaction in which $O_2$ replaces $CO_2$ as a substrate. The products of this oxygenase reaction are 3-phosphoglycerate and a two-carbon acid, 2-phosphoglycolate (fig. 15.27). Phosphoglycolate is oxidized to $CO_2$ by $O_2$ in additional reactions involving enzymes in the cytosol, mitochondria, and another organelle, the peroxisome. This photorespiration is not coupled to oxidative phosphorylation, and it appears to constitute a severe

## Figure 15.27

Photorespiration results from the oxygenation reaction catalyzed by ribulose bisphosphate carboxylase/oxygenase. 2-Phosphoglycolate generated by the reaction moves from the chloroplast to the cytosol, where other enzymes break it down to $CO_2$, $H_2O$, and $P_i$. The oxygenation reaction, like carboxylation, does not require light directly. It occurs mainly during illumination, however, because the formation of the substrate, ribulose-1,5-bisphosphate, requires ATP and NADPH (see fig. 15.25).

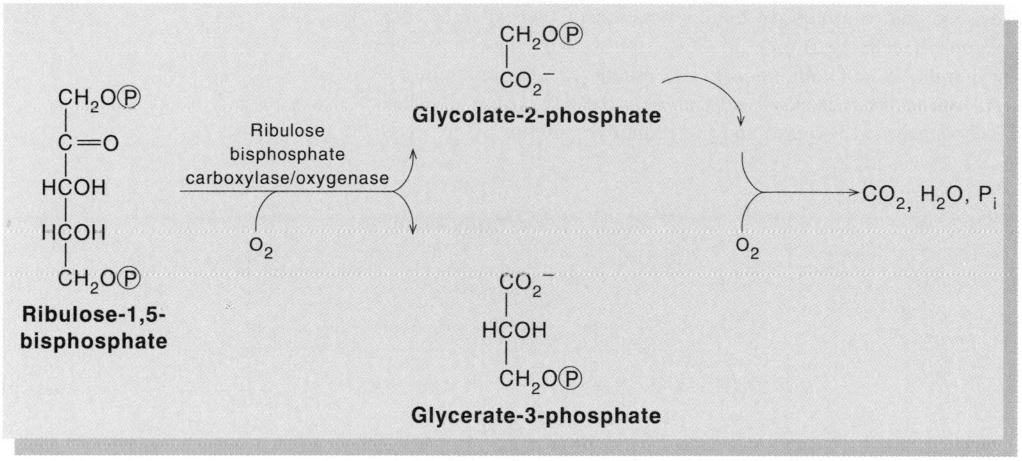

drain on chloroplast metabolism. In some plants, as much as one-third of the $CO_2$ that is fixed is released again by photorespiration. If photorespiration has any benefit to the plant, it is not obvious.

The outcome of the competition between the carboxylase and oxygenase reactions depends on the concentrations of $CO_2$ and $O_2$. The $K_m$ of the activated enzyme for $CO_2$ is about 20 $\mu$M, and that for $O_2$ is about 200 $\mu$M, so $CO_2$ is the preferred substrate. In illuminated chloroplasts, however, the concentration of $O_2$ is elevated as a result of the photosynthetic oxidation of $H_2O$. The concentrations of the two gases are frequently on the order of the $K_m$ values, making $O_2$ a serious competitor. Plants that live in dry climates face the additional dilemma that opening the stomata in the leaves to improve the exchange of $CO_2$ and $O_2$ with the atmosphere can result in dehydration.

If photorespiration is only a liability, natural selection presumably has favored evolution of a ribulose bisphosphate carboxylase with minimal activity as an oxygenase. Nonetheless, the oxygenase activity of the enzyme in higher plants is only slightly below that of the enzymes in photosynthetic bacteria. Some species of plants have evolved an alternative strategy for favoring the carboxylase over the oxygenase: They increase the concentration of $CO_2$ in the region of the enzyme. These plants, which include corn, sugar cane, and numerous tropical species, have a layer of specialized mesophyll cells at the outer surface of the leaf.

Mesophyll cells use $CO_2$ from the air to convert phosphoenolpyruvate to oxaloacetate (fig. 15.28). Oxaloacetate is reduced to malate, which then moves to the bundle sheath cells that surround the vascular structures in the interior of the leaf. Here malate is decarboxylated to pyruvate in an oxidative reaction that reduces $NADP^+$ to NADPH. The pyruvate returns to the mesophyll cells, where it is phosphorylated to phosphoenolpyruvate. This phosphorylation is driven by splitting of ATP to AMP and pyrophosphate and subsequent hydrolysis of the pyrophosphate to phosphate.

This cyclic series of reactions delivers $CO_2$ and reducing power (NADPH) to the bundle sheath cells, at a cost of hydrolyzing one ATP to one AMP (see fig. 15.28). The bundle sheath cells fix the $CO_2$ by the ribulose bisphosphate carboxylase reaction and the reductive pentose cycle. Plants that have this mechanism for concentrating $CO_2$ can grow considerably more rapidly than species that do not, particularly in hot, dry climates. This auxiliary cycle is called the C-4 cycle because it involves the four-carbon acids malate and oxaloacetate. Currently interest has arisen in the possibility of increasing agricultural productivity by using genetic engineering techniques to introduce the C-4 pathway into plant species that lack it.

**Figure 15.28**

The C-4 pathway requires the cooperation of two types of cells. Mesophyll cells (*left*) take up $CO_2$ from the air and export malate to the bundle sheath cells (*right*). The bundle sheath cells return pyruvate to the mesophyll cells and fix the $CO_2$ using ribulose bisphosphate carboxylase and the reductive pentose cycle.

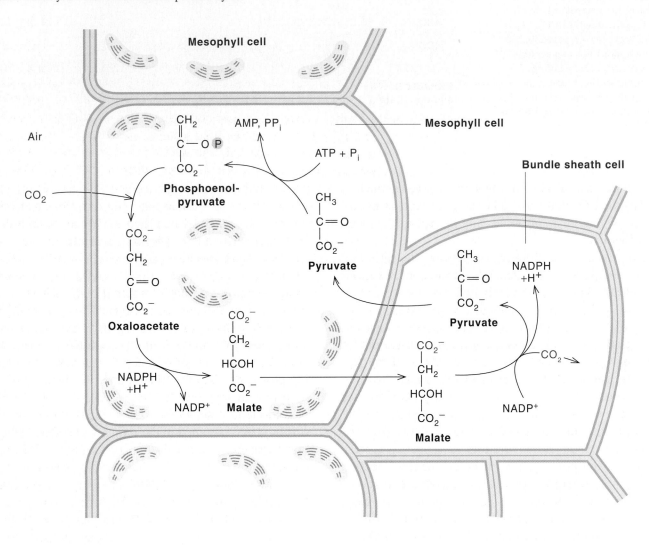

## Summary

1. Photosynthesis, the conversion of sunlight into chemical energy, occurs in algae and some bacteria, in addition to higher plants. In plants, the photochemical reactions of photosynthesis take place in the chloroplast thylakoid membrane. In bacteria, they occur in the plasma membrane.

2. Photosynthetic organisms take advantage of the fact that chlorophyll becomes a strong reductant when it is excited with light. Photooxidation of chlorophyll or bacteriochlorophyll occurs in pigment-protein reaction centers. Chloroplasts have two types of reaction cen-

ters: Photosystem I, which contains the reactive chlorophyll complex P700, and photosystem II, which has P680. The reactive complex in purple photosynthetic bacteria (P870) is a bacteriochlorophyll dimer.

3. Photosynthetic membranes also contain pigment-protein complexes that serve as antennas. When the antenna absorbs a photon, energy hops rapidly from complex to complex by resonance energy transfer until it is trapped in a reaction center.

4. When P870 is excited, it transfers an electron to a bacteriopheophytin. The bacteriopheophytin then re-

duces a quinone, which in turn reduces another quinone, $Q_B$. A cyctochrome replaces the electron released by P870. When $Q_B$ receives a second electron, it picks up protons from the cytosol. Electrons return from $Q_BH_2$ to the cytochrome via a cytochrome $bc_1$ complex, and protons are released outside the cell. This pumping of protons creates an electrochemical potential gradient across the plasma membrane. The return of protons to the cell through an ATP-synthase drives synthesis of ATP.

5. Photosystems I and II of chloroplasts operate in series. Photooxidation of P680 in photosystem II generates a strong oxidant that can oxidize $H_2O$ to $O_2$. In this process, a manganese complex progresses through five different oxidation states ($S_0$–$S_4$). When the complex reaches state $S_4$, $O_2$ is given off. The electron released by P680 goes to a pheophytin and then to two plastoquinones. When the second plastoquinone has been doubly reduced, it picks up protons from the chloroplast stromal space. Electrons move from the reduced plastoquinone to P700 of photosystem I via the cytochrome $b_6f$ complex and a copper protein (plastocyanin).

6. Photooxidation of P700 in photosystem I reduces a chlorophyll, which transfers electrons to a series of membrane-bound iron–sulfur centers, probably by way of a quinone. From the iron–sulfur centers, electrons move to the soluble iron–sulfur protein, ferre-

doxin, and then to a flavoprotein that reduces $NADP^+$. Cyclic electron flow may also occur from the iron–sulfur proteins to carriers between the two photosystems.

7. Electron flow through the cytochrome $b_6f$ complex results in proton translocation from the stroma to the thylakoid lumen. In addition, protons are released in the lumen when $H_2O$ is oxidized and are taken up from the stromal space when $NADP^+$ is reduced. Protons move from the thylakoid lumen back to the stroma through an ATP-synthase, driving the formation of ATP.

8. ATP and NADPH are used for incorporation of $CO_2$ into carbohydrates. $CO_2$ reacts with ribulose-1,5-bisphosphate to give two molccules of 3-phosphoglycerate, under the influence of ribulose bisphosphate carboxylase. 3-Phosphoglycerate can be converted back to ribulose-1,5-bisphosphate at the expense of ATP and NADPH in the reductive pentose cycle. For every 3 molecules of $CO_2$ that enter the cycle, there is a gain of one glyceraldehyde-3-phosphate. Ribulose bisphosphate carboxylase also catalyzes a wasteful oxygenation reaction of ribulose-1,5-bisphosphate. Some species of plants use an auxiliary cycle of reactions (the C-4 cycle) to concentrate $CO_2$ in the cells that carry out the reductive pentose cycle, thus allowing the plants to fix $CO_2$ at elevated rates.

## Selected Readings

Amesz, J. (ed.), *Photosynthesis*. Amsterdam: Elsevier, 1987. Review articles on many aspects of photosynthesis.

Deisenhofer, J., and H. Michel, The photosynthetic reaction centre from the purple bacterium *Rhodopseudomonas viridis. Science* 245:1463, 1989. The structure of a bacterial reaction center is revealed by x-ray crystallography.

Duysens, L. N. M., and J. Amesz, Function and identification of two functional systems in photosynthesis. *Biochim. Biophys. Acta* 64:243, 1962. Evidence for the Z scheme is obtained by measuring oxidation and reduction of cytochrome *f* in algae illuminated with light of various colors.

Forbush, B., B. Kok, and M. McGloin, Cooperation of charges in photosynthetic $O_2$ evolution: II. Damping of flash yield, oscillation, deactivation. *Photochem. Photobiol.* 14:307, 1971. Oscillations in $O_2$ yield during a train of flashes provide evidence for the four S states.

Govindjee (ed.), *Photosynthesis: Energy Conversion by Plants and Bacteria*. New York: Academic Press, 1982. A collection of review articles.

Govindjee and, W. J. Coleman, How plants make oxygen. *Sci. Am.* 262:50–58, 1990.

Jagendorf, A. T., and E. Uribe, ATP formation caused by acid–base transition of spinach chloroplasts. *Proc. Natl. Acad. Sci. USA* 55:197, 1966. Chloroplasts can form ATP in the dark if an electrochemical potential gradient for protons is set up across the thylakoid membrane.

Lorimer, G. H., The carboxylation and oxygenation of ribulose-1,5-bisphosphate: The primary events in photosynthesis and photorespiration. *Ann. Rev. Plant Physiol.* 32:349, 1981. Mechanisms and biochemical significance of the carboxylase and oxygenase reactions.

Rutherford, A. W., Photosystem II, the water-splitting enzyme. *Trends Biochem. Sci.* 14:227, 1989. A readable account of current results and speculations on the oxygen-evolving complex.

Woodbury, N. W., M. Becker, D. Middendorf, and W. W. Parson, Picosecond kinetics of the initial photochemical electron transfer reaction in bacterial photosynthetic reaction centers. *Biochem.* 24:7516, 1985. Fast spectrophotometric techniques are used to follow the initial steps in reaction centers purified from photosynthetic bacteria.

## Problems

1. The word "fixed" is often used to describe what happens to $CO_2$ in photosynthesis or to $N_2$ in nitrogen "fixation." How do you explain the meaning of "to fix" a gas or "a fixed gas" to an English major?

2. If the conversion of light energy into chemical energy occurred with 100% efficiency, how many moles of ATP could be produced per mole of photons of blue light (450 nm) and red light (700 nm)?

3. What arguments can you make for or against the idea that the chlorophylls (fig. 15.4) are aromatic compounds?

4. The lowest energy absorption band of P870 occurs at 870 nm.
   (a) Calculate the energy of an einstein of light at this wavelength.
   (b) Estimate the effective standard redox potential ($E^{\circ\prime}$) of P870 in its first excited singlet state, given that the $E^{\circ\prime}$ for oxidation in the ground state is +0.45 V.

5. The traces shown here are measurements of optical absorbance changes at 870 and 550 nm when a suspension of membrane vesicles from photosynthetic bacteria was excited with a short flash of light. Downward deflection of the traces represent absorbance decreases. Explain the observations. (Absorption spectra of a $c$-type cytochrome in its reduced and oxidized forms are described in the previous chapter.)

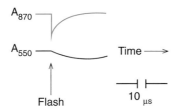

6. You add a nonphysiological electron donor to a suspension of chloroplasts. When you illuminate the chloroplasts, the donor becomes oxidized. How can you determine whether this process involves both photosystems I and II? (In principle, the donor can transfer electrons either to some component on the $O_2$ side of photosystem II or to a component between the two photosystems.)

7. Ubiquinone has an absorbance band at 275 nm. This band bleaches when the quinone is reduced to either the semiquinone or the dihydroquinone. The anionic semiquinone has an absorption band at 450 nm, but neither the quinone nor the dihydroquinone absorbs at this wavelength. A suspension of purified bacterial reaction centers was supplemented with extra ubiquinone and reduced cytochrome $c$ and was then illuminated with a series of short flashes of light. The absorbance at 275 nm decreased on the odd-numbered flashes as shown on the first curve. The absorbance at 450 nm increased on the odd-numbered flashes but returned to the original level on the even-numbered flashes as shown in the second curve.

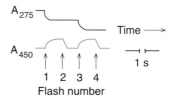

   (a) Explain the patterns of absorbance changes at the two wavelengths.
   (b) Why is it necessary to have reduced cytochrome $c$ present in order to see these effects?

8. Explain why in green plants, the quantum efficiency of photosynthesis drops sharply at wavelengths longer than 680 nm.

9. Do you predict that plant cells lacking carotenoids are more or less susceptible to damage by photooxidation? Explain.

10. Molecular oxygen is an alternative substrate for the ribulose bisphosphate carboxylase–oxygenase and is also a competitive inhibitor with respect to $CO_2$ fixation. Explain.

11. In the formation of phosphoenolpyruvate from pyruvate in the mesophyll cells (fig. 15.28), notice the energetics of the reaction. How does this process compare energetically with the formation of phosphoenolpyruvate from pyruvate in the start of gluconeogenesis?

12. Provide a mechanism for the conversion of pyruvate to phosphoenolpyruvate in the mesophyll cells (fig. 15.28), and account for the consumption of an ATP and $P_i$ and the production of AMP, $PP_i$, and phosphoenolpyruvate.

13. Propose a mechanism for the "oxygenase" activity of ribulose bisphosphate carboxylase.

14. When examining the information described in figures 15.13 or 15.17, one is struck by the similarities with the electron-transport scheme (chapter 14). What types of components do these two schemes have in common?

15. What differences can you find between the pentose phosphate pathway (figs. 12.31 through 12.34 and associated text) and the Calvin cycle?

16. Reduction of 3 moles of $CO_2$ to form 1 mole of triose phosphate requires 9 moles of ATP and 6 moles of NADPH.

(a) What is the source of NADPH in the reduction of $CO_2$?

(b) Account for the ATP consumed in the formation of triose phosphate.

(c) Assume that the $CO_2$ initially is added to PEP in the C-4 pathway. What is the additional cost in ATP per mole of $CO_2$ added?

(d) Is additional NADPH required for $CO_2$ fixation in the C-4 plant?

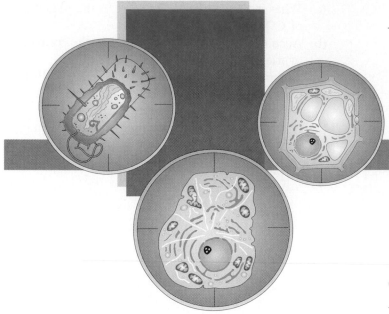

# Structures and Metabolism of Oligosaccharides and Polysaccharides

*Polysaccharides and oligosaccharides are formed from a wide variety of hexoses.*

Thus far we have focused on the roles that sugars and carbohydrates play in energy metabolism. In this chapter we consider the structures, functions, and metabolism of carbohydrates involved in other roles.

Polymeric carbohydrates are major structural components in plant and bacterial cell walls and in the extracellular matrix of vertebrates. Branched-chain polymeric carbohydrates covalently linked to proteins are used to give proteins unique signatures that guide them to their final destination and that facilitate specific interactions between free proteins or proteins attached to cells.

**Figure 16.1**

Some of the sugar building blocks found in polysaccharides.

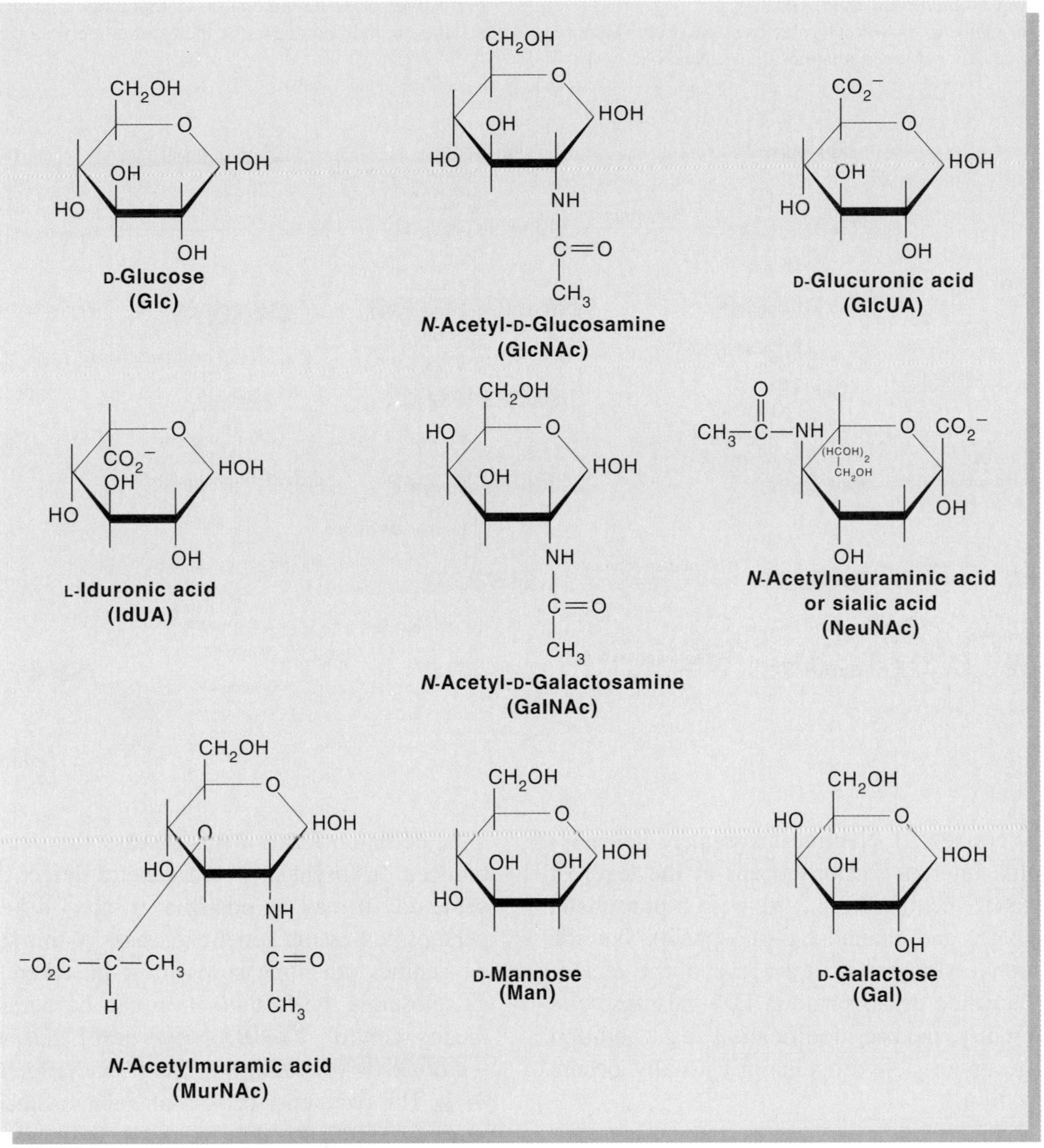

D-Glucose (Glc)

N-Acetyl-D-Glucosamine (GlcNAc)

D-Glucuronic acid (GlcUA)

L-Iduronic acid (IdUA)

N-Acetyl-D-Galactosamine (GalNAc)

N-Acetylneuraminic acid or sialic acid (NeuNAc)

N-Acetylmuramic acid (MurNAc)

D-Mannose (Man)

D-Galactose (Gal)

## Many Types of Monosaccharides Become Integrated into Polysaccharides

Whereas the energy-storage polysaccharides and cellulose are homopolymers composed exclusively of glucose, a large number of monosaccharides are used to build other polysaccharides. This list includes D-mannose, D- and L-galactose, D-xylose, L-arabinose, D-glucuronic acid, L-iduronic acid, N-acetyl-D-glucosamine, N-acetyl-D-galactosamine and N-acetylneuraminic acid. Many of these sugar building blocks are considered in this chapter (fig. 16.1).

## Monosaccharides Are Related through a Network of Metabolic Interconversions

We discussed the biosynthesis of glucose from simpler starting materials in chapter 12. The biosynthesis of other hexoses is linked to glucose by a complex network of reactions (fig. 16.2). In fact, glucose serves as the precursor for the synthesis of many other hexoses without any rearrangement of the six carbon core.

Although the list of hexoses is large, we can make certain generalizations about the pathways by which they

## Figure 16.2

Monosaccharide interconversions. The following abbreviations are used here and throughout the chapter: Gal = galactose, Xyl = xylose, Glc = glucose, GlcUA = glucuronic acid, GlcNAc = N-acetylglucosamine, GalNAc = N-acetylgalactosamine, Fuc = fucose, Man = mannose, ManNAc = N-acetylmannosamine, NeuNAc =

N-acetylneuraminic acid (the most common type of sialic acid = Sia). Sialic acid is a more general term including N- and O-substituted derivatives of neuraminic acid. Compounds in capital boldface type are the unmodified sugars or their amine derivatives.

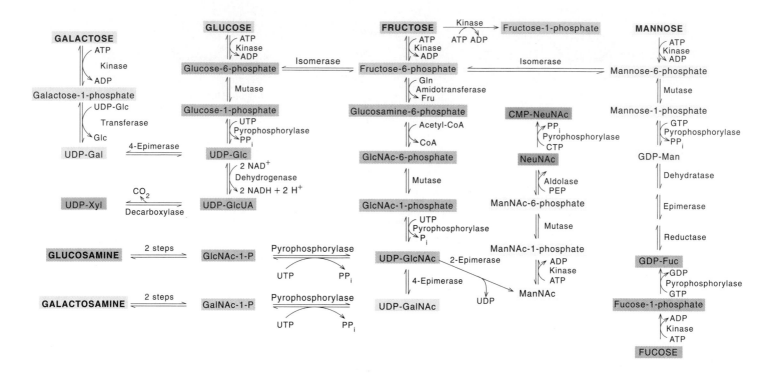

are metabolically connected. Neutral hexoses are never interconverted; some interconversions occur at the level of monophosphorylated hexose (e.g., glucose-6-phosphate, fructose-6-phosphate, and mannose-6-phosphate), but the majority of interconversions occur at the level of the nucleotide sugars (e.g., uridine diphosphate (UDP)-galactose and UDP-glucose). Finally, hexose modification (e.g., addition of an N-acetyl group or dehydrogenation) usually occurs before polymerization.

## Galactosemia Is a Genetically Inherited Disease That Results from the Inability to Convert Galactose into Glucose

Serious diseases often result from the inability of the organism to metabolize certain substances that are ingested as nutrients. In such cases the substance accumulates and often causes irreversible harm to the liver or other vital organs. Galactosemia is a disease that results from the inability to convert galactose into glucose. Reference to figure 16.2 shows that three enzymes, a kinase, a transferase, and an epimerase are required for this conversion. In the

most common form of this disease the transferase is either reduced or absent due to a genetic defect. It is hoped that some day it may be possible to inject a healthy gene into persons possessing genetic diseases of this sort but until that day comes other measures must be taken. In the case of galactosemia the lost function can be compensated for by dietary control. The main source of galactose in the diet (especially of the infant) is the disaccharide lactose found in milk. The damaging effects of galactose accumulation can be minimized by feeding infants with lactose-free milk from the time of birth. Galactose needed for membrane and glycoprotein synthesis can be formed from glucose-1-phosphate (see fig. 16.2).

## Structural Polysaccharides Include Homopolymers and Heteropolymers

The most abundant structural polysaccharide is plant cellulose, a straight-chain homopolymer of glucose with a $\beta(1,4)$ linkage (see chapter 12). Cellulose is formed by the same general mechanism as glycogen, using nucleoside diphosphate sugars. Chitin, a structural polysaccharide found in insects, is also a $\beta(1,4)$ homopolymer (fig. 16.3). The

**Figure 16.3**

Structure of the repeating unit of chitin: a $\beta(1,4)$ homopolymer of *N*-acetyl-D-glucosamine.

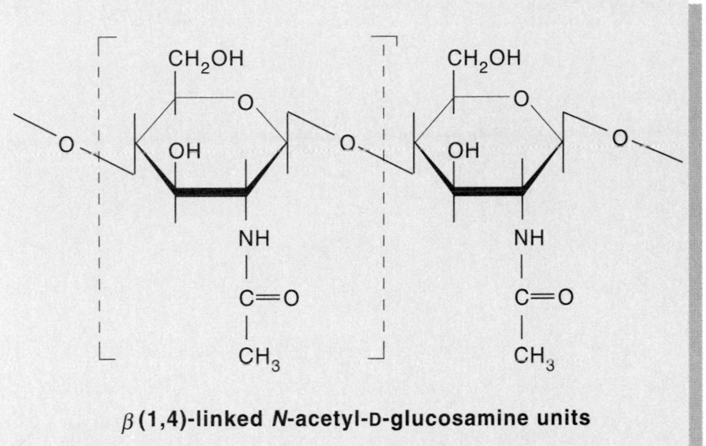

$\beta$(1,4)-linked *N*-acetyl-D-glucosamine units

**Figure 16.4**

Structure of the repeating unit of hyaluronic acid: GlcUA$\beta$(1,3)GlcNAc$\beta$(1,4).

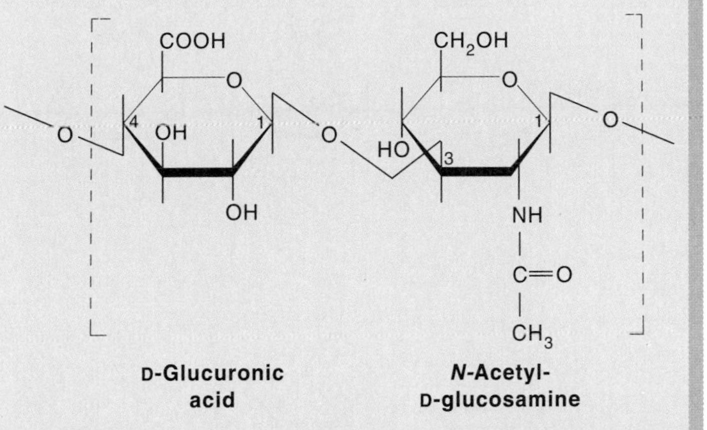

D-Glucuronic
acid

*N*-Acetyl-
D-glucosamine

*N*-acetylglucosamine (GlcNAc) monomers are linked in a similar reaction starting from UDP-*N*-acetylglucosamine monomeric substrates (UDP-GlcNAc).

Many carbohydrates contain modified sugars, like the *N*-acetylglucosamine found in chitin, or two or more different sugars in straight-chain or branched-chain linkages.

Structurally, the simplest and best known of the heteropolysaccharides that contain more than one type of monomer are the glycosaminoglycans. These are long, unbranched polysaccharides composed of repeating disaccharide subunits in which one of the two sugars is either *N*-acetylglucosamine or *N*-acetylgalactosamine (table 16.1). Glycosaminoglycans are highly negatively charged due to the presence of carboxyl or sulfate groups on many of the sugar residues. The negative charge causes the polymeric chains to adopt a stretched or extended conformation, which increases their sphere of influence and generates a high viscosity in the surrounding region. Glycosaminoglycans are usually found in extracellular space in multicellular organisms, where they produce a viscous matrix that resists compression. Such an environment can be beneficial to organisms in many ways: It can provide a passageway for cell migration, lubrication between joints, or rigidity to structures such as cartilage or the ball of the eye.

Hyaluronic acid is a heteropolymer of D-glucuronic acid and *N*-acetyl-D-glucosamine (fig. 16.4). It is considerably larger than most glycosaminoglycans, reaching molecular weights in excess of $10^6$. Hyaluronic acid is formed by the successive addition of glucuronic acid and *N*-acetylglucosamine residues to the non-reducing ends of a growing chain. This process is catalyzed by two enzymes, one specific for the addition of each type of monomer from the

appropriate UDP-sugar. Hyaluronic acid is usually found in a complex with protein (fig. 16.5).

## Oligosaccharides Are Found in Glycoconjugates

One of the most rapidly advancing areas in biochemistry is that concerned with the structure and metabolism of oligosaccharides that are covalently attached to proteins (in glycoproteins) and lipids (in glycolipids). The carbohydrates in these complexes vary in composition, in the way they are linked, their branching patterns and in the sugars that terminate each branch.

A change in the linkage of only one sugar of a complex branched structure may significantly affect its interaction with other biomolecules. For example, influenza virus has a glycoprotein called hemagglutinin which causes agglutination or clumping of red blood cells. The viral hemaglutinin binds to cells via a sialic acid (see fig. 16.1) $\alpha(2,3)$-linked to a galactose residue, whereas other hemaglutinins bind only to sialic acids that are $\alpha(2,6)$-linked to galactose. Such an exquisitely specific method of binding suggests the functional consequences of changes in carbohydrate structure. Other important roles of carbohydrates as recognition markers include the binding of cholera toxin to a specific glycolipid, $GM_1$; the binding of lysosomal enzymes by the mannose-6-phosphate receptor (discussed later on); the binding of terminal sialylated, fucosylated lactosamine sugar sequences by cellular adhesion molecules (CAMs); and the recognition of mammalian eggs by sperm. It is likely that the species specificity of egg fertilization is due to the inability of sperm from one species

## Table 16.1

Structure of Glycosaminoglycans

| Polysaccharide | Monosaccharide Units[a] | | Substituents | Repeating Unit |
|---|---|---|---|---|
| | *A* | *B* | | |
| Hyaluronate | β-D-GlcUA | β-D-GlcN | $R = -C\overset{O}{\underset{CH_3}{}}$ | |
| Chondroitin sulfates Dermatan sulfate | β-D-GlcUA α-L-IdUA | β-D-GalN | $R = -C\overset{O}{\underset{CH_3}{}}$  $R' = -H$ or $-SO_3^-$ | |
| Heparan sulfate and heparin | β-D-GlcUA α-L-IdUA | α-D-GlcN | $R = -C\overset{O}{\underset{CH_3}{}}$  $R' = -H$ or $-SO_3^-$ | |
| Keratan sulfate | β-D-Gal | β-D-GlcN | $R = -C\overset{O}{\underset{CH_3}{}}$  $R' = -H$ or $-SO_3^-$ | |

[a] The polysaccharides are depicted as linear polymers of alternating A and B monosaccharide units.
Abbreviations GlcUA, glucuronic acid; IdUA, iduronic acid; GalN, galactosamine; GlcN, glucosamine; Gal, galactose.

## Figure 16.5

A hyaluronic acid–proteoglycan complex. The individual proteoglycans contain a core protein by which are linked various glycosaminoglycans. The core protein is noncovalently associated with a single hyaluronic acid molecule via two link proteins. Electron micrograph below shows a large aggregate (A) and small aggregate (B) the sizes of which are determined mainly by the length of the hyaluronate central filament of the individual aggregates. Protein is represented in green. Carbohydrate is represented in red. (Electron micrograph from J. A. Buckwalter, and L. Rosenberg. *Coli. Rel. Res.* 3:489–504, 1983.)

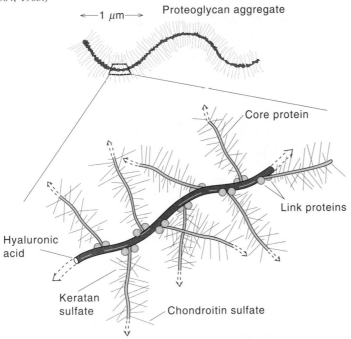

Proteoglycan aggregate

←— 1 μm —→

Core protein

Link proteins

Hyaluronic acid

Keratan sulfate

Chondroitin sulfate

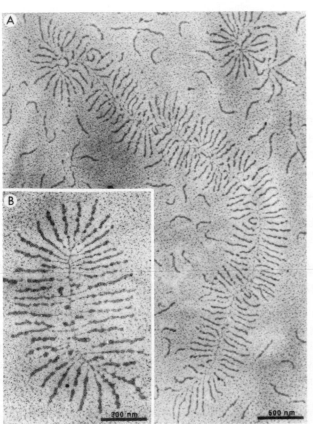

## Figure 16.6

Linkages found between oligosaccharides and proteins in glycoproteins: (a) is an O-glycosidic linkage found in many glycoproteins and mucins; an analogous linkage occurs between N-acetylglucosamine and serine in glycoproteins; (b) is an N-glycosidic linkage found in many glycoproteins; and (c) is an O-glycosidic linkage found only in collagen.

(a) The *N*-acetylgalactosamine-serine linkage

(b) The *N*-acetylglucosamine-asparagine linkage

(c) The galactose-hydroxylysine linkage

to recognize the oligosaccharides of glycoproteins that surround the eggs of other species.

The carbohydrate moiety in a glycoprotein is covalently attached to specific amino acid side chains in a polypeptide chain. Most commonly, two types of glycosidic linkages are found: Carbohydrates linked to the hydroxyl groups of serine, threonine, or hydroxylysine (O-glycosidic linkages) and carbohydrates linked to the amide group of asparagine (N-glycosidic linkages) (fig. 16.6). In the case of O-linked sugars, clusters of serines or threonines are the preferred substrate. For N-glycosidic linkage, only asparagine residues in the sequence Asn-X-Ser(Thr) accept carbohydrates (where "X" is any amino acid except proline).

## *Oligosaccharides Are Synthesized by Specific Glycosyltransferases*

All oligosaccharide structures found in glycoproteins or glycolipids are synthesized by highly specific glycosyltransferases. Most of these enzymes require a divalent cation (usually $Mn^{2+}$), and most are highly specific for the acceptor substrate, a nucleotide sugar, and the linkage by which the sugar is attached to its acceptor. As a rule there is a unique glycosyltransferase for each type of glycosidic bond. For example, two different enzymes attach GlcNAc in $\beta(1,2)$ linkage to mannose for the synthesis of a simple N-linked carbohydrate (fig. 16.7) because the carbohydrate acceptor to which the GlcNAc is added differs markedly in each case.

In addition to the high specificity of glycosyltransferases, a major reason for the ordered addition in oligosaccharide synthesis is that these enzymes are found in different membrane-bound compartments. Nascent glycoproteins come in contact with particular glycosyltransferases only when they are in the cellular compartment where the glycosyltransferase is located.

The early stages of oligosaccharide attachment to protein often begin before protein synthesis is completed. Proteins destined to become glycoproteins due to the addition of glycans are usually synthesized on ribosomes attached to the endoplasmic reticulum (fig. 16.8). Other proteins are synthesized on free ribosomes. Proteins and glycoproteins made in association with the endoplasmic reticulum harbor a short (about 20) hydrophobic amino acid stretch termed a signal sequence at their $NH_2$-terminus. During early translation, the signal sequence is recognized by a soluble protein complex termed the signal recognition particle (SRP). Binding to the SRP blocks further translation until the SRP attaches to a specific receptor on the endoplasmic reticulum membrane. Following this, translation resumes and the growing polypeptide is continuously extruded into the lumen of the endoplasmic reticulum.

Whereas protein synthesis is completed before a protein leaves the endoplasmic reticulum, at the time of exit the synthesis of the oligosaccharide portion of a glycoprotein has only just begun. All nonresident proteins of the endoplasmic reticulum are transferred by a membrane-budding process to the Golgi complex (fig. 16.9). In the Golgi complex oligosaccharide biosynthesis continues with the sequential removal and addition of sugar residues. After exit-

### Figure 16.7

Glycosyltransferases are specific for the acceptor, the sugar transferred, and the linkage whose formation they catalyze. The N-acetylglucosaminyltransferases termed GlcNAc-TI and GlcNAc-TII both transfer N-acetylglucosamine to mannose in $\beta(1,2)$ linkage, but their acceptor substrates are completely different. This sequence of biosynthetic steps was deduced from analyses of a mammalian cell mutant that lacks GlcNAc-TI activity.

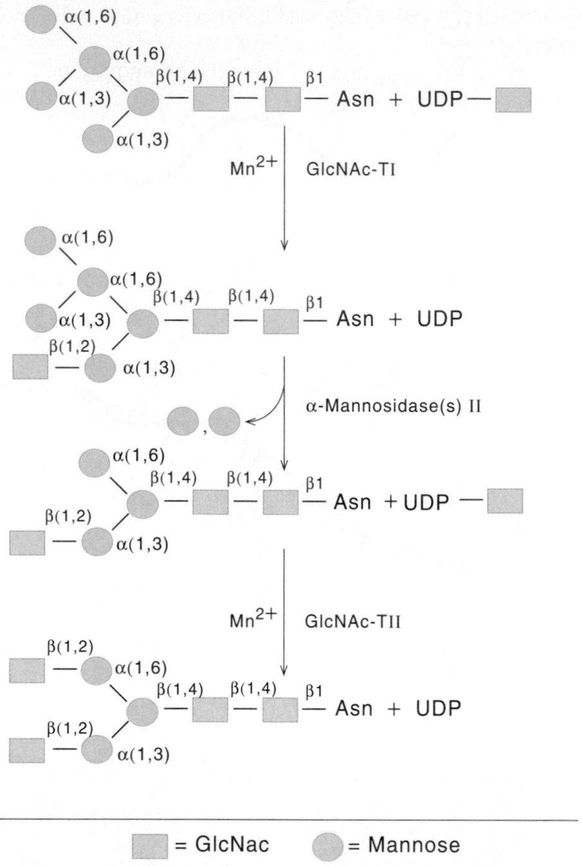

= GlcNac    = Mannose

ing the Golgi complex by further budding, glycoproteins are delivered to one of three locations: the plasma membrane to become a resident integral membrane glycoprotein, the extracellular milieu as a secreted glycoprotein, or the lysosomes.

Because the biosynthesis of O- and N-linked oligosaccharides follows quite different routes, it is convenient to discuss them separately.

## *N-Linked Oligosaccharides Utilize a Lipid Carrier in the Early Stages of Synthesis*

All N-linked oligosaccharides include a common core of five sugars linked to Asn: $Man\alpha(1,3)Man\alpha(1,6)Man\beta(1,4)GlcNAc\beta(1,4)GlcNAc\beta1,Asn$. This reflects a common biosynthetic pathway in which sugars are added to a membrane-associated polyprenol lipid called dolichol phosphate

## Figure 16.8

Diagram of events involved in the targeting and translocation of proteins destined to be completed in the endoplasmic reticulum. Proteins that include a signal sequence of approximately 20 hydrophobic amino acids at their $NH_2$ terminus are recognized by a signal recognition particle (SRP). This complex of proteins and 7S RNA binds to the signal sequence of the nascent protein and the large ribosomal subunit. SRP binding causes translation arrest and the targeting of the translocation complex to the endoplasmic reticulum (ER) where it is bound by the SRP receptor as well as a ribosome receptor protein. As protein synthesis resumes, SRP is released from the signal sequence, and the growing chain traverses the membrane of the endoplasmic reticulum entering the lumen. Signal peptidase cleaves the signal sequence, creating a new $NH_2$ terminus. When protein synthesis is complete, soluble proteins are released into the lumen of the ER. Proteins with additional hydrophobic regions may span the ER membrane so that their C-terminus or N-terminus or both are on the cytoplasmic face of the ER membrane.

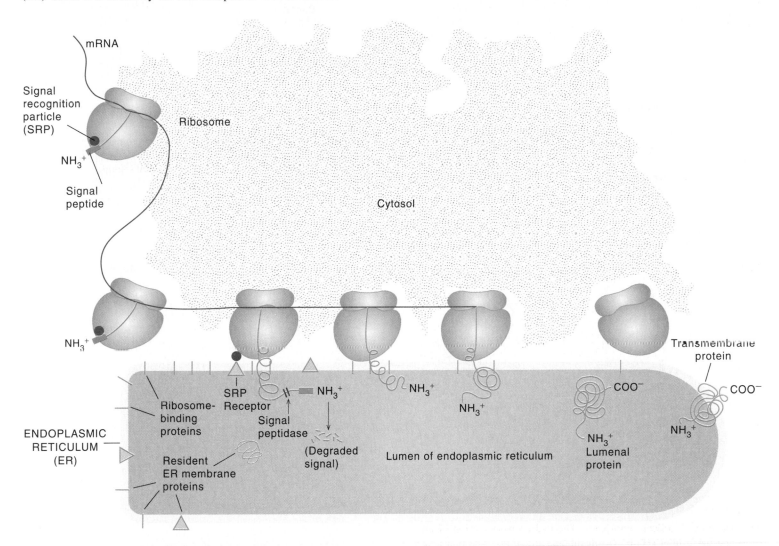

(Dol-P) (fig. 16.10). The Dol-P serves as a substrate for a glycosyltransferase that adds phospho-*N*-acetylglucosamine from UDP-GlcNAc to form Dol-P-P-GlcNAc. Six additional sugars are added one at a time on the cytosolic side of the endoplasmic reticulum (fig. 16.11). Substrates for these additions are activated nucleotide sugars as in polysaccharide synthesis. After the additions are completed, the Dol-P-P-oligosaccharide complex is translocated to the lumenal side of the endoplasmic reticulum. Here more sugars are added; this time the activated sugars are Dol-P-mannose and Dol-P-glucose, which are assembled on the cytosolic side of the endoplasmic reticulum and then translocated to the lumen. The final structure assembled on a dolichol phosphate contains two *N*-acetylglucosamines, nine mannoses, and three glucoses linked as shown in figure 16.11. This oligosaccharide moiety is translocated as a unit to select Asn residues on a growing polypeptide chain. The transfer and subsequent modification of the N-linked carbohydrates is shown schematically in figure 16.12.

Once covalently attached, the oligosaccharide is processed by specific exoglycosidases that sequentially remove some of the terminal sugars (see fig. 16.12, top). Subse-

## Figure 16.9

Schematic diagram showing the relative locations of nucleus, endoplasmic reticulum (ER), Golgi complex, *trans*-Golgi network, and plasma membrane. Glycoproteins synthesized in the lumen of the ER pass to the *cis* cisterna of the Golgi complex by a sequential membrane budding and fusion mechanism. The Golgi cisternae are classified into *cis, medial,* and *trans* in the order of increasing distance from the nucleus. Mature glycoproteins exit from the *trans*-Golgi network in membrane-bound secretory vesicles. Depending on the nature of the membranes, the vesicles have one of several different fates. Following fusion with other membranes, their contents are delivered outside the cell, to the plasma membrane or to other organelles inside the cell, such as lysosomes (see fig. 16.13).

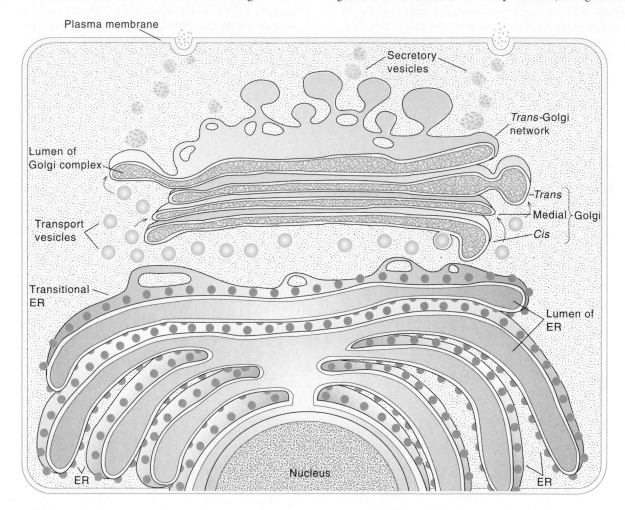

## Figure 16.10

The structure of dolichol phosphate. The hydrocarbon portion of the molecule has a high affinity for membrane structures. The phosphate end forms an activated complex with oligosaccharide intermediates.

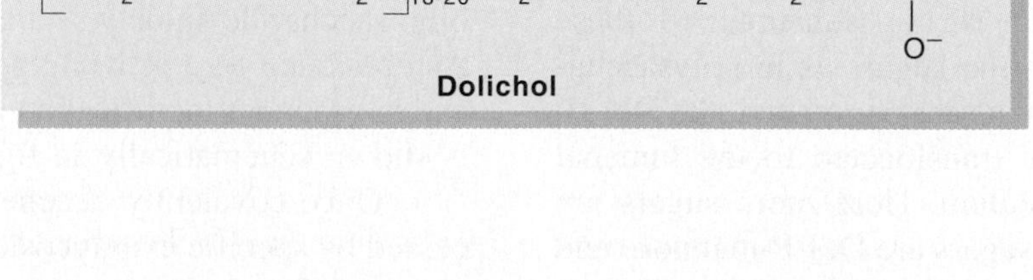

**Dolichol**

## Figure 16.11

Synthesis of the dolichol-linked oligosaccharide. The reaction starts on the cytosolic side of the endoplasmic reticulum. Seven single-step reactions lead to the heptasaccharide that is linked to dolichol phosphate. The heptasaccharide is translocated across the membrane of the endoplasmic reticulum to the lumen, where seven additional one-step additions take place. The activated sugars for these additions are formed on the cytosolic side and translocated to the lumen as hexose-P-dolichol complexes. The initial step in the pathway is inhibited by tunicamycin. The synthesis of Dol-P-Man is inhibited by amphomycin.

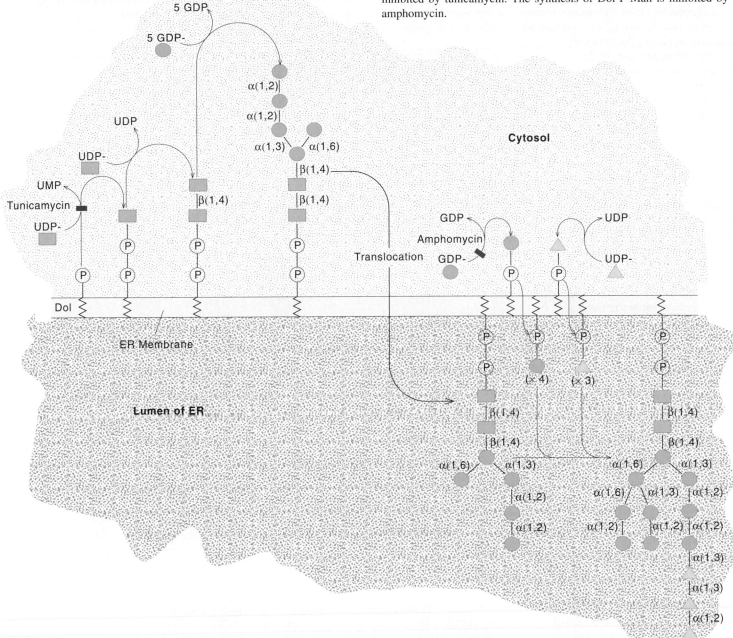

quently many glycoproteins are translocated by budding to the Golgi, where the oligosaccharide undergoes further modification.

Each protein that traverses the three Golgi compartments (*cis, medial,* and *trans*) does so in its own way, usually losing some sugar residues and gaining others (see remainder of fig. 16.12). Finally, in the *trans*-Golgi, proteins are sorted according to the types of oligosaccharides they bear. Some remain in the Golgi, but most continue on to the plasma membrane, the lysosomes, or the extracellular matrix (fig. 16.13).

## Carbohydrate Modification Is Crucial to Targeting Lysosomal Enzymes

The process whereby newly synthesized proteins are directed to specific locations is referred to as protein targeting. Soluble lysosomal hydrolases are targeted to lysosomes

# Figure 16.12

Schematic diagram illustrating the transfer of oligosaccharide to protein in the endoplasmic reticulum and the subsequent processing and maturation of oligosaccharides in the Golgi complex. In the endoplasmic reticulum the oligosaccharide synthesized on Dol-P is transferred to the Asn(N) of an Asn-X-Ser(Thr) sequence in the growing peptide chain. Subsequent trimming of the oligosaccharide occurs by specific exoglycosidases that sequentially remove terminal glucoses and a specific mannose residue. When synthesis of the protein portion is complete, the glycoprotein is transferred to the *cis* compartment of the Golgi complex. For most glycoproteins further processing of the $Man_8GlcNAc_2$ oligosaccharide occurs in this compartment.

In the *medial* Golgi the synthesis of a biantennary structure terminating in GlcNAc is achieved (see fig. 16.7). Also in this compartment additional GlcNAc residues may be added to the core mannose residues to give as many as seven branches. Fucose, galactose, and sialic acid residues are added in the *trans*-Golgi and *trans*-Golgi network compartments. The membranes of the Golgi compartments contain nucleotide-sugar translocases specific for transferring each nucleotide sugar across the membrane.

Inhibitors can block the pathway at specific steps, causing the accumulation of biosynthetic intermediates. Castanospermine inhibits the α-glucosidase that removes the first glucose; deoxymannojirimycin inhibits the α-mannosidase I that removes mannose from the $Man_8$ intermediate, and swainsonine blocks the α-mannosidase II that removes mannose from the $Man_5$ intermediate. In the presence of swainsonine, hybrid N-linked carbohydrates are synthesized.

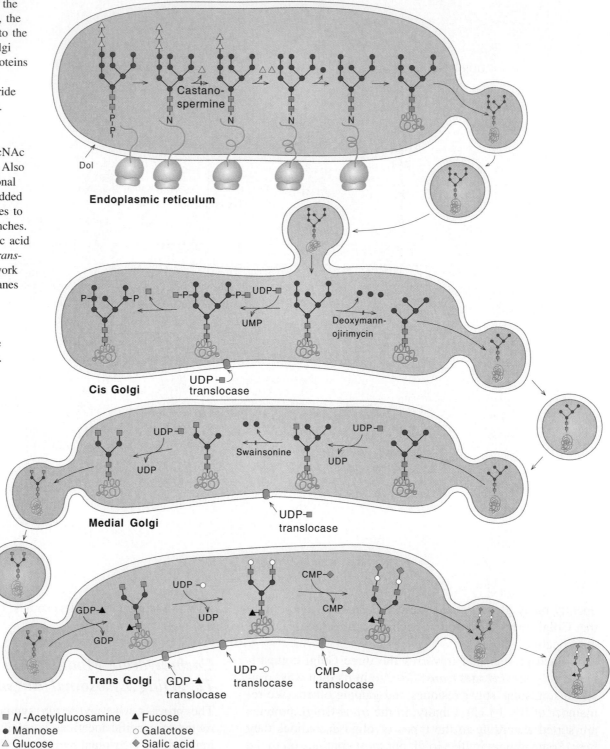

## Figure 16.13

Sorting of proteins for their final destination occurs in the *trans*-Golgi network (TGN). At 20°C secreted proteins accumulate in the TGN. By raising the temperature to 37°C, the sorting and secretion of synchronized populations of marker proteins can be studied. Two major routes to the plasma membrane have been identified: the constitutive pathway and the regulated pathway. Proteins in the latter pathway (including many prohormones) accumulate in secretory granules that fuse with the plasma membrane on receiving a biochemical stimulatory signal. As the name suggests, passage of integral membrane proteins and soluble proteins through the constitutive pathway occurs constantly, with no requirement for a stimulatory signal. Soluble lysosomal hydrolases bound to the Man-6-P receptor exit from the TGN in clathrin-coated vesicles. Clathrin is a fibrous protein that coats certain vesicles. These vesicles fuse with an acidic, prelysosomal compartment where the lysosomal hydrolases are released from the Man-6-P receptor and are ultimately delivered to lysosomes. The Man-6-P receptor does not go to lysosomes but recycles between the plasma membrane, Golgi membranes, and prelysosomes.

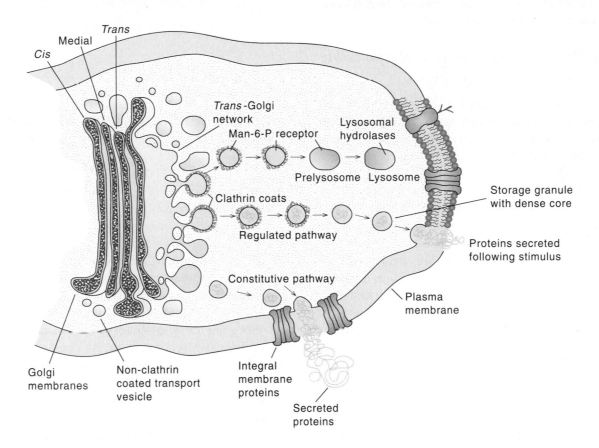

by a specific carbohydrate marker, which they acquire in the Golgi. This is one of the best understood examples of the role of carbohydrate structure in targeting.

Oligomannosyl carbohydrates on soluble enzymes destined to become lysosomal enzymes carry one or two phosphate residues at the 6 position of mannose (Man-6-P). These phosphorylated mannose residues are recognized by a glycoprotein called the Man-6-P receptor, which binds and transports the prelysosomal enzymes to prelysosomal vesicles (see fig. 16.13). This binding ensures that the prelysosomal enzyme enters a vesicle destined to fuse to and thereby deliver its contents to the lysosome. In an acidic, prelysosomal compartment the binding between the Man-6-P receptor and the lysosomal enzyme is disrupted so that the receptor can be recycled to the Golgi apparatus.

Fibroblasts from patients with a lysosomal storage disease, called I-cell disease, cannot add the carbohydrate marker, and consequently their lysosomal hydrolases are largely secreted instead of being targeted to the lysosome. As a result many molecules that are normally degraded by lysosomal hydrolases accumulate in the lysosomes. Morphologists have termed these dense lysosomes inclusion bodies, hence the name I-cell disease.

## Biosynthesis of O-Linked Oligosaccharides in the Cis-Golgi

Glycoproteins in the secretory pathway receive GalNac in O-glycosidic linkage from UDP-GalNAc via a transferase that acts directly on certain Ser or Thr residues of proteins.

No dolichol-linked intermediates are involved. Kinetic-labeling studies indicate that GalNAc is added to many glycoproteins in the *cis*-Golgi. The second sugar in O-linked oligosaccharides is usually Gal; this is added in the *trans*-Golgi compartment. The remainder of the O-linked oligosaccharide is synthesized in the trans Golgi and the trans Golgi network by sequential sugar additions from nucleotide sugars. As usual each of these additions is catalyzed by a specific glycosyltransferase. Each sugar is added to the protein in a specific order. Cytoplasmic and nuclear proteins found with single O-GlcNAc residues pick up these residues from a glycosyltransferase in the cytosol.

## O-Linked Oligosaccharides Are Responsible for Different Blood Group Types

The O-linked oligosaccharides of membrane glycoproteins are usually not very large although they may exhibit branching and various terminal sugar sequences. However, in body fluids very large O-linked structures, termed mucins, are found on proteins. These oligosaccharides often terminate with sugars, which are highly immunogenic and form the basis of the human blood types, giving them the title of blood group substances. Because blood-group-specific sugars are expressed on many O-linked oligosaccharides of cells and fluids and are highly immunogenic, individual persons mount an immune response against blood of a different type. Whereas individuals of similar blood groups can accept blood from one another, individuals of different types often cannot due to immunological rejection.

One of the better known blood-grouping schemes is the ABO system. Blood is considered to be of type A, B, or O. All individuals contain two genes (alleles) for blood type. The possible combinations give rise to six different gene types and four different combinations of blood group antigens in different individuals (table 16.2). Cells of an individual carry A antigens (in gene types *AA* or *AO*), B antigens (in gene types *BB* or *BO*), a mixture of A and B antigens (in gene type *AB*), or neither of these antigens (in gene type *OO*). The explanation for these correlations between gene types and blood group antigens is that the *A* gene and the *B* gene encode different glycosyltransferases. The *A* gene encodes a glycosyltransferase that catalyzes the addition of a terminal *N*-acetylgalactosamine (GalNAc) residue onto a core oligosaccharide, whereas the *B* gene encodes a similar enzyme that adds a galactose (Gal) residue at that site (fig. 16.14). When *A* and *B* genes are present together, both structures are found, but when only *O* genes are present, the site on the oligosaccharide is left unsubstituted. Fortunately for diagnostic purposes the presence or absence of these glycosyltransferases is readily detectable in the milk of the lactating female.

### Table 16.2

The Human ABO Blood Group Scheme

| Gene Type | Antigen |
|-----------|---------|
| *AA* | A |
| *AB* | A,B |
| *BB* | B |
| *AO* | A |
| *BO* | B |
| *OO* | — |

## Specific Inhibitors and Mutants Are Used to Explore the Roles of Glycoprotein Carbohydrates

The functions of the carbohydrate moieties of glycoproteins are difficult to study because of the intimate physical association between oligosaccharides and the protein backbone. However, because several methods enable us to produce glycoproteins with carbohydrates that are severely truncated or missing entirely, we can deduce the function of the affected carbohydrates.

For glycoproteins that contain N-linked carbohydrates, we can completely inhibit oligosaccharide addition by synthesizing the glycoprotein in the presence of tunicamycin (see fig. 16.11). The protein portion of many glycoproteins is synthesized and translocated through the secretory pathway normally in the presence of tunicamycin and can be studied by functional assays. Alternatively, we can eliminate each glycosylation sequence by site-directed mutagenesis of the cloned glycoprotein gene. These artificially constructed genes can be added to cells by a process known as transfection (see chapter 27). This approach has the advantage that the altered protein is synthesized in a normal cell in contrast to a drug-treated cell, in which all glycoproteins are affected, as happens with the tunicamycin treatment.

Although several glycoproteins have been efficiently produced without their oligosaccharides, many glycoproteins cannot be synthesized devoid of carbohydrates. Apparently they require oligosaccharides for proper folding and translocation. In such cases, glycoproteins with altered or truncated carbohydrates are produced to test the effect of oligosaccharide modification on the biological activity of the glycoprotein. One way to do this is to use inhibitors that stop oligosaccharide processing (table 16.3). The glycoproteins so produced have immature N-linked carbohydrates that are quite distinct in structure from mature, complex

## Figure 16.14

The structure and reactions at the oligosaccharide termini of the ABO human blood group antigens. Two enzymes add different hexoses to the termini of O-linked glycoproteins. Individuals may carry both or neither or only one of these enzymes. Individual differences are reflected in the structures of their blood group antigens. These differences are genetically inherited. The *A, B,* and *O* glycosyltransferases have been cloned, and it seems they are all derived from a single gene. Only four nucleotide differences were detected between *A* and *B* genes, and these must be responsible for the different specificities of the A and B transferases. The *O* gene produces a nonfunctional protein.

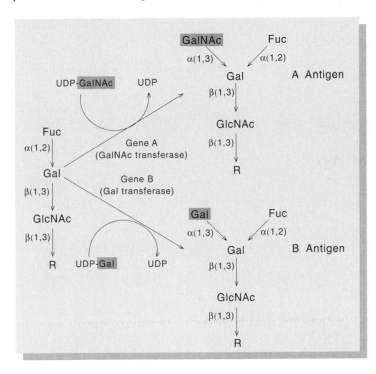

## Figure 16.15

Structure of the peptidoglycan of the cell wall of *Staphylococcus aureus.* (*a*) In this representation X (*N*-acetylglucosamine) and Y (*N*-acetylmuramic acid) are the two sugars in the peptidoglycan. Light green circles represent the four amino acids of the tetrapeptide L-alanyl-D-glutamyl-L-lysyl-D-alanine. Dark green circles are pentaglycine bridges that interconnect peptidoglycan strands. The nascent peptidoglycan units bearing open pentaglycine chains are shown at the left of each strand. (*b*) The structure of X (*N*-acetylglucosamine) and Y (*N*-acetylmuramic acid), which are linked by a $\beta(1,4)$ linkage and alternate in the glycan strand.

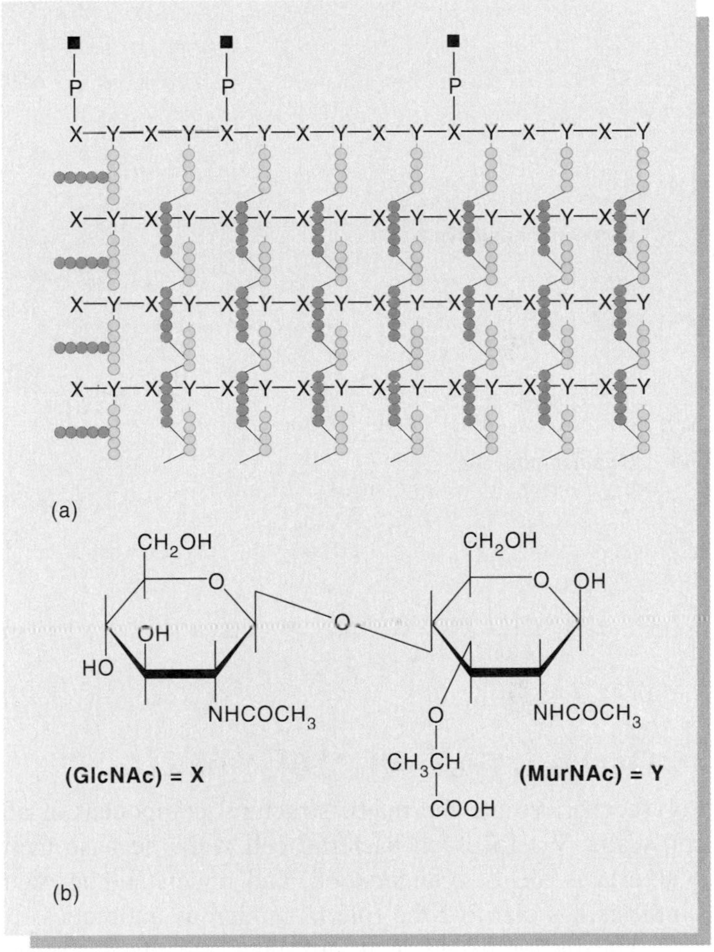

oligosaccharides. A complementary approach is to use mutants of mammalian cells or yeast that are unable to complete the synthesis of mature N-linked carbohydrates. When cloned glycoprotein genes are transfected into such mutants, the carbohydrates produced lack particular structural features.

There are many glycosylation mutants of cultured mammalian cells and yeast. They have been selected as rare survivors of treatments that kill cells expressing a particular carbohydrate or glycoprotein at the cell surface. For example, plants produce a variety of proteins, called lectins, which bind to cell surface oligosaccharides. Lectins are toxic to mammalian cells. They can be used to select for mutants that no longer bind the lectin because they lack a particular carbohydrate at the cell surface. Such glycosylation mutants have low amounts of glycoproteins that require carbohydrates for stable expression.

Some glycosylation mutants that are particularly useful for producing glycoproteins with modified carbohy-

drates are summarized in table 16.4. In some mutants both N- and O-linked carbohydrates are modified, whereas in others only N-linked carbohydrates are affected. We could also obtain glycoproteins with altered carbohydrates by treatment with glycosidases that remove particular sugars. But sugars in oligosaccharides are often sterically protected by the protein, and therefore complete deglycosylation often requires treatment of the unfolded (often denatured) protein. For this reason, the use of glycosylation mutants frequently provides a superior approach for making altered carbohydrates.

**Table 16.3**

Effects of Certain Inhibitors on Glycoprotein Synthesis

| Glycosylation Inhibitor | Major N-linked Carbohydrates Found on Glycoproteins in Presence of Inhibitor |
|---|---|
| Castanospermine | $(Glc_3Man_9GlcNAc_2Asn)$ |
| Deoxymannojirimycin | $(Man_8GlcNAc_2Asn)$ |
| Swainsonine | (Hybrid) |

= Mannose;   ■ = GlcNAc;   ○ = Galactose;   ◇ = Fucose;   △ = NeuNAc;   △ = Glucose

## Bacterial Cell Wall Synthesis

Polysaccharides are the major structural component of all cell walls. We focus on bacterial cell walls because their synthesis is the best understood, and understanding their synthesis has clarified the role of numerous antibiotics.

The bacterial cell wall surrounds the plasma membrane and provides the mechanical strength that enables bacteria to resist shear and osmotic shock. The cell wall is composed of a network of linear heteropolysaccharides cross-linked by peptides (fig. 16.15).

Biosynthesis of bacterial cell wall is remarkable in two respects: (1) It entails the synthesis of a regularly cross-linked polymer; and (2) Part of the synthesis takes place inside the cell and part outside the cell. The synthesis of cell wall is divided into three stages, which occur at different locations: (1) synthesis of UDP-N-acetylmuramyl-pentapeptide, (2) polymerization of N-acetylglucosamine and N-acetylmuramyl-pentapeptide to form linear peptidoglycan strands, and (3) cross-linking of the peptidoglycan strands.

## Synthesis of Activated Monomers Occurs in the Cytoplasm

The first stage in cell wall synthesis (fig. 16.16) entails the synthesis of UDP-N-acetylmuramyl-pentapeptide. First, N-acetylglucosamine-1-phosphate condenses with UTP to form UDP-N-acetylglucosamine. A specific transferase catalyzes a reaction with phosphoenolpyruvate to give the 3-enolpyruvylether of UDP-N-acetylglucosamine. The pyruvyl group is then reduced to a lactyl group by an NADPH-linked reductase, thus forming the 3-O-D-lactylether, or N-acetylglucosamine. This compound is known as UDP-N-acetylmuramic acid (UDP-MurNAc).

Conversion of UDP-N-acetylmuramic acid to its pentapeptide form occurs by the sequential addition of the necessary amino acids. Each step requires ATP and a specific enzyme that ensures the addition of amino acids in the proper sequence, L-alanine is added first, followed by D-glutamic acid, L-lysine (attached by its α-amino group to the γ-carboxyl group of the glutamic acid), and finally the

**Table 16.4**

Glycosylation Mutants that Are Useful for Producing Glycoproteins with Modified Carbohydrates

| Mutant[a] Cell Lines | Glycosylation Enzyme Deficiency | Carbohydrates on Glycoproteins | |
|---|---|---|---|
| | | N-Linked | O-Linked |
| Lec1 Clone 15 B | GlcNAc-TI | | |
| IdID | UDP-Glc-4-epimerase | | Ser |
| Lec8 Clone 13 | UDP-Gal Golgi translocase | | □ — Ser |
| Lec2 Clone 1021 | CMP-NeuNAc Golgi translocase | | ◯ — □ — Ser |

● = Mannose;  ■ = GlcNAc;  ◯ = Galactose;  □ = GalNAc;  △ = NeuNAc

[a] All mutants were derived from Chinese hamster ovary (CHO) cells.

dipeptide D-alanyl-D-alanine is added as a unit. The terminal dipeptide is formed by two enzymatic reactions, conversion of L-alanine to D-alanine by a racemase, followed by the linking of the two alanine residues in an ATP-requiring reaction to form D-alanyl-D-alanine. All of the reactions leading to the pentapeptide formation occur in the cytosol.

## Formation of Linear Polymers Is Membrane-Associated

The next stage in peptidoglycan synthesis results from polymerization of N-acetylglucosamine and N-acetylmuramyl-pentapeptide containing residues into peptidoglycan strands. The relevant reactions take place on the inner surface of the plasma membrane (fig. 16.17).

In step 1 (see fig. 16.17), UDP-N-acetylmuramyl-pentapeptide reacts with a 55-carbon isoprenyl alcohol known as undecaprenol phosphate. This lipid is similar in structure (fig. 16.18) and function to the dolichol phos-

phates in glycoprotein synthesis in eukaryotes. A pyrophosphate linkage is formed with the lipid and UMP is released.

In step 2 (see fig. 16.17) N-acetylglucosamine is added to the lipid intermediate by means of a typical transglycosylation from UDP-N-acetylglucosamine, and UDP is released. In step 3, five glycine residues are sequentially added to the ε-amino group of lysine. Curiously, these glycines are activated by ester formation to a transfer RNA molecule. This form of amino acid activation is rarely seen except in protein synthesis (see chapter 29). In step 4, the disaccharide–oligopeptide unit is transferred from the lipid intermediate to the growing peptidoglycan, and lipid pyrophosphate is generated. It is in this step that we appreciate the dual function of the undecaprenol lipid as activator and transporter of monomer to the extracellular side of the membrane, where cell wall assembly takes place.

In the fifth and final step, one phosphate is hydrolyzed to regenerate the phospholipid, which then can react once again with UDP-N-acetylmuramyl-pentapeptide and partici-

**Figure 16.16**

The first stage of cell wall synthesis: formation of UDP-*N*-acetylmuramyl-pentapeptide (full structure shown at bottom). Point of inhibition by the antibiotic phosphonomycin are indicated.

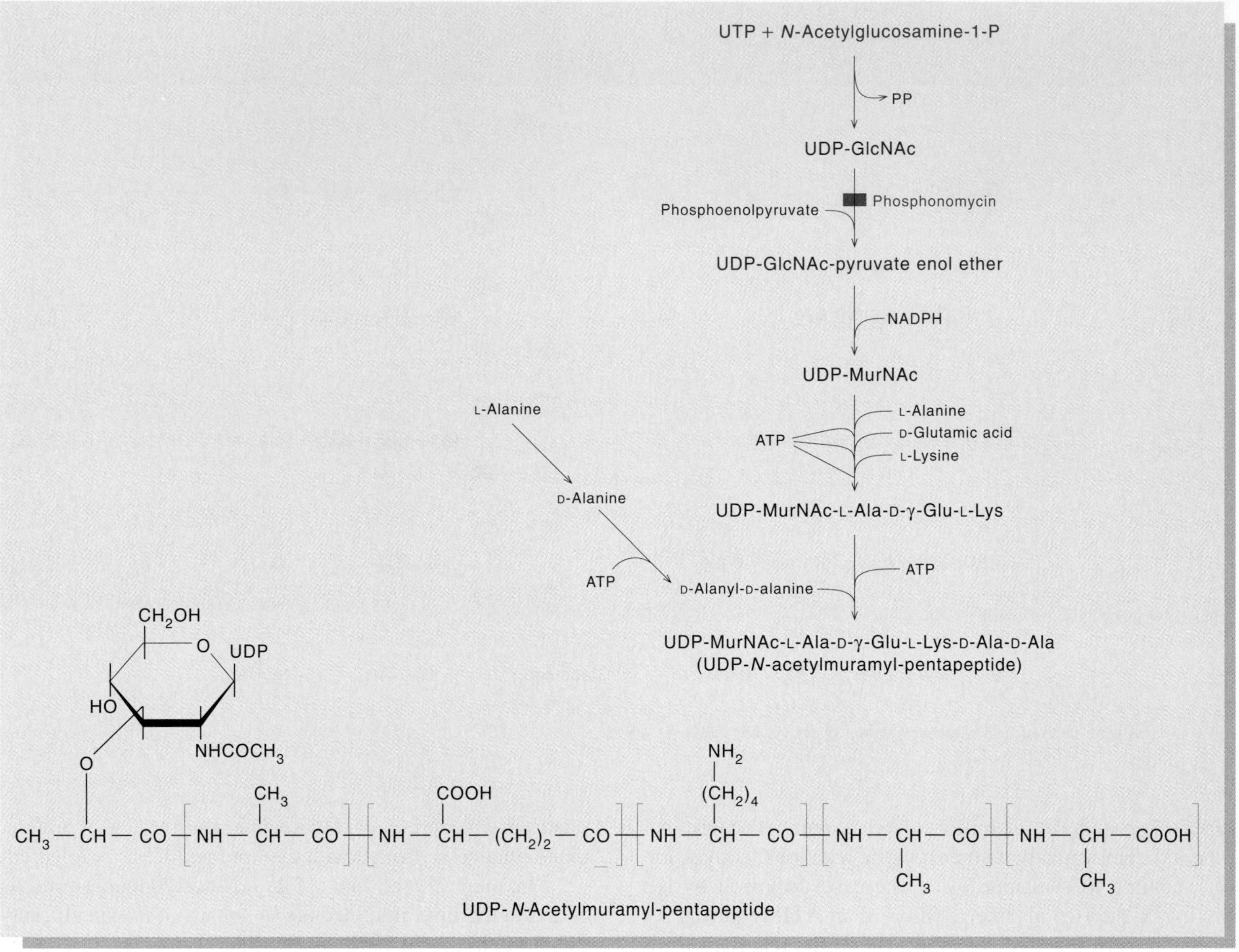
UTP + N-Acetylglucosamine-1-P
PP
UDP-GlcNAc
Phosphoenolpyruvate
Phosphonomycin
UDP-GlcNAc-pyruvate enol ether
NADPH
UDP-MurNAc
L-Alanine
ATP
L-Alanine
D-Glutamic acid
L-Lysine
D-Alanine
UDP-MurNAc-L-Ala-D-γ-Glu-L-Lys
ATP
ATP
D-Alanyl-D-alanine
UDP-MurNAc-L-Ala-D-γ-Glu-L-Lys-D-Ala-D-Ala
(UDP-N-acetylmuramyl-pentapeptide)
CH2OH
O
UDP
HO
NHCOCH3
O
NH2
CH3
COOH
(CH2)4
CH3 — CH — CO — NH — CH — CO — NH — CH — (CH2)2 — CO — NH — CH — CO — NH — CH — CO — NH — CH — COOH
CH3
CH3
UDP-N-Acetylmuramyl-pentapeptide

pate in another cycle, resulting in the addition of a new unit to the growing peptidoglycan strand. The antibiotic bacitracin is a specific inhibitor of the dephosphorylation of the pyrophosphate form of the lipid.

## *Cross-Linking of Linear Polymers Occurs Outside the Plasma Membrane*

The cross-linking of peptidoglycan strands takes place outside the cell membrane at the site of the preexisting wall. Because neither ATP nor any other obvious energy source is available there, a mechanism independent of any external energy source has evolved for this reaction. The reaction is a transpeptidation in which the terminal amino end of an open cross-bridge attacks the terminal peptide bond in an adjacent strand to form a cross-link. The terminal alanine residue is thus eliminated from the strand that becomes cross-linked (fig. 16.19).

## Figure 16.17

The second stage of cell wall synthesis. An ATP-requiring amidation of glutamic acid that occurs between steps 2 and 3 has been omitted. Points of action of the antibiotic inhibitors bacitracin and vancomycin are indicated.

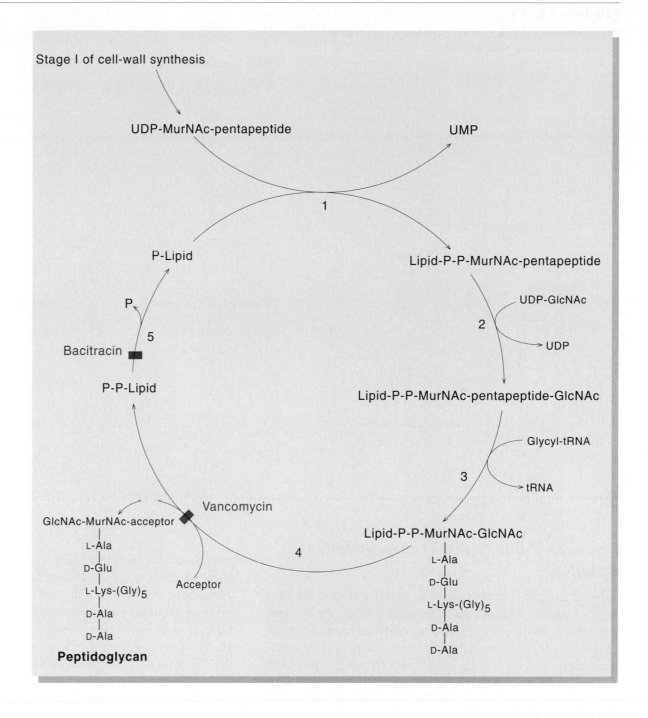

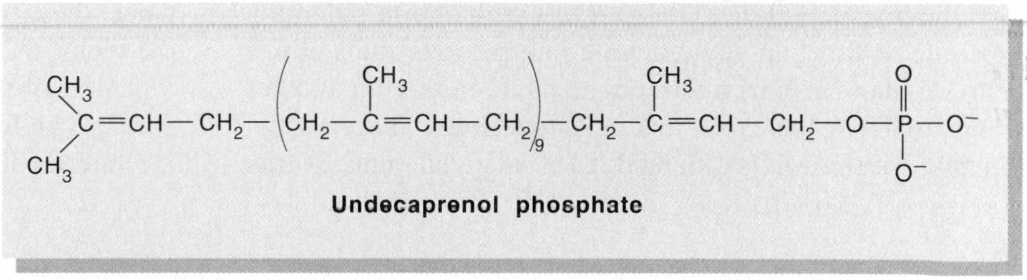

## Figure 16.18

The structure of undecaprenol phosphate. Isoprene phosphates such as undecaprenol phosphate are important carriers and activators in the synthesis of oligosaccharides. Note the similarities to dolichol phosphate.

**Figure  16.19**

The structure of a segment of the peptidoglycan before and
after the final cross-linking reaction.

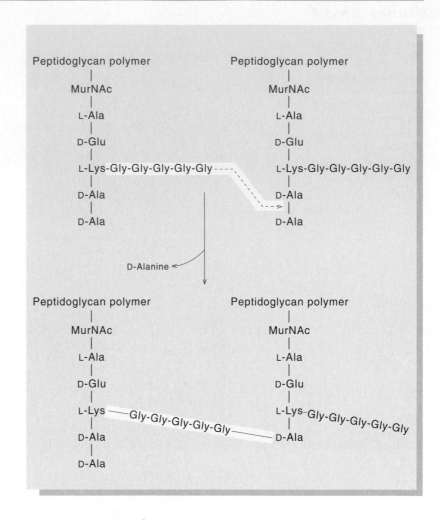

## Penicillin Inhibits the Transpeptidation Reaction

Penicillin is unequaled for usefulness in combating
bacterial diseases and infections. During the 50 years
since Alexander Fleming brought penicillin to the attention
of microbiologists, many biochemists and pharmacologists
have been interested in the mechanism by which this potent
antibiotic kills bacteria. Why do we bring this up here? Simply because penicillin inhibits the cross-linking reaction in
cell-wall synthesis by mocking the terminal D-alanyl-D-
alanine residue of the peptidoglycan strand.

The inhibitory mechanism is shown in figure 16.20.
First the transpeptidase (TPase) reacts with one strand of the
substrate to form an acyl enzyme intermediate, thus eliminating D-alanine. This intermediate then reacts with another
strand to form the cross-link and regenerate the enzyme.
Because penicillin is an analog of alanylalanine, it fits
snugly into the substrate-binding site, with the highly reactive CO—N bond in the $\beta$-lactam ring in the same position
as the bond that is split in the transpeptidation reaction. This
gives it the opportunity to acylate the enzyme, thereby
forming a penicilloyl enzyme, which is inactive (see fig.
16.20). This proposed mechanism is supported by the fact
that the penicilloyl fragment of the antibiotic is found in the
inhibited enzyme.

In addition to penicillin several other antibiotics
(phosphonomycin, bacitracin, and vancomycin) block cell
wall synthesis at different locations (see figs. 16.16 and
16.17). In addition to their biological and medical importance, these antibiotics have been very useful in elucidating
the biosynthetic pathway. This is because they cause accumulation of the intermediate before the blocked step. This
species can frequently be isolated and confirmed as a genu-
ine intermediate in the pathway.

**Figure 16.20**

The third stage of cell wall synthesis. This diagram shows the cross-linking reaction and the mechanism of inhibition by penicillin in the bacterium *Staphylococcus aureus*. (*a*) The end of the peptide side chain of a glycan strand; (*b*) the end of the pentaglycine substituent from an adjacent strand. You may wish to refer to figure 16.19 for details of the structure of the peptidoglycan.

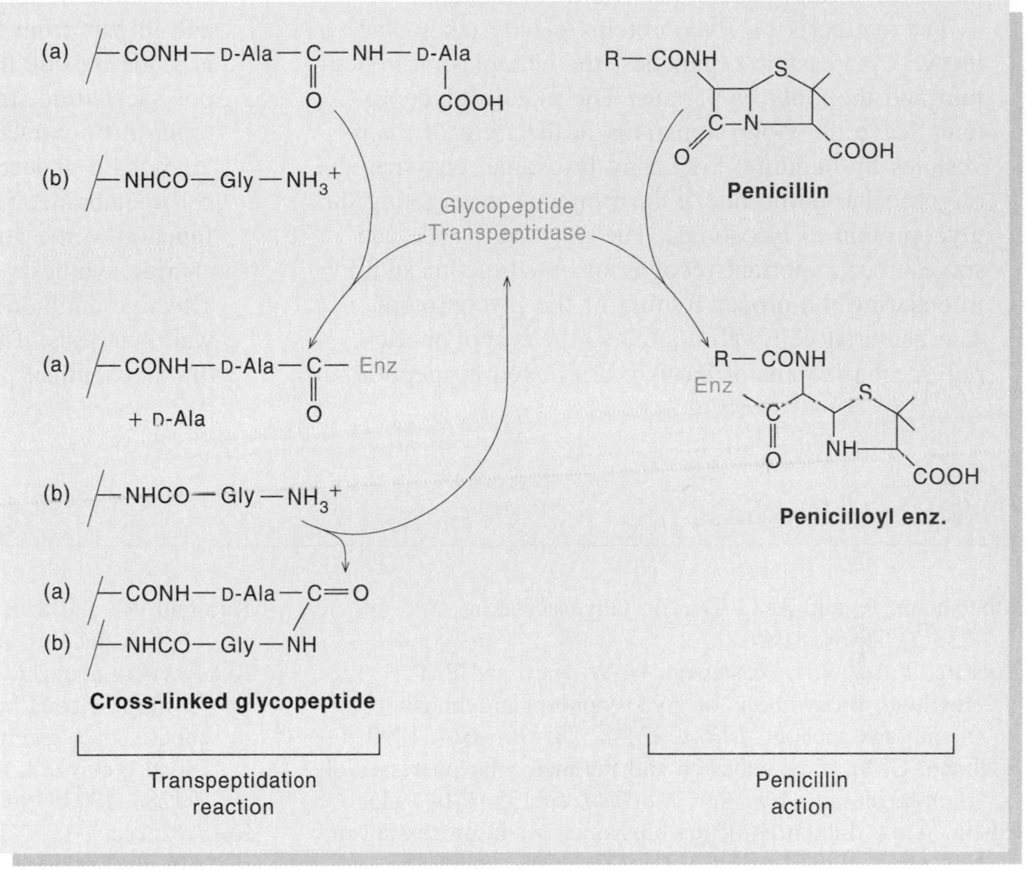

## Summary

This chapter focuses on the synthesis of complex carbohydrates. It begins with a consideration of the hexoses that are the building blocks of complex carbohydrates. Then some aspects of the synthesis of simple homopolysaccharides are examined followed by a brief consideration of heteropolymers that contain more than one hexose. A major portion of the chapter is concerned with glycoproteins that contain complex linear and branched carbohydrates attached to proteins. Finally the synthesis of the bacterial wall is examined.

1. Hexoses, which are the primary building blocks of oligosaccharides and polysaccharides, come in a large variety of types. All hexoses can be derived from glucose through a series of conversions. These conversions usually occur at the level of the monophosphor-ylated sugar or the nucleoside diphosphate sugar. The nucleotide sugar is also the activated substrate for formation of disaccharides, oligosaccharides, and polysaccharides.

2. In higher animals a great variety of branched-chain oligosaccharides are found as conjugates in glycolipids and glycoproteins. A great number of sugars and specific glycosyltransferases are involved in oligosaccharide synthesis. Two types of oligosaccharides are distinguished according to their mode of attachment to the protein in the glycoprotein. The O-linked oligosaccharides are synthesized directly on the amino acid side chain hydroxyl group of a serine or a threonine. The N-linked oligosaccharides are synthesized first on a long-chain dolichol phosphate and then transferred to the asparagine side chain of a receptor protein.

protein-attached oligosaccharide is processed by removal of certain sugars and addition of others.

The synthesis of glycoproteins mostly takes place in two cytoplasmic organelles: the endoplasmic reticulum and the Golgi apparatus. The mature glycoproteins leave the Golgi apparatus in the form of microvesicles by budding. For many lysosomal enzymes the oligosaccharide portion is instrumental in targeting the glycoprotein to lysosomes. The oligosaccharide can serve other important recognition functions in addition to ensuring the proper folding of the glycoprotein.

3. The bacterial cell wall contains a heteropolymeric polysaccharide chain that is cross-linked by peptide linkages. The complexity of the resulting peptidoglycan results in part from the complex repeating units and in part from the fact that a cross-linked polymer is made outside the cell. The partially completed polysaccharide structures are transferred from the cytoplasm to extracellular space by attachment to a long-chain undecaprenol lipid that can traverse the cell membrane. This lipid is similar in structure and function to the dolichol phosphate used in oligosaccharide synthesis in animals. Various compounds are effective antibiotics by blocking specific steps in cell wall synthesis. They are very helpful in elucidating the biochemical pathway.

## Selected Readings

Albersheim, P., and A. G. Darvill, Oligosaccharins. *Sci. Am.* 253(3):58–64, 1985.

Doering, T. L., W. J. Masterson, G. W. Hart, and P. T. Englund, Biosynthesis of glycosylphosphatidylinositol membrane anchors. *J. Biol. Chem.* 265:611–614, 1990.

Edelman, G. M., Cell adhesion and the molecular processes of morphogenesis. *Ann. Rev. Biochem.* 54:135–170, 1985.

Elbein, A. D., Inhibitors of the biosynthesis and processing of N-linked oligosaccharides. *CRC Crit. Rev. Biochem.* 16:21–49, 1984.

Elbein, A. D., Inhibitors of the biosynthesis and processing of N-linked oligosaccharide chains. *Ann. Rev. Biochem.* 56:497–534, 1987.

Fransson, L.-A., Structure and function of cell-associated proteoglycans. *Trends Biochem. Sci.* 12:406–411, 1987. A recent account on a fast moving subject.

Fukuda, M. N., Hempas disease: Genetic defect of glycosylation. *Glycobiology* 1:9–15, 1990.

Fukuda, M. N., K. A. Masri, A. Dell, L. Luzzatto, and K. W. Moremen, Incomplete synthesis of N-glycans in congenital dyserythropoietic anemia type II caused by a defect in the gene encoding $\alpha$-mannosidase II. *Proc. Natl. Acad. Sci. USA* 87:7443–7447, 1990.

Hirschberg, C. B., and M. D. Snider, Topography of glycosylation in the rough endoplasmic reticulum and the Golgi apparatus. *Ann. Rev. Biochem.* 56:63–87, 1987.

Homans, S. W., M. A. J. Ferguson, R. A. Dwek, T. W. Rademacher, R. Anand, and A. F. Williams, Complete structure of the glycosylphosphatidylinositol membrane anchor of rat brain Thy-1 glycoprotein. *Nature* 333:269–272, 1988.

Kochetkov, N. K., and V. N. Shibaev, Glycosyl esters of nucleoside pyrophosphates. *Adv. Carbohydr. Chem. Biochem.* 28:307–325, 1973. A concise review of the chemistry and biochemistry of nucleoside pyrophosphate sugars and derivatives.

Kornfeld, R., and S. Kornfeld, Assembly of asparagine-linked oligosaccharides, *Ann. Rev. Biochem.* 54:631–664, 1985.

Kukowsaka-Latallo, J. F., R. D. Larsen, R. P. Nair, and J. B. Lowe, A cloned human cDNA determines expression of a mouse stage-specific embryonic antigen and the Lewis blood group $\alpha$(1,3/1,4)fucosyltransferase. *Genes Dev.* 4:1288–1303, 1990.

Kukuruzinska, M. A., M. L. E. Bergh, and B. J. Jackson, Protein glycosylation in yeast. *Ann. Rev. Biochem.* 56:915–944, 1987.

Larsen, R. D., L. K. Ernest, R. P. Nair, and J. B. Lowe, Molecular cloning; sequence and expression of a human GDP-L-fucose: $\beta$-D-galactosidase 2-$\alpha$-L-fucosyltransferase cDNA that can form the H blood group antigen. *Proc. Natl. Acad. Sci. USA* 87:6674–6678, 1990.

Maley, F., R. B. Trimble, A. L. Tarentino, and T. H. Plummer, Jr., Characterization of glycoproteins and their associated oligosaccharides through the use of endoglycosidases. *Anal. Biochem.* 180:195–204, 1989.

Paulson, J. C., and K. J. Colley, Glycosyltransferases. Structure, localization, and control of cell type-specific glycosylation *J. Biol. Chem.* 264:17615–17618, 1989.

Pfeffer, S. R., and J. E. Rothman, Biosynthetic protein transport and sorting by the endoplasmic reticulum and Golgi. *Ann. Rev. Biochem.* 56:829–852, 1987.

Phillips, M. L., E. Nudelman, F. C. A. Gaeta, M. Perez, A. K. Singhal, S. Hakomori, and J. C. Paulson, ELAM-1 mediates cell adhesion by recognition of a carbohydrate ligand, sialosyl-Le$^{X}$. *Science* 250:1130–1132, 1990.

Rademacher, T. W., R. B. Parekh, and R. A. Dwek, Glycobiology. *Ann. Rev. Biochem.* 57:785–838, 1988.

Rothman, J. E., The compartmental organization of the Golgi apparatus, *Sci. Am.* 253(3):74–89, 1985.

Stanley, P., Glycosylation mutants of animal cells. *Ann. Rev. Genet.* 18:525–552, 1984.

Stanley, P., Glycosylation mutants and the functions of mammalian carbohydrates. *Trends Genet.* 3:77–81, 1987.

Vliegenthart, J. F. G., L. Dorland, and H. van Halbeek, High-resolution, [1]H-nuclear magnetic resonance spectroscopy as a tool in the structural analysis of carbohydrates related to glycoproteins. *Adv. Carbohydr. Chem. Biochem.* 41:209–374, 1983.

Yamamoto, F., H. Clausen, T. White, J. Marken, and S. Haromori, Molecular genetic basis of the histo-blood group ABO system. *Nature* 345:229–233, 1990.

Yamamoto, F., and S. Hakomori, Sugar-nucleotide donor specificity of histo-blood group A and B transferases is based on amino acid substitutions. *J. Biol. Chem.* 265:19257–19262, 1990.

## Problems

1. The sequence of amino acids in a protein is determined by the base sequence in DNA. What determines the sequence of sugar residues in oligosaccharides and polysaccharides?

2. The interconversion of glucose and galactose occurs at the UDP-hexose level (fig. 16.2) with a 4-epimerase. The epimerase is unusual in that it has a covalently bound $NAD^+$. What function do you propose for the $NAD^+$, and what intermediate do you expect in the reaction?

3. Many amino sugars are found in nature, but glucosamine and galactosamine (figs. 16.1 and 16.2) are by far the most common. By examining their biosynthesis, can you explain why the most common amino sugars in nature are 2-amino sugars?

4. How is lactose made in mothers' milk? What is unusual about the subunit structure of lactose synthase?

5. A great variety of different oligosaccharides result from a limited number of sugars. Explain.

6. How can two different genes for different glycosyltransferases determine ABO blood group types in humans? Explain why blood group O is considered a universal donor. If you have AB type blood, why can you accept any blood type? A small number of people lack the H antigen (the glycosyltransferase that adds Fucα 1) and have Bombay type blood. What blood type can you give to a person with Bombay type blood and why?

7. Explain why fibroblasts from a patient with I-cell disease secrete lysosomal enzymes when grown in tissue culture.

8. Throughout this chapter notice the occurrence of *N*-acetylglucosamine and *N*-acetylgalactosamine in oligo- or polysaccharides. Why are these sugars present in the *N*-acetyl form?

9. Gluconic acid, as its phosphate derivative, is a metabolite in the pentose phosphate pathway. Why isn't gluconic acid, like glucuronic acid, listed in figure 16.1 as a common component of oligo- or polysaccharides?

10. The hydrophobic amino acid sequence that serves as the binding site for the SRP (fig. 16.8) is on the N terminus of the protein. Why isn't it on the C terminus?

11. Figure 16.17 shows a series of reactions in a cycle. Is this a series of reactions like the tricarboxylic acid cycle or does something actually "go" around the cycle?

12. Compare the pentapeptide portion of the peptidoglycans (fig. 16.17) with a normal sequence of amino acids in a protein. What differences can you identify?

13. What complications have to be overcome to synthesize complex carbohydrates (such as cell wall components and O antigens) outside the cell?

14. Bacteria starved for an essential nutrient are not affected by penicillin. Can you explain this?

# Metabolism of Lipids

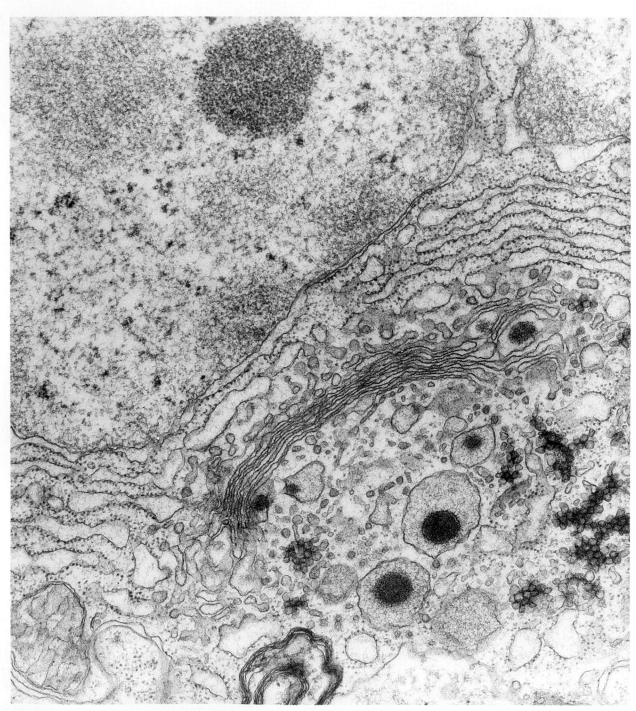

A transmission electron micrograph of an acidic proteoglycan secreting cell from a copepod. The nucleus is at upper left with rough endoplasmic reticulum closest to the nuclear membrane. The smooth endoplasmic reticulum are closest to the Golgi apparatus, which is centrally located in this micrograph. (Courtesy of Dr. Brij L. Gupta, 1964, Ph.D. Thesis, Cambridge University, England.)

# 17

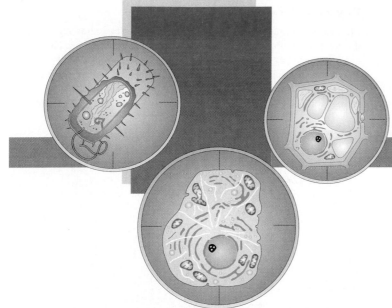

# Structure and Functions of Biological Membranes

*Membranes contain amphipathic lipid molecules that form bilayers with the hydrophilic groups exposed.*

E very cell is surrounded by a plasma membrane which creates a compartment where the functions of life can proceed in relative isolation from the outside world (fig. 17.1). The plasma membrane keeps proteins and other essential materials inside the cell. But the plasma membrane is not simply an inert barrier. Proteins embedded in it work to bring nutrients into the cell and extrude waste products. Other membrane-bound proteins may sense the cell's surroundings, communicate with other cells, or act to move the cell to a new location. In aerobic bacteria, plasma membranes house the electron-transfer reactions that provide the cell with energy.

## Figure 17.1

Cross section of microvilli of cat intestinal epithelial cells, showing the trilaminar (three-layered) structure of the cytoplasmic membranes (165,000×). (Courtesy of Dr. S. Ito.)

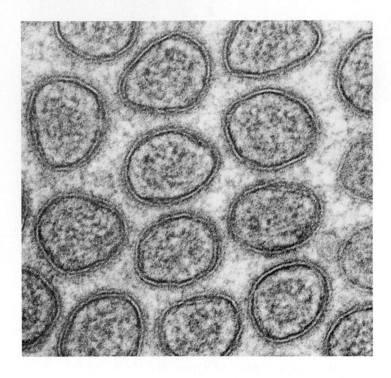

## Figure 17.2

Electron micrograph of a cell from the rat pancreas, showing several different intracellular organelles. (PM = plasma membrane; NE = nuclear envelope; Nu = nucleolus; M = mitochondrion; ER = endoplasmic reticulum; Go = Golgi apparatus; arrows show pore complexes in the nuclear envelope; 24,000×.) (From S. L. Wolfe, *Biology of the Cell,* 2d ed., Copyright © 1981, Wadsworth Publishing Co.)

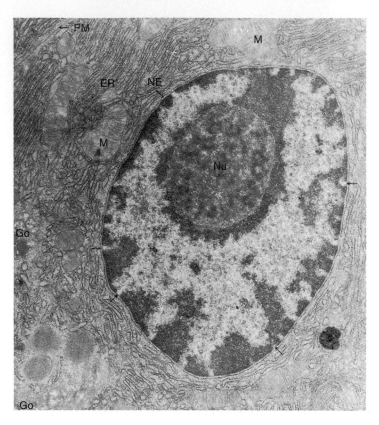

In addition to their plasma membrane eukaryotic cells also contain internal membranes that define a variety of organelles (fig. 17.2). Each of these organelles is specialized for particular functions: The nucleus synthesizes nucleic acids, mitochondria oxidize carbohydrates and lipids and make ATP, chloroplasts carry out photosynthesis, the endoplasmic reticulum and the Golgi apparatus synthesize and secrete proteins, and lysosomes digest proteins. Additional membranes divide mitochondria and chloroplasts into even finer, more specialized subcompartments. Like the plasma membrane, organellar membranes act as barriers to the leakage of proteins, metabolites, and ions; they contain transport systems for import and export of materials, and they are the sites of enzymatic activities as diverse as cholesterol biosynthesis and oxidative phosphorylation.

Although membranes from different organelles or cells may have very different activities, they share the following basic properties:

1. As a rule, biological membranes are impermeable to polar molecules or ions. The nuclear membrane and the outer mitochondrial membrane do have pores that admit relatively large molecules (see fig. 17.2). More commonly, however, an ion or polar molecule can cross a biological membrane only if the membrane has a protein that is a specific transporter for that molecule.

2. Membranes are not rigid but rather adapt flexibly to changes in cellular or organellar shape and size.

3. They are durable. The plasma membrane of an erythrocyte, for example, experiences constant buffeting as the blood courses through the capillaries, and yet it survives for the lifetime of the cell, a matter of months. The membrane either must be remarkably resistant to damage or must reseal very quickly if it is breached.

4. When viewed with an electron microscope after thin-sectioning and staining, membranes typically have a trilaminar appearance of two dark lines separated by a lighter space, with a total thickness on the order of 40 Å (see fig. 17.1).

5. Membranes contain proteins that are not simply structural in nature, but have a variety of enzymatic activities. The identities and amounts of these proteins vary in different cells and may change with time in response to changing conditions.

Our first goal in this chapter is to explain how biological membranes can be, on the one hand, thin and flexible, and on the other, durable and functional. We start by examining the constituents of membranes with the aim of developing a general model for membrane structure. Then we turn to the question of how cells transport polar materials across membranes.

## The Structure of Biological Membranes

Biological membranes consist primarily of proteins and lipids. The relative amounts of these materials vary considerably, depending on the source of the membrane. At one extreme, the inner mitochondrial membrane is about 80% protein and 20% lipid by weight; at the other, the myelin sheath membrane is about 80% lipid and 20% protein. The plasma membrane of human erythrocytes contains about equal amounts of protein and lipid. Many membranes also contain small amounts of carbohydrates. These almost always are covalently attached to either proteins (as glycoproteins) or lipids (as glycolipids or lipopolysaccharides). The mitochondrial inner membrane has little or no carbohydrate, but the myelin membrane has about 3% carbohydrate by weight, and the erythrocyte plasma membrane about 8%.

### Membranes Contain Complex Mixtures of Lipids

Two main types of lipids occur in biological membranes: phospholipids and sterols. The predominant phospholipids in most membranes are phosphoglycerides, which are phosphate esters of the three-carbon alcohol, glycerol. A typical structure is that of phosphatidylcholine (lecithin):

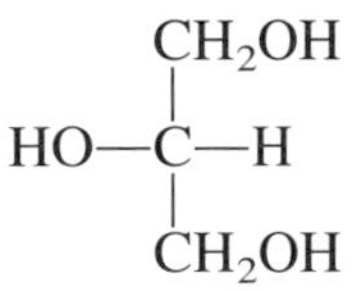

glycerol 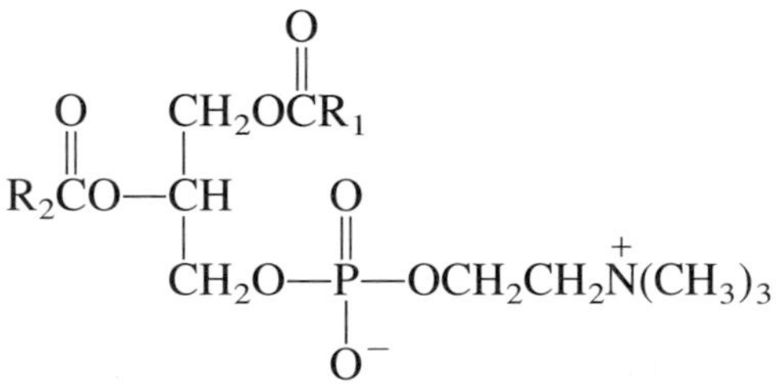                                    phosphatidylcholine

Here $R_1$ and $R_2$ are long, fatty acid side chains. The parent fatty acids $R_1CO_2H$ and $R_2CO_2H$ usually have an even number of carbon atoms; 16- and 18-carbon acids are the most common. The acid esterified to the hydroxyl group on C-1 of the glycerol (that at the top of phosphatidylcholine is drawn above) usually has a fully saturated chain, whereas the acid attached at C-2 often has one or more double bonds, which are almost always *cis* double bonds. Table 17.1 lists some of the fatty acids commonly found in these positions. A phosphatidylcholine that has palmitic acid esterified at both the C-1 and C-2 positions of the glycerol is known by the name dipalmitoylphosphatidylcholine. One with palmitic acid at C-1 and oleic acid at C-2 is called 1-palmitoyl-2-oleoylphosphatidylcholine.

The phosphate group in phosphatidylcholine forms ester linkages both with the hydroxyl group on C-3 of glycerol and also with a second alcohol, choline ($HOCH_2CH_2N(CH_3)_3^+$). In other phospholipids, a variety of other alcohols occupy this second position, and like choline, these all contain polar or electrically charged substituents. Table 17.2 shows the most common of these alcohols and indicates how the phosphoglycerides containing them are named by appending the name of the alcohol to the prefix "phosphatidyl." Free phosphatidic acid, in which the phosphate group is not esterified in this position (fig. 17.3), is an intermediate in phospholipid biosynthesis and metabolism but is not a major constituent of most biological membranes.

In addition to phosphoglycerides, membranes from animal cells usually contain a second group of phospholipids, the sphingolipids. Sphingomyelin, which is representative of this group, has the structure

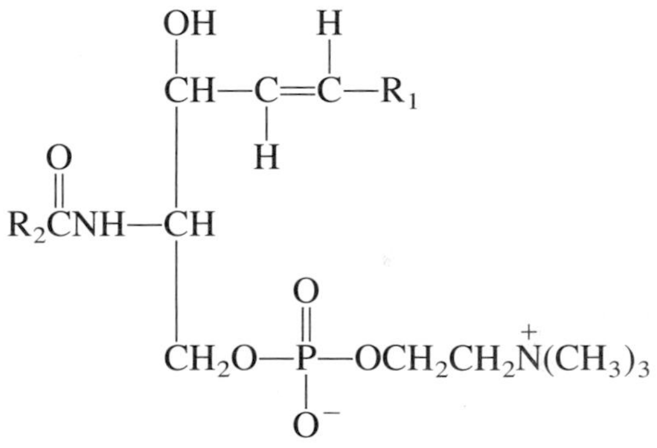

sphingomyelin

Here $R_1 = -(CH_2)_{12}CH_3$, and $R_2$ is of variable length but is typically $-(CH_2)_{16}CH_3$. Although sphingomyelin does not contain glycerol, its structural similarity to phosphatidylcholine is evident (fig. 17.4). Both molecules have two long hydrocarbon side chains and a negatively charged phosphate group esterified to the positively charged choline.

Other sphingolipids have a carbohydrate such as galactose, or a short oligosaccharide chain esterified to the phosphate in place of the choline of sphingomyelin. These compounds are called glycosphingolipids, or more generally, glycolipids. The prototype, in which the substituent is a single galactosyl residue, is called galactosylceramide.

The second major type of lipid found in some biological membranes is cholesterol. Cholesterol (fig. 17.5) is an isoprenoid compound with four fused rings, a short aliphatic chain, and a single hydroxyl group. It occurs in membranes both in its free form and esterified with long-chain fatty acids. Table 17.3 compares the lipid compositions of mem-

**Table 17.1**

Fatty Acids Frequently Found in Membrane Phospholipids

| Name | Structure | Abbreviation[a] |
|------|-----------|------------------|
| **Saturated Fatty Acids[b]** | | |
| Myristic acid | $CH_3(CH_2)_{12}CO_2H$ | 14:0 |
| Palmitic acid | $CH_3(CH_2)_{14}CO_2H$ | 16:0 |
| Stearic acid | $CH_3(CH_2)_{16}CO_2H$ | 18:0 |
| **Unsaturated Fatty Acids[c]** | | |
| Palmitoleic acid | $CH_3(CH_2)_5\overset{H}{\underset{}{C}}=\overset{H}{\underset{}{C}}(CH_2)_7CO_2H$ | $16:1^{\Delta 9}$ |
| Oleic acid | $CH_3(CH_2)_7\overset{H}{\underset{}{C}}=\overset{H}{\underset{}{C}}(CH_2)_7CO_2H$ | $18:1^{\Delta 9}$ |
| Linoleic acid | $CH_3(CH_2)_4\overset{H}{\underset{}{C}}=\overset{H}{\underset{}{C}}CH_2\overset{H}{\underset{}{C}}=\overset{H}{\underset{}{C}}(CH_2)_7CO_2H$ | $18:2^{\Delta 9,12}$ |
| Arachidonic acid | $CH_3(CH_2)_3(CH_2-\overset{H}{\underset{}{C}}=\overset{H}{\underset{}{C}})_4(CH_2)_3CO_2H$ | $20:4^{\Delta 5,8,11,14}$ |

[a] The abbreviation indicates the total number of carbon atoms and, following the colon, the number of double bonds and their positions relative to the carboxyl group. The $\Delta$ superscript indicates the start location of the double bond(s). Numbering of carbons starts from the carboxyl carbon. Thus $18:2^{\Delta 9,12}$ indicates two double bonds, one between carbons 9 and 10 and one between carbons 12 and 13.

[b] A saturated fatty acid usually is esterified at C-1 of the glycerol.

[c] An unsaturated fatty acid usually is present at the C-2 position. Some bacterial phospholipids have a fatty acid with an internal cyclopropane ring here.

**Table 17.2**

Major Phosphoglycerides

| Alcohol (HO—X) | Formula | Phospholipid |
|----------------|---------|--------------|
| Choline | $HO-CH_2CH_2N(CH_3)_3{}^+$ | phosphatidylcholine (lecithin) |
| Ethanolamine | $HO-CH_2CH_2NH_3{}^+$ | phosphatidylethanolamine |
| Serine | $HO-CH_2\underset{\overset{\mid}{CO_2{}^-}}{CHNH_3{}^+}$ | phosphatidylserine |
| Glycerol | $HO-CH_2\underset{\overset{\mid}{OH}}{CHCH_2OH}$ | phosphatidylglycerol |
| Phosphatidylglycerol | 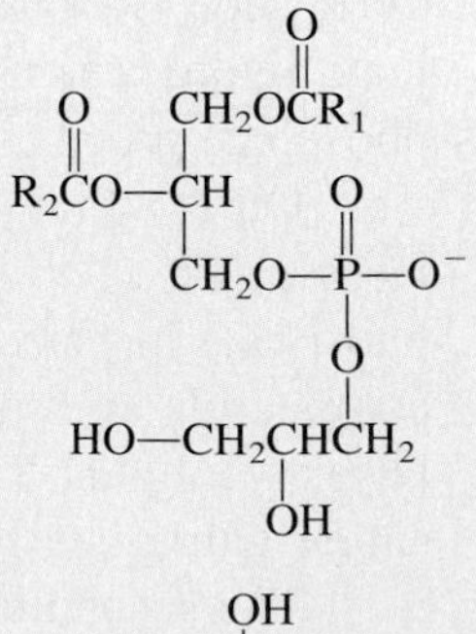 $HO-CH_2\underset{\overset{\mid}{OH}}{CHCH_2}$ | diphosphatidylglycerol (cardiolipin) |
| *myo*-Inositol | 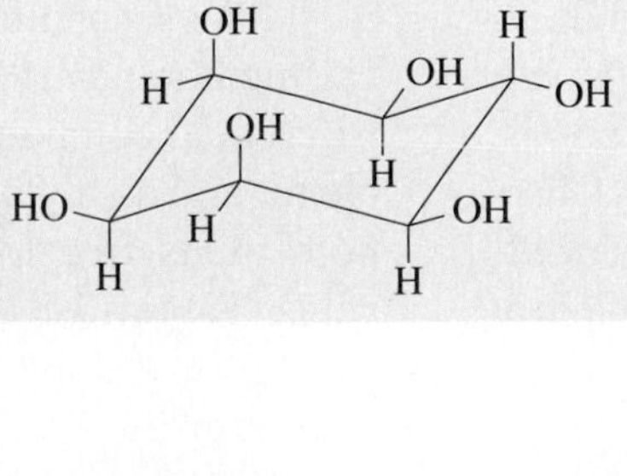 | phosphatidylinositol |

$$R_2\overset{O}{\overset{\|}{C}}O-CH\underset{\underset{\overset{\mid}{O}}{\underset{\overset{\mid}{CH_2O-\overset{O^-}{\underset{\overset{\|}{O}}{P}}-O^-}}{}}}{\overset{\overset{\overset{O}{\|}}{CH_2O\overset{}{C}R_1}}{\underset{}{}}}$$

**Figure 17.3**

Structure of phosphatidic acid, a phosphoglyceride. A saturated fatty acid is esterified at carbon 1 of the glycerol, and a *cis*-unsaturated fatty acid at carbon 2. The cluster of polar and charged oxygens makes the head-group hydrophilic, in marked contrast to the hydrophobic fatty acid chains.

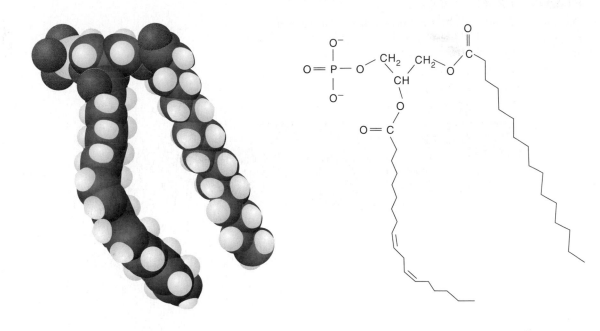

**Figure 17.4**

Structure of a sphingomyelin.

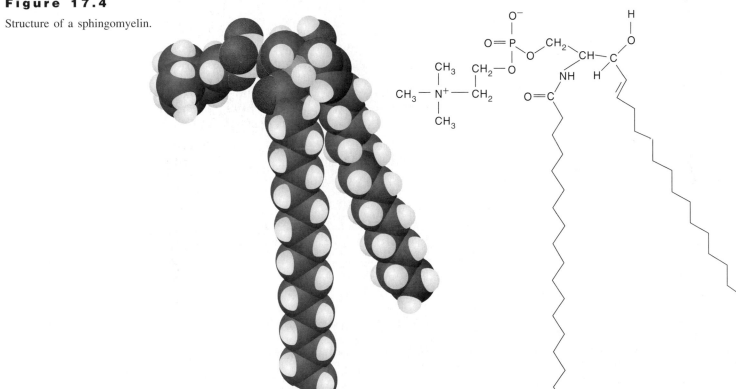

branes from several biological sources. As a rule, phosphatidylcholine is the major phospholipid of animal cell membranes, whereas phosphatidylethanolamine predominates in bacteria. Most bacterial membranes have no cholesterol or sphingolipids. Cholesterol also is low in some organellar membranes of animal cells, such as the inner mitochondrial membrane, but can account for almost a third of the total lipid in cytoplasmic membranes. Plant cell membranes have no cholesterol but some contain relatively large amounts of glycolipids.

## Figure 17.5

Structure of cholesterol, in three different views. The conventional projection is shown at the top right. The more realistic space-filling and conformational models are shown at the lower left and the lower right, respectively.

## Table 17.3

### Lipid Compositions of Membranes

| Source | Lipid Composition[a] (% of total lipids) | | | | | | | | |
| --- | --- | --- | --- | --- | --- | --- | --- | --- | --- |
| | Cholesterol | PC | SM | PE | PI | PS | PG | DPG | Glycolipids |
| **Rat Liver** | | | | | | | | | |
| Plasma membrane | 30 | 18 | 14 | 11 | 4 | 9 | — | — | — |
| Rough endoplasmic reticulum | 6 | 55 | 3 | 16 | 8 | 3 | — | — | — |
| Inner mitochondrial membrane | 3 | 45 | 3 | 25 | 6 | 1 | 2 | 18 | — |
| Nuclear membrane | 10 | 55 | 3 | 20 | 7 | 3 | — | — | — |
| Golgi | 8 | 40 | 10 | 15 | 6 | 4 | — | — | — |
| Lysosomes | 14 | 25 | 24 | 13 | 7 | — | — | 5 | — |
| **Rat Brain Myelin** | 22 | 11 | 6 | 14 | — | 7 | — | — | 21 |
| **Rat Erythrocyte** | 24 | 31 | 9 | 15 | 2 | 7 | — | — | 3 |
| ***E. coli* Plasma Membrane** | 0 | 0 | — | 80 | — | — | 15 | 5 | — |

[a] PC = phosphatidylcholine; SM = sphingomyelin; PE = phosphatidylethanolamine; PI = phosphatidylinositol; PS = phosphatidylserine; PG = phosphatidylglycerol; DPG = diphosphatidylglycerol (cardiolipin).

Source: Adapted from M. K. Jain, and R. C. Wagner, *Introduction to Biological Membranes,* Wiley: New York, 1980.

## Phospholipids Spontaneously Form Ordered Structures in Water

Phospholipid molecules are said to be amphipathic, a term derived from the Greek, meaning having ambivalent feelings. The polar head group of the molecule is intrinsically soluble in water; the fatty acid tails are hydrophobic. The space-filling models shown in figures 17.3 and 17.4 emphasize the segregation of these polar and nonpolar groups into two distinct domains. In this regard, phospholipids resemble detergents such as sodium dodecylsulfate (SDS), which has an ionic sulfate head and a single, long hydrocarbon tail:

$$CH_3(CH_2)_{11}O-\overset{\displaystyle O}{\underset{\displaystyle O}{\overset{\displaystyle \|}{\underset{\displaystyle \|}{S}}}}-O^-Na^+$$

sodium dodecylsulfate

**Figure 17.6**

Structures formed by (*a*) detergents and (*b*) phospholipids in aqueous solution. Each molecule is depicted schematically as a polar head-group (●) attached to one or two long, nonpolar chains. Most detergents have one nonpolar chain; phospholipids have two. At very low concentrations, detergents or phospholipids form monolayers at the air–water interface. At higher concentrations, when this interface is saturated, further molecules form micelles or bilayer vesicles (liposomes).

**Figure 17.7**

Multilayered vesicles (liposomes) formed from sonically dispersed phosphatidylcholine in the presence of 10% diacetylphosphate and 2% potassium phosphotungstate. Each vesicle has a trilaminar structure consisting of two dark layers separated by a light layer. The dark layers contain the electron-dense phosphotungstate ion; the light layer corresponds to the hydrophobic interior of the bilayer.

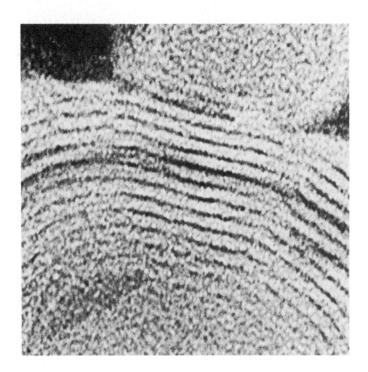

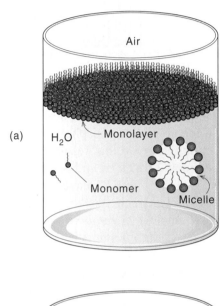

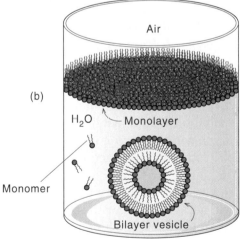

When a detergent is mixed with water, it aggregates spontaneously into spherical or ellipsoidal micelles, in which the polar head-groups of the molecules are exposed to water, but the hydrophobic tails are sequestered (fig. 17.6*a*). Surprisingly, this process is driven, not by a decrease in energy, but rather by an increase in entropy associated with removing the hydrocarbon chains from water. If a hydrocarbon is dissolved in water, the water molecules surrounding it adopt a netlike structure that is more highly ordered than the structure of pure liquid water (see fig. 1.10). Burying the hydrocarbon tails of the detergent molecules in the center of a micelle frees many water molecules from these nets and increases the overall amount of disorder in the system.

Phospholipids themselves generally do not form micelles, because, instead of having a single hydrophobic tail, they have two. The two tails are too bulky to pack together in a spherical or ellipsoidal micelle. They hold the individual molecules apart enough so that portions of the tails remain in contact with water. There is, however, another solution to the problem: The phospholipids can form a bilayer. In this type of aggregate, the hydrocarbon tails of two monolayers of phospholipids pack together while the polar head-groups face outward and remain in contact with water on either side (fig. 17.6*b*). A sheetlike planar bilayer removes the hydrophobic tails from water everywhere except around the edges of the sheet, and by curling up into a spherical vesicle, the bilayer can get rid of its edges. Such a vesicle is called a liposome. Multilayered liposomes resembling onions can be formed by enclosing one vesicle inside another (fig. 17.7).

## Figure 17.8

(*a*) Relative electron densities of myelin membranes from rabbit optic nerve (red line) and sciatic nerve (green line) as a function of distance from the center, measured by x-ray diffraction. The density profile is due mainly to lipids because of the high lipid–protein ratio in myelin. (*b*) Structural interpretation of the electron-density profile, based on the approximate ratios of lipids in mammalian nerve myelin membranes. Side chains from the opposite leaflets of the bilayer do not appear to interdigitate to any great extent.

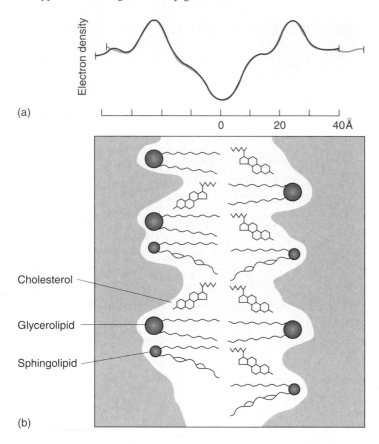

Cholesterol

Glycerolipid

Sphingolipid

(a)

(b)

Although other types of aggregates can form under special conditions, when phospholipids are agitated in the presence of excess water, they tend to aggregate spontaneously to form bilayers. Electron micrographs of bilayers (see fig. 17.7) immediately call to mind the trilaminar appearance of biological membranes (see figs. 17.1 and 17.2). Furthermore, the idea that biological membranes might be built of phospholipid bilayers suggests an explanation for the observation that membranes are largely impermeable to ions. To cross a phospholipid bilayer, an ion would have to traverse the apolar region of hydrocarbon tails, where it would not be well solvated. Small, lipid-soluble molecules, on the other hand, are expected to pass across such a membrane relatively easily, and this is indeed observed to occur. Finally, the presence of a phospholipid bilayer explains the flexibility of natural membranes and the

## Figure 17.9

In an early inaccurate model for the structure of biological membranes, a phospholipid bilayer was coated on both sides by protein in an unfolded or β-pleated sheet conformation. This model reflected the prevailing view of membrane structure from about 1940 until the early 1970s.

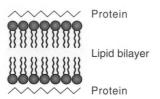

Protein

Lipid bilayer

Protein

ability of membranes to reseal if they are punctured. A bilayer is held together, not by bonds or electrostatic attractions between individual phospholipid molecules, but rather by the entropy increase that results when the hydrocarbon side chains come together and shed their coats of water.

The idea of a lipid bilayer was first proposed by E. Gorter and F. Grendel, who showed in 1925 that the phospholipid content of the erythrocyte plasma membrane is approximately the amount needed to enclose the cell with a bilayer. Subsequent x-ray diffraction measurements confirmed this picture (fig. 17.8).

## Membranes Have Both Integral and Peripheral Proteins

In spite of the arguments presented in the previous section, it is reasonable to ask whether biological membranes are held together simply by phospholipid bilayers. Is it not necessary to reinforce their structure by cross-linking the proteins in the membrane with covalent bonds or at least by favorable electrostatic interactions among the proteins? Considering that some membranes contain even more protein than lipids, any model for membrane structure needs to confront the questions of how these proteins are positioned and of what role, if any, proteins play in holding the membrane together. A possible answer to these questions is that a layer of protein coats each side of the phospholipid bilayer, as shown in figure 17.9. This model was suggested by H. Davson and J. F. Danielli in 1935 and was elaborated by them and others over the course of the next 30 years. It was suggested that the protein coats could take the form of β-pleated sheets. The Davson-Danielli model was attractive because it seemed to account for the trilaminar appearance, impermeability to ions, and durability of membranes. It did not account for the fact that membrane proteins exhibit a variety of enzymatic activities, because the protein in the model played a purely structural role. But at the time the model was proposed, the enzymatic activities of membrane

**Figure 17.10**

Several of the synthetic detergents used for dissolving membranes and solubilizing integral membrane proteins. Triton X-100 and octylglucoside are nonionic detergents; cetyltrimethylammonium bromide and sodium dodecylsulfate (SDS) are ionic. SDS is also an effective denaturant of proteins and is used in polyacrylamide-gel electrophoresis (see chapter 6).

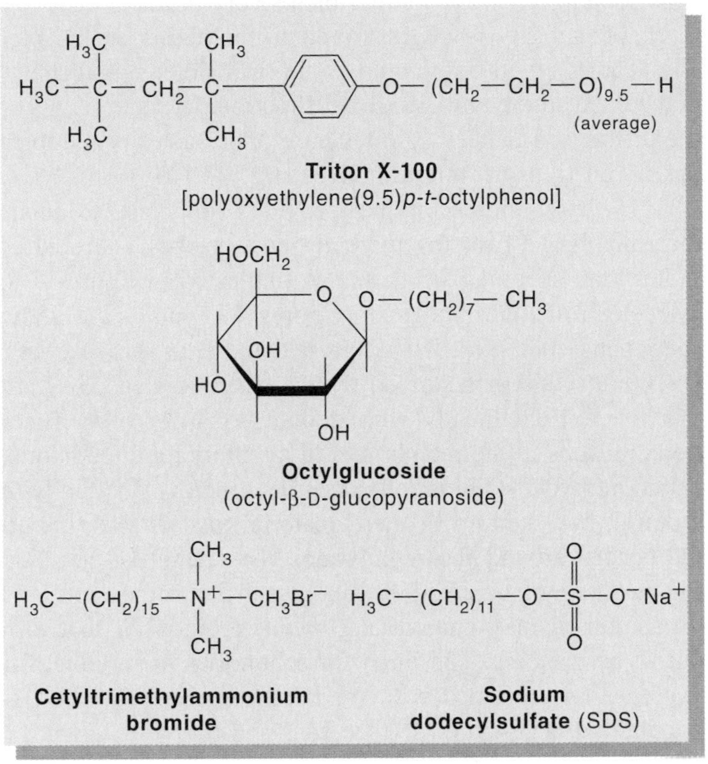

**Triton X-100**
[polyoxyethylene(9.5)*p-t*-octylphenol]

**Octylglucoside**
(octyl-β-D-glucopyranoside)

**Cetyltrimethylammonium bromide**

**Sodium dodecylsulfate** (SDS)

to examine the middle of the phospholipid bilayer, no protein should be found there. On each of these predictions, the model ran into trouble.

Washing preparations of biological membranes with salt solutions usually does remove some proteins, but in most cases they are only a relatively minor fraction of the total protein. Such experiments have led to the realization that biological membranes contain two classes of proteins: peripheral and integral. It is the peripheral proteins that can be removed by washing with salts. In addition to high salt concentrations, EDTA (ethylenediaminetetraacetic acid, a chelator of $Ca^{2+}$ or $Mg^{2+}$ ions) or urea often is used to solubilize these proteins. As a group, peripheral membrane proteins have amino acid compositions similar to those of soluble proteins. On the order of 30% of their residues may be hydrophobic and 70% hydrophilic or neutral. They exhibit the full range of secondary structures and, again, are not remarkable in this regard.

Integral membrane proteins are much more difficult to extract. To solubilize them, it usually is necessary to resort to the use of detergents. Detergents disrupt the phospholipid bilayer of the membrane and incorporate lipids and the hydrophobic portions of proteins into their micelles. Figure 17.10 shows some of the detergents that are used for this purpose. After a membrane has been disrupted, integral membrane proteins can be purified by column chromatography and other conventional techniques, but they almost always require the continued presence of a detergent to remain in solution.

## Integral Membrane Proteins Contain Transmembrane α Helices

As we might expect from the fact that they are soluble only in the presence of detergents, integral membrane proteins tend to have comparatively high contents of hydrophobic amino acid residues. Between 40% and 60% of their side chains may be hydrophobic. In addition, hydrophobic residues often show a curious distribution in the amino acid sequence. They frequently appear in strings of approximately 20 residues, separated by stretches of hydrophilic residues. To describe this aspect of protein structure quantitatively, it is useful to assign each of the 20 natural amino acid residues a number that expresses the side chain's relative hydrophobicity or hydrophilicity. This number can be obtained by measuring the free energy change associated with transfer from an organic solvent to water. On a hydropathy scale, residues with strongly hydrophobic side chains such as isoleucine usually are given positive numbers, and residues with hydrophilic side chains such as arginine, negative numbers. Figure 17.11 shows the average hydropathy

proteins were not recognized as fully as they are today.

One asset of the Davson-Danielli model was its amenability to experimental tests. The model placed all the proteins on the surface of the membrane, where they would be exposed to water and presumably would interact electrostatically with the polar head-groups of the phospholipids. It thus made several predictions. First, to have a stable, sheet-like conformation in contact with water, the proteins found in membranes should contain amino acids with predominantly hydrophilic side chains. Second, it should be possible to wash the proteins off the surface of the phospholipid bilayer by treating membranes with solutions of salts at high ionic strengths. Salt solutions would shield electrostatic interactions holding the proteins to the phospholipid heads. Third, physical probes of protein conformation should show that membrane proteins are unfolded or have high complements of β structure and relatively little α helix. Finally, high-resolution electron-microscopic images should show that membranes have smooth surfaces; and if it is possible

**Figure 17.11**

Hydropathy index for the erythrocyte protein glycophorin, as a function of position in the amino acid sequence. The hydropathy index was averaged over a "window" of seven amino acids, and the window was advanced along the sequence. Glycophorin has a single membrane-spanning segment in the region of residues 75–94. (See fig. 17.18 for a model of how the protein sits in the membrane.) (From J. Kyte and R. F. Doolittle, A simple method for displaying the hydropathic character of a protein, *J. Mol. Biol.* 157:105, 1982. Reprinted by permission.)

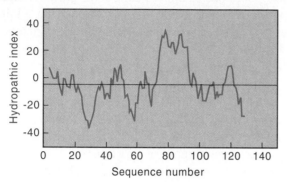

index of the amino acids in glycophorin, a small protein purified from erythrocyte membranes, as a function of location in the amino acid sequence. This protein has one region of predominantly hydrophobic residues extending for about 20 residues. Such regions are uncommon in water-soluble proteins, but larger integral membrane proteins may contain 10 or more of them.

An $\alpha$ helix of 20 amino acid residues is approximately 30 Å long (1.5 Å per residue). This is about the right length to reach across the hydrocarbon region of a bilayer of phospholipids containing 16- and 18-carbon fatty-acids. Hydropathy plots like that of figure 17.11 therefore suggest that integral membrane proteins contain one or more $\alpha$-helical regions extending completely across the bilayer. Physical measurements of secondary structure in integral membrane proteins, though subject to greater technical difficulties than measurements on soluble proteins, support this view. Integral membrane proteins typically have substantial amounts of $\alpha$-helical structure oriented perpendicular to the plane of the membrane.

Very few integral membrane proteins have been crystallized. The reaction-center proteins purified from membranes of photosynthetic bacteria are a notable exception. These proteins were discussed in chapter 15. Before their crystal structures were elucidated, analysis of hydropathy plots suggested that each of the two main protein subunits is folded into five transmembrane $\alpha$ helices, and one such helix was predicted to occur in another subunit. The crystal structures provided a beautiful confirmation of these predictions (see fig. 15.11a). Successful crystallization of the reaction-center proteins was achieved by including small,

amphipathic molecules such as heptane-1,2,3-triol in the solution. These molecules evidently help to fill up the crevices that are occupied by water in crystals of water-soluble proteins.

A three-dimensional structure also has been elucidated for bacteriorhodopsin, an integral membrane protein of the halophilic (salt-loving) bacterium *Halobacterium halobium*. This protein has been studied intensively because of its remarkable activity as a light-driven proton pump (see chapter 14). It forms well-ordered arrays in two-dimensional sheets that can be studied by electron diffraction. Measurements of the diffraction patterns show clearly that bacteriorhodopsin has seven transmembrane helices (fig. 17.12).

Electron microscopy has provided additional evidence that membrane proteins typically have globular structures rather than sheetlike structures as suggested in figure 17.9. The electron micrographs in figures 17.1 and 17.2 show specimens that were sliced into extremely thin sections ($\approx$100 nm) and then stained with an electron-dense reagent that reacted chemically with double bonds or other functional groups in the membrane. In an alternative technique, called negative staining, a thicker specimen is flooded with a dense, but chemically inert, material that simply fills all the spaces around the membrane. With negative staining, the surfaces of unsliced membranes from some cells and organelles display numerous globular projections that turn out to be proteins. The inner mitochondrial membrane, for example, is decorated with the head-pieces of the proton-conducting ATP-synthase (see fig. 14.23).

Another microscopic technique is to freeze the specimen and then fracture it with a knife. A knife cutting through the frozen specimen splits the membrane down the middle, exposing the inside of the bilayer (fig. 17.13a). If the Davson-Danielli model for membrane structure were correct, the two exposed surfaces would be featureless. However, electron micrographs of metallic casts of such samples reveal surfaces studded with particles of various sizes (fig. 17.13b). Additional studies indicate that these particles are proteins that are deeply embedded in the membrane. The particles seen on the inner and outer leaflets of the bilayer usually differ in size and distribution because of an asymmetrical disposition of the proteins across the bilayer.

## Proteins and Lipids Can Move around within Membranes

During the 1960s, various alternatives to the Davson-Danielli model were proposed. Some investigators abandoned the idea of a phospholipid bilayer and suggested instead that membranes consist of aggregates of lipid–protein complexes. However, in 1972, Jon Singer and Garth Nicol-

**Figure 17.12**

A model for the structure of bacteriorhodopsin, a membrane protein from *Halobacterium halobium*. The protein has seven membrane-spanning segments connected by shorter stretches of hydrophilic amino acid residues.

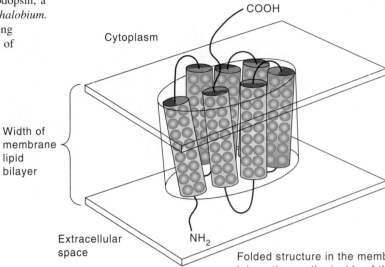

Folded structure in the membrane with charged R-groups interacting on the inside of the protein and hydrophobic surface exposed to membrane lipid side chains.

**Figure 17.13**

Freeze-fracture electron microscopy. (*a*) When struck with a sharp knife, membranes embedded in ice usually fracture between the monolayer leaflets of the lipid bilayer. (*b*) Freeze-fracture electron micrograph of the plasma membrane of *Streptococcus faecalis*, showing a large number of protrusions (presumably proteins) on the outer fracture face and the relative lack of such particles on the inner fracture face (inset). (From H. C. Tsien and M. L. Higgins, Effect of temperature on the distribution of membrane particles in *Streptococcus faecalis* as seen by the freeze-fracture technique, *J. Bacteriol.* 118:725, 1974. Reprinted with permission from American Society for Microbiology.)

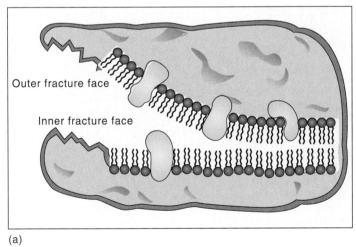

(a)

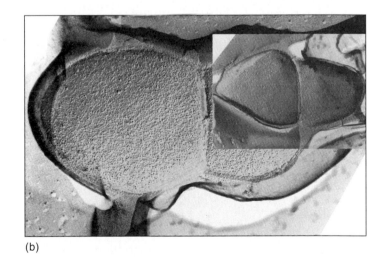

(b)

son incorporated all of the available information into a model they called the fluid-mosaic model. This model, which is now generally accepted, retains the idea that the phospholipid bilayer is the primary structural element of biological membranes. But unlike the Davson-Danielli model, it proposes that integral membrane proteins are embedded in the bilayer, in some cases only partially, but in other cases extending all the way across (fig. 17.14). Peripheral proteins are attached more loosely by ionic interactions with protruding portions of integral proteins or with phospholipid head-groups. A key feature of the fluid-mosaic model is that the integral proteins are, in most cases, not

## Figure 17.14

The fluid-mosaic model for biological membranes as envisioned by Singer and Nicolson. Integral membrane proteins are embedded in the lipid bilayer; peripheral proteins are attached more loosely to protruding regions of the integral proteins. The proteins are free to diffuse laterally or to rotate about an axis perpendicular to the plane of the membrane. For further information, see S. J. Singer and G. L. Nicolson, The fluid mosaic model of the structure of cell membranes, *Science* 175:720, 1972.

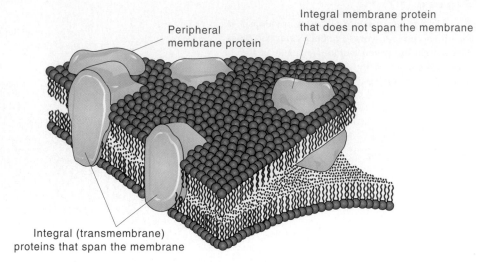

Peripheral membrane protein

Integral membrane protein that does not span the membrane

Integral (transmembrane) proteins that span the membrane

## Figure 17.15

Frye and Edidin's experiment demonstrating lateral diffusion of membrane proteins. (*a*) Proteins on the plasma membranes of human and mouse cells were labeled with dyes that fluoresced at different wavelengths. (*b*) The two populations of cells were mixed and infected with a virus that causes cells to fuse. At short times after mixing, red and green fluorescence from the original cells was seen in separate parts of the membranes of the fused cells. (*c*) Within about 30 min, the two populations of proteins had become intermingled over the entire surface.

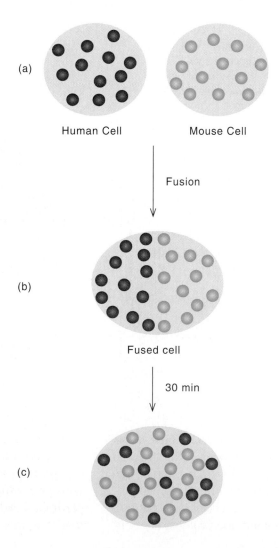

(a)

Human Cell          Mouse Cell

Fusion

(b)

Fused cell

30 min

(c)

## Figure 17.16

Lateral diffusion rates can be studied by measuring the recovery of fluorescence after photobleaching. (*a*) A phospholipid or other component of the membrane is labeled with a fluorescent dye, and the fluorescence from a small region of the membrane is measured through a microscope. The observation region is indicated here by the dotted circle. (*b*) When a laser flash is focused on the observation region, the dye molecules here are destroyed and the amplitude of the fluorescence decreases. (*c*) The fluorescence signal increases again as fresh molecules diffuse into the observation region. (*d*) The diffusion rate is obtained from a plot of the fluorescence as a function of time.

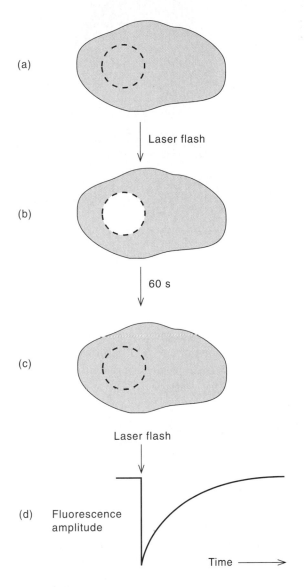

linked together by protein–protein interactions. They are free to diffuse laterally in the bilayer or to rotate about an axis perpendicular to the plane of the membrane. The entire structure thus has the potential of being dynamic rather than static.

Frye and Edidin provided a striking visual demonstration of the dynamic nature of membrane structure. They labeled proteins on the plasma membranes of two samples of cells with fluorescent dyes, human cells with a dye that emitted red light, and mouse cells with a dye that emitted green light (fig. 17.15). The two populations of cells then were mixed and treated with Sendai virus, which causes individual cells to fuse. Immediately after the fusion, red fluorescence from the human proteins could be seen on one half of the hybrid membrane, and green fluorescence from the mouse proteins on the other half. But within a few min, the two types of proteins were intermingled over the entire surface.

If integral membrane proteins are free to diffuse in the membrane, we expect the same to be true of the individual phospholipid molecules that make up the bilayer. To study the dynamics of these motions, phospholipids were labeled with a fluorescent dye that decomposed irreversibly when it was illuminated by a strong laser. When a laser flash was focused to a small spot on the surface of a cell, the labeled phospholipids in this region abruptly ceased fluorescing (fig. 17.16). Fluorescence then rapidly reappeared as the bleached molecules diffused out of the illuminated region and fresh phospholipids diffused in from outside.

Phospholipid molecules in the plasma membrane diffuse rapidly enough to go from one end of an average-sized animal cell to the other in a few minutes. In a bacterial cell, such a trip would take only a few seconds. Integral membrane proteins move more slowly than phospholipids, as we expect in view of their greater mass. Diffusion of membrane proteins plays essential roles in many biochemical processes, including the cellular uptake of lipoproteins (chapter 18), responses of cells to hormones (chapter 24), immunological reactions (supplement 3), vision (supplement 2), and the transport of nutrients and ions. As we see in a later section, however, some membrane proteins cannot move about rapidly because they are attached to cytoskeletal scaffolds.

## Biological Membranes Are Asymmetrical

Although phospholipids diffuse laterally in the plane of the bilayer and rotate more or less freely about an axis perpendicular to this plane, movements from one side of the bilayer to the other are a different matter. Diffusion across the membrane, a transverse, or flip-flop, motion, requires getting the polar head-group of the phospholipid through the

**Figure  17.17**

(*a*) Phospholipids can rotate relatively freely about an axis perpendicular to the plane of the bilayer. (*b*) Rotation about an axis parallel to the plane of the bilayer (flip-flop) is much slower because it requires moving the polar head-group across the bilayer.

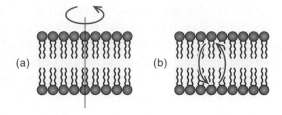

**Figure  17.18**

Topography of glycophorin in the mammalian erythrocyte membrane. Carbohydrate residues (small blue hexagons) are attached to the hydroxyl groups of threonine and serine residues in the N-terminal domain of the protein. The N-terminus and all of the carbohydrates are outside the cell; the C-terminal domain of the protein is inside. The hydrophobic, membrane-spanning domain is flanked by charged amino acid residues that may interact electrostatically with the polar head-groups of the phospholipids.

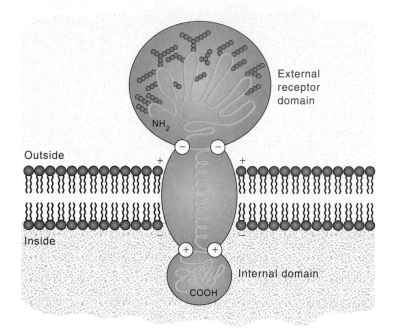

hydrocarbon region in the center of the bilayer (fig. 17.17). Flip-flop motions of phospholipids do occur, and they can be catalyzed enzymatically, but they are much slower than the other types of motions we have described. The same arguments apply even more forcefully to membrane proteins. For a protein to invert its orientation in the membrane, the hydrophilic domain that sticks out into the solution on one side of the membrane has to pass through the center of the bilayer. This rarely occurs. Once a protein has been inserted into a membrane in a particular orientation, it usually retains that orientation indefinitely. An important consequence of these considerations is that biological membranes are both structurally and functionally asymmetrical. An enzyme or receptor embedded in the membrane usually provides a binding site for its substrate or effector only on one side of the membrane.

The orientations of membrane proteins can be probed by examining their reactions with reagents that do not leak across the membrane rapidly and that are introduced only on one side of the membrane or the other. Proteases are useful reagents here. For example, carboxypeptidase, which digests proteins from the C-terminal end, can work on a membrane protein only if this end of the protein is exposed on the side of the membrane on which the enzyme is presented. Erythrocyte membranes are ideal for experiments of this sort, because they can be turned inside out experimentally. In the case of glycophorin, the erythrocyte protein that we mentioned previously in connection with hydropathy plots, the C-terminal end of the protein is inside the cell, and the N-terminal end outside (fig. 17.18). This agrees with the conclusion drawn from the hydropathy plot (see fig. 17.11) that glycophorin makes one pass through the membrane.

Glycophorin is one of the numerous glycoproteins found in the plasma membranes of erythrocytes and other cells. These proteins provide good illustrations of membrane structural asymmetry, because the oligosaccharides

attached to them are almost always on the outside of the cell. Glycophorin has about 100 carbohydrate residues attached to the protein at 16 sites between residues 2 and 50; these are all on the outside of the erythrocyte (see fig. 17.18).

Lipids also show asymmetrical distributions between the inner and outer leaflets of the bilayer. In the erythrocyte plasma membrane, most of the phosphatidylethanolamine and phosphatidylserine are in the inner leaflet, whereas the phosphatidylcholine and sphingomyelin are located mainly in the outer leaflet. A similar asymmetry is seen even in artificial liposomes prepared from mixtures of phospholipids. In liposomes containing a mixture of phosphatidylethanolamine and phosphatidylcholine, phosphatidylethanolamine localizes preferentially in the inner leaflet, and phosphatidylcholine in the outer. For the most part, the asymmetrical distributions of lipids probably reflect packing forces determined by the different curvatures of the inner and outer surfaces of the bilayer. By contrast, the disposition of membrane proteins reflects the mechanism of protein synthesis and insertion into the membrane. We return to this topic in chapter 29.

**Figure 17.19**

(*Top*) Differential scanning calorimetry of various phospholipids dispersed in water. Heat absorption is indicated by a trough in the plot relating differential heat flow to temperature. The lowest point in the trough is the phase transition temperature ($T_m$): (*a*) dipalmitoyl phosphatidylethanolamine; (*b*) dimyristoyl lecithin; (*c*) dipalmitoyl lecithin; (*d*) egg lecithin (plus ethylene glycol to prevent freezing). (From D. L. Melchior and J. M. Steim, Thermotropic transitions in biomembranes, *Ann. Rev. Biophys. Bioeng.* 5:205, 1976. Reproduced, with permission, from the *Annual Review of Biophysics and Bioengineering,* Volume 5, © 1976 by Annual Reviews, Inc.) (*Bottom*) Molecular interpretation of the heat-absorbing reaction during the phase transition.

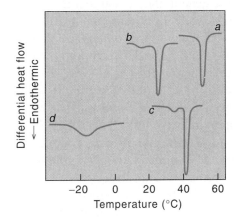

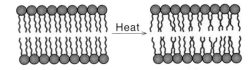

**Figure 17.20**

The amount of disorder in the fatty acid side chains in a phospholipid bilayer, as a function of temperature. The side chains become much more disordered as the temperature increases through the $T_m$. The *solid curve* represents a bilayer that does not contain cholesterol; the *dotted curve,* a bilayer with the same phospholipid plus about 25% cholesterol. The amount of random, disorderly motion of the fatty acid chains can be measured quantitatively by deuterium- or $^{13}$C-NMR. At any given temperature, the disorder is greater near the tips of the chains (toward the middle of the bilayer) than it is close to the head-groups.

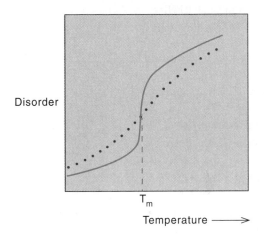

## Membrane Fluidity Is Sensitive to Temperature and Lipid Composition

The degree of fluidity of biological membranes is a function of both temperature and lipid composition. Bilayers containing a single type of phospholipid show an abrupt increase in fluidity (melting) as the temperature is raised through a characteristic and narrow range ($T_m$). The transition can be measured by a variety of physical techniques. One technique that is frequently applied here is differential scanning calorimetry. In this technique, a sample of interest and an inert, reference material are warmed slowly, side by side. The calorimeter records differences between the amounts of heat that must be applied to the two samples to keep their temperatures the same. If the sample undergoes an endothermic phase transition such as melting at a characteristic temperature, extra heat must be applied. Phospholipid bilayers exhibit such a transition at their $T_m$ (fig. 17.19).

The phase transition of a phospholipid bilayer is not as sharp as the melting of a crystaline solid. It extends over several degrees, rather than just a fraction of a degree. In addition, although the fluidity of the bilayer increases substantially at the $T_m$, it also increases gradually with temperature both below and above this point (fig. 17.20). In this regard, the behavior of a bilayer on warming calls to mind the gradual softening of butter.

The temperature of the melting transition ($T_m$) depends strongly on the phospholipid composition in the bilayer. Increasing the length of the fatty acid chains increases the $T_m$; *cis* double bonds decrease the $T_m$ (table 17.4). Increasing the length of the chains makes the formation of kinks more difficult because each kink must disrupt the packing of a larger number of atoms. Double bonds introduce rigid kinks that interfere with the packing of neighboring chains even below the $T_m$. The $T_m$ also depends on the nature of the polar head-group of the phospholipid; phosphatidylethanolamines have higher $T_m$s than phosphatidylcholines or phosphatidylglycerols containing the same fatty acids (see table 17.4). Membranes that contain mixtures of phospholipids may show partial melting transitions at sev-

## Table 17.4

Transition Temperatures for Aqueous Suspensions of Phospholipids

| Phospholipid[a] | $T_m$(°C) |
|---|---|
| Di-14:0 phosphatidylcholine | 24 |
| Di-16:0 phosphatidylcholine | 41 |
| Di-18:0 phosphatidylcholine | 58 |
| Di-22:0 phosphatidylcholine | 75 |
| Di-18:1$^{\Delta 9}$ phosphatidylcholine | −22 |
| Di-14:0 phosphatidylethanolamine | 51 |
| Di-16:0 phosphatidylethanolamine | 63 |
| 1-18:0, 2-18:1$^{\Delta 9}$ phosphatidylethanolamine | 3 |
| Di-14:0 phosphatidylglycerol | 23 |
| Di-16:0 phosphatidylglycerol | 41 |

[a] Phospholipid abbreviations are as in tables 17.1 and 17.3; Di-14:0 phosphatidylcholine, for example, refers to dimyristoyl (14 carbons, 0 double bonds) phosphatidylcholine.

Source: Adapted from M. K. Jain, and R. C. Wagner, *Introduction to Biological Membranes,* Wiley: New York, 1980.

eral temperatures, or (depending on their composition) they may have a single, broadened transition (see fig. 17.19).

We have not yet said much about the second major constituent of eukaryotic membrane lipids, cholesterol. Cholesterol broadens the melting transition of the phospholipid bilayer (see fig. 17.20). Below the $T_m$, cholesterol disorders the membrane because it is too bulky to fit well into the neatly packed arrangement of the fatty acid chains that is favored at low temperatures. Above the $T_m$, cholesterol restricts further disordering because it is too large and inflexible to join in the rapid fluctuations of the chains. If the amount of cholesterol in a phospholipid bilayer is increased to about 30%, roughly the amount in the plasma membranes of typical animal cells, the melting transition becomes so broad as to be almost undetectable.

Cells regulate the lipid compositions of their plasma membrane so that a reasonable membrane fluidity is maintained. They do this by controlling fatty acid biosynthesis so as to vary the lengths of the fatty acid chains and the ratio of unsaturated to saturated fatty acids (see chapter 19). If cells are grown at low temperatures, their phospholipids contain more unsaturated fatty acids or fatty acids with shorter chains or both. These adjustments shift the $T_m$ to lower temperatures, with the result that (to the extent that the melting transition is sharp enough to be measureable) the $T_m$ remains close to or slightly below the ambient temperature. This has been shown in bacteria, plants, and even in animal cells from different parts of the same organism. For example, the ratio of unsaturated to saturated fatty acids is higher in winter wheat than in summer wheat of the same species, and higher in tissue near a reindeer's foot than in the thigh.

## Some Proteins of Eukaryotic Plasma Membranes Are Connected to the Cytoskeleton

Contrary to the general theme that membrane structures are highly dynamic, some of the proteins in eukaryotic plasma membranes are linked to other proteins so that their motions are severely restricted. These linked proteins are attached to the cytoskeleton that is situated in the cytosol of nearly all eukaryotic cells. In the erythrocyte, the cytoskeleton is constituted mainly of spectrin, a protein that has two chains, each with approximately 100 subunits (fig. 17.21). The integral membrane protein glycophorin is linked indirectly to the mesh formed by spectrin through another protein called the band-4.1 protein. The spectrin cytoskeleton is important for maintaining the biconcave shape of the erythrocyte. Cells with a genetic deficiency of spectrin round up into spheres and are more fragile than normal erythrocytes.

One of the major integral proteins of the erythrocyte membrane is the anion channel, or band-3 protein, which moves $Cl^-$ and $HCO_3^-$ anions across the membrane. The anion transporter has two identical subunits with molecular weights of about 95,000, and each subunit probably has 10 or 11 transmembrane helices. The band-3 protein is attached to the spectrin cytoskeleton through a smaller protein, ankyrin. The cytosolic domain of the anion transporter also binds the glycolytic enzyme glyceraldehyde-3-phosphate dehydrogenase.

The cytoskeletons of other eukaryotic cells typically include both microtubules and microfilaments, which consist of long, chainlike oligomers of the proteins tubulin and actin, respectively. Bundles of microfilaments often lie just underneath the plasma membrane (fig. 17.22). They participate in processes that require changes in the shape of the cell, such as locomotion and phagocytosis. In some cells, cytoskeletal microfilaments appear to be linked indirectly through the plasma membrane to peripheral proteins on the outer surface of the cell (fig. 17.23). Among the cell surface proteins connected to this network is fibronectin, a glycoprotein believed to play a role in cell–cell interactions. The lateral diffusion of fibronectin is at least 5,000 times slower than that of freely diffusible membrane proteins.

## Figure 17.21

The structure of spectrin and the location of spectrin in the cytoskeleton. (*a*) An $\alpha\beta$ dimer of spectrin. Both $\alpha$ and $\beta$ subunits are extended structures consisting of end-to-end domains of 106 amino-acyl residues folded into three $\alpha$ helices; the subunits twist about one another loosely as shown. (*b*) The erythrocyte membrane skeleton. Spectrin tetramers ($\alpha_2\beta_2$), shown in yellow, are linked to the cytoplasmic domain of the anion channel (blue) by the protein ankyrin (red), and to glycophorin and actin filaments by protein 4.1. This structure lends stability to the red cell membrane while maintaining sufficient flexibility to allow erythrocytes to withstand substantial shear forces in the peripheral circulation.

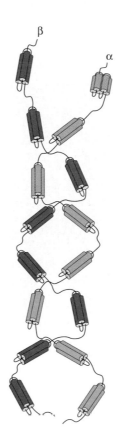

(a)

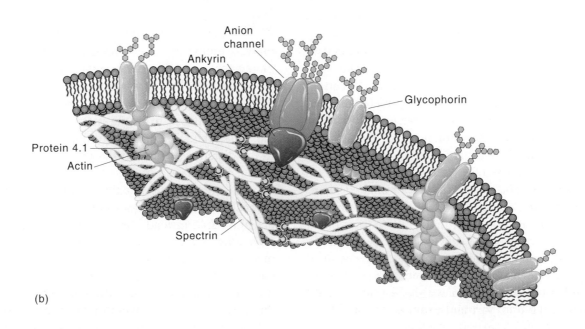

(b)

**Figure 17.22**

Membrane-associated cytoskeletal components of cultured mouse cells. (PM = plasma membrane, MF = microfilaments, MT = microtubules; 54,000×.) (From G. L. Nicolson, Transmembrane control of the receptors on normal and tumor cells. I. Cytoplasmic influence over cell surface components, *Biochem. Biophys. Acta* 457:57, 1976. Reprinted with permission from Elsevier Science Publishers.)

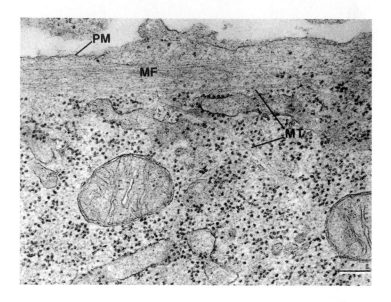

**Figure 17.23**

Schematic illustration of a so-called focal contact, showing how extracellular fibronectin is believed to be indirectly attached to the intracellular cytoskeleton through a transmembrane fibronectin receptor and several other peripheral membrane proteins.

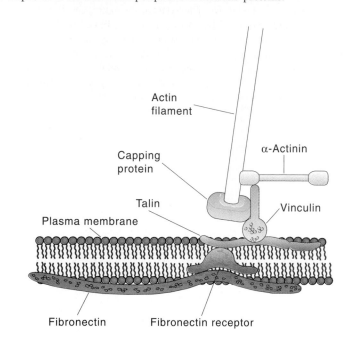

# Transport of Materials across Membranes

Whereas a major function of biological membranes is to maintain the status quo by preventing loss of vital materials and entry of harmful substances, membranes must also engage in selective transport processes. Living cells depend on an influx of phosphate and other ions, and of nutrients such as carbohydrates and amino acids. They extrude certain ions, such as $Na^+$, and rid themselves of metabolic end products. How do these ionic or polar species traverse the phospholipid bilayer of the plasma membrane? How do pyruvate, malate, the tricarboxylic acid citrate and even ATP move between the cytosol and the mitochondrial matrix (see figs. 13.15 and 14.1)? The answer is that biological membranes contain proteins that act as specific <u>transporters</u>, or <u>permeases</u>. These proteins behave much like conventional enzymes: They bind substrates and they release products. Their primary function, however, is not to catalyze chemical reactions but to move materials from one side of a membrane to the other. In this section we discuss the general features of membrane transport and examine the structures and activities of several transport proteins.

# Most Solutes Are Transported by Specific Carriers

For the sake of comparison, let us consider movement of glucose across an inert, porous membrane such as a piece of dialysis tubing (fig. 17.24). The pores in dialysis tubing are large enough so that glucose and water molecules can diffuse from either side of the membrane to the other with little hindrance. If the solutions on the two sides initially contain different concentrations of glucose, molecules diffuse from the more concentrated solution to the more dilute. The net rate of this diffusion is proportional to the difference between the two concentrations $[\Delta C]$, as shown in figure 17.24*c*.

By contrast, the kinetics of transport across a biological membrane usually does not exhibit such a linear dependence on $\Delta C$. Instead, the rate approaches an asymptote at high concentration differences, much as the rate of an enzymatic reaction approaches a maximum rate at high substrate concentrations (fig. 17.24*d*). This behavior suggests that the transport protein has a specific binding site for the material that it transports. The overall rate of transport is limited by the number of these sites in the membrane, and thus fits an

## Figure 17.24

(a) Glucose can diffuse in either direction through the pores in dialysis tubing. (b) If the glucose concentration inside the dialysis bag is initially higher than the concentration outside, glucose diffuses out spontaneously until the two concentrations are equal. (c) At any given time, the net rate of diffusion is proportional to the difference between the two concentrations: $v_{in \rightarrow out} = k[C_{in}] - k[C_{out}] = k[\Delta C]$. (d) The measured rate of transport across a biological membrane usually is not simply proportional to the concentration difference across the membrane but rather approaches a maximum at high values of $[\Delta C]$.

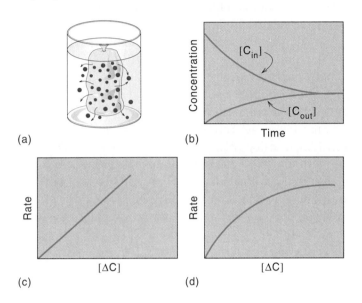

## Figure 17.25

The bacterial lactose-transport protein (lactose permease) transports $\beta$-galactosides, such as lactose, o-nitrophenyl-$\beta$-galactoside, and isopropyl-$\beta$-thiogalactoside. It does not transport galactosides with an $\alpha$-glycosidic linkage.

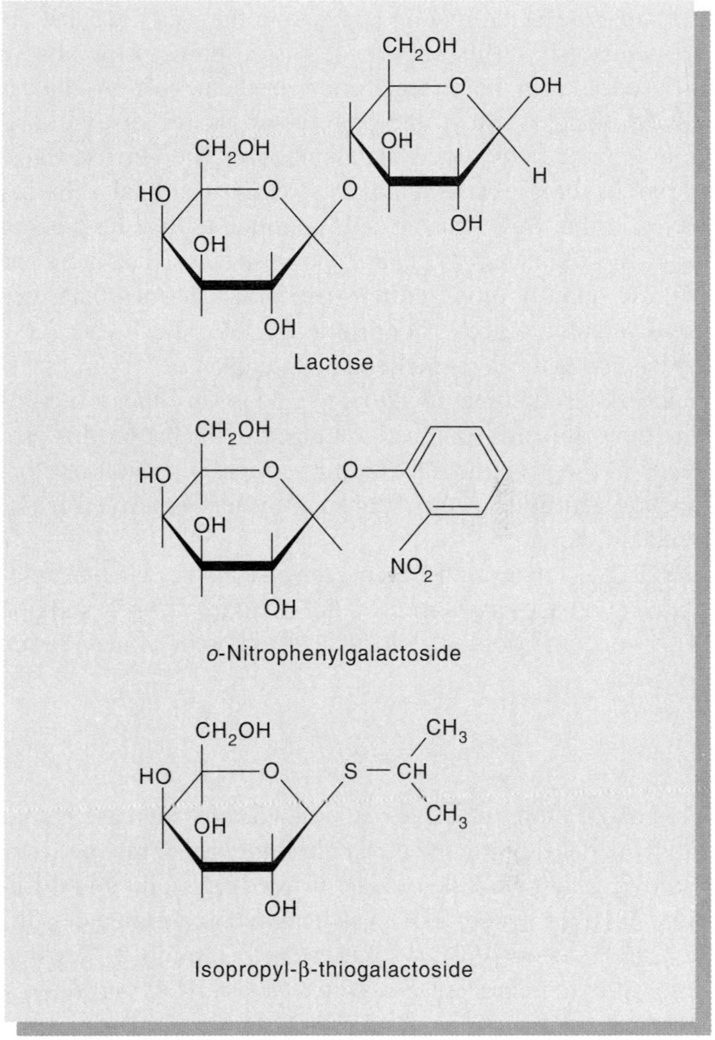

Lactose

o-Nitrophenylgalactoside

Isopropyl-β-thiogalactoside

expression analogous to the Michaelis-Menten equation for enzyme kinetics:

$$v = \frac{V_{max}[\Delta C]}{K_m + [\Delta C]} \tag{1}$$

Just as in enzyme kinetics (see chapter 7), $K_m$ here is an algebraic function of the microscopic rate constants for binding, dissociation, and translocation of the substrate in either direction.

The high specificity of biological transport systems provides compelling support for this interpretation. For example, whereas dialysis tubing is equally permeable to D- and L-glucose, most eukaryotic cells take up D-glucose rapidly but not L-glucose. In bacteria, the synthesis of a specific transport system for a particular nutrient often can be induced by adding a small amount of the nutrient to the growth medium. Thus, if E. coli is grown on medium without lactose, transporters for lactose usually are not found in the plasma membrane. When lactose is added to the culture medium, however, synthesis of a protein that transports lactose is switched on, and the cells soon begin to demonstrate

this activity. The lactose-transport system also catalyzes the uptake of a variety of galactosides that resemble lactose in having a $\beta(1,4)$-glycosidic linkage (fig. 17.25). It does not transport $\alpha$-galactosides.

Like conventional enzymes, transport proteins often are sensitive to specific inhibitors. For example, the anion transporter of the erythrocyte plasma membrane, which moves $HCO_3^-$ and $Cl^-$ ions across the membrane, is inhibited by 1,2-cyclohexadione or derivatives of stilbenedisulfonate. Because these inhibitors react covalently with amino acid residues of the protein, they afford useful probes for the location and structure of the anion-binding site.

## Some Transporters Facilitate Diffusion of a Solute down an Electrochemical Potential Gradient

The presence of a specific transport system in the plasma membrane does not necessarily imply that a cell can pump the substrate in a direction opposite to the way the substrate spontaneously diffuses. In some cases, the net flow always proceeds down the concentration gradient, just as glucose always diffuses out of the more concentrated of two solutions separated by a dialysis membrane. The glucose transporter in the plasma membranes of most animal cells behaves in this way. Muscle cells continue to take up glucose from the blood only because the molecules that enter the cell are quickly modified by metabolic reactions that keep the cytosolic glucose concentration low. And yet, for a given concentration gradient, the uptake of D-glucose is much faster than the uptake of L-glucose or other materials that the cell is not equipped to transport. Transport that proceeds in the same net direction as simple diffusion but that is catalyzed by a specific transport system is termed facilitated diffusion.

If the solute is uncharged, the free energy change associated with moving 1 mol of the material from a solution with concentration $[C_1]$ to a solution with concentration $[C_2]$ is

$$\Delta G_{1 \to 2} = 2.3RT \log \frac{[C_2]}{[C_1]} \qquad (2)$$

This expression, which reflects the entropy increase resulting from distributing the molecules more randomly between the two solutions, holds for both free and facilitated diffusion. If $[C_1] > [C_2]$, $\Delta G_{1 \to 2}$ is negative and molecules diffuse spontaneously from solution 1 to solution 2. Net flux across the membrane ceases when $[C_1] = [C_2]$ and $\Delta G_{1 \to 2}$ is zero.

If the solute carries a net charge, there is an additional thermodynamic effect of moving the charge across any difference in electric potential that exists between the solutions on the two sides of the membrane. The free energy change then is

$$\Delta G_{1 \to 2} = 2.3RT \log \frac{[C_2]}{[C_1]} + z \mathscr{F} \Delta \psi \qquad (3)$$

where $z$ is the charge, $\mathscr{F}$ is the Faraday constant, and $\Delta \psi$ is the electric potential difference ($\Delta \psi = \psi_2 - \psi_1$). The equilibrium condition here is

$$\log \frac{[C_2]}{[C_1]} = -\frac{z \mathscr{F} \Delta \psi}{2.3RT}$$

We used this expression in chapter 14 to evaluate $\Delta \psi$ across the mitochondrial inner membrane.

Among the transporters that facilitate diffusion of charged species down electrochemical potential gradients are the $Na^+$ and $K^+$ channels of nerve cells. These channels are closed in the resting cell but open transiently when the cell is stimulated electrically. The action potential that sweeps down the cell reflects a sudden electric depolarization of the membrane caused by $Na^+$ flowing into the cell, followed by a repolarization as $K^+$ flows out. $Na^+$ influx is thermodynamically downhill because the intracellular $Na^+$ concentration (about 12 mM) is much lower than the extracellular concentration (145 mM) and the internal electric potential is negative relative to the external potential by about 60 mV. These conditions are set up by the membrane $Na^+$–$K^+$ pump, which we discuss shortly. $K^+$ ions, on the other hand, are close to electrochemical equilibrium in the resting cell: The tendency of the negative $\Delta \psi$ to pull $K^+$ into the cell is nearly balanced by a higher intracellular concentration that favors $K^+$ efflux. (The $K^+$ concentrations are about 140 mM inside the cell and 4 mM outside.) When the sudden influx of $Na^+$ causes $\Delta \psi$ to shoot through zero and even change sign, this balance is upset and $K^+$ starts flowing out. The $Na^+$ and $K^+$ channels both close again after 1 to 2 ms, and the $Na^+$–$K^+$ pump soon reestablishes the original conditions.

The anion transporter of the erythrocyte provides another important, though less dramatic, illustration of facilitated diffusion. $CO_2$ produced in respiring tissues diffuses spontaneously into the red blood cells, where carbonic anhydrase catalyzes its reaction with $H_2O$ to form $H_2CO_3$. The $H_2CO_3$ then ionizes to $HCO_3^-$. Some of the $HCO_3^-$ so formed is bound by hemoglobin in the erythrocytes, but a substantial amount is moved back out to the blood plasma by the anion transporter. These processes are reversed when the blood reaches the lungs and $CO_2$ is exhaled. Transport of $HCO_3^-$ in either direction across the erythrocyte membrane is coupled to the movement of $Cl^-$ in the opposite direction. Because an exchange of one anion for another results in no net flow of charge across the membrane, the transport of $HCO_3^-$ and $Cl^-$ is independent of $\Delta \psi$. The $\Delta G$ for moving 1 mol of $HCO_3^-$ into the cell in exchange for 1 mol of $Cl^-$ is

$$\Delta G_{HCO_3^-/Cl^-} = 2.3RT \log \frac{[HCO_3^-]_{in}}{[HCO_3^-]_{out}} +$$

$$2.3RT \log \frac{[Cl^-]_{out}}{[Cl^-]_{in}}$$

$$= 2.3RT \log \left\{ \frac{[HCO_3^-]_{in}[Cl^-]_{out}}{[HCO_3^-]_{out}[Cl^-]_{in}} \right\} \qquad (4)$$

Such coupled transport of two solutes in opposite directions is termed antiport.

**Figure 17.26**

Cells drive active transport in a variety of ways. The plasma-membrane $Na^+$–$K^+$ pump of animal cells (a) and the plasma-membrane $H^+$ pump of anaerobic bacteria (b) are driven by the hydrolysis of ATP. Eukaryotic cells couple the uptake of neutral amino acids to the inward flow of $Na^+$ (c). Uptake of β-galactosides by some bacteria is coupled to inward flow of protons (d). Electron-transfer reactions drive proton extrusion from mitochondria and aerobic bacteria (e). In halophilic bacteria, bacteriorhodopsin uses the energy of sunlight to pump protons (f). E. coli and some other bacteria phosphorylate glucose as it moves into the cell and thus couple the transport to hydrolysis of phosphoenolpyruvate (g).

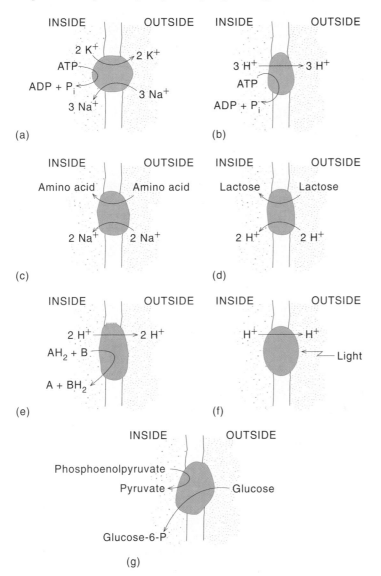

## Active Transport against an Electrochemical Potential Gradient Requires Energy

Cells can transport some materials against gradients of concentration or electric potential or both. As we mentioned in the previous section, eukaryotic cells pump out $Na^+$ ions even though the external $Na^+$ concentration is greater than the cytosolic concentration, and even though the electric potential inside is more negative than the potential outside. Many cells take up amino acids even though the cytosolic concentration exceeds the external concentration, and cells that line the stomach pump out protons against a concentration gradient of more than a million to one. Transport against such concentration and electric gradients is called active transport.

To drive active transport, a cell must couple transport to another process that is thermodynamically favorable, so that the total $\Delta G$ is negative. Cells have developed a variety of schemes for this coupling (fig. 17.26). Some transport systems, including the $Na^+$–$K^+$ pump of animal cells, drive active transport by the hydrolysis of ATP (see fig. 17.26a). The $Na^+$–$K^+$ pump may consume as much as 70% of the ATP that nerve cells synthesize, and even nonexcitable cells spend a substantial portion of their energy resources in this way. Table 17.5 lists some of the other active-transport systems energized by ATP. Many cells contain ATPases that pump protons into lysozomes or intracellular vacuoles. ATPases that pump protons outward are found in the epithelial cells that line the stomach and in the plasma membranes of anaerobic bacteria (see fig. 17.26b). The bacterial enzymes are similar to the proton-conducting $F_1/F_0$ ATP-synthases of mitochondria and chloroplasts (see chapters 14 and 15).

Active transport of a solute against a concentration gradient also can be driven by a flow of an ion down *its* concentration gradient. Table 17.6 lists some of the active-transport systems that operate in this way. In some cases, the ion moves across the membrane in the opposite direction to the primary substrate (antiport); in others, the two species move in the same direction (symport). Many eukaryotic cells take up neutral amino acids by coupling this uptake to the inward movement of $Na^+$ (see fig. 17.26c). As we discussed previously, $Na^+$ influx is downhill thermodynamically because the $Na^+$–$K^+$ pump keeps the intracellular concentration of $Na^+$ lower than the extracellular concentration and sets up a favorable electric potential difference across the membrane. Another example is the β-galactoside transport system of E. coli, which couples uptake of lactose to the inward flow of protons (see fig. 17.26d). Proton influx is downhill because electron-transfer reactions (or,

**Table 17.5**

Some Systems of Active Transport Driven by ATP Hydrolysis

| Substrate | Direction of Pump | Organism, Tissue, or Organelle |
|---|---|---|
| $Na^+/K^+$ | $Na^+$ out, $K^+$ in | Eukaryotic cell plasma membranes |
| $Ca^{2+}$ | In | Muscle sarcoplasmic reticulum |
| $Ca^{2+}$ | Out | Eukaryotic cell plasma membranes |
| $H^+/K^+$ | $H^+$ out, $K^+$ in | Stomach epithelial cell plasma membranes |
| $H^+$ | In | Lysosomes and secretory vesicles; vacuoles in plants and fungi |
| $H^+$ | Out | Anaerobic bacteria ($F_1/F_0$ ATPase) |

**Table 17.6**

Some Systems of Active Transport Driven by Ion Symport or Antiport

| Substrate | Cotransported Ion | Organism or Tissue[a] |
|---|---|---|
| Neutral amino acids | $Na^+$ (symport) | Eukaryotic cells and bacteria |
| Glucose | $Na^+$ (symport) | Intestine and kidney |
| Glutamate | $Na^+$ (symport) | *E. coli* |
| Lactose | $H^+$ (symport) | *E. coli* |
| Xylose | $H^+$ (symport) | *E. coli* |
| Glucose | $H^+$ (symport) | Yeast |
| Galactose | $H^+$ (symport) | Yeast |
| Serotonin, other amines | $H^+$ (antiport) | Adrenal medulla (secretory vesicles) |

[a] Except for the amine transporter of adrenal secretory vesicles, all the transport systems listed here operate in the plasma membrane.

under anaerobic conditions, the proton-conducting ATPase) set up gradients of pH and electric potential across the membrane.

Proton pumps driven by electron-transfer reactions occur in mitochondria, bacteria, and chloroplasts (see fig. 17.26*e* and chapters 14 and 15). Bacteriorhodopsin, which was mentioned above in connection with the structure of membrane proteins, uses light energy to pump protons out of *Halobacterium halobium* (see fig. 17.26*f*). Some bacterial cells can energize sugar transport by still another strategy, group translocation. They modify the sugar by phosphorylation as they transport it (see fig. 17.26*g*). The overall free energy change thus includes the negative $\Delta G$ of the phosphorylation reaction, which in this case is provided ultimately by hydrolysis of phosphoenolpyruvate.

## Isotopes, Substrate Analogs, Membrane Vesicles, and Bacterial Mutants Are Used to Study Transport

To measure rates of transport, methods must be available for identifying molecules that have crossed the membrane. Radioactively labeled substrates are commonly used for this purpose. In a typical experiment, the substrate is added to a suspension of cells, time is allowed for transport to occur, the cells are collected and rinsed rapidly by centrifugation or filtration, and the amount of labeled substrate in the cells is measured. Control measurements usually are made with a labeled substrate that the cells do not transport.

**Figure 17.27**

Facilitated diffusion carried out by the glucose transporter from human erythrocytes. The solubilized transporter was introduced into phospholipid vesicles, and transport was measured with D-glucose and L-glucose as substrates. The initial transport rate is much greater with D-glucose. L-glucose enters the vesicles slowly by simple diffusion but eventually reaches the same internal concentration as D-glucose. (Source: Adapted from M. Kasahara and P. C. Hinkle, Reconstitution and purification of the D-glucose transporter from human erythrocytes, *J. Biol. Chem.* 252:7384, 1977.)

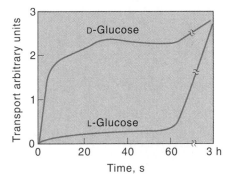

In some cases, rapid spectrophotometric assays can be used to measure transport rates. These assays often employ analogs of the normal, physiological substrate of the transporter, and take advantage of metabolic reactions that alter the spectroscopic properties of the solute after it enters the cell. The bacterial β-galactoside transporter, for example, can be studied by using *o*-nitrophenyl-β-galactoside in place of the physiological substrate, lactose (see fig. 17.25). Once *o*-nitrophenyl-β-galactoside is inside the cell, β-galactosidase hydrolyzes it to galactose and *o*-nitrophenol, which is measurable by its yellow color:

$$o\text{-nitrophenyl-}\beta\text{-galactoside} \xrightarrow{\text{transport}} o\text{-nitrophenyl-}\beta\text{-}$$
(colorless)

$$\text{galactoside} \xrightarrow[\text{H}_2\text{O}]{\substack{\beta\text{-galactosidase} \\ \text{(inside cell)}}} \text{galactose} + o\text{-nitrophenol}$$
(yellow)

If the ability of a transporter to work against a concentration gradient is to be determined, a means must be found to prevent metabolic removal of the transported solute. Here, nonmetabolizable analogs of the substrate can be useful. Glucose transport, for example, can be studied with the analog 2-deoxyglucose, which most cells cannot metabolize beyond formation of the hexose phosphate. In bacteria, mutant strains that cannot metabolize the normal substrate are often used. For example, accumulation of lactose against a concentration gradient can be studied in mutants that lack β-galactosidase, or by using the nonmetabolizable substrate analog isopropyl-β-thiogalactoside (see fig. 17.25).

To study transport in the absence of complicating metabolic processes, it often is advantageous to work with isolated membrane vesicles rather than with whole cells. Cytoplasmic membrane vesicles can be obtained from either eukaryotic or bacterial cells after homogenization or osmotic lysis. Transport proteins that have been solubilized with detergents also can be reincorporated into synthetic phospholipid vesicles (fig. 17.27).

The amino acid sequences of numerous transport proteins have been elucidated during the last 10 years. As with conventional enzymes, molecular biological techniques now make it possible to modify individual amino acid residues in the proteins and thus to probe the roles these residues play in catalysis of facilitated diffusion or active transport.

## Molecular Models of Transport Mechanisms

Judging from their hydropathy profiles, transport proteins typically contain multiple α helices stretching across the membrane. The sequence of the β-galactoside transporter (lactose permease) of *E. coli* illustrates this point (fig. 17.28). This protein probably has 12 transmembrane α helices linked by short runs of more hydrophilic amino acid residues. Amino acid residues that are implicated in binding of substrates (lactose or other β-galactosides) or in the coupling of lactose transport to proton movements are distributed among five of the helices (see fig. 17.28). Although the three-dimensional structure of the protein is not known, it seems likely that these helices pack together in a parallel bundle. Such a bundle could enclose a tubular channel that stretches across the membrane. However, the channel probably is partly occluded by amino acid side chains, because it does not render the membrane freely permeable to ions. Even protons can negotiate the channel only in the presence of an appropriate galactoside.

Figure 17.29 shows a schematic model of how lactose permease could couple the movement of protons and lactose across the membrane. In this model, the transmembrane channel is never fully open but is always plugged at one end or the other. The lactose binding site is centrally located and is accessible to either the extracellular solution or the intracellular solution, depending on where the channel is open. Transitions between these two states might occur by relatively minor conformational changes in the protein. The lactose binding site itself also is presumed to switch between two states with different dissociation constants for lactose, depending on whether or not a nearby amino acid residue is protonated. Binding of a proton from the solution on either side favors the binding of lactose from the same side. The transporter thus would tend to pick up both a proton and

**Figure 17.28**

Secondary structure of lactose permease, based on the protein's hydropathy profile. Amino acid residues are designated by the one-letter code. The boxes indicate hydrophobic segments that probably form transmembrane α helices. Residues implicated in the binding of β-galactosides or in coupling galactoside transport to proton movements are circled. These residues were identified largely by studies of genetically engineered bacterial mutants. For example, replacing the circled Glu (E) residue in helix X (Glu-325) by Ala renders the protein unable to catalyze lactose/H$^+$ symport, although the mutant protein still catalyzes facilitated diffusion of lactose. (Source: From H. R. Kaback, Use of site-directed mutagenesis to study the mechanism of a membrane transport protein, *Biochem.* 26:2071, 1987; and J. C. Collins, S. F. Permuth, and R. J. Brooker, Isolation and characterization of lactose permease mutants with an enhanced recognition of maltose and diminished recognition of cellobiose, *J. Biol. Chem.* 264:14698, 1989.)

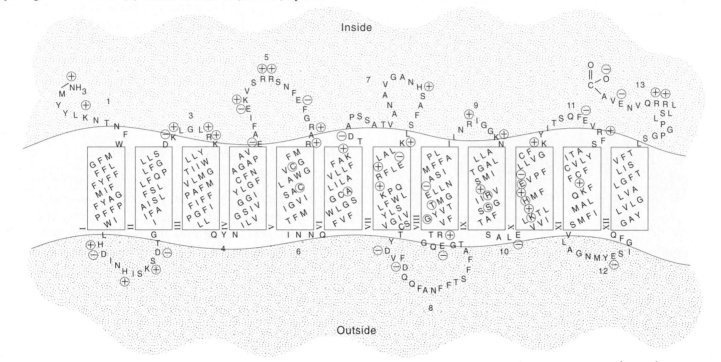

lactose from the solution with the lower pH or the more positive electrical potential and to release them on the other side. A critical feature of the model is that, if the proton-binding site is occupied, conformational changes that expose this site to the opposite side of the membrane can occur only if the lactose-binding site also is occupied. This restriction prevents the protein from catalyzing proton transport in the absence of galactoside transport.

Similar models can rationalize the mechanisms of transport systems that facilitate the diffusion of a single species such as glucose. Again, although the three-dimensional structures of these proteins are not known, hydropathy profiles of the amino acid sequences indicate the presence of numerous α helices that most likely form partially blocked channels across the membrane. The erythrocyte glucose transporter probably has 12 such helices. Facilitated diffusion could result from conformational changes that open the channel alternately on one side of the binding site for glucose and then on the other side. In agreement with the notion that the protein can switch between two conformational states, certain substrate analogs bind to the glucose site best from the extracellular solution, whereas others bind best from inside the cell.

## The Catalytic Cycle of the Na$^+$–K$^+$ Pump Includes Two Phosphorylated Forms of the Enzyme

The eukaryotic Na$^+$–K$^+$ pump, or Na$^+$–K$^+$ ATPase, has two polypeptide subunits. The larger subunit, with a molecular weight of 110,000, contains the cation-binding sites and the catalytic site for ATP hydrolysis; the smaller subunit ($M_r \approx 55,000$) is a glycoprotein the function of which is unknown. ATP binds to the enzyme only from the intracellular side of the membrane, and the enzyme releases ADP and P$_i$ on the same side. Na$^+$ ions also bind most strongly from the intracellular side, but K$^+$ ions bind best from the extracellular side. Ouabain, a specific inhibitor long used as an arrow poison, attaches to the enzyme on the extracellular side. The ion-pumping and ATPase activities of the enzyme are tightly coupled in the sense that ATP hydrolysis occurs at an appreciable rate only if Na$^+$ and K$^+$ both are present on appropriate sides of the membrane.

For each ATP that it splits, the Na$^+$–K$^+$ pump moves three Na$^+$ ions out of the cell and brings in two K$^+$ ions. This means that the pump is intrinsically electrogenic: It creates an electric potential gradient across the membrane

## Figure 17.29

A schematic model for lactose/H$^+$ symport catalyzed by the lactose permease. In the model, the protein forms a channel with gates that can open to expose lactose- and proton-binding sites to the solution on one side of the membrane or the other. In step 1, binding of a proton from the extracellular solution increases the affinity of the lactose-binding site for lactose. Binding of lactose (L) from the extracellular solution (step 2) results in a conformational change that switches the states of the gates (step 3). Lactose and a proton dissociate to the intracellular solution (steps 4 and 5), and the protein relaxes to a conformation that has a low affinity for both lactose and protons. In step 6, the gates switch back to their original state, exposing the unloaded binding sites to the external solution. In living cells, the cycle is driven in the indicated direction by the electrochemical potential gradient for protons (the intracellular solution has a higher pH and a negative electric potential relative to the extracellular solution). The protein can transport a proton into the cell only if a galactoside moves along with it. (Source: J. C. Collins, S. F. Permuth, and R. J. Brooker, Isolation and characterization of lactose permease mutants with an enhanced recognition of maltose and diminished recognition of cellobiose, *J. Biol. Chem.* 264:14698, 1989.)

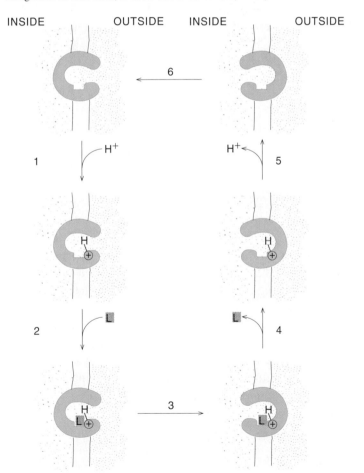

because it moves more positive charges out than it brings in. However, the $\Delta\psi$ of about $-60$ mV that is built up across the plasma membrane of a typical nerve cell probably owes more to leakage of K$^+$ back out of the cell through other types of channels. K$^+$ tends to leak out because the Na$^+$–

K$^+$ pump elevates the internal K$^+$ concentration above the extracellular concentration, and each K$^+$ that exits carries a positive charge. As we discussed previously, K$^+$ approaches an equilibrium when

$$\log \frac{[K^+]_{in}}{[K^+]_{out}} = -\frac{\mathscr{F}\,\Delta\psi}{2.3RT}$$

In the course of the catalytic cycle of the Na$^+$–K$^+$ ATPase, the terminal phosphate group of ATP is transferred to the enzyme, where it forms a carboxylic-phosphoryl anhydride with an aspartyl residue:

$$\text{ATP} + \text{enzyme—CH}_2\text{CO}_2^- \overset{Na^+}{\rightleftharpoons}$$

$$\text{ADP} + \text{enzyme—CH}_2\overset{\text{O}}{\overset{\|}{\text{C}}}\text{—O—}\overset{\text{O}^-}{\underset{\text{O}}{\overset{\|}{\text{P}}}}\text{—O}^- \quad \textbf{(5)}$$

This step of the reaction is promoted by Na$^+$ ions. K$^+$, on the other hand, promotes hydrolysis of the phosphorylated enzyme to yield P$_i$:

$$\text{enzyme—CH}_2\overset{\text{O}}{\overset{\|}{\text{C}}}\text{—O—}\overset{\text{O}^-}{\underset{\text{O}}{\overset{\|}{\text{P}}}}\text{—O}^- \overset{K^+}{\rightleftharpoons}$$

$$\text{enzyme—CH}_2\text{CO}_2^- + \text{P}_i \quad \textbf{(6)}$$

The curious thing about these reactions is that both of them are readily reversible. The reversibility of reaction (5) is not remarkable, because the $\Delta G°$ of hydrolysis of a carboxylic-phosphoryl anhydride is comparable to that of ATP. However, we do not ordinarily expect the reversal of reaction (6) to occur readily, because it requires forming a carboxylic-phosphoryl anhydride without using ATP or any other external source of energy. The explanation must be that the enzyme

$$\text{—CH}_2\overset{\text{O}}{\overset{\|}{\text{C}}}\text{—O—}\overset{\text{O}^-}{\underset{\text{O}}{\overset{\|}{\text{P}}}}\text{—O}^-$$

species that forms from P$_i$ in the reversal of reaction (6) is not identical with the species that forms from ATP in reaction (5). The conformations of the protein evidently differ significantly in two phosphorylated species, as a result of differences in the interactions of the enzyme with Na$^+$ and K$^+$.

These observations suggest a model like that shown in figure 17.30. The enzyme is depicted in two conformations, E$_1$ and E$_2$, both of which can exist in phosphorylated states. E$_1 \sim$ P has a "high-energy" phosphoryl bond that

**Figure 17.30**

A model for operation of the Na$^+$–K$^+$ pump. The enzyme exists in two conformational states, E$_1$ and E$_2$. In state E$_1$, the ion-binding site is open to the intracellular solution and preferentially binds Na$^+$ ions; in E$_2$, the ion-binding site faces the extracellular solution and preferentially binds K$^+$. ATP phosphorylates the enzyme in the E$_1$ conformation to give a "high-energy" intermediate, E$_1 \sim$ P. When the enzyme relaxes to the "low-energy" conformation E$_2$—P, the Na$^+$ ions dissociate on the extracellular side of the membrane and K$^+$ ions are bound. Hydrolysis of E$_2$—P releases P$_i$ inside the cell. This returns the protein to the E$_1$ conformation so that the K$^+$ ions are released inside the cell.

**Figure 17.31**

A model of the structure of porin from the outer membrane of *Rhodobacter capsulatus*. Part (*a*) shows the α-carbon backbones of a trimer of porin molecules viewed along an axis approximately perpendicular to the plane of the membrane. Each molecule forms a tube that passes across the membrane. Part (*b*) shows an individual porin monomer, enlarged slightly from (*a*) and viewed along an axis approximately in the plane of the membrane. The molecule folds as a β-barrel with 16 antiparallel β strands. (From: M. S. Weiss, et al., The three-dimensional structure of porin from *Rhodobacter capsulatus* at 3 Å resolution, *FEBS Lett. 267:268, 1990. Copyright © 1990 Elsevier Science Publishers BV, Amsterdam, Netherlands. Reprinted by permission.*)

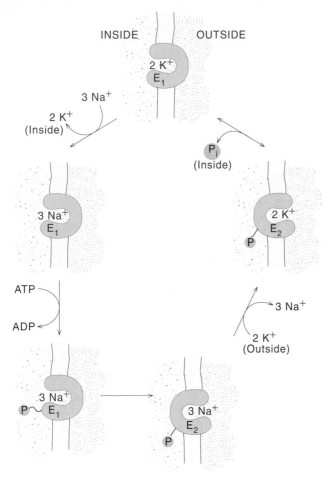

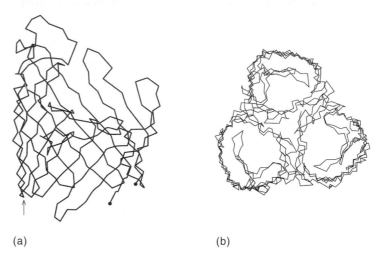

(a)                              (b)

The Ca$^{2+}$-pumping ATPases (see table 17.4) have phosphorylated intermediates similar to those of the Na$^+$–K$^+$ ATPase, and so does the H$^+$–K$^+$ ATPase of stomach epithelial cells. These enzymes are all closely related structurally and probably all operate by similar mechanisms, although the Ca$^{2+}$ pump moves only one type of ion. (The Ca$^{2+}$ ATPase moves two Ca$^{2+}$ ions across the membrane for each ATP that it splits; the gastric enzyme pushes one H$^+$ out and brings one K$^+$ in.) But phosphorylated intermediates are not a hallmark of all ion-pumping ATPases. As we discussed in chapter 14, the F$_1$–F$_0$ ATPases of mitochondria and bacterial plasma membranes do not form such intermediates. The H$^+$-pumping ATPases of lysosomes and vacuoles constitute a third, structurally distinct group of enzymes that remain less thoroughly studied.

## Some Membranes Have Relatively Large Pores

Mitochondria, Gram negative bacteria, and chloroplasts have double envelopes of membranes. As we discussed, their inner membranes are impermeable to ions and polar molecules other than those that are transported by specific

is in equilibrium with the β-γ-pyrophosphoryl bond of ATP; E$_2$—P has a "low-energy" bond in equilibrium with P$_i$. Binding of Na$^+$ from inside the cell occurs in conjunction with phosphorylation of the enzyme, in the E$_1$—conformation, to form E$_1 \sim$ P. In the model, it is the spontaneous relaxation from E$_1 \sim$ P to E$_2$—P that drives the cycle in the direction of Na$^+$ uptake. K$^+$ binds preferentially to E$_2$—P from the extracellular solution and is imported when the dephosphorylated enzyme returns from conformation E$_2$ to E$_1$.

**Figure 17.32**

Gap junctions are formed from hexagonal arrays of a rod-shaped protein. Arrays in the membranes of adjacent cells line up to form intercellular channels that can transmit small molecules from cell to cell in some tissues. (Source: Adapted from L. Makowski, D. L. D. Caspar, W. C. Phillips, and D. A. Goodenough, Gap junction structures. II. Analysis of the x-ray diffraction data, *J. Cell Biol.* 74:629, 1977.)

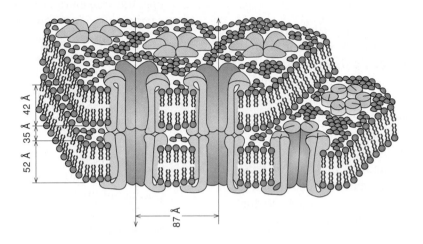

proteins. The outer membranes, by contrast, are indiscriminately permeable to molecules with molecular weights below about 600. The permeability of the outer membranes reflects the presence of pores with diameters of about 10 Å, which are formed from proteins called porins.

Figure 17.31 shows a model of porin from the outer membrane of the bacterium *Rhodobacter capsulatus*. Unlike most integral membrane proteins, porins have a β-stranded secondary structure. The *R. capsulatus* protein appears to form a 16-stranded β barrel that crosses the membrane as a tube. The tubular molecules aggregate as trimers with three parallel pores.

Pores called gap junctions occur in the plasma membranes of cells in some eukaryotic tissues, including liver, brain, and cardiac muscle. These channels allow molecules with molecular weights up to about 1,000 to diffuse between adjacent cells of the same type. Flow of ions through gap junctions provides intercellular communication that enables the heart to beat synchronously. In some tissues, gap junctions participate in distributing nutrients and coordinating embryological development. Gap junctions are formed from hexagonal arrays of a rod-shaped protein that stretches across the plasma membrane. At the center of the hexagon, the cluster of rods evidently leaves a channel through the membrane (fig. 17.32). The hexagonal arrays in the membranes of adjacent cells line up so that small molecules can flow through them from the cytosol of one cell to the cytosol of the other without leaking out to the interstitial space.

Gap junctions close if the $Ca^{2+}$ concentration of the cell rises above normal physiological levels. Because the $Ca^{2+}$-ATPase ordinarily keeps the cytosolic $Ca^{2+}$ concentration below $10^{-7}$ M, a large increase in $[Ca^{2+}]$ could be a signal that a cell has exhausted its energy supplies, and the closing of gap junctions under these conditions could serve to cut off the drain of materials from living cells to neighboring cells that have died.

## Other Specific Interactions Mediated by Membrane Proteins

A variety of signals can be transmitted across membranes without the actual flow of a substance from one side of the membrane to the other. We saw in chapter 12 that some hormones bind to specific receptor sites on the outer surface of the plasma membrane, thereby triggering metabolic changes on the cytosolic side of the membrane. Hormonal systems that function in this way are discussed in greater detail in chapter 24. Other membrane proteins mediate specific cell–cell interactions. Sometimes these interactions merely stimulate particular types of cells to bind to one another, but often they also trigger reactions that result in proliferation or differentiation of the interacting cells. We discuss signals of this type when we consider the interaction between the B and T cells of the immune system (see Supplement 3).

## Summary

In this chapter we discussed the structure and function of biological membranes. First we considered their structure starting with an examination of the constituents of membranes. Then we turned to questions concerning the transport of materials across membranes.

1. Biological membranes consist primarily of proteins and lipids whose relative amounts vary considerably.

2. There are two main types of lipids in biological membranes: Phospholipids and sterols. The predominant phospholipids in most membranes are phosphoglycerides, which are phosphate esters of the three-carbon alcohol glycerol.

3. Due to their amphipathic nature, phospholipids spontaneously form ordered structures in water. When phospholipids are agitated in the presence of excess water, they tend to aggregate spontaneously to form bilayers, which strongly resemble the types of structures they form in biological membranes.

4. Membranes contain proteins that merely bind to their surface (peripheral proteins) and those that are embedded in the lipid matrix (integral proteins). Integral membrane proteins contain transmembrane $\alpha$-helices.

5. Proteins and lipids have considerable lateral mobility within membranes.

6. Biological membranes are asymmetric. Consistent with this asymmetry, a protein that has been inserted into a membrane in a particular orientation usually retains that orientation indefinitely.

7. On heating, phospholipid bilayers undergo a phase transition (melting) from an ordered to a disordered state. The melting temperature ($T_m$) depends strongly on the phospholipid composition in the bilayer. Increasing the length of the fatty-acid chains increases the $T_m$; *cis* double bonds decrease the $T_m$. Cholesterol broadens the melting transition of the phospholipid bilayer. Cells regulate the lipid compositions of their plasma membrane so that a reasonable membrane fluidity is maintained.

8. Some proteins of eukaryotic plasma membranes are connected to the cytoskeleton; this connection inhibits their lateral mobility with the membrane.

9. Biological membranes contain proteins that act as specific transporters of small molecules into and out of the cell. Most solutes are transported by specific carriers that are invariably proteins.

10. Some transporters facilitate diffusion of a solute from a region of relatively high concentration or down a favorable electrochemical potential gradient. Such transporters do not require energy since the transport is in the thermodynamically favorable direction.

11. Other transporters move solutes against an electrochemical potential gradient and require an energy producing process to make them functional. Cells drive such active transport processes in a variety of ways. The transport can be coupled to the hydrolysis of a high energy phosphate, to the cotransport of another molecule down an electrochemical potential gradient, or to the modification of the transported molecule soon after it crosses the membrane.

12. Proteins involved in active transport frequently change their structure during the transport process. For example the $Na^+-K^+$ pump includes two phosphorylated forms of the enzyme involved in the transport of these two ions.

13. Some membranes contain relatively large pores which allow for the free passage of molecules with molecular weights up to about 600. For example the outer membranes of gram negative bacteria contain pores with diameters of about 10 Å which are formed from proteins called porins.

14. Pores called gap junctions occur in the plasma membranes of cells in some eukaryotic tissues. These channels allow molecules with molecular weights up to about 1000 to diffuse between adjacent cells of the same types.

## Selected Readings

Bennett, V., The membrane skeleton of human erythrocytes and its implications for more complex cells. *Ann. Rev. Biochem.* 54:273, 1985.

Deisenhofer, J., O. Epp, N. Miki, R. Huber, and H. Michel, X-ray structure analysis of a membrane protein complex. Electron density map at 3-Å resolution and a model of the chromophores of the photosynthetic reaction center from *Rhodopseudomonas viridis. J. Mol. Biol.* 180:385, 1984.

Gennis, R. B., *Biomembranes: Molecular Structure and Function.* New York: Springer-Verlag, 1989.

Henderson, R., J. M. Baldwin, T. A. Ceska, F. Zemlin, E. Beckmann and K. H. Downing, Model for the structure of bacteriorhodopsin based on high resolution electron cryo-microscopy. *J. Mol. Biol.* 213:899, 1990.

Inesi, G., D. Lewis, D. Nikic, A. Hussain and M. Kirtley, Long-range intramolecular linked functions in the calcium transport ATPase. *Adv. Enzymol.* 65:182, 1992.

Jain, M. K., *Introduction to Biological Membranes,* 2nd ed., New York: Wiley, 1988.

Jennings, M. L., Topography of membrane proteins. *Ann. Rev. Biochem.* 58:999, 1989.

Kaback, H. R., E. Bibi, and P. D. Roepe, β-Galactoside transport in *E. coli:* A functional dissection of lac permease. *Trends Biochem. Sci.* 15:309, 1990.

Silverman, M., Structure and function of hexose transporters. *Ann. Rev. Biochem.* 60:757, 1991.

Singer, S. J., The molecular organization of membranes. *Ann. Rev. Biochem.* 43:805, 1974.

Vance, D. E., and J. E. Vance, (eds.), *Biochemistry of Lipids, Lipoproteins and Membranes.* Amsterdam: Elsevier, 1991.

## Problems

1. Examine the list of ingredients on containers of salad dressing. Without much searching you should be able to find lecithin listed. Remember that ingredients are ordered by weight of the item. (The materials that contribute the most are listed first.) After identifying the major components of salad dressing, propose a function for the lecithin.

2. Both triacylglycerol and phospholipids have fatty acid ester components, but only one can be considered amphipathic. Indicate which is amphipathic, and explain why.

3. Phosphatidylcholine usually has a saturated fatty acyl chain on the C-1 of glycerol and an unsaturated fatty acyl chain on the C-2 position. Can you explain how this happens?

4. Why does the Davson-Danielli membrane model predict that the exposed inside of the lipid bilayer is featureless?

5. The relative orientation of polar and nonpolar amino acid side chains in integral membrane proteins is ''inside-out'' relative to that of the amino acid side chains of water-soluble globular proteins. Explain.

6. What physical properties are conferred on biological membranes by phospholipids? How can the charge characteristics of the phospholipids affect binding of peripheral proteins to the membrane? What role might divalent metal ions play in the interaction of peripheral membrane proteins with phospholipids?

7. Differentiate between peripheral and integral membrane proteins with respect to location, orientation, and interactions that bind the protein to the membrane. What are some strategies used to differentiate between peripheral and integral proteins by means of detergents or chelating agents?

8. Frequently, integral membrane proteins are glycosylated with complex carbohydrate arrays. Explain how glycosylation further enhances the asymmetrical orientation of integral proteins.

9. Integral transmembrane proteins often contain helical segments of the appropriate length to span the membrane. These helices are composed of hydrophobic amino acid residues. In transmembranous proteins with multiple segments that span the membrane, you may find some hydrophilic residue side chains. Why are hydrophilic side chains not favored in single-span membrane proteins? How may the hydrophilic side chains be accommodated in multiple-span proteins?

10. Besides the comments made in the text (page 402) on the free energy changes associated with phosphorylation during sugar transport, what more can be said on the unidirectional aspect of this type of transport system?

11. Elementary portrayals of transport systems have often utilized "revolving doors" as analogies. After reading this chapter what objection(s) do you have with the revolving door idea?

12. Membrane vesicles of *E. coli* that possess the lactose permease are preloaded with KCl and are suspended in an equal concentration of NaCl. It is observed that these vesicles actively, although transiently, accumulate lactose if valinomycin is added to the vesicle suspension. No such active uptake is observed if KCl replaces NaCl in the suspending medium. Explain these results in light of what you know about the mechanism of lactose transport and the properties of valinomycin.

13. The Nernst equation relates the electric potential $\Delta\Psi$ resulting from an unequal distribution of a charged solute across a membrane permeable to that solute, to the ratio between the concentration of solute on one side and on the other:

$$m\,\Delta\Psi - \frac{-2.3RT}{F}\log\frac{[So]_1}{[So]_2}$$

where $m$ is the charge on the solute, $2.3RT/F$ has a value of about 60 mV at 37°C, and $[So]_1$ and $[So]_2$ refer to the concentrations of solute on either side of the membrane. Consider a planar phospholipid bilayer separating two compartments of equal volume. Side 1 contains 50 mM KCl and 50 mM NaCl, whereas side 2 contains 100 mM KCl.

(a) If the membrane is made permeable only to $K^+$, for example, by addition of valinomycin, what is the magnitude of $\Delta\Psi$?

(b) If the membrane is made permeable to $H^+$ and $K^+$, in which direction does $H^+$ initially flow?

(c) If the membrane can be made selectively permeable to both $K^+$ and $Cl^-$, what is the value of $\Delta\Psi$ and the ion concentrations on both sides of the membrane at equilibrium? (*Hint:* Initially, $K^+$ diffuses down its concentration gradient accompanied by an equivalent amount of $Cl^-$. Equilibrium is established when the potentials due to $K^+$ and $Cl^-$ equal each other and the overall membrane potential.)

14. Predict the effects of the following on the initial rate of glucose transport into vesicles derived from animal cells that accumulate this sugar by means of $Na^+$ symport. Assume that initially $\Delta\Psi = 0$, $\Delta pH = 0$ (pH = 7), and the outside medium contains 0.2 M $Na^+$, whereas the vesicle interior contains an equivalent amount of $K^+$.
(a) Valinomycin.
(b) Gramicidin A.
(c) Nigericin.
(d) Preparing the membrane vesicles at pH 5 (in 0.2 M KCl), resuspending them at pH 7 (in 0.2 M NaCl), and adding 2,4-dinitrophenol.

15. In *E. coli,* lactose is taken up by means of proton symport, maltose by means of a binding (ABC-type) protein-dependent system, melibiose by means of $Na^+$ symport, and glucose by means of the phosphotransferase system (PTS). Although this bacterium normally does not transport sucrose, suppose that you isolated a strain that does. How do you determine whether one of the four mechanisms just listed is responsible for sucrose transport in this mutant strain?

16. You are growing some mammalian cells in culture and measure the uptake of D-glucose and L-glucose (see data in the chart). What type of transport is observed with these sugars? (*Hint:* Plot $V$ versus [sugar] and $1/V$ versus $1/$[sugar].) Explain the significance of these data.

| [Sugar](mM) | $V$(mM·cm s$^{-1}$) × 10$^7$ | |
| --- | --- | --- |
| | D-glucose | L-glucose |
| 0.100 | 166 | 4.8 |
| 0.167 | 252 | 8.0 |
| 0.333 | 408 | 16 |
| 1.000 | 717 | 50 |

CHAPTER

# 18

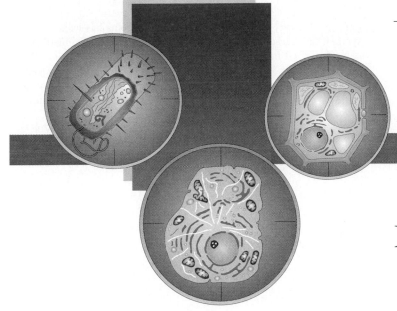

# Metabolism of Fatty Acids

*Fatty acid synthesis and breakdown both occur in steps involving 2 carbon units but by totally different mechanisms.*

F atty acids, as components of phospholipids, are important structural elements of biological membranes and, like carbohydrates, are important sources of energy. In this chapter we focus on the role of fatty acids in energy metabolism. We see that fatty acids are both assembled and de-

411

**Figure 18.1**

Synthesis and degradation of fatty acids. Both synthesis and degradation of fatty acids occur by multistep pathways involving totally different enzymes located in different parts of the cell. Both pathways are similar in that two carbon atoms are either added (in synthesis) or deleted (in degradation) in each round of a repetitious, cyclical process.

The acetyl-CoA used as substrate for fatty acid synthases in the cytosol originates from the mitochondrion. This acetyl-CoA condenses with $CO_2$ to form malonyl-CoA, which eliminates $CO_2$ after an initial condensation reaction. After three more steps a two-carbon unit is added to a growing fatty acid chain. This cycle repeats itself many

times until a fatty acid chain with 16 carbon atoms has been synthesized. Variations of this pathway give rise to unsaturated and polyunsaturated fatty acids.

The degradation of fatty acids also involves a repetitious process yielding a two-carbon unit (acetyl-CoA) in every cycle. This acetyl-CoA can be fed into the TCA cycle, or, alternatively, it can be used to make ketone bodies.

Elaborate controls prevent synthesis and degradation of fatty acids from occurring simultaneously.

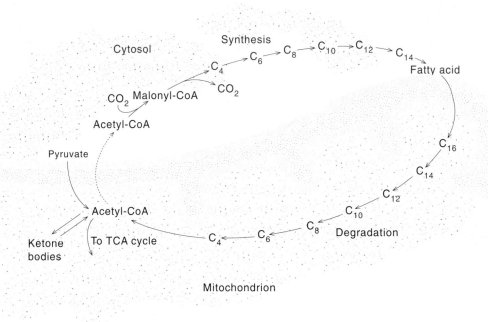

graded in blocks of two carbon atoms (fig. 18.1). Despite this obvious similarity, the processes of synthesis and breakdown are very different. In contrast to the biosynthesis and catabolism of carbohydrates, the two processes for metabolism of fatty acids use completely different sets of enzymes, and they occur in different cellular compartments. These differences facilitate regulatory mechanisms that ensure that biosynthesis and catabolism do not occur simultaneously.

In the first section we deal with reactions associated with fatty acid breakdown. The second section covers the pathway for fatty acid biosynthesis. Finally, we consider the regulatory mechanisms that determine the conditions under which each of these processes occurs.

## Fatty Acid Degradation

### *Fatty Acids Originate from Three Sources: Diet, Adipocytes, and de novo Synthesis*

Fatty acids are mainly derived from the diet. In fact, 30%–40% of the calories ingested each day in the average human diet are provided by the fatty acid components of triacylglycerols and phospholipids. Because of the relationship between a diet high in fat and heart disease, Heart Associations throughout the world recommend that no more than 30% of the calories in our diet be derived from fatty acids. Dietary lipids are degraded by lipases and phospholipases, which are enzymes secreted into the intestinal lumen. Like other digestive enzymes, many of the lipases and phospholipases are derived from the pancreas (fig. 18.2*a*). The fatty acids released by this process are absorbed by cells in the intestinal wall, and triacylglycerols are resynthesized (fig.

## Figure 18.2

(a) Triacylglycerols are digested in the intestine by pancreatic lipase. (b) The fatty acids and monoacylglycerol are absorbed into the intestinal wall. The cells in the wall, enterocytes, resynthesize triacylglycerol and package the lipid with proteins to give lipoprotein particles called chylomicrons, which are secreted from the enterocytes into the lymph.

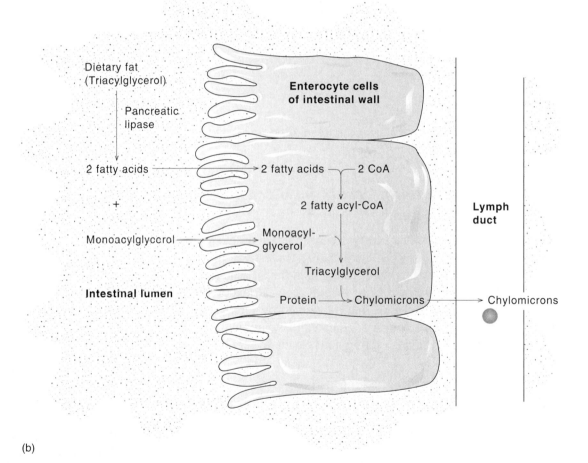

18.2b). These lipids are then packaged into spherical lipoproteins, particles of lipids and proteins, known as chylomicrons, which are secreted into lymphatic vessels and subsequently enter the blood stream. Once in the circulatory system the triacylglycerol components of the chylomicrons are degraded to fatty acids and glycerol by the enzyme lipoprotein lipase, which is attached to the luminal (inner) side of capillary vessels in heart, muscle, adipose (commonly known as fat), and other tissues (fig. 18.3). The released fatty acids are transported into these tissues for generation of energy, storage as an energy reserve in the form of triacylglycerols, or biosynthesis of membrane phospholipids.

If the diet does not provide sufficient lipid to satisfy immediate needs, the fatty acids stored in the adipose tissue in the form of triacylglycerols may be required (discussed later, see fig. 18.20). In addition, most tissues have the ca-

**Figure 18.3**

The chylomicrons are delivered to other tissues in the body via the bloodstream. In the tissues the chylomicrons bind to the endothelial cells of the capillaries and are degraded to fatty acids and glycerol by lipoprotein lipase. This enzyme is secreted by cells in the tissue and is bound to the endothelial cells.

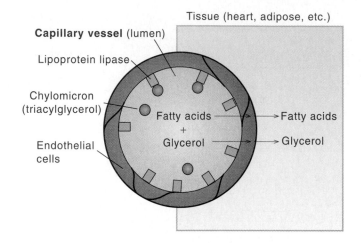

pacity to convert carbohydrates and some amino acids into fatty acids.

## Fatty Acid Breakdown Occurs in Blocks of Two Carbon Atoms

Recall that fatty acid chains nearly always contain an even number of carbon atoms (see chapter 17). Investigations into the mechanisms of fatty acid catabolism began around the turn of the century when Fritz Knoop reported experiments that indicated fatty acids were degraded by removal of two carbons at a time. The data indicated that carbon 3 of a fatty acid was oxidized with subsequent cleavage between carbons 2 and 3.

$$\underset{\beta}{R}\overset{3}{C}H_2\underset{\alpha}{\overset{2}{C}}H_2\overset{1}{C}OOH \longrightarrow \underset{\beta}{R}\overset{O}{\overset{\|}{C}}\underset{\alpha}{\overset{2}{C}}H_2\overset{1}{C}OOH \longrightarrow$$

$$\overset{O}{\overset{\|}{R}}\overset{}{C}OH + \underset{\alpha}{\overset{2}{C}}H_3\overset{1}{C}OOH \quad \textbf{(1)}$$

Carbon 2 is also known as the alpha ($\alpha$) carbon and carbon 3 as the beta ($\beta$) carbon. Hence the term $\beta$ oxidation was coined.

## The Oxidation of Saturated Fatty Acids Occurs in the Mitochondria

We now explore the remarkable process by which a long-chain saturated fatty acid is converted into two-carbon units (acetate), which can be oxidized to $CO_2$ and $H_2O$ via the tricarboxylic acid cycle and the electron-transport chain. Fatty acids that enter cells are activated to their CoA derivatives by the enzyme acyl-CoA ligase and transported into the mitochondria for $\beta$ oxidation as we discuss later in this chapter.

The outline for the $\beta$ oxidation of a saturated fatty acid is shown in figure 18.4. In the first reaction, the acyl-CoA is dehydrogenated by acyl-CoA dehydrogenase to yield the $\alpha,\beta$ (or 2,3)*trans*-enoyl-CoA and $FADH_2$. The electrons from $FADH_2$ are channeled into the electron-transport chain (see chapter 14). Enoyl-CoA is subsequently hydrated stereospecifically by enoyl-CoA hydrase to yield the 3-L-hydroxyacyl-CoA. The hydroxyl group is oxidized by 3-hydroxyacyl-CoA dehydrogenase with $NAD^+$ as coenzyme to yield $\beta$-ketoacyl-CoA and NADH. The final step in the sequence is catalyzed by thiolase to form acetyl-CoA and an acyl-CoA that is two carbons shorter than the initial substrate for $\beta$ oxidation. This acyl-CoA can undergo another cycle of $\beta$ oxidation to yield $FADH_2$, NADH, acetyl-CoA, and an acyl-CoA with two fewer carbons (fig. 18.4*b*). The enzymatic steps are cycled until, in the last sequence of reactions, the four carbon unit,

$$\text{butyryl-CoA } (CH_3CH_2CH_2\overset{O}{\overset{\|}{C}}\text{-CoA})$$

is degraded to two acetyl-CoAs.

## Fatty Acid Oxidation Yields Large Amounts of ATP

$\beta$ Oxidation, in combination with the tricarboxylic acid cycle and respiratory chain, provides more energy per carbon atom than any other energy source, such as glucose and amino acids. The equations for the complete oxidation of palmitoyl-CoA are shown in table 18.1. Equation (1) in the table shows the oxidation of palmitoyl-CoA by the enzymes of $\beta$ oxidation. Each of the products of equation (1) is further oxidized by the respiratory chain, equations (2) and (3), or by the tricarboxylic acid cycle and the respiratory chain, equation (4). When the reactions of equations (1)–(4) are summed, the result is equation (5). Hence, complete oxidation of one molecule of palmitoyl-CoA yields 108 ATP + 16 $CO_2$ + 123 $H_2O$ and CoA. The water generated by $\beta$ oxidation seems almost incidental to this process, but is crucial in several animal species. For example, the oxidation of fatty acids is used as a major source of $H_2O$ by the killer

## Figure 18.4

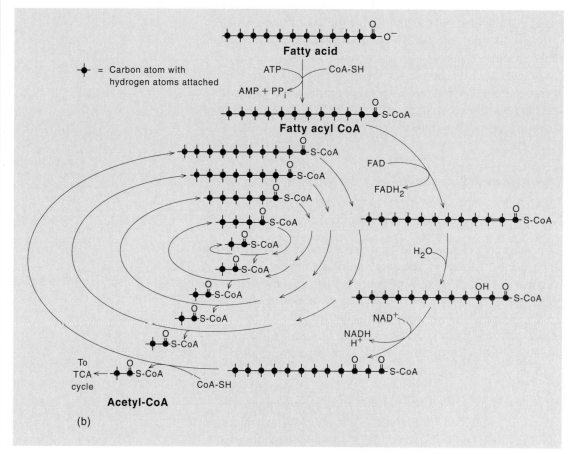

(a) The β oxidation of fatty acids consists of four reactions in the matrix of the mitochondrion. Each cycle of reactions results in the formation of acetyl-CoA and an acyl-CoA with two fewer carbons. (b) The cyclic nature of the β oxidation cycle. In some illustrations the —SH group of CoA is indicated for emphasis. Likewise in some illustrations the —S— group in an acyl-CoA structure is indicated.

---

## Table 18.1

Equations for the Complete Oxidation of Palmitoyl-CoA to $CO_2$ and $H_2O$

1. $CH_3(CH_2)_{14}CO\text{-}CoA + 7\,FAD + 7\,H_2O + 7\,CoA + 7\,NAD^+ \longrightarrow 8\,CH_3CO\text{-}CoA + 7\,FADH_2 + 7\,NADH + 7\,H^+$

2. $7\,FADH_2{}^a + 10\,P_i{}^b + 10\,ADP + 3\frac{1}{2}\,O_2 + 10\,H^+ \longrightarrow 7\,FAD + 17\,H_2O + 10\,ATP$

3. $7\,NADH^a + 18\,P_i + 18\,ADP + 3\frac{1}{2}\,O_2 + 25\,H^+ \longrightarrow 7\,NAD^+ + 25\,H_2O + 18\,ATP$

4. $8\,CH_3CO\text{-}CoA + 16\,O_2 + 80\,P_i + 80\,ADP + 80\,H^+ \longrightarrow 8\,CoA + 88\,H_2O + 16\,CO_2 + 80\,ATP$

5. $CH_3(CH_2)_{14}CO\text{-}CoA^c + 108\,P_i + 108\,ADP + 23\,O_2 + 108\,H^+ \longrightarrow 108\,ATP + 16\,CO_2 + CoA + 123\,H_2O$

[a] The yield of ATP per oxidation of $FADH_2$ and NADH is assumed to be 1.5 and 2.5, respectively.

[b] Remember that condensation of ADP and $P_i$ yields a molecule of $H_2O$ as well as ATP.

[c] See figure 13.4 when accounting for the production of $H_2O$ from the oxidation of acetyl-CoA because each TCA cycle uses 2 molecules of $H_2O$.

whale. This animal lives in the sea but cannot drink the salt water. Similarly, the fat stored in the hump of the camel serves as a source of energy and water in the desert.

The yield of ATP from the oxidation of palmitoyl-CoA can be compared with the yield from glucose. Both are major sources of energy in our bodies. Perhaps the most meaningful comparison is in terms of the number of ATPs produced per carbon atom. For glucose the yield is 30/6 = 5 ATPs per carbon atom, and for palmitoyl-CoA it is 108/16 = almost 7 ATPs per carbon. Thus, the oxidation of palmitoyl-CoA yields more ATP per carbon. The chemical reason for this difference is that 15 of the 16 carbons is palmitate are in the completely reduced state (i.e., no oxygen substitution), whereas all 6 carbons of glucose are partially oxidized (each carbon has a hydroxyl group). The more a carbon atom is substituted with oxygen, the less energy can be harnessed by oxidation to $CO_2$, which is the derivative of carbon that is completely oxidized.

## Additional Enzymes Are Required for Oxidation of Unsaturated Fatty Acids in Mitochondria

We learned in chapter 17 that unsaturated fatty acids are commonly found in triacylglycerols and phospholipids. Nature has also evolved mechanisms for oxidation of these fatty acids. The process is based on $\beta$ oxidation, but additional enzymes are required. As seen in figure 18.5, oleoyl-CoA ($18:1\Delta^9$) can be degraded in the same manner as stearoyl-CoA ($18:0$) through the first three cycles of $\beta$ oxidation. The resulting product, $\Delta^3$-cis-dodecenoyl-CoA ($12:1\Delta^3$), is not a substrate for acyl-CoA dehydrogenase because of its cis double bond. This step is bypassed by a reaction, catalyzed by enoyl-CoA isomerase, in which the double bond is isomerized to $\Delta^2$-trans-dodecenoyl-CoA. This intermediate is a normal substrate for the $\beta$ oxidation enzyme, enoyl-CoA hydrase. Subsequently, the normal route for $\beta$ oxidation resumes. Since acyl-CoA dehydrogenase is not used in one of the rounds of $\beta$ oxidation of oleoyl-CoA, one fewer $FADH_2$ and two fewer ATPs are made than in the $\beta$ oxidation of the corresponding saturated fatty acyl-CoA (i.e., stearoyl-CoA).

Polyunsaturated fatty acids are also degraded by $\beta$ oxidation, but the process requires enoyl-CoA isomerase and an additional enzyme, 2,4-dienoyl-CoA reductase (fig. 18.6). For example, the degradation of linoleoyl-CoA ($18:2\Delta^{9,12}$) begins, like that of oleoyl-CoA, with three rounds of $\beta$ oxidation and results in a $\Delta^3$-cis unsaturated fatty acyl-CoA that is not a substrate for acyl-CoA dehydrogenase. Isomerization of the double bond to the $\Delta^2$-trans position by enoyl-CoA isomerase allows the resumption of

**Figure 18.5**

Unsaturated fatty acids such as oleoyl-CoA can also be degraded to acetyl-CoA in mitochondria. Three cycles of $\beta$ oxidation result in an acyl-CoA ($\Delta^3$-cis-dodecenoyl-CoA) that is not a substrate for acyl-CoA dehydrogenase. This problem is circumvented by isomerization of the double bond to the $\Delta^2$-trans position by enoyl-CoA isomerase. Complete oxidation of the remainder of the molecule by the enzymes of $\beta$ oxidation is now possible.

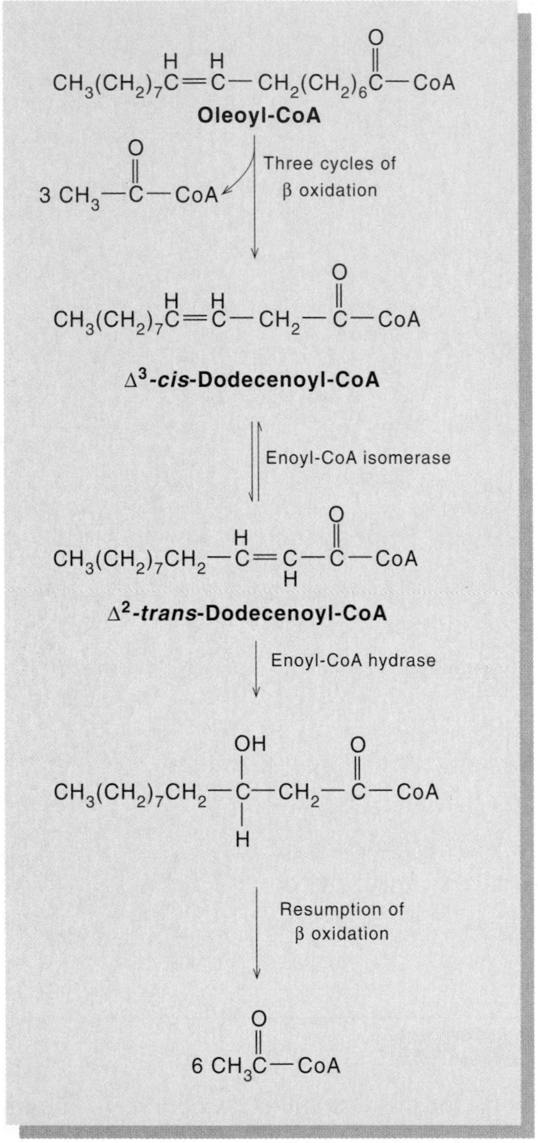

one cycle of $\beta$ oxidation and generation of a $\Delta^4$-cis-enoyl-CoA. Subsequently, acyl-CoA dehydrogenase produces $\Delta^2$-trans,$\Delta^4$-cis-dienoyl-CoA (see fig. 18.6), which has two conjugated double bonds that are not interrupted by a methylene group. This derivative is not a substrate for what would normally be the next enzyme in $\beta$ oxidation, enoyl-CoA hydrase. Instead, the intermediate is converted by 2,4-dienoyl-CoA reductase to $\Delta^3$-trans enoyl-CoA, then to the

**Figure 18.6**

Polyunsaturated fatty acids such as linoleoyl-CoA are also oxidized in the mitochondria. In addition to the enzymes of $\beta$ oxidation and enoyl-CoA isomerase, the process requires 2,4-dienoyl-CoA reductase.

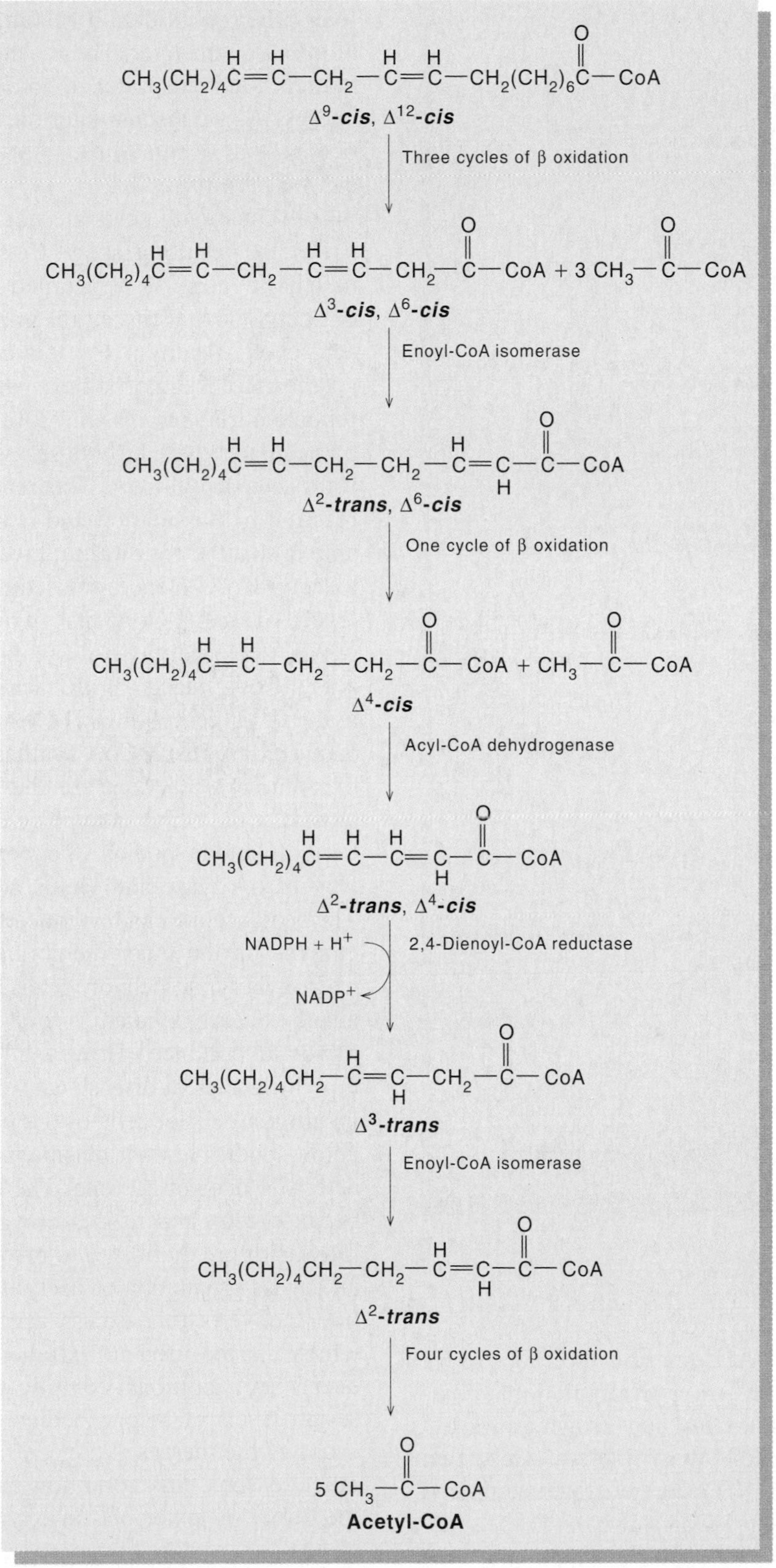

## Figure 18.7

The biosynthesis of ketone bodies (acetoacetate, hydroxybutyrate, and acetone) occurs in the mitochondria of the liver.

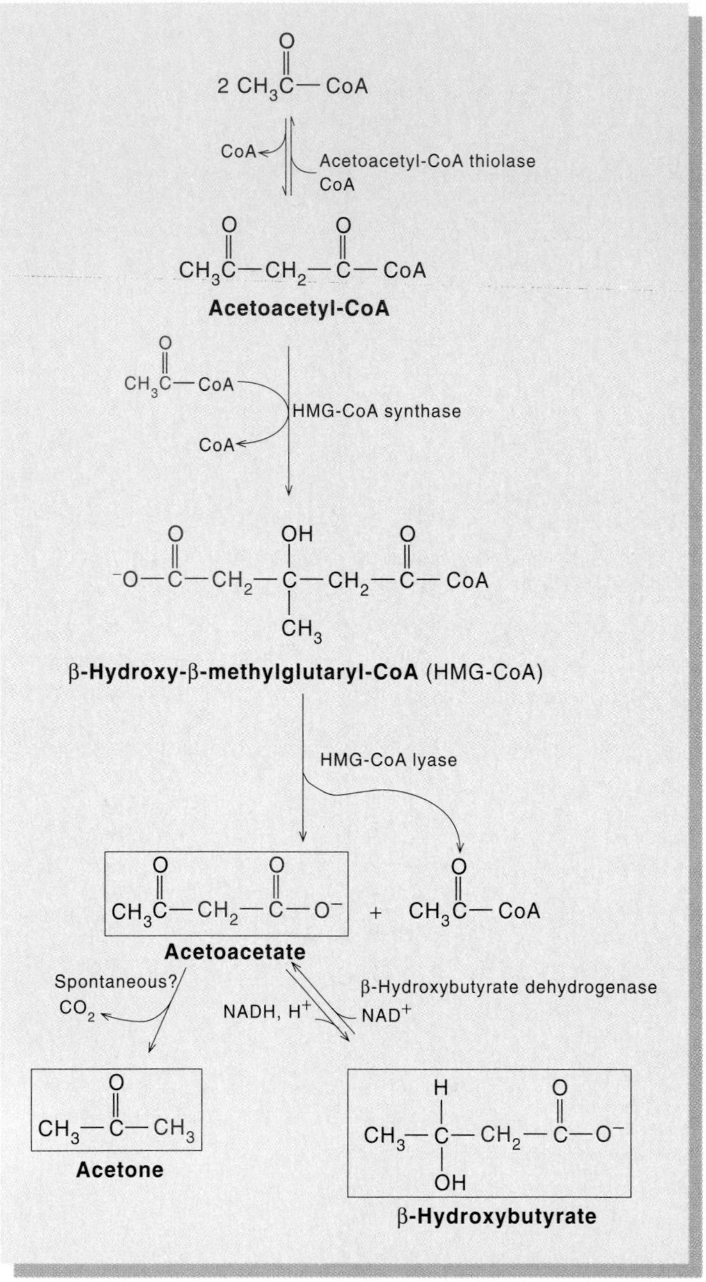

## Ketone Bodies Formed in the Liver Are Used for Energy in Other Tissues

Ketone bodies (acetoacetate, $\beta$-hydroxybutyrate, and acetone structures are presented in fig. 18.7) are made in the liver when $\beta$ oxidation of fatty acids is in excess of that required by the liver. These water-soluble, energy-rich compounds are transported to other tissues for generation of energy. As we discuss later on, excess production of ketone bodies, that occurs during starvation or untreated diabetes, can be harmful.

Once fatty acids are degraded in liver mitochondria, the resulting acetyl-CoA can undergo a number of metabolic fates. As we learned in chapter 13, the utilization of acetyl-CoA is of central importance in the tricarboxylic acid cycle. Alternatively this coenzyme is involved in the synthesis of ketone bodies, which takes place only in the mitochondria (see fig. 18.7). In the first reaction, catalyzed by acetoacetyl-CoA thiolase, two acetyl-CoAs condense to form acetoacetyl-CoA. This reaction is a reversal of the last reaction in $\beta$ oxidation and is thermodynamically unfavorable in that the equilibrium favors thiolytic cleavage of acetoacetyl-CoA. Hence, this compound is formed when the levels of acetyl-CoA rise, which pushes the reaction towards acetoacetyl-CoA synthesis. A third molecule of acetyl-CoA reacts with acetoacetyl-CoA to yield $\beta$-hydroxy-$\beta$-methylglutaryl-CoA (HMG-CoA) in a reaction catalyzed by HMG-CoA synthase. (As we see in chapter 20, the same two reactions are also the first steps in cholesterol biosynthesis, which take place in the cytosol.) In the formation of ketone bodies, the next reaction is catalyzed by HMG-CoA lyase and yields acetoacetate and acetyl-CoA. The acetoacetate can be reduced to $\beta$-hydroxybutyrate by an enzyme on the inner membrane of the mitochondrion, $\beta$-hydroxybutyrate dehydrogenase. Although acetoacetate can also be decarboxylated to form acetone, this is normally of minor importance. However, patients with uncontrolled type I diabetes (a disease caused by a lack of production of insulin in the islet cells of the pancreas) have high levels of ketone bodies in their plasma, and their breath has the characteristic odor of acetone. The rise in ketone bodies occurs because cells and tissues are glucose starved. As a result fatty acids are mobilized to provide energy, which results in an excess production of acetyl-CoA in the liver. Hence, the liver converts this excess acetyl-CoA into ketone bodies which can be used in extrahepatic tissues as a source of energy. An additional complication is that a large rise in the serum levels of ketone bodies lowers the pH of blood because of the increased concentration of these acids. Among other reasons, this condition can be hazardous because of the effect of lower pH on oxygen binding to hemoglobin.

$\Delta^2$-trans-isomer by enoyl-CoA isomerase (see fig. 18.6). With the double bond in the $\Delta^2$-trans-configuration, $\beta$ oxidation can resume. Since acyl-CoA dehydrogenase is bypassed twice during the $\beta$ oxidation of linoleoyl-CoA, two fewer FADH$_2$ and four fewer ATPs are generated compared with the $\beta$ oxidation of stearoyl-CoA (18:0).

**Figure 18.8**

Ketone bodies are converted back to acetyl-CoA in the mitochondria of nonhepatic tissues and used as a source of energy in these tissues.

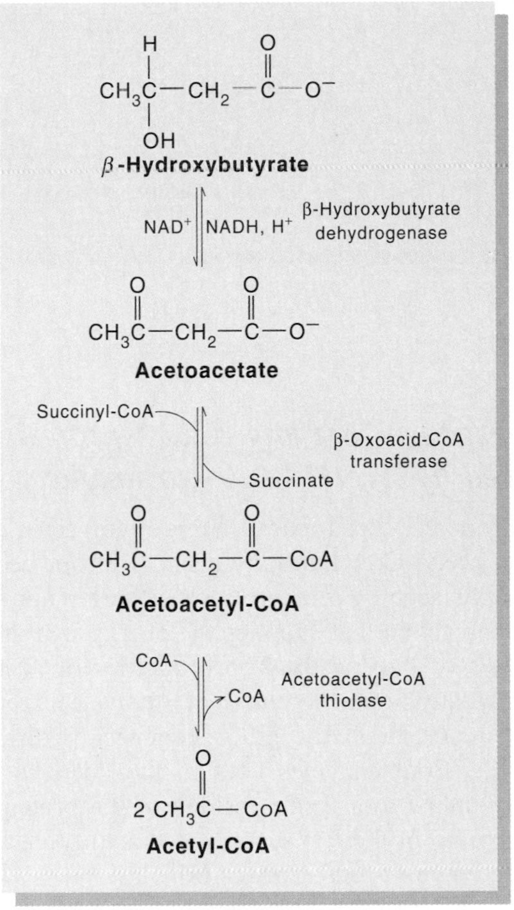

## Summary of Fatty Acid Degradation

In this section we have seen that fatty acids are oxidized in units of two carbon atoms. The immediate end products of this oxidation are $FADH_2$ and NADH, which supply energy through the respiratory chain, and acetyl-CoA, which has multiple possible uses in addition to the generation of energy via the tricarboxylic acid cycle and respiratory chain. Unsaturated fatty acids can also be oxidized in the mitochondria with the help of auxiliary enzymes. Ketone body synthesis from acetyl-CoA is an important liver function for transfer of energy to other tissues, especially brain, when glucose levels are decreased as in diabetes or starvation.

## Biosynthesis of Saturated Fatty Acids

When energy occurs in excess, for example after a meal, organisms make fatty acids to store this energy as components of triacylglycerol. Reserach on the mechanism of fatty acid biosynthesis goes back to the very early years of biochemical investigations. The finding that most fatty acids contain an even number of carbon atoms led Rapier to postulate in 1907 that fatty acids were produced by condensation of an activated two-carbon compound. In some of the first isotopic tracer experiments performed in the late 1930s and early 1940s, David Rittenberg and Konrad Bloch used the newly developed heavy isotopes of carbon (carbon 13) and hydrogen (deuterium) and implicated acetate as the two-carbon compound. Subsequently, Feodor Lynen showed a central role for acetyl-CoA in fatty acid biosynthesis. Precisely how acetyl-CoA was converted into fatty acids eluded workers until the late 1950s when Salih Wakil discovered the involvement of malonyl-CoA. Subsequent progress was rapid, and the scheme for fatty acid biosynthesis as we know it today was elucidated.

The synthesis of saturated fatty acids is very similar in all organisms. The overall reaction for the formation of palmitic acid is

$$\underset{\text{Acetyl-CoA}}{CH_3\overset{O}{\overset{\|}{C}}CoA} + \underset{\text{Malonyl-CoA}}{7\ ^-O\overset{O}{\overset{\|}{C}}CH_2\overset{O}{\overset{\|}{C}}CoA}$$

$$+\ 14\ NADPH + 14\ H^+ \longrightarrow$$

$$\underset{\text{Palmitic acid}}{CH_3(CH_2)_{14}COO^-} + 7\ CO_2 + 8\ CoA$$

$$+\ 14\ NADP^+ + 7\ H_2O \quad \textbf{(2)}$$

Ketone body synthesis is primarily a liver function, since HMG-CoA synthase is present in large quantities only in this tissue. Acetoacetate and $\beta$-hydroxybutyrate are secreted into the blood and carried to other tissues where they are converted into acetyl-CoA as described in figure 18.8. The reactions catalyzed by $\beta$-hydroxybutyrate dehydrogenase and thiolase are common to both the synthesis and degradation of the ketone bodies. However, the second enzyme in the sequence for degradation shown in figure 18.8, $\beta$-oxoacid-CoA transferase, is present in all tissues but liver. Hence, ketone bodies are made in the liver and metabolized to $CO_2$ and energy in nonhepatic (nonliver) tissues. Ketone bodies can also be used to supply these tissues with acetyl-CoA for fatty acid and cholesterol biosynthesis. Notably, ketone bodies are important sources of energy for the brain during starvation. Normally, glucose is the major source of energy in the brain, and the brain does not use fatty acids as a major source of energy.

**Figure 18.9**

Reactions catalyzed by acetyl-CoA carboxylase. In *E. coli*, BCCP and the two enzymatic activities (biotin carboxylase and carboxyltransferase) can be separated from each other. In contrast, in the liver all three components exist on a single multifunctional polypeptide.

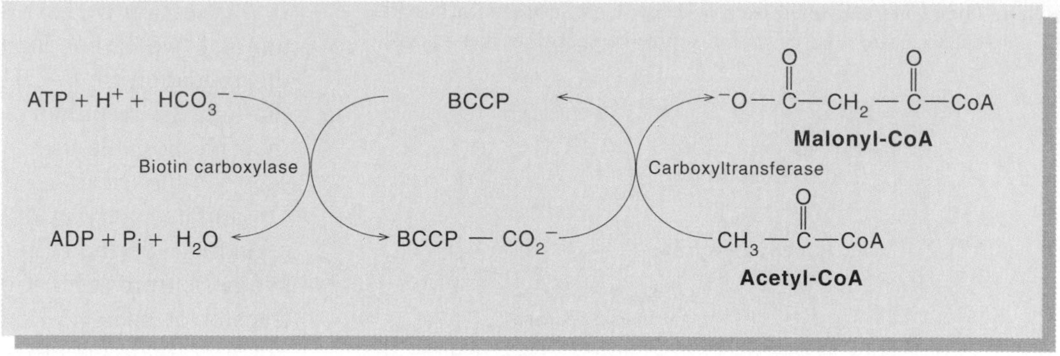

**Figure 18.10**

Structure of $N'$-carboxybiotin linked to biotin carboxyl carrier protein (BCCP). BCCP is one of the components of acetyl-CoA carboxylase isolated from *E. coli*.

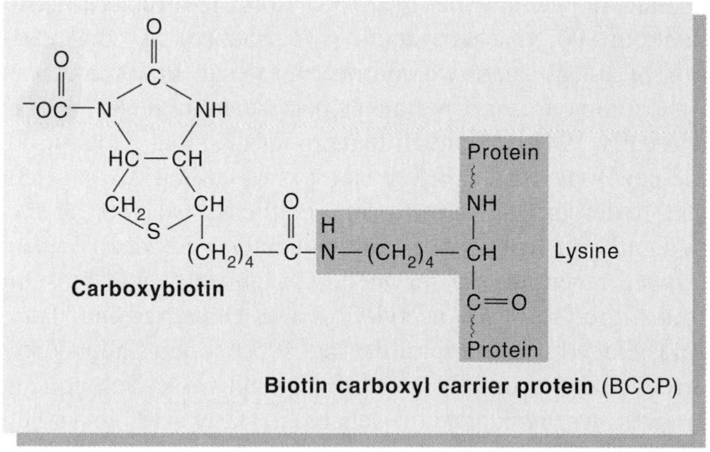

Comparing this equation with the equation for the complete oxidation of palmitoyl-CoA (see table 18.1, equation 1), we find major differences in carriers and intermediates. The principal electron carrier in the anabolic pathway is the NADPH–NADP$^+$ system; in the catabolic pathway, $\beta$ oxidation, the principal electron carriers are FAD–FADH$_2$ and NAD$^+$–NADH. The second striking difference between the two pathways is that malonyl-CoA is the principal substrate in the anabolic pathway but plays no role in the catabolic pathway. These differences reflect the fact that the two pathways do not share common enzymes. Indeed, in animal cells the reactions occur in separate cell compartments; biosynthesis takes place in the cytosol, whereas catabolism occurs in the mitochondria.

## The First Step in Fatty Acid Synthesis Is Catalyzed by Acetyl-CoA Carboxylase

Eight enzyme-catalyzed reactions are involved in the conversion of acetyl-CoA into fatty acids. The first reaction is catalyzed by acetyl-CoA carboxylase and requires ATP. This is the reaction that supplies the energy that drives the biosynthesis of fatty acids. The properties of acetyl-CoA carboxylase are similar to those of pyruvate carboxylase, which is important in the gluconeogenesis pathway (see chapter 12). Both enzymes contain the coenzyme biotin covalently linked to a lysine residue of the protein via its $\epsilon$-amino group. In the last section of this chapter we show that the activity of acetyl-CoA carboxylase plays an important role in the control of fatty acid biosynthesis in animals. Regulation of the first enzyme in a biosynthetic pathway is a strategy widely used in metabolism.

Acetyl-CoA carboxylase of *E. coli* is a multienzyme complex that consists of three protein components that can be isolated individually: Biotin carboxyl carrier protein (BCCP), biotin carboxylase, and carboxyltransferase (fig. 18.9). The reaction sequence involves an initial carboxylation of BCCP, catalyzed by biotin carboxylase. The CO$_2$ is covalently linked to one of the nitrogen atoms of biotin (fig. 18.10). Subsequently, the CO$_2$ is transferred from BCCP to acetyl-CoA in a reaction catalyzed by carboxyltransferase, which yields malonyl-CoA.

A distinctly different form of acetyl-CoA carboxylase is found in the cytosol of animal tissues. The rat liver enzyme is a dimer composed of two identical subunits ($M_r$ of each $= 265,000$) with one biotin per subunit. In contrast to the multienzyme complex in *E. coli*, the three functional parts of acetyl-CoA carboxylase in rat liver occur in a single multifunctional polypeptide. The enzyme, as a dimer, has very low activity. However, in the presence of citrate the enzyme oligomerizes to an active form, a polymer with a

**Figure 18.11**

Activation of acetyl-CoA carboxylase from rat liver *in vitro*. Citrate activates the enzyme and converts it to a polymer. Either palmitoyl-CoA or malonyl-CoA can reverse this process. Citrate does not activate or polymerize the *E. coli* enzyme.

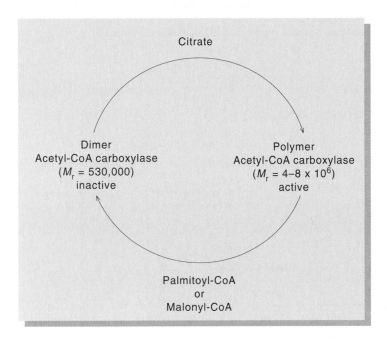

molecular weight between 4 and 8 million (fig. 18.11). The use of citrate to activate acetyl-CoA carboxylase makes good biochemical sense. When there is an abundance of carbohydrate, production of citrate in the mitochondrion occurs in excess. This citrate is transported to the cytosol. Acetyl-CoA carboxylase becomes deactivated and depolymerized when incubated with malonyl-CoA or palmitoyl-CoA. The significance of the concentration of palmitoyl-CoA on the rate of fatty acid biosynthesis is discussed when we look at the regulation of fatty acid metabolism later in the chapter. Incidentally, citrate does not activate or polymerize the *E. coli* enzyme.

## Seven Reactions Are Catalyzed by the Fatty Acid Synthase

After malonyl-CoA synthesis, the remaining steps in fatty acid synthesis occur on fatty acid synthase, which exists as a multienzyme complex. In the initial reactions acetyl-CoA and malonyl-CoA are transferred onto the protein complex by acetyl-CoA transacylase and malonyl-CoA transacylase (step 1 and step 2 in fig. 18.12*a*). The acceptor for the acetyl and malonyl groups is acyl carrier protein (ACP). ACP also carries all of the intermediates during fatty acid biosynthesis. The prosthetic group that binds these intermediates is

the coenzyme phosphopantetheine that is bound to ACP by an ester linkage between the phosphate of the coenzyme and a serine hydroxyl side-chain of the protein (fig. 18.13). The sulfhydryl group of the phosphopantetheine is the attachment site for the intermediates during fatty acid synthesis. Note that CoA also contains this phosphopantetheine moiety which also utilizes its sulfhydryl group to bind fatty acyl residues.

***The Condensation Reaction.*** In the condensation reaction the acetyl group is initially transferred from ACP on to a SH group of 3-ketoacyl-ACP synthase. This acetyl moiety then reacts with malonyl-ACP (step 3 in fig. 18.12*a*) so that the acetyl component becomes the methyl terminal two carbon unit of the acetoacetyl-ACP. The release of $CO_2$ in this condensation reaction provides the extra thermodynamic push to make the reaction highly favorable.

***The Reduction Reactions.*** The object of the next three reactions (steps 4 to 6 in fig. 18.12*a*) is to reduce the 3-carbonyl group to a methylene group. The carbonyl is first reduced to a hydroxyl by 3-ketoacyl-ACP reductase. Next, the hydroxyl is removed by a dehydration reaction catalyzed by 3-hydroxyacyl-ACP dehydrase with the formation of a *trans* double bond. This double bond is reduced by NADPH catalyzed by 2,3-*trans*-enoyl-ACP reductase. Chemically, these reactions are nearly the same as the reverse of three steps in the β-oxidation pathway except that the hydroxyl group is in the D-configuration for fatty acid synthesis and in the L-configuration for β oxidation (compare figs. 18.4*a* and 18.12*a*). Also remember that different cofactors, enzymes and cellular compartments are used in the reactions of fatty acid biosynthesis and degradation.

***Continuation Reactions.*** At this point we have seen one full round of reactions catalyzed by the fatty acid synthase. Each enzyme activity of the complex that we have discussed has been used precisely once. The resulting acyl group on ACP (formed during step 6, fig. 18.12*a*) is transferred to the SH group of the active site of 3-ketoacyl-ACP synthase (the condensing enzyme activity) as was the acetyl group in the first cycle. The acyl group then reacts with another molecule of malonyl-ACP catalyzed by 3-ketoacyl-ACP synthase to yield a 6 carbon intermediate with a ketone on the 3 carbon (fig. 18.12*b*). The reduction reactions (steps 4–6 in fig. 18.12*a*) occur again and a 6 carbon intermediate, hexanoyl-ACP, is formed (fig. 18.12*b*). The hexanoyl group is transferred to the 3-ketoacyl-ACP synthase and condenses with another malonyl-ACP followed by the 3 reduction reactions (fig. 18.12*b*). The biosynthetic process continues to recycle until palmitoyl-ACP is made. At this point, in ani-

## Figure 18.12

Outline of the reactions for fatty acid biosynthesis. Fatty acids grow in steps of two-carbon units and take place on a multienzyme complex. (*a*) The initial reactions of fatty acid biosynthesis are shown. In the first reaction, acetyl-CoA reacts with ACP (acyl carrier protein) to form acetyl-ACP (step 1). ACP is shown with its SH group emphasized (see fig. 18.13) to remind readers that the acyl derivatives are linked to ACP via a thioester bond. Malonyl-CoA, derived from the carboxylation of acetyl-CoA (see fig. 18.9), reacts

with ACP to yield malonyl-ACP (step 2). These two ACP derivatives then condense to form 3-ketoacyl-ACP with the release of ACP and $CO_2$ (step 3). Step 4 involves the reduction of the 3-keto group with NADPH and step 5 the dehydration of 3-hydroxyacyl-ACP to form 2,3-*trans*-enoyl-ACP. The final reaction in the cycle is reduction of the double bond with NADPH (step 6) to give an acyl-ACP with 4 carbons. Screens permit tracking of different molecular groupings.

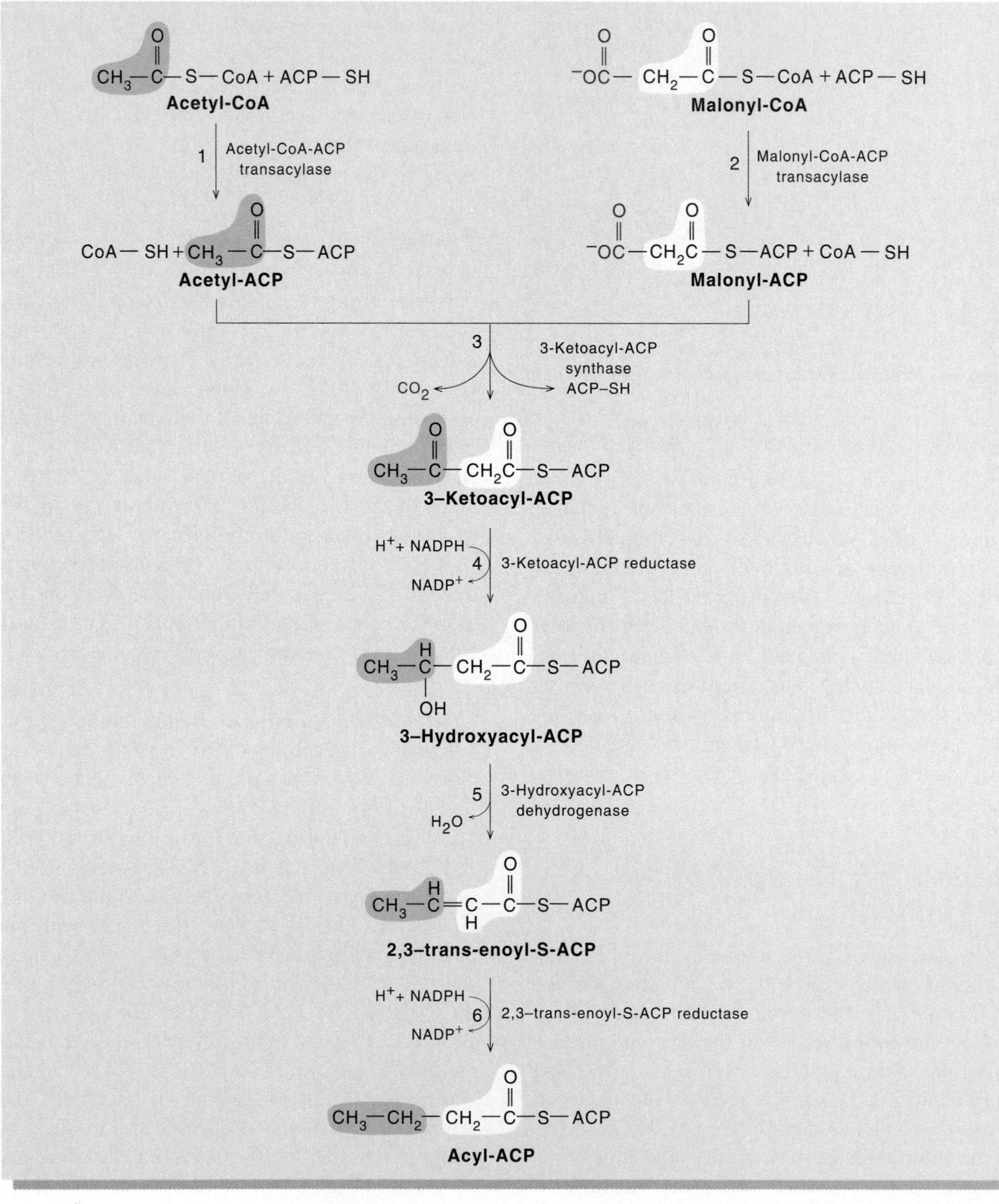

(b) The initial and continuation reactions of fatty acid synthesis are depicted in this part of the figure. The acyl-ACP generated from the first cycle of fatty acid synthesis is condensed with another molecule of malonyl-ACP to yield the 3-keto, 6 carbon intermediate which is reduced, dehydrated and reduced to give an acyl-ACP with 6 carbons. The cycle is then repeated 5 times with the final yield of an acyl-ACP that contains 16 carbons (palmitoyl-ACP).

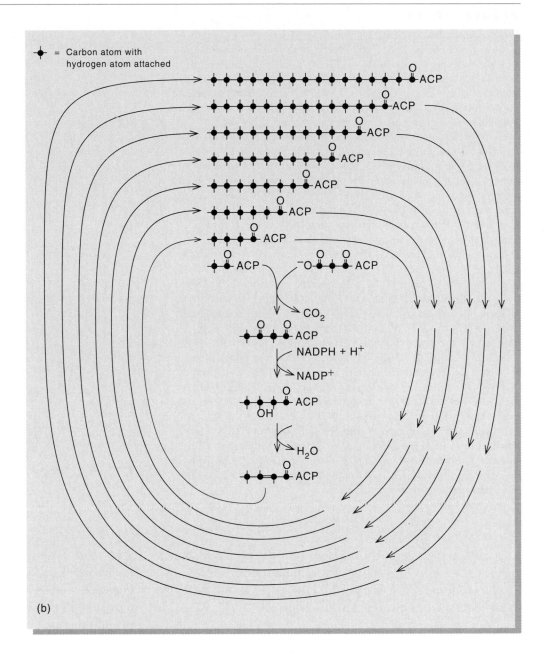

(b)

## Figure 18.13

The phosphopantetheine group in acyl carrier protein (ACP) and in CoA.

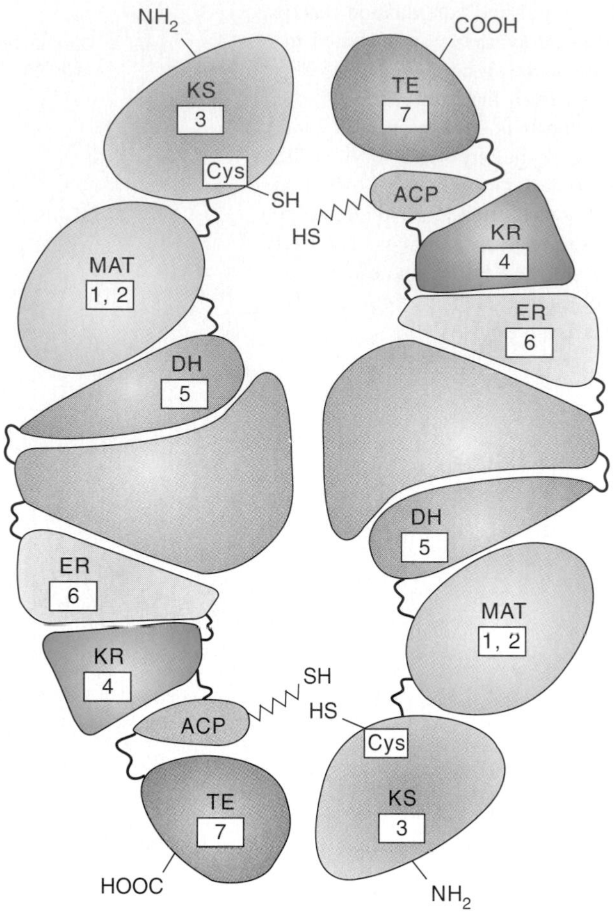

**Figure 18.14**

Proposed organization of the enzymatic activities of fatty acid synthase from animal liver. Fatty acid synthase exists as a dimer of two giant identical peptides ($M_r$ = 272,000). Each subunit has one copy of acyl carrier protein (ACP) and each of the enzyme activities involved in fatty acid synthesis is covalently linked. The two peptides are organized in a head-to-tail configuration in such a way that it is possible to make two fatty acid molecules at the same time.

Fatty acid synthesis begins when the substrates, acetyl-CoA and malonyl-CoA, are transferred onto the protein by malonyl-CoA: acetyl-CoA-ACP transacylase (MAT, steps 1 and 2 in fig. 18.12a). The numbers in parentheses below the abbreviation of the enzyme in this figure refer to the reactions shown in fig. 18.12. (Whereas *E. coli* has separate enzymes that catalyze the transfer of acetyl- and malonyl-CoA to ACP, both reactions are catalyzed by the same enzymatic activity (MAT) on the animal fatty acid synthase.) Subsequently, $\beta$-ketobutyryl-ACP and $CO_2$ are formed in a condensation reaction catalyzed by $\beta$-ketoacyl-ACP synthase (KS, step 3 in fig. 18.12a). The active site cysteine of KS is represented in the figure by Cys-SH. The $\beta$-ketobutyryl-ACP is reduced to the $\beta$-hydroxy derivative by $\beta$-ketoacyl-ACP reductase (KR, step 4 in fig. 18.12a), dehydrated to enoyl-ACP by $\beta$-hydroxylacyl-ACP dehydrase (DH, step 5 in fig. 18.12a) and reduced to butyryl-ACP by enoyl-ACP reductase (ER, step 6 in fig. 18.12a). The butyryl group is then transferred to the cysteine residue of KS and another malonyl group is transferred to ACP by MAT. Another condensation occurs (step 3 in fig. 18.12a) and the $\beta$-ketohexanoyl group is converted to hexanoyl-ACP by the enzymes KR, DH and ER (steps 4, 5, 6 in fig. 18.12a). This enzymatic process continues to recycle until palmitoyl-ACP is made. The palmitate is released from the complex by a thioesterase (TE, reaction 7 in fig. 18.14). (From A. K. Joshi and S. Smith, Mutagenesis of the dehydrase domain of fatty acid synthase, *J. Biol. Chem.* 268:22508–22513, 1993. Reprinted by permission.)

mal cells, palmitate is hydrolyzed from the phosphopantetheine of ACP by the activity of a thioesterase whereas in *E. coli* palmitoyl-ACP is used directly for phospholipid biosynthesis as will be discussed in chapter 19.

## The Organization of the Fatty Acid Synthase Is Different in E. coli and Animals

In *E. coli* and plants the enzymes of fatty acid synthase are believed to occur as a multienzyme complex composed of the individual enzyme activities and ACP grouped together as an aggregate. These proteins are not covalently linked and can be isolated from each other. In contrast the enzyme activities and ACP of the fatty acid synthase of animals are covalently linked in a giant multifunctional polypeptide with a molecular weight of 272,000 (2,505 amino acids). The active form of the enzyme consists of a dimer of this peptide as illustrated in figure 18.14. The legend to figure 18.14 explains how this multifunctional protein makes fatty

acids. The evolution of a multifunctional protein facilitates the channeling of the intermediates of fatty acid synthesis. For each step in the pathway, the next enzyme is always near at hand, and the dilution of intermediates is minimized. In addition, the expression of the seven enzymatic activities of fatty acid biosynthesis is coordinated because only one gene product, as opposed to seven, is required.

## Biosynthesis of Monounsaturated Fatty Acids Follows Distinct Routes in E. coli and Animal Cells

As mentioned in chapter 17, unsaturated fatty acids are abundant in all living organisms. Alternative mechanisms for the biosynthesis of unsaturated fatty acids have evolved. Two chemically distinct pathways exist for the introduction of a *cis* double bond into saturated fatty acids: The anaerobic pathway as typified in *E. coli*, and the aerobic pathway found in eukaryotes.

## Figure 18.15

Anaerobic pathway for biosynthesis of monounsaturated fatty acids in *E. coli*. Synthesis of monounsaturated fatty acids follows the pathway described previously for saturated fatty acids until the intermediate β-hydroxydecanoyl-ACP is reached. At this point an apparent competition arises between the enzymes involved in saturated and unsaturated fatty acid synthesis.

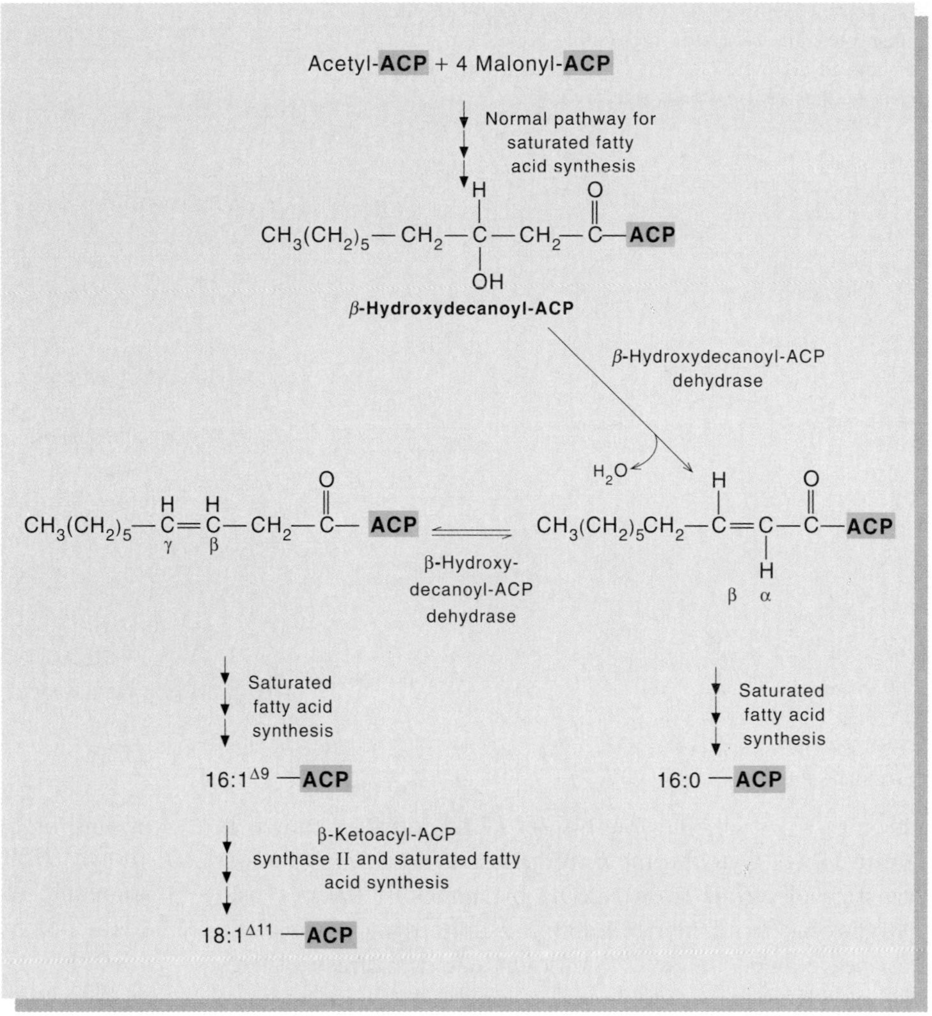

As the name "anaerobic" implies, the double bond of the fatty acid is inserted in the absence of oxygen. Biosynthesis of monounsaturated fatty acids follows the pathway described previously for saturated fatty acids until the intermediate β-hydroxydecanoyl-ACP is reached (fig. 18.15). At this point, a new enzyme, β-hydroxydecanoyl-ACP dehydrase, becomes involved. This dehydrase can form the α-β *trans* double bond, and saturated fatty acid synthesis can occur as previously discussed. In addition, this dehydrase is capable of isomerization of the double bond to a *cis* β-γ double bond as shown in figure 18.15. The β-γ unsaturated fatty acyl-ACP is subsequently elongated by the normal enzymes of fatty acid synthesis to yield palmitoleoyl-ACP ($16:1^{\Delta 9}$). The conversion of this compound to the major unsaturated fatty acid of *E. coli*, *cis*-vaccenic acid ($18:1^{\Delta 11}$), requires a condensing enzyme that we have not previously discussed, β-ketoacyl-ACP synthase II, which shows a preference for palmitoleoyl-ACP as a substrate. The subsequent conversion to vaccenyl-ACP is cata-

lyzed by the usual enzymes of saturated fatty acid biosynthesis.

In contrast to the anaerobic pathway found in *E. coli*, the aerobic pathway in eukaryotic cells introduces double bonds after the saturated fatty acid has been synthesized. Stearoyl-CoA (18:0) is the major substrate for desaturation. Stearic acid is made by the fatty acid synthase as a minor product, the major product being palmitic acid, and is activated to its CoA derivative by acyl-CoA synthase. In eukaryotic cells an enzyme complex associated with the endoplasmic reticulum desaturates stearoyl-CoA to oleoyl-CoA ($18:1^{\Delta 9}$). This remarkable reaction requires NADH and $O_2$ and results in the formation of a double bond in the middle of an acyl chain with no activating groups nearby. The chemical mechanism for desaturation of long-chain acyl-CoAs remains unclear.

Desaturation requires the cooperative action of two enzymes: Cytochrome $b_5$ reductase and stearoyl-CoA desaturase, in addition to an electron carrier protein, cyto-

## Figure 18.16

The aerobic pathway for formation of oleoyl-CoA in eukaryotes. In eukaryotes the double bonds are introduced after the $C_{16}$ and $C_{18}$ saturated fatty acid have been synthesized. F stands for flavin.

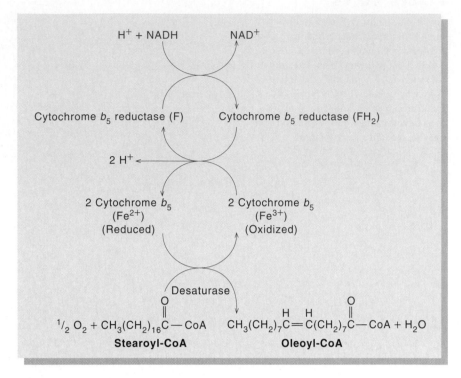

chrome $b_5$. A scheme for this set of reactions is shown in figure 18.16. Cytochrome $b_5$ reductase is a flavoprotein that transfers electrons from NADH by means of flavin (F) to cytochrome $b_5$, a heme-containing protein in which $Fe^{3+}$ can be reduced to $Fe^{2+}$. Stearoyl-CoA desaturase utilizes two electrons from cytochrome $b_5$ coupled with an atom of oxygen to form a *cis* double bond in the $\Delta^9$ position of stearoyl-CoA.

## Biosynthesis of Polyunsaturated Fatty Acids Occurs Mainly in Eukaryotes

*E. coli* does not have polyunsaturated fatty acids, whereas eukaryotes produce a large variety of polyunsaturated fatty acids. Mammals can only desaturate between the $\Delta^9$ position and the carboxyl end of an acyl chain, whereas plants have the enzymes to desaturate at positions $\Delta^9$, $\Delta^{12}$, and $\Delta^{15}$. Mammals need linoleic acid ($18:2^{\Delta9,12}$) and linolenic acid ($18:3^{\Delta9,12,15}$) and are unable to make them. Hence, these are considered essential fatty acids in our diet. One derivative of linoleic acid, arachidonic acid ($20:4^{\Delta5,8,11,14}$), is the biosynthetic precursor of the eicosanoids, which are discussed in chapter 19. A derivative of linolenic acid that has six double bonds, docosahexaenoic acid ($22:6^{\Delta4,7,11,14,17,20}$), is found in abundance in the phospholipids of the retinal membrane of the eye. The function of this highly unsaturated fatty acid in the retina is unknown. Both arachidonic acid and docosahexaenoic acid are made in the liver from the respective essential fatty acid precursors, linoleic ($18:20^{\Delta9,12}$) and linolenic ($18:20^{\Delta9,12,15}$) acids.

Enzyme complexes occur in the endoplasmic reticulum of animal cells that desaturate at $\Delta^5$ if there is a double bond at the $\Delta^8$ position, or at $\Delta^6$ if there is a double bond at the $\Delta^9$ position. These enzymes are different from each other and from the $\Delta^9$-desaturase discussed in the previous section, but the $\Delta^5$ and $\Delta^6$ desaturases do appear to utilize the same cytochrome $b_5$ reductase and cytochrome $b_5$ mentioned previously. Also present in the endoplasmic reticulum are enzymes that elongate saturated and unsaturated fatty acids by two carbons. As in the biosynthesis of palmitic acid, the fatty acid elongation system uses malonyl-CoA as a donor of the two-carbon unit. A combination of the desaturation and elongation enzymes allows for the biosynthesis of arachidonic acid and docosahexaenoic acid in the mammalian liver. As an example, the pathway by which linoleic acid is converted to arachidonic acid is shown in figure 18.17. Interestingly, cats are unable to synthesize arachidonic acid from linoleic acid. This may be why cats are carnivores and depend on other animals to make arachidonic acid for them. Also note that the elongation system in the endoplasmic reticulum is important for the conversion of palmitoyl-CoA to stearoyl-CoA.

## Figure 18.17

Synthesis in mammalian tissues of arachidonic acid from linoleic acid. The $\Delta^5$ and $\Delta^6$ desaturases are separate enzymes and are also different from the $\Delta^9$ desaturase (fig. 18.16). The mechanisms, however, seem to be the same, involving cytochrome $b_5$ and cytochrome $b_5$ reductase. The enzymes for elongation of unsaturated fatty acid such as 18:3 to 20:3 occur on the endoplasmic reticulum.

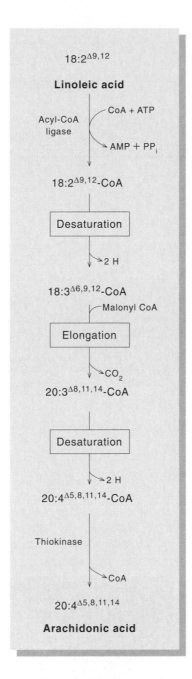

$18:2^{\Delta 9,12}$

**Linoleic acid**

Acyl-CoA ligase — CoA + ATP

→ AMP + PP$_i$

$18:2^{\Delta 9,12}$-CoA

Desaturation

↓ 2 H

$18:3^{\Delta 6,9,12}$-CoA

Malonyl CoA

Elongation

↓ CO$_2$

$20:3^{\Delta 8,11,14}$-CoA

Desaturation

↓ 2 H

$20:4^{\Delta 5,8,11,14}$-CoA

Thiokinase

↓ CoA

$20:4^{\Delta 5,8,11,14}$

**Arachidonic acid**

## Summary of Pathways for Synthesis and Degradation

Before discussing the specific aspects of regulation of fatty acid metabolism, let us review the main steps in fatty acid synthesis and degradation. Figure 18.18 illustrates these processes in a way that emphasizes the parallels and differences. In both cases, two-carbon units are involved. However, different enzymes and coenzymes are utilized in the biosynthetic and degradative processes. Moreover, the processes take place in different compartments of the cell. The differences in the location of the two processes and in the

enzymes used make it possible to regulate the two pathways independently.

# Regulation of Fatty Acid Metabolism

In light of the principles of pathway regulation discussed in chapter 11, you can anticipate that controls must exist to ensure that fatty acid synthesis and breakdown do not occur at the same time. When an organism has satisfied its immediate energy needs, and the limited storage space for glycogen has been filled, most nutrients are directed toward fatty acid synthesis. A virtually unlimited amount of fat can be stored in an animal's body. The last section of this chapter considers the mechanisms employed by animals to regulate the utilization of fatty acids as an energy source.

## The Release of Fatty Acids from Adipose Tissue Is Regulated

There are a number of circumstances in which animals call on their energy reserve in adipose (fat) tissue. For example, during fasting or starvation, mobilization of fatty acids from adipose tissue is an important source of energy. Prolonged work or exercise also promotes the release of fatty acids from fat tissue. The fatty acids are stored in adipose tissue as components of triacylglycerols. Although triacylglycerols are found in liver, intestine, and other tissues, they are primarily found in adipose (fat) tissue, which functions as the main storage depot of triacylglycerols. In a 70-kg human, 135,000 kcal of energy is stored as triacylglycerols in adipose compared with 450 kcal stored in the liver. The specialized cell in adipose tissue for lipid storage is the adipocyte (fig. 18.19). The cytoplasm of the adipocyte is packed with vesicles that are rich in triacylglycerols and that serve as the long-term energy reserves in mammals. When the energy supply from the diet becomes inadequate, the animal responds to the deficiency by release of a hormonal signal, especially epinephrine and glucagon, that is transmitted to the target tissue. These hormones work on the adipocytes similarly to the way they work on cells containing glycogen storage granules. First, the hormones bind to the plasma membrane of the target cells, which stimulates the synthesis of cyclic AMP (cAMP). As shown in figure 18.20 (also see fig. 24.15), the cAMP activates a protein kinase that phosphorylates a key enzyme, triacylglycerol lipase. The lipase, which is active in the phosphorylated form, hydrolyzes triacylglycerol to diacylglycerol with release of a fatty acid. This reaction is rate-limiting for the complete hydrolysis of triacylglycerols. The diacylglycerols and monoacylglycerols are rapidly hydrolyzed the rest of the way to fatty acids and glycerol.

## Figure 18.18

Comparison between synthesis and degradation of fatty acids in the liver. Both processes involve two carbons at a time and very similar intermediates, even though they go in opposite directions. CoA is also heavily involved in both processes. Here the similarities end. The enzymes used in the two processes are totally different, and the coenzymes are also different. In the degradative direction FAD and $NAD^+$ are used, whereas in the synthetic direction the coenzyme NADPH is used. Degradation occurs in the mitochondrial matrix, and synthesis occurs in the cytosol.

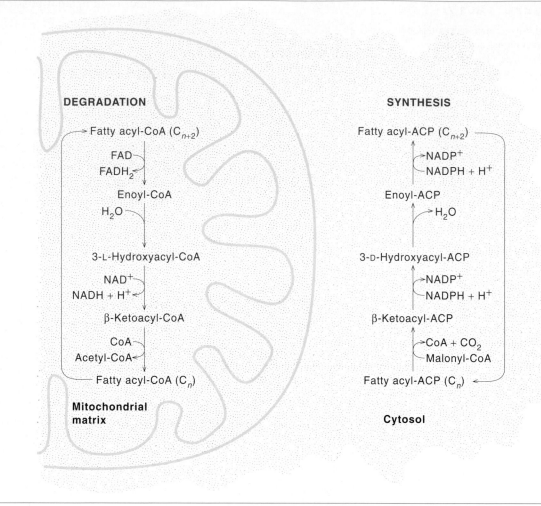

**DEGRADATION**

Fatty acyl-CoA ($C_{n+2}$)

FAD — FADH$_2$

Enoyl-CoA

H$_2$O

3-L-Hydroxyacyl-CoA

NAD$^+$ — NADH + H$^+$

β-Ketoacyl-CoA

CoA — Acetyl-CoA

Fatty acyl-CoA ($C_n$)

**Mitochondrial matrix**

**SYNTHESIS**

Fatty acyl-ACP ($C_{n+2}$)

NADP$^+$ — NADPH + H$^+$

Enoyl-ACP

H$_2$O

3-D-Hydroxyacyl-ACP

NADP$^+$ — NADPH + H$^+$

β-Ketoacyl-ACP

CoA + CO$_2$ — Malonyl-CoA

Fatty acyl-ACP ($C_n$)

**Cytosol**

## Figure 18.19

Scanning electron micrograph of white adipocytes from rat adipose tissue ($600\times$). (Courtesy of Dr. A. Angel, University of Manitoba, and Dr. M. J. Hollenberg of the University of Toronto.)

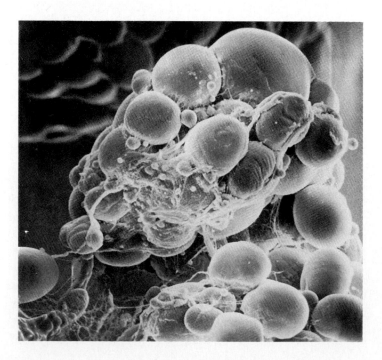

**Figure 18.20**

When certain hormones (e.g., epinephrine) bind to their receptors in adipose tissue, adenylate cyclase is activated. The cAMP that is formed activates protein kinase A, which phosphorylates triacylglycerol lipase. The phosphorylated form of this enzyme is the active species, and triacylglycerols are degraded to fatty acids. The fatty acids are released into the bloodstream, bound by albumin, and delivered to energy-deprived tissues.

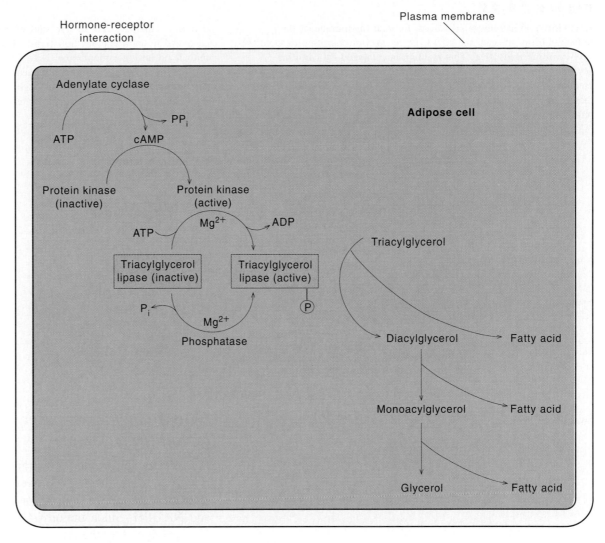

The unesterfied fatty acids move through the plasma membranes of the adipocytes into the bloodstream, where they become bound to the major plasma protein, albumin. The water-soluble product, glycerol, is also released into plasma and removed by the liver for glucose production. Albumin carries the fatty acids to energy-deficient tissues, where fatty acids move from the plasma into the tissues. Cardiac muscle utilizes fatty acids as the major oxidative source of energy for ATP synthesis and, therefore, removes large amounts of fatty acids from the circulation. At the other extreme, the brain does not use fatty acids as a major source of energy but depends mostly on glucose and, to a lesser extent, ketone bodies.

## Transport of Fatty Acids into Mitochondria Is Regulated

The level of fatty acids in the bloodstream is one factor that dictates how much fatty acid is delivered to a tissue. Further controls are necessary in the cell to regulate the utilization of the fatty acid and this depends on the needs of a particular cell type. The heart requires a great deal of energy, so the fatty acids delivered to this organ are largely directed to oxidation, whereas the liver may require fatty acids primarily for membrane lipid and triacylglycerol synthesis. A major control point, therefore, is entry of the fatty acids into the mitochondrion where degradation occurs.

Fatty acids taken into cells are first activated in the cytosol by reaction with CoA and ATP to yield fatty acyl-CoA in a reaction catalyzed by acyl-CoA ligase:

$$RCOO^- + ATP + CoA \xrightarrow{Mg^{2+}}$$

$$RCOCoA + PP_i + AMP \quad (3)$$

The acyl-CoAs can penetrate the outer membrane of mitochondria through a pore but cannot be transported into mitochondria where $\beta$ oxidation occurs. They must first be converted to their carnitine derivatives, which can be transported across the inner membrane of the mitochondria.

## Figure 18.21

Acyl-CoA is not transported across the inner membrane of the mitochondrion. Instead, the acyl-CoA reacts with carnitine to yield the acyl-carnitine derivative. This reaction is catalyzed by carnitine acyltransferase I, which is located on the outer mitochondrial membrane. The acyl-carnitine is transported across the inner membrane by a specific carrier protein. Once inside the matrix of the mitochondrion, the acyl-carnitine is converted back to its acyl-CoA

derivative, the substrate for the start of $\beta$ oxidation. This reaction is catalyzed by carnitine acyltransferase II, which is located on the mitochondrial inner membrane. Note that acyltransferases I and II are oriented in their respective membranes so that the reactions they catalyze occur in intermembrane space and the mitochondrial matrix, respectively. The carnitine is also transferred by the carrier protein.

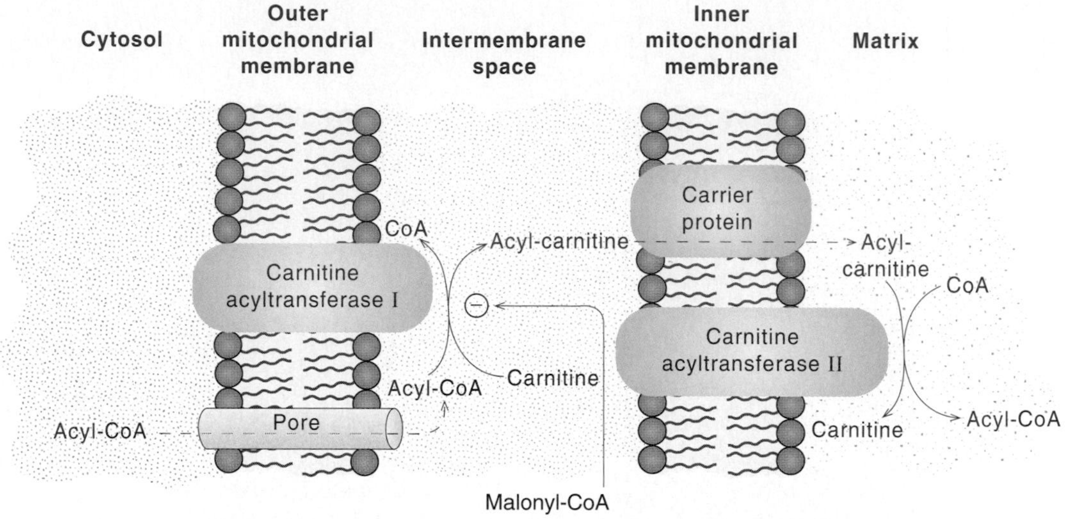

$$RCOCoA + (CH_3)_3N^+\!\!-\!\!CH_2CHCH_2COO^- \rightleftharpoons$$
$$\underset{\displaystyle OH}{\big|}$$

Carnitine

$$(CH_3)_3N^+\!\!-\!\!CH_2CHCH_2COO^- + CoA \quad (4)$$
$$\underset{\displaystyle \underset{\displaystyle RC=O}{\overset{\displaystyle |}{O}}}{\big|}$$

Acyl-carnitine

This reaction is catalyzed by carnitine acyltransferase I on the outer membrane (fig. 18.21). A protein carrier in the inner mitochondrial membrane transfers the acyl-carnitine derivatives across the membrane. Once inside the mitochondria, the reaction is reversed by carnitine acyltransferase II to yield a fatty acyl-CoA (see fig. 18.21). Thus, at least two distinct pools of acyl-CoA occur in the cell, one in the cytosol and the other in the mitochondrion.

This elaborate chain of reactions provides a number of possible points for regulation of the supply of acyl-CoAs for oxidation. The major control point is carnitine acyltransferase I, which is strongly inhibited by malonyl-CoA. Recall that malonyl-CoA is a major substrate for fatty acid biosynthesis. Thus, high levels of malonyl-CoA, which indicate fatty acid synthesis is in progress, prevent fatty acid catabolism, thus avoiding a futile cycle.

## Fatty Acid Biosynthesis Is Limited by Substrate Supply

Just as fatty acid oxidation is limited by substrate supply, so is fatty acid synthesis. The overall equation for fatty acid biosynthesis indicates a need for acetyl-CoA, malonyl-CoA, and NADPH. Since malonyl-CoA is derived from acetyl-CoA catalyzed by the acetyl-CoA carboxylase reaction, we can think of acetyl-CoA as the main substrate for fatty acid synthesis. Acetyl-CoA and NADPH are generated from substrates that originate in the glycolytic pathway and TCA cycle. Thus, glycolysis generates pyruvate, and in the mitochondria the pyruvate is converted to acetyl-CoA as well as oxaloacetate (fig. 18.22). The mitochondrial membrane is impermeable to acetyl-CoA, which is, therefore, converted to citrate and transported into the cytosol. There citrate is converted to acetyl-CoA and oxaloacetate by ATP:citrate lyase. The end result is that an intermediate in the glycolytic degradation of glucose, pyruvate, is converted to acetyl-CoA in the cytosol and can be used for fatty acid biosynthesis.

## Fatty Acid Synthesis Is Regulated by the First Step in the Pathway

An abundance of acetyl-CoA, which would be produced after a carbohydrate-rich meal, creates the possibility for fatty acid synthesis. Since acetyl-CoA can be used in a vari-

## Figure 18.22

Overview of the conversion of carbohydrate to lipid in rat liver cells and its regulation. Colored arrows with pluses and minuses indicate points of activation and inhibition, respectively. Glucagon and epinephrine are the main hormones involved in regulation. Citrate and palmitoyl-CoA are the main substrates involved in regulation. Dashed black arrow indicates compound that passes across inner mitochondrial membrane.

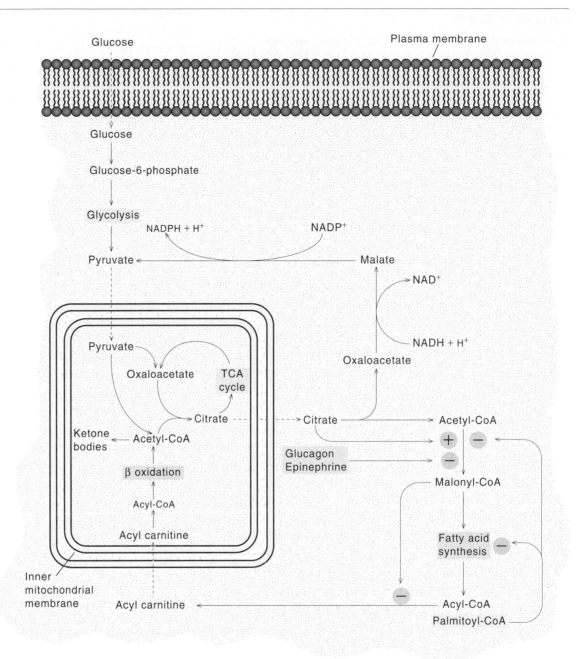

ety of ways (e.g., ketone body synthesis and, as we examine in chapter 20, cholesterol biosynthesis), it is essential that specific controls exist to regulate the synthesis of fatty acids. In fact, multiple types of control have been discovered for the initial step in fatty acid biosynthesis catalyzed by acetyl-CoA carboxylase.

We already learned that citrate activates the liver acetyl-CoA carboxylase and converts it to a high-molecular-weight polymer (see fig. 18.11). The activation by citrate is appropriate because excess cytosolic citrate is a good indication that the energy needs of the cell are satisfied. Furthermore, citrate is the most important source of cytosolic acetyl-CoA as described above and in figure 18.22.

The acetyl-CoA carboxylase is inhibited by palmitoyl-CoA (see figs. 18.11 and 18.22). This control is a classic example of end-product inhibition since palmitoyl-CoA is the final product in fatty acid synthesis that is initiated by acetyl-CoA carboxylase.

Acetyl-CoA carboxylase is also regulated by a chain of reactions promoted by the hormones glucagon and epinephrine. Remember that high levels of these hormones serve as a general signal that energy is in short supply or soon may be needed in large amounts (see chapters 12 and 24). Under these conditions it would be inappropriate to divert energy to the synthesis of fatty acids. These hormones bind to receptors on the plasma membrane and stim-

**Figure 18.23**

Regulation of acetyl-CoA carboxylase by phosphorylation and dephosphorylation. Glucagon is known to activate cAMP-dependent protein kinase; this kinase phosphorylates both serine 77 and serine 1200 of rat acetyl-CoA carboxylase, which inactivates the enzyme. However, there is also an AMP-dependent kinase that phosphorylates serine 79 and serine 1200 and inactivates the rat acetyl-CoA carboxylase. The relative importance of these two kinases in regulating the carboxylase *in vivo* is still unclear. Likewise, the phosphorylated enzyme is a substrate for several different protein phosphate phosphatases, and the physiologically relevant phosphatases are not known. Epinephrine may inhibit the carboxylase via a $Ca^{2+}$-dependent protein kinase.

The dephosphorylated form of the carboxylase does not require citrate for activity, but the phosphorylated form of the enzyme can be activated by citrate *in vitro*. This reaction is reminiscent of the effect of glucose-6-phosphate on glycogen synthase. The active, dephosphorylated form of glycogen synthase has only a small requirement for glucose-6-phosphate, whereas high concentrations of this activator are required to activate the phosphorylated form of glycogen synthase.

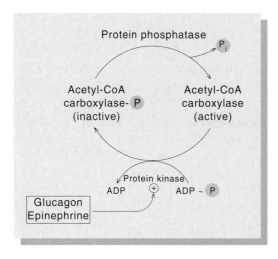

ulate a kinase that phosphorylates and inactivates acetyl-CoA carboxylase (fig. 18.23). Insulin, which often opposes the action of glucagon, has the opposite effect on the carboxylase, possibly by activation of a phosphatase.

## The Controls for Fatty Acid Metabolism Discourage Simultaneous Synthesis and Breakdown

The controls for fatty acid metabolism satisfy most of our expectations for what control systems should do. One of the most important aspects of control is that simultaneous syn-

thesis and breakdown should be discouraged, otherwise energy would be wasted. Two major hormones are implicated in fatty acid breakdown, glucagon and epinephrine. They encourage mobilization of fatty acids from adipose tissue and simultaneously discourage fatty acid synthesis by inhibiting the formation of malonyl-CoA from acetyl-CoA (see fig. 18.22). Hence, the main effect of glucagon and epinephrine is to stimulate breakdown while inhibiting synthesis.

Similarly, factors that stimulate acetyl-CoA carboxylase, the first enzyme in the pathway for fatty acid synthesis, also discourage fatty acid catabolism. This dual effect occurs because the first enzyme in the pathway leads to the formation of malonyl-CoA, which is a potent inhibitor of carnitine acyltransferase I. This inhibition prevents the transport of fatty acids into the mitochondrion, thereby, preventing fatty acid breakdown.

Another point to note about the controls of fatty acid metabolism is that they are designed to meet the metabolic needs of the organism. Thus, an excess of palmitoyl-CoA in the cytosol signals a shutdown of fatty acid synthesis by inhibiting acetyl-CoA carboxylase. Similarly, an abundance of citrate indicates that the energy needs of the organism are being met, and that it is an appropriate time to synthesize fatty acids for energy storage. Citrate accomplishes this by supplying the substrate for fatty acid synthesis, acetyl-CoA, and activates the first enzyme in the pathway, acetyl-CoA carboxylase.

## Long-Term Dietary Changes Lead to Adjustments in the Levels of Enzymes

Before closing we should point out that, over an extended period, dietary conditions can alter the levels of enzymes involved in fatty acid metabolism. For example, the concentrations of fatty acid synthase and acetyl-CoA carboxylase in rat liver are reduced four- to fivefold after fasting. When a rat is fed a fat-free diet, the concentration of fatty acid synthase is 14-fold higher than in a rat maintained on standard rat chow diet. Current evidence indicates that the levels of these enzymes are governed by the rate of enzyme synthesis, not degradation. It appears that synthesis of the enzyme, in turn, is controlled by the rate of transcription of DNA into mRNA. A question of current interest is how this transcription of DNA is regulated.

## Summary

In this chapter we focused on the synthesis and degradation of long-chain fatty acids and on how these processes are regulated. Most of our discussion was concerned with how these reactions take place in the mammalian liver, although occasionally we referred to other animal tissues and to *E. coli*. The following points are the most important.

1. Fatty acids originate from three sources: diet, adipocytes, and *de novo* synthesis.

2. The degradation of fatty acids occurs by an oxidation process in the mitochondria. The breakdown of the 16-carbon saturated fatty acid, palmitate, occurs in blocks of two carbon atoms by a cyclical process. The active substrate is the acyl-CoA derivative of the fatty acid. Each cycle involves four discrete enzymatic steps. In the process of oxidation the energy is sequestered in the form of reduced coenzymes of FAD and $NAD^+$. These reduced coenzymes lead to ATP production through the respiratory chain. The oxidation of fatty acids yields more energy per carbon than the oxidation of glucose because saturated fatty acids are in the fully reduced state.

3. Unsaturated fatty acids are also oxidized in mitochondria with the help of certain additional enzymes that facilitate a continuous flow of the oxidation process.

4. The main end product of fatty acid oxidation is acetyl-CoA, which can be used by the tricarboxylic acid cycle for generation of energy. Alternatively, ketone bodies may be formed from condensation of acetyl-CoAs. Ketone bodies are made in the liver and subsequently diffuse into the blood to be carried to other tissues, where they are converted into acetyl-CoA for various metabolic purposes.

5. Fatty acid biosynthesis also occurs in steps of two carbon atoms. Biosynthesis takes place in the cytosol. In addition to occurring in a different cellular compartment from degradation, biosynthesis involves totally different enzymes and different coenzymes.

6. Fatty acid synthesis takes place in eight steps. All except the first step take place on a multienzyme complex. The intermediates on this complex are carried by attachment of the acid group in thioester linkage to phosphopantetheine of the acyl carrier protein (ACP). The multienzyme complex greatly increases the efficiency of fatty acid synthesis, because for each step in the pathway the next enzyme is always near at hand, and the dilution of intermediates is minimized.

7. Regulation of fatty acid metabolism takes place in such a way that simultaneous synthesis and degradation are minimized. Control factors ensure that synthesis occurs primarily when energy is in excess, and degradation when energy is needed. When energy is abundant, the synthesis of malonyl-CoA by acetyl-CoA carboxylase stimulates fatty acid synthesis and inhibits carnitine acyltransferase I and, as a result, $\beta$ oxidation. The hormones epinephrine and glucagon stimulate degradation and inhibit synthesis. They stimulate degradation by facilitating release of fatty acids stored in the adipocytes. They inhibit synthesis by inactivating the first enzyme in the pathway for synthesis, acetyl-CoA carboxylase. Most other control factors interfere with the supply of substrate for either of the two processes.

## Selected Readings

Cook, H. W., Fatty acid denaturation and chain elongation in eucaryotes. In D. E. Vance, and J. E. Vance (eds.), *Biochemistry of Lipids, Lipoproteins and Membranes*. Amsterdam: Elsevier Science Publishers, 1991. Provides an advanced and current treatment of fatty acid desaturation and its regulation, and cites other key references to this field.

Deuel, H. J., *The Lipids, Biochemistry,* vol. 3. New York: Interscience, 1957. A comprehensive and classical treatise on the biochemistry of lipids until the mid-1950s.

Goodridge, A. G., Fatty acid synthesis in eucaryotes. In D. E. Vance, and J. E. Vance (eds.), *Biochemistry of Lipids, Lipoproteins and Membranes*. Amsterdam: Elsevier Science Publishers, 1991. Provides an advanced treatment of the regulation of fatty acid synthesis and cites other key references related to this topic.

Jackowski, S., J. E. Cronan, and C. O. Rock, Lipid metabolism in procaryotes. In D. E. Vance, and J. E. Vance (eds.), *Biochemistry of Lipids, Lipoproteins and Membranes.*

Amsterdam: Elsevier Science Publishers, 1991. Contains current and advanced information on the metabolism of fatty acids in *E. coli* and other prokaryotes.

McGarry, J. D., and D. W. Foster, Regulation of hepatic fatty acid oxidation and ketone body production. *Ann. Rev. Biochem.* 49:395, 1980. A now classic review article that summarizes the evidence for the regulation of $\beta$ oxidation by malonyl-CoA.

Schulz, H., Oxidation of fatty acids. In D. E. Vance, and J. E. Vance (eds.), *Biochemistry of Lipids, Lipoproteins and Membranes*. Amsterdam: Elsevier Science Publishers, 1991. Provides an advanced and current summary of fatty acid oxidation in prokaryotes and eukaryotes.

Wakil, S. J., Fatty acid synthase, a proficient multifunctional enzyme. *Biochemistry* 28:4523, 1989. Reviews the evidence for the current model of the mammalian fatty acid synthase as depicted in figure 18.14.

## Problems

1. The consensus is that fats containing unsaturated fatty acids are "better for you" than the corresponding saturated forms. Can this statement be explained by the ATP yield that results on complete oxidation (which in turn reflects the caloric content)? Calculate the number of ATPs produced for the complete oxidation of arachidic ($C_{20:0}$) and arachidonic ($C_{20:4}$) acids to assess any differences in energy value of saturated versus polyunsaturated fatty acids.

2. Examine the chemistry of the reactions presented in figure 18.4a. Where else in metabolism have you seen a similar sequence of chemical events? What feature(s) is identical, what different?

3. (a) Calculate the number of moles of ATP produced during the catabolism of a mole of glucose.

   (b) Calculate the number of moles of ATP produced during the catabolism of a mole of decanoic acid [$CH_3(CH_2)_8CO_2H$]. Although this is not a typical fatty acid, it was selected because it has a molecular mass comparable to that of glucose.

   (c) If the heat of combustion of glucose is $-669.9$ kcal/mole and decanoic acid is $-1452$ kcal/mole, and if $-7.3$ kcal are preserved per mole of ATP, what fraction (%) of the energy available is preserved in each case?

   (d) Consider glucose and decanoic acid to represent a typical carbohydrate and fat, respectively. A typical carbohydrate contains 4 kcal/g, and fats contain 9 kcal/g. Calculate the ratio of the nutritional calories for fats to carbohydrates and compare this number with the ratio of the ATPs produced for fats to carbohydrates. Are the numbers consistent?

4. Order the following substances from the least to most oxidized carbon: Formaldehyde, carbon dioxide, methane, formic acid, methanol. Use the open-chain form of glucose and the structural formula for decanoic acid and this scheme to determine the "aver-

age" oxidation state of the carbons in glucose and decanoic acid. How does this oxidation state compare with the data obtained in problems 3a and 3b above? Can you explain your observation?

5. (a) Consider the complete oxidation of glucose ($M_r = 180$) via glycolysis and the TCA cycle, and calculate the moles ATP generated during the oxidation of 1 mole of glucose to $CO_2$ and $H_2O$. Assume that the free energy of ATP hydrolysis under physiological conditions is $-11$ kcal/mole, and assume mitochondrial P/O ratios of 2.5 for NADH oxidation and 1.5 for succinate (or equivalent) oxidation. Estimate the free energy conserved as ATP during the oxidation. Calculate the free energy conserved per gram of glucose oxidized.

   (b) Repeat the calculations for ATP formation, but consider the complete oxidation of 1 mole of palmitic acid ($M_r = 256$) via $\beta$ oxidation and the TCA cycle. Estimate the free energy conserved as ATP energy from palmitate oxidation. Calculate the free energy conserved per gram of palmitic acid oxidized.

   (c) Bearing in mind that respiratory metabolism is an oxidative process, how do you explain the differences in energy content of fat and carbohydrate on a weight basis?

   (d) Explain the rationale behind the use of fat rather than carbohydrate as an energy reserve in plants and animals.

6. Explain the role of carnitine acyltransferases in fatty acid oxidation.

7. Late-night TV ads are often quite educational. One advertisement touted the merits of eating grapefruit to lose weight, but taking them in your lunch presents problems. At that point an individual in a lab coat standing in front of a blackboard covered with molecular formulas proceeds to explain how it is the

citric acid in the grapefruit that burns up the fat and for only $XX you can get a month's supply of their pills. Use what you know about the citric acid cycle and fatty acid metabolism to deduce if added citric acid causes weight loss.

8. After working through problem 7 and having read the chapter, you want to cash in on a get-rich-quick diet scam. You start thinking, if only the body could be tricked into converting some of its fatty acids into acetyl-CoA and then resynthesize fatty acids, there would be a weight loss. If you could actually make this work, would there be a weight loss?

9. Is $\beta$ oxidation (fig. 18.4) best described as a spiral or cyclic process? Why?

10. Acetoacetate (fig. 18.7) is shown to give rise to acetone by a spontaneous reaction. Can you explain how this might occur?

11. $\beta$-Oxoacyl-CoA transferase (see fig. 18.8) is involved in the transfer of a CoASH from succinyl-CoA to acetoacetate to produce succinate and acetoacetyl-CoA. A cursory examination of this reaction suggests a simple transfer of the CoA moiety. However, it is soon realized that the loss of an oxygen by the acetoacetate and the gain of an oxygen by the succinyl group present a dilemma. Produce a rational mechanism that explains the preservation of the thioester energy and solves this dilemma. *Hint:* Consider a succinyl phosphate intermediate.

12. Carnitine deficiency in liver is correlated with hypoglycemia. Suggest a plausible explanation for hypoglycemia in the carnitine-deficient human.

13. (a) Liver mitochondria convert long-chain fatty acids to ketone bodies (acetoacetate and $\beta$-hydroxybutyrate) that are subsequently transported in the plasma to nonhepatic tissues. Suggest some metabolic advantages of supplying ketone bodies to nonhepatic tissues.

    (b) In what way is $\beta$-hydroxybutyrate a better energy source than acetoacetate for nonhepatic tissues?

    (c) Outline the oxidation of $\beta$-hydroxybutyrate to acetyl-CoA in heart mitochondria.

14. Predict the effect on oxidation of ketone bodies and of glucose in nonhepatic tissue of individuals with markedly diminished $\beta$-oxyacid-CoA-transferase activity. Predict the effect if the activity was absent.

15. (a) For an *in vitro* synthesis of fatty acids with purified fatty acid synthase, the acetyl-CoA was supplied as the $^{14}$C-labeled derivative

$$^{14}CH_3-\overset{\overset{\displaystyle O}{\|}}{C}-S-CoA$$

The other reactants, including the malonyl-CoA, were not radioactive. Where is the $^{14}$C-label found in palmitic acid?

   (b) If the malonyl-CoA was supplied as the only labeled compound deuterated as shown in the structure below, how many deuterium atoms would be incorporated in palmitate? On which carbon(s) would these deuterium atoms reside?

$$^-O-\overset{\overset{\displaystyle O}{\|}}{C}-CD_2-\overset{\overset{\displaystyle O}{\|}}{C}-S-CoA$$

   (c) If [3-$^{14}$C]malonyl-CoA (shown below) was used in the reaction, which atoms in palmitate would be labeled? Why?

$$^-O-\overset{\overset{\displaystyle O}{\|}}{^{14}C}-CH_2-\overset{\overset{\displaystyle O}{\|}}{C}-S-CoA$$

16. Except for malonyl-CoA formation, all the individual reactions for palmitate synthesis reside on a single multifunctional protein (fatty acid synthase) in animal cells. It has been shown that a dimer of the multifunctional protein is required to catalyze palmitate synthesis. Explain the molecular basis of this observation.

17. What are the metabolic sources of NADPH used in fatty acid biosynthesis? How many moles of NADPH are required for the synthesis of 1 mole of palmitic acid from acetyl-CoA?

18. Citrate is both a lipogenic substrate and a regulatory molecule in mammalian fatty acid synthesis.

    (a) Explain each function of citrate in fatty acid synthesis.

    (b) Write reactions (including structures) outlining the role of citrate as a lipogenic substrate.

19. Which catalytic activity of the mammalian fatty acid synthase determines the chain length of the fatty acid product?

20. (a) Why is the location of biosynthesis and $\beta$ oxidation of fatty acids in separate metabolic compartments essential to regulation of fatty acid metabolism in the hepatocyte?

    (b) Would you expect an inhibitor of the extramitochondrial carnitine acyltransferase to mimic the effect of malonyl-CoA on $\beta$ oxidation? (Assume that the inhibitor can penetrate the cell membrane.) Explain the rationale for your answer.

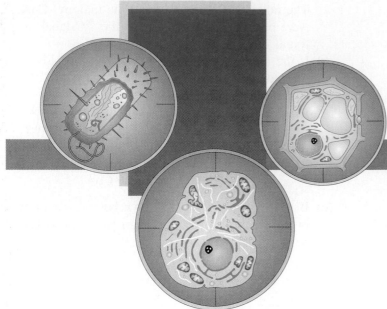

# Biosynthesis of Membrane Lipids

*Fatty acids are linked to glycerol or serine to form complex lipid structures that are used for energy storage, membrane formation, hormones and second messengers.*

Thus far we have been concerned with the metabolism of fatty acids in relationship to storage and release of energy (chapter 18). In this chapter we focus on the metabolism of lipids that serve other roles (fig. 19.1). Most fatty acids that

## Figure 19.1

Outline of pathways for the biosynthesis of major cellular lipids (other than cholesterol) in a mammalian cell. Most of the metabolism of these lipids occurs on membrane surfaces because of the insoluble nature of the substrates and products. These lipids play three major roles: (1) they act as a storehouse of chemical energy, as with triacylglycerols; (2) they are structural components of membranes (boxed compounds); and (3) they act as regulatory compounds (underlined), either as eicosanoids, which act as local hormones, or as phosphorylated inositols and diacylglycerols, which function as second messengers.

Synthesis of most phospholipids starts from glycerol-3-phosphate, which is formed in one step from the central metabolic pathways, and acyl-CoA, which arises in one step from activation of a fatty acid. In two acylation steps the key compound phosphatidic acid is formed. This can be converted to many other lipid compounds as well as CDP-diacylglycerol, which is a key branchpoint intermediate that can be converted to other lipids. Distinct routes to phosphatidylethanolamine and phosphatidylcholine are found in prokaryotes and eukaryotes. The pathway found in eukaryotes starts with transport across the plasma membrane of ethanolamine and/or choline. The modified derivatives of these compounds are directly condensed with diacylglycerol to form the corresponding membrane lipids. Modification of the head-groups or tail-groups on preformed lipids is a common reaction. For example, the ethanolamine of the head-group in phosphatidylethanolamine can be replaced in one step by serine or modified in 3 steps to choline.

Phospholipids containing arachidonic acid serve as the precursor, through liberation of arachidonic acid, of a wide variety of eicosanoids. Similarly, the membrane lipid phosphatidylinositol-4,5-$P_2$ can undergo hydrolysis to inositol-$P_3$ and diacylglycerol. Both of these hydrolysis products are regulatory molecules that serve as second messengers (so called because they are formed in response to hormone binding to the plasma membrane).

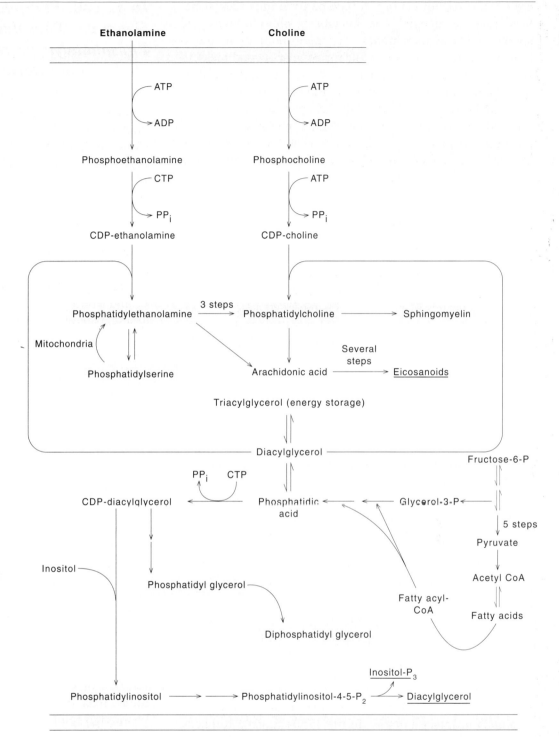

are not utilized for energy storage perform important structural roles as integral components of phospholipids and sphingolipids in membranes. The phospholipids of membranes also serve as a reservoir for cellular second messengers. One fatty acyl component of phospholipids, arachidonic acid, is the precursor of eicosanoids, which trigger a wide range of responses. In this chapter we describe the biosynthesis of the main membrane phospholipids and sphingolipids, and explain how the second messengers that are stored as components of phospholipids are released in response to extracellular stimuli.

# Phospholipids

Phospholipids are ideal compounds for making membranes because of their amphipathic nature (see chapter 17). The polar head-groups of phospholipids prefer an aqueous environment, whereas the nonpolar acyl substituents do not. As a result, phospholipids spontaneously form bilayer structures (see fig. 17.6), which are a dominant feature of most membranes. The phospholipid bilayer is the barrier of the cell membrane that prevents the unrestricted transport of most molecules other than water into the cell. Entry of other molecules is allowed if a specific transport protein is present in the cell membrane. Similarly, the phospholipid bilayer prevents leakage of metabolites from the cell. The amphipathic nature of phospholipids has a great influence on the mode of their biosynthesis. Thus, most of the reactions involved in lipid synthesis occur on the surface of membrane structures catalyzed by enzymes that are themselves amphipathic.

As in other areas of biochemistry, research on phospholipid biosynthesis was initiated only after the structures of the major phospholipids were elucidated. Beginning in 1927 the structure of phosphatidylcholine was confirmed by chemical synthesis. Subsequently in 1932 Charles Best (a codiscoverer of insulin) demonstrated that choline was an essential dietary component. Experiments by Don Zilversmit and Irving Chaikoff in the 1940s with radioactive phosphorus ($^{32}$P) provided insight into the metabolism of phospholipids in intact animals. In the 1950s Eugene Kennedy and co-workers established the role of glycerol-3-phosphate as a precursor of phospholipids and defined the CDP-choline and CDP-ethanolamine pathways. The groundwork laid by these researchers has led to more recent studies that have provided information on the enzymes involved in phospholipid metabolism and the regulation of these pathways.

## In E. coli, *Phospholipid Synthesis Generates Phosphatidylethanolamine, Phosphatidylglycerol, and Diphosphatidylglycerol*

*Escherichia coli* contains three important classes of phospholipids: phosphatidylethanolamine (75%–85%), phosphatidylglycerol (10%–20%), and diphosphatidylglycerol (5%–15%). All three of these phospholipids share the same biosynthetic pathway up to the formation of CDP-diacylglycerol (fig. 19.2), after which the pathways branch (fig. 19.3).

Most of the enzymes for phospholipid synthesis are located on the inner plasma membrane of *E. coli*. Glycerol-3-phosphate acyltransferase, the first enzyme in the pathway, preferentially utilizes saturated fatty acyl derivatives (palmitoyl-CoA or palmitoyl-ACP) for the initial acylation of glycerol-3-phosphate. The second enzyme (see fig. 19.2) 1-acylglycerol-3-phosphate acyltransferase, catalyzes phosphatidic acid formation; this enzyme shows a preference for acyl residues with a double bond. The substrate specificities of these two acyltransferases account for there usually being saturated fatty acids in the SN-1 position and unsaturated fatty acids in the SN-2 position of phospholipids. A third transferase reaction converts phosphatidic acid to CDP-diacylglycerol with CTP as a cosubstrate. Interestingly, CTP is the only high-energy nucleotide used for the synthesis of phospholipids in all organisms. CDP-diacylglycerol is a branchpoint intermediate. It is either converted to phosphatidylserine en route to phosphatidylethanolamine or to phosphatidylglycerol phosphate enroute to phosphatidylglycerol and diphosphatidylglycerol (see fig. 19.3). The hydroxyl group of glycerol-3-phosphate or serine reacts with the high-energy pyrophosphate bond of CDP-diacylglycerol.

Phosphatidylglycerol and diphosphatidylglycerol are also synthesized in the mitochondria of eukaryotes by a pathway similar to that in prokaryotes. The only difference is that diphosphatidylglycerol in eukaryotes is made by the reaction of CDP-diacylglycerol with phosphatidylglycerol rather than the condensation of two molecules of phosphatidylglycerol as occurs in *E. coli*.

## Phospholipid Synthesis in Eukaryotes Is More Complex

Phospholipid synthesis in eukaryotes is more complex than in *E. coli*. This relates to the other roles phospholipids play in membranes aside from their structural role. Eukaryotes

# Figure 19.2

The first phase of phospholipid synthesis in *E. coli* and eukaryotes. Additional routes to and from phosphatidic acid, found predominantly in eukaryotes, are shown in brackets.

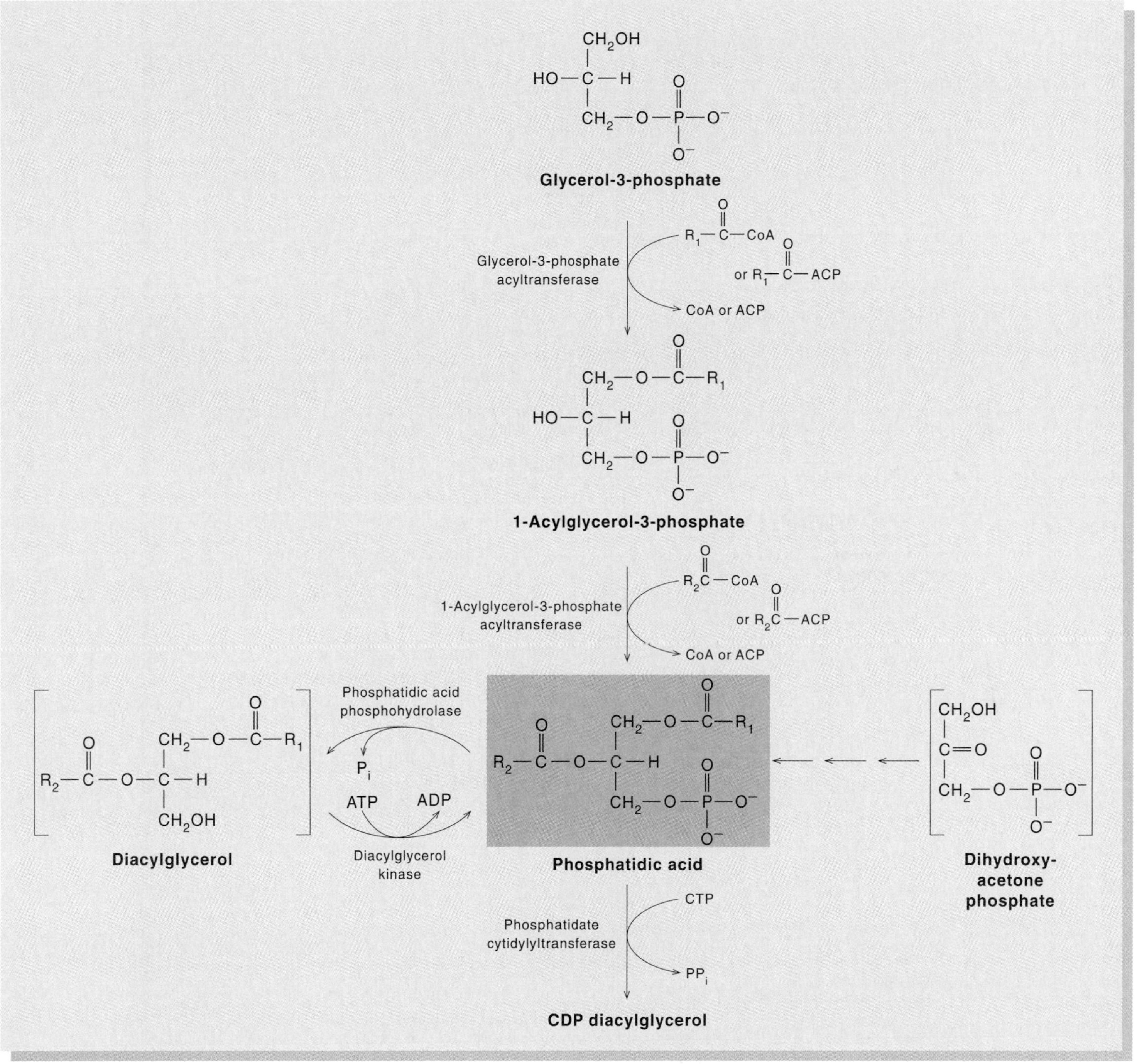

# Figure 19.3

The second phase of phospholipid synthesis in *E. coli,* from CDP-diacylglycerol to the end products.

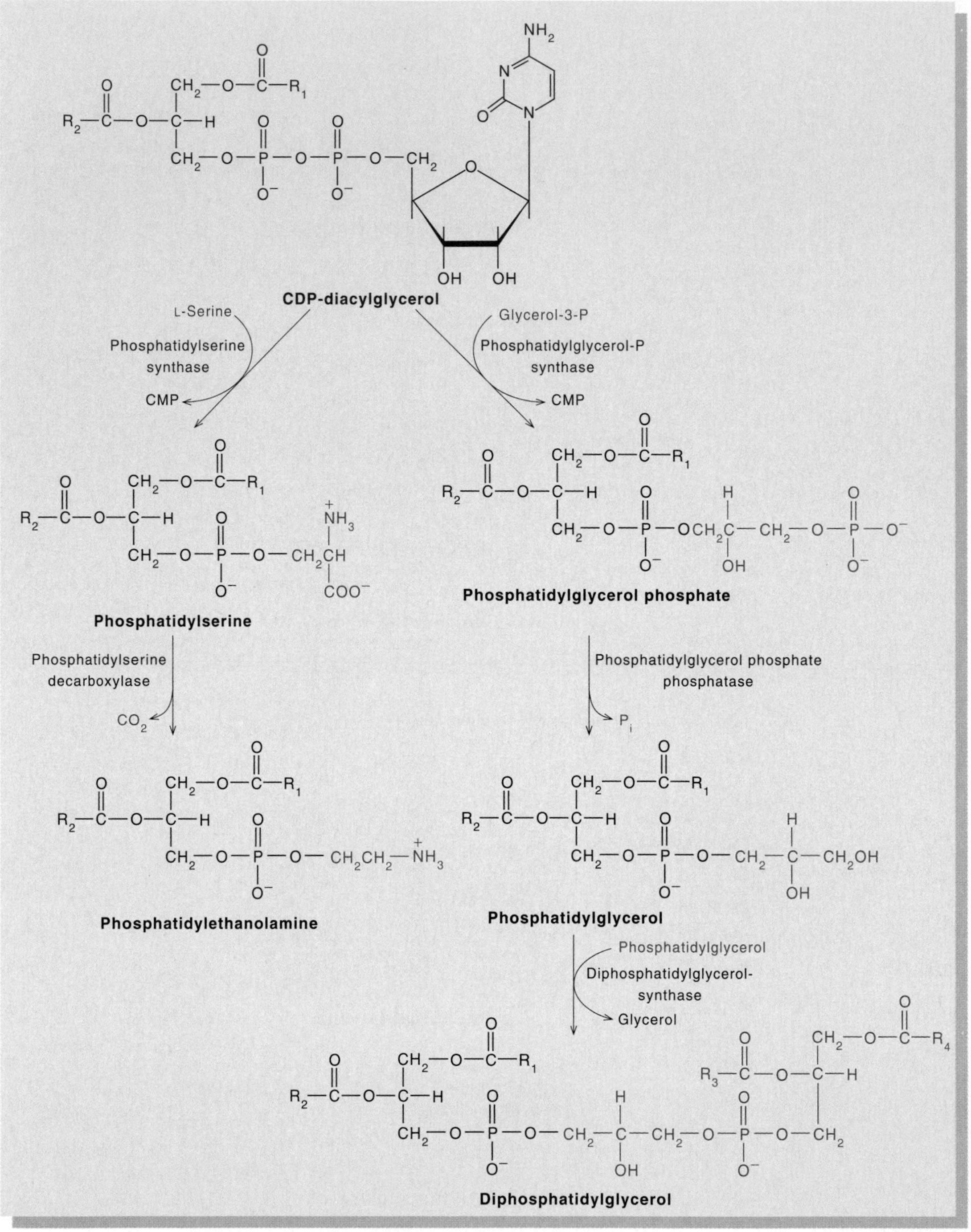

have organelle membranes as well as plasma membranes. In addition, as we will see, the regulation of phospholipid biosynthesis is more complicated than in prokaryotes, partly because a central intermediate in phospholipid biosynthesis, diacylglycerol, is also a precursor of triacylglycerol, the main energy storage lipid.

In the first phase of phospholipid synthesis from glycerol-3-phosphate to phosphatidic acid, the pathways in *E. coli* and eukaryotes are very similar (see fig. 19.2). The major difference is that one additional pathway exists for generation of phosphatidic acid from dihydroxyacetone phosphate, an intermediate in glycolysis. Once phosphatidic acid is made, it is rapidly converted to diacylglycerol or CDP-diacylglycerol (see fig. 19.2) both of which are intermediates for the biosynthesis of eukaryotic phospholipids.

## Diacylglycerol Is the Key Intermediate in the Biosynthesis of Phosphatidylcholine and Phosphatidylethanolamine

Phosphatidylcholine and phosphatidylethanolamine, which are quantitatively the most important phospholipids in eukaryotic cells (see table 17.3), are derived from diacylglycerol as shown in figure 19.4 (also see fig. 19.1). Alternatively, when there is an abundance of fatty acids, diacylglycerol is converted into triacylglycerol, by diacylglycerol acyltransferase (see fig. 19.4). The biosynthesis of phosphatidylcholine begins with choline which is transported into the cell. It is noteworthy that choline is an essential ingredient in the human diet. Once inside the cell, the choline is rapidly phosphorylated to phosphocholine by choline (ethanolamine) kinase (see fig. 19.4). Phosphocholine reacts with CTP to form CDP-choline in a reaction catalyzed by CTP:phosphocholine cytidylyltransferase. The activity of this enzyme is usually rate-limiting for phosphatidylcholine biosynthesis. In the last reaction, diacylglycerol reacts with CDP-choline to yield phosphatidylcholine, catalyzed by CDP-choline:1,2-diacylglycerol phosphocholine-transferase. Similar to the reactions involving CDP-diacylglycerol, the hydroxyl group of diacylglycerol reacts with the pyrophosphate bond in CDP-choline. The biosynthesis of phosphatidylethanolamine proceeds from ethanolamine in a comparable series of reactions (see fig. 19.4).

## Fatty Acid Substituents at sn-1 and sn-2 Positions Are Replaceable

Lung tissue manufactures a specialized species of phosphatidylcholine, dipalmitoyl-phosphatidylcholine, in which palmitic acid is attached to both positions 1 and 2 of the glycerol backbone. This species of phosphatidylcholine is the major component of lung surfactant, which functions to maintain surface tension in the lung alveoli so that they do not collapse when air is expelled. Premature babies do not secrete sufficient dipalmitoyl-phosphatidylcholine-enriched surfactant and, therefore, have major difficulties in breathing. Understanding the biochemistry of lung surfactant phospholipids has led to the development of synthetic surfactant mixtures which can be sprayed into the lung. This treatment decreases the need for respirators prior to maturation of the surfactant biosynthetic machinery of the infant.

The pathway for the synthesis of dipalmitoyl-phosphatidylcholine is illustrated in figure 19.5. The starting species of phosphatidylcholine is made by the CDP-choline pathway (see fig. 19.4). The fatty acid at the sn-2 position, which is usually unsaturated, is hydrolyzed by phospholipase $A_2$, and the lysophosphatidylcholine is reacylated with palmitoyl-CoA. This modification permits alteration of the properties of the phospholipid without resynthesis of the entire molecule, a strategy called remodeling. Deacylation–reacylation of phosphatidylcholine occurs in other tissues and provides an important route for alteration of the fatty acid substituents at both the sn-1 and sn-2 positions. For example, fatty acids at the sn-2 position can be replaced by arachidonic acid, which is stored there until needed for eicosanoid biosynthesis, as we discuss later in this chapter.

## Phosphatidylinositol-4,5-Bisphosphate, a Precursor of Second Messengers, Is Synthesized via CDP-Diacylglycerol

CDP-diacylglycerol is also a precursor of phosphatidylinositol (fig. 19.6), a lipid that is unique to eukaryotes. Phosphatidylinositol accounts for approximately 5% of the lipids present in animal cell membranes (see table 17.3). Also present, at much lower concentrations, are phosphatidylinositol-4-phosphate and phosphatidylino-

# Figure 19.4

The second phase of phospholipid synthesis in eukaryotes. Choline or ethanolamine enters the cell via active transport mechanisms and is immediately phosphorylated by the enzyme, choline (ethanolamine) kinase. The phosphorylated derivatives of choline and ethanolamine are activated to their CDP derivatives by separate enzymes. The last reaction occurs on the endoplasmic reticulum. The diacylglycerol used as a substrate in this reaction may alternatively be converted to storage lipid (triacylglycerol).

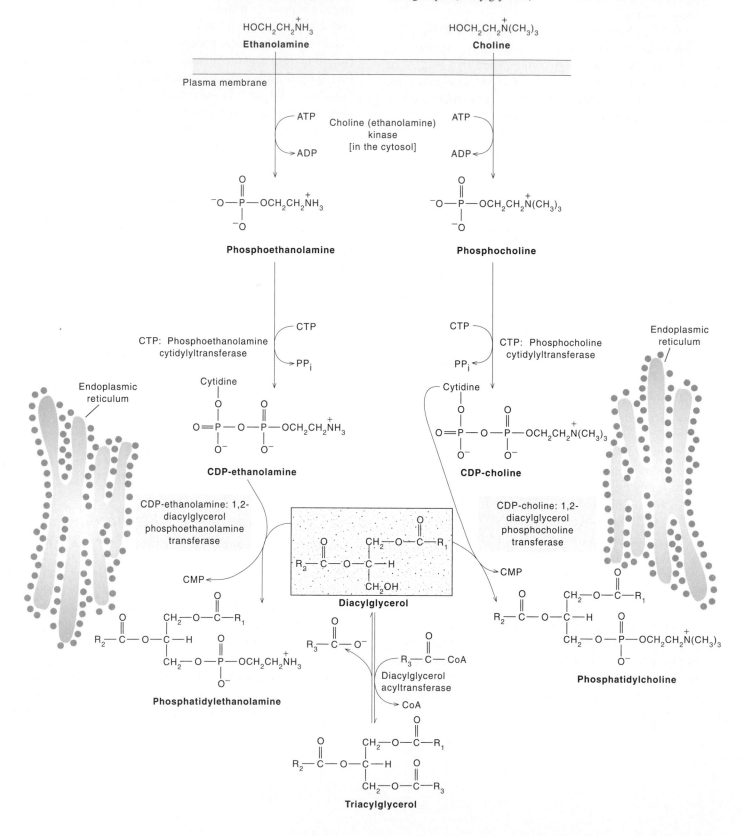

**Figure 19.5**

Biosynthesis of dipalmitoylphosphatidylcholine. $R_2$ is usually an unsaturated fatty acid. Thus this two-step reaction results in the replacement of an unsaturated by a saturated fatty acid at the C-2 position on the glycerol backbone. Dipalmitoylphosphatidylcholine is the major component in lung surfactant, a substance that maintains surface tension in the lung alveoli so that they do not collapse when air is expelled.

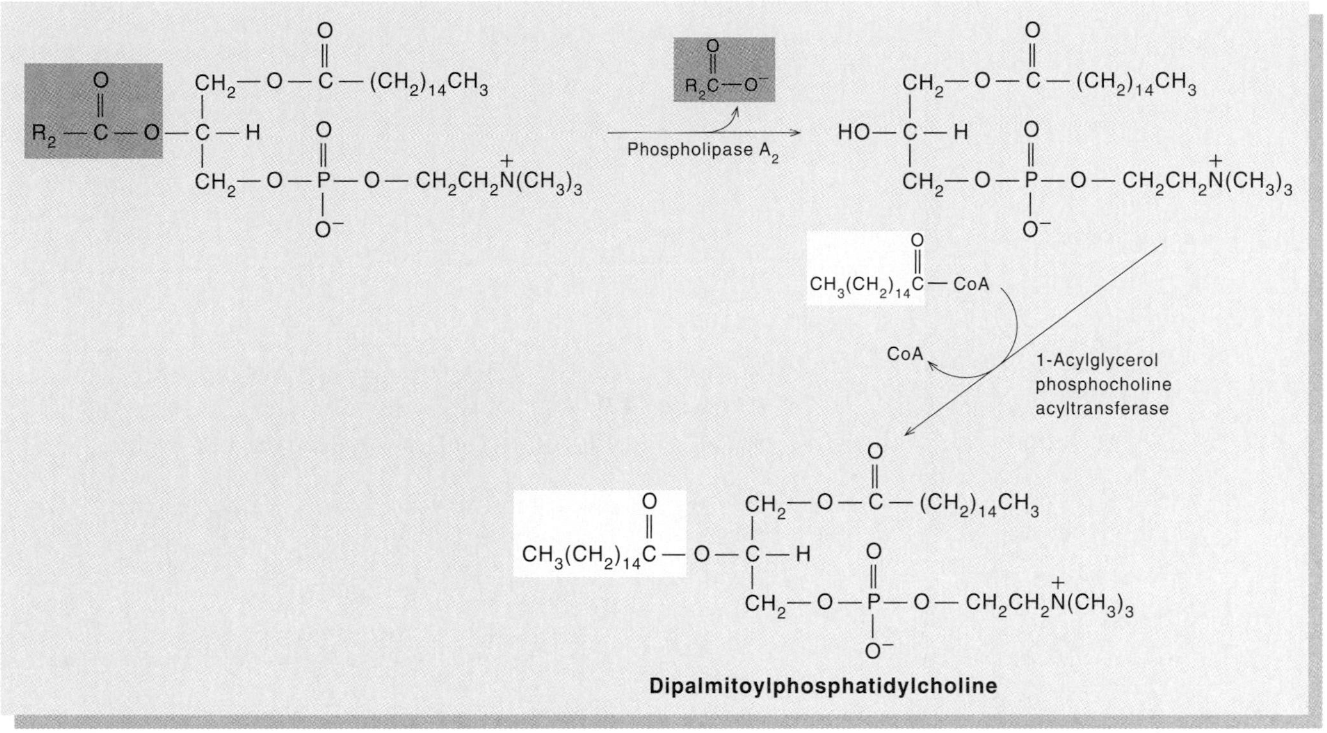

**Dipalmitoylphosphatidylcholine**

sitol-4,5-bisphosphate. These two derivatives are made from phosphatidylinositol by separate kinases that use ATP as the phosphate donor. When certain hormones bind to the cell surface (e.g., vasopressin binds to hepatocytes), the phosphatidylinositol-4,5-bisphosphate, which is localized to the plasma membrane, is attacked by phospholipase C between the glycerol and phosphate moieties. This reaction yields two cellular second messengers: Diacylglycerol and inositol-1,4,5-$P_3$ (fig. 19.7). Each of these compounds has important regulatory functions. The inositol-1,4,5-$P_3$ mobilizes calcium from intracellular stores (endoplasmic reticulum); the rise in cytosolic calcium activates a host of different enzymes (see chapter 24). The rise in diacylglycerol, activates protein kinase C, which plays a major regulatory role as a phosphorylating agent of an extremely diverse group of proteins (see chapter 24 and fig. 24.18).

## The Metabolism of Phosphatidylserine and Phosphatidylethanolamine Is Closely Linked

In prokaryotes, phosphatidylserine is made from CDP-diacylglycerol (see fig. 19.3). The enzyme for this reaction is absent in animal cells, which rely on a base exchange reaction in which serine and ethanolamine are interchanged (fig. 19.8). Although the reaction is reversible, it usually proceeds in the direction of phosphatidylserine synthesis. Phosphatidylserine can be converted back to phosphatidylethanolamine by a decarboxylation reaction in the mitochondria. This may be the preferred route for phosphatidylethanolamine biosynthesis in some animal cells. Furthermore these two reactions (see fig. 19.8) establish a cycle that has the net effect of converting serine into ethanolamine. This is the main route for ethanolamine synthesis

## Figure 19.6

Reaction that converts CDP-diacylglycerol to phosphatidylinositol in eukaryotic cells.

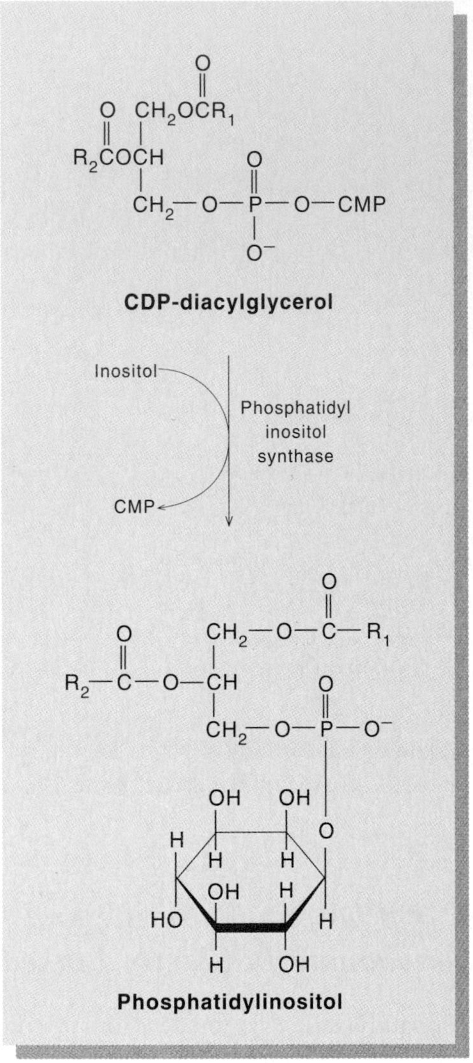

**CDP-diacylglycerol**

**Phosphatidylinositol**

## Figure 19.7

Phospholipase C degradation of phosphatidylinositol-4,5-$P_2$.

**Phosphatidylinositol-4,5-$P_2$**

**Inositol-1,4,5-$P_3$**

**Diacylglycerol**

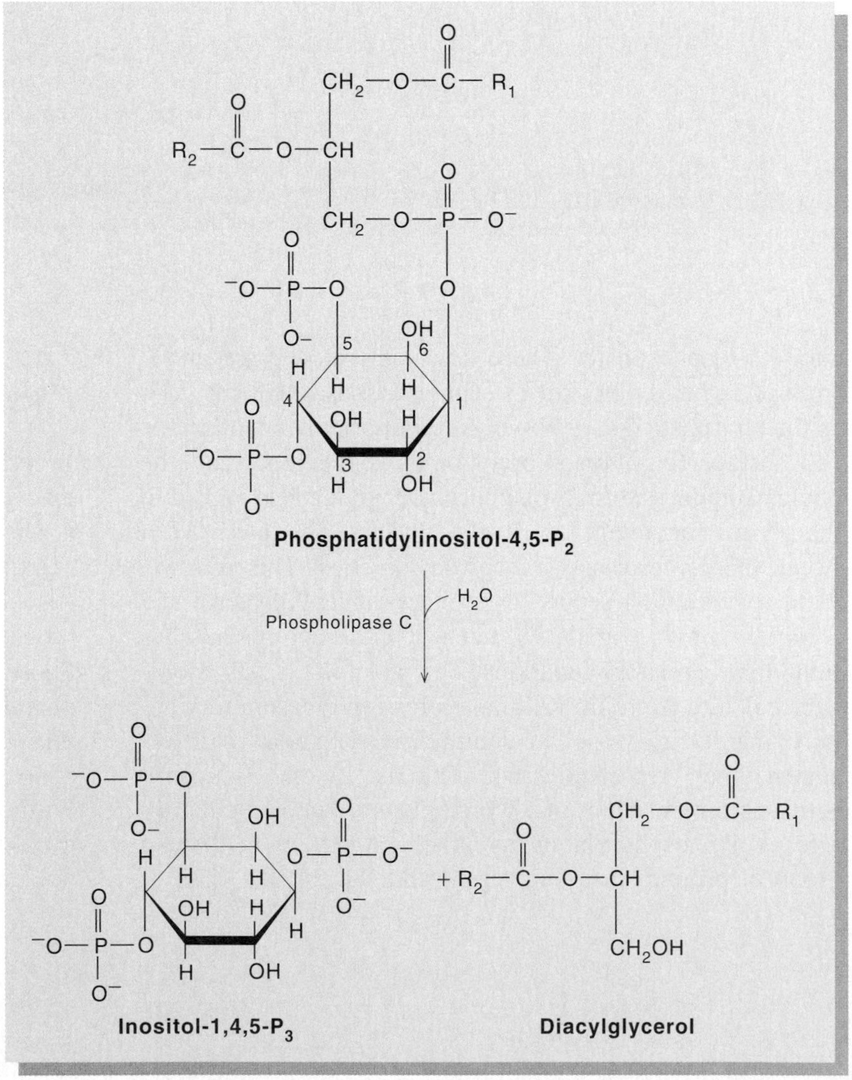

**Figure 19.8**

Phosphatidylserine biosynthesis in animals is catalyzed by a base exchange enzyme on the endoplasmic reticulum. Decarboxylation of phosphatidylserine occurs in mitochondria. The cyclic process of phosphatidylserine formation from phosphatidylethanolamine and the reformation of phosphatidylethanolamine by decarboxylation has the net effect of converting serine to ethanolamine. This is a major mechanism for the synthesis of ethanolamine in many eukaryotes.

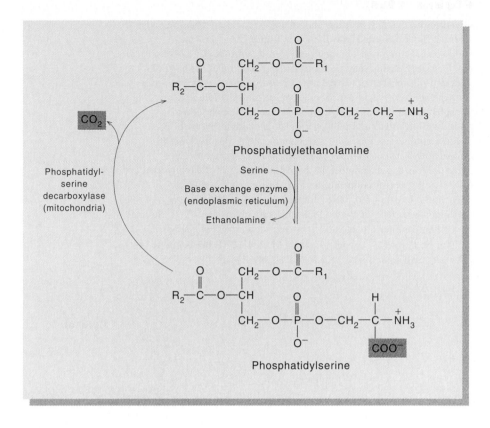

in animal cells. Ethanolamine also enters our bodies as a result of digestion of dietary phosphatidylethanolamine. The primary fate of ethanolamine is for use in the biosynthesis of phosphatidylethanolamine via the CDP-ethanolamine pathway (see fig. 19.4).

## The Final Reactions for Phospholipid Biosynthesis Occur on the Cytosolic Surface of the Endoplasmic Reticulum

The final reactions for the biosynthesis of phosphatidylcholine, phosphatidylethanolamine, phosphatidylserine, and phosphatidylinositol all occur on the cytosolic surface of the endoplasmic reticulum and Golgi apparatus (fig. 19.9). By contrast, phosphatidylglycerol and diphosphatidylglycerol are synthesized on the mitochondrial membrane where they remain for the most part.

Two questions concerning the distribution of lipids in membranes remain unanswered:

1. Because phospholipid synthesis occurs on the cytosolic side of the endoplasmic reticulum and phospholipids are found on both leaflets of the bilayer, how do these lipids reach the inner leaflet of the bilayer? Possibly proteins called flipases catalyze the movement of phospholipids from one side of the bilayer to the other. More research is

required to determine how important these flipases are in movement of lipids between the two leaflets of the bilayer.

2. How does the cell sort and transport phospholipids from the site of synthesis to other membranes in the cell? One view is that phospholipid vesicles that bud from the endoplasmic reticulum are targeted to another membrane where the vesicles fuse with the membrane. Alternatively, phospholipid transfer proteins may be involved. Proteins that transfer phospholipids between membranes *in vitro* have been known for over 25 years but it has not been demonstrated that they function in this way *in vivo*.

## In the Liver, Regulation Gives Priority to Formation of Structural Lipids Over Energy-Storage Lipids

The energy state of the cell dictates the relative rates of phosphatidylcholine, phosphatidylethanolamine, and triacylglycerol biosynthesis. When energy is in short supply, the level of cAMP rises leading to inhibition of fatty acid biosynthesis (see chapter 18). This in turn decreases the supply of diacylglycerol, which limits the synthesis of phosphatidylcholine, phosphatidylethanolamine, and triacylglycerol. When sufficient diacylglycerol is pres-

## Figure 19.9

Glycerolipid synthesis on the endoplasmic reticulum from rat liver.
Fatty acids are inserted into the cytoplasmic surface of the
endoplasmic reticulum (1), and activated to form acyl-CoA thioesters
(2). The acyl chains may be elongated or desaturated or both (3).
Glycerol-phosphate undergoes acyl-CoA-dependent esterification to
form phosphatidic acid (4). The action of phosphatidic acid
phosphatase (5) forms diacylglycerols that are converted to
phosphatidylcholine and phosphatidylethanolamine by acquisition of
phosphocholine and phosphoethanolamine polar head-groups (6).
Phosphatidylserine synthesis occurs by base exchange (7).
Triacylglycerol synthesis occurs by esterification of diacylglycerol (8).
CDP-diacylglycerol is an intermediate in the synthesis of
phosphatidylinositol (9). Once formed, the glycerolipids may move to
the lumenal surface of the endoplasmic reticulum (10). (C = choline;
E = ethanolamine; I = inositol; S = serine; X = polar head-group C,
E, I, or S; P = PO$_4$.) (Source: From R. M. Bell, L. M. Ballas, and R. A.
Coleman, Lipid topogenesis, *J. Lipid Res.* 22:391, 1981.)

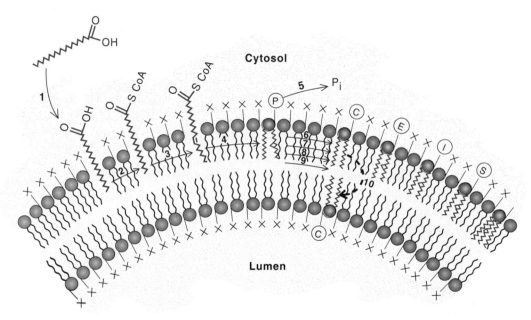

ent, the requirements for the synthesis of the essential membrane components, phosphatidylethanolamine and phosphatidylcholine, are met before an appreciable amount of
energy-storage lipid (triacylglycerol) is made. Any excess
diacylglycerol and fatty acyl-CoA is funneled into triacylglycerol.

Additional regulation of phosphatidylcholine and
phosphatidylethanolamine biosynthesis occurs at the second
step in the biosynthetic sequence (see fig. 19.4) where either
CDP-choline or CDP-ethanolamine are made. For phosphatidylcholine biosynthesis, the activity of CTP:phos-
phocholine cytidylyltransferase (which makes CDP-
choline) is governed by an unusual mechanism. The enzyme

exists in a soluble form as an inactive reservoir and is translocated to the cell membranes where it is activated by membrane lipids (fig. 19.10). This facilitates a rapid response to
a sudden requirement for biosynthesis of phosphatidylcholine, which could be vital to maintaining the integrity of the
cell membrane. The binding of the cytidylyltransferase to
membranes is enhanced by a decrease in phosphatidylcholine, dephosphorylation of the enzyme, or a rise in the concentration of fatty acids or diacylglycerol in the membrane
(see fig. 19.10). The regulation by phosphatidylcholine and
diacylglycerol levels are good examples of metabolic regulation by feedback and feedforward mechanisms, respectively.

**Figure 19.10**

CTP-phosphocholine cytidylyltransferase (CT) is found in two forms, an active form bound to cellular membranes and an inactive form when in a soluble form. The active form of CT is favored by low concentrations of phosphatidylcholine, high concentrations of diacylglycerol and fatty acids and the unphosphorylated state of the enzyme.

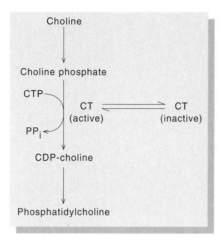

## Phospholipases Degrade Phospholipids

Enzymes that degrade phospholipids are called phospholipases. They are classified according to the bond cleaved in a phospholipid (fig. 19.11). Phospholipases $A_1$ and $A_2$ selectively remove fatty acids from the sn-1 and sn-2 positions, respectively. Phospholipase C cleaves between glycerol and the phosphate moieties; phospholipase D hydrolyzes the head-group moiety X from the phospholipid. Lysophospholipids, which lack a fatty acid at the sn-1 or sn-2 position, are degraded by lysophospholipases.

Phospholipases are found in all types of cells and in various subcellular locations within eukaryotic cells. Some of these enzymes are specific for particular polar head-groups; others are nonspecific. Phospholipase $A_2$ is a major component of snake venom (cobra and rattlesnake) and is partially involved in the deadly effects of these venoms. Because of the high concentration of phospholipase $A_2$ in these venoms, this enzyme has been studied intensively. The pancreas is also rich in phospholipase $A_2$, which is secreted into the intestine for digestion of dietary phospholipids.

In some cases the functions of phospholipases in cells are purely degradative and result in the release of the phospholipid components (fatty acids, glycerol, phosphate, and head-groups). But in many cases phospholipases have important roles in synthesis and regulation. For example, we have seen how phospholipase $A_2$ catalyzes the first step in the "remodeling" of phosphatidylcholine to the surfactant

lipid, dipalmitoylphosphatidylcholine (see fig. 19.5). Phospholipase $A_2$ also plays an important regulatory role in the release of arachidonic acid from the sn-2 position. This molecule is a precursor of hormones known as eicosanoids, which are discussed later in this chapter. The regulatory role of phospholipase C in release of cellular second messengers, diacylglycerol and inositol-1,4,5-$P_3$, has already been mentioned (see fig. 19.7).

## Sphingolipids

Sphingolipids are found in the membranes in animals, plants, and some lower forms of life but are absent from most bacteria such as *E. coli*. The common structural feature of all sphingolipids is a long-chain hydroxylated secondary amine (see fig. 17.4). Sphingolipids are particularly abundant in myelin sheath, a multilayered membranous structure that protects and insulates nerve fibers. Recently several sphingolipids have been implicated as second messengers in cells.

Sphingolipids were first described in a remarkable treatise on the chemical constitution of the brain by Johann L. W. Thudichum, a physician–scientist in London, who published his findings more than 100 years ago. A major impetus for the study of the chemistry and metabolism of the sphingolipids was the discovery of several rare human diseases that could be attributed to the abnormal accumulation of sphingolipids. This accumulation has been shown to result from a defect in catabolism that normally occurs in lysosomes. It is now known that many different kinds of sphingolipids exist, and more than 300 structures have been reported to occur in nature.

## Sphingomyelin Is Formed Directly from Ceramide

Sphingenine is the major long-chain base present in sphingolipids and only differs from sphinganine by having a *trans* double bond between carbons 4 and 5. The biosynthesis of sphinganine occurs on the endoplasmic reticulum and involves a condensation of palmitoyl-CoA with serine and a subsequent reduction of the 3-ketone (fig. 19.12). The sphingolipids are built on a serine-derived core rather than a glycerol core, as is the case with phospholipids. Once the long-chain base is formed, the amino group of serine reacts with acyl-CoA to form *N*-acyl-sphinganine (see fig. 19.12). This derivative is desaturated to *N*-acyl-sphingenine. The *N*-acyl derivatives of sphingenine or sphinganine are usually referred to as ceramide. Sphingomyelin is synthesized by the transfer of the phosphocholine moiety of phosphati-

**Figure 19.11**

Reactions catalyzed by phospholipases. X can be any of the head-groups: choline, ethanolamine, serine, glycerol, or inositol.

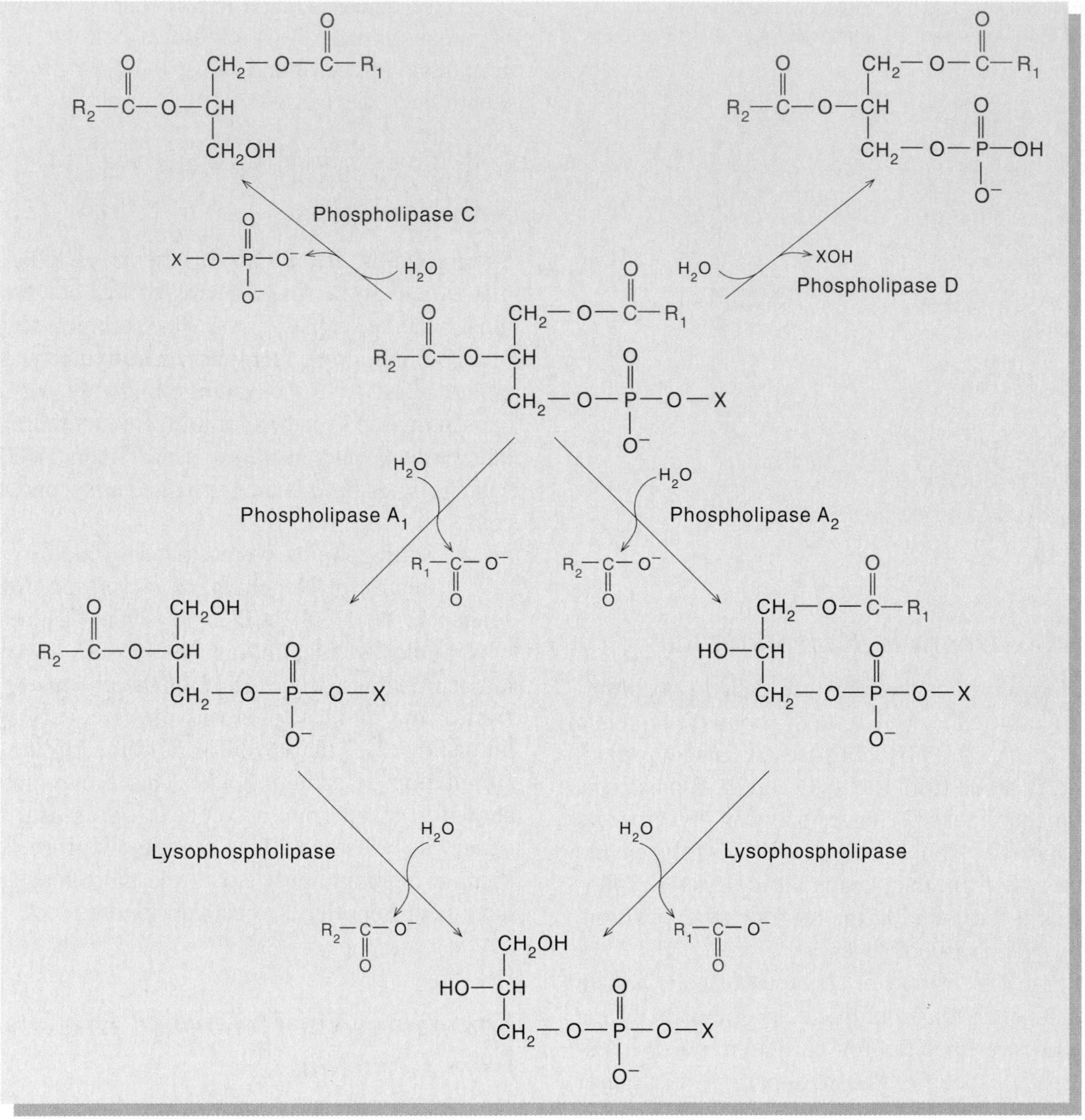

dylcholine to ceramide (see fig. 19.12). Thus, the polar head-groups of phosphatidylcholine and sphingomyelin are identical.

## Glycosphingolipid Synthesis Also Starts from Ceramide

As shown in figure 19.12, ceramide is also the precursor of glucosyl-ceramide, the simplest member of the glycosphingolipids. The biosynthesis of glycosphingolipids shows striking parallels to the synthesis of the O-linked glycoproteins (see chapter 16). Thus, the carbohydrates are added to the acceptor lipid by transfer from a sugar nucleotide (e.g., UDP-glucose). The enzymes involved in these reactions are called glycosyltransferases and are specific for each reaction. Most of the glycosyltransferases involved in the biosynthesis of the glycosphingolipids are located on the lumenal side of the Golgi apparatus. The nucleotide sugars used in biosynthesis are transferred across the membrane into the lumen of the Golgi apparatus by a transporter lo-

## Figure 19.12

The pathways to sphingomyelin and glycosphingolipids. Figure 19.13 elaborates on some pathways from ceramide to glycosphingolipids.

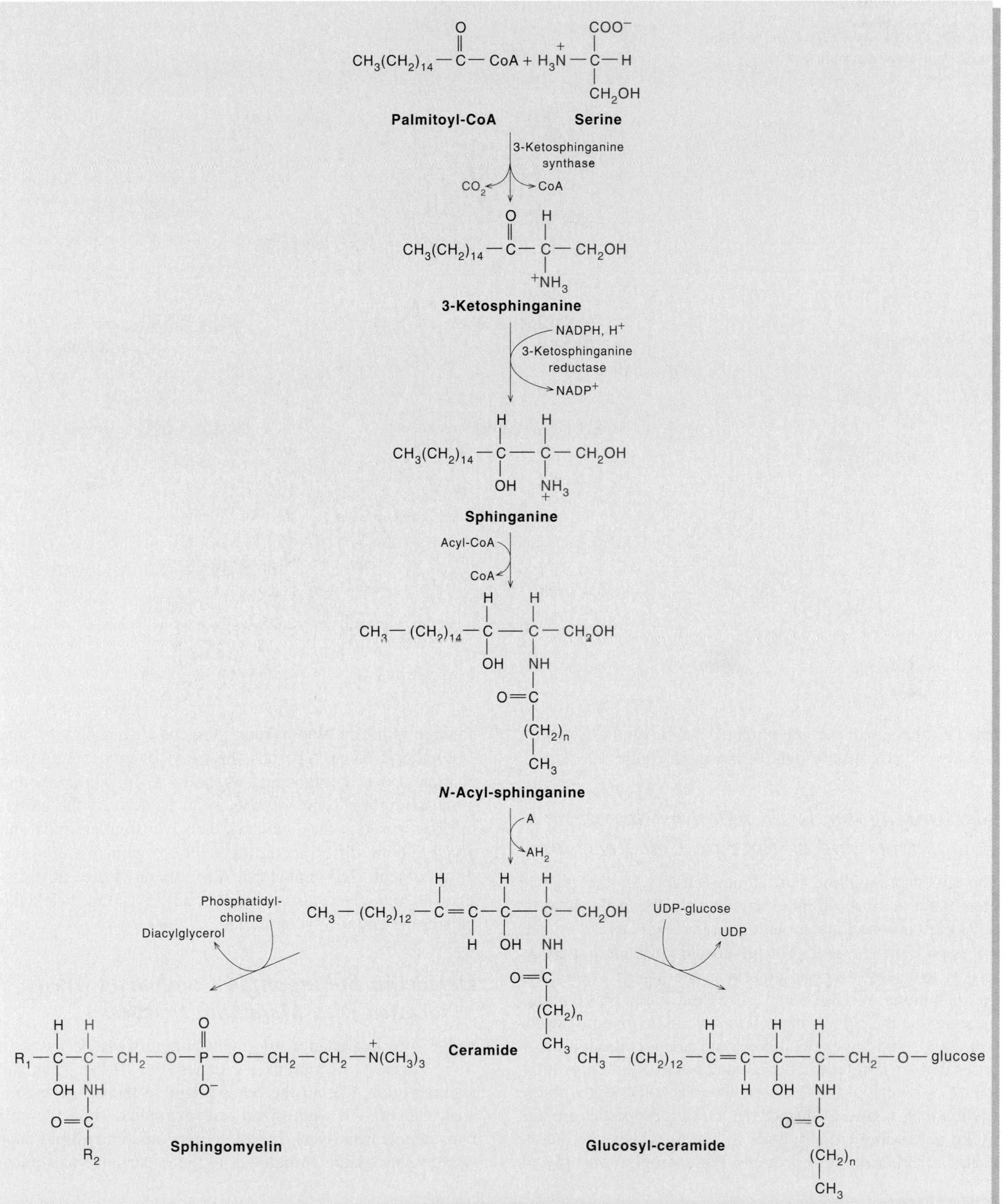

**Figure 19.13**

Outline of biosynthesis of some glycosphingolipids. (Glc = glucose, Gal = galactose, GalNAc = N-acetylgalactosamine, NeuAc = n-acetylneuraminic acid.

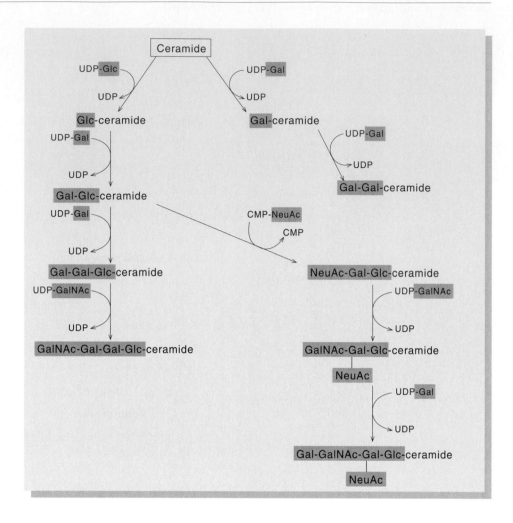

## Glycosphingolipids Function as Structural Components and as Specific Cell Receptors

The glycosphingolipids are found largely in the plasma membrane and are oriented asymmetrically in the bilayer, with the carbohydrate moieties facing exclusively toward the outside of the cell. The glycosphingolipids appear to have a structural role particularly in the myelin sheath and the membrane of the red cell. However, many observations indicate that the glycosphingolipids are more than just structural lipids and have specific cell surface recognition properties that are similar to plasma-membrane-bound glycoproteins (see chapter 16). This conclusion is consistent with the finding that many cell surface recognition properties are due to the protruding carbohydrate moieties, which may be attached to either lipids or proteins. For example, some glyco-

sphingolipids are blood group antigens. A person who displays the B blood type has the same B antigen oligosaccharide as a component of both a membrane-bound glycosphingolipid and glycoprotein. Moreover, the carbohydrate moieties displayed by the glycosphingolipids and glycoproteins differ according to the organism's stage of development. Cells transformed by tumor viruses also display altered glycosphingolipids, just as is the case with membrane-bound glycoproteins.

## Defects in Sphingolipid Catabolism Are Associated with Metabolic Diseases

The degradation of glycosphingolipids occurs in a stepwise fashion in lysosomes (fig. 19.14). Thus, the sphingolipids, like all the components in the cell, are constantly being biosynthesized and degraded, (i.e., the cell constituents turn over). Investigations on sphingolipid degradation have been stimulated by the occurrence of a num-

**Figure 19.14**

Catabolism of some glycosphingolipids. The glycosyl hydrolase
enzymes involved in these reactions are localized in the lysosomes.

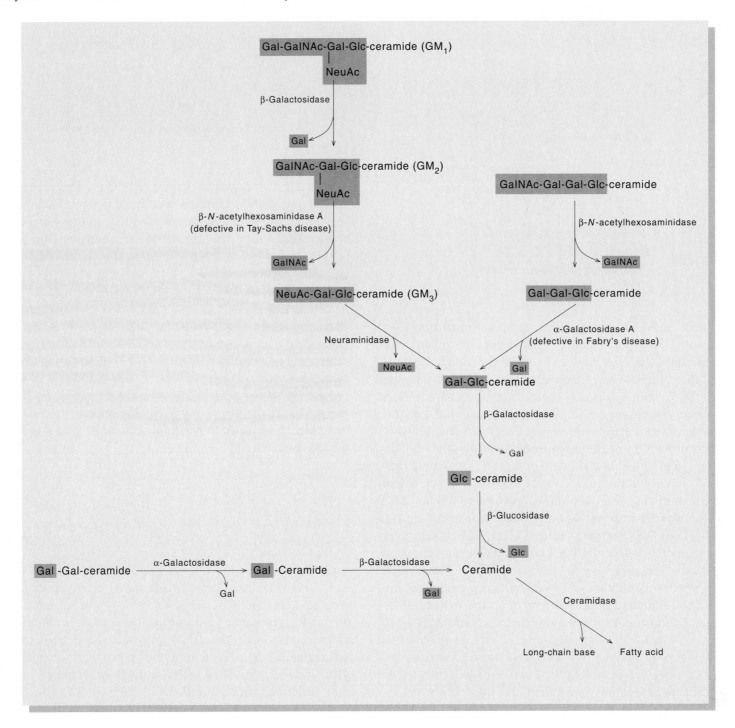

ber of human genetic diseases, the sphingolipidoses, each of
which results from the accumulation of one of these lipids.
Each of these diseases is caused by a deficiency in the activity of an enzyme involved in the catabolism of one of the
sphingolipids. While there is a human disease that results

from a metabolic defect in each one of the degradation reactions shown in figure 19.14, herein we will limit the discussion to two of the better-known disorders.

Fabry's disease was described independently in 1898
by J. Fabry and W. Anderson. Characteristic symptoms are

**Figure 19.15**

Structure of prostaglandin $E_2$. All prostaglandins are derived from $C_{20}$ fatty acids. The PG stands for prostaglandin. The E specifies the position of oxygen substituents, and the subscript refers to the number of double bonds. From the structure it can be seen that the double bonds of $PGE_2$ are $trans\Delta^{13}$ and $cis\Delta^5$.

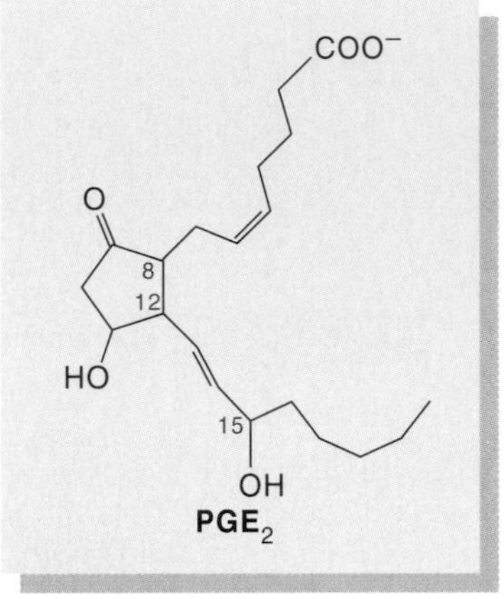

PGE$_2$

**Figure 19.16**

The structure of the thromboxane $A_2$. Thromboxanes differ from prostaglandins in having a cyclic ether (oxane) ring structure.

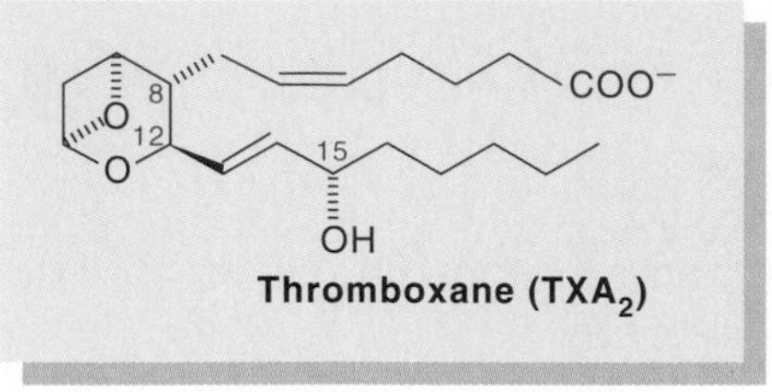

Thromboxane (TXA$_2$)

skin rash, pain in the extremities, and renal impairment accompanied by hypertension. Patients lead a reasonably normal life until their fourth decade, when the kidneys usually fail. Little progress was made in understanding this disease until 1963 when Charles Sweeley and Bernard Klionsky described the structure of Gal-Gal-Glc-ceramide as the major lipid that accumulates in the Fabry-affected kidneys. The accumulation of this lipid throughout the body results in the symptoms described above. In 1967, Roscoe Brady demonstrated that the enzymatic defect was a deficiency of the enzyme that degrades this trihexosyl-ceramide. This enzyme was later shown to be $\alpha$-galactosidase A (see fig. 19.14). The cDNA for the enzyme and the gene that encodes the enzyme have recently been cloned and sequenced. It is now possible to diagnose the disease with biochemical tests based on these discoveries. For example, in prenatal diagnosis, cells are obtained from the amniotic fluid by amniocentesis and culturing. It is then possible to screen the DNA of these cells to see if a normal gene is present or, alternatively, assay the activity of $\alpha$-galactosidase A. There is presently no effective treatment for Fabry's disease other than such drastic procedures as a kidney transplant. Fortunately, Fabry's disease is rare, with only a few hundred cases being reported throughout the world.

The disease described by Warren Tay in 1881, now known as Tay-Sachs disease, occurs much more frequently. It is estimated that 30–50 children with Tay-Sachs disease

are conceived each year in the United States. This disease is devastating, and the children usually do not survive beyond the age of 3. In Tay-Sachs disease the glycosphingolipid abbreviated as GM$_2$ (see fig. 19.14) accumulates, especially in the brain, as a result of the deficiency of $\beta$-$N$-acetylhexosaminidase A.

There is no way of treating Tay-Sachs disease. Enzyme replacement is not considered a likely therapy because infused enzyme cannot penetrate the blood–brain barrier. However, the incidence of the disease has been dramatically decreased by prenatal diagnosis. Tay-Sachs disease is an autosomal recessive disease and so can arise only if both parents are carriers, i.e., if each parent carries a single defective gene for the hexosaminidase A enzyme. In that case there is a 25% chance that a child of these parents will have the disease.

## Eicosanoids Are Hormones Derived from Arachidonic Acid

Eicosanoids are a diverse group of hormones, most of which are derived from the $C_{20}$ polyunsaturated fatty acid, arachidonic acid ($20{:}4^{\Delta 5,8,11,14}$). Most prominent among the group is a series of cyclopentanoic acids known as prostaglandins (PG) (fig. 19.15). Closely related to the prostaglandins are the eicosanoids known as thromboxanes (TX) (fig. 19.16). They differ from prostaglandins in the ring structure, which for thromboxanes is a cyclic ether (oxane ring) rather than the cyclopentane ring of the prostaglandins.

Oxygenated eicosanoids appear to be unique to animal cells. The first oxygenated eicosanoids to be discovered, the prostaglandins, were extracted from human semen in the

## Figure 19.17

Reaction catalyzed by prostaglandin endoperoxide synthase. This enzyme has two catalytic activities, as indicated next to the reaction arrows.

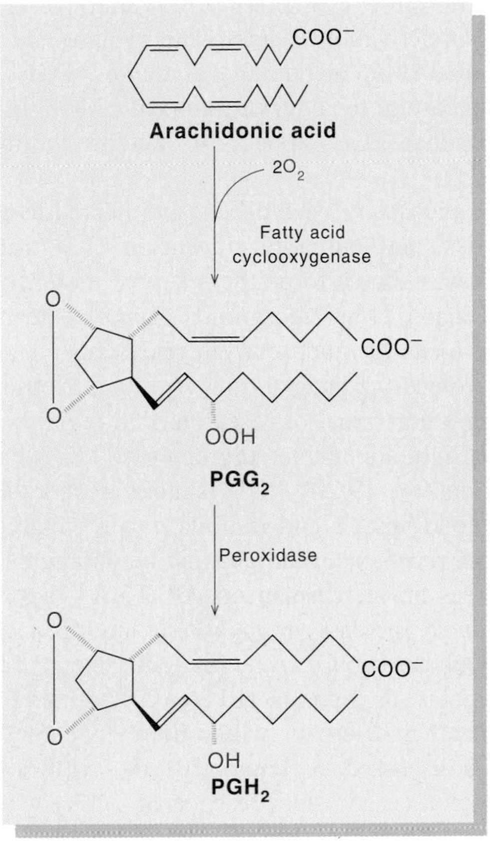

**Arachidonic acid**

$2O_2$

Fatty acid cyclooxygenase

**PGG$_2$**

OOH

Peroxidase

OH
**PGH$_2$**

were awarded the 1982 Nobel Prize in Medicine for their important discoveries.

## Eicosanoid Biosynthesis

Arachidonic acid is not present in significant amounts in tissues as the free acid but is stored as a fatty acid at the SN-2 position of phospholipids. Prostaglandin biosynthesis is initiated by the interaction of a stimulus with the cell surface. Depending on the cell type, the stimulus can take the form of a hormone, such as angiotensin II or antidiuretic hormone, or a protease such as thrombin (involved in blood clotting), or both hormone and protease. These agents bind to a specific receptor that activates a phospholipase A$_2$ that specifically releases the arachidonic acid from a phospholipid such as phosphatidylcholine. The release of arachidonic acid by phospholipase A$_2$ is believed to be the rate-limiting step for the biosynthesis of eicosanoids.

The initial steps for the synthesis of prostaglandins and thromboxanes are the oxidation and cyclization of arachidonic acid to yield the prostaglandins PGG$_2$ and PGH$_2$ (fig. 19.17). (In the abbreviation, PG stands for prostaglandin. The third letter, G or H, refers to the particular structure, and the subscript 2 notes the number of double bonds.) This reaction is catalyzed by a single bifunctional enzyme, prostaglandin endoperoxide synthase, that is present on the endoplasmic reticulum. The formation of PGG$_2$ is catalyzed by the cyclooxygenase component of the enzyme (see fig. 19.17), and the subsequent formation of PGH$_2$ is catalyzed by the peroxidase component. The cyclooxygenase activity has the unusual property of catalyzing its own destruction. Approximately once in every 1,400 substrate turnovers, the cyclooxygenase activity is irreversibly inactivated. This "suicide reaction" occurs both *in vivo* and *in vitro*. The suicide process is an unusual regulatory mechanism, which places an upper limit on cellular prostaglandin biosynthetic activity.

The anti-inflammatory effect of aspirin appears to be due to the acetylation of a serine residue at the active site of the cyclooxygenase, which irreversibly inactivates the enzyme (fig. 19.18). Aspirin has no effect on the peroxidase activity of the enzyme.

PGH$_2$ is the precursor of various prostaglandins and thromboxanes. Each reaction is catalyzed by a separate tissue-specific enzyme as indicated in figure 19.19. The tissue localization of the enzyme relates to the role of the

1930s by Ulf von Euler in Sweden. These compounds, which he believed to arise from the prostate gland (hence the name), were injected into animals and caused the uterus to contract and lowered blood pressure. It is now known that Euler's prostaglandins were from the seminal gland, not the prostate, and that they are present in most tissues of both male and female animals. Since the prostaglandins are present only transiently and at very low concentrations, the structures were not elucidated until the 1960s and later, after the development of sensitive instruments such as the gas chromatograph–mass spectrometer. Sune Bergström and Bengt Samuelsson, also from Sweden, were pioneers in studies on the structure and metabolism of the eicosanoids. Pharmacological interest in prostaglandins was awakened in 1971 by John Vane's discovery that aspirin blocked the synthesis of prostaglandins. Bergström, Samuelsson, and Vane

Figure 19.18

Aspirin inactivates cyclooxygenase by acetylation of a serine, probably at or near the active site.

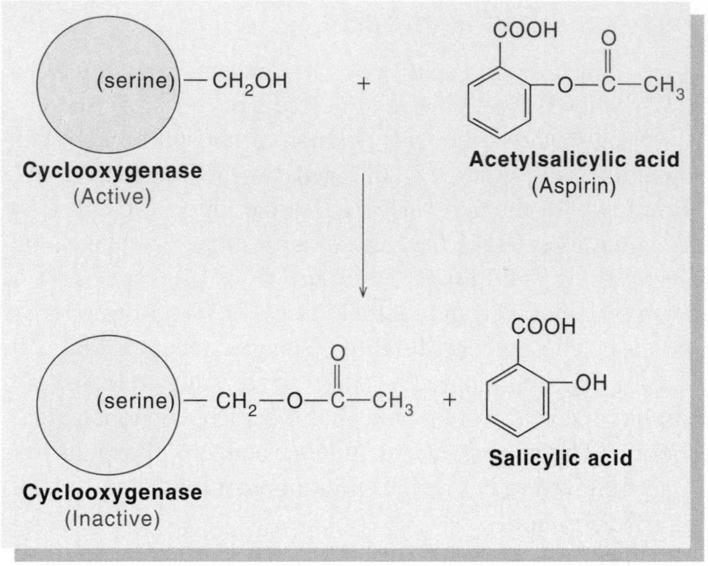

eicosanoid; for example, thromboxane $A_2$ (TXA$_2$) is made in platelets and causes them to aggregate, as occurs in blood clotting, whereas prostaglandin $I_2$ (PGI$_2$) is made in arterial walls and inhibits platelet aggregation.

If the eicosanoids are to serve a useful hormonal function, there must be a rapid means for their removal when their action is no longer required. Consistent with this principle, it has been found that eicosanoids have very short half-lives *in vivo;* injected prostaglandins do not survive a single pass through the circulatory system.

## Eicosanoids Exert Their Action Locally

The diversity of eicosanoid structures is paralleled by a diversity of biological effects. They are generally considered to be hormones that exert their main effect locally. Their target sites include both the cells in which they are formed and neighboring, different cell types. In all cases, the effects of prostaglandins are mediated through specific cell surface receptors, as is the case for other hormones that bind to the outside of the plasma membrane.

Specific binding of PGE$_2$ to a variety of cells correlates with an activation of adenylate cyclase and the accumulation of cAMP. In human adipocytes, PGE$_2$ causes a 15-fold increase in the concentration of cAMP. In platelets, PGI$_2$ rather than PGE$_2$ appears to mediate the increase in cAMP.

The eicosanoids have been implicated as mediators of tissue inflammation since aspirin's anti-inflammatory effect was shown to be the result of its inactivation of cyclooxygenase. How eicosanoids cause tissue inflammation is the focus of much current research.

The role of prostaglandins in blood clotting is also stimulating a great deal of research. PGI$_2$ relaxes coronary arteries and inhibits platelet aggregation. TXA$_2$ has the opposite effects (fig. 19.20). PGI$_2$ is made in endothelial cells that line blood vessels and inhibits platelet aggregation by binding to a receptor on the plasma membrane of platelets, which causes an increase in cAMP. TXA$_2$ suppresses the PGI$_2$-mediated increase in cAMP. It has been speculated that the synthesis of PGI$_2$ may prevent platelets from binding to arterial walls. In damaged areas of arteries (caused by the disease atherosclerosis, which involves cholesterol metabolism as discussed in chapter 20), the synthesis of PGI$_2$ may be decreased and the presence of TXA$_2$ would cause platelets to aggregate, leading to the formation of a blood clot (thrombosis, a major cause of heart attack and stroke). Interestingly, besides the anti-inflammatory effect of aspirin, a low-dose aspirin (one "baby" aspirin daily, or one regular aspirin every three days) has proved useful in the prevention of heart attacks. The low dose of aspirin leads to selective inhibition of platelet thromboxane formation (thus, decreased platelet aggregation) without affecting the synthesis of other eicosanoids in other cells.

# Figure 19.19

Formation of prostaglandins and thromboxanes.

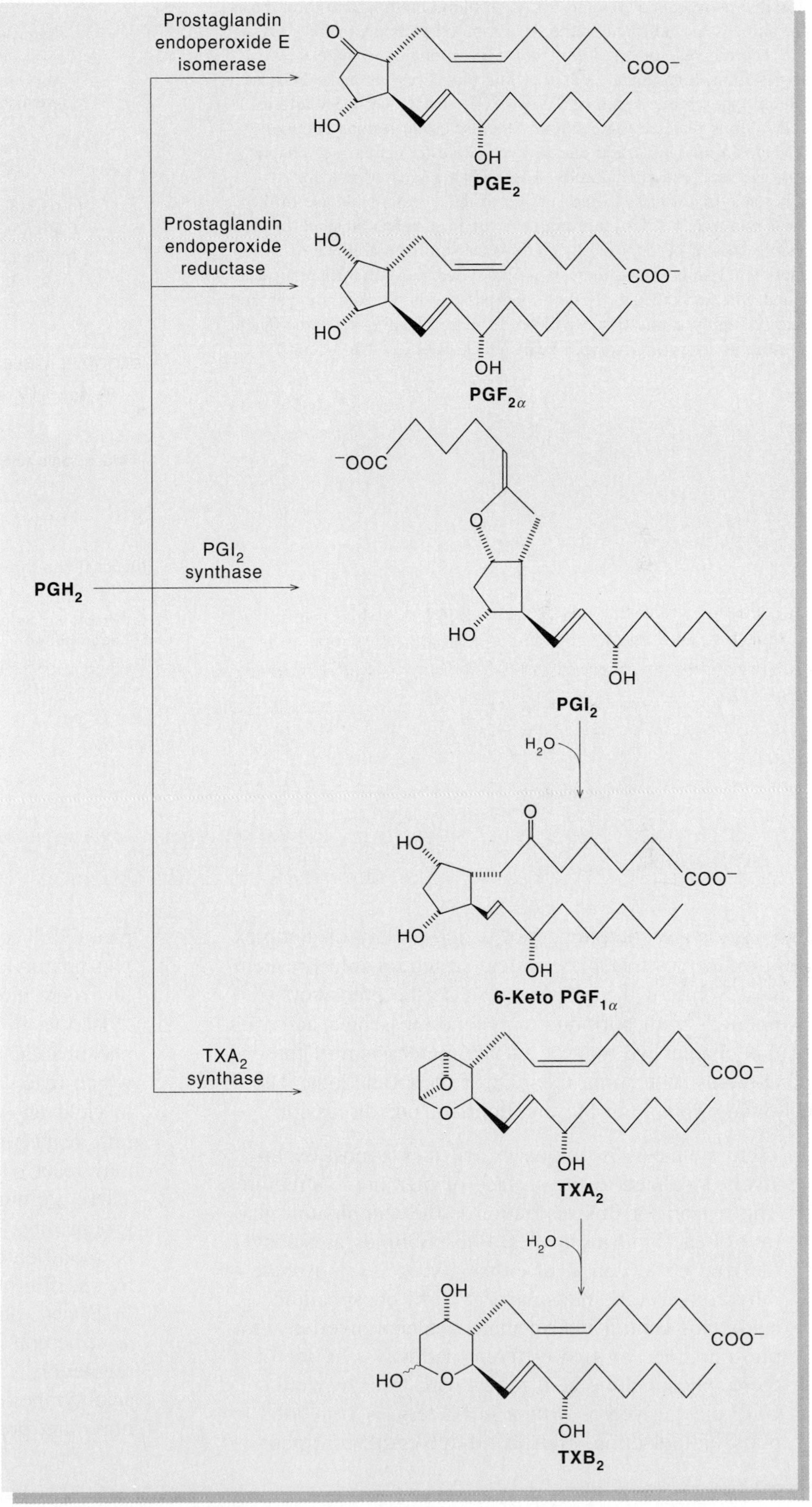

**Figure 19.20**

Possible actions of PGI$_2$ and TXA$_2$ in blood clotting and thrombosis are shown. As indicated in the figure, blood platelets make TXA$_2$, which promotes constriction of coronary arteries and platelet aggregation. Endothelial cells that line blood vessels make PGI$_2$ which opposes the action of TXA$_2$; PGI$_2$ relaxes coronary arteries and inhibits platelet aggregation. Atherosclerotic lesions commonly develop in humans, often due to hypercholesterolemia (see chapter 20), and lack endothelial cells. There is an increased risk for adherence of platelets to the surface of these lesions. A rise in the concentration of TXA$_2$ that might occur in platelets attached to these lesions cannot be opposed by PGI$_2$ because of the absence of endothelial cells. Thus, there is an increased risk for formation of a blood clot (thrombosis). If the blood passage is blocked, oxygen and energy supply to the tissue will be decreased leading to tissue death. If a major artery is involved, heart attack or stroke can occur.

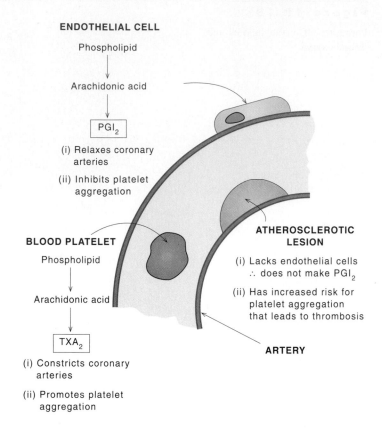

ENDOTHELIAL CELL

Phospholipid
↓
Arachidonic acid
↓
PGI$_2$

(i) Relaxes coronary arteries

(ii) Inhibits platelet aggregation

BLOOD PLATELET

Phospholipid
↓
Arachidonic acid
↓
TXA$_2$

(i) Constricts coronary arteries

(ii) Promotes platelet aggregation

ATHEROSCLEROTIC LESION

(i) Lacks endothelial cells ∴ does not make PGI$_2$

(ii) Has increased risk for platelet aggregation that leads to thrombosis

ARTERY

## Summary

Fatty acids are components of a rich variety of complex lipid molecules that play critical structural roles in membranes. Some of these lipids are also the precursors of compounds with hormone or second-messenger activities. In this chapter we focused on the metabolism of these compounds with some mention of their functions. The following points are the highlights of our discussion.

1. Lipid synthesis is unique in that it is almost exclusively localized to the surface of membrane structures. The reason for this restriction is the amphipathic nature of the lipid molecules. Phospholipids are biosynthesized by acylation of either glycerol-3-phosphate or dihydroxyacetone phosphate to form phosphatidic acid. This central intermediate can be converted into phospholipids by two different pathways. In one of these, phosphatidic acid reacts with CTP to yield CDP-diacylglycerol, which in bacteria is converted to phosphatidylserine, phosphatidylglycerol, or diphos-

phatidylglycerol. In *E. coli*, the major phospholipid, phosphatidylethanolamine, is synthesized by means of this route through the decarboxylation of phosphatidylserine. In the second pathway, found in eukaryotes, phosphatidic acid is hydrolyzed to diacylglycerol, which reacts with CDP-ethanolamine or CDP-choline to yield phosphatidylethanolamine or phosphatidylcholine, respectively. Alternatively, the diacylglycerol may react with acyl-CoA to form triacylglycerol.

2. There are numerous reactions by which the acyl groups or polar head-groups of phospholipids might be modified or exchanged. Phospholipids are degraded by specific phospholipases.

3. In *E. coli*, phospholipids are used almost exclusively as structural components of the cell membranes, and regulation is known to occur at an early stage in fatty acid synthesis. In the mammalian liver, fatty acids are important precursors of both the structural phospholip-

ids and also of the energy-storage lipid, triacylglycerol. The cell's need for structural lipids is satisfied before fatty acids are shunted into energy storage. The rate of triacylglycerol synthesis is regulated by the availability of fatty acid from both diet and biosynthesis. The regulation of phosphatidylcholine biosynthesis occurs primarily at the reaction catalyzed by CTP: phosphocholine cytidylyltransferase. This enzyme is activated by translocation from a soluble form, where it is inactive, to cellular membranes, where it is activated. The regulation of the biosynthesis of other phospholipids in eukaryotic cells is less well understood.

4. The sphingolipids are important structural lipids found in eukaryotic membranes. The acylation of sphingenine produces ceramide, which reacts with phosphatidylcholine to give sphingomyelin. Ceramide also can react with activated carbohydrates (e.g., UDP-glucose) to form the glycosphingolipids. Studies on the catabolism of the sphingolipids have revolved around inherited diseases, sphingolipidoses, that result from a defect in lysosomal enzymes that degrade sphingolipids.

5. Prostaglandins and thromboxanes are hormonelike substances that affect the function of a cell by binding to a receptor on the cell surface. Prostaglandins and thromboxanes are biosynthesized from $C_{20}$ polyunsaturated fatty acids, primarily arachidonic acid. In the initial reaction, arachidonic acid is converted to $PGH_2$ by prostaglandin endoperoxide synthase. $PGH_2$ is then converted to $PGE_2$, $PGF_2$, $TXA_2$, and $PGI_2$.

## Selected Readings

Brindley, D. N., Metabolism of triacylglycerols. In D. E. Vance, and J. E. Vance (eds.), *Biochemistry of Lipids, Lipoproteins and Membranes.* Amsterdam: Elsevier Science Publishers, 1991. This chapter (6) provides an advanced discussion of phosphatidic acid and triacylglycerol metabolism.

Dennis, E. A., Phospholipases. In *Enzymes XVI.* New York: Academic Press, 1983. This article provides a thorough review of phospholipases.

Jackowski, S., J. E. Cronan, and C. O. Rock, Lipid metabolism in procaryotes. In D. E. Vance, and J. E. Vance (eds.), *Biochemistry of Lipids, Lipoproteins and Membranes.* Amsterdam: Elsevier Science Publishers, 1991. This chapter (2) provides an advanced treatment of genetics and metabolism of phospholipids in *E. coli.*

Scriver, C. R., A. I. Beaudet, W. S. Sly, and D. Valle (eds.), *The Metabolic Basis of Inherited Disease,* 6th ed., vol. 2, New York: McGraw Hill, 1989. This book contains in-depth chapters on sphingolipid metabolism and many of the sphingolipidoses.

Smith, W. I., P. Borgeat, and F. A. Fitzpatrick, The eicosanoids: Cyclooxygenase, lipoxygenase and epoxygenase pathways. In D. E. Vance, and J. E. Vance (eds.), *Biochemistry of Lipids, Lipoproteins and Membranes.* Amsterdam: Elsevier Science Publishers, 1991. This chapter (10) provides advanced information on the biochemistry, metabolism, and functions of the eicasanoids.

Sweeley, C. C., Sphingolipids. In D. E. Vance, and J. E. Vance (eds.), *Biochemistry of Lipids, Lipoproteins and Membranes.* Amsterdam: Elsevier Science Publishers, 1991. This is an advanced chapter (11) on the chemistry, metabolism, and function of the sphingolipids.

Vance, D. E. (ed.), *Phosphatidylcholine Metabolism,* Boca Raton, Fla.: CRC Press, 1989. This advanced monograph contains 13 chapters on all aspects of phosphatidylcholine biosynthesis and catabolism.

Vance, D. E., Phospholipid metabolism and cell signalling in eucaryotes. In D. E. Vance, and J. E. Vance (eds.), *Biochemistry of Lipids, Lipoproteins and Membranes.* Amsterdam: Elsevier Science Publishers, 1991. This chapter (7) covers phospholipid metabolism at an advanced level and discusses the role of phosphatidylinositol and other phospholipids in the generation of second messengers in the cell.

Waite, M., *The Phospholipases.* New York: Plenum Press, 1987. This book provides a complete coverage of phospholipases from the technicalities of assay of phospholipases to the proposed mechanism of catalysis.

Waite, M., Phospholipases. In D. E. Vance, and J. E. Vance (eds.), *Biochemistry of Lipids, Lipoproteins and Membranes.* Amsterdam: Elsevier Science Publishers, 1991. This chapter (9) provides advanced knowledge about phospholipases and the hydrolysis of phospholipids.

Weissmann, G., Aspirin. *Sci. Am.* 264:84–90, 1991. An interesting article that discusses the mechanism of action of aspirin in connection with eicosanoid metabolism.

## Problems

1. Would you expect phosphatidylserine decarboxylase (fig. 19.3) to be a pyridoxal phosphate enzyme?
2. Phosphatidylcholine biosynthesis appears to be regulated principally at the step catalyzed by CTP: phosphocholine cytidylyltransferase. Does the type of regulation observed make biochemical sense? Draw a chemical reaction mechanism for this enzyme.
3. Explain why the reactions from choline to phosphatidylcholine in eukaryotic cells are thermodynamically feasible.
4. A patient accumulated the lipid Gal-$\beta$(1,4)-Glc-$\beta$(1,4)-ceramide. What enzymatic reaction might be defective?
5. Sphingolipids as well as other membrane components are constantly degraded and resynthesized. Why does the cell waste so much energy for what appears to be a futile effort?
6. If both parents carry a single defective gene for the hexosaminidase A enzyme, there is a 25% chance that their child will have Tay-sachs disease. What chance is there that their child would be a carrier of a defective gene?
7. Why don't carriers of a defective gene for the hexosaminidase A enzyme suffer from Tay-Sachs disease (or at least some of the symptoms)?
8. One of the pathways for the biosynthesis of phosphatidic acid in eukaryotes originates with dihydroxyacetone phosphate (from glycolysis). This pathway is shown only as three unlabeled arrows in figure 19.2. Given that the three enzymes starting with dihydroxyacetone phosphate are (1) an acyl transferase, (2) a dehydrogenase, and (3) a second acyltransferase, complete the pathway including any missing substrates.
9. This chapter discusses many of the biosynthetic interconversions that are known to occur with the different phosphatidyl components (e.g., phosphatidylethanolamine, phosphatidylcholine, and phosphatidylinositol) of membranes. Only considering the bioenergetics, what is the major difference between the biosynthetic pathways shown in figures 19.4 and 19.6?
10. Low doses of aspirin (one aspirin every other day) are recommended to prevent heart attacks and strokes. Why do three or four tablets per day not work better? (*Hint:* Remember that $TXA_2$ is made in platelets and that $PGI_2$ is made in the arterial walls.)
11. Snake venoms contain many types of lipases, including phospholipase $A_2$. Why do small amounts of this enzyme contribute to some of the toxic effects of snake venom? (Bee venom contains a protein that stimulates phospholipase $A_2$.)

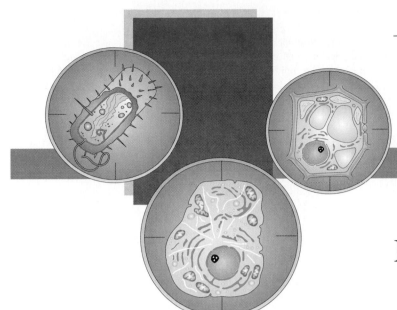

# Metabolism of Cholesterol

*The tetracyclic molecule of cholesterol synthesized in blocks of 5 carbon units is a major component of membranes and also a precursor of steroids and bile acids.*

Steroids are tetracyclic hydrocarbons that are much more common in eukaryotes than in prokaryotes. Cholesterol, the most prominent member of the steroid family, is an important component of many eukaryotic membranes (see chapter 17). In addition, it is the precursor of the other two major classes of steroids: The steroid hormones and the bile acids (fig. 20.1).

Steroid hormones play a key role in the regulation of metabolism. These hormones come in a rich variety, each interacting in a highly specific manner with a receptor protein to promote gene expression in the appropriate target tissue. Bile acids are the primary degradation products of cholesterol. The bile acids are made in the liver, stored in the gall bladder, and secreted into the small intestine, where they aid in the solubilization of lipids, facilitating their digestion by intestinal lipases.

## Figure 20.1

Formation of cholesterol and some of its derivatives. All of the carbon atoms of cholesterol are derived from acetyl-CoA by way of mevalonate in a pathway with 33 reaction steps. From cholesterol a wide variety of steroids and bile acids and bile salts are formed. Many of the reactions leading to cholesterol derivatives are organ-specific. Numbers associated with the arrows indicate the approximate number of steps in the sequence, where that is relevant. Most of the reactions are unidirectional. Excess cholesterol is disposed of by conversion to bile acids and excreted as bile salts (as shown).

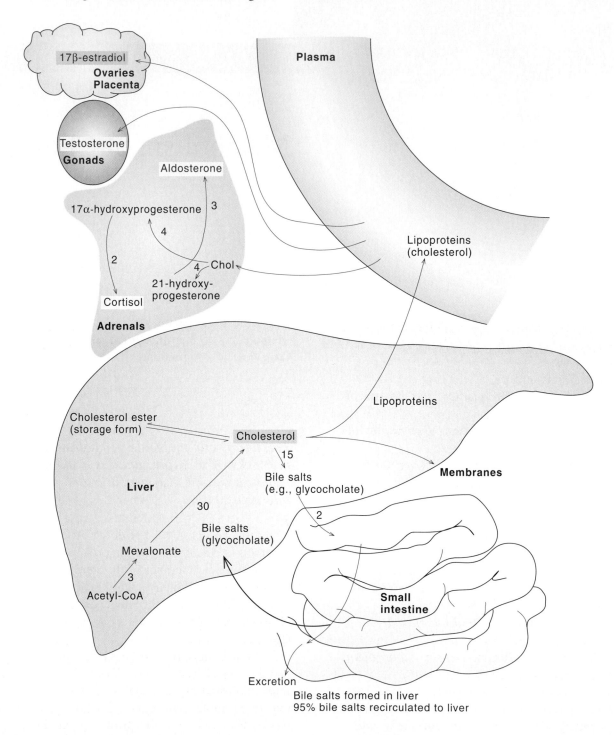

**Figure 20.2**

Basic scheme for cholesterol biosynthesis proposed by Bloch in 1952 (*left*) and scheme proposed for the cyclization of squalene by Woodward and Bloch in 1953 (*right*).

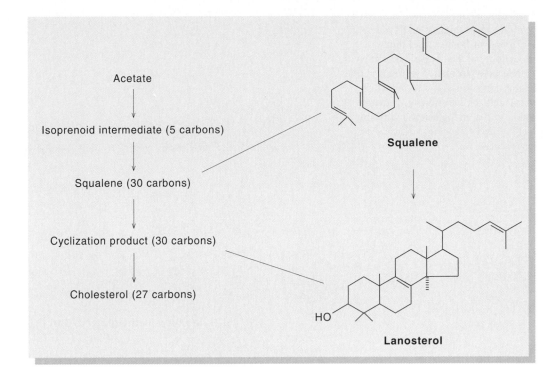

All of these biological roles of the steroids figure prominently in human well-being. Defects in cholesterol metabolism are major causes of cardiovascular disease. It is no wonder that steroids are a central concern in medical biochemistry. In this chapter we discuss the metabolism of these complex lipids and the plasma lipoproteins in which they and other complex lipids are transported to various tissues.

## Biosynthesis of Cholesterol

Early in the 1930s the structure of cholesterol was determined, an achievement that concluded a brilliant chapter in structural organic chemistry. At that time it was not clear how such a complex structure could be assembled from small molecules.

Work on the biosynthesis of cholesterol began in earnest after Rudolf Schoenheimer and David Rittenberg, at Columbia University, developed isotopic tracer techniques for the analysis of biochemical pathways. In 1941, Rittenberg and Konrad Bloch were able to show that deuterium-labeled acetate ($C^2H_3COO^-$) was a precursor of cholesterol in rats and mice. In 1949, James Bonner and Barbarin Arreguin postulated that three acetates could combine to form a single five-carbon unit called isoprene.

This proposal supported with an earlier prediction of Sir Robert Robinson that cholesterol was a cyclization product of squalene, a 30-carbon polymer of isoprene units. In 1953, Robert Burns Woodward and Bloch postulated a cyclization scheme for squalene (fig. 20.2) that was later shown to be correct. In 1956, the unknown isoprenoid precursor was identified as mevalonic acid by Karl Folkers and others at Merck, Sharpe, and Dohme Laboratories. The discovery of mevalonate provided the missing link in the basic outline of cholesterol biosynthesis. Since that time, the sequence and the stereochemical course for the biosynthesis of cholesterol have been defined in detail.

## Mevalonate Is a Key Intermediate in Cholesterol Biosynthesis

The sequence of cholesterol biosynthesis begins with a condensation in the cytosol of two molecules of acetyl-CoA in a reaction catalyzed by thiolase (fig. 20.3). The next step requires the enzyme β-hydroxy-β-methylglutaryl-CoA (HMG-CoA) synthase. This enzyme catalyzes the condensation of a third acetyl-CoA with β-ketobutyryl-CoA to yield HMG-CoA. HMG-CoA is then reduced to mevalonate by HMG-CoA reductase. The activity of this reductase is primarily responsible for control of the rate of cholesterol biosynthesis.

HMG-CoA is an important intermediate for the biosynthesis of both cholesterol and ketone bodies (discussed

## Figure 20.3

Formation of mevalonate. The first two enzymes, thiolase and synthase, are found in both cytosol and mitochondria. The lyase that catalyzes ketone body formation is found only in the mitochondria. The reductase that catalyzes mevalonate formation is found in the endoplasmic reticulum. β-Ketobutyryl-CoA is also known as acetoacetyl-CoA.

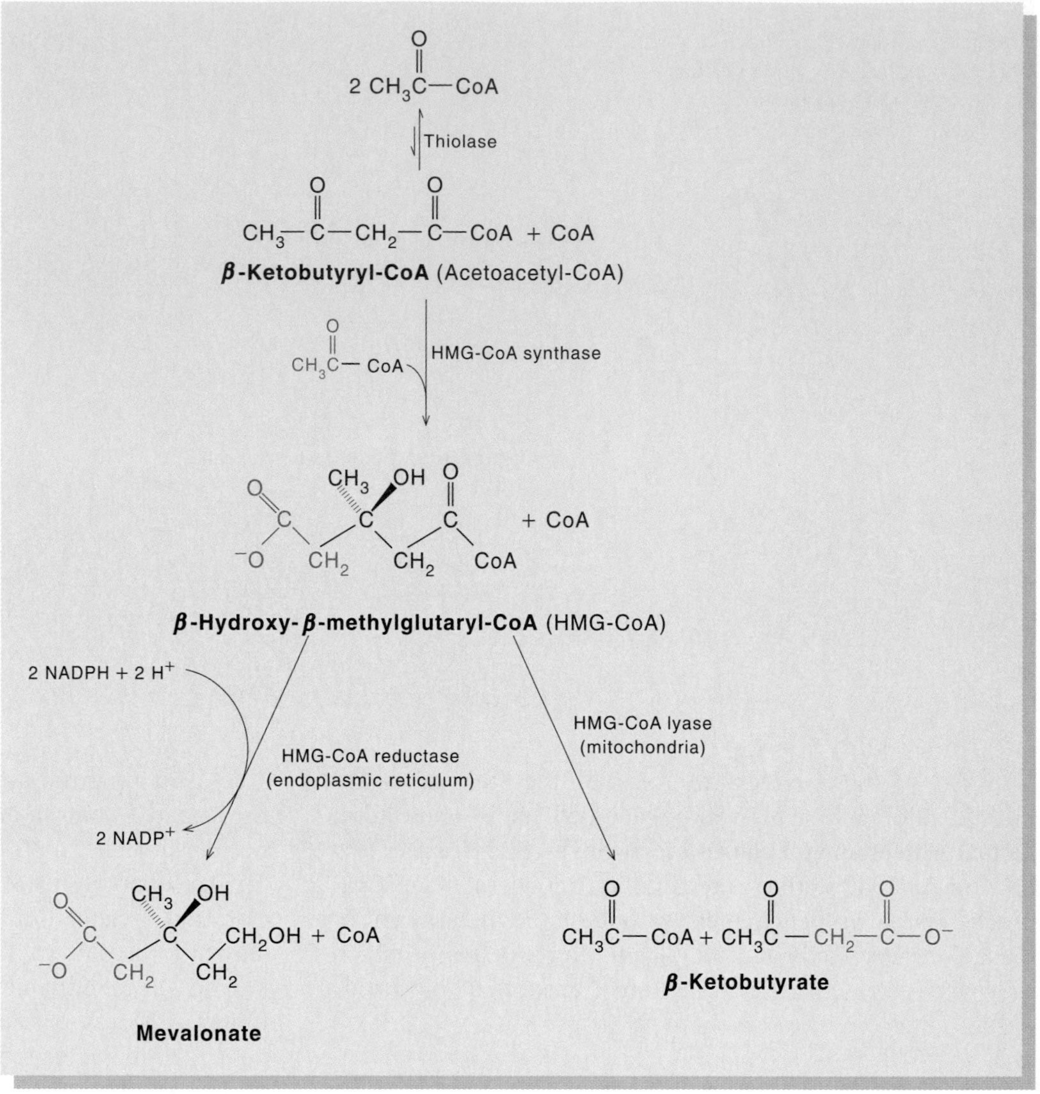

## The Rate of Mevalonate Synthesis Determines the Rate of Cholesterol Biosynthesis

in chapter 18). The biosynthesis of cholesterol is catalyzed by enzymes in the cytosol and enzymes bound to the endoplasmic reticulum. The synthesis of ketone bodies, however, is restricted to the mitochondrial matrix. Thus, thiolase and HMG-CoA synthase are found in both mitochondria and cytosol of rat liver. In contrast, HMG-CoA lyase, which cleaves HMG-CoA to ketone bodies (see chapter 18), is located only in mitochondria. HMG-CoA reductase is bound to the endoplasmic reticulum.

The thiolase and HMG-CoA synthase exhibit some regulatory properties in rat liver (cholesterol feeding causes a decrease in these enzyme activities in the cytosol but not in the mitochondria). However, the primary regulation of cholesterol biosynthesis appears to be centered on the HMG-CoA reductase reaction. HMG-CoA reductase is found on the endoplasmic reticulum, has a molecular weight of 97,092, and consists of 887 amino acids in a single polypeptide chain. The sequence of the enzyme was deduced by Michael Brown and Joseph Goldstein from the sequence of a piece of complimentary DNA (cDNA) derived from mRNA that codes for the reductase. The enzyme

## Figure 20.4

Proposed structure of HMG-CoA reductase derived from studies of recombinant DNA that codes for the enzyme. The enzyme is attached to the endoplasmic reticulum membrane and consists of two domains: the hydrophobic domain, embedded in the membrane, and the catalytic domain, which protrudes into the cytosol. (Source: Adapted from L. Liscum, J. Finer-Moore, R. M. Stroud, K. L. Luskey, M. S. Brown, and J. L. Goldstein, Domain structure of 3-hydroxy-3-methylglutaryl coenzyme A reductase, a glycoprotein of the endoplasmic reticulum, *J. Biol. Chem.* 260:522–530, 1985.)

## Figure 20.5

Structure of lovastatin acid, a potent competitive inhibitor of HMG-CoA reductase. Note the similarity in structure of the red portion of the molecule with mevalonate, the product of the HMG-CoA reductase reaction.

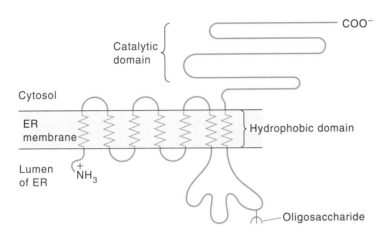

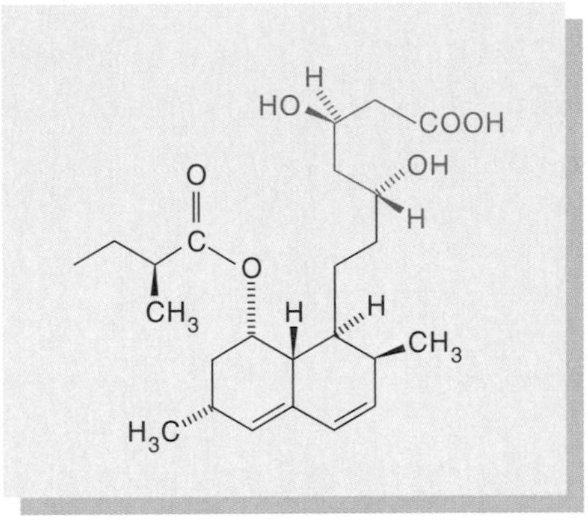

has two domains (fig. 20.4). The amino-terminal domain has seven hydrophobic segments that are thought to cross the membrane as shown in figure 20.4. The carboxyl-terminal domain contains the catalytic site of the enzyme and is thought to protrude into the cytosol.

The activity of the reductase is regulated by three distinct mechanisms. The first control point is at the level of gene expression. The amount of the reductase mRNA produced is modulated by the supply of cholesterol. When cholesterol is in excess, the amount of mRNA for HMG-CoA reductase is reduced; hence, less enzyme is made. In contrast, depletion of cholesterol enhances synthesis of the reductase mRNA.

The second regulatory mechanism involves the degradation of HMG-CoA reductase. As stated in chapter 29 the amount of an enzyme in a cell is determined by both its rate of synthesis and its rate of degradation. The rate of degradation of the reductase appears to be modulated by the supply of cholesterol. Thus, when cholesterol is abundant, the rate of enzyme degradation is twice as fast as when there is a limited supply of cholesterol. The effect of cholesterol on enzyme degradation is mediated by the membrane domain of the enzyme.

The third regulatory mechanism is phosphorylation–dephosphorylation of the reductase, which causes inactivation and activation, respectively. The kinases and phospha-

tases mediate short-term changes in the activity of HMG-CoA reductase.

Cholesterol biosynthesis is also controlled by the plasma levels of low-density lipoproteins, which we discuss in the context of lipoprotein metabolism later in this chapter.

It has long been recognized that a correlation exists between a high level of serum cholesterol and cardiovascular disease (e.g., most heart diseases, stroke). Most serum cholesterol originates from the liver; thus a drug that specifically inhibits cholesterol biosynthesis has been sought. It is logical that such a drug would inactivate HMG-CoA reductase because this enzyme catalyzes the key regulatory step in the pathway. Several fungal metabolites have been isolated that are competitive inhibitors of HMG-CoA reductase. One of the most active compounds is lovastatin (fig. 20.5), which competes favorably ($K_I = 0.6$ nM) with HMG-CoA for the reductase and lowers serum cholesterol in humans. This drug is used for treatment of patients with hypercholesterolemia.

## Six Mevalonates and 10 Steps Are Required to Make Lanosterol, the First Tetracyclic Intermediate

In the next segment of the pathway, mevalonate is converted to squalene, which is cyclized to form lanosterol. The first stage in this sequence of reactions is the synthesis of the five-carbon isoprenoid intermediates, isopentenyl pyro-

## Figure 20.6

The conversion of mevalonate to isopentenyl pyrophosphate and dimethylallyl pyrophosphate. Mevalonate is converted to isopentenyl pyrophosphate in three steps. Each of these steps requires one ATP

cleavage. The conversion of isopentenyl pyrophosphate to dimethylallyl pyrophosphate is readily reversible and does not require any further expenditure of ATP.

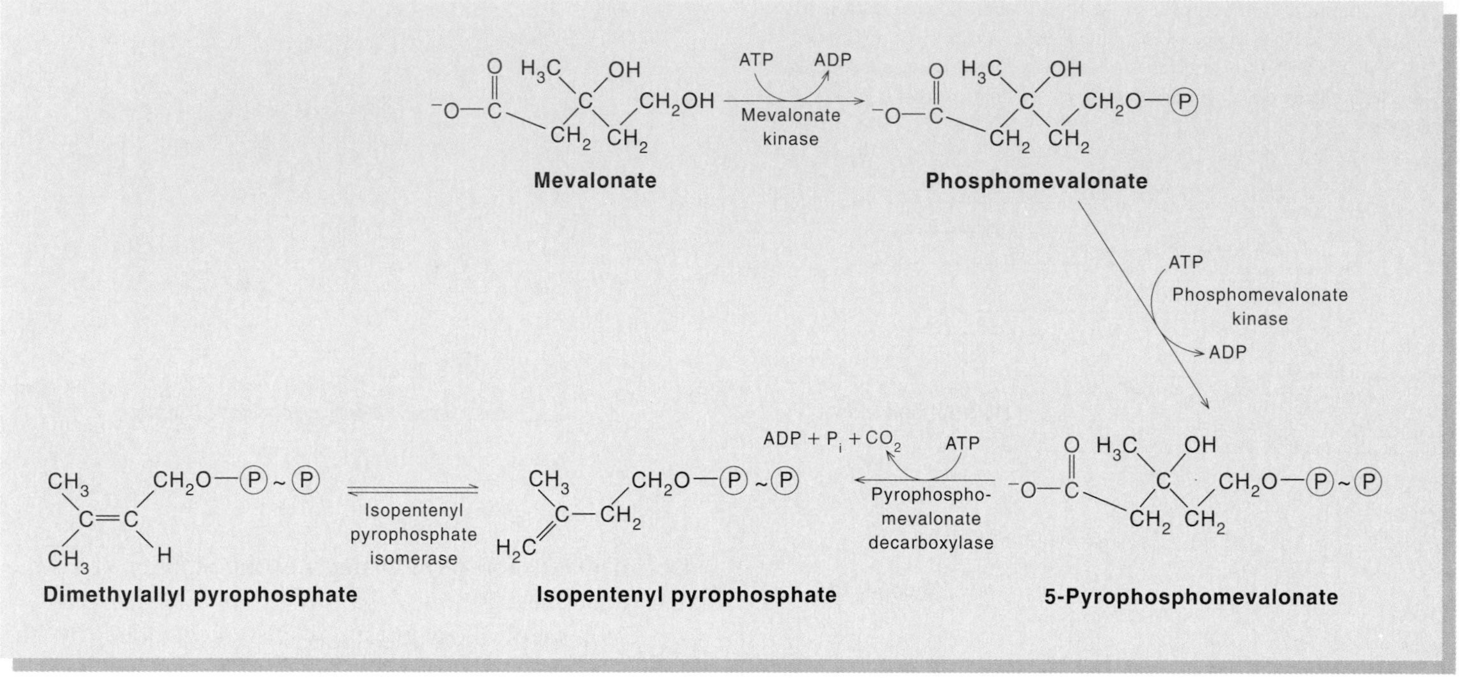

phosphate and dimethylallyl pyrophosphate. The synthesis involves four enzymes (fig. 20.6). One molecule of each of the two $C_5$ isoprenoid pyrophosphates reacts to produce the $C_{10}$ intermediate, geranyl pyrophosphate (fig. 20.7). Subsequently, a $C_{15}$ intermediate, farnesyl pyrophosphate, is formed by the reaction of another molecule of isopentenyl pyrophosphate with geranyl pyrophosphate. Two molecules of farnesyl pyrophosphate react to form presqualene pyrophosphate, which rearranges with the elimination of $PP_i$ to yield the $C_{30}$ intermediate, squalene (fig. 20.8).

The two remaining reactions in the biosynthesis of lanosterol are shown in figure 20.9. In the first of these reactions, squalene-2,3-oxide is formed from squalene. As can be seen in figure 20.8, squalene is a symmetrical molecule, hence the formation of squalene oxide can be initiated from either end of the molecule. The oxide is converted into lanosterol. The reaction can be formulated as proceeding by means of a protonated intermediate that undergoes a concerted series of *trans*-1,2 shifts of methyl groups and hydride ions to produce lanosterol (see fig. 20.9).

## From Lanosterol to Cholesterol Takes Approximately 20 Steps

The last sequence of reactions in the biosynthesis of cholesterol involves approximately 20 enzymatic steps, starting with lanosterol. In mammals the major route involves a series of double-bond reductions and demethylations (fig. 20.10). The sequence of reactions involves reduction of the $\Delta^{24}$ double bond, the oxidation and removal of the $14\alpha$ methyl group followed by the oxidation and removal of the two methyl groups at position 4 in the sterol. The final reaction is a reduction of the $\Delta^7$ double bond in 7-dehydrocholesterol. An alternative pathway from lanosterol to cholesterol also exists. The enzymes involved in the transformation of lanosterol to cholesterol are all located on the endoplasmic reticulum.

## Summary of Cholesterol Biosynthesis

In this section we discussed the remarkable set of reactions that converts a two-carbon precursor, acetyl-CoA, into cho-

## Figure 20.7

Biosynthesis of farnesyl pyrophosphate. Farnesyl pyrophosphate is a $C_{15}$ intermediate containing three $C_5$ isoprenoid subunits. The two transferase reactions involved in the formation of farnesyl pyrophosphate occur by virtually identical mechanisms as shown.

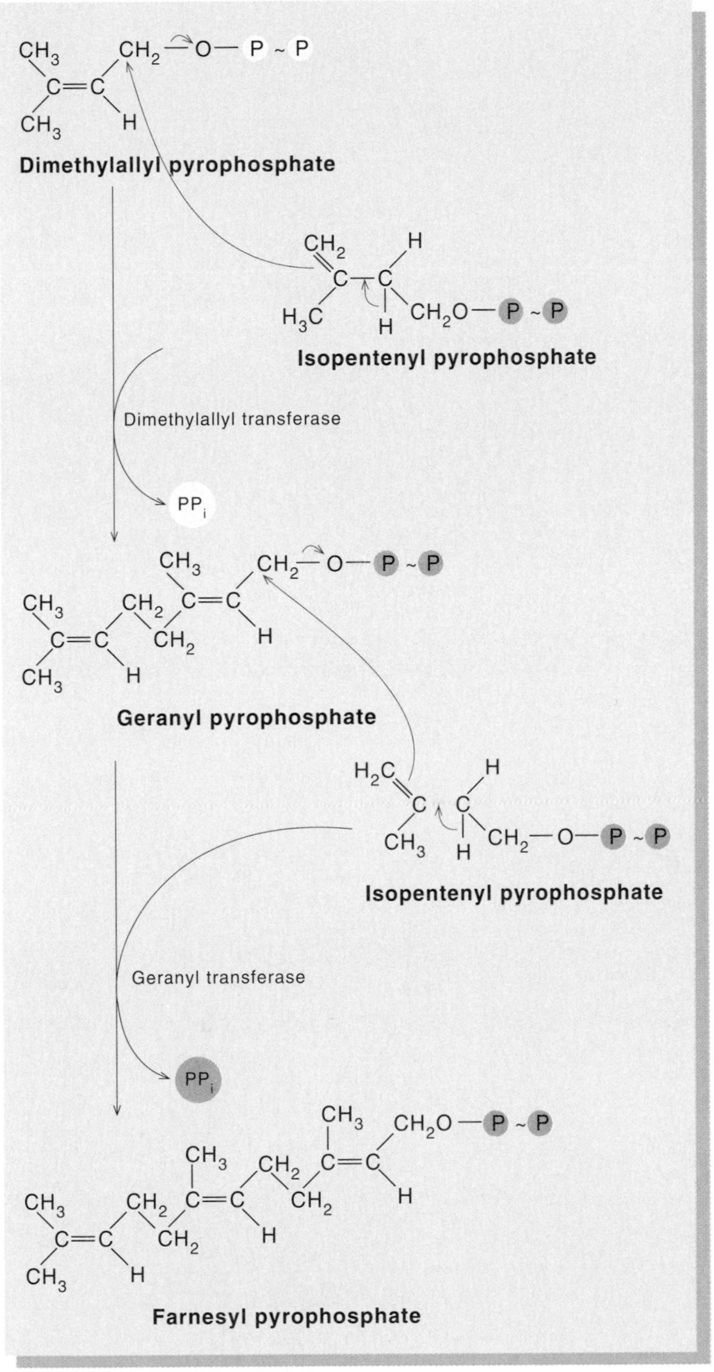

lesterol, a tetracyclic hydrocarbon with a rigid ring structure and a single hydroxyl substituent. The reactions occur in the cytosol and on the endoplasmic reticulum. The key enzyme in the sequence is HMG-CoA reductase, which is regulated by the level of gene expression of the enzyme, by enzyme degradation, and by phosphorylation–dephosphorylation. We now examine how cholesterol and other lipids, particularly triacylglycerols, are transported in the plasma.

# Lipoprotein Metabolism

Unesterified fatty acids are carried in plasma by albumin (chapter 18). The plasma also transports more complex lipids (cholesterol, triacylglycerols) among the various tissues as components of lipoproteins (spherical particles composed of lipids and proteins). Because cholesterol and triacylglycerol are insoluble in an aqueous medium such as the plasma, these lipoproteins (which are soluble in plasma) have evolved for the purpose of transporting complex lipids among tissues. In this section we are concerned with the structure and metabolism of these lipoproteins.

## There Are Five Types of Lipoproteins in Human Plasma

The structures of the various lipoproteins appear to be similar (figs. 20.11 and 20.12). Each of the lipoprotein classes contains a neutral lipid core composed of triacylglycerol and/or cholesteryl ester. Around this core is a coat of protein, phospholipid, and cholesterol, with the polar portions oriented toward the surface of the lipoprotein and the hydrophobic parts associated with the neutral lipid core. The hydrophilic surface interacts with water in plasma, promoting the solubility of the lipoprotein.

The amounts and types of lipids found in human plasma fluctuate according to the dietary habits and metabolic states of the individual. The normal ranges for the lipid levels in plasma are shown in table 20.1. In plasma these lipids are associated with proteins in the form of lipoproteins, which are classified into five major types on the basis of their density (table 20.2, see fig. 20.11). The lipoproteins of lowest density, the triacylglycerol-rich chylomicrons, are the largest in size and contain the most lipid and smallest percentage of protein. At the other extreme are the cholesteryl ester-rich high-density lipoprotein (HDLs), which are the smallest particles and contain the highest percentage by weight of protein and lowest percentage of lipid.

# Figure 20.8

Formation of squalene from farnesyl pyrophosphate. Farnesyl transferase is tightly complexed to the endoplasmic reticulum. Three carbons are labeled ●, ▲, ★) in different structures for purposes of tracking them as the reaction proceeds.

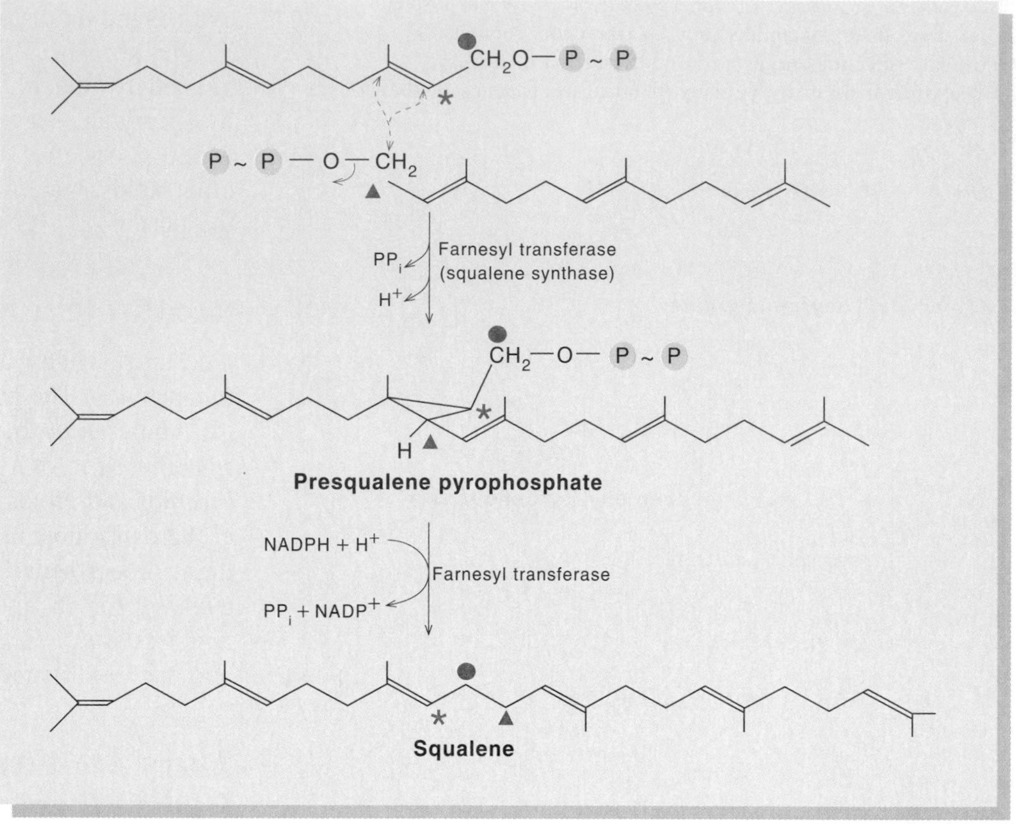

# Figure 20.9

The transformation of squalene into lanosterol. The squalene monooxygenase reaction requires $O_2$, NADPH, FAD, phospholipid, and a cytosolic protein. The cyclase reaction has no known cofactor requirements. The reaction proceeds by means of a protonated intermediate that undergoes a concerted series of *trans*-1,2 shifts of methyl groups and hydride ions to produce lanosterol.

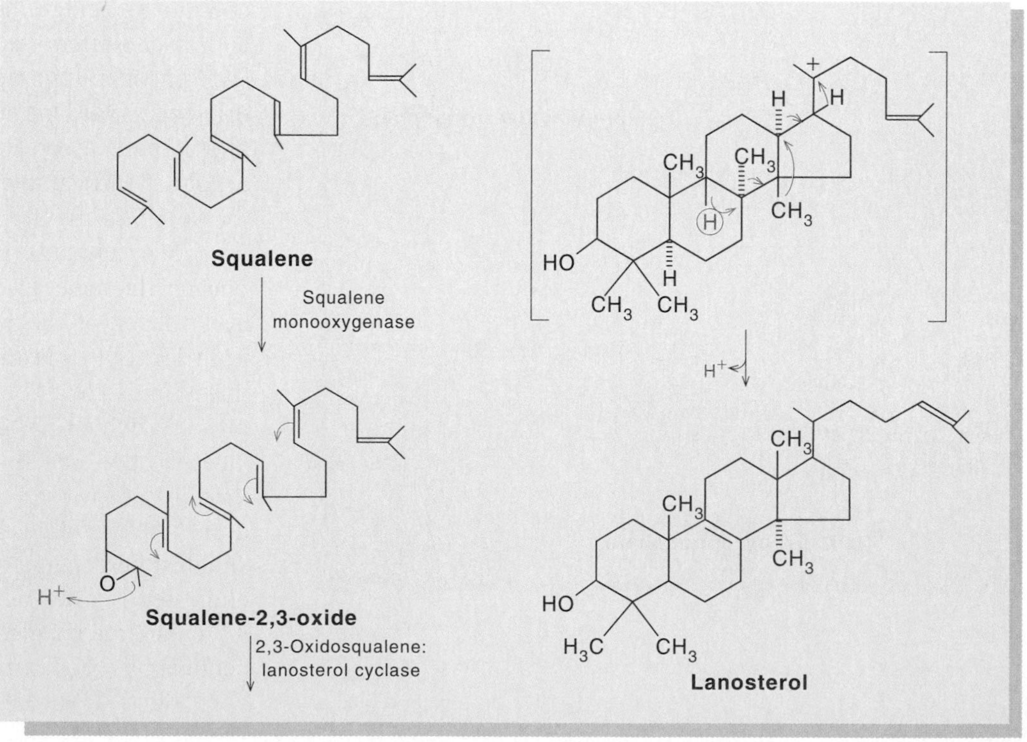

**Figure 20.10**

Two pathways for the conversion of lanosterol to cholesterol. The major route in mammals proceeds through 7-dehydrocholesterol.

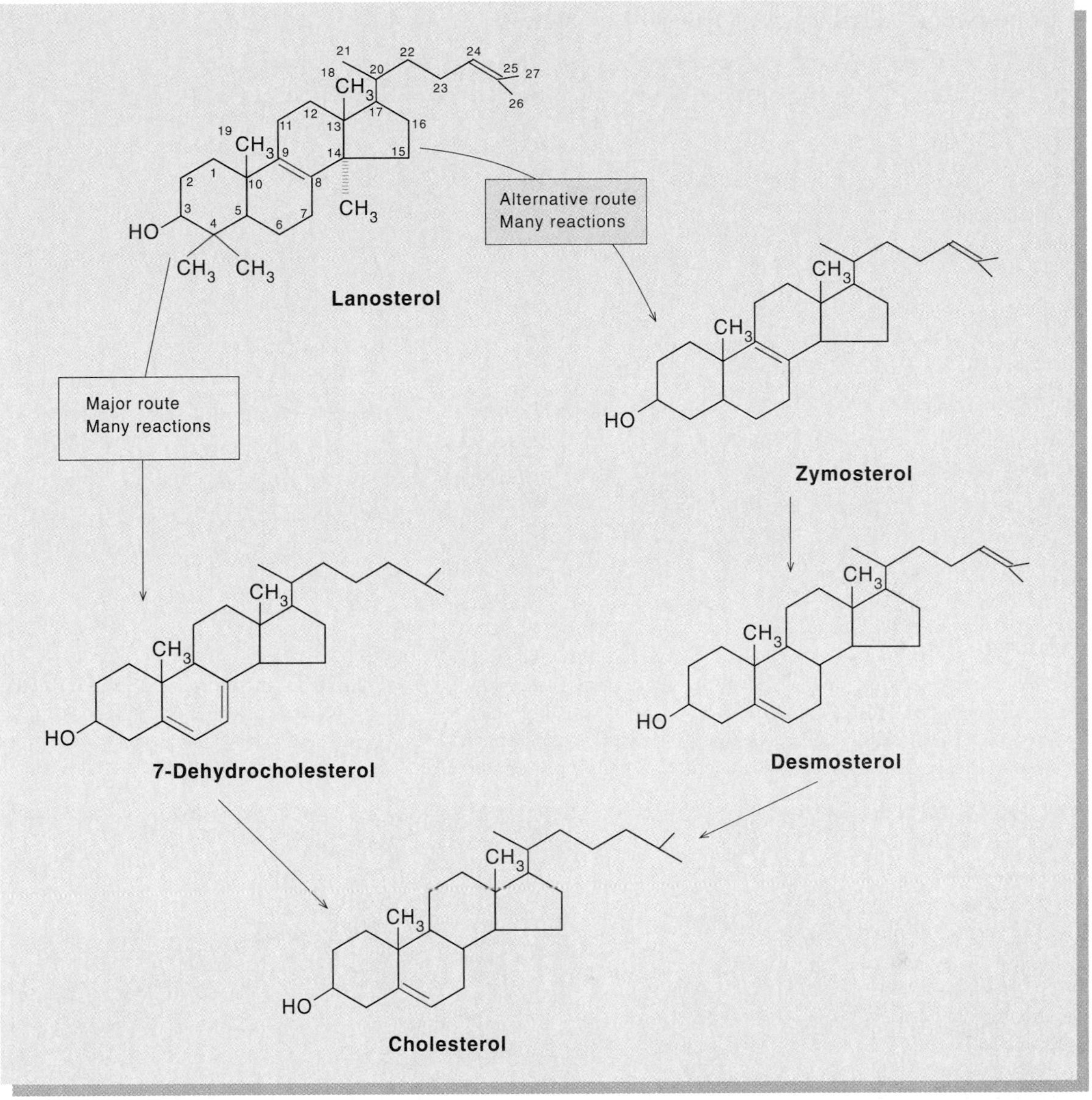

Lanosterol

Alternative route
Many reactions

Major route
Many reactions

Zymosterol

7-Dehydrocholesterol

Desmosterol

Cholesterol

**Table 20.1**

Normal Concentrations of the Major Lipid Classes in Plasma in Humans

| Lipid | Concentration (mg/100 ml) |
|---|---|
| Total lipid | 360–680 |
| Cholesterol and cholesteryl ester | 130–260 |
| Triacylglycerol | 80–240 |
| Phospholipid | 150–240 |

Between these two classes, in both size and composition, are the cholesteryl ester-rich low-density lipoproteins (LDLs), the intermediate-density lipoproteins (IDLs), and the triacylglycerol-rich very-low-density lipoproteins (VLDLs).

There are at least nine apoproteins associated with the lipoproteins, as well as several enzymes and a cholesteryl ester transfer protein. There are two major types of apoproteins. Two apoproteins (apo B100 and apo B48) are tightly integrated into the phospholipid monolayer. The other seven proteins are less tightly associated with the phospholipid and exchange among the lipoproteins. The apoproteins have three major functions. (1) They are important structural components. (2) Some of the apoproteins modulate the ac-

**Table 20.2**

Composition and Density of Human Lipoproteins

|  | Chylomicron | VLDL | IDL | LDL | HDL |
|---|---|---|---|---|---|
| Density (g/ml) | <0.95 | 0.95–1.006 | 1.006–1.019 | 1.019–1.063 | 1.063–1.210 |
| Diameter (nm) | 75–1200 | 30–80 | 25–35 | 18–25 | 5–12 |
| Components (% dry weight) |  |  |  |  |  |
| Protein | 1–2 | 10 | 18 | 25 | 33 |
| Triacylglycerol | 83 | 50 | 31 | 10 | 8 |
| Cholesterol and cholesteryl esters | 8 | 22 | 29 | 46 | 30 |
| Phospholipids | 7 | 18 | 22 | 22 | 29 |
| Apoprotein Composition | A-I, A-II B48 C-I, C-II, C-III | B100 C-I, C-II, C-III E | B100 C-I, C-II, C-III E | B100 | A-I, A-II C-I, C-II, C-III D E |

**Figure 20.11**

Electron micrographs of low-density lipoproteins (LDL), very-low-density lipoproteins (VLDL), chylomicrons (Chylo), and high-density lipoproteins (HDL). Lipids are transported in plasma as components of these particles. The larger particles contain a higher percentage of lipid and therefore are less dense, whereas the smaller particles have less lipid, a higher percentage of protein, and are more dense. (Electron micrographs courtesy of Dr. Robert Hamilton, University of California, San Francisco.)

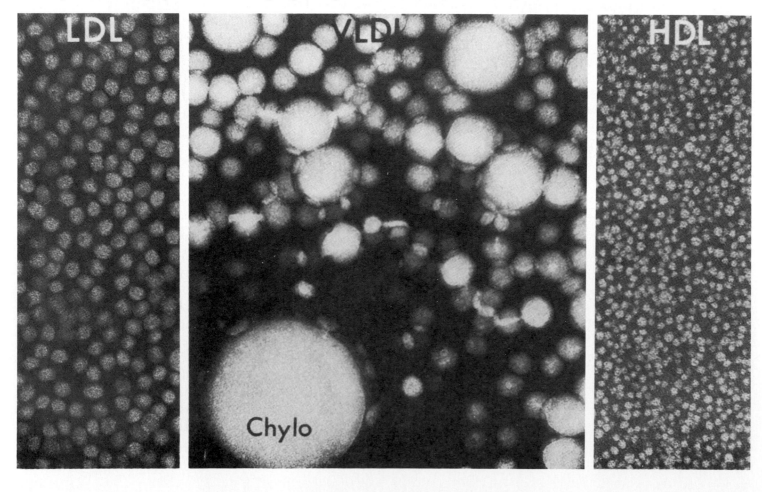

**Table 20.3**

Properties of the Apoproteins of the Major Human Lipoprotein Classes

| Apoprotein | $M_r$ | Plasma Concentration (mg/100 ml) | Miscellaneous |
|---|---|---|---|
| A-I | 29,016 | 9–120 | Major protein in HDL (64%); activates LCAT[a] |
| A-II | 17,400 | 30–50 | Mainly in HDL (20% of dry mass); two identical chains joined by a disulfide at residue 6 |
| B-100 | 513,000 | 80–100 | A single polypeptide that contains 4,536 amino acids; major protein in LDL; is made in the liver and binds to the LDL receptor |
| B-48 | 241,000 | <5 | A protein found exclusively in chylomicrons in humans; made in the intestine |
| C-I | 7,000 | 4–7 | Function unknown |
| C-II | 9,000 | 3–8 | Activates lipoprotein lipase |
| C-III | 9,300 | 8–15 | Inhibits lipoprotein lipase |
| D | 19,000 | 8–10 | Associated with HDL; function unknown |
| E | 33,000 | 3–6 | Found on VLDL and chylomicrons and binds to the LDL receptor |

[a] LCAT = Lecithin-cholesterol acyltransferase.

**Figure 20.12**

Generalized structure of human lipoproteins. The lipoproteins are spherical and vary in diameter from 10 nm to as much as 1,000 nm, depending on the particular proteins and lipids. Each of the lipoprotein classes contains a neutral lipid core composed of triacylglycerol or cholesterol ester or both. Around the core is a layer of protein, phospholipid, and cholesterol that is oriented with the polar portions exposed to the surface of the lipoprotein. Note micellar-like construction.

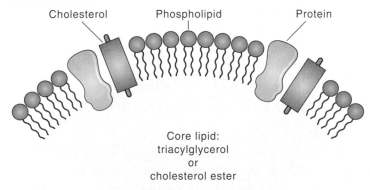

Cholesterol        Phospholipid        Protein

Core lipid:
triacylglycerol
or
cholesterol ester

tivities of enzymes in plasma that are associated with lipoprotein metabolism. (3) Several of the apoproteins interact with receptors on the cell surface.

The structure and function of these apoproteins has been intensely studied in the past decade, and some of the properties of these apoproteins are summarized in table 20.3. All of the apoproteins from human plasma have been sequenced and contain regions that are rich in hydrophobic amino acids, which facilitate binding of lipid.

## Lipoproteins Are Made in the Endoplasmic Reticulum of the Liver and Intestine

Of the various lipid components of the lipoproteins, only the biosynthesis of cholesteryl esters has not yet been mentioned. Cholesteryl ester is the storage form of cholesterol in cells. It is synthesized from cholesterol and acyl-CoA by acyl-CoA:cholesterol acyltransferase (ACAT) (fig. 20.13), which is located on the cytosolic surface of hepatic endoplasmic reticulum. Acylation of the 3 hydroxyl group of cholesterol eliminates the polarity of cholesterol and facilitates the packing of cholesterol as its ester in the core of the lipoprotein or for storage in lipid droplets within cells.

The synthesis of the apoproteins takes place on ribosomes that are bound to the endoplasmic reticulum. As we mentioned previously, the biosynthesis of the other lipids in lipoproteins (cholesterol, triacylglycerols, and phospholipids) also occurs on the endoplasmic reticulum.

Figure 20.13

Biosynthesis of cholesteryl esters. The acyl-CoA: cholesterol acyltransferase involved in cholesteryl ester synthesis is located on the cytosolic surface of liver endoplasmic reticulum.

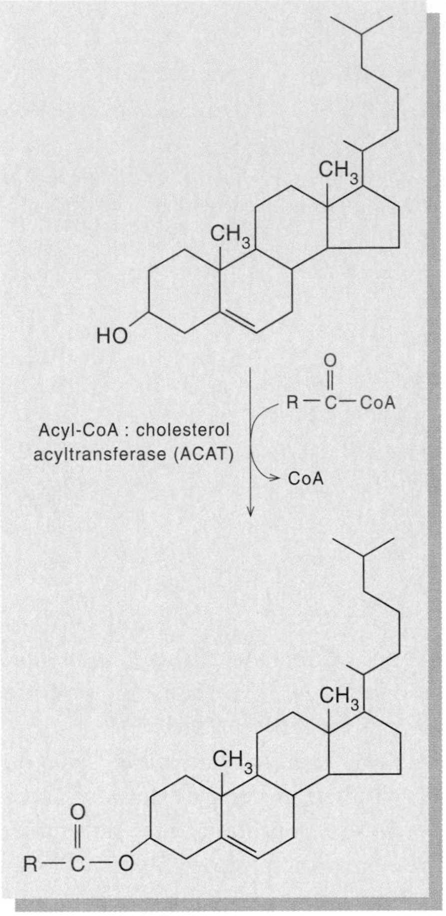

Figure 20.14

Postulated scheme for the synthesis, assembly, and secretion of VLDL by a hepatocyte (liver cell). (1) Synthesis: The apoproteins, phospholipid, triacylglycerol, cholesterol, and cholesteryl esters are synthesized in the endoplasmic reticulum. (2) Assembly: These components are assembled into a prelipoprotein particle in the lumen of the endoplasmic reticulum. (3) Processing: The particle moves to the Golgi apparatus, where modification of the apoproteins occurs. (4) Vesicle formation: A secretory vesicle containing the lipoprotein particles is formed and fuses with the plasma membrane. (5) Secretion: The VLDL is released into the circulation.

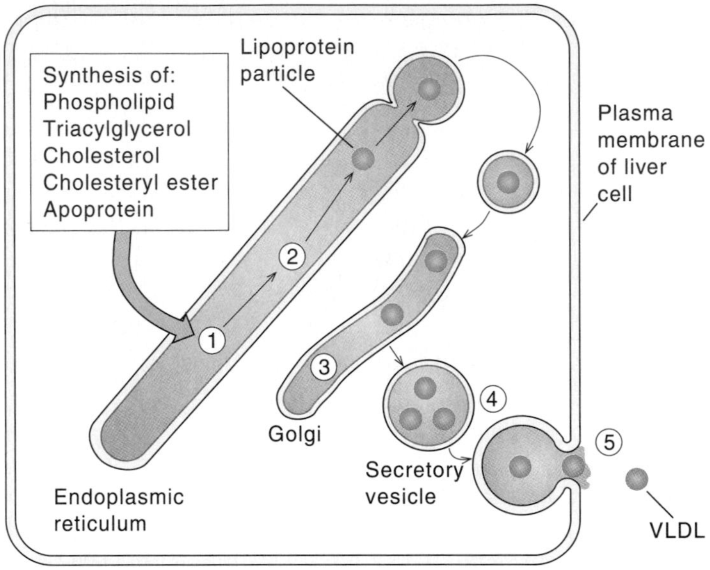

Low-density lipoproteins (LDLs) are produced in plasma from VLDLs, as we shall see later.

How the various components of the lipoproteins are assembled and secreted into the plasma is not known. Current ideas for this process suggest the assembly of VLDL from its components occurs in the endoplasmic reticulum, VLDL is transferred to the Golgi apparatus and is subsequently secreted via secretory (membrane-encapsulated) vesicles. These vesicles fuse with the plasma membrane and release their lipoprotein contents into plasma (fig. 20.14).

The plasma lipoproteins are made mainly in the liver and intestine. In the rat, approximately 80% of the plasma apoproteins originate from the liver; the rest are derived from the intestine. The components of chylomicrons, including apoproteins A-I, A-IV, and B-48; phospholipid; cholesterol; cholesteryl ester; and triacylglycerols, are products of the intestinal cells. Chylomicrons are secreted into lymphatic capillaries, which eventually enter the bloodstream. The liver is the major source of VLDLs and HDLs.

## Chylomicrons and Very-Low-Density Lipoproteins (VLDLs) Transport Cholesterol and Triacylglycerol to Other Tissues

Chylomicrons transport dietary triacylglycerol and cholesteryl ester from the intestine to other tissues in the body. Very-low-density lipoprotein functions in a manner similar to the transport of endogenously made lipid from the liver to other tissues. These two types of triacylglycerol-rich particles are initially degraded by the action of lipoprotein lipase, an extracellular enzyme that is most active within the capillaries of adipose tissue, cardiac and skeletal muscle, and the lactating mammary gland. Lipoprotein lipase catalyzes the hydrolysis of triacylglycerols (see fig. 18.3). The enzyme is specifically activated by apoprotein C-II, which is associated with triacylglycerol-rich chylomicrons and VLDLs, as well as IDL and HDL. As a result, this lipase supplies the heart and adipose tissue with fatty acids, derived from chylomicrons and VLDLs in plasma. The fatty acids produced by the action of lipoprotein lipase on circulating lipoproteins can be taken into the heart, adipose, and muscle tissues and be used for energy or stored as a component of triacylglycerols. Alternatively, the fatty acids can be bound by albumin and transported to other tissues.

As the lipoproteins are depleted of triacylglycerol, the particles become smaller. Some of the surface molecules (apoproteins, phospholipids) are transferred to HDL. In the rat, ''remnants'' that result from chylomicron catabolism are removed by the liver. The uptake of remnant VLDL also occurs, but much of the triacylglycerol is further degraded by lipoprotein lipase to give the intermediate-density lipoprotein (IDL). This particle is converted into LDL via the action of lipoprotein lipase and enriched in cholesteryl ester via transfer from HDL by the cholesteryl ester transfer protein. The half-life for clearance of chylomicrons from plasma of humans is 4–5 min. Patients with the inherited disease, lipoprotein lipase deficiency, clear chylomicrons from the plasma very slowly. When on a normal diet, the blood from these patients looks like tomato soup. A very-low-fat diet greatly relieves this problem.

## Low-Density Lipoproteins (LDLs) Are Removed from the Plasma by the Liver, Adrenals, and Adipose Tissue

Each day approximately 45% of the LDL in human plasma is removed by both the liver and extrahepatic tissues (particularly the adrenals and adipose tissue). The mechanism for the uptake of LDL has been extensively described by Michael Brown and Joseph Goldstein. The apo B100 in LDL particles binds to the cell surface by specific receptors that congregate in areas of the plasma membrane called coated regions, or pits (figs. 20.15, 20.16). These areas of plasma membranes engulf the LDL particles (in a process called endocytosis) to form ''coated vesicles,'' which are somehow directed toward and fuse with lysosomes. The LDL particles are degraded within the lysosomes by the action of proteases and lysosomal acid lipases (lipid degradative enzymes). The cholesterol, or a derivative of cholesterol, diffuses from the lysosomes, suppresses the synthesis of the mRNA for HMG-CoA reductase, and stimulates the activity of acyl-CoA:cholesterol acyltransferase (ACAT). ACAT catalyzes the synthesis of cholesteryl esters (see fig. 20.13), which are then stored within the cell. The cholesterol (or a derivative) also suppresses the synthesis of the LDL receptors and thereby limits the uptake of LDL.

**Figure 20.15**

Receptor-mediated uptake of LDL by human skin fibroblasts. Specific LDL receptors are located in coated regions of the plasma membrane. LDL binding results in uptake by endocytosis and formation of a coated vesicle. This vesicle fuses with a lysosome containing many hydrolytic enzymes that degrade the lipoprotein, releasing cholesterol.

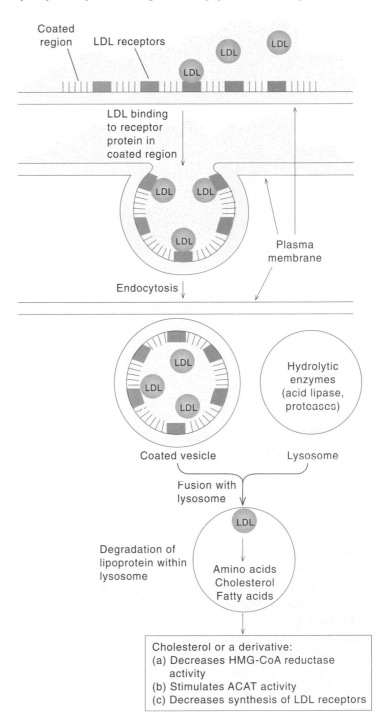

**Figure 20.16**

Electron micrograph of LDL particles (made electron-dense with covalently bound ferritin) bound to coated regions of a human skin fibroblast (97,000 × ). (From R. G. W. Anderson, M. S. Brown, and J. L. Goldstein, Role of the coated and endocytic vesicle in the uptake of receptor-bound low-density lipoprotein in human fibroblasts. *Cell* 10:351, 1977. © Cell Press.)

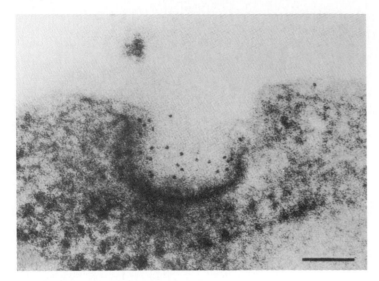

## Serious Diseases Result from Cholesterol Deposits

Brown and Goldstein were able to deduce the pathway for the uptake and catabolism of LDL largely as a result of their studies on the inherited disease familial hypercholestrolemia. Patients with the homozygous (two defective genes) form of this disease have grossly elevated levels of plasma cholesterol (650–1,000 mg/100 ml) (see table 20.1 for normal values), which is largely carried by an elevated concentration of LDL. One result is the formation of cholesterol deposits in the skin (xanthomas) in various areas of the body. Of greater consequence is the deposition of cholesterol in arteries, which results in atherosclerosis, a condition that is the underlying cause of most cardiovascular diseases. Sadly, patients with homozygous familial hypercholesterolemia have symptoms of heart disease by the early teens and usually die from cardiovascular disease before the age of 20. Heterozygous individuals (who have one normal and one defective gene) manifest similar but less severe symptoms. Their plasma cholesterol is in the range of 250–550 mg/100 ml of plasma, and they generally do not have a heart attack before the age of 40. The frequency of the heterozygous form of familial hypercholesterolemia has been estimated at 1 in 500, and that of the homozygous form is about 1 in 1 million.

Familial hypercholesterolemia results from a defective LDL receptor. As a result, the cholesteryl ester in the LDL of the bloodstream cannot be cleared by the normal process, hence, high cholesterol levels occur in the plasma of these patients. Somehow, the high levels of cholesterol lead to the formation of the atherosclerotic lesions, the underlying cause of premature heart disease in these patients.

The importance of the work done on the LDL receptor and familial hypercholesterolemia was recognized in 1985 when Joseph Goldstein and Michael Brown were awarded the Nobel Prize in Physiology or Medicine.

## *High-Density Lipoproteins (HDLs) May Reduce Cholesterol Deposits*

Unlike fatty acids, cholesterol is not degraded to yield energy. Instead excess cholesterol is removed from tissues by HDL for delivery to the liver from which it is excreted in the form of bile salts into the intestine. The transfer of cholesterol from extrahepatic tissues to the liver is called reverse cholesterol transport. When HDL is secreted into the plasma from the liver, it has a discoidal shape and is almost devoid of cholesteryl ester. These newly formed HDL particles are good acceptors for cholesterol in the plasma membranes of cells and are converted into spherical particles by the accumulation of cholesteryl ester. The cholesteryl ester is derived from a reaction between cholesterol and phosphatidylcholine on the surface of the HDL particle catalyzed by lecithin:cholesterol acyltransferase (LCAT) (fig. 20.17). LCAT is associated with HDL in plasma and is activated by apoprotein A-I, a component of HDL (see table 20.3). Associated with the LCAT-HDL complex is cholesteryl ester transfer protein, which catalyzes the transfer of cholesteryl esters from HDL to VLDL or LDL. In the steady state, cholesteryl esters that are synthesized by LCAT are transferred to LDL and VLDL and are catabolized as noted earlier. The HDL particles themselves turn over, but how they are degraded is not firmly established.

Although elevated levels of cholesterol and LDL in human plasma are linked to an increased incidence of cardiovascular disease, recent data have shown that an increase in concentration of HDL in plasma is correlated with a lowered risk of coronary artery disease. Why does an elevated HDL level in plasma appear to protect against cardiovascular disease, whereas an elevated LDL level seems to cause this disease? The answer to this question is not known. An explanation currently favored is that HDL functions in the removal of cholesterol from nonhepatic tissues and the return of cholesterol to the liver, where it is metabolized and secreted. The net effect would be a decrease in the amount of plasma cholesterol available for deposit in arteries (see

## Figure 20.17

Reaction catalyzed by lecithin: cholesterol acyltransferase (LCAT). The resulting cholesteryl ester is transferred to VLDL and LDL particles by a lipid transfer protein.

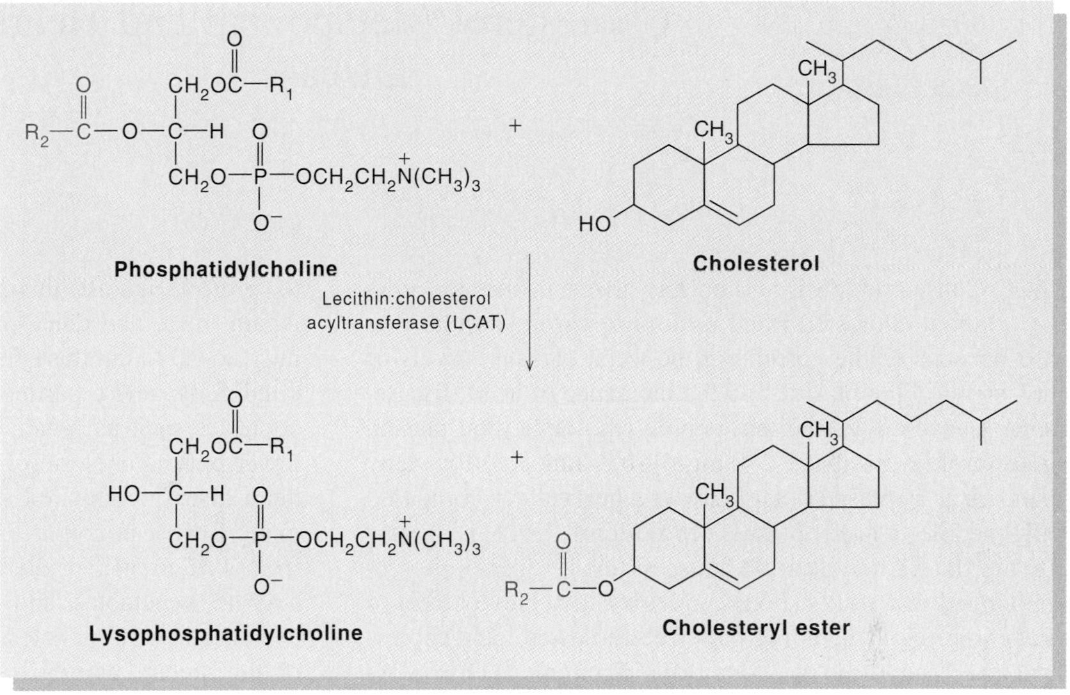

## Figure 20.18

Formation of 7-hydroxycholesterol. The committed and rate-limiting reaction for bile acid synthesis is catalyzed by the endoplasmic reticulum enzyme 7α-hydroxylase.

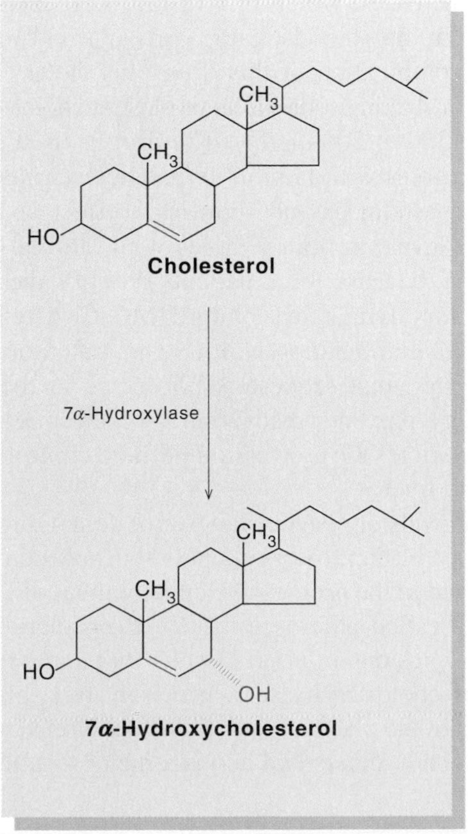

box 20A for more information on cholesterol and heart disease).

## Bile Acid Metabolism

Bile acids are sterol derivatives derived from cholesterol that have two major functions. (1) The cholesterol delivered back to the liver by reverse cholesterol transport is converted into bile acids, which are excreted from the body via the intestine. (2) The bile acids secreted into the intestine are required for the solubilization of dietary lipids so that they can be degraded by lipases and absorbed into the intestinal wall (see fig. 18.2).

The conversion of cholesterol to bile acids is quantitatively the most important mechanism for degradation of cholesterol. In a normal human adult approximately 0.5 g of cholesterol is converted to bile acids each day. The regulation of this process operates at the initial biosynthetic step catalyzed by an enzyme in the endoplasmic reticulum, 7α-hydroxylase (fig. 20.18). The 7α-hydroxylase is one of a group of enzymes called mixed-function oxidases, which are involved in the hydroxylation of the sterol molecule at numerous specific sites. A mixed-function oxidase is an enzyme complex that catalyzes hydroxylation of a substrate with a concomitant production of $H_2O$ from a single molecule of $O_2$. The 7α-hydroxylase is one of several enzymes referred to as cytochrome P450.

# 20A

# Cholesterol Metabolism and Heart Disease

Considerable discussion has arisen in the lay press about cholesterol and its link to cardiovascular disease because of the correlation between elevated levels of cholesterol in the plasma and the incidence of heart disease. Experts generally agree that people who have total plasma cholesterol levels above 240 mg/dl (6.2 mmole/l) for many years are at increased risk of having a heart attack compared with people whose plasma cholesterol level is below 200 mg/dl (5.2 mmole/l). Because of this it is generally recommended that adults should endeavor to achieve levels of total cholesterol (including both free cholesterol and cholesteryl ester) in plasma of 200 mg/dl (5.2 mmole/l) or less. As discussed in the text, plasma cholesterol is largely carried in LDL as cholesteryl ester. The cholesteryl ester carried by LDL is sometimes referred to in the lay press as ''bad'' cholesterol.

However, the relationship between cholesterol levels in plasma and cardiovascular disease is not so simple, because high levels of HDL (which contains 30% cholesterol by weight—see table 20.2) appear to protect against heart attack. Not surprisingly, the cholesterol carried by HDL is referred to as ''good'' cholesterol in the lay press. In addition, many other factors have been linked to increased incidences of heart attack in epidemiological studies. Among these factors are smoking, high blood pressure, genetic background (recall the discussion of familial hypercholesterolemia), diabetes, and obesity.

How can people achieve total cholesterol levels of 200 mg/dl or less? If they were lucky and chose their parents carefully, they have genes that protect them from high cholesterol levels. Such people can eat a diet that is high in fat and cholesterol and still remain well below the 200 mg/dl threshold. Most adults are not in this category and therefore may have to use dietary or drug treatments to reduce plasma cholesterol levels. Cholesterol is abundant in animal meats and dairy products and is generally absent in vegetables. Thus, many people can reduce their plasma cholesterol levels by eating smaller amounts of meat and dairy products. Another complication is that the serum levels of cholesterol are affected by the amount of saturated and polyunsaturated fats in the diet. Saturated fats tend to increase plasma cholesterol levels, whereas polyunsaturated fats tend

to protect against increased cholesterol in the plasma. Again, meat and dairy products are abundant in saturated fatty acids. In contrast, fish have high levels of polyunsaturated fatty acids, particularly $C_{22:5}$ and $C_{22:6}$. People in countries such as Japan, who have diets rich in fish, have lower plasma cholesterol levels and a reduced incidence of heart disease compared with people in North America and many European countries. Thus, reducing the intake of fat from 40% to 30% of dietary calories by eating more carbohydrate, vegetables, and fish, often results in a 15%–20% decrease in serum cholesterol. However, the successfulness of the dietary approach varies widely among individuals because genetic factors again come into play.

Recently a new drug called lovastatin (see fig. 20.5) has become available by prescription for treatment of hypercholesterolemia. This compound, a competitive inhibitor of HMG-CoA reductase, sharply reduces the rate of cholesterol biosynthesis. Equally important, in response to the reduction of cellular cholesterol levels the expression of LDL receptors increases, which in turn allows more LDL to be cleared from the bloodstream, particularly by the liver. Thus, the combination of this drug plus dietary restriction can result in striking reductions in plasma cholesterol levels (e.g., from 300 to 200 mg/dl). The drug is used effectively by heterozygotes with familial hypercholesterolemia. Interestingly, lovastatin has no significant effect on the serum cholesterol levels in homozygotes with familial hypercholesterolemia, because these patients do not have functional LDL receptors. Drugs that inhibit HMG-CoA reductase are often used in combination with a resin, taken orally with a glass of water, that sequesters bile acids in the intestine. This strategy prevents reabsorption of the bile acids and, thereby, promotes the conversion of more cholesterol to bile acids in the liver.

High levels of plasma cholesterol do not directly cause heart attacks. Rather, over long periods cholesterol is somehow involved in the progressive development of a disease of the arteries called atherosclerosis. Atherosclerotic plaques are complex lesions in arterial walls that contain abnormal deposits of cholesteryl esters. Precisely how high cholesterol levels in the plasma relate to the development of atherosclerosis is not understood and is a major frontier of medical research today. The actual sudden onset of heart attack often results from the adherence of platelets to the atherosclerotic lesion and the subsequent formation of a clot that occludes the blood flow in a coronary artery. The heart tissue is therefore deprived of blood supply, a condition that leads to tissue death. Similarly, stroke in the brain can be caused by platelet aggregation on an atherosclerotic lesion in a brain artery.

The subsequent conversion of 7-hydroxycholesterol to cholic acid is outlined in figure 20.19. These reactions involve oxidation of the 3 $\beta$-hydroxyl group, isomerization of the ethenyl double bond, 12$\alpha$-hydroxylation, reduction of the ethenyl (carbon-carbon) double bond, and reduction of the 3 keto group to a 3 $\alpha$-hydroxyl group. Additional hydroxylations and oxidation reactions on the side chain lead to cholic acid, one of the two major human bile acids. The mechanism by which bile acids emulsify triacylglycerols is illustrated in figure 20.20. The bile acid has a hydrophobic side and, by virtue of the three hydroxyl groups and the carboxyl group, a hydrophilic side. The nonpolar surface interacts with triacylglycerol, whereas the polar face is on the surface of the micelle, thus solubilizing (emulsifying) the lipid. The principle involved has similarities to that discussed above for the packaging of triacylglycerol in lipoproteins. The bile acids in micelles, in essence, function as the phospholipid and apoproteins do in lipoproteins.

The bile acids are usually conjugated to the amino acids, glycine or taurine, as shown for the formation of glycocholate in figure 20.21. It is in this form that the bile acids are secreted via the gallbladder into the intestine where they are critically important for the solubilization of dietary lipids. At pH 7, these acids exist as the sodium salts and are often referred to as bile salts. Of course in the acidic environment of the small intestine, they occur largely in their acidic form.

## Metabolism of Steroid Hormones

Steroid hormones have a variety of important functions, including regulation of carbohydrate and mineral metabolism and development of secondary sexual characteristics (Table 20.4). Steroids are often used in the treatment of diseases such as arthritis. Some athletes use steroids to promote muscle growth.

The initial reaction in steroid hormone biosynthesis is catalyzed by desmolase (side-chain cleavage complex), which is found in the mitochondria of steroid-producing tissues (e.g., adrenals, gonads). The reaction is shown in figure 20.22. The product, pregnenolone, is subsequently transferred to the endoplasmic reticulum, where an oxidation of the hydroxyl group and isomerization of the ethenyl double bond produces progesterone. Progesterone is a steroid hormone, and it or pregnenolone is the biosynthetic precursor of all other steroid hormones. The formation of the major steroid hormones involves a series of hydroxylation reactions and other modifications that give rise to the steroid hormones listed in table 20.4. These modifications also improve the aqueous solubility of the steroids.

No known pathway exists in mammals by which the steroid ring nucleus can be degraded to smaller molecules such as acetate. The carbon atoms of the ring system cannot, therefore, be used as a source of metabolic energy. However, many reactions occur, particularly in the liver, for the partial catabolism and inactivation of the steroid hormones. Frequently, these reactions involve reduction of ketone groups or double bonds. These inactive steroids are conjugated to glucuronic acid (fig. 20.23) or sulfate. The excretion of such derivatives results in a very large number of steroid metabolites in the urine. For example, 20 different metabolites of estrogen, conjugated to sulfate or glucuronic acid, have been identified in human urine. The illegal use of steroids by athletes is often detected by analysis of steroid derivatives found in the urine.

Like eicosanoids (chapter 19), steroid hormones are not stored for subsequent secretion. Instead, they are made in response to a stimulus such as corticotropin, a pituitary hormone, which stimulates the synthesis of aldosterone and cortisol in the adrenals. The steroids act by binding to specific receptors as discussed in chapters 24 and 31.

Severe consequences can occur from defective synthesis of a steroid hormone, as is the case in several metabolic diseases. Patients who have 21-hydroxylase deficiency are unable to convert progesterone to aldosterone and cortisol. Instead the progesterone is directed to an excess production of testosterone. In the female, this results in masculinization and hirsutism (growth of hair). In the male, premature masculinization occurs. If the disease is diagnosed before the first birthday, the patient can be treated with the missing steroid hormones, which in turn suppress the synthesis of excess progesterone and, as a consequence, testosterone via feedback mechanisms.

# Figure 20.19

Conversion of 7α-hydroxycholesterol to cholic acid. The increase in the number of polar groups in the conversion to cholic acid increases the water solubility. Bile acids possess a 5 hydrogen: Thus the A and B rings are no longer coplanar but have an A/B *cis*-configuration. This configuration improves the detergent properties of the bile acids so that they are better able to solubilize lipids in the intestine and aid digestion.

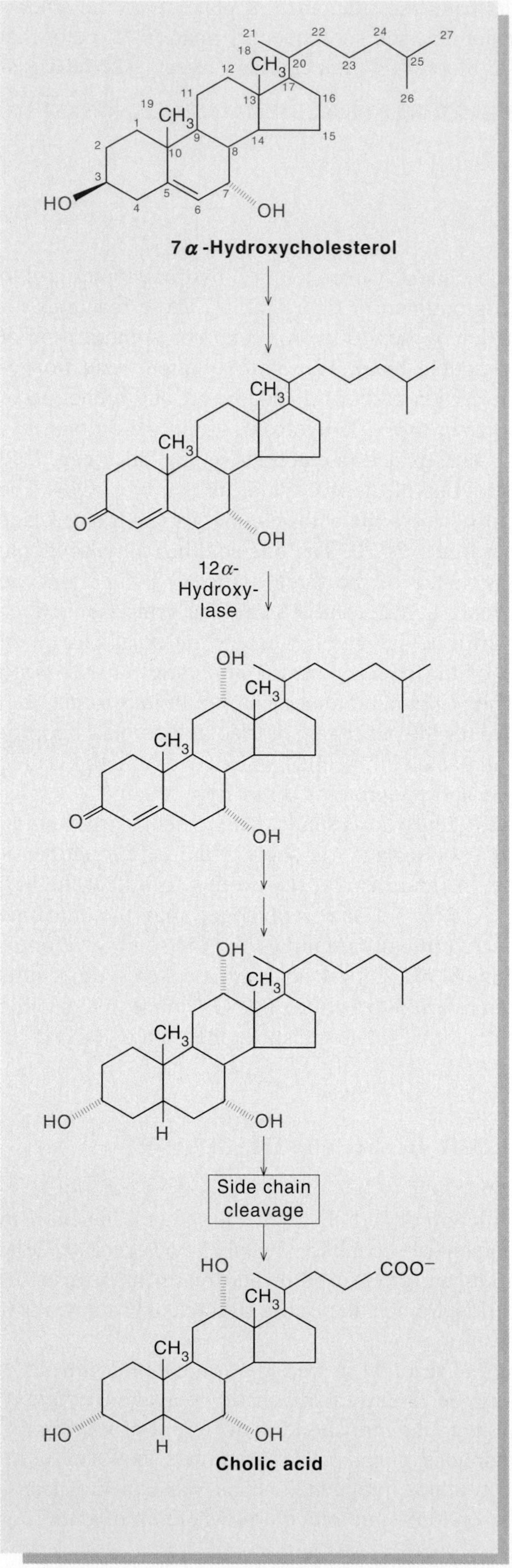

**Figure 20.20**

Bile salts emulsify fats in the intestine. The hydrophobic side or surface of the bile salt associates with triacylglycerols to form a complex. These complexes aggregate to form a micelle, with the hydrophilic side of the bile salt facing outward. Pancreatic lipase also associates with this micelle. The action of the lipase releases free fatty acids that form a much smaller micelle, which can be absorbed by the intestine.

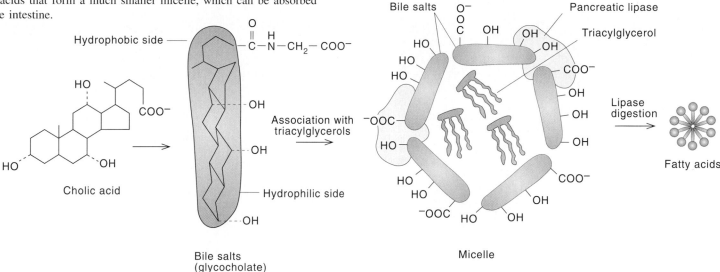

## Overview of Mammalian Cholesterol Metabolism

The metabolism of cholesterol in mammals is extremely complex. A summary sketch (fig. 20.24) helps to draw the major metabolic interrelationships together. Cholesterol is biosynthesized from acetate largely in the liver (fig. 20.24a) or taken in through the diet (fig. 20.24b). From the intestine, dietary cholesterol is secreted into the plasma mainly as a component of chylomicrons. The triacylglycerol components of chylomicrons are quickly degraded by lipoprotein lipase, and the remnant particles are removed by the liver. Apoproteins and lipid components of the chylomicrons and remnants appear to exchange with HDL. Cholesterol made in the liver (fig. 20.24a) has several alternative fates. It can be (1) secreted into plasma as a component of VLDL, (2) stored in droplets as cholesteryl ester, (3) used as a structural component of cell membranes, or (4) converted into bile salts. In plasma, VLDL is converted to IDL and LDL by the action of lipoprotein lipase and through cholesteryl ester exchange reactions with HDL. The LDL serves as a major carrier of cholesterol to extrahepatic cells, which include the adrenals and gonads. LDL and HDL are also returned to the liver. The steroid hormones made in the adrenals and gonads are delivered to various target tissues and promote a wide range of metabolic effects. The bile salts made in the liver are delivered to the upper intestine, where they aid in the solubilization of dietary lipid. Most of the bile salts are resorbed in the lower intestine and returned to the liver by the portal vein. Nevertheless, conversion of cholesterol to bile salts is the major mechanism for excretion of cholesterol from mammals.

## Figure 20.21

Conversion of a cholic acid into glycocholate. Glycocholate is an example of a bile salt. The bile salts solubilize lipids in the small intestine so they can be degraded by lipases.

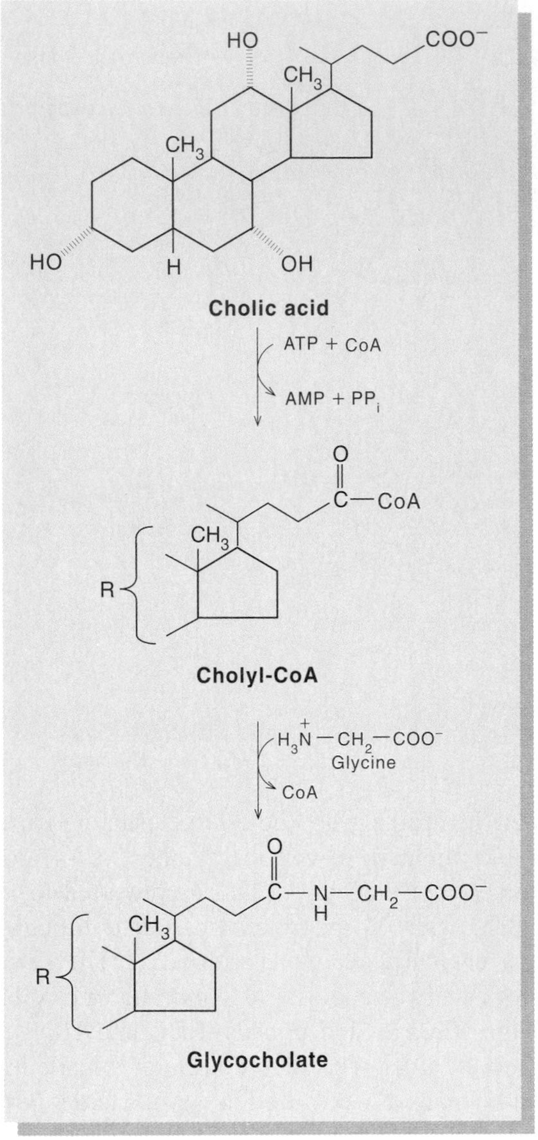

**Cholic acid**

ATP + CoA

AMP + PP$_i$

**Cholyl-CoA**

$H_3\overset{+}{N}$—CH$_2$—COO$^-$
Glycine

CoA

**Glycocholate**

## Figure 20.22

Biosynthesis of progesterone. Desmolase is found in the mitochondria of steroid-producing tissues. Cholesterol is converted to pregnenolone in the mitochondria and then transferred to the endoplasmic reticulum, where it is converted to progesterone. Progesterone is a hormone as well as the precursor of several other hormones.

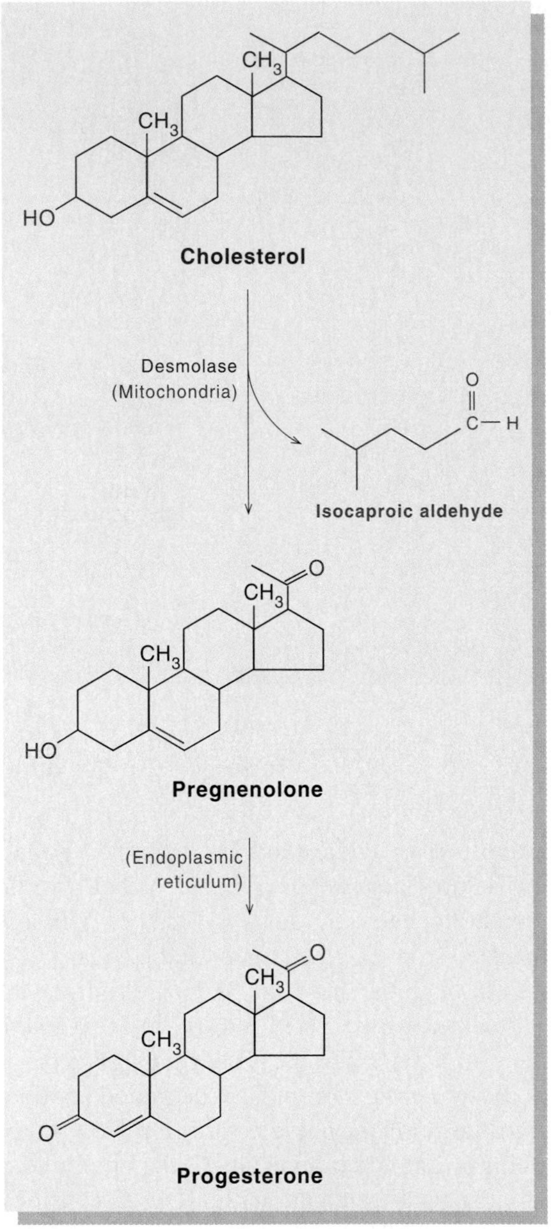

**Cholesterol**

Desmolase
(Mitochondria)

**Isocaproic aldehyde**

**Pregnenolone**

(Endoplasmic
reticulum)

**Progesterone**

## Table 20.4

### Major Steroid Hormones

| Steroid | Function |
|---|---|
| Progesterone | Precursor of other steroids; prepares uterus for implantation of an egg; prevents ovulation during pregnancy |
| Aldosterone (a mineralocorticoid) | Increases retention of sodium ions by the renal tubules |
| Cortisol (a glucocorticoid) | Promotes gluconeogenesis; suppresses inflammatory reactions |
| Testosterone (an androgen) | Promotes male sexual development; promotes and maintains male sex characteristics |
| Estradiol (an estrogen) | Responsible for sexual development in the female; promotes and maintains female sex characteristics |

## Figure 20.23

Inactivated steroid hormones are conjugated to glucuronic acid prior to excretion in the urine.

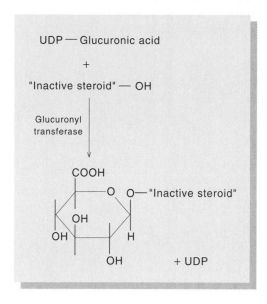

**Figure  20.24**

Fate of cholesterol. (*a*) Cholesterol
biosynthesized in the liver has several
alternative fates. (*b*) Cholesterol obtained
from the diet can enter the plasma and
subsequently the liver.

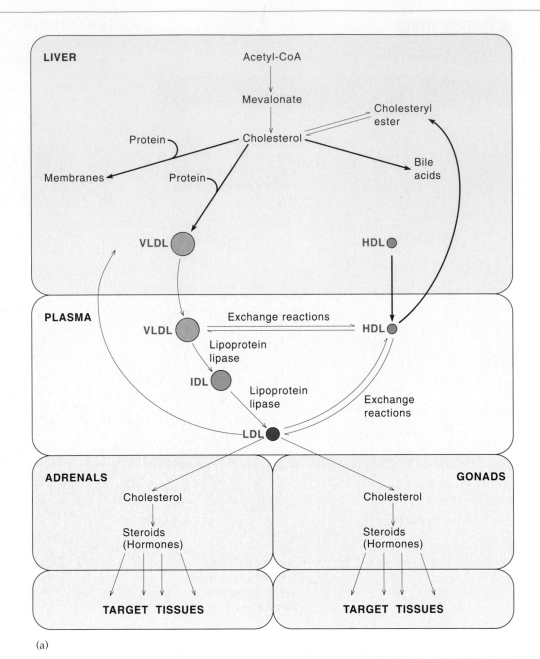

(a)

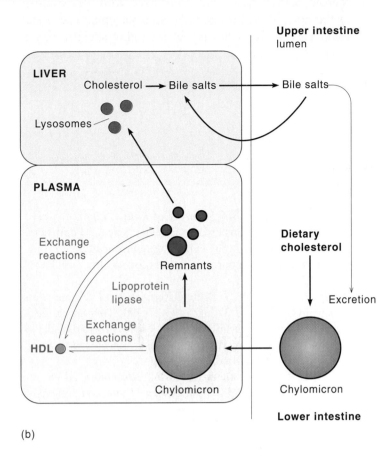

**Upper intestine**
lumen

**LIVER**

Cholesterol → Bile salts → Bile salts

Lysosomes

**PLASMA**

Exchange reactions

Remnants

**Dietary cholesterol**

Lipoprotein lipase

Excretion

Exchange reactions

HDL

Chylomicron

Chylomicron

**Lower intestine**

(b)

## Summary

In this chapter we dealt primarily with the metabolism of cholesterol, the most prominent member of the steroid family of lipids, and with the associated plasma lipoproteins. The chief points in our discussion are as follows:

1. The biosynthesis of steroids begins with the conversion of three molecules of acetyl-CoA into mevalonate, the decarboxylation of mevalonate, and its conversion to isopentenyl pyrophosphate. Six molecules of isopentenyl pyrophosphate are polymerized into squalene, which is cyclized to yield lanosterol. Lanosterol is converted to cholesterol, which is the precursor of bile acids and steroid hormones.

2. The rate of cholesterol biosynthesis appears to be regulated primarily by the activity of HMG-CoA reductase. This key enzyme is controlled by the rate of enzyme synthesis and degradation and by phosphorylation–dephosphorylation reactions. Synthesis of the mRNA for the reductase is inhibited by cholesterol delivered to cells by means of low-density lipoproteins (LDLs).

3. Cholesterol, triacylglycerols, and phospholipids are carried in plasma by lipoproteins, which are synthesized and secreted by the intestine and liver. The major lipoproteins are chylomicrons, very-low-density lipoproteins (VLDLs), low-density lipoproteins (LDLs), and high-density lipoproteins (HDLs). The triacylglycerols in chylomicrons and VLDLs are degraded in plasma by lipoprotein lipase, and the fatty acids are absorbed primarily by heart, skeletal muscle, and adipose tissue. LDLs are removed from plasma by an endocytotic process after binding to specific LDL receptors on the plasma membrane. The LDLs are enzymatically degraded in the lysosomes. In familial hypercholesterolemia, the specific receptors for LDL uptake are defective. High levels of LDL are associated with an increased risk of cardiovascular disease, whereas high levels of HDL seem to protect against this disease.

4. Bile acids are $C_{24}$ acids that are biosynthetically derived from cholesterol. The $7\alpha$-hydroxylation of cho-

lesterol is the committed and rate-limiting reaction in the synthesis of bile acids. Salts formed from the bile acids are secreted into the small intestine and aid the solubilization and digestion of lipids. Formation and excretion of the bile salts is the major route for elimination of cholesterol from the body.

5. Steroid hormones are biosynthesized from cholesterol in the adrenal cortex, gonads, and placenta. These steroids are important hormones for many specific physiological processes.

## Selected Readings

Bloch, K., Cholesterol: Evolution of structure and function. In D. E. Vance, and J. E. Vance (eds.), *Biochemistry of Lipids, Lipoproteins and Membranes*. Amsterdam: Elsevier Science Publishers, 1991. This chapter (12) provides an interesting view of how the structure of cholesterol evolved to optimize its function in cells.

Brown, M. S., and J. L. Goldstein, A receptor mediated pathway for cholesterol homeostasis. *Science* 232:34–47, 1986. An article describing the research on the LDL receptor and familial hypercholesterolemia that won them the Nobel Prize.

Davis, R., Lipoprotein structure and secretion. In D. E. Vance, and J. E. Vance (eds.), *Biochemistry of Lipids, Lipoproteins and Membranes*. Amsterdam: Elsevier Science Publishers, 1991. This chapter (14) provides an advanced discussion on the assembly and secretion of very-low-density lipoproteins.

Edwards, P. A., Regulation of sterol biosynthesis and isoprenylation of proteins. In D. E. Vance, and J. E. Vance (eds.), *Biochemistry of Lipids, Lipoproteins and Membranes*. Amsterdam: Elsevier Science Publishers, 1991. The complexities of the regulation of cholesterol biosynthesis are explained in this chapter (13).

Fielding, P. E., and C. J. Fielding, Dynamics of lipoprotein transport in the circulatory system. In D. E. Vance, and

J. E. Vance (eds.), *Biochemistry of Lipids, Lipoproteins and Membranes*. Amsterdam: Elsevier Science Publishers, 1991. This chapter (15) reviews the current literature on the intricacies of lipoprotein metabolism in the circulatory system.

Goldstein, J. L., and M. S. Brown, Regulation of the mevalonate pathway. *Nature* 343:425–430, 1990. This article describes regulatory mechanisms for cholesterol biosynthesis within the context of the regulation of the biosynthesis of other isoprenoid derivatives.

Makin, H. L. J., *Biochemistry of Steroid Hormones*, 2d ed. Oxford: Blackwell, 1984. An advanced and comprehensive treatment of steroid hormones.

Schneider, W. J., Removal of lipoproteins from plasma. In D. E. Vance, and J. E. Vance (eds.), *Biochemistry of Lipids, Lipoproteins and Membranes*. Amsterdam: Elsevier Science Publishers, 1991. This chapter (16) provides a clear explanation of the current literature on the uptake of lipoproteins into cells and tissues.

Scriver, C. R., A. L. Beaudet, W. S. Sly, and D. Valle, *The Metabolic Basis of Inherited Disease,* 7th ed., vol. 1, New York: McGraw-Hill, 1995. This book has an introductory chapter on lipoprotein structure and metabolism followed by many excellent chapters on disorders of cholesterol and lipoprotein metabolism.

## Problems

1. The conversion of cholyl-CoA to glycocholate (fig. 20.21) is an example of amide bond formation utilizing a CoASH (thioester) derivative. Are all amide bonds made from CoASH derivatives?

2. The conversion of cholic acid to cholyl-CoA (fig. 20.21) also involves the conversion of ATP to AMP and PP$_i$. What intermediate would you expect to be involved in this reaction?

3. Explain the frequencies of the heterozygous and homozygous forms of familial hypercholesterolemia

given in the text (page 472). Do you find anything inconsistent with these frequencies? If so, can you provide an explanation for this?

4. Do you expect the equilibrium position of the reaction catalyzed by CoA: cholesterol acyltransferase (fig. 20.13) to be any different from that of lecithin: cholesterol acyltransferase (fig. 20.17)?

5. A patient homozygous for familial hypercholesterolemia (FH) was treated with lovastatin to lower LDL levels in the blood. This treatment did not have any

effect on LDL levels. Why? After a number of heart attacks, a heart and liver transplant were done, and LDL levels were dramatically lowered. Why were both organs replaced?

6. During routine investigations, the plasma from a family of rats (group 1) was found to have very low concentrations of cholesterol. When the microsomal HMG-CoA reductase from liver was assayed, extremely low activities were found. When the cytosol from normal rats (group 2) was added to the microsomal fraction from group 1 rats, the HMG-CoA reductase activity was gradually restored to normal values. What enzyme activity (or activities) might be deficient in the group 1 rats?

7. Liver cells in culture are given 2-[$^{14}$C]-acetate. Where does this label appear in HMG-CoA?

8. How would a dietary resin that absorbs bile salts reduce plasma cholesterol levels?

9. Notice that the metabolic sequences described in this and previous chapters often involve multiple locations. These locations can be at the organ (fig. 20.1) or the subcellular level (fig. 20.22). This is in contrast, for example, with glycolysis, which occurs in the cytosol, or the tricarboxylic acid cycle, which is completely mitochondrial. Can you provide a possible explanation for this multiple location phenomenon?

# Metabolism of Nitrogen-Containing Compounds

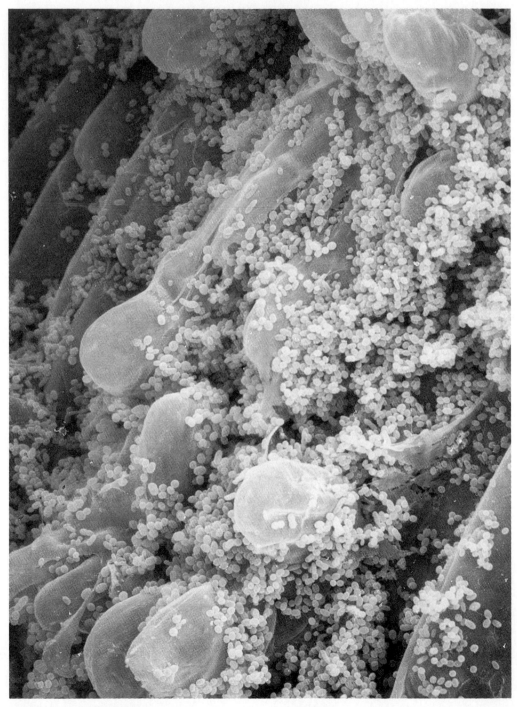

Scanning electron micrograph of part of a sorghum root covered with nitrogen-fixing bacterial cells *(Azospirillum brasilense).* The magnification is 1540 ×. Courtesy of Dr. R. Howard Berg, University of Memphis.

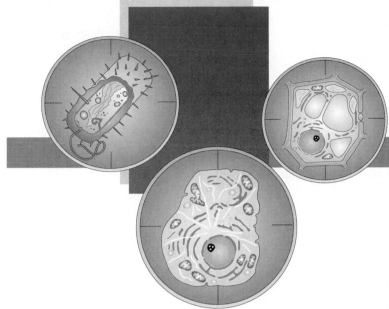

# Amino Acid Biosynthesis and Nitrogen Fixation in Plants and Microorganisms

*The carbon skeletons for amino acid biosynthesis originate from the central metabolic pathways; the first step in amino acid synthesis usually involves addition of the α amino group.*

Amino acids are best known as the building blocks of protein, and indeed that is a main function of the 20 L-amino acids that are commonly found in proteins (see chapter 3). In addition to this role, amino acids serve as precursors to

**Figure 21.1**

Outline of the biosynthesis of the 20 amino acids found in proteins. The *de novo* biosynthesis of amino acids starts with carbon compounds found in the central metabolic pathways. The central metabolic pathways are drawn in black, and the additional pathways are drawn in red. Some key intermediates are illustrated, and the number of steps in each pathway is indicated alongside the conversion arrow. All common amino acids are emphasized by boxes. Dashed arrows from pyruvate to both diaminopimelate and isoleucine reflect the fact that pyruvate contributes some of the side-chain carbon atoms for each of these amino acids. Note that lysine is unique in that two completely different pathways exist for its biosynthesis. The six amino acid families are screened.

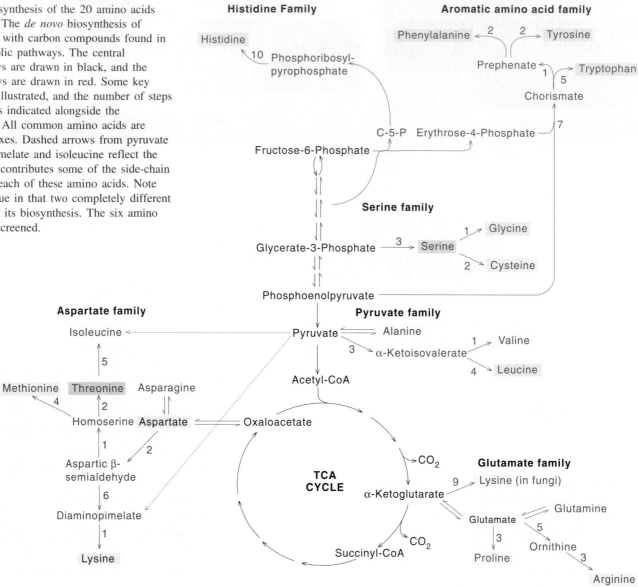

many important small molecule compounds, including nucleotides, porphyrins, parts of lipid molecules, and precursors for several coenzymes. Amino acids also provide the vehicles for converting nitrogen and sulfur from inorganic to organic forms. As an alternative energy source amino acid catabolism can be coupled to the regeneration of ATP from ADP or AMP.

In this chapter we focus on the biosynthesis of amino acids (fig. 21.1) and their role in bringing inorganic nitrogen and sulfur into the biological world. In addition nonprotein amino acids are discussed briefly.

## The Pathways to Amino Acids Arise as Branchpoints from the Central Metabolic Pathways

Inspection of the amino acid biosynthetic pathways shows that all amino acids arise from a few intermediates in the central metabolic pathways (see fig. 21.1). Amino acids derived from a common intermediate are said to be in the same family. For example, the serine family of amino acids, which includes serine, glycine, and cysteine, all arise from glycerate-3-phosphate (see fig. 21.1). The carbon flow from the central metabolic pathways to amino acids is a regulated process that provides amino acids in the amounts needed for maintenance and growth. The flow is regulated mostly by end-product inhibition, so that the amino acid (end product) inhibits the first enzyme in the pathway for its synthesis.

## Our Understanding of Amino Acid Biosynthesis Has Resulted from Genetic and Biochemical Investigations

*Escherichia coli* has served as the organism of choice for examining the pathways for amino acid biosynthesis for two reasons: (1) *E. coli* synthesizes all 20 amino acids commonly found in proteins, and (2) *E. coli* is an organism ideally suited for genetic and biochemical studies.

### The Number of Proteins Participating in a Pathway Is Known through Genetic Complementation Analysis

The investigation of an amino acid biosynthetic pathway in *E. coli* begins with the accumulation of mutants that are deficient in the capacity to synthesize that amino acid. Mutants of this sort are known as auxotrophs. Each auxotroph bears a mutation in one of the genes that encodes an enzyme required for a step in the biosynthetic pathway of the amino acid. It is possible to determine how many steps are in a particular pathway by the process of complementation analysis (see fig. 11.14).

### Biochemists Use the Auxotrophs Isolated by Geneticists

Complementation group analysis sets the stage for the biochemist. First it tells him or her how many proteins must be isolated and characterized. Additional aid to the biochemist is provided by the fact that mutations belonging to complementation groups block specific steps in a pathway. As a result no amino acid is synthesized, and specific intermediates accumulate (fig. 21.2). Owing to their abundance, intermediates accumulating in such mutants can frequently be isolated and their structure determined. By piecing together information of this type from studies on different mutants, an overall picture of the pathway emerges.

Finding intermediates in a pathway is only half the problem. The biochemist must also isolate and characterize the enzymes of the pathway. To this end, extracts are made from either normal or mutant cells. These extracts are tested in conjunction with different intermediates to see whether the extracts can carry out the conversion of one intermediate

**Figure 21.2**

Immediate consequences of a point mutation in a biosynthetic pathway. When the mutation leads to an inactive enzyme, the chain of reactions leading to the end product in the pathway is broken, and frequently large amounts of intermediate are produced, accumulate in the cell, and may leak to the environment.

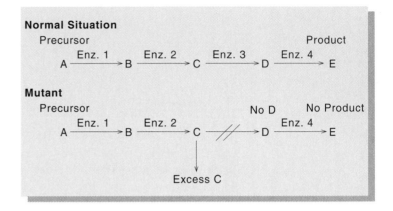

to the next in the pathway. This conversion is used as an assay for isolating the enzyme by progressive subfractionation of the extract to the point at which the enzyme of interest is reasonably pure (see chapter 6 for protein purification techniques). In the final stages of biochemical analysis the structure and catalytic activities of the enzyme are determined.

Sometimes the investigation of a pathway is complicated by the fact that common intermediates are used in the synthesis of more than one amino acid. In such cases intermediates may be converted into more than one product by different enzymes, each product serving as an intermediate for a particular amino acid. Mutants containing defects in enzymes required for the synthesis of more than one amino acid usually require the addition of two or more amino acids to reestablish cell growth.

## The Glutamate Family of Amino Acids and Nitrogen Fixation

A common element in all protein-bound amino acids is the $\alpha$-amino group. Directly or indirectly, the $\alpha$-amino groups of most amino acids are derived from ammonia by way of the amino groups of L-glutamine. Ammonia itself can be incorporated directly into $\alpha$-ketoglutarate to make glutamate, or into glutamate to make glutamine. The uptake of $NH_3$ by glutamate represents a key step in the process, whereby reduced nitrogen becomes incorporated into organic molecules.

# Figure 21.3

The conversion of ammonia into the $\alpha$-amino group of glutamate and into the amide group of glutamine. The direct amination of $\alpha$-ketoglutarate by $NH_4^+$ occurs only under conditions of high $NH_4^+$ concentrations, which are rarely found in nature.

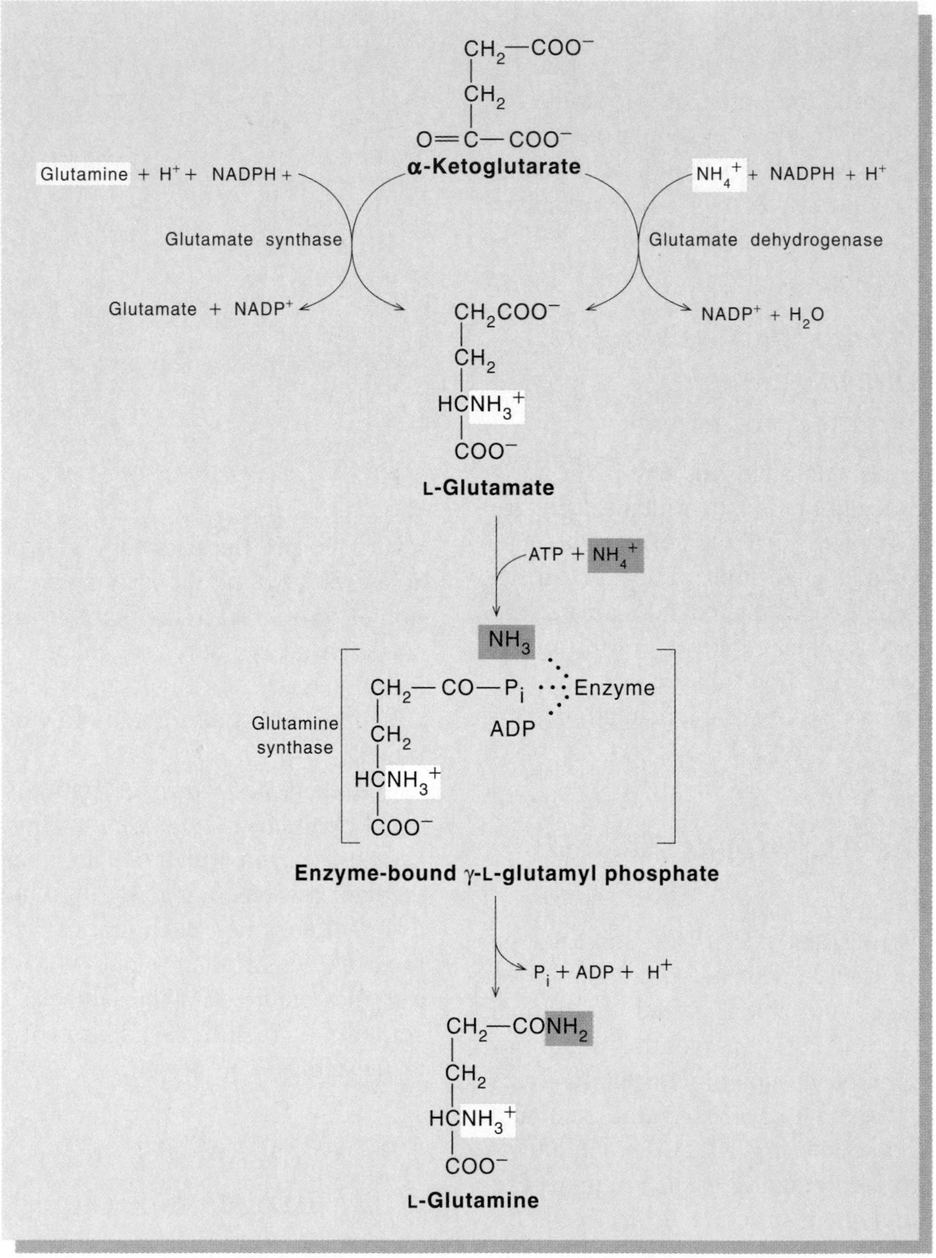

**Figure 21.4**

The structure of glutamine synthase from *Salmonella typhimurium*. The enzyme consists of twelve identical subunits arranged like a hexagonal prism. (*a*) View down the sixfold axis of symmetry. The top ring of monomers are alternately colored light and dark blue and the bottom ring of monomers light and dark red. The active sites of each monomer are marked by pairs of $Mn^{2+}$ ions (white spheres).

(b) Side view of the enzyme along one of the twofold axes of symmetry, showing only the six nearest subunits. (Copyright 1994 by the Scripps Research Institute/Molecular Graphics Images by Michael Pique using software by Yng Chen, Michael Connolly, Michael Carson, Alex Shah, and AVS, Inc. Visualization advice by Holly Miller, Wake Forest University Medical Center.)

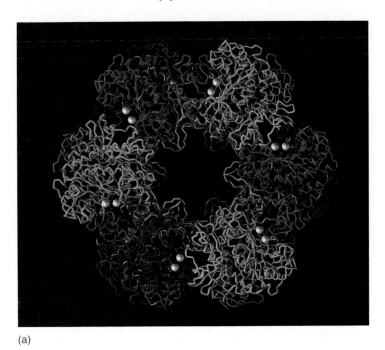

(a)

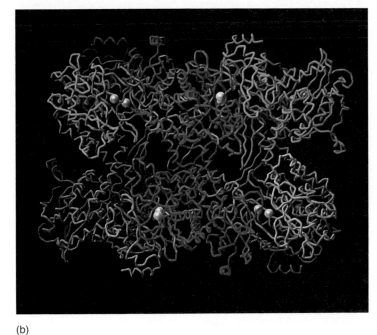

(b)

Glutamate serves as the starting material for the synthesis of glutamine, proline, and arginine (see fig. 21.1). Appropriately, these four amino acids are described as belonging to the glutamate family. In fungi, lysine is also included in this family (see fig. 21.1).

In this section we discuss the pathways for glutamate and glutamine synthesis. We also discuss some other aspects of the nitrogen cycle. Consideration of the arginine pathway is deferred until the next chapter.

### The Direct Amination of α-Ketoglutarate Leads to Glutamate

The most direct route to glutamate (and therefore, amino group formation) is that exhibited by many bacteria when grown in a medium containing an ammonium salt as the sole nitrogen source. The reaction entails a reductive amination catalyzed by glutamate dehydrogenase (fig. 21.3). In *E. coli* this enzyme is specific for NADPH as the hydrogen donor, as might be expected for a biosynthetic reaction involving a reductive step.

Studies of the glutamate dehydrogenases from green plants indicate that they require a very high concentration of $NH_3$ to be effective. In fact an alternative pathway is re-

sponsible for glutamate biosynthesis in most plants. Thus, in green plants, as well as many bacteria and fungi, the formation of the amino group of glutamate usually originates from the amide group of glutamine. The reaction is catalyzed by glutamate synthase (see fig. 21.3).

### Amidation of Glutamate to Glutamine Is an Elaborately Regulated Process

Glutamine, which is so commonly used in amination reactions, is regenerated by a reaction in which ammonia is directly added to glutamate; the reaction is catalyzed by glutamine synthase. This two-step reaction starts with activation of the γ-carboxyl group of glutamate and yields a γ-glutamyl-enzyme complex (fig. 21.3); this is followed by γ-glutamyl transfer of $NH_3$.

Since glutamine synthase is the main enzyme initiating the flow of ammonia nitrogen into organic components, it is not surprising to find that it is a highly regulated enzyme. The *E. coli* enzyme has been studied extensively by Stadtman and Ginsberg. They have shown that the enzyme is composed of 12 identical subunits ($M_r$ 50,000) arranged in two hexameric rings (fig. 21.4). Activity of the enzyme is

**Figure 21.5**

Regulation of glutamine synthase in *E. coli*. The activity of glutamine synthase is inhibited by adenylylation. Both adenylylation and deadenylylation are regulated by a cascade of controls that are responsive to the concentrations of glutamine and $\alpha$-ketoglutarate. The concentrations of these two compounds are an indication of the nitrogen supply. Thus, under conditions of nitrogen excess (*top part of figure*) the concentration of glutamine is high, and the ratio of glutamine to $\alpha$-ketoglutarate is high. This leads to inactivation of the glutamine synthase. Under conditions of nitrogen limitation (*bottom part of figure*) the $\alpha$-ketoglutarate concentration is high, and the ratio of $\alpha$-ketoglutarate to glutamine is high, so the glutamine synthase is activated.

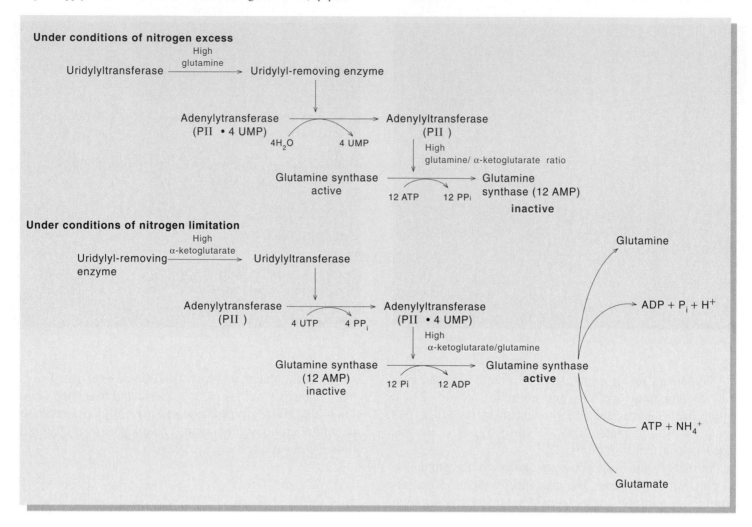

negatively regulated by eight nitrogenous compounds: Carbamyl phosphate, glucosamine-6-phosphate, tryptophan, alanine, glycine, histidine, cytidine triphosphate, and AMP. With the exception of glycine and alanine these compounds receive the amide nitrogen directly from glutamine during their biosynthesis. Therefore most of these compounds can be thought of as end products originating from glutamine. Although glycine and alanine are not direct end products, they are "indicators" of the sufficiency of the nitrogen supply of the cell. Thus, the regulation of glutamine synthase can be regarded as an example of end-product inhibition.

Even more important in the regulation of *E. coli* glutamine synthase activity is the reversible ATP-dependent adenylylation of a specific tyrosyl residue on each subunit. As the enzyme becomes progressively more adenylylated (up to the fully adenylylated form which contains 12 AMP groups per enzyme molecule), the enzyme becomes progressively less active.

The adenylylation reaction and the deadenylylation reaction are regulated by the nitrogen supply in the cell. The immediate small-molecule effectors of this regulatory system are glutamine and $\alpha$-ketoglutarate. A high-glutamine–

**Figure 21.6**

The nitrogen cycle depicts the flow of nitrogen in the biological world. The proteins of animals and plants are cleaved by many microorganisms to free amino acids from which ammonia (or ammonium ion) is released by deamination. Urea, the main nitrogen excretion product of animals, is hydrolyzed to $NH_3$ and $CO_2$. *Nitrosomonas* soil bacteria obtain their energy by oxidizing $NH_3$ to nitrite, $NO_2^-$. *Nitrobacter* obtain their energy by oxidizing nitrite, $NO_2^-$, to nitrate, $NO_3^-$. Plants and many microorganisms reduce nitrate for incorporation into amino acids, completing the cycle. Other microorganisms reduce nitrate partly to $NH_3$ and partly to $N_2$, which is lost to the atmosphere. Atmospheric nitrogen, $N_2$, can be recaptured, reduced, and converted into organic substances by a limited number of nitrogen-fixing bacteria and algae.

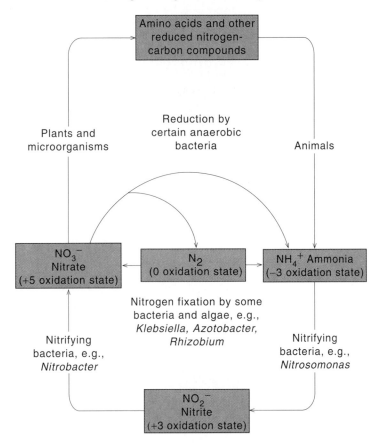

α-ketoglutarate ratio signals nitrogen excess, whereas a high-α-ketoglutarate–glutamine ratio signals nitrogen limitation. Nitrogen excess leads to inactivation of glutamine synthase, just as nitrogen limitation leads to its activation (fig. 21.5). The activation of glutamine synthase favors fixation of ammonia in a condensation with glutamate to form glutamine. Both activation (deadenylylation) and the inacti-

vation (adenylylation) are controlled by a cascade of regulatory interactions illustrated in figure 21.5. The cascades involved here are unusual in that the same two regulatory proteins, uridylyltransferase and adenylyltransferase, function in both cascades. Both of these regulatory proteins have two enzymatically distinct sites arranged so that only one site is active at any given time.

Let us first consider the situation under conditions of nitrogen excess (see fig. 21.5). The first regulatory protein in the cascade is converted into a uridylyl-removing enzyme. This enzyme hydrolyzes UMP from a PII · 4 UMP protein that acts in concert with adenylyltransferase to form adenylate glutamine synthase. The resulting adenylylated enzyme is inactive.

Under conditions of nitrogen limitation, the first regulatory enzyme in the cascade is converted into a uridylyltransferase. This enzyme uridylates the PII protein. The PII · 4 UMP protein and adenylyltransferase deadenylylates the glutamine synthase, 12 AMP.

The control of glutamine synthase by adenylylation and deadenylylation may be limited to certain Gram negative bacteria. However, physiologically analogous modifications of other glutamine synthases probably exist. For example, glutamine synthases in eukaryotic cells are octameric structures that respond to regulation by dissociation to a tetrameric form under conditions in which the $NH_4^+$ supply is restricted.

## The Nitrogen Cycle Encompasses a Series of Reactions in Which Nitrogen Passes through Many Forms

Glutamine synthase is just one of the links in the nitrogen cycle whereby nitrogen passes through several chemical forms. The passage of nitrogen from one form to another involves a chain of widely distributed organisms (fig. 21.6).

$NH_4^+$ is the form in which nitrogen is incorporated into organic materials, but it is often less available to plants or bacteria for biosynthesis than other forms of nitrogen. When present for any length of time in the free state $NH_4^+$ is likely to be oxidized by nitrifying bacteria (such as *Nitrosomonas* and *Nitrobacter*) to nitrite ($NO_2^-$) and nitrate ($NO_3^-$). Reduction of nitrate by plants and bacteria to $NH_4^+$ is vital to maintaining the nitrogen cycle. Nitrogen fixation whereby atmospheric nitrogen is reduced to $NH_3$ occurs in a limited number of microorganisms. Some of these microorganisms exist in the free state, and some exist as symbionts in the root nodules of certain plants.

## Figure 21.7

The eight-step nitrogenase cycle. Nitrogenase contains two complex protein components: component I and component II. The nitrogenase cycle starts when component II (an Fe—protein) binds to component I (usually an Fe—Mo—protein). The binding requires two ATP molecules. During the bound state component II transfers an electron to component I. This is followed by the hydrolysis of the two ATPs and the separation of the two components that begin a new cycle. In each step of the cycle the Fe—protein component II picks up a single electron from reduced ferrodoxin or flavodoxin and binds to two ATP molecules. After three single electron ($e^-$) transfers, an $H_2$ molecule is released from component II, and a $N_2$ molecule binds in its place. Following five more electron transfers, the $N_2$ molecule is fully reduced to 2 $NH_3$'s, which are released.

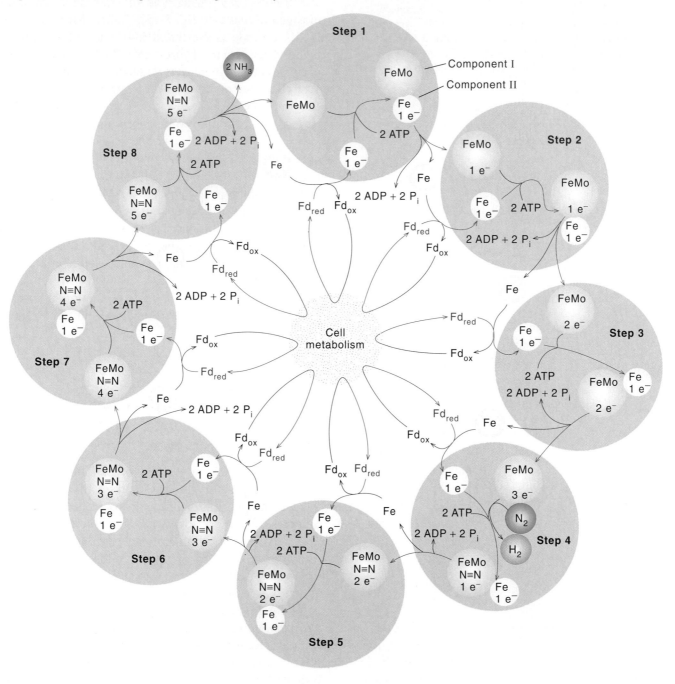

***Nitrogen Fixation Involves an Enzyme Complex Called Nitrogenase.*** The conversion of gaseous nitrogen ($N_2$) into ammonia ($NH_3$) is a most intricate reaction catalyzed by nitrogenase, a two-component enzyme (fig. 21.7). In *Clostridium pasteurianum* component I is a tetrameric protein with a molecular weight of 220,000. The four subunits in component I are of two kinds: Two with molecular weights of 50,000 and two with molecular weights of 60,000. In addition component I has four iron–sulfur centers, each containing four iron atoms and four sulfur atoms; and two extractable iron–molybdenum–sulfur "cofactors" that contain eight iron atoms, six labile sulfur atoms, and one molybdenum atom. Component II from *C. pasteurianum* contains two identical subunits, each with a molecular weight of 29,000. These proteins surround an iron–sulfur center that contains four iron atoms and four sulfur atoms.

Although it might be expected that only six reducing equivalents would be required for each mole of $N_2$ (valence state 0) reduced to $2 NH_3$ (valence state $-3$), in fact, the reaction catalyzed by nitrogenase converts two protons ($H^+$) to hydrogen ($H_2$) for each molecule of $N_2$ reduced. As a result the overall equation for the conversion of nitrogen to ammonia requires eight reducing equivalents.

$$N_2 + 8\,H^+ + 8\,e^- + 16\,ATP \longrightarrow$$
$$2\,NH_3 + H_2 + 16\,ADP + 16\,P_i \quad \textbf{(1)}$$

Why is hydrogen production a necessary step in the reduction of dinitrogen? It appears that it is necessary for the binding of nitrogen. The reduction of nitrogen by nitrogenase can be viewed as a cycle of eight electron transfers, each of which occurs in the cyclic transfer of a single electron from reduced ferrodoxin or flavodoxin to the Fe protein (component II) and the regeneration of the reduced electron donor by cell metabolism (e.g., pyruvate oxidation). On reduction, the Fe—protein binds two molecules of ATP, after which it associates with the Fe—Mo—protein (component I). The transfer of an electron to the Fe—Mo—protein is accompanied by the hydrolysis of the two bound ATP molecules and the dissociation of components I and II. Overall, the sequential functioning of eight Fe—protein reduction–oxidation cycles is required for one Fe—Mo protein cycle. The first three transfers, which result in the production of one $H_2$ molecule, somehow pave the way for the binding of dinitrogen at the active site. The remaining transfers are necessary to reduce a single dinitrogen to two ammonias.

***Nitrate and Nitrite Reduction Play Two Roles.*** As pointed out above, $NH_3$ tends to be oxidized to nitrites or nitrates, if it is not immediately incorporated into organic matter. Nitrate and nitrite reduction play two physiological roles. One, exhibited primarily by bacteria, is a dissimilatory role in which nitrate and nitrite serve as terminal electron acceptors. The reduction serves to oxidize reducing equivalents such as NADH that are generated during oxidation of substrates. The other role played by these reductions is an assimilatory one in which nitrate is first converted to nitrite and thence to $NH_3$. These reactions are observed in bacteria, fungi, and plants.

Generally, the assimilatory nitrate and nitrite reductases are soluble enzymes that utilize reduced pyridine nucleotides or reduced ferrodoxin. In contrast, the dissimilatory nitrate reductases are membrane-bound terminal electron acceptors that are tightly linked to cytochrome $b_1$ pigments. Such complexes allow one or more sites of energy conservation (ATP generation) coupled with electron transport.

The ammonium ion ($NH_4^+$), produced by fermentative bacteria that use nitrite as an oxidant, or produced during the decomposition of organic materials, is an important source of nitrogen for many plants and bacteria. Nevertheless, under vigorous aerobic conditions much of the $NH_4^+$ so produced is converted back to nitrite and nitrate by nitrifying bacteria.

## The Serine Family and Sulfur Fixation

As already noted the serine family includes three amino acids: Serine, glycine, and cysteine (see fig. 21.1). We focus on cysteine synthesis, which funnels sulfur into the biochemical world and supplies the cysteine needed for biosynthesis.

## Cysteine Biosynthesis Occurs by Sulfhydryl Transfer to Activated Serine

The biosynthesis of L-cysteine entails the sulfhydryl transfer to an activated form of serine. This pathway to L-cysteine has been most thoroughly studied in *E. coli*. In the first step an acetyl group is transferred from acetyl-CoA to serine to yield *O*-acetylserine (fig. 21.8*a*). The reaction is catalyzed by serine transacetylase. The formation of cysteine itself is catalyzed by *O*-acetylserine sulfhydrylase.

In some organisms sulfur incorporation involves homocysteine as an intermediate. In such cases cysteine formation occurs by a transsulfuration reaction, with the intermediate formation of L,L-cystathionine (fig. 21.8*b*). Cystathionine is formed in a simple condensation reaction from serine and homocysteine by cystathionine-$\beta$ synthase.

## Figure 21.8

The biosynthesis of cysteine by direct sulfhydrylation and by a transsulfuration route in which the sulfur is derived from homocysteine. The direct sulfhydrylation pathway (*a*) is indicated as occurring with $H_2S$ as the source of sulfur. The transsulfuration pathway (*b*) passes through homocysteine. Two routes leading to homocysteine are shown. One starts with L-methionine. This is the sole route in animals. The second route involves the homocysteine-synthase-catalyzed conversion of *O*-acetyl-L-homoserine as shown.

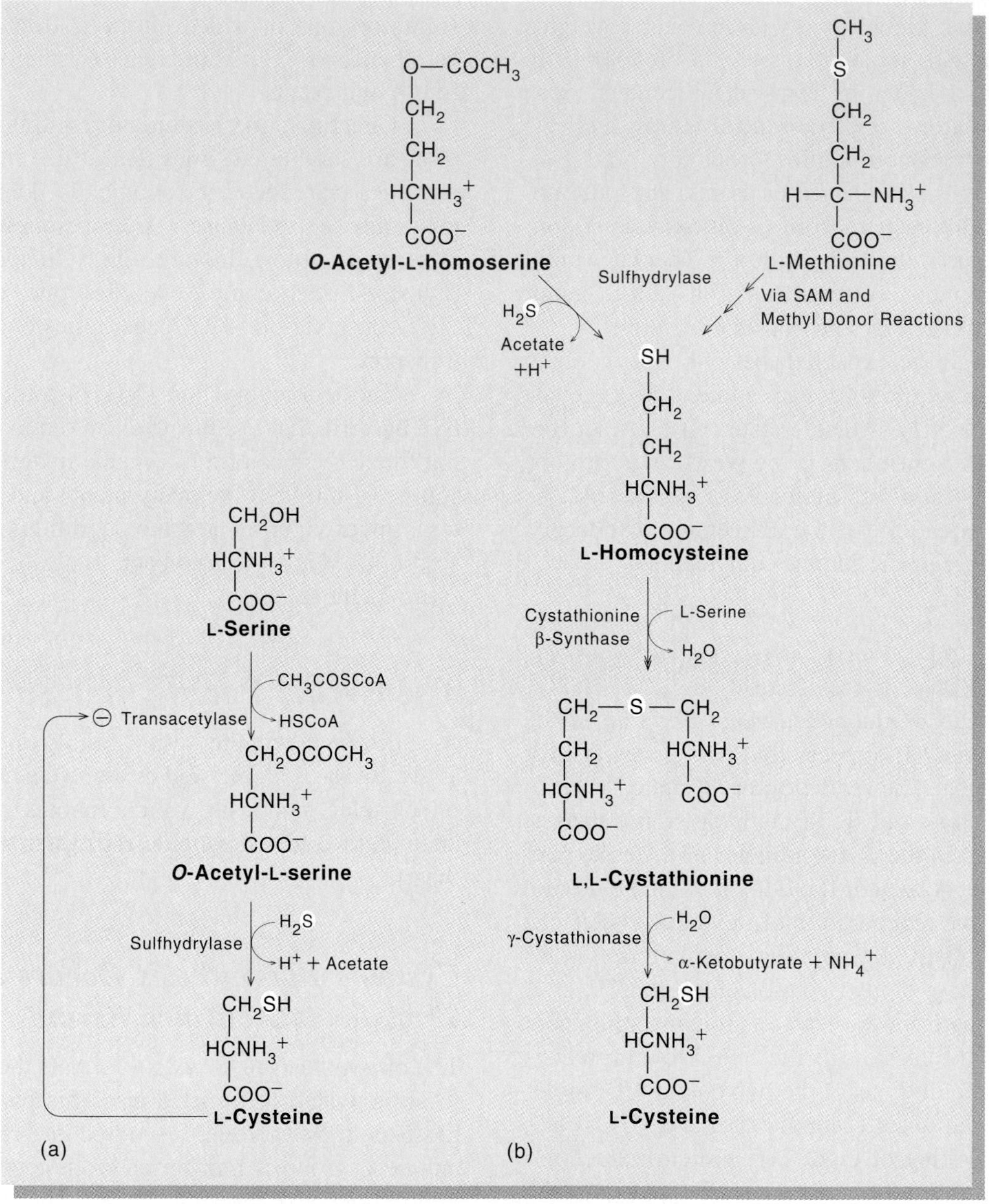

## Figure 21.9

Formation of 3'-phosphoadenosine-5'-phosphosulfate (PAPS), an active intermediate involved in sulfate reduction. The eight-electron reduction of $SO_4^{2-}$ to $H_2S$ is poorly understood except for the initial steps in the activation of sulfate (shown in yellow). Reduction in yeast and plants involves the APS derivative shown. In *E. coli* a PAPS derivative is used.

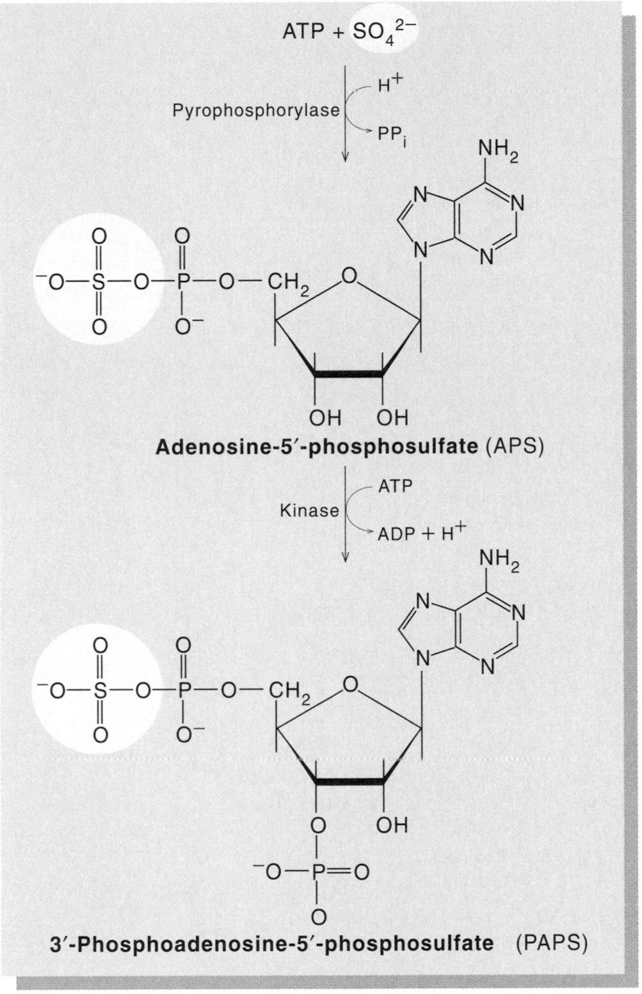

**Adenosine-5'-phosphosulfate** (APS)

**3'-Phosphoadenosine-5'-phosphosulfate** (PAPS)

The subsequent cleavage of cystathionine to yield cysteine, $\alpha$-ketobutyrate and $NH_4^+$ is catalyzed by $\gamma$-cystathionase, a pyridoxal-phosphate-containing enzyme. This transsulfuration pathway is one of the routes used for methionine catabolism.

## Sulfate Must Be Reduced to Sulfide before Incorporation into Amino Acids

Most sulfur in nature exists in the inorganic form of sulfate ion, $SO_4^{2-}$. Sulfate must be reduced to $H_2S$ before it can be incorporated into amino acids. The eight-electron reduction

of sulfur from $SO_4^{2-}$ to $H_2S$ starts with activation of sulfate (fig. 21.9). In a reaction catalyzed by adenylylsulfate pyrophosphorylase, inorganic sulfate is converted to adenosine-5'-phosphosulfate (APS). In *E. coli* and certain other bacteria APS is phosphorylated at the 3'—OH by APS kinase to yield 3'-phosphoadenosine-5'-phosphosulfate (PAPS). The reduction of the sulfonyl moiety of APS (in yeast and plants) or of PAPS (in *E. coli*) to sulfite occurs by transfer of the sulfonyl group to a thiol acceptor, such as thioredoxin to yield an —S—$SO_3^-$ derivative. Reaction with a second thiol group (on thioredoxin or a molecule of glutathione) yields sulfite and an oxidized acceptor.

The six-electron reduction of sulfite to sulfide is catalyzed by sulfite reductase, a multisubunit complex composed of a flavoprotein and a heme iron–sulfur protein.

# The Aspartate and Pyruvate Families Both Make Contributions to the Synthesis of Isoleucine

The aspartate (oxaloacetate) family of amino acids includes aspartate, asparagine, methionine, lysine, threonine, and isoleucine (see fig. 21.1). The pyruvate family includes alanine, valine, leucine, and also lysine and isoleucine (see fig. 21.1). Threonine is a precursor of isoleucine. It is converted into isoleucine by a group of enzymes that are also used in the synthesis of valine (fig. 21.10).

## Isoleucine and Valine Biosynthesis Share Four Enzymes

The first step in valine biosynthesis is a condensation between pyruvate and "active" acetaldehyde (probably hydroxyethyl thiamine pyrophosphate) to yield $\alpha$-acetolactate. The enzyme acetohydroxy acid synthase usually has a requirement for FAD, which, in contrast to most flavoproteins, is rather loosely bound to the protein. The very same enzyme transfers the acetaldehyde group to $\alpha$-ketobutyrate to yield $\alpha$-aceto-$\alpha$-hydroxybutyrate, an isoleucine precursor. Unlike pyruvate, the $\alpha$-ketobutyrate is not a key intermediate of the central metabolic routes; rather it is produced for a highly specific purpose by the action of a deaminase on L-threonine as shown in figure 21.10.

Conversion of the acetohydroxy acids to the $\alpha,\beta$-dihydroxyacid precursors of valine and isoleucine is catalyzed by acetohydroxy acid isomeroreductase. The $\alpha,\beta$-dihydroxy acids are both converted to the $\alpha$-keto acid precursors of valine and isoleucine by a dihydroxy acid dehydrase. Finally, the two amino acids are formed in trans-

## Figure 21.10

The biosynthesis of isoleucine and valine. The reactions leading to valine are catalyzed by the same enzymes that catalyze the corresponding reactions in isoleucine biosynthesis. Common enzymes are screened in yellow.

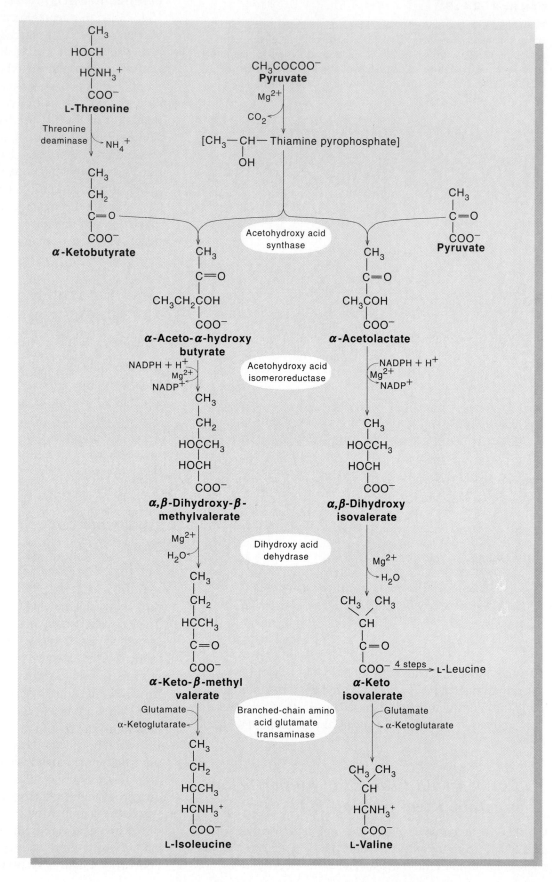

Figure 21.11

Three herbicides considered to be ''safe.'' These compounds all act on an enzyme required for branched amino acid biosynthesis. Since this enzyme is not present in mammals, the herbicides are considered to be harmless to humans.

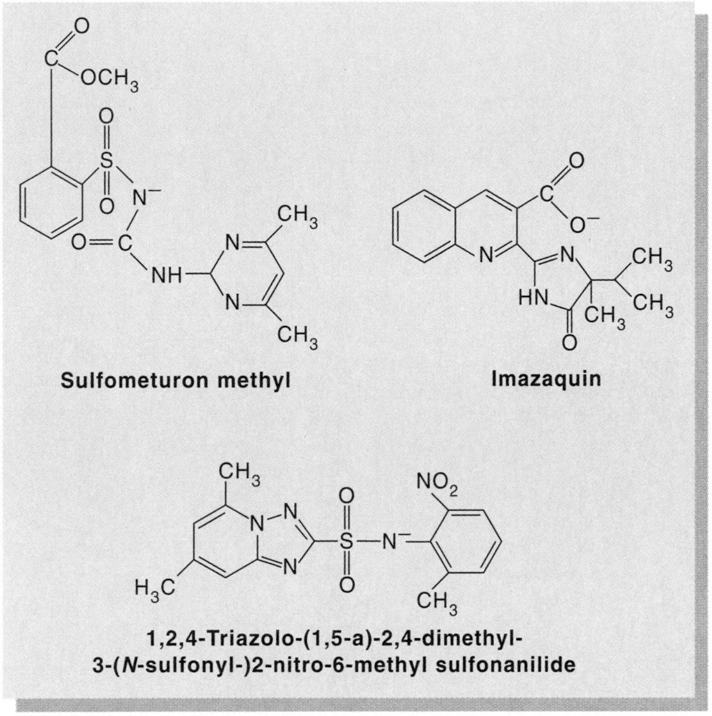

**Sulfometuron methyl**

**Imazaquin**

**1,2,4-Triazolo-(1,5-a)-2,4-dimethyl-3-(N-sulfonyl-)2-nitro-6-methyl sulfonanilide**

amination reactions in which glutamate is the amino donor (branched-chain amino acid glutamate transaminase).

## Amino Acid Pathways Absent in Mammals Offer Targets for Safe Herbicides

In recent years, agribusiness firms have developed empirically several compounds that inhibit essential steps in the biosynthesis of amino acids found in plants but missing in animals. One of these compounds, glyphosate, is a highly specific inhibitor of 5-enol pyruvyl-shikimate-3-phosphate synthase (an enzyme needed for aromatic amino acid biosynthesis). Glyphosate is the active ingredient in the widely used herbicide Roundup.

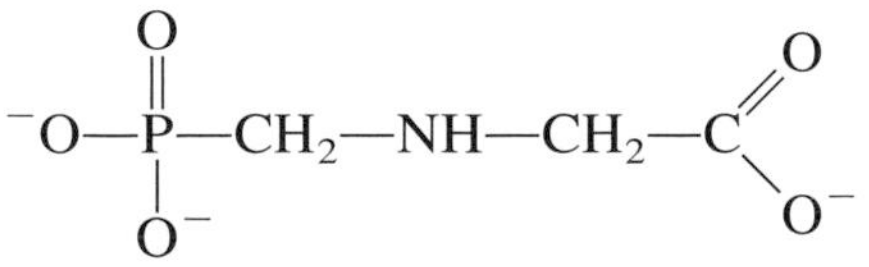

Glyphosate

Three other classes of compounds, although quite different from each other, are all inhibitors of acetohydroxy acid synthase (an enzyme required for branched-chain amino acid biosynthesis (see fig. 21.10). These three classes are sulfonylureas, imidazolinones, and triazolpyrimidines, which are the active ingredients in, respectively, Oust, Sceptor, and a third commercial herbicide still under development (fig. 21.11).

Because animals do not synthesize either the aromatic or branched-chain amino acids, these materials can be applied to kill unwanted vegetation without causing harm to domestic animals or humans. Certain derivatives can often be selectively applied to combat noxious plants without appreciable harm to crops. More promising, however, is the prospect of using biotechnology to incorporate genes for enzymes specifically resistant to one of the herbicides into seeds used for crop production.

## Chorismate Is a Common Precursor of the Aromatic Amino Acid Family

Aromatic amino acid biosynthesis proceeds via a long series of reactions, most of them concerned with the formation of the aromatic ring before branching into the specific routes to phenylalanine, tyrosine, and tryptophan. Chorismate, the common intermediate of the three aromatic amino acids, (see fig. 21.1) is derived in eight steps from erythrose-4-phosphate and phosphoenolpyruvate. We focus on the biosynthesis of tryptophan, which has been intensively studied by both geneticists and biochemists.

## Tryptophan Is Synthesized in Five Steps from Chorismate

The pathway leading from chorismate to L-tryptophan (fig. 21.12) has been thoroughly investigated. In E. coli the details of the enzymatic steps, the correlation between DNA sequence and the protein products, and the factors controlling the transcription of the structural genes exceed those known for most other sets of related genes. The extensive gene–enzyme analysis of Charles Yanofsky laid the foundation for others to explore details of some of the enzymatic steps by physical and kinetic approaches. Comparative studies of other bacteria and fungi have revealed a variation on the themes found in E. coli, particularly with respect to the distribution on one protein or another of the sequence of enzyme activities, which are identical in all forms. In addition, these studies have also revealed differences in the way the genes are arranged in the DNA and in the way expression of those genes is controlled.

## Figure 21.12

The biosynthesis of tryptophan from the branchpoint compound, chorismate in *E. coli*. The first step involves the conversion of chorismate to the aromatic compound anthranilate. The anthranilate is transferred to a ribose phosphate chain. The product is cyclized to indoleglycerol phosphate by the removal of water and loss of the ring carboxyl by indoleglycerol phosphate synthase. Finally, in a replacement reaction catalyzed by tryptophan synthase, glyceraldehyde-3-phosphate is removed from indoleglycerol phosphate, and the enzyme-bound indole is condensed with serine. The structure of phosphoribosyl pyrophosphate is shown in figure 21.13. The red arrow indicates that the end product of this pathway inhibits the first enzyme in the pathway.

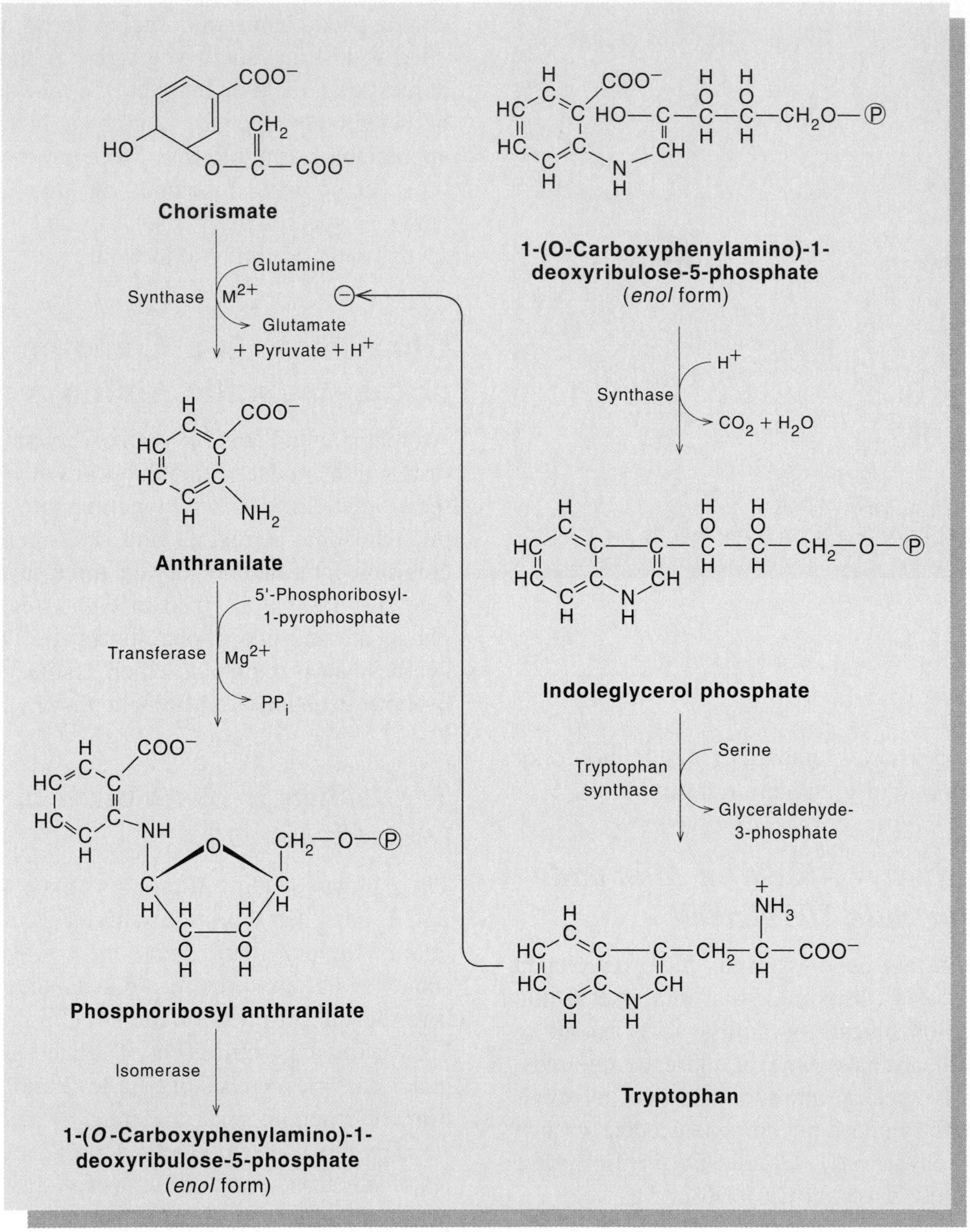

## Table 21.1

Distribution of Tryptophan Biosynthetic Enzyme Activities on Different Proteins in Bacteria and Fungi

| Organisms | Anthranilate Synthase | | Phosphoribosyl Anthranilate Transferase | Phosphoribosyl Anthranilate Isomerase | Indole-Glycerol-P Synthase | Tryptophan Synthase | |
|---|---|---|---|---|---|---|---|
| | *CoI* | *CoII* | | | | α | β |
| *E. coli, Salmonella typhimurium* | | | | | | | |
| *Serratia marcescens* | | | | | | | |
| *Pseudomonas putida* | | | | | | | |
| *Acinetobacter calcoaceticus* | | | | | | | |
| *Bacillus subtillis* | | | | | | | |
| *Neurospora crassa* | | | | | | | |
| *Sacaromyces cerevisiae* | | | | | | | |

Note: ⌐‾‾‾‾⌐ = covalent linkage; ⌐------⌐ = obligatory association required for full activity.
A single band covering two activities indicates a single polypeptide.

The first specific step in tryptophan biosynthesis is the glutamine-dependent conversion of chorismate to the simple aromatic compound anthranilate. Like most other glutamine-dependent reactions, the reaction can also occur with ammonia as the source of the amino group. However, high concentrations of ammonia are required. Thus far, almost all the anthranilate synthases examined have the glutamine amidotransferase activity (component II) and the chorismate-to-anthranilate activity (component I) on separate proteins.

Anthranilate is transferred to a ribose phosphate chain in a phosphoribosyl-pyrophosphate-dependent reaction catalyzed by anthranilate phosphoribosyltransferase. Phosphoribosylanthranilate undergoes a complex reaction known as the Amadori rearrangement, in which the ribosyl moiety becomes a ribulosyl moiety. The product, 1-(*O*-carboxyphenylamino) - 1 - deoxyribulose - 5' - phosphate, is cyclized to indoleglycerol phosphate by the removal of water and loss of the ring carboxyl by indole-glycerol phos-

phate synthase. The final step in tryptophan biosynthesis is a replacement reaction, catalyzed by tryptophan synthase, in which glyceraldehyde-3-phosphate is removed from indoleglycerol phosphate and the enzyme-bound indole undergoes a β-replacement reaction with serine. Isotopes and mutants were instrumental in demonstrating that indole is not a true intermediate in the tryptophan biosynthetic pathway (see Methods of Biochemical Analysis 21A).

Among different organisms the five enzyme activities required for tryptophan synthesis are distributed on different proteins (table 21.1). For example, in *E. coli,* indoleglycerol phosphate synthase catalyzes both the isomerization of phosphoribosylanthranilate and the cyclization step. Of particular interest is the occurrence of a single protein of the catalytic activities for nonconsecutive reactions in some cases. If in such cases the proteins were separate from each other in the cell, this arrangement, for example, in *Neurospora,* would necessitate the product of one reaction leaving the product site of one enzyme to be acted on by another enzyme and then returning to the substrate site of a third enzyme on the same protein that exhibited the first enzyme activity. The persistence of this arrangement during evolution makes attractive the idea that all the tryptophan biosynthetic enzymes exist in the cell as a single multienzyme (and multiprotein) complex. However, if so, the complex must be quite labile because individual gene products are so readily separated.

The evolution of the gene fusion that resulted in the conversion of the α and β peptides of tryptophan synthase as found in E. coli to the single peptide with α and β domains as found in fungi and yeast has been studied experimentally. Fusion of the trpB and trpA genes in the order in which they occur on the chromosome to yield a single β,α chain results in a protein of only limited activity. In contrast, when the gene fusion is done in the reverse of the genetic order (i.e., A to B), the resulting α,β chain yields a protein that is much more active. In addition, a short interdomain peptide region is required for full activity in a fusion protein.

## Amino Acids Act as Negative Regulators of Their Own Synthesis

Metabolite flow to tryptophan is controlled by inhibition of anthranilate synthase by tryptophan (see fig. 21.12). This is consistent with the principle that the end product inhibits the first enzyme which is unique to the pathway. This is a general phenomenon in amino acid biosynthetic pathways to ensure against the buildup of excess end product which is always a waste and oftentimes can cause problems.

In many cases the amino acid pathway branches so that two or more amino acids are formed. Aspartate is the precursor of four other amino acids found in proteins: Isoleucine, threonine, methionine, and lysine (see fig. 21.2). The first step in this overall pathway entails the conversion of aspartate to β-aspartyl-phosphate by aspartokinase. One might imagine that all four of the amino acid end products of this pathway would act together to inhibit this enzyme. However, in E. coli a different solution has been found. In this bacterium there are three aspartokinases which appear to be parts of different multienzyme complexes leading to threonine and leucine for aspartokinase I, methionine for aspartokinase II and lysine for aspartokinase III. As might be expected threonine and isoleucine inhibit aspartokinase I,

methionine inhibits aspartokinase II and lysine inhibits aspartokinase III. The evolution of three different aspartokinases to carry out the same reaction in this situation underscores the tremendous importance of efficient regulation of these pathways.

## Histidine Constitutes a Family of One

Histidine is unusual in two respects: It is in a family by itself, and both its structure and its pathway show a strong interplay with the purine pathway. The starting point for histidine biosynthesis is phosphoribosyl pyrophosphate (PRPP) as in the purine pathway (see chapter 23). The first specific step in histidine biosynthesis entails a condensation reaction between PRPP and ATP leading to phosphoribosyl ATP (see fig. 21.13). In the fifth step most of the purine nucleotide donated in the first step is returned to the purine pathway while the histidine precursor, imidazole glycerol phosphate, which now contains a newly synthesized imidazole ring, undergoes additional conversions leading to the formation of histidine. Not only does interplay occur between the histidine pathway and the purine pathway, but the final products of the two pathways, histidine and purine nucleotide, both contain imidazole rings.

## Nonprotein Amino Acids Are Derived from Protein Amino Acids

In addition to the 20 amino acids most frequently found in proteins a large group of amino acids occur in plants, bacteria, and animals that are not found in proteins. Some are found in peptide linkages in compounds that are important as cell wall or capsular structures in bacteria or as antibiotic substances produced by bacteria and fungi. Others are found as free amino acids in seeds and other plant structures. Some amino acids are never found in proteins. These nonprotein amino acids, numbering in the hundreds, include precursors of normal amino acids, such as homoserine and diaminopimelate; intermediates in catabolic pathways, such as pipecolic acid; D enantiomers of "normal" amino acids; and amino acid analogs, such as azetidine-2-carboxylic acid and canavanine, that might be formed by unique pathways or by modification of normal amino acid biosynthetic pathways.

**Table 21.2**

Some D-Amino Acids Found in Peptide Antibiotics

| Antibiotic | D-Amino Acids Present | Produced by |
|---|---|---|
| Actinomycin $C_1$ (D) | D-Valine | *Streptomyces parralus* and others |
| Bacitracin A | D-Asparagine, D-glutamate, D-ornithine, D-phenylalanine | *Bacillus subtilis* |
| Circulin A | D-Leucine | *Bacillus circulans* |
| Fungisporin | D-Phenylalanine, D-valine | *Penicillium spp.* |
| Gramicidin S | D-Phenylalanine | *Bacillus brevis* |
| Malformin $A_1$, C | D-Cysteine, D-leucine | *Aspergillus niger* |
| Mycobacillin | D-Aspartate, D-glutamate | *Bacillus subtilis* |
| Polymixin $B_1$ | D-Phenylalanine | *Bacillus polymyxa* |
| Tyrocidine A, B | D-Phenylalanine | *Bacillus brevis* |
| Valinomycin | D-Valine | *Streptomyces fulrissimus* |

## A Wide Variety of D-Amino Acids Are Found in Microbes

Certain D-amino acids with some L enantiomers are commonly found both in microbial cell walls and in many peptide antibiotics. For example, the peptidoglycans of bacteria contain both D-alanine and D-glutamate. The latter is present in a $\gamma$-glutamyl linkage. In some forms, the $\alpha$ carboxyl of the D-glutamyl residue is either amidated or in peptide linkage with glycine. D-Lysine or D-ornithine is found in the glycopeptide of some Gram positive organisms. The capsule of the anthrax bacillus is composed of a nearly pure homopolymer of D-glutamate in $\gamma$ linkage. Other bacilli also produce $\gamma$-linked polyglutamates, some of which form separate D-glutamate and L-glutamate chains, whereas others form a copolymer of D- and L-glutamate. A wide variety of D-amino acids have been found in antibiotics; some of these are listed in table 21.2.

D-Alanine is found in bacterial cell wall peptidoglycan. L-Alanine is converted to D-alanine by a racemase that contains pyridoxal phosphate as a cofactor. The racemization is followed by the formation of a D-alanyl-D-alanine dipeptide, which is accompanied by the conversion of ATP to ADP. The dipeptide is subsequently incorporated into the glycopeptide (see fig. 16.16).

In most cases of formation of peptides containing D-amino acid, the L form of the amino acid is the substrate for the incorporating enzyme. In contrast, the free D-amino acid is ordinarily a poor substrate for the incorporation reaction. Whether the racemization occurs on the enzyme or afterwards remains to be determined in most cases. In the case of the D-valyl residue formed in penicillin, a tripeptide derivative containing L-valine is an intermediate, and the conversion is thought to occur by way of an $\alpha,\beta$-dehydro form of the valyl residue.

**Figure 21.13**

The biosynthesis of histidine. The 5-aminoimidazole-4-carboxamide ribotide formed during the course of histidine biosynthesis is also an intermediate in purine nucleotide biosynthesis. Therefore it can be readily regenerated to an ATP, thus replenishing the ATP consumed in the first step in the histidine biosynthetic pathway (see fig. 23.13).

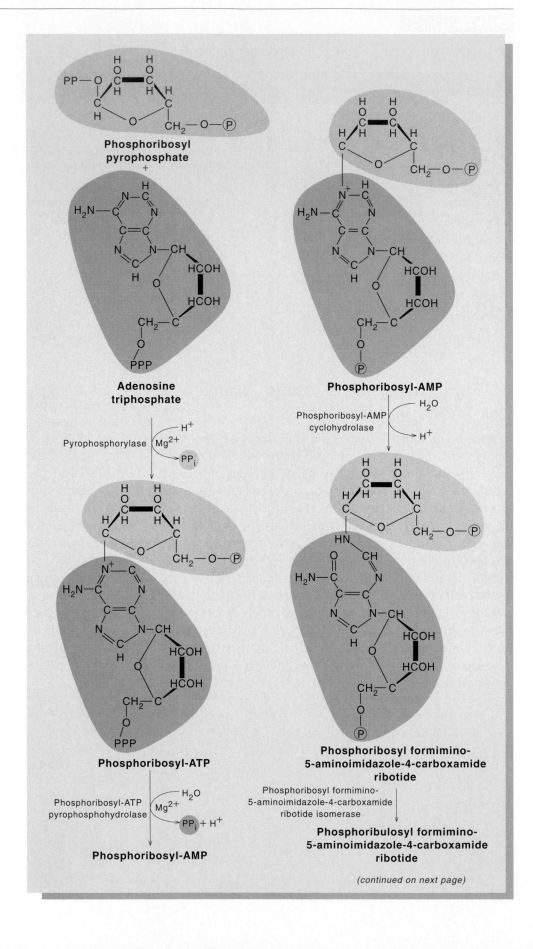

Phosphoribosyl pyrophosphate
+
Adenosine triphosphate

Pyrophosphorylase   Mg²⁺   H⁺   PPᵢ

Phosphoribosyl-ATP

Phosphoribosyl-ATP pyrophosphohydrolase   Mg²⁺   H₂O   PPᵢ + H⁺

Phosphoribosyl-AMP

Phosphoribosyl-AMP

Phosphoribosyl-AMP cyclohydrolase   H₂O   H⁺

Phosphoribosyl formimino-5-aminoimidazole-4-carboxamide ribotide

Phosphoribosyl formimino-5-aminoimidazole-4-carboxamide ribotide isomerase

Phosphoribulosyl formimino-5-aminoimidazole-4-carboxamide ribotide

(continued on next page)

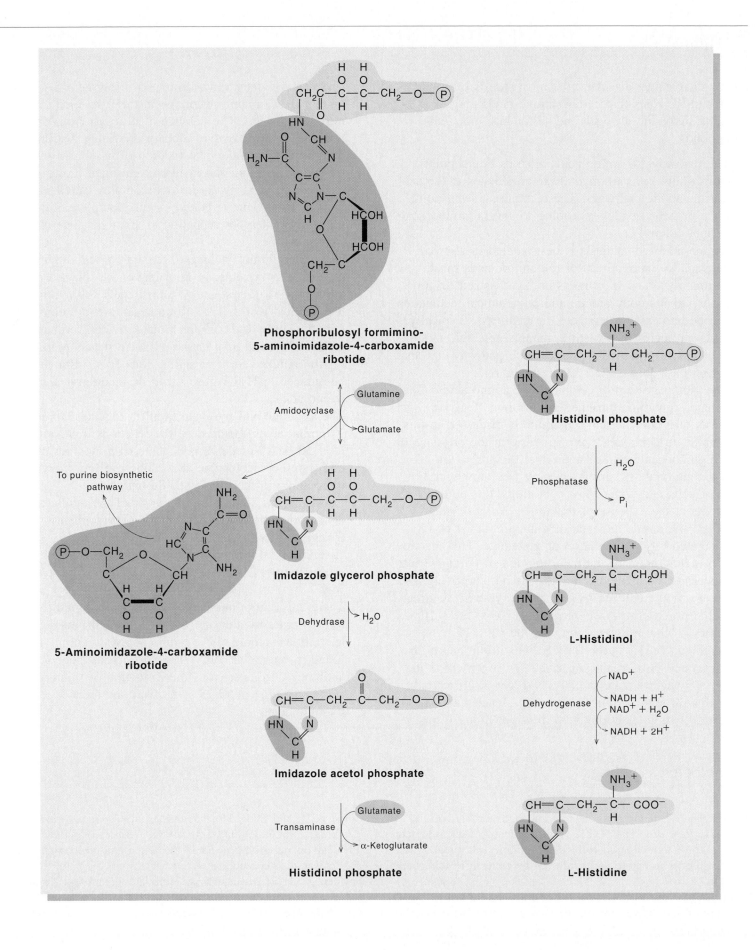

**Phosphoribulosyl formimino-5-aminoimidazole-4-carboxamide ribotide**

Amidocyclase

Glutamine

Glutamate

To purine biosynthetic pathway

**5-Aminoimidazole-4-carboxamide ribotide**

**Imidazole glycerol phosphate**

Dehydrase $\rightarrow H_2O$

**Imidazole acetol phosphate**

Transaminase

Glutamate

$\alpha$-Ketoglutarate

**Histidinol phosphate**

**Histidinol phosphate**

Phosphatase

$H_2O$

$P_i$

**L-Histidinol**

Dehydrogenase

$NAD^+$

$NADH + H^+$

$NAD^+ + H_2O$

$NADH + 2H^+$

**L-Histidine**

## Summary

In this chapter we discussed the biosynthesis of amino acids and the roles that certain amino acids play in bringing inorganic nitrogen and sulfur into bioorganic compounds.

1. The pathways to amino acids arise as branchpoints from a few key carbohydrate intermediates in the central metabolic pathways. The common starting point for a branched pathway leading to several amino acids defines a family.

2. Amino acid biosynthesis is best studied in microorganisms in which all 20 of the amino acids most commonly found in proteins are synthesized. In microorganisms both genetic and biochemical techniques can be harnessed to analyze the pathways. Typically, research begins by isolating mutants defective in single steps in a particular biosynthetic pathway and analyzing the consequences of the mutation.

3. The glutamate family contains four amino acids: Glutamate, glutamine, proline, and arginine. Only the synthesis of glutamate and glutamine is discussed in this chapter. In some cell types growing in the presence of a high concentration of ammonia, the amination of $\alpha$-ketoglutarate occurs directly by free ammonia. For most cells and under most conditions, this amination occurs at the expense of the amide group of glutamine. The amide nitrogen of glutamine must be regenerated by the amidation of glutamate. This reaction is a major route for the incorporation of ammonia into bioorganic compounds, and so it is not surprising that the reaction catalyzed by glutamine synthase is a regulated process, which is most sensitive to the sufficiency of the nitrogen supply of the cell.

4. Ammonia ($NH_3$) is the form in which nitrogen is incorporated into organic materials. Nitrogen exists in the $-3$ valence state in $NH_3$. Nitrogen itself actually passes through various forms and valence states as a result of its interactions with different living forms.

The valence states range from $-5$ in nitrates to $-3$ in ammonia or organic materials. In the 0 valence state, nitrogen is a gas. The passage of nitrogen from one form to another involves a chain of widely distributed organisms. The biological fixation of gaseous nitrogen by both free-living and symbiotic nitrogen-fixing bacteria is catalyzed by an enzyme complex called nitrogenase. Dinitrogen is bound by this complex as it gradually undergoes reduction to ammonia, one electron at a time.

5. The serine family includes three amino acids: Serine, glycine, and cysteine. In this chapter we focused on the synthesis of cysteine, which funnels sulfur into the biochemical world. The biosynthesis of L-cysteine entails the sulfhydryl transfer to an activated form of serine. Most sulfur in nature exists in the inorganic, highly oxidized form of sulfate ion. This sulfur must be reduced to $H_2S$ before it can be incorporated into amino acids.

6. The aspartate and pyruvate families together contain 11 amino acids. Because of the reactions involved in its synthesis, isoleucine is considered a member of both families. Isoleucine and valine use four enzymes in common in their biosynthetic pathways.

7. Chorismate is a common precursor of the amino acids of the aromatic amino acid family. Tryptophan is synthesized in five steps from chorismate.

8. Histidine is in a family of one. There are nine steps in this pathway which interacts with the purine pathway.

9. There are a large number of amino acids found in different organisms that are not incorporated into proteins. For example, the D-amino acids are commonly found in microbial cell walls and in many peptide antibiotics. In most cases the formation of D-amino acid containing peptides starts from the related L-amino acid.

## Selected Readings

Barker, H. A., Amino acid degradation by anaerobic bacteria. *Ann. Rev. Biochem.* 50:23, 1981. A review of an important group of fermentation pathways of amino acid breakdown that occur in nature and could not be covered in this chapter.

Bishop, P. E., and R. D. Joerger, Genetics and molecular biology of alternative nitrogen fixation systems. *Annu. Rev. Plant Physiol. Plant Mol. Biol.* 41:109–125, 1990.

Christen, P., and D. E. Metzler, (eds.), *Transaminases.* New York: John Wiley and Sons, 1985. A series of review chapters describing in detail the scope and mechanisms of transamination reactions.

Fowden, L., P. J. Lea, and E. A. Bell, The nonprotein amino acids of plants. *Adv. Enzymol.* 50:117, 1979. A discussion of the occurrence and biosynthesis of naturally occurring amino acid analogs in plants.

Katz, E., and A. L. Demain, The peptide antibiotics of *Bacillus:* Chemistry, biogenesis and possible functions. *Bacteriol. Rev.* 41:449, 1977. A description of several peptide antibiotics showing the distribution of D-amino acid in these compounds.

Ledley, F. D., H. E. Grenett, M. McGinnis-Shelnutt, and S. L. C. Woo, Retroviral-mediated gene transfer of human phenylalanine hydroxylase into NIH 3T3 and hepatoma cells. *Proc. Natl. Acad. Sci. USA* 83:409, 1986.

Mazelis, M., Amino acid catabolism. In B. J. Mifflin (ed.), *The Biochemistry of Plants,* vol. 5. New York: Academic Press, 1980, pp. 541–567. A survey of some of the amino acid catabolic pathways that have been found in plants.

Meister, A., *Biochemistry of the Amino Acids,* vol. 2, 2d ed. New York: Academic Press, 1965, pp. 593–1084. A very complete survey of amino acid catabolic pathways as they were known up to that time.

Meister, A., *Biochemistry of the Amino Acids,* vols. 1 and 2. New York: Academic Press, 1965. The two-volume classic provides a thorough discussion of amino acid literature, occurrence, properties, and metabolism of amino acids up to that time.

Meister, A., Glutathione metabolism and its selective modification. *J. Biol. Chem.* 263:17205, 1988. A minireview describing the many important metabolic roles for glutathione.

Miflin, B. J. (ed.), *The Biochemistry of Plants: A Comprehensive Treatise,* vol. 5, *Amino Acids and Derivatives.* New York: Academic Press, 1980. This volume contains 10 chapters by several authors detailing amino acid biosynthesis pathways in plants.

Neidhardt, F. C., J. L. Ingraham, K. B. Low, B. Magasanik, M. Schaechter, and H. E. Umbarger, (eds.), *Escherichia coli* and *Salmonella typhimurium: Cellular and Molecular Biology,* vol. 1. Washington: American Society for Microbiology, 1987. This volume contains seven chapters by several authors describing in detail the pathways of amino acid biosynthesis in bacteria with particular emphasis on enzymatic and genetic control mechanisms.

Warren, M. J., and A. I. Scott, Tetrapyrrole assembly and modification into the ligands of biologically functional cofactors. *Trends Biol. Sci.* 51:486–491, 1990.

Wellner, D., and A. Meister, A survey of inborn errors of metabolism and transport in man. *Ann. Rev. Biochem.* 50:911, 1981. This review documents the importance of the pathways that break down amino acids in humans.

Yamada, K. S., T. Kinoshita, T. Tsunoda, and K. Aida (eds.), *The Microbial Production of Amino Acids.* New York: John Wiley and Sons, 1972. A collection of essays describing microbial processes used in Japanese industry for the production of amino acids. Includes examples in which the regulatory mechanisms functioning in most cells have been modified or bypassed.

## Problems

1. Why did nature ''waste'' an ATP in glutamine biosynthesis? The lone pair of electrons on an ammonia could have attacked the $\gamma$-carbonyl group of a glutamate. Subsequent elimination of an oxide ion and release of a proton from the nitrogen would produce a glutamine without the consumption of an ATP.

2. Does it appear to be a paradox that L-glutamate is both the product and an initial reactant in the glutamine biosynthetic pathway? Assuming that glutamate synthase (fig. 21.3) is utilized, how can you explain this paradox?

3. Examine the bioenergetics of the synthesis of glutamine synthesis from $\alpha$-ketoglutarate via glutamate synthase or glutamate dehydrogenase (fig. 21.3). Is there a difference?

4. What is the function of NADPH in the reactions catalyzed by glutamate dehydrogenase and glutamate synthase?

5. *E. coli* strains have been isolated that are unable to grow in a medium containing L-valine but lacking L-isoleucine and L-leucine. The same organism can grow on a medium lacking all three amino acids. Provide an explanation.

6. The amino acid 2-aminobutanoate is a product of some bacteria (not a protein component). Predict how the bacteria produces this amino acid.

7. A certain bacteria that was a tryptophan auxotroph was observed to grow well when it was supplied with tryptophan, but as soon as the tryptophan in the environment was exhausted it started to excrete a metabolite on the tryptophan biosynthetic pathway. Why didn't it excrete the metabolite before it exhausted the environmental tryptophan?

8. This chapter categorizes the amino acids into families based on the origin of their carbon skeleton. Is this an absolute pattern? Take a closer look at the information in this chapter to produce an answer.

9. What is the function of the acetylation of serine with acetyl-CoA during the biosynthesis of cysteine (fig. 21.8)?

10. Which ribose carbons are incorporated into tryptophan?

11. Molecules with structures as diverse as carbamoyl-phosphate, tryptophan, and cytidine triphosphate are feedback inhibitors of the *E. coli* glutamine synthase. The feedback inhibition is cumulative, with each metabolite exerting a partial inhibition on the enzyme. Why would complete inhibition of the glutamine synthase by a single metabolite be metabolically unsound?

12. Given the structural diversity of the compounds that feedback-inhibit glutamine synthase, would you predict that they interact at a common regulatory site? Why or why not?

13. How does increased synthesis of aspartate and glutamate affect the TCA cycle? How does the cell accommodate this effect?

14. In what sense may indole be viewed as an "intermediate" in L-tryptophan biosynthesis?

15. When $^{14}$C-labeled 4-hydroxyproline was administered to rats, the 4-hydroxyproline in newly synthesized collagen was not radiolabeled. Explain.

16. The accumulation of biosynthetic intermediates, or of metabolites derived from these intermediates, has proven to be valuable in the analysis of biosynthetic pathways in microorganisms. It was found that these accumulations occurred only after the required amino acid had been consumed and growth had stopped. How might you account for this observation?

# Demonstration That Indole Is Not a True Intermediate in the Tryptophan Biosynthetic Pathway

The tryptophan pathway provides another example in which nutritional studies with mutants of *Neurospora* and *E. coli,* isotope incorporation studies, and enzymatic analyses have been exploited to reveal the steps in a biosynthetic pathway. For example, early studies with tryptophan-requiring organisms found in nature revealed that some could use indole (a compound known to be formed by the microbial degradation of tryptophan), and others could use anthranilate. Later, after Beadle and Tatum introduced the approach of studying metabolism with mutants of the bread mold *Neurospora,* tryptophan-requiring mutants of this organism were found that could use indole or either anthranilate or indole. Still later, similar mutants of *E. coli* were found, and mutants of both organisms were described that accumulated one or the other of these compounds. Clearly, those mutants that grew on anthranilate or indole were blocked in some step before these compounds, and those that accumulated them were blocked in the step after them.

Incorporation studies with isotopes showed that when anthranilate was converted to tryptophan, the carboxyl group of anthranilate was lost as carbon dioxide, but the nitrogen was retained. Because the enzymes in the tryptophan biosynthetic pathway have only a limited specificity, it was possible to substitute 4-methyl-anthranilate in *E. coli* extracts that could convert anthranilate to indole. This "nonisotope" label was conserved during the conversion to yield 6-methyl indole.

It was thus clear that some two-carbon unit replaced the carboxyl carbon of anthranilate. Further studies with such *E. coli* extracts indicated that phosphoribosyl pyrophosphate was a good cosubstrate for the formation of indole from anthranilate. Fractionation of these extracts, as well as examination of mutants blocked between anthranilate and indole, revealed that an intermediate in this conversion was indole-3-glycerol phosphate. Extracts from one group of such mutants could not form indole-3-glycerol phosphate, whereas the other group could not convert it to indole and glyceraldehyde-3-phosphate. The latter group was found to accumulate the dephosphorylated derivative, indole-3-glycerol, in culture fluids.

The two intermediates in the conversion of anthranilate to indole-3-glycerol phosphate, phosphoribosylanthranilate and 1-(*O*-carboxyphenylamino)-1-deoxyribulose-5'-phosphate, were originally postulated to account for the involvement of phosphoribosyl pyrophosphate in indole-3-glycerol phosphate formation. Support for the postulate was obtained when the dephosphorylated derivative of the second of these intermediates was found in the culture fluids of certain tryptophan-requiring bacterial mutants. The corresponding derivative of the first intermediate has not been found, probably because of its instability. Indeed, this compound, when formed in extracts, is rapidly broken down to anthranilate and ribose-5-phosphate.

For several years, indole, which was accumulated by some mutants and used to satisfy the tryptophan requirement by others, was considered an intermediate in tryptophan biosynthesis. Such a role for indole would have been of interest, because it appeared to be an exception to the generalization that biosynthetic intermediates had to bear a charge. It was found that extracts of cells that utilized indole did indeed catalyze the condensation of indole with serine, and extracts of cells that accumulated indole catalyzed the

4-Methylanthranilate      6-Methyl indole

cleavage of indole-3-glycerol phosphate to indole and glyceraldehyde-3-phosphate. Furthermore, *E. coli* mutants of these two classes clearly were affected in separate genes. However, the two products of these genes catalyzed their corresponding reactions faster when they were associated in a complex of the form $\alpha_2\beta_2$, where $\alpha$ and $\beta$ stand for different protein subunits. The complex itself catalyzes the overall reaction

Indole-3-glycerol phosphate + serine $\longrightarrow$

tryptophan + glyceraldehyde-3-phosphate

faster than either of the separate reactions. The same was found with extracts of *Neurospora* in which the two partial reactions were catalyzed by the same protein and with which no evidence for indole as a free intermediate could be found. Thus, it became clear that indole, although historically important in deciphering the pathway to tryptophan, occurs only as a bound intermediate.

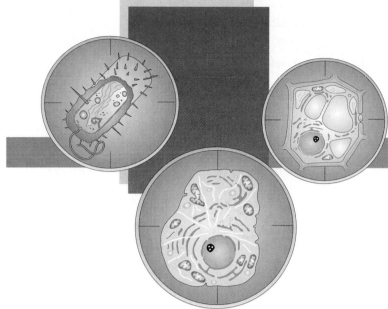

# Amino Acid Metabolism in Vertebrates

*Vertebrates obtain most of their amino acids by
nutrition. Amino acid degradation in vertebrates
usually starts with the loss of the α amino nitrogen
and finishes with the return of the carbon skeletons
to the central metabolic pathways.*

Amino acid metabolism in vertebrates contrasts sharply
with amino acid metabolism in plants and microorganisms.
Most striking is the fact that plants and microorganisms can
synthesize all 20 amino acids required for protein synthesis
whereas vertebrates can only synthesize about half this
number. This inability leads to complex nutritional needs
for vertebrates. We discuss these needs in light of the pathways for biosynthesis that still exist.

**Figure 22.1**

Amino acids that are synthesized *de novo* in mammals. All such amino acids are related by a small number of steps to glycolysis or TCA cycle intermediates. (Also see fig. 21.1.)

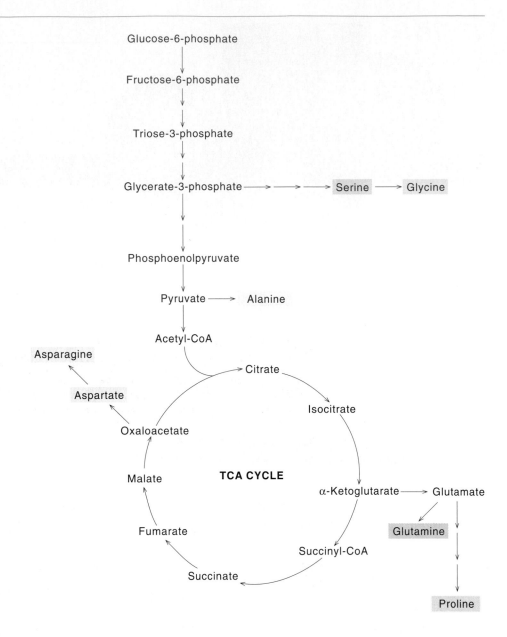

By contrast with the biosynthetic pathways, all degradation pathways for commonly occurring amino acids are found in vertebrates. This makes it possible for vertebrates to dispose of potentially harmful excesses of amino acids and their partial degradation products. The need for efficient degradation pathways is highlighted by the existence of many pathological states that result from deficiencies in the degradative pathways. We discuss some of these pathological states and the mutational deficiencies that lead to them in light of the degradation pathways.

In addition to serving in their role as precursors of proteins, amino acids are essential precursors of other biomolecules, of which we discuss two examples: The synthesis of porphyrin and the synthesis of glutathione.

## Humans and Rodents Synthesize Less Than Half of the Amino Acids They Need for Protein Synthesis

In view of the central importance of amino acids in proteins, we might expect that all organisms would possess the necessary enzymes to synthesize the protein amino acids. Surprisingly, only eight protein amino acids can be synthesized by the standard *de novo* pathways (fig. 22.1). The remainder are synthesized by alternative pathways (sometimes referred to as salvage pathways) or supplied by nutrients.

Why did the evolutionary process not favor the preservation of pathways for synthesizing all of the amino acids needed in higher animals? Can this be a case of too great a

**Table 22.1**

The Essential Amino Acids

| For Weight Gain in Protein-Starved Adult Rats | | For Positive Nitrogen Balance in Adult Humans | | For Mouse L Cells in Culture | |
|---|---|---|---|---|---|
| *Essential* | *Nonessential* | *Essential* | *Nonessential* | *Essential*[a] | *Nonessential* |
| Histidine | Alanine | Isoleucine | Alanine | Arginine | Alanine |
| Isoleucine | Arginine[b] | Leucine | Arginine | Cysteine | Asparagine |
| Leucine | Asparagine | Lysine | Asparagine | Glutamine | Aspartate |
| Lysine | Aspartate | Methionine | Aspartate | Histidine | Glutamate |
| Methionine | Cysteine | Phenylalanine | Cysteine | Isoleucine | Glycine |
| Phenylalanine | Glutamate | Threonine | Glutamate | Leucine | Proline |
| Threonine | Glutamine | Tryptophan | Glutamine | Lysine | Serine |
| Tryptophan | Glycine | Valine | Glycine | Methionine | |
| Valine | Proline | | Histidine[c] | Phenylalanine | |
| | Serine | | Proline | Threonine | |
| | Tyrosine | | Serine | Tryptophan | |
| | | | Tyrosine | Tyrosine | |
| | | | | Valine | |

[a] The medium also contained 0.25–1% dialyzed horse serum.

[b] Arginine is required in the diet of young rats.

[c] Histidine is required in infant humans.

genetic load? Not likely since these mammals are believed to have in excess of 50,000 genes and only 84 genes are required to encode all of the enzymes needed to make the 20 major amino acids found in proteins. A more likely reason for this loss of major pathways is that some of the intermediates in amino acid biosynthesis may have toxic effects on other biochemical processes occurring in higher eukaryotes. Perhaps the most probable reason of all is that these amino acids were not needed because these vertebrates eat organisms that contain an abundance of amino acids.

## Many Amino Acids Are Required in the Diet for Good Nutrition

The inability of mammals to synthesize all of the amino acids they require has led to the classification of amino acids as essential and nonessential. Conceptually this distinction seems clear. Practically speaking, it turns out to be more complex to designate essential amino acids. An "essential" amino acid, in this classification, means one that must be supplied in the diet if the organism is to maintain a positive nitrogen balance. As we will see, the absence of a *de novo* pathway for the biosynthesis of an amino acid does

not necessarily mean that the amino acid must be obtained from the diet. For one thing, it is frequently possible for the organism to make one amino acid from another. Furthermore, an alternative pathway for synthesis may supply sufficient amino acid for maintenance if not for growth.

The classical differentiation between essential and nonessential amino acids was made by W. C. Rose. His identification scheme was based on the weight gain of growing white rats that were fed diets containing 19 of the 20 amino acids found in proteins. These results are shown in the first two columns of table 22.1. For humans, the results were based not on weight gain or loss but on short-term maintenance of a positive nitrogen balance (see table 22.1 middle columns). If the nitrogen balance was negative (total nitrogen excretion exceeding total nitrogen intake) during the period when a single amino acid was excluded from the diet, it was concluded that tissue protein was being degraded to supply the missing amino acid. For comparison, the amino acids found to be essential and nonessential for a strain (L) of mouse fibroblasts grown in cell culture are also included in table 22.1 (last two columns).

The results in table 22.1 indicate that given 19 other amino acids, the particular amino acid can be formed either

## Figure 22.2

Synthesis of arginine by the salvage pathway found in vertebrates and by the *de novo* pathway found in plants and bacteria. The final steps from ornithine to arginine are also part of the urea cycle (see fig. 22.7).

by the *de novo* pathway used by plants or by a "salvage" pathway at the expense of some other amino acid.

As an example, arginine can be synthesized by a circuitous route that starts from the amino acid proline (fig. 22.2). The enzymes needed for *de novo* synthesis of proline are still formed (or found) in animal tissues. In addition, animals contain a proline oxidase that yields $\Delta'$-pyrroline-5-carboxylic acid, which is in equilibrium with glutamic-$\gamma$-semialdehyde. A transaminase converts the semialdehyde to ornithine. From ornithine to arginine, enzymes of the urea cycle are used (see fig. 22.7). Thus, although the pathway by which bacteria and plants form arginine is not present in animal tissues, arginine can be formed, but in some organisms (e.g., rats) at a rate insufficient for growth. Two other amino acids normally present in the diet also can be formed from other amino acids: Cysteine can be formed from dietary methionine, and tyrosine can be formed by hydroxylation of phenylalanine. These routes and the corresponding enzymes are listed in table 22.2.

## Essential Amino Acids Must Be Obtained by Degradation of Ingested Proteins

Amino acids that originate from catabolism are derived from three sources: Dietary proteins, storage proteins, and metabolic turnover of endogenous proteins. Catabolism of dietary proteins is a characteristic of higher animals, whereas the catabolism of storage protein is best illustrated by the germination of protein-storing seeds, such as beans or peas. Turnover of endogenous proteins occurs in all cells. The amino acids to which they are degraded can be recycled to make new proteins or derivatives that involve amino acids as precursors.

Protein catabolism begins with hydrolysis of the covalent peptide bonds that link successive amino acid residues in a polypeptide chain (fig. 22.3). This process is termed proteolysis, and the enzymes responsible for the action are called proteases. In humans and many other animals, proteolysis occurs in the gastrointestinal tract; this type of proteolysis results from proteases secreted by the stomach, pancreas, and small intestine.

The initial products of proteolytic digestion are free amino acids and small peptides. Further digestion of pep-

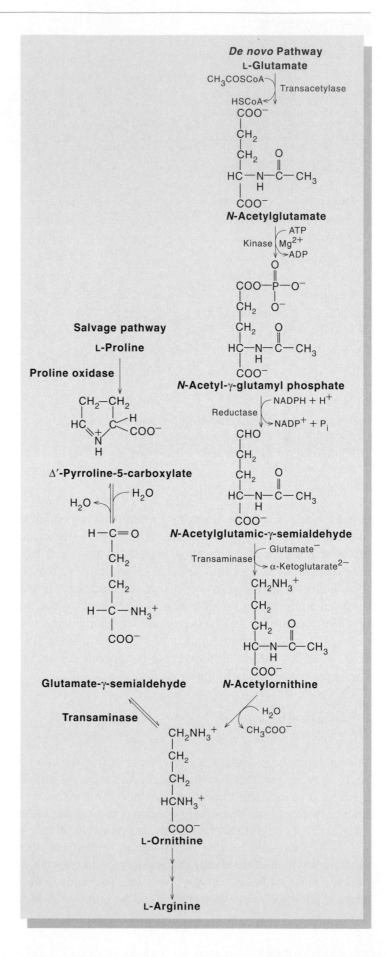

**Table 22.2**

Salvage Pathways Allowing the Formation of Certain Nonessential Amino Acids from Other Amino Acids

| Amino Acid Formed | Formed From | Enzymes Required |
|---|---|---|
| Arginine | Proline | Proline oxidase<br>Ornithine-glutamate transaminase<br>Ornithine transcarbamoylase<br>Argininosuccinate synthase<br>Argininosuccinate lyase |
| Cysteine | Methionine | S-Adenosylmethionine synthase<br>α-Methyltransferase<br>S-Adenosylhomocysteinase<br>Cystathionine-β-synthase<br>Cystathionine-γ-lyase |
| Tyrosine | Phenylalanine | Phenylalanine-4-monooxygenase |

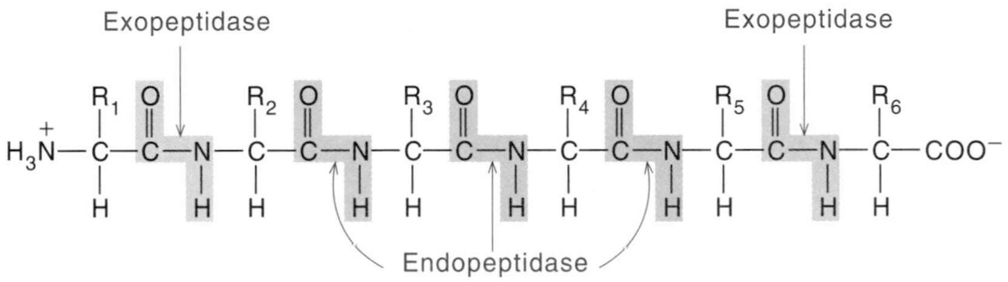

**Figure 22.3**

A protease hydrolyzes a peptide bond. Proteases have varying degrees of specificity, depending on the chemical nature of the R group and the location of the peptide linkage. Exopeptidases attack one or both ends of a polypeptide chain, and endopeptidases attack interior linkages.

tides results from the action of peptidases secreted (or formed) by the intestinal mucosa. Some peptidases act on their substrates by hydrolyzing internal peptide bonds (endopeptidases); others act by removing amino acids one at a time from the end of the peptide (exopeptidases). A sign of the competitiveness of microorganism for the available food supply is that they make proteases and peptidases, which can be secreted into the surrounding medium so as to break down potential nutrient proteins to a suitable size for absorption.

## Amino Acids May Be Reutilized or They May Be Degraded When Present in Excess

Degradative pathways for most amino acids begin by removal of the α-amino nitrogen. There are two major routes of deamination: Transamination and oxidative deamination.

## Transamination Is the Most Widespread Form of Nitrogen Transfer

The process of transamination is illustrated in figure 22.4 for an undesignated amino acid donating its amino group to the TCA cycle intermediate α-ketoglutarate. The reaction leads to an α-keto acid and glutamate. Most transaminases involved in amino acid catabolism exhibit a fairly broad specificity for the α-amino acid. The amino group acceptor is usually α-ketoglutarate. The mechanism for this reaction was discussed in chapter 10 (see fig. 10.4).

## Net Deamination via Transamination Requires Oxidative Deamination

Transamination does not result in net deamination because one amino acid is replaced by another amino acid. The main function of transamination in catabolism is to funnel the amino nitrogen into one or a few amino acids. For glutamate

**Figure 22.4**

Transamination and deamination. Glutamate transaminase catalyzes the transfer of the α-amino group of an amino acid to α-ketoglutarate. The reaction is highly reversible because the reacting functional groups of the products are identical to those of the reactants. Transamination is not deamination. Transamination yields ammonia only if it is linked to another type of deamination process. Here net deamination results from the combined action of glutamate transaminase and glutamate dehydrogenase. In this process the α-ketoglutarate is recycled.

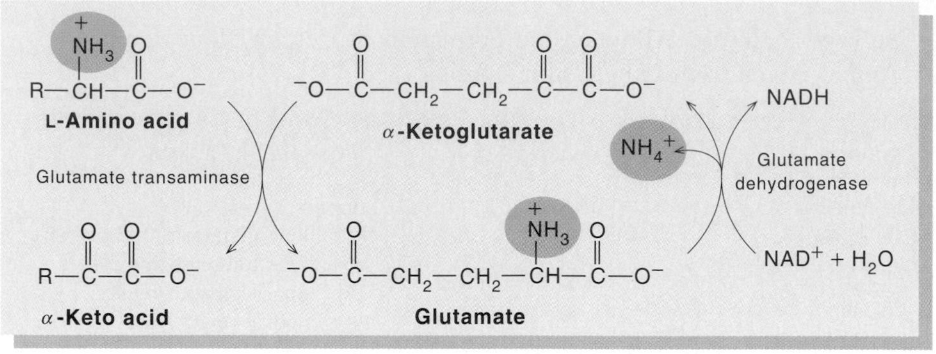

to play a role in the net conversion of amino groups to ammonia, a mechanism for glutamate deamination is needed so that α-ketoglutarate can be regenerated for further transamination. Most often this entails the oxidative deamination of glutamate in a reaction catalyzed by an NAD$^+$-linked enzyme, glutamate dehydrogenase. This broadly distributed enzyme is located in the mitochondria of eukaryotic cells. It catalyzes release of the amino group of glutamate, leading to the regeneration of α-ketoglutarate.

$$\text{Glutamate} + \text{NAD}^+ + \text{H}_2\text{O} \longrightarrow$$
$$\alpha\text{-ketoglutarate} + \text{NH}_4^+ + \text{NADH} \quad \textbf{(1)}$$

The overall process of transamination of α-ketoglutarate and regeneration of the α-ketoglutarate is shown in figure 22.4.

Catalysis of glutamate dehydrogenase starts with a hydride transfer from the α carbon of the amino acid to NAD$^+$ (fig. 22.5). The resulting α-iminoglutarate hydrolyzes to α-ketoglutarate and ammonia.

We discussed a glutamate dehydrogenase reaction in the previous chapter in conjunction with the biosynthesis of glutamate (see fig. 21.3). The deamination of glutamate to α-ketoglutarate involves the same compounds, but the reaction is reversed. This reaction is close enough to equilibrium that it can occur in either direction, but the glutamate dehydrogenases involved in catalyzing the forward and backward reactions are usually different and located in different places. The glutamate dehydrogenase involved in deamination is located in the mitochondria, and it uses NAD$^+$ as a cosubstrate. The amination reaction occurs in the cytosol, and it usually is specific for NADPH as cosubstrate.

Some other amino acids also undergo deamination directly by oxidative reactions. Whether transamination or direct deamination is more important as an initial step in amino acid breakdown depends on the organism or tissue under investigation.

## In Many Vertebrates Ammonia Resulting from Deamination Must Be Detoxified Prior to Elimination

The NH$_4^+$ resulting from deamination of amino acids is converted to ammonia either directly or indirectly (e.g., by means of a transamination to yield a readily deaminated product such as glutamate). In microorganisms using a single amino acid as a nitrogen source, the ammonia so liberated is assimilated and used to form other nitrogen-containing cellular components. When the amino acid is a carbon source, much more ammonia is liberated than is needed for biosynthesis, and it is disposed of by excretion to the surrounding medium. This simple disposal mechanism is adequate for free-living microorganisms because the ammonia is carried away in the surrounding medium or escapes into the atmosphere.

Ammonia is also the major nitrogenous end product in some of the simpler aquatic and marine animal forms, such as protozoa, nematodes, and even bony fishes, aquatic amphibia, and amphibian larvae. Such animals are called ammonotelic. But in many animals, NH$_3$ is toxic, and its removal by simple diffusion is difficult. Thus, in terrestrial snails and amphibia, as well as in other animals living in environments in which water is limited, urea is the principal end product (fig. 22.6). Urea formation also helps to maintain osmotic balance with seawater in cartilagenous fishes. In such animals, most of the urea secreted by the kidney glomerulus is reabsorbed by the tubules. Indeed, the amount of nitrogen excreted by the kidneys of fishes is small com-

## Figure 22.5

The glutamate-dehydrogenase-catalyzed reaction involves hydride transfer from glutamate to NAD$^+$, leading to $\alpha$-iminoglutarate imine followed by hydrolysis to $\alpha$-ketoglutarate. The amine group in glutamate is written in the uncharged form ($-NH_2$) because this is believed to be the reactive species.

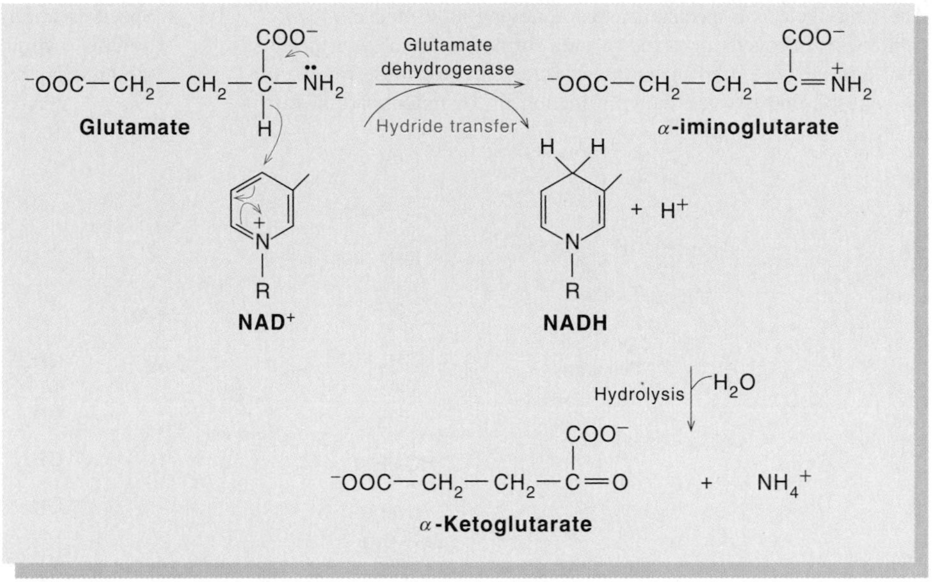

## Figure 22.6

Excretory forms of nitrogen in different organisms. NH$_3$ is the most common end product of nitrogen metabolism. In many organisms NH$_3$ is toxic. To prevent the harmful excess of ammonia, it is converted to urea or uric acid before excretion.

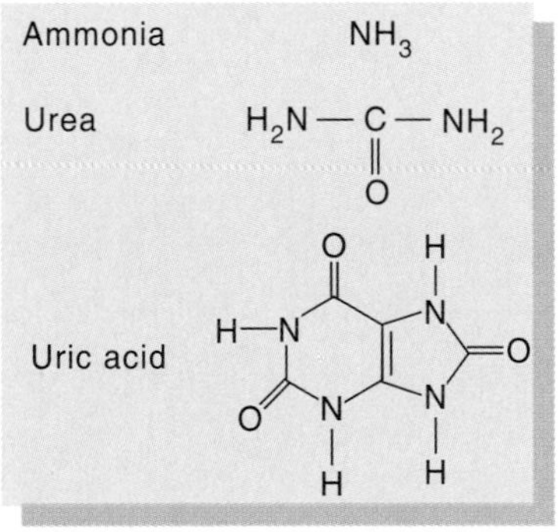

pared with that excreted by the gills, and in most fishes, ammonia is the major form of excreted nitrogen.

Another form of "detoxified" ammonia that is used in nitrogen excretion is uric acid. Uric acid is the predominant nitrogen excretory product in birds and terrestrial reptiles (turtles excrete urea, whereas alligators excrete ammonia unless they are dehydrated, in which case they, too, excrete uric acid). Uric acid formed as a product of amino acid catabolism involves the *de novo* pathway of purine biosynthesis; therefore, its formation from NH$_3$ liberated in amino acid catabolism is described elsewhere (see chapter 23). In mammals, uric acid is exclusively an intermediate in purine catabolism, and in most mammals (primates excluded), it is further converted by uricase to allantoin.

## Urea Formation Is a Complex and Costly Mode of Ammonia Detoxification

Urea formation in the liver starts with the multistep conversion of ornithine to arginine (fig. 22.7). This is followed by the breakdown of arginine into ornithine and urea. The cyclic nature of this pathway was first appreciated by Hans Krebs and Henseleit in 1932. In subsequent years the impor-

# Figure 22.7

The urea cycle is a mechanism for removing unwanted nitrogen. Sources of nitrogens involved in urea formation are shown in red. Five enzymes are used in the urea cycle. Three of these function in the cytosol, and two, as shown, function in the mitochondrial matrix.

Specific carriers in the inner mitochondrial membrane transport ornithine, citrulline, ammonium ion, and $HCO_3^-$ ($CO_2$) into and out of the mitochondrial matrix.

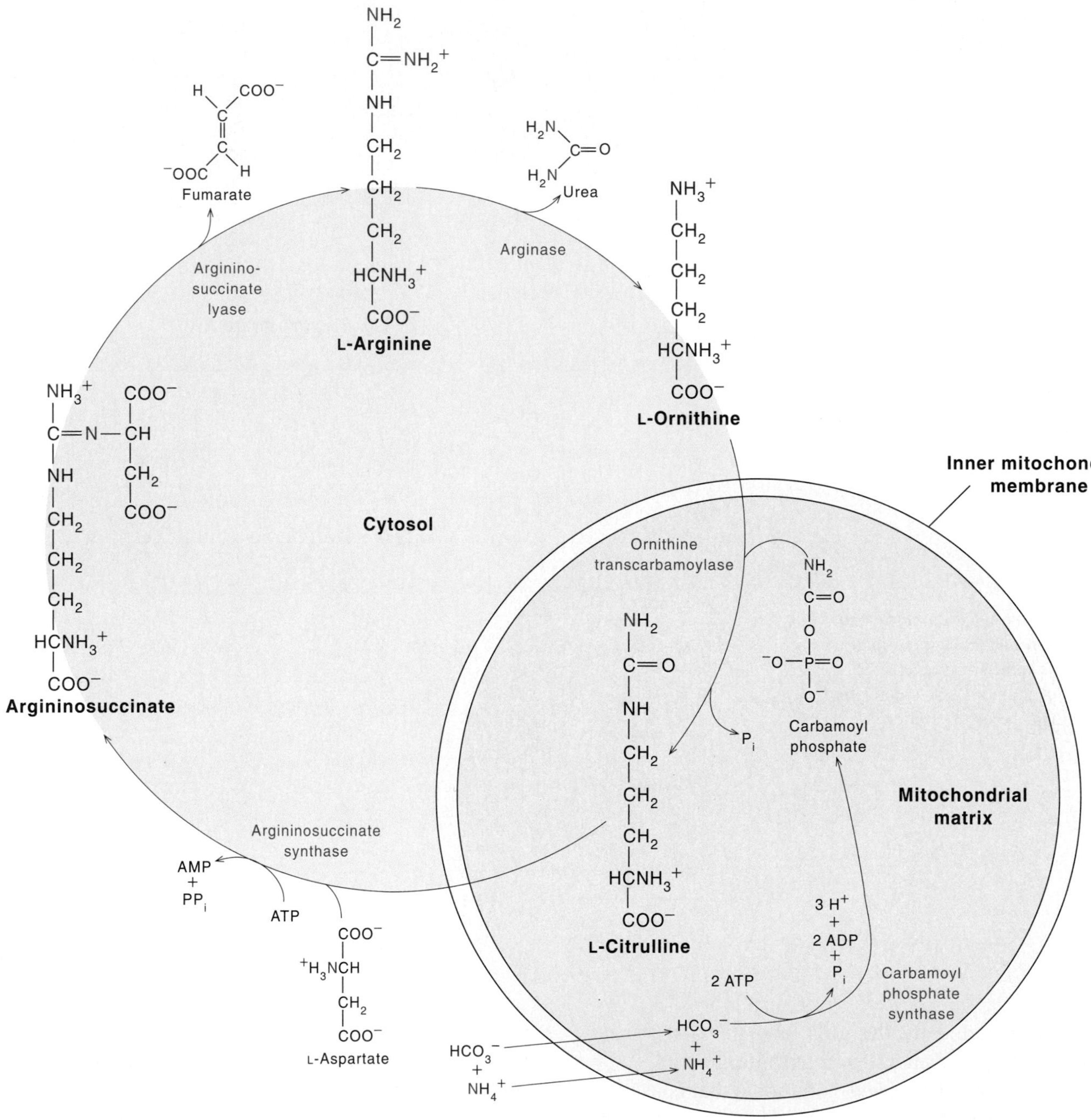

**Figure 22.8**

The mechanism of formation of carbamoyl phosphate. The reaction involves three steps, all of which take place on the same enzyme, carbamoyl phosphate synthase.

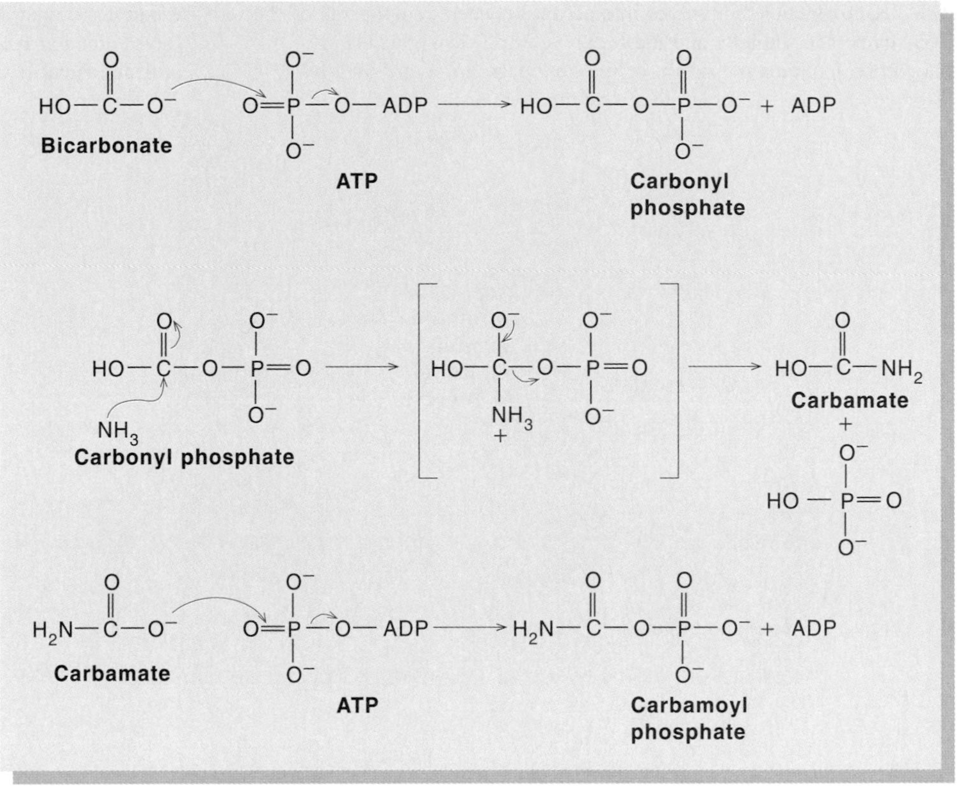

tant details of the pathway were determined by many workers, including P. P. Cohen, S. Grisolia, and S. Ratner.

The complete urea cycle as it occurs in the mammalian liver requires five enzymes: Argininosuccinate synthase, arginase, and argininosuccinate lyase (which function in the cytosol), and ornithine transcarbamoylase, and carbamoyl phosphate synthase (which function in the mitochondria). Additional specific transport proteins are required for the mitochondrial uptake of L-ornithine, $NH_3$, and $HCO_3^-$ and for the release of L-citrulline.

The free ammonia formed by oxidative deamination of glutamate is converted into carbamoyl phosphate in a three-step reaction requiring two ATP molecules (fig. 22.8).

$$NH_4^+ + HCO_3^- + 2\,ATP \longrightarrow$$
$$\text{carbamoyl phosphate} + HPO_4^{2-} + 2\,ADP + 2\,H^+ \quad \textbf{(2)}$$

First the bicarbonate ion is activated (step 1). The activated carbon is subject to nucleophilic attack by ammonia, leading to a carbamate intermediate (step 2). Finally, in a reaction similar to step 1, a second phosphoryl group is transferred to carbamate to form carbamoyl phosphate (step 3).

The carbamoyl group of carbamoyl phosphate has a high group-transfer potential, which is displayed by its transfer to the terminal amino group of ornithine to form L-citrulline (see fig. 22.7). In the process, inorganic phosphate is released. Before further reaction can occur the citrulline must be transported across the mitochondrial membrane to the cytosol, where the remaining reactions leading to urea formation occur. Citrulline reacts with L-aspartate in an ATP-dependent reaction to form argininosuccinate, AMP, and $PP_i$. The $PP_i$ is subsequently hydrolyzed to inorganic phosphate, so in effect the cost of this step is two ATP molecules. Argininosuccinate is cleaved to fumarate and L-arginine. The fumarate returns to the pool of TCA cycle intermediates, whereas the arginine becomes hydrolyzed to urea and ornithine. The ornithine is reutilized in further rounds of the urea cycle. Urea diffuses into the bloodstream and is ultimately eliminated through the kidneys in the urine. The stoichiometry for the urea cycle is

$$CO_2 + NH_4^+ + 3\,ATP + \text{aspartate} + 2\,H_2O \longrightarrow$$
$$\text{urea} + 2\,ADP + 2\,P_i + AMP + PP_i$$
$$+ \text{fumarate} + 6\,H^+ \quad \textbf{(3)}$$

**Figure 22.9**

The "Krebs bicycle" involves interaction between components of the TCA cycle (on the left) and the urea cycle (on the right). This interaction explains the origin of the amino group contributed by aspartate to urea formation. The amino group originates from a transamination reaction involving oxaloacetate. The resulting aspartate is deaminated to fumarate, which can be recycled to oxaloacetate.

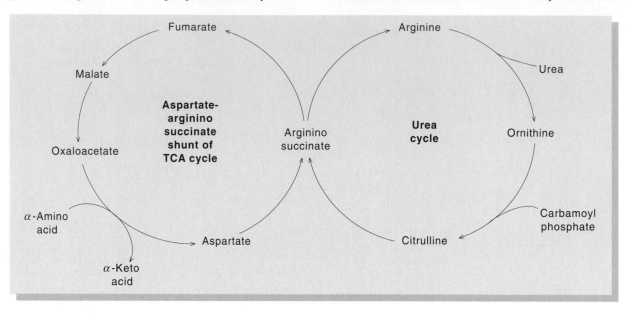

In each turning of the urea cycle two nitrogens are eliminated, one originating from the oxidative deamination of glutamate and the other coming from the $\alpha$-amino group of aspartate. Because the $PP_i$ produced in the urea cycle is subsequently hydrolyzed, it takes the equivalent of four high-energy phosphates to form a single molecule of urea. Thus, the cost of this form of detoxification of ammonia is quite high.

## The Urea Cycle and the TCA Cycle Are Linked by the Krebs Bicycle

The fumarate released in the urea cycle links the urea cycle with the TCA cycle. This fumarate is hydrated to malate, which is oxidized to oxaloacetate. The carbons of oxaloacetate can stay in the TCA cycle by condensation with acetyl-CoA to form citrate, or they can leave the TCA cycle either by gluconeogenesis to form glucose or by transamination to form aspartate as shown in figure 22.9. Because Krebs was involved in the discoveries of both the urea cycle and the TCA cycle, the interaction between the two cycles shown in figure 22.9 is sometimes referred to as the Krebs bicycle.

## More Than One Carrier Exists for Transporting Ammonia from the Muscle to the Liver

The urea cycle is a unique function of the liver. Excess ammonia formed in other tissues must be carried in a nontoxic form to the liver. In many tissues glutamine serves as the carrier of excess nitrogen. The glutamine is formed in the tissues in a reaction, catalyzed by glutamine synthase, that combines $NH_3$ with glutamate.

$$ATP + NH_4^+ + glutamate \xrightleftharpoons{\text{Glutamine synthase}} ADP + P_i + glutamine + H^+ \quad \textbf{(4)}$$

This reaction involves activation of the $\gamma$-carboxyl group of glutamate to yield a $\gamma$-glutamyl enzyme complex, together with the cleavage of ATP to ADP and $P_i$. In a second step, the $\gamma$-glutamyl group is transferred to $NH_4^+$.

After the glutamine reaches the liver, the enzyme glutaminase releases the ammonia from the glutamine by the reaction

$$Glutamine + H_2O \longrightarrow glutamate + NH_4^+ \quad \textbf{(5)}$$

## Figure 22.10

The glucose–alanine cycle. Active muscle functions anaerobically and synthesizes alanine by a transamination reaction between glutamate and pyruvate. The alanine is transported to the liver, where the pyruvate is regenerated and converted to glucose by gluconeogenesis. The glucose then is transported back to the muscle tissue, where it is used for energy production in glycolysis.

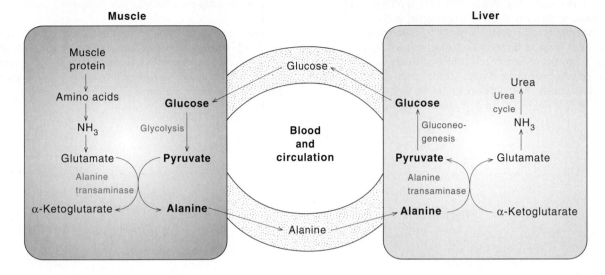

Ammonia is also transported from skeletal muscle to the liver in the form of the amino acid alanine (fig. 22.10). This alanine is formed in the muscle tissue by a transamination reaction between pyruvate and glutamate. Then the alanine is transported by the bloodstream to the liver, where it reacts with $\alpha$-ketoglutarate to reform pyruvate and glutamate. This reaction is catalyzed by alanine transaminase. The nitrogen originating from the glutamate is processed by the urea cycle. When the blood glucose concentration is low, the pyruvate resulting from alanine transamination is used to make glucose via the gluconeogenesis pathway. The glucose can be returned to the skeletal muscle to supply immediately available energy to keep the muscle going. Thus, the transport of alanine from muscle to liver results in a reciprocal transfer of glucose to muscle. The entire cyclical process is referred to as the glucose–alanine cycle (see fig. 22.10).

## Amino Acid Catabolism Can Serve as a Major Source of Carbon Skeletons and Energy

Thus far we have considered the deamination of amino acids and the fate of the resulting ammonium ion. The carbon skeleton remaining after deamination can be used in various biosynthetic pathways, or it can be degraded by reactions coupled to the production of energy.

Catabolism of amino acids usually entails their conversion to intermediates in the central metabolic pathways. All amino acids can be degraded to carbon dioxide and water by appropriate enzyme systems. In every case, the pathways involve the formation, directly or indirectly, of a dicarboxylic acid intermediate of the tricarboxylic acid cycle, of pyruvate, or of acetyl-CoA (fig. 22.11).

Acetyl-CoA so formed can be oxidized to carbon dioxide by means of the TCA cycle or, when cycle function is restricted, can be converted to acetoacetate and lipid. Amino acids metabolized to acetoacetate and acetate are termed ketogenic. At one time, it was thought that the ketogenic property was readily explained by the absence in animal tissues of a mechanism for net conversion of acetate residues into glucose (see glyoxylate cycle, chapter 12). There have been claims, however, that animal tissues, particularly the liver, do contain the enzymes of the glyoxylate cycle although we do not know the level to which they function under different circumstances. The ketogenic effect could be due to the limited function of this pathway.

In contrast to amino acids that lead to C-2 carbon units such as acetate, amino acids that give rise to C-4 or C-5 carbon units such as $\alpha$-ketoglutarate or any of the four-

## Figure 22.11

Pathways for the degradation of the 20 amino acids (highlighted in colored boxes) found in proteins. The strategy followed for amino acid degradation in gross respects is similar (except for direction) to the strategy for amino acid biosynthesis. Thus, the $\alpha$-amino groups are usually removed at an early stage in degradation, and the carbon skeletons filter into the central metabolic pathways. However, enormous differences exist in the specific pathways used in the two processes. Only a handful of amino acids involve similar sequences in both directions, and most of these involve highly reversible reactions, indicated by double arrows. The differences can be seen by comparing figures 22.11 and 21.1. In degradation pathways the carbon skeletons are funneled almost exclusively to intermediates in the TCA cycle. The dashed arrows in this figure associated with tyrosine and isoleucine reflect the fact that the carbon skeletons of these amino acids are split into two components, which are separately processed.

Amino acid degradation serves three purposes: (1) supplying energy, (2) supplying intermediates for the synthesis of other compounds, and (3) removing harmful excesses of certain amino acids. The number of steps in each pathway is indicated by a number alongside the conversion arrow.

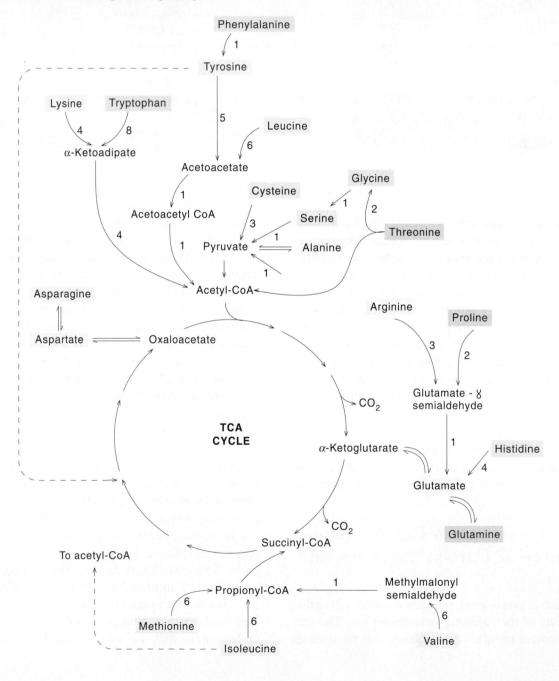

carbon dicarboxylic acids (see fig. 22.11) can stimulate TCA cycle function. This is because they are intermediates in the cycle. For their further metabolism they must leave the cycle by one of two routes (see fig. 13.1). By one route, the conversion of oxaloacetate to phosphoenolpyruvate results in gluconeogenesis when carbohydrate utilization is restricted. For this reason, such amino acids are considered glycogenic. By the other route, pyruvate is formed and, after conversion of the latter to acetyl-CoA, can be oxidized completely to carbon dioxide and water, provided there is ample TCA cycle function.

## For Many Genetic Diseases the Defect Is in Amino Acid Catabolism

One of the first indications that genes affect phenotypes by the nature of the proteins they encode came from the work of a London pediatrician named Garrod (1908). Garrod was analyzing a disease in humans known as alkaptonuria, in which the cartilaginous tissues are dark and the urine turns black on exposure to air. He suggested that this was due to an abnormality in the metabolism of the amino acid phenylalanine because feeding phenylalanine to patients with this syndrome resulted in increased secretions of homogentisic acid; it is the homogentisic acid that turns black on air oxidation. Long before the full explanation for alkaptonuria was known, it had been found that it followed a recessive pattern of inheritance, and Garrod proposed that the condition was due to a defective enzyme. It was not until 1958 that the full biochemistry leading to this problem was appreciated. Humans have developed a multistep metabolic pathway for disposing of the excess phenylalanine (or tyrosine), and one of the intermediates in this pathway is homogentisate (fig. 22.12). Alkaptonuria is due to a deficiency of the homogentisic acid oxidase enzyme. When this enzyme is not present in sufficient amounts, homogentisate accumulates, resulting in the darkening of the urine and other problems associated with this condition.

Garrod was too far ahead of his time to be fully appreciated when he made his great discovery; his incredible insight was noted by George Beadle, who shared the Nobel Prize with E. L. Tatum in Physiology and Medicine for their contributions to biochemical genetics. Beadle pointed out that the one-gene–one-enzyme hypothesis was implicit in Garrod's work and was actually formulated by Garrod in almost the same terms.

## Most Human Genetic Diseases Associated with Amino Acid Metabolism Are Due to Defects in Their Catabolism

Since Garrod's work researchers have described many metabolic diseases that are due to the inability of the affected individual to dispose of specific dietary components. The diseases may be difficult to treat in the cases of errors in amino acid catabolism, because the culprit amino acid is one of the normal constituents of protein and is required for growth and development as well as for replacement of those body proteins that undergo rapid turnover. Since the affected fetus is usually carried by a mother, who is heterozygous for the deficiency (carrying one normal and one defective gene) and whose own metabolism is essentially normal, the development of the fetus is essentially normal. Thus, management of these diseases is possible, but it is dependent on prenatal diagnosis or diagnosis soon after birth. Treatment consists of a low-protein diet, carefully selected to supply enough of the culprit amino acid for protein formation but not enough to allow high plasma levels of it or of the offending metabolites. Supplements of nonoffending amino acids, prepared by synthesis or by fermentation processes, could be employed to compensate in part for the low-protein diet. Indeed, the chemical industries have made such preparations available.

Some of the diseases of amino acid catabolism are listed in table 22.3. These naturally occurring defects have been invaluable in demonstrating the obligatory nature of some of the steps in amino acid breakdown. A mutation that causes a defective enzyme usually leads to (1) a substantial accumulation of the intermediate that is a substrate for that enzyme, and (2) a drastic lowering of all the intermediates below that step in the pathway.

The fact that these genetic diseases are due to single (recessive) gene mutations gives hope on two fronts. It should soon be possible to identify carriers of any of these traits so that, with the aid of genetic counseling, it will be possible for parents to avoid giving birth to homozygous children that carry two defective genes. For those rare individuals affected, it may one day be possible to provide, through transplant, cells capable of metabolizing the offending amino acids. It has already been demonstrated, for example, that fibroblasts from patients with human "maple syrup" urine disease (see table 22.3) can be transfected with a cDNA for the missing component of branched-chain $\alpha$-keto acid dehydrogenase to yield cells in which the missing activity is restored.

**Figure 22.12**

The conversion of phenylalanine and tyrosine to fumarate and acetoacetate.

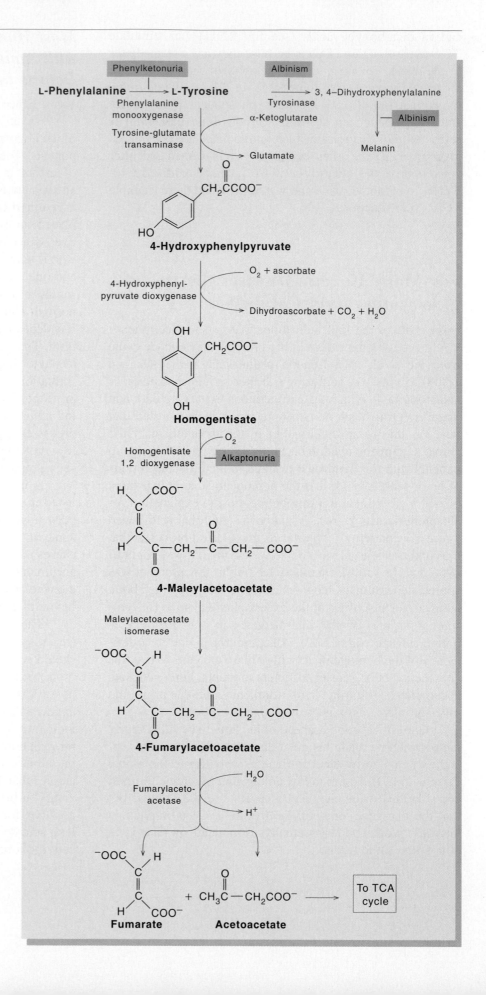

**Table 22.3**

Some Inborn Errors of Amino Acid Metabolism in Humans

| Amino Acid Catabolic Pathway Involved | Condition | Distinctive Clinical Manifestation | Enzymatic Block or Deficiency |
|---|---|---|---|
| Arginine and the urea cycle | Argininemia and hyperammonemia | Mental retardation | Arginase |
| | Hyperammonemia | Neonatal death, lethargy, convulsions | Carbamoyl phosphate synthase |
| | Ornithinemia | Mental retardation | Ornithine decarboxylase |
| Glycine | Hyperglycinemia | Severe mental retardation | Glycine-cleavage system |
| Histidine | Histidinemia | Speech defects, mental retardation in some; in others, none | Histidase |
| Isoleucine, leucine, and valine | Branched-chain ketoaciduria ("maple syrup" urine disease) | Neonatal vomiting, convulsions, and death; mental retardation in survivors | Branched-chain keto acid dehydrogenase complex |
| Isoleucine, methionine, threonine, and valine | Methylmalonic acidemia | Similar to the preceding except that methylmalonate accumulates | Methylmalonyl-CoA mutase (some patients respond to vitamin $B_{12}$ therapy) |
| Leucine | Isovaleric acidemia | Neonatal vomiting, acidosis, lethargy, and coma; survivors mentally retarded | Isovaleryl-CoA dehydrogenase |
| Lysine | Hyperlysinemia | Mental retardation and some noncentral nervous system abnormalities | Lysine-ketoglutarate reductase |
| Methionine | Homocystinuria | Mental retardation common: several eye diseases and thromboembolism common; osteoporosis and faulty bone structures | Cystathionine-$\beta$-synthase |
| Phenylalanine | Phenylketonuria and hyperphenyl-alaninemia | Vomiting is an early neonatal symptom, but mental retardation and other neurologic disorders develop in the absence of dietary treatment. | Phenylalanine-L-monoxygenase |
| Proline | Hyperprolinemia, type I | Probably not etiologically associated with any disease; proline excreted | Proline oxidase |
| Tyrosine | Alkaptonuria | Homogentisic acid in urine darkens on standing; in adult years, pigment deposits cause darkening of skin, cartilage; arthritis develops | Homogentisic acid oxidase |
| | Albinism | The most common type, oculocutaneous albinism, results in white hair, pink skin, and an extreme photophobia owing to lack of pigment in the eye | Tyrosinase of the melanocyte is absent |

# Amino Acids Serve as the Precursors for Compounds Other Than Proteins

As stated at the outset of this chapter, the primary fate of amino acids is their incorporation into protein. However, amino acids also serve as precursors for a number of other important molecules. These include processes leading to the formation of the porphyrin nucleus found in many oxygen- and electron-carrying proteins, of biologically active amines, and of glutathione. These two topics are examined below.

## *Porphyrin Biosynthesis Starts with the Condensation of Glycine and Succinyl-CoA*

Early isotope tracer experiments by David Shemin permitted the elucidation of the formation of the immediate precursor of the porphyrin needed for the cytochromes and for hemoglobin. These studies indicated that the glycine methylene carbon and nitrogen were incorporated along with both carbons of acetate. Subsequent enzymatic studies in both bacteria and animals revealed a condensation reaction between succinyl-CoA and glycine to yield δ-aminolevulinate and $CO_2$ (presumably by way of an enzyme-bound β-keto acid, α-amino-β-ketoadipate) (fig. 22.13).

δ-Aminolevulinate is also the precursor to porphobilinogen in plants, blue-green algae, and most eubacteria, but it is not formed by a condensation of glycine and succinyl-CoA. Rather it is formed from glutamate by reduction of the α-carboxyl group to yield α-glutamyl semialdehyde. As has been emphasized in describing reduction of other carboxyl groups, an "activation" of the carboxyl group is required. The reaction in the δ-aminolevulinate pathway is unique in that the glutamate is transferred to a tRNA acceptor. Hence, the reaction requires the expenditure of two high-energy phosphate bonds. The glutamyl tRNA is the substrate for a specific reductase. The reduced product, α-glutamyl semialdehyde, undergoes an unusual intramolecular transamination reaction in which the amino group on C-2 of the semialdehyde is transferred to the C-1 position (which becomes C-5 in δ-aminolevulinate).

The pyrrole monomer porphobilinogen arises from the condensation of two molecules of δ-aminolevulinate with the ions of two water molecules. This reaction is catalyzed by δ-aminolevulinate dehydrase. Condensation of four porphobilinogen molecules yields the branchpoint compound in tetrapyrrole synthesis, uroporphyrinogen III. This is a complex reaction requiring two enzymes: Uroporphyrinogen I synthase, which catalyzes a head-to-tail condensation

of four porphobilinogen molecules (fig. 22.14), and uroporphyrinogen cosynthase, which inverts one of the units and closes the ring.

Derivatives of porphyrins are coenzymes in a number of oxidation–reduction reactions (see chapters 10, 14, and 15).

**Figure 22.13**

Tetrapyrrole biosynthesis. The sequence by which four porphobilinogen residues are converted to uroporphyrinogen III is the sequential head-to-tail condensation of the four residues by uroporphyrinogen I synthase (porphobilinogen deaminase) to yield the unrearranged hydroxymethylbilane. This unstable intermediate is rearranged and cyclized by uroporphyrinogen III cosynthase to yield uroporphyrinogen III. Hydroxymethylbilane in the absence of cosynthase spontaneously cyclizes to yield the unrearranged urobilinogen I, which is not an intermediate in the pathway (hence the name uroporphyrinogen I synthase for the deaminase).

## *Glutathione Is γ-Glutamylcysteinylglycine*

The tripeptide γ-glutamylcysteinylglycine, or glutathione, is found in nearly all cells and plays a variety of roles. It is formed in two steps, each requiring ATP. In the first step glutamate condenses with a cysteine. The γ- rather than the α-carboxyl of glutamate makes a peptide bond with the α-amino group of cysteine.

$$\text{Glutamate + cysteine + ATP} \xrightarrow{\underset{\text{synthase}}{\gamma\text{-Glutamylcysteine}}} \text{γ-glutamylcysteine + ADP + P}_i \quad \textbf{(6)}$$

In the second step, the condensation of the dipeptide with glycine, a normal peptide linkage is made.

$$\text{γ-Glutamylcysteine + glycine + ATP} \xrightarrow{\underset{\text{synthase}}{\text{Glutathione}}} \text{glutathione + ADP + P}_i \quad \textbf{(7)}$$

Glutathione helps to maintain the sulfhydryl groups of proteins in a reduced state. An enzyme, protein-disulfide reductase, catalyzes sulfhydryl disulfide interchanges between glutathione and proteins. The reductase is important in insulin breakdown and may catalyze the reassortment of disulfide bonds during polypeptide chain folding.

Mutants of *E. coli* have been isolated that are essentially devoid of glutathione owing to the loss of one or the other of the two synthases. Such cells have normal growth rates, but they are more sensitive to such sulfhydryl reagents as mercurials.

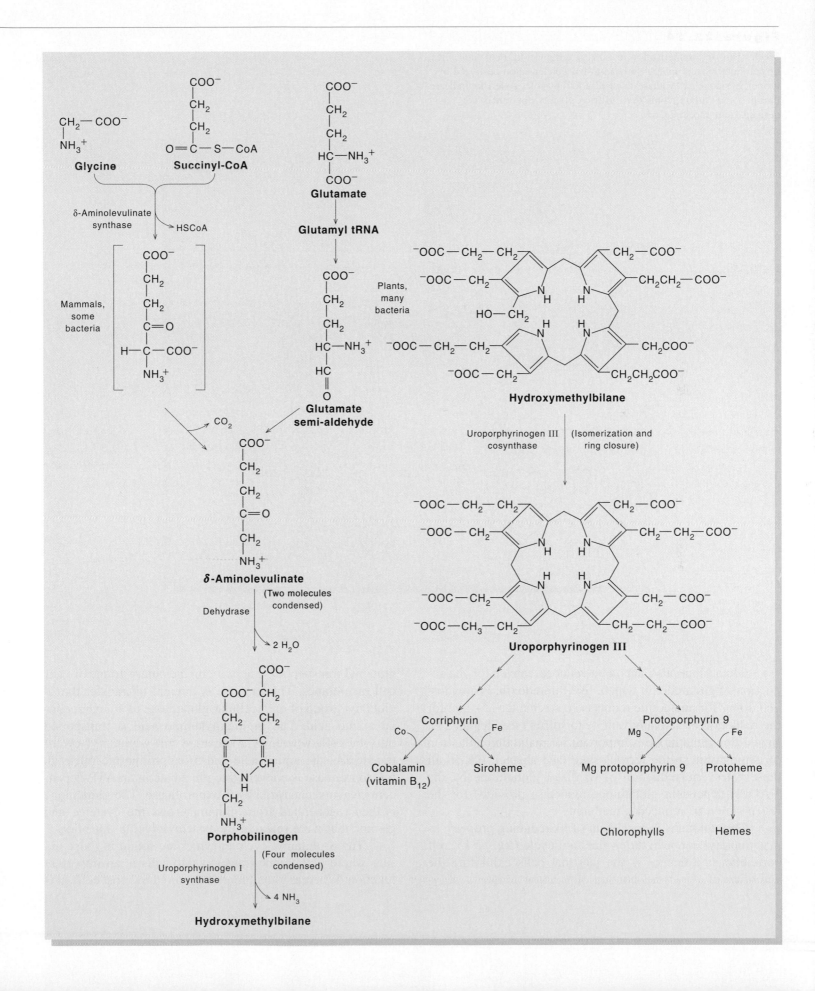

**Figure 22.14**

Mechanism of polymerization in a linear tetrapyrrole. Four molecules of porphobilinogen undergo a head-to-tail condensation catalyzed by uroporphyrinogen I synthase to yield a tetrapyrrole. Asterisks indicate nitrogen and carbon atoms derived from glycine; the others are derived from succinyl-CoA.

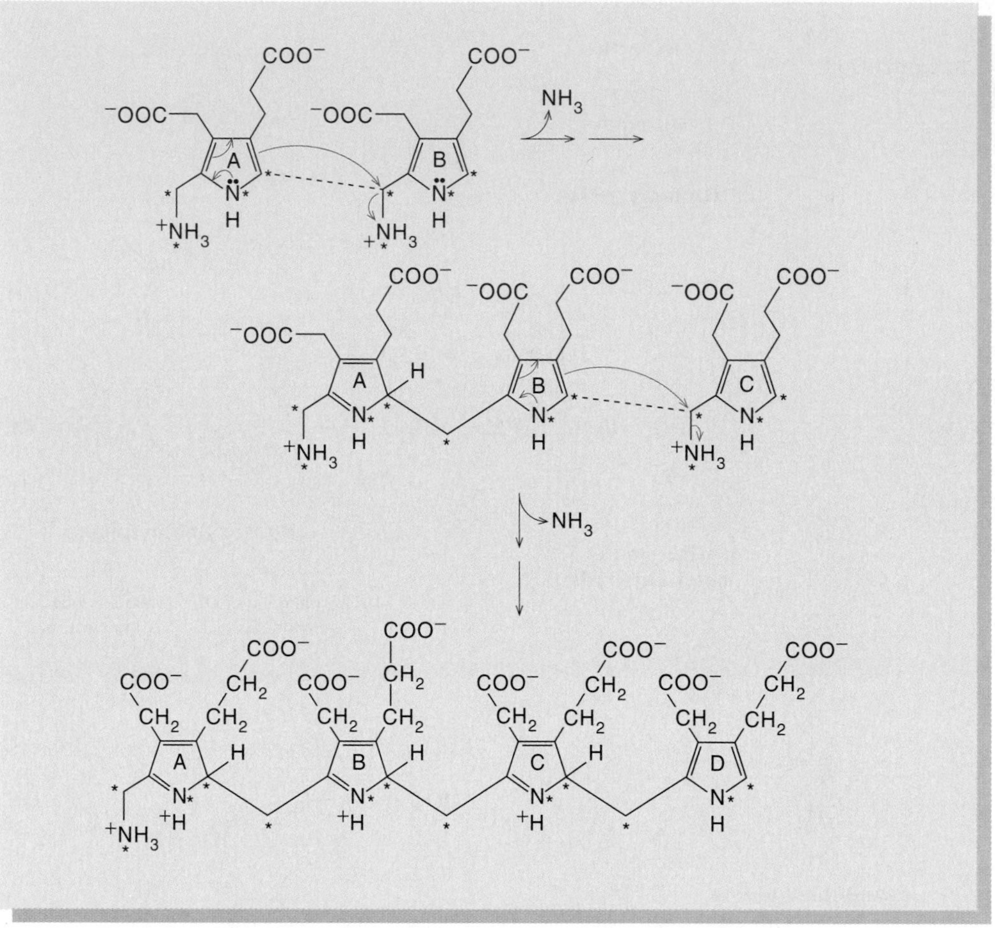

Glutathione also acts as a reduced carrier for the reduction of glutaredoxin which, like thioredoxin, is a hydrogen donor for nucleotide reductase (see chapter 23), and for the reduction of activated sulfate to sulfite (see chapter 21). In addition glutathione is important for maintaining the iron of hemoglobin in the ferrous state (see chapter 12). In all these roles, the reduction of oxidized glutathione by the NADPH-dependent glutathione reductase provides for the regeneration of reduced glutathione.

Glutathione is independent of its reducing property as a $\gamma$-glutamyl donor in the $\gamma$-glutamyl cycle (fig. 22.15). Of particular significance is the fact that cells exhibiting the activities of the cycle contain substantial amounts of $\gamma$-glutamyl transpeptidase activity on the outer surface of their cell membranes. This enzyme is thought to transfer the $\gamma$-glutamyl group of extracellular glutathione to an extracellular amino acid. The $\gamma$-glutamylamino acid is transported into the cell, where, as a substrate for $\gamma$-glutamyl cyclotransferase, the amino acid and 5-oxoproline are released. 5-Oxoprolinase is converted to glutamate in an ATP-dependent cleavage catalyzed by 5-oxoprolinase. The glutathione is then regenerated from glutamate and the cysteine and glycine that were released by cysteinylglycine dipeptidase.

The $\gamma$-glutamyl cycle enzymes are found in those tissues where glutathione transport into cells is an important function. Whereas glutathione is exported by most cells, it is

**Figure 22.15**

The γ-glutamyl cycle proposed by Alton Meister. The cycle involves enzymes forming glutathione, the excretion of glutathione (1) and, in cells that import glutathione, the γ-glutamyl transpeptidase-dependent transport (2) of another amino acid (or peptide), the cleavage of the intracellular γ-glutamyl amino acid, and the ATP-dependent conversion of 5-oxoproline to glutamate. The cysteinylglycine (Cys-Gly) formed in the reaction is transported by an uncharacterized transport system (3) and cleaved by an intracellular protease (4) or cleaved by a membrane-bound protease and the free amino acids transported. The γ-glutamyl cycle enzymes are found in those tissues for which the transport of glutathione into cells is an important function. This includes liver and kidney cells in animals.

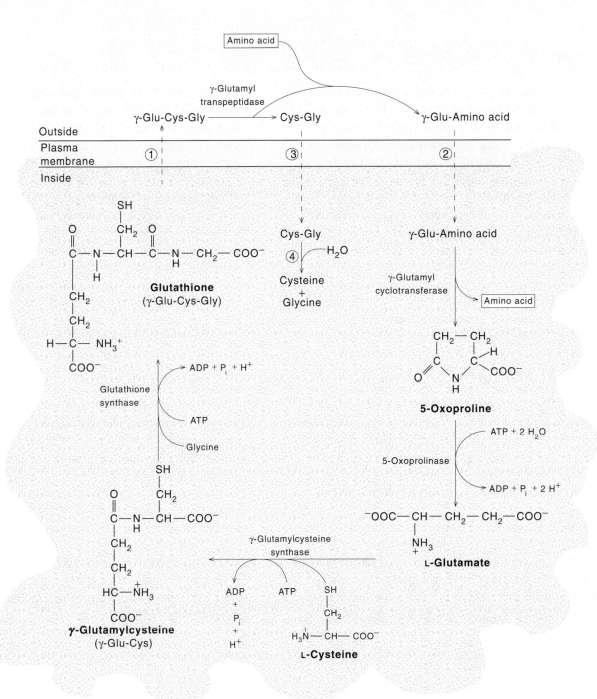

efficiently transported into cells that contain the membrane-bound $\gamma$-glutamyl transpeptidase. Thus, the $\gamma$-glutamyl transpeptidase appears to facilitate salvage of glutathione secreted by some tissues into the bloodstream, as well as to permit an energy-driven transport system for amino acids. Either oxidized or reduced glutathione is a substrate for the transpeptidase. The enzyme also catalyzes the hydrolysis of glutathione to glutamate and cysteinylglycine and glutamine to glutamate and ammonia. Many amino acids can serve as acceptors for the $\gamma$-glutamyl group, including $\gamma$-glutamyl amino acids (to yield $\gamma$-glutamyl-$\gamma$-glutamyl amino acids) and even glutathione itself. Such a transport across cell membranes is probably especially important for cysteine and methionine, as well as for glutathione.

## Summary

1. Only eight of the *de novo* pathways for amino acid biosynthesis can be found in humans. These amino acids are all related by a small number of steps to glycolytic or TCA cycle intermediates. A number of additional amino acids can be formed from these amino acids. Essential amino acids are those that must be supplied in the diet.

2. For most amino acids the $\alpha$-amino group is removed at an early stage in catabolism, usually in the first step. Transaminases are specific for different amino acids. Frequently $\alpha$-ketoglutarate is the acceptor for the amino group, in which case it is converted into glutamate. The $\alpha$-ketoglutarate can be regenerated from the glutamate by oxidative deamination.

3. A great deal of excess $NH_3$ frequently results from amino acid catabolism. This excess ammonia must be eliminated. In bacteria and lower eukaryotes the ammonia can usually be removed by simple diffusion, but in higher eukaryotes this is not feasible. Since the ammonia is frequently quite toxic, it is detoxified before removal by conversion to urea or uric acid. An intricate pathway resulting in the conversion of ammonia into urea involves five enzymes: Three located in the cytoplasm and the remaining two in the mitochondrial matrix.

4. All amino acids can be degraded to $CO_2$ and water via the TCA cycle by the appropriate enzymes: The pathways often contain branchpoints to useful biosynthetic products. In every case, the pathways involve the formation of a dicarboxylic acid intermediate of the TCA cycle, of pyruvate, or of acetyl-CoA.

5. The discussion of amino acid catabolism is organized according to the common intermediates formed during degradation. Alanine, glycine, threonine, serine, and cysteine are degraded to acetyl-CoA by way of pyruvate. Threonine is also degraded to acetyl-CoA via pyruvate, but it also yields acetyl-CoA directly. Phenylalanine, tyrosine, tryptophan, lysine, and leucine also lead to acetyl-CoA, but they go by way of acetoacetyl-CoA rather than pyruvate. Arginine, histidine, proline, glutamic acid, and glutamine are all degraded to $\alpha$-ketoglutarate. Catabolism of methionine, valine, and isoleucine leads to succinyl-CoA. Aspartate and asparagine are converted to oxaloacetate on degradation.

6. The importance of catabolic pathways is underscored by a broad spectrum of human metabolic diseases in each of which one enzyme for normal amino acid catabolism is either missing or defective.

7. Many biologically important routes of amino acid utilization, other than those leading to incorporation into proteins, are known. Some of these routes are distinctly anabolic pathways in which the amino acids serve as an initial substrate in an independent biosynthetic pathway. Other simple pathways involve the conversion of one amino acid to another, such as the formation of tyrosine from phenylalanine. The utilization of glycine in the formation of porphyrin derivatives occurs by very complex highly branched pathways. Some other biologically important pathways lead to the biosynthesis of small peptides as in the biosynthesis of glutathione.

## Selected Readings

Battersby, A. R., C. J. R. Fookes, G. W. J. Matcham, and E. McDonald, Biosynthesis of the pigments of life: Formation of the macrocycle. *Nature* 285:17, 1980. This paper discusses the steps in tetrapyrrole biosynthesis and the pathways diverting this nucleus to chlorophylls, hemes, cytochromes, and other macrocyclic pigments.

Bender, D. A., *Amino Acid Metabolism*. New York: John Wiley and Sons, 1985.

Fowden, L., P. J. Lea, and E. A. Bell, The nonprotein amino acids of plants. *Adv. Enzymol.* 50:117, 1979. A discussion of the occurrence and biosynthesis of naturally occurring amino acid analogs in plants.

Katz, E., and A. L. Demain, The peptide antibiotics of *Bacillus:* Chemistry, biogenesis and possible functions. *Bacteriol. Rev.* 41:449, 1977. A description of several peptide antibiotics showing the distribution of D-amino acid in these compounds.

Kishore, G. M., and D. M. Shah, Amino acid biosynthesis inhibitors as herbicides. *Ann. Rev. Biochem.* 57:627–663, 1988. The focus is on the biosynthesis of essential amino acids.

Meister, A., *Biochemistry of the Amino Acids*, vols. 1 and 2. New York: Academic Press, 1965. The two-volume classic provides a thorough discussion of amino acid literature, occurrence, properties, and metabolism of amino acids up to that time.

Torchinsky, Y. M., Transamination, its discovery, biological and chemical aspects (1937–1987). *Trends Biochem. Sci.* 12:115–117, 1987.

Yamada, K., S. Kinoshita, T. Tsunoda, and K. Aida (eds.), *The Microbial Production of Amino Acids*. New York: John Wiley and Sons, 1972. A collection of essays describing microbial processes used in Japanese industry for the production of amino acids. Includes examples in which the regulatory mechanisms functioning in most cells have been modified or bypassed.

Also see readings at the end of chapter 21.

## Problems

1. The pathway for the biosynthesis of arginine (fig. 22.2b) has several intermediates that have an *N*-acetyl group. Compare these intermediates with those in the salvage pathway (fig. 22.2a) and propose a reason for the acetyl groups.

2. In the conversion of the *N*-acetyl-$\gamma$-glutamyl phosphate to *N*-acetylglutamic-$\gamma$-semialdehyde (fig. 22.2), two processes occur: An elimination of a phosphate and a reduction. Which step occurs first? Also can you propose a reason for the use of the phosphate group in the first place?

3. Often ammonia is portrayed as "feeding into" the urea cycle, for example, see figure 22.10. Do ammonia molecules "feed into" the urea cycle?

4. What effect does deprivation of dietary pyridoxal phosphate have on the capacity to metabolize amino acids?

5. The *de novo* biosynthetic pathway for the biosynthesis of arginine has been lost by higher animals. However the biosynthesis of arginine via a salvage pathway from proline can occur.
   What arguments can you make that it is unlikely that mammals would completely lose all capabilities to produce arginine?

6. Reduce the names "ketogenic" and "glucogenic" to the simplest possible terms.

7. Use figure 22.11 to deduce which amino acids are glucogenic, ketogenic, or both.

8. Many of the inborn errors in amino acid metabolism appear to result in mental retardation (table 22.3). The majority of these individuals appear normal at birth, but their mental capabilities fail to develop. Can you provide an explanation for this observation?

9. Why have many inborn errors been found in the metabolism of amino acids in humans but "none" in glycolysis, the citric acid cycle, or electron transport?

10. Fumarate is a product of both argininosuccinate lyase (fig. 22.7) and fumarylacetoacetase (fig. 22.12), even though the reactions are quite different. If you were describing these two reactions to an organic chemistry student using one word per reaction, what words would you pick?

11. The biosynthesis of δ-aminolevulinate is known to occur with the loss of one glycine carbon-bound hydrogen, producing the intermediate shown in figure 22.13. What coenzyme would you expect to participate in this process? How does the same coenzyme stabilize the carbanion formed in the decarboxylation part of the reaction?

12. What are the functions of the two water molecules that are utilized by the enzyme 5-oxoprolinase (fig. 22.15)?

13. Would you anticipate elevated arginase activity in the liver of an untreated diabetic animal? Why or why not?

14. The concentration of phenylalanine in the blood of neonates is used to screen for phenylketonuria (PKU). Explain the biochemical basis for the correlation of elevated blood phenylalanine concentration and PKU. Explain why restriction of dietary phenylalanine is critically important for youngsters with PKU.

15. (a) L-Glutathione is not a primary gene product as are proteins. What "information" is used to direct the synthesis of L-glutathione?

    (b) Predict the effect of a glutathione synthase inhibitor on cells exposed to oxidative stress.

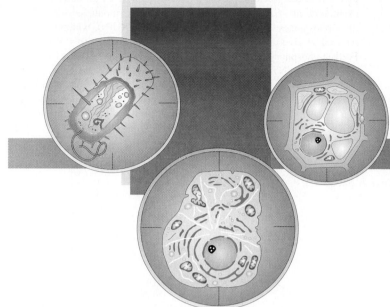

# Nucleotides

*Nucleotides are synthesized from many simple precursors and the pentose, phosphoribosyl pyrophosphate. Nucleotides function as energy carriers and regulatory molecules and as precursors to coenzymes and nucleic acids.*

This chapter deals with the biosynthesis of ribonucleotides and deoxyribonucleotides, their role in metabolic processes, and the pathways for their degradation (fig. 23.1). The biosynthesis of nucleotides is a vital process because these compounds are indispensable precursors for the synthesis of both RNA and DNA. Without RNA synthesis, protein synthesis is halted; and unless cells can synthesize DNA, they

## Figure 23.1

Synthesis and utilization of nucleotides. Numbers associated with arrows indicate the number of enzymatic steps involved. The starting materials are ribose-5-phosphate, glutamine, glycine, formyltetrahydrofolate, and aspartate. Serine also provides methylene tetrahydrofolate for converting dUMP to dTMP. Before it is used, ribose-5-phosphate is activated by conversion to phosphoribosylpyrophosphate (PRPP). PRPP is the starting material onto which the purine ring is built to give IMP, the precursor of AMP and GMP. In contrast to that strategy the pyrimidine ring is built from $HCO_3^-$, glutamine and aspartate before PRPP is added to give OMP, the precursor of UMP and ultimately of all other pyrimidine nucleotides. Phosphorylation of the ribonucleoside monophosphates (NMPs) gives diphosphates (NDPs), which can be reduced to give the deoxyribonucleoside diphosphates (dNDPs). Further phosphorylation of the NDPs gives ribonucleoside triphosphates (NTPs), which are the building blocks for RNA. But each NTP has another very important role in the cell, shown in the boxes. dNDPs are phosphorylated to dNTPs, three of which directly contribute to DNA synthesis. The fourth building block required, dTTP, must be derived indirectly from dCDP and dUDP.

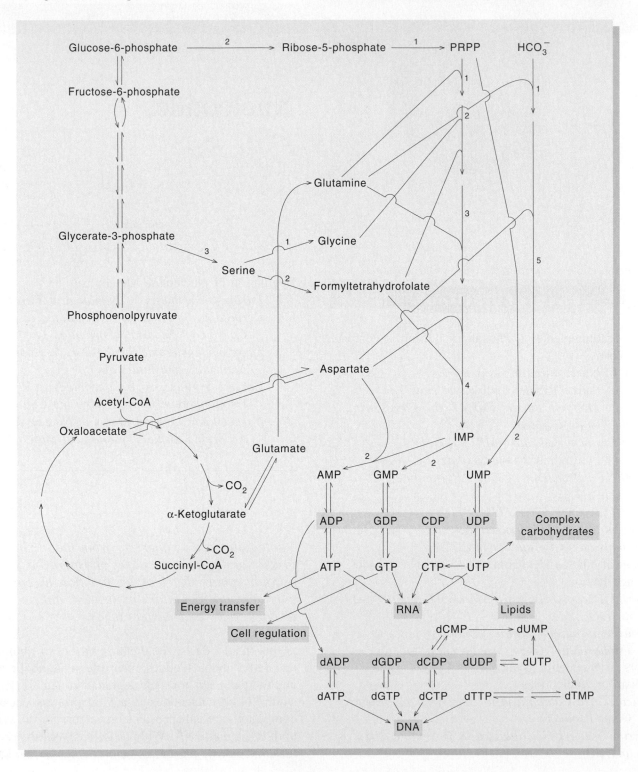

cannot divide. Nucleotides are also necessary for constant repair of DNA, a process necessary for cell survival.

It is not surprising that inhibitors of nucleotide biosynthesis are very toxic to cells. As we will see, their toxicity has been used to advantage in the treatment of cancer as well as in the treatment of certain diseases resulting from infections by viruses, bacteria, or protozoans.

Nucleotides play important roles in all major aspects of metabolism. ATP, an adenine nucleotide, is the major substance used by all organisms for the transfer of chemical energy from energy-yielding reactions to energy-requiring reactions such as biosynthesis. Other nucleotides are activated intermediates in the synthesis of carbohydrates, lipids, proteins, and nucleic acids. Adenine nucleotides are components of many major coenzymes, such as $NAD^+$, $NADP^+$, FAD, and CoA. (See chapter 10 for structures of these coenzymes.)

Pathways for the metabolic degradation of nucleotides are very important to the organism, as demonstrated by the fact that several genetic defects causing blocks in these pathways have serious consequences for the health of the organism.

The physiological importance of nucleotides is reflected in the careful regulation of their intracellular levels as well as intra- and extracellular levels of nucleosides and nucleobases. Levels of ATP, ADP, and AMP are tightly controlled in a variety of cells, as are the levels of dATP, dCTP, dGTP, and dTTP. This regulation is determined by the concentration and location of the enzymes of nucleotide metabolism and by levels of substrates, products, and effectors.

## Nucleotide Components: A Phosphoryl Group, a Pentose, and a Base

Each nucleotide is composed of three parts: (1) a heterocyclic nitrogenous base, (2) a pentose, and (3) a phosphoryl group (fig. 23.1). Nucleotides differ from one another through differences in each of these components.

Nucleotide bases (or nucleobases) belong to two classes: Pyrimidines and purines (table 23.1). The former have a single six-member ring containing two nitrogen atoms. In pyrimidine nucleotides, N-1 of the pyrimidine base is attached to the pentose C-1′. The common pyrimidine bases in nucleotides are uracil, cytosine, and thymine (fig. 23.2). Purine bases have a six-member pyrimidine ring fused to a five-member imidazole ring, the fused system containing four nitrogen atoms. In purine nucleotides, the glycosidic linkage is between N-9 of the purine and C-1′ of the pentose. The common purine bases in nucleotides are adenine, guanine, and hypoxanthine (see fig. 23.2). Nucleo-

### Table 23.1

Names of Common Bases, Nucleosides, and Nucleotides

| Base | Nucleoside[a] | Nucleotide |
|---|---|---|
| *Purines* | | |
| Adenine | Adenosine (A) | AMP[b] or adenylate |
| Guanine | Guanosine (G) | GMP or guanylate |
| Hypoxanthine | Inosine (I) | IMP or inosinate |
| | | |
| *Pyrimidines* | | |
| Uracil | Uridine (U) | UMP or uridylate |
| Thymine | Thymidine (T) | TMP or thymidylate |
| Cytosine | Cytidine (C) | CMP or cytidylate |

[a]With the exception of thymidine, these are the names of the ribonucleosides. Deoxyribonucleosides are indicated by the prefix *deoxy* in front of the nucleoside name; thus deoxyadenosine contains deoxyribose instead of ribose. Thymidine indicates the deoxyribose derivative of thymine. Standard one-letter abbreviations are shown in parentheses.

[b]AMP is the abbreviation for the most commonly used name for this nucleotide, adenosine-5′-monophosphate, or adenosine-5′-phosphate. Similarly, ADP and ATP designate the 5′-diphosphate and -triphosphate of adenosine, respectively. Adenylate (or adenylic acid) is an alternative nomenclature.

tide catabolism leads to the purines, xanthine and uric acid (see fig. 23.20), and xanthosine monophosphate is a nucleotide intermediate of metabolism (see fig. 23.11).

The pentose component of naturally occurring nucleotides is ribose or 2-deoxyribose (i.e., ribose with a hydrogen instead of a C-2′—OH). In nucleotides the purine or pyrimidine is attached to C-1′ of the pentose in the $\beta$ configuration. This means that the base is *cis* relative to C-5′—OH and *trans* relative to the C-3′—OH. The major function of deoxyribonucleotides (those that have 2-deoxyribose as the pentose) is to serve as building blocks for DNA. Although ribonucleotides similarly serve as the units for RNA synthesis, they also have a multitude of other functions in cell metabolism. In some synthetic nucleosides with therapeutic properties, other pentose components such as arabinose are present. (See fig. 12.2 for the structure of arabinose.)

The phosphoryl group of nucleotides is most commonly substituted on the C-5′—OH of the pentose. However, in cyclic nucleotides a single phosphoryl group is esterified to both the C-5′—OH and the C-3′—OH (e.g., see fig. 12.28). In nucleic acids each nucleotide unit has one phosphoryl group esterified to the C-5′—OH and another at the C-3′—OH, and these phosphoryl groups link the nucleotide units.

A nucleoside has no phosphoryl group; it consists of a purine or pyrimidine linked to ribose or deoxyribose. The

## Figure 23.2

Structures of three common ribonucleotides (a) and four common deoxyribonucleotides (b). See Table 23.1 for alternative names and for names of the corresponding bases and nucleosides. The ribonucleotides contain a ribose sugar, whereas the deoxyribonucleotides have a deoxyribose that lacks a hydroxyl group at C-2′ of the pentose. In all cases the phosphoryl group is attached as an ester of one of the pentose hydroxyls, most commonly at C-5′. The bases are attached by a glycosidic bond from a ring nitrogen to C-1′ of the pentose. When the base is a purine (e.g., adenine, guanine, or hypoxanthine) and the pentose is deoxyribose, this glycosidic bond is acid-labile. A nucleoside has no phosphoryl group, only a hydroxyl group at C-5′.

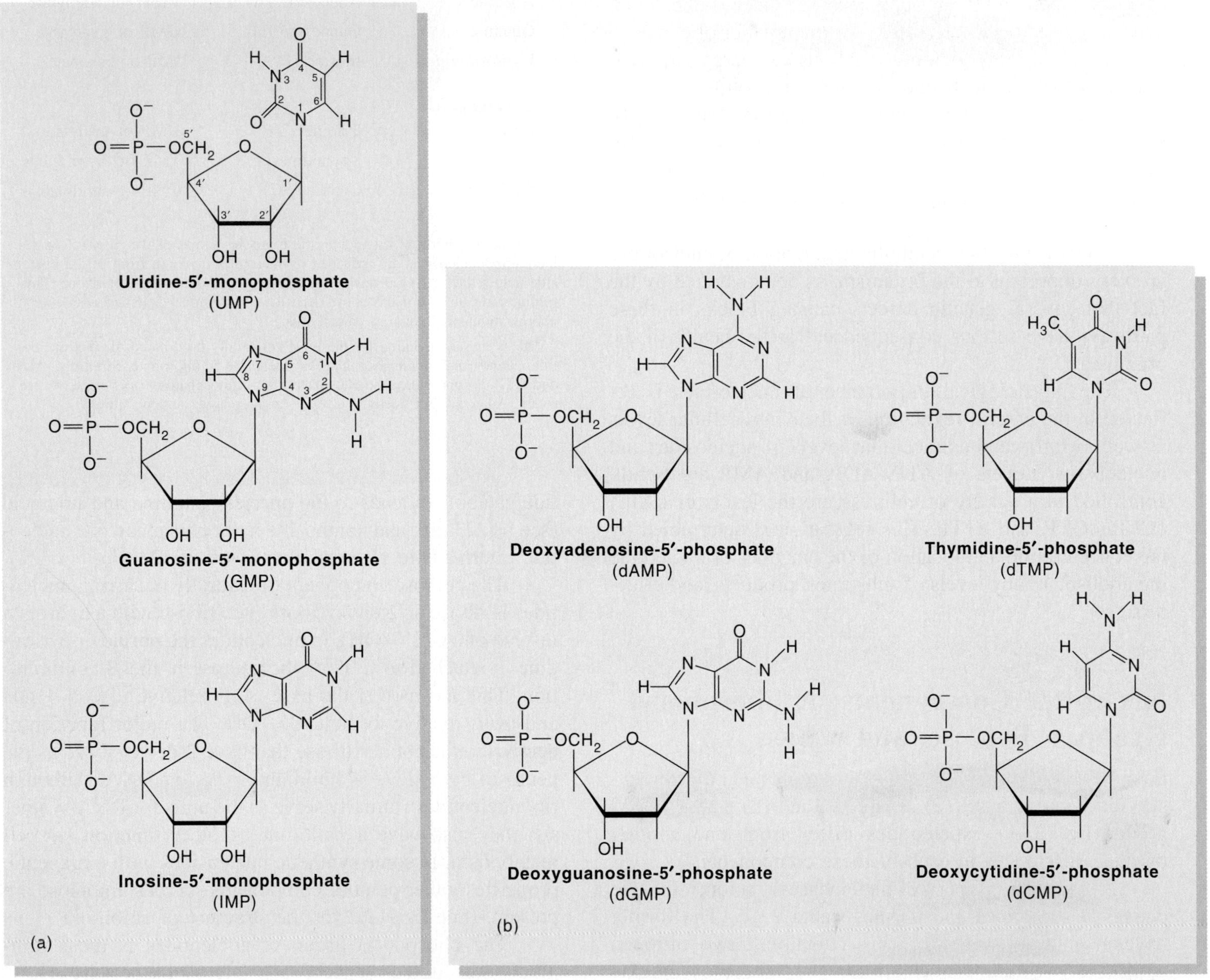

**Uridine-5′-monophosphate**
(UMP)

**Guanosine-5′-monophosphate**
(GMP)

**Inosine-5′-monophosphate**
(IMP)

(a)

**Deoxyadenosine-5′-phosphate**
(dAMP)

**Thymidine-5′-phosphate**
(dTMP)

**Deoxyguanosine-5′-phosphate**
(dGMP)

**Deoxycytidine-5′-phosphate**
(dCMP)

(b)

**Figure 23.3**

Tautomeric equilibrium of guanine and adenine. The keto and amino forms are strongly favored for guanine and adenine, respectively. Comparable tautomeric equilibria exist for thymine, uracil, and cytosine.

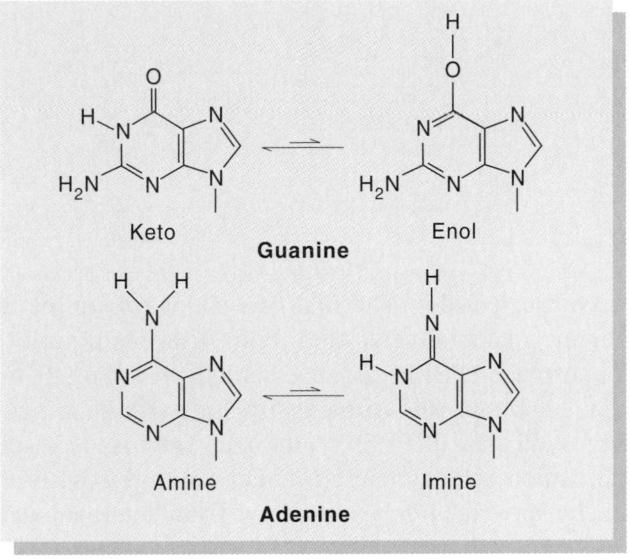

**Table 23.2**

Ionization Constants of the Ribonucleotides (Presented as p$K$ Values)

|  | Base | Secondary Phosphate | Primary Phosphate |
|---|---|---|---|
| Adenosine-5'-phosphate (AMP) | 3.8 | 6.1 | 0.9 |
| Uridine-5'-phosphate (UMP) | 9.5 | 6.4 | 1.0 |
| Cytidine-5'-phosphate (CMP) | 4.5 | 6.3 | 0.8 |
| Guanosine-5'-phosphate (GMP) | 2.4, 9.4 | 6.1 | 0.7 |

**Figure 23.4**

Uncharged and protonated forms of adenosine. The charged base resonates between the two structures shown on the right. Cytosine protonates in a similar way.

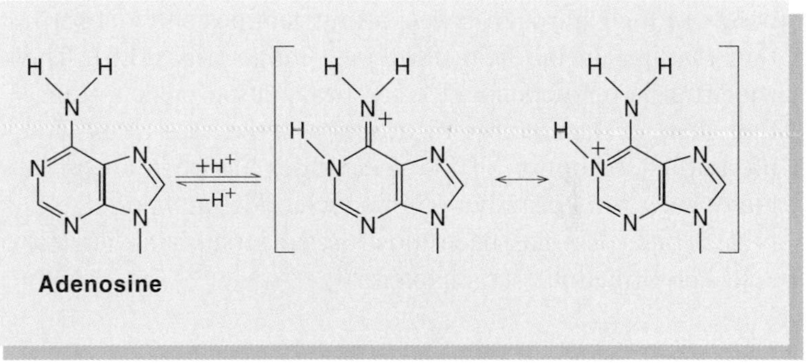

simplest nucleotides are therefore nucleoside-5'-monophosphates. However, nucleoside-5'-diphosphates and nucleoside-5'-triphosphates are nucleotide forms that are also extremely important in cell metabolism. In diphosphates a second phosphoryl group is added to the first in acid anhydride linkage (also called pyrophosphate linkage), and in triphosphates a second anhydride bond links another phosphoryl group. The nucleoside diphosphates (NDPs) and the nucleoside triphosphates (NTPs) dissociate three and four protons, respectively, from their phosphate groups. In the triphosphates, the phosphate immediately attached in ester linkage to the 5'-carbon is designated $\alpha$; the middle phosphate, in pyrophosphate linkage, is called $\beta$; and the terminal phosphate, also in pyrophosphate linkage, is called $\gamma$. The NDPs and NTPs can form complexes with $Mg^{2+}$ or $Ca^{2+}$ and probably exist in these complexes in the cell. The NDPs and NTPs have a number of important metabolic functions in the cell: They serve as energy-carrying enzyme cofactors (e.g., see chapters 12–15) and as substrates for the biosynthesis of nucleic acids (see chapters 26 and 28).

All commonly occurring bases in nucleotides are capable of existing in two tautomeric forms, which differ by the placement of a proton and some electrons. For example, guanosine can undergo a change from a keto form to an enol form as shown in figure 23.3. The keto form is so strongly favored that it is difficult to detect even trace amounts of the enol form at equilibrium. Similarly, the keto forms of thy-

midine or uridine are strongly preferred. Adenosine and cytidine can isomerize to imino forms, but the amino forms are strongly preferred (see fig. 23.3). Even though the unusual tautomers are present in very small amounts, it is conceivable that when present in DNA they contribute to the mutation process.

Some nucleotides undergo protonation in acid and some undergo deprotonation in base; the relative p$K$ values are listed in table 23.2. At neutrality there is no charge on any of the bases. Three of the bases, A, C, and G, undergo protonation as the pH is lowered. The adenine moiety in AMP (adenosine monophosphate) protonates on the N-1 position of the purine rather than on the amino group (fig. 23.4). The charged form is stabilized by the resonance hy-

**Figure 23.5**

Uncharged and protonated forms of guanosine.

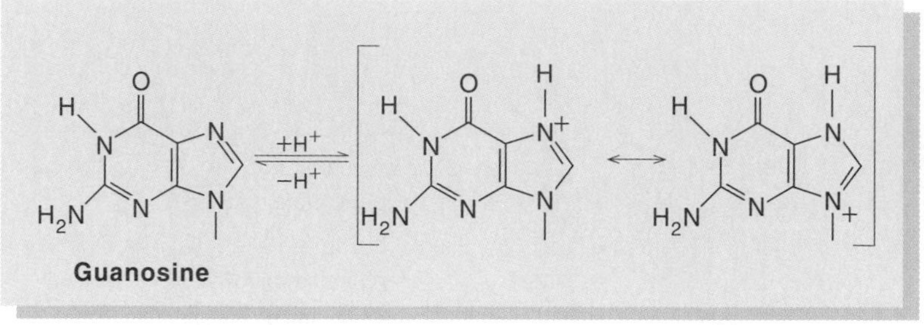

brids shown. In CMP the proton adds to the comparable N-3 ring nitrogen. In guanosine a proton adds to N-7 rather than the amino group (fig. 23.5), again indicating the unusually low basicity of the amino groups on the nucleotides compared with primary aliphatic amines. On the basic side of neutrality both UMP and GMP lose a proton from the imino nitrogens at positions 3 and 1, respectively. As you might expect, the ionization constants for the primary and secondary dissociations of the phosphate group do not differ appreciably for the various nucleotides (see table 23.2).

Owing to the large number of conjugated double bonds in their nitrogen bases, all nucleotides show absorption maxima in the near-ultraviolet range (fig. 23.6). The spectrum is pH-dependent since protonation or deprotonation changes the electron distribution in the base rings. The ultraviolet absorption of the nucleotides has been useful in many ways for the study of mononucleotides and polynucleotides. Indeed, it has been most useful in studying nucleic acid conformation (see chapter 25).

## Overview of Nucleotide Metabolism

All organisms synthesize, interconvert, and catabolize various purine and pyrimidine nucleotides. However, cells of different types, or even the same cells in different stages of development, differ greatly in their ability to carry out some of the reactions involved, with some cells favoring one set of reactions and others another. In the rest of the chapter we deal with the details of these reactions.

## Synthesis of Purine Ribonucleotides *de Novo*

Purine nucleotides can be synthesized in three ways: By *de novo* synthesis, by reconstruction from purine bases through the addition of a ribose phosphate moiety, or by phosphory-

lation of nucleosides. The first two pathways are the more important quantitatively, and both use phosphoribosylpyrophosphate (PRPP) as an essential precursor. It has a similar important role in pathways for pyrimidine nucleotide biosynthesis. PRPP is synthesized from ribose-5-phosphate, which cells synthesize either from glucose-6-phosphate by an oxidative pathway or from intermediates of glycolysis by a nonoxidative pathway (see chapter 12).

The formation of PRPP from ribose-5-phosphate and ATP is catalyzed by ribose-5-phosphate pyrophosphokinase (fig. 23.7). This is an unusual kinase because the pyrophosphoryl group is transferred rather than the phosphoryl group.

The ultimate precursors of the purine ring were established by administering isotopically labeled compounds to pigeons and tracing the incorporation of labeled atoms into the purine ring of uric acid. Birds were used in these experiments because they excrete waste nitrogen largely as uric acid, a purine derivative that is easily isolated in pure form. Chemical degradation of the uric acid revealed the origins of the atoms as depicted in figure 23.8. A flow scheme showing the successive incorporation of atoms from these precursors is shown in figure 23.9.

## *Inosine Monophosphate (IMP) Is the First Purine Nucleotide Formed*

The pathway from PRPP to the first complete purine nucleotide, inosine monophosphate (IMP), involves 10 steps and is shown in figure 23.10. It would seem logical that the purine should be built up first, followed by addition of ribose-5-phosphate, but this is not the case. The starting point is PRPP, to which the imidazole ring is added; the six-member ring is built up afterward.

Step 1, in which phosphoribosylamine is formed, is catalyzed by glutamine phosphoribosylpyrophosphate ami-

## Figure 23.6

Absorption spectra of the common ribonucleoside-5'-monophosphates in protonated and unprotonated forms. The lower pH gives absorbance for the protonated form. Absorbance values are relative to the value at the maximum for each compound. Molar absorbances (that is, the absorbance of a 1-M solution) at the wavelength and pH giving the maximum absorbance are as follows: AMP, $15.4 \times 10^3$; UMP, $10.0 \times 10^3$; CMP, $13.2 \times 10^3$; GMP, $13.7 \times 10^3$. The comparable deoxyribonucleotides and the nucleoside di- and triphosphates have similar absorbance curves and molar absorbances.

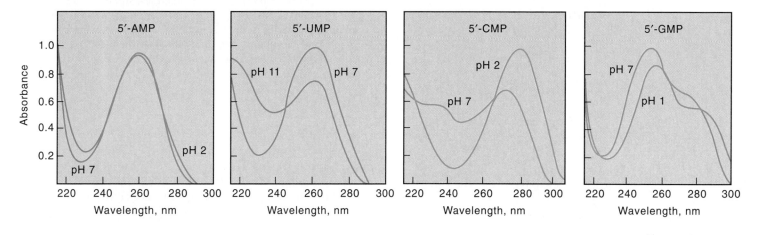

## Figure 23.7

Synthesis of phosphoribosylpyrophosphate (PRPP). This is an unusual kinase-catalyzed reaction because the group transferred is the pyrophosphate group rather than the phosphate group.

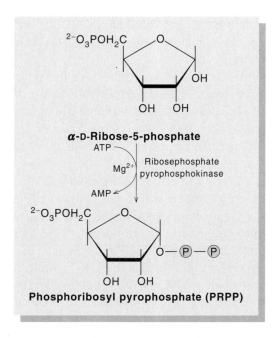

## Figure 23.8

Precursors of the purine ring of uric acid in pigeons as determined by isotope labeling experiments. The indicated precursor substances were administered one at a time to pigeons. Each precursor was labeled with isotopic nitrogen or carbon, and in each case the excreted uric acid was purified and degraded chemically. The isotope content of the various degradation products indicated that precursors contributed the specific atoms indicated. Glycine contributes the two bridge carbons (4) and (5) as well as N7.

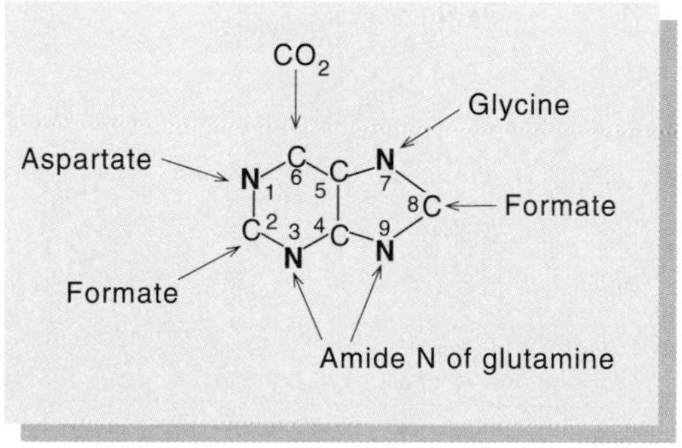

**Figure 23.9**

Summary of incorporation of precursors into the purine ring of IMP. In steps 3 and 9 formate or other indirect donors of a one-carbon unit, such as serine, donate their atoms in the form of a formyl group attached to reduced folate.

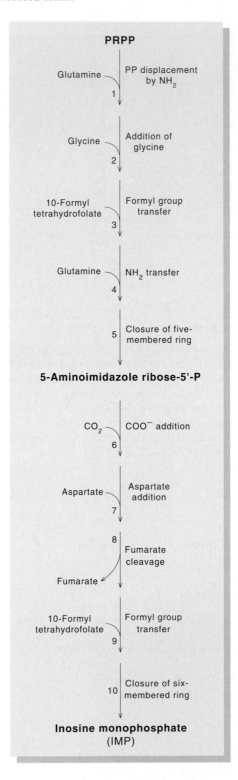

**Figure 23.10**

Biosynthetic pathway to inosine monophosphate. The group or atom introduced at each step is shown in red. The first step, in which phosphoribosylamine is formed, involves inversion of the configuration at C-1 of the ribose. In step 2 an amide bond is formed by the synthase between the carboxyl group of glycine and the amino group of phosphoribosylamine. An additional formyl group is added to the glycine amino group in the next step. An additional amide group is donated by a glutamine, and next the five-member imidazole ring is formed. Further groups are introduced through successive attachments to the imidazole, and finally the purine ring is completed to form inosine-5′-monophosphate.

dotransferase, an enzyme containing nonheme iron. The reaction involves inversion of the configuration at C-1 of the ribose and leads to the $\beta$ configuration that is characteristic of naturally occurring nucleotides. This step involves commitment of PRPP to the purine biosynthetic pathway and, as you might expect, is subject to important feedback inhibitory effects by purine nucleotides. We examine these effects a little later.

In step 2, an amide bond is formed by the synthase between the carboxyl group of glycine and the amino group of phosphoribosylamine, with ATP supplying energy and being hydrolyzed to ADP and inorganic phosphate. After these two steps have introduced atoms 4, 5, 7, and 9 of the purine ring, the remaining atoms are introduced one by one (steps 3, 4, 6, 7, and 9).

In step 3, carbon 8 is introduced as a formyl group that is transferred from 10-formyltetrahydrofolate (10-formyl-$H_4$ folate). (To review the way tetrahydrofolate derivatives accept a one-carbon unit from donors, such as serine, glycine, or formate, and transfer it to a suitable acceptor in biosynthetic reactions, see chapter 10; in chapter 10 10-formyl $H_4$ folate is written as $N^{10}$-formyl $H_4$ folate, an alternative terminology that is still in use.) Now the five components that will constitute the imidazole part of the purine are present, but before ring closure occurs, N-3 of the purine ring is introduced (step 4) by transfer of another amino group from glutamine to phosphoribosyl-formylglycinamide. ATP provides energy for the amido group transfer, being itself hydrolyzed to ADP and phosphate.

The imidazole ring is closed in an essentially irreversible cyclization requiring the presence of $Mg^{2+}$ and $K^+$ (step 5). Then C-6 of the purine ring is introduced by addition of bicarbonate in the presence of a specific carboxylase (step 6). This carboxylation is unusual in that it does not

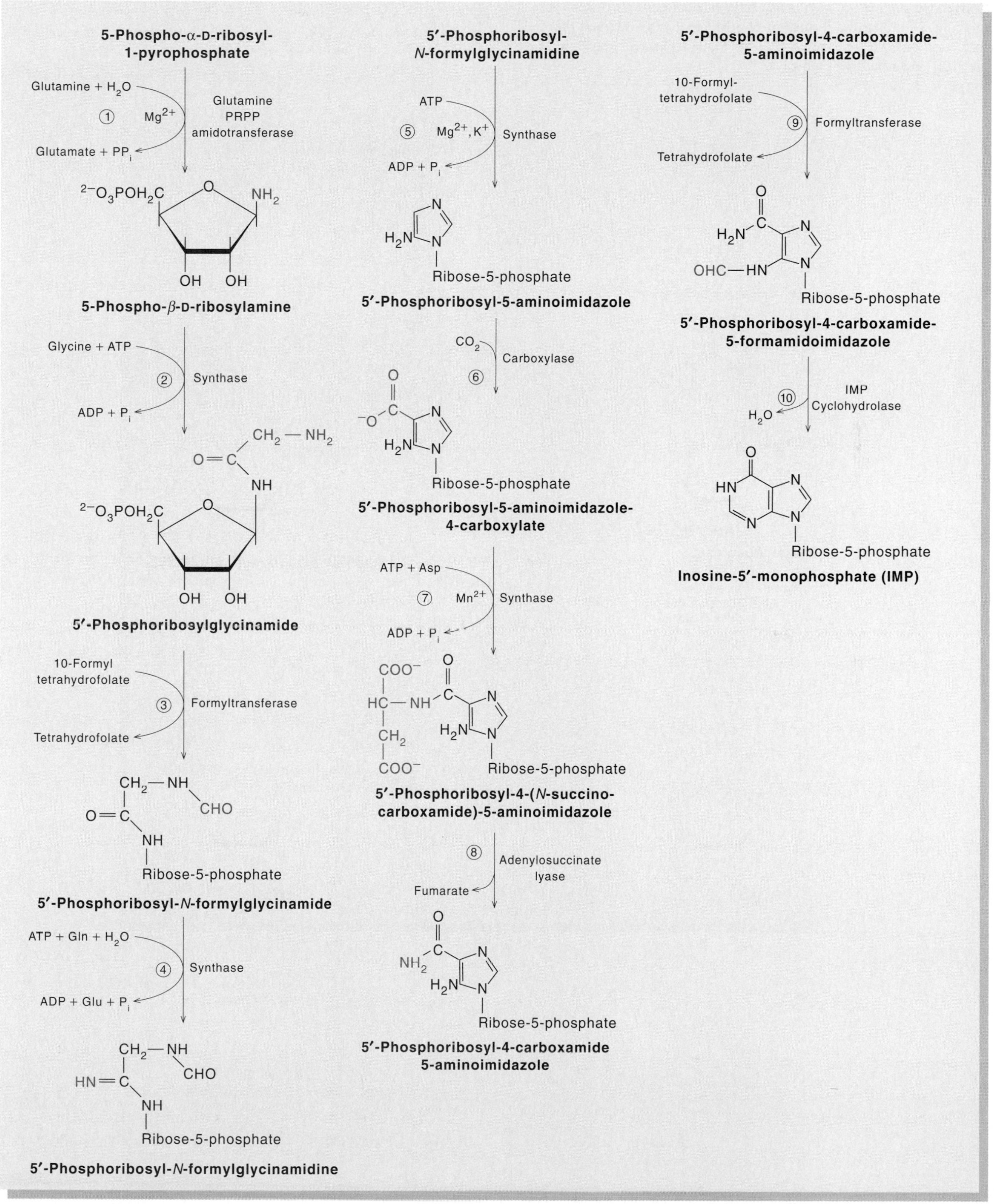

**5-Phospho-α-D-ribosyl-1-pyrophosphate**

Glutamine + H₂O

① Mg²⁺   Glutamine PRPP amidotransferase

Glutamate + PP_i

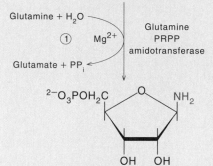

**5-Phospho-β-D-ribosylamine**

Glycine + ATP

② Synthase

ADP + P_i

**5′-Phosphoribosylglycinamide**

10-Formyl tetrahydrofolate

③ Formyltransferase

Tetrahydrofolate

**5′-Phosphoribosyl-N-formylglycinamide**

ATP + Gln + H₂O

④ Synthase

ADP + Glu + P_i

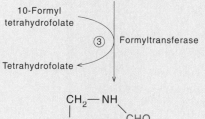

**5′-Phosphoribosyl-N-formylglycinamidine**

---

**5′-Phosphoribosyl-N-formylglycinamidine**

ATP

⑤ Mg²⁺, K⁺   Synthase

ADP + P_i

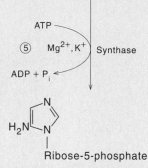

Ribose-5-phosphate

**5′-Phosphoribosyl-5-aminoimidazole**

CO₂

⑥ Carboxylase

Ribose-5-phosphate

**5′-Phosphoribosyl-5-aminoimidazole-4-carboxylate**

ATP + Asp

⑦ Mn²⁺   Synthase

ADP + P_i

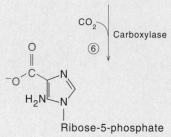

Ribose-5-phosphate

**5′-Phosphoribosyl-4-(N-succino-carboxamide)-5-aminoimidazole**

⑧ Adenylosuccinate lyase

Fumarate

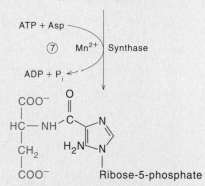

Ribose-5-phosphate

**5′-Phosphoribosyl-4-carboxamide 5-aminoimidazole**

---

**5′-Phosphoribosyl-4-carboxamide-5-aminoimidazole**

10-Formyl-tetrahydrofolate

⑨ Formyltransferase

Tetrahydrofolate

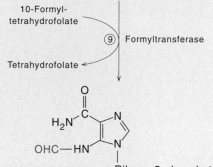

Ribose-5-phosphate

**5′-Phosphoribosyl-4-carboxamide-5-formamidoimidazole**

⑩ IMP Cyclohydrolase

H₂O

Ribose-5-phosphate

**Inosine-5′-monophosphate (IMP)**

## Figure 23.11

Conversion of IMP to AMP and GMP. In both cases two steps are required. Note that the formation of AMP requires GTP, and the formation of GMP requires ATP. This tends to balance the flow of the IMP down the two pathways.

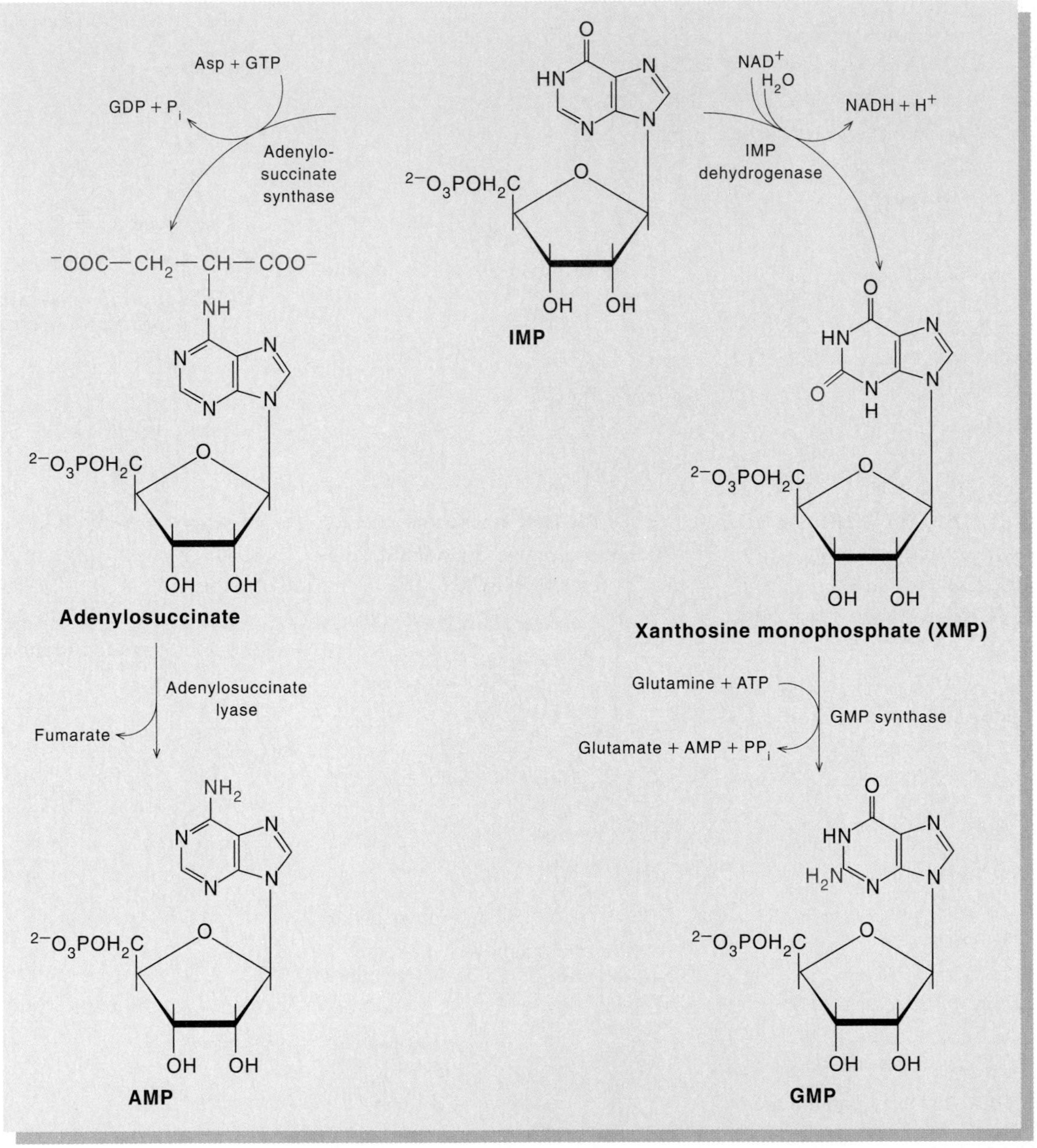

seem to involve biotin and is not coupled with any energy-yielding process such as ATP hydrolysis. The equilibrium is unfavorable for the formation of the carboxylate. *In vivo* the reaction proceeds at physiological concentrations of bicarbonate because of coupling with subsequent steps that are thermodynamically favorable.

Next, in steps 7 and 8, N-1 of the purine ring is contributed by aspartate. Aspartate forms an amide with the 4-carboxyl group, and the succinocarboxamide so formed is then cleaved with release of fumarate. Energy for carboxamide formation is provided by ATP hydrolysis to ADP and phosphate. These reactions resemble the conversion of citrulline to arginine in the urea cycle (chapter 22) and the conversion of IMP to AMP (see fig. 23.11).

The final atom of the purine ring is provided in step 9 by donation of a formyl group from 10-formyltetrahydrofolate to the 5-amino group of the almost completed ribonucleotide. In the final step, ring closure is effected by elimination of water to form IMP (inosine monophosphate), the first product with a complete purine ring. Although this final ring closure does not require energy from ATP, closure of the imidazole ring does, and the synthesis of IMP from ribose-5-phosphate requires a total of six high-energy phosphate groups from ATP (assuming hydrolysis of pyrophosphate released during phosphoribosylamine synthesis, step 1 of figure 23.10).

## IMP Is Converted into AMP and GMP

IMP does not accumulate in the cell but is converted to AMP, GMP, and the corresponding diphosphates and triphosphates. The two steps of the pathway from IMP to AMP (fig. 23.11) are typical reactions by which the amino group from aspartate is introduced into a product. The 6-hydroxyl group of IMP (tautomeric with the 6-keto group) is first displaced by the amino of aspartate to give adenylosuccinate, and the latter is then cleaved nonhydrolytically by adenylosuccinate lyase to yield fumarate and AMP. In the condensation of aspartate with IMP, cleavage of GTP to GDP and phosphate provides energy to drive the reaction.

Conversion of IMP to GMP also proceeds by a two-step pathway (see fig. 23.11): First, dehydrogenation of IMP to xanthosine-5′-phosphate (XMP), and second, transfer of an amino group from glutamine to C-2 of the xanthine ring to yield GMP. The second reaction also involves the cleavage of ATP to AMP and inorganic pyrophosphate. The inorganic pyrophosphate is in turn hydrolyzed to inorganic phosphate by the ubiquitous inorganic pyrophosphatase in a reaction with a very favorable equilibrium. This hydrolysis

**Figure 23.12**

The structure of orotic acid. Loss of proton leads to orotate. Deposits of sodium orotate cause a painful condition.

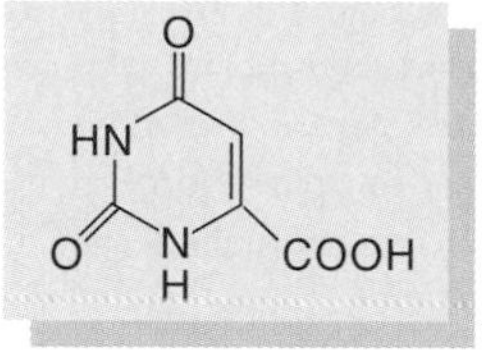

is coupled with the GMP synthase reaction because pyrophosphate is a product of the latter and a substrate of the pyrophosphatase. The net result is that the release of two high-energy phosphate groups is used to drive the GMP synthase reaction to completion.

## Synthesis of Pyrimidine Ribonucleotides *de Novo*

Like purine nucleotides, pyrimidine nucleotides can be synthesized either *de novo* or by the "salvage" pathways from nucleobases or nucleosides. However, salvage is less efficient because, except in the case of utilization of uracil by bacteria, and to some extent by mammalian cells, pyrimidine nucleobases are not converted to nucleotides directly but only via nucleosides.

The biosynthetic pathway to pyrimidine nucleotides is simpler than that for purine nucleotides, reflecting the simpler structure of the base. In contrast to the biosynthetic pathway for purine nucleotides, in the pyrimidine pathway the pyrimidine ring is constructed before ribose-5-phosphate is incorporated into the nucleotide. The first pyrimidine mononucleotide to be synthesized is orotidine-5′-monophosphate (OMP), and from this compound, pathways lead to nucleotides of uracil, cytosine, and thymine. OMP thus occupies a central role in pyrimidine nucleotide biosynthesis, somewhat analogous to the position of IMP in purine nucleotide biosynthesis. Like IMP, OMP is found only in low concentrations in cells and is not a constituent of RNA.

Early clues to the nature of the pyrimidine pathway were provided by the observations that orotic acid (6-carboxyuracil, fig. 23.12) can satisfy the growth requirement of mutants of the fungus *Neurospora* that are unable to make pyrimidines, and that isotopically labeled orotate is an immediate precursor of pyrimidines in *Neurospora* and a number of bacteria.

## Figure 23.13

Biosynthesis of UMP. The parts of the intermediates derived from aspartate are shown in red. Bold type indicates atoms derived from carbamoyl phosphate. In contrast to purine nucleotide synthesis, where ring formation starts on the sugar, in pyrimidine biosynthesis the pyrimidine ring is completed before being attached to the ribose.

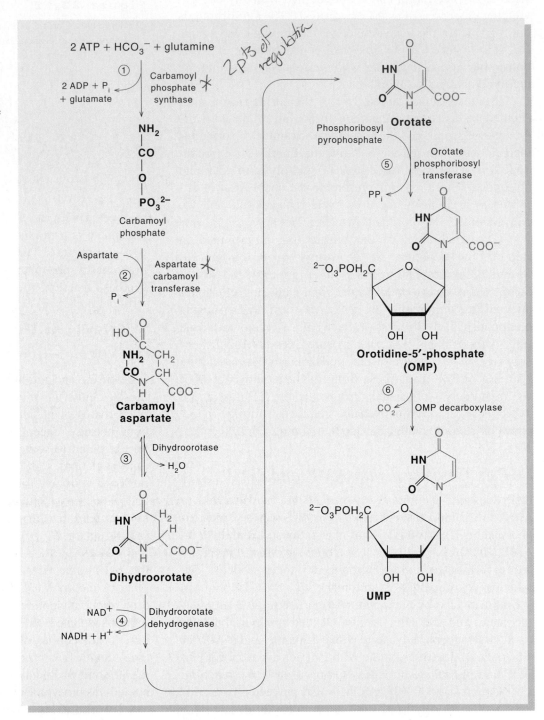

### UMP Is a Precursor of Other Pyrimidine Nucleotides

The pathway for UMP synthesis is shown in figure 23.13. It starts with the synthesis of carbamoyl phosphate, catalyzed by carbamoyl phosphate synthase. This enzyme is present in microorganisms and in the cytosol of all eukaryotic cells that are capable of forming pyrimidine nucleotides. Eukaryotes also have another carbamoyl phosphate synthase, a distinct enzyme that uses ammonia as a substrate instead of glutamine. It is associated with citrulline formation in the pathway for arginine biosynthesis, and in mammals it is a mitochondrial enzyme present predominantly in the liver, where it catalyzes a step in the urea cycle (chapter 22). In

**Figure 23.14**

The production of CTP by amination of UTP. The conversion of the pyrimidine ring of uracil to cytosine occurs at the level of the nucleotide triphosphate. The enzyme responsible for this conversion is known as cytidine triphosphate synthase.

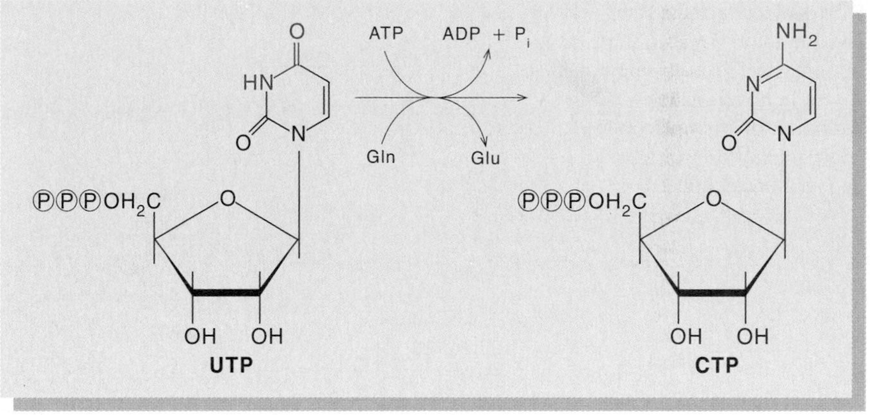

the cytosol, the enzyme product, carbamoyl phosphate, next reacts with aspartate to form carbamoyl aspartate. The reaction is catalyzed by aspartate carbamoyltransferase (aspartate transcarbamoylase or aspartate transcarbamylase), and the equilibrium greatly favors carbamoyl aspartate synthesis.

In the third step, the pyrimidine ring is closed by dihydroorotase to form 1-dihydroorotate. Dihydroorotate is then oxidized to orotate by dihydroorotate dehydrogenase. This flavoprotein in some organisms contains FMN and in others both FMN and FAD. It also contains nonheme iron and sulfur. In eukaryotes it is a lipoprotein associated with the inner membrane of the mitochondria. In the final two steps of the pathway, orotate phosphoribosyltransferase yields orotidine-5′-phosphate (OMP), and a specific decarboxylase then produces UMP.

Low activities of orotidine phosphate decarboxylase and (usually) orotate phosphoribosyltransferase are associated with a genetic disease in children that is characterized by abnormal growth, megaloblastic anemia, and the excretion of large amounts of orotate. When affected children are fed a pyrimidine nucleoside, usually uridine, the anemia decreases and the excretion of orotate diminishes. A likely explanation for the improvement is that the ingested uridine is phosphorylated to UMP, which is then converted to other pyrimidine nucleotides so that nucleic acid and protein synthesis can resume. In addition, the increased intracellular concentrations of pyrimidine nucleotides inhibit carbamoyl phosphate synthase, the first enzyme in the pathway of orotate synthesis.

## CTP Is Formed from UTP

After UMP is synthesized, it is phosphorylated to UTP. Then CTP is formed by reaction of UTP with glutamine in a reaction driven by the concomitant hydrolysis of ATP to ADP and inorganic phosphate (fig. 23.14). Both the mammalian and bacterial cytidine triphosphate synthases can use ammonia as a donor in place of glutamine, but this reaction is of no physiological significance because the $K_m$ for ammonia is very high and the reaction rate low.

## Biosynthesis of Deoxyribonucleotides

Tracer studies with isotopically labeled precursors have shown that both in mammalian tissues and in microorganisms, deoxyribonucleotides are formed from corresponding ribonucleotides by replacement of the 2′—OH group with hydrogen.

Three types of ribonucleotide reductase catalyze this reduction of the ribose ring. The most widely distributed in nature occurs in mammalian and plant cells, in yeast, and in some prokaryotes. This type of reductase contains a tyrosyl radical closely associated with nonheme iron, as in the reductase from E. coli. The E. coli reductase is composed of two nonidentical subunits, both contributing to the active site; it is specific for the reduction of diphosphates (ADP, GDP, CDP, and UDP).

An unusual feature of ribonucleotide reductase is that the reaction it catalyzes involves a radical mechanism. The mammalian type of reductase initiates this reaction by the tyrosyl radical-nonheme iron. Hydroxyurea and related compounds inhibit the mammalian reductase by abolishing the radical state of the tyrosine residue. Inhibition of DNA synthesis by such compounds is secondary to this effect.

The physiological reducing substrate is a low-molecular-weight (13,000) electron-transport protein, thioredoxin. Thioredoxin has two half-cystine residues that are separated in the polypeptide chain by two other residues. The oxidized form of thioredoxin, with a disulfide bridge between the

**Figure 23.15**

Ribonucleotide reductase and the thioredoxin system. In some lactobacilli, vitamin $B_{12}$ is involved in the reduction of ribonucleotide triphosphate to deoxyribonucleotide.

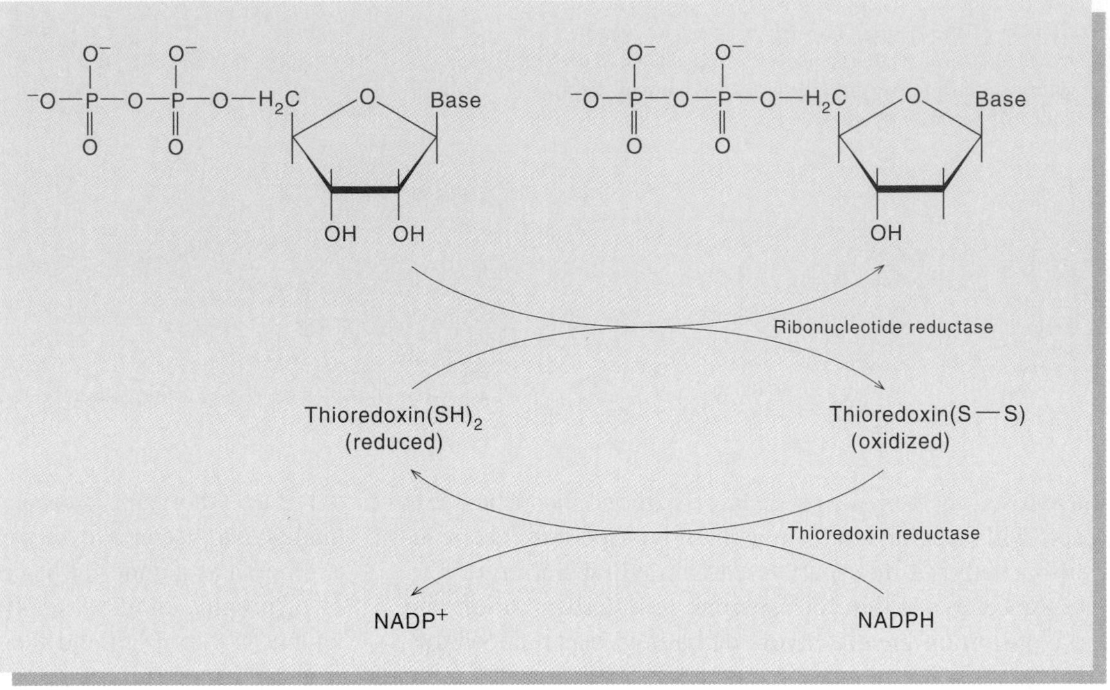

half-cystines, is reduced by NADPH in the presence of a flavoprotein, thioredoxin reductase. The reduced form of thioredoxin, with two cysteine residues present, is the reducing substrate for ribonucleotide reduction. The flow of electrons from NADPH to ribose is shown in figure 23.15.

## Thymidylate Is Formed from dUMP

Since DNA contains thymine (5-methyluracil) as a major base instead of uracil, the synthesis of thymidine monophosphate (dTMP, or thymidylate) is essential to provide dTTP (thymidine triphosphate), which is needed for DNA replication together with dATP, dGTP, and dCTP.

Thymidylate is synthesized from dUMP, which may be formed by two pathways in cells. The major precursor of dUMP is dCMP, which is converted to dUMP by deoxycytidylate deaminase:

$$\text{dCMP} + \text{H}_2\text{O} \longrightarrow \text{dUMP} + \text{NH}_3 \qquad (1)$$

This enzyme is widely distributed in animal tissues, and the enzyme produced by yeast, by bacteriophage T2-infected *E. coli*, and by animal cells has been studied extensively. Deamination of 5-methyldeoxycytidylate and 5-hydroxymethyldeoxycytidylate is also catalyzed by this enzyme.

Another route to dUMP is the reduction of UDP to dUDP, followed by phosphorylation of dUDP to dUTP (or direct reduction of UTP to dUTP in some microorganisms). The dUTP is then hydrolyzed to dUMP. This circuitous route to dUMP is dictated by two considerations. First, the ribonucleotide reductase in most cells acts only on ribonucleoside diphosphates, probably because this permits better regulation of its activity. Second, cells contain a highly active deoxyuridine triphosphate diphosphohydrolase (dUTPase). It prevents the incorporation of dUTP into DNA by keeping intracellular levels of dUTP low by means of the reaction

$$\text{dUTP} + \text{H}_2\text{O} \longrightarrow \text{dUMP} + \text{PP}_i \qquad (2)$$

Methylation of dUMP to give thymidylate is catalyzed by thymidylate synthase and utilizes 5,10-methylenetetrahydrofolate as the source of the methyl group. This reaction is unique in the metabolism of folate derivatives because the folate derivative acts both as a donor of the one-carbon group and as its reductant, using the reduced pteridine ring as the source of reducing potential. Consequently, in this reaction, unlike any other in folate metabolism, dihydrofolate is a product (fig. 23.16). Since folate derivatives are present in cells at very low concentrations, continued syn-

## Figure 23.16

Thymidylate biosynthesis. (R = p-aminobenzoyl-L-glutamate.) Red symbols indicate precursors of the methyl group of thymidylate. The dTMP is formed from dUMP. 5,10-Methylenetetrahydrofolate donates the methyl group in this reaction. This methyl group donor is regenerated by a two-step process involving first the reduction of 7,8-dihydrofolate to tetrahydrofolate and thence to 5,10-methylenetetrahydrofolate, in which serine supplies the methylene group.

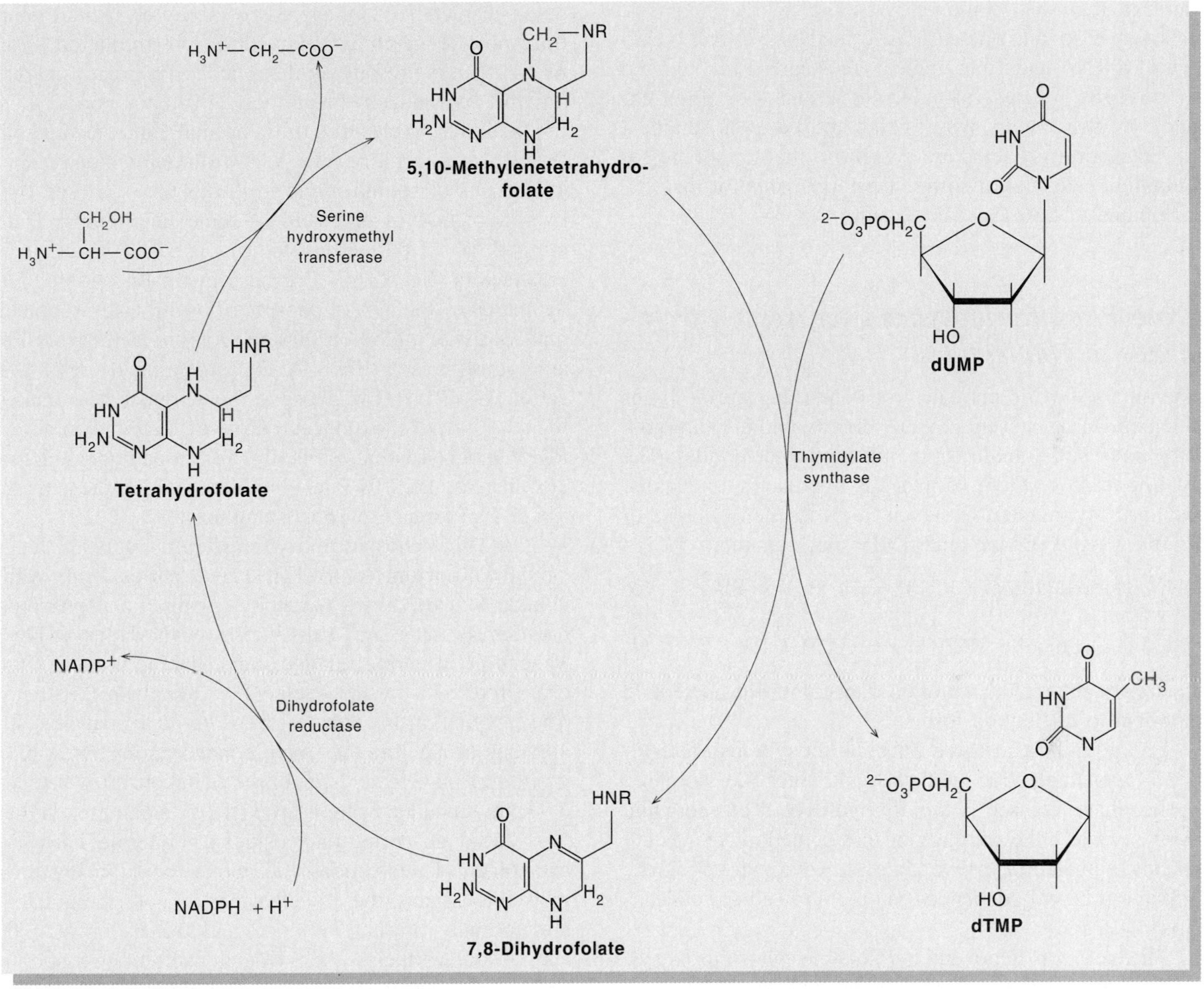

thesis of thymidylate requires regeneration of 5,10-methylenetetrahydrofolate from dihydrofolate. As shown in figure 23.16, this occurs in two steps catalyzed by the enzymes dihydrofolate reductase, an NADP$^+$-linked dehydrogenase, and serine hydroxymethyltransferase. Thymidylate synthesis can be interrupted, and consequently the synthesis of DNA arrested, by the inhibition of either thymidylate synthase or dihydrofolate reductase. Many potent inhibitors are known for each of these enzymes, the best known being 5-fluoro-dUMP for thymidylate synthase and methotrexate for dihydrofolate reductase. We discuss these and other inhibitors of nucleotide synthesis in a later section.

# Formation of Nucleotides from Bases and Nucleosides (Salvage Pathways)

In addition to the pathways for synthesis *de novo,* mammalian cells and microorganisms can readily form mononucleotides from purine bases and their nucleosides and to a lesser extent from pyrimidine bases and their nucleosides. In this way bases and nucleosides formed by constant breakdown of mRNA and other nucleic acids can be reconverted (or "salvaged") to useful nucleotides, and the energy expended by the cell in synthesizing the bases is retained.

Transporter systems are present in the membranes of mammalian cells and of some microorganisms for the efficient uptake of bases and nucleosides.

## *Purine Phosphoribosyltransferases Convert Purines to Nucleotides*

In mammals specific enzymes for converting purine bases to nucleotides are present in many organs, and in heart muscle this may be the main source of purine nucleotides. The most important of these enzymes is hypoxanthine-guanine phosphoribosyltransferase, which catalyzes the formation of IMP from hypoxanthine and GMP from guanine:

$$\text{Hypoxanthine} + \text{PRPP} \rightleftharpoons \text{IMP} + \text{PP}_i \qquad (3)$$

$$\text{Guanine} + \text{PRPP} \rightleftharpoons \text{GMP} + \text{PP}_i \qquad (4)$$

This enzyme returns the major products of purine nucleotide catabolism to nucleotide forms.

The equilibria in these phosphoribosyltransferase reactions favor nucleotide synthesis, and since the inorganic pyrophosphate released is rapidly hydrolyzed by inorganic pyrophosphatase, the coupling of these reactions makes the synthesis of nucleotide irreversible. However, the efficiency of salvage is heavily dependent on the intracellular concentration of PRPP.

The role of hypoxanthine-guanine phosphoribosyltransferase in purine salvage has been confirmed by the abnormally high excretion of purines (as uric acid) in humans who lack hypoxanthine-guanine phosphoribosyltransferase. Studies of purine metabolism in cultures of cells from patients with this hereditary disorder also support this conclusion.

Besides this salvage role, hypoxanthine-guanine phosphoribosyltransferase is probably important also for the transfer of purines from liver to other tissues. Purine biosynthesis *de novo* is especially active in the liver, and extrahepatic cells that have a low capacity for the synthesis of purines *de novo,* such as erythrocytes and bone marrow cells, depend on uptake of hypoxanthine and xanthine from the blood to meet their needs for purine nucleotides. Blood levels of xanthine and hypoxanthine, normally about 0.04 mM, are maintained by release of these bases from the liver.

The neurologic disorder of children called Lesch-Nyhan syndrome is due to a congenital lack of hypoxanthine-guanine phosphoribosyltransferase. The disorder is characterized by aggressive behavior, mental retardation, spastic cerebral palsy, and self-mutilation. Purine metabolism is profoundly disturbed, with greatly increased *de novo* biosynthesis of purines (200 times normal), overproduction of uric acid (6 times normal), and elevated blood levels of uric acid. Severe gout, an inflammatory response to deposits of sodium urate in the joints, is caused by the overproduction of uric acid in some individuals. The increased rate of purine biosynthesis is probably due to several factors (fig. 23.17). Because nucleotide production is depressed by the lack of phosphoribosyltransferase, the normal feedback inhibition that controls the activity of PRPP amidotransferase is lifted. At the same time, decreased utilization of PRPP by phosphoribosyltransferase makes a higher intracellular concentration of PRPP available for PRPP amidotransferase. Finally, decreased nucleotide pools also increase the PRPP level by deregulating the activity of ribose-5-phosphate pyrophosphokinase.

Adenine phosphoribosyltransferase catalyzes the conversion of adenine to AMP in many tissues, by a reaction similar to that of hypoxanthine-guanine phosphoribosyltransferase, but is quite distinct from the latter. It plays a minor role in purine salvage since adenine is not a significant product of purine nucleotide catabolism (see below). The function of this enzyme seems to be to scavenge small amounts of adenine that are produced during intestinal digestion of nucleic acids or in the metabolism of 5'-deoxy-5'-methylthioadenosine, a product of polyamine synthesis.

Although purine nucleosides are intermediates in the catabolism of nucleotides and nucleic acids in higher animals and humans, these nucleosides do not accumulate and are normally present in blood and tissues only in trace amounts. Nevertheless, cells of many vertebrate tissues contain kinases capable of converting purine nucleosides to nucleotides. Typical of these is adenosine kinase, which catalyzes the reaction

$$\text{Adenosine} + \text{ATP} \longrightarrow \text{AMP} + \text{ADP} \qquad (5)$$

2'-Deoxyadenosine is also a substrate, though a relatively poor one. In some mammalian cells deoxycytidine kinase also phosphorylates deoxyadenosine and deoxyguanosine (see next section). In addition, indirect evidence suggests the existence of a specific deoxyadenosine kinase in human cells. Deoxyadenosine has received special attention as a substrate for kinases because congenital inability to catabo-

**Figure 23.17**

Mechanism of overproduction of purine nucleotides in the congenital deficiency of hypoxanthine-guanine phosphoribosyltransferase. The loss of the transferase prevents the recycling of hypoxanthine and guanine. This increases uric acid production as well as the *de novo* synthesis of purine nucleotides.

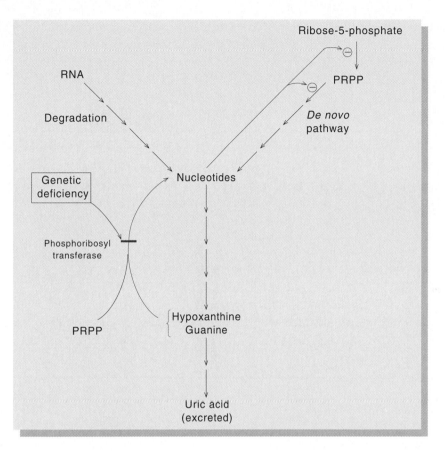

## Inhibitors of Nucleotide Synthesis

There are several distinct types of inhibitors of nucleotide biosynthesis, each type acting at different points in the pathways to purine or pyrimidine nucleotides. All these inhibitors are very toxic to cells, especially rapidly growing cells, such as those of tumors or bacteria, because interruption of the supply of nucleotides seriously limits the cell's capacity to synthesize the nucleic acids necessary for protein synthesis and cell replication. In some cases, the toxic effect of such inhibitors makes them useful in cancer chemotherapy or in the treatment of bacterial infections. However, some of these agents can also damage the rapidly replicating cells of the intestinal tract and bone marrow. This danger imposes limits on the doses that can be used safely.

6-Mercaptopurine (table 23.3) and related thiopurines are potent inhibitors of purine nucleotide biosynthesis, but they are inactive until they are converted to the correspond-

lize this nucleoside results in immune deficiency disease. A mitochondrial kinase specific for deoxyguanosine has been reported. None of these kinases has been studied extensively, and their role in various tissues and organs is uncertain.

In contrast to the extensive salvage and reutilization of purine bases, pyrimidine bases are poorly utilized by mammalian tissues.

## Conversion of Nucleoside Monophosphates to Triphosphates Goes through Diphosphates

The products of the biosynthetic pathways discussed in the preceding sections are, in most cases, mononucleotides. In cells, a series of kinases (phosphotransferases) converts these mononucleotides to their metabolically active diphosphate and triphosphate forms.

## Table 23.3

Some Inhibitors of Nucleotide Metabolism Together with Enzymes Inhibited

| Inhibitor | Enzymes Inhibited | Inhibitor | Enzymes Inhibited |
|---|---|---|---|
| **Inhibitor of purine nucleotide biosynthesis** | | **Inhibitors of both purine and pyrimidine nucleotide biosynthesis** | |
| 6-Mercaptopurine | Figure 23.11, adenylosuccinate synthase, IMP dehydrogenase; Figure 23.10, reaction 1 | $H_2N-C-NHOH$, with =O on carbon — Hydroxyurea | Figure 23.15, ribonucleotide reductase |
| **Inhibitors of pyrimidine nucleotide biosynthesis** | | Acivicin | Figure 23.10, reactions 1 and 4; Figure 23.11, GMP synthase; Figure 23.13, reaction 1 |
| $COO^-$ $CH-NH-CO-CH_2-PO_3^{2-}$ $CH_2$ $COO^-$ — $N$-(Phosphonacetyl)-L-aspartate (PALA) | Figure 23.13, aspartate carbamoyltransferase | **Indirect inhibitors of nucleotide biosynthesis** | |
| 5-Fluorouracil | Figure 23.16, thymidylate synthase | Methotrexate (amethopterin) | Figure 23.16 dihydrofolate reductase |
| | | Sulfonamide | Dihydropteroate synthase |

**Figure 23.18**

Activated forms of two inhibitors of nucleotide metabolism.

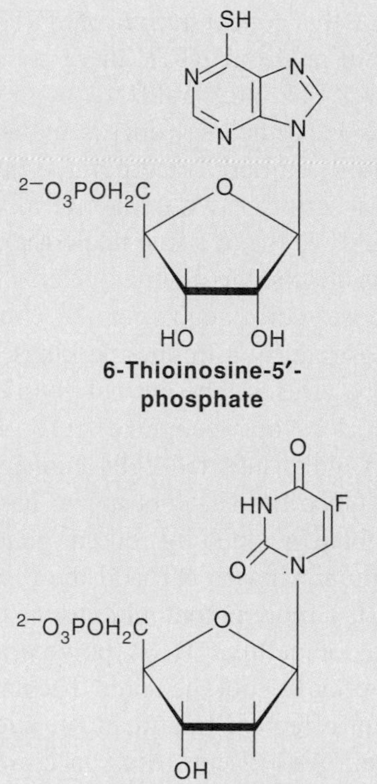

6-Thioinosine-5'-
phosphate

5-Fluoro-2'-deoxyuridine-5'-
phosphate (F-dUMP)

ing ribonucleoside 5'-phosphates. 6-Mercaptopurine is converted by the action of hypoxanthine-guanine phosphoribosyltransferase to the nucleotide 6-thioinosine-5'-monophosphate (T-IMP) (fig. 23.18). The latter inhibits several enzymes of purine biosynthesis, and it is uncertain which effect is primarily responsible for the toxicity. It blocks conversion of IMP to adenylosuccinate and to XMP, key reactions in the formation of AMP and GMP (see fig. 23.11). T-IMP is also capable of "pseudofeedback" inhibition of glutamine PRPP amidotransferase (see fig. 23.10), the first committed step in the purine nucleotide pathway. This enzyme is highly responsive to intracellular concentrations of both normal ribonucleoside 5'-monophosphates and analogs. 6-Mercaptopurine is used clinically in the treatment of leukemia.

One inhibitor of those shown in table 23.3 specifically interferes with a step in pyrimidine nucleotide biosynthesis. N-(Phosphonacetyl)-L-aspartate (PALA) is a powerful inhibitor of the carbamoyl transferase reaction. PALA was synthesized to act as an analog of the transition state intermediate (see chapter 9) postulated to be formed in the aspar-

tate carbamoyltransferase reaction, and its tight binding to the enzyme is probably due to its resemblance to the intermediate. PALA is also under trial as an anticancer agent.

5-Fluorouracil, an agent widely used in treating solid tumors, interferes with thymidylate synthesis (see fig. 23.16). Its major inhibitory effect occurs after its conversion in the cell to 5-fluoro-2'-deoxyuridine-5'-monophosphate (see fig. 23.18). The latter acts as an analog of dUMP and binds very tightly to thymidylate synthase (see fig. 23.16) in the presence of methylenetetrahydrofolate, forming a covalent complex with the enzyme that is unable to undergo the normal catalytic reaction. 5-Fluorouracil may also exert a cytotoxic effect through incorporation of 5-fluorouridine phosphate into RNA.

Hydroxyurea interferes with the synthesis of both pyrimidine and purine nucleotides (see table 23.3). It interferes with the synthesis of deoxyribonucleotides by inhibiting ribonucleotide reductase of mammalian cells, an enzyme that is crucial and probably rate-limiting in the biosynthesis of DNA. It probably acts by disrupting the iron-tyrosyl radical structure at the active site of the reductase. Hydroxyurea is in clinical use as an anticancer agent.

Acivicin is a potent inhibitor of several steps in purine nucleotide biosynthesis that utilize glutamine. The enzymes it inhibits are glutamine PRPP amidotransferase (step 1, fig. 23.10), phosphoribosyl-N-formylglycinamidine synthase (step 4, fig. 23.10), and GMP synthase (see fig. 23.11). In pyrimidine nucleotide biosynthesis the enzymes inhibited are carbamoyl synthase (step 1, fig. 23.13) and CTP synthase (see fig. 23.14). Acivicin is under trial for the treatment of some forms of cancer.

Another group of inhibitors prevents nucleotide biosynthesis indirectly by depleting the level of intracellular tetrahydrofolate derivatives. Sulfonamides are structural analogs of p-aminobenzoic acid (fig. 23.19), and they competitively inhibit the bacterial biosynthesis of folic acid at a step in which p-aminobenzoic acid is incorporated into folic acid. Sulfonamides are widely used in medicine because they inhibit growth of many bacteria. When cultures of susceptible bacteria are treated with sulfonamides, they accumulate 4-carboxamide-5-aminoimidazole in the medium, because of a lack of 10-formyltetrahydrofolate for the penultimate step in the pathway to IMP (see fig. 23.10). Methotrexate, and a number of related compounds inhibit the reduction of dihydrofolate to tetrahydrofolate, a reaction catalyzed by dihydrofolate reductase. These inhibitors are structural analogs of folic acid (see fig. 23.19) and bind at the catalytic site of dihydrofolate reductase, an enzyme catalyzing one of the steps in the cycle of reactions involved in thymidylate synthesis (see fig. 23.16). These inhibitors therefore prevent synthesis of thymidylate in replicating

**Figure 23.19**

Normal metabolites with which indirect inhibitors of nucleotide metabolism compete. Methotrexate and trimethoprim compete with folate and, more importantly, with dihydrofolate (fig. 23.16); sulfonamides (sulfa drugs) compete with p-aminobenzoate.

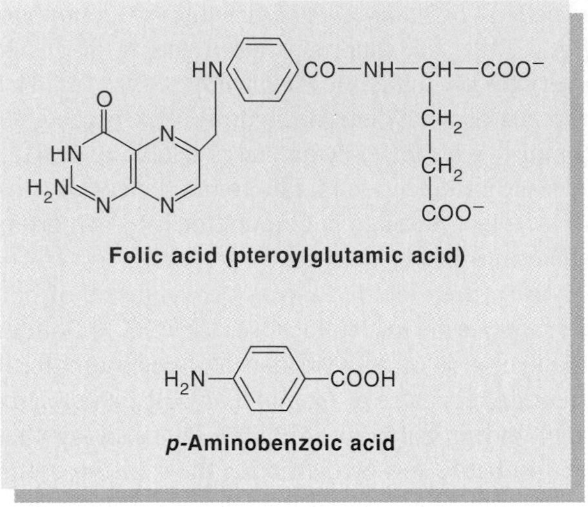

**Folic acid (pteroylglutamic acid)**

**p-Aminobenzoic acid**

cells, and as a secondary effect synthesis of purine nucleotides is also decreased because of 10-formyltetrahydrofolate depletion. Methotrexate is used as an anticancer drug.

A number of drugs that are of great importance for the treatment of viral infections and of certain types of cancer are nucleosides or have closely related structures. Examples are shown in table 23.4.

3′-Azido-2′-deoxythymidine (AZT) has received much attention because it is the only drug with proven effectiveness in arresting the course of acquired immune deficiency syndrome (AIDS). It has been shown to be particularly effective in preventing the spread of the disease from the pregnant mother to the fetus. The virus that causes this disease, human immunodeficiency virus (HIV), contains an RNA-directed DNA polymerase. Once the virus enters a host cell, this reverse transcriptase catalyzes the synthesis of a double-stranded DNA copy of the viral RNA as the first step in virus replication. AZT is converted to its monophosphate (AZTMP) by thymidine kinase; thymidylate kinase then converts the monophosphate to the triphosphate (AZTTP). AZTTP is a powerful inhibitor of the reverse transcriptase. More importantly, it is also a good substrate for this enzyme and, when incorporated into the growing DNA chain, promptly terminates chain extension, because there is no 3′—OH group on the AZT residue to which the next nucleotide unit can become attached. The effect of

AZTTP on DNA polymerase of the host cell is much weaker. The effect of AZT is further increased by the accumulation of considerable amounts of AZTMP in cells. This monophosphate inhibits thymidylate kinase, with a consequent decrease in the concentration of dTTP in the cell. With less competition from dTTP there is more efficient incorporation of AZTTP into viral DNA by the reverse transcriptase. 2′,3′-Dideoxyinosine (ddI) is under trial for use against AIDS and is similarly incorporated into viral DNA by the reverse transcriptase with chain termination.

Acyclovir (ACV) is not a true nucleoside, because the guanine residue is attached to an open-chain structure, but it mimics deoxyribose well enough for the compound to be accepted as a substrate by a thymidine kinase specified by certain herpes-type viruses. The normal thymidine kinase in mammalian cells does not recognize ACV as a substrate, however, so only virus-infected cells convert ACV to its monophosphate. Once the first phosphate has been added, the second phosphate is added by cellular guanylate kinase; several other cellular kinases can add the third phosphate. The triphosphate is a more potent inhibitor of the viral DNA polymerases than of cellular DNA polymerases and also inactivates the former but not the latter. The net result is that ACV has been an effective treatment of, and prophylaxis for, genital herpes. Also it can result in dramatic relief of pain associated with "shingles" caused by reactivation of latent varicella-zoster virus, and has been successful in many patients with herpes encephalitis.

Gancyclovir (GCV), like ACV, is an open-chain analog of deoxyguanosine. It is more effective than ACV in the treatment of infections by human cytomegalovirus (HCMV). These infections can cause birth defects in newborns and can be fatal or cause blindness in immunosuppressed individuals, such as transplant recipients, cancer patients receiving chemotherapy, or AIDS patients. HCMV-infected cells are more efficient at forming the triphosphate of the drug, though the mechanism is unclear. Furthermore, the viral DNA polymerase is much more sensitive to inhibition (and chain termination) by the triphosphate than are cellular DNA polymerases.

Cytosine arabinose (araC) is taken up by cells and converted to its triphosphate (araCTP), which is a substrate for cell DNA polymerases, and is a relative chain terminator. This means that the cell DNA polymerases have difficulty in adding the next nucleotide after araCMP has been added to the chain. Incorporation of two or more successive araCMP residues causes chain termination. The araC is a component of a combination of drugs with proven effectiveness against some forms of leukemia.

**Table 23.4**

Other Nucleoside Analogs Used in Therapy

| Analog | Enzyme Inhibited by Active Form | Clinical Condition Treated | Analog | Enzyme Inhibited by Active Form | Clinical Condition Treated |
|---|---|---|---|---|---|
| Azidothymidine (AZT) | Viral DNA polymerase | AIDS (Human immuno-deficiency virus infection) | Acyclovir (ACV) | Viral DNA polymerase | Herpes simplex virus infections |
| 2′,3′-Dideoxyinosine (ddI) | Viral DNA polymerase | AIDS (Human immuno-deficiency virus infection) | Gancyclovir (GCV) | Viral DNA polymerase | Human cytomegalovirus infections |
| | | | Cytosine arabinoside (araC) | DNA polymerase | Leukemia |

## Catabolism of Nucleotides

In the small intestine, ribonuclease and deoxyribonuclease I, which are secreted in the pancreatic juice, hydrolyze nucleic acids mainly to oligonucleotides. The oligonucleotides are further hydrolyzed by phosphodiesterases, also secreted by the pancreas, to yield 5′- and 3′-mononucleotides. Most of the mononucleotides are then hydrolyzed to nucleosides by various group-specific nucleotidases or by a variety of nonspecific phosphatases. The resulting nucleosides may be absorbed intact by the intestinal mucosa, or they may undergo phosphorolysis by nucleoside phosphorylases and hydrolysis by nucleosidases to free bases:

$$\text{Nucleoside} + P_i \rightleftharpoons \text{base} + \text{ribose-1-phosphate} \quad (6)$$

$$\text{Nucleoside} + H_2O \rightleftharpoons \text{base} + \text{ribose} \quad (7)$$

Experiments with labeled nucleic acid indicate that the free bases of ingested nucleic acids are used only to a small extent for synthesis of tissue nucleic acids, and in the case

## Figure 23.20

Major pathways of purine degradation in animals. Primates excrete uric acid. Mammals other than primates catabolize uric acid to other end products. In contrast to the catabolism of carbohydrates, lipids, or amino acids, the catabolism of nucleotides results in no energy production in the form of ATP. In both GMP and AMP catabolism ribose-1-phosphate is released.

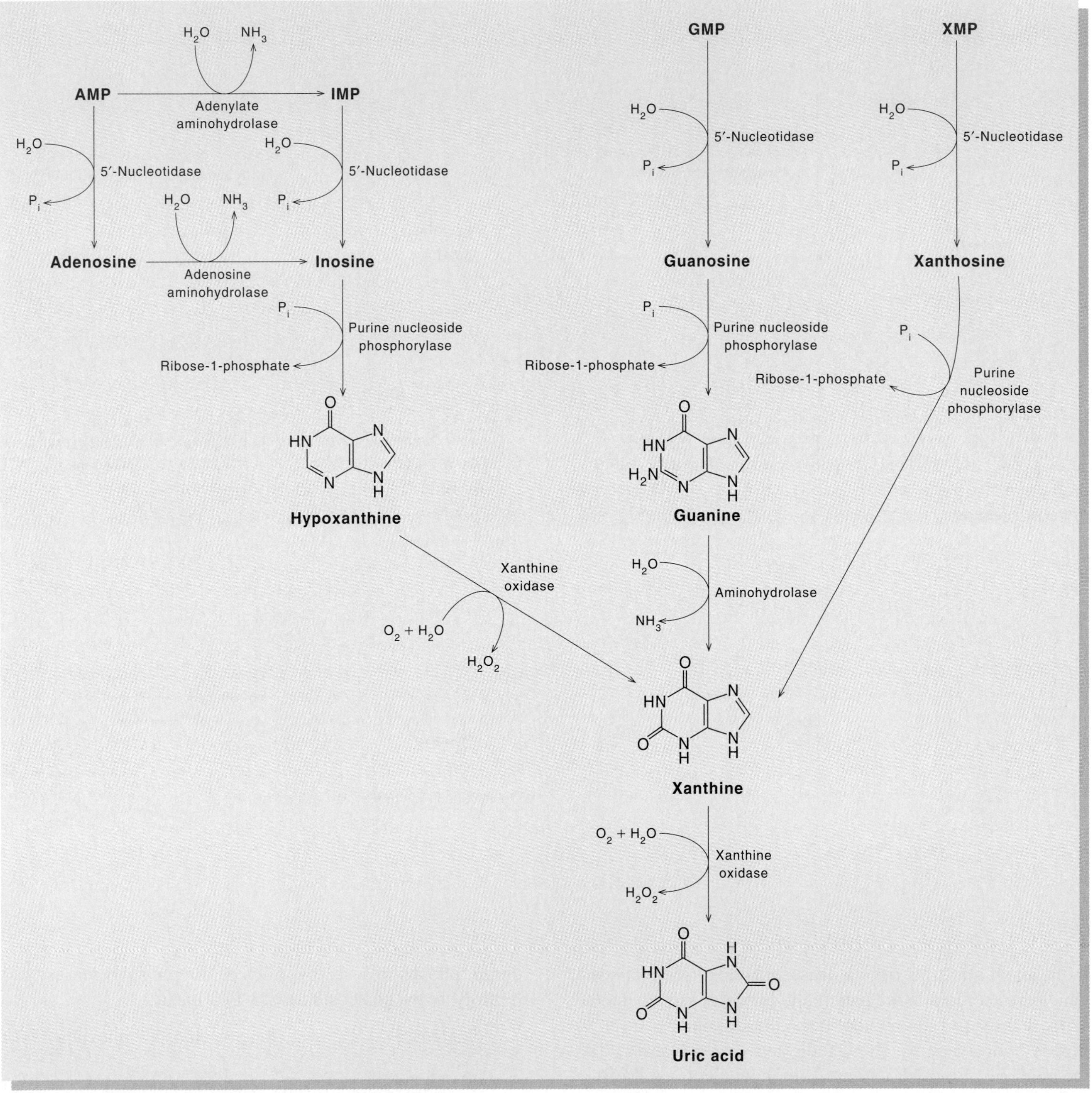

of purines, most of the bases are catabolized due to the presence in intestinal mucosa of a high level of xanthine oxidase, a catabolic enzyme.

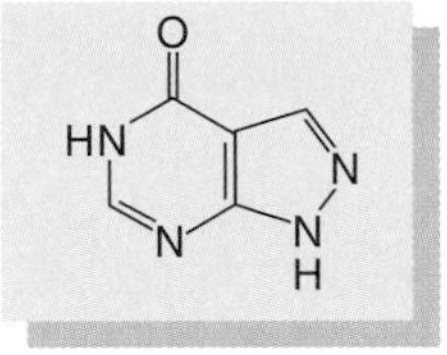

## Purines Are Catabolized to Uric Acid and Then to Other Products

After purine nucleotides have been converted to the corresponding nucleosides by 5'-nucleotidases and by phosphatases, inosine and guanosine are readily cleaved to the nucleobase and ribose-1-phosphate by the widely distributed purine nucleoside phosphorylase. The corresponding deoxynucleosides yield deoxyribose-1-phosphate and base with the phosphorylase from most sources. Adenosine and deoxyadenosine are not attacked by the phosphorylase of mammalian tissue, but much AMP is converted to IMP by an aminohydrolase (deaminase), which is very active in muscle and other tissues (fig. 23.20). An inherited deficiency of purine nucleoside phosphorylase is associated with a deficiency in the cellular type of immunity.

An adenosine aminohydrolase (deaminase) is also present in many mammalian tissues. This enzyme is of great interest because hereditary deficiency of the enzyme is linked to a severe (usually fatal) defect in the immune system, marked by a serious deficiency in lymphocytes and consequent inability to combat infections.

Inosine formed by either route is then phosphorolyzed to yield hypoxanthine. Although, as we have previously seen, much of the hypoxanthine and guanine produced in the mammalian body is converted to IMP and GMP by a phosphoribosyltransferase, about 10% is catabolized. Xanthine oxidase, an enzyme present in large amounts in liver and intestinal mucosa and in traces in other tissues, oxidizes hypoxanthine to xanthine, and xanthine to uric acid (see fig. 23.20). Xanthine oxidase contains FAD, molybdenum, iron, and acid-labile sulfur in the ratio $1:1:4:4$, and in addition to forming hydrogen peroxide, it is also a strong producer of the superoxide anion $\cdot O_2$, a very reactive species. The enzyme oxidizes a wide variety of purines, aldehydes, and pteridines.

Guanine aminohydrolase (guanine deaminase or guanase), present in liver, brain, and other mammalian tissues, provides another pathway to xanthine, this time from guanine. Subsequent oxidation of xanthine to uric acid then occurs.

Gout is a relatively common ($\approx 3$ per 1,000 persons) derangement of purine metabolism that is associated with elevated plasma levels of uric acid. The excessive uric acid leads to painful deposits of monosodium urate in the

**Figure 23.21**

The structure of allopurinol, an analog of hypoxanthine. Allopurinol inhibits xanthine oxidase and is used in the treatment of gout.

cartilage of joints, especially of the big toe. Uric acid deposits also may occur as calculi in the kidney, with resultant renal damage. The genetics are complex and incompletely understood. Individuals suffering from gout and other metabolic disorders producing elevation of serum uric acid may be treated with the xanthine oxidase inhibitor, allopurinol (fig. 23.21), an analog of hypoxanthine. Allopurinol is also a substrate of xanthine oxidase, but the product, oxypurinol, binds very tightly to the reduced form of xanthine oxidase and inactivates the enzyme. Allopurinol is nontoxic, and administration causes a marked decrease in the uric acid concentration in serum and in urinary excretion of uric acid. During allopurinol treatment, serum levels of hypoxanthine and xanthine are prevented from accumulating by operation of the salvage pathway and because the consequent accumulation of IMP, GMP, and AMP causes feedback inhibition of purine biosynthesis (see fig. 23.24).

Mammals other than primates further oxidize urate by a liver enzyme, urate oxidase. The product, allantoin, is excreted. Humans and other primates, as well as birds, lack urate oxidase and hence excrete uric acid as the final product of purine catabolism. In many animals other than mammals, allantoin is metabolized further to other products that are excreted: Allantoic acid (some teleost fish), urea (most fishes, amphibians, some mollusks), and ammonia (some marine invertebrates, crustaceans, etc.). This pathway of further purine breakdown is shown in figure 23.22.

## Pyrimidines Are Catabolized to β-Alanine, $NH_3$, and $CO_2$

Deaminases present in many cells are able to deaminate cytosine or its nucleosides or nucleotides to the corresponding uracil derivatives. Cytosine aminohydrolase (deaminase) appears to occur only in microorganisms (yeast and bacteria), but cytidine aminohydrolase is widely distributed in bacteria, plants, and mammalian tissues. A distinct deox-

**Figure 23.22**

Degradation of uric acid to excretory products. Mammals other than primates oxidize uric acid further to allantoin. Humans and other primates as well as birds lack urate oxidase and hence excrete uric acid as the final product of purine catabolism. In many animals other than mammals, allantoin is metabolized further to urea or ammonia and $CO_2$ as shown.

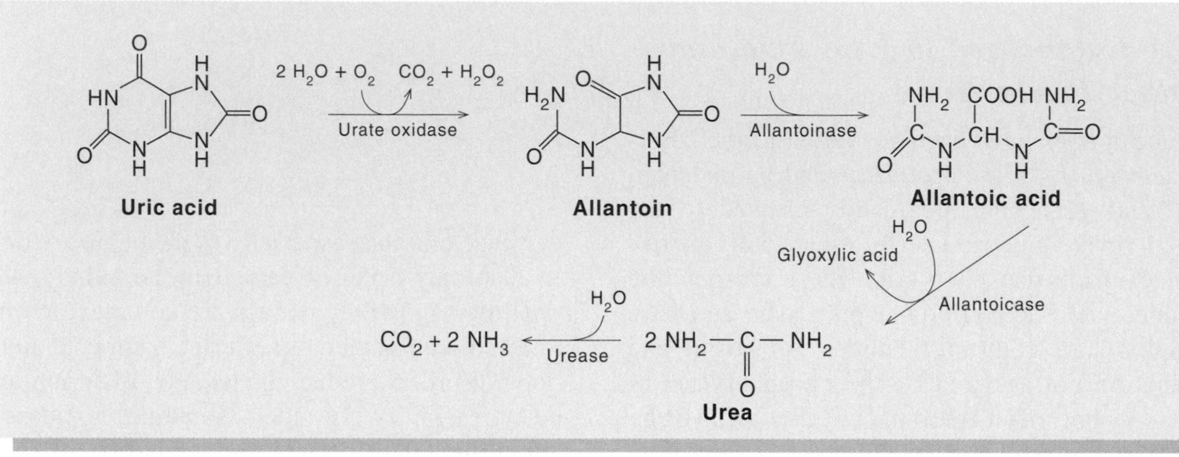

## Regulation of Nucleotide Metabolism

ycytidine aminohydrolase is present in various mammalian tissues and tumors, in plants, and in bacteria. A deoxycytidylate aminohydrolase that is similarly distributed produces dUMP, which is susceptible to attack by 5'-nucleotidase to give deoxyuridine. Although the physiological function of these aminohydrolases is not completely understood, the uridine and deoxyuridine formed can be further degraded by uridine phosphorylase to uracil as previously discussed, so that these reactions provide a pathway for converting nucleotides of uracil and cytosine to uracil and ribose-1-phosphate or deoxyribose-1-phosphate (fig. 23.23). Similarly, thymine nucleosides and nucleotides can be converted by 5'-nucleotidase and phosphorylase to thymine.

Enzymes present in mammalian liver are capable of the catabolism of both uracil and thymine. The first reduces uracil and thymine to the corresponding 5,6-dihydro derivatives. This hepatic enzyme uses NADPH as the reductant, whereas a similar bacterial enzyme is specific for NADH. Similar enzymes are apparently present in yeast and plants. Hydropyrimidine hydrase then opens the reduced pyrimidine ring, and finally the carbamoyl group is hydrolyzed off from the product to yield $\beta$-alanine or $\beta$-aminoisobutyric acid, respectively, from uracil and thymine (see fig. 23.23).

## Regulation of Nucleotide Metabolism

Although much remains to be discovered about the details of the regulation of nucleotide metabolism, a number of important control points are rather well understood. We discuss some of these here, together with their known effects on intracellular nucleotide pools.

## Purine Biosynthesis Is Regulated at Two Levels

Many lines of evidence indicate that the first committed step in *de novo* purine nucleotide biosynthesis, production of 5-phosphoribosylamine by glutamine PRPP amidotransferase, is rate-limiting for the entire sequence. Consequently, regulation of this enzyme is probably the most important factor in control of purine synthesis *de novo* (fig. 23.24). The enzyme is inhibited by purine-5'-nucleotides, but the most inhibitory nucleotides vary with the source of the enzyme. Inhibition constants ($K_I$) are usually in the range $10^{-3}$–$10^{-5}$ M. The maximum effect of this end-product inhibition is produced by certain combinations of nucleotides (e.g., AMP and GMP) in optimum concentrations and ratios, indicating two kinds of inhibitor binding sites. This is an example of a concerted feedback inhibition.

The rate of the amidotransferase reaction is also governed by intracellular concentrations of the substrates L-glutamine and PRPP. Competing metabolic reactions or drugs that alter the supply of these substrates also affect the rate of IMP synthesis.

The second important level of regulation of purine nucleotide synthesis is in the branch pathways from IMP to AMP and to GMP (see figs. 23.11 and 23.24). The first of the two reactions leading from IMP to AMP is the irreversible synthesis of adenylosuccinate. This requires GTP as a source of energy and is inhibited by AMP. Of the two reactions required to convert IMP to GMP, the first is irreversible and is inhibited by GMP, and the second requires ATP as a source of energy. Thus, two types of regulation occur at this level of purine nucleotide synthesis: (1) a "forward"

## Figure 23.23

Degradation of pyrimidine bases. Parts of this pathway are widely distributed in nature. The entire pathway is found in mammalian liver. As in purine nucleotide catabolism, no ATP results from catabolism, and the ribose-1-phosphate is released during catabolism before destruction of the base.

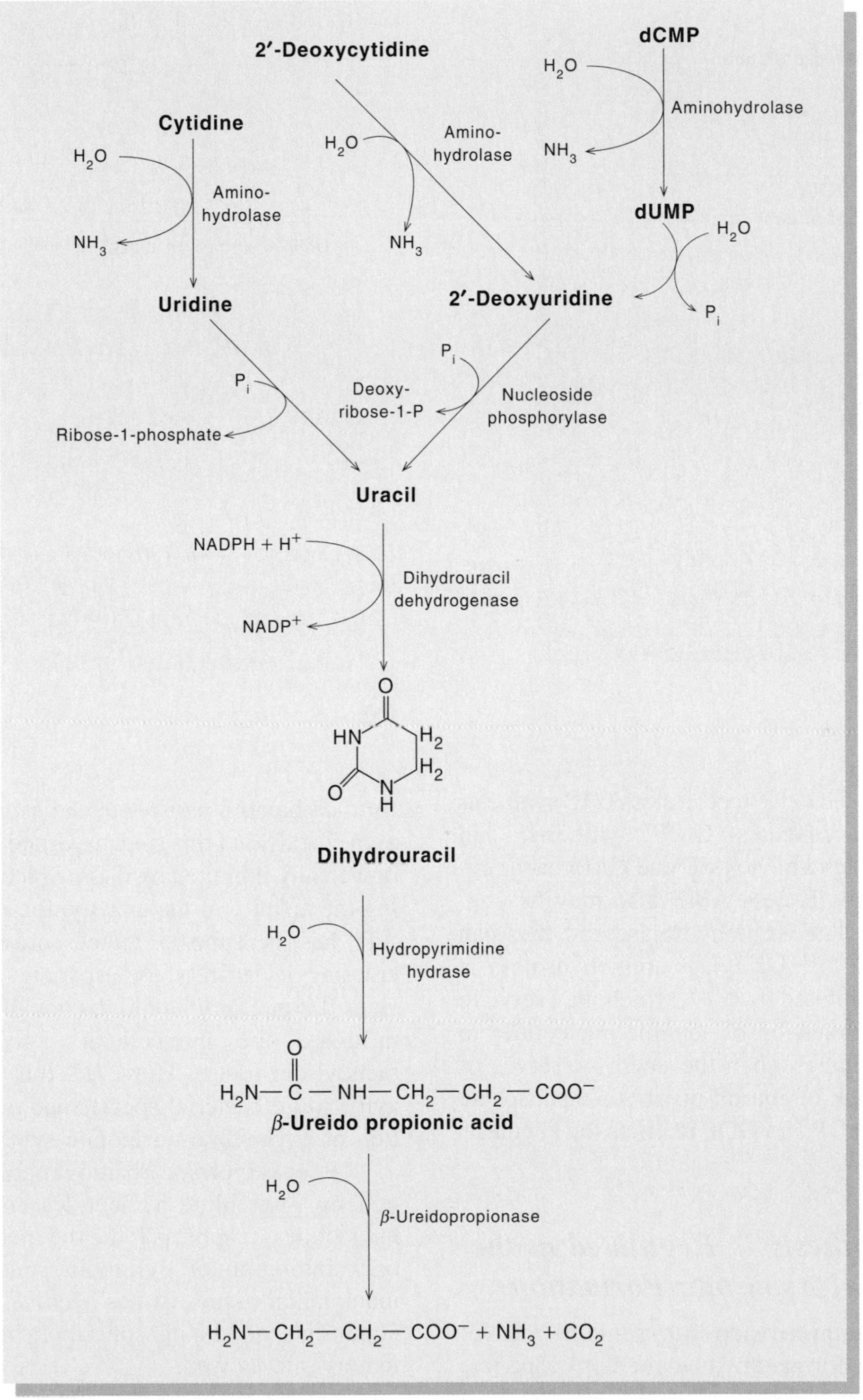

## Figure 23.24

Regulation of purine biosynthesis. Red arrows show points of inhibition ⊖ or activation ⊕. In addition to the feedback inhibition, GTP stimulates ATP synthesis, and ATP stimulates GTP synthesis, thus helping to ensure a balance between the pools of the two nucleoside triphosphates. The full biosynthetic pathways are shown in figures 23.10 and 23.11.

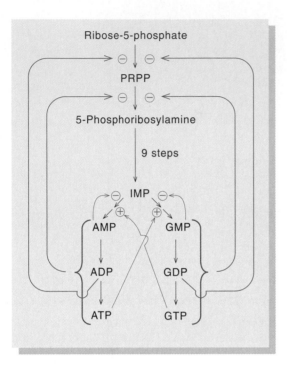

## Figure 23.25

Proposed scheme for the regulation of deoxyribonucleotide synthesis in *E. coli* and mammalian cells. Red arrows indicate points of activation ⊕ and inhibition ⊖. (Source: Adapted from L. Thelander and P. Reichard, Reduction of ribonucleotides, *Ann. Rev. Biochem.* 48:133, 1979.)

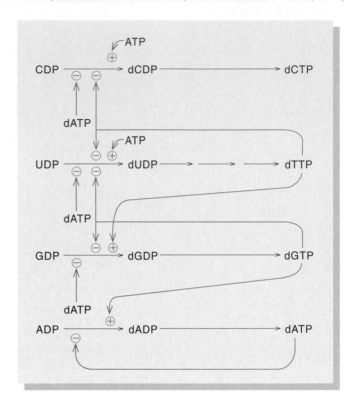

control, by which increased GTP accelerates AMP synthesis and increased ATP accelerates GMP synthesis, and (2) feedback inhibition, by which AMP and GMP each regulate their own synthesis. Excess AMP also may be converted to IMP by adenylate aminohydrolase and thus can serve as a source of GMP. Adenylate aminohydrolase is activated by ATP and inhibited by GTP, which may serve to control this potential conversion of adenine nucleotides to guanine nucleotides. Finally, when the energy reserves of the cell are low, feedback inhibition of ribose-5-phosphate pyrophosphokinase by ADP and GDP restricts the synthesis of PRPP (see fig. 23.24).

## *Pyrimidine Biosynthesis Is Regulated at the Level of Carbamoyl Aspartate Formation*

In bacteria, the first committed step in pyrimidine nucleotide biosynthesis is the formation of carbamoyl aspartate

from carbamoyl phosphate and aspartate. In *E. coli,* the enzyme catalyzing this step, aspartate carbamoyltransferase, is powerfully inhibited by CTP, which acts chiefly by decreasing the affinity of the enzyme for aspartate (see chapter 9). ATP has the opposite effect, activating the enzyme by increasing its affinity for aspartate. Concentrations of ATP and CTP in *E. coli* are high enough for these nucleotides to influence the intracellular activity of aspartate carbamoyltransferase. However, this is not a regulatory enzyme in all bacterial species and is not involved in regulation of pyrimidine nucleotide synthesis in animal cells.

In eukaryotes, carbamoyl phosphate synthase is inhibited by pyrimidine nucleotides and stimulated by purine nucleotides; it appears to be the most important site of feedback inhibition of pyrimidine nucleotide biosynthesis in mammalian tissues. It has been suggested that under some conditions, orotate phosphoribosyltransferase may be a regulatory site as well.

**Figure 23.26**

Phases in the life cycle of a typical eukaryotic cell. The cell cycle is divided into the resting stage ($G_0$), the prereplication stage ($G_1$), the synthesis or replication stage (S), the postreplication stage ($G_2$), and the mitotic stage (M).

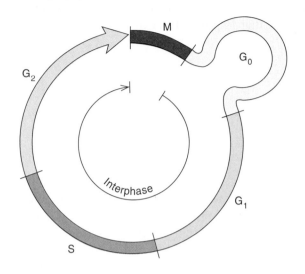

## Deoxyribonucleotide Synthesis Is Regulated by Both Activators and Inhibitors

The manner in which the reduction of ribonucleotides to deoxyribonucleotides is regulated has been studied with reductases from relatively few species. The enzymes from *E. coli* and from Novikoff's rat liver tumor have a complex pattern of inhibition and activation (fig. 23.25). ATP activates the reduction of both CDP and UDP. As dTTP is formed by metabolism of both dCDP and dUDP, it activates GDP reduction, and as dGTP accumulates, it activates ADP reduction. Finally, accumulation of dATP causes inhibition of the reduction of all substrates. This regulation is reinforced by dGTP inhibition of the reduction of GDP, UDP, and CDP and by dTTP inhibition of the reduction of the pyrimidine substrates. Because evidence suggests that ribonucleotide reductase may be the rate-limiting step in deoxyribonucleotide synthesis in at least some animal cells, these allosteric effects may be important in controlling deoxyribonucleotide synthesis.

A second enzyme on the pathway to dTTP that is subject to allosteric control is deoxycytidylate deaminase, which supplies dUMP for thymidylate synthesis. The enzyme in mammalian cells, yeast, and bacteriophage T2-infected *E. coli.* is allosterically activated by dCTP (hydroxymethyl dCTP for the phage enzyme) and inhibited by dTTP.

## Enzyme Synthesis Also Contributes to Regulation of Deoxyribonucleotides during the Cell Cycle

Many of the enzymes participating in *de novo* synthesis of deoxyribonucleotide triphosphates, as well as those responsible for interconversion of deoxyribonucleotides, increase in activity when cells prepare for DNA synthesis. The need for increased DNA synthesis occurs under three circumstances: (1) when the cell proceeds from the $G_0$, or resting, stage of the cell cycle to the S, or synthetic or replication, stage (fig. 23.26); (2) when it performs repair after extensive DNA damage; and (3) after infection of quiescent cells with virus. When cells leave $G_0$, for example, enzymes such as thymidylate synthase and ribonucleotide reductase, increase as well as the corresponding mRNAs. These increases in enzyme amount supplement allosteric controls that increase the activity of each enzyme molecule. Corresponding decreases in amounts of these enzymes and their mRNAs occur when DNA synthesis is completed.

## Intracellular Concentrations of Nucleotides Vary According to the Physiological State of the Cell

As the cell progresses through its life cycle (fig. 23.26) marked changes occur in the concentration of some nucleotides, particularly an increase of the deoxyribonucleoside triphosphates as the cell enters S phase. Conditions that damage cells, such as starvation, lack of oxygen, or presence of toxic materials, also affect nucleotide levels. Consequently, there are no "correct" values of nucleotide levels, even for a specific type of cell. Some general statements are possible, however. ATP is generally present in normal cells at a concentration of 2–10 mM, but other ribonucleoside triphosphates are at lower concentrations (0.05–2 mM), depending on conditions and the cell type. The ribonucleoside mono- and diphosphates have lower concentrations than the corresponding triphosphates, but ADP is typically comparable with or even higher than GTP, CTP, or UTP. Deoxyribonucleoside triphosphates (dNTPs) are present in much lower concentrations, 2–60 $\mu$M. The concentration of the dNTP present at lowest concentration, dCTP in some cells and dGTP in others, is sufficient for only a few minutes of DNA synthesis, so that during DNA replication, synthesis of dNTPs must keep pace with utilization for DNA synthesis.

In prokaryotes, the levels of deoxyribonucleotides vary from undetectable to 200 picomoles per $10^6$ cells (compared with 3–30 picomoles of dATP, dCTP, and dGTP per $10^6$ eukaryotic cells). As in eukaryotic cells, dTTP is usually present at higher concentrations than the other deoxyribonucleoside triphosphates.

The arrest of cell growth by many agents is associated with depletion of one or more of the deoxyribonucleotide pools. Thus, thymidine at millimolar concentrations arrests cell growth and decreases the concentration of dCTP dramatically, whereas the concentrations of dATP, dGTP, and especially dTTP increase. This effect is considered to be mediated by the allosteric inhibition of ribonucleotide reductase, referred to earlier. Hydroxyurea, another agent that arrests cell growth by blocking DNA synthesis, depletes the pools of dATP and dGTP, and in some cells it is the effect on the latter, also brought about by ribonucleotide reductase inhibition, that is probably critical.

## Summary

Nucleotides are the building blocks for nucleic acids; they are also involved in a wide variety of metabolic processes. They serve as the carriers of high-energy phosphate and as the precursors of several coenzymes and regulatory small molecules. Nucleotides can be synthesized *de novo* from small-molecule precursors or, through salvage pathways, from the partial breakdown products of nucleic acids. The highlights of our discussion in this chapter are as follows.

1. The ribose for nucleotide synthesis comes from glucose, either by means of the pentose phosphate pathway or from glycolytic intermediates through transketolase-transaldolase reactions. Ribose-5-phosphate is converted to phosphoribosylpyrophosphate (PRPP), the starting point for purine synthesis. This pathway also incorporates into purines atoms from glycine, aspartate, glutamate, $CO_2$, and one-carbon fragments carried by folates. IMP synthesized by this route is converted by two-step pathways to AMP and GMP, respectively.

2. The biosynthetic pathway to UMP starts from carbamoyl phosphate and results in the synthesis of the pyrimidine orotate, to which ribose phosphate is subsequently attached. CTP is subsequently formed from UTP. Deoxyribonucleotides are formed by reduction of ribonucleotides (diphosphates in most cells). Thymidylate is formed from dUMP.

3. Several inhibitors of nucleotide biosynthesis are known. Each is extremely toxic, especially to rapidly growing cells, where the need for nucleic acid synthesis is greatest. In limited amounts, some of the inhibitors have chemotherapeutic value in the treatment of cancer and other illnesses. Some analogs of normal nucleosides are proving to be useful in the treatment of AIDS and certain other viral infections.

4. Nucleic acids and nucleotides are degraded to nucleosides or free bases before they are ingested. Nucleotides or their partial degradation products may be reutilized for nucleic acid synthesis, or they may be further catabolized for excretion or for use in the synthesis of other products. Purine nucleotides are degraded via guanine, hypoxanthine, and xanthine to uric acid, which in some species is degraded further before excretion. Inherited deficiencies in some of the enzymes involved in nucleotide degradation and salvage cause severe impairment of health, a fact testifying to the importance of the degradative pathways. Nucleotides of uracil and cytosine are degraded via uridine and uracil to simpler substances such as $\beta$-alanine.

5. All biosynthetic pathways are under regulatory control by key allosteric enzymes that are influenced by the end products of the pathways. For example, the first step in the pathway for purine biosynthesis is inhibited in a concerted fashion by nucleotides of either adenine or guanine. In addition, the nucleoside monophosphate of each of these bases inhibits its own formation from inosine monophosphate (IMP). On the other hand, adenine nucleotides stimulate the conversion of IMP into GMP, and GTP is needed for AMP formation.

## Selected Readings

Andersson, M., L. Lewan, and U. Stenram. Compartmentation of purine and pyrimidine nucleotides in animal cells. *Int. J. Biochem.* 20:1039–1050, 1988. A critical review of the evidence for compartmentation, and the significance of the phenomenon.

Jones, M. E., Pyrimidine nucleotide biosynthesis in animals: Genes, enzymes and regulation of UMP biosynthesis. *Ann. Rev. Biochem.* 49:253–279, 1980. Authoritative outline of the regulatory properties of the two multifunctional proteins responsible for pyrimidine nucleotide synthesis in animals.

Kornberg, A., *DNA Replication,* 2d ed. San Francisco: Freeman, 1989. See especially chapter 1.

Manfredi, J. P., and E. W. Holmes, Purine salvage pathways in myocardium. *Ann. Rev. Physiol.* 47:691–705, 1985. Although this review applies specifically to salvage in heart muscle, it is an excellent summary of purine salvage pathways and the metabolism of the purine nucleotides formed.

Mathews, C. K., L. K. Moen, and R. G. Sargent, Enzyme interactions in deoxyribonucleotide synthesis. *Trends Biochem. Sci.* 13:394–397, 1988.

Plageman, P. G. W., R. M. Woehlheuter, and C. Woffendin, Nucleoside and nucleobase transport in animal cells. *Biochem. Biophys. Acta* 947:405–443, 1988. Reviews the field in detail.

Reichard P., Interactions between deoxyribonucleotide and DNA synthesis. *Ann. Rev. Biochem.* 57:349–374, 1988. Comprehensive review of deoxyribonucleotide synthesis and its relation to DNA synthesis.

Weber, G., (ed.), Enzyme pattern-targeted chemotherapy. *Adv. Enzyme Regul.* 24, 1985. This volume contains several chapters about inhibitors of nucleotide metabolism and their mechanism of actions.

## Problems

1. How do you explain the observation that pyrimidine biosynthesis in bacteria is regulated at the level of aspartate carbamoyltransferase, whereas most of the regulation in humans is at the level of carbamoyl phosphate synthase?

2. Compare and contrast the pathways for the biosynthesis of IMP (fig. 23.10) with that of UMP (fig. 23.13). What are the two most striking features of this comparison?

3. The first enzyme in the biosynthesis of IMP (fig. 23.10) utilizes a glutamine to provide an amino group. We have seen the same phenomenon in the two previous chapters. Why doesn't nature use an ammonium ion rather than the more expensive (in the expenditures of ATP) glutamine as an amine donor?

4. Examine the reactants and products of the second enzyme on the IMP biosynthetic pathway. What reaction intermediate would you expect?

5. What function can you propose for the ATP in the fourth reaction in figure 23.10?

6. What function can you give ATP (in a mechanistic sense) in the fifth reaction (fig. 23.10) in the formation of IMP?

7. Examine figure 23.15 and deduce the fate of the oxygen on carbon two of the ribose residue.

8. What happens if you give allopurinol to a chicken?

9. Explain why Lesch-Nyhan patients suffer from severe gout. Although these patients can be treated with allopurinol to relieve the symptoms of gout, this treatment has no effect on the severe mental retardation. Suggest a possible explanation.

10. Why can the toxic effect of sulfanilamide on bacteria be reversed by *p*-aminobenzoate? Why are sulfa drugs not very toxic to humans?

11. Substitute thymine for uracil in figure 23.23 and determine the metabolites that would follow in the pathway?

12. Explain how antifolates like methotrexate selectively kill cancer cells. Why do cancer patients lose their hair, intestinal mucosa, cells of the immune system, and so forth when treated with antifolates?

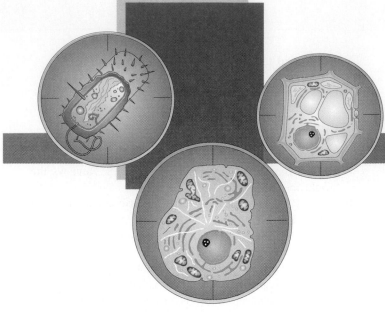

# Integration of Metabolism and Hormone Action

*The metabolic processes of vertebrates in different tissues are regulated by hormones.*

In chapter 11 we described strategies used to organize biochemical conversions so that all essential conversions are thermodynamically feasible and kinetically regulated. In the twelve chapters that followed, we witnessed countless examples employing these basic strategies. This is an appro-

**Table 24·1**

Fuel reserves in a normal 70-kg human (kcal)

| Organ | Glucose or Glycogen | Triacylglycerols | Mobilizable Proteins |
|---|---|---|---|
| Liver | 400 | 450 | 400 |
| Brain | 8 | 0 | 0 |
| Muscle | 1,200 | 450 | 24,000 |
| Adipose tissue | 80 | 135,000 | 40 |

(Source: Data from G. T. Cahill, *Clin. Endocrinol. Metab.* 5:598, 1976.)

priate time to see how various organizational strategies are integrated in the metabolism of a multicellular organism. The integration process is complicated by the fact that specific tissues assume unique roles in the overall metabolism of the organism. The best understood multicellular organisms are vertebrates, and the best understood aspect of their metabolism is that dealing with energy production and consumption. Consequently, we examine the integration of energy metabolism for several vertebrate tissues. First we consider how energy metabolism is manipulated between these tissues. Then we explore the underlying mechanisms that permit a smooth flow between energy-supplying and energy-consuming tissues. Following this we expand the discussion to the topic of hormonal regulation of metabolism.

## Tissues Store Biochemical Energy in Three Major Forms

In addition to relying on nutrition to supply the chemical fuel for their energy needs, vertebrates maintain fuel reserves in various tissues. These reserves are of three types: Glycogen, triacylglycerols, and proteins. Greater than 90% of the fuel reserves in an average human exist in the form of triacylglycerols stored in the adipose tissue (table 24.1). In obese individuals this fuel reserve can be severalfold higher. Despite their abundance, triacylglycerols in large measure are held in check until the more readily utilizable glycogen reserves located in liver and muscle tissues are close to exhaustion. Generally, glycogen reserves of the liver play the most widely useful role despite the greater abundance of glycogen often found in muscle tissue. This is because, of the two tissues, only the liver is capable of degrading glycogen to (neutral) glucose that can be secreted into the bloodstream. Once the glycogen reserves have been depleted from the liver, the energy-hungry organism turns to the lipid (triacylglycerol) reserves stored in adipose tissue. The first step in mobilizing lipid for energy consumption involves the

hydrolysis of triacylglycerols to fatty acids and glycerol. The fatty acids so produced are transported to other tissues, where they are degraded to the activated two-carbon unit, acetyl-CoA, in a repetitious multistep process (see chapter 18). A third major fuel reserve, protein, mostly from muscle tissues, can be mobilized by breakdown to amino acids that are transported to the liver. In the liver the amino acids are deaminated and converted into TCA cycle intermediates. The use of muscle tissue proteins to satisfy energy needs is bound to physically weaken the organism, so it comes as no surprise that muscle tissue is used for energy purposes only as a last resort. The best known exception to this rule occurs during periods of prolonged muscle inactivity, as in the case of a temporarily incapacitated limb.

There must be a way in which the energy-requiring tissues can send signals to the energy-producing tissues which reflect their needs. The signals are supplied by circulating hormones, which regulate the metabolic activities of different tissues in the organism (fig. 24.1).

## Each Tissue Makes Characteristic Demands and Contributions to the Energy Pool

Some tissues are mainly energy suppliers, and others are mainly energy consumers. Still other tissues are important both as consumers and as suppliers. In this section we survey energy metabolism in five well-characterized vertebrate tissues: Liver, adipocytes, striated muscle, smooth heart muscle, and brain.

### Brain Tissue Makes No Contributions to the Fuel Needs of the Organism

Brain tissue does not contribute to the energy needs of other tissues. Rather, it makes major demands on the glucose supply, accounting for upwards of 50% of the glucose con-

**Figure 24.1**

Some aspects of the regulation of energy metabolism of cells in a typical mammal. Each cell type serves a particular function and has specific requirements for maintenance and growth. Some cells are primarily energy producers, and others are primarily energy consumers. The activities of different cell types are regulated by an intricate hierarchy of hormone-secreting cells. To be susceptible to a specific hormone, a cell must possess a specific hormone receptor. On contact with the receptor the hormone causes a structural change in the hormone receptor, which in turn triggers a chain of reactions in the cell.

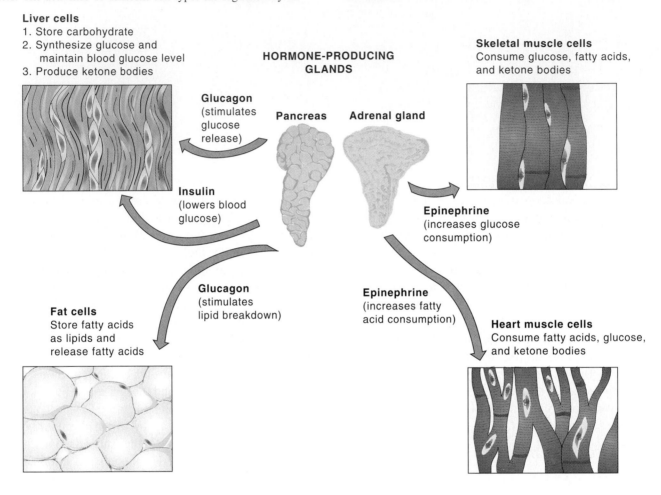

**Liver cells**
1. Store carbohydrate
2. Synthesize glucose and maintain blood glucose level
3. Produce ketone bodies

**HORMONE-PRODUCING GLANDS**

**Skeletal muscle cells**
Consume glucose, fatty acids, and ketone bodies

**Glucagon**
(stimulates glucose release)

**Pancreas**

**Adrenal gland**

**Insulin**
(lowers blood glucose)

**Epinephrine**
(increases glucose consumption)

**Glucagon**
(stimulates lipid breakdown)

**Epinephrine**
(increases fatty acid consumption)

**Fat cells**
Store fatty acids as lipids and release fatty acids

**Heart muscle cells**
Consume fatty acids, glucose, and ketone bodies

sumption in the resting human. To satisfy its needs, the brain normally absorbs glucose from the bloodstream. It metabolizes the glucose to $CO_2$ and $H_2O$ by a combination of glycolysis and the TCA cycle. Energy consumption by the brain is approximately constant: A sleeping brain and a wide-awake, mentally alert brain burn about the same amount of glucose. Because of its vital importance and delicate constitution, brain tissue is somehow given priority in meeting demands on the available supplies of blood glucose in times of starvation or other forms of stress, when the glucose supply is limited. Under conditions of prolonged starvation the blood glucose level drops somewhat. As a consequence, the energy needs of brain tissue are met by a combination of glucose and ketone bodies, which are metabolized via the TCA cycle.

## Heart Muscle Utilizes Fatty Acids in Preference to Glucose to Fulfill Its Energy Needs

Continuous operation is an obvious necessity for the heart muscle. Its energy needs vary considerably, depending on the physical activity of the organism. For reasons that are unclear, the heart relies mainly on fatty acids supplied from the bloodstream to fulfill its energy needs. Perhaps this is because the fatty acid supply is more reliable than the fluctuating carbohydrate supply. The fatty acids are metabolized by $\beta$-oxidation in the densely packed mitochondria of the heart muscle cells. Most organisms have a very extensive supply of fatty acids; thus the functioning of the heart muscle is ensured. Under conditions of vigorous activity the

**Figure 24.2**

The cycling of lactate and glucose between muscle and liver in the Cori cycle. Under conditions of intense activity the muscle operates anaerobically so that the end product of glucose breakdown is lactate. This product can pass through the bloodstream to the liver and be converted to glucose, which can be returned to the muscle.

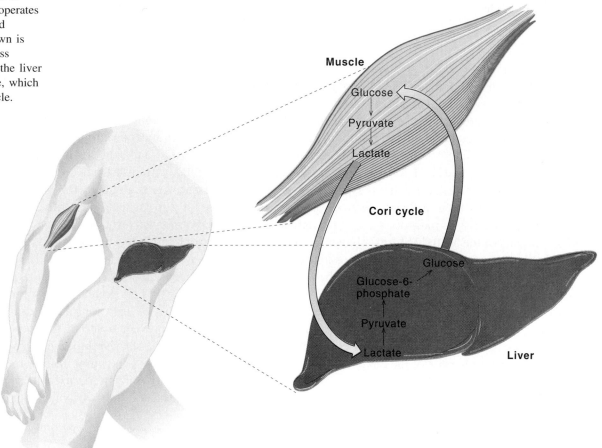

heart muscle also uses glucose as a source of energy. Under starvation conditions the heart muscle can switch to using ketone bodies supplied from the liver via the bloodstream. The ketone bodies are degraded in the mitochondria of the heart muscle by way of the TCA cycle. Like the brain, heart muscle makes no energy contributions to other tissues of the organism.

## Skeletal Muscle Can Function Aerobically or Anaerobically

Whereas the heart muscle operates under strictly aerobic conditions, skeletal muscle can function either aerobically or anaerobically. The energy consumption of skeletal muscle varies enormously with the extent of muscle activity. During periods of mild exertion, muscle tissue requires moderate levels of glucose, which can be supplied by the blood glucose or the breakdown of the glycogen reserves present in the muscle tissue. This glucose is metabolized aerobically all the way to $CO_2$ and $H_2O$. During times of great exertion, muscle tissue uses oxygen faster than it can be supplied by the bloodstream, so that the muscle must operate anaerobically. Under these conditions, glucose breakdown stops at the 3-carbon acid, lactate. The lactate so produced is secreted into the bloodstream and picked up by the liver, which converts the lactate back to glucose. This glucose can be returned to the muscle for further glycolysis. The cycling of glucose and lactate between skeletal muscle and liver is known as the Cori cycle (fig. 24.2). When the blood glucose level is low, muscle tissue can utilize fatty acids as an alternative supply of energy. The fatty acids are mobilized via the β-oxidation pathway in the muscle mitochondria. Muscle lactate can also be harnessed by heart muscle, which catabolizes it for energy purposes (box 24A).

BOX

# 24A

# The Lactate Dehydrogenase of Heart Muscle

Most vertebrates possess at least two genes for lactate dehydrogenase (LDH) that make similar but nonidentical polypeptides called M and H. In embryonic tissue, both genes are equally active, resulting in equimolar amounts of the two gene products and a statistical array of tetramers ($M_4$, $M_3H$, $M_2H_2$, $H_3M_1$, and $H_4$ in the ratios of $1:4:6:4:1$). These so-called isoenzymes, or isozymes, can usually be detected by their differing electrophoretic mobilities. As embryonic tissue multiplies and differentiates, the relative amounts of the M and H forms change. In pure heart tissue, the $H_4$ tetramer predominates. In skeletal muscle, which functions anaerobically under stress, the $M_4$ isozyme predominates. It seems likely that the M and H forms were designed to serve different functions. A clue to these functions may be revealed by the inhibiting effect of pyruvate on the dehydrogenase. Pyruvate can form a covalent complex with $NAD^+$ at the active site of the enzyme according to the following reaction:

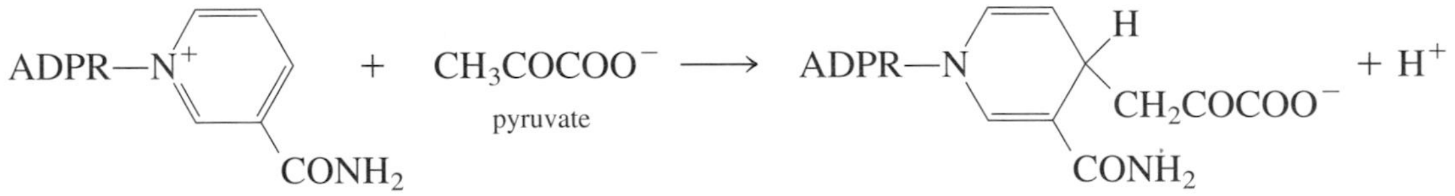

The $H_4$ tetramer shows a much greater inhibition by this compound than the $M_4$ tetramer. Active muscle tissue is anaerobic and produces a good deal of pyruvate. Inhibition of LDH under anaerobic conditions would shrink the supply of $NAD^+$ and shut down the glycolytic pathway (see chapter 12) with disastrous consequences. In fact, active muscle tissue has augmented levels of pyruvate but converts this readily to lactate. This conversion is possible because the predominant form of LDH in muscle is $M_4$, which is only poorly inhibited by excess pyruvate. The function of LDH in aerobic tissue is less clear. Heart muscle is aerobic tissue, and consequently, most of its pyruvate is funneled into the Krebs cycle for greater energy production (see chapter 13). The LDH of heart muscle, $H_4$, might be inhibited by pyruvate to prevent waste of this potential high-energy carbon source or excessive buildup of pyruvate resulting from the conversion of incoming lactate to pyruvate. It seems likely that the $H_4$ enzyme is used in such tissues to convert absorbed lactate into pyruvate.

$$ADPR-N^+ \text{(pyridine-CONH}_2\text{)} + CH_3COCOO^- \text{ (pyruvate)} \longrightarrow ADPR-N \text{(dihydropyridine, H, CH}_2COCOO^-, CONH_2\text{)} + H^+$$

In times of starvation, muscle tissue is usually quite inactive, so its energy demands are low. In fact, under such conditions the muscles tend to supply energy to the rest of the organism via degradation of muscle protein to amino acids. These amino acids are transported to the liver where they are converted to glucose or used for biosynthesis of other proteins essential for maintenance of the organism's vital processes.

## Adipose Tissue Maintains Vast Fuel Reserves in the Form of Triacylglycerols

Under conditions of nutritional excess, fatty acids are absorbed by adipose tissue where they are converted to storage lipids in the form of triacylglycerols. The triacylglycerols can be mobilized at a later time, when the carbohydrate

energy reserves are low. The process of mobilization starts with conversion of triacylglycerols to fatty acids, which are transported by the bloodstream to the liver for further processing (fig. 24.3).

## The Liver Is the Central Clearing House for All Energy-Related Metabolism

Glucose is the most readily utilizable energy source of the organism, and it is a primary function of the liver to maintain blood glucose at a reasonable level for absorption into most tissues. Thus, in times of glucose excess the liver absorbs glucose and converts it to glycogen, which it stores for future energy needs. At other times, when the glucose concentration in the bloodstream drops, the liver converts glycogen into glucose, which it secretes into the bloodstream.

**Figure 24.3**

Synthesis and degradation of triacylglycerols in adipose tissue. Fatty acids are delivered to adipose tissue. In times of energy excess these are converted to triacylglycerols and stored until needed, at which point the triacylglycerols are converted back to fatty acids.

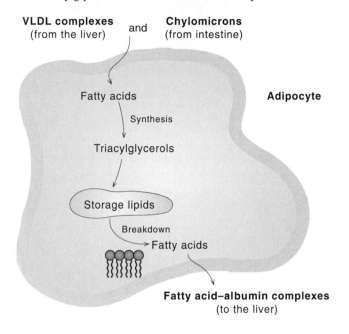

**Figure 24.4**

The plasma levels of fatty acids and ketone bodies increase during starvation, whereas the glucose levels decrease. Due to the dramatic rise in ketone bodies there are more ATP equivalents in the plasma of the starving animal than in the normal animal.

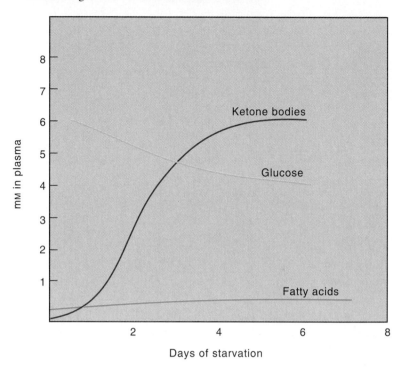

When the blood glucose level falls and the liver's glycogen reserves are also exhausted, the liver still has the capacity to synthesize glucose via gluconeogenesis from amino acids that are supplied from protein breakdown. Under starvation conditions the liver forms increasing amounts of ketone bodies (see fig. 18.7). This is due to elevated concentrations of acetyl-CoA, which favor the formation of ketone bodies. The ketone bodies are secreted and used as a source of energy by other tissues, especially those tissues like the brain that cannot catabolize fatty acids directly.

The major events in fuel storage and energy utilization are summarized in table 24.2. In this table, energy metabolism is considered under conditions of glucose excess, such as just after a meal; under conditions when the glucose supply is scant, as after prolonged periods between meals; and under starvation conditions, when the glucose supply is severely limited.

Most often the metabolic state of an organism is reflected by the small molecules present in plasma. Figure 24.4 illustrates the plasma levels of glucose, ketone bodies, and fatty acids as a function of the number of days of starvation. As you can see, the glucose level drops about 30% as the period of starvation becomes prolonged; the fatty acid level rises about twofold, while the level of ketone bodies rises severalfold.

## Pancreatic Hormones Play a Major Role in Maintaining Blood Glucose Levels

Thus far we have argued that it is the liver's job to maintain normal blood glucose levels. But the liver cannot by itself sense the blood glucose levels. It relies on hormonal signals transmitted by the pancreas (fig. 24.5). The pancreas sends two quite different signals. When the blood glucose level is high, the pancreas secretes insulin, which binds to specific receptors located on the outer plasma membranes of insulin-responsive cells (target cells). The effect of insulin on liver cells is to stimulate uptake of excess glucose, which is converted to glycogen. When the blood glucose level is low, the pancreas secretes the protein hormone, glucagon, which binds to different receptors on liver cells. This hormone stimulates the pathway leading to glycogen breakdown and subsequent secretion of glucose by the liver.

Glucagon and insulin bind to specific receptors on the outer plasma membrane of a target cell. In the case of glucagon, this binding indirectly stimulates the enzyme adenylate cyclase, on the inner surface of the membrane, to catalyze the production of cyclic AMP. Depending on the cell type,

## Table 24.2

Main Energy-Related Reactions in Five Vertebrate Tissues*

| Tissue | Just after Eating | When Hungry | Starvation Conditions |
|---|---|---|---|
| Brain | Glucose → $CO_2 + H_2O$ | Glucose → $CO_2 + H_2O$ | Ketone bodies → $CO_2 + H_2O$ ；Glucose → $CO_2 + H_2O$ |
| Heart muscle | Fatty acids → $CO_2 + H_2O$ ；Glucose → $CO_2 + H_2O$ | Fatty acids → $CO_2 + H_2O$ ；Glucose → $CO_2 + H_2O$ | Ketone bodies → $CO_2 + H_2O$ |
| Skeletal muscle | Glucose → $CO_2 + H_2O$ ；Glucose → Glycogen ；Glucose → Lactate | Fatty acids ；Glycogen → $CO_2 + H_2O$ ；→ $CO_2 + H_2O$ ；→ Lactate | Fatty acids → $CO_2 + H_2O$ ；Ketone bodies → $CO_2 + H_2O$ ；Proteins → Amino acids |
| Adipose tissue | Fatty acids → Triacylglycerols | Triacylglycerols → Fatty acids | Triacylglycerols → Fatty acids |
| Liver | Glucose → Glycogen ；Fatty acids → Triacylglycerols | Fatty acids → $CO_2 + H_2O$ ；Glycogen → Glucose | Amino acids → $CO_2 + H_2O$ → Glucose ；Fatty acids → $CO_2 + H_2O$ → Ketone bodies |

* Pink indicates imports; blue indicates exports

**Figure 24.5**

The opposing effects of insulin and glucagon on the
blood glucose level. Insulin and glucagon are both
secreted by the pancreas. In many cases they act on
the same tissues, but their effects are opposite. Insulin
promotes the storage of glucose as glycogen, the use
of glucose as an energy source, and the synthesis of
proteins and fats. As a result, insulin tends to lower
the blood glucose level. Glucagon acts in the opposite
direction, and its action therefore tends to raise the
blood glucose level. The secretion of these hormones
is regulated by the blood glucose level. A low blood
glucose level favors the secretion of glucagon, and a
high blood glucose level favors the secretion of
insulin.

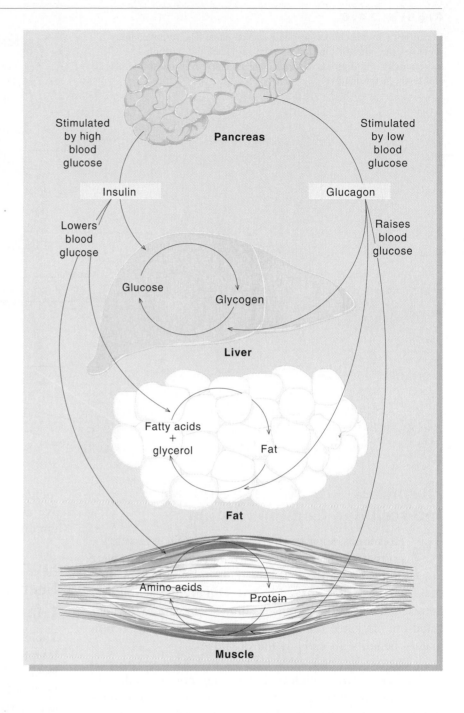

the cAMP exerts different effects inside the cell. In liver
cells cAMP sets off a chain of reactions that results in the
breakdown of glycogen to glucose-1-phosphate and subsequent conversion of glucose-1-phosphate to glucose (see
fig. 12.29). In adipocytes the cAMP also leads to glycogen
breakdown, but the main effect in the case of adipocytes is
to activate the first enzyme in the pathway for lipid breakdown, triacylglycerol lipase. This enzyme hydrolyzes the
triacylglycerol to diacylglycerol with release of fatty acid,
the rate-limiting step in the complete hydrolysis of the triacylglycerols. Fatty acids are transported to other tissues (see

chapter 20), where they can be metabolized as a source of
energy.

Glucagon and insulin are only two of the many hormones that regulate vertebrate metabolism, and cAMP is
only one of several so-called second messengers that trans-
duce the hormone signal to the inside of the cell. In the
remainder of this chapter we take a comprehensive look at
the hormone systems that provide the main mechanisms for
regulating energy metabolism and other metabolic activities
as well.

**Figure 24.6**

Location of major endocrine glands in humans. The hypothalamus regulates the anterior pituitary, which regulates the hormonal secretions of the thyroid, adrenals, and gonads (ovary in the female and testis in the male).

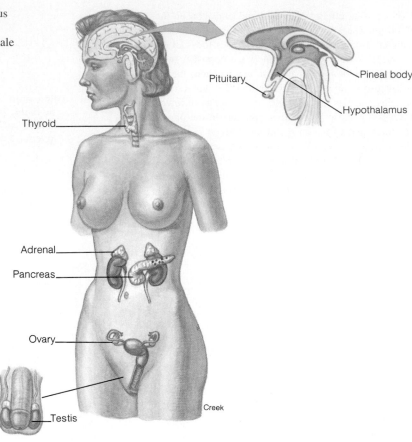

## Hormones Are Major Vehicles for Intercellular Communication

Hormones, such as glucagon and insulin, override the normal cellular controls.[a] Whatever the cells were doing before they received their hormonal instruction, their activities are redirected. Usually the action triggered by the hormone brings no immediate gain to the cells acted on, but since it is beneficial for the organism, it ultimately benefits all cells of the organism.

The two hormones that we have just discussed are part of a larger picture, which includes many hormones that directly affect most tissues in the vertebrate. Each hormone is synthesized in a specialized endocrine gland (fig. 24.6 and table 24.3) and is secreted under special circumstances. Once secreted, the hormone diffuses throughout the entire organism, but it triggers reactions only in those cells that carry specific receptors for the hormone.

In the following sections we discuss how hormones are synthesized, transported, and degraded. Then we explore basic aspects of hormone–receptor interaction and the more direct biochemical consequences of these interactions. We also indicate some of the ways in which the hormonal circuits themselves are regulated.

## Hormones Are Synthesized and Secreted by Specialized Endocrine Glands

Table 24.3 lists many of the better known hormones of vertebrates. Although many different classes of compounds are used as hormones by animals and plants, most vertebrate hormones fall into one of three classes: Polypeptides, amino acid derivatives, and steroids.

### Polypeptide Hormones Are Stored in Secretory Granules after Synthesis

Insulin, the first polypeptide hormone to be identified, was discovered by Frederick G. Banting and Charles Best in 1922. They found that this substance, which they isolated from the pancreas, restored normal glucose utilization in experimental animals lacking a pancreas. Insulin was also the first protein to be sequenced, a landmark accomplishment achieved by Fred Sanger in 1955. About 20 years

---

[a]One of the clearest examples of how a hormonal signal can override an intracellular signal is presented for glycogen phosphorylase in figure 9.20.

**Figure 24.7**

Processing pathway of preproopiomelanocortin. This precursor polypeptide is cleaved into a variety of active peptides. With the exception of the signal peptidase cleavage site, the cleavage sites are generally pairs of basic amino acids, although one site contains four. Which active peptides are produced depends on the processing pathway, which varies in different cell types. Thus, in the anterior and intermediate lobes of the pituitary gland, the precursor polypeptide is cleaved to yield corticotropin and β-lipoprotein. In the intermediate lobe only, these polypeptide hormones are further cleaved to yield the melanocyte-stimulating hormone, γ-MSH and α-MSH; γ-lipotropin; and β-endorphin.

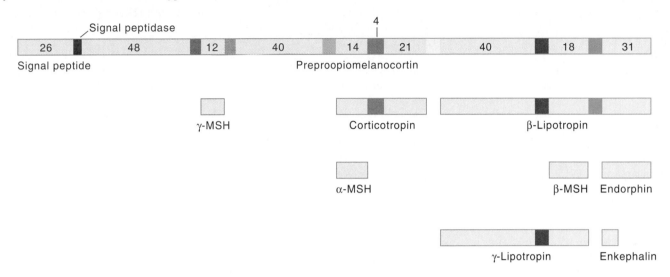

later, Steiner discovered that the two polypeptide chains of insulin are synthesized as a single polypeptide, proinsulin, which folds and is cross-linked by disulfide bonds (see fig. 29.21). An internal peptide is then removed by the concerted action of specific proteases. All of these events occur within the pancreatic β cells that synthesize insulin.

With the advent of techniques to isolate and translate mRNA in cell-free systems, it was discovered that the primary translation product of insulin mRNA—called preproinsulin—is even larger than proinsulin (see chapter 29). Like all secreted polypeptide hormones, proinsulin is synthesized with a hydrophobic signal sequence at the amino terminus that directs the nascent polypeptide into the endoplasmic reticulum and is then removed. Surprisingly, in some cases several different peptide hormones can be liberated from the same precursor by proteolytic processing. A striking example is preproopiomelanocortin (fig. 24.7), which is a precursor for corticotropin (ACTH), β-lipotropin, three melanocyte-stimulating hormones (MSH), endorphin, and an enkephalin. When several different polypeptide hormones are cleaved from a common precursor, the cleavage pattern can vary to yield a different spectrum of peptides, depending on the cell type. Proteolytic processing of these precursors often occurs at the site of dibasic amino acids.

All polypeptide hormones are synthesized from mRNA as precursors, which contain signal peptides that direct them into the lumen of the endoplasmic reticulum. In a few cases (e.g., growth hormone and prolactin) no further processing occurs. However, in most cases further processing is required. The peptides synthesized by neurosecretory cells of the hypothalamus, for example, are often much larger than the final hormones. Oxytocin and vasopressin, each nine residues long, represent the amino terminals of precursors called proneurophysins, which are 160 and 215 amino acids long, respectively. After oxytocin or vasopressin is cleaved from proneurophysin in the hypothalamus, where it is synthesized, the hormone remains associated with the neurophysin as it passes down the axons to the posterior pituitary, where it is secreted. The neurophysin may serve to protect the hormone from degradation prior to secretion. Somatostatin, another hypothalamic hormone, is a 14-amino-acid product cleaved from the carboxyl terminus of a precursor that contains 121 amino acids. Thyrotropin-releasing hormone (TRH) is the smallest known polypeptide hormone (pyroGlu-His-ProNH$_2$). In the synthesis of TRH, the glutamyl residue is cyclized and the carboxyl-terminal amide originates from an adjacent glycine. The precursor polypeptide (255 amino acids) contains five copies of the sequence Lys-Arg-Gln-His-Pro-Gly-Arg-Arg within it (fig. 24.8).

Polypeptide hormones are usually stored in secretory granules after their passage through the endoplasmic reticulum and Golgi apparatus. Release of these hormones into the bloodstream is accomplished by fusing the secretory granule membranes with the plasma membrane. This event is often regulated by other hormones.

**Table 24.3**

Vertebrate Hormones[a]

| Hormone | Structure | Function |
|---|---|---|
| **Pineal** | | |
| Melatonin | *N*-Acetyl-5-methoxytryptamine | Regulates circadian rhythms |
| **Hypothalamus**[b] | | |
| Corticotropin-releasing factor (CRF or CRH) | Polypeptide (41 residues) | Stimulates ACTH and $\beta$-endorphin secretion |
| Gonadotropin-releasing factor (GnRF) or (GnRH) | Polypeptide (10 residues) | Stimulates LH and FSH secretion |
| Prolactin-releasing factor (PRF) | (May be TRH) | Stimulates prolactin secretion |
| Prolactin-release inhibiting factor (PIF) | (May be 56-residue peptide from GnRH precursor) | Inhibits prolactin secretion |
| Growth hormone-releasing factor (GRF or GRH) | Polypeptide (40 and 44 residues) | Stimulates GH secretion |
| Somatostatin (Growth hormone-release inhibiting factor, SIF) | Polypeptide (14 and 28 residues) | Inhibits GH and TSH secretion |
| Thyrotropin-releasing factor (TRF or TRH) | Polypeptide (3 residues) | Stimulates TSH and prolactin secretion |
| **Pituitary** | | |
| Oxytocin (ocytocin) | Polypeptide (9 residues) | Uterine contraction, milk ejection |
| Vasopressin (antidiuretic hormone, ADH) | Polypeptide (9 residues) | Blood pressure, water balance |
| Melanocyte-stimulating hormones (MSH) | $\alpha$ Polypeptide (13 residues) $\beta$ Polypeptide (18 residues) $\gamma$ Polypeptide (12 residues) | Pigmentation |
| Lipotropin (LPH) | $\beta$ Polypeptide (93 residues) $\gamma$ Polypeptide (60 residues) | Fatty acid release from adipocytes |
| Corticotropin (adrenocorticotropic hormone, ACTH) | Polypeptide (39 residues) | Stimulates adrenal steroid synthesis |
| Thyrotropin (thyroid-stimulating hormone, TSH) | 2 Polypeptides ($\alpha$, 96 residues; $\beta$, 112 residues) | Stimulates thyroid hormone synthesis |
| Growth hormone (GH) or somatotropin | Polypeptide (191 residues) | General anabolic effects; stimulates release of insulinlike growth factor-I |
| Prolactin | Polypeptide (197 residues) | Stimulates milk synthesis |
| Luteinizing hormone (LH) | 2 Polypeptides ($\alpha$, 96 residues; $\beta$, 121 residues) | Ovary: luteinization, progesterone synthesis; testis: interstitial cell development, androgen synthesis |
| Follicle-stimulating hormone (FSH) | 2 Polypeptides ($\alpha$, 96 residues; $\beta$, 120 residues) | Ovary: follicle development, ovulation, estrogen synthesis; testis: spermatogenesis |
| **Thyroid** | | |
| Thyroxine and triiodothyronine | Iodinated dityrosine derivatives (see fig. 24.9) | General stimulation of many cellular reactions |
| Calcitonin | Polypeptide (32 residues) | $Ca^{2+}$ and $P_i$ metabolism |
| Calcitonin gene-related peptide (CGRP) | Polypeptide (37 residues) | Vasodilator |
| **Parathyroid** | | |
| Parathyroid hormone (PTH) | Polypeptide (84 residues) | $Ca^{2+}$ and $P_i$ metabolism |

[a] Only the more common hormones of known structure are listed.

[b] Most of the hypothalamic releasing factors are also called hypothalamic regulatory hormones.

**Table 24.3**

Vertebrate Hormones[a] (Continued)

| Hormone | Structure | Function |
|---|---|---|
| **Alimentary tract**[c] | | |
| Gastrin | Polypeptide (17 residues) | Stimulates acid secretion from stomach and pancreatic secretion |
| Secretin | Polypeptide (27 residues) | Regulates pancreas secretion of water and bicarbonate |
| Cholecystokinin | Polypeptide (33 residues) | Secretion of digestive enzymes |
| Motilin | Polypeptide (22 residues) | Controls gastrointestinal muscles |
| Vasoactive intestinal peptide (VIP) | Polypeptide (28 residues) | Gastrointestinal relaxation; inhibits acid and pepsin secretion |
| Gastric inhibitory peptide (GIP) | Polypeptide (43 residues) | Inhibits gastrin secretion |
| Somatostatin | Polypeptide (14 residues) | Inhibits gastrin secretion; inhibits glucagon secretion |
| **Heart** | | |
| Atrial natriuretic peptide (ANP) | Several active peptides cleaved from precursor polypeptide of 126 residues | Smooth muscle relaxation; diuretic activity |
| **Pancreas** | | |
| Insulin | 2 Polypeptides (21 and 30 residues) | Glucose uptake, lipogenesis, general anabolic effects |
| Glucagon | Polypeptide (29 residues) | Glycogenolysis, release of lipid |
| Pancreatic polypeptide | Polypeptide (36 residues) | Glycogenolysis, gastrointestinal regulation |
| Somatostatin | Polypeptide (14 residues) | Inhibition of somatotropin and glucagon release |
| **Adrenal cortex** | | |
| Glucocorticoids | Steroids (cortisol, corticosterone) | Many diverse effects on protein synthesis and inflammation |
| Mineralocorticoids | Steroids (aldosterone) | Maintains salt balance |
| **Adrenal medulla** | | |
| Epinephrine (adrenalin) | Tyrosine derivative (see fig. 24.10) | Smooth muscle contraction, heart function, glycogenolysis, lipid release |
| Norepinephrine (noradrenalin) | Tyrosine derivative (see fig. 24.10) | Arteriole contraction, lipid release |
| **Gonads** | | |
| Estrogens (ovary) | Steroids (estradiol, estrone) | Maturation and function of secondary sex organs |
| Progestins (ovary) | Steroids (progesterone) | Ovum implantation, maintenance of pregnancy |
| Androgens (testes) | Steroids (testosterone) | Maturation and function of secondary sex organs |
| Inhibins A and B | 1 Polypeptide ($\alpha$, 134 residues; $\beta$, 115 and 116 residues) | Inhibit FSH secretion |
| **Placenta** | | |
| Estrogens | Steroids | Maintenance of pregnancy |
| Progestins | Steroids | |
| Choriogonadotropin | 2 Polypeptides ($\alpha$, 96 residues; $\beta$, 147 residues) | Similar to LH |
| Placental lactogen | Polypeptide (191 residues) | Similar to prolactin |
| Relaxin | 2 Polypeptides (22 and 32 residues) | Muscle tone |
| **Liver** | | |
| Angiotensin[d] | Polypeptide (8 residues) | Responsible for essential hypertension |
| **Kidney** | | |
| 1,25-dihydroxyvitamin $D_3$ | Steroid | Calcium uptake, bone formation |

[c] Many of these peptides are also found in the brain, where they may modulate neural activity.

[d] The liver secretes $\alpha_2$-globulin, which is cleaved by renin, a kidney enzyme, to give a decapeptide, proangiotensin, from which the carboxyl-terminal dipeptide is removed to give angiotensin.

**Figure 24.8**

Biosynthesis of thyrotropin-releasing hormone (TRH). Five copies of TRH are contained within a 255-amino-acid precursor polypeptide (pre-pro-TRH). The precursor has a hydrophobic signal peptide (dark green) and five copies of the sequence Lys-Arg-Gln-His-Pro-Gly-Arg-Arg (light green). The dibasic amino acids are recognized by specific proteases to liberate Gln-His-Pro-Gly, which is subsequently converted into TRH. The amino-terminal glutamine is converted into pyroglutamine, and the carboxyl-terminal amide is derived from the neighboring glycine, which is removed by a specific enzyme.

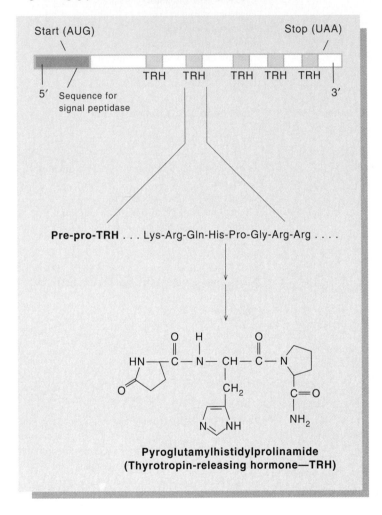

one or two positions by a special peroxidase; then two iodinated residues condense as shown in figure 24.9.

The secretion of thyroid hormones starts with endocytosis of the modified thyroglobulin, followed by fusion of the endocytotic vesicles with lysosomes. The lysosomal enzymes then degrade the thyroglobulin, liberating triiodothyronine and thyroxine into the circulation. Only about five molecules of $T_3$ and $T_4$ are generated from each molecule of thyroglobulin. Thyroid hormone secretion is stimulated by thyrotropin (TSH), a pituitary hormone that activates adenylate cyclase in its target cells.

Epinephrine, originally called adrenalin, was the first hormone to be isolated, characterized, and synthesized. Epinephrine and its precursor, norepinephrine, are synthesized from tyrosine in the chromaffin cells of the adrenal medulla. They are also synthesized by neurons of the central and peripheral nervous system. The biosynthetic pathway is shown in figure 24.10. The first step, which involves oxidation of tyrosine to 3,4-dihydroxyphenylalanine (dopa), is catalyzed by tyrosine hydroxylase. This is the rate-limiting enzyme in the pathway. The amount and activity of tyrosine hydroxylase are regulated by cAMP-dependent mechanisms that are responsive to the neurotransmitter acetylcholine. This neurotransmitter is liberated by special neurons that impinge on the chromaffin cells. Second, dopa is decarboxylated to dopamine, which is then $\beta$-hydroxylated to produce norepinephrine. Finally, the N-methylation of norepinephrine ($S$-adenosylmethionine is the methyl donor) produces epinephrine. Both of these catecholamines (norepinephrine and epinephrine) are stored in chromaffin granules, where they are complexed with ATP and proteins called chromogranins. Neural stimulation of the medulla is mediated by acetylcholine, which binds to receptors on the membranes of medullary cells (see supplement 1). This event leads to a local depolarization and an influx of calcium. As a result, the chromaffin granules fuse with the cell membrane, and a packet of catecholamines, ATP, and protein is extruded into the extracellular fluid.

## Thyroid Hormones and Epinephrine Are Amino Acid Derivatives

Thyroxine ($T_4$) and the more potent triiodothyronine ($T_3$) are cleaved from a large precursor protein called thyroglobulin. Thyroglobulin exists as a dimer of two identical polypeptides ($M_r \approx 330,000$). It is a storage protein for iodine and can be considered a prohormone of the circulating thyroid hormones. Thyroglobulin is secreted into the lumen of the thyroid gland, where specific residues are iodinated in

## Steroid Hormones Are Derived from Cholesterol

Steroid hormones are derived from cholesterol by a stepwise removal of carbon atoms and hydroxylation. The steroid hormones are synthesized by cells of the adrenal cortex (in the case of glucocorticoids and mineralocorticoids) and the gonads (in the case of estrogens, progestins, and androgens). The hormone names just mentioned are generic names for entire classes of compounds that interact with specific receptors; for example, estrogen is the generic

**Figure 24.9**

Pathway of thyroxine (T$_4$) and triiodothyronine (T$_3$) synthesis.
Thyroid cells actively transport iodine (I$^-$), which is incorporated into
a few tyrosine residues of thyroglobulin by the enzyme
iodoperoxidase. After condensation of iodinated tyrosine residues, the
thyroglobulin is proteolytically degraded liberating thyroxine and
triiodothyronine.

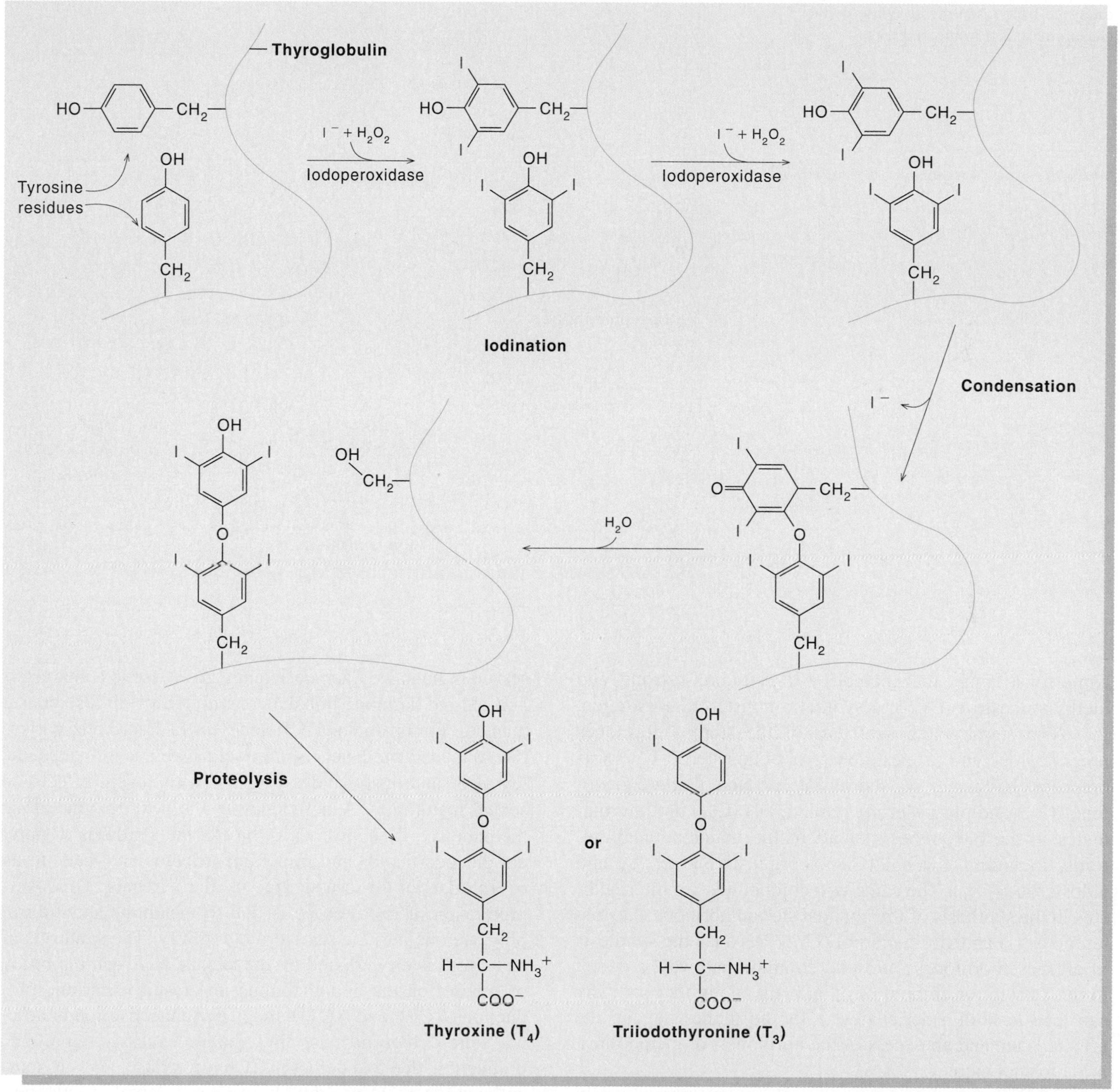

## Figure 24.10

Pathway of epinephrine synthesis. Epinephrine and its precursor, norepinephrine, are synthesized from tyrosine. The synthesis occurs in the chromaffin cells of the adrenal medulla and in neurons of the central and peripheral nervous system. The first step, which is catalyzed by tyrosine hydroxylase, is the rate-limiting step in the pathway.

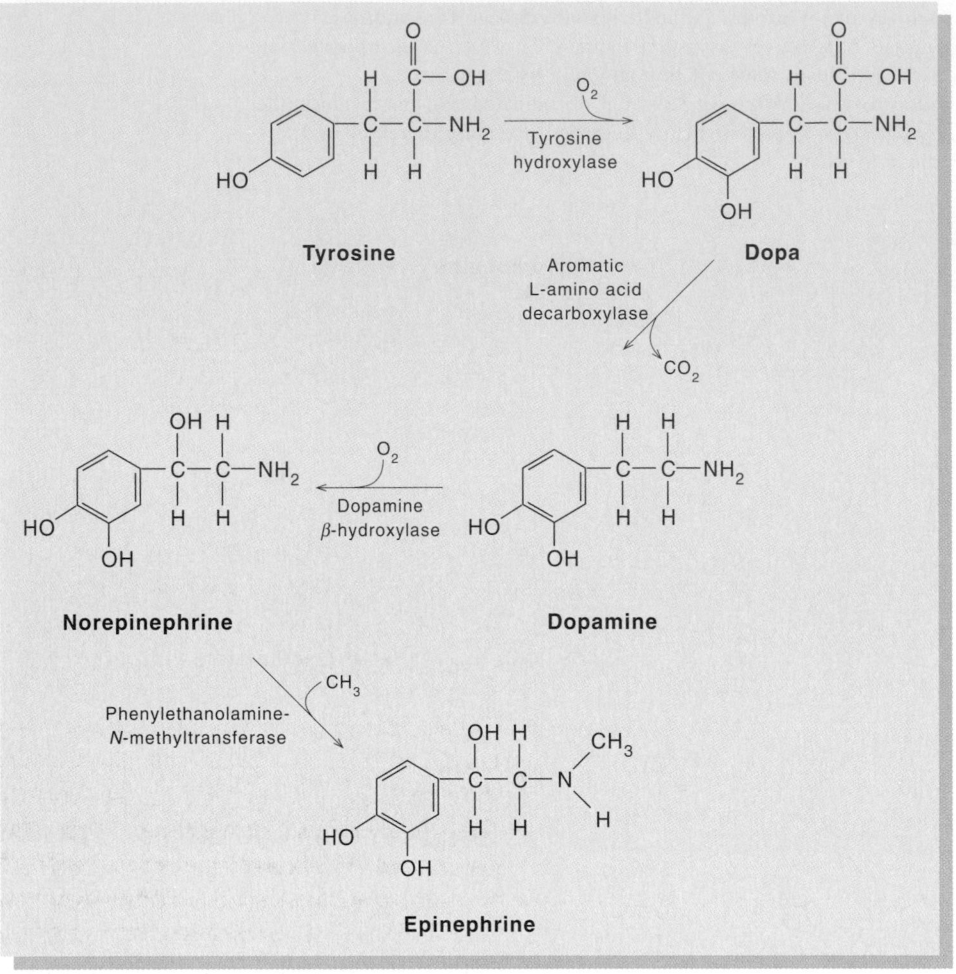

name for a family that includes $17\beta$-estradiol, estrone, and diethylstilbestrol (DES), a synthetic nonsteroidal estrogen.

The step-by-step synthesis of the steroid hormones pregnenolone and progesterone from cholesterol ($C_{27}$) was presented in chapter 20 (see fig 20.22). Note that pregnenolone ($C_{21}$) and progesterone (table 20.4) ($C_{21}$) are intermediates in the biosynthesis of all of the major adrenal steroids, including cortisol ($C_{21}$), corticosterone ($C_{21}$), and aldosterone ($C_{21}$). The same two compounds are intermediates in the synthesis of the gonadal steroid hormones, testosterone ($C_{19}$) and $17\beta$-estradiol ($C_{18}$). Because the synthesis of all these hormones follows a common pathway, a defect in the activity or amount of an enzyme along that pathway can lead to both a deficiency in the hormones beyond the affected step and an excess of the hormones, or metabolites, prior to that step.

Deficiencies in each of the six enzymes involved in the conversion of cholesterol to aldosterone have been ob-

served in humans. Each deficiency gives rise to a characteristic steroid hormone imbalance, with telling clinical consequences. For example, a deficiency in 17-hydroxylase gives rise to inadequate levels of cortisol as well as to inadequate levels of androgens and estrogens, with severe effects on sexual maturation. A deficiency in the next enzyme along the pathway, 21-hydroxylase, blocks the synthesis of adrenal glucocorticoids and mineralocorticoids and leads to an overproduction of testosterone by the adrenals. This overproduction of testosterone is due to metabolic shunting of progesterone into the sex steroid pathway. The synthesis of androgens is exacerbated by the lack of feedback inhibition of cortisol on the hypothalamus; the result is chronic production of CRF and ACTH and perpetual activation of adrenal steroid biosynthesis. In females, excessive androgen production, due to a defect in 21-hydroxylase, leads to masculinization or, to use the clinical term, female pseudohermaphrodism—that is, a genetic female and male appear-

## Figure 24.11

Conversion of testosterone to 5α-dihydrotestosterone (5α-DHT). Receptor-binding studies indicate that 5α-DHT has a higher affinity for the androgen receptor than testosterone does.

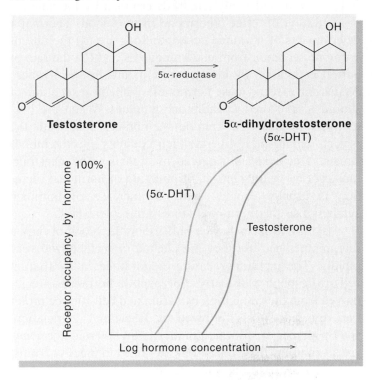

Testosterone is also a prohormone of metabolites that bind to different receptors. For example, some of the effects of androgens on the production of red blood cells (erythropoiesis) are due to the reduction of testosterone to 5β-dihydrotestosterone within precursor cells. 5β-Dihydrotestosterone binds to a receptor that is distinct from the one that binds testosterone and 5α-dihydrotestosterone (see fig. 24.11). Testosterone also influences erythropoiesis by stimulating the kidney to produce a specific growth factor, erythropoietin. Another striking example of testosterone as a prohormone occurs in the brain, where testosterone influences neural development and activity (e.g., male-specific mating behavior and bird songs) by being converted into 17β-estradiol, which interacts with estrogen receptors. Testosterone is metabolized to 17β-estradiol by aromatase, the same enzyme involved in 17β-estradiol synthesis in the ovary (fig. 24.12).

Cell specificity of mineralocorticoid action is achieved in a different manner. Aldosterone, cortisol, and corticosterone bind with similar affinities to mineralocorticoid and glucocorticoid receptors. However, aldosterone activates only its own receptor in target tissues such as the kidney because of an enzyme, 11β-hydroxysteroid dehydrogenase, that converts the prevalent glucocorticoids into inactive 11-keto derivatives but does not affect aldosterone.

Vitamin $D_3$ is a precursor of the hormone 1,25-dihydroxyvitamin $D_3$. Vitamin $D_3$ is essential for normal calcium and phosphorus metabolism. It is formed from 7-dehydrocholesterol by ultraviolet photolysis in the skin. Insufficient exposure to sunlight and absence of vitamin $D_3$ in the diet leads to rickets, a condition characterized by weak, malformed bones. Vitamin $D_3$ is inactive, but it is converted into an active compound by two hydroxylation reactions that occur in different organs. The first hydroxylation occurs in the liver, which produces 25-hydroxyvitamin $D_3$, abbreviated 25(OH)$D_3$; the second hydroxylation occurs in the kidney and gives rise to the active product 1,25-dihydroxyvitamin $D_3$ 24,25(OH)$_2D_3$ (fig. 24.13). The hydroxylation at position 1 that occurs in the kidney is stimulated by parathyroid hormone (PTH), which is secreted from the parathyroid gland in response to low circulating levels of calcium. In the presence of adequate calcium, 25(OH)$D_3$ is converted into an inactive metabolite, 24,25(OH)$_2D_3$. The active derivative of vitamin $D_3$ is considered a hormone because it is transported from the kidneys to target cells, where it binds to nuclear receptors that are analogous to those of typical steroid hormones. 1,25(OH)$_2D_3$ stimulates calcium transport by intestinal cells and increases calcium uptake by osteoblasts (precursors of bone cells).

ance. This reversal of sexual phenotype is explained by the fact that during embryonic development of mammals, the external genitalia develop from common precursor cells. In the absence of hormonal stimulation, they develop into female structures, but androgens direct their development into male structures.

Testosterone is both a hormone and a prohormone. The high levels of testosterone normally produced in the male by the testes play a major role in the growth and function of many tissues in addition to reproductive organs. Essentially all the sexual differences in nonreproductive tissues, such as muscle, liver, and brain, are a consequence of androgen action. Although testosterone is the major circulating androgen, many target cells reduce this steroid to 5α-dihydrotestosterone, a steroid that binds to the androgen receptor with higher affinity than testosterone (fig. 24.11). When 5α-reductase is defective, the androgen receptors are only partially activated, and a full androgen response is not obtained. A deficiency in this enzyme therefore leads to abnormal development of male genitalia—that is, they are of female phenotype (a clinical condition referred to as male pseudohermaphrodism, type 2). In this example, testosterone can be considered a prohormone of a more active androgen.

**Figure 24.12**

Metabolic conversion of testosterone by target cells. Testosterone (T) is the predominant androgen in the bloodstream. When testosterone enters target cells, it can be metabolized in a variety of different ways. It can either (*a*) bind directly to androgen receptors ($R^a$) or (*b*) be reduced to 5α-dihydrotestosterone (5α-DHT), which then binds to $R^a$ with higher affinity. (*c*) Other target cells reduce testosterone to 5β-dihydrotestosterone (and other 5β metabolites), which bind to a distinct receptor ($R^β$). (*d*) Yet other cells convert testosterone into an estrogen, 17β-estradiol, which binds to estrogen receptors ($R^e$).

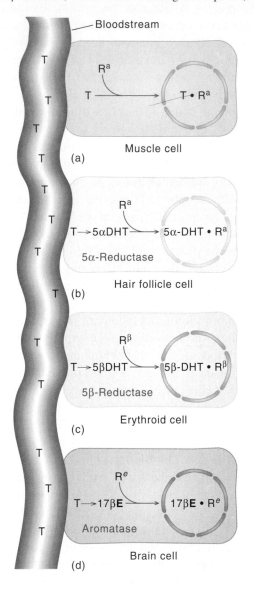

## The Circulating Hormone Concentration Is Regulated

The occupancy of hormone receptors can fluctuate greatly and is ultimately determined by the concentration of "free" hormone in the blood. The major determinants of hormone concentrations are (1) the rate of hormone secretion from endocrine cells and (2) the rate of hormone removal by clearance or metabolic inactivation. As we have seen, most hormones (with the exception of steroids) are stored in secretory granules. When the hormone is needed, the granule membranes fuse with the plasma membrane to liberate their contents into the bloodstream. This event is triggered by signals from other hormones or by neural signals. Stimulation of hormonal secretion is usually coupled with an increase of hormone synthesis, so that hormonal stores are replenished.

Most hormones have a half-life in the blood of only a few minutes because they are cleared or metabolized very rapidly. The rapid degradation of hormones allows target cells to respond transiently. Polypeptide hormones are removed from the circulation by serum and cell surface proteases, by endocytosis followed by lysosomal degradation, and by glomerular filtration in the kidney. Steroid hormones are taken up by the liver and metabolized to inactive forms, which are excreted into the bile duct or back into the blood for removal by the kidneys. Catecholamines are metabolically inactivated by O-methylation, by deamination, and by conjugation with sulfate or glucuronic acid.

Thyroid hormones and most steroid hormones are associated with carrier proteins in the serum. The carrier proteins are called, appropriately, thyroxine-binding globulin, transcortin (for cortisol), and sex-steroid-binding protein. These proteins have a high affinity ($K_d \approx 10^{-9}-10^{-8}$ M) for their respective hormones. They buffer the concentration of "free" hormone and retard hormone degradation and excretion. The carrier proteins are distinguishable from the intracellular receptors for these hormones.

## Hormone Action Is Mediated by Receptors

Hormone action begins with the binding of the hormone to a receptor on (or in) a target cell. Binding of hormone molecule induces a conformational change in its receptor, and this change is detected by other macromolecules. Hence the hormone–receptor interaction can be transduced from one molecule to another. In the simplest scheme, a hormone receptor might be a rate-limiting en-

**Figure 24.13**

The conversion of vitamin $D_3$ to an active compound. Vitamin $D_3$ is formed from 7-dehydrocholesterol by ultraviolet photolysis in the skin. Vitamin $D_3$ is inactive, but it is converted into an active compound by two hydroxylation reactions that occur in different organs. The first reaction occurs in the liver and results in 25-hydroxyvitamin $D_3$. The second hydroxylation occurs in the kidney and results in 1,25-dihydroxyvitamin $D_3$.

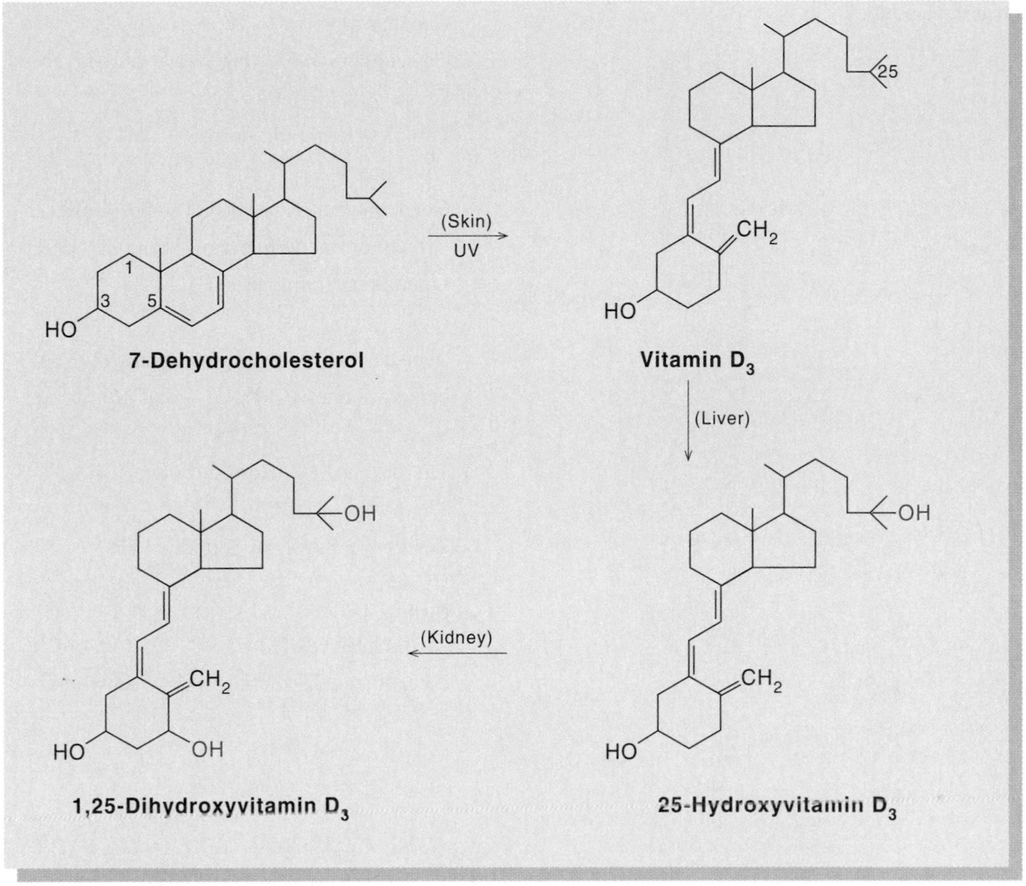

zyme or be coupled to a rate-limiting enzyme. In cases in which the receptors regulate adenylate cyclase, another protein is interposed between the receptor and the adenylate cyclase. The protein directly activated by hormone–receptor interaction is sometimes referred to as an acceptor protein. GTP-binding proteins (G proteins) are acceptors for all receptors that activate or inhibit adenylate cyclase (described in more detail a little later). For many membrane receptors, the molecular intermediates are still unknown. Generalized schemes for hormone–receptor activation of physiological responses are shown in figure 24.14.

Receptors for steroid hormones are located within the cell and bind to specific DNA sequences when they are activated by the appropriate hormones. The binding of these receptors to DNA in the promoter region of a gene activates transcription, in most cases by helping to assemble an efficient initiation complex. The acceptor proteins in these cases may be other transcription factors.

Overall, it appears that all hormones act by binding to receptors whether they are located in the cell membrane or inside the cell. Binding of hormone induces a conformational change in the receptor that is transmitted to its active site (if it is an enzyme) or to other macromolecules. In either case, a chain of events is elicited that ultimately affects a vast array of metabolic processes, ranging from alterations in enzyme activities to changes in gene expression. These changes may lead to profound alterations in cell growth, morphology, and function.

## Figure 24.14

Possible mechanisms of hormone action. (*a*) The hormone (H) theoretically activates an enzyme (E) directly as an allosteric effector. (*b*) Alternatively, a separate binding protein for the hormone, called a receptor (R), may then activate an enzyme. (*c*) Another possibility interposes an acceptor protein (A) between the receptor and the enzyme. Each interaction is reversible.

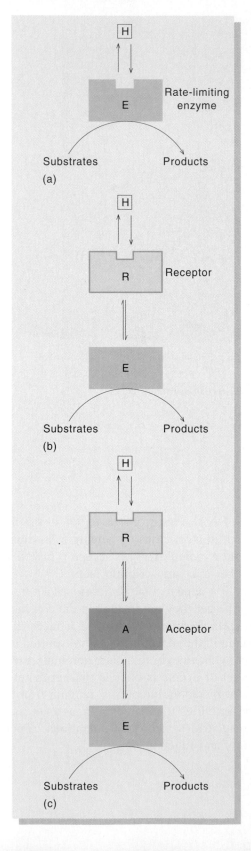

### Table 24.4

Hormones that Activate or Inhibit Adenylate Cyclase

**Activators**
Corticotropin (ACTH)
Calcitonin
Catecholamines (acting on $\beta_1$ and $\beta_2$ receptors)
Choriogonadotropin
Follicle-stimulating hormone (FSH)
Glucagon
Gonadotropin-releasing hormone (GnRH)
Growth hormone-releasing hormone (GRH)
Luteinizing hormone (LH)
Lipotropin (LPH)
Melanocyte-stimulating hormones (MSH)
Parathormone (PTH)
Secretin
Thyrotropin regulatory hormone (TRH)
Thyrotropin (TSH)
Vasoactive intestinal peptide (VIP)
Vasopressin
**Inhibitors**
Angiotensin
Catecholamines (acting on $\alpha_2$ receptors)

## Many Plasma Membrane Receptors Generate a Diffusible Intracellular Signal

Activation of many membrane-associated hormone receptors generates a diffusible intracellular signal called a second messenger. Many different receptors generate the same second messenger. Five intracellular messengers are currently known: Cyclic AMP, cyclic GMP, inositol trisphosphate, diacylglycerol, and calcium. There are many membrane receptors for which second messengers do not exist or have not yet been identified (table 24.4).

## The Adenylate Cyclase Pathway Is Triggered by a Membrane-Bound Receptor

Many hormones bind to receptors that act through an intermediary protein to either activate or inhibit adenylate cyclase (table 24.4). Cyclic AMP is formed from ATP by adenylate cyclase, which is bound to the inside of the cell membrane (fig. 24.15).

## Figure 24.15

The adenylate cyclase pathway of hormone receptor action. When the receptor is unoccupied, the G protein α subunit has GDP bound, and it is complexed with the subunits; in this form it cannot activate adenylate cyclase. Binding of hormone activates the receptor, which leads to replacement of GDP by GTP, and the activation subunit then interacts productively with adenylate cyclase to stimulate the synthesis of cAMP. The intrinsic GTPase activity of the α subunit leads to GDP production and the inactivation of the G protein. If hormone remains bound, the G protein can be activated again. Meanwhile, the cAMP produced binds to the R subunits of cAMP-dependent protein kinase, leading to the dissociation of two catalytic subunits, which can then phosphorylate critical proteins; in this case the activation of phosphorylase kinase is shown.

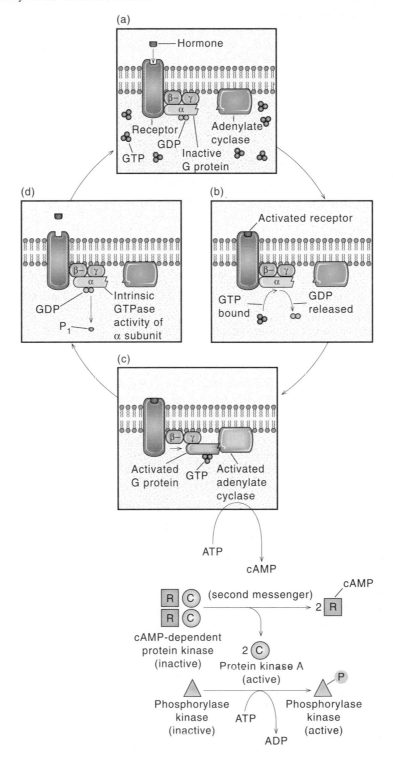

**Table 24.5**

Some G Protein Receptors and Effectors[a]

| G Protein | Receptors for | Effectors | Signaling pathways |
|---|---|---|---|
| $G_s$ | Epinephrine, norepinephrine, histamine, glucagon, ACTH, luteinizing hormone, follicle-stimulating hormone, thyroid-stimulating hormone, and others | Adenylate cyclase<br>$Ca^{2+}$ channels | ↑ cAMP<br>↑ $Ca^{2+}$ influx |
| $G_{olf}$ | Odorants | Adenylate cyclase | ↑ cAMP (olfaction) |
| $G_{t1}$ (rods) | Photons | cGMP phosphodiesterase | ↓ cGMP (vision) |
| $G_{t2}$ (cones) | Photons | cGMP phosphodiesterase | ↓ cGMP (color vision) |
| $G_{i1}$, $G_{i2}$, $G_{i3}$ | Norepinephrine, prostaglandins, opiates, angiotensin, many peptides | Adenylate cyclase<br>Phospholipase C<br>Phospholipase $A_2$<br>$K^+$ channels | ↓ cAMP<br>↑ Inositol trisphosphate, diacylglycerol, $Ca^{2+}$<br>Arachidonate release<br>Membrane polarization |
| $G_0$ | Probably many, but not yet defined | Phospholipase C<br>$Ca^{2+}$ channels | ↑ Inositol trisphosphate, diacylglycerol, $Ca^{2+}$<br>↓ $Ca^{2+}$ influx |
| Gq | Norepinephrine, vasopressin | Phospholipase C | ↑ Inositol bisphosphate, diacylglycerol, $Ca^{2+}$ |

[a] H. R. Bourne et al., ''The GTPase superfamily: A conserved switch of diverse cell functions'' in *Nature,* 348:126, 1990. Copyright © 1990 Macmillan Magazines Ltd., London, England.

As we indicated previously, the interaction between the hormone receptor and the adenylate cyclase is mediated by a G protein. Most G proteins are heterotrimers of $\alpha$, $\beta$, and $\gamma$ subunits that are associated with the membrane and can make contact with both the membrane receptor and an effector molecule such as adenylate cyclase as shown in figure 24.15. There are several classes of G proteins with different specificities (table 24.5). In the absence of hormone activation, the $\alpha$ subunit of the G protein binds GDP at an allosteric site. On binding of hormone, the conformation of the $\alpha$ subunit changes so that the GDP is displaced and replaced by GTP. The GTP-bound $\alpha$ subunit may dissociate from the $\beta$, $\gamma$ subunits, allowing it to activate an effector molecule. In the case of receptors coupled with $G_s$, the $\alpha_s$-GTP complex activates adenylate cyclase, which proceeds to synthesize a burst of cAMP. The period of activation is brief, as the $\alpha$ subunit also contains a GTPase activity that hydrolyzes a phosphate residue from the GTP, thereby returning the $\alpha$ subunit to its original inactive state. Further cAMP synthesis is contingent on the same chain of reactions that began with hormone binding to the hormone receptor protein.

The cAMP formed on the inner side of the plasma membrane diffuses into the cytoplasm and binds to regulatory subunits of cAMP-dependent protein kinase A. This enzyme is composed of four subunits: Two catalytic subunits, C; and two regulatory subunits, R. Binding of cAMP to the regulatory subunits promotes the dissociation of the two catalytic subunits, which then become active (see fig. 24.15). The catalytic subunits phosphorylate a wide variety of proteins on serine or threonine residues, inducing conformational changes that alter their function. Many substrates of the cAMP-dependent kinase A are rate-limiting enzymes in key metabolic pathways. The activity of the kinase is determined by the intracellular cAMP concentration, which is a function of its rate of synthesis by adenylate cyclase and its rate of degradation by cAMP phosphodiesterases. There are several phosphodiesterases, and their activities can be hormonally regulated.

Another similar class of receptors are coupled to $G_i$ proteins. When activated, these G proteins in turn activate a variety of effectors as indicated in table 24.5. One of these effectors is adenylate cyclase again, but in this case the $\alpha_i$-GTP subunit inhibits enzyme activity, thus lowering cAMP levels. The list of known G proteins is increasing at a rapid pace.

**Figure 24.16**

The reversible conversion of the $\alpha_s$ subunit of the $G_s$ protein between its active and inactive forms and inhibitory effects of cholera bacterial toxin.

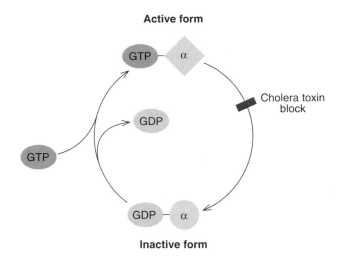

**The G Protein Cycle Is a Target for Certain Bacterial Toxins**

Cholera is caused by infestation of the bacterial species *Vibrio cholerae*. These bacteria secrete a toxin that binds to the plasma membranes of the epithelial cells that line the gut. A portion of the toxin penetrates the cells and catalyzes covalent attachment of an ADP-ribose group to the $\alpha_s$ subunit of the $G_s$-protein. This modification blocks hydrolysis of GTP to GDP on the $\alpha_s$ complex, which locks the complex in its active form (fig. 24.16). The resulting persistent activation of adenylate cyclase leads to a continuing loss of fluid from the intestinal cells, which is lethal unless the patient receives adequate liquids to replace the lost water and salts. In pertussis (whooping cough) the invading bacteria (*Bordetella pertussis*) secrete a toxin that causes a similar ADP-ribosylation of the $\alpha_i$ subunit in the $G_i$-protein. In this case the ADP-ribosylation prevents the exchange of GTP for the bound GDP, which blocks the normal inhibition of adenylate cyclase by this G protein.

**Multicomponent Hormonal Systems Facilitate a Great Variety of Responses**

Before considering other types of secondary messengers associated with hormonal systems, we might reflect on why such systems are often complex. It would certainly be simpler to have just one transmembrane protein with a site for binding the hormone on the extracellular side and a site for synthesizing cAMP on the cytosolic side (see fig. 24.14a). A multicomponent system, however, has many advantages, the foremost of which is that it presents many points for control. Such a system can be used in various ways in the same or in different cell types, leading to a rich variety of possible responses. For example, there is more than one type of hormone system in the liver cell that activates the same G protein that leads to activation of the adenylate cyclase enzyme (these two hormonal systems involve glucagon and epinephrine). In addition, a rich variety of other hormonal systems operate through completely different G proteins to either activate or inhibit the adenylate cyclase enzyme (see tables 24.4 and 24.5).

Variability in hormonal response patterns does not stop at the level of second-messenger synthesis. Thus, cyclic AMP can activate the well-known cAMP-dependent protein kinase A, but the possibility of other cAMP-responsive enzymes or cAMP-activated regulatory proteins should not be ruled out. The protein kinase activated by cAMP can activate a number of other enzymes. For example, in the liver, phosphorylase kinase (see fig. 24.15) is activated and catalyzes the breakdown of glycogen. In adipocytes, triacylglycerol lipase is activated and catalyzes the breakdown of triacylglycerols.

One should think of the individual proteins in a multicomponent hormonal response system like the parts of a machine which may play similar roles in different machines where the overall function is quite different. Each cell type may contain some but not all elements, permitting cells to respond differently to the same hormonal signal or in a similar way to different hormonal signals.

**The Guanylate Cyclase Pathway**

Another second-messenger system involves receptors that either are coupled to guanylate cyclase or guanylate cyclase itself. One class of receptors, those that are activated by atrial, cardiac, or brain natriuretic factors (a natriuretic factor increases the urinary secretion of sodium), is bifunctional in that the external domain of these membrane receptors binds the peptide hormones, resulting in the activation of an internal cytoplasmic domain that has guanylate cyclase activity and produces cGMP from GTP. This is the best example of the simplest receptor–effector coupling scheme illustrated in figure 24.14a. Another class of receptors activates intracellular guanylate cyclases by an unknown mechanism. The cGMP that is produced can either

**Figure 24.17**

Phosphatidylinositol-4,5-bisphosphate (PIP$_2$) and the two second messengers, diacylglycerol and inositol trisphosphate, that are derived from it.

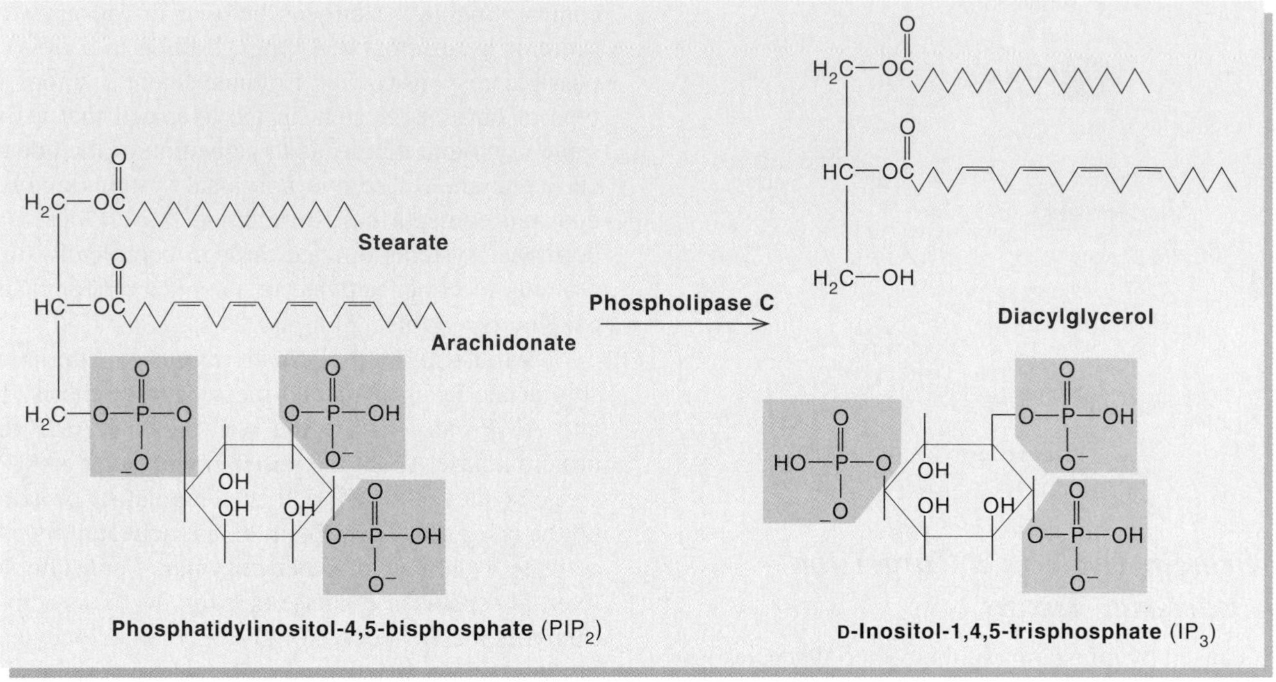

Phosphatidylinositol-4,5-bisphosphate (PIP$_2$)          D-Inositol-1,4,5-trisphosphate (IP$_3$)

act directly on target molecules such as ion channels, or it can activate a cGMP protein kinase G that can phosphorylate many proteins, resulting in a variety of metabolic changes, depending on the cell type. For example, a cAMP phosphodiesterase can be activated by cGMP-dependent phosphorylation, resulting in lowering of cAMP levels; this effect helps explain why cAMP and cGMP sometimes have opposite effects on cells. The cGMP kinases differ from the cAMP kinases described previously in that the regulatory and catalytic domains are part of the same molecule rather than being encoded by separate genes.

## Calcium and the Inositol Trisphosphate Pathway

The outline of another important second-messenger system was elucidated during the last few years. Chemical messengers that act via this system include a variety of hormones (e.g., catecholamines, vasopressin, and angiotensin) as well as some neurotransmitters (e.g., acetylcholine acting on pancreatic acinar cells to stimulate secretion of digestive

enzymes or acting on pancreatic $\beta$ cells to stimulate insulin secretion). The receptors for these hormones and neurotransmitters are membrane-bound proteins with their hormone binding sites facing the outside of the cell. Their activation by an appropriate signal is transmitted by a G protein (G$_0$, or Gq, table 24.6), which then activates a phosphodiesterase (phospholipase C) that cleaves the polar inositol trisphosphate (IP$_3$) from phosphatidylinositol-4,5-bisphosphate (PIP$_2$). As a result, IP$_3$ enters the cytoplasm, whereas the diacylglycerol moiety remains in the membrane (fig. 24.17). Both breakdown products of PIP$_2$ play important second-messenger roles (fig. 24.18). IP$_3$ stimulates the release of calcium from intracellular stores (residing in the endoplasmic reticulum) into the cytoplasm. The calcium is bound by the protein calmodulin, which then activates one or more calcium-dependent protein kinases. Meanwhile, the diacylglycerol, along with phosphatidylserine, activates a membrane-associated protein kinase C that phosphorylates serine and threonine residues (see fig. 24.18). Thus, several kinases are activated by this complex system, and they, like the cAMP-dependent protein kinases, modify the function of rate-limiting enzymes and regulatory proteins involved in

## Figure 24.18

The phosphatidylinositol ($PIP_2$) pathway. Binding of certain hormones and some other ligands to a variety of membrane receptors (R) leads to activation of GTP-binding proteins, which then activate a phospholipase C that cleaves the inositol trisphosphate ($IP_3$) moiety from $PIP_2$ and leaves diacylglycerol (DG) in the membrane. Diacylglycerol, in conjunction with phosphatidylserine (PS), activates a protein kinase (C) that phosphorylates enzymes associated with key metabolic pathways, thereby activating or inactivating them.

Meanwhile, the polar $IP_3$ moiety binds to intracellular receptors on the endoplasmic reticulum, resulting in the liberation of calcium into the cytosol. The calcium binds to calmodulin, which then activates another group of protein kinases (the $Ca^{2+}$/CaM-dependent protein kinases). Hormonal stimulation can be short-lived because $IP_3$ and DG are rapidly degraded to inactive forms that are ultimately recycled to $PIP_2$; $Ca^{2+}$ is pumped back into the endoplasmic reticulum, where it is sequestered.

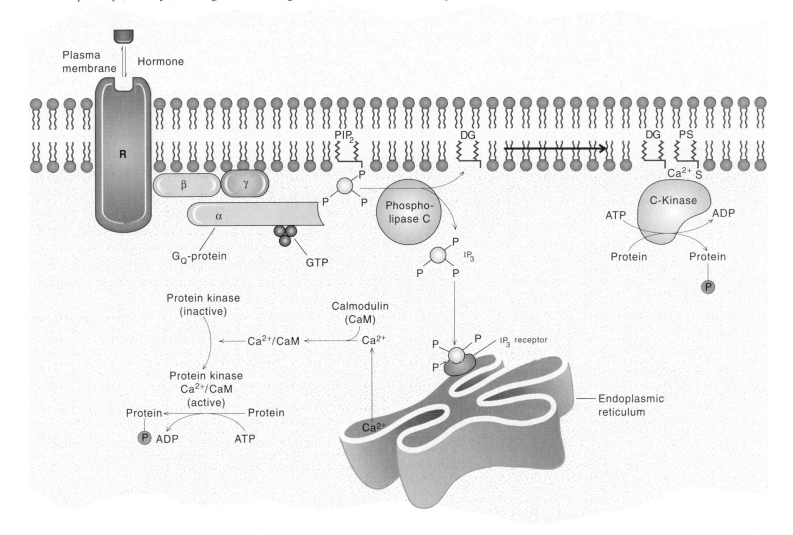

a variety of metabolic pathways. In addition to intracellular messengers, the breakdown of $PIP_2$ also stimulates the production of extracellular modulators of hormone activity. Thus, following the breakdown of $PIP_2$, arachidonic acid (one of the main fatty acids found in the diacylglycerol moiety of $PIP_2$) is metabolically converted into eicosanoids of different sorts. Formation of eicosanoids, as well as their function as local hormones were discussed in chapter 19.)

$IP_3$ is rapidly degraded to inactive $IP_2$ and then on to inositol. Meanwhile, diacylglycerol is phosphorylated and then converted to CDP-diacylglycerol, which combines with inositol to form phosphatidylinositol. The latter is subsequently phosphorylated in two steps to $PIP_2$. The degradation and resynthesis of $PIP_2$ completes the so-called phosphatidylinositol cycle.

**Figure 24.19**

Activation of steroid hormone receptors by the hormone. In the absence of the hormone, the steroid receptors are complexed through the hormone-binding domain to another protein known as heat shock protein 90 (hsp90). Both the hormone-binding domain and the hsp90 prevent functional interaction of the receptor with DNA. Binding of the hormone frees the receptor from hsp90 and promotes dimerization of the receptor, which can then bind to the palindromic hormone response element (HRE) and activate transcription.

**Without hormone**

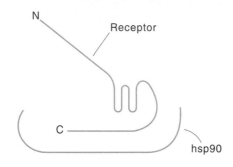

**With hormone**

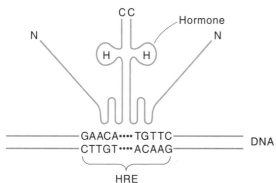

### Steroid Receptors Modulate the Rate of Transcription

The receptors for all classes of steroid hormones, including the receptors for 1,25-dihydroxyvitamin $D_3$, and thyroid hormones are intracellular proteins that are not very abundant, usually only $10^2-10^5$ molecules per cell.

All of these classes of receptors have a similar structure composed of several functional domains, including regions that bind the hormone, bind to DNA, activate transcription and allow dimerization. In the absence of hormone these receptors are usually sequestered by other proteins located either in the cytoplasm or nucleus. For example,

glucocorticoid receptors are sequestered by hsp90 in the cytoplasm. Binding of a hormone such as corticosterone to the COOH-terminal domain of the receptor causes a conformational change that liberates the receptor and allows it to interact with another activated receptor, forming a homodimer (fig. 24.19). The receptor dimers recognize specific DNA sequences that are located in the promoter region of the genes that these receptors regulate (table 24.6). The positioning of receptors at the promoter allows their activation domains to stimulate transcription. Transcriptional regulation is discussed in detail in chapter 31.

Thyroid hormone receptors differ from steroid hormone receptors in that they are always bound to DNA. In the absence of thyroid hormones, these receptors inhibit the expression of genes to which they bind; the addition of hormone converts them into transcriptional activators.

### Hormones Are Organized into a Hierarchy

The synthesis of many hormones is regulated by a cascade of hormones (fig. 24.20). Frequently, a hypothalamic hormone impinges on the pituitary to stimulate synthesis of a hormone that activates hormone synthesis in yet another organ. The end products of these cascades generally feedback-inhibit the production of hormones at the beginning of the cascade (fig. 24.21).

Another common mechanism for modulating hormonal response involves two (or more) hormonal inputs with both positive and negative effects (see fig. 24.21). The hypothalamic peptides, somatostatin and GRF, have opposite effects on GH synthesis and secretion. Similarly, glucagon and insulin have opposite effects on gluconeogenesis in the liver (see the discussion earlier in this chapter), and some of the effects of ecdysone on gene expression in insects are blocked by juvenile hormone (a terpene derivative; fig. 24.22).

Considerable "cross-talk" also occurs between different hormones at the level of receptor function. For instance, the diacylglycerol-activated protein kinase C phosphorylates and thereby inhibits the activity of insulin and epidermal growth factor receptor kinases (epidermal growth factor is discussed in a later section). Likewise, when cAMP-dependent protein kinases are active they can inhibit receptors for epinephrine by phosphorylating them.

Cells also have mechanisms that tend to prevent chronic stimulation. Exposure of cells to epinephrine leads to an initial sharp rise in cAMP levels; however, cAMP levels fall nearly to basal levels within an hour or so, despite the continuous presence of saturating amounts of epinephrine. Furthermore, if the hormone is removed and the cells

**Table 24.6**

Regulation of Specific Genes by Steroid and Thyroid Hormones

| **Glucocorticoids** | | **Androgens** | |
|---|---|---|---|
| Tyrosine aminotransferase | Liver | β-Glucuronidase | Kidney |
| Tryptophan oxygenase | Liver | Aldolase | Prostate |
| Glutamine synthase | Liver, retina | Prostate-binding proteins | Prostate |
| Phosphoenolpyruvate carboxykinase | Kidney | Ovomucoid | Oviduct |
| | | Ovalbumin | Oviduct |
| Ovalbumin | Oviduct | | |
| Conalbumin | Oviduct (liver)[a] | **1,25-Dihydroxyvitamin D₃** | |
| α-Fetoprotein (↓) | Liver | Calcium-binding protein | Intestine |
| α₂-Globulin | Liver | Ostercalcin | Bone |
| Metallothionein | Liver | | |
| Proopiomelanocortin (↓) | Pituitary | **Ecdysone** | |
| Mammary tumor virus | Mammary gland | Dopa-decarboxylase | Epidermis |
| | | Vitellogenin | Fat body |
| **Estrogens** | | Larval scrum protein I | Fat body |
| Ovalbumin | Oviduct | | |
| Conalbumin | Oviduct (liver) | **Thyroid Hormones** | |
| Ovomucoid | Oviduct | Carbamyl phosphate synthase | Liver |
| Lysozyme | Oviduct | Growth hormone | Pituitary |
| Vitellogenin | Liver | Prolactin (↓) | Pituitary |
| apo-VLDL | Liver | α-Glycerophosphate dehydrogenase | Liver (mitochondria) |
| Glucose-6-P-dehydrogenase | Uterus | Malic enzyme | Liver |
| **Progestins** | | | |
| Avidin | Oviduct | | |
| Ovalbumin | Oviduct | | |
| Conalbumin | Oviduct | | |
| Uteroglobin | Uterus | | |

[a] In the liver, the product of the conalbumin gene is called *transferrin*.
Note: ( ↓ ) means that mRNA levels are decreased by hormone.

(Reprinted with permission from *Nature* (H. R. Bourne et al., The GTPase superfamily: A conserved switch of diverse cell functions, *Nature*, 348:126, 1990). Copyright 1990 Macmillan Magazines Limited.)

are challenged within a few hours, the secondary response is lower. This phenomenon, called desensitization, is associated with both an uncoupling of receptors from adenylate cyclase activation and a decrease in the number of receptors accessible to hormone binding. It occurs only after productive receptor function. The main consequence of desensitization is that exposure to hormone results in transient activation of cellular events, rather than chronic activation.

The response to many polypeptide hormones also diminishes with chronic stimulation, but the time course is much longer (many hours to days) and the mechanism is different from that involved in desensitization. Binding of insulin, calcitonin, LH, TRH, and EGF to their respective receptors promotes a physiological response but ultimately leads to the clearance of these receptors from the surface and a blunting of that response. The hormone-mediated loss of receptors is often referred to as down-regulation. In this case, the receptors are internalized and degraded. Hormone binding leads to a clustering of receptors, as if they were cross-linked. These clusters, or patches, aggregate in membrane structures called coated pits (the intracellular side of the pits is coated with a scaffolding protein called clathrin). Small endosome vesicles that engulf the receptors and their ligands bud off from the coated pits, migrate within the cell, and associate with other membranous structures known collectively as GERL (Golgi–endoplasmic-reticulum lysosomes). After a few hours they fuse with lysosomes, at which point the lysosomal enzymes degrade both the receptor and the hormone (fig. 24.23). Replacement of the hormone receptors requires protein synthesis.

Most hormones are released in a cyclic manner (examples include insulin, glucagon, growth hormone, and many

## Figure 24.20

Growth hormone cascade. Neurosecretory cells in the hypothalamic region of the brain are activated by neurotransmitters from other neurons to secrete growth hormone releasing factor (GRF), a 44-amino-acid peptide that travels through the portal circulation to the anterior pituitary, where it binds to membrane receptors on somatotroph cells. GRF binding stimulates the production of cAMP, which activates growth hormone (GH) synthesis and secretion. GH, a 191-amino-acid polypeptide, passes into the bloodstream and travels to the liver, where it binds to membrane receptors that probably produce a second messenger (as yet unknown) that activates transcription of insulinlike growth factor-1 (IGF-1) gene, which codes for a 71-amino-acid peptide. IGF-1 is secreted from the liver into the bloodstream and ultimately binds to membrane receptors (which are tyrosine kinases similar to the insulin receptor) located on many peripheral cells. Activation of these receptors leads to cellular proliferation under appropriate conditions.

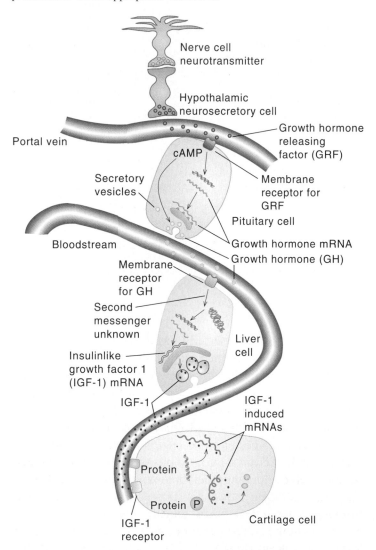

## Figure 24.21

Control of hormone synthesis and secretion in the anterior pituitary. Neurosecretory neurons in the hypothalamus liberate polypeptides that either stimulate ⊕ or inhibit ⊖ hormone synthesis by specialized pituitary cells containing the appropriate receptors. For example, GnRH stimulates gonadotroph cells in the pituitary to synthesize and secrete LH and FSH. These pituitary hormones then impinge on target cells, typically stimulating them to make other low-molecular-weight hormones. The end products of these cascades feedback-inhibit hormone production at either or both hypothalamic and pituitary levels.

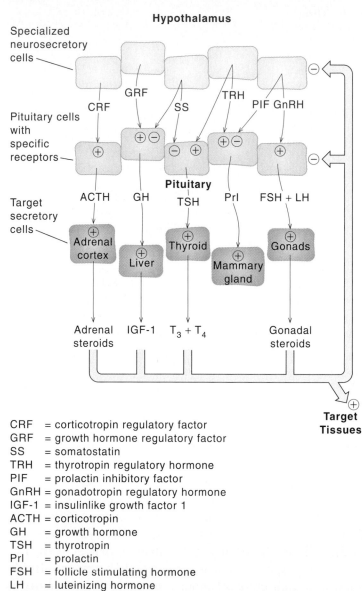

CRF = corticotropin regulatory factor
GRF = growth hormone regulatory factor
SS = somatostatin
TRH = thyrotropin regulatory hormone
PIF = prolactin inhibitory factor
GnRH = gonadotropin regulatory hormone
IGF-1 = insulinlike growth factor 1
ACTH = corticotropin
GH = growth hormone
TSH = thyrotropin
Prl = prolactin
FSH = follicle stimulating hormone
LH = luteinizing hormone
$T_3$, $T_4$ = thyroid hormones

**Figure 24.22**

Structure of β-ecdysone and juvenile hormone. These hormones play major roles in the growth and maturation of insects by controlling the timing for molting of the insect exoskeleton.

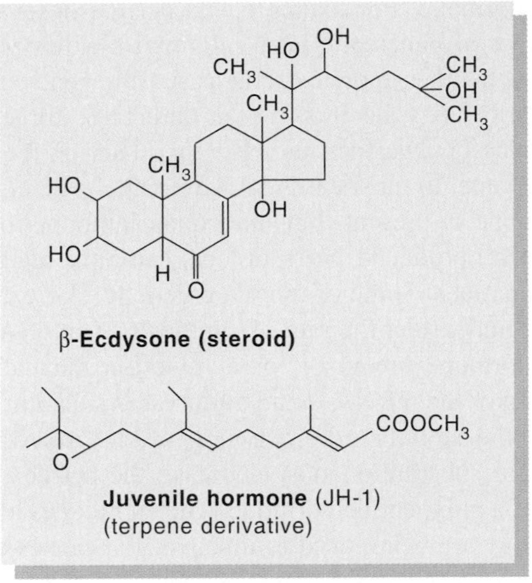

**β-Ecdysone (steroid)**

**Juvenile hormone** (JH-1)
(terpene derivative)

**Figure 24.23**

The down-regulation of receptors by endocytosis. The hormone-mediated loss of receptors is often referred to as down-regulation. Continuous activation of receptors by hormone often leads to patching, a clustering of receptors as if they were cross-linked. Endocytosis of the patches removes them from the cell surface. The endocytotic vesicles, sometimes called receptosomes, fuse with lysosomes, where the contents are degraded by the lysosomal enzymes. Receptosomes also may allow entry of receptors into other cell compartments, such as the nucleus, by fusing with these organelles.

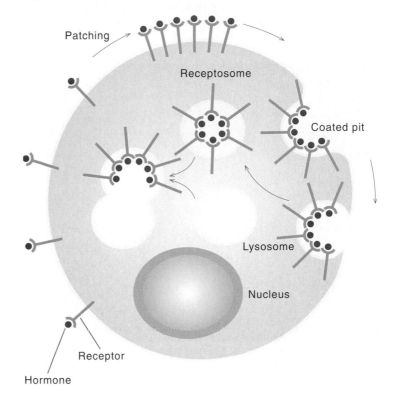

of the hypothalamic releasing hormones). In some cases the cycles appear to be autonomously regulated, but in others they are clearly entrained by neural and hormonal signals that may vary, depending on the developmental stage or physiological condition. The periodic release of peptide hormone can promote distinctly different responses than would be achieved by chronic release.

## Diseases Associated with the Endocrine System

Human diseases related to endocrine dysfunction can be broadly grouped into (1) overproduction of a particular hormone, (2) underproduction of a hormone, and (3) target-cell insensitivity to a hormone.

## Overproduction of Hormones Is Commonly Caused by Tumor Formation

Most cases of hormonal overproduction are associated with enlargement of the normal endocrine organ, frequently owing to a tumor. Pituitary neoplasms usually affect the production of only one pituitary hormone as a result of the cancerous proliferation of the cell type that normally synthesizes that hormone. Examples include giantism (acromegaly), which is associated with proliferation of the somatotroph cells that synthesize growth hormone, and Cushing's syndrome, which is usually due to overproduction of ACTH. Adrenal and parathyroid tumors that lead to the overproduction of various adrenal steroids and parathyroid hormones have also been described.

Occasionally, tumors of organs that do not usually produce a given hormone (ectopic tumors) synthesize and secrete peptide hormones, as in ACTH synthesis by certain lung tumors and GRH production from a pancreatic islet

## Figure 24.24

A transgenic mouse that synthesizes rat growth hormone ectopically and a normal littermate. A chimeric gene with the mouse metallothionein promoter fused to the structural gene of rat growth hormone (rGH) was introduced into the germline of mice by microinjection of cloned DNA into fertilized eggs. Because of the metallothionein promoter, these genes were expressed in many large organs (such as liver, kidney, heart, intestine) that do not normally make GH. As a consequence, the circulating GH level was elevated several hundredfold, an effect that led to increased growth of those mice that inherited the gene. (From R. D. Palmiter, R. L. Brinster, R. E. Hammer, and R. M. Evans. Reprinted by permission from *Nature* 300:611–615. © 1982 Macmillan Magazines Limited. Photo courtesy of Ralph Brinster.)

tumor. Expression of hormones by these ectopic tumors represents a curious activation of gene expression in an inappropriate tissue. The same result can be obtained experimentally by producing animals in which a structural gene codes for a hormone that is under the transcriptional control of a promoter from a gene expressed in another tissue. This type of overproduction is illustrated in experiments with metallothionein-growth hormone (fig. 24.24). In this situation the growth hormone gene has been engineered so that it is not under normal feedback control.

The excessive production of thyroid hormone in Graves' disease is associated with an enlarged thyroid gland, but in this case a circulating immunoglobulin that mimics the activity of thyroid stimulating hormone (TSH) is implicated. Finally, we saw in an earlier section how an inappropriate steroid hormone may be secreted in excessive amounts when specific enzymes in the adrenal steroid biosynthetic pathway are present at inadequate levels.

## Underproduction of Hormones Has Multiple Causes

A wide variety of defects can lead to inadequate production of hormones. In some conditions, not enough cells produce a given hormone. For example, in juvenile-onset diabetes the number of pancreatic islet cells that synthesize insulin decreases. In other, more extreme cases, the gene coding for the hormone may be missing or defective; for example, some forms of dwarfism are due to a lack of the growth hormone gene. In most cases, however, the gene coding for the hormone is present, but inadequate amounts of active hormone are produced. Such defects can have many possible explanations. Some of them are genetic. For example, a mutation may affect the rate of synthesis of mRNA coding for the hormone precursor, or a mutation may affect the processing of the mRNA, or an amino acid substitution may decrease the activity or processing of the mRNA, or an amino acid substitution may decrease the activity or processing of a polypeptide hormone. The techniques of molecular biology are being used to discover the causes of many of these genetic disorders. Other defects in hormone synthesis are the consequence of an inadequate supply of precursors. For example, iodine is essential for thyroid hormone synthesis, and its absence leads to goiter. This condition, involving enlargement of the thyroid, is caused by high concentrations of TSH that result from lack of normal feedback by $T_3$ and $T_4$ on the hypothalamus. Likewise, synthesis of vitamin D requires ultraviolet irradiation of 7-dehydrocholesterol, without which the formation of $1,25(OH)_2D_3$ cannot occur, so that rickets ensues. A rare cause of rickets involves a defect in the enzyme that converts $25(OH)D_3$ to $1,25(OH)_2D_3$, the active form of vitamin D. Similarly, defects in enzymes involved in adrenal steroid biosynthesis lead to Addison's disease, and a defect in $5\alpha$-reductase leads to one form of testicular feminization, the result of inadequate production of dihydrotestosterone.

## Target-Cell Insensitivity Results from a Lack of Functional Receptors

The most dramatic examples of target-cell insensitivity are those in which the correct receptors are lacking. In complete testicular feminization, androgen receptors are missing from all cells. The consequence is phenotypic expression of female characteristics in genotypic males. A rare form of dwarfism (Laron dwarfs) is associated with high plasma levels of growth hormone (GH) but low levels of insulinlike growth factor-1 (IGF-1); these individuals have a defect in the GH receptor. Occasionally, antibodies are directed

Table 24.7

## Table 24.7

Vertebrate Growth Factors

| Factor[a] | $M_r$ | Cell Types Affected |
|---|---|---|
| Epidermal growth factor (EGF)[b] | 6,400 | Epithelial and mesodermal cells |
| Tumor growth factor-$\alpha$ (TGF-$\alpha$) | 7,000 | Same |
| Insulinlike growth factor-1 (IGF-1) | 7,000 | Same |
| Insulinlike growth factor-2 (IGF-2) | 7,000 | Same |
| Fibroblast growth factor (FGF)[c] | 26,000 (dimer) | Same |
| Platelet-derived growth factor (PDGF)[b] | 31,000 | Same |
| Nerve growth factor (NGF)[c] | 13,000 | Sensory and sympathetic neurons |
| Erythropoietin | 23,000 | Erythroid cell precursors |
| Macrophage colony-stimulating factor (M-CSF or CSF-1)[b] | 70,000 (dimer) | Macrophage precursors |
| Granulocyte colony-stimulating factor (G-CSF) | 25,000 | Granulocyte precursors |
| Granulocyte–macrophage colony-stimulating factor (GM-CSF) | 23,000 | Granulocytes and macrophage precursors |
| Multicolony-stimulating factor (multi-CSF) or Interleukin-3 (IL-3) | 25,000 | Precursors of most hematopoietic cells |
| Interleukin-2 (IL-2)[c] | 13,000 | T lymphocytes |

[a] The genes for most of these growth factors and many of their receptors have been cloned.

[b] The EGF receptor is homologous to the *erb*B and *neu* oncogenes, and the M-CSF receptor is homologous to the *fms* oncogene. PDGF is homologous to the *sis* oncogene.

[c] There are many members of the FGF, NGF, and interleukin families of growth factors.

against receptors and interfere with normal hormone binding, such as occurs in one form of adult-onset diabetes. Remarkably, the antibody itself promotes insulin effects.

## Growth Factors

In addition to the many hormones that have been discussed, a large number of growth factors exist that have hormonelike activities. As their name implies, growth factors often stimulate the proliferation of particular cells. Unlike hormones, most growth factors are synthesized by a variety of cell types rather than in a specialized endocrine gland. However, the mechanisms of action of growth factors and hormones are probably very similar.

All of the growth factors characterized so far are proteins (table 24.7). Many of them are cleaved from larger precursors. The 53-amino-acid epidermal growth factor (EGF) is cleaved from a precursor of 1,168 amino acids. This precursor is a membrane-spanning protein, with the EGF moiety and nine related sequences in the extracellular domain. EGF is homologous to several other growth factors,

including $\alpha$-tumor growth factor and a protein secreted by vaccinia virus. These factors bind to membrane receptors and are thought to trigger intracellular events the same way as hormones.

The EGF receptor, for example, shows striking similarity to the insulin receptor in overall organization and amino acid sequence; however, instead of separate $\alpha$ and $\beta$ chains, the EGF receptor is composed of a single polypeptide that appears to be a composite of the $\alpha$ and $\beta$ chains of the insulin receptor. As is true with the insulin receptor, the intracellular domain of the EGF receptor has tyrosine kinase activity.

The receptor for insulinlike growth factor-1(IGF-1) is also similar to the insulin receptor in structure and tyrosine kinase activity. Thus, in this case it is likely that the hormones (insulin and IGF-1) and their receptors have each evolved from common ancestral genes by duplication and divergence.

The receptors for one of the colony-stimulating factors (M-CSF) and for platelet-derived growth factor (PDGF) are also membrane proteins with tyrosine kinase activity. Although the mechanisms by which these receptors regulate cellular activities are not clear yet, a major hypothesis is that they phosphorylate (and thereby either activate or inhibit) regulatory enzymes, including some serine–threonine protein kinases. PDGF also appears to activate the inositol tris-phosphate/diacylglycerol pathway by phosphorylating phospholipase C on tyrosine residucs (see fig. 24.17). This would also lead to activation of several serine/threonine kinases.

Normal cells require growth factors (mitogens) for proliferation. In the absence of these factors they reversibly withdraw from the cell cycle and become arrested. Transformed cells (tumor cells) have a relaxed cell cycle control and can traverse the cell cycle in the absence of added growth factors. Cell transformation can be achieved by activation or inappropriate expression of a variety of cellular or viral genes. The isolation and characterization of many of these transforming genes (oncogenes) have revealed striking homology with growth factors or their receptors. For example, the oncogene product *sis* resembles PDGF in its sequence and its affinity for specific receptors, and the EGF receptor is homologous to the *erb*B oncogene. Thus, a prevalent idea is that many oncogenes lead to abnormal cell proliferation by directly or indirectly providing a rate-limiting factor, be it a growth factor, its receptor, or an intracellular intermediate (see supplement 4).

## Plant Hormones

Plant hormones (also called growth regulators) coordinate the growth and development of plants. The major hormones discovered to date fall into six classes: Auxins, cytokinins, gibberellins, abscisic acid, ethylene and oligosaccharides. The structures of representative members of each class are shown in figure 24.25. All of these hormones are low-molecular-weight compounds; indeed, one hormone, ethylene, is a gas. The polypeptide hormone of higher plants, systemin, is an 18-amino-acid peptide that is released on wounding and activates the plant defense system. Polypeptide and steroid hormones have not been described for higher plants, although some yeasts and fungi use these compounds as mating factors. Each class of compounds elicits many diverse responses, and considerable interaction occurs among different plant hormones in the control of physiological processes. Furthermore, the same process is controlled by different hormones (or combinations of hormones) in different species. These considerations make it difficult to analyze and generalize about the mechanism of plant hormone action.

Auxins are synthesized in the apical buds of growing shoots. They stimulate growth of the main shoot, but inhibit the development of lateral shoots; this effect has led to the horticultural practice of pinching off the apical buds to stimulate the formation of bushy plants. The curvature of plants toward the light (phototropism) is thought to be due to transport of auxins away from the light, which stimulates more rapid growth of cells on the darker side of the shoot. Auxins bind to specific membrane proteins. The affinity of auxins for these proteins, coupled with their location and abundance on target cells, supports the view that the membrane proteins may be the receptors that mediate auxin action.

Auxin stimulation of growth occurs in two phases. The earliest response (called the rapid response) is an increase in proton transport out of the cell, which occurs after a lag of a few minutes. This hydrogen ion pump is thought to be coupled with a membrane ATPase, but it is not clear whether the receptor interacts directly with the ATPase or whether other intermediates are involved. It is thought that lowering the extracellular pH activates enzymes that partially degrade the cell wall, thereby loosening it and allowing for cell expansion. A subsequent effect of auxin (the slow response) is to increase the synthesis of proteins and nucleic acids, resulting in sustained growth. Auxin has been shown, for example, to increase the amount of cellulase mRNA during pea cell expansion. A 20-kd membrane protein that binds auxins has been characterized that appears to be the auxin receptor in that an antibody against it blocks auxin-stimulated events.

Cytokinins are adenine derivatives that are produced in the roots and that promote growth and differentiation of numerous tissues. Cytokinins can overcome the auxin-mediated inhibition of lateral shoots; in other tissues both hormones act synergistically. Plant cells can be grown in culture if auxins and cytokinins are provided. With relatively balanced concentrations of both hormones, cells proliferate but remain unorganized. If the ratio of auxins to cytokinins is high, shoots develop; if the ratio is low, root development is favored. Intermediate concentrations of both hormones promote undifferentiated growth, with neither roots nor shoots. As these results show, plant cells can display a wide range of physiological responses. They are much more plastic in their developmental potential than animal cells. Indeed, normal plants can be grown from a single tissue culture cell. Although this diversity of responses is fascinating, it has also made it difficult to define the mechanism of action of plant hormones.

Gibberellins also promote shoot elongation, and frequently they act synergistically with auxins. Gibberellins stimulate the accumulation of specific mRNAs such as amylase mRNA in germinating seeds. This fact suggests that their receptors (or a second messenger) act at the genetic level.

## Figure 24.25

Some common plant hormones. All of these hormones are low-molecular-weight compounds. One hormone, ethylene, is a gas.

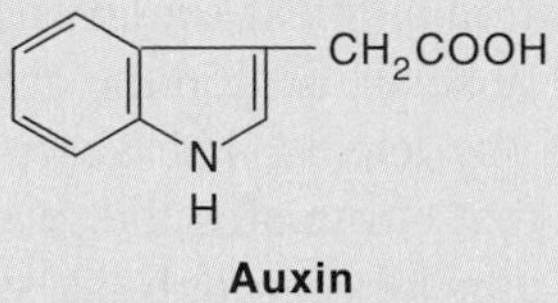

**Auxin**
(indole acetic acid)

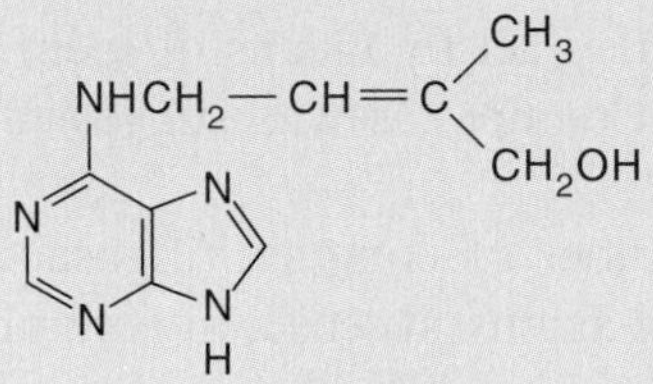

**Cytokinin**
(zeatin)

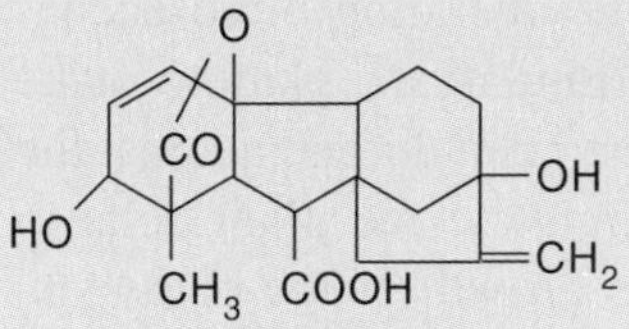

**Gibberellin**
(gibberellic acid)

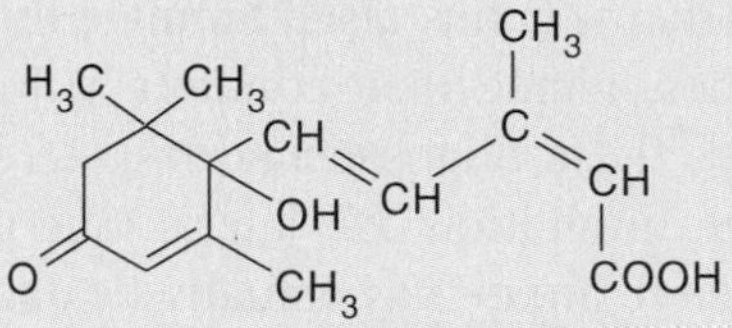

**Abscisic acid**

$H_2C=CH_2$
**Ethylene**

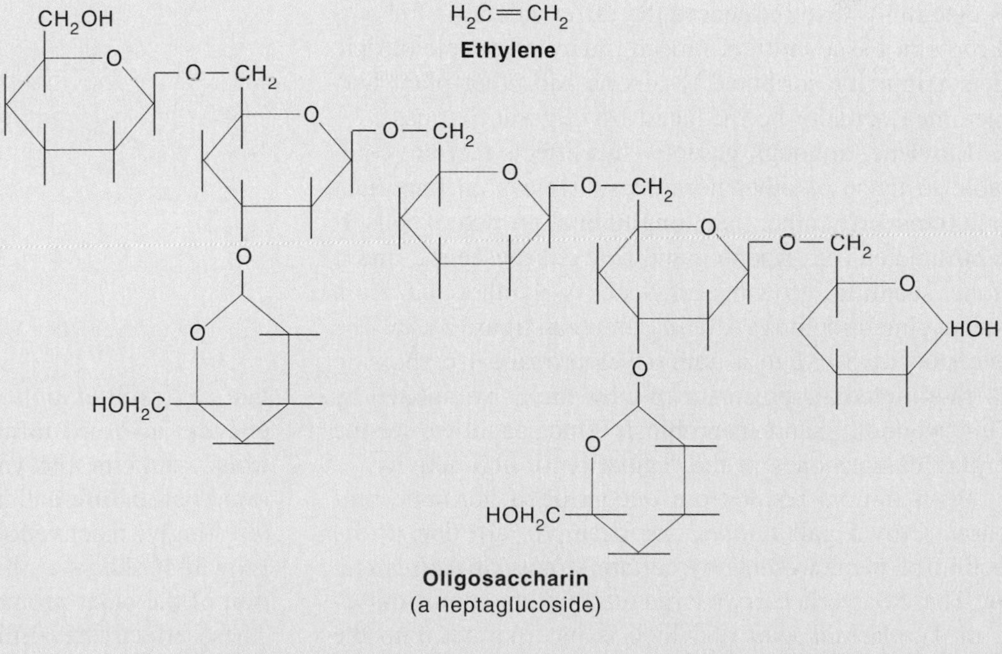

**Oligosaccharin**
(a heptaglucoside)

Abscisic acid is antagonistic to many other plant hormones. It inhibits germination of seeds and shoot growth while promoting resting bud formation and leaf senescence. Wilting stimulates the synthesis of abscisic acid in the chloroplasts within mesophyll cells of the leaves. The abscisic acid in this case is a stress signal that stimulates the guard cells to close and thus minimize water loss through the stomata. Abscisic acid stimulates $K^+$ efflux from guard cells and into adjacent cells; thus it may act by means of membrane receptors to modulate ion pumps, as was suggested for auxins.

Oligosaccharin is one member of complex oligosaccharides that function in defense against disease, in control of plant growth, and in differentiation. They are released from the cell wall by specific degradative enzymes. For example, when bacteria or fungi infect plants, they produce enzymes that degrade the cell wall, liberating oligosaccharides. Some of these compounds stimulate the plant to make antibiotics or proteases that help the plant defend against the pathogen or insect. There they stimulate the plant cells to produce antibiotics that inhibit the growth of the pathogen. When certain plant cells are damaged they can produce the enzymes required to liberate oligosaccharides, thus allowing a hormonal response even when the pathogen does not produce the appropriate enzymes. Auxins also stimulate the production of oligosaccharides, which then counteract the auxin-stimulated growth; thus these oligosaccharides serve as feedback inhibitors. The inhibition of lateral shoot growth that has been attributed to auxin may actually be due to the production of oligosaccharides.

Addition of oligosaccharides to combinations of auxin plus cytokinins also influences the differentiation of shoots and roots in tissue culture. Indeed, many of the pleiotropic effects originally attributed to auxins and other plant hormones may actually be mediated by oligosaccharins.

Ethylene, although gaseous, has effects that are comparable to those of other hormones. It plays an important role in transverse rather than longitudinal growth of cells. It also stimulates fruit ripening and flower senescence, and it inhibits seedling growth. Ethylene is synthesized from S-adenosylmethionine (SAM), as shown in figure 24.26. The conversion of SAM to 1-aminocyclopropane-1-carboxylic acid, the immediate precursor of ethylene, is stimulated by auxins, wounding, and anaerobiosis. Once again we see the interplay of hormones in the regulation of cell activity.

Plant tumors result from uncontrolled hormone production. Crown gall tumors, for example, are due to the infection of plant wounds by certain strains of *Agrobacterium*. These bacteria carry a large plasmid, the tumor-inducing, or $T_1$ plasmid, part of which is incorporated into the plant genome. This DNA encodes several genes that stimu-

late cell proliferation. One gene product is a rate-limiting enzyme involved in auxin biosynthesis, and another controls cytokinin biosynthesis. Together they promote the rapid but undifferentiated growth that is the crown gall. Interestingly, inactivation of one of the genes promotes the growth of shoots at the site of infection, whereas inactivation of the other promotes the growth of roots (fig. 24.27). These effects are reminiscent of the action of auxins and cytokinins in tissue culture.

### Figure 24.26

Synthesis of ethylene, a gaseous plant hormone involved in fruit ripening and flower senescence.

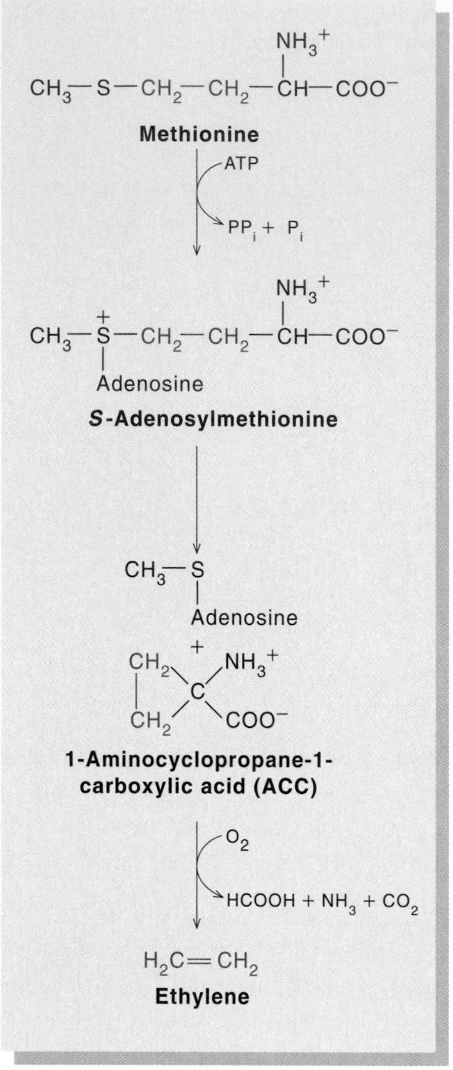

**Figure 24.27**

Crown gall tumors of plants are caused by certain strains of the bacterium *Agrobacterium* that can infect plant cells and introduce new genetic information coding for plant hormones. The transforming DNA of these bacterial strains is carried on a large plasmid; it contains genes coding for rate-limiting enzymes in hormone biosynthesis. (*a*) The transforming DNA of wild-type bacterial plasmid codes for rate-limiting enzymes involved in both auxin and cytokinin biosynthesis, resulting in undifferentiated growth (called callus). (*b*) Mutations that disrupt the gene involved in auxin biosynthesis give rise to tumors that produce only cytokinins, which leads to a proliferation of leaves and shoots. (*c*) Alternatively, mutations that disrupt the gene involved in cytokinin synthesis give rise to tumors that produce only auxin. In these tumors, only roots differentiate.

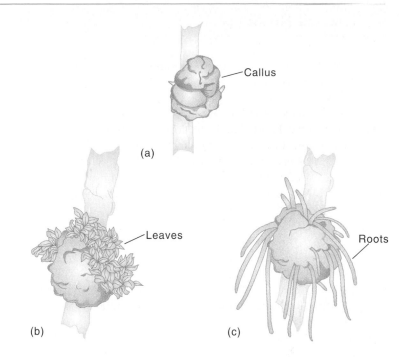

## Summary

In this chapter we focused on the ways in which various metabolic activities are integrated, with special attention to the nature and functioning of hormones. The following points are the highlights of our discussion.

1. Tissues store biochemically useful energy in three major forms: Carbohydrates, lipids, and proteins. Each tissue makes characteristic demands on and contributions to the energy supply of the organism.

2. Hormones are chemical messengers formed in specific tissues. They circulate between tissues of multicellular organisms and serve to coordinate metabolic activities, maintain homeostasis of essential nutrients, and prepare the organism for reproduction.

3. Most hormones fall into three classes: Polypeptides, steroids, and amino acid derivatives. Polypeptide hormones are synthesized from large precursors. Steroid hormones are derivatives of cholesterol. Thyroid hormones and epinephrine are amino acid derivatives.

4. A number of factors—synthesis, rate of release, and rate of elimination—determine the concentration of circulating hormone.

5. Hormones act by reversibly binding to proteins called receptors, an event that results in a conformational change that is detected by other macromole-cules (acceptors) and that eventually leads to activation of rate-limiting enzymes. Each class of hormones binds to specific receptors that activate other membrane proteins.

6. Most membrane receptors generate a diffusible intracellular signal called a second messenger. Five intracellular messengers are currently known: Cyclic AMP, cyclic GMP, inositol triphosphate, diacylglycerol, and calcium. Second messengers usually activate or inhibit the action of one or more enzymes.

7. Steroid hormones penetrate the cell and bind to receptors in the nucleus, and activate (or sometimes repress) transcription of specific genes. Thyroid hormones act similarly.

8. A large number of diseases are due to either overproduction or underproduction of hormones, or to insensitivity of target tissues to circulating hormones. Knowledge of hormone biosynthesis, secretion, and interaction with target cells is essential to an understanding of the biochemical basis of these disorders.

9. In addition to classical hormones, other chemical messengers called growth factors serve to coordinate growth of tissues during development.

10. Plants make several hormones that regulate growth and differentiation.

## Selected Readings

*Annual Reviews of Biochemistry* and *Annual Reviews of Physiology.* Over 40 volumes in each series with many relevant reviews in the area of hormone receptors and hormone action. Provides good access to primary literature.

Baxter, J. D., and K. M. MacLoed, Molecular basis for hormone action. In P. K. Bondy and L. E. Rosenberg (eds.), *Metabolic Control and Disease.* Philadelphia: Saunders, 1980. A useful summary.

Berridge, M. J., Inositol trisphosphate and calcium signalling. *Nature* 301:315–325, 1993.

Bredt, D. S., and S. H. Snyder, Nitric oxide, a novel neuronal messenger. *Neuron* 8:3–11, 1992. A review.

Carafoli, E., and J. T. Penniston, The calcium signal. *Sci. Amer.* 253(5):70–78, 1985.

Carpenter, G., Receptors for epidermal growth factor and other polypeptide mitogens. *Ann. Rev. Biochem.* 56:881–914, 1987:

Clapham, E. D., and E. J. Neer, New roles for G-protein $\beta\gamma$-dimers in transmembrane signalling. *Nature* 365:403–406, 1993.

Collins, S., M. G. Caron, and R. J. Lefkowitz, From ligand binding to gene expression: New insights into the regulation of G-protein-coupled receptors. *Trends Biochem. Sci.* 17:37–39, 1992.

Czech, M. P., J. K. Klarlund, K. A. Yagaloff, A. P. Bradford, and R. E. Lewis, Insulin receptor signaling. *J. Biol. Chem.* 263:11017–11020, 1988.

deGroot, L. J., et al. (eds.), *Endocrinology,* 3 vols. New York: Grune and Stratton, 1979. Over 2,000 pages of comprehensive treatment, primarily from a medical point of view.

DeVos, A. M., M. Ultsch, and A. A. Kossiakoff, Human growth hormone and extracellular domain of its receptor: Crystal structure of the complex. *Science* 255:306–312, 1992.

Fantl, W. J., D. E. Johnson, and L. T. Williams, Signalling by receptor tyrosine kinases. *Ann. Rev. Biochem.* 62:453–481, 1993.

Feig. L. A., The many roads that lead to RAS. *Science* 260:767–768, 1993.

Funder, J. W., Mineralcorticoids, glucocorticoids, receptors and response elements. *Science* 259:1132–1133, 1993.

Gerisch, G., Cyclic AMP and other signals controlling cell development and differentiation in *Dictyostelium. Ann. Rev. Biochem.* 56:853–879, 1987.

Gilman, A. G., G proteins: Transducers of receptor-generated signals. *Ann. Rev. Biochem.* 56:615–650, 1987.

Guillemin, R., Peptides in the brain: The new endocrinology of the neuron. *Science* 202:390, 1978. Nobel laureate speech.

Jones, A. M., Surprising signals in plant cells. *Science* 263:183–184, 1994.

Kikkawa, U., A. Kishimoto, and Y. Nishizuku, The protein kinase C family: Hctrogeneity and its implications. *Ann. Rev. Biochem.* 58:31–44, 1989.

Lehmann, J. M., L. Jong, A. Fanjul, J. F. Cameron, X. P. Lu, P. Haefner, M. I. Dawson, and M. Pfahl, Retinoids Selective for Retinoid X Receptor Response Pathways. *Science* 258:1944–1946, 1992.

Lowy, D. R., and B. M. Willumsen, Function and regulation of RAS. *Ann. Rev. Biochem.* 62:851–891, 1993.

Marx, J., Two major signal pathways linked. *Science* 262:988–990, 1993.

Napier, R. M., and M. A. Venis, From auxin-binding protein to plant hormone receptor? *Trends Biochem. Sci.* 16:72–75, 1991.

O'Malley, B. W., and L. Birnbaumer, *Receptors and Hormone Action,* 3 vols. New York: Academic Press, 1977. Review articles by many authors cover most aspects of hormone action.

Pelech, S. L., and D. E. Vance, Signal transduction via phosphatidylcholine cycles. *Trends Biochem. Sci.* 14:28–30, 1989.

Pifkis, S. J., M. R. El-Maghrabi, and T. H. Claus, Hormonal regulation of hepatic gluconeogenesis and glycolysis. *Ann. Rev. Biochem.* 57:755–784, 1987.

Rasmussen, H., The cycling of calcium as an intracellular messenger. *Sci. Am.* 261(4):66–73, 1989.

*Recent Progress in Hormone Research.* New York: Academic Press. An annual publication with nearly 40 volumes. A good place to find a recent summary.

Riddiford, L. M., and J. W. Truman, Biochemistry of insect hormones and insect growth regulators. In *Biochemistry of Insects.* New York: Academic Press, 1978.

Simon, M. I., P. Strathmann, and N. Gautam, Diversity of G proteins in signal transduction. *Science* 252:802–808, 1991.

Simpson, I. A., and S. W. Cushman, Hormonal regulation of mammalian glucose transport. *Ann. Rev. Biochem.* 55:1059–1089, 1986.

Sutherland, E. W., Studies on the mechanism of hormone action. *Science* 177:401, 1972. Nobel laureate speech related to discovery of cAMP as second messenger.

Wittinghofer, A., and E. F. Pai, The structure of Ras protein: a model for a universal molecular switch. *Trends Biochem. Sci.* 16:382–387, 1991.

Yarden Y., and A. Ullrich, Growth factor receptor tyrosine kinases. *Ann. Rev. Biochem.* 57:443–478, 1988.

## Problems

1. Most metabolic conversions can occur in either direction by using different pathways. Eukaryotic cells frequently take advantage of subcellular compartments to separate oppositely directed pathways. Use fatty acid synthesis and degradation as an example, and discuss the design of the paths from the point of view that they are thermodynamically favorable and kinetically regulated (be sure to consider subcellular compartments). Discuss the design of glycolysis and gluconeogenesis in the liver.

2. If a starving person (one who has gone a number of weeks with no food) is given a shot of insulin, what happens? Explain your answer.

3. Why are all known hormone receptors proteins? Can other macromolecules serve as receptors?

4. What is the advantage of the fact that muscle tissue can use an anaerobic metabolism?

5. List several reasons why polypeptide hormones are synthesized as precursors.

6. A patient has a hypothyroid condition. He has low serum $T_3$ and $T_4$ levels and elevated levels of serum thyroid-stimulating hormone (TSH). After injection of thyrotropin-releasing hormone (TRH), his serum TSH goes even higher. Is his defect primary (thyroid), secondary (pituitary), or tertiary (hypothalamus)?

7. Thyrotropin-releasing hormone (TRH) contains an unusual amino acid, pyroglutamate (fig. 24.8). Have we encountered this amino acid before under a different name? Where do you think the name pyroglutamate came from?

8. Activation of most membrane-associated hormone receptors generates a second messenger. What is a second messenger and what are the five second messengers currently known? What role do G proteins play in second-messenger formation?

9. Is vitamin D a hormone or a vitamin? Explain your answer?

10. Inhibitors of protein synthesis have been shown to block both the rapid and slow auxin-mediated growth responses. How do you explain these observations?

# Notes on Nutrition

Water accounts for over half the body mass (55%) of the average human. Of the remaining 45%, 19% is protein, 19% is lipid, less than 1% is carbohydrate, and 7% is inorganic material. Nutrients must contain the raw materials that go into the construction of the components of the human body. In addition, nutrients must supply the necessary chemical energy and enzyme cofactors (vitamins and trace metal elements) that are required for the maintenance and growth of the human body. The human body requires nutrients such as water, amino acids, fats, carbohydrates, and major minerals in large amounts. Vitamins and trace metal elements are required in smaller amounts.

In the ''notes'' that follow, we will first consider some of the human body's energy requirements vis-á-vis nutrients. Then we will consider the nutritional value of different types of food.

## Energy Requirements

Gross energy requirements for the human body vary over a wide range according to whether an individual is in an active or a resting state. An average 154-lb man requires somewhere between 2,000 and 3,000 kcal per day to maintain his basal metabolic rate (BMR), the energy required by an awake individual during physical, digestive, and emo-

tional rest, and moderate activity. The additional energy expenditure required for an average adult to engage in various activities has been estimated (table 1). The metabolic rate required for daily activities is measured directly by the heat evolved, or indirectly by the volumes of oxygen consumed and carbon dioxide evolved per unit time.

Humans also differ in their energy intake requirements according to their stage in life, sex, and weight (table 2). In general, the energy intake requirements of infants are lower than those of adults, despite the fact that they are in a growing state. This is because infants weigh less. Females generally require less energy intake than males because they have less muscle tissue.

If an individual consumes more food than is needed to supply his or her energy intake requirements, the individual will gain weight; if the individual consumes less food than is needed, the individual will lose weight. Clearly, a happy medium must be struck to maintain a reasonably normal body weight.

## Nutritional Value of Different Foods

Foods differ in their energy value, which is usually expressed in kilocalories or kilojoules, with proteins and carbohydrates having less caloric value than fats (lipids). Fats

## Table 1

Additional Energy Expenditure for a 154-lb Adult during Various Activities

| Activity | Extra Energy | |
| | Kilojoules per Hour | Kilocalories per Hour |
| --- | --- | --- |
| Lying still, awake | 29 | 7 |
| Eating | 117 | 28 |
| Driving car | 264 | 63 |
| Dishwashing | 293 | 70 |
| Typing (at high speed, electric typewriter) | 293 | 70 |
| Walking 3 mph | 586 | 140 |
| Bicycling ~8 mph | 732 | 175 |
| Walking 4 mph | 996 | 238 |
| Running | 2051 | 490 |
| Swimming 2 mph | 2314 | 553 |

## Table 2

Recommended Dietary Allowances for Energy Intake[*]

| Age (year) and Sex Group | Weight (lb) | Height (in) | Energy Needs (kcal) |
| --- | --- | --- | --- |
| Infants | | | |
| 0.0–0.5 | 13 | 24 | 650 |
| 0.5–1.0 | 20 | 28 | 850 |
| Children | | | |
| 1–3 | 29 | 35 | 1300 |
| 4–6 | 44 | 44 | 1800 |
| 7–10 | 62 | 52 | 2000 |
| Males | | | |
| 11–14 | 99 | 62 | 2500 |
| 15–18 | 145 | 69 | 3000 |
| 19–22 | 160 | 70 | 2900 |
| 23–50 | 174 | 70 | 2900 |
| 51+ | 170 | 68 | 2300 |
| Females | | | |
| 11–14 | 101 | 62 | 2200 |
| 15–18 | 120 | 64 | 2200 |
| 19–22 | 128 | 65 | 2200 |
| 23–50 | 138 | 64 | 2200 |
| 51+ | 143 | 63 | 1900 |
| Pregnancy 1st trimester | | | +0 |
| 2nd trimester | | | +300 |
| 3rd trimester | | | +300 |
| Lactation | | | +500 |

Source: From Recommended Dietary Allowances, Revised 1989, Food and Nutrition Board, National Academy of Sciences–National Research Council, Washington, D.C.

[*] The data in this table have been assembled from the observed median heights and weights as surveyed in the U.S. population (NHANES II data).

are the most efficient source of chemical energy for ATP production.

The ideal diet not only supplies essential calories, it provides the organism with the specific raw materials needed for maintenance and balanced growth. The recommended daily percentages of fats, proteins, and carbohydrates are indicated in figure 1 and table 3.

Since the human body is composed of only about 1% carbohydrate by mass, it may seem surprising that the recommended diet contains more than 50% carbohydrate (see figure 1). This discrepancy arises because carbohydrates are the ideal nutrients for satisfying immediate energy needs, and they also furnish the carbon skeletons for the synthesis of other types of biomolecules.

Understanding the total nutritional needs of the human body requires a more detailed consideration of the chemical composition of different nutrients. As we have seen in Chapter 22, the body cannot synthesize many of the amino acids needed to build proteins. For this reason, the body requires proteins that contain the essential amino acids in sufficient quantities. Too little animal protein will lead to nutritional deficiencies because plant proteins do not contain an adequate supply of all the amino acids that active humans must receive from their diets.

Fruits and vegetables are very important in the diet in order to assure the proper supply of vitamins needed for normal metabolic processes. Vegetables are also an important source of linoleic acid, an essential fatty acid that is in

## Figure 1

**United States dietary goals.** Graphical comparison of the composition of the current U.S. diet and the dietary goals for the U.S. population suggested by the Senate Select Committee on Human Nutrition. (From *Dietary goals for the United States,* 2nd ed., U.S. Government Printing Office, Washington, D.C. 1977.)

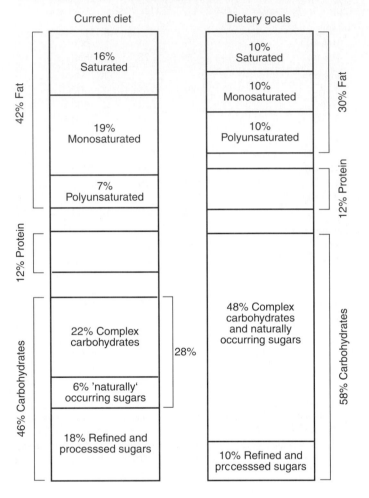

## Table 3

### Dietary Recommendations of Major Health Organizations

|  | Senate Select Committee on Human Nutrition[a] | USDA[b] and Surgeon General[c] |
|---|---|---|
| Obesity | Reduce to ideal weight | Maintain ideal weight |
| Fat | 30% of total calories | Avoid too much fat |
| Type of fat |  | Avoid too much saturated fat |
|  | 10% MUFA[d] |  |
|  | 10% PUFA |  |
| Cholesterol | 300 mg day$^{-1}$ | Avoid too much cholesterol |
| Carbohydrate | 58% of total calories |  |
| Type of carbohydrate | 48% complex CHO | Eat foods with starch and fiber |
|  | 10% simple sugars | Avoid too much sugar |
| Vitamins/minerals |  | Eat a variety of foods |
| Alcohol |  | If you drink do so in moderation |
| Sodium | Limit sodium intake to 5 g day$^{-1}$ | Avoid too much sodium |

[a] Senate Select Committee on Human Nutrition. *Dietary goals for the United States,* 2nd ed. U.S. Government Printing Office, 1977.

[b] U.S. Department of Agriculture. *Nutrition and your health.* U.S. Government Printing Office, 1980.

[c] The Surgeon General's Report on Nutrition and Health, U.S. Government Printing Office, 1988.

[d] MUFA = monounsaturated fatty acids; PUFA = polyunsaturated fatty acids.

short supply in animal tissues (table 4). Finally, vegetables are an important source of fiber, which serves to aid in the digestive processes.

In Chapter 20, we discussed the health-threatening situation that often arises from excess cholesterol. In this regard, vegetable fats and oils are better for humans than animal fats because they contain no cholesterol and less of the saturated fatty acids that lead to elevated levels of blood cholesterol (see tables 4 and 5).

## Table 4

### Lipid Composition of Foods

| Food Group | Percent Linoleic Acid | P/S Ratio[a] |
|---|---|---|
| Milk | | |
| Butterfat | 2 | 1/34 |
| Meat | | |
| Lard | 10 | 1/34 |
| Beef fat | 2 | 1/24 |
| Vegetables and fruits | | |
| Corn oil margarine | 32 | 1/0.45 |
| Corn oil | 57 | 1/0.2 |
| Safflower oil | 77 | 1/0.1 |
| Soybean oil | 52 | 1/0.3 |
| Peanut oil | 28 | 1/0.5 |
| Bread and cereals[b] | | |
| Wheat flour, white | | 1/0.3 |
| Rolled oats | | 1/0.5 |

[a] Total polyunsaturated fatty acids/total saturated fatty acids.

[b] Composition of extracted lipid.

## Table 5

### Cholesterol Content of Foods

| Food Group | Cholesterol (mg/100 g) |
|---|---|
| Cow's milk | |
| Skimmed | 1 |
| Whole | 11 |
| Butter | 250 |
| Cheese | 100 |
| Meat | |
| Beef, pork, chicken, bacon, fish | 60 to 70 |
| Brain | 2000 |
| Kidney | 375 |
| Liver | 300 |
| Sweetbreads | 250 |
| Egg yolk | 1500 |
| Egg (whole) | 550 |
| Oysters | 200 |
| Vegetables, fruits (all) | 0 |
| Bread, cereal (all) | Trace (from additives) |

# References

Montgomery, R., T. W. Conway, and A. A. Spector, *Biochemistry, A Case-Oriented Approach,* 5th ed. St. Louis: C. V. Mosby, 1990.

Devlin, T. M. (ed.), *Textbook of Biochemistry with Clinical Correlations,* 3rd ed. New York: Wiley-Liss, 1992.

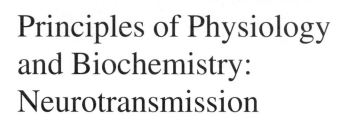

# Principles of Physiology and Biochemistry: Neurotransmission

In chapter 24 we considered some of the ways in which multicellular organisms coordinate the metabolic activities of different cells and tissues. We saw that hormones released in one part of the body travel through the blood to remote target tissues, where they bind to specific receptors. Many of the hormone receptors are membrane proteins that are designed to regulate the activities of other membrane-bound enzymes. In this chapter we turn to another mechanism that higher animals use for communication among different cells and organs, neurotransmission. Neurotransmission also is mediated by membrane-bound proteins, and it hinges critically on movements of ions across membranes, a topic we encountered before in chapters 14 and 17.

The cell body of a mammalian motor neuron has an extended process termed an axon that branches at its tip to make multiple contacts (synapses) with a muscle cell (fig. S1.1). Some axons actually reach lengths of several meters. Much of the axon may be encased in myelin sheaths, which are multilayered membranes formed by other cells that wrap themselves around the neuron. Shorter processes (dendrites) extend from the cell body to make contacts with other neurons. If a motor neuron is stimulated electrically or is triggered by its connecting neurons, an electric signal called an action potential sweeps down the axon and is transmitted to the muscle cell, which then proceeds to contract. Other

## Figure S1.1

Schematic diagram of a typical motor neuron (a nerve cell conducting impulses to muscle cells).

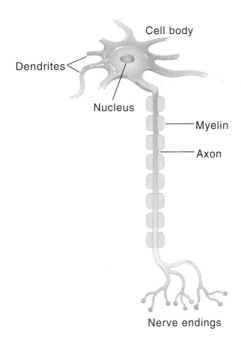

Cell body

Dendrites

Nucleus

Myelin

Axon

Nerve endings

types of neurons transmit electric signals to the brain from sensory organs, such as the eye, nose, or skin, or from place to place within the central nervous system. Our goals in this chapter are to examine the biochemical nature of such signals and to consider how they are transmitted from cell to cell.

## Pumping and Diffusion of Ions Creates an Electric Potential Difference across the Plasma Membrane

We saw in chapter 17 that the plasma membranes of most eukaryotic cells contain ATP-driven pumps that transport $Na^+$ out of the cell and bring $K^+$ in. In addition, the plasma membrane contains specific ion channels that facilitate $K^+$ diffusion out of the cell, down its electrochemical potential gradient. Outward diffusion of $K^+$ moves positive charge across the membrane, making the inside of the cell negatively charged relative to the extracellular solution. The steady-state electric potential difference ($\Delta\psi$) across the plasma membrane varies among different types of cells, depending on the permeability of the membrane to $K^+$ and other ions, but is particularly large in electrically excitable cells, such as nerve and muscle cells. In a

motor neuron that is not excited electrically, $\Delta\psi$ ($\psi_{in} - \psi_{out}$) is about $-70$ mV. In a muscle cell, it is about $-90$ mV.

A mammalian motor neuron has an internal $K^+$ concentration of about 150 mM. The external $K^+$ concentration is about 5.5 mM. $K^+$ ions are almost at equilibrium in this situation because, although the concentration gradient set up by the $Na^+$–$K^+$ ATPase favors diffusion of $K^+$ out of the cell, $\Delta\psi$ favors $K^+$ influx. As we discussed in chapters 14 and 17, the net change in free energy for diffusion of $K^+$ in either direction can be calculated by summing the contributions of these opposing terms:

$$\Delta G_{out \to in}^{K^+} = 2.3RT \log\left(\frac{[K^+]_{in}}{[K^+]_{out}}\right) + \mathscr{F}(\psi_{in} - \psi_{out})$$

$$\approx (1.42 \text{ kcal/mol}) \times \log\left(\frac{150}{5.5}\right)$$

$$+ (23 \text{ kcal/mol} \cdot \text{V}) \times (-0.07 \text{ V})$$

$$\approx 2.0 - 1.6 = 0.4 \text{ kcal/mol} \qquad \textbf{(1)}$$

$$\Delta G_{in \to out}^{K^+} = -\Delta G_{out \to in}^{K^+} \approx -0.4 \text{ kcal/mol} \qquad \textbf{(2)}$$

The situation for $Na^+$ is quite different because both the concentration gradient and $\Delta\psi$ favor $Na^+$ influx. The intracellular $Na^+$ concentration in a mammalian neuron is about 15 mM, and the extracellular concentration is about 150 mM. Combining the concentration and electrical terms in this case gives

$$\Delta G_{out \to in}^{Na^+} = 2.3RT \log\left(\frac{[Na^+]_{in}}{[Na^+]_{out}}\right) + \mathscr{F}(\psi_{in} - \psi_{out})$$

$$\approx (1.42 \text{ kcal/mol}) \times \log\left(\frac{15}{150}\right)$$

$$+ (23 \text{ kcal/mol} \cdot \text{V}) \times (-0.07 \text{ V})$$

$$\approx -1.4 - 1.6 = -3.0 \text{ kcal/mol} \qquad \textbf{(3)}$$

$Na^+$ ions are substantially out of equilibrium because, although the plasma membrane does have specific channels for $Na^+$ diffusion, in the resting nerve cell almost all of these channels are in an inactive, nonconducting state.

Note that according to equations (1) and (2), even $K^+$ is not really at equilibrium across the membrane: $\Delta G_{in \to out}^{K^+}$ is small, but not zero. This is because the rate of flow of $K^+$ out of the cell is limited by the number of $K^+$-conducting channels and by the intrinsic resistance that these channels offer to the flow of ions. Although the overall conductivity of the resting membrane to $K^+$ is about 100 times greater than the conductivity to $Na^+$, it still is much lower than the conductivity of an equal thickness of water.

**Figure S1.2**

A device for eliciting and recording action potentials along a nerve axon. Brief closure of the switch connected to the stimulating electrode causes a current pulse into the axoplasm. If an impulse is generated, resultant potential changes can be detected by the recording electrode, which is connected to an oscilloscope or other recording device.

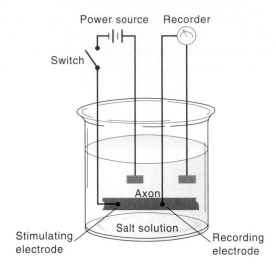

## An Action Potential Is a Wave of Local Depolarization and Repolarization of the Plasma Membrane

In a classic series of studies beginning in the 1930s, A. L. Hodgkin and A. F. Huxley showed that the action potential that sweeps along a stimulated nerve results from changes in the conductivity of the plasma membrane to $Na^+$ and $K^+$. In a typical experiment, the membrane potential of a squid nerve axon was probed by an intracellular electrode made by pulling a glass pipet to a fine tip (fig. S1.2). A brief voltage spike was applied to stimulate the nerve some distance away from the recording electrode, and signals appearing at the recording electrode were measured as a function of time.

To initiate an action potential, the electrical stimulus must depolarize the membrane locally by about 20 mV, that is, change $\Delta \psi$ from $-70$ to about $-50$ mV. When this threshold is reached, an action potential is launched down the axon. After a delay that depends on the distance between the stimulating and recording electrodes, the membrane potential registered by the recording electrode drops abruptly through zero and overshoots to about $+40$ mV (fig. S1.3). The depolarization of the membrane lasts for about 1 ms, after which $\Delta \psi$ rebounds to a negative value somewhat beyond the original resting potential. This wave of local depo-

**Figure S1.3**

A typical action potential $V$ (blue) that might be recorded by the instrument in figure S1.2 if the stimulating current is sufficient to depolarize the membrane by at least 20 mV. The membrane potential returns to its resting value of about $-60$ mV within about 5 ms. Accompanying changes in relative $Na^+$ conductance, $g_{Na^+}$ (red), and $K^+$ conductance, $g_{K^+}$ (green), are also shown. Depolarization is correlated with an increase in $g_{Na^+}$, whereas repolarization is accompanied by a decrease in $g_{Na^+}$ and a transient increase in $g_{K^+}$. Note that the membrane potential transiently becomes more negative than the resting value (hyperpolarizes) until $g_{K^+}$ returns to its normal value. The time of appearance of the action potential after stimulation at $T = 0$ depends on the distance between the recording and stimulating electrodes. (*Note:* A convenient measure of membrane permeability is the conductance ($g$) of the membrane; this is the reciprocal of electrical resistance (ohms) and is measured in reciprocal ohms (mhos). mmho refers to millimhos or mhos $\times 10^{-3}$.) (Source: S. W. Kuffler and J. G. Nicholls, *From Neuron to Brain.* Copyright © 1976 Sinauer Associates, Sunderland, Mass.)

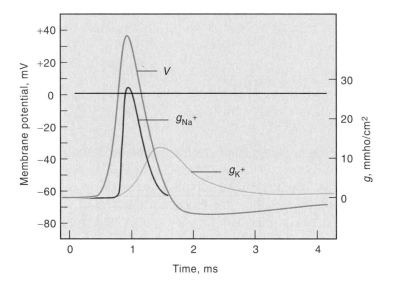

larization and recovery propagates along the axon at a rate that, with myelinated nerve fibers, can exceed 100 m/s.

Experiments in which the salts in the extracellular and intracellular solutions were varied showed that the initial, rapid rise of $\Delta \psi$ to a positive value results from a transient increase in the conductivity of the plasma membrane to $Na^+$. The increased conductivity allows $Na^+$ to flow rapidly into the cell, down its gradients of both concentration and electric potential (see equation (3)). However, $Na^+$ influx shuts off abruptly in less than 1 ms, when the conductivity decreases again. As the conductivity to $Na^+$ drops, the conductivity to $K^+$ increases, and $K^+$ begins to flow rapidly out of the cell. $K^+$ efflux is favored, not only by the high ratio of the intracellular to extracellular $K^+$ concentrations, but now also by $\Delta \psi$, which at this time is opposite in sign to the resting potential. (See equations (1) and (2) but with $\Delta\psi$ changed from $-0.07$ to $+0.04$ V.) It is the flow of $K^+$ out of the cell that returns $\Delta\psi$ to a negative value. Hyperpolarization of the membrane to about $-80$ mV occurs for a period of several milliseconds after the action potential, while the $K^+$ conductivity remains somewhat elevated above its original level. Meanwhile, the $Na^+$–$K^+$ ATPase works to restore the original $Na^+$ and $K^+$ concentrations by pumping $Na^+$ out of the cell and bringing $K^+$ in, and the membrane returns to its resting condition.

These observations raise intriguing questions. What is the nature of the channels that allow $Na^+$ or $K^+$ to flow across the membrane? How are the conductivities of the cation channels gated on or off? Does this gating account for the triggering of an action potential by a local depolarization of the membrane and for propagation of the action potential along the axon? A partial answer to these questions is that the cation channels are membrane proteins that undergo conformational changes in response to the membrane potential itself. These conformational changes can cause the channels to flip from "closed" (nonconductive) states to "open" (conductive) states, or *vice versa*. Suppose that a local change of $\Delta\psi$ from the resting value of $-70$ mV to $-50$ mV causes a $Na^+$ channel to open. The resulting flow of $Na^+$ into the cell through this channel causes a further change of the local $\Delta\psi$ in the same direction, which acts to open other nearby $Na^+$ channels. A wave of depolarization thus would spread out from the region of the stimulus. If the opening of $K^+$ channels requires a larger threshold change in $\Delta\psi$, a phase of $K^+$ efflux would follow after a delay.

## Action Potentials Are Mediated by Voltage-Gated Ion Channels

Ion channels have been purified from several types of excitable cells. The proteins that make up the voltage-gated $Na^+$ channels of brain neurons were first identified by labeling them with reactive derivatives of neurotoxins obtained from scorpions. Intact channels were purified from both brain and muscle after solubilization with detergents. When the purified proteins were incorporated into phospholipid vesicles or planar bilayer membranes, they were found to conduct $Na^+$ across the membrane. The ion specificity of the reconstituted channels and the alterations of the conductance in response to changes in $\Delta\psi$ or to various neurotoxins were similar to the properties of the original nerve or muscle membranes. In addition to scorpion toxins, a variety of other specific neurotoxins bind to the purified channels and inhibit their activities. These include tetrodotoxin (a poison obtained from puffer fish) and saxitoxin (from the marine dinoflagellates that cause the poisonous "red tide" contamination of shellfish), both of which block the flow of $Na^+$ through the channel. Some other toxins (*e.g.*, veratridine and batrachotoxin) bind only after the $Na^+$ channel has switched to its conductive state and prevent the normal return to the nonconductive state.

Electrical measurements with purified, reconstituted $Na^+$ channels have demonstrated that an individual channel can exist in either conductive or nonconductive states. As shown in figure S1.4, the fraction of time that a channel spends in the conductive state depends strongly on $\Delta\psi$, increasing abruptly as $\Delta\psi$ is made less negative. These results agree well with the notion that a local depolarization of the plasma membrane switches the nearby $Na^+$ channels to their conductive states and thus launches an action potential that propagates down the cell.

$Na^+$ channels have one major subunit ($\alpha$) with a molecular weight of about 260,000 and a variable number of smaller subunits. Binding sites for tetrodotoxin, saxitoxin, and scorpion toxins are found on the $\alpha$ subunit. The protein purified from mammalian brain has two of the smaller subunits ($\beta_1$ and $\beta_2$); that from skeletal muscle has only one ($\beta_1$). The protein obtained from the electric organ of the electric eel has only the $\alpha$ subunit. Where it is present, $\beta_2$ is linked to $\alpha$ by a disulfide bond but can be removed without altering the ion conductance of the channel or the sensitivity to $\Delta\psi$ or neurotoxins. All of the subunits have complex carbohydrates attached on their extracellular surfaces, and $\alpha$ also can be phosphorylated at multiple sites on its intracellular surface. Phosphorylation appears to raise the threshhold change in $\Delta\psi$ needed for opening of the channel and may play a physiological role in regulating the channel's activity.

Once the major subunits of the $Na^+$ channel proteins had been purified, it was possible to isolate and sequence cDNA clones that encoded them. This was done for the $\alpha$ subunits of the $Na^+$ channels from the electric eel and rat brain and also for a related $Ca^{2+}$ channel from skeletal muscle. In addition, studies of fruit flies (*Drosophila*) with a genetic neurological disorder called "Shaker" led to the cloning and sequencing of the DNA encoding of a voltage-gated $K^+$ channel from brain cells. These studies showed that portions of the $Na^+$, $Ca^{2+}$, and $K^+$ channels have homologous amino acid sequences and suggested structural models for the organization of the channels in the membrane.

The $K^+$ channel from *Drosophilia* consists of a single polypeptide that probably has six transmembrane $\alpha$ helices (fig. S1.5). The pore for $K^+$ diffusion probably is formed from a tetramer of this polypeptide. In the $\alpha$ polypeptides of

## Figure S1.4

Ionic conductivity of purified $Na^+$ channels reincorporated into phospholipid bilayers. (*a*) The current through a single channel is measured as a function of time. Downward deflections of the traces indicate openings of the channel (increased current of $Na^+$ across the membrane). The numbers on the right indicate the electric potential difference across the membrane ($\Delta\psi$). When $\Delta\psi$ is made less negative, the channel opens more frequently and stays open longer. (*b*) The percentage of time the channel is in the open state is plotted as a function of $\Delta\psi$. (From W. A. Catterall, Structure and function of voltage-sensitive ion channels, *Science* 242:50, Oct. 7, 1988. Copyright 1988 by the AAAS. Reprinted by permission.)

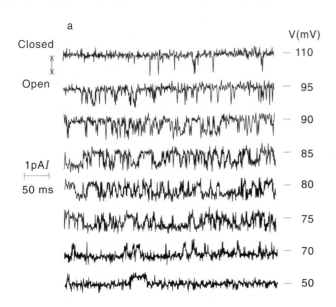

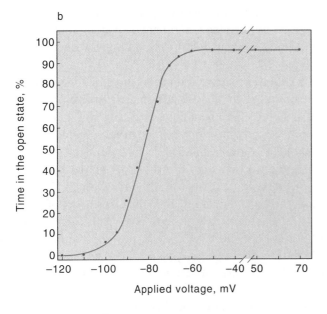

## Figure S1.5

Proposed transmembrane topology of the voltage-gated $K^+$ channel from *Drosophila*. The carboxyl- and amino-terminal ends are labeled C and N, respectively, and the putative transmembrane $\alpha$ helices are represented below labels 1–6. The effects of site-directed mutations in the protein suggest that part of the pore of the channel is lined by amino acids in the loop labeled H5, where alternate residues are denoted by filled circles. Mutations in this region change the ionic selectivity of the channel. (Reprinted with permission from *Nature*. A. J. Yool and T. L. Schwarz, Alteration of ionic selectivity of a $K^+$ channel by mutation of the H5 region, *Nature* 349:700, 1991. Copyright 1991 Macmillan Magazines Limited.)

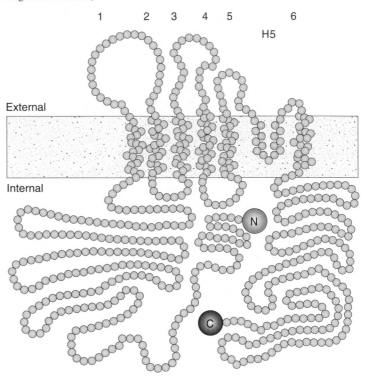

the $Na^+$ and $Ca^{2+}$ channels, there are four domains, each of which probably has six transmembrane $\alpha$ helices (fig. S1.6). These domains are structurally homologous to each other and to the $K^+$-channel protein. Within each domain, the transmembrane helices labeled 1, 2, 5, and 6 in figure S1.5 are composed almost exclusively of hydrophobic amino acid residues. Helices 3 and 4, on the other hand, also include numerous acidic and basic residues. Helix 4 has a stretch of residues in which every third amino acid is basic (either lysine or arginine); helix 3 has an excess of acidic residues (aspartates and glutamates).

In current structural models of the $Na^+$ channel, the four homologous $\alpha$-helix domains pack together around a central pore formed by the four copies of the amphipathic helix 3. These helices probably are oriented so that most of their ionizable amino acid side chains face the aqueous space in the pore, whereas most of their nonpolar side chains face outward and interact with hydrophobic residues of other helices (fig. S1.7). The diameter of such a pore is

## Figure S1.6

A functional map of the $Na^+$ channel $\alpha$ subunits. The transmembrane folding model of the $\alpha$ subunit is depicted with experimentally demonstrated sites of cAMP-dependent phosphorylation (P), interaction of site-directed antibodies that define transmembrane orientation ( $\succ$ ), covalent attachment of $\alpha$-scorpion toxins (ScTx), glycosylation ( $\psi\psi$ ), and modulation of channel inactivation (h). (From W. A. Catterall, Structure and function of voltage-sensitive ion channels, *Science* 242:50, Oct. 7, 1988. Copyright 1988 by the AAAS. Reprinted by permission.)

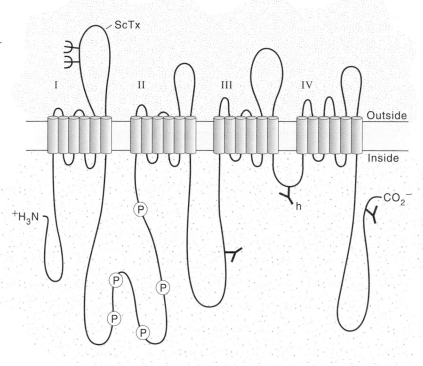

## Figure S1.7

End view of energy-optimized parallel tetramer of the sodium channel helix 3. Colors: light blue, $\alpha$ carbon backbone; red, acidic residues; blue, basic residues; yellow, polar and neutral residues; and purple, lipophilic residues. This view is from the intracellular face of the membrane, at which the N terminus of each segment is predicted to be located. (From M. Montal, *Proteins: Structure, Function, and Genetics,* vol. 8, University of California Press at San Diego, La Jolla, Calif., p. 226.)

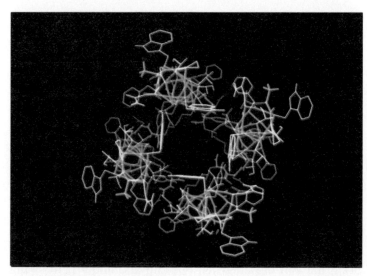

consistent with the observation that the $Na^+$ channel conducts $Na^+$, $Li^+$, and some other small cations across the membrane, but excludes cations larger than $Cs^+$. The net negative charge of the residues lining the pore explains why the channel conducts cations but not anions. In agreement with this model, the $Na^+$ conductivity decreases at low pH, when one or more of the aspartate and glutamate side chains are likely to be protonated. Also, synthetic peptides containing the amino acid sequence of helix 3 have been shown to form cation-selective channels in phospholipid bilayer membranes.

The switching of the $Na^+$ and $K^+$ channels from their closed, nonconductive conformation to an open conformation seems likely to involve movements of helix 4 relative to helix 3. The resting potential of about $-70$ mV across the membrane (negative inside) would exert forces that tend to pull the positively charged helix 4 inward and push the negatively charged helix 3 outward. These forces presumably are balanced by other intramolecular forces that maintain the structure of the resting channel. Depolarization of the membrane would upset this balance and thus could cause one of the helices to slip and perhaps rotate with respect to the other. Indications of such a movement can be seen by electrophysiological measurements. When the membrane is first depolarized, there is a small, net movement of positive charge outward (fig. S1.8). This gating current precedes the

## Figure S1.8

Illustration of the gating current that precedes the inward-directed $Na^+$ current associated with depolarization during the action potential. The gating current is believed to be related to the voltage-dependent opening of the $Na^+$ channels and perhaps may reflect rearrangement of a dipolar "voltage sensor" associated with the channel gate. (Source: Adapted from C. M. Armstrong and F. Bezanilla, Charge movement associated with the opening and closing of the activation gates of the Na channels, *J. Gen. Physiol.* 63:533, 1974.)

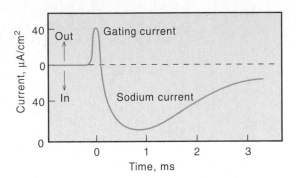

## Figure S1.9

"Ball-and-chain" model for inactivation of the $K^+$ channel. (*a*) Open conformation, (*b*) inactive conformation. (From Research News, *Science* 250:507, 1990. Reprinted by permission of Suzanne Black.)

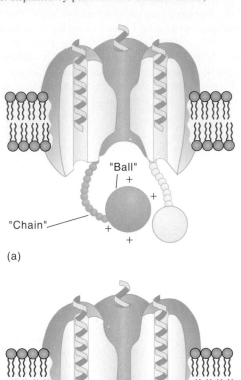

(a)

(b)

large inward flow of $Na^+$ that ensues when the channel opens.

A novel technique has been used to examine the effects of site-directed mutations of the $K^+$ channel protein. mRNA encoding the *Drosophila* $K^+$ channel was injected into oocytes of the South American frog, *Xenopus laevis.* The channel protein was expressed in functional form in the oocyte membrane, which lacks a voltage-gated $K^+$ channel of its own. By introducing mRNA coding for proteins with mutations in various regions, it was found that amino acid residues in the region between helices 5 and 6 determine the specificity of the channel for $K^+$ and other small cations. This section of the protein may form a hairpin loop that lines part of the channel (see fig. S1.5). The amino-terminal end of the protein was shown to play a critical role in return of the channel from the open to the closed state. When the first 20 amino acid residues at this end of the protein were deleted, the channel still switched normally to the open state in response to depolarization of the membrane. However, the channel then failed to inactivate, and remained in the conductive state indefinitely. Adding a synthetic peptide with the amino acid sequence of the 20 N-terminal residues restored the inactivation.

The first 11 residues at the N-terminal end of the $K^+$ channel are predominantly hydrophobic in nature. When individual residues in this region were replaced by amino acids with polar side chains, inactivation of the channel was slowed relative to the rate in the wild-type protein. Following the hydrophobic residues, the protein has a cluster of five positively charged residues. Mutations that deleted

these charged residues or replaced them by noncharged residues also slowed inactivation of the channel. However, deletion mutations in a region just beyond here (residues 23–27) made inactivation occur more rapidly than in the wild-type protein! Conversely, insertion mutations in this last region slowed inactivation.

These results suggest that the first 20 residues constitute a positively charged, amphipathic "ball" that can plug the channel, perhaps by interacting with negatively charged groups near the opening of the pore. The next 20 residues or so could form a "chain" that flexes to allow the ball to swing into or out of the opening (fig. S1.9). Shortening the chain by deletions in the region of residues 23–27 might bring the ball closer to the opening and allow it to swing into place more quickly; lengthening the chain by insertion mutations could have the opposite effect.

## Figure S1.10

Structures of some of the major neurotransmitters.

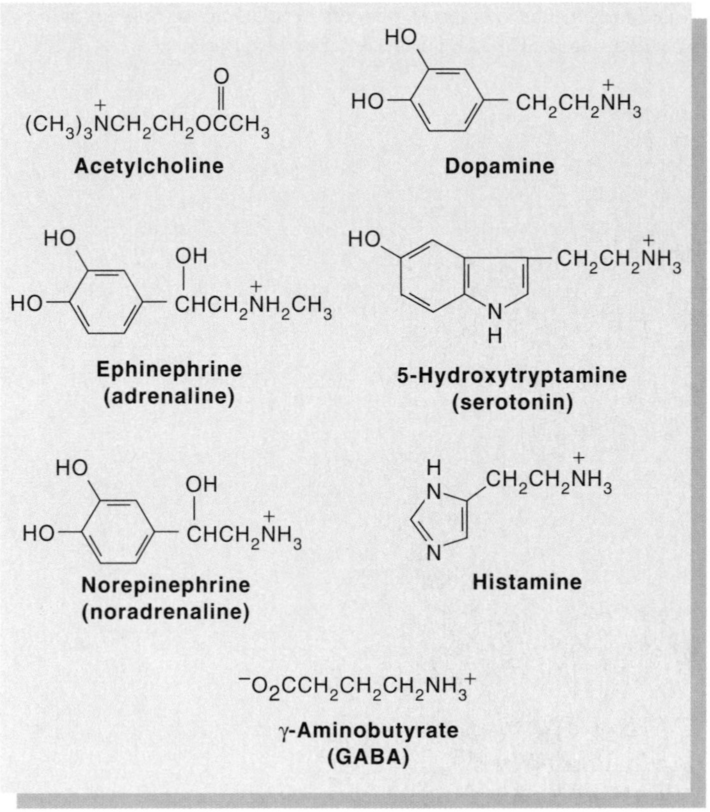

**Acetylcholine**
$(CH_3)_3\overset{+}{N}CH_2CH_2OCCH_3$

**Dopamine**

**Ephinephrine (adrenaline)**

**5-Hydroxytryptamine (serotonin)**

**Norepinephrine (noradrenaline)**

**Histamine**

$^-O_2CCH_2CH_2CH_2\overset{+}{N}H_3$

**$\gamma$-Aminobutyrate (GABA)**

## Synaptic Transmission Is Mediated by Ligand-Gated Ion Channels

Nerve impulses must be transmitted from one nerve cell to another and also from neurons to muscle and glandular cells. This transmission is mediated by chemical messengers, neurotransmitters, that diffuse from one cell to the next. A large variety of molecules serve as such neurotransmitters, including acetylcholine, epinephrine, norepinephrine, dihydroxyphenylalanine, 5-hydroxytryptamine (serotonin), $\gamma$-aminobutyrate, and histamine (fig. S1.10). Some neurotransmitters, including acetylcholine, trigger an action potential in the second cell; others, such as $\gamma$-aminobutyrate, inhibit the receiving cell's response to other signals.

Figure S1.11 depicts the process of synaptic transmission for acetylcholine, which acts as an excitatory neurotransmitter at mammalian neuromuscular junctions and between some types of neurons in the central nervous system. The transmitting (presynaptic) cell synthesizes acetylcholine from choline and acetyl-CoA.

It stores the acetylcholine near the synapse in intracellular synaptic vesicles. When an action potential arrives, the depolarization of the presynaptic cell's plasma membrane triggers an influx of $Ca^{2+}$ through voltage-gated $Ca^{2+}$ channels. $Ca^{2+}$ ions cause synaptic vesicles to fuse with the plasma membrane and thus to release their contents of acetylcholine into the synaptic cleft. Acetylcholine diffuses through the synaptic cleft to the plasma membrane of the receiving (postsynaptic) cell, where it binds to a receptor. The acetylcholine receptor is itself a ligand-gated cation channel. Binding of acetylcholine causes the receptor to undergo a conformational change that opens a pore for cations to flow across the plasma membrane. Although both $Na^+$ and $K^+$ can flow through this pore, the major effect is an influx of $Na^+$ because the driving force for this is considerably greater than the driving force for movements of $K^+$ (see equations (1), (2), and (3)). $Na^+$ influx depolarizes the plasma membrane of the postsynaptic cell and thus initiates an action potential as we discussed previously.

Similar events occur at synapses where the presynaptic cell releases an inhibitory neurotransmitter such as $\gamma$-aminobutyrate, except that here the receptor typically provides a pore that is specific for $K^+$ or $Cl^-$ ions. $K^+$ efflux or $Cl^-$ influx hyperpolarizes the plasma membrane to a potential more negative than the resting potential, opposing the depolarization induced by stimulatory neurotransmitters.

To prepare a synapse to respond to another signal, the neurotransmitter must be removed quickly from the synaptic cleft. In some cases, the transmitter is taken up by the presynaptic neuron and repackaged in synaptic vesicles; in other cases, it is broken down by extracellular enzymes. Acetylcholine, for example, is hydrolyzed rapidly to choline and acetate by the enzyme acetylcholinesterase.

## Figure S1.11

Schematic diagram of a synaptic junction in which acetylcholine is the chemical transmitter. Arrival of an action potential at the terminus of the presynaptic cell (*top*) stimulates $Ca^{2+}$ uptake, which triggers release of acetylcholine (ACh) from vesicles near the terminus of the presynaptic cell. Release is accomplished by vesicle fusion with the plasma membrane. The interaction of acetylcholine with its receptors in the postsynaptic membrane triggers depolarization, thus initiating an action potential in the postsynaptic cell (*bottom*).

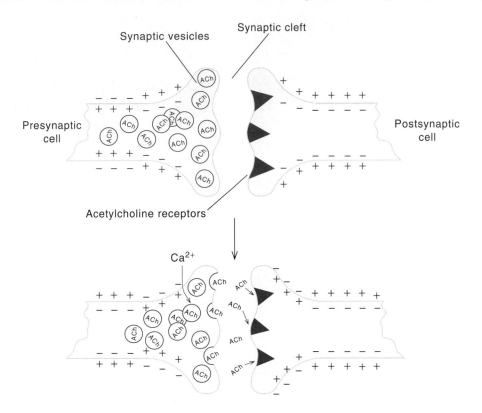

Dopamine is oxidized to 3,4-dihydroxyphenylacetaldehyde by the enzyme monoamine oxidase. Inhibitors of acetylcholinesterase are widely used as insecticides; inhibitors of monoamine oxidase are helpful in treating some neurological disorders.

The acetylcholine receptor was purified for the first time from membranes of the electric organ of the electric ray fish, *Torpedo californica.* When the purified receptor was incorporated into phospholipid vesicles, the vesicles became permeable to $Na^+$ in the presence of acetylcholine analogs. This permeability was blocked by agents that were known to inhibit synaptic transmission, including protein neurotoxins from snake venoms ($\alpha$-bungarotoxin, from snakes of the genus *Bungarus,* or cobratoxin from cobras).

The acetylcholine receptor has four subunits ($\alpha$, $\beta$, $\gamma$, and $\delta$, with $M_r \approx 52$, 56, 63, and $66 \times 10^3$, respectively). The complete receptor includes two copies of the $\alpha$ subunit, and one of each of the others. The overall structure, as visualized by electron microscopy, resembles a cylindrical bundle of five, approximately parallel rods, with a water-filled channel along the axis of the cylinder (fig. S1.12). This assembly projects about 70 Å into the synaptic cleft on one side of the membrane and about 40 Å into the intracellular space on the other. The binding site for acetylcholine is in the extracellular region. The central channel has an opening with a diameter of about 25 Å in the synaptic cleft but narrows to only a few angstroms near the center of the membrane.

The individual subunits of the acetylcholine receptor have homologous amino acid sequences. Each of them has four regions of hydrophobic amino acid residues that probably are embedded in the membrane, although only one transmembrane helix (the M2 helix) from each subunit is clearly resolved in the structural maps currently provided by electron microscopy. These helices also include many uncharged, polar residues, notably serines and threonines, which probably line the receptor's central channel (fig. S1.13). The M2 helices are bent near the center of the phospholipid bilayer, at the site of a leucine residue that is conserved in the amino acid sequences of all the subunits (fig. S1.14). The leucine side chains appear to form a tight, hydrophobic ring that blocks the flow of hydrated ions at this point. Exactly how the binding of acetylcholine in the protein's extracellular domain causes this ring to expand remains to be discovered.

## Figure S1.12

Images of the acetylcholine receptor from a view looking down on the plane of the membrane. This picture was made by mathematically filtering electron micrographs of ordered, two-dimensional arrays of the receptor. Each of the four pentameric structures shown represents an individual acetylcholine receptor with its central transmembrane pore. The dark areas indicate the five subunits ($\alpha_2\beta\gamma\delta$). (From R. M. Stroud, M. P. McCarthy, and M. Schuster, Nicotinic acetylcholine receptor superfamily of ligand-gated ion channels, *Biochem.* 29:11009, 1990.)

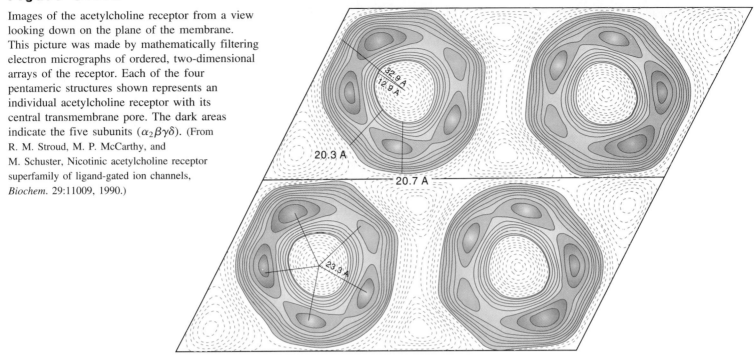

## Figure S1.13

A model of the channel formed by the M2 regions of the five subunits of the acetylcholine receptor. Polar amino acid side chains, shown in red, line the channel. (Model of the channel formed by the M2 regions of the five subunits of the acetylcholine receptor, modelled by A. Kamb, J. Newdoll and R. M. Stroud, using the software MidasPlus of the UCSF Computer Graphics Lab.)

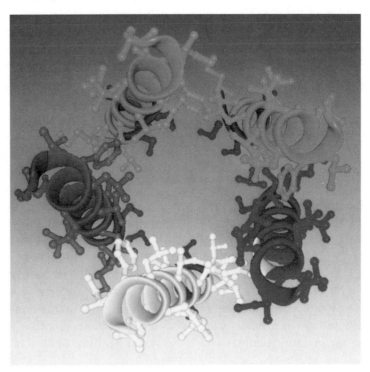

## Figure S1.14

A model of the gate in the closed acetylcholine channel. The M2 helices are bent about halfway across the phospholipid bilayer. A leucine side chain (L) from each of the five helices projects into the channel here, forming a hydrophobic barrier to the diffusion of ions. (From N. Unwin, Nicotinic acetylcholine receptor at 9 Å resolution, *J. Mol. Biol.* 229:1101, 1993. Reprinted by permission.)

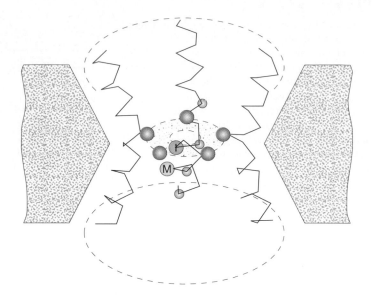

## Summary

In this chapter we examined the way in which signals are propagated along neurons and transmitted to other nerve or muscle cells. The following points are central to our discussion.

1. Action potentials are waves of depolarization and repolarization of the plasma membrane. In a resting nerve cell, the electric potential gradient ($\Delta\psi$) across the plasma membrane is about $-70$ mV, inside negative. This potential difference is generated mainly by the unequal rates of diffusion of $K^+$ and $Na^+$ ions down concentration gradients maintained by the $Na^+$–$K^+$ ATPase.

2. The plasma membranes of electrically excitable cells contain specific ion channels that can switch between closed (nonconductive) and open (conductive) states. This switching is controlled by changes in $\Delta\psi$. When a neuron receives an electric stimulus that depolarizes the membrane locally by about 20 mV, $Na^+$ channels in this region switch to the open state. $Na^+$ flows into the cell through these channels, down its electrochemical gradient. This flow causes $\Delta\psi$ to drop rapidly to zero and to overshoot briefly to a positive value. A wave of depolarization spreads along the membrane as the changing $\Delta\psi$ triggers additional $Na^+$ channels to open. The $Na^+$ channels close again spontaneously in about 1 ms, and channels that are specific for $K^+$ open. $K^+$ efflux then returns $\Delta\psi$ to a value somewhat more negative than its original value.

3. The voltage-gated $Na^+$ channel has one major subunit ($\alpha$) and several smaller subunits. In the $\alpha$ subunit there are four homologous domains, each of which probably has six transmembrane $\alpha$ helices. One of the transmembrane helices from each domain has negatively charged residues that probably line the pore across the membrane; another helix has an abundance of positively charged residues. Changes in $\Delta\psi$ could cause these helices to move with respect to each other and thus to open or close the pore.

4. The voltage-gated $K^+$-channel protein consists of a single subunit with six putative transmembrane $\alpha$ helices that are homologous to the $\alpha$ helices in the individual domains of the $Na^+$ channel. The intact channel probably is composed of four copies of the protein. Mutations of amino acid residues in a region between two of the transmembrane helices alter the ion specificity of the channel. The amino-terminal region of the protein is critical for the return of the channel to a closed state. This could involve a "ball" domain that swings on a "chain" to plug the channel.

5. Synaptic transmission between excitable cells is mediated by ligand-gated ion channels. The presynaptic

neuron releases a neurotransmitter that diffuses to the postsynaptic cell. Binding of the neurotransmitter to a receptor in the plasma membrane of the postsynaptic cell causes a pore for ions to open, which can result either in depolarization of the membrane (initiating an action potential) or in hyperpolarization (inhibiting the firing of the cell), depending on the ion specificity of the channel.

6. The acetylcholine receptor, which mediates transmission at neuromuscular junctions and at some synapses, has the subunit composition $\alpha_2\beta\gamma\delta$. The five subunits form a pentagonal structure with a transmembrane channel at the center. The channel has a wide aperture in the synaptic cleft but narrows tightly in the center of the phospholipid bilayer. The side chains of a ring of leucine residues in the central region could form the gate that opens or closes to control the flow of ions across the membrane. Binding of acetylcholine to the extracellular region of the protein causes conformational changes that must be relayed to the gate.

## Selected Readings

Catterall, W. A., Structure and function of voltage-sensitive ion channels. *Science* 242:50, 1988.

Changeux, J. P., Chemical signaling in the brain. *Sci. Am.* 269:58–62, 1993.

Hoshi, T., W. N. Zagotta, and R. W. Aldrich, Biophysical and molecular mechanisms of *Shaker* potassium channel inactivation. *Science* 250:533, 1990.

Rehm, H., and B. L. Temple, Voltage-gated $K^+$ channels of the mammalian brain. *FASEB J.* 5:164, 1991.

Stroud, R. M., M. P. McCarthy, and M. Shuster, Nicotinic acetylcholine receptor superfamily of ligand-gated ion channels. *Biochemistry* 29:11009, 1990.

Unwin, N., Nicotinic acetylcholine receptor at 9 Å resolution. *J. Mol. Biol.* 229:1101, 1993.

Yool, A. J., and T. L. Schwarz, Alteration of ionic selectivity of a $K^+$ channel by mutation of the H5 region. *Nature* 349:700, 1991.

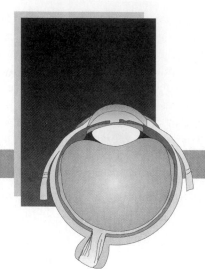

# Principles of Physiology and Biochemistry: Vision

Perhaps it is not surprising that the photochemistry of vision is different from that of photosynthesis. Animals that can see use light to obtain information; photosynthetic organisms use light as a source of energy. However, vision and photosynthesis both start with excitation of an electron from one molecular orbital to another orbital of higher energy. The excited molecule must then undergo a transformation to a metastable product. Since chlorophyll can undergo such a transformation with a quantum yield near 1.0, it is not hard to imagine an eye that uses electron-transfer reactions like those that work so well in photosynthesis. But the eyes of multicellular organisms are different. Instead of chlorophyll, their light-sensitive cells contain a complex called rhodopsin, which consists of a protein, opsin, and a linear polyene, 11-*cis*-retinal. Instead of undergoing oxidation when it is excited with light, the retinal isomerizes. In this chapter we examine the biochemistry of this reaction to light.

## The Visual Pigments Are Found in Rod and Cone Cells

 The eyes of vertebrates are marvelously complex organs, with many different types of specialized cells (fig. S2.1). Light rays entering the eye are refracted by the cornea, the clear tissue at the front of the eye.

## Figure S2.1

Structure of the human eye.

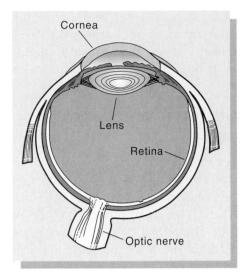

The light traverses an aqueous chamber and reaches the lens, which is densely packed with proteins called crystallins. Adjustments in the shape of the lens focus a sharp optical image onto the retina, a thin layer of tissue that lines the back of the eye. The retina is a neural tissue with several different layers of cells. Some of these cells, the rod and cone cells, contain the visual pigments. Other cells make synaptic connections to the rods or cones and to additional neural cells that carry impulses to the brain.

In both rods and cones, the light-sensitive molecules reside in a layered system of membranes at one end of the cell (figs. S2.2 and S2.3). The membranes form by invaginations of the plasma membrane near the middle of the cell. In cone cells, the membranes remain contiguous with the plasma membrane. In rods, they pinch off to form a stack of autonomous flattened vesicles, or disks, in the outer segment of the cell. This region of the cell is connected by a thin cilium to the inner segment, which is packed with mitochondria and ribosomes. The basal part of the cell ends in a synaptic junction with another neural cell called a bipolar cell.

Rod cells are specially adapted for vision in dim light. Cones provide visual acuity in bright light and also serve for perception of color. Animals such as owls, which have very high visual sensitivity in dim light but cannot distinguish colors, have only rod cells. Some animals such as pigeons have only cones and are skilled at distinguishing colors in bright light but have poor night vision. Primates have both rods and cones, with rods considerably outnumbering cones in all but the central portion of the retina.

## Figure S2.2

Thin-section electron micrograph (59,200 ×) of a portion of a rod cell in a rabbit retina. Part of the outer segment is shown at the top and part of the inner segment at the bottom. (Courtesy Dr. Ann Bunt-Milam.)

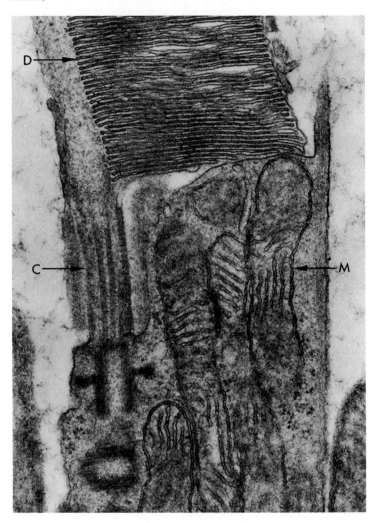

## Rhodopsin Consists of 11-*cis*-Retinal Bound to the Protein, Opsin

Figure S2.4 shows the structures of 11-*cis*-retinal and its more stable isomer all-*trans*-retinal. The retinals are related to the alcohol retinol, or vitamin $A_1$. Mammals cannot synthesize these compounds *de novo* but can form them from dietary carotenoids such as $\beta$-carotene. A deficiency of vitamin A causes night blindness, along with serious deterioration of the eyes and other tissues.

Opsin is an intrinsic membrane protein with a molecular weight of about 38,000. It accounts for about 95% of the protein in the disk membranes. The carboxyl-terminal end

Schematic diagram of a rod cell. The orientation of the cell is the same as in figure S2.2. In the retina, many rod cells are stacked side by side, with the outer segments all pointing out to the periphery of the eye. Light enters the cells end-on through the inner segment after passing through several layers of other neural cells. In cone cells, the outer segments are shorter and are conical rather than cylindrical in shape.

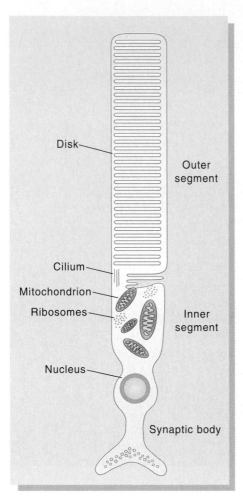

ception of color by cones depends on the fact that there are three different types of cones with different absorption spectra. One of these absorbs blue light (440 nm) maximally, the second absorbs green (530 nm), and the third absorbs yellow (570 nm). In some species of birds, the third component absorbs maximally at 630 nm. However, all cones contain 11-*cis*-retinal bound to proteins that are similar to opsin.

## Light Isomerizes the Retinal of Rhodopsin to All-**trans**

The discovery that rhodopsin contains the 11-*cis* isomer of retinal came as a surprise. In the early 1950s, Ruth Hubbard, George Wald, and their co-workers found that rhodopsin decomposes into retinal and opsin when the retina is exposed to light. The retinal that was released was the all-*trans* isomer. When opsin was mixed with a crude preparation of retinal, rhodopsin was regenerated. Crystalline all-*trans*-retinal, however, did not bind to opsin. The regeneration achieved with the crude preparation proved to be due to a small amount of the 11-*cis* isomer in the mixture. Because rhodopsin binds 11-*cis*-retinal in the dark but releases all-*trans*-retinal after illumination, Hubbard and Kropf concluded that the action of light in vision is to isomerize the chromophore about its C-11–C-12 double bond. This change in structure must somehow be translated into an electrophysiological signal that is transmitted to the brain.

## Transformations of Rhodopsin Can Be Detected by Changes in Its Absorption Spectrum

The release of all-*trans*-retinal from opsin takes several minutes and is too slow to be an obligatory step in visual perception. It appears to be a step in the regeneration of the active form of rhodopsin. The all-*trans*-retinal leaves the rod cell and is isomerized back to the 11-*cis* form before it returns.

To explore the changes in rhodopsin that precede release of all-*trans*-retinal from the protein, Toru Yoshizawa and Wald measured the optical absorbance changes that occurred when they illuminated rhodopsin at low temperatures. At 77 K, illumination caused the absorption band of the rhodopsin to shift from 500 to 543 nm. The product of this transformation is now called bathorhodopsin. Bathorhodopsin is stable indefinitely in the dark at 77 K, but if it is warmed above about 130 K it decays spontaneously to a species that absorbs maximally at 497 nm. This is called lumirhodopsin. If the sample is warmed further to about

of rhodopsin is exposed to the cytoplasmic space outside the disk, and the amino end to the space inside the disk. Seven hydrophobic α-helical stretches of the protein probably lace back and forth across the phospholipid bilayer. The retinal is bound to a lysyl-$NH_2$ group by a Schiff's base linkage (see fig. S2.4) in a region that is buried in the bilayer.

In solution, 11-*cis*-retinal absorbs maximally near 380 nm, but in rhodopsin the peak is at 500 nm (fig. S2.5). The absorption spectrum of rhodopsin is essentially identical to the spectrum of the sensitivity of the rod cells (after correction for absorption in the cornea and lens). The per-

## Figure S2.4

Structures of retinals, retinols, and β-carotene. The structure of 11-*cis*-retinal (*top*) indicates the numbering system used for the carbons. In rhodopsin, 11-*cis*-retinal is bound by a protonated Schiff's base linkage to a lysine of opsin.

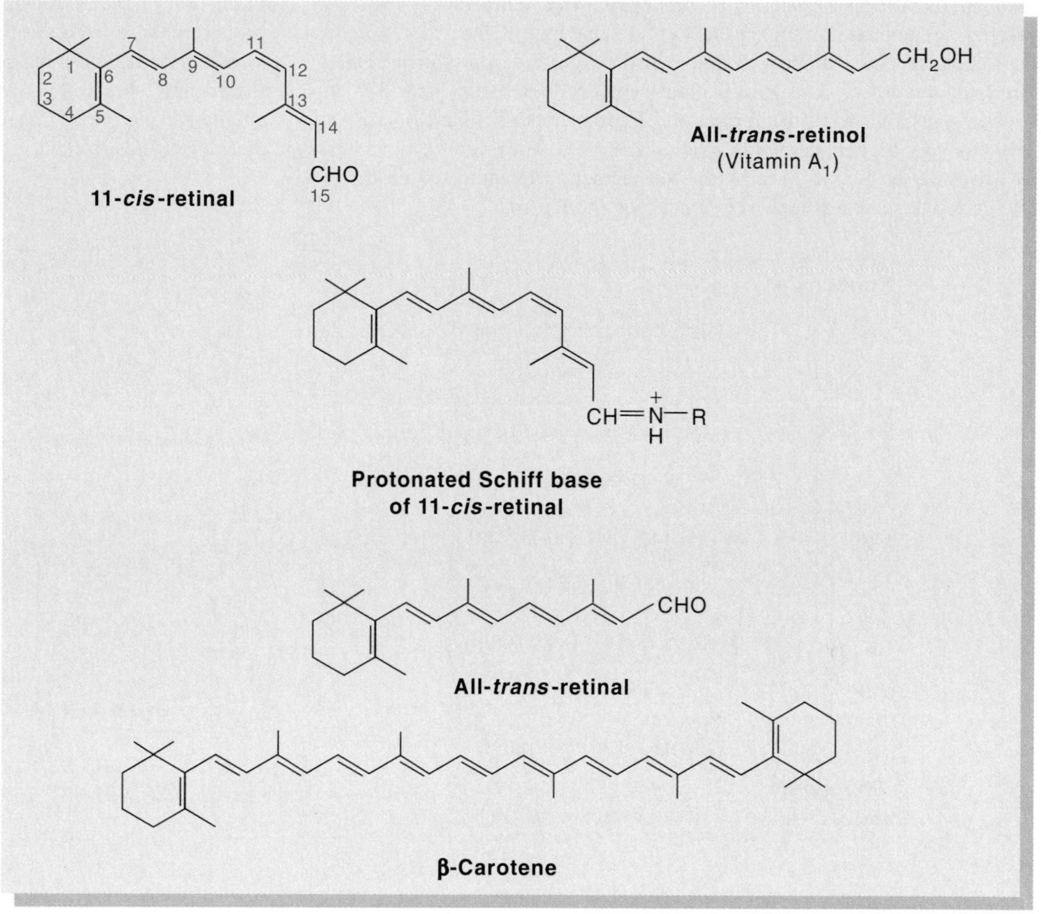

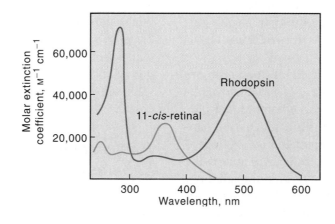

**11-*cis*-retinal**

**All-*trans*-retinol**
(Vitamin A₁)

**Protonated Schiff base
of 11-*cis*-retinal**

**All-*trans*-retinal**

**β-Carotene**

## Figure S2.5

Absorption spectra of rhodopsin and of 11-*cis*-retinal in hexane solution. The absorption of rhodopsin in the 280-nm region is due mainly to the opsin.

## Figure S2.6

Photochemical transformations of rhodopsin. The numbers in parentheses indicate the optical absorption maxima of the intermediates. The numbers in color on the right are approximate half-times for the conversions in rod outer segment membranes near 37° C. The steps following the formation of bathorhodopsin have progressively higher thermal activation energies. They can be blocked by lowering the temperature below the temperatures indicated on the left. Asterisk indicates the excited state of rhodopsin.

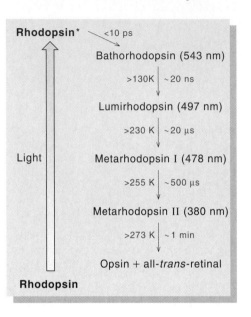

## Figure S2.7

Excitation of the protonated Schiff's base causes a movement of positive charge from the N toward the ring. The C-11–C-12 bond loses much of its double-bond character. The valence bond diagrams shown here should not be taken too literally, but they give a good qualitative picture of the redistribution of electrons that occurs when the molecule is excited.

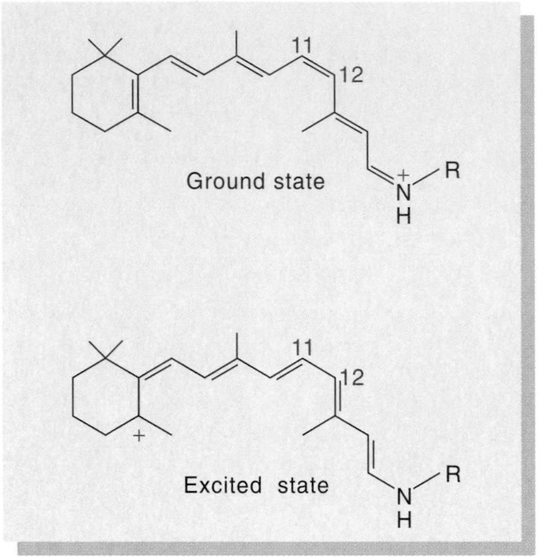

230 K, lumirhodopsin decays to metarhodopsin I, which absorbs at 478 nm. Above about 255 K, metarhodopsin I decays to metarhodopsin II, which absorbs at 380 nm. These transformations are outlined in figure S2.6. Subsequent kinetic measurements by other investigators showed that rhodopsin passes sequentially through the same series of states if it is excited with a short flash of light at physiological temperatures. The numbers on the right side of figure S2.6 give approximate half-times for the transformations at 37°C.

How can the absorption spectrum of rhodopsin go through such wild changes from 500 to 543 nm and eventually to 380 nm? In thinking about this, remember that rhodopsin's initial absorption spectrum is already shifted substantially compared with the spectrum of free 11-*cis*-retinal (see fig.S2.5). Furthermore, the absorption maxima of the cone pigments vary from 450 to 630 nm, even though these pigments all contain 11-*cis*-retinal. The explanation for these spectral differences depends partly on the fact that the nitrogen atom of the retinylidine Schiff's base linkage in rhodopsin is protonated and is therefore positively charged (fig. S2.7). The protonation state of the nitrogen has been demonstrated by NMR and resonance Raman spectroscopy. When the retinal absorbs light and is raised to an excited state, the electron density on the nitrogen increases, and the positive charge moves to the opposite end of the molecule, as represented roughly in figure S2.7. Because of the redistribution of charge, the relative energies of the excited and ground states are extremely sensitive to the positions of other charged or polar groups nearby. The structure of the protein thus can affect the absorption spectrum of the pigment profoundly. An arrangement of charged or polar groups that stabilizes the excited state relative to the ground state shifts the absorption spectrum to longer wavelengths.

## Isomerization of the Retinal Causes Other Structural Changes in the Protein

Let us now consider bathorhodopsin, which appears to be the first metastable product of the photochemical reaction. Resonance Raman measurements indicate that the retinal in bathorhodopsin has isomerized to the all-*trans* form but is still held as a protonated Schiff's base.

**Figure S2.8**

Potential energy surfaces of ground and excited states in rhodopsin as functions of the angle of rotation of the bond between carbons 11 and 12. A rotational angle of 0° means that the retinyl group is 11-*cis* (rhodopsin); 180° means that it is all-*trans* (bathorhodopsin). In the ground state (*lower curve*), rhodopsin is stabilized with respect to bathorhodopsin, possibly because the positive charge of the protonated Schiff's base interacts more favorably with a negatively charged group of the protein (see fig. S2.9). The energy of the excited state (*upper curve*) is less sensitive to the position of the Schiff's base than the energy of the ground state is, because the positive charge has moved to the opposite end of the retinal molecule (see fig. S2.7). The energy of the excited state is minimal when the rotation angle is about 90°. Rhodopsin is raised from the ground state to the excited state by light (*vertical arrow*). It relaxes back to the ground state along a path like that indicated by the dashed arrow, usually ending up as bathorhodopsin but sometimes returning to the original 11-*cis* isomer.

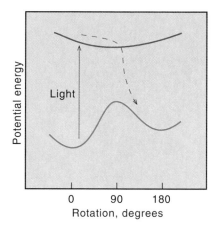

Isomerization of the retinal Schiff's base can occur when the molecule is excited with light, because the C-11–C-12 bond loses much of its double-bond character in the excited state. The valence bond diagrams of figure S2.7 illustrate this point. In the ground state of rhodopsin, the potential energy barrier to rotation about the C-11–C-12 bond is on the order of 30 kcal/mol. This barrier essentially vanishes in the excited state. In fact, the energy of the excited molecule probably is minimal when the C-11–C-12 bond is twisted by about 90° (fig. S2.8). The excited molecule oscillates briefly about this intermediate conformation, and when it decays back to a ground state it usually settles into the all-*trans* isomer, bathorhodopsin.

The barrier to isomerization of rhodopsin in the dark is physiologically important because it limits the "noise" in our perception of light. If the barrier were low enough to be overcome thermally, discrimination of light from dark would be more difficult.

Because opsin binds 11-*cis*-retinal but not all-*trans*-retinal, the all-*trans* molecule that is created in bathorhodopsin must find itself initially in a binding site that is tailored for the 11-*cis* structure. In solution, 11-*cis*-retinal is less stable than all-*trans*-retinal because of steric repulsion between the methyl group on C-13 and the hydrogen on C-10 (see fig. S2.4). On the protein, compensating interactions must stabilize the 11-*cis* complex relative to the all-*trans* complex. These interactions are disrupted when the pigment is isomerized. The pronounced shift of the absorption spectrum to longer wavelengths in bathorhodopsin could be explained if the isomerization results in a movement of the Schiff's base nitrogen, increasing the distance between the positively charged nitrogen and a nearby negatively charged group (fig. S2.9). Such a movement could destabilize the ground state of bathorhodopsin relative to the excited state. A repositioning of the Schiff's base is also likely to lead to changes in the structure of the protein surrounding the binding site.

The reorganization of rhodopsin set in motion by isomerization of the retinyl group continues as the system relaxes through the states lumirhodopsin, metarhodopsin I, and metarhodopsin II. Physical measurements indicate that the protein undergoes particularly substantial changes in conformation during the transition to metarhodopsin II. The nitrogen of the retinyl Schiff's base becomes exposed to the aqueous solution and loses its proton, and another group takes up a proton from the solution. The loss of the positive charge on the nitrogen accounts for the shift of the absorption maximum to 380 nm, where unprotonated Schiff's bases of all-*trans*-retinal absorb.

The formation of metarhodopsin II is fast enough to be an obligatory step in visual transduction. It clearly is associated with changes in the interactions between rhodopsin and its surroundings. A reasonable hypothesis, therefore, is that the changes in protein structure allow metarhodopsin II to initiate an interaction with some other component of the disk membrane. We explore the nature of this component in the following sections.

## The Conductivity Change That Results from Absorption of a Photon

The human eye is amazingly sensitive. After being in the dark for a time, one can perceive continuous light that is so weak that an individual rod cell absorbs a photon, on the average, only once every 38 min. A flash of light is detectable if approximately six rods each absorb one photon. This means that absorption of a single photon by any one of the approximately $3 \times 10^7$ molecules of rhodopsin in a rod

## Figure S2.9

A model for the photochemical conversion of rhodopsin to bathorhodopsin. In rhodopsin (*top*), the positively charged nitrogen of the Schiff's base is stabilized by electrostatic interactions with one or more negatively charged groups of the protein ($A_1^-$). When rhodopsin is converted to bathorhodopsin (*bottom*) the isomerization about the C-11 C-12 double bond moves the nitrogen, increasing the distance between the positive and negative charges. The separation of electric charge could account for the shift of the absorption spectrum to longer wavelengths. The isomerization also can lead to further structural changes, including proton movements. The proton of the Schiff's base could be passed to another basic group ($A_2^-$), and $A_1^-$ might pick up a proton from another acidic group ($A_3H$). Other groups of the protein also must adjust to the new geometry of the retinyl Schiff's base.

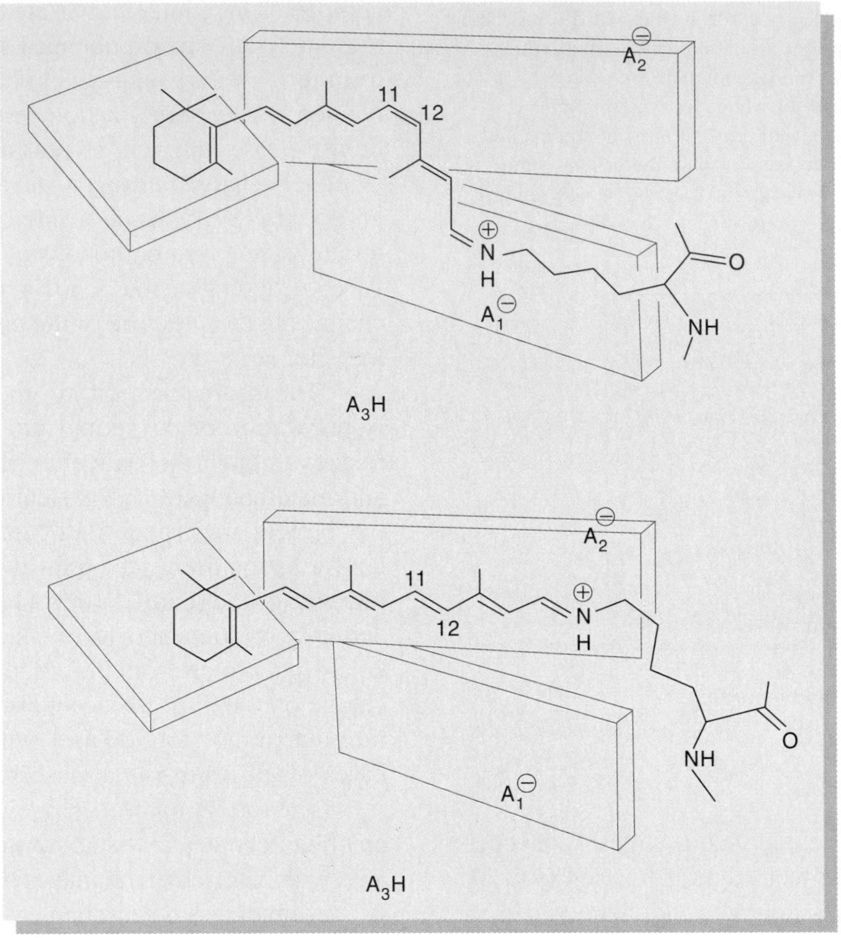

must be sufficient to excite the cell and trigger a neuronal response. The combined responses of six rods can elicit the sensation of seeing.

The electrophysiological response of a rod or cone to light involves a change in the permeability of the plasma membrane to cations. In the inner segment and basal parts of the cell, the plasma membrane contains a $Na^+$–$K^+$ pump, which uses ATP to move $Na^+$ out of the cell and $K^+$ in. (Recall our discussion of ion pumps in chapter 17.) $K^+$ diffuses back out of the cell relatively freely, and its efflux causes the membrane to become negatively charged by about 20 mV on the inside relative to the outside. $Na^+$, on the other hand, cannot readily pass back across the membrane into the inner segment. Instead, $Na^+$ diffuses in the extracellular space to the region of the outer segment, where it reenters the cell through channels that are selectively permeable to cations (fig. S2.10). Once back inside, the $Na^+$ flows to the inner segment to be pumped out again. When a

**Figure S2.10**

Na$^+$ is pumped out of the inner segment (*solid arrows*) and diffuses back into the cell through channels in the outer segment (*dashed arrows*). In rods of vertebrates, the Na$^+$ channels are held open in the dark, and they close in the light. (In invertebrates, the channels open in the light.)

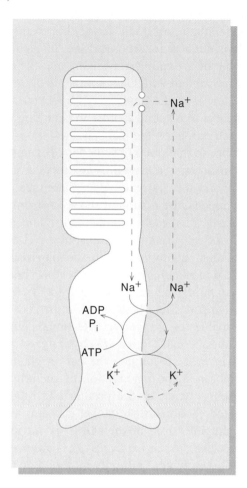

rod cell of a vertebrate's retina is excited with light, some of the cation channels in the outer segment suddenly close, decreasing the inward movement of Na$^+$. The interruption of the Na$^+$ current results in an increase in the electric potential across the plasma membrane. This hyperpolarization causes a change in the movement of an unidentified chemical transmitter to the bipolar cell that makes a synaptic junction with the rod. The bipolar cell then sends a signal to a ganglion cell, the third component in the hierarchy of retinal neurons.

The hyperpolarization of the rod and the response of the bipolar cell are graded effects. Absorption of a single

photon causes the membrane potential to increase by about 1 mV, with the effect peaking about 1 s after the excitation. Up to about 100 photons per rod, the more light the cell absorbs, the larger the amplitude of the hyperpolarization and the faster the hyperpolarization occurs. The reaction of the ganglion cell, however, is all or none. When the ganglion cell is triggered, it responds with an action potential that proceeds to the brain. Sensitivity to weak light is enhanced (with a sacrifice in spatial resolution) by having multiple rod cells connected to each bipolar cell and multiple bipolar cells connected to each ganglion cell. The output of cone cells is not summed in this way. This difference partly explains why the cones provide better visual acuity but lower sensitivity than the rods.

In the rods of vertebrates, absorption of a single photon decreases the current of Na$^+$ into the outer segment by about 3%. Exactly how many Na$^+$ channels have to close in order to achieve this effect is uncertain, but estimates have ranged from 25 to 1,000. Because a photon excites only one molecule of rhodopsin, the cell must have a mechanism for amplifying the effect of light by a substantial factor. Recall also that the plasma membrane of the rod outer segment is not contiguous with the disk membranes that hold the rhodopsin (see fig. S2.3). This fact suggests that excitation of rhodopsin causes a change in the concentration of a diffusible molecule that moves from the disk to the plasma membrane.

## The Effect of Light Is Mediated by Guanine Nucleotides

There is strong evidence that the diffusible molecule that moves between the disk and the plasma membrane is 3′,5′-cyclic-GMP (cGMP). cGMP causes an increase in the permeability of the plasma membrane to Na$^+$ (fig. S2.11). This appears to be a direct effect of cGMP on the cation channels, rather than an indirect effect mediated by a kinase, because it can be seen in the absence of ATP.

The disk membranes of the rod outer segment contain a phosphodiesterase that hydrolyzes cGMP to 5′-GMP. If the membranes are kept in the dark, the phosphodiesterase remains in a relatively inactive state and the cGMP content of the cell is high. When the cell is illuminated and rhodopsin is converted to metarhodopsin II, the activity of the phosphodiesterase increases, initiating a drop in the cGMP content. The reduced cGMP concentration decreases the Na$^+$ permeability of the plasma membrane (see fig. S2.11)

**Figure S2.11**

cGMP increases the electric conductivity of the rod cell cytoplasmic membrane. The conductivity was measured with a patch of membrane held at the tip of a micropipet. Experiments with solutions containing various ions showed that cGMP affects the movement of $Na^+$ and other cations across the membrane. Note the sigmoidal concentration dependence: the Hill coefficient is 1.6. (For more details, see E. E. Fesenko et al. *Nature* 313:310, 1985.)

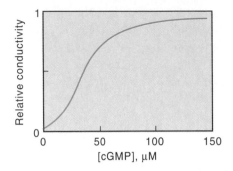

and thus results in hyperpolarization of the membrane. In agreement with this scheme, injecting extra cGMP into rod outer segments temporarily prevents the hyperpolarization caused by light.

Activation of the phosphodiesterase by light requires the presence of GTP and is associated with the binding of GTP to another protein, transducin. Transducin is a member of the family of G proteins that participate in the activation or inhibition of adenylate cyclase by hormones in other tissues (see chapter 24). Like other G proteins, transducin consists of three subunits, $\alpha$, $\beta$, and $\gamma$ (fig. S2.12). In the resting state, the $\alpha$ subunit contains a molecule of bound GDP. When rhodopsin is transformed to metarhodopsin II by light, it interacts with transducin, causing GTP to displace the bound GDP. Once GTP is attached, the $\alpha$ subunit probably separates from the $\beta$ and $\gamma$ subunits and binds to an inhibitory subunit of the phosphodiesterase. The removal of the inhibitory component activates the phosphodiesterase.

The molecules of metarhodopsin II that are generated by light diffuse rapidly in the plane of the membrane, and each of them typically triggers about 500 molecules of transducin to bind GTP in place of GDP. Each of the transducin $\alpha$-subunit-GTP complexes can activate a molecule of phosphodiesterase, which goes on to hydrolyze many molecules of cGMP. These effects account for the large amplification in the response to light. They can be demonstrated with purified preparations of rhodopsin, transducin, and the phosphodiesterase. For example, illuminated rhodopsin that is incorporated into phospholipid vesicles is capable of activating transducin, and the complex of transducin's $\alpha$ subunit with GTP is capable of activating the phosphodiesterase in the absence of rhodopsin.

For the eye to respond rapidly to changing light intensities, the activation of the phosphodiesterase must be a transient response that quickly switches off again. The activation of transducin is reversed by hydrolysis of the bound GTP to GDP and free inorganic phosphate (see fig. S2.12). When this happens, the $\alpha$ subunit separates from the inhibitory polypeptide of the phosphodiesterase and recombines with the $\beta$ and $\gamma$ subunits of transducin. The inhibitory polypeptide then recombines with the phosphodiesterase, returning the phosphodiesterase to its resting, inactive state.

To switch off the response to light, two additional things must happen: Metarhodopsin II must decay to a form that is incapable of activating additional molecules of transducin, and the cGMP that was hydrolyzed must be re-synthesized in order to reopen the cation channels in the plasma membrane. The inactivation of metarhodopsin involves phosphorylation of the protein and binding of the phosphorylated rhodopsin to another protein, arrestin (see fig. S2.12). Regeneration of the cGMP is carried out by a guanylate cyclase, which catalyzes the synthesis of cGMP from GTP. This enzyme is strongly inhibited by $Ca^{2+}$. $Ca^{2+}$ enters the rod outer segment through the same cation channels that admit $Na^+$, and it is exported by a transport protein that exchanges $Ca^{2+}$ for $Na^+$. The $Ca^{2+}$ concentration in the cell thus decreases on illumination, when the cation channels close. The drop in the $Ca^{2+}$ concentration results in activation of the guanylate cyclase.

## Figure S2.12

Steps in the activation of the cGMP phosphodiesterase by light. Light converts rhodopsin (R) to metarhodopsin II ($M_{II}$). $M_{II}$ reacts catalytically with transducin, which has GDP bound to its $\alpha$ subunit. The interaction with $M_{II}$ changes the specificity of the nucleotide binding site so that GDP is released and GTP is bound. In the dark, the phosphodiesterase (PDE) is inactivated by an inhibitory polypeptide (I). When the I polypeptide is removed by binding to the complex of the transducin $\alpha$ subunit and GTP, the active phosphodiesterase can hydrolyze cGMP to 5'-GMP. $M_{II}$ goes on to activate several hundred additional molecules of transducin before slower enzymatic reactions (*thin arrows*) convert it to an inactive form. The inactivation involves phosphorylation of rhodopsin and binding to another protein, arrestin (A). The GTP on transducin's $\alpha$ subunit eventually is hydrolyzed to GDP. When this happens, I returns to the phosphodiesterase, the three subunits of transducin reassemble, and the system returns to its resting state.

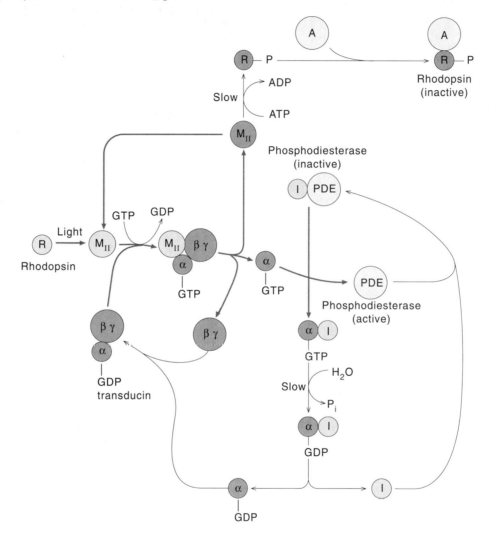

## Summary

In this supplement we described the biochemical mechanisms involved in vision. The main points of our discussion are as follows.

1. Light rays entering the eye of a vertebrate are refracted by the cornea and focused on the retina. Rod and cone cells in the retina contain the visual pigments that are responsible for the initial response to light.
2. The light-sensitive protein complex in the rods, rhodopsin, consists of 11-*cis*-retinal bound as a Schiff's base to a protein, opsin. Rhodopsin is an integral constituent of membranes that form a stack of disks at one end of the cell. Cone cells, which are responsible for the perception of color, contain similar complexes in infoldings of the plasma membrane.

3. When rhodopsin absorbs light, the retinal isomerizes to the all-*trans*-isomer. This structural change initiates a series of transformations of the pigment–protein complex that result in an interaction with another protein, transducin. The interaction with rhodopsin causes one of transducin's subunits to take up a molecule of GTP in exchange for bound GDP and to react with a phosphodiesterase that hydrolyzes cGMP. This reaction activates the phosphodiesterase. The resulting drop in cGMP concentration leads to a decrease in the $Na^+$ permeability of the plasma membrane. A hyperpolarization of the membrane then triggers the transmission of a signal at the rod's synaptic junction to an adjacent neural cell.

## Selected Readings

Birge, R. R., Nature of the primary photochemical events in rhodopsin and bacteriorhodopsin. *Biochim. Biophys. Acta* 1016:293, 1990. A review covering rhodopsin's structure, spectroscopic properties and responses to light. This article also discusses the closely related protein, bacteriorhodopsin, which serves as a light-driven proton pump in halophilic bacteria.

Chabre, M., Trigger and amplification mechanisms in visual phototransduction. *Ann. Rev. Biophysics Biophysical Chem.* 14:331, 1985. A review covering the structure of rhodopsin and the roles of cGMP, GTP, the phosphodiesterase, and transducin in the response of the rod cell to light.

Fesenko, E. E., S. S. Kolesnikov, and A. L. Lyubarsky, Induction by cyclic GMP of cation conductance in plasma membrane of retinal rod outer segment. *Nature* 313:310, 1985. cGMP increases the $Na^+$ conductance of the cytoplasmic membrane.

Nathans, J., Rhodopsin: Structure, function and genetics. *Biochemistry* 31:4923, 1992. A review of recent work on the structure and reactions of rhodopsin.

# Storage and Utilization of Genetic Information

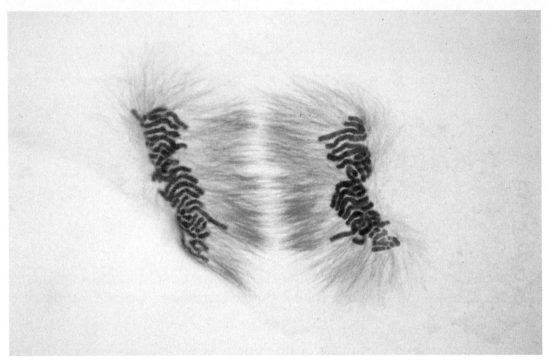

Light micrograph of late mitosis in a plant.
Microtubules are stained red and
chromosomes are counterstained blue.
(Courtesy of Andrew Bajer.)

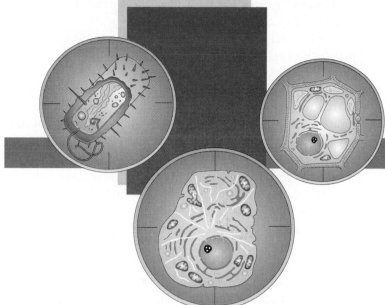

# Structures of Nucleic Acids and Nucleoproteins

*Most DNAs form a double stranded helical structure in which nitrogen bases from opposing chains form hydrogen bonds with one another.*

Nucleic acids occupy a unique position in the biochemical world. Not only are they involved in many important reactions, but they carry genetic information, which must be faithfully duplicated so that it can be passed from one cell generation to the next and from one organism to the next. Nonetheless this information must be mutable to produce the variability on which the evolutionary selection process feeds. Finally the DNA must be selectively transcribed so that each cell can synthesize the proteins it needs.

In the following five chapters we focus on the biochemistry of DNA, RNA, and protein, with occasional reminders of the genetic significance of these closely related processes. Appropriately, this section on nucleic acids and protein metabolism begins with a description of key experiments that demonstrated the genetic significance of nucleic acids. Following this we turn to a consideration of the structural properties of DNA and chromosomes. Next, reactions involving DNA are discussed, first purely biochemical reactions (chapter 26) and then *in vitro* and *in vivo* reactions in which DNA is manipulated (chapter 27). Chapter 28 deals with information transfer from DNA to RNA, and, finally, chapter 29 covers the mechanism of information transfer from RNA to protein.

# The Genetic Significance of Nucleic Acids

After the discovery around the turn of the century that genes are carried by chromosomes, a great deal of effort went into characterizing the sizes and shapes of chromosomes. But it was not until much later that significant progress was made in elucidating the chemical nature of the gene. Biologists were aware, as early as 1900, that chromosomes are composed of both nucleic acids and proteins. However, the seemingly simple chemical composition of nucleic acids misled early investigators into believing that nucleic acids were a purely structural component of the chromosome. The favored theory was that the arrangement of specific proteins along the chromosome accounted for gene specificity. This notion was dispelled in 1944 when Oswald T. Avery and his colleagues at the Rockefeller Institute (now University) demonstrated that purified deoxyribonucleic acid (DNA) contains the genetic determinants of the bacterium *Diplococcus pneumoniae* (now called *Streptococcus pneumoniae*). In this section we discuss Avery's results, as well as some other historically important observations that led up to his experiments.

## *Transformation Is DNA-Mediated*

In 1928, Fred Griffith was experimenting with two different strains of pneumococcus. Type S bacteria (S for smooth, from the appearance of bacterial colonies on agar plates) are encapsulated by polysaccharide. The capsules protect them from the host immune system, making the S bacteria pathogenic; even when small numbers of S bacteria are injected into mice, death results. By contrast, R bacteria (R for rough colonies) are nonencapsulated; they are readily attacked by the mouse's immune system and consequently are nonpathogenic. Although heat-killed S bacteria by themselves are nonvirulent, Griffith found that when heat-killed S bacteria were mixed with live R bacteria and introduced into a susceptible laboratory mouse, death of the animal frequently occurred (fig. 25.1*a*). S bacteria could then be detected in the blood. Apparently the genetic factor required for encapsulation was transferred from killed S cells to live R cells.

Griffith's result was duplicated outside of the animal by Dawson and Sia in 1930. Both bacterial strains were grown in liquid growth medium and distinguished by the distinctive appearance of their colonies on plates (see fig. 25.1*b*). In parallel with the results obtained in the mouse, heat-treated extracts of S cells transformed R cells into S cells.

More than 10 years later, Avery provided convincing proof that the active agent in the S cell extracts was the cellular DNA. He and his co-workers did this by purifying the DNA from S cells and showing that it had the capacity to transform R cells into S cells *in vitro* (see fig. 25.1*b*). The transforming activity in the purified extract was destroyed if the extract was first incubated with the enzyme DNase which specifically degrades DNA. By contrast the transforming activity was not affected by RNase, proteases, or enzymes that degrade capsular polysaccharides. It was subsequently found that DNA could be used to transfer many other genetic traits between the appropriate pairs of donor and recipient bacterial strains. For example, resistance to the antibiotics streptomycin or penicillin could be transferred, with the DNA of resistant cells, to sensitive cells. The transformation studies with different traits conferred by the donor DNA showed that the active genetic material being transferred faithfully reflected the genetic patterns of the donor strains. The procedure of altering the genetic composition of one cell strain by exposing it to DNA of another strain is termed transformation.

Transformation occurs naturally in only a small number of bacterial species other than *Streptococcus pneumoniae*. In organisms that are not transformed naturally, laboratory procedures are available to make the cell envelope partially permeable (e.g., of *E. coli*) and permit uptake of DNA. This is termed artificial transformation. It can be brought about by $Ca^{2+}$ treatment and temperature shock; but the most efficient method found to be useful with most organisms is electroporation, or electric shock. For instance, by exposing yeast to electric field pulses it is possible to obtain transformation efficiencies that are $10^4$ times greater than those obtained by conventional methods. Transformation is an essential step in the cloning of genes (see chapter 27). Transformation with self-replicating plasmids that may be carrying different genes has made it possible to select cells carrying the gene(s) of interest from a large population of cells.

Avery's experiments on cells were followed by additional experiments on viruses, demonstrating that the genetic information of a virus is carried by nucleic acids. In the case of a cell the genetic information is always carried by DNA. In the case of viruses the genetic information can be carried by either DNA or RNA, depending on the virus.

## Structural Properties of DNA

The early work equating genetic material with DNA plunged genetics into an entirely new vocabulary of chemical terms. The genetic consequences of these early studies

## Figure 25.1

*In vivo* and *in vitro* evidence that DNA causes transformation of *Streptococcus pneumoniae*. (*a*) Transformation experiment by Griffith. R bacteria are nonvirulent. S bacteria are virulent. A mixture of R bacteria and heat-killed S bacteria is also virulent if transformation has occurred. (*b*) Transformation experiment by Avery and co-workers. When bacteria from a liquid culture are spread on a semisolid medium, each cell adheres to the medium at random. As time passes, the cells and their offspring grow and divide, leading to visible clones or colonies, each of which arose from a single cell. R and S cells each have a distinct clonal morphology. Whereas R cells produce small rough colonies, S colonies are smooth. R cells exposed to DNA from S cells produce a mixture of both types of colonies: untransformed R colonies and transformed S colonies.

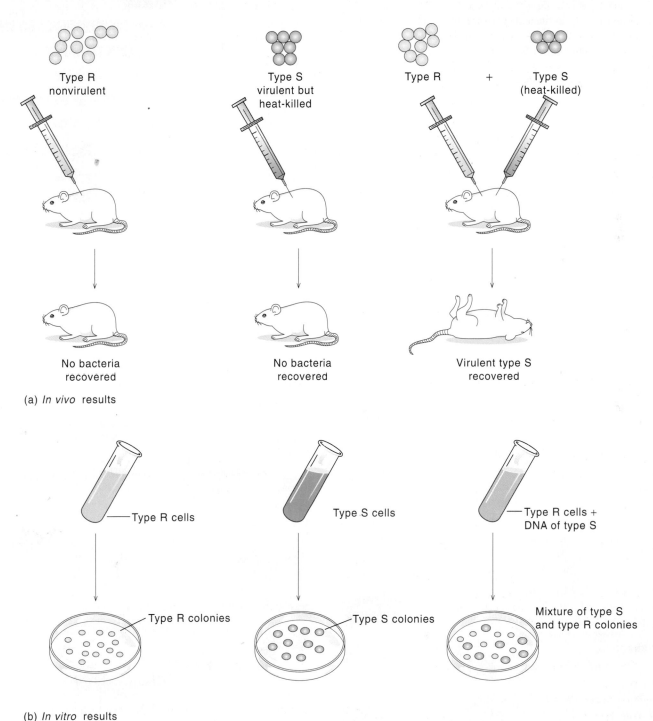

## Figure 25.2

Genome size in different cells, viruses, and plasmids. Plasmids are small circular DNA molecules that replicate autonomously in cells harboring them. Unlike viruses, they do not form any complex nucleoprotein structures. In the case of plasmids, most viruses, and bacteria, the genome size is equivalent to the size of the chromosomal DNA because there is only one chromosome. For all of the remaining eukaryotic organisms listed here, the genome is subdivided into two or more chromosomes. Some organisms contain more than 100 chromosomes. All chromosomes are believed to contain a single DNA molecule.

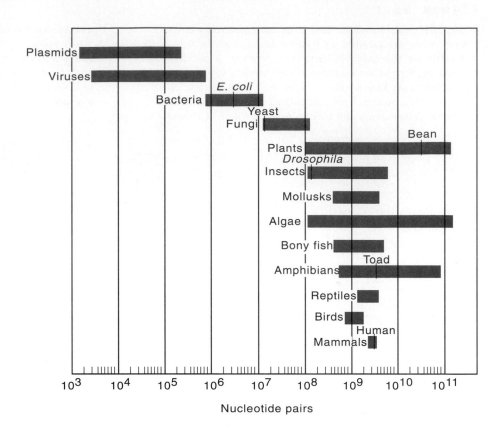

could be understood, but the chemistry was new. The remainder of this chapter focuses on the chemical and structural properties of DNA.

The amount of DNA per cell differs widely among different organisms (fig. 25.2). Mammalian cells contain about 1,000-fold more DNA than bacterial cells. Bacterial viruses such as the T type (T1–T7) bacteriophages (phages) that infect *E. coli* contain 10- to 20-fold less DNA than the bacterial host chromosome. The DNA of the smallest viruses is about 1/10 the size of the smallest T phage DNA, containing barely enough genetic material to accommodate about 10 genes. This finding is consistent with the fact that viruses do not contain sufficient genetic information for independent growth, but can only grow parasitically in the host cells they infect. On the other hand, the amount of DNA per cell is not always directly proportional to the amount of genetic information an organism carries. This is because complex eukaryotes contain a great deal of noninformational DNA in their chromosomes.

In addition to the main DNA associated with the cell or the virus, informational DNA is found in special organelles, such as chloroplasts and mitochondria. This DNA carries genes whose products are exclusively associated with organelle function.

## The Polynucleotide Chain Contains Mononucleotides Linked by Phosphodiester Bonds

Nucleotides are the building blocks of nucleic acids; their structures and biochemistry were discussed in chapter 23. When a 5′-phosphomononucleotide is joined by a phosphodiester bond to the 3′-OH group of another mononucleotide, a dinucleotide is formed. The 3′-5′-linked phosphodiester internucleotide structure of nucleic acids was firmly established by Lord Alexander Todd in 1951. Repetition of this linkage leads to the formation of polydeoxyribonucleotides in DNA or polyribonucleotides in RNA. The structure of a short polydeoxyribonucleotide is shown in figure 25.3. The polymeric structure consists of a sugar phosphate diester backbone with bases attached as distinctive side chains to the sugars.

The polynucleotide chain has a directional sense with 5′ and 3′ ends. Either of these ends may contain a free hydroxyl group or a phosphorylated hydroxyl group. The structure shown in figure 25.3 contains a phosphate group on the 5′ end but none on the 3′ end. By convention, one writes a nucleic acid sequence from the 5′ to the 3′ end so that a comparable structure is written pTpApCpG. With no

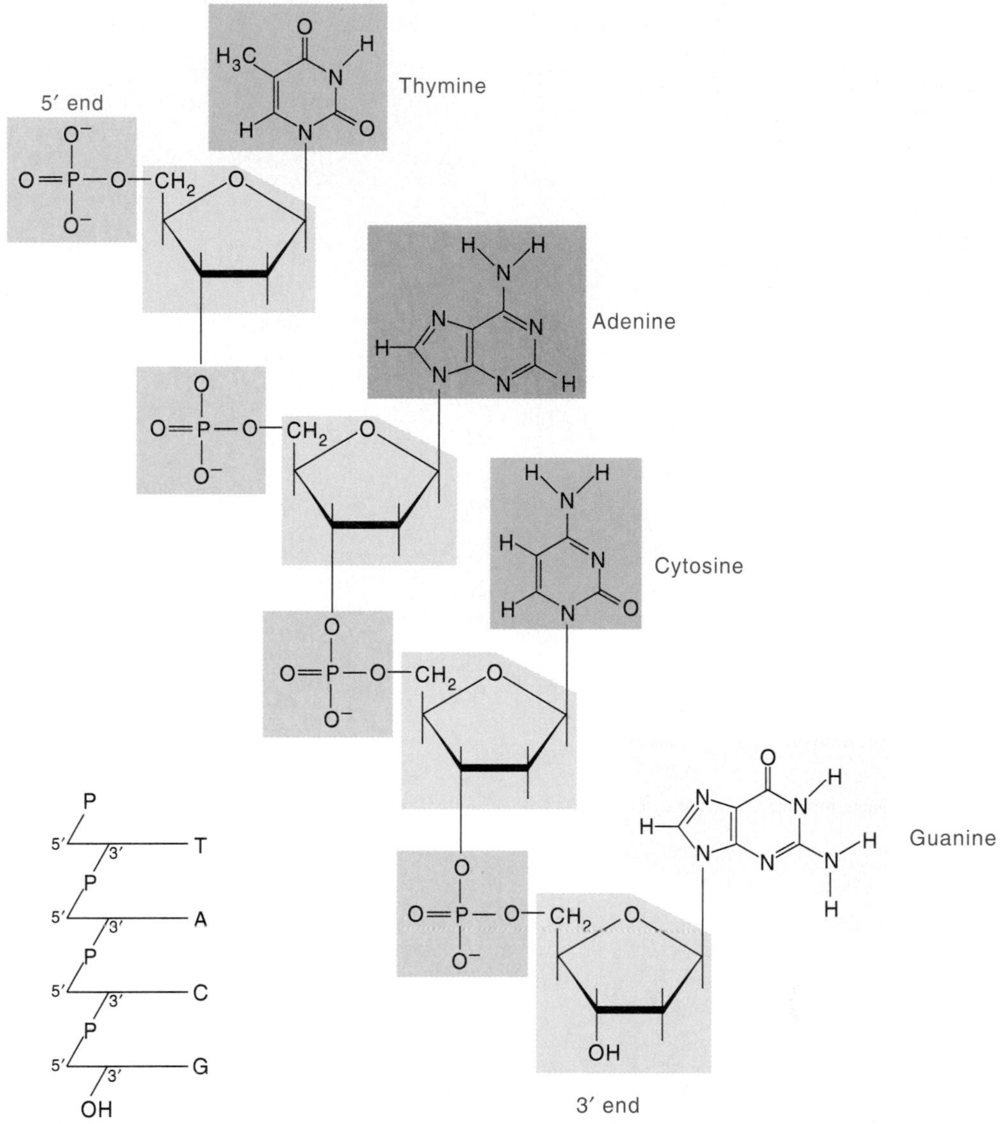

**Figure 25.3**

The structure of a deoxyribonucleotide. Drawn in abbreviated form at lower left. The illustrated structure is written pTpApCpG.

phosphate on the 5′ end, the structure is designated TpApCpG; alternatively, if the terminal phosphate is on the 3′ end rather than the 5′ end, the structure is written TpApCpGp. When the phosphates are not indicated, the structure is indicated by dashes on either end: -TACG-. The letters "d" or "r" sometimes precede the capital letter of the nucleotide to indicate a deoxyribo- or a riboderivative.

## Most DNAs Exist as Double-Helix (Duplex) Structures

Like most other types of biological macromolecules, nucleic acids adopt highly organized three-dimensional structures. The dominant factors that determine nucleic acid conformations are the limitations imposed by the stereochemistry of the polynucleotide chains, the high negative charge resulting from the regularly repeating phosphate groups, and the noncovalent affinities between purine and pyrimidine bases.

A body of chemical information that proved vital to understanding DNA structure came from Erwin Chargaff's analyses of the nucleotide composition of duplex DNAs from various sources (table 25.1). Although the base compositions varied over a wide range, Chargaff found that within the DNA of each source that he examined, the amount of A was very nearly equal to the amount of T, and the amount of G was very nearly equal to the amount of C. The C is present as both unmodified C and, to a lesser extent, 5-methyl-cytosine, which results from postreplicative

## Table 25.1

Base Composition of DNAs from Different Sources

|  | (A) Adenine | (G) Guanine | (C) Cytosine | (5-MC) 5-Methyl-cytosine | (T) Thymine | $\dfrac{A + T}{G + C + 5\text{-MC}}$ |
|---|---|---|---|---|---|---|
| Human | 30.4 | 19.6 | 19.9 | 0.7 | 30.1 | 1.53 |
| Sheep | 29.3 | 21.1 | 20.9 | 1.0 | 28.7 | 1.38 |
| Ox | 29.0 | 21.2 | 21.2 | 1.3 | 28.7 | 1.36 |
| Rat | 28.6 | 21.4 | 20.4 | 1.1 | 28.4 | 1.33 |
| Hen | 28.0 | 22.0 | 21.6 |  | 28.4 | 1.29 |
| Turtle | 28.7 | 22.0 | 21.3 |  | 27.9 | 1.31 |
| Trout | 29.7 | 22.2 | 20.5 |  | 27.5 | 1.34 |
| Salmon | 28.9 | 22.4 | 21.6 |  | 27.1 | 1.27 |
| Locust | 29.3 | 20.5 | 20.7 | 0.2 | 29.3 | 1.41 |
| Sea urchin | 28.4 | 19.5 | 19.3 |  | 32.8 | 1.58 |
| Carrot | 26.7 | 23.1 | 17.3 | 5.9 | 26.9 | 1.16 |
| Clover | 29.9 | 21.0 | 15.6 | 4.8 | 28.6 | 1.41 |
| Neurospora crassa | 23.0 | 27.1 | 26.6 |  | 23.3 | 0.86 |
| Escherichia coli | 24.7 | 26.0 | 25.7 |  | 23.6 | 0.93 |
| T4 Bacteriophage | 32.3 | 17.6 |  | 16.7[a] | 33.4 | 1.91 |

[a] In the T4 bacteriophage all of the cytosine exists in the 5-hydroxymethyl form 5-HMC.

## Figure 25.4

Dimensions and hydrogen bonding of (a) thymine to adenine and (b) cytosine to guanine. Note that two hydrogen bonds are formed in the A-T base pair and three in the G-C base pair. The overall dimensions of the base pairs are the same. Consequently they fit at any position in an otherwise regular polymeric structure. (Source: Adapted from S. Arnott, M. H. F. Wilkins, L. D. Hamilton, and R. Langridge, Fourier synthesis studies of lithium DNA, part III: Hoogsteen models, J. Mol. Biol. 11:391, 1965.)

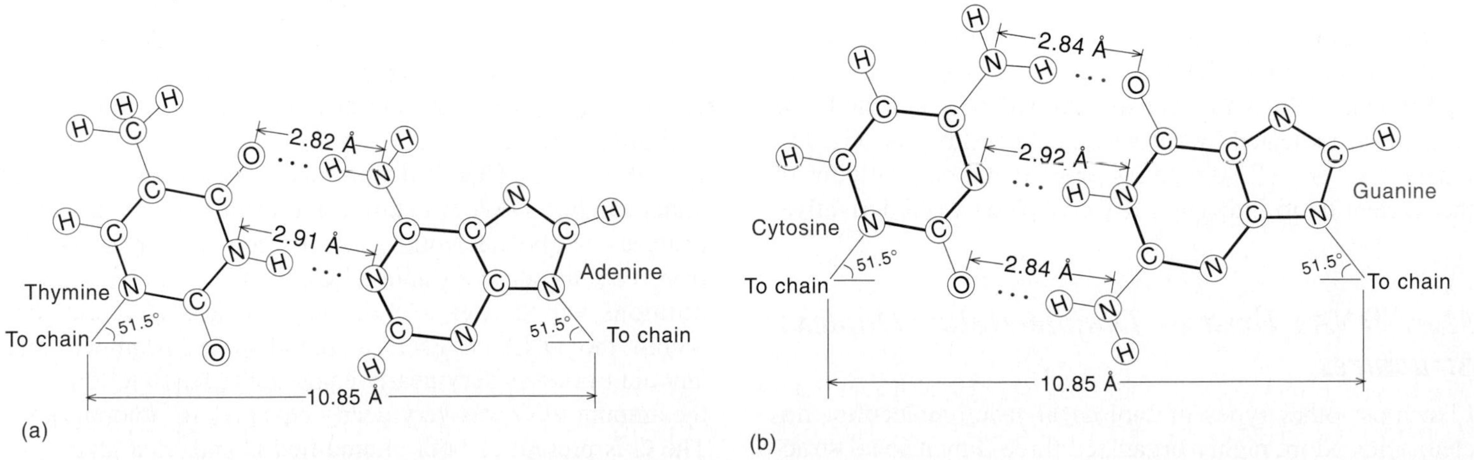

**Figure 25.5**

Segment of DNA, drawn to emphasize the hydrogen bonds formed between opposing chains. Each type of base is represented by a different color, with the sugar-phosphate backbones in blue and yellow. Note the three hydrogen bonds in the G-C pairs and the two in the A-T pairs (A, red; T, green; G, yellow; C, blue). The two strands are antiparallel: One strand (left side) runs 5′ to 3′ from top to bottom, and the other strand (right side) runs 5′ to 3′ from bottom to top. The planes of the base pairs are turned 90° to show the hydrogen bonds between the base pairs. (Source: Adapted from A. Kornberg, The Synthesis of DNA, *Scientific American,* October 1968.)

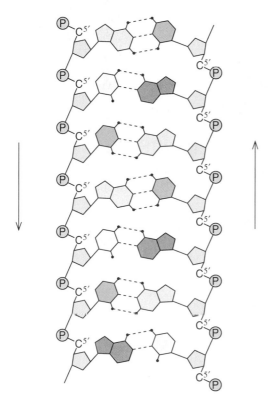

modification. The two equalities were the first indication that regular complexes occur between A and T and between G and C in DNA.

While searching for the meaning of these equalities, James Watson noted that hydrogen-bonded base pairs with the same overall dimensions could be formed between A and T and between G and C (fig. 25.4). The A-T base-paired structure has two hydrogen bonds, whereas the G-C base pair has three. The hydrogen-bonded pairs are formed between bases of opposing strands and can only arise if the directional senses of the two interacting chains are opposite or antiparallel (fig. 25.5). With this notion in mind Francis Crick took a closer look at the x-ray diffraction pattern produced by DNA and was able to interpret the diffraction pattern in terms of a helix (see Methods of Biochemical

Analysis 25A) composed of two polynucleotide strands. In this structure the planes of the base pairs are perpendicular to the helix axis, and the distance between adjacent pairs along the helix axis is 3.4 Å, bringing them into close contact (fig. 25.6). The structure repeats itself after 10 residues, or once every 34 Å along the helix axis; the repeating distance is referred to as the pitch length or just the pitch. An average-sized bacterial gene, which encodes the information to make a single protein, is about 1,000 bp in length, equivalent to 100 helical turns. As we see (chapters 26 and 28) the complementary structure of duplex DNA hints at how the genetic material is faithfully replicated as well as how it is expressed.

An important feature of the helical structure is the grooved nature of the surface resulting from the helical twist. Alternating wide (major) and narrow (minor) grooves are displayed in a side view of the helix structure (see fig. 25.6*a* and *b*). Different sections of the purine and pyrimidine bases are exposed in these two grooves as indicated in figure 25.6*c*. Many different proteins interact with DNA; most of the interactions occur with the phosphoryl groups on the outer surface of the structure and with the purine and pyrimidine bases in the major groove because of its greater accessibility. Specific instances of DNA-protein interactions are considered in later chapters (chapters 30 and 31).

## Hydrogen Bonds and Stacking Forces Stabilize the Double Helix

Several factors account for the stability of the double-helix structure. The negatively charged phosphoryl groups are all located on the outer surface, where they have a minimum effect on one another. The repulsive electrostatic interactions generated by these charged groups are often partly neutralized by interaction with cations such as $Mg^{2+}$, basic polyamines (such as putrescine and spermidine), and the positively charged side chains of chromosomal proteins. The core of the helix is composed of the base pairs held together by the specific hydrogen bonds and also by favorable stacking interactions between the planes of adjacent base pairs. These stacking interactions are complex, involving dipole–dipole interactions and van der Waals forces; this results in a stacking energy comparable in magnitude to the stabilizing energy generated by the hydrogen bonds between the base pairs. The result is that stacking is maximized in most nucleic acid structures. In this connection it is noteworthy that two fully extended polynucleotide strands can form a hydrogen-bonded base-paired complex, leading to a stepladderlike structure (fig. 25.7). In this structure the chains do not form a helix but lie straight, with a distance of

# Figure 25.6

The most common form of the double-helix DNA. The base-paired structure shown in figure 25.5 forms the helix structure shown in (*a*) and (*b*) by a right-handed twist. The two strands are antiparallel as indicated by the curved arrows in (*a*). (Reprinted with permission from *Nature* (171:737, 1953) Copyright 1953 Macmillan Magazines Limited.) In (*b*) a space-filling model depicts the sugar-phosphate backbones as strings of mostly gray, red, white, and yellow spheres, and the base pairs are rendered as horizontal flat plates composed of dark blue spheres. (Reprinted with permission from *Nature* (175:834, 1955) Copyright 1955 Macmillan Magazines Limited.) In (*c*) the orientation of the groups in the base pairs with respect to the major and minor grooves is indicated.

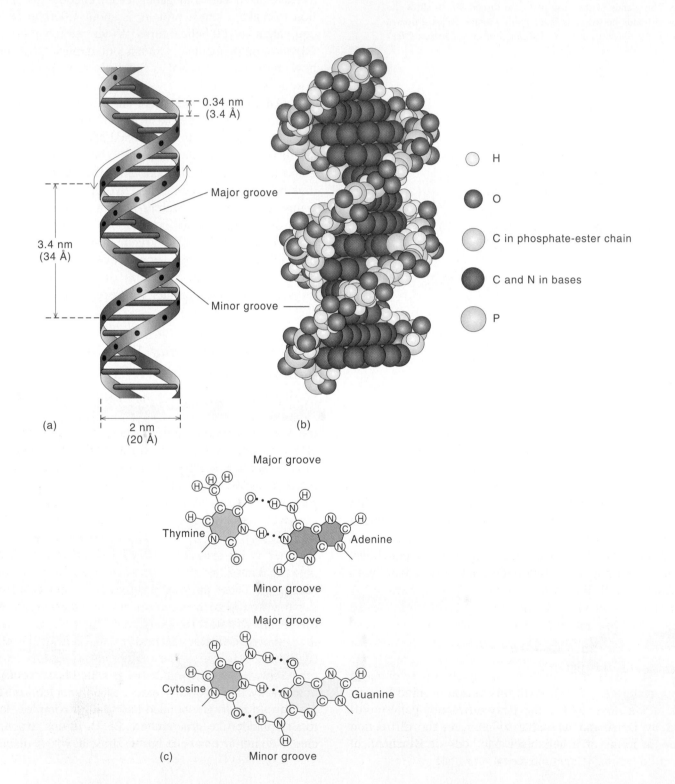

**Figure 25.7**

Different conformations of base-paired DNA: (*a*) the untwisted straight ladder, (*b*) the normal spiral ladder. The stepladder structure is unstable; it can be converted into a spiral ladder by a right-handed twist, a change that permits the planes of the base pairs to come into close contact.

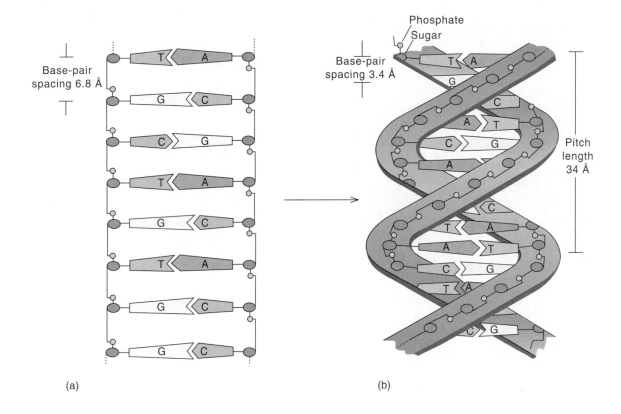

(a)     (b)

6.8 Å between adjacent nucleotides in the direction of the long axis. This 6.8-Å distance between adjacent base pairs produces a gap that would presumably be filled by water. Such a conformation is unstable because the planes of the bases prefer close contacts with one another as does water. The stepladder structure is related to the helix structure by a simple right-handed (clockwise) twist (see fig. 25.7*a*). Following this operation the distance between base pairs decreases until they are in close contact, with a spacing of 3.4 Å.

## *Conformational Variants of the Double-Helix Structure*

The same base-pairing arrangement is found in all naturally occurring double-helix structures. However, the inherent flexibility in the ribose sugar and the degrees of freedom generated by several rotatable single bonds per residue, six in the sugar-phosphate backbone and one in the C-1′–

N-glycosidic linkage, lead to considerable variation in the conformations adopted by double-helix structures.

The most striking conformational variant observed for a DNA double helix with Watson-Crick base pairing is referred to as the Z form. In Z DNA the backbone is twisted in the left-handed (counterclockwise) direction. This structure was first detected by Alex Rich and his co-workers (fig. 25.8). The Z form is a considerably slimmer helix than the B form and contains 12 bp/turn rather than 10. In the Z form, the planes of the base pairs are rotated approximately 180° with respect to the helix axis from their orientation in the B form (fig. 25.9).

Because of the different orientations of the bases in Z DNA, this DNA conformation requires that the sequence of purine and pyrimidine bases in each chain strictly alternate. An alternating sequence of G and C, or T and G or A and C residues can adopt a Z conformation. Of course, in all cases the opposing strand must contain a sequence of bases that is complementary as in all DNA duplex structures. An alternating A-T DNA sequence cannot adopt the Z confor-

## Figure 25.8

Space-filling models of (*a*) B DNA and (*b*) Z DNA. The irregularity of the Z DNA backbone is illustrated by the heavy lines that go from phosphate to phosphate residue along the chain. In contrast, B DNA has a smooth line that connects the phosphate groups and the two grooves, neither one of which extends into the helix axis of the molecule. The space-filling model is excellent for displaying the volume occupied by molecular constituents and the shape of the outer surface. (From A. Wang et al., Left-handed double helical DNA: Variations in the backbone conformation, *Science,* 211, 1981. Copyright 1981 by the AAAS. Reprinted by permission.)

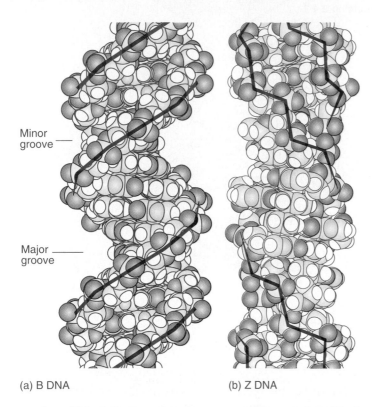

Minor groove

Major groove

(a) B DNA    (b) Z DNA

mation. This is believed to have something to do with the way that water molecules orient around an A-T base pair in the Z helix.

The biologic significance of Z DNA is currently unclear. However, several cellular proteins that bind specifically to Z DNA have been isolated from the nuclei of *Drosophila* fruit flies. The mere existence of such proteins suggests that they may function in some specific role when they encounter stretches of DNA that can adopt a Z conformation.

## Duplex Structures Can Form Supercoils

The detection of different conformations of DNA underscores the inherent flexibility built into the DNA duplex. All the conformations discussed thus far involve regular linear duplexes. Energetically favorable interactions with other molecules, particularly proteins, can induce additional conformations that do not result in major changes in either pair-

## Figure 25.9

The change in topological relationship if a four-base-pair segment of B DNA is converted into Z DNA. Such a conversion could be accomplished by rotation of the bases relative to those in B DNA. This rotation is shown diagrammatically by coloring one surface of the bases. All of the colored areas are at the bottom in B DNA. In the segment of Z DNA, however, four of them are turned upward. The turning is indicated by the curved arrows.

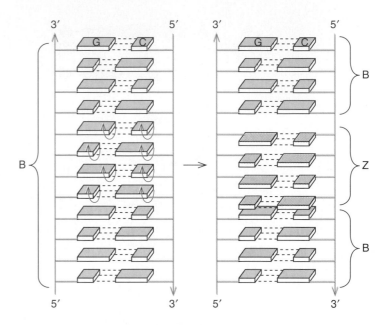

ing or stacking. Several conformations are believed to play important roles in different situations. Bends are known to be important in structures formed by chromosomes (discussed later in this chapter). Cruciforms, in which a single chain folds back on itself into a hairpinlike duplex, are formed at the ends of eukaryotic chromosomes (see the discussion of telomeres in chapter 26). Supercoiled DNA is a very common type of tertiary structure in which the double-helix segments twist around each other. Supercoiled DNA is topologically constrained by being covalently closed and circular, or by being complexed to proteins so that the ends of the DNA cannot rotate freely.

DNA can form right-handed (negatively supercoiled) or left-handed (positively supercoiled) supercoils. A right-handed supercoiled structure is shown in the middle of figure 25.10. Negative supercoiling imparts a torsional stress to the DNA that favors unwinding, whereas positive supercoiling favors tighter winding of the double helix.

Supercoiling of circular duplex DNA is quantitatively considered in terms of the linking number ($L$), an integer that specifies the number of complete turns made by one strand around the other. The linking number can change only if a covalent linkage in the DNA backbone is broken and reformed. Enzymes called topoisomerases (see chapter

## Figure 25.10

A circular duplex molecule in different topological states. (Adapted from a diagram supplied by M. Gellert.) The linking number can be changed only by breakage and re-formation of the phosphodiester linkages, as shown in the conversion of the relaxed circular form (1) to the strained negatively supercoiled form (2). The strain in the negatively supercoiled form can be partitioned in different ways between twist ($T$) and supercoiling ($S$) as shown in the interconversion between (2) and (3). No phosphodiester linkages are broken in making this interconversion, and consequently no change occurs in the linking number ($L$) as indicated.

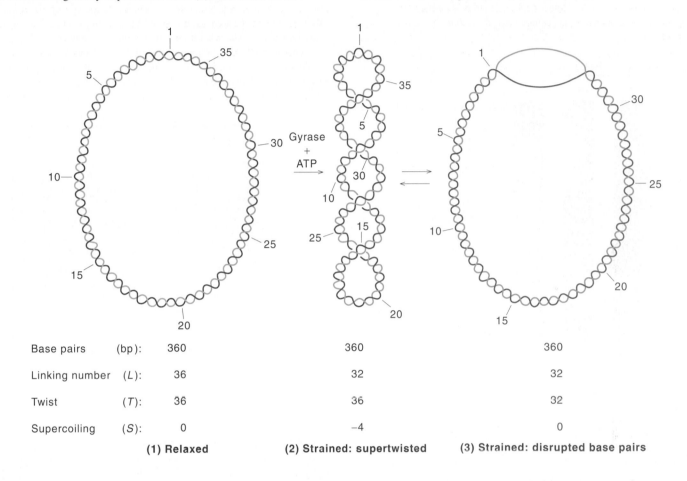

| | | (1) Relaxed | (2) Strained: supertwisted | (3) Strained: disrupted base pairs |
|---|---|---|---|---|
| Base pairs | (bp): | 360 | 360 | 360 |
| Linking number | ($L$): | 36 | 32 | 32 |
| Twist | ($T$): | 36 | 36 | 32 |
| Supercoiling | ($S$): | 0 | −4 | 0 |

26) catalyze changes in the linking number. If a molecule of DNA is projected onto a two-dimensional surface, the linking number is defined as the excess of right-handed over left-handed crossings of one strand over the other. Linear duplex DNA with free ends adopts a conformation in solution close to the B form, with about 10 bp/turn. Therefore, a closed circular duplex with this extent of twist is presumed to be under no torsional strain, and is said to be relaxed. Because B DNA is a right-handed helix, the linking number is normally positive by the sign convention we have adopted. The values of the linking number of relaxed DNA, $L°$, are distributed over a narrow range of integral values centered around 1 per 10 base pairs. DNA with a mean linking number smaller than this is termed negatively supercoiled, or underwound; DNA with a larger linking number is termed positively supercoiled, or overwound. The deviation of the linking number from its relaxed value, $\Delta L = L - L°$, can be partitioned between twist (altered double-helix coiling) and supercoiling.

$$L = \text{twist } (T) + \text{supercoiling } (S)$$

At present we do not know precisely how $L$ partitions between twist and supercoiling for helices under torsional stress. For example, consider the hypothetical situation illustrated in figure 25.10. A 360-bp structure in the circular relaxed form ($L = +36$, $T = +36$, $S = 0$) is indicated on the left. Exposure to the bacterial topoisomerase known as DNA gyrase introduces negative supercoils into such a structure in a reaction that requires ATP. If four negative supertwists are introduced, $L$ is reduced to +32. Barring other changes, $T$ remains fixed and $S$ becomes −4. In fact, the torsional strain introduced by the four negative supertwists tends to reduce T, causing a partial unwinding of the duplex. At one extreme, this effect can lead to the unwind-

## Figure 25.11

Electrophoretic patterns of highly supercoiled or partially supercoiled DNA. Strip A represents a sample of circular duplex DNA obtained by deproteinization of the animal virus SV40. In strips B and C the DNA has been exposed for increasing times to an enzyme (topoisomerase) that catalyzes relaxation. Adjacent bands differ by 1 in linking number. (From W. Keller, Characterization of purified DNA-relaxing enzyme from human tissue culture cells, *Proc. Natl. Acad. Sci. USA* 72:2553, 1975.)

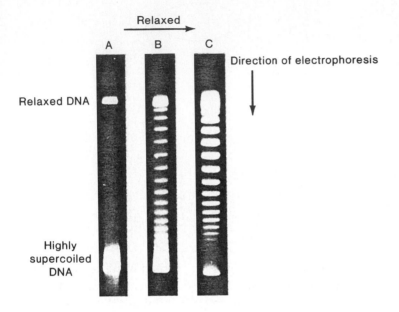

## Figure 25.12

Effect of temperature on the relative absorbance of native, renatured, and denatured DNA. When native DNA is heated in aqueous solution, its absorbance does not change until a temperature of about 80°C is reached, after which the absorbance rises sharply, by about 40% (curve *a*). On cooling the absorbance falls, but along a different curve, and it does not return to its original value (curve *b*). Renatured DNA, in which the two strands have been brought back into perfect register, shows a sharp melting curve similar to native DNA (curve *c*). Renatured DNA is prepared from denatured DNA by holding the temperature at about 25°C below the denaturation temperature for an extended time. This subject is discussed in detail later in the text. The temperature at which the native DNA is half denatured is labeled $T_m$.

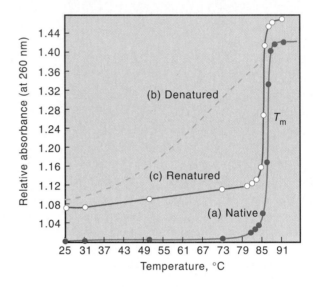

ing of four helical turns, in which case the supercoiling disappears entirely ($L = +32$, $T = +32$, $S = 0$). The actual situation probably leads to a reduction in the negative value of $S$ and a concomitant reduction in $T$ that is spread over the entire duplex as pictured. Note that the linkage number $L$ changes only when covalent bonds are broken, as in the case of gyrase treatment.

For DNA with a molecular weight of less than $10^7$, agarose gel electrophoresis is a most effective method for assessing the extent of supercoiling. DNA isomers differing by 1 in linking number form separate bands in the gel (fig. 25.11). The more highly supercoiled molecule migrates more rapidly through the gel as a result of its more compact structure.

## DNA Denaturation Involves Separation of Complementary Strands

In the laboratory, double-stranded DNA can be separated into single strands. The process of separating the polynucleotide strands of duplex nucleic acid structures is called denaturation. Denaturation disrupts the secondary binding forces that hold the strands together. Recall that the second-

ary binding forces include the edge-to-edge hydrogen bonds between the base pairs of opposing strands and the face-to-face stacking forces between the planes of adjacent base pairs. Individually, these secondary forces are weak, but when they act cooperatively they give rise to a DNA duplex that is highly stable in aqueous solution. The conditions required to denature DNA provide us with a measure of the strength of these interactions.

One of the simplest ways to denature DNA is by heating. The extent of denaturation at any temperature can be measured by the change in ultraviolet absorbance of a solution of DNA. A rise in absorbance coincides with the disruption of the regular base-paired structure, and the separation of the two polynucleotide strands from one another. The sharpness of the disruption of the regularly hydrogen-bonded base-paired native structure may be likened to the melting of a pure organic compound. It is customary to refer to this ultraviolet absorption temperature profile as a melting curve (curve *a* in fig. 25.12).

The melting temperature, $T_m$, of DNA is defined as the temperature at the midpoint of the absorption increase. This is about 85°C for the example shown in figure 25.12. Rapid cooling of the denatured DNA solution leads to re-formation of intrastrand hydrogen bonds but in a nonspecific, irregular manner. The absorbance decreases, but only by about three-fourths of the total original increase, and the decrease occurs over a much broader range of temperatures, as shown by curve *b* in figure 25.12. On subsequent reheating and cooling, the absorbance follows this cooling curve in the appropriate direction, indicating that denaturation results in an irreversible change.

The melting curve is an excellent tool for detecting DNAs that occur naturally in the single-stranded conformation. For example, in certain bacteriophages that infect *E. coli,* such as $\phi$X174, fd, or M13, the DNA exists as a single, circular strand. The ultraviolet absorption temperature curve for the DNA of these phages is similar in shape to that observed for denatured DNA (curve *b* in fig. 25.12). Broad melting curves also are characteristic of most RNAs, which rarely have regions of regular base pairing that extend for more than 10 or 20 residues.

Melting curves also have provided evidence that the stability of the double-helix structure is a function of its base composition. The midpoint of denaturation ($T_m$) of naturally occurring DNAs is precisely correlated with the average base composition of the DNA: The higher the mole percent of G-C base pairs, the higher the $T_m$ (fig. 25.13). This seems reasonable, because the G-C base pair contains three hydrogen bonds, whereas the A-T base pair contains only two (see fig. 25.4); thus DNA with a greater G-C content is expected to be more stable. As indicated above, base stacking is also believed to contribute to the stability of the duplex structure. In general the interaction energy gained by stacking between adjacent G-C base pairs is greater than that gained by interaction between A-T base pairs.

Other factors present in aqueous solution can affect the stability of the double-helix structure in a positive or a negative way. For example, salt has a stabilizing effect, which is mainly due to the repulsive electrostatic interactions between the negatively charged phosphate groups. Salt shields this charge interaction and therefore stabilizes the duplex structure. Thus, DNA in 0.15 M NaCl denatures at a $T_m$ about 20°C higher than DNA in 0.01 M phosphate. In pure water (no salt present) DNA denatures at room temperature. Extremes of pH also have a destabilizing effect on the double-helix structure. When the pH is above 11.5 or below 2.3, extensive deprotonization or protonization, respectively, occurs in the hydrogen-bonding groups of the bases, which in turn disrupts the hydrogen-bonded structure. Alkali is an excellent DNA denaturant; it permits rapid separation of the strands without degradation. Alkali both denatures and degrades RNA to 2'(3')-mononucleotides.

Many solutes that can form hydrogen bonds also lower the melting temperature (decrease the stability) of double-

**Figure 25.13**

Dependence of the temperature midpoint ($T_m$) of DNA on the content of guanine and cytosine. As the percentage of G + C increases, the $T_m$ increases. Two curves are shown to illustrate the point that the denaturation temperature is shifted to lower values when the ionic strength is lowered.

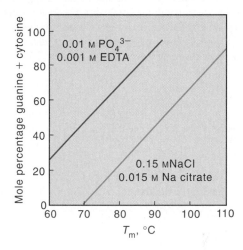

helix structures. The organic compounds formamide and urea are frequently used to lower the denaturation temperature as well as to prevent reaggregation in DNA manipulations, when it is important to avoid nonspecific aggregation. Reagents that increase the solubility of the DNA bases (e.g., methanol) or disrupt the water shell around them (e.g., trifluoroacetate) reduce the hydrophobic interactions between the bases and lower the $T_m$. Most proteins that bind to DNA inhibit denaturation. However, some DNA-binding proteins destabilize the native state. Proteins of this class usually bind preferentially to single-stranded DNA, thereby favoring separation of the double strands. We discuss so-called single-strand binding proteins in the next chapter because they play an important role in DNA synthesis.

## DNA Renaturation Involves Duplex Formation from Single Strands

We have seen that when a solution of heat-denatured DNA is allowed to cool rapidly, the regularly hydrogen-bonded structure does not reform. However, reassembly of the two separated polynucleotide strands into the native structure, called renaturation, is possible under certain specialized conditions.

The first indication that renaturation was possible came from Paul Doty's laboratory in experiments performed by Julius Marmur. They observed that when transforming DNA was heated and rapidly cooled it

**Figure 25.14**

Steps in denaturation and renaturation of a DNA duplex. In step 1 the temperature is raised to the point where the two strands of the duplex separate. If denatured DNA is slowly cooled, the events depicted as steps 2 and 3 follow. In step 2 a second-order reaction occurs in which two complementary strands of DNA must collide and form interstrand hydrogen bonds over a limited region. Step 3 is a first-order reaction in which additional hydrogen bonds form between the complementary strands that are partially hydrogen-bonded (zippering). Once complementary strands are partially bonded, the zippering reaction occurs rapidly. In the overall process, step 2 is rate-limiting.

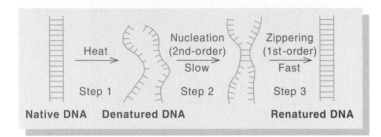

was biologically inactive; however, when denatured DNA was slowly cooled, a small percentage of the initial transforming activity was recovered. Further experiments showed that the optimal temperature for this recovery of activity, or renaturation, was about 25°C below the $T_m$. Similar to denaturation, renaturation can be followed spectrophotometrically. If the temperature of a denatured DNA solution is maintained at $T_m - 25°C$ for a long time, the absorbance of the solution gradually decreases until it approaches a value close to that of native DNA (see curve c in fig. 25.12). The optimum temperature for renaturation is called the annealing temperature. At the annealing temperature, irregularly hydrogen-bonded structures are unstable, but regularly hydrogen-bonded structures are stable. Consequently, prolonged exposure of denatured DNA at this temperature favors the formation of regularly base-paired structures.

Kinetic analysis indicates that renaturation is a two-step process. In the slow step effective contact is made between two complementary regions of DNA originated from separate strands. This rate-limiting step called nucleation is a function of the concentration of complementary strands. Nucleation is followed by a relatively rapid zippering up of adjoining base residues into a duplex structure. The steps involved in denaturation and renaturation are depicted in figure 25.14.

Since nucleation involves interaction between two molecules, it should occur at a rate proportional to the square of the concentration of single strands. If c is the concentration of single-stranded DNA at time $t$, then the rate equation for the loss of single-stranded DNA is

$$-\frac{dc}{dt} = k_2 c^2 \tag{1}$$

where $k_2$ is the rate constant for a second-order reaction. Starting with a concentration $c_0$ of completely denatured DNA, the amount of single-stranded DNA left after renaturation for time $t$ is given by

$$\frac{c}{c_0} = \frac{1}{1 + k_2 c_0 t} \tag{2}$$

At time $t_{1/2}$, when half of the DNA is renatured, $c/c_0 = 0.5$ and $t = t_{1/2}$, from which it follows that

$$c_0 t_{1/2} = \frac{1}{k_2} \tag{3}$$

The rate of renaturation is also a function of chain length, but this effect is usually eliminated as a variable by shearing the starting DNA down to a uniform size. For a typical renaturation experiment the values of $c/c_0$ are plotted as a function of $c_0 t$, and the resulting curve is referred to as a "cot" curve.

In figure 25.15 cot curves for four DNA samples and one synthetic polyribonucleotide sample are presented. The mouse satellite DNA is a fraction of the DNA from the mouse that contains highly repetitious sequences. The T4 DNA is the total DNA isolated from the bacteriophage T4. Similarly the E. coli DNA is the total DNA isolated from E. coli bacteria. The calf nonrepetitive fraction represents a fraction of calf nuclear DNA that contains mostly sequences that are represented only one time per haploid genome. At the top of this semilogarithmic plot is an additional scale indicating the nucleotide complexity, N, which is defined as the number of nucleotides in a nonrepeating sequence. If no sequences in the cellular DNA repeat, then N is equal to the number of nucleotides in the genome. It can be seen that $c_0 t_{1/2}$ is proportional to N for these samples.

This family of curves has been used to calibrate more complex situations in which the test DNA is a mixture of unique sequence and repetitive DNA. In such cases the repetitive DNA fraction tends to anneal more rapidly. Bacterial DNA (E. coli) contains very little repetitive DNA (0.3%). This is mainly accounted for by the eight genes for E. coli ribosomal RNA that have nearly identical sequences. In complex eukaryotes single-copy DNA (i.e., nonrepetitive DNA) accounts for 40%–70% of the DNA, most of the remainder being roughly divided between middle repetitive ($<10^4$ copies/genome) and highly repetitive ($>5 \times 10^4$ copies/genome). Further analyses of eukaryotic gene struc-

**Figure 25.15**

Reassociation of double-stranded nucleic acids from various sources. The genome size is indicated by arrows near the upper nomographic scale. Over a factor of $10^{10}$, this value is proportional to the $c_0t$ (the "cot") required for half-reaction. All DNAs were sheared so that they have approximately the same fragment size (about 400 nucleotides, single-stranded). Correction has been made to give the rate that would be observed at 0.18 M sodium ion concentration. No correction for temperature has been applied because it was approximately optimum in all cases. The labels for the different DNAs should not concern the average reader. Mouse satellite and calf (nonrepetitive fraction) are fractions of the genome obtained from the indicated animals. (Source: Adapted from R. J. Britten and D. E. Kohne, Repeated sequences in DNA, *Science*, 161:529, 1968.)

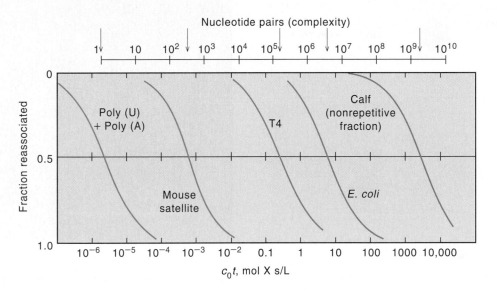

## Chromosome Structure

All types of nucleic acids interact with proteins. Chromosomal DNA forms stable nonspecific complexes with structural proteins that stabilize their tertiary structure; it also forms transient complexes with enzymes and regulatory proteins that modulate DNA and RNA metabolism. The gross tertiary structure of DNA in *E. coli* and a typical eukaryotic chromosome is described in the next section.

ture by a variety of other techniques have shown that most genes encode proteins that belong to the unique (single-copy) class. The middle repetitive class includes transfer RNA genes and ribosomal RNA genes that form part of the biochemical machinery for protein synthesis (chapter 29), as well as the genes encoding histones, the main chromosomal proteins. Some highly repetitive sequences occur in tandem (see mouse satellite DNA in fig. 25.15), and still other repetitive DNA elements are distributed at random throughout the genome. It has been argued that some families of repetitive DNA, referred to as "selfish DNA," represent "parasitic" sequences that replicate together with the genome without conferring any positive or negative characteristics to the host cells that have these sequences.

## *Physical Structure of the Bacterial Chromosome*

A single chromosome of *E. coli* contains about $3 \times 10^9$ daltons or about $4.5 \times 10^6$ bp of DNA. If all of this DNA were in a duplex structure stretched end to end, it would be 1.5 mm long, which is about 75 cell diameters. In fact, the chromosome is circular, centrally located in the cell, and highly folded, so that it is only about 2 $\mu$m across. No dramatic change in chromosome morphology is seen prior to cell division as in eukaryotes. Clearly the degree of compaction observed throughout the cell cycle does not interfere with transcription to any great extent because all the genes in *E. coli* are readily expressible.

Electron micrographs of the *E. coli* chromosome suggest a folded circular structure containing about 40–100 supercoiled loops (diagrammatically indicated in fig. 25.16). It is believed that the folded structure is held together by an RNA-protein core, although the manner in which this is done is not well understood. The structure is further stabilized because the core forms a complex with positively charged polyamines and certain basic proteins. D. E. Pettijohn and his co-workers have provided evidence of such a core. They first showed that the individual super-

## Figure 25.16

The *E. coli* chromosome exists as a circular, folded, supercoiled duplex (*a*). This can be converted to a partially unfolded structure by brief treatment with RNase (*c*). There are 50–100 loops in the structure; supercoiling may be selectively eliminated from individual loops by single-strand nicking of the DNA within the loop (*b*).

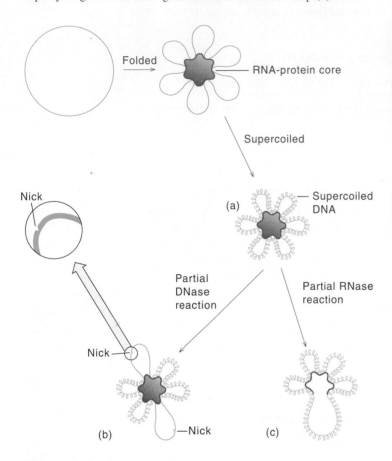

## Figure 25.17

Swollen fibers of chromatin from the nucleus of the chicken red blood cell. The electron micrograph is enlarged about 325,000× and negatively stained with uranyl acetate. (Micrograph courtesy of A. L. Olins and D. E. Olins.)

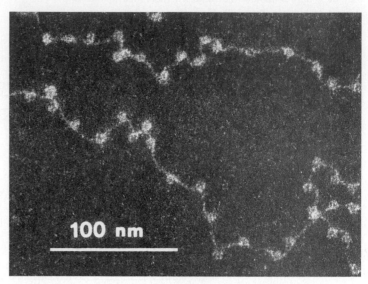

evidence for significant stretches of DNA with no function. Frequently, genes with a related function are clustered. These clustered genes are usually transcribed into single expression units (messenger RNAs) containing the information for the synthesis of several functionally related proteins (see chapters 28 and 29).

## *Eukaryotic DNA Is Complexed with Histones*

DNA in eukaryotic chromosomes exists in a highly compacted form known as chromatin, a complex of DNA with a great variety of proteins. Five proteins called histones are present in large amounts (see table 25.2) and are believed to form a regularly repeating structural motif. The remaining proteins are present in smaller amounts and are irregularly distributed.

Electron-microscopic and x-ray diffraction studies on chromatin suggest that the DNA forms a coiled-coil structure with the histones of chromatin. A breakthrough in our understanding of the nucleohistone complex came when D. E. Olins and A. L. Olins observed that chromatin viewed after sudden swelling in water had a beaded structure (fig. 25.17). The beads, called nucleosomes, contain four of the five histones; they are about 10 nm in diameter, and the spacing between the beads is about 14 nm. Brief enzymatic digestion of chromatin with micrococcal nuclease fragments this structure. The DNA–histone frag-

coiled loops maintain their supercoiling independently of one another. Thus, if a single nick is introduced into one of the loops by limited DNase action, that loop adopts an expanded relaxed conformation, but supercoiling in the other loops is maintained (see fig. 25.16). Limited RNase or protease treatment causes the partial breakdown of the looped structures without interfering with the supercoiling (see fig. 25.16). These results have led to the conclusion that each of the loops is a domain, the lateral motion of which is restricted by an RNA-protein core complex.

## *The Genetic Map of* Escherichia coli

The circular *E. coli* chromosome contains enough base pairs to make about 3,000 average-sized genes. The relative positions of over half of these genes are known. In regions where our understanding is reasonably complete the impression is obtained of a tightly organized genome. Coding regions are interspersed with regulatory regions; there is no

### Table 25.2

Characteristics of Histones

| Name | Ratio of Lysine to Arginine | $M_r$ | Copies per Nucleosome |
|------|------|------|------|
| Histone H1[a] | 20 | 21,000 | 1 (not in bead) |
| Histone H2a | 1.2 | 14,500 | 2 (in bead) |
| Histone H2b | 2.5 | 13,700 | 2 (in bead) |
| Histone H3 | 0.7 | 15,300 | 2 (in bead) |
| Histone H4 | 0.8 | 11,300 | 2 (in bead) |

[a] Not found in lower eukaryotes such as yeast.

ments from this partial digestion give rise to a banded pattern on agarose gel electrophoresis that suggests nucleoprotein structures containing 200 bp of DNA or multiples thereof (400, 600, 800 bp, etc.). Electron-microscopic examination of the individual fractions isolated after gel electrophoresis confirms the suggested correlation between size of the DNA estimated on gels and the number of nucleosomes. Thus, the most rapidly moving DNA band seen on gels was derived from a structure containing one nucleosome, and the second fastest migrating species contains nucleosome dimers, and so forth. Evidently, the brief treatment with endonuclease preferentially cleaves DNA in the internucleosomal region, where the DNA is least likely to be protected from enzyme attack. More extensive nuclease treatment results in a single band on gels that contain a single nucleosome with 140 bp of DNA. It appears that exhaustive nuclease digestion has removed all of the DNA that is not in direct contact with the nucleosome. From these results it was deduced that nucleosomes contain a core of histone with 140 base pairs of DNA wrapping; an additional 60 bp of more exposed DNA connects adjacent nucleosomes.

The histones present in chromatin are of five major types: H1, H2a, H2b, H3, and H4 (table 25.2). The lysine-rich histone H1 is not present in the nucleosome core particles, as evidenced by its release on extensive nuclease treatment and the finding that H1 is the only histone that readily exchanges between free and DNA-bound histone. H1 may play a key role in the conversion of chromatin to the highly compacted chromosome that occurs immediately before cell division. The other eight histones, two each of the other four histones, form the protein core of the nucleosome. These protein octamers do not come apart even when chromosomes duplicate.

An illustration of the coiled-coil structure of the nucleosome is presented in figure 25.18. The 140 bp of DNA

make about one and three-quarters superhelical turns about the histone octamer. An additional 60 bp of spacer DNA (not shown) connect adjacent nucleosomes. A survey of chromatins from different species and in different tissues of the same species has shown that the spacer DNA actually varies in length from about 20–95 bp.

Salt bridges between positively charged basic amino acid side chains of histones and the negatively charged DNA phosphates play a major role in stabilizing the DNA–histone complex. Indeed, treatment of chromatin with concentrated NaCl (1–2 M), which is known to disrupt electrostatic bonds, causes a complete dissociation of DNA and histone in the nucleohistone complex.

Higher order structures beyond that of the nucleosome are less well understood. Electron-microscopic investigations indicate two types of fibers with diameters of 10 nm and 30 nm. To account for the 10 nm fiber the nucleosomes can be arranged edge to edge in a zigzag fashion to produce a fibril that is 10 nm wide. When the ionic strength is raised on an isolated preparation of chromatin, the fibrils reversibly condense into an irregularly supercoiled fiber about 30 nm in diameter. The nucleosome particles are thought to have their cylindrical axes approximately perpendicular to the long axis of the 30 nm fiber with six to seven nucleosomes per turn (fig. 25.19). Further coiling of these structures is necessary to explain the much larger structures seen in mitotic chromosomes.

## Organization of Genes within Eukaryotic Chromosomes

The organization of genes within a typical eukaryotic chromosome is far more complex and less well understood than in prokaryotes. It is highly likely that a much lower percentage of the DNA is informational in complex eukaryotes than in prokaryotes. E. coli contains about 3,000 genes; the

## Figure 25.18

(a) Path of DNA that can account for the bipartite structure of the nucleosome core is a superhelix with an external diameter of 110 Å and a pitch of 27 Å; the turns of the 20-Å-wide DNA helix are nearly in contact. About 80 bp of DNA occur per turn; the nucleosome core, an enzymatically reduced form of the nucleosome consisting of some 140 bp, has about one and three-quarter turns wrapped on it. The histone octamer complex, containing two each of histones H2a, H2b, H3, and H4, is packed on the inside of the DNA coiled-coil structure. (© Scientific American, Inc., George V. Kelvin. Reprinted with permission.) In (b) we see this histone octamer inserted into the nucleosome core. The H3-H4 tetramer is shown in yellow, and an H2a-H2b dimer is shown at each end in purple.

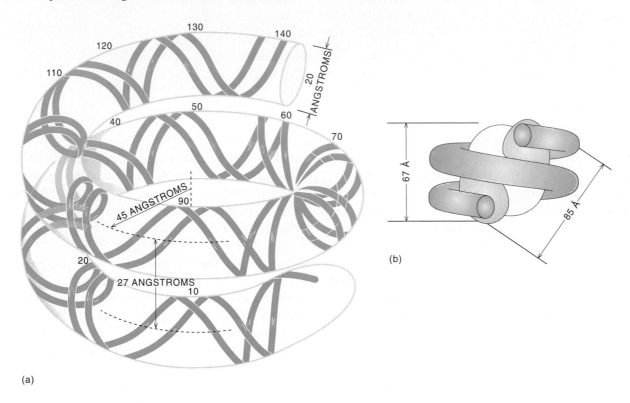

(a)

(b)

human genome is 1,000 times larger but it is very unlikely to contain 1,000 times the number of functional genes. Most estimates hover around 50,000. Even if there were 200,000 human genes, this would still leave approximately 90% of the DNA with no coding function. Measurements on the fruit fly *Drosophila melanogaster* indicate that the coding information for proteins in most genes probably accounts for no more than one-tenth of the base pairs within the gene. The discovery that noncoding regions (introns) occur between coding regions (exons) helps to explain the excessive amounts of DNA that appear to be present in eukaryotes (see chapter 28), although it is probably not the whole story. Noncoding regions serving as control loci may be larger on the average in eukaryotes. Nonfunctioning genes that have lost their initiation sites for being expressed (promoters) and repetitive genetic elements with no apparent coding function also help to account for the large amount of noncoding DNA.

As we have already discussed, not all repetitive DNA in eukaryotes is noncoding. Thus, histone genes, for example, are typically reiterated many times. The five different histone genes are usually clustered, and this cluster is then tandemly repeated many (up to 100 or more) times. Ribosomal RNA genes are also tandemly clustered. Other nonidentical but functionally related genes that show clustering include the globin genes and the immunoglobulin genes.

**Figure 25.19**

Helical superstructures might be formed with increasing salt concentration (*bottom to top*) as is suggested here. The zigzag pattern of nucleosome (1, 2, 3, 4) closes up, eventually to form a solenoid, a helix with about six nucleosomes per turn. (The helix is probably more irregular than it is in this drawing.) Cross-linking data indicate that H1 molecules on adjacent nucleosomes make contact. Extrapolation from the zigzag form to the solenoid suggests (but does not prove) that the aggregation of H1 at higher ionic strengths gives rise to a helical H1 polymer (not shown) running down the center of the solenoid. In the absence of H1 (*bottom*) no ordered structures are formed. The details of H1 associations are not known at this time; the drawing is meant to indicate only that H1 molecules contact one another and link DNA. (© Scientific American, Inc., George V. Kelvin. Reprinted with permission.)

30 nm

H1

DNA

## Summary

1. The genetic material of cells and viruses consists of DNA or RNA. That DNA bears genetic information was first shown when the heritable transfer of various traits from one bacterial strain to another was found to be mediated by purified DNA.

2. All nucleic acids consist of covalently linked nucleotides. Each nucleotide has three characteristic components: (1) a purine or pyrimidine base; (2) a pentose; and (3) a phosphate group. The purine or pyrimidine bases are linked to the C-1' carbon of a deoxyribose sugar in DNA or a ribose sugar in RNA. The phosphate groups are linked to the sugar at the C-5' and C-3' positions. The purine bases in both DNA and RNA are always adenine (A) and guanine (G). The pyrimidine bases in DNA are thymine (T) and cyto-

sine (C); in RNA they are uracil (U) and cytosine. The bases may be post-replicatively or posttranscriptionally modified by methylation or other reactions in certain circumstances.

3. DNA exists most typically as a double-stranded molecule, but in rare instances it exists (in some phages and viruses) in a single-stranded form. The continuity of the strands is maintained by repeating 3',5'-phosphodiester linkages formed between the sugar and the phosphate groups; they constitute the covalent backbone of the macromolecule. The side chains of the covalent backbone consist of the purine or pyrimidine bases. In double-stranded, or duplex, DNA the two chains are held together in an antiparallel arrangement. The base composition of DNA varies character-

istically from one species to another in the range of 25%–75% guanine plus cytosine. Specific pairing occurs between bases on one strand and bases on the other strand. The complementary base pairs are either A and T, which can form two hydrogen bonds, or G and C, which can form three hydrogen bonds. The duplex is stabilized by the edge-to-edge hydrogen bonds formed between these planar base pairs and face-to-face interactions (stacking) between adjacent base pairs. Twisting of the duplex structure into a helix makes stacking interactions possible.

4. The right-handed helical structure, known as B DNA, is the most commonly occurring conformation of linear duplex DNA in nature. In this structure, the distance between stacked base pairs is 3.4 Å, with approximately 10 base pairs per helical turn. The inherent flexibility of the structure, however, makes a variety of conformations possible under different conditions. In some instances, nucleotide sequence and degree of hydration dictate which conformations are favored. DNA interacts with a variety of proteins inside the cell, and these proteins can also have a significant influence on its secondary and tertiary structure.

5. Circular DNA molecules, which are topologically confined so that their ends are not free to rotate, can form supercoils that are either right-handed (negative) or left-handed (positive). Negative supercoiling exerts a torsional tension favoring the untwisting of the primary right-handed double helix, whereas positive supercoiling has the opposite effect. Negatively supercoiled DNAs are most commonly observed in prokaryotes, which contain an enzyme that generates the supercoiled structure.

6. When duplex DNA (or RNA) is heated, it dissociates (denatures) into single strands. The temperature at which denaturation occurs (the melting temperature) is a measure of the stability of the duplex and is a function of the G-C content of the DNA. A preparation of denatured DNA may be renatured in the native duplex structure by maintaining the temperature about 25°C below the melting temperature. The rate of renaturation is a measure of the sequence complexity of the DNA. In prokaryotes, which consist predominantly of unique sequences, the complexity (the number of base pairs) is approximately equal to the genome size. However, eukaryotic cells contain DNAs of varying sequence complexity that renature at quite different rates. The fastest renaturing fractions are present in many copies per nucleus, whereas the slowest renaturing fractions are present in single copies. Analyses by other techniques have shown that some of the repetitive DNA sequences exist as tandemly repeated structures, while other types of repetitive sequences are dispersed throughout the genome.

7. In the chromatin of eukaryotic cells DNA forms a coiled-coil structure with an approximately equal weight of a mixture of five basic proteins known as histones. Four of these histones in pairs form an octamer around which the DNA duplex occurs in a left-handed helix. The DNA octamer complex is called a nucleosome. Each nucleosome contains about 140 base pairs of DNA in a nuclease-resistant "nucleosome core" and approximately 60 base pairs of spacer between core particles. Histone H1 binds to the chromatin independently of the octamer and is the first histone to dissociate from the chromatin when the ionic strength is raised. Beyond the nucleosome the higher order structure of the chromosome involves coiled-coil structures with varying degrees of regularity.

## Selected Readings

Avery, O. T., C. M. MacLeod, and C. McCarthy, Studies on the chemical nature of the substance inducing transformation of pneumococcal types. *J. Exp. Med.* 79:137–158, 1944.

Camerini-Otero, R. D., and P. Hsieh, Parallel DNA triplexes, homologous recombination and other homology-dependent DNA interaction. *Cell* 73:217–224, 1993.

Cantor, C. R., C. L. Smith, and M. K. Mathew, Pulsed-field gel electrophoresis of very large molecules. *Ann. Rev. Biophys. Chem.* 17:287–304, 1988.

Daniels, D. L., G. Plunkett III, V. Burland, and F. R. Blattner, Analysis of the *Escherichia coli* genome: DNA sequence of the region from 84.5 to 86.5 minutes. *Science* 257:771–778, 1992.

Dervan, P. B., Reagents for the site-specific cleavage of megabase DNA. *Nature* 359:87–88, 1992.

Dickerson, R. E., The DNA helix and how it is read. *Sci. Am.* 249(6)d:94–111, 1983.

Hershey, A. D., and M. Chase, Independent functions of viral proteins and nucleic acid in growth of bacteriophage. *J. Gen. Physiol.* 36:39–56, 1952.

Hillary, C. M., J. T. Finch, B. F. Luisi, and A. Klug, The structure of an oligo(dA), oligo(dT) tract and its biological implications. *Nature* 330:221–236, 1987.

Kang, C., X. Zhang, R. Ratliff, R. Moyzis, and A. Rich, Crystal structure of four-stranded Oxytricha telomeric DNA. *Nature* 356:126–131, 1992.

Kim, S. H., Three-dimensional structure of transfer RNA. *Prog. Nuc. Acid Res. Mol. Biol.* 17:181–216, 1973.

Kornberg, R. D., and A. Klug, The nucleosome. *Sci. Am.* 244(2):52–64, 1981.

Lerman, L. S., S. G. Fischer, I. Hurley, K. Silverstein, and N. Lumelsky, Sequence-determined DNA separations. *Ann. Rev. Biophys. Bioeng.* 13:399–423, 1983.

Morse, R. H., and R. T. Simpson, DNA in the nucleosome. *Cell* 54:285–287, 1988.

Nadeau, J. G., and D. M. Crothers, Structural basis for DNA bending. *Proc. Natl. Acad. Sci.* 86:2622–2626, 1989.

Ramakrishnan, V., J. T. Finch, V. Graziano, P. L. Lee, and R. M. Sweet, Crystal structure of globular domain of histone H5 and its implications for nucleosome binding. *Nature* 362:219–223, 1993.

Rich, A., A. Nordheim, and A. H.-J. Wang, The chemistry and biology of left-handed Z DNA. *Ann. Rev. Biochem.* 53:791–846, 1984.

Richmond, T. J., J. T. Finch, B. Rushton, D. Rhoades, and A. Klug, Structure of the nucleosome core particle at 7 Å resolution. *Nature* 311:532–537, 1984.

Roberts, R. W., and D. M. Crothers, Stability and properties of double and triple helices: Dramatic effects of RNA or DNA backbone composition. *Science* 258:1463–1466, 1992.

Saenger, W., *Principles of Nucleic Acid Structure.* New York: Springer-Verlag, 1984.

Schmid, M. B., Structure and function of the bacterial chromosome. *Trends Biochem. Sci.* 13:131–135, 1988.

Schwartz, D. C., and C. R. Cantor, Separation of yeast chromosome-sized DNAs by pulsed field gradient gel electrophoresis. *Cell* 37:67–75, 1984.

Strobel, S. A., L. A. Doucette-Stamm, L. Riba, D. E. Housman, P. B. Dervan, Site-specific cleavage of human chromosome 4 mediated by triple-helix formation. *Science* 254:1639–1642, 1991.

Structures of DNA, *Cold Spring Harbor Symp. Quant. Biol.* 47, 1983.

van Holde, K. E., *Chromatin.* New York: Springer-Verlag, 1988.

Watson, J. D., and F. H. C. Crick, Molecular structure of nucleic acids. *Nature* 171:737–738, 1953.

Wells, R. D., D. A. Collier, J. C. Hanvey, M. Shimizu, and F. Wohlrab, The chemistry and biology of unusual DNA structures adopted by oligopurine · oligopyrimidine sequences. *Faseb J.* 2:2939–2949, 1988.

## Problems

1. Briefly describe how Avery was able to show that DNA is the genetic material in cells.

2. Why is the bond holding nucleotides together in nucleic acids called a phosphodiester bond?

3. Give the structure of the DNA strand complementary to pTpApCpG (see structure shown in fig. 25.3) in the abbreviated form.

4. How many different base-paired structures with two hydrogen bonds can be made using guanine (G) and thymine (T)? Which one of these is most similar to the standard Watson-Crick (G-C) base pair?

5. (a) List the hydrogen bond donors and acceptors available in the major and minor grooves of the DNA double helix.

   (b) The intermolecular forces that stabilize DNA-protein complexes often involve hydrogen bonds between specific amino acids and the exposed surfaces of the bases. Explain why interaction with the major groove is more common among DNA-binding proteins than interaction with the minor groove.

6. How many base pairs are found in the DNA pictured at the right in figure 25.10 if two negative supercoils are introduced without breaking the backbone? Describe the effect of a short stretch of Z DNA on the overall conformation of this DNA.

7. Describe a physical method that can be used to estimate the base composition of DNA. Describe the data obtained with two DNA samples: One with high G-C content and another with high A-T content? (Assume that the concentration of the samples is equal.)

8. Why is DNA denatured at either low pH (pH 2) or high pH (pH 11), and why is DNA stable at pH 7? (*Hint:* See p$K$ values in table 23.2.)

9. You are given a sample of nucleic acid extracted from a virus. How would you determine whether the virus has an RNA or DNA genome and whether it is single- or double-stranded?

10. A column packed with the material hydroxyapatite preferentially binds double-stranded DNA over single-stranded DNA. The double-stranded DNA can

be eluted by changing the salt concentration. Using a hydroxyapatite column and the information in figure 25.15, propose a method for separating a mixture of equal amounts of T4 bacteriophage and *E. coli* DNAs into the respective components.

11. Why can't RNA duplexes or RNA–DNA hybrids adopt the B conformation?

12. Some DNA-binding proteins specifically bind to Z DNA. How could these proteins help stabilize DNA in the Z configuration? (*Hint:* How do single-stranded DNA-binding proteins destabilize duplex structures?)

13. Linear duplex DNA can bind more ethidium bromide than covalently closed circular DNA of the same molecular weight. Why? (*Hint:* Ethidium bromide molecules bind between adjacent base pairs of DNA, causing the duplex to unwind in the region of binding.)

14. Give the relative times for 50% renaturation of the following pairs of denatured DNAs, starting with the same initial DNA concentrations.
(a) T4 DNA and *E. coli* DNA, each sheared to an average single-strand length of 400 nucleotides.
(b) Unsheared T4 DNA and sheared T4 DNA.

15. When histone proteins are isolated from chromatin their mass is equal to the DNA, and the molar ratio of four of the histones is 1:1:1:1 (H2a:H2b:H3:H4), while H1 is found in half the amount (0.5). Discuss whether or not these data fit the bead-and-string model for nucleosomes.

16. Your supervisor shows you two test tubes and announces: "The labels fell off of these tubes in the freezer! One was supposed to contain DNA from *E. coli* and one was DNA from *Mycobacterium tuberculosis*. I can't figure out which is which, so you will just have to grow cells and prepare more DNA." Because *M. tuberculosis* is a dangerous pathogen, you wish to avoid culturing the cells. Can you devise a simple strategy to determine which sample is from which organism, based on the fact that *E. coli* DNA contains 52% G + C and *M. tuberculosis* DNA is 70% G + C? Show an example of the expected results.

# X-Ray Diffraction of DNA

An x-ray pattern of DNA (fig. 1) is obtained by holding a stretched fiber containing many DNA molecules in a vertical direction and exposing it to a collimated monochromatic beam of x-rays. Only a small percentage of the x-ray beam is diffracted. Most of the beam travels through the specimen with no change in direction. A photographic film is held in back of the specimen; a hole in the center of the film allows the incident undiffracted beam to pass through (fig. 2). Coherent diffraction occurs only in certain directions, specified by Bragg's law: $2d \sin \theta = n\lambda$. Here $d$ is the distance between identical repeating structural elements: $\theta$ is the angle between the incident beam and the regularly spaced diffract-

ing planes. $\lambda$ is the wavelength of x-rays used, and $n$ is the order of diffraction, which may equal any integer but is usually strongest for $n = 1$. The most important point is that $\sin \theta \approx \theta$ and $d \approx 1/\theta$, so that a spot far out on the photographic film is indicative of a repeating element of small dimension and vice versa.

Watson and Crick were the first to appreciate the significance of strong 3.4-Å and 34-Å spacings and the central crosslike pattern, which reflects a helix structure in the x-ray diffraction pattern of DNA. They interpreted this as arising from the hydrogen-bonded antiparallel double-helix structure.

## Figure 1

Diffraction pattern of a fibrous sample of DNA. (© M. H. F. Wilkins.)

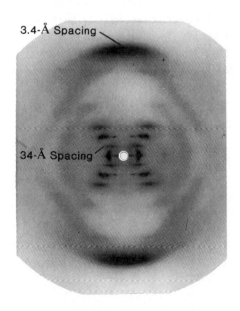

## Figure 2

Camera setup for obtaining DNA diffraction pattern.

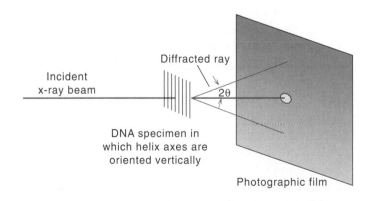

# DNA Replication, Repair, and Recombination

*The genetic message carried by the sequence of bases in the DNA is replicated by a template mechanism.*

From the complementary duplex structure of DNA it is a short intuitive hop to a model for replication that satisfies the requirement for one round of DNA duplication for every cell division. Such a proposal was made by Watson and

## Figure 26.1

Watson-Crick model for DNA replication. The double helix unwinds at one end. New strand synthesis begins by absorption of mononucleotides to complementary bases on the old strands. These ordered nucleotides are then covalently linked into a polynucleotide chain, a process resulting ultimately in two daughter DNA duplexes.

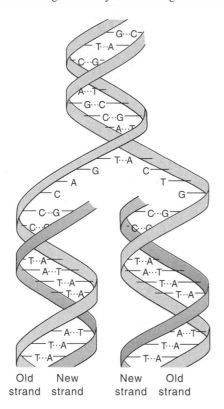

Old     New         New     Old
strand  strand      strand  strand

Crick when they proposed the duplex structure for DNA (fig. 26.1). First, the double helix unwinds; next, mononucleotides are absorbed into complementary sites on each polynucleotide strand; and finally these mononucleotides become linked to yield two identical daughter DNA duplexes. What could be simpler! Subsequent biochemical investigations showed that in many respects this model for DNA replication was correct, but they also indicated a much greater complexity than was initially suspected. Part of the reason for the complications is that replication must be very fast to keep up with the cell division rate, and it must be very accurate to ensure faithful transfer of information from one cell generation to the next.

In addition to the replication process there are two other major areas of DNA metabolism. One is concerned with repair of damaged DNA, and the other is concerned with recombination between DNA molecules. In this chapter we deal mainly with the replication process, but we also consider chromosome repair and recombination.

# The Universality of Semiconservative Replication

The Watson-Crick model for DNA replication is called semiconservative because the daughter duplexes arising from replication each contain one old (conserved) strand and one new strand.

Matthew Meselson and Frank Stahl conceived a way of demonstrating the semiconservative mode of replication that involved the use of isotopes that result in DNAs of different densities after replication. For this purpose *E. coli* bacteria were grown for several generations on a medium in which all the nitrogen was of the heavy $^{15}$N isotope type (normal nitrogen is $^{14}$N). The resulting DNA had a greater than normal density because the $^{15}$N was incorporated into the bases of the DNA. Following this step, the bacteria were transferred to a growth medium containing normal ($^{14}$N) nitrogen, and the cells were allowed to go through one or more rounds of duplication. The DNA from these cells was then isolated and analyzed by density-gradient centrifugation. Pure $^{15}$N DNA produces a single band of DNA (fig. 26.2, frame 1). The same is true for $^{14}$N DNA (see fig. 26.2, frame 2). The only difference is that denser DNA produces a band further down the tube. Thus, the location of the DNA in the centrifuge tube can be used as a measure of the density of the DNA. Cells containing pure $^{15}$N DNA were allowed to grow in $^{14}$N medium for one or more generations, and the DNA from them was similarly analyzed. After precisely one generation in $^{14}$N medium, the only band visible in the isolated DNA was that corresponding to $^{15}$N–$^{14}$N-hybrid DNA (see fig. 26.2, frame 4). These data argue strongly against the conservative mode of replication, but they do not discriminate between the semiconservative and other modes of replication in which both strands might be labeled approximately equally. For this purpose the results in further generations were examined. Only in the semiconservative mode would equal amounts of DNA with the hybrid density and the light density be expected after two generations. This result is shown in figure 26.2, frame 5. In the third generation we still see a band with the hybrid density, but the amount of pure light DNA has increased (see fig. 26.2, frame 6). These results strongly support the semiconservative mode for DNA replication. Indeed it is hard to think of another mode of replication that could give rise to these results.

Subsequent studies initiated by others have shown that the DNA in all chromosomes regardless of their size replicate in this semiconservative fashion. Thus, the semiconservative mode of DNA replication is well established and essentially universal. The genetic information implanted in

## Figure 26.2

The Meselson-Stahl experiment demonstrating semiconservative replication for *E. coli* chromosomal DNA. Cesium chloride (CsCl) density-gradient centrifugation is used to discriminate between DNAs of different densities. When a concentrated solution of CsCl is centrifuged at high speed (50,000–100,000 times gravity), a stable concentration gradient of CsCl develops, with the concentration increasing along the direction of the centrifugal force. Since CsCl is much denser than water, this concentration gradient produces a density gradient. Macromolecules of DNA present in the solution are driven by the centrifugal field into the region where the solution density is equal to their own density. If the DNA has a uniform density, then after many hours of centrifugation (about 36 h) equilibrium is established, with all of the DNA concentrated in a single band. *E. coli* DNA has different densities when cells are grown in $^{14}$N or $^{15}$N medium (frames 1 and 2). When cells containing pure heavy DNA ($^{15}$N–$^{15}$N DNA) are grown in $^{14}$N medium for one generation, all of the DNA is of intermediate density ($^{14}$N–$^{15}$N). After two generations of growth in $^{14}$N medium, the cells contain equal amounts of light ($^{14}$N–$^{14}$N) DNA and intermediate density DNA (frame 5). In subsequent generations the hybrid DNA reappears in constant amounts, but the amount of light DNA increases. These results support the model of a semiconservative mode of DNA replication.

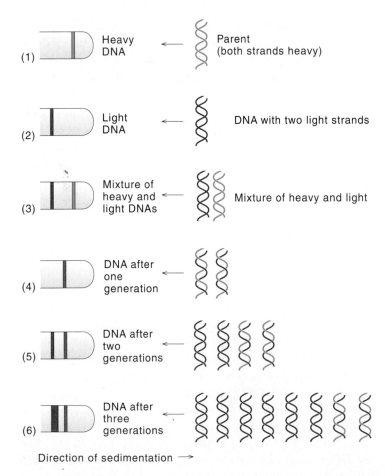

(1) Heavy DNA ← Parent (both strands heavy)

(2) Light DNA ← DNA with two light strands

(3) Mixture of heavy and light DNAs ← Mixture of heavy and light

(4) DNA after one generation

(5) DNA after two generations

(6) DNA after three generations

Direction of sedimentation →

the base sequence is directly transferred from one generation of DNA to the next by the ability of DNA single strands to serve as templates for the absorption of complementary mononucleotides.

## Overview of DNA Replication in Bacteria

More is known about the replication of DNA in *E. coli* than in any other system. The *E. coli* bacterium contains a single circular chromosome with about $4.5 \times 10^6$ bp. Within this chromosome there are about 2,000 genes, each gene being represented by a unique sequence of bases and encoding the genetic information essential for the synthesis of a specific protein. Exactly how many of these genes are required for DNA synthesis is not known; so far about 30 have been found, and in many cases the gene-encoded proteins have been extensively characterized. First we consider the overall strategy for bacterial DNA replication; then we discuss some of the proteins involved in the synthesis.

### *Growth during Replication Is Bidirectional*

Replication of the *E. coli* chromosome can be made visible by very delicate techniques developed by John Cairns and Ric Davern for isolating intact $^3$H-labeled chromosomes and subjecting them to autoradiography. After one round of replication in labeled medium, chromosomes appear circular and uniformly labeled.

Initiation of a second round of replication leads to a replication "eye" at the initiation site of replication (fig. 26.3). As synthesis proceeds the size of the replication eye becomes larger; at this stage the replicating chromosome is referred to as a theta structure because it has the appearance of the Greek letter $\theta$. Semiconservative replication is consistent with the density of the autoradiographic tracks made by different parts of the chromosome after one and two rounds of replication in [$^3$H]thymidine (see fig. 26.3).

It seems clear that the replication eye must contain two partially separated parental DNA strands that are base-paired with short strands of newly synthesized DNA. Not resolved by these autoradiographs is the question of whether replication occurs in one or both directions about the origin of replication. If growth is unidirectional we expect one growth point (fig. 26.4*a*), called a growth fork; if growth is bidirectional we expect two growth points or two growth forks (see fig. 26.4*b*). Although examples of both types of replication occur, it is believed that most bacterial chromosomes including *E. coli* replicate bidirectionally.

Evidence for bidirectional replication is supported by additional experiments. Growing bacteria were briefly ex-

## Figure 26.3

Simulated autoradiographs of the *E. coli* chromosome after one or more replications in the presence of [³H]thymidine. After one round of replication, the autoradiograph shows a circular structure that is uniformly labeled. The second round of replication begins with the formation of a replication "eye." One branch in the replication eye is twice as strongly labeled as the remainder of the chromosome, indicating that this branch contains two labeled strands. This structure is consistent with semiconservative replication for the *E. coli* chromosome.

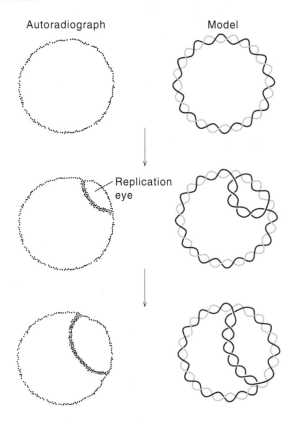

Autoradiograph    Model

Replication eye

## Figure 26.4

Schematic diagrams of two different modes of DNA synthesis at the growth fork(s). In unidirectional replication (*a*) one growth fork occurs; in bidirectional replication (*b*) two occur. Red indicates regions containing newly synthesized DNA.

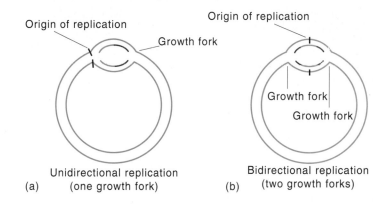

Origin of replication

Growth fork

Origin of replication

Growth fork
Growth fork

(a)    Unidirectional replication (one growth fork)

(b)    Bidirectional replication (two growth forks)

posed to radioactive thymidine after which the replicating chromosomes were examined by autoradiography; both of the forks in the replicating structures were intensely labeled (see fig. 26.4*b*). These results show that both forks must be active in replication, a finding consistent with bidirectional growth.

## Growth at the Replication Forks Is Discontinuous

Continuous synthesis on both strands of the replication fork would require synthesis in the $5' \rightarrow 3'$ direction on one strand and in the $3' \rightarrow 5'$ direction on the other because of the antiparallel nature of the DNA duplex (fig. 26.5*a*). Continuous synthesis on both strands seems unlikely since the only known enzymes that catalyze DNA synthesis add bases

to the growing chain in the $5' \rightarrow 3'$ direction (see below). For this reason it was hypothesized that replication is discontinuous on one of the branches at the replication fork (see fig. 26.5*b*). Meticulous electron-microscopic examination of replication forks in bacterial viruses did in fact show that transient gaps sometimes are apparent on one of the DNA strands near the replicating fork. Observations such as this led to the notion of leading-strand and lagging-strand synthesis (see fig. 26.5*b*). The $5' \rightarrow 3'$ synthesis of the leading strand can occur continuously in the direction of unwinding at the replication fork. But synthesis of the lagging strand in the $5' \rightarrow 3'$ direction only occurs in discontinuous spurts in the opposite direction.

This concept of discontinuous synthesis was supported by the finding that about half of the newly synthesized DNA is first made in small pieces that subsequently become incorporated into larger units of DNA. Small replication fragments were first detected in the laboratory of Okazaki. This was done by exposing growing cells to ³H-labeled thymidine for a very short time (2–10 s), followed by rapid isolation of the radioactively labeled DNA. After longer labeling times, 1–2 min, most of the labeled DNA was found in much larger segments of DNA.

A closer examination of the Okazaki fragments led to detection of short stretches of ribonucleotides at the $5'$ ends. From this and many other observations on different systems it was determined that a new DNA chain can be initiated only by attaching the first deoxynucleotide through its $5'$-phosphate to the $3'$-OH of a short RNA polynucleotide. An RNA strand that functions in this capacity is called a primer. Primers are formed at points along the chromosome; they base-pair with the single-stranded template DNA in the regions where they are formed.

## Figure 26.5

Models for synthesis at the replication fork. (*a*) Continuous synthesis on both strands. Note that both growth arrows are pointing in the same direction, which would require growth in the $5' \rightarrow 3'$ direction on one strand and in the $3' \rightarrow 5'$ direction on the other. If growth occurs only in the $5' \rightarrow 3'$ direction, synthesis would have to be discontinuous on one strand, as in (*b*). Alternatively, it could be discontinuous on both strands (*c*).

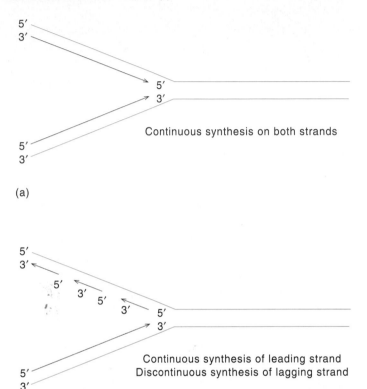

Continuous synthesis on both strands

(a)

Continuous synthesis of leading strand
Discontinuous synthesis of lagging strand

(b)

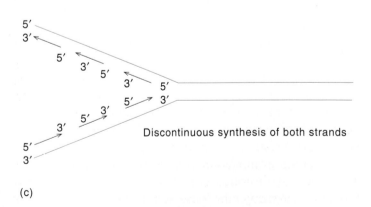

Discontinuous synthesis of both strands

(c)

A detailed model for discontinuous synthesis could now be proposed (fig. 26.6). First, RNA primers are made on the single-strand region of the template; then DNA is synthesized. Finally, the RNA is removed from the fragments, and the gaps are filled in and ligated. An understand-

## Figure 26.6

A model for discontinuous DNA synthesis. Synthesis occurs in a region that has been partially single-stranded. First, RNA primers are formed at various points on the single-stranded region (1). DNA synthesis starts at the 3' ends of the primers (2). Primers are removed (3). Gaps between DNA fragments are filled in by further DNA synthesis (4). Fragments are ligated to make one long, continuous piece of DNA (5). Newly synthesized RNA and DNA are indicated in red.

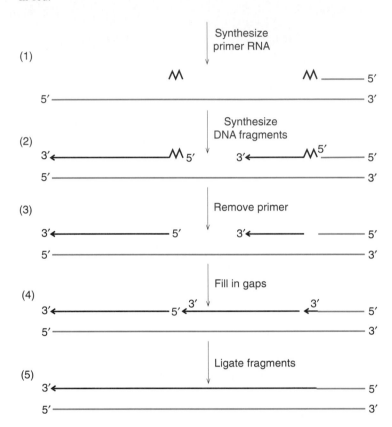

ing of the detailed steps of this process has come largely from analysis of simpler viral replicating systems and more recently from examination of a system intended to resemble the *E. coli* chromosome (see below).

## *Proteins Involved in DNA Replication*

Table 26.1 contains a list of some proteins involved in DNA replication and their functions. How does one go about analyzing such a complex situation? Historically, three general methods have been used for the identification and characterization of the proteins involved in DNA replication: Purification, reconstitution, and mutation. Insofar as possible, all three methods are used together.

The first method involves isolation of proteins with enzymatic activities that are logically related to the replication process, such as DNA polymerases and ligases. This is

## Table 26-1

Proteins Involved in DNA Replication

| Protein | Gene(s) | Function |
|---|---|---|
| DnaA | *dnaA* | Initiator protein: Binds *oriC;* promotes double-helix opening; DnaB loading |
| DnaC | *dnaC* | Complexes with DnaB; delivers DnaB to DNA |
| DNA polymerase III holoenzyme | | |
| $\alpha$ | *dnaE* | Polymerase |
| $\epsilon$ | *dnaQ* | 3' to 5' exonuclease |
| $\theta$ | *holE* | Unknown |
| $\tau$ | *dnaX* | |
| $\gamma$ | *dnaX* | |
| $\delta$ | *holA* | |
| $\delta'$ | *holB* | |
| $\chi$ | *holC* | |
| $\psi$ | *holD* | |
| $\beta$ | *dnaN* | Processivity factor: "Sliding clamp" |
| PolI | *polA* | Prime removal and gap filling; Initial leading strand synthesis on ColE1 |
| Ligase | *lig* | Joins nascent DNA fragments |
| Gyrase | *gyrA* *gyrB* | Type II topoisomerase; replication swivel, DNA supercoiling, decatenation |
| DnaB | *dnaB* | 5' to 3' helicase and activator of primase |
| Primase | *dnaG* | Primer synthesis |
| SSB | *ssb* | Binds single-stranded DNA |

Source: Adapted from T. A. Baker and S. H. Wickner, Genetics and enzymology of DNA replication in *Escherichia coli. Ann. Rev. Genetics,* 26:447, 1992.

the classical biochemical approach and can be applied to any biological system. After isolation and characterization, several approaches may be used to demonstrate that the purified enzyme is active in the replication process *in vivo.* Sometimes this can be done by using inhibitors that act on both the purified protein and the cellular process. The concentration of inhibitor required to inhibit DNA replication *in vivo* should be approximately the same as that required to inhibit the purified enzyme *in vitro.* In prokaryotes, mutations have been very useful for confirming the functions of isolated proteins in the replication process. The induction of a new enzyme activity associated with a biological process, such as virus infection or cell proliferation, also provides useful evidence.

A second method used to identify proteins needed for replication involves reconstitution. Whole-cell lysates containing all of the components necessary for replication are fractionated, and the DNA replication process is then reconstituted with various combinations of the purified or partially purified proteins. Components of the replication system are recognized on the basis of their ability to restore overall activity *in vitro.* This procedure can be applied to any organism, even when relevant genetic mutants are not available.

The third method uses genetic mutation as the primary tool. This method requires the isolation of conditional mutants, that is, mutants that behave normally under one set of conditions but abnormally under another. Most commonly, temperature-sensitive DNA replication mutants are used. Such mutants have been isolated in *E. coli,* and they grow normally at a low (permissive) temperature (33°C) but poorly or not at all at a high (nonpermissive) temperature (41°C). Preliminary analysis of the temperature-sensitive step provides clues to the stage of replication affected. For example, the length of time required for DNA synthesis to stop after cells are shifted from permissive to nonpermissive temperatures can indicate whether the mutation occurs in a protein involved in initiation or elongation. If the mutation is in the gene for a protein required for elongation, most DNA synthesis stops immediately at the nonpermissive temperature because the majority of cells are usually at some stage in the elongation process. If the mutation is in a gene required only for initiation of replication, most DNA synthesis continues for some time and stops when the rounds of replication in progress are completed. *In vitro* assays are then used to aid in purifying the corresponding proteins from wild-type cells. Extracts from cells with the temperature-dependent defect are not active in DNA synthesis at the elevated temperature, but activity can be restored by adding the corresponding protein from normal wild-type cells. This complementation assay can be used to aid in the purification of particular replication proteins. To prove that the correct protein has been purified from wild-type cells, the proteins from the temperature-sensitive mutant also must be purified and shown to be abnormal, frequently exhibiting unusual instability at elevated temperatures.

## Characterization of DNA Polymerase I in Vitro

We look at *E. coli* DNA polymerase I as an example of how biochemical and genetic studies are used to characterize a DNA replication enzyme. Then we consider the other major proteins involved in DNA replication in *E. coli* before examining the proteins that participate in the synthesis of DNA in different types of organisms.

The Watson-Crick proposal of a complementary duplex structure for DNA stimulated a search for a DNA polymerase enzyme with certain implied properties. The enzyme should require an intact DNA chain to serve as a template for the absorption of complementary bases, and the newly synthesized DNA should be a complement of one of the template DNA chains. Arthur Kornberg and his co-workers isolated a DNA-synthesizing enzyme from cells of *E. coli* that satisfied these requirements and named it DNA polymerase; it is now known as DNA polymerase I (PolI), or the Kornberg enzyme. This enzyme requires a DNA template, the four commonly occurring deoxynucleotide triphosphates, and $Mg^{2+}$ ions for making DNA. The enzyme catalyzes the addition of mononucleotides to the 3'-OH end of a growing chain (fig. 26.7). Simultaneously with the formation of this linkage, the linkage between the two phosphates is broken, releasing a pyrophosphate group. The energy produced by the cleavage provides the energy necessary for linking the mononucleotide to the growing DNA chain. Subsequent cleavage of the pyrophosphate to orthophos-

phates ensures the irreversibility of the reaction. Bases added to the growing chain are determined by the sequence of bases in the DNA template. As nucleotides complementary to those on the template are added, the single-stranded DNA template gradually becomes converted to a double helix. Structure studies have shown that the enzyme has a complex surface with specific attachment sites for the template chain, the growing chain, and monomer nucleoside triphosphate. The enzyme is highly selective, because the only nucleotides it links to the growing chain are those that form Watson-Crick base pairs with the template strand. As the new chain lengthens by synthesis, the enzyme moves along the template one base at a time.

In addition to the characteristic template-directed polymerization activity, PolI contains an activity that results in the removal of mononucleotides from the 3' end of a polynucleotide strand. This enzyme activity leads to cleavage of the bond between the 5'-phosphate of the terminal residue and the 3'-OH group of the penultimate residue. Since this is the same linkage that is made during synthesis, the degradation reaction may be thought of as a reversal of the polymerization process. For net chain elongation beyond the 3'-OH end of the DNA strand, the polymerization rate must exceed the depolymerization rate. Polymerization is much faster than depolymerization if the correct base-paired nucleotide is inserted into the growing chain. If a mismatched base is accidentally inserted, the opposite is true, and the mismatched base is usually removed. The combined polymerization–depolymerization reaction has been viewed as a ''proofreading,'' error-reducing mechanism in DNA synthesis.

*E. coli* DNA polymerase I has a second associated activity catalyzing $5' \rightarrow 3'$ degradation of DNA. Whereas $3' \rightarrow 5'$ degradation activity of the enzyme is much more effective on unpaired or mispaired bases, the $5' \rightarrow 3'$ activity cleaves preferentially at base-paired regions. The ability of PolI to degrade DNA (or RNA) in the $5' \rightarrow 3'$ direction as well as to carry out a polymerization reaction suggests a role in removing RNA primer during gap filling.

## Crystallography Combined with Genetics to Produce a Detailed Picture of DNA PolI Function

The question arises as to whether the three activities of PolI just described all originate from a single active site or from more than one site. Cleavage of PolI by the protease subtilisin leads to a small fragment ($M_r = 30,000$) with $5' \rightarrow 3'$-nuclease activity and a large fragment ($M_r = 70,000$), called the Klenow fragment, exhibiting the polymerization and $3' \rightarrow 5'$ depolymerization activities. A bril-

## Figure 26.7

Template and growing strands of DNA. (*a*) Nucleotides are added one at a time to the 3'-OH end of the growing chain. Only residues that form Watson-Crick H-bonded base pairs with the template strand are added. (*b*) Covalent bond formation between the 3'-OH end of the growing chain and the 5'-phosphate of the mononucleotide is accompanied by pyrophosphate removal from the substrate nucleoside triphosphate.

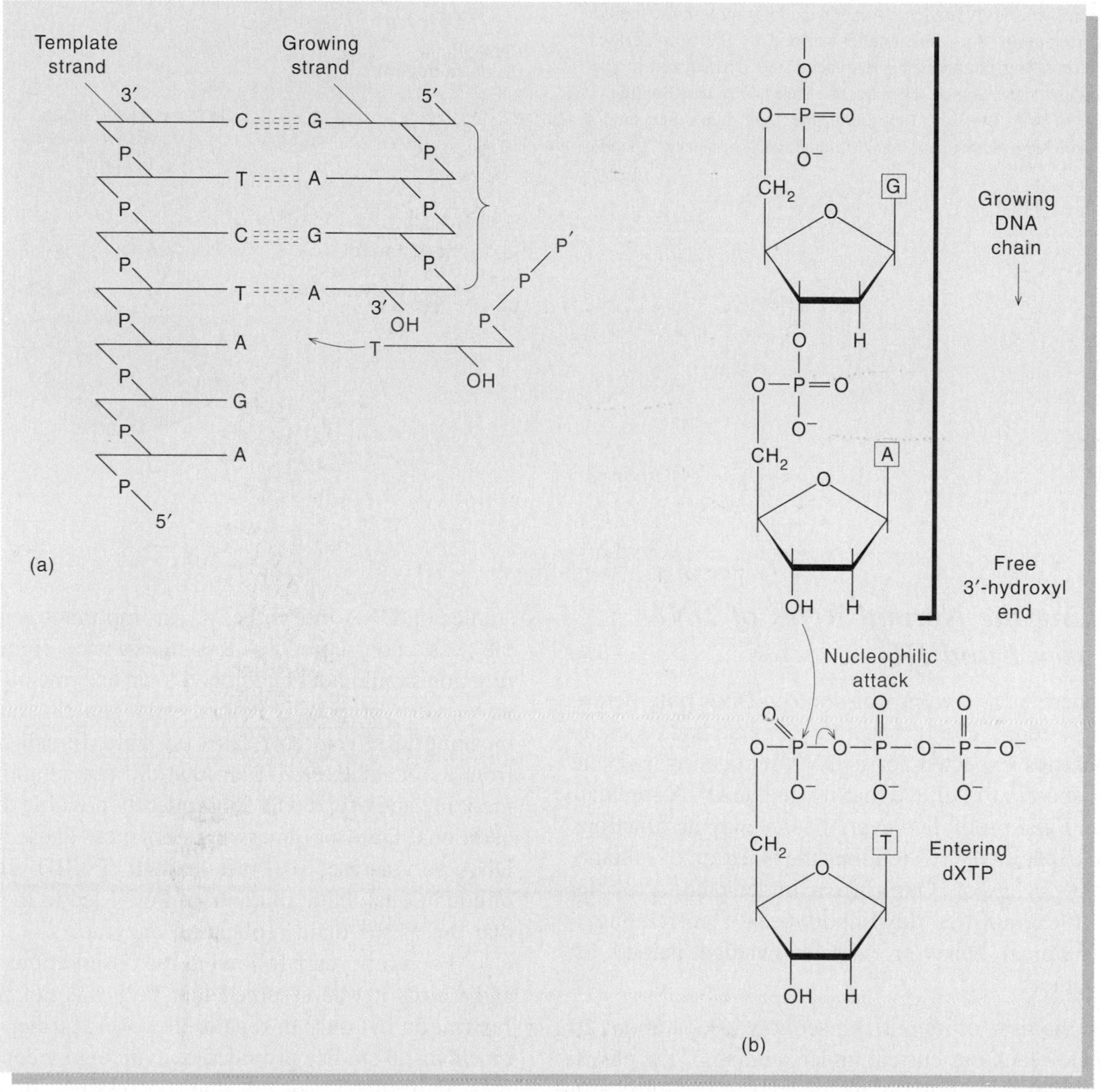

liant series of investigations by Tom Steitz and his co-workers led to a detailed understanding of the Klenow fragment. These studies were done on wild-type and mutant enzyme lacking the nuclease activity. From their studies Steitz and his colleagues concluded that within the Klenow fragment the polymerase and 3' → 5' exonuclease activities were located in different regions. Thus, the polymerase has three enzyme activities all resulting from different parts of the same protein molecule. The approximate locations of these activities are indicated relative to a chain undergoing synthesis in figure 26.8. Recent investigations on cocrystals of DNA and PolI indicate that the DNA makes a sharp bend between the 3' → 5' exo and the polymerase active sites. This somewhat clumsy-looking structure may facilitate the detection of imperfections in newly synthesized DNA and the backing up process that must accompany proofreading excision by the 3' → 5' exo.

**Figure 26.8**

A schematic drawing of the possible structure of the DNA-PolI complex in an elongating complex of DNA. The Klenow fragment contains the 5′–3′ polymerase site and the error correcting 3′–5′ exonuclease site. The fragment cleaved from the PolI enzyme by limited proteolysis contains the 5′–3′ exonuclease site. The 5′–3′ exonuclease removes RNA leaders from Okazaki fragments to make way for polymerization. Any mismatches formed by the polymerase result in the elongating strand being displaced and hydrolyzed at the 3′–5′ exonuclease site. (Source: This drawing is primarily based on the structural studies of L. S. Beese, V. Derbyshire, and T. A. Steitz, Structure of DNA polymerase I Klenow fragment bound to duplex DNA, *Science* 260:352–355, 1993.)

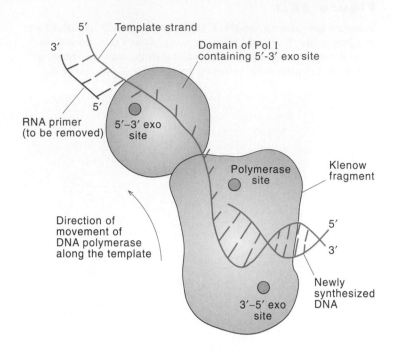

## Establishing the Normal Roles of DNA Polymerases I and III

All of Kornberg's early work was done on DNA polymerase in cell-free systems. The fact that the enzyme had so many of the properties expected for a DNA-replicating enzyme led most observers to believe that it was the DNA-replicating enzyme. Final proof, however, that an enzyme functions in the same capacity *in vivo* requires the isolation of mutants that affect its behavior. Our current understanding of the role of PolI exemplifies the importance of correlating a given biochemical behavior with knowledge gained by studying mutants.

From the time of its initial discovery it took about 20 years to reach our current understanding of the physiological role of PolI. A major step in this direction was taken by Cairns and DeLucia, who laboriously scanned several thousand strains of mutagenized *E. coli* to find one that contained almost no PolI polymerizing activity (1%–2% of normal). This mutant grew well under normal conditions, suggesting that PolI was not involved in replication. However, the mutant in question was hypersensitive to ultraviolet irradiation. Since ultraviolet irradiation was known to damage DNA, the hypersensitivity of the mutant suggested that PolI might be involved in repairing chromosome damage. The discovery of the polymerase mutant had a profound effect on thought and experimental design in further studies on DNA biosynthesis. An important general principle was underscored by this unexpected finding—that a function should not be assigned to an enzyme on the basis of its *in vitro* properties alone. Only genetic mutants make meaningful *in vivo* correlates possible. In cell-free extracts from a mutant *E. coli* strain that did not contain PolI polymerizing activity, it was subsequently possible to detect two additional DNA polymerizing enzymes. These were named DNA polymerases II (PolII) and III (PolIII). The behavior of conditional lethal mutants of PolIII led to the conclusion that this is the main replication enzyme.

For a few years following the observations of DeLucia and Cairns it was assumed that PolI was not important in replication but only in repair. However, further genetic and biochemical studies provided convincing evidence that PolI is a multifunctional enzyme possessing, in addition to 5′ → 3′ polymerizing activity, the 3′ → 5′ and 5′ → 3′ degradation activities described earlier. The original mutant of PolI isolated by Cairns and DeLucia was inactivated only in its polymerizing function. Subsequently, mutants in PolI that affect the 5′ → 3′ degradation function were found to be conditionally lethal and did not permit elongation of DNA synthesis under nonpermissive conditions; thereby, it was demonstrated that this activity of the PolI enzyme is indispensable for chromosome replication. In this regard it should be mentioned that of the three known *E. coli* DNA polymerases, only PolI has a 5′ → 3′ exonuclease activity.

## Figure 26.9

Steps in the sealing of a DNA nick, catalyzed by DNA ligase. The bacterial ligase uses $NAD^+$ to make an enzyme–AMP intermediate. Mammalian DNA ligases and bacteriophage T4 ligase use ATP for the same purpose.

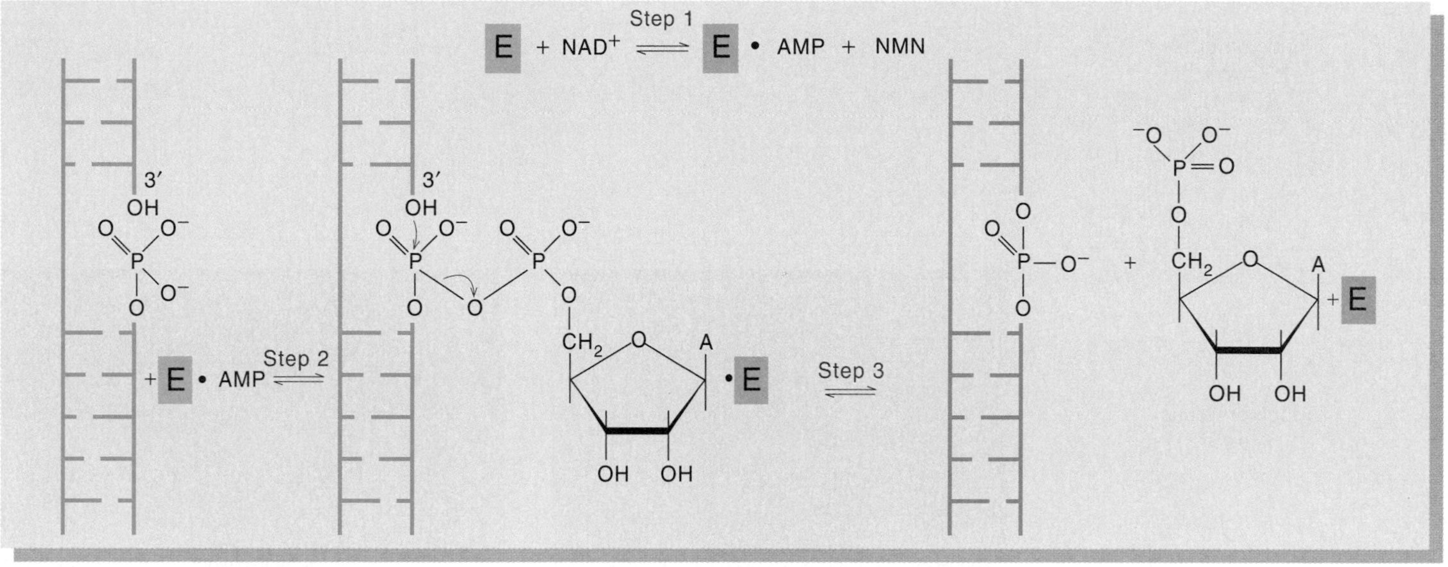

Step 1

$E + NAD^+ \rightleftharpoons E \cdot AMP + NMN$

Step 2

Step 3

## Figure 26.10

Type I and type II topoisomerases relax negatively supercoiled DNA in steps of one and steps of two, respectively. Type II topoisomerases can also add additional negative supercoils (as indicated by the double arrow). The latter reaction requires energy input, which is encoded by ATP cleavage.

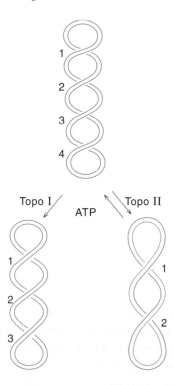

Topo I   ATP   Topo II

## Other Proteins Required for DNA Synthesis in Escherichia coli

As mentioned earlier, discontinuous DNA synthesis necessitates the existence of an enzyme for joining the newly synthesized segments (see fig. 26.6). Such an enzyme has been found in a variety of cell types and is called polynucleotide ligase (fig. 26.9).

Topoisomerases can introduce negative supercoils into DNA or relax negatively supercoiled DNA. They can also catalyze the linking together (catenation) of double-stranded circular DNAs or the decatenation of linked circular molecules (figs. 26.10 and 26.11). Topoisomerases must serve vital functions because mutants carrying defective topoisomerases of one sort or another show severely impaired growth properties. It seems likely that decatenation and modulation of supercoiling could both be very important to cell survival.

At the growth fork it is necessary that the parental double helix be unwound to present further single-stranded regions to serve as templates for continued replication. The dnaB protein of E. coli is believed to be a helicase that directly catalyzes this process.

Single-strand binding protein (SSB) is found in abundance in E. coli, and it is believed to be bound in mass at the replication fork. This fact and other evidence described later on indicate that it plays an important role at the growth fork.

## Figure 26.11

Catenation by topoisomerases. (*a*) Two circular DNAs can be catenated by type I topoisomerase only if one of the DNAs is nicked. This is not necessary when using a type II topoisomerase. (*b*) Electron micrographs of catenated DNA before (i) and after (ii) incubation with DNA gyrase. The catenate contains one large circular DNA and one small circular pBNP66 plasmid DNA. (Source: Adapted from M. Gellert, L. M. Fisher, H. Ohmori, M. H. O'Dea, and K. Mizuchi, DNA gyrase: Site-specific interactions and transient double-strand breakage of DNA, *Cold Spring Harbor Symp. Quant. Biol.* 45:301, 1981.)

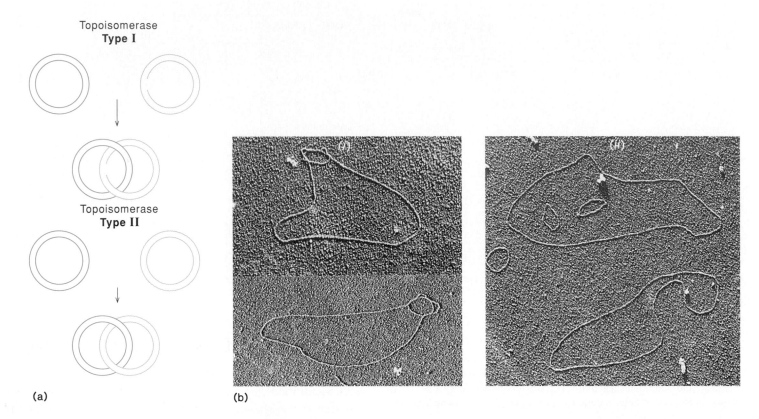

Topoisomerase
**Type I**

Topoisomerase
**Type II**

(a)                    (b)

Two functions can be suggested, both inferred from its preference for binding to single-stranded DNA. It could protect single-stranded DNA from nucleases, or it could facilitate unwinding by inhibiting rewinding.

There are two enzymes that can catalyze the synthesis of RNA with the help of a DNA template. One of these, called primase is encoded by the *dnaG* gene. Primase catalyzes the synthesis of small primer RNAs that are required for DNA synthesis. The other enzyme called RNA polymerase catalyzes the synthesis of all the other RNAs found in *E. coli*. For some time it was thought that RNA polymerase catalyzed the synthesis of the primers that initiate replication of the chromosome. Presently, it is believed that all primer synthesis is catalyzed by primase.

### Replication of the Escherichia coli Chromosome

The elongation step in *E. coli* chromosomal DNA synthesis is depicted in figure 26.12. As the DNA gradually unwinds new synthesis takes place on the two strands. At the heart of

the unwinding process we find a DnaB helicase, and just behind it we find primase. Single strand binding protein (SSB) coats the newly unwound DNA strands. The primase travels in the same direction as the DnaB helicase, making periodic pauses to synthesize short RNA primers on the lagging strand. PolIII extends the leading strand and also extends the lagging strand. In the latter case the first DNA bases are added to the primers. Following the action of PolIII on the lagging strand, PolI replaces the RNA bases with DNA bases, and DNA ligase then links the short DNAs on the lagging strand.

### Initiation and Termination of Escherichia coli Chromosomal Replication

Two aspects of *E. coli* chromosomal replication still to be considered are initiation and termination. From what has been said we conjecture that replication initiates at a unique site, proceeds bidirectionally, and terminates at a point where the two oppositely advancing growth forks meet. To study initiation it was first necessary to isolate that segment

**Figure 26.12**

Model for the replication fork in the *E. coli* chromosome. The DnaB protein unwinds the duplex while the primase directly behind it synthesizes RNA primer on the lagging strand. SSB protein binds to single-stranded regions wherever they are. DNA polymerase III progressively adds nucleotides to the leading strand and also adds nucleotides to the lagging strand starting at the RNA primers. After this, DNA polymerase I replaces the RNA primer with the DNA equivalent. Finally DNA ligase knits the short DNA segments together on the lagging strand. (Source: Adapted from K. Arai, R. L. Low, and A. Kornberg, movement and site selection for priming by the primosome in phage φX174 DNA replication, *Proc. Natl. Acad. Sci. USA* 78:711, 1981.)

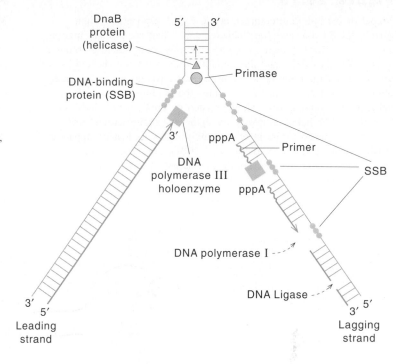

of the chromosome that carries the unique origin of replication. This was done by chopping the bacterial chromosome into small pieces, circularizing the small pieces, and then transfecting them into cells to see which pieces could sustain autonomous replication. In this way a unique segment of chromosome was discovered which contains 245 bp required to initiate replication.

With circular minichromosomes containing this unique segment termed *oriC*, it has been possible to study the initiation and replication process *in vitro*. First, in crude extracts it was possible to show that replication initiates within or near the *oriC* sequence and proceeds bidirectionally. About 13 different proteins participate in the *oriC*-directed DNA replication as judged by their activity in a reconstituted replication reaction.

Initial complex formation entails binding of the dnaA protein to several sites (probably four) of the *oriC* region (fig. 26.13). This is followed by the cooperative binding of more dnaA proteins. In all, 20–40 dnaA proteins bind to a region of about 200 bp. This binding is climaxed by a structural change in the dnaA–*oriC* complex, which results in localized melting of the DNA at one end of the *oriC* region. The melting reaction, which reproduces the local opening of the duplex requires that the DNA be negatively supercoiled. As we have already discussed, negative supercoiling favors unwinding of the duplex (see chapter 25).

The dnaB protein is transferred to this complex from a dnaB–dnaC complex. The dnaC dissociates, leaving the dnaB protein bound to the template. Melting of the DNA duplex proceeds bidirectionally from *oriC* as the dnaB helicase migrates from the dnaA–*oriC* complex to provide a template for the priming and replication enzymes. Addition of the priming and replication enzymes results in immediate initiation and elongation. Continued elongation requires gyrase and SSB. Gyrase provides a swivel to permit continued unwinding of parental strands while SSB stabilizes the transiently single-stranded regions. DNA gyrase proves useful in the late stages of replication as well, because many circular dimers are formed that require decatenation and negative supercoils so they can function again as active templates.

## DNA Replication in Eukaryotic Cells

Mutant studies have not been that helpful in finding proteins with known functions in vertebrate DNA replication. The major approach in vertebrate investigations has been isolation of proteins having activities logically related to DNA replication and isolation of replicative intermediates. Different forms of DNA polymerases, DNA ligases, topoisomerases, single-strand binding proteins and unwinding enzymes have all been found.

## Figure 26.13

Model of the initiation complex at the *E. coli* replication origin (*oriC*). R1-R4 indicates four 9 bp sequences that are primary binding sites for the DnaA protein. The consensus sequence for these binding sites is TTATCCACA and their relative orientation is indicated by the arrows. After the binding of about 20 dnaA proteins at *oriC*, a small region of the DNA melts. This permits the binding of the dnaB helicase. The binding of dnaB is assisted by dnaC, which does not bind itself. The helicase unwinds a significant region around *oriC*, creating the necessary room for assembly of the replication apparatus. (Source: Adapted from T. A. Baker and S. H. Wickner, Genetics and enzymology of DNA replication in *Escherichia coli, Ann. Rev. Genetics* 26:447, 1992).

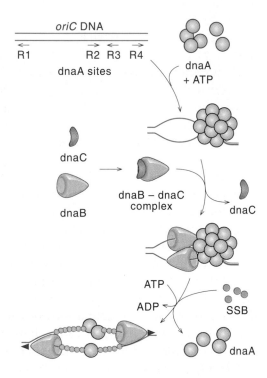

## Figure 26.14

Multiple-origin model for eukaryotic chromosomal DNA replication. (*a*) Autoradiograph of short-term labeling of a eukaryotic chromosome during replication and its interpretation. (*b*) Overall replication scheme for a eukaryotic chromosome. Only a short region of the chromosome is shown. It is believed that replication origins are relatively free of proteins.

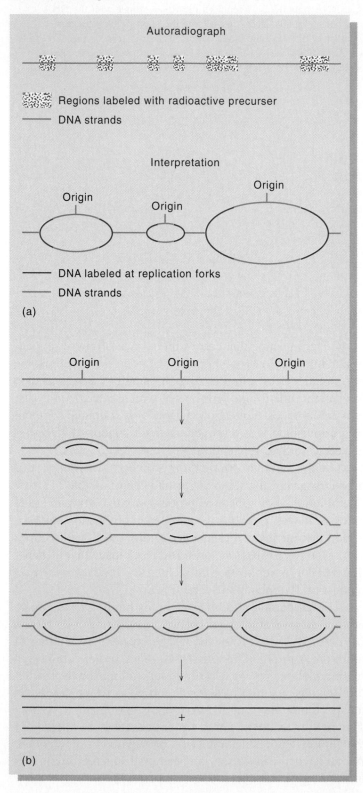

## *Eukaryotic Chromosomal DNA*

Although general features of DNA replication in eukaryotes are thought to be similar to those of prokaryotes, there are some interesting differences. The chromosomes of higher eukaryotic organisms are quite large, in some cases 1,000 times larger than their bacterial counterparts. For these larger DNA molecules to replicate in a reasonable time, they have multiple origins of replication. The simultaneous synthesis of DNA at several points along the chromosome has been demonstrated by incorporating radioactive nucleotides into the replicating chromosomes for a short period and then observing the distribution of radioactive DNA by autoradiography (fig. 26.14). Multiple regions of incorporated label are observed, and replication proceeds bidirectionally from these regions. Termination of replication occurs at the point where the growth forks from two adjacent replication units meet (see fig. 26.14). DNA on the lagging strand of a fork is made discontinuously. The Okazaki fragments are much shorter than those found in prokaryotes, averaging only between 100 and 200 nucleotides in length. Synthesis of these DNA fragments is initiated on primers that arc found covalently attached to the 5′ ends of newly synthesized fragments.

Remember that DNA in eukaryotic chromosomes is associated with histones in complexes called nucleosomes (see fig. 25.18). The disassembly of the DNA–histone complex presumably occurs directly in front of the replication fork. Two observations suggest that nucleosome disassembly may be a rate-limiting step in the migration of the replication fork in chromatin: (1) The rate of migration of replication forks in eukaryotes is slower than in prokaryotes; and (2) the length of the replication fragments on the lagging strand is similar to the length of DNA between adjacent nucleosomes (about 200 bp). Before the newly replicated DNA is reassembled into nucleosomes, the RNA primers must be removed, and the gaps must be filled by enzymes unknown at the present time. Finally, replication fragments must be linked together by DNA ligase.

## SV40 Is Similar to Its Host in Its Mode of Replication

Much has been learned about the replication of chromosomal DNA in eukaryotes by studying a simple monkey virus known as SV40. The viral genome of SV40 consists of a circular duplex DNA molecule of about 5,200 bp with one origin for replication. The SV40 viral genome is complexed with histones to form a nucleoprotein structure very similar to that observed for chromatin. Since SV40 encodes only a single replication protein (T antigen), the virus must make extensive use of the host's cellular replication machinery. As a result many similarities exist between viral and cellular DNA replication. In both cases initiation of DNA synthesis involves two nascent strands. The leading strand grows continuously while the lagging strand grows discontinuously by joining together small (about 200 bp) segments of DNA that are independently initiated with RNA primers. Completion of replication occurs when two oppositely moving forks meet. In linear cellular chromosomes the two merging forks originate from adjacent origins, but in circular SV40 chromosomes they have a single origin.

The development of an efficient cell-free replication system has greatly accelerated progress in understanding the molecular mechanisms involved in SV40 DNA replication. The origin of replication is recognized by the viral T anti-

gen. In addition to its specific binding activity, the T antigen has helicase activity. Once it is bound to the origin, T antigen enters the duplex and catalyzes the unwinding of the two DNA strands. Unwinding appears to be a critical step that establishes the replication forks and generates the single-stranded DNA regions required for the priming and elongation of nascent strands.

In addition to specific nucleotide sequence elements, the T-antigen-mediated unwinding reaction requires accessory proteins contributed by the host cell. For example, a single-stranded DNA-binding protein is required to prevent reassociation of the single strands exposed during unwinding. Such a protein has been found, and it binds specifically to single-stranded DNA.

Of four distinguishable DNA polymerase activities $\alpha$, $\beta$, $\gamma$, and $\delta$, it appears that $\alpha$ and $\delta$ are required for SV40 DNA replication. DNA polymerase $\alpha$ has long been considered the major replicative polymerase in animal cells. The enzyme is composed of four subunits. The largest subunit contains the polymerase active site and a $3′ \rightarrow 5′$ exonuclease that serves a proofreading function during polymerization. The smallest subunit of DNA polymerase $\alpha$ contains a primase capable of synthesizing short RNA transcripts that can serve as primers for subsequent DNA chain elongation by the polymerase-carrying subunit. Each time a polymerase binds to the template primer it adds a certain number of nucleotides to the growing chain before it dissociates. A highly processive enzyme adds a great number of residues before it dissociates. DNA polymerase $\alpha$ is not a highly processive enzyme because fewer than 100 nucleotides are polymerized per binding event.

The properties of DNA polymerase $\delta$, which is also required for SV40 DNA replication, contrast with those of DNA polymerase $\alpha$. First $\delta$ has no primase activity, and second it is a highly processive enzyme capable of catalyzing the polymerization of more than 1,000 nucleotides per binding event.

These differences between the two polymerases support the suggestion that DNA polymerase $\alpha$ might be best suited to serve as the lagging-strand polymerase and that DNA polymerase $\delta$ would be best suited to serve as the leading-strand polymerase. This is because the leading-strand polymerase would be expected to be highly processive and would derive little benefit from an associated primase activity. By contrast, the lagging-strand polymerase would require only moderate processivity, and it would benefit from a tightly associated primase activity (fig. 26.15). Most recent evidence (cited in Waga and Stillman reference) suggests that DNA polymerase $\delta$ is involved in the replication of both strands while DNA polymerase $\alpha$ and primase are involved in the synthesis of RNA-DNA primers.

## Figure 26.15

Hypothetical scheme for the concurrent replication of leading and lagging strands at the replication fork on the SV40 chromosome. The scheme follows the same general notion as one proposed by Kornberg for *E. coli* chromosomal replication. In this scheme polymerase δ functions on the leading strand, and polymerase α functions in conjunction with primase on the lagging strand. The 37-kdα protein, PCNA, greatly augments the activity of polymerase δ but has no effect on polymerase α. (Source: Adapted from B. Stillman, Initiation of eukaryotic DNA replication *in vitro, BioEssays* 9:56, 1988.)

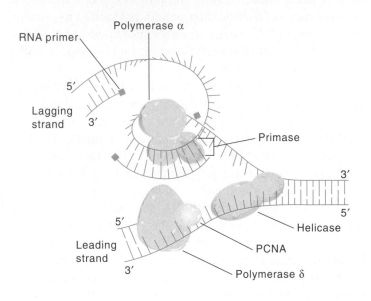

## Figure 26.16

Structure of a thymine dimer formed in DNA by exposure to short-wavelength ultraviolet light.

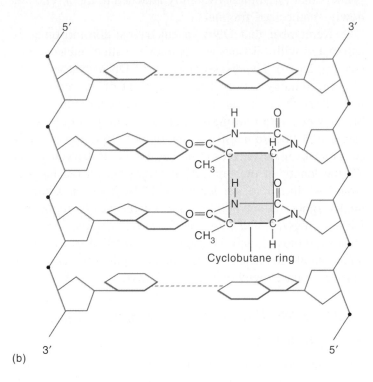

## Several Systems Exist for DNA Repair

We have seen that chromosomes are usually formed by a single DNA molecule regardless of their size. Such a large molecule makes an easy target to attack and damage. In fact, limited damage occurs quite frequently. A single lesion in a DNA molecule left unrepaired could interfere with the replication process and probably would cause cell death. Clearly it is highly advantageous to have a repair system and indeed most cells have more than one.

DNA damage is caused by a variety of physical, chemical, and biological agents. Physical agents include ultraviolet light and ionizing radiation. Damage can be in the form of a missing, incorrect, or modified base or an alteration in the structural integrity of the DNA strands by breaks, cross-links, or dimerization of bases, usually pyrimidines.

Adjacent pyrimidine bases in a DNA strand form dimers with high efficiency after absorbing ultraviolet light (fig. 26.16). By contrast, purines are quite resistant to damage by ultraviolet. Pyrimidine dimers formed within an otherwise intact DNA duplex have provided a useful substrate to assay for DNA repair. These dimers can be repaired directly by enzymatic photoreactivation (fig. 26.17). The

## Figure 26.17

Thymine dimers may be monomerized from DNA by enzymatic photoreactivation. In this case no nucleotides are removed in the repair reaction.

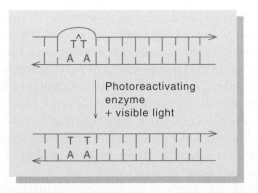

## Figure 26.18

Pyrimidine dimers and other forms of DNA damage can be removed by a general excision repair mechanism. The first reaction in this form of repair involves forming nicks about the damaged region of the DNA. In (*a*) we see the mode of incision of UV-irradiated DNA by the pyrimidine-dimer-specific glycosylase and AP endonuclease activities of *Micrococcus luteus* and bacteriophage T4. In (*b*) we see the mode of incision of the uvrABC endonuclease of *E. coli*. (Source: Adapted from G. Walker, Inducible DNA repair systems, *Ann. Rev. Biochem.* 54:425, 1985.)

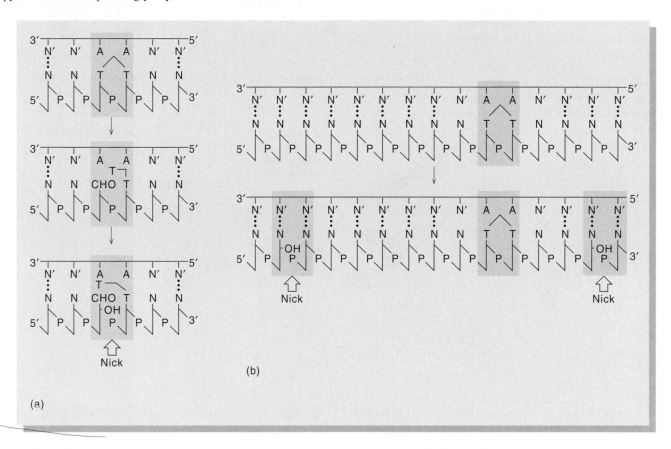

(a)

(b)

photoreactivation enzyme binds to the DNA containing the pyrimidine dimer and uses visible light to cleave the dimer without breaking any phosphodiester bonds.

Systems that function with the help of visible light are quite common. This relates to the fact that sunlight is a mixture of ultraviolet and visible light. The potentially harmful effects of moderate doses of ultraviolet are overcome by the repair processes triggered by the visible light. Ultraviolet tanning lamps usually eliminate most of the visible light and consequently could be quite harmful if used excessively or even in moderation.

Pyrimidine dimers and other forms of DNA damage can also be removed by other mechanisms that do not rely on light (fig. 26.18). This type of repair entails removal of the damaged region and resynthesis of the excised region. Special glycosylases that recognize abnormal or incorrectly paired bases can cleave N-glycosidic bonds to generate an apurinic or apyrimidinic site (see fig. 26.18a). Alternatively, a double incision is made in the strand that carries the lesion by a complex of three proteins encoded jointly by the *uvrA, uvrB,* and *uvrC* genes (see fig. 26.18*b*). In the case of pyrimidine dimers, an incision is made seven nucleotides 5′ to the pyrimidine dimer, followed by a second incision of three or four nucleotides 3′ to the same dimer. The *uvrD* gene product, a helicase, together with PolI and possibly a single-strand binding protein, releases the 12–13 nucleotide oligomer generated by the incision of the uvrABC enzyme complex. After release of this oligomer, PolI and DNA ligase resynthesize the excised region and ligate the nicks.

## Synthesis of Repair Proteins Is Regulated

In *E. coli* the synthesis of many enzymes involved in repair is regulated by the so-called SOS system. Two proteins, lexA and recA, form the working machinery of this regulatory system (fig. 26.19). Under normal conditions the lexA protein inhibits the expression of about 17 genes (the *din* genes), the encoded proteins of which are

**Figure 26.19**

Model for the SOS regulatory system. In normally growing cells the SOS functions associated with DNA repair are not expressed. This is because lexA repressor inhibits their transcription. LexA repressor also inhibits its own expression and that of *recA*. SOS functions are turned on by a series of reactions that starts with DNA damage. DNA damage results in an inducing signal that activates the protease function of recA. This protease cleaves lexA protein, so that all genes that were formerly inhibited by lexA can be expressed. Once the damage is repaired, the level of lexA repressor builds up again, and the SOS genes return to their usual repressed state.

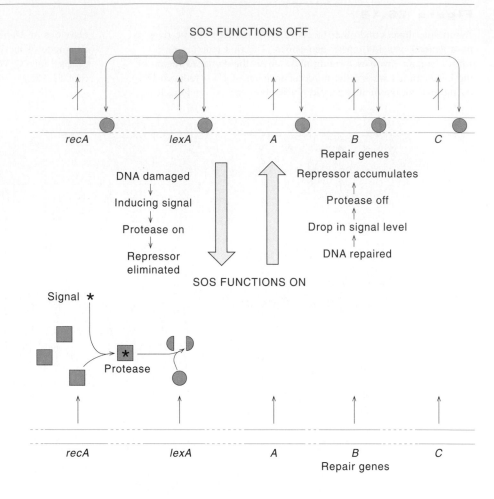

exclusively involved in DNA repair. The lexA protein does this by binding tightly to the control regions of these genes.

The recA protein has two functions: One is to catalyze DNA recombination (see next section), the other is to inactivate the lexA protein. It does the latter by mediating the cleavage of the lexA protein. Damage to the chromosome usually results in DNA fragments. A DNA fragment that binds the recA protein activates a highly specific protease function that cleaves the lexA protein at a specific -AlaGly-bond. The fragmented lexA protein falls off the DNA, thereby permitting the *din* genes to express at a high rate. As soon as the DNA damage has been corrected, the recA protease returns to its dormant form, and newly synthesized lexA protein is free to bind to the *din* genes again.

## DNA Recombination

Individuals belonging to the same species carry approximately the same number of genes positioned in the same relative locations on their homologous chromosomes; each gene is represented by more than one variant which frequently results in different visible characteristics (phenotypes) in different individuals. The different representations

of a gene are referred to as alleles for that gene. Alleles that give rise to different phenotypes serve as useful markers for detecting recombination between homologous chromosomes. Most cells (diploid cells) of an organism contain pairs of homologous chromosomes, one from each parent. During formation of the sex cells (haploid cells) a reduction takes place so that each sex cell only contains one chromosome of each type. The process of chromosome segregation, which takes place during formation of the sex cells, is called meiosis (fig. 26.20). Homologous chromosomes segregate at random during the first phase of meiosis (see fig. 26.20a). This results in the production of sex cells with a very large number of possible combinations of chromosomes. Some combinations of alleles undoubtedly have selective advantages over others. Indeed the primary object of mating seems to be to produce new combinations of genes. But what of the alleles on the same chromosomes? Nature has provided another mechanism for reshuffling alleles on the same chromosome, a process called recombination. Although recombination can take place in most cells of the organism, meaningful genetic recombination that can be passed from one generation to the next only takes place during the first phase of meiosis (see fig. 26.20c).

**Figure 26.20**

Meiosis. The process of meiosis involves two cell divisions with two segregation cycles, meiosis I and meiosis II. These cycles are pictured for a hypothetical cell containing four chromosomes. (*a*) During the first cycle homologous chromosomes pair and then segregate. (*b*) During the second cycle the sister chromatids from each chromosome segregate. The second cycle is very much like mitosis (see fig. 1.21). Each diploid cell that enters meiosis ultimately yields four haploid cells. These haploid cells are the sex cells for the next round of mating. (*c*) Recombination between homologous chromosomes during meiosis I leads to different combinations of alleles. In this illustration alleles for the same gene are represented by corresponding capital or lower-case letters.

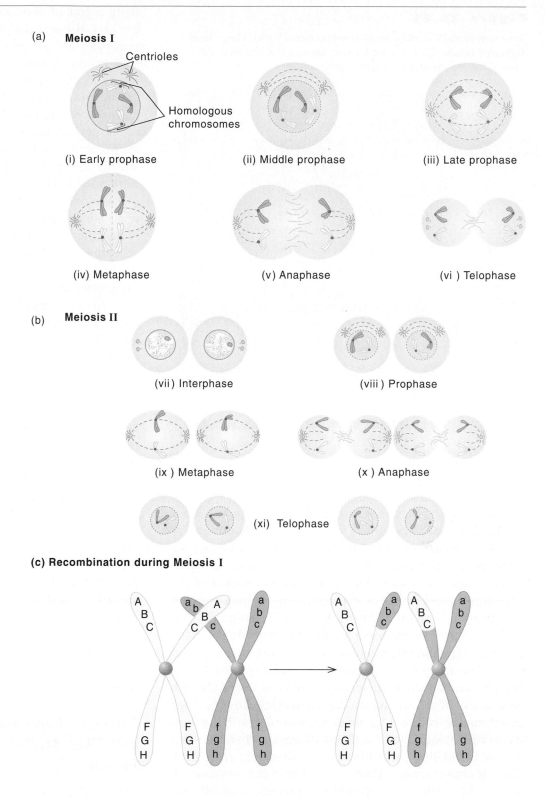

(a) **Meiosis I**

(i) Early prophase  (ii) Middle prophase  (iii) Late prophase

(iv) Metaphase  (v) Anaphase  (vi) Telophase

(b) **Meiosis II**

(vii) Interphase  (viii) Prophase

(ix) Metaphase  (x) Anaphase

(xi) Telophase

(c) **Recombination during Meiosis I**

In the 1960s Matthew Meselson demonstrated that homologous recombination involves breakage and rejoining between existing chromosomes. He made this demonstration on bacteriophages carrying different alleles that infect the same cell. With the help of isotopic labeling he showed that recombinant chromosomes can form without DNA replication, thus demonstrating the breakage–rejoining mechanism. Only a small fraction of new DNA appears in the recombinant chromosome, and this can be accounted for by the repair processes that accompany breakage and rejoining.

## Figure 26.21

Meselson model for phage recombination (circa 1964). Phage form staggered breaks. Base pairing leads to an annealed complex with gaps. Gaps are mended by repair synthesis.

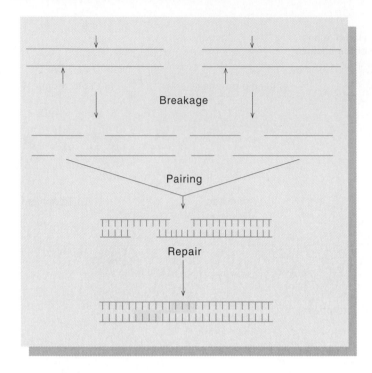

From his studies Meselson proposed a three-step model for homologous recombination (fig. 26.21): (1) Formation of staggered breaks in two parental DNAs; (2) base pairing between single-stranded regions of the two parental types; and (3) repair synthesis. The most important aspect of the Meselson model is that it explained why homologous recombination is so common—a key step in the process involves interaction between complementary single-strand regions originating from homologous chromosomes.

Meselson's results spearheaded investigations of the mechanism of recombination in eukaryotes. Here also the mechanism of recombination can be explained by breakage and rejoining with only a small amount of new DNA synthesis being required in the region of the join. In eukaryotes it was found that recombination is invariably reciprocal. Thus, if chromosome AB recombines with chromosome a,b so that Ab forms as a recombinant product, then aB also forms. R. Holliday proposed a model to explain how recombination occurs in eukaryotes (fig. 26.22). Depending on how the recombination complex finally breaks up, the genes on either side of the region of interaction may or may not show recombination.

The Holliday model is depicted in figure 26.22. Ini-

tially a pair of homologous chromosomes containing two markers, A,a and B,b, is aligned so that homologous DNA sequences are adjacent to one another (frame *a*). The first step entails making single-strand breaks at identical or nearly identical points in the two duplexes (frame *b*). The cut strands migrate and base-pair with homologous regions on the opposing unbroken DNA strands. The pairing need not be perfect, but it should be regular enough to give transient stability to the complex. Subsequently the crossover strands become covalently linked to the opposing strands to form a stable bridge structure. This bridge between the duplexes is known as the Holliday junction. Further exchange of the single strands between the interacting chromosomes is mediated by bridge migration (frame *f*), which can be accompanied by further heteroduplex formation. The more the bridge moves from the initial position of crossing over, the greater the region of heteroduplex formation.

Another way of displaying frame *f* is shown in *g*; to go from (*f*) to (*g*) the ends of the duplexes are merely pulled apart. In (*h*) we have the same structure as in (*g*), but an arrow is shown indicating a twisting of one of the duplexes. The structure in (*h*) leads to the structure in (*i*), called the chi form because it is shaped like the Greek letter chi ($\chi$). The chi form divides by two single-strand cuts in either the horizontal direction (*i*) or the vertical direction (*j*), leading to quite different products, (*k*) or (*n*). In both structures some DNA exchange has occurred in the region of heteroduplex formation. In (*k*), the outside markers (i.e., the markers outside the region of heteroduplex formation) A and B and a and b have not recombined; that is, they are on the same DNA molecules (chromosomes) as they were before the recombination event. The structure in (*n*), however, is recombinant for the parental genetic markers. The repair process that follows exchange closes the gaps in the severed strands, (*l*) and (*o*).

Genetic investigations have supported the Holliday model and at the same time indicate that different organisms can only be explained by variations of the model.

## *Enzymes Have Been Found in* Escherichia coli *That Mediate the Recombination Process*

Cursory inspection of both the Meselson and Holliday models suggests the need for special enzymes to mediate the recombination process. First an endonuclease is needed to make the initial strand breaks that are required for strand invasion. Second an enzyme is required to mediate formation of the recombination complex. Finally an enzyme is probably required to resolve the recombinant duplexes.

**Figure 26.22**

Holliday model of genetic recombination. In (a) we see two paired homologous chromosomes with different marker alleles, A, B, and a, b. In (b), single-strand cuts have been made at identical loci on the two chromosomes. In (c) and (d) we see these strands beginning to invade the duplexes of the homologous strands. The invading strands base-pair at homologous sites. In (e) the 3' ends of the invading strand (indicated by the arrowheads in (d) become covalently linked to the 5' end of the cut strand in place. This produces a bridge commonly called a *Holliday junction*. The bridge can migrate to the right as shown (or to the left, not shown) in (f) to form more heteroduplex structures. In (g) we see another way of displaying (f), and (h) is the same as (g), except for the arrow indicating the way in which the lower half of the structure is twisted to form (i). Then (i) may be cut in one of two ways to yield either (j) or (m). Parts (k) and (n) are alternative ways of displaying (j) and (m), respectively. Finally, in (l) and (o) the duplexes resulting from (k) and (n) are mended.

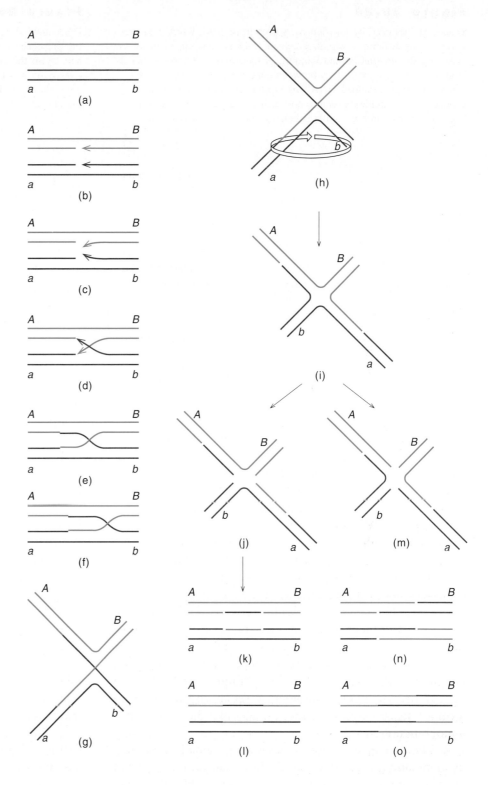

Genetic and biochemical studies on *E. coli* have led to the discovery of these three types of enzymes.

The search for enzyme activity associated with recombination has been based on analysis of organisms carrying mutations that affect general recombination. Such mutants, known as *rec* mutants, were first discovered more than 20 years ago, and new ones are still being found. *E. coli rec* mutations, *recA, recB, recC,* and *recD* reduce recombination efficiency to different extents. The *recA* mutants are totally deficient in homologous recombination, whereas *recB* or *recC* mutants can recombine but only about $10^{-4}$ times as well as wild-type cells.

**Figure 26.23**

Reactions catalyzed by purified recA protein *in vitro*. RecA catalyzes a number of different reactions between DNA strands, all of them involving the unwinding and winding of base-paired structures. (*a*) D-loop formation by interaction between supercoiled circular duplex DNA and single-stranded DNA. (*b*) Strand exchange between a gapped circular duplex structure and a linear duplex structure. (*c*) Complex formation between two helices, one of which is gapped.

*recA enzyme* → D-loop formation

(a)

*recA enzyme* → Strand exchange and linear duplex formation

(b)

*recA enzyme* → Complex formation between two helices, one of which is gapped

(c)

We have discussed the repair role played by the protease function of *recA* (see fig. 26.19). The second class of reactions catalyzed by recA protein indicates a prominent role in the initiation of homologous pairing during genetic recombination. Thus, the purified recA protein catalyzes various forms of complex formation and exchange between duplex and single-stranded DNAs in reactions that entail the cleavage of ATP (see fig. 26.23). D-loop formation can result when the recipient is a circular duplex and the donor is a single strand (see fig. 26.23*a*). Strand exchange can occur when a single-stranded fragment is removed in the presence of a double-stranded fragment (see fig. 26.23*b*). RecA protein also causes a complex to form between two circular helices, provided that at least one input helix is gapped on one strand (see fig. 26.23*c*). Interestingly, while the two helices must contain homologous sequences in order to form a four-stranded structure, the gap may lie in a nonhomologous region.

The fact that mutations in *recA* completely eliminate recombination, as well as the finding that the recA-related protein catalyzes strand exchange reactions like those required in the recombination models we have been discussing, makes it highly likely that *recA* plays a key role in these processes *in vivo*. If this is so, then recA protein is clearly situated at the hub of activities in the recombination complex.

The recA enzyme does not catalyze any recombination event unless at least one free end of a single strand or a

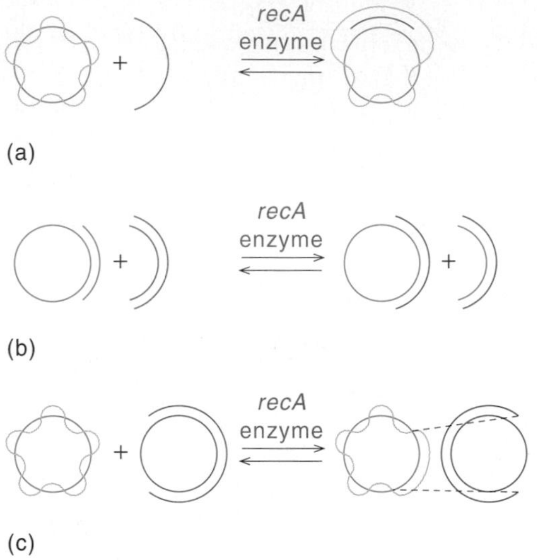

**Figure 26.24**

DNA unwinding by exoV enzymes. This is best observed *in vitro* in the presence of ATP and $Ca^{2+}$. The $Ca^{2+}$ inhibits the exonucleolytic activity of the enzyme. The ATP provides the energy that drives the enzyme in a concerted manner into the duplex structure. The duplex unwinds in front of the path of the enzyme and rewinds in back of the enzyme.

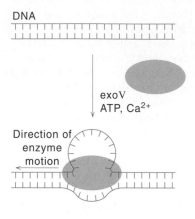

DNA

exoV
ATP, $Ca^{2+}$

Direction of enzyme motion

DNA duplex is available. Thus, it seems unlikely that the recA protein could be involved in making the initial incision(s) in the duplex structure that is(are) required to initiate recombination. There are reasons for believing that the *recB* and *recC* genes are involved in this capacity. First, a mutation in either of these genes reduces recombination by a factor of about $10^{-4}$. Second, these two genes together with *recD* encode the subunits of a nuclease known as exoV. ExoV is both a nuclease and a helicase. The helicase activity is best demonstrated *in vitro* under conditions where the nuclease activity is blocked. This can be done by adding $Ca^{2+}$. When linear duplex DNA is incubated with exoV in the presence of ATP and $Ca^{2+}$, it begins to unwind at one end. As the unwinding progresses, the single-stranded regions collapse back on each other to reform a duplex. A double-loop structure is maintained in the vicinity of the migrating enzyme (fig. 26.24). What is the point of this migrating behavior? Further experiments indicate that exoV scans the duplex looking for preferred sequences where it makes single-strand incisions.

The recombination process triggered by the initial scissions made by exoV and mediated by the recA enzyme are not sufficient for recombination between interacting chromosomes. An additional enzyme, resolvase, is required for efficient separation of the recombining chromosomes in the recA–chromosome complex. This protein is encoded by the *ruvC* gene in *E. coli* and is absolutely required for homologous recombination in the bacterium. Its effectiveness has also been demonstrated in a cell-free *in vitro* system.

Incidentally, the fact that the recA enzyme carries two enzyme activities, one for homologous recombination and one for regulating the concentration of DNA repair genes, including itself (the recA protease, which cleaves the lexA repressor), leads to the suspicion that recA is also more directly involved in DNA repair. Recombination between two homologous chromosomes badly damaged in different regions could lead to one normal chromosome. Although, technically a haploid organism, *E. coli* carries two copies of the bacterial chromosome for a considerable time prior to cell division. Thus, the necessary identical pairs of chromosomes are present for homologous recombination.

## Other Types of Recombination

There are two other types of recombination, on which we will not elaborate. Site-specific recombination, is limited to highly select regions of the genome and to very specific functions. One of the best understood examples of site-specific recombination is the integration of bacteriophage λ DNA into the *E. coli* host chromosome (see chapter 30). Finally nonspecific recombination occurs between nonhomologous regions of chromosomes with little or no sequence homology. In the next section we see that at one stage in its life cycle a retrovirus recombines in this way.

## RNA-Directed DNA Polymerases

It is a commonly held belief that RNA preceded DNA in the early evolution of living systems. If this is the case then the first DNA polymerases must have been capable of transferring sequence information from RNA to DNA. Enzymes of this sort are called reverse transcriptases because they do the reverse of common transcriptases (see chapter 28). Reverse transcriptases no longer play the central role in genetic information transfer, but they are still found in all species and function in a number of capacities in both cellular and viral metabolism.

## Retroviruses Are RNA Viruses That Replicate through a DNA Intermediate

RNA viruses that cause tumors (oncogenic RNA viruses) are called retroviruses because their life cycle involves a DNA intermediate. The ability of retroviruses to use such a route for replication hinges on a viral-encoded enzyme called reverse transcriptase, an enzyme with three discrete activities: (1) It catalyzes the synthesis of DNA from the viral plus-strand; (2) it catalyzes the synthesis of DNA plus-strand from the viral minus-strand DNA; and (3) it catalyzes the degradation of the viral RNA from an RNA–DNA heteroduplex.

Following viral infection the virus-plus strand first functions as an mRNA for the synthesis of the viral proteins. Once the viral reverse transcriptase has been synthe-

sized, it is used in conjunction with the viral RNA template and a tRNA primer for the synthesis of minus-strand DNA. Next the RNase H activity of the viral enzyme removes extensive sections of RNA from the DNA–RNA heteroduplex so that the newly synthesized minus-strand DNA can be used as a template for the synthesis of viral plus-strand DNA. The complete process of duplex DNA synthesis is depicted in figure 26.25. The duplex DNA is subsequently integrated more or less indiscriminately at many sites in the host genome. There it can remain indefinitely or until the cell dies. The integrated viral DNA serves as a template for the synthesis of viral plus strand, completing the nucleic acid cycle that began with virus infection. Additional properties of retroviruses, including their association with malignant transformation, are taken up in Supplement 4: Carcinogenesis.

## Hepatitis B Virus Is a DNA Virus That Replicates through an RNA Intermediate

Hepatitis B is an animal virus with a small circular DNA genome. A gap in one strand is bridged by an incomplete complementary strand. The longer strand has a protein bound at its 5' end. A DNA polymerase present in the mature virus particle can elongate the 3' ends of the incomplete strands. An unusual feature of this virus is that replication involves an RNA intermediate that must be reverse transcribed.

## Some Transposable Genetic Elements Encode a Reverse Transcriptase That Is Crucial to the Transposition Process

A widely distributed group of genetic elements exist that are capable of integrating at random sites in the host chromosome. These transposable genetic elements usually carry genes to facilitate their own transposition in addition to conferring specific properties on the host cells in which they are located. Many of these transposable elements encode a reverse transcriptase that functions in the transposition process. The first step in transposition of elements of this type involves transcription of the genetic element. The resulting RNA, like the retroviral RNA, has two functions: One as a messenger RNA for the synthesis of element-encoded proteins and one as a template for the synthesis of the transposable element DNA. The transposable element integrates at other sites on the host chromosome without any apparent regard for sequence homology at the sites of integration. Transposable genetic elements that use a reverse transcriptase in this way have been found in yeast, *Drosophila*, and mice, and they are presumed to be present in a very wide range of organisms.

## Figure 26.25

A model for the generation of double-stranded DNA
carrying two copies of LTR. The sequence X is a marker
for the plus strand; the complementary sequence X' occurs
on the minus strand. (*a-c*) Synthesis of minus-strand DNA
from the genomic RNA template using a tRNA primer
(represented by inverted J in step *a*). (*d*) Degradation of the
RNA portion of the resulting DNA:RNA hybrid by
RNAseH. (*e*) Bridge between the newly synthesized segment
of minus-strand and repeated sequences at the 3' end of
genomic RNA. (*f*) Extension of minus-strand DNA
(*leftward*) and initiation of plus-strand DNA (*heavy arrow
moving rightward*). (*g*) Completion of 300-nucleotide
fragment of plus-strand DNA. (*h*) Bridge between sequences
repeated in minus-strand DNA and in the 300-nucleotide
fragment. (*i*) Completion of synthesis and tidying up: ▭▪▫
= LTR, with sequences from the 5' and 3' ends of the viral
genome separated by the terminal repeat. In this figure red
stands for RNA and blue stands for DNA.

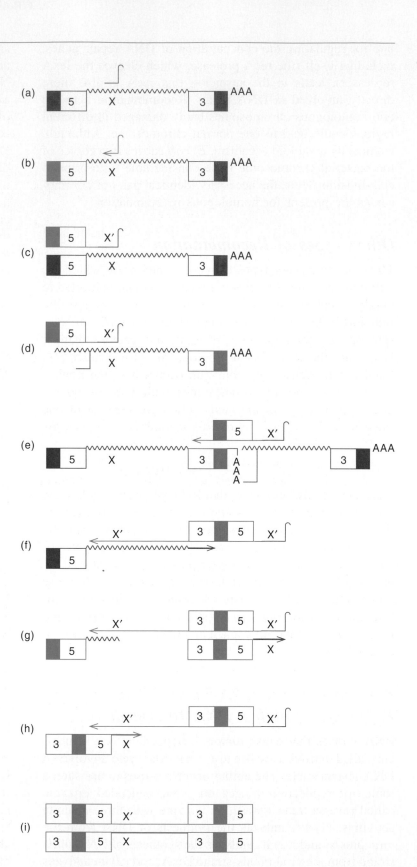

## Bacterial Reverse Transcriptase Catalyzes Synthesis of a DNA–RNA Molecule

Except as intermediates, molecules in which DNA and RNA are covalently linked to each other have only been found in certain bacteria. For example, some strains of *E. coli* carry a genetic element that encodes a reverse transcriptase. This genetic element produces one long transcript that folds into a complex three-dimensional structure. Part of this molecule serves as a template for the synthesis of a reverse transcriptase, which recognizes a specific G residue on the folded RNA as a primer start site for DNA synthesis. Synthesis starts from the 2′-OH group of this G residue using another part of the RNA as a template. DNA synthesis arrests at a specific point on the RNA template. An RNase H removes that region of the RNA in which DNA–RNA heteroduplex has been formed, leaving a covalently linked DNA–RNA molecule. No function has been found for this unusual nucleic acid (see Linn and Maas reference).

## Telomerase Facilitates Replication at the Ends of Eukaryotic Chromosomes

Circular bacterial chromosomes are initiated by RNA primers. At some stage the RNA primers must be eliminated and replaced by DNA. Due to the circular nature of the chromosome an upstream DNA molecule can always serve as a primer for regions from which RNA primers are eventually removed. This guarantees that the primer requirement does not interfere with complete replication of the chromosome.

Since eukaryotic chromosomes are linear, the ends of these chromosomes require a special solution to ensure complete replication. This can be seen in figure 26.26. At the very end of a linear duplex a primer is necessary to initiate DNA replication. After RNA primer removal there is bound to be a gap at the 5′ end of the newly synthesized DNA chains. Since DNA synthesis always requires a primer the usual way of filling this gap is not going to solve the problem. This dilemma is overcome by a special structure at the ends (telomeres) of eukaryotic chromosomes and a special type of reverse transcriptase (telomerase) that synthesizes telomeric DNA. In many eukaryotes the telomeres contain short sequences (frequently hexamers) that are tandemly repeated many times. Telomerase contains an RNA that binds to the 3′ ends and also serves as a template for the extension of these ends. Prior to replication, the 3′ ends of the chromosome are extended with additional tandemly repeated hexamers. The 3′ ends are extended sufficiently so that there is room to accommodate an RNA primer. In this way there is no net loss of DNA from the 5′ ends as a result of replication. After replication the 3′ end is somewhat

### Figure 26.26

Synthesis at the ends of a eukaryotic chromosome. One end of the linear DNA of a eukaryotic chromosome is diagrammed. A flush-ended DNA duplex presents a problem for completing synthesis at the 5′ end (*a*). This is because of the RNA primer requirement for DNA synthesis. When the primer at the 5′ end is removed there is no conventional way to fill the gap. A solution to this problem is shown in (*b*). The ends of eukaryotic chromosomal DNAs consist of highly repetitious tandem repeats (telomeres). These repeats on the 3′ end serve as both primer and template for extending the 3′ end. The extended 3′ end can accommodate a primer RNA, so after chromosomal DNA replication no loss occurs from the 5′ end of the DNA. Another process is needed to remove the extension from the 3′ end. New synthesis is indicated in red. The zigzag represents primer.

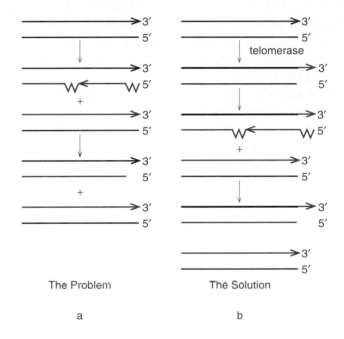

longer than it was before because of the extra added tandem repeats. An additional mechanism is necessary to keep the telomeres from growing indefinitely by this effect. Possibly related to this requirement, it has been observed that the telomeres become shorter in somatic tissues with aging. This is believed to be due to a deficiency of telomerase in most somatic tissues.

## Other Enzymes That Act on DNA

Many other enzymes modify DNA and degrade it in various ways. In the next chapter we discuss some of those enzymes that have been particularly useful in DNA manipulations. In Supplement 3 (Immunobiology) we discuss DNA splicing operations that are essential for assembling complex antibody genes prior to their expression.

## Summary

This chapter deals with reactions involved in DNA synthesis, degradation, repair, and recombination. The chief points to remember are as follows.

1. DNA replication proceeds by the synthesis of one new strand on each of the parental strands. This mode of replication is called semiconservative, and it appears to be universal. DNA synthesis initiates from a primer at a unique point on a prokaryotic template such as the *E. coli* chromosome. From the initiation point, DNA synthesis proceeds bidirectionally on the circular bacterial chromosome. The bidirectional mode of synthesis is not followed by all chromosomes. For some chromosomes, usually small in size, replication is unidirectional.

2. In eukaryotic systems, replication can start at several points (still not well defined) along the chromosome. Replication is usually bidirectional about each initiation site. The termination points of replication are interspersed between initiation sites. In most cases of unidirectional or bidirectional replication, synthesis occurs nearly (but not exactly) simultaneously on both strands of the parent DNA template. Because synthesis can occur only in the $5' \rightarrow 3'$ direction on the growing chain, and the two strands in the parent duplex are oriented in opposite directions, synthesis can occur continuously on only the leading strand. On the other (lagging) strand it must pause for the template to unwind. Synthesis on the lagging strand does not occur continuously but rather in small discontinuous spurts, generating Okazaki fragments.

3. Many proteins are required for DNA synthesis and chromosomal replication. These include polymerases; helicases, which unwind the parental duplex; enzymes that fill in the gaps and join the ends in the case of lagging-strand synthesis; enzymes that synthesize RNA primers at various points along the DNA template; topoisomerases, which permit rotation and supercoiling; and single-strand DNA-binding proteins, which stabilize single-stranded regions that are transiently formed during replication. Most of these proteins have been isolated from whole cells and studied in cell-free systems.

4. In *E. coli,* mutations have been isolated in the genes encoding a number of these enzymes. Many of these mutations are conditional because the functional enzymes involved are required for DNA synthesis and cell viability. Mutants carrying mutationally altered proteins have been important in confirming their roles predicted from cell-free studies.

5. In eukaryotes, considerable progress has been made in studying the *in vitro* replication of animal viruses, such as SV40. The importance ascribed to the enzymes that have been characterized is largely based on a comparison of their properties with similar prokaryotic enzymes whose functions are better understood.

6. Many enzymes that act on DNA are involved in processes other than DNA synthesis. They include DNA repair enzymes, DNA degradation enzymes, and DNA recombination enzymes.

7. Enzymes that catalyze the synthesis of DNA using an RNA template are known as reverse transcriptases. The first reverse transcriptase discovered was encoded by an RNA retrovirus. This enzyme is needed in the virus replication cycle. Some animal viruses pass through an RNA intermediate and also require a reverse transcriptase to replicate the viral DNA. Similarly, a number of transposable elements found in cellular chromosomes replicate through RNA intermediates; they usually encode a reverse transcriptase. A unique reverse transcriptase called telomerase is used to synthesize the DNA at the ends of linear eukaryotic chromosomes.

# Selected Readings

Baker, T. A., and S. H. Wickner, Genetics and enzymology of DNA replication in *Escherichia coli. Ann. Rev. Genet.* 26:447–469, 1992.

Bauer, W. R., F. H. C. Crick, and J. H. White, Supercoiled DNA. *Sci. Am.* 243(4):118–133, 1980.

Beese, L. S., V. Derbyshire, and T. A. Steitz, Structure of DNA polymerase I Klenow fragment bound to duplex DNA. *Science* 260:352–355, 1993.

Beese, L. S., and T. A. Steitz, Structural basis for the 3' → 5'-exonuclease activity of *E. coli* DNA polymerase I: A two metal ion mechanism. *EMBO J.* 10:25–33, 1991.

Bell, S. P., and B. Stillman, ATP-dependent recognition of eukaryotic origins of DNA replication by a multiprotein complex. *Nature* 357:128–134, 1992.

Blackburn, E. H., Telomerases. *Ann. Rev. Biochem.* 61:113–129, 1992.

Bohr, V. A., and K. Wasserman, DNA repair at the level of the gene. *Trends Biochem. Sci.* 13:429–432, 1988.

Bramhill, D., and A. Kornberg, A model for initiation at origins of DNA replication. *Cell* 54:915–918, 1988.

Campbell, J. L., Eukaryotic DNA replication. *Ann. Rev. Biochem.* 55:733, 1986.

Challberg, M. D., and T. J. Kelly, Animal viruses and DNA replication. *Ann. Rev. Biochem.* 58:671–717, 1989.

Cox, M. M., and I. R. Lehman, Enzymes of general recombination. *Ann. Rev. Biochem.* 56:229–262, 1987.

Demple, B., and P. Karran, Death of an enzyme: Suicide repair of DNA. *Trends Biochem. Sci.* 8:137–139, 1983.

Dunderdale, H. J., F. E. Benson, C. A. Parsons, G. J. Sharples, R. G. Hoyd, and S. C. West, Formation and resolution of recombination intermediates by *E. coli* RecA and RunC protein. *Nature* 354:506–510, 1991.

Fangman, W. F., and B. J. Brewer, Activation of replication origins with yeast chromosomes. *Ann. Rev. Cell. Biol.* 7:375–402, 1991.

Fink, G. R., J. D. Boeke, and D. J. Garfinkel, The mechanisms and consequences of retrotransposition. *Trends Genet.* 2:118–123, 1986.

Holliday, R., A different kind of inheritance. *Sci. Am.* 260(4):60–73, 1989. On DNA methylation.

Itoh, T., and J. Tomizawa, Antisense RNA. *Ann. Rev. Biochem.* 60:631–652, 1991. Includes an excellent description of the initiation of DNA synthesis for Col E1 plasmid.

Kohlstaedt, L. A., J. Wang, J. M. Friedman, P. A. Rice, and T. A. Steitz, Crystal structure at 3.5 Å resolution of HIV-1 reverse transcriptase complexed with an inhibitor. *Science* 256:1783–1790, 1992.

Kong, X-P, R. Onrust, M. O'Donnell, and J. Kuriyan, Three-dimensional structure of the β subunit of *E. coli* DNA polymerase III holoenzyme: A sliding clamp. *Cell* 69:425–437, 1992.

Kornberg, A., and T. A. Baker, *DNA replication,* 2d ed. New York: Freeman, 1991. A magnificent up-to-date, clearly written and thorough treatment.

Landy, A., Dynamic, structural and regulatory aspects of site-specific recombination. *Ann. Rev. Biochem.* 58:913–950, 1989.

Linn, D., and W. Maas, Reverse transcriptase-dependent synthesis of a covalently linked branched DNA–RNA compound in *E. coli* B. *Cell* 56:891–904, 1989.

Lohman, T. M., W. Bujalowski, and L. B. Overman, *E. coli* single strand binding protein. *Trends Biochem. Sci.* 13:250–255, 1988.

Maxwell, A., and M. Gellert, Mechanistic aspects of DNA topoisomerases. *Adv. Prot. Chem.* 38:69–107, 1986.

McHenry, C. S., DNA polymerase III holoenzyme of *E. coli. Ann. Rev. Biochem.* 57:519–550, 1988.

Meselson, M., and F. W. Stahl, The replication of DNA in *Escherichia coli. Proc. Natl. Acad. Sci. USA* 44:671–682, 1958. A classic paper.

Messer, W., and W. Noyer-Weidner, Timing and targeting: The biological function of Dam methylation in *E. coli. Cell* 54:734–737, 1988.

Modrich, P., DNA mismatch correction. *Ann. Rev. Biochem.* 56:435–466, 1987.

Newlin, C. S., Yeast chromosome replication and segregation. *Microbiol. Rev.* 52:568–601, 1988.

Ogawa, T., and T. Okazaki, Discontinuous DNA replication. *Ann. Rev. Biochem.* 57:519–550, 1988.

Radman, M. and R. Wagner, The high fidelity of DNA replication. *Sci. Am.* 259(2):40–46, 1988.

Sancar, A., and G. B. Sancar, DNA repair enzymes. *Ann. Rev. Biochem.* 57:29–67, 1988.

Sancar, G. B., and A. Sancar, Structure and function of DNA photolyases. *Trends Biochem. Sci.* 12:259–261, 1987.

Shapiro, J. A., *Mobile Genetic Elements.* New York: Academic Press, 1983.

Smith, G. R., Homologous recombination in *E. coli:* Multiple pathways for multiple reasons. *Cell* 58:807–809, 1989.

Stahl, F. W., Genetic recombination. *Sci. Am.* 256(2):90–101, 1987.

Varmus, H., Reverse transcription. *Sci. Am.* 257(3):56–64, 1987.

Waga, S., and Stillman, B., Anatomy of a DNA replication fork revealed by reconstitution of SV40 DNA replication in vitro. *Nature.* 369:207–212, 1994.

Wang, J. C., DNA topoisomerases. *Ann. Rev. Biochem.* 54:665–697, 1985.

Wang, T. S. F., Eukaryotic DNA polymerases. *Ann. Rev. Biochem.* 60:513–553, 1991.

## Problems

1. In the Meselson-Stahl experiment illustrated in figure 26.2, a sample of the DNA shown in tube 4 (labeled with $^{15}N$ followed by one generation in $^{14}N$) was heat-denatured prior to being subjected to centrifugation in a CsCl density gradient. This gradient showed two peaks of single-stranded DNAs of different densities. How did this experiment further support the idea of semiconservative replication?

2. Explain why Cairns and coworkers used $[^3H]$-thymidine to label replicating *E. coli* DNA in the experiments shown in figure 26.3.

3. Draw the chemical reaction mechanism for the formation of a phosphodiester linkage during DNA synthesis. Discuss the significance of the pyrophosphate product that is formed. What is the significance of the $Mg^{2+}$ requirement?

4. Cairns and De Lucia isolated a mutant strain of *E. coli* that had only about 1% of the DNA PolI activity found in wild-type cells, yet the strain replicated its DNA at a normal rate. Explain how this discovery was important in understanding the role of the different DNA polymerases in replication and repair.

5. All enzymes that make DNA in a template-dependent fashion require a primer. How does the use of a primer increase the fidelity of DNA synthesis, and why is this primer usually RNA?

6. Even with its proofreading activity, *E. coli* DNA polymerase III still exhibits a measurable rate of nucleotide misincorporation (about one mistake per $10^{10}$ nucleotides incorporated). Mutants of *E. coli* DNA polymerase III can be isolated that have a lower than normal rate of misincorporation. Why might such mutants, which can be said to have hyperaccurate DNA replication, be evolutionarily unfavorable?

7. *E. coli* has a genomic complexity of about $4 \times 10^6$ bp, and each replication fork can move at a rate of about $10^3$ bp/s (base pairs per second). How long does it take to replicate the *E. coli* chromosome? With an ample carbon source and ideal growth conditions, cells of *E. coli* can divide in about 20 min. How can this shorter division time occur if the rate of fork migration remains constant at $10^3$ bp/s?

8. Draw the chemical reaction mechanism for DNA ligase in *E. coli* (uses $NAD^+$ as the source of energy and forms a covalent intermediate with an $\epsilon$-amino group of lysine). Why are ligation reactions that require ATP more thermodynamically favorable?

9. Humans have about $2.9 \times 10^9$ bp of DNA in their genome, and the replication fork migrates at the rate of approximately 30 bp/s. How long would it take to replicate the entire genome if it was a single continuous piece of DNA? How many replication origins are required to replicate this DNA in one hour?

10. Why does the reaction catalyzed by the *dnaB* gene product require ATP hydrolysis? Does this protein alter the linking number of DNA? Explain.

11. Eukaryotic DNA is replicated at a slower rate than prokaryotic DNA. One reason may be the requirement for the deposition of histone proteins on DNA (histone synthesis and DNA replication are coupled). Describe a model for the replication of eukaryotic DNA and nucleosome formation.

12. The graph shows *E. coli* labeled with radioactive thymidine for a short pulse (10 s) followed by a chase with an excess of nonradioactive thymidine. The DNA is then extracted and centrifuged in alkaline sucrose gradients (under high pH conditions the DNA denatures). Explain what these data imply, and interpret these results in light of our current model for DNA replication.

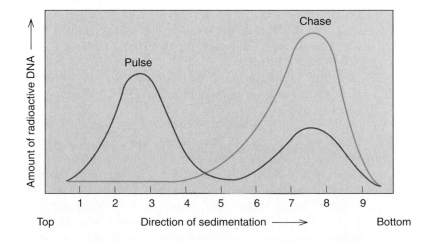

13. How is the SOS response reversed following repair of DNA?

14. Compare and contrast the excision repair and photo-reactivation mechanisms for correction of ultraviolet-induced thymine dimers.

15. What is the role of the recA protein in *E. coli* DNA repair and in recombination? What use are *recA* mutants in biochemical and genetic research?

16. Normal human fibroblasts are grown in culture and then exposed to UV light. A short time later the DNA is extracted and applied to an alkaline sucrose gradient, and the data in graph (*a*) are observed. Another sample of cells is also exposed to UV light, but about 12 h are allowed to pass before the DNA is extracted and applied to an alkaline sucrose gradient (graph *b*). Explain these data from what you know about DNA repair (*Hint:* see fig. 26.18.). An-

other sample of fibroblast cells, isolated from a patient with xeroderma pigmentosum (a disease resulting from the inability to repair DNA damage caused by UV light) was exposed to UV light and then applied to a gradient after a short time. Would you expect the data to resemble those in graph (*a*) or (*b*)? Why?

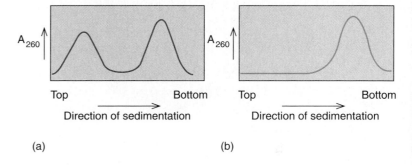

(a)　　　　　　　　　　(b)

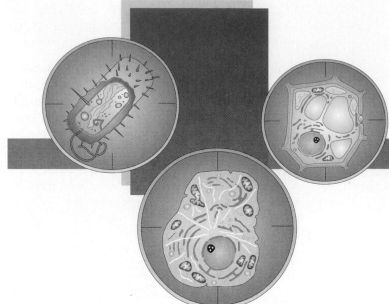

# DNA Manipulation and Its Applications

*DNAs from different sources can be knitted together to produce new genetic alignments.*

A broad array of techniques has made it possible to investigate the fine structure of DNA as well as to redesign existing genes. The impact of these techniques cannot be overestimated. It is possible to map and isolate genes from complex eukaryotes—feats that could never have been accomplished by classical genetic methods. Genes can be redesigned to achieve a variety of practical goals. Newly designed genes may be inserted into the same species from which the original unmodified genes came or into other species.

In this chapter we first consider DNA sequencing. Then we explore the different approaches for amplifying and isolating specific genes or gene segments. Following this, we examine the methods currently available for restructuring existing DNA sequences. Finally, we look at some of the major advances in our understanding of gene structure and location that have resulted from the new technology. This part of the discussion focuses on two examples: The mapping of the human globin gene family and the mapping of the gene responsible for the genetically inherited disease, cystic fibrosis.

## Sequencing DNA

If we are going to manipulate a segment of DNA, it is most useful to know something about its primary structure. The most complete information comes from total sequence analysis.

Initial efforts at sequencing nucleic acids were confined to RNA molecules that could be readily isolated in pure form. The first sequence to be determined was that for tyrosine tRNA from yeast. From roughly 100 pounds of yeast Robert Holley was able to isolate enough of the tyrosine tRNA to carry out a sequence analysis. This historically significant effort required a number of enzymes and chromatographic techniques. We shall not elaborate on this accomplishment because sequencing of RNA is no longer done directly. In fact, the sequencing of RNA has been replaced by the sequencing of "cDNA," which results from the reverse transcription of RNA into DNA.

Two quite different methods have been developed for sequencing DNA. One method, developed by Walter Gilbert and Alan Maxam and involving cleavage of preexisting DNA, uses a chemical approach. A second method, involving premature termination of newly synthesized DNA, uses an enzymatic approach that was developed by Fred Sanger.

For pure sequencing, Sanger's is the method of choice. It employs chain-terminating dideoxynucleoside triphosphates to produce a continuous series of fragments in reactions catalyzed by DNA polymerase. Dideoxynucleoside triphosphates (ddXTPs) resemble deoxynucleoside triphosphates except that they lack a 3′ —OH group. They can add to a growing chain during polymerization, but they cannot be added onto, and as a result they act as chain terminators.

The DNA being sequenced is mixed with a suitable primer, radioactive dXTPs, DNA polymerase I (PolI), and a small amount of one ddXTP. The primers determine where DNA synthesis starts, and the ddXTP determines the base type where elongation stops. The products of four separate reaction mixtures, each differing only by the ddXTP it contains, are analyzed in figure 27.1. As depicted, reaction 1, using ddATP, contains all fragments with an A terminus; reaction 2, using ddCTP, contains all C terminations; and so on. After the newly synthesized oligonucleotides are separated from the template by denaturation they are fractionated by electrophoresis for a limited time on polyacrylamide gels. The positions of the fragments on the gel are detected by autoradiography. The sequence is read directly from the composite autoradiogram, starting with the fastest moving (smallest) band at the bottom of the gel and moving up. If the first band is in reaction 3, it is a G residue; the next band up, appearing from reaction 4, would be T; and so on. Up to 800 residues can be read from a single gel.

## Methods for Amplification of Select Segments of DNA

Cellular genomes are very large; even *E. coli* contains more than a million base pairs, and eukaryotic genomes frequently contain a billion or more base pairs in one complete genome. Because of their large size it is impractical to fractionate the cellular genome and expect to obtain enough of a particular DNA segment for sequencing or other investigations. Two methods of amplifying defined segments of DNA have been developed. The first of these takes the desired segment and amplifies it *in vitro* with DNA PolI using DNA primers that bind to the ends of the region of interest. This method involves repeated cycles of synthesis and is appropriately named the polymerase chain reaction (PCR) method. The second method inserts the DNA segment of interest into a plasmid or virus that can be amplified *in vivo*. We discuss both of these methods because they are both useful in many ways.

## Amplification by the Polymerase Chain Reaction

PCR entails enzymatic amplification of specific DNA sequences using two oligonucleotide primers that flank the DNA segment to be amplified (fig. 27.2). The primers must complement opposite strands so that after annealing, their 3′ ends in effect face each other (see fig. 27.2b).

The PCR procedure has three steps, which are usually repeated many times in a cyclical manner:

1. Denaturation of the original double-stranded DNA sample at high temperature

2. Annealing of the oligonucleotide primers to the DNA template at low temperature (37°C)

3. Extension of the primers using DNA polymerase

## Figure 27.1

The Sanger dideoxynucleoside method of sequencing DNA. (*a*) A suitable template is chosen, and the primer is chosen so that DNA synthesis begins at the point of interest. The primer is radioactively labeled. In addition to the template–primer complex the reaction mixture contains all four radioactive deoxyribonucleoside triphosphates and small amounts of a single dideoxynucleoside triphosphate. The dideoxy compound serves as a chain terminator. (*b*) After synthesis in the presence of DNA polymerase I, the products of the reaction mixture are separated by gel electrophoresis and analyzed by autoradiography. The gel is run under denaturing conditions in warm urea so that single-stranded fragments separate strictly according to size. For a given dideoxy compound all fragments terminating with that particular base should give rise to bands on the gel. The interpretation of the gel pattern is given in (*c*). The smallest labeled fragment moves the fastest and appears at the bottom of the gel. (*d*) A typical sequencing film. The sequence begins CAAAAAACGG. (Courtesy of GIBCO-BRL, Life Technologies, Inc., Gaithersburg, Md.)

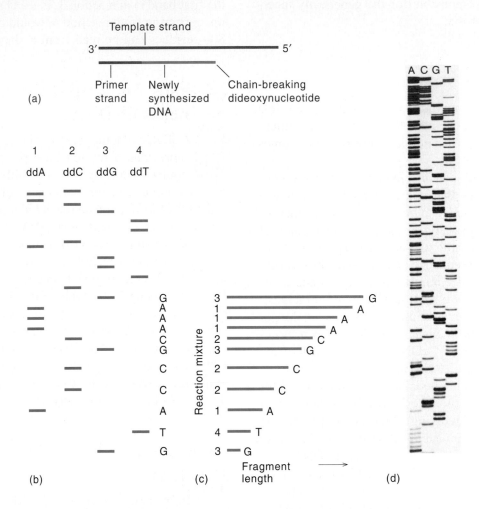

These steps are illustrated in figure 27.2. Each set of three steps comprises a cycle. The extension products of one primer provide a template for the other primer in a subsequent cycle so that each successive cycle essentially doubles the amount of DNA. This results in the exponential accumulation of the specific target fragment by approximately $2^n$, where $n$ is the number of cycles. The specific target fragment is also referred to as the "short product" and is defined as the region between the 5′ ends of the extension primers. Each primer is physically incorporated into one strand of the short product.

Other products are also synthesized during the succession of cycles, such as the "long product" of indefinite length, which is derived from the template molecules. However, the amount of long product only increases arithmetically during each cycle of the amplification process because the quantity of original template remains constant.

At the end of the PCR process the short product is so overwhelmingly abundant compared with the long product that its purification is not required for most purposes.

## Figure 27.2

Steps in the polymerase chain reaction (PCR). The DNA to be amplified is denatured and annealed with two oligonucleotides that flank the region of interest. These oligonucleotides (or primers) are extended. Extension continues to the ends of the DNA strands. The products are again denatured and annealed to primers for a second round of extension. This process of denaturation, annealing, and primer extension is repeated many times. The primary product of the reaction is duplex DNA, bounded by the sequences of the primers. (From J. L. Marx, Multiplying genes by leaps and bounds, *Science* 240:1408–1410, June 10, 1988. Copyright 1988 by the AAAS. Reprinted by permission.)

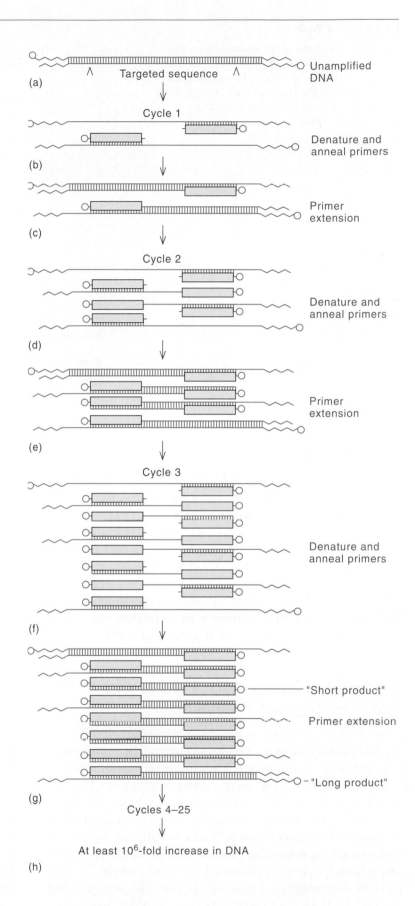

(a) Unamplified DNA

Targeted sequence

Cycle 1

(b) Denature and anneal primers

(c) Primer extension

Cycle 2

(d) Denature and anneal primers

(e) Primer extension

Cycle 3

(f) Denature and anneal primers

(g) "Short product"

Primer extension

"Long product"

Cycles 4–25

(h) At least $10^6$-fold increase in DNA

## Figure 27.3

Cleavage map of the SV40 genome. The zero point of the map is the unique *Eco*R1 site. For clarity, the circular genome is shown opened at the R1 site, and the cleavage sites (and resulting fragments) for each restriction enzyme are indicated on a separate line.

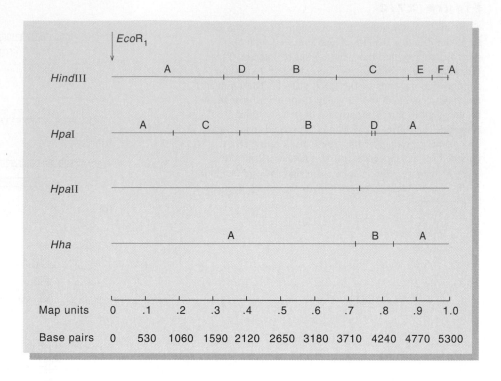

## DNA Cloning

The second method for DNA amplification is more complicated than PCR, but it has several advantages. DNA to be amplified by cloning is linked to a plasmid or a virus that can be replicated indefinitely in the appropriate host cell. After amplification the DNA of interest can be cut from the plasmid or virus and reisolated by gel electrophoresis. Cloning is not only useful for amplifying a segment of DNA, it can be adapted to the isolation of a DNA segment of interest from a large mixture such as is obtained from the isolation of the entire genome.

### *Restriction Enzymes Are Used to Cut DNA into Well-Defined Fragments*

Most of the enzymes that are absolutely essential for cloning were discussed in the previous chapter. The most important enzymes that have not been discussed yet are the restriction enzymes. Systematic cleavages of duplex DNA at specific sites requires restriction enzymes. Each species of bacteria harbors a unique restriction enzyme, and hundreds of restriction enzymes with different specificities have been isolated, giving researchers a great deal of choice as to how and where DNA is cut. Some of the most commonly used restriction enzymes and their recognition sites are indicated in table 27.1. Most of these enzymes recognize a sequence of either four or six contiguous base pairs. The cleavage sites are situated so that a blunt-ended or staggered-ended DNA results from the cleavage reaction. As a rule the recognition sites are located on an axis of symmetry so that the freshly cleaved segments have identical structures at their ends.

A viral genome cleaved exhaustively with a particular restriction enzyme usually yields several fragments. Some restriction enzyme cleavage sites for the 5,300 bp (5.3 kb) SV40 virus genome are shown in figure 27.3. The duplex fragments obtained after cleavage can be separated according to size by gel electrophoresis. Nondenaturing conditions are used so that the duplex strands stay together. The larger a fragment is, the slower it migrates on the gel. After electrophoresis for a time sufficient to separate the fragments, the gel is stained with a fluorescent dye such as ethidium bromide and viewed under long-wavelength ultraviolet light (long-wavelength UV is used because it does not damage the DNA). Individual fragments may be extracted from the gel for sequencing, PCR amplification, or cloning (described later on).

The problem of determining how a set of restriction fragments are normally connected is resolved by determining the sequences by a second set of fragments cut with a different restriction enzyme. The overlapping information obtained from the two sets of fragments permits a determination of the complete sequence of the intact genome. The

**Table 27.1**

Recognition Sequences and Cutting Sites of Selected Restriction Enzymes

| Enzyme | Recognition Sequences | Enzyme | Recognition Sequences |
|---|---|---|---|
| AluI | ↓<br>AGCT<br>TCGA<br>↑ | HpaII | ↓<br>CCGG<br>GGCC<br>↑ |
| BamHI | ↓<br>GGATCC<br>CCTAGG<br>↑ | KpaI | ↓<br>GGTACC<br>CCATGG<br>↑ |
| BglII | ↓<br>AGATCT<br>TCTAGA<br>↑ | MboI | ↓<br>GATC<br>CTAG<br>↑ |
| ClaI | ↓<br>ATCGAT<br>TAGCTA<br>↑ | PstI | ↓<br>CTGCAG<br>GACGTC<br>↑ |
| EcoRI | ↓<br>GAATTC<br>CTTAAG<br>↑ | PvuI | ↓<br>CGATCG<br>GCTAGC<br>↑ |
| HaeII | ↓<br>GGCC<br>CCGG<br>↑ | SalI | ↓<br>GTCGAC<br>CAGCTG<br>↑ |
| HindII | ↓<br>GTPyPuAC<br>CAPuPyTG<br>↑ | SmaI | ↓<br>CCCGGG<br>GGGCCC<br>↑ |
| HindIII | ↓<br>AAGCTT<br>TTCGAA<br>↑ | XmaI | ↓<br>CCCGGG<br>GGGCCC<br>↑ |

strategy of sequencing overlapping fragments is identical to that used in primary structure determination of proteins (see chapter 3).

## Plasmids Are Used to Clone Small Pieces of DNA

In a simple procedure for "DNA cloning," an autonomously replicating plasmid and insert DNA are cut with a restriction enzyme and then the pieces are annealed and covalently joined by the action of DNA ligase. The resulting recombinant molecules are then transfected into *E. coli,* where they replicate. When plasmid vectors are used, a population of permeabilized cells is bathed in the plasmid DNA containing the inserted DNA. Because only a small number of cells become transfected by this procedure, a way to se-

lect cells that carry the desired hybrid plasmids is needed.

A particularly useful plasmid vector for selecting transfected cells called pBR322 is itself a hybrid plasmid (fig. 27.4). This plasmid contains two genes, *amp*[r] and *tet*[r], which confer resistance to penicillin and tetracycline, respectively. *Pst*I restriction fragments of foreign DNA may be inserted into the unique *Pst*I restriction site on pBR322 (see fig. 27.4). This is done by digesting pBR322 with *Pst*I, mixing with the restriction fragments to be cloned at low temperatures to permit annealing to take place between the two DNAs, and finally ligating the annealed fragments with DNA ligase. The product contains some of the original pBR322 and some pBR322 with inserted foreign DNA. When this mixture is used in transfection, most cells are not transfected, some are transfected with pBR322, and some are transfected with the desired hybrid plasmid. The three types of cells may be readily distinguished by their drug-

## Figure 27.4

Structure of the pBR322 plasmid (*a*) and construction of a hybrid plasmid containing the pBR322 vector and a segment of foreign DNA (*b*). For pBR322 the unique sites for various restriction enzymes are indicated. Also indicated are the locations of the tetracycline (*tet$^r$*) and the ampicillin (*amp$^r$*) resistance genes and the origin for DNA replication. The hybrid plasmid is constructed by treating the plasmid and the foreign DNA with the *Pst*I restriction enzyme and mixing the two DNAs together in the presence of DNA ligase.

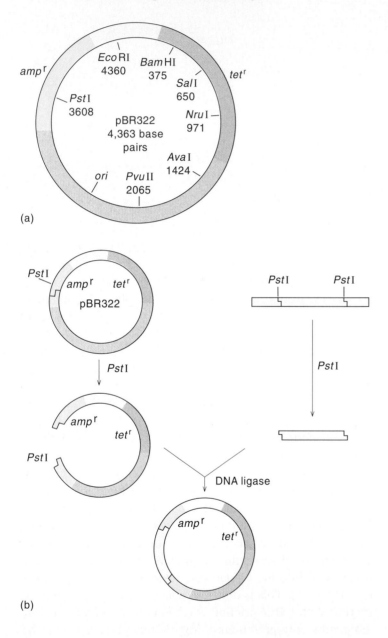

(a)

(b)

## Figure 27.5

Application of the replica plating technique to the detection of hybrid plasmid-containing cells. About $10^7$ bacteria are spread on a plate. After overnight growth the plate (master plate) appears as a uniform "lawn" of bacteria, but in reality it consists of very small colonies that have merged to give a uniform appearance. A piece of velvet is lightly pressed against the surface of this lawn, and some cells stick to the velvet. Several essentially identical impressions of the lawn are transferred to fresh plates, which contain normal medium or normal medium supplemented with antibiotics. The results on the replica plates after overnight growth are indicated. The plate in normal medium again gives rise to a lawn of cells as virtually all of the cells transferred grow into colonies. The plates containing antibiotics only give rise to a few colonies, each of which is derived from a single cell that carries the plasmid-conferred drug resistance(s).

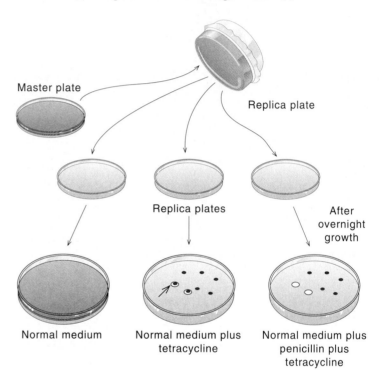

plating" (fig. 27.5). The first step when using this approach is to spread a large population of treated bacteria on an agarose plate containing growth medium. Within 12 h a seemingly homogenous "lawn" of cells develops on the surface of the agarose. Actually, the lawn results from the growth of many microcolonies to the point of confluency. At this point a piece of velvet is lightly pressed against the surface of the plate, and this impression is transferred to other agarose plates containing growth medium with tetracycline or penicillin plus tetracycline. Only the transfected cells produce colonies on the plates containing the antibiotics, and because of their small number, each of these gives rise to readily detectable clones. The clones present on the tetracycline-containing plates, which are missing on the penicillin plus tetracycline plates, most likely contain the

resistant properties. Normal cells do not grow in the presence of tetracycline or penicillin. Transfected cells with the DNA inserted in the plasmid are tetracycline-resistant but penicillin-sensitive, because the insert has disrupted the *amp$^r$* gene. Cells containing the desired plasmids can be distinguished from those containing pBR322 by "replica

## Figure 27.6

Electrophoretogram of restriction enzyme digests of pBR322 and pBR322 with a DNA insert at the *Pst*I site. The insert is assumed to have no internal *Bam*HI restriction sites. In channels A and B the pBR322 is predigested with *Pst*I and *Bam*HI, respectively. The resulting DNA migrates with the same mobility because the plasmid has one site for each of these enzymes and therefore has the same molecular weight. In C and D the hybrid plasmid containing a DNA insert is treated with *Bam*HI and *Pst*I, respectively. In C the hybrid plasmid has been linearized by one cut at the *Bam*HI site in the *tet*^r gene. It runs more slowly than the pBR322 because it is larger due to the DNA insert. In D the plasmid cuts at two *Pst*I sites located between the pBR322 sequences and the insert sequences. Consequently, one segment migrates at the rate of a linearized pBR322 plasmid. The other segment, also linearized, migrates at a rate characteristic of the size of the DNA insert. The electrophoresis is run from left to right; fragments are stained with ethidium bromide and photographed with UV light.

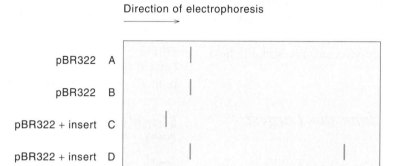

## Figure 27.7

The nutrient agar plate contains a continuous lawn of *E. coli* bacteria except for circular clearings that represent phage plaques. Each plaque was originally derived from a single phage infecting a single *E. coli* bacterium. After infection the phage multiplies, ultimately producing about 100 mature viruses. The phages also produce an enzyme that causes the harboring cell to lyse. When this happens the phages are released, and each of them infects a neighboring cell and goes through the same infectious cycle. The process continues. Each cycle takes about 30 min. Eventually a visible clearing can be seen on the plate.

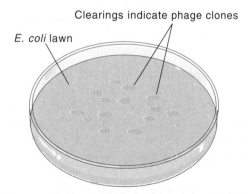

desired hybrid plasmids (see fig. 27.5). These clones are usually plucked from the tetracycline plates and retested to eliminate any uncertainty about the original drug testing. Once this is confirmed, the appropriate plasmid-containing cells are grown in liquid culture. After a moderate density of growth is achieved, the plasmid DNA is selectively amplified by overnight growth. Plasmid DNA replication continues for several hours, until each cell contains 1,000 to 2,000 copies of the small circular plasmid DNA. This DNA is readily separable from the host DNA and can be characterized by its rate of migration on gel electrophoresis (fig. 27.6) or other more specific tests to determine if it contains the inserted DNA sequence. If desired, the inserted sequence may be removed from the plasmid vector by digestion with *Pst*I, the restriction enzyme used in the initial construction of the hybrid plasmid. The cleaved fragments can be separated readily by gel electrophoresis. Many plasmid vectors, other than pBR322, have specific advantages for other purposes.

## *Bacteriophage λ Vectors Are Useful for Cloning Larger DNA Segments*

Bacteriophage λ possesses a number of advantages as a cloning vector. DNA fragments as large as 24 kb can be propagated using such vectors. The primary pool of clones can be amplified by limited phage growth as plaques, and the entire collection of phage clones (recognized as clear plaques) can be stored for long periods in a small volume (fig. 27.7).

Because it does not accommodate molecules of DNA that are much longer than the viral genome, the use of λ as a vector for cloning substantial DNA fragments requires the removal of a significant portion of the viral DNA beforehand. Fortunately, the central third of the genome contains genes that are not essential for phage production and can therefore be deleted.

Bacteriophage λ vectors that accommodate foreign DNA fragments generated by a variety of restriction endonucleases have been constructed. The recombinant DNA molecules that incorporate some of these vectors can be introduced directly into *E. coli* by transfection. Alternatively, recombinant DNA molecules can be packaged into phage particles and subsequently infected into suitable host cells.

## Cosmids Are Used to Clone the Largest Segments of DNA[a]

Although plasmids and bacteriophage λ are both highly useful vectors, the size of the DNA fragments that can be cloned in them is limited. With plasmids, the larger the fragment of foreign DNA inserted, the lower the efficiency of ligation and transfection, making the cloning of DNA fragments larger than 15 kp experimentally difficult. In λ vectors, the length of the nonessential region of λ DNA limits fragment size to 24 kb or less. Also, the original λ vectors do not allow propagation of viable bacterial cells that carry the inserted DNA fragment; the insert is propagated as part of a virus that lyses the cell.

Cosmids were developed as vectors for cloning large DNA fragments. The first part of their name, "cos," comes from the fact that cosmids contain the cohesive ends, or *cos,* sites of normal λ. These ends are essential for packaging the DNA into λ phage heads. The last part of their name, "mid," indicates that cosmids carry a plasmid origin of replication like the one found in the pBR322 plasmid. Such cosmids can be used for cloning in the same way as any other plasmid vector. However, because cosmids also contain the *cos* sites, cosmid DNA along with an inserted DNA fragment can be packaged as a λ phage. The result after packaging is a defective but nevertheless infectious phage particle. Once the cosmid and the inserted DNA fragment are introduced by infection into a λ-sensitive cell, the plasmid replicates. Since cosmids lack the entire bacteriophage genome except for the region adjacent to the *cos* sites, these vectors can propagate exogenously derived DNA fragments of up to 40–50 kb in length.

[a] Yeast artificial chromosomes have been developed that can clone hundreds of kilobases of DNA (see Watson, Gilman, Wilkowski, and Zoller reference).

## Shuttle Vectors Can Be Cloned into Cells of Different Species

Vectors that include replication systems derived from more than one host species are known as shuttle vectors. Such vectors commonly include a replication system able to function in *E. coli* and one that works in a second host, which may be bacterial or eukaryotic. Initial cloning and amplification of the DNA segment to be studied is often carried out in *E. coli* because it is easier to make large quantities when culturing in *E. coli*. The recombinant DNA molecule, consisting of the "bifunctional vector" plus the cloned segment of DNA, is then introduced into the second host, where the purpose is usually to measure the expression of the genes carried by the vector. Shuttle vectors that can replicate in both *E. coli* and yeast are the most common.

## Constructing a "Library"

Cloning can involve a single vector-linked DNA fragment or a collection of independently isolated vector-linked DNA fragments derived from a single organism. Such a collection is termed a "library" and may serve as the source of well-defined sequences from a given organism. Each clone of a library harbors a particular DNA segment from the desired organism. Within the entire library a sequence may be repeated, but other sequences may be missing. The ideal library, which can only be approached, represents all of the sequences with the smallest possible number of clones.

A library from the same cell or organism can be prepared in two ways. The genome may be fragmented and ligated to the appropriate vector to produce a genomic DNA library. An alternative approach is to construct a cDNA library in which the DNA fragments to be cloned are obtained by reverse transcription from the cellular RNA. Each of these libraries has advantages and disadvantages, and for a specific purpose, one library is usually preferred over the other.

The vast majority of DNAs within a library are uncharacterized. As a rule, the task of finding the desired genes or sequences within a library greatly exceeds the task of constructing the library.

## A Genomic DNA Library Contains Clones with Different Genomic Fragments

A major concern in constructing a genomic DNA library is to maximize the probability that all segments of the genome are represented. If the genomic DNA is prepared by cutting

## Table 27.2

Theoretical Number of Clones Required to Fully Represent the Entire Genome of Various Organisms

| Size of Cloned DNA Fragment (bp) | Genome Size (bp) | | |
|---|---|---|---|
| | $2 \times 10^6$ (e.g., bacteria) | $2 \times 10^7$ (e.g., fungi) | $3 \times 10^9$ (e.g., mammals) |
| $5 \times 10^3$ | 400 | 4,000 | 600,000 |
| $10 \times 10^3$ | 200 | 2,000 | 300,000 |
| $20 \times 10^3$ | 100 | 1,000 | 150,000 |
| $40 \times 10^3$ | 50 | 500 | 75,000 |

with a restriction enzyme, an added concern is that the enzyme cleaves genes of interest at one or more sites. To increase the likelihood of isolating desired genes in one piece, different restriction enzymes can be used on different parallel preparations. But even if the genes of interest are not cut by the enzyme(s) chosen, the DNA fragments produced may be inconveniently small to work with. An enzyme that recognizes a sequence of six bases (a six-cutter) gives an average fragment size of 4,096 bp[b], which is a reasonable size for making a plasmid library but much smaller than the size desirable for cloning in λ or a cosmid vector. Therefore, when large, randomly generated fragments are desired, the method of choice usually entails making an incomplete digest with a four-cutter restriction enzyme, which produces overlapping ends that can be readily cloned into the chosen vector as described earlier. The extent of digestion is controlled so that cleavage occurs at only some of the restriction enzyme recognition sites and the average size of the fragments produced is in the desired range. The conditions used thus depend on whether the product is going to be cloned in a plasmid, a λ phage, or a cosmid vector. Table 27.2 gives the minimum number of clones (i.e., the size of the library) required to fully represent the entire genome in a genomic DNA library, as a function of the average size of the cloned fragments and the size of the genome. Since DNA fragments in a population are cloned on a random basis, the chance of finding a given single-copy gene in a library of the indicated size is 50%. A clone bank should be 3 to 10 times the minimum size to give a high probability that a particular segment is represented.

[b] $(\frac{1}{4})^6 = 1/4,096$

## A cDNA Library Contains Clones Reflecting the mRNA Sequences

A cDNA library consists of a collection of clones that contain DNA copies of the cellular or organismic RNA. If the RNA is obtained from a differentiated multicellular organism, then the library varies in composition according to the type of cell used as the RNA source and to the physiological state of the cell. This variation is a reflection of the relative abundances of particular mRNAs made by different cell types. If a cDNA species corresponding to a particular gene product is desired, it is often possible to select a cell type suspected to synthesize a large amount of the corresponding mRNA or mRNA-related protein. Thus, pituitary cells can be used if cDNA encoding growth hormone is desired, whereas liver cells can be used if a serum albumin cDNA is the goal. The mRNAs present in low amounts clearly require the screening of a larger library than the mRNAs present in medium or high abundance.

Once the crude mRNA fraction has been isolated from the chosen cells or tissue, it is converted to duplex DNA molecules with the help of reverse transcriptase. This duplex DNA does not have "sticky ends" for insertion into a vector. For this purpose DNA linkers are attached to the ends. Linkers are synthetic single-stranded oligonucleotide segments (6, 8, 10, or 12 bases in length) that self-associate to form symmetrical, blunt-ended, double-stranded molecules containing the recognition sequence for a particular restriction enzyme. Figure 27.8 shows an eight-base linker (CCTGCAGG) containing a PstI recognition site. This linker self-associates to produce an eight-base, blunt-ended, duplex structure that adds to the double-stranded cDNA in the presence of T4 ligase. The resulting product is treated

**Figure 27.8**

Insertion of cDNA into pBR322 plasmid by
the linker method. The strategy here is to
open up the plasmid with a restriction enzyme
that makes staggered cuts and to attach
linkers that contain the same recognition site
to the cDNA. After the linkers are attached to
the cDNA, the duplex is treated with the
same restriction enyzme (*Pst*I) to expose the
overhangs. The two DNAs are mixed together
and ligated. After transfection, cells
containing the hybrid plasmids are recognized
by tetracycline resistance and ampicillin
sensitivity. Identification of the insert is
discussed in the text.

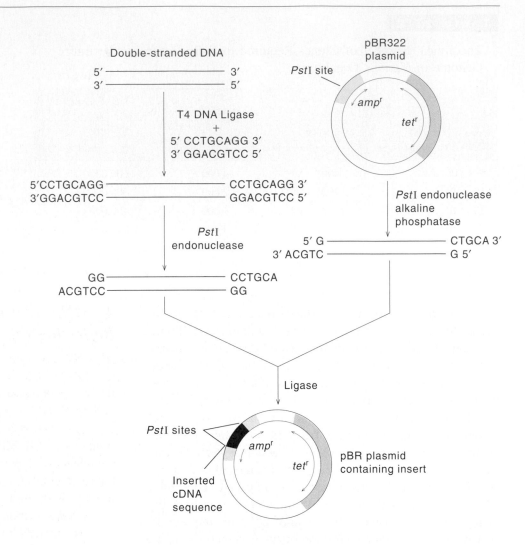

## Numerous Approaches Can Be Used to Pick the Correct Clone from a Library

A library can contain thousands or even tens of thousands of different kinds of clones (see table 27.2), making it a challenging endeavor to isolate a clone with the DNA of interest. Most currently used procedures for screening large numbers of colonies for plasmids or phage that contain specific DNA inserts are variants of the colony hybridization method developed by Grunstein and Hogness. This procedure makes use of a specific radioactive probe that contains some sequences complementary to those in the DNA of interest. The colonies to be screened are first grown on agar petri plates (fig. 27.9). A replica of each plate is made on another agar plate, which is stored for reference. A replica is

with *Pst*I to produce the characteristic 3′ overhang. The plasmid, linearized with *Pst*I, and the two DNAs are mixed and reacted with ligase to produce plasmid with the insert.

also made on a nitrocellulose filter. The colonies formed on the filter are lysed and the contents denatured simultaneously by treatment with sodium hydroxide. After heating, the denatured DNA is fixed on the filter at each site where a colony was located. The DNA on the filter is then hybridized with a radioactively labeled nucleic acid probe complementary to the specific DNA sequence to be selected. The presence of hybridized probe at sites occupied by DNA derived from colonies that include the DNA fragment of interest is detected by autoradiography. The colony whose DNA hybridizes with the nucleic acid probe can then be picked from the reference plate, which contains a viable bacterial colony at a corresponding location.

## Cloning in Systems Other than Escherichia coli

Despite the success and broad applications of *E. coli* cloning systems, instances occur in which gene products cannot be made in this bacterium. Either they are not synthesized in

## Figure 27.9

Colony hybridization procedure used to identify bacterial clones harboring a plasmid containing a specific DNA. Step 1: Replica-plate the colonies containing plasmids onto nitrocellulose paper. Step 2: Lyse cells with NaOH and fix denatured DNA to paper. Step 3: Hybridize to $^{32}$P-labeled DNA carrying the desired sequence and autoradiograph the product. Locations of desired DNA should be emphasized in the autoradiograph. Clones carrying desired plasmids (circled) may then be isolated from a corresponding agar replica plate carrying untreated colonies.

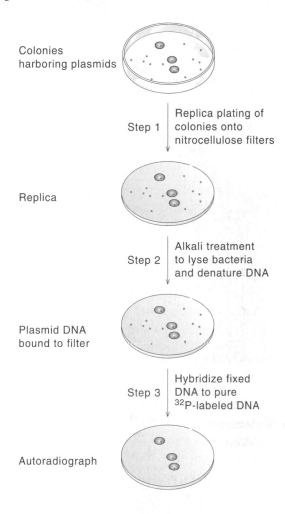

Colonies harboring plasmids

Step 1 — Replica plating of colonies onto nitrocellulose filters

Replica

Step 2 — Alkali treatment to lyse bacteria and denature DNA

Plasmid DNA bound to filter

Step 3 — Hybridize fixed DNA to pure $^{32}$P-labeled DNA

Autoradiograph

their entirety, or they are rapidly broken down after synthesis. In addition, the study of certain processes indigenous to other species (e.g., photosynthesis, antibiotic production) often requires the use of a host bacterial species that naturally carries out the process, which may exclude *E. coli.*

Effective cloning systems are available for a variety of bacterial hosts, including *Bacillus subtilis, Streptomyces spp.,* and *Agrobacter tumefaciens.* Cloning systems have also been developed for eukaryotic hosts such as the yeast *Saccharomyces cerevisiae,* mammalian cells in tissue culture, and plant cells.

## Site-Directed Mutagenesis Permits the Restructuring of Existing Genes

By combining different procedures of manipulation, it is now possible to make discrete changes in genes. This technique, called site-directed mutagenesis, is one of the most important in modern genetics and biochemistry. The first site-directed mutagenesis studies were carried out by David Shortle and Daniel Nathans in 1978 with the help of the mutagen sodium bisulfite, which deaminates C residues so that they become converted into U residues.

Directed mutagenesis as it is practiced today is based on the chemical synthesis of a deoxyoligonucleotide that contains discrete changes in its sequence from that normally observed in the genome under investigation. These changes may be single-base or multibase; they may involve base changes, base deletions, or base additions.

Many variations of site-directed mutagenesis exist. One can start out with a circular, single-stranded DNA and anneal it to a synthetic primer DNA carrying the desired changes (fig. 27.10). This primer can be extended, and the resulting product can be transfected. Finally, one selects clones of cells containing the plasmid with the desired changes.

The polymerase chain reaction (PCR) will probably be the method of choice in the future for carrying out site-directed mutagenesis. In its most general form the use of PCR for this purpose requires a piece of duplex starting DNA, two outside flanking primers that are perfect complements to segments of opposing strands, and two complementary primers with the desired changes in their sequence as diagrammed in figure 27.11. The steps followed parallel the steps of PCR amplification.

PCR amplification can be coupled with classical cloning methods using cloning vectors. For this purpose the PCR product should contain restriction sites at its ends that are suitable for cloning. Thus, PCR amplification and cloning need not be thought of as alternatives for particular purposes but as complementing each other to give a greater variety of approaches.

## Recombinant DNA Techniques Were Used to Characterize the Globin Gene Family

The human globin family is a paradigm for studying differential gene activity during development and the molecular basis of genetic disorders in gene expression. Hemoglobin is a tetramer containing two $\alpha$-like and two $\beta$-like subunits (see chapter 5). These proteins are en-

## Figure 27.10

Scheme for oligonucleotide-directed mutagenesis of double-stranded circular plasmid DNA. Supercoiled plasmid circles are nicked in one strand and rendered partially single-stranded by treatment with exonuclease. The gapped circles are hybridized with a homologous oligodeoxynucleotide carrying, by design, some mismatches. *In vitro* DNA synthesis, primed in part by the oligodeoxynucleotide, leads to heteroduplex plasmid circles. (Source: After G. Dalbadie-McFarland, L. W. Cohen, A. D. Riggs, C. Morin, K. Itakura, and J. H. Richards, Oligonucleotide-directed mutagenesis as a general and powerful method for studies of protein function, *Proc. Natl. Acad. Sci. USA* 79:6408–6412, 1982.)

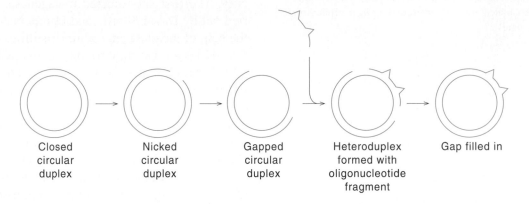

Closed circular duplex → Nicked circular duplex → Gapped circular duplex → Heteroduplex formed with oligonucleotide fragment → Gap filled in

## Figure 27.11

Illustration of a general method of mutagenesis using PCR. Primers are represented as short lines with arrowheads pointing in the 3′ direction. The bump in primers 2 and 3 and their products represent a mismatched base, a deliberate alteration in base sequence from that present in the starting DNA. Of the four major products resulting from step 3 only D is extendable by DNA polymerase.

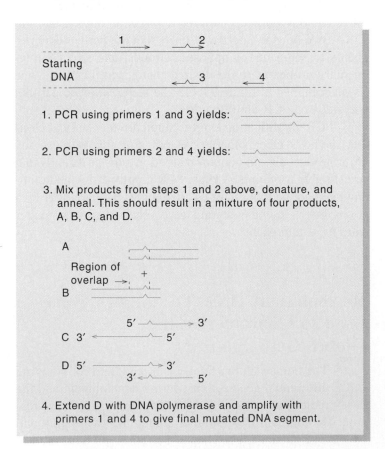

1. PCR using primers 1 and 3 yields:

2. PCR using primers 2 and 4 yields:

3. Mix products from steps 1 and 2 above, denature, and anneal. This should result in a mixture of four products, A, B, C, and D.

4. Extend D with DNA polymerase and amplify with primers 1 and 4 to give final mutated DNA segment.

coded by a small number of genes that are expressed sequentially during development. The information summarized in figure 27.12 indicates that the $\alpha$-like and $\beta$-like globin gene families have coordinated programs for expression: Two switches exist for the $\beta$-like genes, whereas a single switch results in activation of adult $\alpha$-globin production early in fetal life. A combination of classical and recombinant DNA techniques has been used to show that the $\alpha$-like genes are located in a single cluster on chromosome 16, and the $\beta$-genes are located in a single cluster on chromosome 11 (fig. 27.13). We focus on the contributions to our understanding of the globin genes that have resulted from investigations using the recombinant DNA approach.

## *DNA Sequence Differences Were Used to Detect Defective Hemoglobin Genes*

All of the hemoglobin genes are represented by two or more alleles within the human population; the genes are said to be polymorphic. This polymorphism frequently shows up in readily detectable phenotypes when the differences occur in vital areas of the polypeptide chains. Polymorphisms show up in the DNA even more frequently because the DNA contains sequence differences in both the coding and the noncoding regions of a gene, and these differences can be detected even when no visible effect is apparent in the organism. Because of their frequency and ease of detection by recombinant DNA methods, DNA polymorphisms have become extremely useful in mapping the human genome.

## Figure 27.12

Changes in types of hemoglobin observed in early development. A single switch in gene expression is observed for $\alpha$-like chains. Two switches in gene expression are observed for $\beta$-like chains. The corresponding tetrameric hemoglobin molecules observed at different stages in development are also indicated.

| | Early embryo | 8-week gestation | 8-month gestation |
|---|---|---|---|
| $\alpha$-like chains | $\zeta$ ⟶ | $\alpha$ ⟶ | |
| $\beta$-like chains | $\varepsilon$ ⟶ | $\gamma$ ($\gamma_G$ and $\gamma_A$) ⟶ | $\beta$ and $\delta$ ⟶ |
| Tetramers | $\zeta_2\varepsilon_2$ ⟶ | $\alpha_2\varepsilon_2$, $\zeta_2\gamma_2$ ⟶ $\alpha_2\gamma_2$ ⟶ | $\alpha_2\beta_2$, $\alpha_2\delta_2$ ⟶ |

## Figure 27.13

The chromosomal localization and genomic organization of the human globin genes. The $\alpha$- and $\beta$-globin gene complexes are positioned on chromosomes 16 and 11, respectively. For each complex, the arrangement of genes on the chromosome is depicted above, and the general structure of the major gene is shown below, together with the location of the intervening sequences, or introns (IVS), and codon numbers. Coding regions are shown by solid boxes and IVS regions by open boxes. Genes with the $\psi$ symbol in front are called pseudogenes because they are sequence-related but not expressed.

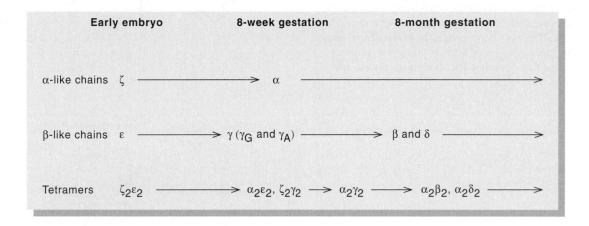

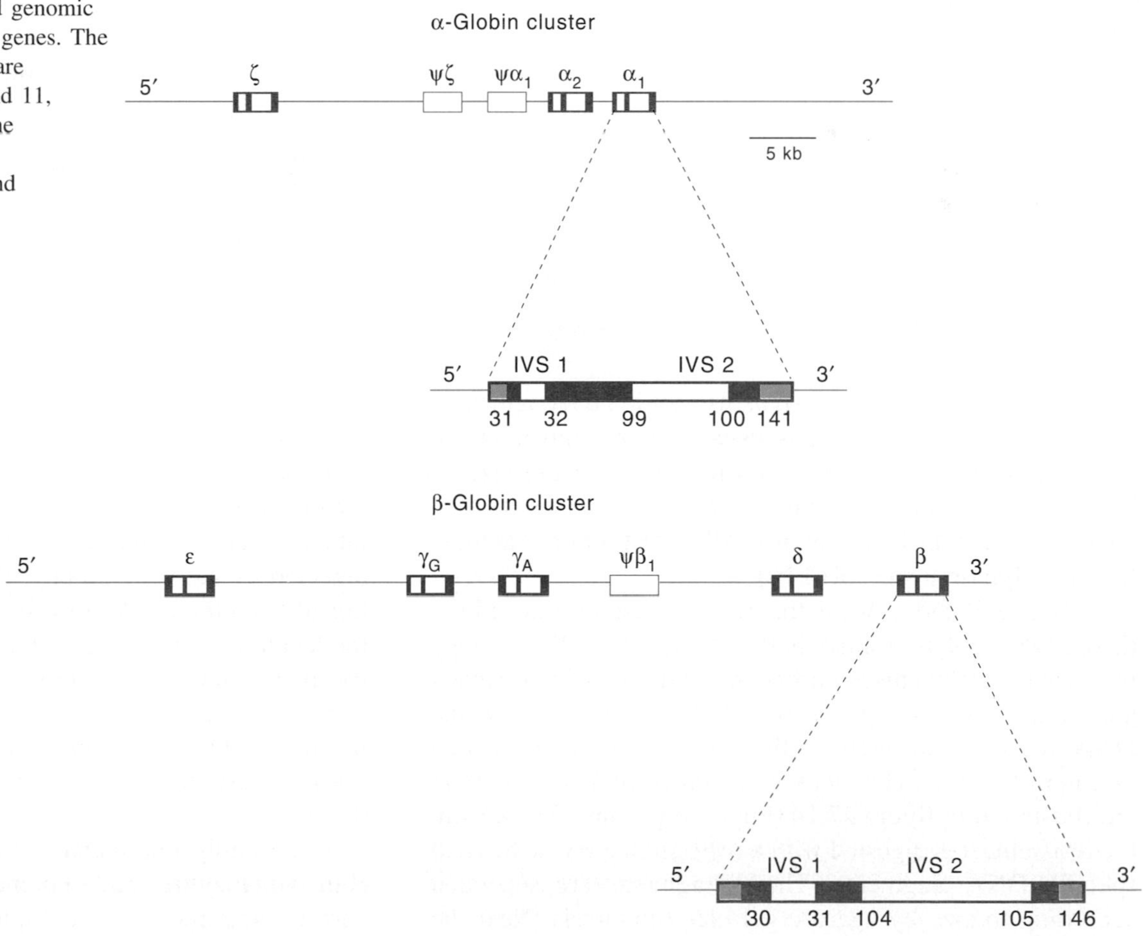

**Figure 27.14**

The steps involved in assaying by Southern blotting. The DNA to be analyzed is digested with a restriction enzyme (1). The resulting fragments are electrophoresed on an agarose gel (2). The DNA fragments on the gel are transferred to a cellulose nitrate sheet by placing the cellulose nitrate sheet next to the gel and passing solvent through the gel into the sheet. Flow of the solvent is maintained by blotting the far side of the cellulose nitrate sheet with paper towels. The DNA, first denatured with alkali, flows with the solvent but gets stuck in the sheet (3). The sheet is hybridized to radioactively labeled DNA containing the gene sequence of interest (4). The hybridized sheet is autoradiographed to determine the location of the labeled restriction fragment on the gel (5).

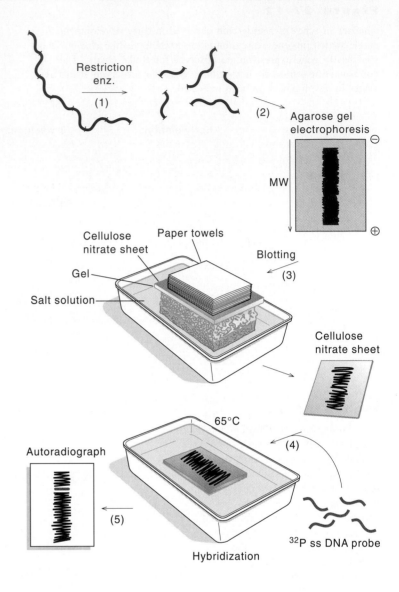

Whereas DNA polymorphisms should be recognizable by sequence differences, it is usually more convenient to detect these polymorphisms by differences in the size of DNA fragments obtained with restriction enzymes. Differences observed in this way are called restriction fragment length polymorphisms (RFLPs).

Kan and Dozy were the first to discover an allele-linked DNA polymorphism in the globin genes. With it they predicted which fetuses carried normal and which carried abnormal sickle-cell genes for $\beta$-globin. To analyze the DNA for these differences they used a technique called Southern blotting. The steps involved in Southern blotting are illustrated in figure 27.14. First the genomic DNA from the test subject is digested with a restriction enzyme to yield specific DNA fragments. These fragments are separated according to size by agarose gel electrophoresis. Next the

DNA is denatured and transferred from the agarose gel to a cellulose nitrate sheet. The DNA firmly bound to the sheet is hybridized with a radioactively labeled DNA probe, which carries some of the sequences of interest. The radiolabel, which hybridizes to specific regions of the sheet, is detected by autoradiography. By comparing the results obtained from the DNA of different individuals one can see if the labeled DNAs move with the same or a different mobility. If they move differently, there must be a RFLP difference between the individuals. Detection of an RFLP by this means usually depends on the restriction enzyme used in the initial digestion. Some enzymes show a difference; others do not.

To apply this technique to their hemoglobin studies Kan and Dozy first had to prepare a DNA probe that carried specific sequences in the $\beta$-globin gene. This was a rela-

# Figure 27.15

Inheritance pattern of an RFLP associated with sickle-cell disease. Humans carry two alleles for the same gene, and each offspring inherits one allele from each of its parents in an entirely random fashion. Normal individuals are homozygous for normal *Hb* alleles; individuals with sickle-cell trait are heterozygous, with the one normal *Hb* allele and one *Hb*^s allele; and individuals with sickle-cell disease are homozygous for the *Hb*^s allele. At the top (*a*) we see a three-generation pedigree analysis for a family that carries both the normal and the sickle-cell gene for β-globin. Males are represented by squares and females by circles. A purple circle or square indicates an individual who is homozygous normal. A half-filled circle or square (red/purple) indicates a heterozygous individual with sickle-cell trait. A filled circle or square (red) indicates a homozygous individual

with sickle-cell disease. In (*a*) both sets of grandparents produce a heterozygous individual with sickle-cell trait. Because one of the grandparents is homozygous normal and the other is heterozygous, there is a 50% chance that the grandparent mating will give rise to a heterozygous offspring as shown and also a 50% chance that they will have normal offspring (not shown). The two heterozygous parents have an increased chance of having abnormal offspring because in this mating each parent carries one abnormal gene or allele. There is a 25% chance of a homozygous sickle-cell anemic offspring, a 50% chance of an abnormal offspring with sickle-cell trait, and a 25% chance of a normal offspring. Below the pedigree chart is the electrophoretic pattern of a *Hpa*I digest probed with β-globin cDNA by the Southern blotting technique (*b*). At the bottom we see an interpretation of the normal and abnormal DNAs (*c*).

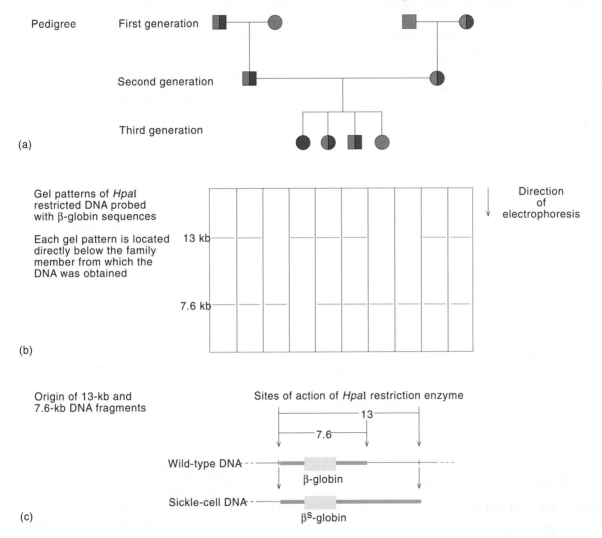

tively simple task because reticulocytes, which contain vast quantities of hemoglobin, are also greatly enriched in the β-globin mRNA. From the purified β-globin mRNA isolated from reticulocytes, a cDNA probe was made with the help of reverse transcriptase. Using a *Hpa*I restriction enzyme digest of the total genomic DNA in conjunction with

Southern blotting, Kan and Dozy showed that the normal β-globin gene was contained within a 7.6-kb *Hpa*I restriction fragment, whereas the β-globin gene of sickle-cell anemia, *Hb*^s, was contained within a 13-kb fragment (fig. 27.15). Further analysis showed that the RFLP resulted from a *Hpa*I restriction site 5 kb to the 3' side of the β-globin gene, that was present in the normal case and absent in $Hb^s$. Subsequent analysis showed sequence differences within the coding regions. One may wonder why the RFLP outside the coding region was so commonly associated with the abnormal gene. A possible explanation for this is that in the distant past a mutation occurred which resulted in the 7.6-kb type and a few 13-kb types before introduction of the sickle-cell gene mutation. After the $Hb^s$ mutation was introduced into the 13-kb type, it became greatly expanded because of the selective advantages of this gene in heterozygotes. In this connection it should be noted that heterozygotes carrying one normal gene and one sickle-cell gene fare far better when infected by malaria. As a result in areas where malaria is prevalent the heterozygote has a selective advantage over the normal homozygote.

Knowledge of this and other polymorphisms has been used for pre- or postnatal diagnosis of the sickle-cell gene. Such information can be of great practical value in genetic counseling. Incidentally, it is now possible to diagnose sickle-cell disease (which occurs in individuals that are homozygous for the sickle-cell gene) with greater certainty because the point mutation leading to the defect in the coding region itself produces a recognizable RFLP.

## The β-Globin cDNA Probe Was Used to Characterize the Normal β-Globin Gene

Detailed mapping with DNA probes was first successfully executed on the human β-globin gene. All members of a human genomic library that annealed to the radioactive cDNA probe for the β-globin gene were isolated, and each of these was sequenced. This analysis resulted in a complete description of the β-globin gene (see fig. 27.13). The β-globin gene is appreciably longer than the β-globin mRNA; in addition to containing regions that are present in the final mRNA the gene contains two "intervening" regions not represented in the mRNA sequences. We have more to say about the significance of these intervening sequences (IVS, also called introns) in the next chapter.

## Chromosome Walking Permitted Identification and Isolation of the Regions around the Adult β-Globin Genes

The original cDNA probe carrying the β-globin mRNA sequences could only detect members in the genomic library that contained sequences homologous to those present in the probe. To explore the region flanking the β-globin gene the genomic library was probed further. For this purpose members of the genomic library that hybridized with the original

cDNA were themselves converted into radioactive probes, and these were used to locate additional members of the library that contained sequences flanking the β-globin gene. A cyclic repetition of this process resulted in a gradual extension of sequence information in and around the β-globin gene. Using the library in this manner to extend the map is known as chromosome walking (fig. 27.16). A parallel approach was used to extend the map around the adult α-globin gene (see fig. 27.13). It can be seen that in both cases several genes occur in a cluster for each of the protein types. Most of these can be correlated with the genes that are expressed at different times during development. In addition, genes, called pseudogenes, occur that have strong sequence similarities to known genes but are never expressed. It is not clear if these pseudogenes have a function or simply represent evolutionary "junk," which has not been removed.

## Walking and Jumping Were Both Used to Map the Cystic Fibrosis Gene

By combining the linkage information obtained from RFLP mapping with other DNA manipulation techniques, it has been possible to locate genes causing serious genetic disorders even when these genes are only known from their inheritance patterns. The list of serious disorders that can be linked to single genetic loci is growing. It includes Huntington's disease, Duchenne's muscular dystrophy, polycystic kidney disease, cystic fibrosis, chronic granulomatous disease, peripheral neurofibromatosis, central neurofibromatosis, familial polyposis coli, and multiple endocrine neoplasia. One of the most spectacular achievements has been the determination of the gene causing cystic fibrosis (CF).

Cystic fibrosis is the most common serious genetic disorder in caucasian populations. The major clinical symptoms include chronic pulmonary disease, pancreatic exocrine insufficiency, and an increase in the concentration of sweat electrolyte. Bearers of this disease are readily diagnosed and often die of congestive lung complications before age 30. Pedigree analysis shows that, as in the case of sickle-cell anemia, a single gene inherited in autosomal recessive fashion results in the disease syndrome. The frequency of the disease is 1 in 2,000, from which it may be calculated that the carrier frequency is about 1 in 20 (the frequency of the heterozygote for a rare allele is twice the square root of the frequency of the homozygote). By classical genetic analysis the CF locus has been assigned to the long arm of chromosome 7 near the *met* locus. A map of the region containing the *met* locus and the CF gene is shown in figure 27.17.

In many genetic disorders, cystic fibrosis included, the analysis begins before the responsible gene and its protein

## Figure 27.16

The linkage map of the human β-globin gene locus as shown by the structural analysis of overlapping λ genomic clones. Both λHβG1 and λHβG3 clones contained the entire β-globin gene. Other clones detected by "walking" led to the discovery of other β-globinlike genes. These included four genes that are expressed and two pseudogenes that are not expressed. The genomic segments of the clones isolated are shown together with the cleavage sites for the enzyme EcoR1. The numbers on the top line indicate the size of the fragments in kilobase pairs.

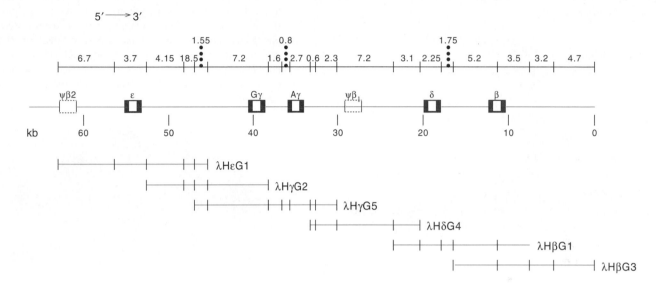

## Figure 27.17

Map of restriction fragment length polymorphisms (RFLPs) closely linked to the cystic fibrosis (CF) gene. The inverted triangle near the right-hand end indicates the location of the ΔF$_{508}$ mutation characteristic of most persons with cystic fibrosis disease. (Source: Adapted from B-S. Kerem, J. M. Rommens, J. A. Buchanan, D. Markiewicz, T. K. Cox, A. Chakravarti, M. Buchwald, and L-C. Tsui, Identification of the cystic fibrosis gene: Genetic analysis, *Science* 245:1075, 1989.)

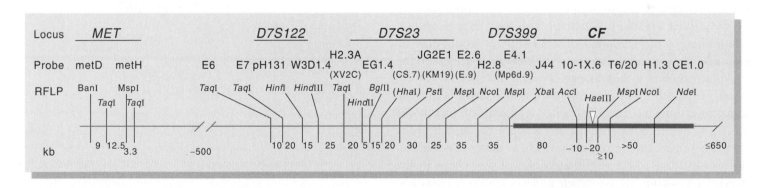

product are known. Thus, one cannot locate the gene directly, as in the case of the β-globin gene and then determine its approximate location in subsequent analysis. For genes such as those responsible for cystic fibrosis the approximate location is determined by conventional genetic analysis, and then one attempts to close in on the gene by a process called "reverse genetics." Conventional genetic mapping by recombination frequency is not practical with human genes below map distances of 1 centimorgan (cM)[c] because of the small number of test recombinant crosses that are ordinarily available for observation. Unfortunately, a centimorgan on the human genome is equivalent to a

[c] Genetic loci 1 cM apart recombine 1% of the time at meiosis.

## Figure 27.18

Jumping and walking to find the cystic fibrosis gene. Following each jump, the locus defined by the jump was used as a starting point for a chromosome walk. Each DNA segment so found was hybridized to a sweat gland cDNA library until a match was found. It seemed likely that the sweat gland cDNA library would have a good representation of the cystic fibrosis gene transcript because the disease involves the sweat glands.

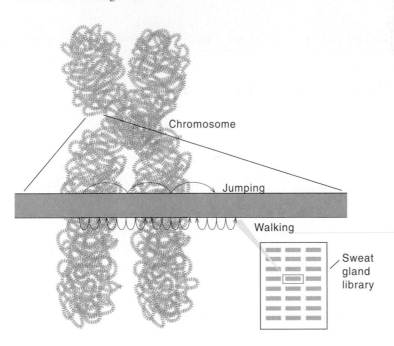

## Figure 27.19

Predicted structure of the CFTR protein. Cylinders represent membrane-spanning helical segments. The cytoplasmically oriented NBFs are shown as blue spheres with slots to indicate the points of entry of nucleotides. R represents the large polar domain, which is linked to two halves of the protein molecule. Charged amino acids are shown as small circles with the charge sign. Net charges on the internal and external loops joining the membrane cylinders and on regions of the NBFs are contained in open squares. Potential sites for phosphorylation by protein kinases A or C (PKA or PKC) and N-glycosylation (N-linked CHO) are indicated. (K = Lys; R = Arg; H = His; D = Asp; E = Glu.) (From J. R. Riordan et al., Identification of the cystic fibrosis gene, *Science* 245:1066, Setp. 8, 1989. Copyright 1989 by the AAAS. Reprinted by permission.)

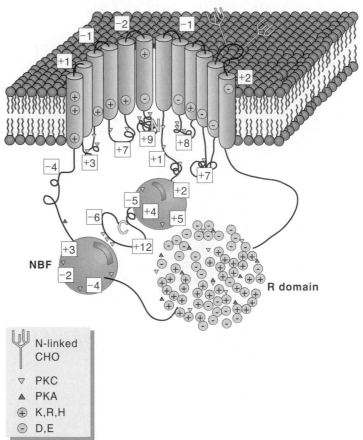

physical distance of about 1,000 kbp, so this presents the problem of "walking along" a chromosome for a million base pairs. Chromosome walking procedures are not suited to such long distances because of the size limitation of probes. Even cosmids, which provide the largest probes, cannot harbor probes larger than 40 kbp'. It would take 25 cosmids end-to-end to span 1,000 kbp. A walk involving this many probes would take a very long time even if it were possible. Another hindrance to such a long walk is that the human genome is sprinkled with segments of repetitious DNA. Those regions "interrupt" a walk because a probe that encounters such a region anneals to many members of the library that could be situated almost anywhere in the genome. Walking breaks down under such conditions and would benefit by an alternative procedure. To traverse long distances of the genome and to skip troublesome regions of repetitious sequences, a procedure called chromosome "jumping" was devised. Probes for jumping carry small segments from the same chromosome, which are ordinarily about 500 kb apart. The precise distance is variable and a function of how the jumping probes are made. We do not go into the complexities here of constructing such a library except to say that there are several ways in which this can be done.

Usually a jumping library and a walking library are prepared from the same genome and used in conjunction with each other. The region around each locus recognized by the jumping library is scrutinized by the walking library as shown in figure 27.18.

This complex analysis would be in vain if one did not have some criterion for knowing when the goal of finding the cystic fibrosis gene was reached. This is where clues from the physiological nature of the condition became useful. In a brilliant strategy a cDNA library was prepared from the mRNA fraction of sweat gland tissue. Recall that in cystic fibrosis sweat glands malfunction; therefore, the mRNA for the CF gene might be well represented in the mRNA fraction of the sweat gland cells.

While the walking and jumping process was in progress each new segment mapping in the general region of interest was tested against the sweat gland cDNA library. Finally, a member of the walking library was found that annealed with a member of the sweat gland cDNA library. Was this match fortuitous or did it mean the CF gene had been found? To answer this question the cDNA discovered in this way was used to probe the genomic library. By this means a gene was mapped that extended over a region of about 250 kb with 23 introns. A unique transcript, approximately 6,500 nucleotides in length was detected in extracts of sweat gland tissue which matched the transcript size expected from this gene. The protein predicted from a sequence analysis of this transcript consists of two similar motifs, each with (1) a domain having properties consistent with membrane association and (2) a domain believed to be involved in ATP binding (fig. 27.19). Finally it was discovered that many CF patients carry a three base deletion in this transcript which should result in the loss of a phenylalanine residue from the protein. This defect correlates with the notion that CF patients have a faulty membrane protein that leads to the secretion problems characteristic of the disease. The fact that the abnormal gene is located as close as one can tell by classical genetics to the CF locus adds additional support to the notion that the CF gene has been found.

Finding the disease gene does not, of course, mean that a cure is in the offing. However, the characterization of the disease gene will be a tremendous aid in diagnosing carriers and fetuses that are homozygous for the disease gene. It also should be a help in focusing approaches to finding a cure for the disease.

## Summary

1. Sequencing DNA uses chemical methods to cleave specific bases in preexisting DNA or carries out the synthesis under conditions where synthesis is interrupted at specific bases.
2. A specific segment of DNA can be synthesized *in vitro* by the polymerase chain reaction. Short segments of DNA bordering the segment of interest are added to a mixture containing the segment of interest, a DNA polymerase, and the deoxyribotriphosphate substrates. The DNAs are first denatured, then annealed, and then synthesized. This cycle is repeated 20 or more times by raising the temperature to stop synthesis and lowering the temperature for annealing and synthesis. The outcome is a mixture in which the vast majority of the DNA is newly synthesized DNA bounded by the sequences of the added primers.
3. Another procedure for amplification cuts DNA containing the segment of interest into small pieces with a restriction enzyme. The cut pieces are incorporated into a plasmid or virus "vector" to be amplified in a suitable host. After growth, the mixture is plated to produce a mixture of bacterial or viral clones. The clone or clones of interest are identified often by hybridization of the clones after replica plating with a radioactive probe, followed by autoradiography to find the clone of interest.
4. Most cloning has been done in *E. coli*. Yeast is the most used eukaryotic host. Cloning is also possible in a number of plant and animal cells.
5. Mapping with recombinant DNA probes was first applied to the human globin genes. Starting probes were obtained by isolating the globin messenger from reticulocytes and converting it into a cDNA, which was used to scan a human genomic library for cross hybridizing members. Once detected and purified, these cross hybridizing members carrying globin messenger sequences were themselves converted to radioactive probes and used to further scan the genomic library for nearby sequences. By repeating this cycle several times, a process known as chromosome walking revealed a region around the adult hemoglobin gene that contained several closely related genes associated with hemoglobin.
6. Frequently, alleles of the same gene can be distinguished by restriction site differences in the genes themselves or in nearby locations. Alleles identified in this way are said to show restriction fragment length polymorphism. The allele responsible for sickle-cell disease was identified in this way.
7. The cystic fibrosis gene has been mapped by chromosome walking and jumping, a newer approach in which the relevant probes contain segments of the

genome that are normally located about 500 kbp from one another. A cDNA library was made from normal sweat gland tissue, chosen because of the disease's association with abnormal release of sweat salt suggested that the sweat gland would contain an abundance of the messenger associated with the gene. By hybridizing the genomic DNA probes with the cDNA sweat gland library, a segment of genome was identi-

fied as a candidate for the cystic fibrosis gene. This gene was characterized in detail and found to encode a complex transmembrane protein that carries a specific amino acid change in over half of the persons with cystic fibrosis. This correlation is overwhelming support that the gene responsible for cystic fibrosis has been mapped and characterized.

## Selected Readings

Caruthers, M. H., A. D. Barone, S. L. Beaucage, D. R. Dodds, E. F. Fisher, L. J. McBride, M. Matteucci, Z. Stabinsky, and J. Y. Tang, Chemical synthesis of deoxyoligonucleotides. *Methods Enzymol.* 154:287–313, 1987.

Cohen, S. N., A. Change, H. Boyer, and R. Helling, Construction of biologically functional bacterial plasmids *in vitro. Proc. Natl. Acad. Sci. USA* 70:3240–3244, 1973.

Gusella, J. F., DNA polymorphism and human disease. *Ann. Rev. Biochem.* 55:831–854, 1986.

Hunkapiller, T., R. J. Kaiser, B. F. Koop, and L. Hood, Large-scale and automated DNA sequence determination. *Science* 254:59–67, 1991. State of the art on the mammoth project to sequence the human genome.

Jackson, D. A., R. H. Symons, and P. Berg, Biochemical method for inserting new genetic information into DNA of Simian Virus 40: Circular SV40 DNA molecules containing lambda phage genes and the galactose operon of *E. coli. Proc. Natl. Acad. Sci. USA* 69:2904–2909, 1972.

Kerem, B., J. M. Rommens, J. A. Buchanan, D. Markiewicz, T. K. Cox, A. Chakravarti, M. Buchwald, and L. C. Tsui, Identification of the cystic fibrosis gene: Genetic analysis. *Science* 245:1073–1079, 1989.

Mansour, S. L., K. R. Thomas, and M. R. Capecchi, Disruption of the proto-oncogene *int-2* in mouse embryo-derived stem cells: A general strategy for targeting mutations to non-selectable genes. *Nature* 336:348–352, 1988.

Maxam, A. M., and W. Gilbert, A new method of sequencing DNA. *Proc. Natl. Acad. Sci. USA* 74:560–564, 1977.

Mullis, K. B., The unusual origin of the polymerase chain reaction. *Sci. Am.* 262:56–65, 1990.

Riordan, J. R., J. M. Rommens, B. S. Kerem, N. Alon, R. Rozmahel, Z. Grezelczak, J. Zielenski, S. Lok, N. Plasvsic, J.-L. Chou, M. T. Drumm, M. C. Iannuzzi, F. S. Collins, and L-C. Tsui, Identification of the cystic fibrosis gene: Cloning and characterization of complementary DNA. *Science* 245:1066–1073, 1989.

Sambrook, J., E. F. Fritsch, and T. Maniatis, *Molecular Cloning: A Laboratory Manual,* 2d ed., Cold Spring Harbor Laboratory Press, Cold Spring Harbor, N.Y, 1989. A three-volume collection that is thorough.

Sanger, F., Sequences, sequences, and sequences. *Ann. Rev. Biochem.* 57:1–28, 1988. A scientific memoir.

Sanger, F., and A. R. Coulson, A rapid method for determining sequences in DNA by primed synthesis with DNA polymerase. *J. Mol. Biol.* 94:444–448, 1975.

Watson, J. D., M. Gilman, J. Witkowski, and M. Zoller, *Recombinant DNA,* 2d ed. Scientific American Books. New York: W. H. Freeman Company, 1992. This text is an excellent elementary text on the subject of recombinant DNA. It contains many exciting chapters on specific applications and is extremely well referenced.

## Problems

1. Read the rest of the sequence in the autoradiogram in figure 27.1d as far as possible.
2. What are the major advantages of the polymerase chain reaction (PCR) method for amplifying defined segments of DNA as opposed to the use of conventional cloning methods? How might the PCR method be used to test for infection with the AIDS virus and how would this be an improvement over the antibody test currently used? (The current ELISA test is an indirect test for the presence of antibodies against the HIV proteins.)
3. Calculate the frequency of occurrence of restriction sites for PstI and HindIII in the DNA from a thermophile (80% G + C) and from *E. coli* (52% G + C).

4. You just isolated a novel recombinant clone and purified the desired insert (a 10,000 bp linear duplex DNA) from the vector. Now you wish to map the recognition sequences for restriction endonucleases A and B. You cleave the DNA with these enzymes and fractionate the digestion products according to size by agarose gel electrophoresis. Comparison of the pattern of DNA fragments with marker DNAs of known sizes yields the following results:
   (a) Digestion with A alone gives two fragments, of lengths 3,000 and 7,000 bp.
   (b) Digestion with B alone generates three fragments, of lengths 500, 1,000, and 8,500 bp.
   (c) Digestion with A and B together gives four fragments, of lengths 500, 1,000, 2,000 and 6,500 bp.
   Draw a restriction map of the insert, showing the relative positions of the cleavage sites with respect to one another.
5. Draw the ends of a DNA fragment digested with the restriction endonuclease *Bam*HI. How do these ends differ from those generated by *Mbo*I? If *Mbo*I and *Bam*HI ends were to be ligated together, would the resulting junction be cleavable by *Bam*HI or *Mbo*I?
6. Describe a procedure for cloning a DNA fragment into the *Bam*HI site of pBR322.
7. How large a genomic library should you construct in order to detect and isolate a 15-kb gene out of a genome containing $3 \times 10^9$ bp?
8. If you were interested in isolating a cDNA for human serum albumin, why would you use a cDNA library established from mRNA isolated from liver? If you wanted to isolate the gene for albumin, why would you use a genomic library established from any human tissue?
9. Which of the *E. coli* vectors on the left (a, b, c) would be used to achieve the cloning objectives on the right (1–5)?
   (a) Plasmid   (1) Genomic library
   (b) Cosmid    (2) DNA sequencing
   (c) Lambda    (3) cDNA library
                 (4) Small inserts
                 (5) Genomic walking

10. Site-directed mutagenesis is one of the most powerful tools available to the biochemist. What are some of the applications of this technique? How can the PCR method be used to do site-directed mutagenesis, and what is the advantage of this method?
11. The Southern blot technique is often used to compare genes from different organisms. For example, one could use the human globin gene probe described in the text to determine the extent of homology between globin genes from different primates. How could one reduce the stringency of the hybridization conditions (step 4 of fig. 27.14) to permit such a "heterologous hybridization"?
12. An unusual feature of the sickle-cell variant of the $\beta$-globin gene is that it directly alters a cleavage site for restriction endonuclease *Mst*II. *Mst*II recognizes the sequence CCTGAGG, which is mutated to CCTGTGG in the sickle-cell gene. How would you use this information and the Southern blot method to analyze fetal cells in amniotic fluid to determine whether the fetus carries sickle-cell anemia? What problems might you encounter in using this method?
13. Describe the procedure called "chromosome jumping." How was this procedure used to map the cystic fibrosis gene?
14. Describe a procedure using the PCR technique that could be used to determine whether a normal individual is a carrier of the cystic fibrosis $\Delta F_{508}$ mutation. What problems could you anticipate with this method?

CHAPTER

28

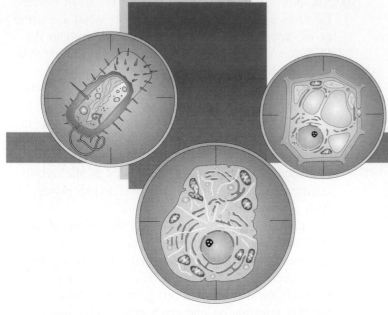

# RNA Synthesis and Processing

*Single stranded RNA molecules are synthesized by a template mechanism from select regions of the DNA genome.*

In prokaryotes DNA, RNA, and protein synthesis all take place in the same cellular compartment. In eukaryotes the DNA is compartmentalized in the cell nucleus, and it became clear long before the biochemistry of these three processes was understood that DNA synthesis takes place in the nucleus, whereas the bulk of protein synthesis takes place in the cytoplasm. From these observations on eukaryotes it was self-evident that DNA cannot be directly involved in the synthesis of protein but must somehow transmit its genetic information for protein synthesis to the cytoplasm. Careful experiments with radioactive labels were used to demonstrate that RNA synthesis takes place in the nucleus; much of this RNA is degraded rather quickly, but the portion that survives is mostly transferred to the cytoplasm (fig. 28.1). From observations of this kind it became clear that RNA was the prime candidate for the carrier of genetic information for the synthesis of proteins.

In this chapter we focus on the structure and metabolism of the major classes of cellular RNA.

## The First RNA Polymerase to Be Discovered Did Not Require a DNA Template

The first enzyme discovered that could catalyze polynucleotide synthesis was a bacterial enzyme called polynucleotide phosphorylase. This enzyme, isolated by Severo Ochoa and Marianne Grunberg-Manago in 1955, could make long chains of 5′-3′-linked polyribonucleotides starting from nucleoside diphosphates. However, there was no template requirement for this synthesis, and the sequence was uncontrollable except in a crude way by adjusting the relative concentrations of different nucleotides in the starting materials.

Shortly after Ochoa's studies, Sam Weiss began a search for a DNA-directed RNA polymerase. His experimental design was influenced by the theory that RNA must be made on a DNA template if it is to carry the genetic message. He was also influenced by Kornberg's discovery that DNA synthesis required nucleoside triphosphates for substrates rather than nucleoside diphosphates. With crude liver extracts, Weiss was able to demonstrate a capacity for RNA synthesis that was severely inhibited by DNase. Weiss's results touched off systematic investigations of RNA metabolism in many laboratories. A continuous expansion of research effort from that time on has yielded a wealth of understanding about the transcription process and related aspects of RNA metabolism.

**Figure 28.1**

In eukaryotic cells the nuclear membrane separates the processes of RNA and protein synthesis. This can be demonstrated with radioactive substrates that are precursors of RNA and protein. Immediately after exposure of cells to labeled precursors, the RNA label becomes fixed in the nucleus, and the protein label becomes fixed in the cytoplasm. Eventually most of the labeled RNA becomes transferred to the cytoplasm, and a fraction of the labeled protein becomes transferred to the nucleus.

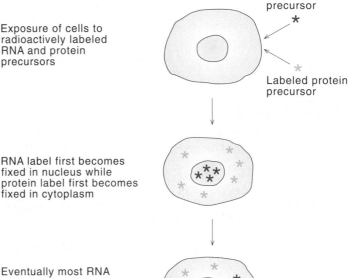

1. Exposure of cells to radioactively labeled RNA and protein precursors

Labeled RNA precursor

Labeled protein precursor

2. RNA label first becomes fixed in nucleus while protein label first becomes fixed in cytoplasm

3. Eventually most RNA label becomes transferred to cytoplasm, and some protein label becomes transferred to nucleus

## DNA–RNA Hybrid Duplexes Suggest That RNA Carries the DNA Sequences

Sol Spiegelman reasoned that if RNA was made on a DNA template, it should be complementary to one of the DNA chains. In this case it should be possible to make a DNA–RNA duplex by annealing, as was done for complementary DNA chains (see chapter 26). Using ³H-labeled T2 bacteriophage DNA and ³²P-labeled RNA that was synthesized after T2 infection of *E. coli* cells, he was able to show that the newly synthesized RNA forms a specific complex with the viral DNA (fig. 28.2). It was presumed that the specific complex must involve Watson-Crick-like base pairing between an RNA and a DNA chain, with uracil playing the role of thymine in the RNA. Complementary interaction was a strong indication that the RNA was a product of DNA-directed synthesis. Although most RNAs are synthesized as a result of transcription from a DNA template, different strategies are used in their synthesis and in their posttranscriptional modification. The strategy used in a given case is strongly related to the function of

**Table 28.1**

Types of RNA in *Escherichia coli*

| Type | Function | Number of Different Kinds | Number of Nucleotides | Percent of Synthesis | Percent of Total RNA in Cell | Stability |
|------|----------|--------------------------|----------------------|---------------------|----------------------------|-----------|
| mRNA | Messenger | Thousands | 500–6000 | 40–50 | 3 | Unstable ($t_{1/2} = 1$–3 min) |
| rRNA | Structure and function of ribosomes | 3 { 23S 16S 5S | 2800 1540 120 | 50 | 90 | Stable |
| tRNA | Adapter | 50–60 | 75–90 | 3 | 7 | Stable |

(Source: G. Zubay, *Biochemistry,* 2d ed., Macmillan, New York, 1988, p. 893.)

**Figure 28.2**

Demonstration that phage RNA has sequences complementary to phage DNA. S. Spiegelman and B. Hall used [3]H-labeled T2 DNA and [32]P-labeled RNA, the latter made after T2 infection of *E. coli* cells. The two nucleic acids were mixed together, annealed, and then centrifuged to equilibrium in a CsCl density gradient. In (*a*) the DNA was first heat-denatured and then mixed with RNA and annealed at 65°C prior to centrifugation. In (*b*) the DNA denaturation step was left out. Most of the RNA goes to the bottom of the centrifuge tube because of its high density. The DNA bands about one-third of the way from the top. In (*a*) some RNA also bands at approximately the same location as the DNA, but in (*b*) this is not the case. The comigration of a fraction of the RNA with the DNA is believed to be due to the formation of a specific DNA–RNA hybrid duplex. The RNA is much smaller than the DNA, so the RNA in the hybrid duplex migrates at the density of the DNA. In (*b*) no hybrid duplex forms because the DNA was not denatured before carrying out the annealing process.

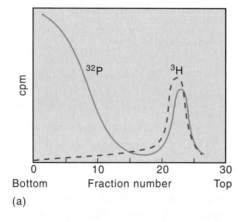

(a)

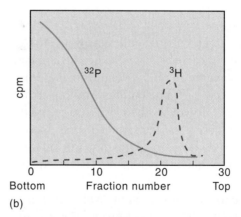

(b)

the RNA. For this reason, it is important that we discuss the different classes of RNA found in the cell before we consider their mode of synthesis or postsynthesis modification.

## There Are Three Major Classes of RNA

There are three major types of RNA that are transcribed from the DNA template: Messenger RNA (mRNA), ribosomal RNA (rRNA), and transfer RNA (tRNA). These three RNAs work together in protein synthesis. Some of the prop-erties of the three RNAs as they are found in *E. coli* are summarized in table 28.1.

## *Messenger RNA Carries the Information for Polypeptide Synthesis*

Messenger RNA carries the message from the DNA for the synthesis of a polypeptide chain with a specific sequence of amino acids. Each protein is unique, resulting in an enor-mous variety of mRNAs. It should be noted (see table 28.1) that although 40%–50% of the RNA synthesized in *E. coli*

## Figure 28.3

The tertiary structure of yeast phenylalanine tRNA. (*a*) The full tertiary structure. Purines are shown as rectangular slabs, pyrimidines as square slabs, and hydrogen bonds as lines between slabs. (Source: From G. J. Quigley and A. Rich, Structural domains of transfer RNA molecules, *Science,* 194:796, 1976.) (*b*) Nucleotide sequence. Residues that appear in most of the yeast tRNAs and residues that appear to be constantly a purine or a pyrimidine are indicated. Residues involved in tertiary base pairing are shown connected by solid lines. Several of the nucleotides are methylated. These are indicated by a small m. (Illustration copyright by Irving Geis. Reprinted by permission.)

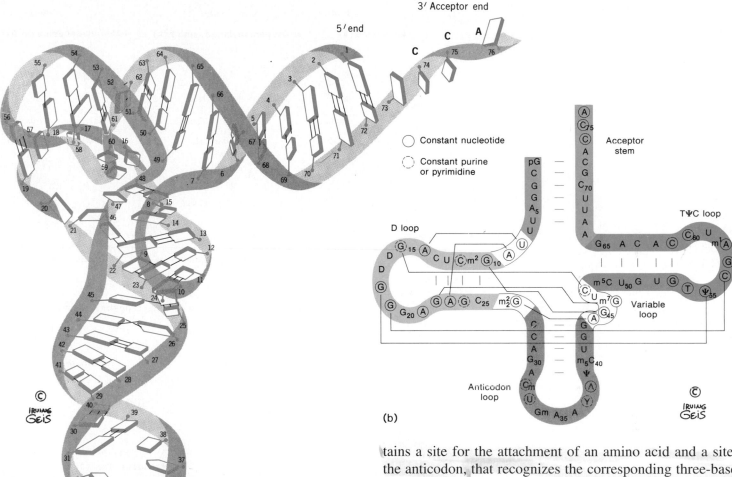

(a)

(b)

is mRNA, at any given time mRNA only accounts for about 3% of the cellular RNA. This is because bacterial mRNA is very unstable, with an average half-life of only 1–3 min. Messenger RNAs are heterogenous in both size and sequence, reflecting the fact that each mRNA encodes the information for the synthesis of a different protein(s).

### *Transfer RNA Carries Amino Acids to the Template for Protein Synthesis*

The function of transfer RNA is to carry amino acids to the mRNA template, where the amino acids are linked into a specific order. For this purpose each tRNA molecule con-

tains a site for the attachment of an amino acid and a site, the anticodon, that recognizes the corresponding three-base codon on the mRNA (see chapter 29). Since there are only 20 amino acids that are incorporated into proteins and more than twice this number of tRNAs in most cells, several amino acids must be represented by more than one tRNA. Each amino acid is enzymatically attached to the 3′ end of one or more tRNA by a specific aminoacyl-tRNA synthase that recognizes both the amino acid and the tRNA.

The primary structure is known for all *E. coli* tRNAs, and the three-dimensional structures of some of them have been determined by x-ray crystallography. The three-dimensional structure of phenylalanine tRNA of yeast is shown in figure 28.3. All tRNAs show similarities in their folded structures with four loops. Except for the variable loop, the loops are usually the same size in different tRNAs. The tRNA[Phe] molecule contains 20 bp that are hydrogen-bonded in Watson-Crick fashion and an additional 40 or so hydrogen bonds formed in other ways. These additional hydrogen bonds and the accompanying base stacking stabi-

## Figure 28.4

The structures of some modified nucleosides found in tRNA. The parent ribonucleosides are shown on the left in yellow screens. The other bases found in RNA result from post-transcriptional modification.

Adenosine (A)    $N^6$-Isopentenyladenosine ($i^6$A)    1-Methyladenosine ($m^1$A)

Guanosine (G)    Inosine (I)    1-Methylinosine ($m^1$I)

$N^2,N^2$-Dimethylguanosine ($m_2^2$G)    1-Methylguanosine ($m^1$G)

Uridine (U)    Ribothymidine (T)    Dihydrouridine (D)

4-Thiouridine ($s^4$U)    Pseudouridine ($\psi$)

Cytidine (C)    2-Thiocytidine ($s^2$C)    5-Methylcytidine ($m^5$C)

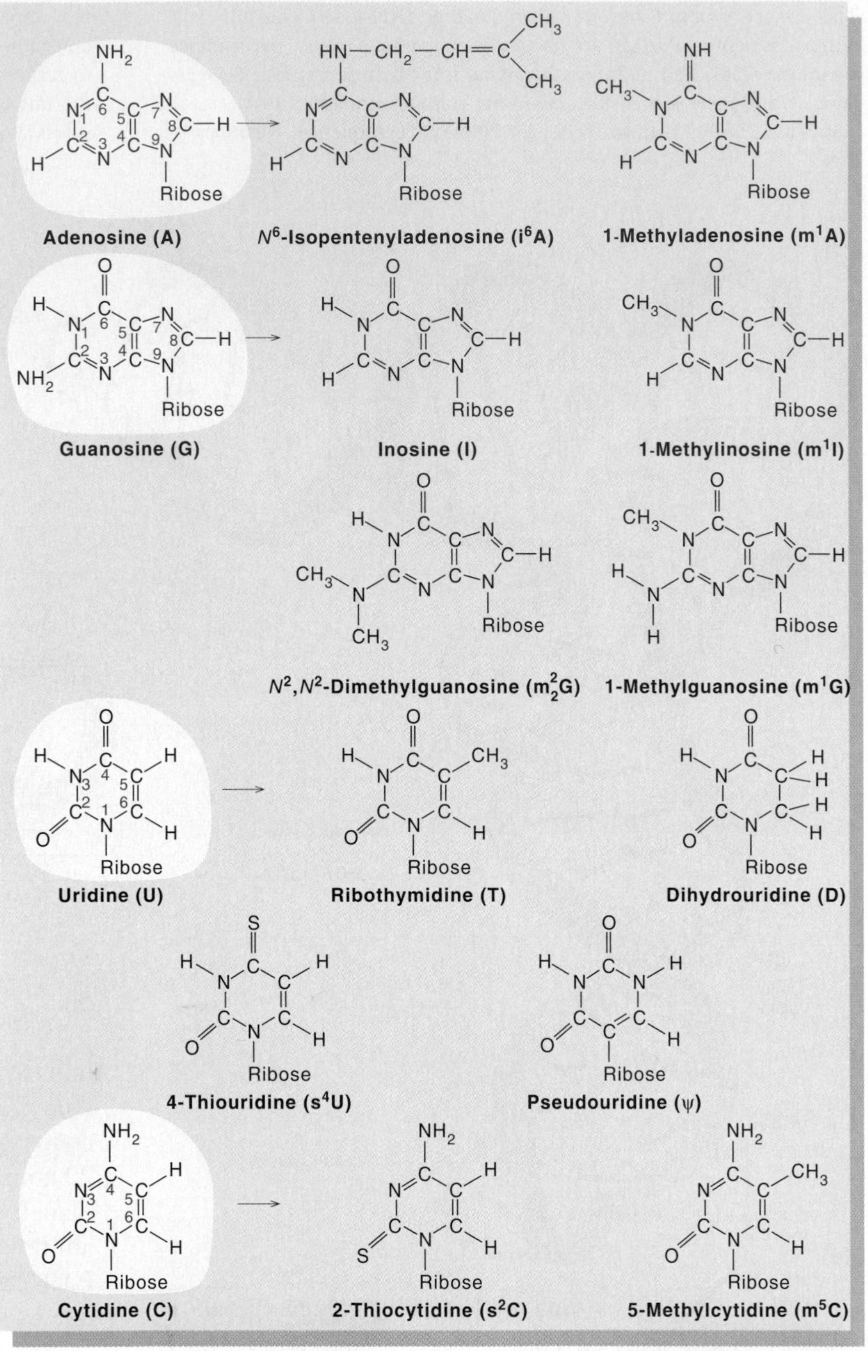

lize the tRNA$^{Phe}$ in the complex folded structure shown in figure 28.3a.

All tRNAs contain several ribonucleotides that differ from the usual four (12 in the case of tRNA$^{Phe}$). The structures for some of these are shown in figure 28.4. Only four ribonucleotides are incorporated into RNA in the transcription process. All of the rare bases found in the mature tRNA result from posttranscriptional modification.

The complex folded structures adopted by tRNAs illustrates the fact that nucleic acids with a properly adjusted primary sequence can adopt complex secondary and tertiary structures. Apropos of this, Francis Crick once said that transfer RNA is an RNA molecule trying to look like a protein.

## Figure 28.5

Composition of the *E. coli* ribosomes. The 70S ribosome can dissociate into a 50S and a 30S subunit. *In vitro* this can be done by lowering the Mg ion concentration. The individual subunits can be dissociated into their constituent RNAs and proteins by exposure to urea denaturant. Molecular weights are given for the subunits and the proteins, and the numbers of nucleotides are given for the RNAs.

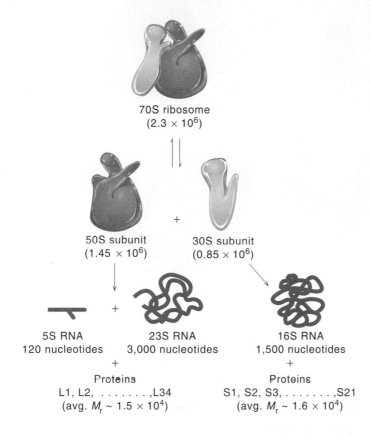

70S ribosome
$(2.3 \times 10^6)$

50S subunit                30S subunit
$(1.45 \times 10^6)$        $(0.85 \times 10^6)$

5S RNA          23S RNA          16S RNA
120 nucleotides   3,000 nucleotides   1,500 nucleotides
+                +
Proteins          Proteins
L1, L2, . . . . . . . ,L34   S1, S2, S3, . . . . . . . ,S21
(avg. $M_r \sim 1.5 \times 10^4$)   (avg. $M_r \sim 1.6 \times 10^4$)

## Ribosomal RNA Is an Integral Part of the Ribosome

The bulk of the cellular RNA is ribosomal RNA. Although seven genes exist in *E. coli* for rRNA, they all lead to essentially the same three ribosomal RNA molecules (see table 28.1) which differ substantially in size. The three rRNAs are always found in a complex with proteins in a functional component known as the ribosome. The ribosome is the site where mRNA and tRNAs meet to engage in protein synthesis. In *E. coli*, ribosomes are referred to as 70S particles, a measure of their rate of sedimentation and hence their size (S refers to Svedberg units, which are defined in chapter 6). A 70S ribosome consists of two dissociable subunits: A 50S subunit and a 30S subunit. Each of these contains both RNA and protein. The 50S subunit contains 23S and 5S rRNAs. The 30S subunit contains a single 16S rRNA (fig. 28.5). Eukaryotic ribosomes are similar in structure, although they are somewhat larger (80S) and con-

tain mostly larger rRNAs (25–28S, 18S, 5S, and an additional 5.5–5.8S; this additional rRNA corresponds in sequence to the first 150 nucleotides of the prokaryotic large rRNA subunit). Chloroplasts and mitochondria have ribosomes and rRNA that are distinctly different from those present in the cytoplasm and strongly resemble those of prokaryotes, testifying to their evolutionary origin from bacteria.

## The Fine Structure of the Ribosome Is Beginning to Emerge

Results from many different experimental approaches are beginning to coalesce to produce a three-dimensional picture of the ribosome that includes the location of its individual structural components and functional sites. The development of this picture has been especially challenging because the ribosome is large, fragile, and structurally complex and has resisted efforts to produce crystals that are capable of giving high-resolution structural information.

The current view of the overall morphology of ribosomes is based largely on electron-micrographic studies of the subunits of *E. coli*. From this analysis it appears that both subunits are asymmetrical (fig. 28.6).

The relative location of individual ribosomal proteins within the two subunits has been examined in two ways. One method involves determining which ribosomal proteins can be chemically cross-linked to each other and has yielded an elaborate grid of spatial relationships based on the frequency of cross-linking. The other method relies on neutron diffraction whereby the individually deuterated ribosomal proteins are located within the ribosomal subunit. The two methods of determining the location of ribosomal proteins have yielded a consistent spatial picture.

Recently, information concerning protein locations within the 30S subunit has been combined with a secondary structure model of 16S rRNA and the location of protein binding sites in rRNA to generate a partial three-dimensional picture of where the rRNA and proteins are situated in this ribosomal subunit (fig. 28.7).

## Overview of the Transcription Process

All DNA-dependent RNA polymerases carry out the following reaction:

$$NTP + (NMP)_n \xrightarrow[\text{DNA}]{Mg^{2+}} (NMP)_{n+1} + PP_i$$

The subsequent breakdown of $PP_i$ ensures the irreversibility of this reaction, as is the case in DNA polymerization reactions; this helps explain why both DNA and RNA polymer-

**Figure 28.6**

Gross shapes of the *E. coli* ribosomal subunits and the ribosome.

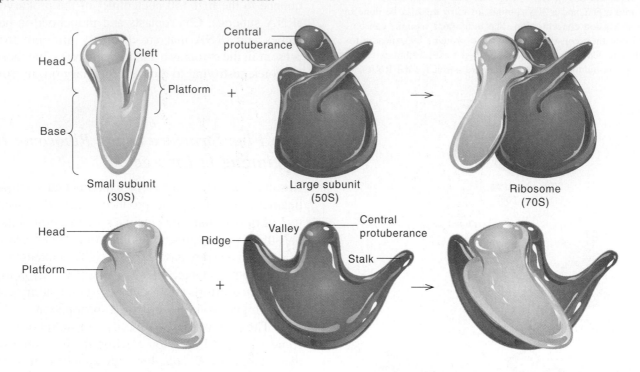

ase utilize NTPs rather than NDPs. The DNA template strand determines which base is added to the growing RNA molecule. For example, a cytosine in the template strand of DNA means that a complementary guanine is incorporated at the corresponding location of the RNA. Synthesis proceeds in a $5' \rightarrow 3'$ direction, with each new nucleotide being added onto the $3'$—OH end of the growing RNA chain.

The overall process for RNA synthesis on a duplex DNA template can be conceptually divided into initiation, elongation, and termination (fig. 28.8). In the initiation phase of the reaction, the RNA polymerase binds at a specific site on the DNA called the promoter. Here it unwinds and unpairs a small region of the DNA. Elongation begins by the base pairing of ribonucleotide triphosphates to one strand of the DNA, followed by the stepwise formation of covalent bonds from one base to the next. As elongation proceeds the DNA unwinds progressively in the direction of synthesis. The short stretch of DNA–RNA hybrid formed during synthesis is prevented from becoming longer than 10–20 bp by a rewinding of the DNA and the simultaneous displacement of the newly formed RNA. Termination occurs at a sequence recognized by the RNA polymerase as a stop signal. At this point the ternary complex of DNA, RNA, and polymerase breaks up.

First we discuss the process of transcription in bacteria and then in eukaryotes. In both cases we consider the properties of the RNA polymerase(s) first.

## Bacterial RNA Polymerase Contains Five Subunits

*Escherichia coli* contains one RNA polymerase, which transcribes all three major types of RNA. The active enzyme is a pentamer containing four different polypeptide chains with a total molecular weight of about 500,000. The subunits of the enzyme can be separated by electrophoresis on polyacrylamide gels. The four different polypeptide chains, termed $\beta'$, $\beta$, $\sigma^{70}$, and $\alpha$, have molecular weights of 155,000, 151,000, 70,000, and 36,500, respectively. Additional $\sigma$-like proteins have been identified, which we discuss later.

A complex with the subunit structure $\alpha_2\beta\beta'\sigma^{70}$ can carry out the functions necessary for synthesis of RNA and is referred to as the holoenzyme. Holoenzyme can be reversibly separated into two components by chromatography on a phosphocellulose column.

$$\alpha_2\beta\beta'\sigma^{70} \rightleftharpoons \alpha_2\beta\beta' + \sigma^{70}$$

Holoenzyme          Core          Sigma-70<br>polymerase

**Figure 28.7**

Arrangement of components in the *E. coli* 30S ribosomal particle. In (*a*) the relative locations of the ribosomal proteins, numbered 1–21, are shown. (Illustration prepared by Dr. Malcolm Capel from data described in M. S. Capel, M. Kjeldguard, D. M. Engelman, *J. Mol. Biol.* 200:66–87, 1988.) In (*b*) the conformation of the rRNA is shown. The location of the proteins is indicated by numbers given in the figure. (Illustration prepared by S. Stern, B. Weiser, and H. F. Noller from data described in *J. Mol. Biol.* 204:447–481, 1988.)

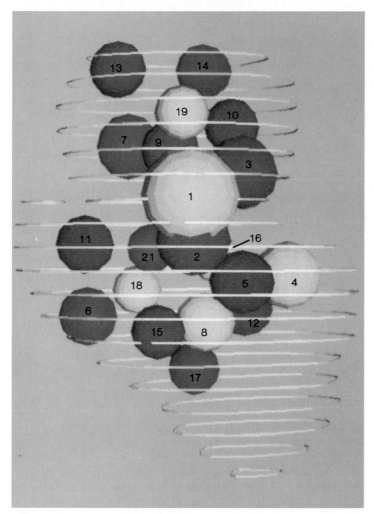

(a)

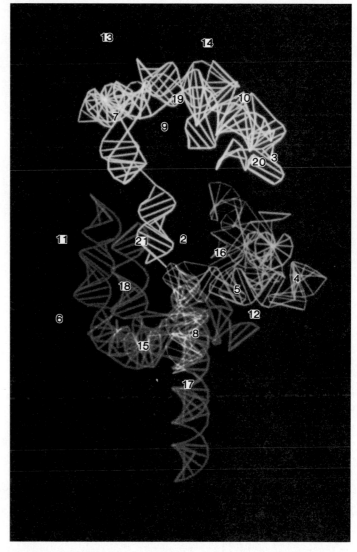

(b)

The enzyme without the sigma factor, called core polymerase, retains the capability to synthesize RNA, but it is defective in the ability to bind and initiate transcription at true initiation sites on the DNA. In fact when RNA polymerase was first purified from crude extracts it was missing the σ factor. The assay for polymerase involved the use of DNA with single-strand nicks. When a DNA template was used that did not have single-strand nicks, this enzyme was not active. This led to a search for a missing factor. When this factor ($\sigma^{70}$) was added back to the purified core enzyme and the uncut DNA template, the enzyme was able to bind tightly and selectively initiate RNA chains. Because of its role in binding and initiation, $\sigma^{70}$ is often referred to as an initiation factor.

The precise functions of the subunits of the core enzyme are not known. $\beta'$ is a basic (positively charged) polypeptide thought to be involved in DNA binding. The $\beta$ subunit is the site of binding of several inhibitors of transcription and is thought to contain most or all of the active sites for phosphodiester bond formation. The $\alpha$ subunit is necessary for reconstituting active enzyme from separated subunits.

**Figure 28.8**

Overview of RNA synthesis.

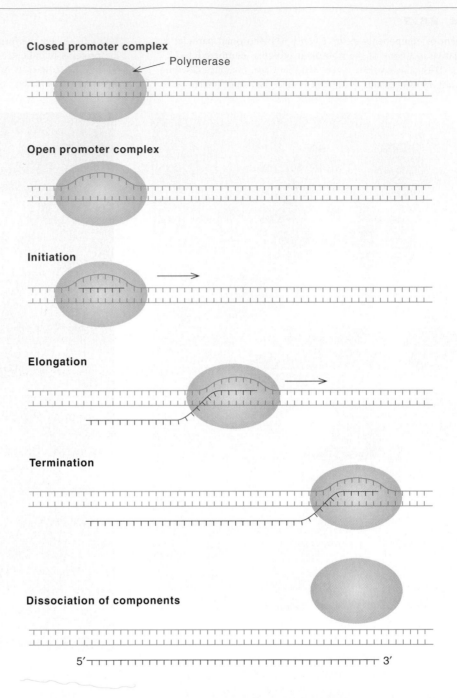

Closed promoter complex

Polymerase

Open promoter complex

Initiation

Elongation

Termination

Dissociation of components

5′ 3′

## Binding at Promoters

In the cell, RNA polymerase transcribes only select regions of the DNA; in these regions it synthesizes RNA that is complementary to one of the DNA strands. This selective action is possible because the holoenzyme is able to recognize and form a stable complex with DNA at specific promoters. RNA polymerase is able to form unstable nonspecific complexes at any place on the template, mainly by an interaction with the DNA phosphates, but it either rapidly dissociates and rebinds or slides along the DNA until it reaches a promoter. It then forms a moderately stable com-

plex with the promoter, most likely interacting with particular nucleotides in the −35 and −10 regions of the promoter. Initially the complex contains DNA in the double-helical or unmelted state; in this state they are referred to as closed promoter complexes (fig. 28.9). The next step involves breaking the H bonds (melting) over a stretch of about 10 bp of DNA, from positions −9 to 2, and a conformational change in the polymerase; this complex is termed the open promoter complex.

Common features have been identified in the sequence of over 250 different *E. coli* promoters at two locations: One 6-bp region centered at −35 (35 bp upstream of the initia-

## Figure 28.9

Various types of RNA polymerase–DNA binding complexes. A nonspecific complex is formed at any point along the DNA. The closed promoter complex is formed at a polymerase-binding site. Following the formation of the closed promoter complex, an open promoter complex is formed at the same site.

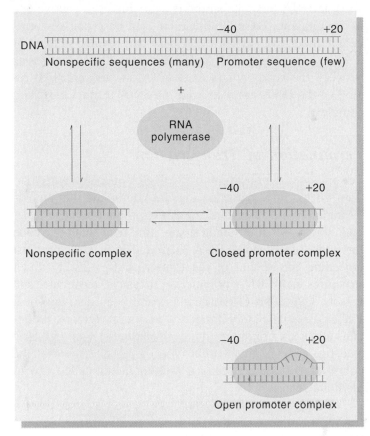

20–40 min. At the other extreme one finds very strong promoters for rRNA genes, which are transcribed at the rate of one per second. This 2,000-fold difference in transcription rate is primarily a function of the base sequence of the promoter.

## Initiation at Promoters

Once the polymerase binds to the promoter and strand separation occurs, initiation usually proceeds rapidly (1–2 s). The first, or initiating, NTP, which is usually ATP or GTP, binds to the enzyme. The binding is directed by the complementary base in the DNA template strand at the start site. A second NTP binds, and initiation occurs on formation of the first phosphodiester bond by a reaction involving the 3'-hydroxyl group of the initiating NTP with the inner phosphorus atom of the second NTP. Inorganic pyrophosphate derived from the second NTP is a product of the reaction. This process is illustrated in figure 28.10.

## Alternative Sigma Factors Trigger Initiation of Transcription at Different Promoters

Although most promoters in *E. coli* appear to utilize $\sigma^{70}$ as an initiation factor, several additional sigma-like initiation factors have been discovered that bind to core polymerase to form holoenzymes that recognize different species of promoters. One such factor is the product of the *rpoH* gene, which is involved in heat shock regulation. This factor is called sigma-32 ($\sigma^{32}$) because of its molecular weight of 32,000. It becomes bound to core polymerase in *E. coli* cells that have been subjected to heat shock and directs the polymerase to bind to and initiate at a class of promoters (the heat shock promoters) responsible for high-level expression of a dozen or so heat shock genes. Another such factor, $\sigma^{54}$, ($M_r = 54,000$) encoded by the *rpoN* gene is required for expression of certain genes involved in nitrogen metabolism. The consensus sequences associated with different classes of promoters are given in table 28.2.

## Elongation of the Transcript

After initiation has occurred, chain elongation proceeds by the successive binding of the nucleoside triphosphate complementary to the base at the growth point in the template strand, bond formation with pyrophosphate release, and translocation of the polymerase one base farther along on the template strand. Transcription proceeds in the 5' → 3' direction, antiparallel to the 3' → 5' strand of the templating DNA strand. Once elongation has produced an RNA chain about 10 bases long, the $\sigma$ subunit dissociates from the holoenzyme, leaving core polymerase to continue the

tion site) and one 6-bp region centered at −10 contain sequences that are similar but not identical in all bacterial promoters. The average sequences in both of these regions (TTGACA at −35 and TATAAT at −10) are termed the consensus sequences. A promoter with the consensus sequences in both regions leads to a very high level of transcription. Remarkably no naturally occurring promoter has the consensus sequence. It seems likely that this is part of a strategy which ensures that regulatory proteins can stimulate promoters and thereby exert an influence over their activity. This issue is taken up in chapters 30 and 31. Promoter strength in the absence of regulatory proteins decreases the more the actual sequences deviate from the consensus sequences, with some bases being more important than others. Naturally occurring promoters differ greatly in strength, which is best defined in terms of the frequency of RNA initiation. Some promoters are very weak, such as that associated with the *lac* repressor gene (discussed in chapter 30); as a result the *lac* repressor gene is transcribed only once in

## Figure 28.10

Details of phosphodiester bond formation. The $\alpha$, $\beta$, and $\gamma$ phosphates are indicated on the initiating NTP, which in this case is ATP. The colored ovals represent NTP-binding sites on the RNA polymerase. The biochemistry of bond formation in RNA synthesis is very similar to that in DNA synthesis.

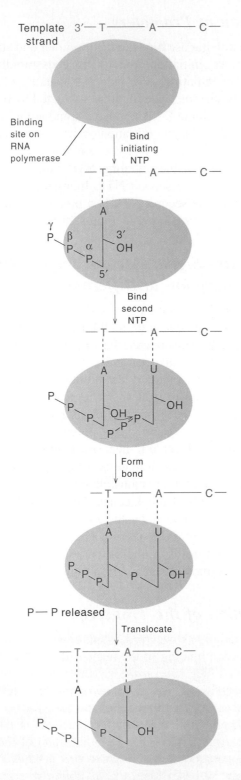

elongation reaction until a terminator signal is reached. The released $\sigma$ is available to bind to a free core polymerase and re-form a holoenzyme capable of binding at the same or other promoters where it initiates additional RNA chains.

As the polymerase traverses the DNA, it must continually cause a melting or strand separation of the DNA so that a single DNA template strand is available at the active site of the enzyme. During elongation, one base pair re-forms behind the active site for every base pair opened in front of it. The short transient RNA–DNA hybrid duplex that forms between the newly synthesized RNA and the unpaired region of the DNA helps to hold the RNA to the elongating complex.

## Termination of Transcription

Termination of transcription involves stopping the elongation process at a region on the DNA template that signals termination and release of the RNA product and the RNA polymerase. Most terminators are similar in that they code for a double-stranded RNA stem-and-loop structure just preceding the 3′ end of the transcript (fig. 28.11). Such structures cause RNA polymerase to pause, terminate, and detach. Two types of terminators have been distinguished. The first is sufficient without any accessory factors; it contains about six uridine residues following the stem and loop (see fig. 28.11). The second type of terminator lacks the polyU stretch and requires a protein factor called rho to facilitate release.

Rho binds to the RNA and is able to traverse RNA in an ATP-requiring reaction in the 5′ → 3′ direction on the RNA. C residues on the RNA are absolutely essential for this to happen. A helicase activity in rho is essential for its termination activity. If rho catches up with the RNA polymerase at a pause site or a termination site, it is highly likely to catalyze termination. This general picture of rho action is supported by the finding that interjection of a translational block that inhibits ribosome movement encourages premature release of the transcript. In *E. coli*, ribosomes normally traverse the nascent transcript before transcription is completed. These ribosomes make it difficult for rho to bind and reach the elongating polymerase. A translational block facilitates rho binding and migration to the elongating RNA polymerase.

## Comparison of Escherichia coli *RNA Polymerase with DNA PolI and PolIII*

In table 28.3 the bacterial RNA polymerase and the bacterial DNA polymerase are compared. Both types of enzymes are DNA-template-directed and require 4 NTPs and a divalent cation. While RNA polymerase makes single-stranded

**Table 28.2**

Summary of *Escherichia coli* Sigma Factors

| Sigma Factor | Gene | Consensus Sequence | | Genes Recognized |
| | | -35 Region | -10 Region | |
|---|---|---|---|---|
| $\sigma^{70}$ | *rpoD* | TTGACA | TATAAT | most genes |
| $\sigma^{32}$ | *rpoH (htpR)* | CTTGAA | CCCCAT-TA | heat-shock regulated |
| $\sigma^{54}$ | *rpoN (ntrA)* | CTGGCACN$_5$TTGCA | | nitrogen regulated |
| $\sigma^{E}$ | not identified | GAACTT | TCTGA | *rpoH, htrA* |
| $\sigma^{F}$ | not identified | TAAA | GCCGATAA | flagellar, chemotaxis |
| $\sigma^{K}$ | *rpoK (katF)* | Unknown | | *KatE* |

**Figure 28.11**

Important features of a typical transcription unit. DNA is shown with promoter and terminator regions expanded below. RNA is transcribed starting in the promoter region at +1 and ending after the stem and loop of the terminator. The protein resulting from translation of this RNA is shown above with its N and C termini indicated.

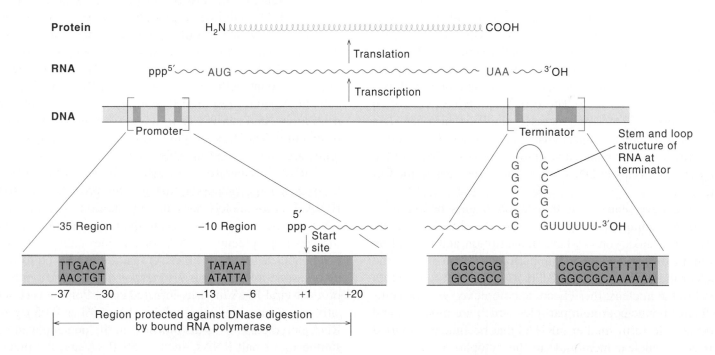

chains involved directly or indirectly in protein synthesis, the DNA polymerases replicate and repair the duplex DNA. Only RNA polymerase recognizes start-and-stop sequences in the duplex DNA. Thusfar only the DNA polymerases have been shown to have a proofreading function, but the possibility of a proofreading function for RNA polymerase has not been outruled.

## Important Differences Exist between Eukaryotic and Prokaryotic Transcription

Although the basic mechanism by which RNA is synthesized is quite similar in prokaryotes and eukaryotes, several important differences occur. Most parts of the bacterial

**Table 28.3**

Comparison of *Escherichia coli* RNA Polymerase with DNA Polymerases I and III

| | RNA Polymerase | DNA Polymerases I and III |
|---|---|---|
| **Similarities** | | |
| DNA-template-directed | Yes | Yes |
| Requires 4 NTPs | Yes (rNTPs) | Yes (dNTPs) |
| Requires divalent cation | Yes | Yes |
| **Differences** | | |
| Function | Transcription | Replication and repair |
| Initiates chains | Yes | No |
| Terminates chains | Yes | No |
| Recognizes sequences | Yes | No |
| Uses intact duplex template | Yes | No |
| Product | Single-strand RNAs | Duplex DNA strands |
| Proofreading | ? | Yes |

DNA are readily accessible to RNA polymerase binding and transcription. By contrast, most DNA in eukaryotic cells exists in a condensed form (chromatin), which is not readily accessible to transcription. The small fraction of DNA accessible to the RNA polymerase in any given cell type is especially sensitive to cleavage by mild treatment with bovine pancreatic DNase I. These regions of the DNA often contain bound RNA polymerase, modified histones, and additional nonhistone proteins. Active regions are often undermethylated compared with the total DNA. Most of the methylated groups in DNA are on the C residues in the CG sequence.

We have stated that translation begins before transcription is completed in prokaryotes. The situation is quite different in eukaryotes, where transcription and translation occur in different cellular compartments separated by the nuclear membrane. Large precursors of mRNA are synthesized in the nucleus; these become complexed with proteins to form ribonucleoprotein particles which are modified and processed to form smaller mRNAs that become transported across the nuclear membrane to the cytoplasm.

## Eukaryotes Have Three Nuclear RNA Polymerases

Unlike prokaryotes, in which all major types of RNA are synthesized by one RNA polymerase, eukaryotic cells contain three nuclear DNA-dependent RNA polymerases, each responsible for synthesizing a different class of RNAs.

Nuclear extracts can be fractionated by chromatography on DEAE-cellulose to give three peaks of RNA polymerase activity (the use of column chromatography is explained in chapter 6). These three peaks correspond to three different RNA polymerases (I, II, and III), which differ in relative amount, cellular location, type of RNA synthesized, subunit structure, response to salt and divalent cation concentrations, and sensitivity to the mushroom-derived toxin α-amanitin. The three polymerases and some of their properties are summarized in table 28.4.

RNA polymerase I is located in the nucleolus and synthesizes a large precursor that is later processed to form rRNA. It is completely resistant to inhibition by α-amanitin. RNA polymerase II is located in the nucleoplasm and synthesizes large precursor RNAs (sometimes called heterogeneous nuclear RNA, or hnRNA) that are processed to form cytoplasmic mRNAs. It is also responsible for the synthesis of most viral RNA in virus-infected cells. PolII is very sensitive to α-amanitin, being inhibited by 50% at 0.05 μg/ml. RNA polymerase III is also located in the nucleoplasm and synthesizes small RNAs, such as 5S RNA and the precursors to tRNAs. This enzyme is somewhat resistant to α-amanitin, requiring about 5 μg/ml to reach 50% inhibition.

## Eukaryotic RNA Polymerases Are Not Fully Functional by Themselves

Crude enzyme preparations of RNA polymerases I, II, and III have been shown to be capable of selective transcription

## Table 28.4

Comparison of Eukaryotic DNA-Dependent RNA Polymerases

| Type | Location | RNAs Synthesized | Sensitivity to α-Amanitin |
|------|----------|------------------|---------------------------|
| RNA polymerase I | Nucleolus | Pre-rRNA | Resistant |
| RNA polymerase II | Nucleoplasm | hnRNA, mRNA | Sensitive |
| RNA polymerase III | Nucleoplasm | Pre-tRNA, 5S RNA | Sensitive to very high levels |
| Mitochondrial | Mitochondria | Mitochondrial | Resistant |
| Chloroplast | Chloroplasts | Chloroplast | Resistant |

of defined DNA templates, initiating at sites known in some cases to be utilized *in vivo*. Fractionation of these polymerase-containing extracts has revealed many additional protein factors that stimulate *in vitro* transcription. These factors are divided into two categories. Basal transcription factors are required for the transcription of virtually all genes of a particular class. In addition to basal transcription factors there are an array of factors required for activated transcription. Transcription studies are presently in a stage of rapid development, and the distinction between factors required for basal transcription and activated transcription is not always clear. The situation is much more complex than for prokaryotes.

First we discuss basal transcription factors (fig. 28.12). With all three polymerases, two or more basal transcription factors are required, and some or most of these factors must bind to the promoter before the polymerase can bind.

Because many eukaryotic genes have been cloned during the last few years, it has become possible to compare the DNA sequences preceding genes that may act as promoter-like signals for RNA polymerase II. One feature that stands out is a common sequence, TATAAA, called the TATA box, found usually 25–30 bp before the transcription start site in many but not all PolII-transcribed genes.

Phil Sharp and Leonard Guarente showed that at least four transcription factors are required in addition to polymerase II for initiation from the major late promoter of adenovirus. *In vitro* studies indicate that these factors assemble in an orderly fashion (see fig. 28.12b). First the TFIID complex binds to the TATA box. Sequential binding of TFIIA, TFIIB, RNA polymerase II, and TFIIE follow. It is believed that this multifactor complex functions for a large number of eukaryotic promoters that contain TATA boxes.

The promoter region for RNA polymerase I also involves the cooperative binding of additional basal transcription factors. In this case two transcription factors bind upstream of the transcription start site (see fig. 28.12a).

In contrast to the initiation complex for RNA polymerases I and II, the region necessary for selective transcription of 5S RNA by RNA polymerase III is located in a region 40–80 bp downstream of the transcription start site (see fig. 28.12c). One of the additional protein complexes needed for selective transcription of the 5S gene (TFIIIA) binds to this site, and in conjunction with two other protein factors (TFIIIB and TFIIIC), directs RNA polymerase III to bind and initiate transcription. The binding site for factor TFIIIA was determined by the DNA "footprinting" technique (Methods of Biochemical Analysis 28A), a technique commonly used to locate binding sites for specific DNA-binding proteins.

PolIII polymerase also transcribes tRNA genes. In this case TFIIIA does not participate in the transcription process (see fig. 28.12c).

## Recent Evidence Suggests That the TATA-Binding Protein May Be Required by All Three Polymerases

Sometimes it is said that the basal transcription factor TFIID is the TATA-binding protein. Actually TFIID is a multiprotein complex of which the TATA-binding protein is but one component. Initially it was thought that the TATA-binding protein was exclusively involved with PolII promoters that contain a TATA element. However, closer examination of the different transcription systems indicates that the TATA-binding protein is part of the transcription initiation complex for PolII promoters that do not contain a TATA element. The TATA-binding protein is also required for transcription of ribosomal RNA genes. PolIII polymerase also may use the TATA-binding protein. In all of those instances where the TATA sequence is missing from the promoter, the exact function of the TATA-binding protein is not clear since it does not bind directly to the DNA in such

## Figure 28.12

Formation of the initiation complex for transcription for the three major classes of eukaryotic RNA polymerase: PolI, PolII, and PolIII. Each polymerase consists of many subunits (not shown). In addition to the firmly bound subunits, a number of protein factors, called transcription factors (TFs), only associate with the polymerases at the initiation site for transcription. For all three polymerases some of these transcription factors must bind to the promoter before the polymerase can bind. In each case we see an orderly progression of binding of factors and polymerase. In the case of PolIII, three transcription factors are involved for 5S rRNA genes, and only two are involved for tRNA genes.

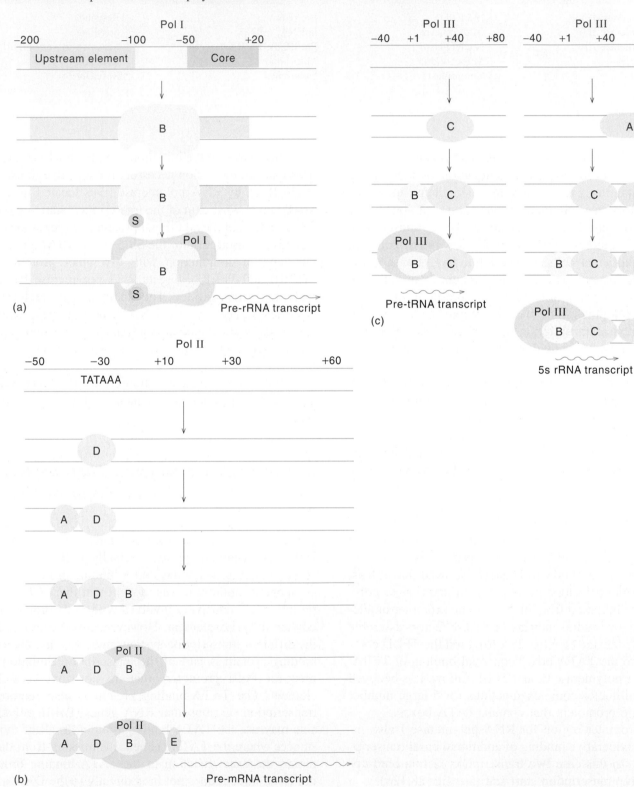

**Figure 28.13**

*Cis* elements involved in transcription in yeast (*a*) and in vertebrates (*b*). Upstream activator sequences (UAS) in yeast are similar in function to upstream enhancers in vertebrates. Yeast has no parallel to downstream enhancers found in vertebrates.

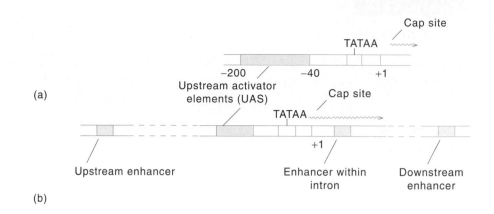

(a)

(b)

complexes. Perhaps it helps to link other transcription factors to the polymerase.

## In Eukaryotes Promoter Elements Are Located at a Considerable Distance from the Polymerase Binding Site

Eukaryotic promoters often contain binding sites adjacent to the polymerase-binding site where additional protein factors may bind. Promoter regions further removed from the polymerase-binding site are also quite common. The transcription factors that bind to these regions tend to be less general and more gene-specific than the factors that bind next to the polymerase. Considerable differences occur between the additional promoter elements found in yeast and vertebrates. In the case of yeast, additional promoter elements called upstream activator sequences (UAS) are usually located 40–200 bp upstream from the transcription start site (see fig. 28.13). The UAS elements bind additional transcription activation factors, which interact with those that bind at the TATA box. In vertebrates activator elements called enhancers are found upstream, downstream, and even in the middle of genes. Remarkably these elements may be located as far as 10 kb from the gene they influence. Like UAS elements, enhancers serve as binding sites for additional transcription activators. We have more to say about transcription activators in chapter 31 when we discuss regulation of gene expression in eukaryotes.

## Many Viruses Encode Their Own RNA Polymerases

Three strategies are used by different DNA viruses to accomplish transcription of viral DNA. The first type utilizes the host RNA polymerase, in some cases modifying it or

synthesizing new promoter-specific factors to direct it to read the viral promoters. Examples of such viruses are bacteriophage $\phi$X174, and T4 (of *E. coli*).

The second type of virus utilizes the host RNA polymerase to transcribe "early" viral genes including a gene for a new RNA polymerase that transcribes exclusively the remaining "late" viral genes. *E. coli* bacteriophages T7 and T3 are the best known examples of this type. T7 RNA polymerase recognizes specifically the T7 late promoters, all of which contain a nearly identical sequence of 18–22 nucleotides immediately upstream of the 5′-triphosphate terminal GTP start site. T7 RNA polymerase also recognizes specific termination points on the template and ignores the ones normally recognized by *E. coli* RNA polymerase.

A third type of virus, exemplified by bacteriophage N4, carries a virus specific RNA polymerase in its virion. This polymerase enters the cell together with the viral DNA and transcribes some early viral genes. Some of these genes code for specificity factors that direct the host RNA polymerase to transcribe late genes. Vaccinia virus is another example of a virus that contains a virion-encapsulated RNA polymerase.

## RNA-Dependent RNA Polymerases of RNA Viruses

The RNA genomes of single-stranded RNA bacterial viruses, such as Q$\beta$, MS2, R17, and f2, are themselves mRNAs. Bacteriophage Q$\beta$ codes for a polypeptide that combines with three host proteins to form an RNA-dependent RNA polymerase (replicase). The three host proteins are ribosomal protein S1 and two elongation factors for protein synthesis: EF-Tu and EF-Ts (see table 28.5). The Q$\beta$ replicase functions exclusively with the Q$\beta$ RNA plus strand template. It first makes a complementary RNA transcript (minus strand) and ultimately uses the minus strand as

## Table 28.5

RNA-Synthesizing Enzymes

| | Template | Primer | Molecular Weight of Subunit(s) | Gene Name | Substrate | Inhibition by Rifampicin |
|---|---|---|---|---|---|---|
| **Template-dependent** | | | | | | |
| Enzymes from bacteria | | | | | | |
| Holoenzyme *(E. coli)* | DNA | — | 155,000<br>151,000<br>70,000<br>36,500 | *rpoC*<br>*rpoB*<br>*rpoD*<br>*rpoA* | 4 NTPs | Yes |
| DNA primase | DNA | — | 65,000 | *dnaG* | 4NTP, 4 dNTP | No |
| Enzymes from phage or phage-infected bacteria | | | | | | |
| T7 RNA polymerase | T7 DNA | — | 99,000 | T7 *gene1* | 4 NTPs | No |
| N4 RNA polymerase | N4 DNA | ? | 350,000 | Viral | 4 NTPs | No |
| Qβ replicase | Qβ RNA | — | 65,000<br>55,000<br>43,000<br>35,000 | *rpsA*<br>Viral<br>*tuf*<br>*tsf* | 4 NTPs | No |
| **Template-independent** | | | | | | |
| CCA enzyme | — | 3′ end tRNA | 45,000 | *cca* | CTP<br>ATP | No |
| Poly(A) polymerase (eukaryotic) | — | 3′ end mRNA | | | ATP | No |
| Polynucleotide phosphorylase | — | 3′ end RNA | 86,000<br>48,000 | *pnp* | 4 NDPs | No |

(Source: G. Zubay, *Biochemistry,* 2d ed., Macmillan, New York, 1988, p. 813)

a template to synthesize multiple copies of viral RNA plus strands. Like the DNA-dependent RNA polymerases, the replicase utilizes rNTPs and transcribes in the $5' \rightarrow 3'$ direction. The phage RNA must first act as an mRNA to direct the synthesis of the aforementioned component of the replicase, since uninfected cells do not have an RNA-dependent RNA polymerase or replicase.

RNA tumor viruses (retroviruses) that infect animal cells exhibit a different replication strategy. In their virions they carry an enzyme that uses the viral RNA as a template to synthesize a DNA copy (see chapters 26 and 27). This DNA becomes integrated into the host genome. Subsequently the viral RNA is transcribed from the integrated viral DNA using host cell RNA polymerase.

## Other Types of RNA Synthesis

In addition to the cellular enzyme(s) that catalyzes DNA-directed RNA synthesis, cellular enzymes are involved in polyribonucleotide synthesis that do not use a template. Some of the properties of these enzymes are summarized in table 28.5. We have already mentioned polynucleotide phosphorylase in this chapter, and in chapter 26 we discussed the importance of DNA primase to DNA synthesis.

Two enzymes are known to add ribonucleotides post-transcriptionally to the 3′ hydroxyl end of specific RNAs. One adds the CCA sequence found in all tRNAs at their 3′ ends. The 3′ terminal adenine in this sequence serves as the amino acid attachment site. The 3′-CCA is relatively unstable and is continually being rejuvenated by this enzyme,

## Figure 28.14

Comparative processing of major transcripts in prokaryotes and eukaryotes.

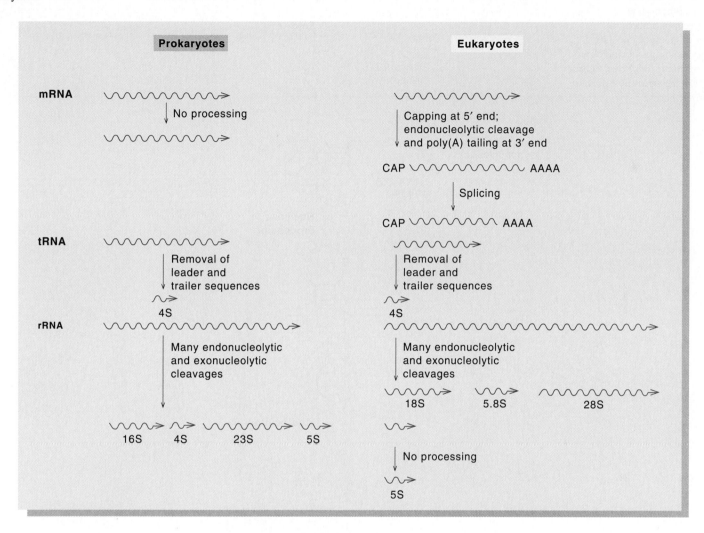

called the CCA enzyme, or tRNA nucleotidyltransferase.

In eukaryotes, 100–200 adenosine residues are added to the 3' ends of most mRNAs by a poly(A) polymerase. This addition occurs in the nucleus before the mRNA is fully processed and transported to the cytoplasm.

## Posttranscriptional Alterations of Transcripts

Most RNAs are not made in their final functional forms as they peel off the DNA template (fig. 28.14). They must undergo backbone phosphodiester bond cleavages into smaller molecules (processing) and individual base changes

(modification). The types of alterations that pre-tRNA and pre-rRNA transcripts undergo are very similar in prokaryotes and eukaryotes. We focus on the situation in *E. coli* because it is the best understood.

## *Processing and Modification of tRNA Requires Several Enzymes*

Transfer RNAs are processed from larger precursors in both prokaryotic and eukaryotic cells. This processing involves two types of nucleases: Endoribonucleases, that cleave at internal sites in the RNA, and exonucleases, that remove nucleotides from the ends of the chains.

**Figure 28.15**

Processing and modification of
*E. coli* tyrosine tRNA.
(T = ribothymidine;
$\psi$ = pseudouridine;
$i^6A$ = isopentyladenosine;
mG = methylguanosine;
$s^4U$ = thiouridine.) See figure
28.4 for the structures of these
modified bases.

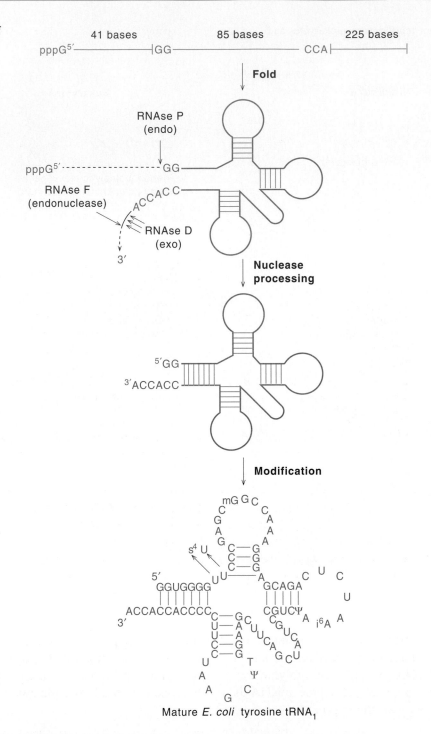

Mature *E. coli* tyrosine tRNA₁

The processing steps of *E. coli* tyrosine tRNA^Tyr are
diagrammed in figure 28.15. The initial transcript has, in
addition to the 85 nucleotide residues of the final product,
41 residues at the 5′ end and 225 residues at the 3′ end; it
probably folds to form the typical cloverleaf structure of the
mature tRNA prior to processing. Processing begins when a
specific endonuclease called RNaseF cleaves the precursor

at a site three nucleotides beyond what will be the 3′ end of
the mature tRNA. Another endonuclease, RNaseP, then
cleaves the remaining RNA to produce the mature 5′ end.
At the 3′ end, exonuclease RNaseD sequentially removes
additional nucleotides usually until it reaches the 3′ terminal
CCA sequence. Some tRNAs encode the CCA terminal sequence; some do not. Following processing individual bases

**Figure 28.16**

Processing of *E. coli* ribosomal RNA. The ribosomal RNA is transcribed as one long RNA molecule, which contains the sequences for the three ribosomal RNAs and one or two tRNA molecules. Many processing sites occur and many different enzymes are involved in the processing, as indicated by the vertical arrows and the symbols associated with these arrows. The various nucleases are described in table 28.7. (Source: Adapted from D. Apirion and P. Gegenheimer, Processing of bacterial RNA, *FEBS Lett.* 125:1, 1981.)

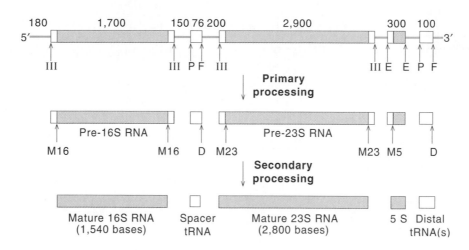

## Processing of Ribosomal Precursor Leads to Three RNAs

Both eukaryotic and prokaryotic cells synthesize large precursors to rRNA that are processed to produce the mature rRNAs. The scheme for processing rRNA in *E. coli* is summarized in figure 28.16.

The initial transcript is over 5,500 nucleotides long and includes, from the 5' end of the RNA: A 16S rRNA, a spacer region with one or two tRNAs, a 23S rRNA, and a 5S rRNA, and in some cases one or two additional tRNAs. Extra bases are found preceding and following each of these RNAs. Primary processing events include the endonucleolytic action by RNaseIII to produce pre-16S and pre-23S RNAs. This is followed by the action of specific ribonucleases to produce the tRNAs and pre-5S rRNA. Secondary processing by endonucleases M16, M23, and M5 results in mature 16S, 23S, and 5S RNAs, respectively. Extra bases on the 3' end of the tRNAs are removed by exonuclease RNaseD. This processing scheme has been deduced by observing the accumulation of intermediates in mutant strains defective in one or more of the nucleases and by cleaving the intermediates *in vitro* with purified or partially purified nucleases.

on the tRNA molecule are modified by a variety of enzymes, including methylases, deaminases, thiolases, pseudouridylating enzymes, and transglycosylases. Some of the modified bases that result from the action of these enzymes are illustrated in figure 28.4.

## Eukaryotic Pre-mRNA Undergoes Extensive Processing

In prokaryotes, most mRNAs function in translation with no prior alterations. Indeed little opportunity arises for any alterations because translation starts on the nascent transcript before transcription has been completed. By contrast, in eukaryotes transcription occurs in the nucleus, and a transcript undergoes extensive changes before being transported to the cytoplasm. First, the 5' end of the message is modified, a process called capping (fig. 28.17). This usually occurs before transcription has been completed. Changes at the 3' end of the transcript are part of the termination mechanism because the PolII enzyme does not appear to recognize any termination signal. The RNA polymerase transcribes well beyond the useful part of the message. These extended transcripts are cleaved about 12 nucleotides downstream of an AAUAA sequence, after which poly(A) polymerase adds about 200 adenylate residues to the 3' end. In many cases, especially in higher eukaryotes, noncoding regions called introns are removed from interior locations, and the remaining message is reunited. This process is called splicing because it is similar to the process of splicing in film editing. We discuss splicing in some detail.

To assist in the modification and processing of mRNAs, eukaryotic cells contain in their nuclei small nuclear RNAs (snRNAs), which are complexed with specific proteins to form small nuclear ribonucleoprotein particles (snRNPs). These RNAs have been named U1, U2, U3, . . . , U13 and range in size from 100 to 220 bases. One, U3, is

**Figure 28.17**

Structure of the 5′ methylated cap of eukaryotic mRNA. A 7-methylguanosine (in red) is attached through a triphosphate linkage formed between its 5′-OH and the 5′-OH of the terminal residue in the initial transcript. Note that the 2′-OH groups on the last two bases of the initial transcript have also been modified by methylation (in red). $N_1$, $N_2$, and $N_3$, can be any purine or pyrimidine bases.

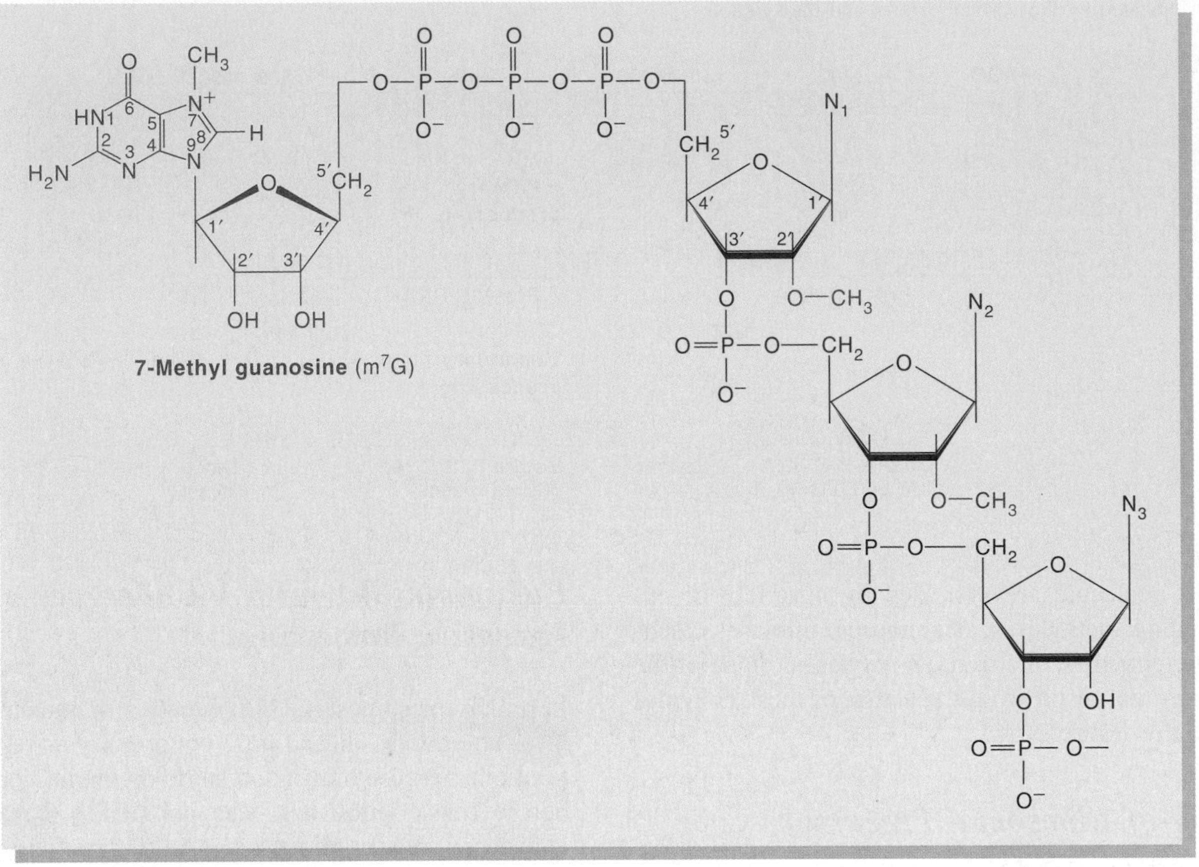

7-Methyl guanosine ($m^7G$)

found in the nucleolus, the site of rRNA synthesis. All are very abundant, and as many as 1 million copies of most of them may occur per nucleus. Their base sequences are highly conserved among organisms, and all seem to contain unusual trimethylguanosine structures at their 5′ ends. Each snRNP contains an snRNA and 6–12 proteins, some of which are common to all snRNPs. The snRNPs appear to be involved in the processing and modification of RNAs, including splicing (U1, U2, U5, U6), polyadenylation of pre-mRNA (U11), formation of 3′ ends of histone mRNAs (U7), and maturation of rRNA (U3).

*Splicing Entails the Removal of Internal Sequences.* It was established in the early 1970s that a great deal of RNA (hnRNA) turns over in the nucleus without ever reaching the cytoplasm. What could this RNA be? Was it a specific type of RNA that never made it to the cyto-

plasm, or was it evidence for processing of mRNA precursors? The big surprise came in 1977 when Ric Robert's and Phil Sharp's laboratories simultaneously discovered that the mRNAs of adenovirus undergo extensive processing in which internal segments are removed from the mRNA precursors. Very soon thereafter it was found that this phenomenon was general and widespread, especially in higher eukaryotes.

One of the early demonstrations of sequence removal resulted from the finding that mouse β-globin precursor mRNA did not form a perfect hybrid with DNA complementary to mature mRNA. When the hybrid was observed with an electron microscope, a loop in the DNA of the heteroduplex appeared, which suggested that the precursor contained internal sequences not present in the mature mRNA (fig. 28.18). These noncoding intervening sequences (introns) are interspersed with coding sequences (exons).

The presence of introns implies a function, but in

## Figure 28.18

Electron micrograph showing mouse β-globin precursor mRNA (nascent transcript) hybridized with DNA (cDNA) complementary to mature mRNA (upper photo). A control experiment (lower photo) shows that mature mRNA forms a perfect hybrid with DNA complementary to mature mRNA (cDNA) as expected. The intron region in the nascent transcript is indicated by a loop in the upper figure. Schematics indicating the RNA and cDNA are shown on the right. A second small intron is present in the precursor mRNA near the 5′ end and is the reason that the 5′ end of the RNA is not hybridized to the cDNA in the upper photo. (From A. Kinniburgh, J. Mertz, and J. Ross, The precursor of mouse β-globin contains two intervening sequences, *Cell* 14:681, 1978. © Cell Press.)

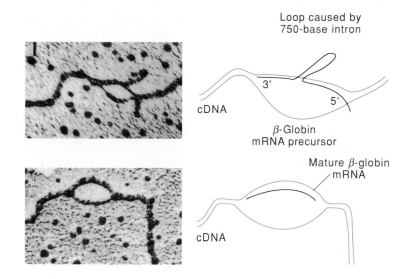

many cases the presence of introns may represent no more than a stage in the evolution of a gene. This argument is supported by the finding that introns are far less common in unicellular eukaryotes such as the yeast *Saccharomyces*, and they are very rare in prokaryotes such as *E. coli*. Frequently, splice points are correlated with "domains" that define protein structural units (see chapter 4). Similar domains are often seen in different proteins. For example, the exons of hemoglobin encode three structural domains of different types, whereas the heavy-chain immunoglobulin exons encode four domains that are quite similar in structure.

***Splicing Is a Two-Step Process.*** By attaching the promoter for the bacteriophage SP6 RNA polymerase to the β-globin gene, it has become possible to transcribe the gene *in vitro* with SP6 RNA polymerase to produce abundant amounts of β-globin pre-mRNA. This source of precursor mRNA was used to determine the steps and factors involved in splicing. The process involves cleavage of the pre-mRNA at the 5′ splice site to generate the 5′-proximal exon and an RNA species containing the intron in a "lariat" configuration connected to the distal exon (fig. 28.19). The lariat is formed via a 2′-5′ phosphodiester bond, which joins the 5′ terminal guanosine of the intron to an adenosine residue within the intron at a spot 18–40 nucleotides upstream of the 3′ splice site. These two RNA species are probably held together in a noncovalent complex until the next step (see fig. 28.19, step 2) in the reaction, which entails cleavage at the 3′ splice site to generate the free intron RNA and ligation of the two exons via a 3′-5′ phosphodiester bond. U1

and U2 snRNAs, as well as several additional protein factors, are necessary for the reactions to occur *in vitro*. These various factors are assembled in a large 40S–60S ribonucleoprotein called a spliceosome.

Removal of internal sequences in eukaryotes is not restricted to mRNA processing. It also occurs in the processing of rRNA and some tRNAs. In tRNAs the mechanism appears to be different in that the signal for splicing originates not from the primary sequence but from the secondary or tertiary structure of the pre-tRNA.

***RNA Editing Involves Changing Some of the Primary Sequence of a Nascent Transcript.*** On some occasions, the nascent transcript requires alterations in its sequence to convert it into a translatable mRNA. For example, in the mitochondrial message for cytochrome c oxidase from the protozoan *Leishmania tarentolae*, several U residues are added at different points to the nascent transcript (fig. 28.20). To make the editing changes a so-called guide RNA must form a complex with the nascent transcript. The changes in sequence are dictated by the complementary sequence in the guide RNA in the region where editing occurs. In addition to the guide RNA a special enzyme system is required, which is capable of inserting and removing bases at various points in the nascent transcript.

Editing seems like a roundabout way to get a functional message. Why not just make the transcript so that it can be translated in the first place. Possibly, editing serves a regulatory function. Alternatively it has been proposed that editing may be a device used to accelerate the evolutionary process for a gene.

## Figure 28.19

Splicing scheme for pre-mRNA. In step 1 the 2′-OH on an adenosine attacks a phosphate that is 5′-linked to a guanine residue. This leads to a lariat configuration connected to the distal exon. The lariat is formed via a 2′-5′ phosphodiester bond, which joins the 5′ terminal guanosine of the intron to an adenosine residue within the intron, 18–40 nucleotides upstream of the 3′ splice site. In the next step a cleavage occurs at the 3′ splice site to generate the free intron RNA, and a ligation occurs of the two exons via a 3′-5′ phosphodiester bond. Bases usually found in the region of the splice sites are indicated. The spliceosome is presumed to hold the reacting components in the proper juxtaposition to facilitate the two-step reaction.

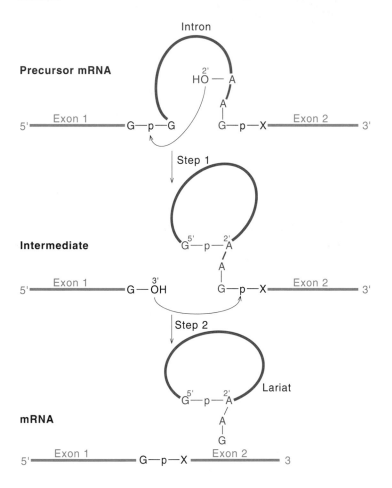

## Figure 28.20

Editing of the cytochrome b message in the flagellated protozoan *Leishmania tarentolae*. Editing, which involves the insertion of 11 U residues, creates a translatable RNA where no translation was possible before. Structure shown for mRNA is after editing.

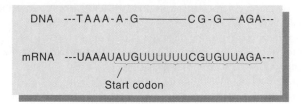

## Some RNAs Function Like Enzymes

For the more than 50 years since Sumner's discovery of urease, evidence has grown for all enzymes being proteins. This isn't quite true, as coenzymes frequently show similar activities to enzymes to which they are normally associated. Moreover, if we accept the broader definition of an enzyme to include the ability to bring reactive groups in close proximity so that they are encouraged to react, then we must regard all nucleic acid templates as enzymes. Nevertheless, it came as a giant surprise in the early 1980s when it was found that a number of biochemical activities normally as-

sociated with protein enzymes were shared by RNAs. In this section we describe some of these results, the new approaches they have led to, and the far-reaching implications of these findings.

## Some RNAs Are Self-Splicing

In 1982, Tom Cech discovered that when the pre-rRNA of the protozoan *Tetrahymena* was incubated with $Mg^{2+}$ and guanosine monophosphate, splicing of the RNA occurred without the involvement of any protein. This ability of RNA to carry out its own splicing has given rise to the term "ribozyme."

The mechanism of self-splicing in this case is somewhat different from that observed in the spliceosome reaction (fig. 28.21). First the 3′ hydroxyl group of the guanosine cofactor attacks the phosphodiester bond at the 5′ splice site. This is followed by another transesterification reaction in which the 3′ hydroxyl group of the upstream RNA attacks the phosphodiester bond at the 3′ splice site, thereby completing the splicing reaction. The final reaction products include the spliced rRNA and the excised oligonucleotide.

Since its initial discovery, self-splicing has been found to occur for RNAs from a wide variety of organisms. Certain precursor RNAs that exhibit self-splicing produce lariats, just like those seen in the commonly observed splicing reactions that are catalyzed by spliceosomes (see fig. 28.19). These findings suggest that at one time all splicing reactions were RNA-catalyzed.

## Some Ribonucleases Are RNAs

Ribozyme activities are not confined to splicing reactions. Among the large number of RNases, the most sophisticated enzymes are probably those involved in processing because they must attack preRNAs at specific sites. In his studies on

## Figure 28.21

Self-splicing of pre-rRNA from the protozoan *Tetrahymena*. The first step is a transesterification reaction in which the 3′ hydroxyl group of a guanosine attacks the phosphodiester bond at the 5′ splice site. The second step involves another transesterification reaction in which the 3′ hydroxyl group of the upstream exon attacks the phosphodiester bond at the 3′ splice site and displaces the 3′ hydroxyl group of the intron.

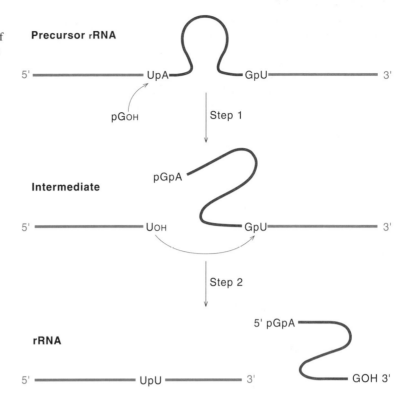

the RNaseP-processing enzyme, Sid Altman discovered that this enzyme is a ribonucleoprotein composed of both a 20 kilodalton protein and a 377-nucleotide RNA molecule. Stimulated by Cech's results, Altman tested the RNA component of RNaseP alone and found that it still had catalytic activity.

Since then other RNA phosphodiesterases have been discovered. Of special interest it was found that single-stranded plant RNAs, called viroids, cleave the viroid RNA at specific linkages. This activity is essential to the life cycle of the viroid.

### Ribosomal RNA Catalyzes Peptide Bond Formation

Recently Harry Noller has extended the range of ribozymes to include peptide bond catalysis. Noller and his colleagues found that the removal of all protein from the ribosomes of certain thermophilic bacteria left the ribosomes with the ability to catalyze peptide bond formation.

### Catalytic RNA May Have Evolutionary Significance

It seems likely that the range of catalytic activities for RNA will be expanded by future discoveries. For many years prior to the discovery of RNA catalysis, molecular evolu-

tionists struggled with the question of which came first, RNA or protein. Proteins make excellent enzymes, but they lack the template activity required for transmission of information and evolution. On the other hand, nucleic acids (RNA) have excellent template activity, like DNA, but they did not have any known catalytic activities. For a period of time it was impossible to imagine how either of these molecules could have evolved first since proteins cannot exist without templates for ordering amino acids in a polypeptide chain and RNAs cannot exist without enzymes. This dilemma appears to have been resolved by the recent discovery of ribozymes. Now a reasonable argument can be made that the first enzymes were RNAs and the first proteins were made under the direction of ribozymes.

Although this argument gives us a theoretically satisfying resolution to a central dilemma in molecular evolution, it still remains to demonstrate that RNAs have the versatility to function as enzymes for a much wider range of reactions. In this regard it seems unlikely that the full range of ribozymes that probably existed in early evolution are still present. More than likely most primitive ribozymes have long since been replaced by more efficient protein enzymes.

This creates a dilemma of a different sort. How are we to demonstrate ribozyme activities that no longer exist? Perhaps we must design our own ribozymes to demonstrate the full potential for RNA to act as an enzyme. This approach is being used by Jack Szostak and Gerald Joyce, who indepen-

# Figure 28.22

Inhibitors of RNA synthesis.

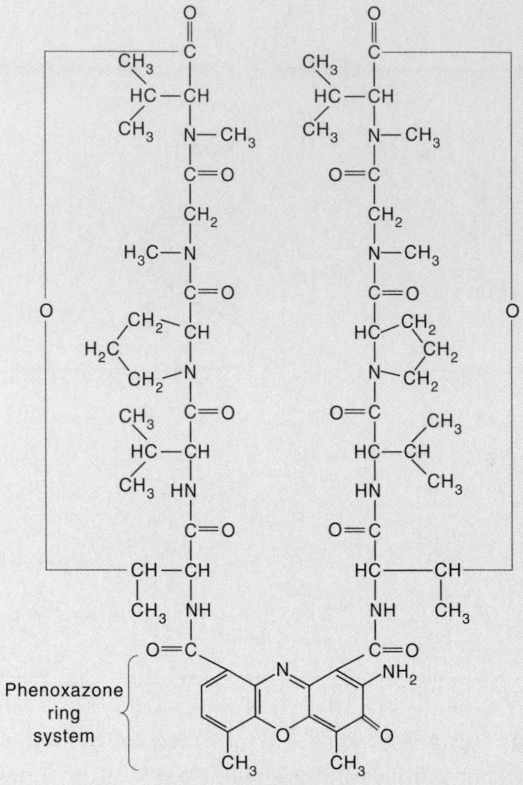

**Actinomycin D**

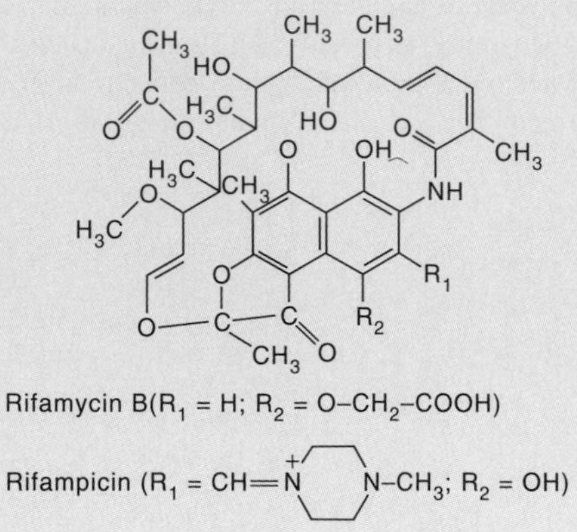

Phenoxazone ring system

Rifamycin B(R$_1$ = H; R$_2$ = O–CH$_2$–COOH)

Rifampicin (R$_1$ = CH=N$^+$—N–CH$_3$; R$_2$ = OH)

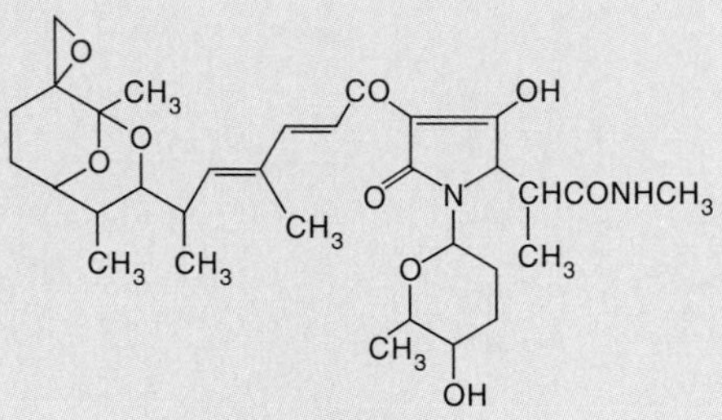

**Streptolydigin**

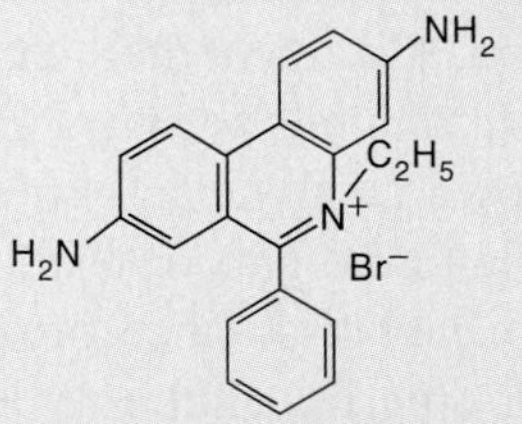

**Ethidium bromide**

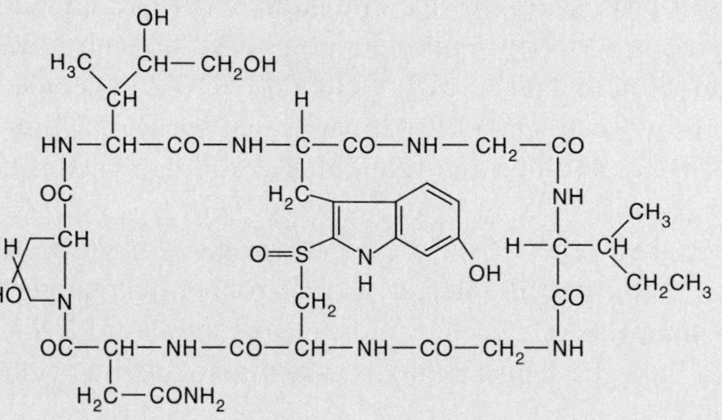

**α-Amanitin**

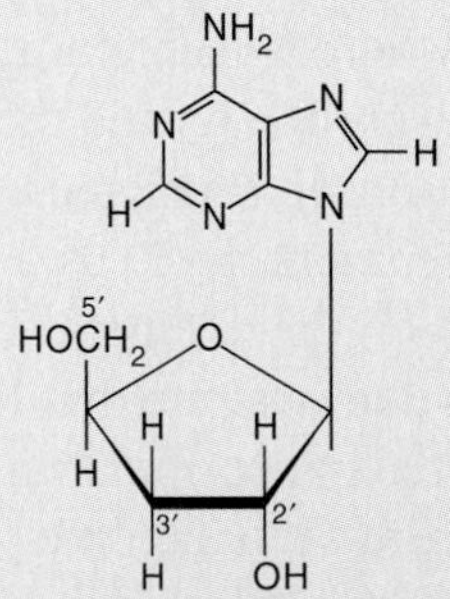

**Cordycepin**

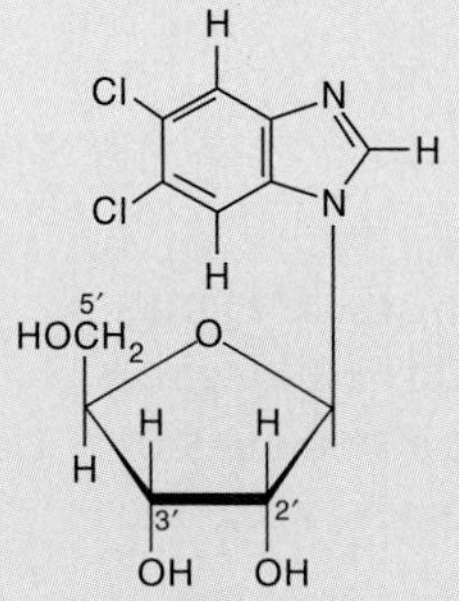

**DRB**

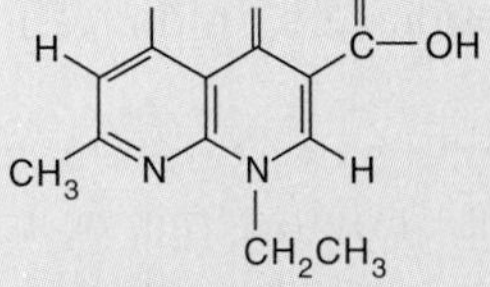

**Nalidixic acid**

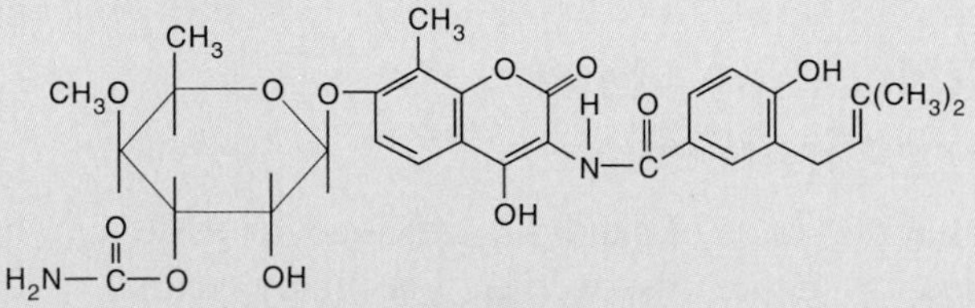

**Novobiocin**

dently have developed *in vitro* systems to select for RNA molecules with predesignated activities. So far, Joyce has selected for a *de novo* synthesized RNA that can cleave DNA, while Szostak has selected for an RNA that catalayzes primer extension by trinucleotides. Rapid progress using this approach to assess the potential of RNA to function as an enzyme is expected.

# Inhibitors of RNA Metabolism

A large variety of inhibitors of RNA synthesis have been identified. Some of these inhibitors have proved useful in elucidating transcription mechanisms, and some have facilitated selection for superior mutant strains with enzymes that are resistant to their inhibition. The inhibitors fall into three classes (fig. 28.22).

## Some Inhibitors Act by Binding to DNA

The best known example of inhibitors that bind to DNA is actinomycin D, an antibiotic produced by *Streptomyces antibioticus*. The inhibition of RNA synthesis is caused by the insertion (intercalation) of its phenoxazone ring between two G-C base pairs, with the side chains projecting into the minor groove of the double helix, hydrogen-bonded to guanine residues. RNA polymerase binding to DNA that contains actinomycin D is only slightly impaired, but RNA chain elongation in both eukaryotes and prokaryotes is blocked. Ethidium bromide also intercalates into DNA and at low concentrations preferentially binds to negatively supercoiled DNA (see chapter 25). It has been used to selectively inhibit transcription in mitochondria, which contains supercoiled DNA.

## Some Inhibitors Bind to RNA Polymerase

Rifampicin is a synthetic derivative of a naturally occurring antibiotic, rifamycin, that inhibits bacterial DNA-dependent RNA polymerase but not T7 RNA polymerase or eukaryotic RNA polymerases. It binds tightly to the $\beta$ subunit. Although it does not prevent promoter binding or formation of the first phosphodiester bond, it effectively prevents synthesis of longer RNA chains. It does not inhibit elongation when added after initiation has occurred. Another antibiotic, streptolydigin, also binds to the $\beta$ subunit; it inhibits all bond formation.

The most useful inhibitor of eukaryotic transcription has been $\alpha$-amanitin, a major toxic substance in the poisonous mushroom *Amanita phalloides*. The toxin preferentially binds to and inhibits RNA polymerase II (see table 28.4). At high concentrations it also can inhibit RNA polymerase III but not RNA polymerase I or bacterial, mitochondrial, or chloroplast RNA polymerases.

## Some Inhibitors Are Incorporated into the Growing RNA Chain

Cordycepin in its 5'-triphosphorylated form is a substrate analog that is incorporated into growing RNA chains by most RNA polymerases. It causes chain termination after incorporation, since it does not contain the 3' hydroxyl group necessary for the formation of the next phosphodiester bond.

# Summary

In this chapter we described the synthesis, transcription, and posttranscriptional reactions undergone by the three major classes of RNA. The main points we covered are as follows:

1. RNA is synthesized in the $5' \rightarrow 3'$ direction by the formation of 3'-5'-phosphodiester linkages between four ribonucleoside triphosphate substrates, analogous to the process of DNA synthesis. The sequence of bases in RNA transcripts catalyzed by DNA-dependent RNA polymerases is specified by the complementary sequences of the DNA template strand.

2. Some newly synthesized RNA transcripts are the functional species, whereas others must be modified or processed into the mature functional species. Modifying enzymes add nucleotides to the 5' or 3' ends or alter bases within the RNA, such as by methylation of specific residues. Specific processing enzymes cleave RNA internally, splice together noncontiguous regions of a transcript, or remove nucleotides from the 5' or 3' ends.

3. The major classes of RNA in both prokaryotes and eukaryotes are messenger RNA, ribosomal RNA, and

transfer RNA. These distinct classes of RNA play specific functional or structural roles in the translation of genetic information into proteins.

4. DNA-dependent synthesis of RNA in *E. coli* is catalyzed by one enzyme, consisting of five polypeptide subunits. The complete holoenzyme is composed of four polypeptides (the core enzyme) and an additional polypeptide that confers specificity for initiation at promoter sequences in the DNA template.

5. The steps involved in transcription include binding of polymerase at the initiation site, initiation, elongation, and termination.

6. In eukaryotes most transcription takes place in the nucleus. Three nuclear RNA polymerases, I, II and III, are responsible for the synthesis of rRNA, mRNA, and small RNA transcripts, respectively. The polymerases contain more subunits than in *E. coli,* and other proteins must bind at the initiation sites or near the

initiation sites before the polymerases can begin transcription.

7. Viruses sometimes use the host RNA polymerase in a modified form and sometimes synthesize their own RNA polymerase.

8. One of the more interesting processing reactions of nascent transcripts involves the removal of internal sequences. This type of processing is referred to as splicing. Most splicing reactions appear to require host proteins; however, some only require RNA, a fact demonstrating that RNA is capable of functioning like an enzyme in making and breaking of phosphodiester linkages in polyribonucleotides. Catalytic RNAs may have evolutionary significance.

9. Inhibitors of RNA synthesis may be classified according to their mechanism of action. Some bind to DNA, some bind to RNA, and some are incorporated into the growing RNA chain during transcription.

## Selected Readings

Ahsen, U., and H. F. Noller, Footprinting the sites of interaction of antibiotics with catalytic group I intron RNA. *Science* 260:1500–1503, 1993.

Altman, S., M. Baer, C. Guerrier-Takada, and A. Vioque, Enzymatic cleavage of RNA by RNA. *Trends Biochem. Sci.* 11:515–518, 1986.

Barinaga, M., Ribozymes: Killing the messenger. *Science* 262:1512–1514, 1993.

Bass, B. L., Splicing: The new edition. *Nature* 352:283–284, 1991.

Bear, D. G., and D. W. Peabody, The *E. coli* rho protein: An ATPase that terminates transcription. *Trends Biochem. Sci.* 13:343–348, 1988.

Beaudry, A. A., and G. R. Joyce, Directed evolution of an RNA enzyme, *Science* 257:635–641, 1992.

Bjork, G. R., J. U. Ericson, C. E. D. Gustafsson, T. G. Hdagervall, Y. H. Josson, and P. M. Wikstrom, Transfer RNA modification. *Ann. Rev. Biochem.* 56:263–287, 1987.

Buratowski, S., S. Hahn, L. Guarente, and P. A. Sharp, Five intermediate complexes in transcription initiation by RNA polymerase II. *Cell* 56:549–561, 1989.

Burtis, K. C., and B. X. Baker, *Drosophila* double sex gene controls somatic sexual differentiation by producing alternatively spliced mRNAs encoding related sex-specific polypeptides. *Cell* 56:997–1010, 1989.

Cattaneo, R., RNA editing: In chloroplast and brain. *Trends Biochem. Sci.* 17:4–6, 1992. A recent review dealing with RNA editing.

Cech, T., RNA editing: World's smallest introns? *Cell* 64:667–669, 1991.

Cech, T. R., RNA as an enzyme. *Sci. Am.* 225(5):64–75, 1986.

Cech, T. R., and B. L. Bass, Biological catalysis by RNA. *Ann. Rev. Biochem.* 55:599–630, 1986.

Chambon, P., Split genes. *Sci. Am.* 244:60–66, 1981.

Chowrira, B. M., A. Berzal-Herranz, and J. M. Burke, Novel guanosine requirement for catalysis by the hairpin ribozyme. *Nature* 354:320–323, 1991. In some cases the guanine amino group serves a catalytic role in self-splicing.

Crick, F., Central dogma of molecular biology. *Nature* 227:561–563, 1970.

Dahlberg, A. E., The functional role of ribosomal RNA in protein synthesis. *Cell* 57:525–529, 1989.

Darnell, J. E., Jr., RNA. *Sci. Am.* 253(4):68–78, 1985.

Deutscher, M. P., The metabolic role of RNases. *Trends Biochem. Sci.* 13:136–139, 1988.

Dorit, R. L., L. Schoenbach, and W. Gilbert, How big is the universe of exons? *Science* 250:1377–1382, 1990.

Forster, A. C., A. C. Jeffries, C. C. Sheldon, and R. H. Symons, Structural and ionic requirements for self-cleavage of virusoid RNAs and trans self-cleavage of viroid RNA. *Cold Spring Harb. Symp. Quant. Biol.* 52:249–259, 1987.

Futcher, B., Supercoiling and transcription, or vice versa? *Trends Genet.* 4:271–272, 1988.

Geiduschek, E. P., and G. P. Tocchini-Valentini, Transcription by RNA polymerase III. *Ann. Rev. Biochem.* 57:873–914, 1988.

Haas, E. S., D. P. Morse, J. W. Brown, F. J. Schmidt, and N. R. Pace, Long-range structure in ribonuclease P RNA. *Science* 254:853–856, 1991.

Hall, B. D., and S. Spiegelman, Sequence complementarity of T2-DNA and T2 specific RNA. *Proc. Natl. Acad. Sci. USA* 47:137–146, 1964. The first use of RNA–DNA hybridization.

Helmann, J. D., and M. J. Chamberlin, Structure and function of bacterial sigma factors. *Ann. Rev. Biochem.* 57:839–872, 1988.

Hou, Y.-M., and P. Schimmel, A simple structural feature is a major determinant of the identity of a transfer RNA. *Nature* 333:144–145, 1988.

Khoury, G., and P. Gruss, Enhancer elements. *Cell* 33:313–314, 1983. The first report on enhancers.

Koleske, A. J., and R. A. Young, An RNA polymerase II holoenzyme responsive to activation. *Nature* 368:466–473, 1994.

Landweber, L. R., and W. Gilbert, RNA editing as a source of genetic variation. *Nature* 363:179–182, 1993.

Leff, S. D., M. G. Rosenfeld, and R. M. Evans, Complex transcriptional units: Diversity in gene expression by alternative RNA processing. *Ann. Rev. Biochem.* 55:1091–1117, 1986.

Lorsch, J. R., and J. W. Szostak, In vitro evolution of new ribozymes with polynucleotide kinase activity. *Nature* 371:31–36, 1994.

Lührmann, R., B. Kastner, and M. Bach, Structure of spliceosomal snRNPs and their role in pre-mRNA splicing. *Biochem. Biophys. Acta* 1087:265–292, 1990.

Nikoliu, D. B., S-H. Hu, J. Lin, A. Gasch, A. Hoffman, M. Horikoshi, N-H. Chua, R. G. Roeder, and S. K. Burly, Crystal structure of TF IID TATA-box binding protein. *Nature* 360:40–45, 1992.

Padgett, R. A., P. J. Grabowski, M. M. Komarska, S. Seller, and P. A. Sharp, Splicing of messenger RNA precursors. *Ann. Rev. Biochem.* 55:1119–1150, 1988.

Patzelt, E., K. L. Perry, and N. Agabian, Mapping of branch sites in trans-spliced pre-mRNAs of *Trypanosoma brucei. Mol. Cell. Biol.* 9:4291–4297, 1989. Cases where the splicing event involves transcripts from different chromosomes.

Petska, S., J. A. Langer, K. C. Zoon, and C. E. Samuel, Interferons and their actions. *Ann. Rev. Biochem.* 56:757–777, 1987.

Peterson, M. G., J. Inostroza, M. E. Maxon, F. Osvaldo, A. Admon, D. Reinberg, and R. Tjian, Structure and functional properties of human general transcription factor IIt. *Nature* 354:369–373, 1991.

Proudfoot, N. J., How RNA polymerase II terminates transcription in higher eukaryotes. *Trends Biochem. Sci.* 114:105–110, 1989.

Steitz, J. A., "Snurps." *Sci. Am.* 258(6):56–63, 1988.

Stuart, K., RNA editing in mitochondrial mRNA of trypanosomatids. *Trends Biochem. Sci.* 16:68–72, 1991. RNA editing is processing that involves the removal, addition, and modification of nucleotides in the coding regions of nascent transcripts.

Thompson, C. C., S. L. McKnight, Anatomy of an enhancer. *Trends in Genetics* 8:232–236, 1992.

Weiner, A. M., mRNA splicing and autocatalytic introns: Distant cousins or the products of chemical determination? *Cell* 72:161–164, 1993.

Xing, Y., C. V. Johnson, P. R. Dobner, and J. B. Lawerence, Higher level organization of individual gene transcription and RNA splicing, *Science* 259:1330, 1993.

## Problems

1. Compare the reactions catalyzed by RNA and DNA polymerases. What are the similarities and differences?

2. One strand of DNA is completely transcribed into RNA by RNA polymerase. The base composition of the DNA template strand is: G = 20%, C = 25%, A = 15%, T = 40%. What would you expect the base composition of the newly synthesized RNA to be?

3. Although 40 to 50% of the RNA being synthesized in *E. coli* at any given time is mRNA, only about 3% of the total RNA in the cell is mRNA. Explain.

4. In *E. coli*, the mRNA fraction is heterogeneous in size, ranging from 500 to 6,000 nucleotides. The largest mRNAs have many more nucleotides than needed to make the largest proteins. Why are some of these mRNAs so large in bacteria?

5. Base-stacking interactions are an important stabilizing force in nucleic acid structures. Describe how base stacking contributes to the tertiary structure of the tRNA molecule.

6. Why is a single-strand binding protein or a DNA helicase not required for transcription as they are for replication?

7. When yeast phenylalanine tRNA is digested with a small amount of RNase (partial digest) such as T2 RNase, almost all of the cleavage sites are found in the anticodon loop. Explain this observation.

8. Speculate on the advantage of having three rRNAs (16S, 23S, and 5S) as part of the same RNA precursor (such as is found in *E. coli*)?

9. In *E. coli* the precise spacing between the −35 and −10 conserved promoter elements has been found to be a critical determinant of promoter strength. What

does this suggest about the interaction between RNA polymerase and these elements? What sorts of evidence could you obtain about this interaction by doing "footprint" experiments? (Also explain how you would do these experiments.)

10. Human RNA polymerase II generates RNA at a rate of approximately 3,000 nucleotides per minute at 37°C. One of the largest mammalian genes known is the 2,000-kbp (1 kbp = one kilo base-pair = 1,000 base-pairs) gene encoding the muscle protein, dystrophin. How long would it take one RNA polymerase II molecule to completely transcribe this gene? How long would it take to transcribe the adult $\beta$-globin gene (1.6 kbp)?

11. In early research with intact eukaryotic mRNA, the RNA appeared to have two 3'-ends and no 5'-terminus. Explain these observations.

12. What is unique about the promoter for RNA polymerase III? Diagram how this promoter works.

13. It has recently been demonstrated that the TATA-binding protein, in addition to flattening and widening the DNA by interacting (atypically) with the minor groove, also introduces a sharp bend (>100°) into its recognition sequence on binding. How might these changes in the structure of the DNA facilitate assembly of the RNA polymerase II transcription initiation complex?

14. Propose a sequence for the guide RNA that directs the editing of precursor RNA to yield the mRNA shown in figure 28.20.

15. A particular eukaryotic DNA virus is found to code for two mRNA transcripts, one shorter than the other, from the same region on the DNA. Analysis of the translation products reveals that the two polypeptides share the same amino acid sequence at their amino-terminal ends but are different at their carboxyl-terminal ends. The longer polypeptide is coded by the shorter mRNA! Suggest an explanation.

16. Although a phosphodiester bond must be formed between the upstream and downstream exons, there is no direct ATP requirement for splicing of mRNA introns or for splicing of the *Tetrahymena* pre-rRNA intron. Explain this observation.

# Use of the Footprinting Technique to Determine the Binding Site of a DNA-binding Protein

To determine the site of action of the transcription factor TFIIIA, the approximate location of its binding site on a 5S gene of *Xenopus borealis* was determined by the so-called "footprinting" technique. DNA fragments containing the 5S gene were labeled at the 5′ end with $^{32}$P mixed with TFIIIA and then digested with DNAseI. The resulting DNA fragments were electrophoresed on a polyacrylamide gel that was subsequently autoradiographed (see fig. 1). Regions of the DNA that were protected from DNase attack by TFIIIA binding appear as a blank spot (footprint) on the autoradiogram. The footprint shows that the protected region is situated between the 45th and the 90th base pair.

## Figure 1

DNaseI protection (footprinting) experiment on 5S DNA of *Xenopus*. The diagram on the left indicates the region on the gel that corresponds to the 5S RNA gene. Arrow points in the direction of transcription. Cross-hatched area indicates the region that binds transcription factor protein. Column labeled Xbs refers to an intact gene containing 160 bp of the 5′ flanking sequence, the 5S RNA gene (120 bp), and the 3′ flanking sequence (138 bp). The various deleted 5S DNAs are preceded by 74 bp of the plasmid pBR322 sequence. Numbers in other columns refer to portions of the 5S gene that have been deleted. All samples were subjected to partial digestion with DNaseI, then were electrophoresed and autoradiographed. In + columns, transcription factor protein was added before DNaseI treatment. (Courtesy Donald D. Brown of the Carnegie Institute of Washington.)

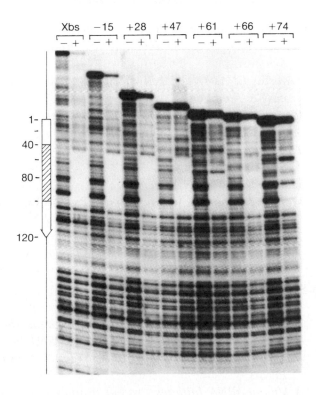

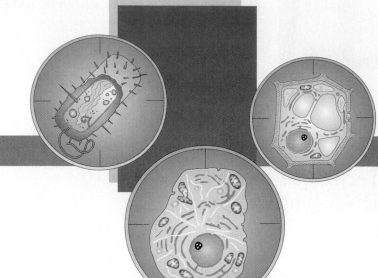

# Protein Synthesis, Targeting, and Turnover

*The arrangement of amino acids in polypeptide chains is determined by the arrangement of codons in messenger RNA molecules.*

Proteins are informational macromolecules, the ultimate heirs of the genetic information encoded in the sequence of nucleotide bases within the chromosomes. Each protein is composed of one or more polypeptide chains, and each peptide chain is a linear polymer of amino acids. The order of the amino acids commonly found in the polypeptide chain is determined by the order of nucleotides in the corresponding messenger RNA template. In this chapter we examine four aspects of protein metabolism (fig. 29.1): (1) The process whereby amino acids are ordered and polymerized into polypeptide chains; (2) posttranslational alterations in polypeptides, which occur after they are assembled on the ribosome; (3) the targeting process whereby proteins move from their site of synthesis to their sites of function; and (4) the proteolytic reactions that result in the return of proteins to their starting material, amino acids.

## The Cellular Machinery of Protein Synthesis

Amino acids are assembled into polypeptides on ribosomes. Prior to their interaction with messenger RNA (mRNA), amino acids are covalently attached to transfer RNA (tRNA) to form aminoacyl-tRNAs. The aminoacyl-tRNAs attach to specific sites on the mRNA. Messenger RNA contains the instructions for translation in the form of the genetic code that specifies the amino acid sequence of the polypeptide to be synthesized. Each ribosome binds to and moves along the messenger RNA while producing a single polypeptide. The direction or polarity of translation is from the 5′ to the 3′ terminus on the message, whereas the polypeptide is synthesized from the amino to the carboxyl terminus. As a rule, the information in a single mRNA molecule is translated simultaneously by a number of ribosomes that form a structure called a polysome. In prokaryotic cells, translation of mRNA begins while it is still being transcribed so that assemblies of nascent polypeptides and mRNAs can be seen on chromosomal DNA (fig. 29.2). In eukaryotic cells, transcription and translation are separate, and polysomes occur either free in the cytosol or bound to membranes in the endoplasmic reticulum (fig. 29.3).

### Messenger RNA Is the Template for Protein Synthesis

The mRNA molecule carries the genetic message in the form of a sequence of nucleotides that determines the order of amino acids in the polypeptide chain. Each amino acid is represented in the mRNA by a sequence of three nucleotides called codons. Codons are arranged in a contiguous reading frame, which is flanked on either side by bases that are not translated. These untranslated regions frequently have roles in regulating the processing and expression of the message. The 5′ end of the reading frame begins with a start codon, usually consisting of the nucleotides AUG, a sequence that codes for the amino acid methionine. Methionine is always used to initiate translation. The 3′ end of the reading frame contains one or more of three stop codons: UAA, UAG, or UGA. Stop codons serve as signals to terminate the polypeptide chain. The 3′ end of the message in eukaryotes usually contains a posttranscriptionally added poly(A) tail. This poly(A) tail has no known role in translation except that it may increase the lifetime of the message.

The 5′ end of the mRNA plays a special role in the selection of the start codon. This selection process occurs in fundamentally different ways in prokaryotes and eukaryotes. In prokaryotes (fig. 29.4) the mRNA contains a specific ribosome-binding site upstream of the initiating start codon. This binding site allows the ribosome to identify the correct initiating AUG. Prokaryotic mRNAs are frequently polycistronic, meaning that they encode more than one polypeptide chain. Internal ribosome-binding sites allow for independent initiation at internal reading frames. Eukaryotic mRNAs, on the other hand (see fig. 29.4), are usually monocistronic, encoding only one polypeptide chain. In keeping with this architecture, eukaryotic ribosomes interact with a ribosome entry site at the 5′ terminus of mRNA and move along it by a scanning mechanism to find the start codon. Identification of the ribosome entry site may be partly facilitated by the cap structure (see fig. 28.17) at the 5′ end of the mRNA. Initiation in eukaryotes usually occurs at the first AUG sequence downstream from the ribosome entry site. In multicellular eukaryotes the initiating AUG sequence is selected by a preferred initiation context of adjacent bases.

### Transfer RNAs Order Activated Amino Acids on the mRNA Template

Transfer RNAs contain two crucial functional sites: A site for attachment of the amino acids and a site that interacts with the mRNA during translation on the ribosome. At least one type of tRNA corresponds to each of the 20 types of amino acids that are incorporated into proteins. For accurate translation to occur, tRNAs must be distinguishable from one another by the molecules that recognize these specific types while still being recognizable to the molecules that interact with all tRNAs. A major challenge has been to understand how the similarities and differences in the structures of different tRNAs are related to the functions that

## Figure 29.1

Overview of reactions in protein synthesis. (aa₁, aa₂, aa₃ = amino acids 1, 2, 3.) Protein synthesis requires transfer RNAs for each amino acid, ribosomes, messenger RNA, and a number of dissociable protein factors in addition to ATP, GTP, and divalent cations. First the transfer RNAs become charged with amino acids, then the initiation complex is formed. Peptide synthesis does not start until the second aminoacyl tRNA becomes bound to the ribosome. Elongation reactions involve peptide bond formation, dissociation of the discharged tRNA, and translocation. The elongation process is repeated many times until the termination codon is reached. Termination is marked by the dissociation of the messenger RNA

from the ribosome and the dissociation of the two ribosomal subunits. The polypeptide chain sometimes folds into its final form without further modifications. Frequently the folded polypeptide chain is modified by removal of part of the polypeptide chain or by addition of various groups to specific amino acid side chains. The completed polypeptide chain migrates to different locations according to its structure. It may remain in the cytosol or it may be transported into one of the cellular organelles or across the plasma membrane into extracellular space. Proteins have different lifetimes. Eventually a protein is degraded, usually down to its component amino acids.

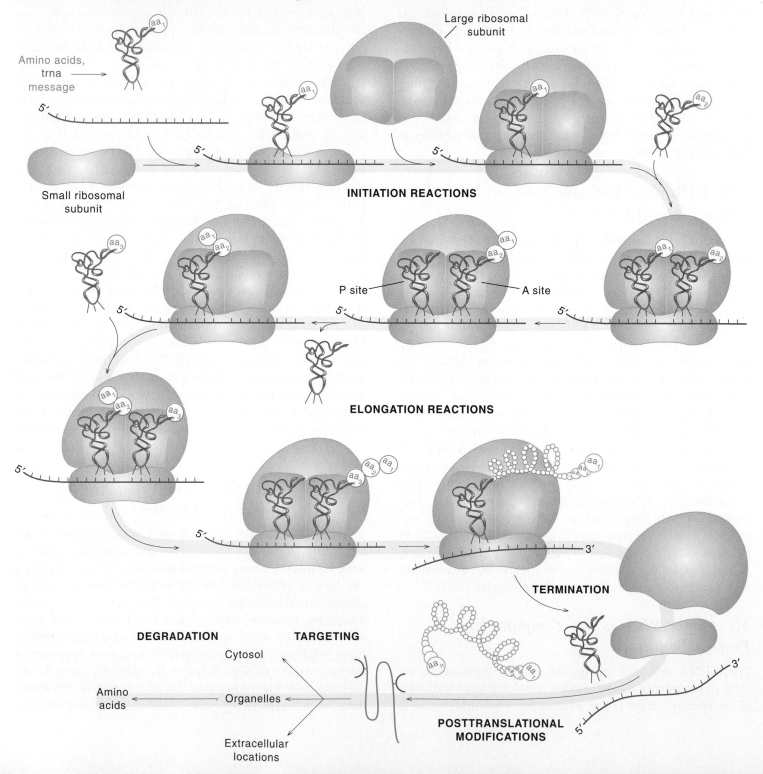

## Figure 29.2

(*a*) Electron micrograph of *E. coli* polysomes. Ribosomes are the dark structures connected by the faintly visible mRNA strand. A DNA strand connecting the polysomes from which the mRNA is being transcribed is visible as a horizontal line. (From O. L. Miller, B. A. Hankalo, and C. A. Thomas, "Visualization of bacterial genes in action," *Science* 169:392, 1970, © 1970 by AAAS.) (*b*) Line drawing for clarification. In bacteria, translation usually begins before transcription is completed, resulting in a DNA–mRNA–ribosome complex as shown. As the mRNA grows, the number of ribosomes associated with it increases. Here the mRNA appears to be growing from left to right. It is not possible to see the growing polypeptide chains on the ribosome because of the staining procedure used and the relatively small size of the polypeptide chains.

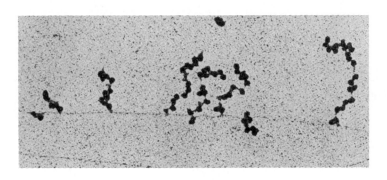

(a)

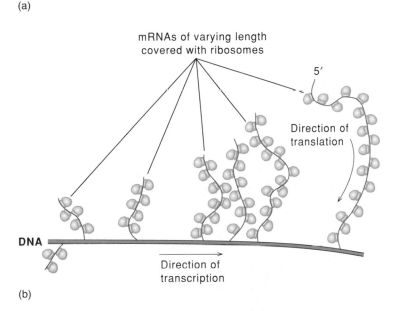

(b)

they play in protein synthesis. Here we summarize the structural features found in all tRNAs that allow them to perform their common functions.

Transfer RNAs that correspond to a single amino acid type are known as cognate tRNAs. In writing, the different cognate tRNAs are designated by a superscript. For example, tRNA$^{Phe}$ and tRNA$^{Ser}$ designate the tRNAs that correspond to the amino acids phenylalanine and serine, respectively. Some amino acids have several different cognate

## Figure 29.3

(*a*) Electron micrograph of mammalian rough endoplasmic reticulum. Continuous sheets of membrane create a compartment distinct from the surrounding cytosol. (From S. L. Wolfe, *Biology of the Cell,* 2d ed., 1981, Wadsworth Publishing Co.) (*b*) Clarifying line drawing shows expanded region of endoplasmic reticulum with ribosomes attached to the surface of the membrane facing the cytosol. The term "rough endoplasmic reticulum" arose because at low magnification the attached ribosomes give the endoplasmic reticulum a rough appearance.

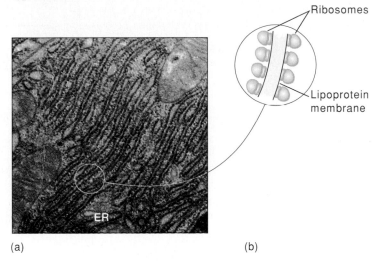

(a)                                    (b)

tRNAs, which are called isoacceptor tRNAs. Isoacceptor tRNAs are distinguished from one another by subscripts. A single cell typically contains a total of 50 or more different tRNAs.

The tRNAs are built on a structural theme so similar that a single figure can describe the common features of their primary, secondary, and tertiary structures (fig. 29.5). All tRNAs are composed of a single polynucleotide chain of from 70–95 residues. This chain folds back on itself to form a cloverleaf structure composed of four double-stranded stems and four single-stranded loops. The overall folding is such that 5′ and 3′ ends of the molecule are brought together to create a stem with seven regular Watson-Crick base pairs. This stem is known as the acceptor stem because during protein synthesis the amino acid is transiently attached to the ribose at this terminus. The 3′ end of the molecule is always terminated with the sequence CCA, which identifies the tRNA to the components that interact with all of these molecules during protein synthesis.

The unpaired loops of tRNA are named according to their unique structural features. Loop I varies in size from 7 to 11 unpaired bases and frequently contains the unusual base dihydrouracil; it is designated the D loop. Loop II contains the three bases known as the anticodon and is therefore designated the anticodon loop. This portion of tRNA plays a

## Figure 29.4

Simplified diagram of mRNA structure. (*a*) Typical eukaryotic mRNA. An AUG start codon is located near the 5' end of the mRNA. The single reading frame ends with one of the three trinucleotide sequences that represents a stop codon. Frequently, but not always, the 5' end of the mRNA is capped, and the 3' end contains a poly(A) tail. The cap structure is described in figure 28.17.

(*b*) Typical bacterial mRNA. Some bacterial mRNAs contain more than one reading frame; some contain only one. Each reading frame contains a start and a stop codon. Recognition of the start codons is facilitated by the presence of a ribosome recognition sequence (Shine-Dalgarno). The space between any two reading frames in the same transcript varies from one transcript to the next.

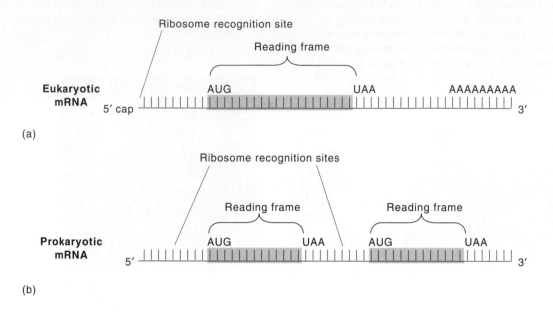

## Figure 29.5

The general structure of a tRNA molecule. (*a*) Representation in the form of a cloverleaf, which is the simplest way of observing the secondary structure. (*b*) A more realistic drawing of the three-dimensional folded structure. Color coding shows how the various loops in the cloverleaf structure correspond to the parts of the folded

structure. For a more detailed drawing showing the types of hydrogen bonds and the structures of the individual bases, see figure 28.3. (From A. Rich and S. H. Kim, The three-dimensional structure of transfer RNA, *Sci. Amer.* 238:52–62, January 1978. Copyright © 1978 by Scientific American, Inc. All rights reserved. Reprinted by permission.)

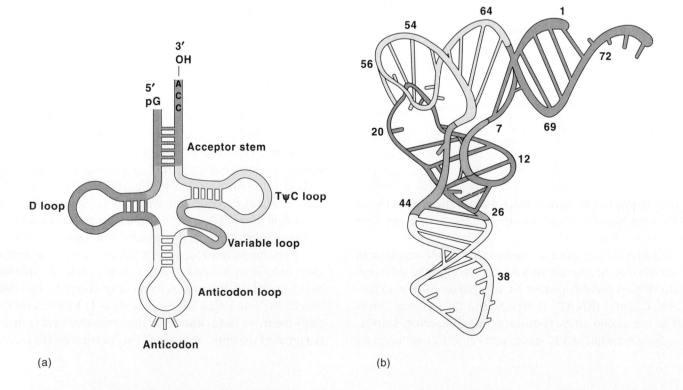

**Figure 29.6**

Functional sites on the prokaryotic ribosome. This figure shows a ribosome in the process of elongating a polypeptide chain. The mRNA and the EF-Tu-aminoacyl-tRNA complex are more closely associated with the 30S subunit. The peptidyl transferase and the EF-G elongation factor are associated with the 50S subunit, as is the nascent polypeptide chain. IF and EF are standard abbreviations for proteins that function as initiation factors and elongation factors, respectively. (From C. Bernabeu and J. A. Lake, Nascent polypeptide chains emerge from the exit domain of the large ribosomal subunit: Immune mapping of the nascent chain, *Proc. Nat'l. Acad. Sci. USA* 79:3111–3115, 1982. Copyright © 1982 National Academy of Sciences, Washington, D.C. Reprinted by permission.)

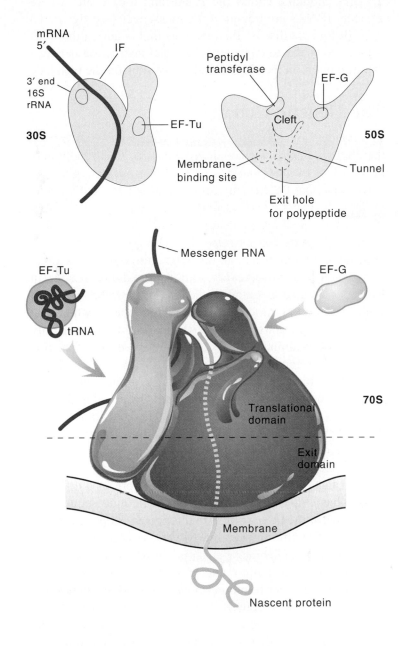

key role in translation by pairing with the complementary codon in mRNA and thus serves to align amino acids in their appropriate sequence. Loop III, the variable loop, may contain as few as 3 or as many as 21 bases, making it the major site of size variation in tRNA. Loop IV contains the unusual ribothymidine and pseudouridine bases as part of an invariant sequence; for this reason it is known as the TψC loop.

All tRNAs in solution fold into a three-dimensional L-shaped structure like that of tRNA$^{Phe}$ (fig. 29.5b). This structure is composed of two helical arms joined at right angles. The ribose moiety to which the amino acid is joined is at the end of one arm, identifying it as the acceptor arm. The anticodon is at the end of the other arm, identifying it as

the anticodon arm. The 3′ terminal ribosome and anticodon are separated by a diagonal distance of about 75 Å.

## Ribosomes Are the Site of Protein Synthesis

We considered the structure of the ribosome in some detail in the previous chapter without referring to those sites that are functionally important in protein synthesis. Here we can localize some of the functional sites on the two ribosomal subunits (fig. 29.6). The mRNA binds to the smaller ribosomal subunit. The peptidyl transferase is an integral part of the 50S subunit, and the elongation factor EF-G binds to the 50S subunit. The nascent polypeptide chain exits through a channel in the 50S subunit. Two functional sites occur on the 50S subunits, called the P site and the A site, where adjacent tRNAs are bound to the messenger (see fig. 29.1).

Before we discuss the reactions that occur on the ribosome, it is useful to consider the crucial interaction between the anticodon on the tRNAs and the codons on the mRNAs.

## The Genetic Code

The concept of a genetic code grew out of the realization that both nucleic acids and proteins are linear polymers made of a limited number of building blocks, plus the knowledge that the structure of proteins is genetically determined. The genetic code is simply the sequence relationship between nucleotides in genes (or mRNA) and the amino acids in the proteins they encode. The coding ratio is the number of nucleotides in an mRNA required to represent an amino acid. Since there are four different nucleotides in RNA and 20 different amino acids in proteins, the minimum acceptable coding ratio is 3:1. A singlet code, in which one nucleotide represents one amino acid, could code for only four different amino acids because there are only four different nucleotides in DNA or RNA. A doublet code, in which two nucleotides code for a single amino acid, could code for 16 amino acids since there are 16 possible doublet sequences in RNA. A triplet code, made from four nucleotides taken three at a time, generates a total of 64 possible triplet sequences. This would be more than enough to represent the 20 amino acids found in proteins.

In fact, a triplet code creates a problem of a different sort. What is the use of the extra trinucleotide sequences? Allowing for the three triplets that represent stop signals, we are still left with 61 possible triplets to code for just 20 amino acids. A nondegenerate code, in which a unique triplet codes for each amino acid, would imply that 41 triplet sequences are never found in translation reading frames. On the other hand, a degenerate code, in which all triplets are used, requires that each amino acid be represented, on the average, by more than one triplet sequence. We return to this point later. For now, we simply note that in most cases, point mutations in which one base pair in the DNA is replaced by another result either in no amino acid change or in a change to another amino acid. This is the result that we expect if most of the triplet sequences do in fact represent amino acids.

Another question relates to the location of codons in the translation reading frame. Codons could be arranged in a close-packed side-by-side arrangement. In this event each nucleotide within the translation reading frame would represent a code letter within one, and only one, codon. Alternatively, codons could be separated by one or more "spacer nucleotides." In this event some nucleotides within the reading frame would represent code letters within codons,

while others would serve as spacer nucleotides. Finally, we can imagine an overlapping codon arrangement, in which all nucleotides represent code letters, and some nucleotides represent code letters in more than one codon. However, biochemical experiments that we describe later in this chapter favor a nonoverlapping triplet code without spacers. Further genetic experiments in support of this type of code (not elaborated on here) indicate that when a single base pair is added or deleted, a change occurs in the reading frame of the message downstream of the change. Such an alteration in the DNA is known as a frameshift mutation. A frameshift mutation produces a new termination point because the frameshift changes the reading of the original stop codon and produces a new stop codon.

## The Code Was Deciphered with the Help of Synthetic Messengers

In 1961, Marshall Nirenberg and Heinrich Mattaei were attempting to synthesize proteins in cell-free extracts of *E. coli*. These extracts, containing the essential cellular components, were supplemented with nucleotides, salts, and radioactive amino acids, which were used so that the expected small amounts of protein could be detected. For mRNA they chose to use viral RNA from tobacco mosaic virus (TMV), because their goal was to make viral protein. As a control for these first "incorporation" experiments, they needed an mRNA that would not be expected to code for a protein. For this purpose, they chose poly(U), which contained only uridine residues. When Nirenberg and Mattaei added this control RNA to their cell-free system, substantial incorporation was evident. They showed that this synthetic messenger specifies the synthesis of a polypeptide containing only phenylalanine residues (polyphenylalanine).

A wave of excitement was set in motion by this discovery. Soon afterwards it was demonstrated that poly(A) promotes polylysine synthesis and poly(C) promotes polyproline synthesis. From these observations it seemed clear that the code words UUU, AAA, and CCC correspond to the amino acids phenylalanine, lysine, and proline, respectively.

Polynucleotide phosphorylase was used to produce "RNA" of random sequence, the composition of which reflected the mixture of nucleoside diphosphates in the reaction mixture. Mixed polynucleotides containing two bases were used in the incorporating system and shown to incorporate a pattern of amino acids consistent with a triplet code, but the observed incorporation could not define the code sequence.

The chemical synthesis of short oligonucleotides of known sequence provided a way out of the dilemma. First,

**Figure 29.7**

Demonstration that poly(UC) codes for a Ser-Leu repeating peptide. When poly(UC) containing a strict alternating sequence of U and C is used as an mRNA, only serine and leucine are incorporated. Analysis of the polypeptide product indicates that it contains an alternating sequence of serine and leucine.

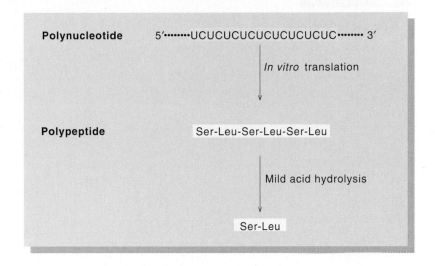

Philip Leder and Nirenberg showed that nucleotide triplets cause the specific binding of aminoacyl-tRNA to ribosomes. They cleverly observed that this binding could be readily measured by simply passing a solution containing radioactive aminoacyl-tRNA, ribosomes, and the trinucleotide through a nitrocellulose filter. Ribosomes, along with bound ligands, were adsorbed to the filter, and the coding specificity of some triplets was quickly established. Thus UUC, UCU, and CUU were shown to specify phenylalanine, serine, and leucine, respectively. However, many triplets did not give clear binding signals and there was concern that the binding reaction might not display the correct specificity.

The code was ultimately defined by the translation of polynucleotides of repeating sequences. These polynucleotides with long repeat sequences were produced by chemically synthesizing short, defined-sequence oligomers and then amplifying them enzymatically. When repeating dinucleotides were translated (fig. 29.7), they yielded repeating dipeptides, and when repeating trinucleotides (not shown) were translated, they yielded up to three individual peptides, each containing only a single amino acid. Eventually the sequence of bases from natural messages was correlated with the sequence of amino acids in the proteins they encoded.

## The Code Is Highly Degenerate

A triplet code, one made from four nucleotides taken three at a time, generates a total of 64 different triplet sequences or codons. Three of these codons, as we will see, are utilized to terminate translation and are not generally used to specify amino acids. The remaining 61 codons and the 20 amino acids can be neatly summarized by grouping codons with the same first and second bases into a grid (table 29.1). The four horizontal sections are composed of codons with the same first base. The four vertical sections are composed of codons with the same second base. The boxes representing the vertical–horizontal intersections contain codon families, the members of which differ only in their 3′ terminal base. For example, the codons UCU, UCC, UCA, and UCG constitute a family encoding serine. Thus, the genetic code is degenerate, that is, one amino acid is generally specified by multiple or synonymous codons.

As we have indicated, the codon AUG is the only one generally used to specify methionine, but it serves a dual function in that it is also used to initiate translation. Occasionally, GUG and UUG are also read as an initiating codon in bacteria, but in internal positions these codons are always read as valine and leucine, respectively. In eukaryotes, initiation at codons other than AUG is much less frequent than in prokaryotes. Weak initiation occasionally occurs at GUG, CUG, and ACG codons in eukaryotic systems. The UGA triplet also serves a dual function; it is usually recognized as a stop, but on occasion it serves as a codon for selenocysteine (box 29A).

Organisms differ in the frequency with which they utilize synonymous codons. Some synonymous codons are used frequently, and some are used infrequently. For *E. coli* and yeast this frequency of use, known as codon usage, correlates with the abundance in the organism of the tRNAs that recognize particular synonymous codons. Also, for the same two species, proteins that occur in the greatest abundance employ high-usage codons more frequently and low-usage codons less frequently than the average proteins. It is not clear that this same trend is followed in multicellular eukaryotes, such as *Drosophila* and humans.

## Table 29.1

The Genetic Code

| First Position (5′ end) | Second Position U | Second Position C | Second Position A | Second Position G | Third Position (3′ end) |
|---|---|---|---|---|---|
| U | UUU ⎫ Phe / UUC ⎭ / UUA ⎫ Leu / UUG ⎭ | UCU ⎫ / UCC / UCA Ser / UCG ⎭ | UAU ⎫ Tyr / UAC ⎭ / UAA ⎫ STOP / UAG ⎭ | UGU ⎫ Cys / UGC ⎭ / UGA STOP / UGG Trp | U C A G |
| C | CUU ⎫ / CUC / CUA Leu / CUG ⎭ | CCU ⎫ / CCC / CCA Pro / CCG ⎭ | CAU ⎫ His / CAC ⎭ / CAA ⎫ Gln / CAG ⎭ | CGU ⎫ / CGC / CGA Arg / CGG ⎭ | U C A G |
| A | AUU ⎫ / AUC Ile / AUA ⎭ / AUG Met | ACU ⎫ / ACC / ACA Thr / ACG ⎭ | AAU ⎫ Asn / AAC ⎭ / AAA ⎫ Lys / AAG ⎭ | AGU ⎫ Ser / AGC ⎭ / AGA ⎫ Arg / AGG ⎭ | U C A G |
| G | GUU ⎫ / GUC / GUA Val / GUG ⎭ | GCU ⎫ / GCC / GCA Ala / GCG ⎭ | GAU ⎫ Asp / GAC ⎭ / GAA ⎫ Glu / GAG ⎭ | GGU ⎫ / GGC / GGA Gly / GGG ⎭ | U C A G |

Source: Robert F. Weaver, and Philip W. Hedrick, *Basic Genetics.* Copyright © 1991 Wm. C. Brown Communications, Inc., Dubuque, Iowa. All Rights Reserved. Reprinted by permission.

## Wobble Introduces Ambiguity into Codon–Anticodon Interactions

For 61 triplets to act as codons, tRNAs must interact specifically with each triplet. Strict Watson-Crick base pairing between codon and anticodon would require 61 different anticodons and, correspondingly, 61 different tRNAs. As the characterization of tRNAs progressed it became clear that in many cases individual tRNAs could recognize more than one codon. In all cases the different codons recognized by the same tRNA were found to contain identical nucleotides in the first two positions and a different nucleotide in the third position (in the 3′ position of the codon). This relationship is the reason that the genetic code can be so neatly

BOX

# 29A

# Site-Specific Variation in Translation Elongation

The unusual amino acid selenocysteine (a derivative of cysteine in which the sulfur atom is replaced by a selenium atom) is an essential component in a small number of proteins. These proteins occur in prokaryotes and eukaryotes ranging from *E. coli* to humans. In all cases, selenocysteine is incorporated into protein during translation in response to the codon UGA. This codon usually serves as a termination codon but occasionally, in some required but unknown context of bases, is used to specify selenocysteine instead.

In *E. coli,* the products of four genes (*selA, selB, selC,* and *selD*) are required for the incorporation of selenocysteine. The product of the *selC* gene is tRNA$^{Ser}$ (a suppressor tRNA the anticodon of which is UCA). The first step in the incorporation of selenocysteine is catalyzed by seryl-tRNA synthase. The products of *selA* and *selD* function in the subsequent conversion of Ser-tRNA$^{Ser}$ to selenocysteyl-tRNA$^{Ser}$. The probable pathway is

tRNA$^{Ser}$ $\longrightarrow$ seryl-tRNA$^{Ser}$ $\longrightarrow$
phosphoseryl-tRNA$^{Ser}$ $\longrightarrow$ selenocysteyl-tRNA$^{Ser}$

Incorporation of selenocysteyl-tRNA$^{Ser}$ into protein in response to the UGA codon requires SELB (the protein product of the *selB* gene in *E. coli*). SELB is homologous in sequence to EF-Tu and probably replaces it in translation by specifically recognizing selenocysteyl-tRNA and UGA in the appropriate sequence context. Selenocysteyl-tRNA$^{Ser}$, in combination with SELB, must be capable of competing with termination factors for the translation of the termination codon when it occurs in the "right" context of bases. This process is known as site-specific variation in translation elongation.

Other proteins of unknown function that are homologous to EF-Tu are known to exist. Conceivably, these proteins might participate in the incorporation of other rare amino acids into unique positions in proteins.

arranged as shown in table 29.1, by codon families that differ in their third base.

The 3′ terminal redundancy of the genetic code and its mechanistic basis were first appreciated by Francis Crick in 1966. He proposed that codons and anticodons interact in an antiparallel manner on the ribosome in such a way as to require strict Watson-Crick pairing (that is, A-U and G-C) in the first two positions of the codon but to allow other pairings in its 3′ terminal position. Nonstandard base pairing between the 3′ terminal position of the codon and the 5′ terminal position of the anticodon alters the geometry between the paired bases; Crick's proposal, labeled the wobble hypothesis, is now viewed as correctly describing the codon–anticodon interactions that underlie the translation of the genetic code.

By inspecting the geometry that would result from different wobble pairings (fig. 29.8) and recognizing that inosine, the deaminated form of adenine, frequently occurs in

tRNAs in the 5′ position of the anticodon, Crick grasped the relationship between codon and anticodon. According to this relationship (table 29.2), when C or A occurs in the 5′ position of an anticodon, it can pair only with G or U, respectively, in the 3′ position of a codon. Transfer RNAs containing either G or U in the 5′, or wobble, position of the anticodon can each pair with two different codons, whereas an inosine (I) in this position produces a tRNA that can pair with three codons differing in the 3′ base. Subsequent sequence analysis of many tRNAs has proved this hypothesis as a feature of the translation of the "universal" genetic code (but see the following discussion).

A careful comparison of the wobble rules with the genetic code indicates that the minimum number of tRNAs required to translate all 61 codons is 31. With the addition of tRNA$_i^{Met}$ the total comes to 32. Most cells contain many more than this minimum number of tRNA types.

## Figure 29.8

Examples of standard (*a*) and wobble (*b* and *c*) base pairs formed between the first base in the anticodon and the third base in the codon.

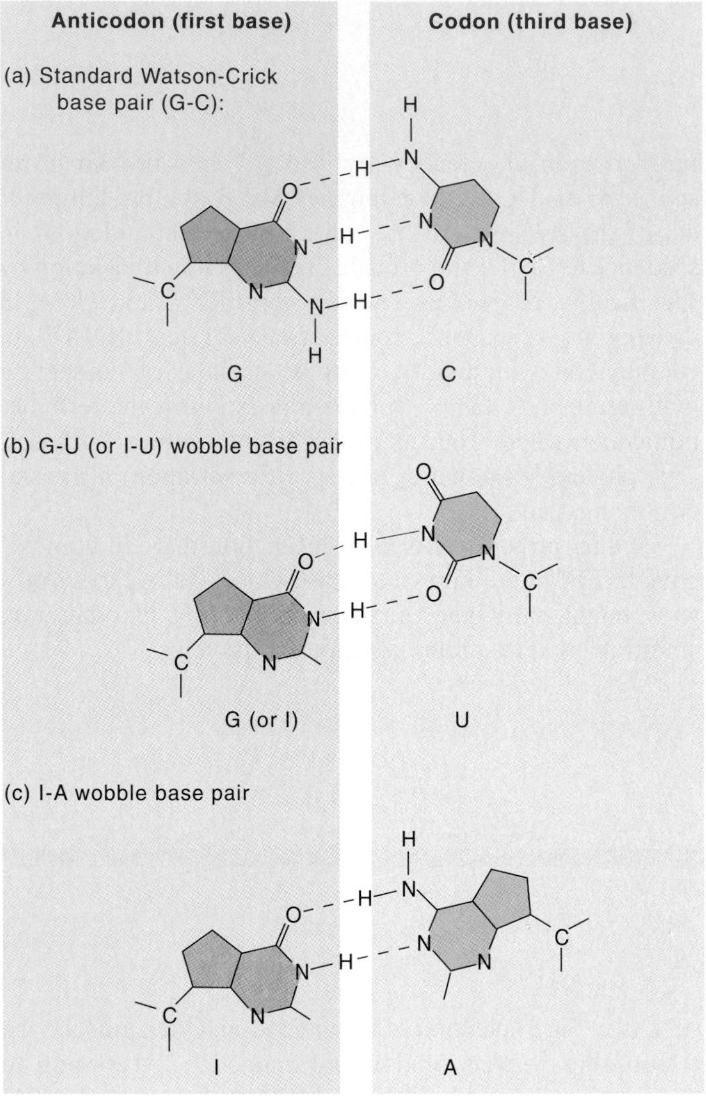

| Anticodon (first base) | Codon (third base) |

(a) Standard Watson-Crick base pair (G-C):

G    C

(b) G-U (or I-U) wobble base pair

G (or I)    U

(c) I-A wobble base pair

I    A

## Table 29.2

The Wobble Rules of Codon–Anticodon Pairing

| 5' Base of Anticodon | 3' Base of Codon |
| --- | --- |
| C | G |
| A | U |
| U | A or G |
| G | C or U |
| I | U, C, or A |

## The Code Is Not Quite Universal

It was originally believed that exactly the same genetic code is utilized by all genetic systems. Initial experiments supported this conclusion, since it was found that some mammalian mRNAs could be faithfully translated in cell-free bacterial systems. We now know, however, that significant variations in the meaning of specific code words occur in many genetic systems. The exact scope and nature of these variations are still being discovered, but the current view is that the variations in genetic meaning reflect divergences from the standard, or "universal," genetic code described

earlier, rather than independent origins of the genetic code.

Many of the known variations in the genetic code are found in genes of mitochondria and chloroplasts. It is easy to see why these genetic systems might be more plastic, since they frequently encode only 10–20 proteins. The remainder of the organellar proteins are derived by importing nuclear gene products.

The tRNAs used to translate mitochondrial mRNAs are entirely derived from mitochondrial chromosomes. The first clue that something was unusual about the mitochondrial genetic code was that only 24 types of tRNA could be found. According to Crick's rules of wobble, 32 tRNAs minimally are required for the translation of all 61 codons. One possible solution to this conundrum was that mitochondrial genes do not utilize all 61 codons. Another possibility was that the wobble rules might be different for mitochondria. In fact, the latter is the case. Crick's original wobble rules stated that at least two different tRNAs are needed to translate four codon families. In all of these cases (with the exception of the codon family specifying Arg; see the footnote to table 29.3), single tRNAs have been found responsible for specifying all four code words, and these tRNAs all contain a U in the "wobble" position of their anticodons. It appears that the mitochondrial ribosome allows these tRNAs to pair with all four members of the codon family. The six mitochondrial tRNAs that pair with the normal two codons contain an altered U in the wobble position, and this modification causes them to conform to the normal "wobble" rules.

Another peculiarity of the mitochondrial code emerged from a study of yeast codon usage. By comparing tables 29.3 and 29.1, you will see that the mitochondrial code has several differences in code word meaning. The codons beginning with CU represent Thr instead of Leu, the AUA codon represents Met instead of Ile, and the UGA codon represents Trp rather than a stop signal.

**Table 29.3**

The Genetic Code of Yeast Mitochondria

| | | Second Position | | | | | | | |
|---|---|---|---|---|---|---|---|---|---|
| **First Position (5′ end)** | | **U** | | **C** | | **A** | | **G** | | **Third Position (3′ end)** |

| First Position (5′ end) | | Second Position U | | Second Position C | | Second Position A | | Second Position G | | Third Position (3′ end) |
|---|---|---|---|---|---|---|---|---|---|---|
| **U** | UUU<br>UUC | Phe | AAG | UCU<br>UCC | | UAU<br>UAC | Tyr | AUG | UGU<br>UGC | Cys | ACG | **U**<br>**C** |
| | UUA<br>UUG | Leu | AAU* | UCA<br>UCG | Ser | AGU | UAA<br>UAG | STOP | | UGA<br>UGG | Trp | ACU* | **A**<br>**G** |
| **C** | CUU<br>CUC<br>CUA<br>CUG | Thr | GAU | CCU<br>CCC<br>CCA<br>CCG | Pro | GGU | CAU<br>CAC | His | GUG | CGU<br>CGC<br>CGA<br>CGG | Arg | GCA[b] | **U**<br>**C**<br>**A**<br>**G** |
| | | | | | | | CAA<br>CAG | Gin | GUU* | | | | |
| **A** | AUU<br>AUC | Ile | UAG | ACU<br>ACC | | AAU<br>AAC | Asn | UUG | AGU<br>AGC | Ser | UCG | **U**<br>**C** |
| | AUA<br>AUG | Met | UAC[a] | ACA<br>ACG | Thr | UGU | AAA<br>AAG | Lys | UUU* | AGA<br>AGG | Arg | UCU* | **A**<br>**G** |
| **G** | GUU<br>GUC<br>GUA<br>GUG | Val | CAU | GCU<br>GCC<br>GCA<br>GCG | Ala | CGU | GAU<br>GAC | Asp | CUG | GGU<br>GGC<br>GGA<br>GGG | Gly | CCU | **U**<br>**C**<br>**A**<br>**G** |
| | | | | | | | GAA<br>GAG | Glu | CUU* | | | | |

*Source:* From S. G. Bonitz et al., Codon recognition rules in yeast mitochondria, in *Proc. Natl. Acad. Sci. USA* 77:3167, 1980.

The codons (5′ → 3′) are at the left and the anticodons (3′ → 5′) are at the right in each box. (* designates U in the 5′ position of the anticodon that carries the —CH$_2$NH$_2$CH$_2$COOH grouping on the 5′ position of the pyrimidine.)

[a]Two tRNAs for methionine have been found. One is used in initiation and one is used for internal methionines.

[b]Although an Arg tRNA has been found in yeast mitochondria, the extent to which the CGN codons are used is not clear.

## The Rules Regarding Codon–Anticodon Pairing Are Species-Specific

The genetic code differs very little between species. By contrast, considerable differences occur between species in the anticodon translation system of tRNA, as evidenced by the mitochondrial tRNA system. In all systems the bases in the anticodon–codon complex run antiparallel, as in standard double-helix pairing, and in all cases only Watson-Crick-like base pairing occurs between the first two bases in the codon and the opposing bases in the anticodon segment of the tRNA. However, for the 3′ base in the codon, the rules for pairing vary with the species and with the base in question. These rules, summarized in table 29.4, are as follows.

When the 5′ base in the anticodon is a G it can pair with either a U or a C, and this is true in all organisms.

When the 5′ base in the anticodon is an A it can pair only with a U in the codon. However, it is rare that an A is found in this position of the anticodon. An A in this position in eukaryotes is usually deaminated to an inosine (I) base, which has an expanded capacity for pairing. Base A can pair only with U, but I can pair with U, C, or A. In eubacteria, deamination is limited to the conversion of the ACG sequence to an ICG anticodon. When C is the 5′ base in the anticodon it pairs with G only. The only known exception to this rule is found in eubacteria, in which the C is covalently modified in the tRNA that recognizes the AUA codon. Thus, in this one instance the modified C can pair with a 3′ A in the codon. A U base in the 5′ position of the anticodon shows the greatest variability. An unmodified U can pair with any of the four bases in the 3′ position of the codon.

Anticodon–Codon Pairing

| Anticodon First Base | Codon Third Base | Examples |
|---|---|---|
| U | U, C, A, G | Mitochondrial code in family boxes |
| *U | A, G | Mitochondrial code in two-codon sets |
| †U | A | Eukaryotes |
| ‡U | U, A, G | Eubacteria in family boxes |
| C | G | All codes |
| *C | A | Bacteria, isoleucine codon AUA |
| G | U, C | All codes |
| A | U | Rare |
| I | U, C, A | Eukaryotes, ICG in eubacteria |

*, †, ‡ = Various modifications of U (Yokoyama et al., 1985)

*C = modified C

(From Thomas Jukes et al., *Cold Spring Harbor Symp. Quant. Biol.* 32:775, 1987. Copyright © 1987 Cold Spring Harbor Laboratory Press, Cold Spring Harbor, N.Y. Reprinted by permission.)

This situation is reflected in the U-family box in mitochondria (table 29.3).

U can also be modified in various ways. In mitochondria one type of modification permits a U to pair with either an A or a G in the two-codon sets. In eukaryotes, another type of U modification limits U to pairing with an A in the codon, and in eubacteria, a third type of U modification permits U pairing with U, A, or G but not C in family boxes.

## The Steps in Translation

Protein synthesis involves more than 100 different proteins and more than 30 kinds of RNA molecules. The process begins by the attachment of amino acids to specific tRNA molecules. Subsequent steps take place on the ribosome; amino acids are transported to the ribosome on their tRNA carriers, and they do not leave the ribosome until they have become an integral part of a polypeptide chain.

### Synthases Attach Amino Acids to tRNAs

A unique class of enzymes, called aminoacyl-tRNA synthases, attach amino acids to their cognate tRNAs. This attachment serves two functions: (1) The linkage between amino acid and tRNA activates the amino acid, making the subsequent formation of a peptide bond energetically favorable. (2) The tRNA directs the amino acid to a designated location on a messenger RNA so that the amino acid is incorporated at the appropriate location in the polypeptide chain.

All synthase reactions proceed in two separate steps (fig. 29.9). In the first step the synthase recognizes its corresponding amino acid and its second substrate, ATP, and forms a mixed anhydride bond between the carboxyl group of the amino acid and the phosphate of AMP with the release of $PP_i$.

$$\text{Amino acid} + \text{ATP} \longrightarrow \text{aminoacyl-AMP} + PP_i \quad (1)$$

The equilibrium constant for this reaction is about 1, so that the energy derived from the cleavage of the phosphate anhydride of ATP is conserved in the mixed anhydride. Aminoacyl-AMP remains tightly bound to the enzyme and, as we will soon show, this fact has allowed researchers to crystallize this important complex and analyze its structure.

The second reaction catalyzed by the aminoacyl-tRNA synthases results in the attachment of the amino acid through an ester linkage to the 3′ terminal ribose of tRNA:

$$\text{Aminoacyl-AMP} + \text{tRNA} \longrightarrow \text{aminoacyl-tRNA} + \text{AMP} \quad (2)$$

Synthases differ with respect to their site of attachment to tRNA. Some synthases form the 2′ ester, some form the 3′ ester, and still others produce a mixture of the two. The specificity of the synthases was determined by analyzing their ability to act on tRNA derivatives lacking one or the other terminal hydroxyl group. Once esterified to the terminal ribose, the aminoacyl group can migrate between the vicinal 2′ and 3′ hydroxyl groups. Thus, in cells, aminoacyl-tRNAs are mixtures of 2′ and 3′ esters. Only the 3′ derivative is a substrate for the subsequent transpeptidation reaction catalyzed by the ribosome.

The sum of the two reactions catalyzed by aminoacyl-tRNA synthases is

$$\text{Amino acid} + \text{ATP} + \text{tRNA} \longrightarrow \text{aminoacyl-tRNA} + \text{AMP} + PP_i \quad (3)$$

This overall reaction is reversible but is driven to completion by the subsequent hydrolysis of $PP_i$ to two equivalents of $P_i$ through the action of ubiquitous pyrophosphatases. Thus, the formation of aminoacyl-tRNA consumes two equivalents of ATP. The energy that ultimately drives the formation of the peptide linkage during protein synthesis is derived from the ester linkage that joins amino acids to tRNA.

## Figure 29.9

Formation of aminoacyl-tRNA. This is a two-step process involving a single enzyme that links a specific amino acid to a specific tRNA molecule. In the first step (1) the amino acid is activated by the formation of an aminoacyl-AMP complex. This complex then reacts with a tRNA molecule to form an aminoacyl-tRNA complex (2).

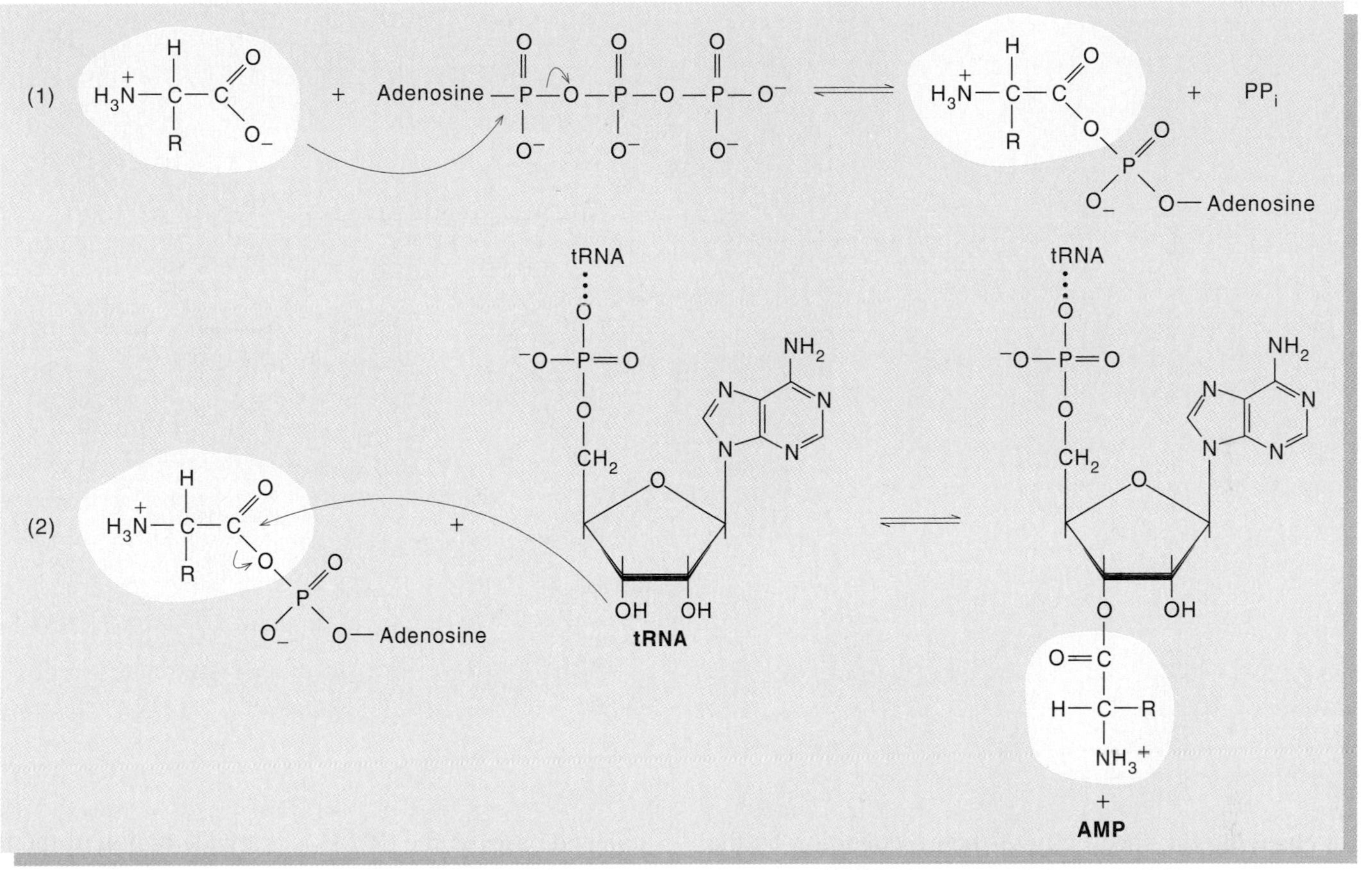

## Each Synthase Recognizes a Specific Amino Acid and Specific Regions on Its Cognate tRNA

Our understanding of synthase reactions and the types of the active sites involved in these reactions was advanced substantially by the crystallization and structural solution of tyrosyl-tRNA synthase complexed with the reaction intermediate tyrosyl-adenylate (fig. 29.10). The reaction intermediate is bound in a deep cleft in the enzyme and interacts with it through 11 hydrogen bonds. Six of these bonds are with the AMP moiety, and five are with the tyrosyl moiety of the intermediate. The amino acid selectivity of tyrosyl-tRNA synthase is thus determined primarily by the formation of specific hydrogen bonds with the amino acid.

Most cells contain only one synthase for each of the 20 amino acids specified by the genetic code. Each enzyme must be capable of recognizing its unique amino acid and one or more cognate tRNAs. Solving the puzzle of how synthases recognize tRNAs has been one of the major challenges in understanding the nature of the translation of the genetic code itself. The identity relationship between the tRNA and its synthase has, in fact, been called the "second genetic code."

Surprisingly, synthases fall into two categories with respect to the importance of the anticodon as a specificity element that they recognize. Some synthases (17 in *E. coli*) recognize the anticodon and some do not (3 in *E. coli*). This distinction has been demonstrated in several ways. First, some tRNAs can be genetically altered in their anticodon

**Figure 29.10**

Recognition of an adenylylated amino acid by the proper synthase. Shown is adenyl tyrosine bound to the tyrosyl synthase. A network of H bonds not only stabilizes the reaction but also serves to discriminate between different amino acid residues. MC designates main chain (backbone) carbonyl or amino groups participating in hydrogen bonding.

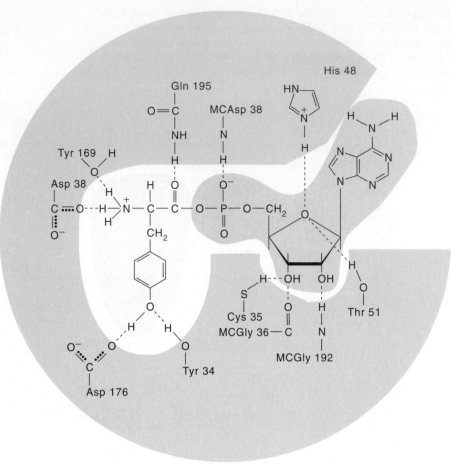

without changing the specificity of their recognition by the cognate synthase. Other synthases apparently require an unaltered anticodon in order to recognize their cognate tRNAs. Four examples are shown in figure 29.11.

The crystal structure of glutaminyl-tRNA synthase, complexed with tRNA and ATP, has been determined by Tom Steitz and his colleagues (fig. 29.12). This accomplishment provided the first structure of a tRNA–protein complex and thus offers important insight into the general nature of the recognition of tRNAs by proteins.

Glutaminyl-tRNA synthase is known to require the integrity of the anticodon loop of tRNA for recognition and is thus presumed to interact with both "ends" of the tRNA. In keeping with this interpretation, the synthase is asymmetrical and longer ($\approx 100$ Å) than the tRNA. Moreover, the crystal structure demonstrates the occurrence of significant contact between the protein and bases in the anticodon loop. The additional contacts between the synthase and tRNA appear to occur along the inside of the L-shaped structure. The recognition of the tRNA appears to arise from direct hydrogen bonding between amino acid residues of the pro-

tein and bases of the tRNA over a wide region of the tRNA structure.

The acceptor stem of the bound tRNA substrate plunges deeply into the active site pocket created by the dinucleotide fold, also the site of ATP binding. Surprisingly, binding in the active site induces a significant conformational change in the 3' terminal CCA acceptor sequence. This conformational change appears to cause the melting of the A-U pair at the end of the acceptor stem in a manner that allows these bases to interact with amino acid side chains of the protein. At this point in our understanding, it seems likely that the interactions that allow synthases to identify their cognate tRNAs are both diverse and complex.

## Aminoacyl-tRNA Synthases Can Correct Acylation Errors

Many aminoacyl-tRNA synthases appear to contain a second catalytic site that serves to correct errors by hydrolyzing incorrectly matched amino acids and tRNAs. These proofreading hydrolysis reactions are best understood in the case

**Figure 29.11**

Identity elements in four tRNAs. Each circle represents one nucleotide. Filled circles indicate nucleotides that serve as recognition elements to the appropriate aminoacyl-tRNA synthase. It is possible that other identity elements occur in these structures that are still to be discovered. (From L. H. Schulman and J. Abelson, Recent excitement in understanding transfer RNA identity, *Science* 240:1591, June 17, 1988. Copyright 1988 by the AAAS. Reprinted by permission.)

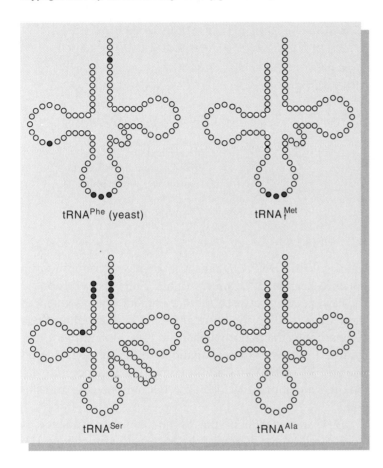

tRNA$^{Phe}$ (yeast)

tRNA$_f^{Met}$

tRNA$^{Ser}$

tRNA$^{Ala}$

**Figure 29.12**

Solvent-accessible surface representation of the GlnRS enzyme complexed with tRNA and ATP. The region of contact between tRNA and protein extends across one side of the entire enzyme surface and includes interactions from all four protein domains. The acceptor end of the tRNA and the ATP are seen in the bottom of the deep cleft. Protein is inserted between the 5′ and 3′ ends of the tRNA and disrupts the expected base pair between U1 and A72. (From M. G. Rould, J. J. Persona, D. Söll, and T. Steitz, "Structure of *E. coli* glutamyl-tRNA synthetics complexed with tRNA$^{Gln}$ and ATP at 2.8-Å resolution, implications for tRNA discrimination," *Science* 246:1135–1142, 1989, © 1989 by the AAAS.)

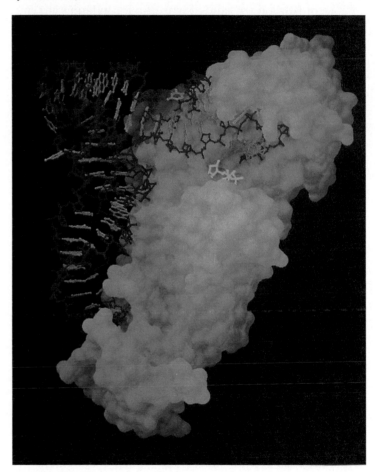

of isoleucyl-tRNA synthase. The amino acids isoleucine and valine, which differ by only a single methylene group, are among the most difficult to discriminate. Isoleucyl-tRNA synthase is capable of distinguishing these two amino acids at its aminoacylation site but occasionally and mistakenly forms valyl-tRNA$^{Ile}$. When isoleucyl-tRNA synthase encounters valyl-tRNA$^{Ile}$ the hydrolytic active site specifically degrades it:

$$\text{Valyl-tRNA}^{Ile} + H_2O \longrightarrow \text{valine} + \text{tRNA}^{Ile} \quad \textbf{(4)}$$

In this way the erroneous incorporation of valine in place of isoleucine is avoided. The sequential action of aminoacylation and proofreading sites contributes to an overall error frequency of translation of less than 1 in 10,000.

## A Unique tRNA Initiates Protein Synthesis

The translation of every protein begins with the incorporation of the amino acid methionine. A unique initiator tRNA, tRNA$_i^{Met}$, is responsible for the incorporation of this initiating methionine in all protein-synthesizing systems, and it also plays an important role in selecting the appropriate translation start site in mRNA. Generally, only two tRNAs occur in cells that specify methionine. We designate the one that is responsible for the incorporation of internal methionines simply as tRNA$^{Met}$. A single methionyl-tRNA synthase is responsible for activating both methionine isoacceptor tRNAs.

The functional discrimination between the two isoacceptor tRNAs is accomplished by protein initiation and elongation factors. Initiation factors recognize tRNA$_i^{Met}$, and elongation factors recognize tRNA$^{Met}$. Clearly, tRNA$_f^{Met}$ and tRNA$^{Met}$ must possess structural features that allow them to be distinguished from all other tRNAs by the single methionyl-tRNA synthase, but they must be sufficiently different that they can be discriminated by the factors.

In prokaryotic systems, a specific formylating enzyme exists that can recognize tRNA$_f^{Met}$ and formylate, its amino terminus utilizing $N^{10}$-formyltetrahydrofolate as the formyl donor. This reaction serves to ensure that the initiator does not participate in the elongation reactions. For unknown reasons this recognition step is not possessed by eukaryotic systems.

## Translation Begins with the Binding of mRNA to the Ribosome

One of the major demands of protein synthesis is to select the appropriate initiator codon, generally AUG, for translation. This is accomplished at the level of the ribosome by the binding of the small ribosomal subunit to mRNA. The recognition of the appropriate start codon occurs in different ways in prokaryotes and eukaryotes.

In eukaryotes, the AUG sequence closest to the 5′ end of the mRNA usually serves as the start codon for the single protein encoded by each mRNA. The small ribosomal subunit binds at the 5′ end of the mRNA and moves along the strand until it encounters an AUG sequence that is recognized by base pairing with the anticodon of Met-tRNA$_i^{Met}$. In higher eukaryotes, but not in yeast, the recognition of this initiating AUG is facilitated by an initiation context of flanking bases. The preferred initiation context is GCCGCCpurCCAUGG. How this sequence is recognized is unknown, but when the initiation context of the first AUG departs from this sequence, the 40S subunit can bypass it and initiate at the next AUG downstream.

In prokaryotic systems, the initiating AUG codon may occur at any point within the mRNA, and more than one start site may occur within the same mRNA. How do prokaryotic ribosomes select the appropriate initiating codon from the much more abundant internal AUG sequences? The answer was suggested in the early 1970s when Shine and Dalgarno noticed that bacterial mRNAs contain a complementary purine-rich region (which has become known as the Shine-Dalgarno sequence) centered approximately 10 bases toward the 5′ side of the initiating AUG sequence and

### Table 29.5

The Sequence of the 3′ End of *E. coli* 16S RNA and Some Shine-Dalgarno Sequences at the 5′ End of Bacterial mRNAs

|  | The pyrimidine-rich complement to the Shine-Dalgarno sequence |
| --- | --- |
| 16S rRNA | 3′ ··· HOAUUCCUCCA CUA ··· 5′ |
| *lacZ* mRNA | 5′ ··· ACACAGGAAAC AGCUAUG ··· 3′ |
| *trpA* mRNA | 5′ ··· ACGAGGGGAAAUCUGAUG ··· 3′ |
| RNA polymerase β mRNA | 5′ ··· GAGCUGAGGAACCCUAUG ··· 3′ |
| r-Protein L10 mRNA | 5′ ··· CCAGGAGCAAA GCUAAUG ··· 3′ |

The purine-rich Shine-Dalgarno sequence    The initiation codon

that *E. coli* 16S RNA contains a seven-base pyrimidine-rich sequence near its 3′ terminus (table 29.5). They proposed that base pairing between these complementary sequences could serve to align the initiating AUG for decoding. Such base pairing is the major mechanism for codon initiator recognition in *E. coli*. Thus, mutations in the Shine-Dalgarno sequence of mRNA that improve pairing enhance translation, and mutations that decrease pairing decrease translation.

The importance of the Shine-Dalgarno sequence is underscored by the action of the bacterial toxin colecin E3. This toxin inactivates the small subunit of the prokaryotic ribosome by cleavage of about 50 residues from the 3′ terminus of 16S rRNA. The cleavage disrupts the sequence that is complementary to the Shine-Dalgarno sequence and thus specifically inhibits the initiation process. Because of the fundamental differences between prokaryotic and eukaryotic initiation just described, colecin E3 does not inhibit the eukaryotic ribosome.

## Dissociable Protein Factors Play Key Roles at the Different Stages in Protein Synthesis on the Ribosome

At each stage in protein synthesis on the ribosome—initiation, elongation, and termination—a different set of protein factors is engaged by the ribosome. Why do such protein factors, which are crucial to the translation, exist separate from the ribosome? Why must they cycle on and

**Figure 29.13**

Formation of the initiation complex for protein synthesis in prokaryotes. *E. coli* has three initiation factors bound to a pool of 30S ribosomal subunits. One of these factors, IF-3, holds the 30S and 50S subunits apart after termination of a previous round of protein synthesis. The other two factors, IF-1 and IF-2, promote the binding of both fMet-tRNA$^{fMet}$ and mRNA to the 30S subunit. The binding of mRNA occurs so that its Shine-Dalgarno sequence pairs with 16S

rRNA and the initiating AUG sequence with the anticodon of the initiator tRNA. The 30S subunit and its associated factors can bind fMet-tRNA$^{fMet}$ and mRNA in either order. Once these ligands are bound, IF-3 dissociates from the 30S subunit, permitting the 50S subunit to join the complex. This releases the remaining initiation factors and hydrolyzes the GTP, which is bound to IF-2.

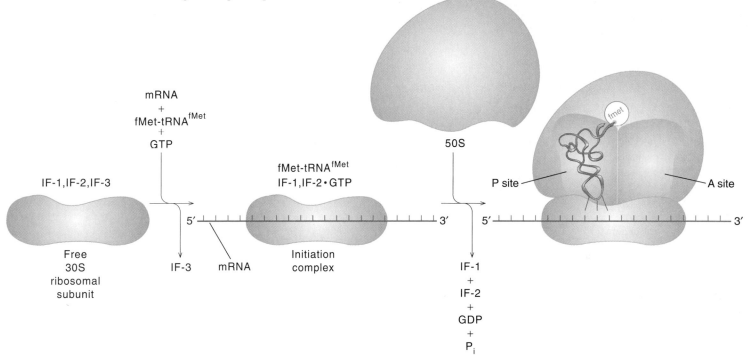

off the ribosome, and why are their functions not performed by firmly bound ribosomal proteins? The answers to these questions appear to lie in the economy of structure that is afforded by the cycling strategy. Since factors cannot interact simultaneously with the ribosome, it makes sense for the ribosome to interact with them sequentially at a single site.

## *Protein Factors Aid Initiation*

Even though specific differences distinguish the initiation process in eukaryotes and prokaryotes, three things must be accomplished to initiate protein synthesis in all systems: (1) The small ribosomal subunit must bind the initiator tRNA; (2) the appropriate initiating codon on mRNA must be located; and (3) the large ribosomal subunit must associate with the complex of the small subunit, the initiating tRNA, and mRNA. Nonribosomal proteins, known as initiation factors (IFs), participate in each of these three processes. IFs interact transiently with a ribosome during initiation and thus differ from ribosomal proteins, which remain continuously associated with the same ribosome.

*E. coli* has three initiation factors (fig. 29.13) bound to a small pool of 30S ribosomal subunits. One of these factors, IF-3, serves to hold the 30S and 50S subunits apart after termination of a previous round of protein synthesis. The other two factors IF-1 and IF-2, promote the binding of both fMet-tRNA$_i^{Met}$ and mRNA to the 30S subunit. As we noted before, the binding of mRNA occurs so that its Shine-Dalgarno sequence pairs with 16S RNA and the initiating AUG sequence with the anticodon of the initiator tRNA. The 30S subunit and its associated factors can bind fMet-tRNA$_i^{Met}$ and mRNA in either order. Once these ligands are found, IF-3 dissociates from the 30S, permitting the 50S to join the complex. This releases the remaining initiation factors and hydrolyzes the GTP that is bound to IF-2. The initiation step in prokaryotes requires the hydrolysis of one equivalent of GTP to GDP and P$_i$.

Initiation of protein synthesis in eukaryotic cells requires a much more complex spectrum of initiation factors, abbreviated eIF. At least nine separate factors have been identified, some of which are composed of as many as 11 different peptide subunits. The exact function of only a few

**Figure 29.14**

Formation of the initiation complex for protein synthesis in eukaryotes. The reaction begins with the small subunit held apart from the large subunit by an antiassociation factor and ends with the hydrolysis of GTP and joining of the large subunit as in prokaryotes. The intervening reactions are different. A much more complex spectrum of initiation factors (eIFs) is involved, and the exact function of only a few of these factors is known with certainty. The

Met-tRNA$^{fMet}$ first binds to the small subunit as a ternary complex with the eIF-2 and GTP. This ternary complex then binds to the 5' end of mRNA with the aid of several factors, one of which, eIF-4F, contains a subunit that specifically binds to the terminal cap structure of the mRNA. Binding to mRNA is followed by a scanning reaction that moves the small subunit along the mRNA, usually to the first AUG, in a reaction driven by the hydrolysis of ATP to ADP and $P_i$.

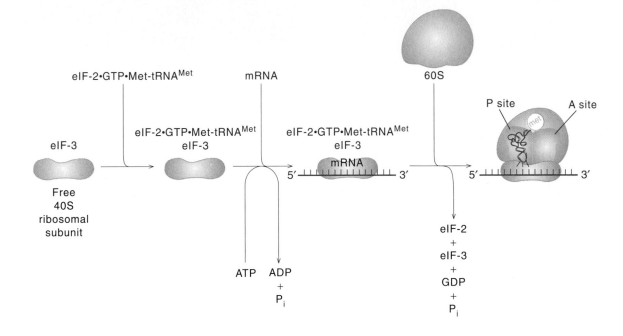

of these factors is known. The main features of the process are outlined in figure 29.14. As in the prokaryotic system, the reaction begins with the small subunit held apart from the large subunit by an antiassociation factor and ends with the hydrolysis of GTP and joining of the large subunit. The intervening reactions are different in prokaryotes and eukaryotes. The Met-tRNA$_i^{Met}$ first binds to the small subunit as a ternary complex with eIF-2 and GTP. The resulting preinitiation complex then binds to the 5' end of mRNA with the aid of several factors, one of which, eIF-4F, contains a subunit that specifically binds to the terminal cap structure of mRNA. Binding to mRNA is followed by a scanning reaction that moves the small subunit along the mRNA, usually to the first AUG, in a reaction driven by the hydrolysis of ATP to ADP and $P_i$. The mRNA binding and scanning reactions, which have no counterpart in the prokaryotic system, position the ribosome on the initiating AUG sequence in the appropriate flanking nucleotide sequence.

## Three Elongation Reactions Are Repeated with the Incorporation of Each Amino Acid

At the conclusion of the initiation process, the ribosome is poised to translate the reading frame associated with the initiator codon. The translation of the contiguous codons in mRNA is accomplished by the sequential repetition of three reactions with each amino acid. These three reactions of elongation are similar in both prokaryotic and eukaryotic systems; two of them require nonribosomal proteins known as elongation factors (EF). Interestingly, the actual formation of the peptide bond does not require a factor and is the only reaction of protein synthesis catalyzed by the ribosome itself.

The elongation reactions begin with the binding of the aminoacyl-tRNA specified by the codon immediately adjacent to the initiator codon. The binding of this aminoacyl-tRNA is catalyzed by an aminoacyl-tRNA binding factor,

**Figure 29.15**

Addition of the second aminoacyl-tRNA to the ribosome complex and the accompanying EF-Tu, EF-Ts cycle in *E. coli*. The purpose of the cycle is to regenerate another protein aminoacyl-tRNA complex suitable for transferring further aminoacyl-tRNAs to the A site on the ribosome.

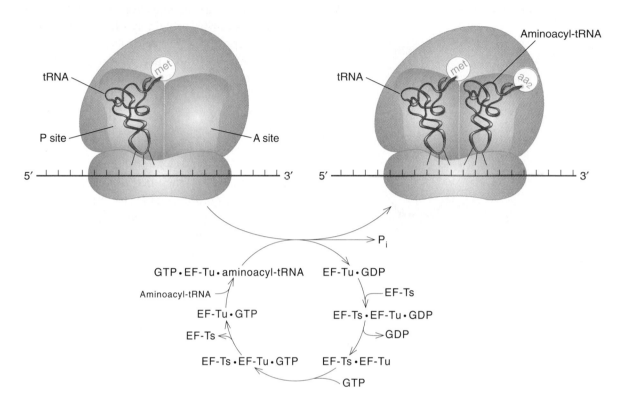

designated EF-Tu in bacteria and EF-1 in eukaryotic systems. The factor interacts with the ribosome as a ternary complex bound to both aminoacyl-tRNA and GTP. The binding of this complex to the ribosome is coupled to GTP hydrolysis. A productive complex forms if, and only if, the anticodon of the tRNA in the complex is complementary to the codon bound to the A site on the ribosome. Following the binding of aminoacyl-tRNA, EF-Tu is released from the ribosome as a complex with GDP. A second elongation factor, EF-Ts, catalyzes the regeneration of the EF-Tu–GTP complex so that it can again bind aminoacyl-tRNA. The series of reactions involved in this regeneration is depicted in figure 29.15. EF-1 is a multisubunit protein that combines the functional properties of EF-Tu and EF-Ts.

Peptide bond formation occurs immediately following the dissociation of the binding factor from the ribosome. This reaction is known as transpeptidation, and the enzymatic center that catalyzes it is known as peptidyl transferase, although it promotes the conversion of an ester to a peptide bond. Transpeptidation is generally assumed to proceed by nucleophilic attack by the amino group of the incoming aminoacyl-tRNA on the carbonyl of the ester of peptidyl-tRNA with the formation of a tetrahedral intermediate (fig. 29.16).[a]

The final step in elongation is known as translocation (fig. 29.17). This reaction, like aminoacyl-tRNA binding, is catalyzed by a factor (the translocation factor, known as EF-G in prokaryotic systems and EF-2 in eukaryotic systems) that cycles on and off the ribosome and hydrolyzes GTP in the process. The overall purpose of translocation is to move the ribosome physically along the mRNA to expose the next codon for translation.

The structural differences between bacterial and eukaryotic elongation factors are highlighted by the selective action that diphtheria has on eukaryotic systems (box 29B).

[a]H. F. Noller, V. Hossarth, and L. Zimniak. Ribosomes stripped of all their protein can still catalyze transpeptidation. *Science* 256:14–16, 1992.

## Figure 29.16

Formation of the first peptide linkage. The formylmethionine group is transferred from its tRNA at the P site to the amino group of the second aminoacyl-tRNA at the A site of the ribosome. This involves nucleophilic attack by the amino group of the second amino acid on the carboxyl carbon of the methionine. The resulting bond formation attaches both amino acids to the tRNA at the A site.

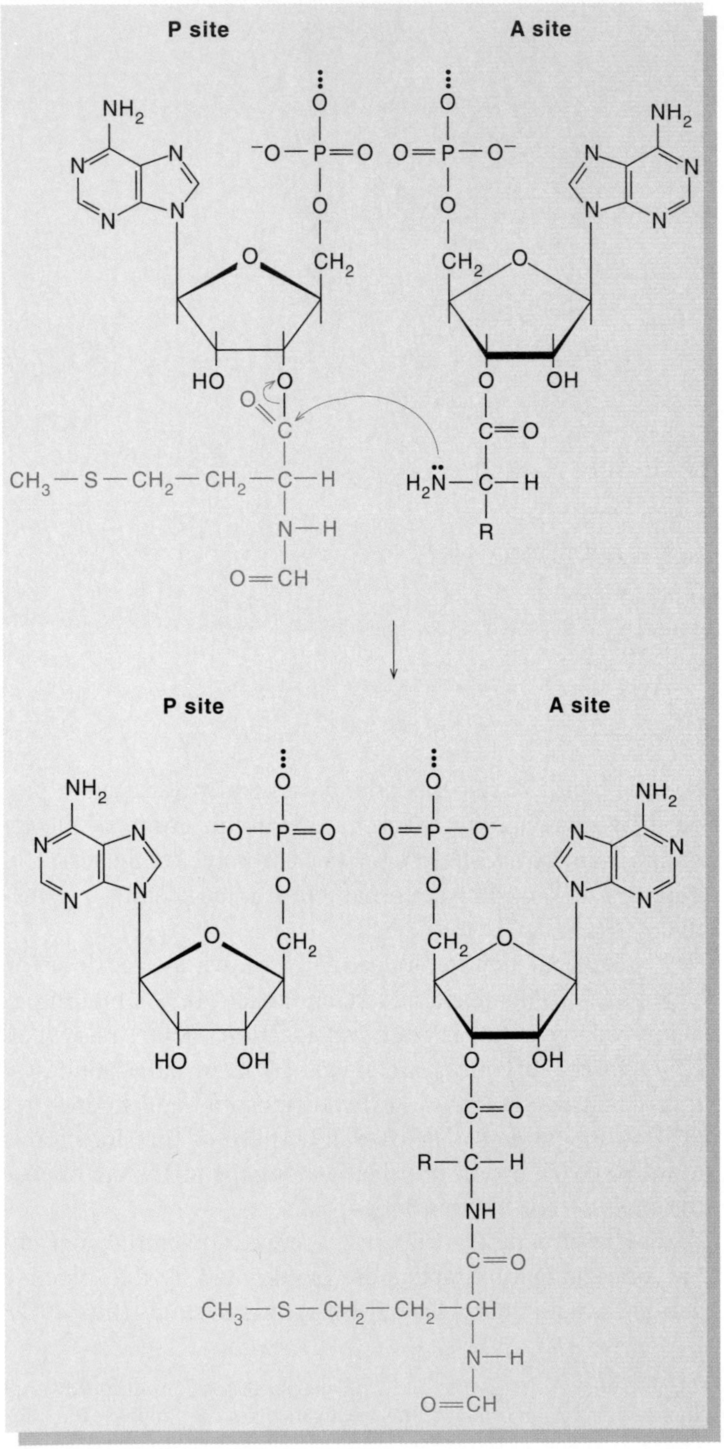

## Two GTPs Are Required for Each Step in Elongation

The hydrolysis of GTP plays a conspicuous role in the translation process. Two equivalents of GTP are hydrolyzed during elongation with the incorporation of each amino acid. This hydrolysis accounts for about half of the total energy consumed during protein synthesis. The chemical and functional purposes of GTP hydrolysis are best understood in the case of *E. coli* EF-Tu and EF-G. The sites of GTP binding and hydrolysis are situated on these factors. The interaction of the factor–GTP complex with the ribosome is believed to activate the hydrolytic site. GTP hydrolysis leads to GDP. No covalent intermediates are formed. Rather, the change of bound ligand to GDP that results from hydrolysis and release of $P_i$ is believed to change the conformation of the factor. The factor when bound to GTP is thought to be in the "on" configuration, able to bind to the ribosome and through binding to cause either aminoacyl-tRNA binding or translocation. The binding of GDP to the factor puts it in the "off" configuration and causes it to dissociate from the ribosome. Hydrolysis itself is not required for these changes to occur. This point is demonstrated by the fact that the nonhydrolyzable analog of GTP, GMPPCP, containing a methylene bridge instead of an oxygen between the $\beta$ and $\gamma$ phosphorus, can substitute for it in the reactions catalyzed by the elongation factors (see fig. 29.18). The use of this analog slows these reactions by delaying the dissociation of the factors from the ribosome.

   The translation factors that interact with GTP are members of the so called G protein superfamily, which includes the signal-transducing G proteins that link membrane receptors with their intracellular targets (see chapter 24) and the Ras proteins that function as growth regulators (see supplement 4: Carcinogenesis and Oncogenes). All G proteins bind and hydrolyze GTP, and it is believed that they all act by the same general mechanism we have just described. Thus, when bound to GTP these proteins are in their "active" configuration, and when bound to GDP as a result of hydrolysis they are converted to their "inactive" configuration.

## Figure 29.17

The translocation reaction in *E. coli*. The translocation reaction occurs immediately after peptide synthesis. It involves displacement of the discharged tRNA from the P site and concerted movement of the peptidyl-tRNA and mRNA so that the peptidyl-tRNA is bound to the P site and the same three nucleotides in the mRNA. The A site is vacated and ready for the addition of another aminoacyl-tRNA. Translocation in eukaryotes is similar except that the EF-2 factor is involved instead of the EF-G factor.

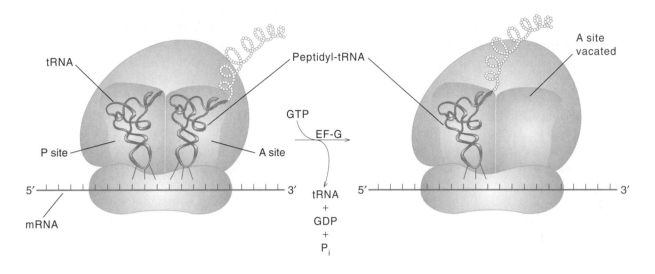

## Figure 29.18

The structure of guanylyl methylene diphosphonate (GMPPCP).

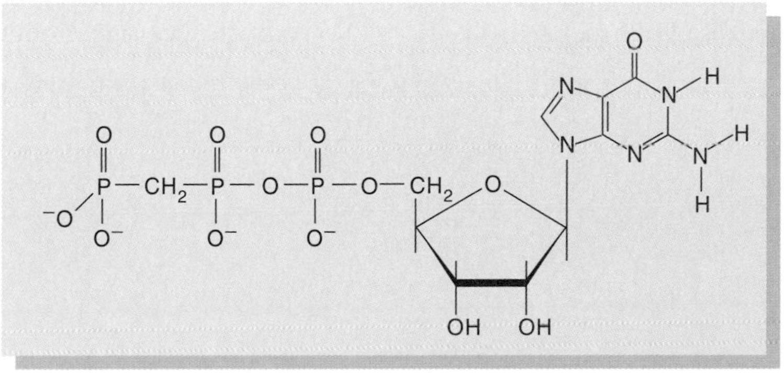

BOX

29B

# Mechanisms of Damage Produced by Certain Toxins

A single exotoxin, diphtheria toxin, is responsible for the pathogenesis of *Corynebacterium diphtheriae* and the disease diphtheria. The pathogenic consequences can be prevented by immunization with toxoid, an inactivated form of the purified toxin. Curiously, the structural gene for the toxin is carried by a bacterial virus, called β phage, that must infect the bacterium to induce toxin production. The widespread immunization against diphtheria employed in the United States has caused β phage, but not *C. diphtheriae,* largely to disappear. A catalytically identical but structurally very different exotoxin is produced by *Pseudomonas aeruginosa.*

In the cytoplasm of the cell, the catalytic portion of diphtheria toxin acts as a very specific protein-modifying enzyme. It catalyzes the ADP-ribosylation and consequent inactivation of EF-2 by the following reaction:

$$\text{EF-2} + \text{NAD}^+ \longrightarrow \text{ADP-ribosyl-EF-2}$$
$$+ \text{ nicotinamide} + \text{H}^+$$

This reaction is reversible when conducted *in vitro,* but under the conditions of pH and nicotinamide concentration that exist in the cell, it is irreversible. Thus, diphtheria toxin kills cells by irreversibly destroying the ability of EF-2 to participate in the translocation step of protein synthesis elongation. A number of other protein toxins have subsequently been found to ADP-ribosylate and inactivate cellular proteins involved in other essential cellular pathways. For example, cholera and pertussis toxins ADP-ribosylate and inactivate proteins important to cAMP metabolism.

The enzymatic specificity of diphtheria toxin deserves special comment. The toxin ADP-ribosylates EF-2 in all eukaryotic cells *in vitro* whether or not they are sensitive to the toxin *in vivo,* but it does not modify any other protein, including the bacterial counterpart of EF-2. This narrow enzymatic specificity has called attention to an unusual posttranslational derivative of histidine, diphthamide, that occurs in EF-2 at the site of ADP-ribosylation (see fig. 1). Although the unique occurrence of diphthamide in EF-2 explains the specificity of the toxin, it raises questions about the functional significance of this modification in translocation. Interestingly, some mutants of eukaryotic cells selected for toxin resistance lack one of several enzymes necessary for the posttranslational synthesis of diphthamide in EF-2 that is necessary for toxin recognition, but these cells seem perfectly competent in protein synthesis. Thus, the *raison d'être* of diphthamide, as well as the biological origin of the toxin that modifies it, remains a mystery.

Some other toxins inhibit protein synthesis by inactivating the ribosome through alterations of rRNA. A fungal toxin, α-sarcin, inactivates the ribosome by specific nuclease cleavage of a single phosphodiester bond in a purine-rich region found in the 23–28S RNA of all ribosomes. A second group of toxins, known as the ribosome-inactivating proteins, is abundant in plants of many species. The best known example of this type of toxin is ricin, the toxic agent in the castor bean. These proteins all act by the curious mechanism of removing a single adenine residue, by N-glycolytic cleavage, from the site in rRNA adjacent to the

**Figure 1**

Inactivation of the EF-2 factor by diphtheria toxin through the ADP-ribosylation of a modified histidine side chain.

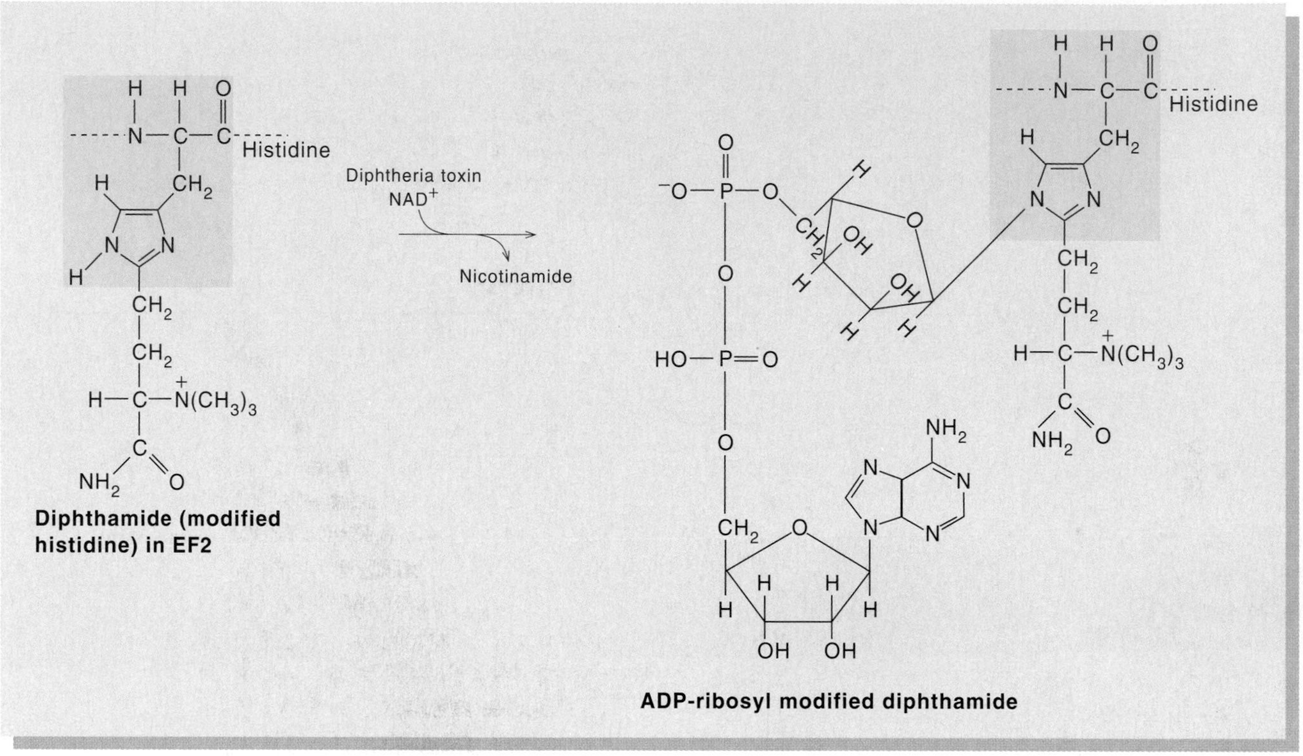

Diphthamide (modified histidine) in EF2

ADP-ribosyl modified diphthamide

one that is cleaved by $\alpha$-sarcin. Both of these modifications of rRNA alter the ribosome's ability to interact with factors, a fact suggesting that this region of rRNA is part of the factor interaction site on the large subunit. One of the ribosome-inactivating proteins, trichosanthin, was originally isolated from a plant tuber as the active principle of an Ori-ental folk medicine used to induce abortions. More recently, trichosanthin has been found to selectively inhibit viral protein synthesis in HIV-infected T cells, and it is now undergoing clinical trials as a therapeutic agent for the treatment of AIDS. The biological basis of this selective inhibition is unknown.

## Figure 29.19

The release reaction in *E. coli*. The release reaction occurs when the codon adjacent to the anticodon–codon complex is one of the stop codons, for example, UAA. The stop is recognized by release factor proteins that cause the peptidyl transferase to transfer the nascent polypeptide to water, forming a free polypeptide. Following the release of the polypeptide, the final tRNA and the mRNA dissociate from the ribosome, and the ribosome dissociates into its constituent subunits. RF is an abbreviation for protein release factor.

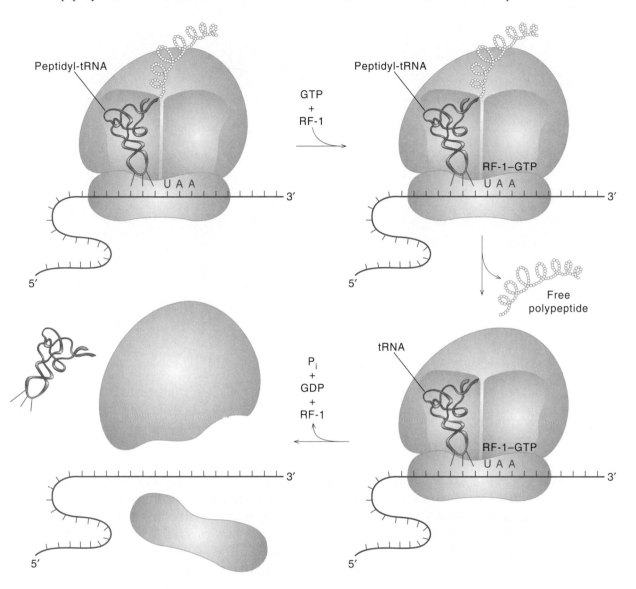

## *Termination of Translation Requires Release Factors and Termination Codons*

The last step in translation involves the cleavage of the ester bond that joins the now complete peptide chain to the tRNA corresponding to its C-terminal amino acid (fig. 29.19). Termination requires a termination codon, mRNA and at least one protein release factor (RF). The freeing of the ribosome from mRNA during this step requires the participation of a protein called ribosome releasing factor (RRF).

In *E. coli,* termination codons that arrive at the A site on the ribosome are recognized by one of three protein release factors, RF-1 recognizes UAA and UAG, and RF-2 recognizes UAA and UGA. The third release factor, RF-3, does not itself recognize termination codons but stimulates the activity of the other two factors.

The consequence of release factor recognition of a termination codon in the A site is to alter the peptidyl transferase center on the large ribosomal subunit so that it can accept water as the attacking nucleophile rather than requiring the normal substrate, aminoacyl-tRNA (fig. 29.20). In other

**Figure 29.20**

An *in vitro* assay for release factors.

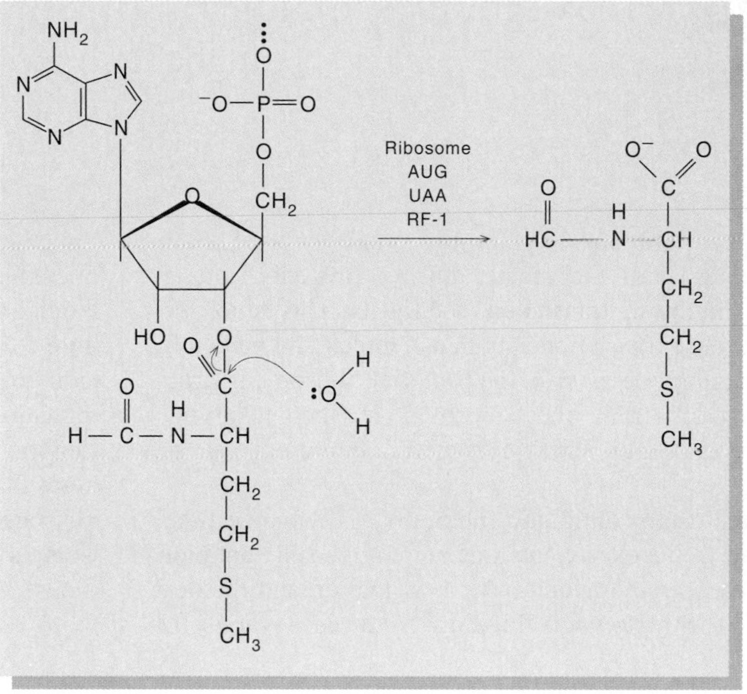

words, the termination reaction serves to convert the peptidyl transferase into an esterase. This feature of the termination reaction is clearly seen in the simple *in vitro* reaction that occurs when *E. coli* ribosomes are combined with fMet-tRNA$_i^{Met}$, RF-1, and two separate nucleotide triplets: AUG and UAA (see fig. 29.20). Formylmethionine is produced by hydrolysis, and this reaction is specifically inhibited by antibiotics such as sparsomycin that inhibit peptidyl transferase.

## *Ribosomes Can Change Reading Frame during Translation*

Normally, ribosomes march in lock step down the mRNA, one codon at a time, beginning at the initiating AUG and ending at a termination codon. In this way translation is rigidly maintained in a single reading frame. In the last few years, a number of specific exceptions to this seemingly rigid rule have been discovered in both prokaryotic and eukaryotic systems. In these cases, the correct expression of specific messages requires the ribosome to violate the rule of lock-step translation. One way that this occurs is known as translational frameshifting. Frameshifting generally occurs one base at a time; this change, or slipping, can be in either the +1 or −1 direction. Another way involves skipping as many as 50 bases downstream in the mRNA. This

process is known as translational jumping. From an informational point of view, translational jumping is analogous to mRNA splicing except that the information is not removed but is simply ignored by the translation system. Both frameshifting and jumping are favored at specific sites on the mRNA. In some cases these sites are known to require the presence of specific sequences or specific secondary structures in the mRNA. Although the nature of these sequences and structures have been identified in several cases, the mechanisms by which they are read have not been determined.

A notable instance of frameshifting occurs in the gene for the release factor RF-2, which recognizes the UGA stop codon in *E. coli*. This gene does not make RF-2 unless a frameshift takes place at an in-frame UGA sequence. When there is an abundance of the release factor, the frameshift is unlikely to occur because release occurs when the ribosome reaches this point on the messenger. However, when the release factor is in short supply, the ribosome is likely to pause at this site and then undergo a frameshift to make functional release factor. In this example, frameshifting results in a negative feedback mechanism for regulating the amount of release factor that is synthesized.

Translational fidelity of bacterial ribosomes is influenced in a more general way by the binding of antibiotics to the ribosome (box 29C).

BOX

## 29C    Antibiotics Inhibit by Binding to Specific Sites on the Ribosome

Many antibiotics (generally, small organic compounds with therapeutic utility) prevent bacterial growth by inhibiting translation (see fig. 1). This is not surprising, because translation is both a complex and metabolically essential process. Also, the bacterial ribosome is structurally distinct from the ribosome in the eukaryotic cytoplasm, and thus specific bacterial inhibitors can be found.

A great many antibiotic inhibitors of ribosome function belong to the class known as aminoglycoside antibiotics. Of these, streptomycin is the best known and the best investigated. Streptomycin binding produces a variety of functional alterations in the ribosome. One of the first to be recognized was the loss of translational fidelity. When bound to the small subunit, streptomycin distorts its structure so as to allow altered codon–anticodon pairing and the consequent incorporation of incorrect amino acids. Indeed, mutations to streptomycin resistance frequently prevent antibiotic binding and involve alterations in ribosomal proteins that are known to play a role in maintaining the fidelity of translation. Streptomycin binding also alters the ribosome's ability to participate properly in the initiation reactions.

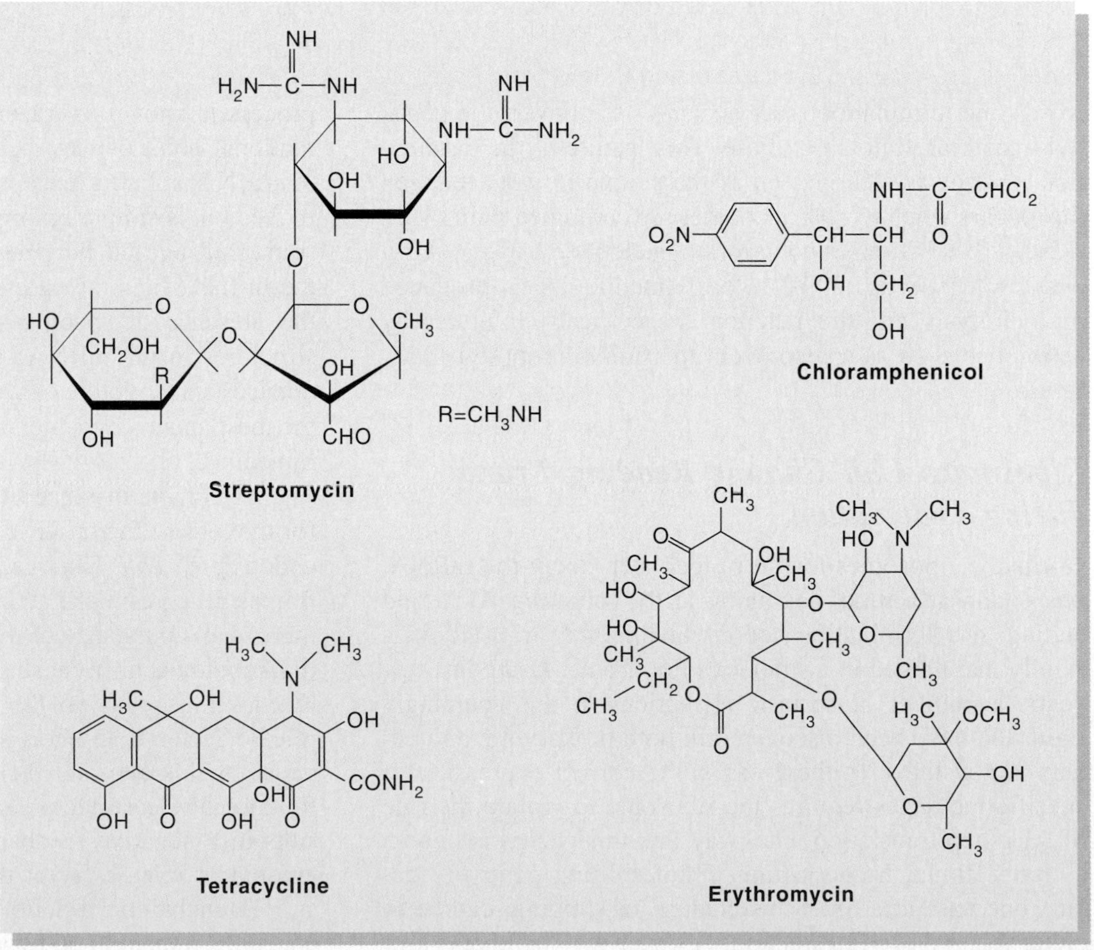

### Figure 1

The structures of some antibiotic inhibitors of protein synthesis. All of the inhibitors shown function by binding to specific sites on the ribosome.

Streptomycin

Chloramphenicol

Tetracycline

Erythromycin

# Targeting and Posttranslational Modification of Proteins

Despite the fact that only 20 amino acids (plus selenocysteine and formylmethionine in prokaryotic systems) are known to be directly specified by the genetic code, chemical analysis of mature proteins has revealed hundreds of different amino acids, all of them structural variants on the original 20. This structural diversity, which greatly expands the chemical lexicon of proteins, results from posttranslational modification of the primary products of translation. Our knowledge of the nature and significance of enzymatic reactions that bring about these important alterations is still very incomplete.

In addition to the modification of amino acid side chains many cases occur in which parts of the originally synthesized polypeptide chain are removed during the process of maturation. The types of modification and processing that polypeptide chains undergo is strongly related to the site of protein synthesis and to the mechanisms that are involved in targeting the polypeptide chain to its final destination. In fact, processing frequently begins during polypeptide synthesis and continues for some time thereafter. A major division involving the types of processing reactions that polypeptide chains undergo is related to the site of protein synthesis and the final destination of the protein. Proteins synthesized on free polysomes either remain in the cytosol or are targeted to the mitochondria, the chloroplasts, or the nucleus. Those that remain in the cytosol are often ready to function as soon as they are released from the ribosome. Those that must be transported to another site usually undergo substantial modification during the transport process. Many proteins, however, are synthesized on membrane-bound polysomes rather than on free polysomes. In bacteria such proteins are synthesized on polysomes associated with the inner plasma membrane; in eukaryotes, they are synthesized on polysomes associated with the endoplasmic reticulum. Proteins that are synthesized on ribosomes bound to the endoplasmic reticulum (ER) either remain in the ER or are targeted to the Golgi apparatus, secretory granules, the plasma membrane, or lysosomes. As a rule proteins that are synthesized on membrane-bound polysomes undergo more extensive modification before they reach their final destination and become fully functional (see chapter 16).

## Proteins Are Targeted to Their Destination by Signal Sequences

Despite the apparent complexity of the problem, protein transport in all systems is accomplished by a single, rather simple underlying mechanism; each polypeptide destined for transport contains an amino acid sequence known as a signal, or leader sequence that identifies the polypeptide to the appropriate transporting system. This generality was first recognized in the middle 1970s by Gunter Blobel, who articulated the underlying signal hypothesis (see fig. 16.8). Frequently, the signal sequence is cleaved from the parent polypeptide during the transport process. A protein containing a signal sequence that is cleaved on transport is known as a preprotein. A protein containing a peptide sequence that must be removed for the protein to be active is known as a proprotein, and a protein that contains both of these sequences is known as a preproprotein.

Many protein hormones are synthesized in a preproprotein form. For example, insulin mRNA is translated into a single polypeptide chain, preproinsulin, that contains 84 residues of proinsulin plus a 23-residue signal peptide (fig. 29.21). Within the islets of Langerhans of the pancreas, the N-terminal sequence is removed cotranslationally in targeting proinsulin to the Golgi apparatus, and the two disulfides joining the ends of the molecule are formed. Following this, the C-peptide region of proinsulin is removed to yield the mature circulating form of insulin with 51 residues in two disulfide-linked peptides.

Signal sequences usually occur at the amino terminus of polypeptides to be transported. These sequences vary in length from 10 to 40 amino acid residues and are generally characterized by the occurrence of a block of hydrophobic residues. The presence of hydrophobic residues reflects the fact that protein transport invariably involves the movement of proteins across lipid membranes. The hydrophobic residues both target the protein to the appropriate compartment and initiate penetration into the membrane.

## Some Mitochondrial Proteins Are Transported after Translation

Some proteins, especially those destined for the eukaryotic mitochondria and chloroplasts, are transported after their synthesis on free polysomes is complete. Such transport is known as posttranslational transport. In the case of posttranslational transport it is believed that the polypeptide to be transported must be unfolded from its native folded configuration by a system of polypeptide-chain-binding proteins (PCBs) before it can pass through the membrane. Posttranslational transport into the mitochondrion requires both ATP and a proton gradient. Presumably the energy from one or both of these sources is used to unfold the protein or separate it from the PCB system so that it can pass through the membrane.

The transport of proteins from the cytoplasm to the mitochondrion is further complicated by the fact that mito-

**Figure 29.21**

Processing of insulin. Insulin is synthesized by membrane-bound polysomes in the β cells of the pancreas. The primary translation product is preproinsulin, which contains a 24-residue signal peptide preceding the 81-residue proinsulin molecule. The signal peptide is removed by signal peptidase, cutting between Ala (−1) and Phe (+1), as the nascent chain is transported into the lumen of the endoplasmic reticulum. Proinsulin folds and two disulfide bonds cross-link the ends of the molecule as shown. Before secretion, a trypsinlike enzyme cleaves after a pair of basic residues 31, 32 and 59, 60; then a carboxypeptidase B-like enzyme removes these basic residues to generate the mature form of insulin.

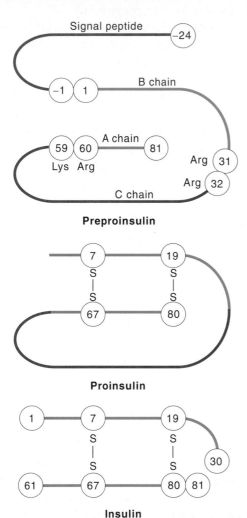

protein attached to the outer membrane of the mitochondria. This receptor protein guides the polypeptide chain to a transport channel, across which it is transported into the matrix. Once in the matrix, the first signal sequence is removed by a specific protease. A second signal sequence, which is exposed by this cleavage reaction, results in the transport of the polypeptide chain by a similar mechanism across a transport channel into the intermembrane space. In intermembrane space the second signal sequence is removed, and the protein folds into its mature configuration, which includes associating with a heme group.

As in the case of mitochondrial proteins, some of the proteins of chloroplasts are synthesized directly in the organelle, whereas others are synthesized in the cytosol and must be transported. The mechanisms for transport are very similar to those observed for transported mitochondrial proteins.

All nuclear proteins are synthesized on free polysomes in the cytosol. In contrast to the transport processes that involve specific amino-terminal sequences that are cleaved, more complex structural features are recognized in nuclear proteins, and they are somehow selectively transported in an intact state into the nucleus.

## Eukaryotic Proteins Targeted for Secretion Are Synthesized in the Endoplasmic Reticulum

The endoplasmic reticulum (ER) is the largest membrane-bounded organelle in a typical eukaryotic cell (see fig. 16.9). The ER consists of a continuous network of tubules and cisternae extending throughout the cytoplasm, with a total surface area many times that of the plasma membrane. Most of the ER is studded with ribosomes to form the rough endoplasmic reticulum (RER). The ribosomes of the RER are the site of synthesis of membrane and secretory proteins and are the starting point for the protein secretory pathway. The membrane and lumen of the ER contain a characteristic set of proteins that function to process secretory and membrane proteins. After only a short time in the RER, secretory proteins are transported, by a process of vesicle budding and fusion, to the Golgi apparatus and from there to the cell surface, secretory granules, or lysosomes. As you will recall, the Golgi apparatus is a flattened stack of membranes that is the primary site of protein targeting as well as a principal site of carbohydrate addition to form glycoproteins (see chapter 16).

Let us review the way in which proteins that are destined for assembly into membranes, for secretion, or for targeting to other areas of the cell enter the lumen of the ER (see fig. 16.8). The initial targeting of nascent polypeptides

chondria themselves are compartmentalized. They have both an outer membrane and an inner membrane (see chapter 14). Some proteins are targeted to either of these membranes, some are targeted to the fluid layer enclosed by the inner membrane, called the matrix, and others such as cytochrome $c_1$ are targeted to the fluid layer bounded by the inner membrane and the outer membrane, known as the intermembrane space (fig. 29.22). Cytochrome $c_1$ has two signal sequences located in series at the amino-terminal end of the preprotein. The first is recognized by a specific receptor

**Figure 29.22**

Two successive translocations are required to target proteins such as cytochrome $c_1$ to the intermembrane space. The precursors of cytochrome $c_1$ have two uptake-targeting sequences at its N terminus. The first targets the polypeptide to the matrix. In the matrix the first target sequence is cleaved by a specific protease. The second target sequence is thereby exposed and targets the polypeptide to intermembrane space, where the second target sequence is removed by another protease. The molecule folds and adds heme to become fully functional. (Source: After F. U. Hartl, J. Ostenmann, B. Guiard, and W. Neupert, *Cell* 51:1021–1027, 1987; and E. C. Hurt and A. P. G. M. van Loon, How proteins find mitochondria and intramitochondrial compartments, *Trends Biochem. Sci.* 11:204–207, 1986.)

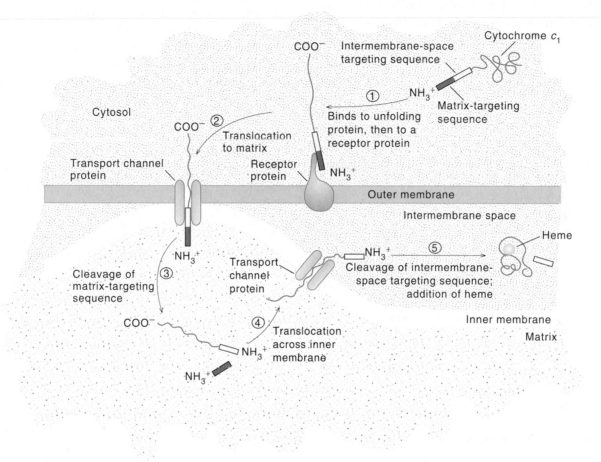

to the ER membrane results from the cotranslational recognition of a signal sequence by a ribonucleoprotein complex called the signal recognition particle, or SRP. The SRP is a complex that consists of six different proteins and a single 300-residue RNA molecule designated 7S RNA. This complex is especially adapted to recognize and bind to the signal sequence when the total nascent chain has reached a length of about 90 amino acid residues. On binding, the SRP arrests translation of the nascent chain while it searches for a specific receptor known as the SRP receptor, or "docking protein." The SRP receptor is an integral membrane protein that protrudes on the inner face of the ER membrane. When docking is complete, the SRP dissociates from the ribosome, which is now bound by the signal sequence of its nascent polypeptide to the SRP receptor. After dissociation of the SRP, the translation resumes, and translocation of the nascent polypeptide across the membrane begins. It has recently been discovered that GTP is essential for the docking maneuver and that its hydrolysis is probably essential for the subsequent release of SRP. Furthermore, sequencing of one of the protein subunits of SRP, SRP54, has revealed that it is homologous to the GTP-binding domain of the G proteins. Thus, signal sequence recognition by SRP may operate in a manner analogous to the GTP-dependent functioning of members of the G protein family.

Numerous nonmembranous proteins are retained by and function in the ER. Retention of these proteins is accomplished by a surprisingly simple mechanism. Soluble ER proteins all share the common C-terminal sequence. Lys-Asp-Glu-Leu (or KDEL in single-letter language). A protein receptor appears to be bound to the ER membrane that recognizes and binds this sequence. The passive nature of this retention mechanism is illustrated by the fact that if the four C-terminal residues are removed from an ER protein, it is no longer retained by the organelle.

## Proteins That Pass through the Golgi Apparatus Become Glycosylated

Glycosylation presents a conspicuous modification of proteins as they pass through the Golgi apparatus (see chapter 16). Many of the cell surface and secretory proteins produced here are glycoproteins. Because of the diversity of the glycosylation reactions that occur in the Golgi and the fact that this organelle is the principal site of protein transport, it seems likely that the carbohydrate moieties attached to proteins are responsible for targeting them to their destinations.

One clear-cut example supporting this conclusion is the role of mannose-6-phosphate residues in targeting proteins to the lysosome. Digestive enzymes destined for the lysosome are identified by the presence of mannose-6-phosphate by binding to a specific mannose-6-phosphate receptor protein. The critical step in targeting glycoproteins from the Golgi to the lysosome is their recognition by the enzymes that catalyze the two-step phosphorylation of terminal mannose residues in their N-linked oligosaccharide side chains. Failure to phosphorylate the mannose residues results in the secretion of lysosomal enzymes. This malfunction, you may remember, is associated with the syndrome known as I-cell disease, a condition that leads to the crowding of lysosomes with damaged proteins that it cannot degrade (see chapter 16).

## Processing of Collagen Does Not End with Secretion

Recall that collagen is an extracellular matrix protein that serves as a major constituent of many connective tissues (see figs. 4.10 to 4.13). Collagen fibrils have a distinctive banded pattern with a periodicity of 680 Å. Individual fibrils are composed of three polypeptide chains wound around one another in a right-handed helix with a total length of 3,000 Å. Each of the polypeptide chains in the triple helix has a repetitious tripeptide sequence, Gly-X-Y, where X is frequently a proline and Y is frequently a hydroxyproline. The latter amino acid is not one of the 20 that are specified genetically, so it must be formed posttranslationally by a modification of some of the prolines.

Since collagen is a secreted protein, we know that it must follow the route of synthesis that starts on a ribosome bound to the endoplasmic reticulum (fig. 29.23). This is where the fun begins, as we find that the nascent collagen polypeptide has extensive N and C termini (150 and 250

amino acids, respectively, for type I collagen) that are not found in mature collagen. The function of these extensions appears to be to facilitate the initial interaction of chains in triplets and to stabilize the triple helices once they have been formed. The first modification reaction to take place in the endoplasmic reticulum is hydroxylation of specific proline residues and some lysine residues by two different enzyme systems. Glycosylation begins soon thereafter with O-glycosylation of certain hydroxylysine residues and N-glycosylation of certain asparagine residues. Next, the modified polypeptide chains, in clusters of three, form interchain disulfides near the C termini. This brings the chains in close proximity, thereby facilitating the winding reaction that leads to triple-helix formation. The winding reaction proceeds in the C to N direction. The triple-helix procollagen molecule is packaged into a secretory vesicle. Following exocytosis, the ends of the procollagen molecule are removed by extracellular proteases. Collagen fibrils form by the spontaneous association of the mature collagen molecules.

## Bacterial Protein Transport Frequently Occurs during Translation

The problem of protein transport in bacterial systems is relatively simple. Polypeptides synthesized in the cytoplasm may function there, may be inserted into the plasma membrane, or may be secreted by being passed through this membrane.

Most noncytoplasmic bacterial proteins are targeted into or across the plasma membrane while they are being synthesized on the ribosome. This process is known as cotranslational transport. Cotranslational secretion of proteins in *E. coli* involves a group of proteins encoded by the *sec* (for secretion) genes. Some of these proteins are membrane proteins that function by recognizing and binding the signal sequence of nascent polypeptide chains as they emerge from the ribosome. As a consequence of cotranslational transport and the action of the sec proteins, most bacterial ribosomes engaged in the synthesis of secreted proteins are tethered to the cytoplasmic face of the plasma membrane by their nascent peptide chains. A leader peptidase that is an integral membrane protein cleaves the leader sequence from the secreted polypeptide.

## Protein Turnover

Cellular proteins are continuously being formed and degraded. At first glance, continuous degradation appears to be wasteful. However, protein degradation is of major biological importance in regulating protein levels, in protecting

**Figure 29.23**

Major events in the posttranslational processing of collagen.

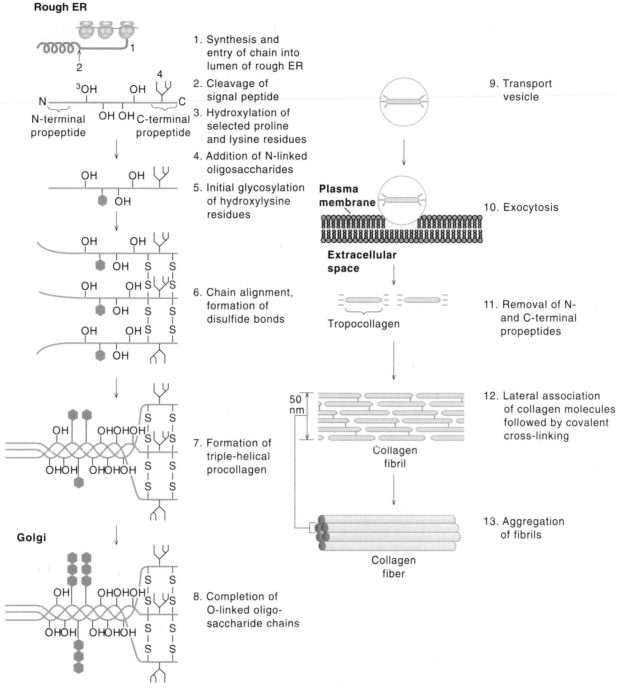

**Rough ER**

1. Synthesis and entry of chain into lumen of rough ER

2. Cleavage of signal peptide

N-terminal propeptide
C-terminal propeptide

3. Hydroxylation of selected proline and lysine residues

4. Addition of N-linked oligosaccharides

5. Initial glycosylation of hydroxylysine residues

6. Chain alignment, formation of disulfide bonds

7. Formation of triple-helical procollagen

**Golgi**

8. Completion of O-linked oligosaccharide chains

9. Transport vesicle

**Plasma membrane**

10. Exocytosis

**Extracellular space**

Tropocollagen

11. Removal of N- and C-terminal propeptides

50 nm

Collagen fibril

12. Lateral association of collagen molecules followed by covalent cross-linking

13. Aggregation of fibrils

Collagen fiber

against the accumulation of abnormal proteins, in controlling growth and development, and in allowing adaptation to changing environmental conditions.

Although many proteolytic enzymes are known to exist, we are just beginning to understand how they operate to govern the levels and types of proteins within cells—to distinguish those proteins that must be degraded from those that must be preserved.

## The Lifetimes of Proteins Differ

The level of a protein within a cell is determined by the balance between its rates of synthesis and degradation. As a consequence, changes in protein levels can be brought about by changes either in synthetic or degradative rates. Moreover, a rapid rate of degradation ensures that the concentration of a protein rises or falls rapidly when its synthetic rate changes.

## Table 29.6

Half-Lives of Some Proteins in Mammalian Cells

| Enzyme | Half-Life (h) |
|---|---|
| **Rapidly degraded** | |
| 1. c-myc, c-fos, p53 oncogenes | 0.5 |
| 2. Ornithine decarboxylase | 0.5 |
| 3. δ-Aminolevulinate synthase | 1.1 |
| 4. RNA polymerase I | 1.3 |
| 5. Tyrosine aminotransferase | 2.0 |
| 6. Tryptophan oxygenase | 2.0 |
| 7. β-Hydroxyl-β-methylglutaryl coenzyme A reductase | 2.0 |
| 8. Deoxythymidine kinase | 2.6 |
| 9. Phosphoenolpyruvate carboxykinase | 5.0 |
| **Slowly degraded** | |
| 1. Arginase | 96 |
| 2. Aldolase | 118 |
| 3. Cytochrome $b_5$ | 122 |
| 4. Glyceraldehyde-3-phosphate dehydrogenase | 130 |
| 5. Cytochrome $b$ | 130 |
| 6. Lactic dehydrogenase (isoenzyme 5) | 144 |
| 7. Cytochrome $c$ | 150 |

## Table 29.7

Correlation between Half-Lives of Cytosolic Proteins and Amino Acid Residue at the N Terminal

| Amino-Terminal Residue | Half-Life |
|---|---|
| **Stabilizing** | |
| Methionine | |
| Glycine | |
| Alanine | > 20 h |
| Serine | |
| Threonine | |
| Valine | |
| **Destabilizing** | |
| Isoleucine | ≈30 min |
| Glutamate | |
| Tyrosine | |
| Glutamine | ≈10 min |
| Proline | ≈7 min |
| **Highly Destabilizing** | |
| Leucine | |
| Phenylalanine | |
| Aspartate | ≈3 min |
| Lysine | |
| Arginine | ≈2 min |

(*Source:* From A. Bachmjuir et al., *In vivo* half-life of a protein is a function of its amino-terminal residue, *Science* 234:179, 1986. Copyright © 1986.)

The half-lives of eukaryotic proteins are different and distinct. Representative examples are listed in table 29.6. It is apparent that degradative rates of individual proteins within a single cell can vary over a wide range. At least in the mammalian liver, enzymes that occupy important metabolic control points are degraded most rapidly, whereas especially long-lived proteins are rarely the sites of metabolic control. In addition, it has been established that degradative rates of specific proteins can vary with changes in physiological conditions. Thus, the protein degradative mechanism is in some way tuned to metabolic control.

How are individual proteins selected for hydrolysis by the proteolytic machinery? Two sequence characteristics have been identified as correlating with rates of protein degradation. Several years ago it was observed through sequence analysis that rapidly degraded liver proteins, those with half-lives of less than 2 h, nearly all contain regions of their sequence that are rich in the amino acids proline, glutamate, serine, and threonine. These regions, involving from 10 to 60 amino acid residues, have been designated PEST sequences on the basis of the single-letter designations of

their constituent amino acids. Presumably, the PEST sequences create structural domains that are recognized by proteolytic enzymes.

A second structural feature that has been correlated with protein degradative rates is the N-terminal residue of the mature form of cytoplasmic proteins. This relationship, summarized in table 29.7, has become known as the N-terminal rule. Here the presumption is that the amino-terminal residue is at least partly responsible for recognition by the degradative machinery. Proteins that are degraded according to the N-terminal rule are believed to be recognized by the ubiquitin-ATP-dependent pathway, which we describe shortly.

## Abnormal Proteins Are Selectively Degraded

A very important function of protein degradation is to protect the organism against the consequences of intracellular accumulation of abnormal proteins. All cells possess a degradative system that is capable of recognizing "abnormal" proteins.

The features of this degradative system are best seen in *E. coli,* in which altered intracellular proteins can be readily manipulated. For example, it has been known for many years that incomplete chains of $\beta$-galactosidase are rapidly degraded even though the completed protein is very stable. In a similar way, many protein alterations that result from miscoding induced by streptomycin, or by the incorporation of amino acid analogs, also fail to accumulate because the protein products that contain them are rapidly degraded. This system presumably limits expression of heterologous proteins in bacteria that are degraded because they cannot form their native structure.

Another manifestation of the defense against abnormal proteins is seen in the heat shock response. This defense reaction, involving programmed changes in gene transcription and translation, is exhibited by essentially all cells under stressful conditions. The cells reduce their overall rates of gene transcription and translation and for a brief time produce a small repertoire of proteins called heat shock proteins (hsp). Some hsp play a role in the folding and transport of proteins during normal cellular function. In the heat shock response it is believed that these proteins protect against the presence of damaged proteins by binding to them and promoting either their refolding or their proteolytic degradation.

## Proteolytic Hydrolysis Occurs in Mammalian Lysosomes

The classic studies of Christian DeDuve in the 1960s established that mammalian cells contain a degradative organelle, the lysosome, that is produced in the Golgi apparatus and contains a large number of proteases and other hydrolytic enzymes. A similar vacuole containing hydrolytic enzymes is also present in yeast and higher plants. Together, the hydrolytic enzymes in these organelles are capable of completely degrading many macromolecules and delivering their monomeric units to the cytoplasm for further metabolism. The metabolic importance of the lysosome is demonstrated by the existence of various lysosomal storage diseases, in which one or more hydrolytic enzymes is missing. An accumulation of undegraded molecules results from these genetic diseases, frequently with devastating consequences.

A primary function of the lysosome is to digest protein-containing particles derived from the extracellular space. One mechanism of delivery is the process of endocytosis. Endocytosis is the invagination of a group of occupied receptors on the plasma membrane. Most mammalian cells can also engulf large extracellular particles by the less specific processes of pinocytosis and phagocytosis. The endocytic vesicles formed by these processes fuse with lysosomes to form secondary lysosomes where hydrolysis occurs. Lysosomes also function to degrade intracellular proteins, especially under conditions of nutritional deprivation. Under these circumstances, lysosomes can engulf cytoplasmic contents to form autophagic vacuoles and thus recycle cellular proteins.

Lysosomal proteases are called cathepsins, a name derived from the Greek term meaning "to digest." The interior of the lysosome is acidic and the cathepsins, like all lysosomal hydrolases, possess acidic pH optima and exhibit little enzymatic activity at neutral pH. This characteristic protects the cell from autolytic breakdown that might result from leakage of lysosomal contents into the neutral cytoplasm.

## Ubiquitin Tags Proteins for Proteolysis

A number of different proteolytic systems are thought to be responsible for the degradation of soluble proteins in the cytoplasm of eukaryotic cells. One of the best understood is that which involves ATP and the protein ubiquitin. Ubiquitin is a small protein of only 76 residues. It occurs universally in eukaryotic cells and is highly conserved in sequence; only three residues distinguish the ubiquitin in yeast and humans. The covalent attachment of ubiquitin to proteins is thought to "tag" them for subsequent hydrolysis by cellular proteases.

Three enzymes participate in the ATP–ubiquitin system that prepares proteins for proteolysis (fig. 29.24). In the first step of this series of reactions, the C-terminal glycine residue of ubiquitin is activated by forming a thiol ester with a specific activating enzyme designated E1. This reaction is driven by the hydrolysis of ATP to AMP and $PP_i$. Ubiquitin is then transferred to a sulfhydryl group of a second protein (E2) that is a substrate for a group of ubiquitin-targeting proteins designated E3. The E3 proteins are responsible for identifying proteins for degradation (some on the basis of the identity of their N-terminal residue and some because they are "abnormal") and catalyzing the covalent transfer of ubiquitin from E2 to these proteins. This attachment is via isopeptide bonds that join the carboxyl group of the previously activated C-terminal glycine residue of ubiquitin with $\epsilon$-amino groups of lysine side chains of the targeted proteins. Proteins tagged in this way are then recognized and degraded by specific proteases that release ubiquitin so that it can recycle.

## ATP Plays Multiple Roles in Protein Degradation

ATP plays a conspicuous and rather surprising role in the degradation of proteins. Clearly, the hydrolysis of peptide bonds is a reaction that in and of itself does not require the input of energy. Nonetheless, ATP is required for the action of many proteolytic enzymes, independent of its role in ubiquitination that we have just seen.

One particularly well-understood ATP-dependent protease is the La protease that is the product of the *lon* gene of *E. coli*. This protease, like many ATP-dependent proteases, is a large protein, and its ability to hydrolyze proteins is tightly coupled to its ability to hydrolyze ATP. Approximately two ATPs are hydrolyzed to ADP and $P_i$ for each peptide bond that is hydrolyzed. It appears that the hydrolysis of ATP is required to activate the proteolytic active site of La. Other proteases seem to require ATP as an allosteric effecter to activate their hydrolytic sites, but this ATP is not hydrolyzed.

Thus, ATP serves at least three roles in intracellular proteolysis: (1) It functions in the tagging of proteins through covalent attachment of ubiquitin, (2) it activates proteases such as La through hydrolysis, and (3) it serves as a positive allosteric effecter of other proteases without being hydrolyzed. The presumed function of this energetic requirement is to ensure the fidelity of protein degradation. It is, after all, nearly as important to cellular function to degrade proteins correctly as it is to synthesize them correctly.

**Figure 29.24**

The ubiquitin marking system targets certain proteins for degradation. At least three enzymes, E1, E2, and E3, are involved in addition to ubiquitin-specific proteases.

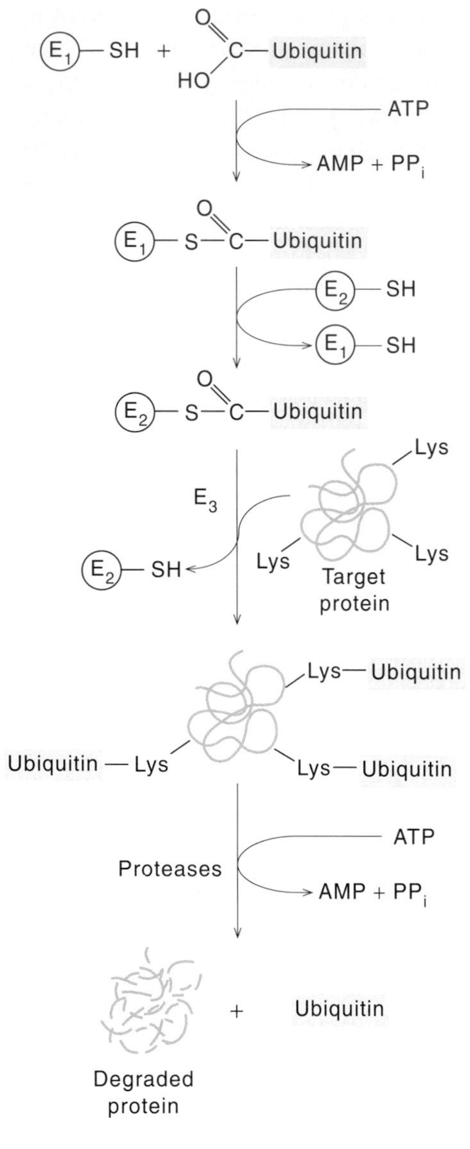

## Summary

We focused in this chapter on the complex mechanisms of protein synthesis. The following points are central to this subject.

1. Three types of RNA carry out protein synthesis: Ribosomal RNA, transfer RNA, and messenger RNA. Ribosomal RNA is invariably complexed with many proteins to form ribosomes, on which amino acids are assembled into polypeptides. The amino acids are brought to the ribosomes attached to transfer RNAs. The messenger RNA contains the instructions for translation in the form of the genetic code. Messenger RNAs form transient complexes with ribosomes. Individual aminoacyl-tRNAs bind to specific sites on the messenger RNAs. The interacting site on the messenger is the codon; the interacting site on the tRNA is the anticodon.

2. The part of the messenger that is translated is the reading frame. Eukaryotic messages carry only one reading frame, whereas prokaryotic messengers may carry more than one. In prokaryotes the initiation codons are recognized by a ribosome-binding site upstream of the start codon.

3. Most transfer RNAs have common parts and uncommon parts. The common parts facilitate binding of the aminoacyl-tRNAs to common sites on the ribosome. The uncommon sites permit specific reactions with charging enzymes that covalently attach the correct amino acids to the correct tRNA. Another uncommon site on the tRNAs is the anticodon, which leads to specific complex formation with the complementary codon site on the messenger.

4. Attachment of the amino acid to the tRNA is catalyzed by a specific aminoacyl synthase, which recognizes all the cognate tRNAs for a specific amino acid.

5. A unique methionyl-tRNA binds to the initiation codon on all messages.

6. The genetic code is the sequence relationship between nucleotides in the messenger RNA and amino acids in the proteins they encode. Triplet codons are arranged on the messenger in a nonoverlapping manner without spacers.

7. The code was deciphered with the help of synthetic messengers with a defined sequence, by analyzing the types of polypeptide chains that were made when these messengers were used in an *in vitro* protein-synthesizing system.

8. The genetic code is highly degenerate, with most amino acids represented by more than one codon. In many cases the 3′ base in the codon may be altered without changing the amino acid that is encoded.

9. The codon–anticodon interaction is limited to Watson–Crick pairing for the first two bases in the codon but is considerably more flexible in the third position.

10. Translation begins with the binding of the ribosome to mRNA. A number of protein factors transiently associate with the ribosome during different phases of translation: Initiation factors, elongation factors, and termination factors.

11. Initiation factors contribute to the ribosome complex with the messenger RNA and the initiator methionyl-tRNA. Elongation factors assist the binding of all the other tRNAs and the translocation reaction that must occur after each peptide bond is made. Termination factors recognize a stop signal and lead to the termination of polypeptide synthesis and the release of the polypeptide chain and the messenger from the ribosome.

12. A large number of antibiotics have been characterized that inhibit protein synthesis. These antibiotics are usually made by a particular microorganism, and they inhibit protein synthesis in a broad family of other organisms, mostly bacterial.

13. Specific enzymes catalyze folding after polypeptide synthesis.

14. Proteins are targeted to their destination by signal sequences built into the polypeptide chain. These signals are usually located at the N-terminal end of the protein and are generally cleaved during protein maturation.

15. Posttranslational modifications include many covalent alterations: Polypeptide processing, attachment of carbohydrate or lipid groups to specific side chains, and addition of many other low-molecular-weight ligands to side chains.

16. Intracellular protein degradation is not random. Different proteins have quite different half-lives, which are related to specific structural features. Imperfectly folded proteins and polypeptide fragments are frequently degraded most rapidly. In eukaryotes, lysosomes play a major role in protein degradation.

## Selected Readings

Bachmjuir, A., and A. Varshavsky, The degradation signal is a short-lived protein. *Cell* 56:1019–1032, 1989.

Beasley, E. M., and G. Schatz, Import of proteins into mitochondria. *Chemtracts* 2:305–317, 1991.

Bjork, G. R., J. U. Ericson, C. E. D. Gustafsson, T. G. Hagervall, Y. H. Jonsson, and P. M. Wilkstrom, Transfer RNA modification. *Ann. Rev. Biochem.* 56:263–288, 1987.

Böck, A., K. Forchhammer, J. Heider, and C. Baron, Selenoprotein synthesis: An expansion of the genetic code. *Trends Biochem. Sci.* 16:463–467, 1991.

Bond, J. S., and P. E. Butler, Intracellular proteases. *Ann. Rev. Biochem.* 56:333, 1987. An overview of the types of proteolytic enzymes that are found in cells and how they may function in biologically important cleavages of proteins.

Brunori, M., M. C. Silvestrini, and M. Pocchiari, The scrapie agent and the prion hypothesis. *Trends Biochem. Sci.* 13:309–313, 1988.

Burgess, T. L., and R. B. Kelly, Constitutive and regulated secretion of proteins. *Ann. Rev. Cell. Biol.* 3:243–294, 1987.

Craig, E. A., Chaperones: Helpers along the pathways to protein folding. *Science* 260:1902–1903, 1993.

Crick, F. H. C., Codon–anticodon pairing: The wobble hypothesis. *J. Mol. Biol.* 19:548–555, 1966. A classic paper.

Ellis, R. J., and S. M. van der Vies, Molecular chaperones. *Ann. Rev. Biochem.* 60:321–348, 1991.

Englander, S. W., In pursuit of protein folding. *Science* 262:848–850, 1993.

Englesberg-Kulka, H., and R. Schoulakerk-Schwarz, A flexible genetic code, or why does selenocysteine have no unique codon? *Trends Biochem. Sci.* 13(11):419–421, 1988.

Ferguson, M. A. J., and A. F. Williams, Lipids as membrane tethers for proteins. *Ann. Rev. Biochem.* 57:285, 1988.

Fessler, J. H., and L. I. Fessler, Biosynthesis of procollagen. *Ann. Rev. Biochem.* 47:129–162, 1978.

Finley, D., and A. Varshavsky, The ubiquitin system functions and mechanisms. *Trends Biochem. Sci.* 10:343–347, 1985.

Fox, T. D., Natural variation in the genetic code. *Ann. Rev. Gen.* 21:67, 1987. A review of the exceptions to the "universal" genetic code.

Gold, L., Posttranslational regulatory mechanisms in *E. coli. Ann. Rev. Biochem.* 57:199, 1988. A summary of the mechanisms of initiation and regulation of translation in prokaryotic systems.

Hershey, J. W. B., Protein phosphorylation controls translation rates. *J. Biol. Chem.* 264:20823, 1989. Describes how protein kinases are thought to regulate translation in eukaryotic systems.

Hershko, A., Ubiquitin-mediated protein degradation. *J. Biol. Chem.* 263:15237–15240, 1988.

Hoagland, M. B., M. L. Stephenson, J. F. Scott, L. I. Hecht, and P. Zamecnik, A soluble ribonucleic acid intermediate in protein synthesis. *J. Biol. Chem.* 231:241–257, 1958. Describes the pioneering tracer studies that chart the course of amino acid into polypeptide chain.

Kleinkauf, H., and H. Dohren, Nonribosomal polypeptide formation on multifunctional proteins. *Trends Biochem. Sci.* 8:281–283, 1983.

Kozak, M., The scanning model for translation: An update. *Mol. Cell. Biol.* 8:2737, 1989. The current view of the way in which the eukaryotic ribosome selects the initiating codon, by the originator of the scanning model.

Moore, P. B., The ribosome returns. *Nature* 33:223–227, 1988.

Nirenberg, M. W., and J. H. Mattaei, The dependence of cell-free protein synthesis in *E. coli* upon naturally occurring or synthetic polyribonucleotides. *Proc. Natl. Acad. Sci. USA* 47:1588–1602, 1961. The landmark paper reporting the finding that poly(U) stimulates the synthesis of polyphenylalanine.

Noller, H. F., V. Hoffarth, and L. Zimniak, Resistance of peptidyl transferase to protein extraction procedures. *Science* 256:1416–1418, 1992. Demonstration that peptidyl transferase is an RNA.

Normanly, J., and J. Abelson, tRNA identity. *Ann. Rev. Biochem.* 58:1029, 1989. A summary of the chemical features of tRNA.

Pain, V. M., Initiation of protein synthesis in mammalian cells. *Biochem. J.* 235:625, 1986. A comprehensive review of the complex process of translational initiation in mammalian systems, emphasizing the mechanism of the process and its regulation.

Parker, J., Errors and alternatives in reading the universal genetic code. *Microbiol. Rev.* 53:273, 1989. A summary of the current knowledge of alternative mechanisms of translation.

Pelham, H. R. B., Control of protein exit from the endoplasmic reticulum. *Ann. Rev. Cell. Biol.* 5:1, 1989. Describes the current picture of how proteins are sorted and transported through the endoplasmic reticulum.

Pfeffer, S. R., and J. E. Rothman, Biosynthetic protein transport and sorting by the endoplasmic reticulum and Golgi. *Ann. Rev. Biochem.* 56:829, 1987. An excellent overview of the major features of protein targeting in eukaryotic cells.

Proud, C. G., Guanine nucleotides, protein phosphorylation and control of translation. *Trends Biochem. Sci.* 11:73–77, 1986.

Rechsteiner, M., S. Rogers, and K. Rote, Protein structure and intracellular stability. *Trends Biochem. Sci.* 12:390–394, 1987.

Rothman, J. E., Polypeptide chain binding proteins: Catalysts of protein folding and related processes in cells. *Cell* 59:591, 1989. A description of the proteins that are thought to be involved promoting the formation of three-dimensional structure in proteins.

Rould, M. A., J. J. Perona, and T. A. Steitz, Structural basis of anticodon loop recognition by glutaminyl-tRNA synthetase. *Nature* 352:213–218, 1991.

Saks, M. E., J. R. Sampson, and J. N. Abelson, The transfer RNA identity problem: A search for rules. *Science* 263:191–197, 1994.

Siegel, V., and P. Walter, Each of the activities of signal recognition particle (SRP) is contained within a distinct domain. *Cell* 52:39–49, 1988.

Thompson, R. C., EFTu provides an internal kinetic standard for translational accuracy. *Trends Biochem. Sci.* 13:91–93, 1988.

Tobias, J. W., T. E. Shrader, G. Rocap, and A. Varshavsky, The N-end rule in bacteria. *Science* 154:1374–1377, 1991.

Von Figura, K., and A. Hasilik, Lysosomal enzymes and their receptors. *Ann. Rev. Biochem.* 55:167–193, 1986.

Webb, R., and L. A. Sherman, Chaperones classified. *Nature* 359:458–486, 1992.

## Problems

1. Compare the translation initiation signals in prokaryotic and eukaryotic systems, and describe those features of each type of mRNA that determine the frequency with which a particular message is translated. What consequences do these differences have for gene organization in the two systems?

2. The relationship between tRNAs and their synthases is sometimes called the "second genetic code." Explain.

3. A single tRNA can insert serine in response to three different codons: UCC, UCU, or UCA. What is the anticodon sequence of this tRNA?

4. How much energy is required to synthesize a single peptide bond in protein synthesis? How does this compare with the free energy of formation of the peptide linkage, which is about 5 kcal/mole?

5. Explain this statement: "The universal genetic code is not quite universal."

6. Explain why the use of GUG and UUG as initiation codons in place of AUG was not expected, even based on Crick's wobble hypothesis.

7. Assuming that translation begins at the first codon, deduce the amino acid sequence of the polypeptide encoded by the following mRNA template:

AUGGUCGAAAUUCGGGACACCCAUUUGAA–
–GAAACAGAUAGCUUUCUAGUAA

8. Assume that you have a copolymer with a random sequence containing equimolar amounts of A and U. What amino acids would be incorporated and in what ratio, when this copolymer is used as an mRNA?

9. Researchers often design degenerate oligonucleotides based on a protein sequence for use as hybridization probes to isolate the corresponding gene. (A degenerate oligonucleotide is actually a mixture of oligonucleotides, the sequences of which differ at positions corresponding to degeneracies in the genetic code.) The N-terminal amino acid sequence of a protein is:

Met-Val-Asp-Ser-Asn-Trp-Ala-Gln-Cys-Asp-Pro-Ala-Thr

Give the sequence of the least degenerate 20-residue-long oligonucleotide that hybridizes to the gene encoding this protein.

10. The effect of single-point mutations on the amino acid sequence of a protein can provide precise identification of the codon used to specify a particular residue. Assuming a single base change for each step, deduce the wild-type codon in each of the following cases.

(a) Gln ⟶ Arg ⟶ Trp

(b) Glu ⟶ Lys ⟶ Ile

(c)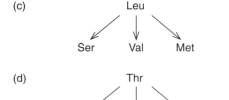
Leu → Ser, Val, Met

(d)
Thr → Ile, Pro, Lys

11. Even though the roles of IF-2, EF-Tu, EF-G, and RF-3 in protein synthesis are quite different, all four of these proteins share a domain with significant amino acid sequence similarity. Suggest a role for this conserved domain.

12. The antibiotic fusidic acid inhibits protein synthesis by preventing EF-G from cycling off of the ribosome. Fusidic-acid-resistant mutants of EF-G have been isolated. Fusidic acid resistance is recessive to sensitivity. In other words, an *E. coli* cell containing two EF-G genes, one resistant and one sensitive, is still sensitive to the antibiotic. Why? (*Hint:* Look at fig. 29.2.)

13. What are the major differences between the mechanisms of protein import into endoplasmic reticulum compared to protein import into mitochondria?

14. Scientists have tried to isolate the peptidyl transferase from ribosomes for many years without success. It is now thought that this activity is part of the ribosome (large subunit). Discuss this point, in view of what you know about other catalytic RNP complexes (RNA-protein complexes).

15. What are the possible amino acid changes that can result from a single nucleotide change in a GAA codon? Knowing the structures of the amino acids, what do you predict are the effects of the altered amino acids?

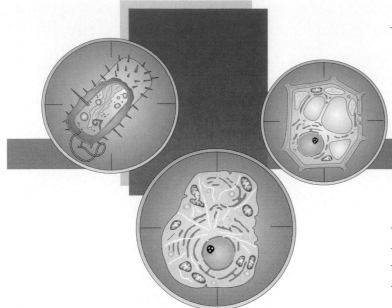

# Regulation of Gene Expression in Prokaryotes

*In bacteria the level of expression of a particular messenger RNA is a function of the affinity of the RNA polymerase for a DNA promoter; this affinity is modulated by regulatory proteins that bind to the DNA or the RNA polymerase.*

In all biological systems, gene expression is regulated so that gene products are produced either before or as they are needed. In this chapter we examine the mechanisms that ensure efficient regulation in the bacterium *Escherichia coli* and the bacteriophage λ.

*E. coli* maintains all of its genes in a state where they can be turned on or turned off on short notice. The short messenger lifetime makes it possible to control gene expression from the transcription level. The lack of separate compartments for RNA and protein synthesis has fostered mechanisms where translation actually exerts a direct role on transcription. These are some of the special features that have influenced the evolution of regulatory systems in *E. coli*.

## Control of Transcription Is the Dominant Mode of Regulation in *Escherichia coli*

The *E. coli* chromosome contains about 3,000 genes. This system is regulated so that under conditions of active growth, only about 5% of the genome is actively transcribed at any given time. The remainder of the genome is either silent or transcribed at a very low rate. When growth conditions change, some active genes are turned off, and other, inactive genes are turned on. The cell always retains its totipotency, so that within a short time (seconds to minutes in most cases) and given appropriate circumstances, any gene can be fully turned on. The fully expressing rRNA gene makes one copy per second, a fully turned-on β-galactosidase gene makes about one copy per minute, and a fully turned-on biotin synthase gene makes about one copy every 10 min. In the maximally repressed state, all these genes express less than one transcript every 10 min.

The level of transcription for any particular gene usually results from a collection of control elements organized into a hierarchy that coordinates all the metabolic activities of the cell. For example, when the rRNA genes are highly active, so are the genes for ribosomal proteins, and the latter are regulated in such a way that stoichiometric amounts of most of the ribosomal proteins are produced. When glucose is abundant, most genes involved in processing more complex carbon sources are turned off by a process called catab-olite repression. If the glucose supply is depleted and lactose is present, then the genes involved in lactose catabolism are expressed. In *E. coli*, the production of most RNAs and proteins is regulated exclusively at the transcriptional level, although notable exceptions occur. Rapid response to changing conditions is ensured partly by a short mRNA lifetime—on the order of 1–3 min for most mRNAs. Some mRNAs have appreciably longer lifetimes (10 min or longer) and the consequent potential for much higher levels of protein synthesis per mRNA subject to translational control. Examples of all these situations are considered later. Finally, the fine-level control for any particular enzyme system is subject to regulation by activators or inhibitors.

## The Initiation Point for Transcription Is a Major Site for Regulating Gene Expression

The rate of initiation of transcription can be regulated in several ways, most of which influence the rate of formation of the RNA polymerase–DNA promoter complex. The primary sequence of nucleotides in the promoter region is the first factor to be considered. The closer this sequence is to the consensus sequence, the greater the affinity of the polymerase for the promoter (fig. 30.1).

The rate of initiation of transcription also can be altered by changes in the RNA polymerase structure. This can occur by subunit replacement, subunit covalent modification, or small-molecule-induced allosteric transition. During a temperature upshift ($30 \rightarrow 42°C$), the usual $\sigma$ subunit ($\sigma^{70}$) is partially replaced by an alternative $\sigma$ factor ($\sigma^{32}$), changing the types of promoters recognized by the polymerase. In bacteriophage T4 infection, the subunits of the polymerase become ribose-adenylated, lowering the affinity of polymerase for bacterial promoters and raising the affinity for phage promoters. Binding of guanosine tetraphosphate (ppGpp) to RNA polymerase changes the structure of the polymerase so that it has a greatly lowered affinity for rRNA, tRNA, and ribosomal protein promoters and at the same time a somewhat greater affinity for some other promoters.

Finally, the rate of initiation of RNA synthesis can be controlled by auxiliary regulatory proteins that affect the rate of formation of the polymerase–promoter complex. According to their positive or negative action on gene expression, regulatory proteins are known as activators or repressors, respectively. Activators augment polymerase binding to the promoter, whereas repressors have the opposite effect.

**Figure 30.1**

Schematic diagram of DNA conformation in the rapid-start complex. Two regions most important in polymerase binding are lettered with most favored sequences. Transcription starts at the +1 base pair. Upstream of the start bases are numbers starting with −1.

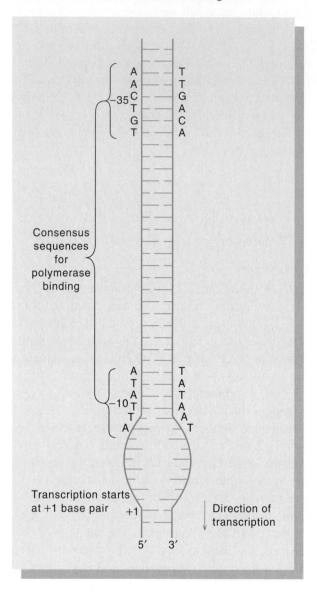

**Figure 30.2**

Different genetic elements of the *lac* operon. The operon contains a control region, the promoter–operator region, and three structural genes, *z*, *y*, and *a*. The *i* gene, a repressor, is also shown. It is not part of the operon, but it is located at an adjacent site on the genome with its own promoter.

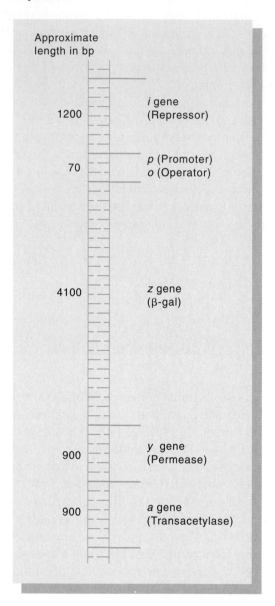

## Regulation of the Three-Gene Cluster Known as the *Lac* Operon Occurs at the Transcription Level

In bacteria it is common to find units of expression that contain clusters of two or more functionally related genes. Such is the case with the *lac* operon, a three-gene cluster associated with the metabolism of the dis-

saccharide lactose (fig. 30.2). The first of these genes, the *z* gene, encodes β-galactosidase, which hydrolyzes β-galactosides, in particular lactose, to produce the monosaccharides, glucose and galactose. The middle gene, *y*, encodes lactose permease, which is associated with the active transport of lactose into the cell. The third gene, *a*, encodes thiogalactoside transacetylase. A useful function has yet to be found for this gene.

## Figure 30.3

Effect of inducer on β-galactosidase synthesis. Differential plot expressing accumulation of β-galactosidase as a function of increase in mass of cells in a growing culture of *E. coli*. Because the abscissa and ordinate are expressed in the same units (micrograms of protein), the slope of the straight line gives galactosidase as the fraction ($P$) of total protein synthesized in the presence of inducer. (Source: After Melvin Cohn, *Bact. Rev.* 21:140, 1957.)

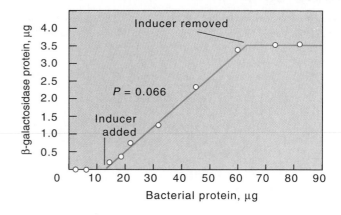

## Figure 30.4

Inducers of the *lac* operon. (*a*) All inducers have the β-galactoside structure shown, in which R can be a variety of substituents. (*b*) Allolactose is the natural inducer when cells are grown on lactose.

(*c*) Isopropyl-β-D-thiogalactoside (IPTG) is a synthetic inducer useful in the laboratory; the β-oxygen is replaced by a β-sulfur atom. This change prevents hydrolysis by β-galactosidase.

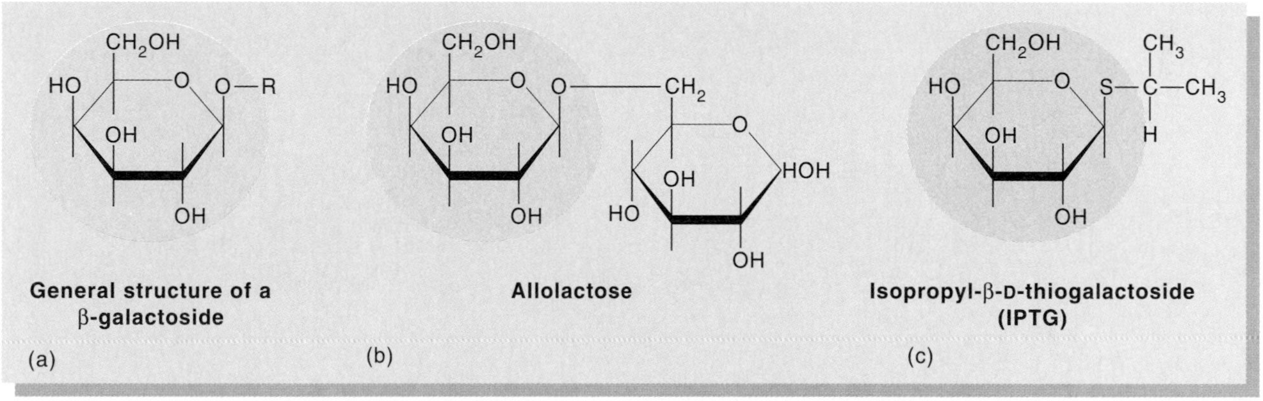

**General structure of a β-galactoside**

(a)

**Allolactose**

(b)

**Isopropyl-β-D-thiogalactoside (IPTG)**

(c)

Expression of the *lac* operon is regulated by controlling elements, which are separate from the structural genes. The controlling elements consist of a promoter locus, which is the site where RNA polymerase binds and initiates transcription; the promoter locus also contains sites for the binding of a repressor and an activator. The *i* gene encodes the repressor, and the *crp* gene encodes the activator.

## β-Galactosidase Synthesis Is Augmented by a Small-Molecule Inducer

Wild-type *E. coli* cells grown in the absence of lactose contain an average of 0.5–5.0 molecules of β-galactosidase per cell, whereas bacteria grown in the presence of an excess of lactose or certain lactose analogs contain 1,000–10,000 molecules per cell. Radioactive amino acid has been used as a tracer to show that the increase in enzyme activity observed on induction results from *de novo* protein synthesis. When excess β-galactoside inducer is added, enzyme activity increases at a rate proportional to the increase in total protein within the culture (fig. 30.3). Enzyme formation reaches its maximum rate within 3 min after inducer is added. Removal of inducer leads to cessation of enzyme synthesis in about the same amount of time.

A large number of compounds have been tested for their capacity to induce β-galactosidase. All inducers contain an intact, unsubstituted galactosidic residue (fig. 30.4). Many compounds that are not themselves substrates for β-galactosidase such as thiogalactosides are good inducers

**Figure 30.5**

Conversion of lactose to allolactose, the natural inducer of the *lac* operon. Ultimately, lactose is broken down to its constituent monosaccharides, galactose and glucose.

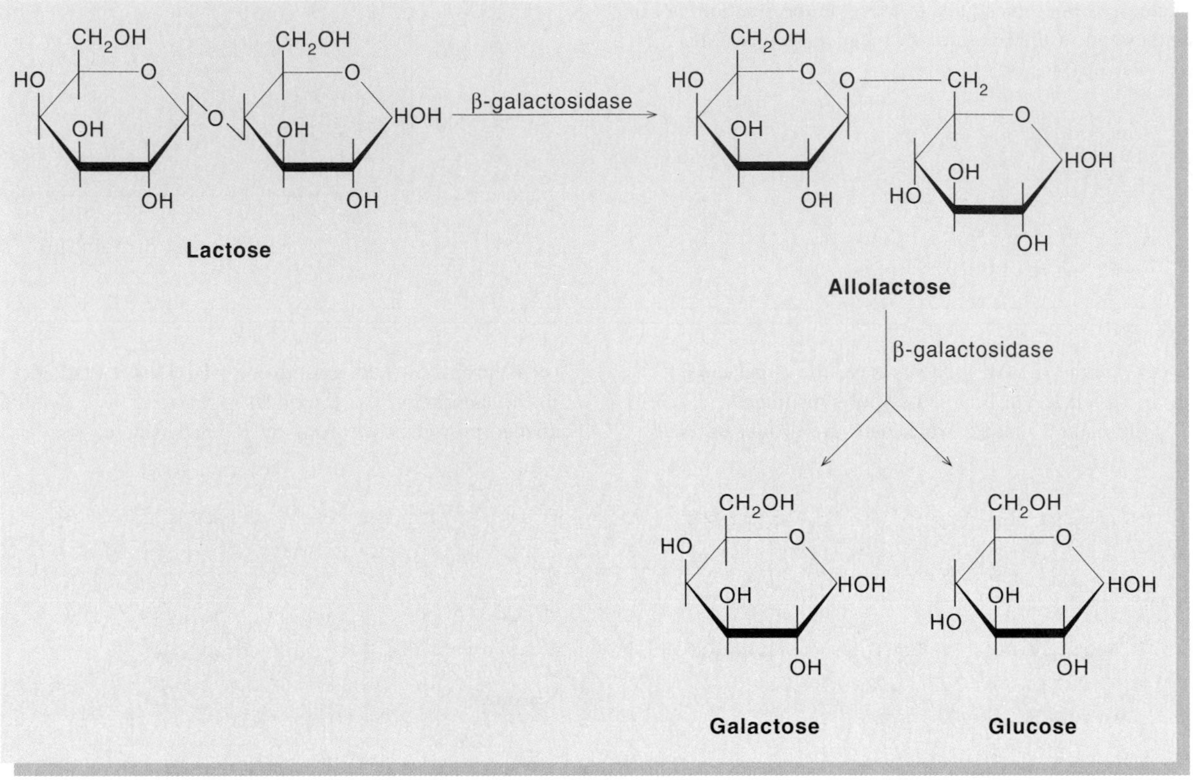

(such compounds are called gratuitous inducers because they are not substrates for the enzyme). No correlation exists between affinity for $\beta$-galactosidase and the capacity to induce. Lactose, the natural substrate of the operon, is not an inducer *in vivo*. Rather, allolactose, which is formed as an intermediate in lactose metabolism in the presence of the very limited amount of $\beta$-galactosidase that exists in uninduced cells, is believed to be the natural inducer (fig. 30.5). The three proteins of the *lac* operon are coordinately induced, that is, they are induced to the same extent by the same inducer. These results suggest that the receptor for the inducer is distinct from the proteins encoded by the operon, and the inducer acts at one site.

## A Gene Was Discovered That Leads to Repression of Synthesis in the Absence of Inducer

Two distinct types of mutations have been observed in the genes associated with the *lac* operon. One class of mutations includes structural gene mutations: (1) $\beta$-galactosidase

mutations ($z^+ \rightarrow z^-$), expressed as the loss of the capacity to synthesize active $\beta$-galactosidase; (2) permease mutations ($y^+ \rightarrow y^-$), expressed as the loss of the capacity to concentrate lactose; and (3) transacetylase mutations ($a^+ \rightarrow a^-$), expressed as the loss of the capacity to form thiogalactoside transacetylase. The other class of mutations involves controlling elements of the operon such as *i* gene mutations ($i^+ \rightarrow i^-$), expressed as the capacity to synthesize large amounts of $\beta$-galactosidase even in the absence of inducer. This type of *i* gene mutation is called a constitutive mutation. Structural mutations usually affect only the enzyme in whose gene the mutation occurs. In contrast, constitutive mutations invariably affect the amounts of all three structural gene products but not their structures (at this point you may wish to refer to box 30A for more information on genetic concepts and notation).

The most informative genetic studies were performed on cells containing two copies of the *lac* operon with point mutations in different genes. Partial diploids (merodiploids) of this sort are constructed by mating experiments in which a second copy of the *lac* region is incorporated into the test cell by mating. For this purpose the F factor plasmid is a

# 30A Genetic Concepts and Genetic Notation

Much of the early work on the *lac* operon was purely genetic. It is essential that certain aspects of genetics be understood. The information in this box should be adequate for those with no prior exposure to genetics except for what they have already encountered in this text.

Genes are specified by one or more small letters in italic. Thus, $z$ indicates the gene for $\beta$-galactosidase, and *lac* indicates the operon. Frequently a superscript is appended to the genetic symbol. The two most common superscripts are $+$, indicating a normal (wild-type) gene, and $-$, indicating a nonfunctioning (mutant) gene. Different representations of the same gene are referred to as alleles. Thus, $z^+$ and $z^-$ are both alleles of the $z$ gene.

Cells that carry a single copy of each gene are referred to as haploids. Cells that carry two copies of each gene are referred to as diploids. Bacteria are haploid cells because they carry a single chromosome with a unique representation for each gene. Bacterial cells that are partial diploids (merodiploids) may occur naturally, or they may be selected for by genetic techniques.

A favored method for constructing merodiploids is to infect the bacterial cell with a virus or a plasmid DNA that carries the extra genes of interest. The F plasmid is commonly used for this purpose. In strict usage, the genetic representation for a cell carrying the *lac* operon on the chromosome and the F plasmid would be $z^+y^+a^+//Fz^+y^+a^+$,

where the diagonal lines separate the host chromosome to the left and the plasmid chromosome to the right. As a rule, however, only one diagonal line is used to separate the genetic symbols. Also, for convenience, if all the alleles for a given gene are wild type, they may not be shown. Thus, $z^+y^+a^+/Fz^-y^+a^+$ and $z^+/Fz^-$ may be taken as representations of the same genetic state in cases in which it is understood that the *lac* operon is present on both the host and the plasmid chromosome.

Two genetic elements located on the same chromosome are said to be in the *cis* orientation. Two genetic elements located on different chromosomes in the same cell are said to be in the *trans* orientation. In the merodiploid $z^-y^-a^+/Fz^+y^+a^+$, the two mutant genes are in the *cis* orientation. In the merodiploid $z^-y^+a^+/Fz^+y^-a^+$, they are in the *trans* orientation.

A major reason for using merodiploids is to study the interaction between different alleles of the same gene. This often tells us a great deal about how a gene or the gene product functions. The two simplest types of interactions are dominant and recessive. A cell that is $z^+/Fz^-$ behaves like a $z^+$ cell as far as the metabolism of $\beta$-galactosidase is concerned. Therefore, the $z^+$ allele is dominant to the $z^-$ allele, or conversely, the $z^-$ allele is recessive to the $z^+$ allele.

favorite. Merodiploids of the type $z^+y^-a^-/Fz^-y^+a^+$ or $z^-y^+a^+/Fz^+y^-a^-$ behave like normal wild-type cells with respect to the expression and metabolism of the *lac* operon. This demonstrates that the distribution of normal genes and mutant genes on the two chromosomes does not influence the phenotype as long as there is at least one functional gene of each type. This tells us that each of the structural genes behave as an independent entity not affected by other genes on the same operon.

The study of merodiploids of the types $i^+z^-/Fi^-z^+$ give the same inducible phenotype. This demonstrates that the $i^+$ inducible allele is dominant to the $i^-$ constitutive

allele, and that it is active on the same chromosome *(cis)*, or on a different chromosome *(trans)* with respect to the structural gene it influences (table 30.1). In fact the $i^+$ gene could be moved to any location on the chromosome and it would still show dominant behavior. The fact that the influence of the $i$ gene is not sensitive to location suggests that $i$ gene action results from a diffusible gene product. The dominance of the inducible $i^+$ allele to the constitutive $i^-$ allele suggests that the former corresponds to the active form of the $i$ gene.

Further understanding of $i$ gene function has come from study of rare mutations designated $i^s$. Mutants bearing

## Table 30.1

Expression of $\beta$-Galactosidase as a Function of Genotype

| Genotype[a] | Phenotype | |
|---|---|---|
| | −Inducer | +Inducer |
| $i^+O^+Z^+$ | − | + |
| $i^-O^+Z^+$ | + | + |
| $i^+O^+Z^-$ | − | − |
| $i^+O^+Z^+/i^-O^+Z^+$ | − | + |
| $i^sO^+Z^+/i^+O^+Z^+$ | − | − |
| $i^+O^cZ^+$ | + | + |
| $i^+O^cZ^+/i^+O^+Z^+$ | + | + |
| $i^+O^cZ^+/i^+O^+Z^-$ | + | + |
| $i^+O^cZ^-/i^+O^+Z^+$ | − | + |

[a] The diagonal indicates that two *lac* operons, including the $i^+$, are present in the same cell. Under phenotype a + indicates a high level expression of $\beta$-galactosidase, a − indicates low level or the absence of expression.

this allele are noninducible, meaning that they have lost their capacity to express the structural gene products of the operon. In merodiploids of the constitution $i^+/i^s$, the $i^s$ allele is dominant, that is, the merodiploids cannot synthesize structural gene products even in the presence of inducer (see table 30.1). The most likely explanation for the $i^s$ mutant is that it is an allele of $i$ in which the repressor is not influenced by the inducer so that it always represses.

## A Locus Adjacent to the Operon Is Found to Be Required for Repressor Action

The vast majority of constitutive mutants result from $i^-$ mutations. However, occasional constitutive mutants have been mapped outside of the $i$ gene in the region of the *lac* operon promoter. Rare mutants of this type designated $O^c$ are much easier to isolate by selection for constitutivity in cells diploid for the $i^+$ gene. This selection procedure minimizes the chance of finding constitutive mutants that result from $i$ gene mutations because both copies of the $i^+$ gene would have to mutate to $i^-$ simultaneously to give a constitutive phenotype. If the probability of an $i^+ \rightarrow i^-$ mutation is $10^{-6}$, the probability of two such events occuring simultaneously in the same cell is $10^{-12}$, which is extremely unlikely. As a result, constitutive mutations obtained under these circumstances are almost always of the $o^c$ type. Like $i^-$ mutations, $o^c$ mutations affect the quantity of $\beta$-galactosidase synthesized but not its structure.

In merodiploids of the type $o^+z^+/o^cz^+$. $\beta$-galactosidase is constitutively expressed, showing that the $o^c$ allele is dominant to $o^+$ in this situation. In merodiploids of the type $o^cz^+/o^+z^-$ the $o^c$-allele is dominant also; but in $o^cz^-/o^+z^+$, it is recessive. Thus, the $o^c$-allele is dominant only when it is located *cis* to the structural genes it influences. From this result, François Jacob and Jacques Monod inferred that $o^+ \rightarrow o^c$ mutations correspond to a modification of the DNA structure that affects the ability of the repressor to bind.

## Genetic Studies on the Repressor Gene and the Operator Locus Lead to a Model for Repressor Action

The behavior of the various mutations we have just discussed led Jacob and Monod to propose a model for the regulation of protein synthesis. The genetic elements of this model consist of a structural gene or genes, a regulator gene, and an operator locus (fig. 30.6).

1. The structural gene produces an mRNA that serves as a template for protein synthesis.
2. The regulator gene (not itself part of the operon) produces a repressor that can interact with the operator locus.
3. The operator is always adjacent to the structural genes it controls.
4. The operator and its associated structural genes are referred to as the operon.
5. The repressor molecule combines with the operator locus to prevent the structural gene(s) from synthesizing mRNA. In induction, the inducer combines with the repressor, changing its structure so that it no longer binds to the operator; this region of the genome is then free to combine with RNA polymerase.

The Jacob-Monod operon hypothesis has provided a tremendous stimulus for investigations directed toward understanding not only the *lac* system but other genetic regulatory systems as well.

## Biochemical Investigations Verify the Operon Hypothesis

Genetic studies led to the operon hypothesis. Biochemical investigations were essential to provide direct evidence for the hypothesized properties of the repressor. The first task was to isolate the repressor.

## Figure 30.6

Schematic model illustrating the operon hypothesis. This diagram is modified from the original proposed by Jacob and Monod, who thought *i* gene repressor was an RNA rather than a protein. (*a*) The *i* gene encodes a repressor that binds tightly to the operator *o* locus, thereby preventing transcription of the mRNA from the *z, y,* and *a* structural genes. (*b*) When inducer is present, it combines with repressor, changing its structure so it can no longer bind to the operator locus. Inducer also can remove repressor already complexed with the *o* locus.

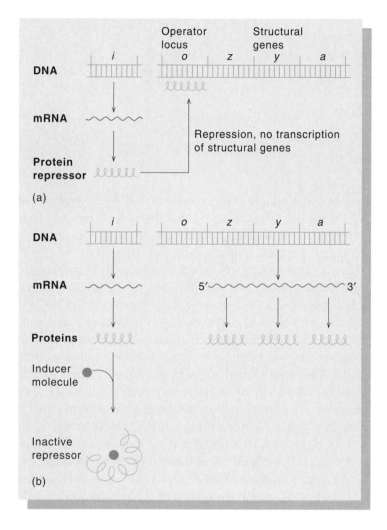

According to the operon hypothesis, inducer is supposed to bind to repressor. Walter Gilbert and Benno Muller-Hill used $^{14}$C-labeled isopropyl-$\beta$-D-thiogalactoside isopropylthiogalactopyranoside (IPTG), the strongest known inducer (see fig. 30.4), to monitor repressor purification from a crude cell extract. IPTG has a further advantage for such studies in that it is completely stable. A crude extract of disrupted cells was fractionated by standard protein purification procedures, and the fraction containing the $^{14}$C-IPTG-labeled product was isolated.

Several properties of normal and abnormal repressor were studied. The repressor was found to bind strongly to the *lac* promoter. This binding was disrupted by adding inducer. Promoter DNA containing an $o^c$ mutation was not effective in binding repressor. Only repressor prepared from $i^+$ cells was effective in binding to DNA.

A more detailed characterization of the operator binding site was made. Eight $o^c$ mutations were found to involve base replacements near the center of this region. The reactivity of wild-type operator to various chemical agents was determined in the presence and in the absence of repressor. Repressor binding substantially decreases the reactivity toward dimethylsulfate of several purine bases in the operator region (fig. 30.7). Dimethylsulfate reacts with N-7 of guanine and the N-3 of adenine in the double helix. Finally, if all the thymines in the DNA are replaced by bromouracil, a number of the bromouracil bases become cross-linked to the bound repressor in the presence of ultraviolet light. Taking into account the normal twist of the double helix, all the groups shown by chemical methods to be in the vicinity of the repressor are situated on one side of the double helix.

The sequence of bases in the operator region shows a remarkable symmetry property. Twenty-eight out of 36 of the base pairs in this region of the promoter are located on a twofold (dyad) axis of symmetry (see fig. 30.7). We shall see that dyad symmetry is a common property for repressor or activator binding sites and that it relates to the way in which DNA and regulatory proteins interact.

Biochemical proof that repressor inhibits operon expression was shown in a cell-free system in which most of the components were prepared from an $i^-$ extract. Addition of repressor to such an extract inhibited the synthesis of $\beta$-galactosidase. This inhibition was reversed by addition of IPTG inducer.

The detailed biochemical studies not only confirmed the Jacob-Monod hypothesis, they gave further details about the nature of repressor interaction and a detailed characterization of the repressor binding site. The repressor binds mainly to one side of the double helix, over a 36-base region covered by the symmetry axis. It inhibits expression because it binds to a site that overlaps the polymerase binding site.

## An Activator Protein Is Discovered That Augments Operon Expression

Although the genetic and biochemical studies on the action of repressor on the *lac* operon answered many questions about gene expression of the *lac* operon, they left equally important questions unanswered. It had been known since the turn of the century that the *lac* operon expresses at a

**Figure  30.7**

The operator locus (the presumptive repressor binding site). Bases are numbered +1 for the first base transcribed and −1 for the base before that. Regions showing dyad symmetry are underlined and overlined. Arrows indicate point mutations leading to the constitutive phenotype (oᶜ). Circled bases are those groups that are strongly protected against reaction with dimethylsulfoxide when *lac* repressor is bound. Shaded circles indicate those groups that become cross-linked to repressor in the presence of ultraviolet light when thymine in the DNA is replaced by 5-bromouracil.

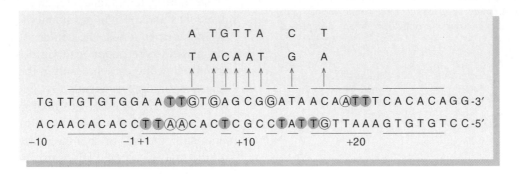

greatly reduced level if lactose and glucose are present simultaneously. Either of these sugars can be used by the bacterium as a source of carbon compounds and energy, but the lactose is not utilized to any appreciable extent until the glucose supply has been exhausted. This effect is called catabolite repression. As long as glucose is available, lactose is underutilized.

A turning point in our understanding of catabolite repression was provided by Earl Sutherland, who found that when glucose was added to growing *E. coli* cells, the level of 3′,5′-cAMP (cAMP) was drastically reduced. Could the lack of cAMP be responsible for the poor expression of the *lac* operon in the presence of glucose? In support of this I. Pastan and R. Perlman found that large quantities of cAMP added to the growth medium could partially reverse the glucose catabolite repression effect. In a cell-free system containing crude extracts from *E. coli* and DNA containing the lac operon, G. Zubay found that the low-level expression of the *lac* operon could be greatly increased by addition of cAMP. This provided support for the notion that cAMP was playing a direct role in activating the *lac* operon. Further investigations were facilitated by Jonathan Beckwith's genetic studies and the isolation of key mutants relating to the action of cAMP.

Beckwith and his colleagues isolated a large family of mutants that were permanently catabolite-repressed. These mutants fell into two categories: Those that could be phenotypically corrected by growing in the presence of cAMP and those that could not. The first class of mutants were believed to be defective in the synthesis of cAMP, and the latter class were presumed to be defective in the protein receptor for cAMP. Cell-free extracts were prepared from both of these mutants. When used for cell-free synthesis of β-galactosidase, it was found that mutants of the first type were greatly stimulated by addition of cAMP, confirming the belief that these mutants were defective in the synthesis of cAMP but nothing else. When extracts from mutants of the second type were used, cAMP had no stimulating effect, suggesting that a protein necessary for cAMP action was missing or defective. Further cell-free studies were performed in which mutants of the second type were used in conjunction with partially purified extracts from a normal strain. Addition of small amounts of extracts from a normal strain reestablished the stimulatory effect of the cAMP. The purification of the cAMP receptor protein was monitored with this system. Ultimately, a single protein called CAP was found to be responsible for the effect. Soon afterwards it was found that CAP is a dimer composed of identical subunits, each with an $M_r$ of 22,000. CAP binds to DNA and this binding is greatly stimulated in the presence of cAMP. The cAMP apparently alters the conformation of CAP so that it can form a strong complex with DNA at the *lac* promoter region.

A series of genetic deletions was used to demonstrate that the site necessary for CAP stimulation of the *lac* operon is in the −50−−80 region of the *lac* operon. A 14-bp segment between −53 and −68 shows dyad symmetry for 12 of the 14 bp (fig. 30.8).

In the absence of bound repressor, the full sequence of reactions leading to initiation of transcription is summarized by the following set of equations. First, the coactivator cAMP combines with the activator CAP, which then binds in the −60 region of the promoter. This complex stimulates the binding of RNA polymerase to an adjacent site on the promoter. CAP stimulates binding of RNA polymerase by an affinity between CAP and polymerase.

$$cAMP + CAP \rightleftharpoons cAMP\text{-}CAP$$

$$cAMP\text{-}CAP + DNA \rightleftharpoons cAMP\text{-}CAP\text{-}DNA$$

$$cAMP\text{-}CAP\text{-}DNA + polymerase \rightleftharpoons cAMP\text{-}CAP\text{-}DNA\text{-}polymerase$$

The overall strategy in creating a promoter sequence responsive to cAMP-CAP activation can be summarized as follows. The nucleotide sequence in the RNA polymerase-binding site is adjusted so that polymerase by itself produces a low level of transcription. An adjacent site for binding CAP is created so that when CAP is binding, the additional affinity contributed by favorable contacts between the CAP and the polymerase convert this into a high-level promoter. The crucial importance of the promoter sequence in eliciting the polymerase response to CAP is shown by the fact that certain mutations that result in producing the consensus sequence for the promoter (see fig. 30.1) eliminate the catabolite repression effect for the *lac* operon *in vivo* as well as the need for cAMP in the cell-free system.

## Enzymes That Catalyze Amino Acid Biosynthesis Are Regulated at the Level of Transcription Initiation

Among the genes concerned with biosynthetic processes, those uniquely involved in synthesis of the amino acid tryptophan are possibly the best understood. This is mostly a result of the efforts of Charles Yanofsky and his colleagues, who have used a wide variety of genetic and biochemical techniques to probe the complexities of this system.

Wild-type *E. coli* can synthesize all 20 types of amino acids from simpler substrates, but for most of them, it does so only when they are not available in adequate amounts in the growth medium. For example, synthesis of enzymes for tryptophan synthesis is sharply reduced when the external tryptophan supply is high. Lowering available L-tryptophan selectively stimulates synthesis of the mRNA and associated enzymes (fig. 30.9). The five contiguous structural genes are transcribed as a single polycistronic mRNA. Initiation of transcription is regulated in part by the interaction of the tryptophan repressor, the protein product of the *trpR* gene, with its target site on the DNA, the *trp* operator, *trpO* (fig. 30.10). Binding of L-tryptophan to the repressor causes a structural alteration essential for strong specific binding to the *trpO* locus. Like *lac* repressor, the *trp* repressor binds at a site that overlaps the RNA polymerase-binding site, and

**Figure 30.8**

CAP-binding region in the *lac* promoter. Regions showing dyad symmetry that are believed to interact strongly with CAP are overlined and underlined. Circled bases are those that are strongly protected from reaction with dimethylsulfate when CAP is bound. Point mutations L8 or L29 result in a promoter that is not stimulated in transcription by cAMP. Red dots indicate those phosphate positions that, if ethylated by ethylnitrosourea, block CAP binding. If the area of interest exists as a normal DNA double helix when binding CAP, then most of the groups implicated in CAP binding appear on one side of the DNA.

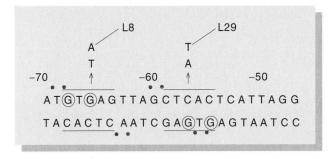

the region where the repressor binds contains a number of bases arranged with dyad symmetry (fig. 30.11).

The most significant difference between the action of the *trp* and *lac* repressors relates to the function of the small-molecule effector. In the case of *lac,* the effector molecule allolactose acts as an antirepressor (inducer), causing release of repressor from the operator; in the case of *trp,* the effector molecule L-tryptophan acts as a corepressor, stimulating the binding of repressor to the operator. It should be obvious that the difference in action of these small-molecule effectors, the concentrations of which dictate the level of operon activity, is well suited to the different metabolic needs of the cell satisfied by the two operons.

## *The* trp *Operon Is Also Regulated after the Initiation Point for Transcription*

From this point on, the close parallel between the regulation of the *trp* operon and the *lac* operon ends. The *trp* operon has no positive control system like cAMP-CAP, but it does have another means for regulating transcription at a site downstream from the initiation point. The existence of a second signal for regulation was first suspected when it was discovered that a genetic deletion of some of the bases between the initiation site for transcription and the first structural gene (*trpE*) raised the level of expression of the operon 8- to 10-fold. This was true even in strains with a defective repressor gene (*trpR⁻* strains). How could this be, and what could be the mechanism of action? An important clue was

**Figure 30.9**

The tryptophan operon, indicating the location of the different genes, the polypeptide chains, the resulting enzyme complexes, and the reactions catalyzed by the enzyme complexes.

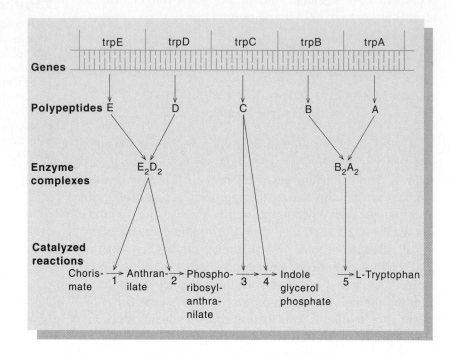

**Figure 30.10**

Schematic diagram of the repressor control of *trp* operon expression. The *trp* promoter (*P*) and *trp* operator (*O*) regions overlap. The *trp* aporepressor is encoded by a distantly located *trpR* gene. L-Tryptophan binding converts the aporepressor to the repressor that binds at the operator locus. This complex prevents the formation of the polymerase–promoter complex and transcription of the operon that begins in the leader region (*trpL*). Only a fraction of the transcripts extends beyond the attenuator locus in the leader region. The regulation of this fraction is discussed in the text.

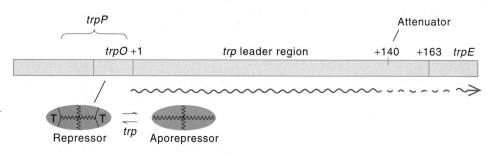

**Figure 30.11**

The promoter–operator region of the tryptophan operon. Two regions, PBS1 and PBS2, where polymerase binds are bracketed. Regions within the repressor-binding site showing dyadic symmetry are underlined and overlined. Single-base changes that lead to operator constitutive (*o*c) mutants are indicated below the duplex.

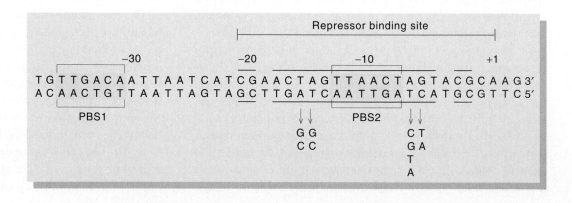

## Figure 30.12

The leader region for the tryptophan operon. The region of the leader RNA containing the hypothesized leader polypeptide is shown. The translation start of the trpE protein is also shown.

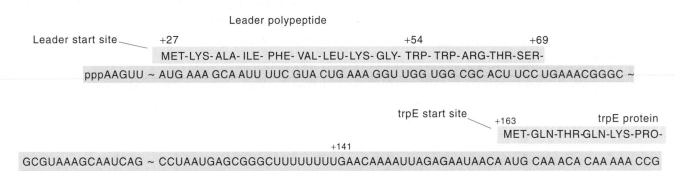

provided by sequence analysis of the 162 bases in the *trp* leader region, that is, the region between the initiation site for transcription and the initiation site for translation of the first structural gene (fig. 30.12). This leader region contains a potential initiation codon (bases 27–29), two tandem *trp* codons (bases 54–59), and a terminator codon (bases 69–71). A so-called leader peptide of 14 amino acids would result from translation of this region.

There are numerous reasons for believing that translation of the leader peptide up to or through the *trp* codons regulates attenuation of mRNA transcription. First of all, selective starvation of cells for tryptophan relieves attenuation and permits most RNA polymerase molecules to read through the leader region. The only other amino acid that relieves attenuation of the *trp* operon when it is lacking is arginine; arginine starvation is about 80% as effective as tryptophan starvation. It should be noticed that an *arg* codon is located adjacent to the two *trp* codons in the leader region. Most telling of all was the finding that a mutation resulting in the replacement of the AUG start codon by AUA, which should eliminate translation of the leader peptide, also prevents transcription beyond the attenuator.

Other experiments indicated that the fraction of tRNA that is charged with an amino acid is a crucial factor in the attenuation response. This has been examined *in vivo* by comparing the *trp* operon enzyme levels in *trpR*⁻ strains that are otherwise normal with strains that are defective in some respect in charged tRNA$^{Trp}$. Such structural defects in tRNA$^{Trp}$ or in the charging enzyme elevates expression, probably by permitting polymerase to transcribe through the attenuator.

These results support the hypothesis that transcription read-through requires partial translation of the leader sequence. However, only if the translation pauses or stops in the region where the *trp* or *arg* codons occur is read-through

favored. A careful examination of the secondary-structure possibilities in the attenuator region suggests why this is so. The leader region RNA between bases 50 and 141 has the potential to form a variety of base-paired conformations. Figure 30.13 illustrates the most likely secondary structures that form in terminated *trp* leader RNA. These are based on analysis of regions of the transcript that show resistance to RNase T1 digestion under mild conditions and the base pairing established by studies of defined oligonucleotides. Four regions of base pairing that can form three stem-and-loop structures have been proposed. Region 1, which includes the tandem *trp* codons and the leader peptide translation stop codon (bases 54–68), can base-pair with region 2 (bases 76–91). Although region 2 (bases 74–85) also should be able to base-pair with region 3 (bases 108–119), stem-and-loop 2 · 3 has not been observed *in vitro*, presumably because stem-and-loop 3 · 4 and stem-and-loop 1 · 2 form preferentially. Region 3 (bases 114–121) can base-pair with region 4 (bases 126–134). The existence of this stem and loop is inferred from the T₁ RNase-resistance of the GC-rich region from residue 107 to the 3′ end of the transcript. The stem-and-loop structure formed between regions 3 and 4, followed by a sequence of U residues, is a common structure for a transcription terminator (see chapter 28). Hence, it is expected that conditions under which this structure is preserved would favor transcription termination. In support of this, a number of single-base replacement mutations have been isolated that lead to mispairing in the 3 · 4 stem; all of these lower the level of transcription termination in the leader to some extent.

The absence of translation of the leader would not perturb this 1 · 2 and 3 · 4 structure (fig. 30.14*a*). Consistent with this, changing the initiator codon of the leader peptide by a single base prevents read-through, as discussed earlier. On the other hand, selective starvation resulting from either

**Figure 30.13**

Proposed secondary structures in the leader RNA. Four regions, labeled 1, 2, 3, and 4 at the left, can base-pair to form three stem-and-loop structures (1.2, 2.3, and 3.4). The arrows in the main figure mark the RNAse T1 cleavage sites.

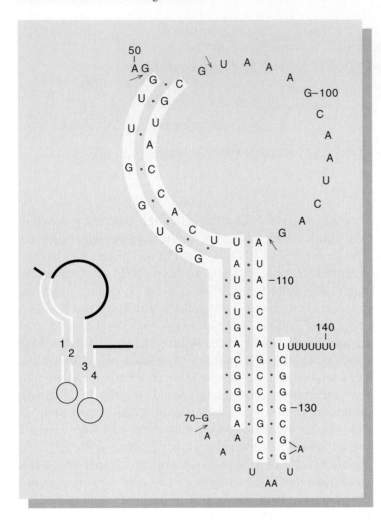

tryptophan or arginine deprivation stimulates transcription read-through. Most likely, this is because the ribosome stalls in the region of the *trp* or *arg* codons (bases 54–62). The resulting rupture of the base-paired 1 · 2 structure would make region 2 available for pairing with region 3. This would encourage disruption of the stem-and-loop 3 · 4 structure, resulting in transcription read-through (see fig. 30.14*b*). In the presence of an adequate supply of all amino acids, translation would proceed beyond this critical region, so that region 2 would not be available for base pairing, and the 3 · 4 loop would be maintained, favoring transcription termination at the attenuator (see fig. 30.14*c*).

This attenuator mechanism of control is amazingly simple because it requires no proteins other than those normally used for transcription and translation. One might expect such a simple and effective mechanism to be used repeatedly for other operons involved in amino acid biosynthesis. Indeed, for several other amino acid biosynthetic pathways in *E. coli* for which tRNA charging is involved in regulation, attenuator mechanisms have been found.

## Genes for Ribosomes Are Coordinately Regulated

Over 100 genes in *E. coli* participate in the synthesis of the RNAs and proteins that constitute the enzymatic machinery for translation. The relevant gene products make up between 20% and 40% of the dry cell mass. In rapidly growing cells about 85% of the RNA is ribosomal, 10% is tRNA, and most of the remainder is mRNA. The various RNAs and proteins are produced according to need. For ribosomes and tRNAs, this results in a synthesis rate that is roughly proportional to the cell growth

**Figure 30.14**

Model for attenuation in the *trp* operon, showing ribosome and leader RNA. (*a*) Where no translation occurs, as when the leader AUG codon is replaced by an AUA codon, stem-and-loop 3.4 is intact, and termination in the leader is favored. (*b*) Cells are selectively starved for tryptophan so that the ribosome stops prematurely at the tandem *trp* codons. Under these conditions, stem-and-loop 2.3 can form, and this is believed to lead to the disruption of stem-and-loop 3.4. (*c*) All amino acids, including excess tryptophan, are present so that stem-and-loop 3.4 is present.

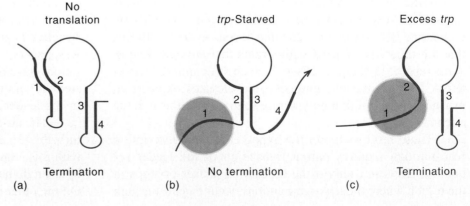

## Figure 30.15

A typical rRNA (*rrn*) operon contains two promoters and genes for 16S, 23S, and 5S rRNA and a single 4S tRNA gene. The four fully processed RNAs are derived from a single intact 30S primary transcript.

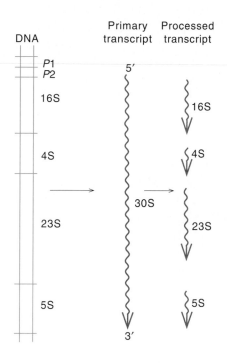

## Figure 30.16

Guanosine tetraphosphate (ppGpp) concentration under normal conditions, after amino acid starvation, and after readdition of amino acids (● = wild-type cells; ▲ = *relA* cells; and ■ = *spoT* cells).

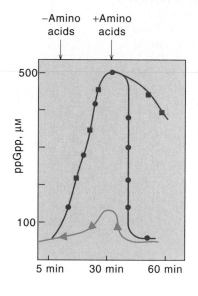

rate; the relative amounts of the three rRNAs (16S, 23S, and 5S), the 60 or so tRNAs, and the 50 ribosomal proteins, are consistent with the stoichiometric needs for making ribosomes.

## *Control of rRNA and tRNA Synthesis by the* rel *Gene*

Under conditions of rapid growth, *E. coli* cells contain about $10^4$ ribosomes. The maximum rate of reinitiation at the ribosomal gene promoter is about one per second. In rapid growth, *E. coli* can duplicate once every 20 min, which would allow for the synthesis of only about 1,200 molecules of rRNA if there were only one gene for ribosomal RNA. In fact, there are seven copies for ribosomal RNA operons in the bacterial chromosome, which makes it possible for rRNA synthesis to maintain the necessary pace under conditions of rapid growth. These operons are dispersed at seven locations around the circular *E. coli* chromosome. Each operon is transcribed into a single transcript, which is processed into four or five RNAs (fig. 30.15). The order of RNAs in the original transcript starting from the 5′ end is 16S, 4S, 23S, and 5S. In some operons additional 4S genes for tRNA are located downstream from the 5S gene. All the

known ribosomal RNA operons appear to have two strong promoters located in tandem; this probably permits a more rapid rate of reinitiation for the genes than would be possible with only one promoter.

As stated earlier, rRNA synthesis is usually maintained at a rate proportional to the gross rate of protein synthesis. In a normal wild-type cell, when protein synthesis is limited (e.g., by amino acid availability), M. Cashel and J. Gallant have shown that the ppGpp concentration rises rapidly from about 50 $\mu$M to 500 $\mu$M (fig. 30.16). Concomitantly, rRNA synthesis ceases abruptly. This is part of the phenotype known as the stringent response: First, amino acid deprivation or other factors that slow down protein synthesis provoke an increased rate of accumulation of ppGpp; this in turn leads to an inhibition of rRNA synthesis.

Observations on different mutants indicate that the concentration of ppGpp is regulated by a careful balance between its rate of synthesis (controlled by the *rel* gene product) and rate of breakdown (controlled by the *spoT* gene product). Correlated observations indicate that the synthesis of rRNA is inversely proportional to the ppGpp concentration. In a *relA* mutant cell, neither the rapid rise in ppGpp concentration nor the cessation of rRNA synthesis is seen when amino acids are removed. The *relA* gene encodes a protein that is required for ppGpp synthesis. If

**Figure 30.17**

Schematic diagram of ppGpp
synthesis and the hypothesized
mechanism for its action.
ppGpp is synthesized on the
ribosome when there is a
peptidyl-tRNA on the P site of
the ribosome and uncharged
tRNA on the A site. The
ppGpp probably inhibits rRNA
synthesis by complexing with
the RNA polymerase.

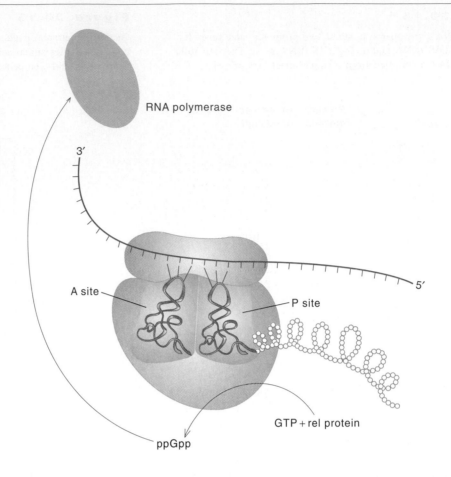

amino acids are reintroduced into the growth medium of a
wild-type culture, the ppGpp concentration falls rapidly
(half-life about 20 s), and the rate of rRNA synthesis rises
rapidly. In a mutant called *spoT⁻*, the normal rise in ppGpp
level is observed on amino acid starvation, but the level of
ppGpp falls much more slowly on readdition of amino acids
to the growth medium. Correlated with this, the rate of
rRNA synthesis in a *spoT⁻* mutant also increases very
slowly on readdition of amino acids.

The synthesis of ppGpp has been studied both in crude
cell-free extracts of *E. coli* and in a partially purified system
to determine what factors influence its rate of synthesis. It
was found that ppGpp is synthesized on the ribosome from
GTP in the presence of the protein encoded by the wild-type
*relA* gene. Maximum ppGpp synthesis occurs in the presence of ribosomes associated with mRNA and uncharged
tRNA with anticodons specified by the mRNA. If the uncharged tRNA bound to the ribosome acceptor site (A site)
is replaced by charged tRNA, the rate of ppGpp synthesis is
greatly lowered. If uncharged tRNA anticodons are not
complementary to the mRNA codons exposed on the ribosome for protein synthesis, ppGpp synthesis does not occur.

Cell-free synthesis studies also strongly support the
notion that ppGpp directly inhibits RNA synthesis. Thus, in
a cell-free system the DNA-directed synthesis of rRNA with
*E. coli* RNA polymerase is strongly inhibited by 100–
200 $\mu$M ppGpp. Such experiments have led to the hypothesis that ppGpp, by binding to RNA polymerase, alters its
structure so that it has a lowered affinity for rRNA promoters. The *in vivo* and *in vitro* studies on ppGpp and rRNA
have resulted in a model for how the level of amino acid
charging of tRNA controls the rate of rRNA synthesis (fig.
30.17). First, uncharged tRNA that is codon-specific for the
exposed codons on the mRNA becomes bound to the ribosome acceptor site, creating a situation unfavorable for protein synthesis but favorable for ppGpp formation. Second,
the ppGpp diffuses and binds to RNA polymerase, thereby
lowering its affinity for rRNA promoters.

This alteration of the RNA polymerase by ppGpp affects the ability of RNA polymerase to interact with promoters in a differential way. For the promoters of the rRNA
operons, the polymerase–promoter interaction is strongly
inhibited by ppGpp. For some other promoters the effect of
ppGpp is actually stimulating. As first observed in the cell-

free system, ppGpp stimulates expression from the *lac* and *trp* operons. In more recent *in vivo* studies Cashel has found that the biosynthetic operons associated with arg, gly, his, leu, lys, phe, ser, thr, and val biosynthetic enzymes have an absolute requirement for ppGpp. This is a most appropriate response because an absence of an amino acid leads to ppGpp buildup, which then stimulates the genes required for biosynthesis of the amino acid.

The observations on ppGpp's role in rRNA synthesis show that this nucleotide is an important control factor regulating rRNA synthesis, but it does not eliminate the possibility that other factors also affect the level of rRNA. *In vivo* and *in vitro* evidence indicates that the inhibitory effect of ppGpp on transcription extends to most tRNA and ribosomal protein genes. Ribosomal protein gene expression also appears to be regulated at the translational level.

## Translational Control of Ribosomal Protein Synthesis

In exponentially growing *E. coli* cells, synthesis rates of most ribosomal proteins are nearly identical and coordinately regulated. M. Nomura and his coworkers have suggested that free ribosomal proteins inhibit the translation of their own mRNA and that as long as the assembly of ribosomes removes ribosomal proteins, the corresponding mRNA escapes this feedback inhibition. This hypothesis has been tested *in vitro* using a protein-synthesizing system with various template DNA molecules carrying ribosomal protein genes; it has been tested *in vivo* by examining the effect of overproduction of certain ribosomal proteins on the synthesis of other ribosomal proteins using various recombinant plasmids. By these means it was found that certain ribosomal proteins selectively inhibit the synthesis of other ribosomal proteins whose genes are part of the same operon; this autogenous, or self-imposed, inhibition occurs at the level of the translation of the mRNA rather than at the level of the transcription of mRNA. Figure 30.18 illustrates how this scheme works for the regulatory protein L1. L1 and L11 are encoded by the $P_{L11}$ operon. L1 can form a complex with either the 5′ end of its own mRNA or with 23S rRNA. It binds more strongly to the 23S rRNA. However, if an excess of L1 occurs over the available 23S rRNA, then L1 binds to the 5′ end of the mRNA, thereby inhibiting the synthesis of both L1 and L11. In this way the amounts of L1 and L11 proteins synthesized are kept in register with the amounts of rRNA synthesized.

Other operons carrying ribosomal protein genes are regulated in a similar manner. Most ribosomal protein genes exist in operons with promoters at one end. In some cases, individual operons are regulated by one of the encoded ribosomal proteins, in other cases, operons appear to be subdivided into units of regulation, and individual units of regulation are regulated at the translation level by one of the translation products.

**Figure 30.18**

Schematic diagram explaining autogenous inhibition by L1. L1 can form a complex either with the 5′ end of its own mRNA or with 23S rRNA. If there is an excess of L1 over the available 23S rRNA, then L1 binds to the 5′ end of the mRNA and inhibits both L1 and L11 synthesis. The inhibition of L1 translation by this binding depends on the fact that the translation of L1 and L11 is somehow coupled.

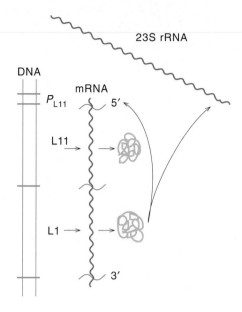

## Regulation of Gene Expression in Bacterial Viruses

Bacterial viruses rarely coexist indefinitely with the host cell. Temperate viruses can adopt either an active, replicating state or a dormant, prophage state. In the prophage state, the viral genome exists at a low copy number, sometimes in a host-integrated form as with λ bacteriophage, and other times in an independent plasmidlike chromosome, as with P1 bacteriophage. In the active, replicating state, temperate viruses duplicate rapidly so that within about an hour they kill the host cell and release infectious particles. Most temperate viruses in the prophage state utilize the unmodified host RNA polymerase. But once they enter the lytic cycle, whether they are temperate or virulent, they either modify or replace the host polymerase to enhance their own transcription. The lytic life cycle of viruses employs multiple regulatory proteins that function in a cascade. The pattern of regulation is unidirectional and irre-

Regulatory Proteins of λ

| Protein | Function |
|---------|----------|
| cI | At low concentrations represses $P_R$ and $P_L$ and activates $P_{RM}$; at high concentrations represses $P_R$, $P_L$, and $P_{RM}$ |
| cII | Activator for $P_{int}$ and $P_{RE}$ |
| cIII | Stabilizes cII |
| cro | At low concentrations represses $P_{RM}$; at high concentrations represses $P_{RM}$, $P_R$, and $P_L$ |
| N | An antiterminator at $t_{L1}$, $t_{R1}$, and $t_{R2}$ |
| Q | An antiterminator at $t_{6S}$ |

versible. A small percentage of virus genes designated as early genes are expressed first. Subsequently, late genes are turned on, and early genes are sometimes turned off, until mature virus particles are assembled and exit the cell. The pattern of gene expression observed for a bacteriophage has been likened to the pattern of expression observed in the development of differentiated cells in multicellular organisms (see chapter 31).

In this chapter we consider the patterns of gene expression found for λ, the best understood of the temperate phages.

## λ Metabolism Is Directed by Six Regulatory Proteins

λ encodes six regulatory protein: cI, cII, cIII, cro, N, and Q (table 30.2). The developmental processes accompanying infection center on the roles of these six regulatory proteins and the factors that govern the level of their expression.

## The Dormant Prophage State of λ Is Maintained by a Phage-Encoded Repressor

About half of the time when λ infects a cell it adopts a dormant lysogenic state in which the virus is linearly integrated into the host genome. This state is maintained by moderate amounts of the λ-encoded cI repressor. The cI repressor prevents the lytic cycle from developing by inhibiting two promoters: the $P_L$ promoter for early leftward transcription and $P_R$ the promoter for early rightward transcrip-

tion (fig. 30.19). It does this by binding to sites that prevent the polymerase from binding, just like the bacterial repressor proteins we have already discussed. The repressor binding site closest to the $P_R$ promoter is actually a composite of three adjacent repressor binding sites, $O_{R1}$, $O_{R2}$, and $O_{R3}$, which have different affinities for repressor. At moderate concentrations (about 100 copies of cI repressor per cell) $O_{R1}$ and $O_{R2}$ are occupied. At higher concentrations of cI the $O_{R3}$ site is also occupied.

Binding of repressor at $O_{R1}$ and $O_{R2}$ has two effects: It inhibits transcription from the $P_R$ promoter while it stimulates the promoter for repressor maintenance $P_{RM}$. Thus, at moderate concentrations the cI repressor functions simultaneously as a repressor and as an activator. The stimulation of leftward transcription from the $P_{RM}$ promoter is due to favorable contacts made between the repressor bound at $O_{R2}$ and the polymerase bound to the $P_{RM}$ promoter. At more elevated concentrations of cI when the $O_{R3}$ binding site is also occupied by repressor, the $P_{RM}$ promoter is inhibited. This prevents overproduction of cI.

As long as a moderate concentration of the cI repressor is present, λ remains dormant. Exposure of lysogenic cells to ultraviolet light or DNA-damaging drugs such as mitomycin C leads to wholesale destruction of the cI protein. This is because DNA damage activates the host recA protease, which destroys the cI repressor (the activated *recA* protease cleaves the cI repressor in much the same way that it cleaves the *lexA* repressor; see chapter 26). With *cI* gone the prophage spontaneously enters the lytic cycle. This mechanism of prophage activation seems to be an instruction to the prophage that when conditions are unfavorable in the cell, it is time to replicate and find a new host cell to infect or lysogenize.

## Events That Follow Infection of **Escherichia coli** by Bacteriophage λ Can Lead to Lysis or Lysogeny

Infection of *E. coli* by λ starts by attachment of the virus to the bacterial membrane and injection of the viral genome (fig. 30.20). This is followed by circularization catalyzed by the cellular DNA ligase enzyme. Only three genes are expressed at this time: The regulatory gene *N* from the $P_L$ promoter and the regulatory genes *cro* and *cII* from the $P_R$ promoter. The N protein combines with the host RNA polymerase, permitting extension of the early right and early left transcripts. This leads to expression of the other regulatory genes: *cII*, *cIII*, and *Q*, and replication and recombination genes. At this point the infection process can proceed in either of two, mutually exclusive directions: The lytic direc-

## Figure 30.19

A segment of the λ genome containing the $P_{RM}$ and nearby $P_L$ and $P_R$ promoters. Transcription from $P_{RM}$ results in cI synthesis. An expanded view is shown of the three cI binding sites flanking the $P_{RM}$ and $P_R$ promoters. Transcripts that originate from the three promoters are shown by the wavy red lines. The $O_{R1}$, $O_{R2}$, and $O_{R3}$ sites are binding sites for the cI repressor.

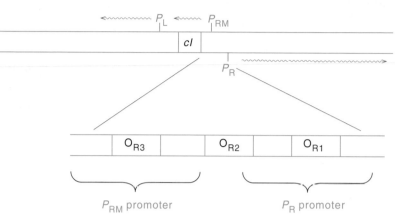

tion or the lysogenic direction. First we consider the events that take place in the lytic direction.

## The N Protein Is an Antiterminator that Results in Extension of Early Transcripts

The N protein produced by early leftward transcription influences the RNA polymerase by causing it to ignore certain termination signals. For N to act it must first bind to the polymerase. Remarkably N can only attach to a transcribing polymerase at two polymerase pause sites, which are designated $nut_L$ and $nut_R$ (fig. 30.21). At these points during transcription the polymerase pauses and can pick up the N protein. The polymerase structure is altered after binding N so that it no longer recognizes the early termination sites, $t_{R1}$, $t_{R2}$, and $t_{L1}$, as stop signals. As a result, polymerase transcribes through these termination sites, producing longer transcripts from which additional gene products are expressed (see figs. 30.21 and 30.22).

## Another Antiterminator, the Q Protein, Is the Key to Late Transcription

Extension of the rightward transcript stimulated by the N protein leads to expression of the $Q$ gene protein, a key regulatory protein for late transcription. The Q protein functions like N as an antiterminator; in this case antitermination occurs at the $t_{6S}$ terminator (see fig. 30.22). This leads to transcription of all of the late genes, including the lysis gene and the genes for the head and tail proteins of the mature phage. At late times other factors encourage rightward transcription from the $P_{R'}$ promoter.

## Cro Protein Prevents Buildup of cI Protein during the Lytic Cycle

As the lytic cycle progresses, the phage DNA replicates. The increase in the number of gene copies can result in overexpression of the cI, which could shut down the lytic cycle. This effect is overcome by the regulatory protein cro, which is specifically designed to inhibit cI synthesis during late infection.

The cro protein and the cI protein bind to exactly the same sites on the DNA. Despite this fact, the cI protein is required for lysogeny, whereas the cro protein is required for lysis. These requirements can be shown with mutants. A $cI^-$ mutant invariably undergoes lysis, whereas a $cro^-$ mutant can lysogenize but cannot complete the lytic cycle. This remarkable difference in the behavior of $cro$ and $cI$ results from the fact that although they bind to the same sites, they do so with totally different relative affinities (fig. 30.23). $Cro$ binds preferentially to $O_{R3}$ and less strongly to $O_{R1}$ and $O_{R2}$.

The binding of $cro$ to $O_{R3}$ turns off $cI$ expression originating from the $P_{RM}$ promoter. At high concentrations $cro$ binds to one or more of the other sites, turning off the $P_R$ promoter. The $P_L$ promoter is also turned off at high concentrations of $cro$. The most important physiologic effect of $cro$ is the turning off of the $P_{RM}$ promoter. In the absence of $cro$, $cI$ expression increases because of the increase of the number of $cI$ gene copies resulting from early replication of viral DNA during the lytic cycle. This buildup shuts down transcription from the $P_R$ and $P_L$ promoters before sufficient transcripts have been made to ensure the lytic pathway. The binding properties of $cI$ and $cro$ to the tripartite operator shared by $P_{RM}$ and $P_R$ are shown in fig. 30.23.

## Figure 30.20

Overview of λ development.

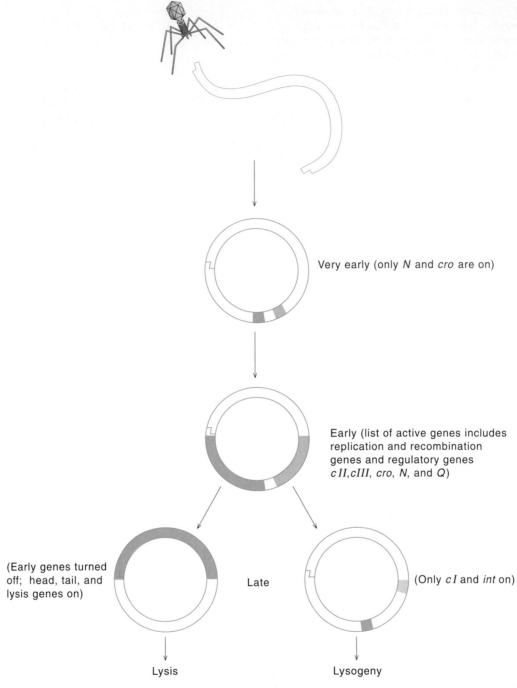

Very early (only *N* and *cro* are on)

Early (list of active genes includes replication and recombination genes and regulatory genes *c*II,*c*III, *cro*, *N*, and *Q*)

(Early genes turned off;  head, tail, and lysis genes on)

Late

(Only *c*I and *int* on)

Lysis

Lysogeny

## Late Expression along the Lysogenic Pathway Requires a Rapid Buildup of the cII Regulatory Protein

Thus far we have considered the events that take place after infection or after activation of the prophage that lead to phage replication and ultimately lysis. As already indicated, when λ infects a cell, the cells can follow this lytic pathway, or alternatively they can follow the lysogenic pathway, in which case the λ genome becomes dormant and integrates into the host genome. The choice appears to depend primarily on the level of cII protein formed during the early stages of infection. If this level is sufficiently high, the lysogenic pathway is followed; otherwise the lytic pathway is followed. Although we do not understand precisely what controls the level of cII expression, it is easy to appreciate the importance of cII to elaboration of the lysogenic pathway. Thus, cII protein is an activator for two promoters: $P_{RE}$ and

## Figure 30.21

Early transcripts. The early left transcript is initiated from the $P_L$ promoter. In the absence of N protein this transcript terminates at $t_{L1}$. In the presence of N protein the polymerase picks up an N protein at the N utilization site, $nut_L$. This makes it possible for the polymerase to transcribe through $t_{L1}$. The early right transcript tends to terminate at $t_{R1}$ unless N protein is present. In this case the N protein becomes bound to the polymerase at the $nut_R$ site.

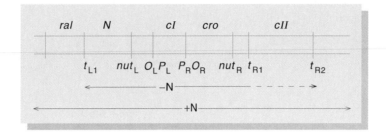

## Figure 30.22

Bacteriophage λ DNA. *Top:* Circular map indicating locations of main control genes and early and late functions. *Bottom:* Expanded region containing control genes and main promoters and terminators. Arrows indicate main transcripts and conditions under which they are active. In the absence of N protein, about half the transcription beginning at $P_R$ reads through $t_{R1}$ to $t_{R2}$ (indicated by dashed portion of arrow). More information on regulatory proteins and promoters is given in table 30.2.

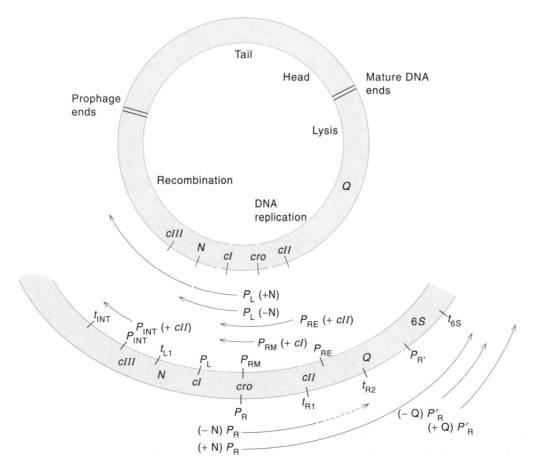

## Figure 30.23

Segment of the λ genome showing the three operators $O_{R1}$, $O_{R2}$, and $O_{R3}$ around the $P_{RM}$ and the $P_R$ promoters. The cI and cro regulatory proteins bind to these operators with different relative affinities. The net result of these differing affinities is that cI is required for lysogeny and cro is required for the lytic cycle.

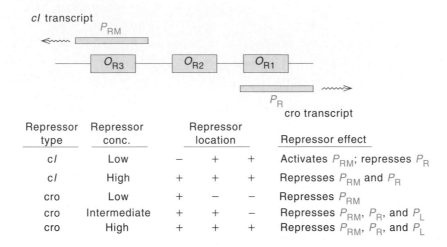

| Repressor type | Repressor conc. | Repressor location | | | Repressor effect |
|---|---|---|---|---|---|
| cI | Low | − | + | + | Activates $P_{RM}$; represses $P_R$ |
| cI | High | + | + | + | Represses $P_{RM}$ and $P_R$ |
| cro | Low | + | − | − | Represses $P_{RM}$ |
| cro | Intermediate | + | + | − | Represses $P_{RM}$, $P_R$, and $P_L$ |
| cro | High | + | + | + | Represses $P_{RM}$, $P_R$, and $P_L$ |

## Figure 30.24

A segment of the λ genome containing the $P_{RM}$ and nearby $P_L$, $P_R$, and $P_{RE}$ promoters. Transcripts that can originate from the three promoters are shown by the wavy lines. The $P_{RE}$ and $P_{RM}$ require different activators for expression. The transcript from $P_{RE}$ is 10 times more effective in cI expression because it has a Shine-Dalgarno sequence for ribosome attachment.

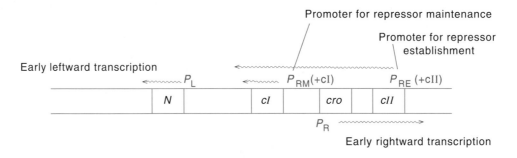

$P_{int}$ (see fig. 30.22). The $P_{RE}$ promoter leads to transcripts that are very efficient in producing high levels of the cI repressor (fig. 30.24). High levels of cI can shut down the $P_L$ and $P_R$ promoters before the lytic cycle gets underway. The other promoter activated by cII is $P_{int}$, which leads to int protein synthesis. This protein is required for the integration of the phage genome into the host genome, a process that completes the lysogenic process. After integration, moderate levels of cI resulting from transcription initiated at the $P_{RM}$ promoter (as described above) keep both the $P_L$ and the $P_R$ promoters in the repressed state.

## Interaction between DNA and DNA-Binding Proteins

So far we have focused on where regulatory proteins bind to the DNA and under what conditions they bind. The ability to isolate and sequence DNA promoters and to isolate and characterize regulatory proteins has greatly increased our appreciation of how specific complexes are formed between DNA and regulatory proteins.

The three-dimensional structures of about 20 regulatory proteins have been determined; in four cases the structures of the specific DNA–regulatory protein complexes have also been determined.

### Recognizing Specific Regions in the DNA Duplex

The problem of regulation in prokaryotes such as *E. coli* is far simpler than in complex eukaryotes because of the smaller number of genes. It seems likely that a system of this complexity (about 3,000 genes) should contain no more than 100–300 regulatory proteins because genes with related functions are often under the control of the same regulatory proteins.

Each member of this diverse family of regulatory proteins must be able to scan the DNA structure and recognize control regions with a high degree of specificity. How is this possible, particularly as the specific regions of DNA are embedded in a core of base pairs that are already hydrogen-bonded to each other? Present indications are that the secondary structure of DNA is left intact when these regions interact with their regulatory proteins. Thus, the base pairs must be identifiable by the groups that are exposed in the core of an intact duplex structure. As in most specific inter-

actions between nucleic acids and proteins, it seems likely that the specificity of interactions between DNA and regulatory proteins must result primarily from hydrogen bond interactions. Inspection of the DNA duplex in the major and minor grooves reveals that many of the hydrogen-bonding groups of the base pairs are available for such interactions (see fig. 25.6). In the major groove the GC and AT base pairs both have three hydrogen-bonding groups accessible for interaction with other molecules. In the minor groove the GC base pair also has three, but the AT base pair has only two. There are other considerations. The major groove can accommodate a larger protein element than the minor groove. An $\alpha$ helix or a two-chain $\beta$ structure can fit comfortably into the major groove so that their polar side chains have no difficulty making hydrogen bonds with a core of base pairs. Such structures cannot be fitted into the minor groove, the contacts of which are probably limited to those that can be made with an extended polypeptide chain. Based on size considerations alone it seems likely that regulatory proteins make most of their specific contacts with DNA in the major groove. Quite remarkably almost all regulatory proteins involve $\alpha$-helical elements interacting in the major groove.

### The Helix-Turn-Helix Is the Most Common Motif Found in Prokaryotic Regulatory Proteins

In prokaryotes such as *E. coli*, the helix segment recognized by the DNA is part of a larger domain known as the helix-turn-helix motif. A protruding recognition helix is supported by a second segment of helix, which stabilizes the recognition helix and fixes its orientation with respect to the remainder of the regulatory protein.

The primary sequence for the helix-turn-helix motif consists of a 20-amino-acid sequence in which 6 of the amino acids tend to be conserved (residues 4, 5, 8, 9, 10, and 15 in fig. 30.25). Four of these residues, 4, 5, 10, and 15, usually have hydrophobic side chains; they make stabilizing contacts between the two helices, ensuring the preservation of their mutual orientation. Two other residues, 8 and 9, are important for making the bend between the two segments of $\alpha$ helix. Residue 9 is frequently a glycine because its small side chain (—H) is often necessary to make a bend. The remaining amino acid residues vary considerably in different proteins. These are the residues that either face the DNA or the remaining portion of the regulatory protein.

**Figure 30.25**

Helix-turn-helix motif. Helix 2 is the recognition helix. Individual amino acids are numbered. Residues 4, 5, 8, 9, 10, and 15 tend to be conserved in different regulatory proteins. (From Carl Branden and John Tooze, *Introduction to Protein Structure*, Garland Publishing, Inc., New York, 1991, p. 102.)

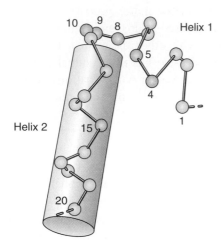

The simplest way for a segment of protein helix to make contact in the major groove is for it to run parallel to the groove. Although it adopts this orientation in some cases, it frequently orients itself in other ways in different regulatory proteins. In all known orientations ample opportunity arises for many of the amino acid side chains to make contact with the hydrogen-bonding groups in the DNA base pairs.

## Helix-Turn-Helix Regulatory Proteins Are Symmetrical

Three helix-turn-helix regulatory proteins are depicted in figure 30.26: cro repressor, λ repressor, and CAP activator. Of these, only in cro is the recognition helix oriented so that it runs parallel to the major groove when bound to the DNA. In all three cases the regulatory proteins are homodimers containing monomers with recognition helices spaced precisely 34 Å apart along the direction of the DNA helix's axis so that they can make identical contacts with adjacent major grooves of the DNA duplex. The strategy seems clear; the regulatory protein contains two identical half-sites for interaction with two virtually identical half-sites in the DNA.

Beyond helix 2 in the helix-turn-helix protein we see a great deal of variability in overall structure in the different regulatory proteins. This variability is not surprising because the remainder of the structure serves quite different functions in different regulatory proteins.

The symmetrical aspect of the regulatory protein-binding sites is mirrored in the DNA sequence of the binding site. Recall that the sequence in this region for the *lac* repressor binding site (fig. 30.7), the CAP binding site (fig. 30.8), and the *trp* repressor binding site (fig. 30.14) all show a dyad axis of symmetry with respect to the arrangement of base pairs.

## DNA–Protein Cocrystals Reveal the Specific Contacts between Base Pairs and Amino Acid Side Chains

Examination of DNA–regulatory protein complexes have permitted reasonable guesses to be made about the precise nature of the contacts between amino acid side chains and DNA in many cases. Figure 30.27 illustrates three examples: One for the 434 phage repressor (fig. 30.27*a*), one for the λ cI repressor (fig. 30.27*b*), and one for the trp repressor (fig. 30.27*c*). In all cases only half-sites are depicted because symmetry considerations dictate that the two half-sites should have virtually identical structures.

In the case of the two phage repressors, several contacts are observed between individual amino acid side chains and individual base pairs. In addition, one or two hydrogen bonds are observed to nearby phosphate groups. In both of these cases the side chain that is contacting a base participates in the hydrogen bond to a phosphate group, either directly or indirectly.

In these two examples we can see that the amino acids participating in direct interaction are frequently glutamine or asparagine, the two amino acids with amide side chains. The presence of both a hydrogen bond donor and a hydrogen bond acceptor gives the amide side chain the advantage of greater versatility than most amino acid side chains for forming hydrogen bonds. Indeed, in one case we can see that glutamine makes two hydrogen bonds with an adenine. In four of these five examples glutamine is used. The greater length of the glutamine side chain gives it more "reaching power" for making contact with the base pairs.

Despite the apparent advantages in the amide-containing amino acids, we do not always find amides used as part of the recognition sequence. For example, the trp repressor does not use any amides. The basic amino acids, lysine and arginine, and the hydroxylic amino acid, threonine, are the main interacting amino acids in the trp repressor. The side

**Figure 30.26**

The structures of three regulatory proteins. They all possess twofold axes of symmetry, and the protruding helical cylinders that interact with adjacent major grooves on the DNA (red) are spaced about 34 Å apart. N and C labels indicate the N and C termini of the polypeptide chains.

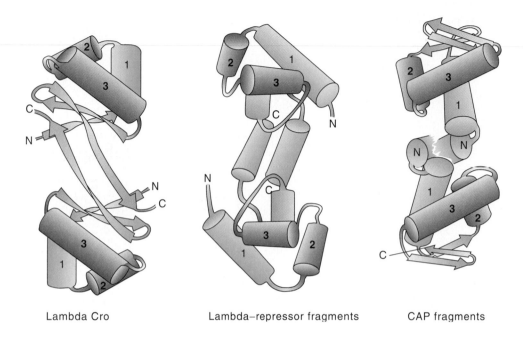

Lambda Cro          Lambda–repressor fragments          CAP fragments

chain in arginine can make a pair of hydrogen bonds with a guanine, which should result in a particularly strong interaction.

The trp repressor presents an unusual variation for interaction with the base pairs in its operator. Except for the arginine–guanine interaction, the interactions are mediated by a water molecule (see fig. 30.27c). For example, a threonine hydroxyl group makes a hydrogen bond with a water molecule, which in turn makes two hydrogen bonds with an adenine.

## Some Regulatory Proteins Use the β-Sheet Motif

Despite the dominance of the α-helical motif as a recognition unit, the β-sheet motif has been used in a limited number of cases. In these cases (three are known) two chains in the anti-parallel orientation interact with half-sites in adjacent large grooves of the DNA (see fig. 30.28 for an example). (The β-sheet motif obviously works in these cases de-

spite the fact that it is more difficult to stabilize than a segment of α helix.)

## Involvement of Small Molecules in Regulatory Protein Interaction

A common but not universal feature of regulatory proteins is their sensitivity to small-molecule effectors. The lambdoid phage repressors are examples of molecules that almost always function as repressors; their action is controlled merely by their concentration. When they are needed, they are synthesized; when they are no longer needed or their presence is undesirable, they are selectively removed by degradation.

For many other bacterial regulatory proteins the action of the regulatory protein depends on specific small-molecule effectors. As we have seen in the case of the lac repressor, the binding of allolactose results in release of the repressor from the operator. By contrast the binding of cAMP

## Figure 30.27

Specific interactions between three different repressors and their operator binding sites. Only half the operator binding site is shown because identical contacts are made with the other half. The numbers associated with the amino acid side chains refer to the distance of amino acids from the amino-terminal end of the protein. Nucleotides are numbered from the central dyad at the operator. (*a*) The 434 phage repressor; (*b*) The λ repressor; (*c*) The trp repressor. (Source: Adapted from T. Steitz, *Q. Rev. Biophys.* 23:236, 1990.)

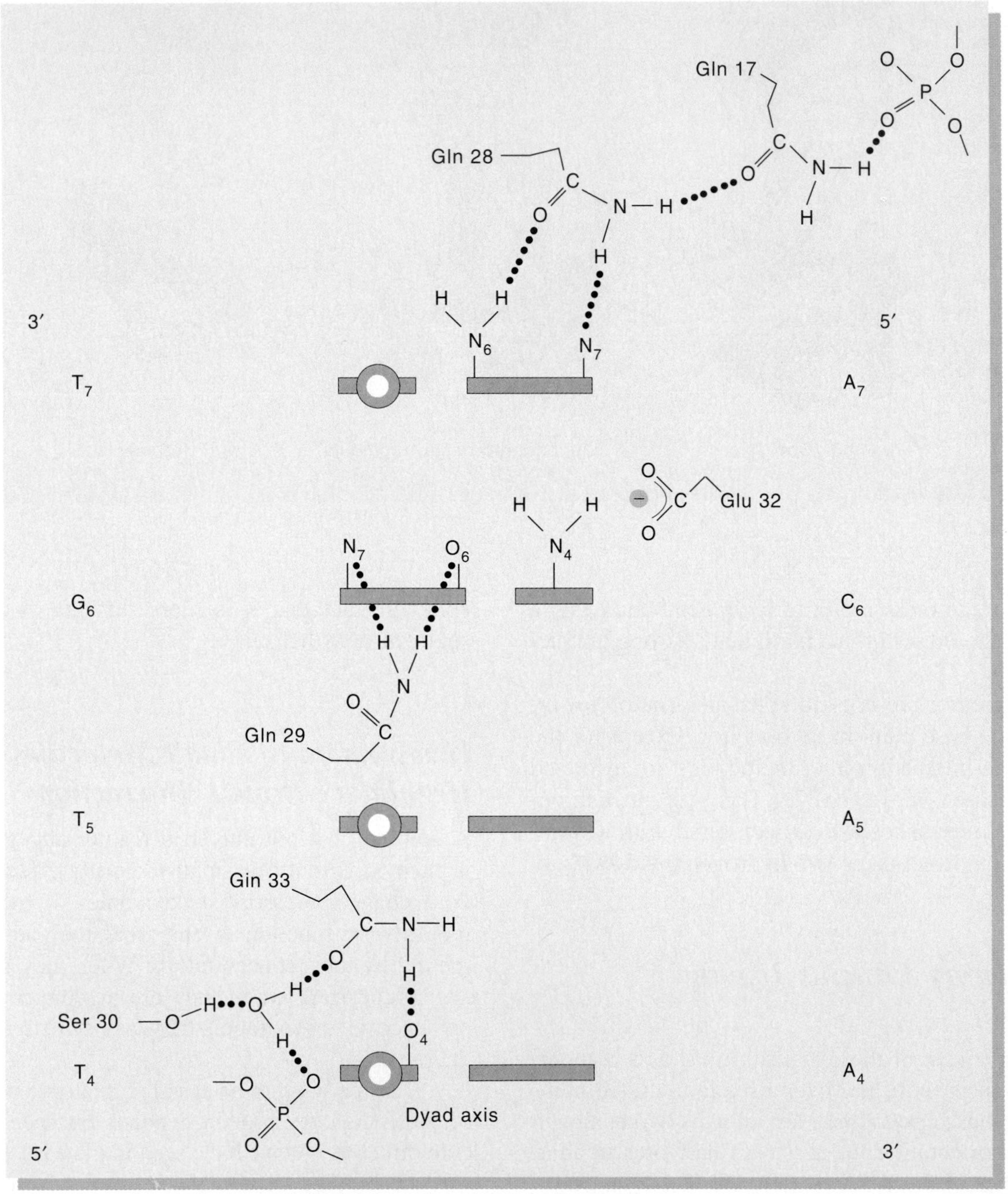

(a)

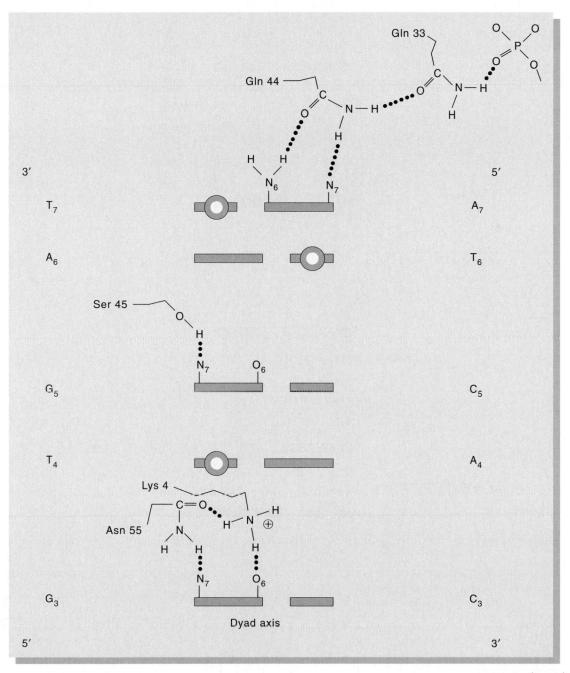

(b)

(cont.)

**Figure 30.27** (*cont.*)

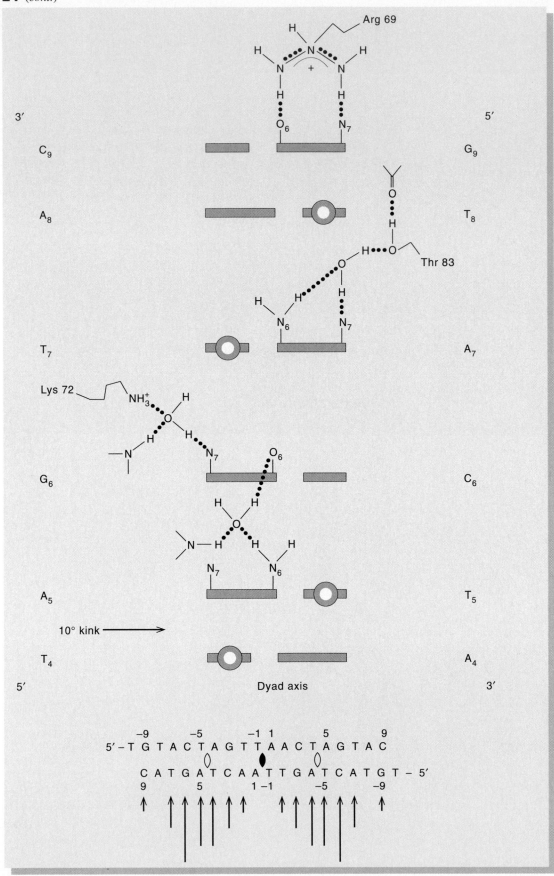

(c)

## Figure 30.28

Diagram of the met repressor–DNA structure illustrating the regions of the dimeric met repressor that contact DNA. The two-stranded $\beta$ sheet of the repressor is bound in the major groove of DNA, where it forms the sequence-specific interactions.

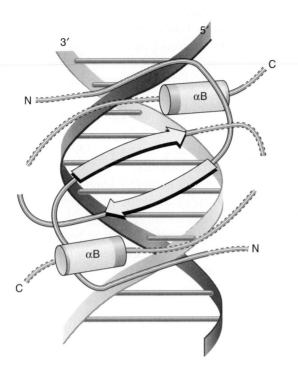

is necessary for the binding of the CAP protein to its DNA-binding site. The small molecule signals a particular metabolic state, and the response of the regulatory protein is appropriate to the metabolic needs of the cell.

It is often difficult to see the change in regulatory protein structure as a result of the binding of its small-molecule effector. In spite of this the theory is that the small-molecule effector changes the structure of the regulatory protein sufficiently to cause release or binding to the appropriate DNA-binding site as the case may be.

In the case of the trp repressor the binding of tryptophan to the repressor converts the inactive repressor (aporepressor) into an active repressor by a slight shift in the orientation of the recognition helices (fig. 30.29).

## Figure 30.29

Docking of the trp repressor to the trp operator. The repressor does not bind unless it first complexes with tryptophan. Note how the angle of the recognition helices changes when tryptophan is bound.

With regard to the trpR/O interaction, the bound tryptophan not only causes a readjustment of the so-called recognition helix, but it also shapes the side chains in its immediate environment in a very special way.

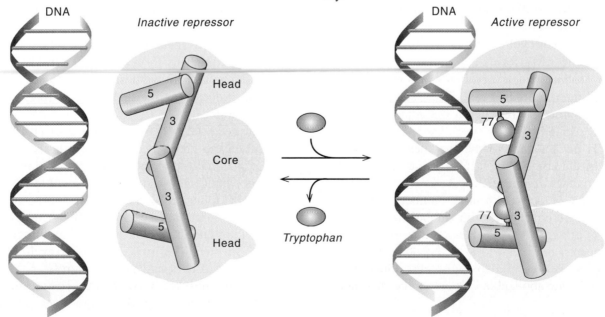

## Summary

In this chapter we discussed the regulatory systems of the *E. coli* bacterium and the λ bacteriophage. The main points in our presentation are as follows.

1. *Escherichia coli* carries about 3,000 genes. Only a small fraction of the genome is actively transcribed at any given time. But all of the genes are in a state where they can be readily turned on or turned off in a reversible fashion. The level of transcription is regulated by a complex hierarchy of control elements.

2. In the most common form of control, expression is regulated at the initiation site of transcription. There are several ways of doing this, all of them revolving around protein or small-molecule factors that influence the binding of RNA polymerase at the transcription start site.

3. The *lac* operon, a cluster of three genes involved in the catabolism of lactose, exemplifies both positive and negative forms of control that influence the rate of initiation of transcription. Jacob and Monod identified the repressor as a negative control element that is *trans*-dominant and the operator as a *cis*-dominant site for binding the repressor. Transcription is initiated by RNA polymerase at the promoter, which overlaps the operator site on one side of the three structural genes of the *lac* operon. The tight complex between repressor and operator prevents initiation, and it is broken when lactose is present. The lactose is readily converted to allolactose, which binds to the lac repressor. This changes the structure of the repressor so that it dissociates from the DNA.

4. Initiation of transcription proceeds at a greatly increased rate when cyclic AMP is present. This is because cyclic AMP forms a complex with the CAP activator protein, which then binds at a site adjacent to the polymerase-binding site. The CAP protein enhances polymerase binding at the adjacent site by cooperative binding.

5. Lactose is the substrate of the enzymes of the *lac* operon. In the absence of lactose, there is no use for enzymes of the *lac* operon.

6. CAP and cAMP activate a large number of genes in *E. coli* that are concerned with catabolism. When glucose is present, the cAMP is greatly lowered and the *lac* operon is expressed at a very low level, even when lactose is present. This is because glucose is a more readily metabolizable carbon source than lactose.

7. The *trp* operon contains a cluster of five structural genes associated with tryptophan biosynthesis. Initiation of transcription of the *trp* operon is regulated by a repressor protein that functions similarly to the lac repressor. The main difference is that the trp repressor action is subject to control by the small-molecule effector, tryptophan. When tryptophan binds the repressor, the repressor binds to the *trp* operator. Thus, the effect of the small-molecule effector here is opposite to its effect on the *lac* operon. When tryptophan is present, there is no need for the enzymes that synthesize tryptophan.

8. The *trp* operon has a control locus called an attenuator about 150 bases after the transcription initiation site. The attenuator is regulated by the level of charged tryptophan tRNA, so that between 10% and 90% of the elongating RNA polymerases transcribe through this site to the end of the operon. Low levels of *trp* tRNA encourage transcription through the attenuator.

9. Ribosomal RNA and protein synthesis are both controlled at the level of initiation of transcription. This is a result of the direct binding of guanosine tetraphosphate, ppGpp, to the RNA polymerase. This binding decreases the affinity of RNA polymerase for the initiation sites of transcription. Guanosine tetraphosphate is synthesized when the general level of amino-acid-charged tRNA is low.

10. The synthesis of ribosomal proteins is regulated at the level of translation. Certain ribosomal proteins bind to specific sites on the ribosomal RNAs or their own mRNAs. In the absence of the ribosomal RNAs, they bind to their own mRNAs, which inhibits their translation. This form of translational control regulates the rate of synthesis of ribosomal proteins so that it does not exceed the rate of ribosomal RNA synthesis.

11. Viruses borrow heavily on the host enzymatic machinery to obtain energy for synthesis, as well as for replication, transcription, and translation. The virus infective cycle is strongly irreversible. Virus infection is followed by the gradual turning on of viral genes. Viral enzymes are the first viral gene products; in late infection, the virus structural proteins are favored. The irreversible lytic cycle of the virus is directed by a cascade of controls.

12. In λ the host RNA polymerase is used throughout. Regulation is achieved through a series of repressors

and activators, as well as two viral proteins that bind directly to the RNA polymerase. The viral proteins that bind to the polymerase modify it so that it can transcribe through provisional stop signals.

13. When λ phage infects an *E. coli* cell, it does not always produce viral progeny. Sometimes it integrates its genome into the host genome and replicates only as the host genome replicates. This so-called lysogenic state can be disrupted by DNA-damaging conditions such as exposure to UV radiation. Under these conditions the dormant viral genome enters the active replication cycle.

14. Bacterial regulatory proteins are controlled by small-molecule effectors; viral regulatory proteins are not. Bacterial genes are regulated in a highly reversible manner; viral genes are usually turned on only once.

15. Proteins that regulate transcription usually bind to specific sites on the DNA. The recognition process involves specific hydrogen bonds formed between amino acid side chains of the protein and the base pairs of the DNA. Most of this interaction takes place in the major groove of the DNA, which is more accessible to the protein. The vast majority of regulatory proteins interact with the DNA from the side chains of a segment of α helix that fits snugly into the major groove. The binding site on the DNA usually consists of two half-sites, which are arranged on a dyad axis of symmetry that matches two half-sites on the regulatory protein. The two half-sites are situated in adjacent major grooves on one side of the DNA.

## Selected Readings

Anderson, J. E., M. Ptashne, and S. C. Harrison, Structure of the repressor-operator complex of bacteriophage 434. *Nature* 306:846–852, 1987.

Brennan, R. G., and B. W. Matthews, The helix-turn-helix DNA binding motif. *J. Biol. Chem.* 264:1903–1906, 1989. A mini review.

Gilbert, W., and B. Muller-Hill, Isolation of the lac repressor. *Proc. Natl. Acad. Sci. USA* 56:1891–1898, 1966. First isolation of a repressor protein.

Gold, L., Posttranscriptional regulatory mechanisms in *Escherichia coli. Ann. Rev. Biochem.* 56:199–234, 1988.

Goodrich, J. A., and W. R. McClure, Competing promoters in prokaryotic transcription. *Trends Biochem. Sci.* 16:394–396, 1991. Two or more bacterial promoters are often found in close proximity and may compete for the binding of RNA polymerase.

Gralla, J. D., Transcriptional control—Lessons from an *E. coli* promoter data base. *Cell* 66:415–418, 1991.

Green, P. J., O. Pines, and M. Inouye, The role of antisense RNA in gene regulation. *Ann. Rev. Biochem.* 55:569–597, 1986.

Helman, J. D., and M. J. Chamberlain, Structure and function of bacterial sigma factors. *Ann. Rev. Biochem.* 57:839–872, 1988.

Jacob, F., and J. Monod, Genetic regulatory mechanisms in the synthesis of proteins. *J. Mol. Biol.* 3:318–356, 1961. A classic paper.

Kaiser, D., and R. Losick, How and why bacteria talk to each other. *Cell* 73:873–886, 1993.

Kustu, S., A. K. North, and D. S. Weiss, Prokaryotic transcriptional enhancers and enhancer-binding proteins. *Trends Biochem. Sci.* 16:397–401, 1991. First discovered in eukaryotes, enhancers have now been found to exist for a number of prokaryotic genes.

Losick, R., and P. Stragier, Crisscross regulation of cell-type-specific gene expression during development in *B. subtilis. Nature* 355:601–604, 1992.

Nomura, M., R. Gourse, and G. Baughman, Regulation of the synthesis of ribosomes and ribosomal components. *Ann. Rev. Biochem.* 53:75–117, 1984.

Pabo, C. O., and R. T. Sauer, Transcription factors: Structural families and principles of DNA recognition. *Ann. Rev. Biochem.* 61:1053–1095, 1992.

Parkinson, J. S., Signal transduction schemes of bacteria. *Cell* 73:857–872, 1993.

Ptashne, M., *A Genetic Switch: Gene Control and Phage λ.* Cambridge, Mass.: Cell Press, and Palo Alto, Calif.: Blackwell Scientific, 1987.

Ptashne, M., A. D. Johnson, and C. O. Pabo, A genetic switch in a bacterial virus. *Sci. Am.* 247(5):128–140, 1982.

Roberts, J. W., RNA and protein elements of *E. coli* and λ transcription antitermination complexes. *Cell* 72:653–656, 1993.

Simons, R. W., and N. Kleckner, Biological regulation by antisense RNA in prokaryotes. *Ann. Rev. Genet.* 22:87–600, 1988.

Steitz, T. A., Structural studies of protein–nucleic acid interaction: The sources of sequence-specific binding. *Quar. Rev.*

*Biophys.* 23:205–280, 1990. A very readable and very thorough review of the subject with excellent illustrations, written by one of the pioneers in the field.

Storz, G., L. A. Tartaglia, and B. N. Ames, Transcriptional regulator of oxidative stress-inducible genes: Direct activation by oxidation. *Science* 248:189–194, 1990.

Stwinowski, Z., R. W. Schevitz, R.-G. Zhang, C. L. Lawson, A. Joachimiak, R. Q. Marmorstein, B. F. Luisi, and P. B. Sigler, Crystal structure of trp repressor/operator complex at atomic resolution. *Nature* 335:321–329, 1988.

Weintraub, H., Antisense RNA and DNA. *Sci. Am.* 262(1):40–46, 1990.

Wolberger, C., Y. Dong, M. Ptashne, and S. C. Harrison, Structure of phage 434 Cro/DNA complex. *Nature* 335:789–795, 1988.

Yanofsky, C., Operon-specific control by transcription attenuation. *Trends Genet.* 3:356–360, 1987.

Zubay, G., M. Lederman, and J. DeVries, DNA-directed peptide synthesis III. Repression of $\beta$-galactosidase synthesis and inhibition of repressor by inducer in a cell-free system. *Proc. Natl. Acad. Sci. USA* 58:1669–1675, 1967. Showing that repressor works in a cell-free system.

Zubay, G., D. Schwartz, and J. Beckwith, Mechanism of activation of catabolite-sensitive genes: A positive control system. *Proc. Natl. Acad. Sci. USA* 66:104–110, 1970. First isolation of an activator protein.

## Problems

1. What set of data originally led Jacob and Monod to suggest the existence of a repressor in *lac* operon regulation?

2. The *lac* promoter–operator region is found in many vectors used by molecular biologists to clone genes. Not infrequently, high levels of transcription driven by the *lac* promoter generate toxic levels of the cloned gene product, causing *E. coli* cells to grow poorly. Faced with such a situation, how could you minimize expression of the *lac* promoter?

3. With respect to $\beta$-galactosidase production in the presence and absence of inducer, what would be the phenotype of the following *E. coli* mutants:
   (a) $i^s o^+ z^+$
   (b) $i^s o^c z^+$
   (c) $i^s o^c z^- / i^+ o^+ z^+$
   (d) $i^+ o^c z^- / i^- o^+ z^+$

4. In a cell that is *lacZ*$^-$, what would be the relative thiogalactoside transacetylase concentration, compared with wild type, under the following conditions?
   (a) After no treatment
   (b) After addition of lactose
   (c) After addition of IPTG

5. Consider a negatively controlled operon with two structural genes (*A* and *B*, for enzymes A and B), an operator gene (*O*), and a regulatory gene (*R*). The first line of data in the table gives the enzyme levels in the wild-type strain after growth in the absence or presence of the inducer. Complete the table for the other cultures.

| | | | Uninduced | | Induced | |
|---|---|---|---|---|---|---|
| | | | *Enz A* | *Enz B* | *Enz A* | *Enz B* |
| | Strains | | | | | |
| **Haploid strains** | | | | | | |
| (1) $R^+ O^+ A^+ B^+$ | | | 1 | 1 | 100 | 100 |
| (2) $R^+ O^c A^+ B^+$ | | | | | | |
| (3) $R^- O^+ A^+ B^+$ | | | | | | |
| **Diploid strains** | | | | | | |
| (4) $R^+ O^+ A^+ B^+ / R^+ O^+ A^+ B^+$ | | | | | | |
| (5) $R^+ O^c A^+ B^+ / R^+ O^+ A^+ B^+$ | | | | | | |
| (6) $R^+ O^+ A^- B^+ / R^+ O^+ A^+ B^+$ | | | | | | |
| (7) $R^- O^+ A^+ B^+ / R^+ O^+ A^+ B^+$ | | | | | | |

6. Explain why, when *E. coli* is grown in the presence of *both* glucose and IPTG, $\beta$-galactosidase protein levels are lower than when grown only in the presence of IPTG.

7. Although *E. coli* promoters generally conform to a rather well-defined consensus sequence, no perfect match to this consensus has ever been observed in a naturally occurring promoter. Suggest an explanation.

8. Do you expect a regulatory mechanism like the prokaryotic attenuator to be found in eukaryotes? Why or why not?

9. The *lac* repressor has an "on" rate constant for the binding of the *lac* operator (when cloned into $\lambda$) of about $5 \times 10^{10} \text{M}^{-1} \text{s}^{-1}$. This value is much greater

than the calculated diffusion-controlled process, which is about $10^8 M^{-1} s^{-1}$ for a molecule the size of the lac repressor. Explain why this repressor binding works better than expected.

10. A mutation in the *trp* leader region is found to result in a reduction in the level of *trp* operon expression when the mutant is grown in rich medium. However, when the mutant is grown in a medium lacking glycine, a stimulation in the level of *trp* enzymes is observed. Explain these observations. What do you anticipate is the effect of growing the mutant in a medium lacking both glycine and tryptophan?

11. In *E. coli* no pools of free rRNAs or ribosomal proteins are floating around in the cell even when the bacteria are grown at different growth rates. Explain how *E. coli* coordinates the biosynthesis of the ribosome.

12. Draw a graph of rRNA gene transcription levels under the conditions shown in figure 30.16.

13. The rRNA genes of *E. coli* are present in multiple copies to facilitate production of many copies of rRNA during periods of rapid growth. If ribosomal proteins and RNAs need to be assembled in a $1:1$ ratio, then why are single-copy genes adequate for ribosomal protein expression?

14. Describe the principal differences between patterns of control of gene expression used by bacterial host and bacteriophage systems.

15. How does the clustering of genes on the bacteriophage λ genome (fig. 30.20) facilitate its genetic regulation?

16. How is the synthesis of the CAP protein regulated? What is unusual about this regulation?

17. Gene regulatory proteins in bacteria were predicted (before their precise structure was known) to interact in the major groove of DNA by a two-site model of binding. Describe the data that showed this model to be correct.

18. Referring to figure 30.27c, draw a detailed structure of the interaction of residue Arg69 of the *trp* repressor with G9 of its recognition sequence. Explain why binding of repressor proteins to DNA does not disrupt the DNA double helix.

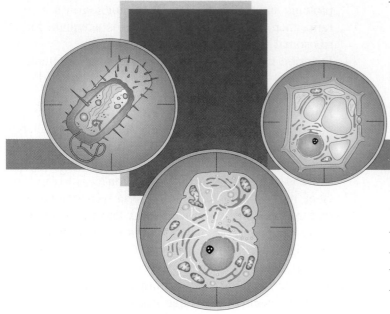

# Regulation of Gene Expression in Eukaryotes

*Gene expression in eukaryotes is usually regulated at the level of transcription initiation by a complex of RNA polymerase and an array of regulatory proteins that bind in and around the DNA promoter.*

The structural and metabolic differences between prokaryotes and eukaryotes are reflected by many differences in the modes used to regulate gene expression (fig. 31.1). The physical separation of transcription and translation in eukaryotes permits more elaborate processing of messenger RNA while eliminating the possibility of regulatory processes that require strict coupling of the transcription and translation processes. The much greater size of the eukaryotic genome necessitates the existence of many more DNA-binding proteins for regulating transcription, and we will see that basically new types of regulatory systems have developed to fulfill this need. Finally, two kinds of regulatory phenomena occur, which are almost unique to multicellular eukaryotes. First, some regulatory processes facilitate communication (including metabolic regulation) between cells; second, other regulatory devices trigger changes during the course of development that lead to differentiated cells.

It was much easier to discuss regulation in prokaryotes, in which so much of what is known comes from work on *E. coli* and related bacteriophages. In eukaryotes the important information comes from many quarters, and the array of processes is wider by far. We, therefore, needed to be brief and had to leave out a great deal. In so doing we pondered what was most important for students to know. To compensate for the necessary omissions we included an extensive reference list. Consult this for topics we left out or discussed only briefly. The supplements on immunobiology and carcinogenesis are also of value on those subjects.

First we look at a unicellular organism, and then we examine mechanisms prevalent in multicellular eukaryotes.

## Gene Regulation in Yeast: A Unicellular Eukaryote

The yeast, *Saccharomyces cerevisiae,* has a cell volume 10 times larger and a genome about 4 times larger than *E. coli.* Yeast cells are nearly spherical in shape, divide by budding, and are bounded by a cytoplasmic membrane that is surrounded by a thick polysaccharide cell wall (fig. 31.2). Structures visible in the electron microscope include a nucleus, mitochondria, and microsomes, as well as ribosomes, a Golgi apparatus with secretory vesicles, and several types of granular and vesicular inclusions. The number of mitochondria, microsomes, and ribosomes fluctuates widely with growth conditions, reflecting the presence of regulatory devices that control their numbers.

## Galactose Metabolism Is Regulated by Specific Positive and Negative Control Factors in Yeast

Yeast mRNAs are usually translated so that only the 5' proximal AUG is recognized as a translation start. This minimizes the value of polycistronic mRNAs. Even when functionally related genes are clustered, they usually give rise to separate transcripts. Three of the four genes (*GAL7, GAL10,* and *GAL1*) associated with galactose utilization are clustered on chromosome XI, whereas the fourth, for galactose transport is specified by a gene (*GAL2*) located on chromosome XII. (Note: In yeast the normal wild-type genotype is capitalized, and the mutant genotype is written in lower case.) Expression of the four structural genes is regulated by specific positive and negative controls. Transcription of the *GAL1, GAL7,* and *GAL10* genes is increased over 1,000-fold when galactose is present, suggesting that galactose is an inducer.

Each of the structural genes is associated with a distinct mRNA. The *GAL7* and *GAL10* genes are transcribed from the same DNA strand, whereas the *GAL1* gene, approximately 600 bp from *GAL10,* is transcribed from the complementary DNA strand (fig. 31.3).

Yeast is ideally suited for genetic analysis because it can be grown and examined in either the haploid or diploid states. Genetic investigations of the GAL system in diploids indicates two *trans*-acting gene products that regulate expression: *GAL4* and *GAL80.* Recall from observations on *E. coli* that *trans*-acting genetic loci usually signify that diffusible gene products are the active agent. Most *gal4* mutants are uninducible for the *GAL* structural genes in the haploid state but inducible in the diploid state when paired with a *GAL4* wild-type gene. This suggests that the active *GAL4* gene product is a positive control protein (an activator) like CAP in the *lac* operon. Rare *GAL4$^c$* mutants result in constitutive expression of the *GAL* structural genes, implying that the structure of the GAL4$^c$ protein in these unusual mutants is modified so that it is no longer repressible. Most *gal80* mutants also give rise to constitutive expression of the *GAL* genes, which suggests that GAL80 normally acts as a negative control protein (a repressor) of *GAL* gene expression. Rare *GAL80$^s$* mutants are uninducible in the haploid or the diploid states where they are paired with wild-type *GAL80.* Based on these results it was proposed that the wild-type GAL80 protein binds the inducer galactose and that this binding converts the protein to an inactive form. The GAL80$^s$ protein appears to have lost the site for galactose binding. As a result it functions as a repressor even in the presence of galactose. This is similar to the situation for *i$^s$* mutants in the *lac* operon of *E. coli* (see

## Figure 31.1

Overview of some of the unique features associated with regulation of gene expression in higher organisms. Eukaryotes may be unicellular, as in the case of yeast, or multicellular, as in the case of vertebrates. Regulation of gene expression in all eukaryotes bears a close resemblance to regulation of gene expression in simple unicellular prokaryotes. Yet some striking differences exist. Thus, multicellular eukaryotes are composed of differentiated cells that can express only a limited number of the total genes in their chromatin. And most of the chromatin in any particular cell type is highly condensed (heterochromatic), but a small fraction of the chromatin is expanded, or swollen (euchromatic). It is the latter portion of the chromatin that has the potential to be expressed. The regions of the chromatin that are euchromatic are different in different cell types. To convert a region of the genome from the potentially active to the active state requires the binding of regulatory proteins to the promoter. The role

of these proteins is to create an environment favorable for the binding of RNA polymerase. Here we come to one of the most striking differences between prokaryotes and eukaryotes. In prokaryotes the regulatory proteins usually bind in the immediate vicinity of the RNA polymerase. In eukaryotes the regulatory proteins may be bound to the DNA at several places, ranging from a point adjacent to the polymerase-binding site to points a thousand or more bases away. Although the explanation for how distantly bound regulatory proteins can influence gene expression is not totally clear, the consensus is that chromosome folding is the most likely way of bringing such proteins into the immediate proximity of the RNA polymerase-binding site. The function of all of the regulatory proteins binding to a single promoter of the DNA is to promote formation of the transcription initiation complex.

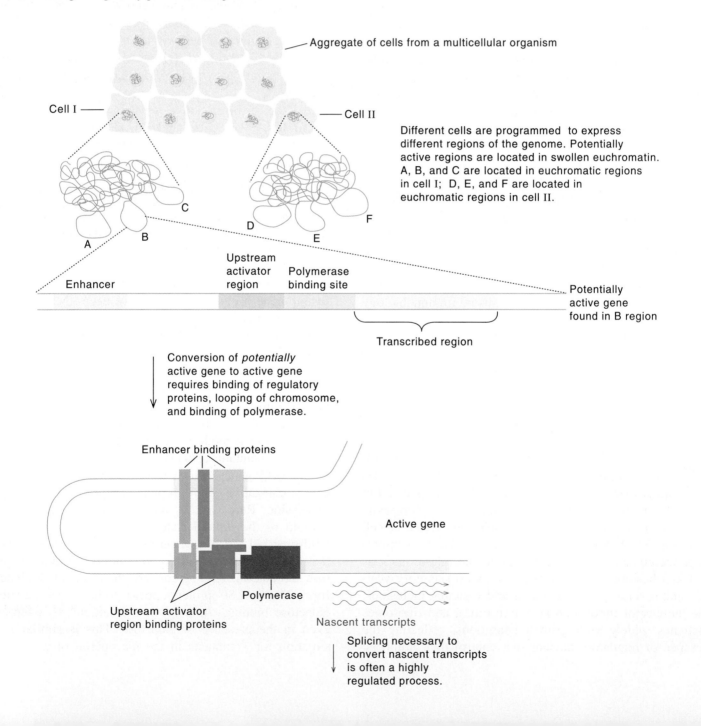

## Figure 31.2

Diagram of a haploid yeast cell. A yeast cell contains many of the organelles characteristic of a typical eukaryotic cell. Duplication occurs by a budding process. The bud gradually grows until, just before pinching off, it contains a nuclear equivalent of chromosomes, as well as some mitochondria and other elements present in the cytoplasm. Mechanical strength of the yeast cell is guaranteed by a thick cell wall.

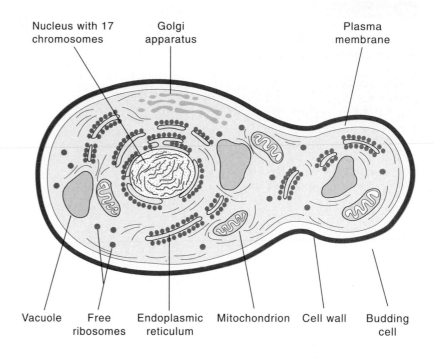

## Figure 31.3

Model for the regulation of enzymes of the *GAL* system. (*a*) Three structural genes synthesize distinct mRNAs and enzymes (transferase, epimerase, and kinase). Arrows next to the genes indicate the direction of transcription. Synthesis requires the GAL4 protein. However, it is inactive in the absence of inducer because of complex formation with the GAL80 protein. The GAL80 protein does not prevent the GAL4 protein from binding to specific sites on the DNA, but it does prevent GAL4 protein from activating transcription once bound. (*b*) GAL4 protein becomes active when the GAL80 protein is removed by adding inducer, which binds to the GAL80 protein. Recent evidence suggests that GAL80 protein may not dissociate from GAL4 on induction.

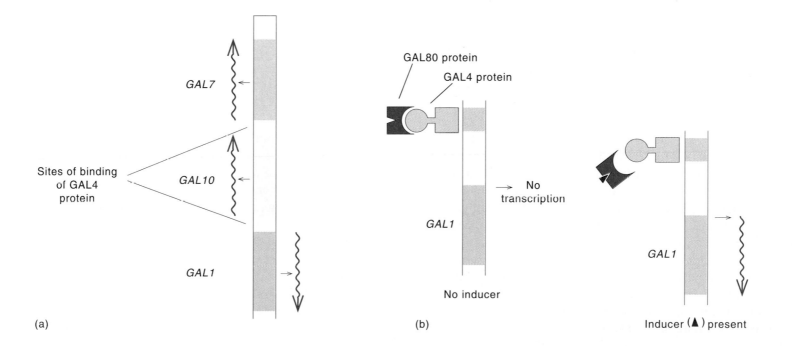

Phenotypes of Haploid and Diploid Yeast with
Regulatory Gene Mutations

| Genotype | GAL10 (epimerase)[a] Noninduced | Induced | GAL1 (kinase) Noninduced | Induced |
|---|---|---|---|---|
| Wild type | − | + | − | + |
| gal4 | − | − | − | − |
| gal4/GAL4 | − | + | − | + |
| gal80 | + | + | + | + |
| GAL4[c] | + | + | + | + |
| GAL80[s] | − | − | − | − |
| GAL80[s]/GAL80 | − | − | − | − |

[a] In the induced state, galactose is added to the growth medium. It should
be noted that GAL10 and GAL1 respond in identical fashion to the different
mutations.

chapter 30). The phenotypes for different regulatory gene
mutants are summarized in table 31.1.

Based on these genetic studies a model was proposed
for the mechanism of regulation of the *GAL* system (see fig.
31.3*b*). In this model the GAL4 protein binds to a site upstream of the gene(s) it regulates and promotes RNA polymerase II-dependent transcription of these genes. The
GAL80 protein prevents GAL4 protein from functioning as
an activator by binding to it. Induction by galactose results
from the binding of the sugar to the GAL80 protein, which
changes its structure so that the GAL4–GAL80 complex is
altered. It is not clear if actual break-up of the complex is
required to relieve the repression.

The proposed model postulates *cis*-acting sequences
upstream of the target genes for GAL4 protein. The GAL4
protein binds to four related 17-bp sequences located between 150 and 200 bases upstream of the *GAL10* gene. The
specific DNA sequences to which the GAL4 protein binds
were determined by reacting the DNA with dimethylsulfate
and then following the methylation pattern by nucleotide
sequencing to determine which guanine bases are protected
by the regulatory protein. The sequences that showed protection in a wild-type *GAL4* strain were not protected in the
*gal4* mutant derivative. Further manipulations of the promoter region showed that only one of the 17-bp sequences is
needed for normal regulation.

## The GAL4 Protein Is Separated into Domains with Different Functions

The model for GAL4 protein specifies a multifunctional
protein with binding sites for DNA and GAL80 protein as
well as a site that somehow activates transcription. Deletion
studies were performed to see if these different sites could
be allocated to different parts of the 881-amino-acid GAL4
protein. A mutant protein containing only the first 98
N-terminal amino acids still binds DNA but cannot bind the
GAL80 protein or activate transcription. Additional mutant
studies show that amino acids 148–196 and 768–881 are
required to activate transcription and that amino acids 851–
881 are required for GAL80 protein building.

Further experiments indicate that different parts of the
GAL4 protein and the lexA bacterial repressor protein (see
chapter 26) are functionally interchangeable. In a "domain
swap" experiment, the DNA-binding domain of the lexA
protein was inserted in place of the GAL4 DNA-binding
domain (fig. 31.4). As might be expected this hybrid protein
could not activate the wild-type *GAL1* gene for transcription
unless other changes were made in the DNA. If the DNA-
binding site for the GAL4 protein was replaced by the
DNA-binding site for the lexA protein, then the same hybrid
protein was able to activate transcription of the *GAL1* gene.

The *GAL* system shows both similarities and differences to typical bacterial regulatory systems. In both systems DNA-binding regulatory proteins play a major role.
However, in the yeast system the binding site for the regulatory proteins is often located some distance upstream from
the RNA polymerase-binding site. In yeast such distant sites
required for activation are referred to as upstream activator
sequences (UASs).

## Mating Type Is Determined by Transposable Elements in Yeast

Yeast has two haploid cell mating types: *MATα* (or simply
α) and *MATa* (or *a*); on contact, haploid cells of opposite
mating types fuse to form a single diploid (*a/α*) cell. Diploid cells can grow and divide indefinitely as diploid cells,
or they can sporulate, a process in which they undergo meiosis and give rise to two α and two *a* cells for each diploid
cell. The haploid mating type is determined by specific sequences at the *MAT* locus (fig. 31.5). These sequences are
found at the *HMLα* or *HMRa* loci, where they are usually
not expressed. When the sequences stored at the *HMLα*
locus are transposed to *MAT*, they express; similarly when
sequences are transposed from *HMRa* to *MAT*, they express.

## Figure 31.4

Ptashne's domain-swap experiment. The GAL4 protein normally binds at a site upstream from the *GAL1* gene. In the absence of GAL80 this activates transcription from the *GAL1* gene. If the DNA-binding domain of GAL4 is replaced by the DNA-binding domain of the lexA protein, transcription from the *GAL1* gene does not occur because the hybrid regulatory protein does not bind at the appropriate site. However, if the DNA-binding site is also changed to that of the DNA-binding site for the lexA protein, then the *GAL1* gene becomes active again.

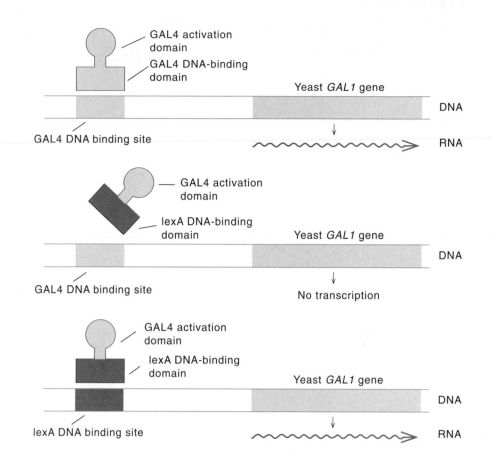

## Figure 31.5

Yeast chromosome III, showing the *HML, MAT,* and *HMR* loci. W is a region common to *MAT* and *HML,* but not *HMR* (≈750 bp). X is a region found at *MAT, HMLα,* and *HMRa* (≈700 bp). $Y_a$ is a specific substitution found at *MATa* and *HMRa* (≈600 bp). $Y_α$ is a specific substitution found at *HMLα* and *MATα* (750 bp). $Z_1$ is a region found at *MAT, HMLα,* and *HMRa* (≈250 bp). $Z_2$ is a region found at *MAT* and *HMLα* (≈70 bp).

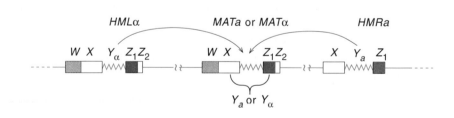

Why are the α and a sequences only expressed when they are present at the *MAT* locus? The explanation for this was suggested by the finding of four unlinked genes: *SIR1–SIR4.* If any of these genes are mutated, the a and α sequences at *HMLα* and *HMRa* express just as they do at *MAT.* Further experiments showed that *trans*-acting regulatory factors, encoded by the *SIR* genes act as repressors at "silencer" sites (*HMLE* and *HMRE*) adjacent to *HMRa* and

*HMLα.* Remarkably, these silencer sites are located more than a kilobase away from the regions they control (fig. 31.6). The silencer sites are not found at or near the mating type locus (*MAT*), explaining why the a and α coding sequences are expressed when transposed to *MAT.*

An important question remaining is how does a single genetic locus determine the haploid yeast cell mating type? Haploid cells that carry *MATα* behave as α cells; similarly

## Figure 31.6

Structure of mating loci determinants in yeast. The genetic regions *HMLα* and *HMRa* are normally silent, whereas *MATa* or *MATα* are active. *MATa*, which contains the $Y_a$ segment, expresses transcript *a*1. *MATα*, which contains the $Y_a$ segment, expresses two transcripts: $\alpha_1$ and $\alpha_2$. The structures in and around the $Y_a$ and $Y_\alpha$ segments are the same at the storage locus and the expression locus. The inactivity at the storage loci results from far upstream *cis*-acting elements not present at *MAT*. These elements, in conjunction with the four *SIR* gene products, negatively regulate expression of mating type genes at the storage loci.

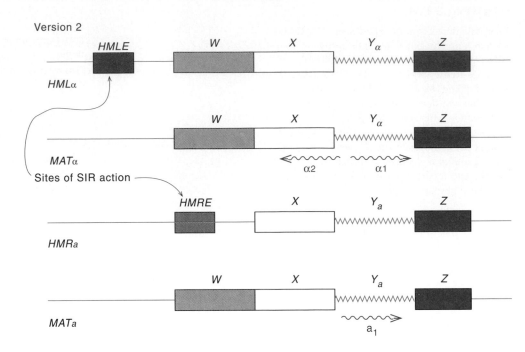

Version 2

## Figure 31.7

Regulation of *a*-specific genes (*a*sg) and *α*-specific genes (*α*sg) in *α* and *a* haploid cells. In an *α* cell, *α*2 inhibits *a*sg and *α*1 activates *α*sg. In an *a* haploid cell, the *a*1 transcript made at *MATa* does not appear to exert any regulatory function. (⊕ = positively regulated; ⊖ = negatively regulated.)

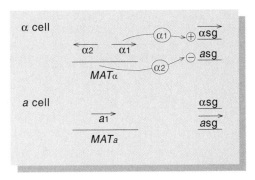

cells that carry *MATa* behave as *a* cells; and, finally, diploid cells that carry both *MATα* and *MATa* do not express any haploid-specific genes. To explain the key role of *MAT* it was proposed that the *MAT* locus encodes regulatory proteins that control unlinked genes that determine mating type. Subsequently it was shown that the *MATα* locus contains two genes: *MATα*1 and *MATα*2, which encode the *α*1 and *α*2 regulators, respectively. Ira Herskowitz and his coworkers found that *α*1 turns on expression of the *α*-

specific genes, and *α*2 turns off expression of *a*-specific genes (fig. 31.7).

In addition to understanding the regulatory mechanisms that give the haploid cells their specific properties, considerable progress has been made in determining the mechanism that regulates meiosis in diploid cells. Diploid *a/α* cells undergo sporulation and meiosis when they are subjected to nutritional starvation. It is believed that the reactions leading to meiosis are triggered by a complex of the *α*2 and *a*1 proteins, which repress the haploid specific gene *RME*1 (regulator of meiosis), allowing the cells to initiate meiosis (fig. 31.8).

The control of mating type in yeast illustrates a specialized role of mobile or transposable genetic elements in which mobile elements hop from one site to another. Mobile genetic elements are commonly found in both prokaryotes and eukaryotes, but most of them show little site specificity. Indeed, their almost random insertion behavior has led to the proposal that their major function in most cases is to promote evolutionary change.

Mating type in yeast is much better understood than mating type of multicellular organisms such as vertebrates where sex type is determined by gene dosage. For example, in humans, males carry one X chromosome whereas females carry two X chromosomes. The mechanism used in yeast and other fungi is more responsive to factors in the external environment, which may be more suitable for a unicellular organism that is directly exposed to its surroundings.

## Figure 31.8

Regulation of meiosis in yeast. Haploid cells are not able to initiate meiosis because they express *RME*1, which is a negative regulator of meiosis. Diploid cells are able to initiate meiosis because they make a complex of α2 and *a*1 that inhibits the synthesis of the *RME*1 product.

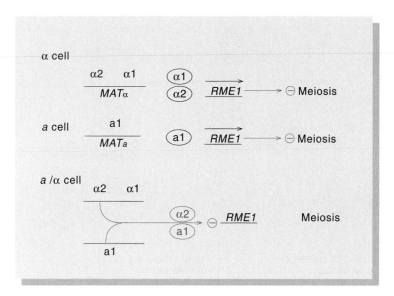

## Figure 31.9

Nuclear transplantation technique. The nucleus of an unfertilized egg is mechanically removed with a micropipette (*left*), and a nucleus from a blastula cell is removed and microinjected into the enucleated egg.

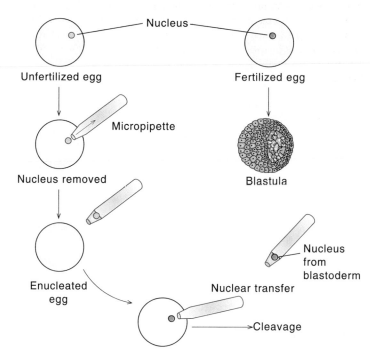

This concludes our discussion of regulatory phenomena in yeast. All of the regulatory processes in yeast are highly reversible, just as they are in *E. coli*. In multicellular eukaryotes additional regulatory mechanisms often show an irreversible character.

# Gene Regulation in Multicellular Eukaryotes

We have seen that yeast differentiates into three cell types: The *a* and α haploid types and the *a*/α diploids. In multicellular eukaryotes we must deal with a much wider range of cell types and a much larger genome. First we consider some aspects of the changing chromosome architecture, which reflects changes in gene activity. Then we examine the underlying changes in the way that expression is controlled at the transcriptional, posttranscriptional, and translational levels. Finally we consider some developmental processes that are triggered by changes in gene expression.

## *Nuclear Differentiation Starts in Early Development*

In the mature adult state, a multicellular eukaryote contains many cells, each with a potential for the constitutive or inducible expression of a unique subset of genes belonging to the total genome.

Whereas most differentiated cells in the adult contain the normal amount of genomic DNA, the expression of only a fraction of the genome in any particular cell type suggests that the average nucleus may have lost its totipotency. The first direct evidence that nuclei irreversibly differentiate during development came from R. Briggs and T. J. King in 1952. Working with frog eggs, they showed that the original nucleus from an unfertilized egg could be replaced by the diploid nucleus from a developing animal (fig. 31.9). If the inserted nucleus came from a frog very early in development (blastula), a mature frog developed. If, however, the nucleus came from a later stage of development, no growth or only limited growth ensued. These nuclear transplantation experiments demonstrated that during development the nucleus assumes a pattern of expression that is difficult or impossible to reverse even when the partially differentiated nucleus is returned to the environment of the egg cytoplasm.

In 1962, J. Gurdon took this type of experiment one step further by showing that adult frogs could be developed by injecting enucleated eggs with nuclei from the intestinal epithelium of tadpoles. The success frequency was much lower than when nuclei from earlier stages of development were used. Nevertheless, such results showed the influence of the cytoplasm on nuclear expression; they also demonstrated that in certain cases nuclei already tentatively com-

**Figure 31.10**

Manipulation of mouse teratoma cells. The cells from a teratocarcinoma may be dispersed and grown in tissue culture. These cells can be injected into an embryo (*a*), in which case the resulting animal is a chimera in which some cells come from the original parents and others arise from the cells injected into the blastocyst. (*b*) Alternatively, these cells may be implanted subcutaneously, in which case the animal develops a tumor at the site of implantation. (From G. Zubay, *Genetics*, Benjamin/Cummings, Menlo Park, Calif., 1987, p. 301.)

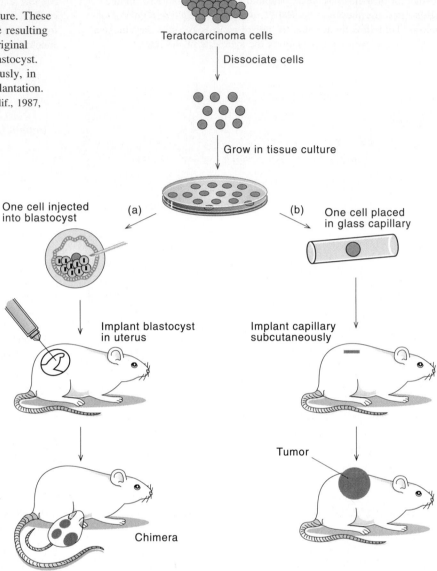

mitted to a specific pathway of development can be reprogrammed by placing them in a different environment. The most important lesson to be gained from such experiments is that the nucleus tends to assume a stable pattern of expression that is ultimately dictated by its surroundings.

The most convincing evidence for the importance of cytoplasm on nuclear development comes from studies on teratoma, a unique type of tumor found in many kinds of mammals. It is composed both of neoplastic cells, like other tumors, and of many kinds of differentiated cells. A typical teratoma may include nerve, muscle, blood, and skin cells, and other differentiated cells all mixed together with undifferentiated neoplastic stem cells. Only the stem cells are neoplastic, producing more stem cells or more differentiated cells, usually both. In many ways the stem cells behave like

embryonic cells. They can be dissociated into single cells and cultured *in vitro* like bacterial cells. Such cells can be reintroduced into the animal by subcutaneous implantations, in which case they produce a tumor. Most remarkably, they can be implanted in an early embryo (blastocyst) to produce a hybrid chimera, in which, in favorable cases, all of the tissues possess some cells from the tumor parent (fig. 31.10). Even egg cells have been isolated that are derived from the tumor parent. These egg cells can be fertilized by normal sperm, resulting in progeny that could truly be said to have a tumor for a mother. This result shows that the neoplastic stem cells are capable of reverting to completely normal behavior when subjected to the embryonic environment.

These pioneering studies demonstrating nuclear differentiation and dedifferentiation are supported by a broad range of genetic and biochemical findings that indicate that chromosomal structural differences often can be correlated with changes in the potential for gene expression.

## Chromosome Structure Varies with Gene Activity

In interphase cells, chromosomes are in a dispersed form called chromatin. Chromatin exists in a highly condensed form known as heterochromatin or a swollen form known as euchromatin. All nuclei contain both types of chromatin. Some chromosomes or parts of chromosomes are always heterochromatic (constitutive heterochromatin); others are heterochromatic only during certain times of the cell cycle or in certain cell types (facultative heterochromatin). Many studies indicate that euchromatin is relatively active in RNA synthesis, and heterochromatin is relatively inactive. Whereas the euchromatic state seems to be necessary for a high level of transcription, for most genes it is not sufficient. This has been demonstrated by autoradiography of cells grown briefly in the presence of $^3$H-labeled RNA precursors. Those regions of the genome that are transcriptionally active incorporate the label; the extent of labeling is proportional to the amount of RNA synthetic activity. Some euchromatic regions appear active by this test, whereas others do not.

## Giant Chromosomes Permit Direct Visualization of Active Genes

Polytene chromosomes are unusually large chromosomes found in the salivary glands of certain insects. Because of their size, they are convenient vehicles for studying the relationship between chromosome structure and activity. Polytene chromosomes are produced by the repeated replication of interphase chromosomes without separation, resulting in a large number of chromosomes that remain laterally aligned. For example, the DNA content in the giant salivary gland cells of *Drosophila melanogaster* may be as much as a thousand times that of other cells in the fruit fly. Microscopically, the stained chromosomes appear as cross-banded extended bodies (fig. 31.11). About 5,000 bands are in *Drosophila*, and it is tempting to associate single bands with single genes. However, the average DNA content found in each band is 30,000 bp, which is considerably more DNA than necessary to encode the average protein (about 1,000 bp).

Transcription in polytene chromosomes usually is associated with local swellings of the bands, called puffs (see

**Figure 31.11**

A segment of giant chromosome from the midge larva of *Rhyncosciara* at different stages during development. The arrows and connecting lines indicate comparable bands. Changes in the extent of swelling of different regions reflect the activity of those regions in transcription. (Source: Drawing based on the results of M. F. Breuer and C. Pavin, *Chromosoma* 7:257–280, 1946.)

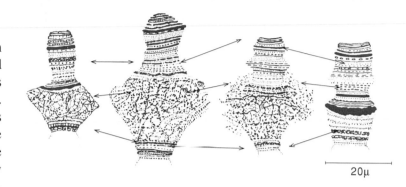

20µ

fig. 31.11). Sometimes puffing results in a broadening and lengthening, and sometimes the extended DNA projects laterally into loops that combine to form a large ringlike structure. As a rule, a puff originates from a single band, but it can result from swelling of one or more bands. The extent of incorporation of labeled RNA precursors into a puff is approximately proportional to the size of the puff. A detailed investigation of the puffing patterns as a function of the physiological state of the organism shows that different bands become activated, swollen, and transcriptionally active in a sequentially related, tissue-specific fashion.

The correlation of puffs with genetic functions has focused on two types of gene products: Secretory proteins and proteins formed in response to heat shock. Certain puffs in *Drosophila* can be artificially induced by the insect hormone, ecdysone, which is instrumental in regulating development. Some puffs are induced directly by exposure to the hormone, whereas others are induced indirectly, depending for puffing on the gene products of the more directly induced puffs. This dependence is shown by adding drugs that inhibit protein synthesis, with the result that secondary puffs do not form.

## In Some Cases Entire Chromosomes Are Heterochromatic

In some cases entire chromosomes can be heterochromatized and stay that way through successive cell divisions. Male and female mammals are distinguished by the fact that females carry two X chromosomes, and males carry only one. Invariably, one of the X chromosomes in the somatic tissue of females is inactivated and condensed into a hetero-

**Figure 31.12**

Diagram of nuclei obtained from cells in the mucous membrane of the human mouth. (*a*) Nucleus from a female, showing one Barr body (arrow). (*b*) Nucleus from a male, with no Barr body. (*c*) Nucleus from an XXX female, showing two Barr bodies. (*d*) Nucleus from an XXXX female, showing three Barr bodies.

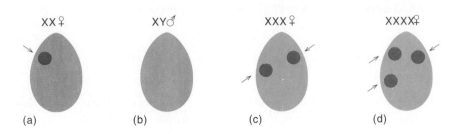

chromatic state known as a Barr body (fig. 31.12). The consequences of X-chromosome inactivation are readily observed in females heterozygous for an X-linked mutation.

For example, the enzyme, glucose-6-phosphate dehydrogenase (G-6-PD), is encoded by an X-linked gene. A female that is heterozygous for this gene may carry two alleles that produce electrophoretically distinct forms of the same enzyme (A and B). When isolated cells from a skin biopsy are cloned, each clone contains either the A or the B form of the G-6-PD enzyme, but never both. If every skin cell of a single organism were to be analyzed for the enzyme, large homogenous patches expressing one or the other of the G-6-PD alleles would be found. This pattern indicates that the decision as to which X chromosome of the female is inactivated is made at the multicellular stage in the embryo. Once the decision is made, that X chromosome remains inactive through successive cell divisions. There is an equal chance that the X chromosome from either parent is inactivated. In genetically abnormal cells that contain three or four X chromosomes, two or three Barr bodies, respectively, can be found (see fig. 31.12).

X-chromosome inactivation provides a simple means of maintaining equal amounts of active X-linked genes in both males and females. This form of so-called dosage compensation is not found in all species. For example, in *D. melanogaster* neither of the X chromosomes in the female cells is condensed into a Barr body. In this species, dosage compensation, which regulates the activity of specific alleles, operates by a different mechanism so that many alleles in the female X are about 50% as active as the corresponding alleles in the male. However, not all alleles in the *Drosophila* X chromosome are regulated in this way, so that for certain genes twice as much gene product is produced in females as in males.

## Biochemical Differences between Active and Inactive Chromatin

Much of the preceding discussion on chromatin structure indicates that active chromatin exists in a more swollen state than inactive chromatin. Several biochemical changes accompany the transition from condensed to swollen chromatin. These changes include a redistribution of nucleosomes along the DNA duplex, chemical modification of histones, alteration in the pattern of nonhistone chromosomal protein binding, and chemical modification of the DNA. Currently, most of these changes are discussed in a general, descriptive manner because their causes and consequences are not known.

***DNA in Active Chromatin Is More Susceptible to DNase Degradation.*** The greater accessibility of transcriptionally active chromatin has been elegantly demonstrated by Groudine and Weintraub for the hemoglobin and ovalbumin genes in the chicken. Chromatin was isolated from chicken erythrocytes, in which the hemoglobin genes were recently very active, and from the oviduct, in which the ovalbumin gene is very active. These two chromatins were degraded by DNaseI using an assay that measured gross DNA degradation as well as that of specific genes. In both cases the rate of DNA degradation from the active gene was much greater than that from average DNA. Thus in erythrocyte chromatin the globin gene DNA was rapidly degraded, whereas in oviduct chromatin the ovalbumin gene DNA was rapidly degraded. Extensive regions surrounding the active genes were also quite susceptible to DNaseI hydrolysis.

***DNA Methylation Is Correlated with Inactive Chromatin.*** Methylation of cytosine in CpG sequences may play a regulatory role in the gene expression. In chapter 25 it was noted

## Figure 31.13

Hemimethylated DNA. The unmethylated C residue in the indicated sequence is destined to be methylated shortly after DNA replication.

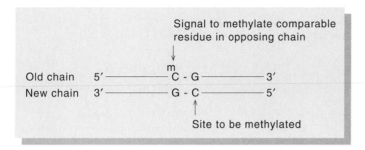

that 5-methylcytosine (5-MC) is the main modified base in vertebrate DNA. Most of the 5-MC occurs in the dinucleotide CpG. Indeed, in mammals and birds approximately 50%–70% of all such dinucleotide sequences are modified. In this connection it is noteworthy that CpG sequences occur much less frequently than would be expected statistically. The mechanism of methylation has been explored by DNA transfection of certain tissue culture grown cells. If the DNA used in transfection is methylated, the pattern of methylation is maintained through many cell duplications. Likewise, when unmethylated DNA is used in transfection, a non-methylated pattern is maintained. These results indicate that under many circumstances methylation is passively maintained by a signal that recognizes hemimethylated DNA. If the C residue in a CpG dinucleotide is methylated, immediately after semi-conservative DNA replication, only the parental DNA chain is methylated; the passive methylation system signals methylation of the corresponding C residue in the new DNA chain (fig. 31.13).

The extent of methylation of a gene is correlated with its ability to transcribe. Given that DNA methylation usually reduces transcription, two important, closely related questions remain unanswered: How is methylation regulated *in vivo?* How does methylation interfere with transcription? Since methylation is known not to interfere with the elongation phase of RNA synthesis, it seems likely that methylation blocks initiation. The binding of polymerase and other regulatory proteins at the initiation locus is sensitive to modification of these nucleotides. The precise inhibition mechanisms, however, await further elucidation.

Before turning to other questions we note that methylation or something like methylation could be at the root of X chromosome inactivation. In this connection it may be rele-

vant that *Drosophila,* which does not show the methylation phenomenon, does not show X-chromosome inactivation either.

***Enhancers Are Promoter Elements that Operate over Great Distances.*** In chapter 28 we saw that several proteins in addition to RNA polymerase II are required for the initiation of transcription at the adenovirus major late promoter; this promoter is prototypical of many eukaryotic promoters. In addition to signals for regulatory protein binding in the immediate vicinity of the polymerase-binding site, other signals function at more distant locations. In yeast, sequences of this type have been called UAS sequences because their action is usually confined to regions located a few hundred bases upstream from the promoters they influence. In vertebrates *cis*-active sequences operating over much greater distances have been discovered. These sequences have been named enhancers because they enhance expression of genes located in their vicinity. Enhancer signals are effective downstream as well as upstream of the promoters they influence. Moreover many enhancers are equally effective in either orientation on the DNA.

The first enhancer was discovered by George Khoury in the SV40 virus (see chapter 26). It should be recalled that the SV40 genome is a circular duplex containing about 5,300 bp. In SV40-infected cells, viral RNA synthesis is divided into early and late phases. Early transcription originates from a block of sites illustrated in figure 31.14. A TATA box, typical of eukaryotic gene promoters (see chapter 28) occurs about 27 bp upstream from the transcription start site. Immediately upstream of the TATA box the SV40 promoter contains a series of three tandemly repeated GC-rich segments that bind a transcription activator protein known as SP1. Further upstream from the transcription start site a tandemly repeated 72-bp sequence occurs (−116 to −188 and −189 to −261 from the 5′ end of the messenger). Removal of one of these sequences has no effect on transcription, but removal of both 72-bp sequences drastically lowers early transcription.

Surprisingly, the precise location or orientation of the 72-bp segment is not critical to the stimulating effect on transcription. Thus, the 72-bp segment remains effective after inversion, or after translocation further upstream or downstream from the transcription start site. Foreign genes inserted into DNA containing the SV40 enhancer are frequently stimulated in the same way as the SV40 early region, demonstrating the general stimulating effect of this enhancer.

## Figure 31.14

Region in and around the early transcription start sites of the SV40 genome. The base pair number on the circular genome is indicated. There are several transcription start sites. One cluster of start sites (early), used initially after infection, is located about 27 bp downstream from the TATA box. The other cluster of start sites (late early) is used at later times. This cluster of start sites is located upstream of the TATA box. One of the products of early transcription is the protein known as large T antigen. This protein binds in and around the core *ori*. T antigen inhibits early transcription. The upstream positive control *cis* elements include the tandem 72-bp enhancer sequences and three tandem 21-bp sites. The latter function in conjunction with host-encoded protein known as SP1.

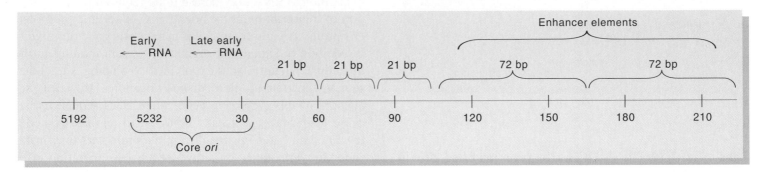

## Figure 31.15

Enhancers are frequently composed of two or more components, or enhansons. Each enhanson contains binding sites for two or more proteins, which can interact with each other when bound to the enhanson if not before. An enhancer that contains two enhansons, each of which can bind two different proteins, can be arranged in $2^4$, or 16, ways with respect to binding sites. From considerations such as this it is clear that a great deal of variety is possible in enhancer construction with a very limited number of DNA-binding sites for different regulatory proteins. The proteins that bind to the enhancer do not all make direct contact with the DNA. Often one finds pyramids of proteins that bind to each other in which only the base of the pyramid makes direct contact with the DNA.

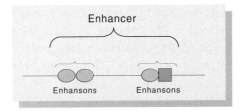

A more detailed analysis of the SV40 enhancer shows that it is composed of elements 15–20 bps in length that bind one or more protein factors specifically. These elements, sometimes called enhansons, are ineffective when separated from one another. It is believed that different enhancers are composed of different combinations of enhansons (fig. 31.15). The emerging picture of the enhancer is of a complex, multicomponent segment of DNA that can bind different combinations of protein activators. This appears to be a major strategy used in multicellular eukaryotes to produce a greater variety of responses. It has the advantage that, given a finite set of regulatory proteins, a much greater variety of combinations can be produced. The use of complexes containing combinations of regulatory proteins has implications for the structures of regulatory proteins that are discussed later on.

It is not certain how proteins that bind to enhancers influence transcription over long distances. However, several proposals have been made: (1) The enhancer element may function as an attachment point to a structural component of the nucleus to stimulate transcription; (2) it may serve as an initial binding site for some factor required for transcription that must subsequently move along the duplex to the initiation point for transcription; or (3) folding or looping of the chromosome may bring the *cis* bound enhancer proteins into close contact with the gene it stimulates (fig. 31.16). Currently, the third possibility is strongly favored, but this does not exclude the other two mechanisms.

## DNA-Binding Proteins That Regulate Transcription in Eukaryotes Are Often Asymmetrical

In the previous chapter we discussed DNA-binding proteins that regulate transcription in prokaryotes. The principles that govern recognition between proteins in eukaryotes show some similarities and some differences. In both cases specific recognition is dominated by interactions that take place in the major groove of the DNA. The specific interactions usually involve H bond formation

## Figure 31.16

Possible mechanisms for enhancer action. (1) The enhancer draws the associated gene to the nuclear matrix, where it is more accessible to the transcription apparatus. (2) The enhancer is the initial binding site for an element that subsequently moves. (3) The enhancer folds or loops, depending on its polarity, to bind with other promoter elements.

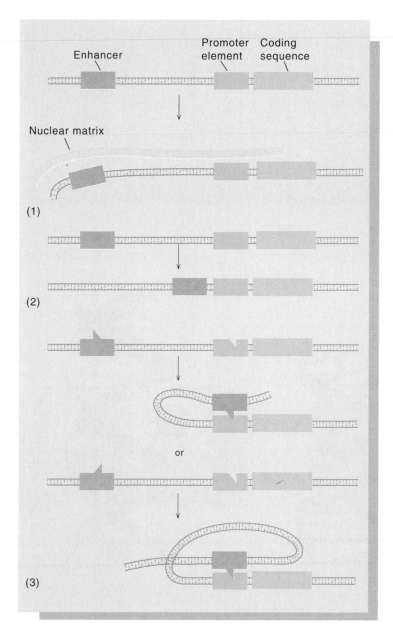

teract in a symmetrical fashion with the DNA. Half-sites in the protein bind to half-sites in the DNA, which are usually spaced one helix turn apart. In eukaryotes most of the cases that have been examined so far involve proteins that interact in an asymmetrical fashion with the DNA. In some cases we find eukaryotic proteins that contain multiple sites for interaction within one polypeptide chain, and in other cases we find two structurally distinct proteins interacting to form heterodimers that interact with the DNA.

## The Homeodomain

The homeodomain has a motif that has been recognized in a large family of eukaryotic regulatory proteins. The name "homeodomain" derives from the fact that many of the mutations that affect the body plan in a developing *Drosophila* embryo were referred to as homeotic gene mutations (this is taken up later). Walter Gehring found that many of these homeotic genes and other genes that regulate development in *Drosphila* encode regulatory proteins that possess a common 180-base segment in the 3' exon. The base homology ranges between 60% and 80%, depending on the specific genes being compared, and the amino acid homology is even higher (up to 87% fig. 31.17). The high basicity of the amino acid sequence in this region—called the homeobox—and other structural features suggest that this region encodes a DNA-binding protein that regulates gene expression by binding to specific sites on the DNA. Indeed, sufficient amino acid homology exists between the homeobox and the helix-turn-helix motif seen in bacterial regulatory proteins to suggest that the gross structures of these very distantly related regulatory proteins are quite similar.

One of the most exciting features relating to the homeobox is that very similar sequences have been identified in many other animals, including frogs, mice, and even humans. This finding suggests that homeobox proteins occur in a wide range of organisms, possibly playing similar roles in regulating developmental processes.

Whereas amino acid sequence comparisons suggested that the homeodomain would contain a helix-turn-helix motif, the 60-residue homeodomain forms a stable structure that can bind to DNA as a monomer. The recognition helix in the homeodomain is longer and makes more contacts with the DNA core than the recognition helices from bacterial regulatory proteins (fig. 31.18).

Although an isolated homeodomain can fold correctly and bind DNA with a specificity similar to that of the intact proteins, it is believed that the precise DNA-binding specificity is modulated by other regions of the protein. Protein–protein interactions may also have a role in modulating many homeodomain–DNA interactions. For example, the

between the base pairs in the DNA and the side chains in the proteins. In both cases the $\alpha$ helix is the most common element used for DNA recognition. The most striking difference between DNA-binding proteins in prokaryotes and eukaryotes has to do with the symmetry of the interaction. In prokaryotes DNA-binding proteins almost always are composed of an equal number of identical subunits that in-

## Figure 31.17

The homeobox sequence found in three *Drosophila* regulatory proteins. The sequence of amino acids along the main, continuous line is that found in the *Antp* homeobox. At points where the sequence of ftz proteins is different, the differences are shown above the corresponding Antp proteins, and at points where the sequence of ubx proteins is different, they are shown below the corresponding Antp proteins. Certain proteins isolated from the human and mouse embryos have segments with similar sequences. Note the high basicity of the sequence, which should favor electrostatic binding to nucleic acid.

|     |     |     | 5 |     |     |     |     |     | 10 |     |     |     |     | 15 |     |     |     |     | 20 |
|-----|-----|-----|-----|-----|-----|-----|-----|-----|-----|-----|-----|-----|-----|-----|-----|-----|-----|-----|-----|
|     |     |     | Thr<br>Gly |     |     |     |     |     |     |     |     |     |     |     |     |     |     |     |     |
| Arg<br>Ser | Lys | Arg | Gly | Arg | Gln | Thr | Tyr | Thr | Arg | Tyr | Gln | Thr | Leu | Glu | Leu | Glu | Lys | Glu | Phe |

|     |     |     | 25 |     |     |     |     |     | 30 |     |     |     |     | 35 |     |     |     |     | 40 |
|-----|-----|-----|-----|-----|-----|-----|-----|-----|-----|-----|-----|-----|-----|-----|-----|-----|-----|-----|-----|
|     |     |     |     |     | Ile |     |     |     |     |     |     | Asp |     |     | Asn |     |     | Ser |     |
| His | Phe<br>Thr | Asn | Arg<br>His | Tyr | Leu | Thr | Arg | Arg | Arg | Arg | Ile | Glu | Ile<br>Met | Ala | His<br>Tyr | Ala | Leu | Cys | Leu |

|     |     |     | 45 |     |     |     |     |     | 50 |     |     |     |     | 55 |     |     |     |     | 60 |
|-----|-----|-----|-----|-----|-----|-----|-----|-----|-----|-----|-----|-----|-----|-----|-----|-----|-----|-----|-----|
| Ser<br>Thr | Glu | Arg | Gln | Ile | Lys | Ile | Trp | Phe | Gln | Asn | Arg | Arg | Met | Lys | Ser<br>Trp<br>Leu | Lys | Lys | Asp<br>Glu | Arg<br>Asn<br>Ile |

## Figure 31.18

(*a*) Complex formed between the *Drosophila* homeobox protein engrailed and DNA seen from two angles. In addition to the main contacts made by the recognition helix in the major groove, additional contacts are made in the minor groove, which can be seen in (*b*). (From C. R. Kissinger, B. Liu, E. Martin-Blanco, T. B. Kornberg, and C. O. Pabo, Crystal structure of an engrailed homeodomain-DNA complex at 2.8 Å resolution: A framework for understanding homeodomain-DNA interactions, *Cell* 63:579–590, November 2, 1990. Copyright © Cell Press. Reprinted by permission.)

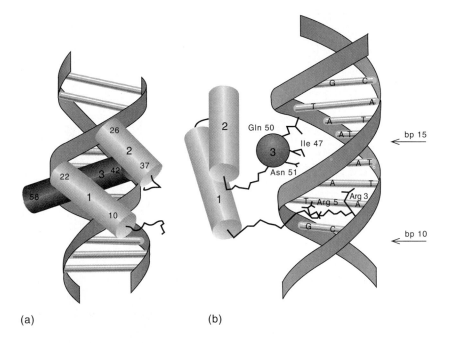

(a)                    (b)

yeast α2 protein forms homodimers but also forms complexes with a related homeodomain protein a1 (see fig. 31.8). Each of these complexes has different site preferences, and it is possible that these accessory proteins in addition to adding new DNA contacts actually alter the way that α2 interacts with DNA.

We tentatively conclude that the asymmetrical interaction of the homeodomain protein with DNA gives it a versatility not displayed by the symmetrical helix-turn-helix homeodimers found in bacteria.

## Zinc Finger

Another DNA-binding motif very common in eukaryotic regulatory proteins is the zinc finger. This motif was first identified as the DNA-binding structure in the RNA poly-

## Figure 31.19

Schematic representation of the $C_2$-$H_2$ zinc finger found in Xfin from *Xenopus laevis*. (Adapted from M. S. Lee et al., *Science* 245:645, 1989.) The recognition helix is stabilized by a complex involving zinc. Cysteine sulfurs are in yellow, and histidine nitrogens are blue.

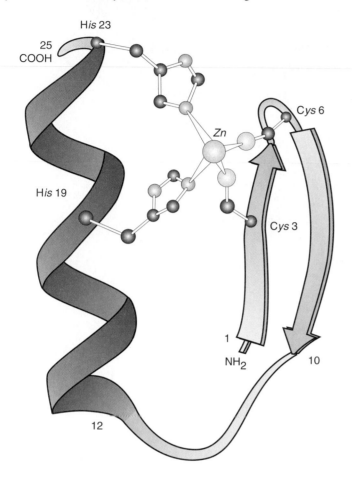

merase III transcription factor TFIIIA which binds to the internal control region of the 5S rRNA gene (discussed in chapter 28). Zinc fingers that resemble TFIIIA are present in the mammalian transcription factor SP1 and in a variety of other regulatory proteins found in eukaryotes. This type of zinc finger motif consists of about 30 amino acids with two cysteine and two histidine residues that stabilize the domain by tetrahedrally coordinating a $Zn^{2+}$ ion (fig. 31–19). A region of about 12 amino acids between the cysteine–histidine pairs is characterized by scattered basic residues and several conserved hydrophobic residues. Proteins in this family usually contain tandem repeats of the 30-residue zinc finger. The crystal structure of a zinc finger–DNA complex containing three fingers from zif268 and a consensus zif-binding site has been reported. The crystal structure shows that the zinc fingers bind in the major groove and wrap part way around the double helix (fig. 31.20). Each finger has a similar way of docking against the DNA and makes base

contacts with a 3-bp subsite arranged in a way that reflects the helical pitch of the DNA and the 3-bp periodicity of the subsites.

## Leucine Zipper

A third type of DNA-binding domain was first described for the mammalian enhancer-binding protein C-EBP. Proteins of this class show a primary sequence similarity consisting of a highly conserved stretch of about 30 amino acids with a substantial net basic charge immediately followed by a region containing four leucine residues positioned at intervals of seven amino acids. The latter segment, named the leucine zipper by Steve McKnight, is required for dimerization and for DNA binding. It is believed that dimerization of proteins in this group is stabilized by hydrophobic interactions between closely apposed $\alpha$-helical leucine repeat regions of the two proteins. The two helical cylinders are believed to be oriented in parallel in a coiled-coil fashion (fig. 31.21). The main function of the leucine motifs is to bring two proteins together so they can form homodimers or heterodimers that bind to dissimilar half-sites on the DNA.

## Helix-Loop-Helix

The helix-loop-helix motif appears to be another way of creating heterodimers that can bind to asymmetric sites on the DNA. Like the leucine zipper proteins, the helix-loop-helix proteins have a basic region that contacts the DNA and a neighboring region that mediates dimer formation. Based on sequence patterns, it has been proposed that this dimerization region forms an $\alpha$ helix, a loop, and a second $\alpha$ helix. Like the leucine zipper protein, the activity of the helix-loop-helix proteins is modulated by heterodimer formation. For example, the MyoD protein, which appears to be the primary signal for differentiation of muscle cells, binds DNA most tightly when it forms a heterodimer with the ubiquitously expressed E2A protein.

This brief description of three new types of regulatory proteins found primarily if not exclusively in eukaryotes by no means includes all of the known types of DNA-binding proteins. Furthermore, the progress in discovering new regulatory proteins is so rapid that anything we say here is bound to need supplementation if the reader wants to be up to date on this subject.

## Transcription Activation Domains of Transcription Factors

Thus far we have focused on the DNA-binding domains of regulatory proteins. Many regulatory proteins have additional domains that are involved in transcription activation.

## Figure 31.20

The complex formed between the zif268 zinc finger protein and DNA. The three fingers fit into the major groove and wrap partway around the duplex. The zinc is not shown in this figure. (From N. P. Pavletich and C. O. Pabo, Zinc finger-DNA recognition: Crystal structure of a Zif268 DNA complex at 2.1 Å, *Science* 252:809–817, 10 May 1991. Copyright 1991 by the AAAS. Reprinted by permission.)

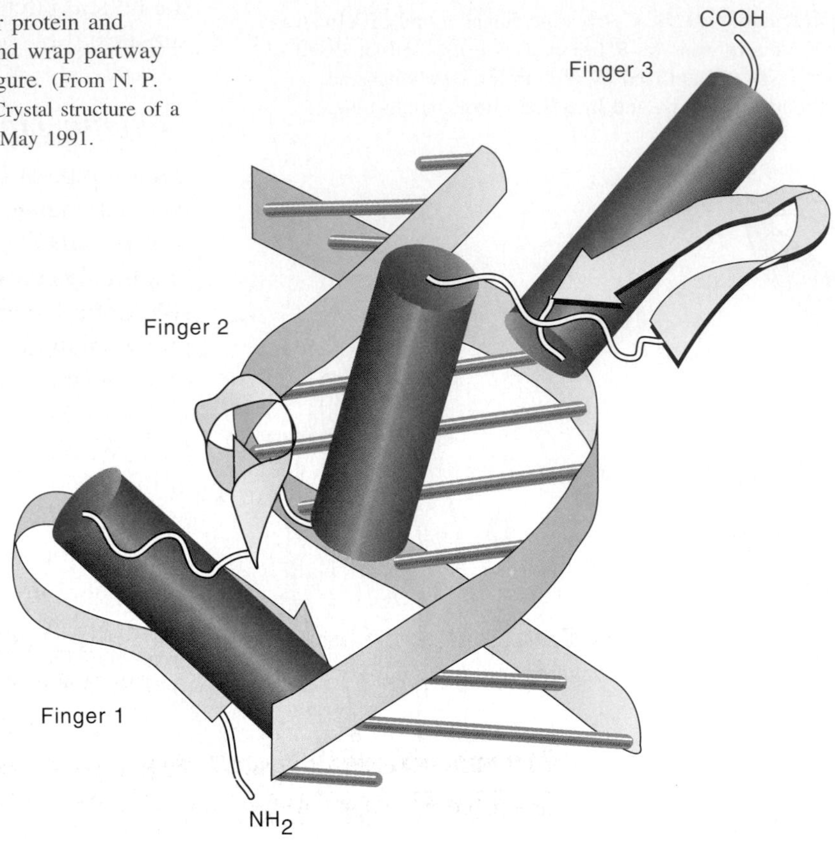

## Figure 31.21

Diagrammatic sketch of a leucine zipper dimer. The protein monomers are held together by interaction between leucine side chains (green knobs in the upper part of the structure). The part of the protein monomers that interacts with the major groove of the DNA is shown (in red). (Adapted from C. R. Vinson et al., *Science* 246:911, 1989.)

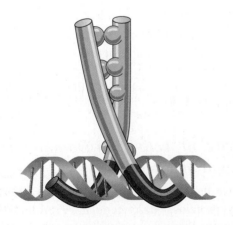

The amino acid sequences of mammalian DNA regulatory proteins suggest the existence of at least three different types of activation domains: Acidic, glutamine-rich, and proline-rich.

Transcriptional activation functions of DNA-binding factors depend on regions of from 30 to 100 amino acids that are separate from the DNA-binding domains. Factors often have more than one activation domain, and several apparently unrelated structural motifs have been identified that confer these functions. The first defined activation regions in eukaryotic transcription factors were identified by studies of the yeast factor, GAL4 (discussed previously). The activation domains of these factors consist of relatively short stretches of amino acids with significant negative charge that can form amphipathic $\alpha$-helical structures.

Deletion analysis of the transcription factor of SP1 has revealed four separate regions that contribute to transcriptional activation; all lie outside the zinc-finger-binding domain. The two strongest activation domains contain about 25% glutamine and very few charged amino acid residues. Several other transcription factors show glutamine-rich regions in the domain required for transcription activation.

These are but two of the types of domains found to be involved in transcription activation. Undoubtedly, this list will grow in the near future, as will our understanding about how these transcription activation domains work. Thus far they are likely to represent regions that function by contacting other regulatory proteins, transcription factors, and the RNA polymerase itself.

## Alternative Modes of mRNA Splicing Present a Potent Mechanism for Posttranscriptional Regulation

The basic mechanisms of RNA splicing were discussed in chapter 28. RNA splicing was first discovered for transcription of adenovirus where different reading frames are connected to the same 5′ end. Thus, we have known about the phenomenon of alternative splicing as long as we have known about splicing itself. Alternative splicing occurs for eukaryotic viruses such as SV40 and polyoma, as well, and it is also a common phenomenon in eukaryotic genes that contain multiple exons in their nascent transcripts. It is clearly a regulatory phenomenon in viruses since we see a shift in the types of splicing as virus infection progresses. For eukaryotic genes it is also a regulatory phenomenon since different modes of splicing are found in different tissues of the same multicellular organism. The type of splicing found usually involves segments from the same transcript, but on occasion transplicing occurs where two independent transcripts participate in a common splicing operation. In the most common splicing situation a promoter is present at one end of the transcript and the combination of exons that is used in the mature mRNA varies according to the pattern of splicing. In some cases one or more exons are excluded from the message by selective splicing. In other cases part of an intron is fused to one of the exons to make the final message. Splicing can also be influenced by the choice of the promoter or the choice of the polyadenylation site. In the latter two situations parts of the upstream or downstream regions of the gene are excluded from the nascent transcript, and so the splicing of the transcript is limited to the RNA remaining.

The possibilities for alternative splicing are enormous. One particularly elaborate example is seen in the gene for tropomyosin in vertebrates. Recall that tropomyosin is a key component of vertebrate striated muscle (see fig. 5.18). The mRNA for tropomyosin found in striated muscle undergoes nine splices in the process of maturation (fig. 31.22). Variants of tropomyosin resulting from alternative splicing are found in other tissues of the same organism. It seems likely that the types of tropomyosins found in different tissues are best suited to the needs of the tissues.

P. Bingham and his coworkers have uncovered another regulatory role for splicing in *Drosophila*. In three different genes they observed that the final splices are subject to regulation. These final splicing operations appear to be regulated by the protein encoded by the mature mRNA. If that protein is present in sufficient concentrations, then it inhibits the final splices, thereby preventing the mRNA from becoming functional and turning out unneeded protein. It is not clear how common this type of feedback inhibition is in *Drosophila* or other species.

## Gene Expression Is Also Regulated at the Levels of Translation and Polypeptide Processing

Following messenger formation, the amount and types of proteins can be modulated in additional ways. The initial polypeptide can be processed in various ways so that different polypeptides or proteins are expressed in different tissues. Such a situation exists for processing the precursor polypeptide preproopiomelanocortin (see fig. 24.7). This polypeptide is processed in different ways in the anterior and intermediate lobes of the pituitary gland to give rise to different hormones in these two tissues.

Because of the longer lifetime of eukaryotic messengers, it seems likely that translation level controls should play a greater role in regulation of eukaryotic gene expression. Despite this belief, few mechanisms have been elucidated.

Inactivation of eukaryotic translation factors by covalent modification is one of the few mechanisms known to regulate the rate of translation. Specific protein kinases have been identified that phosphorylate and inactive both eIF-1 and EF-2. The significance of the phosphorylation of EF-2 as a regulatory mechanism of the elongation rate is still not clear, but the phosphorylation of eIF-2 appears to be a general mechanism for controlling translation initiation in many cells.

The regulation of translation through the phosphorylation of eIF-2 is best understood as it operates in the rabbit reticulocyte. Two protein kinases specific for the a subunit of eIF-2 have been purified from reticulocytes. One of these kinases, termed the heme-regulated inhibitor repressor (HRI), serves to coordinate the rate of hemoglobin synthesis (more than 90% of the total protein synthesized in the reticulocyte is hemoglobin) with the availability of hemin (the

**Figure 31.22**

Alternative modes of splicing of the tropomyosin gene transcript in
different tissues. (Source: Adapted from R. E. Breitbart, A. Andreadis, and
B. Nadal-Ginard, Alternative splicing: A ubiquitous mechanism for the
generation of multiple protein isoforms from single genes, *Ann. Rev. Biochem.*
56:467–495, 1987.)

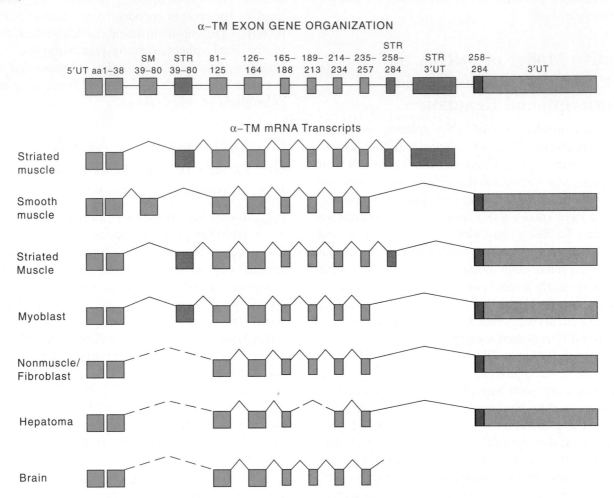

precursor of the heme group in hemoglobin). Hemin binds
to and inhibits the activity of HRI, thereby enhancing the
rate of globin synthesis (fig. 31.23).

The second eIF-2-specific kinase appears to be present
at low levels in most mammalian cells. This kinase is activated by double-stranded RNA and for this reason has been
named the double-stranded RNA-activated inhibitor (DAI).
DAI may play a role in defending cells against invasion by
viruses.

HRI and DAI are different proteins, but both phosphorylate the same amino acid residue in a subunit of eIF-2,
and, in consequence, both kinases inhibit protein synthesis
by the same mechanism. Phosphorylated eIF-2 is capable of

catalyzing the binding of Met-tRNA to the ribosome, but it
does so in a stoichiometric rather than a catalytic manner.
This mechanism of inhibition results from the fact that eIF-2
dissociates from the ribosome as a complex with GDP, and
in order to recycle the bound GDP must exchange for GTP.
Phosphorylated eIF-2 binds to, but is unable to dissociate
from, the guanine nucleotide exchange factor (GEF) that
catalyzes the exchange of GTP for GDP. Because cells contain fewer copies of GEF than eIF-2, all of the GEF can be
sequestered by partial phosphorylation of eIF-2, and protein
synthesis initiation ceases for lack of GEF. This makes the
rate of protein synthesis initiation exquisitely sensitive to
the state of phosphorylation of eFI-2.

**Figure 31.23**

Regulation of protein synthesis in the rabbit reticulocyte. The vast majority of the protein synthesized in the rabbit reticulocyte is hemoglobin. The gross rate of protein synthesis in the reticulocyte is controlled indirectly by the concentration of heme. Heme inactivates a kinase that would otherwise inactivate the initiation complex involving eIF-2 and eIF-2B. The kinase phosphorylates the eIF-2 factor, making it impossible for the eIF-2–eIF-2B complex to exchange GDP for GTP.

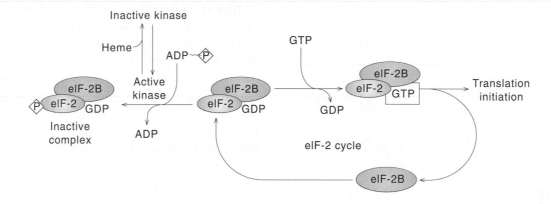

## Patterns of Regulation Associated with Developmental Processes

As cells proliferate within the embryo, they assume different properties; changes are evidenced by the genes they express and the proteins they synthesize. Developmental differences can be maintained or altered during subsequent cell duplication. During embryonic growth, cells that are initially capable of following any pathway of development become committed to a particular pathway. As a rule, pathways have many branchpoints, so that when progeny cells of partially committed cells reach a branchpoint they may differentiate further down a more specialized pathway. This process of gradual commitment, repeating itself many times, gives rise to a complex pattern of cell lineages, all arising from the initially fertilized egg.

It seems likely that the pivotal events in the evolution of a differentiated cell reflect changes in gene expression that result from a complex hierarchy of controls. The key to understanding differentiation, therefore, is to identify the regulatory factors responsible for the controls involved in differentiation and to explain how they act.

We focus on regulatory events that occur in development. First we discuss some of the classical studies of gene expression in the embryogenesis of sea urchins and amphibians and then turn to some of the developmental events in fruit flies, the best characterized developmental system.

## During Embryonic Development in the Amphibian, Specific Gene Products Are Required in Large Amounts

During early development, when cell divisions are occurring rapidly, demand for certain products associated with the translation apparatus (ribosomes) increases. Special regulatory mechanisms ensure that these gene products are produced in the required amounts and that they are not overproduced in the more mature organism.

## Ribosomal RNA in Frog Eggs Is Elevated by DNA Amplification

In eukaryotic organisms ribosomal RNA genes are present in clusters of hundreds to thousands of copies. The nascent transcript, a 45S molecule, undergoes an elaborate series of processing reactions to produce three kinds of ribosomal RNAs: 28S (25–28S), 18S (17–18S), and 5.8S (5.5–5.8S). Multiple copies of ribosomal RNA genes are organized into tandem arrays separated by nontranscribed spacers, which vary in length in different species from about 2,000 to 30,000 bp. *In situ* hybridization studies, using radioactive rRNA and autoradiography, demonstrate that the ribosomal RNA genes are localized around the nucleolus, where rRNA processing and ribosome assembly take place. Germ cells and, in particular, immature eggs (oocytes) often raise the

**Figure 31.24**

Time course of rDNA amplification in the *Xenopus laevis* oocyte. The amount of rDNA per cell is plotted against time. Note that the ordinate is a logarithmic scale. The sharp drop in rDNA at fertilization is thought to be due to dilution of the amplified rDNA copies by cell division, rather than to their destruction. (From A. P. Bird, Gene reiteration and gene amplification, *Cell Biol.* vol. 3, *Gene Expression: The Production of RNAs,* L. Goldstein and D. M. Prescott (eds.), Copyright © 1980 Academic Press, New York.)

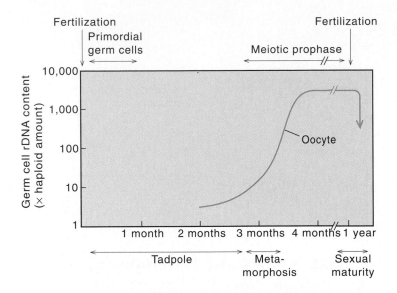

level of rDNA per nucleus by amplification to satisfy the high demand for rRNA during the very rapid early cleavage stages. This type of amplification has been most thoroughly investigated in the frog *Xenopus laevis* (fig. 31.24) in which amplification increases the number of rRNA genes about 1,000-fold. These extrachromosomal genes are transcribed during oogenesis and subsequently discarded.

## 5S rRNA Synthesis in Frogs Requires a Regulatory Protein

An entirely different mechanism is used to elevate the level of 5S rRNA synthesis during oocyte development. Recall that eukaryotic ribosomes contain the small rRNA components (5S and 5.8S). Whereas the 5.8S rRNA is contained in the unprocessed 45S transcript with the large rRNAs, the 5S exists independently. In amphibians the 5S rRNA genes are organized into the different multigene families the expression of which is under developmental control. For instance, in *X. laevis* the normal haploid genome contains about 20,000 copies of the 5S rRNA genes, organized into three multigenic families. The genes of two major families, comprising about 98% of all the 5S rRNA genes, are expressed only in growing oocytes. The genes of the third family, existing in about 400 copies, are active in most (somatic)

cells and growing oocytes. This developmental control enables the oocyte to accumulate 5S rRNA at rates 1,000-fold higher than is possible in somatic cells.

In chapter 28 we noted (see fig. 28.12) that PolIII, which transcribes the 5S genes, requires the binding of three transcription factors to the 5S promoter. These factors are known as TFIIIA, TFIIIB, and TFIIIC. The abundance of TFIIIA is much greater in the oocyte than in most somatic tissue. As a consequence, only in the oocyte is there enough TFIIIA to bind to all of the 5S promoters. Competition binding experiments in which both somatic and oocyte 5S gene DNAs are exposed to limited amounts of the TFIIIA protein show that the regulatory protein binds considerably more firmly to the somatic 5S genes. This fact probably explains why the somatic 5S genes are uniquely active in somatic tissues where the amounts of TFIIIA protein are much lower. Under a variety of conditions, large amounts of 5S rRNA are synthesized in the oocyte but not in somatic cells; this difference is directly related to the concentration of TFIIIA regulatory protein in the two situations.

## Early Development in *Drosophila* Leads to a Segmented Structure that Is Preserved to Adulthood

Whereas frogs have proved valuable in the study of isolated embryonic events, they have not been as useful in understanding the overall pattern of early embryonic developments. For this purpose it has been necessary to turn to other organisms to identify the genes involved in regulating development. Many fundamental aspects of developmental biology have been studied using genetic and molecular biology techniques on investigations of *D. melanogaster*.

Some steps in the development of *Drosophila* are shown in figure 31.25. Beginning shortly after fertilization, the new diploid nucleus undergoes a rapid series of divisions with no segregation of nuclei into separate cells. After the eighth division, when there are 256 nuclei (fig. 31.25*c*), the nuclei begin to migrate to the periphery of the egg cytoplasm. After another nuclear doubling, cell membranes form around a group of cells at the posterior end of the egg; these cells are progenitors of germ cells for the subsequent generation. The remaining nuclei continue to divide until there are about 6,000 nuclei at the periphery (fig. 31.25*e*). At this point, about 2 h after fertilization, membranes are formed separating the nuclei into a monolayer of cells. The resulting structure, known as the blastoderm, is essentially a cell monolayer enclosing the yolk.

The blastoderm divides into 14 different segments: Md, Mx, Lb, T1–T3, and A1–A8 (see fig. 31.25*f*). Three of

## Figure 31.25

Steps in the development of *Drosophila melanogaster*. The fertilized egg contains a single zygotic nucleus (*a*). This divides (*b*) every 10 min. After eight divisions, when there are 256 nuclei, nuclear migration toward the outer cortex structure begins (*c*). Eventually all the nuclei form a monolayer on the cortex surface. The first nuclei to become enclosed are the pole cells, which become germ cells in the adult organism (*d*). The fully formed blastoderm contains about 6,000 cells, which form a monolayer around the cortex (*e*). Even before a cell membrane has formed around the nuclei, the embryo has become functionally divided into a segmented structure (*f*). During gastrulation (*g*), there is continued cell duplication, a folding of sheets of cells, and mass migration of segments. Eventually this structure hatches into the first larval stage (*h*). The remaining stages between the larva and the adult fly (*i*) are not illustrated. In (*f*) through (*i*), various segments are labeled. Three segments, Md, Mx, and Lb, fuse to make the head structure. Thoracic segments T1–T3 and abdominal segments A1–A8 retain their segmental appearance in the adult.

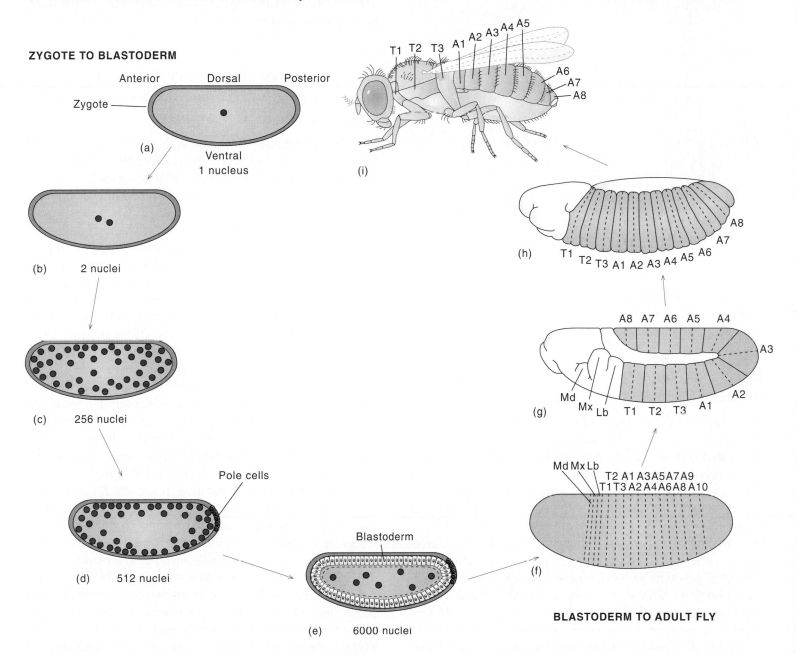

these segments, Md, Mx, and Lb, fuse to become part of the head structure. The remaining segments become subdivided into two compartments, with anterior and posterior parts. During gastrulation there is a continued cell duplication, folding of sheets of cells, and mass migration of segments (see fig. 31.25*g*). The embryo eventually hatches into the first larval stage (see fig. 31.25*h*). Cells within the larva are of two types. About 80% are embryonic precursors to adult tissues rather than larval tissues. These latter cells form packets within the larva, called imaginal disks, that are ar-

**Figure 31.26**

Imaginal disks in a mature larva and the structures they lead to in the adult fly.

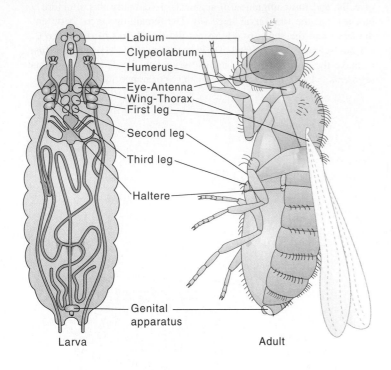

Larva                    Adult

rested in development until pupation (fig. 31.26), when the hormone, ecdysone, signals their differentiation into specific adult structures.

## Early Development in Drosophila Involves a Cascade of Regulatory Events

There are two distinct phases in Drosophila embryogenesis; the first precedes cellularization of the blastoderm and is associated with a cascade of interacting regulators; the second occurs after cellularization and depends on intercellular signals that must be carried, at least in part, by membrane-bound proteins. We focus on some of the events occurring in the first phase because they are better understood.

Recall that in the infectious cycle of the λ bacteriophage, each stage is characterized by the expression of specific gene products under the control of one or more regulatory proteins. A new phase in the development of the bacteriophage results from the synthesis of one or more new regulatory proteins. A hierarchy determines the order in which the different regulatory proteins make their appearance. Early development in Drosophila is very similar except for its greater complexity. In fact, some of the earliest regulatory proteins in the developing oocyte are supplied by surrounding cells. Another difference in Drosophila development is that the large size of the oocyte permits the estab-

lishment of gradients within a single cell. The regulatory protein encoded by a messenger RNA injected at the anterior end of the oocyte is likely to exist in highest concentration at the anterior end and at lowest concentration at the posterior end. Similarly, mRNAs injected into the oocyte at the posterior end establish regulatory protein gradients in the opposite direction. These gradients are crucial for regulating the expression of genes involved in early development. Some gradients are also established in the perpendicular direction, on the dorsal-ventral axis, but we overlook these for our present purposes.

## Three Types of Regulatory Genes Are Involved in Early Segmentation Development in Drosophila

Identifying regulatory mutants in *E. coli* or yeast usually entails detecting phenotypes that affect the amounts of a number of structural gene products that are under the control of the same regulatory gene. In searching for mutants that carry mutations in regulatory genes that affect development, the task of detection is somewhat more complex. In the first place we are always dealing with a diploid organism that carries two alleles for each gene. An organism that carries a mutation in a regulatory gene does not show an abnormal phenotype if the mutant allele is recessive to the wild-type allele. However, on inbreeding two such heterozygotes one would expect one-quarter of the progeny to be homozygous for the mutant allele and therefore show an abnormal phenotype. What does this phenotype look like? It may have misplaced parts or it may have an incorrect number of segments. In many cases development is arrested at the stage when the associated regulatory gene is needed in normal development. The first insight into the genetic system directing *Drosophila* development came from the discovery of bizarre mutations that affect the body plan (fig. 31.27). For example, the mutation *Antennapedia* results in a pair of extra legs sprouting from the head in place of antennae. Another mutation, *Bithorax,* results in an extra pair of wings appearing where normally there should be much smaller appendages called halteres. These mutations occur in regulatory genes known as homeotic genes.

In addition to homeotic genes two other types of genes occur that influence the segmentation pattern; these are called maternal-effect genes and segmentation genes (table 31.2). Maternal-effect genes are so called because they affect the phenotype only according to the information present in the female parent. Maternal-effect mutations occur in genes responsible for establishing the anterior–posterior axes in the young embryo. Segmentation mutations affect the number and polarity of the body segments. Homeotic

## Figure 31.27

Abnormal phenotypes resulting from homeotic mutations.
*Antennapedia* results in a pair of extra legs sprouting from the head in place of antennae. *Bithorax* results in an extra pair of wings appearing where halteres normally appear.

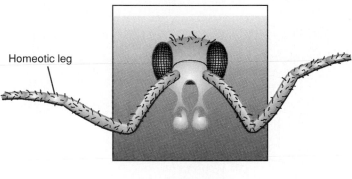

Homeotic leg

### Table 31.2

**Regulatory Genes Involved in Segmentation Development**

**Maternal-effect genes—establish gradients**
Anterior
  bicoid (*bcd*)
Posterior
  *nos*
Dorsal–ventral
  dorsal (*dl*)

**Segmentation genes**
Gap genes—define four broad regions in the egg
  hunchback (*hb*)
  Krüppel (*Kr*)
  Knirps (*kni*)
Pair-rule genes—define seven bands
  runt (*runt*)
  hairy (*h*)
  fushi tarazu (*ftz*)
  even skipped (*eve* )
  paired (*prd*)
  odd paired (*opa*)
Segment-polarity genes—define 14 bands
  engrailed (*en*)
  wingless (*wg*)
  gooseberry (*gb*)

**Homeotic genes—specify structures associated with individual segments**
  *BX-C* locus
  *ANT-C* locus

mutations are more specific than segmentation mutations; they result in changes in structures that are uniquely associated with individual segments or subsegments.

## Analysis of Genes That Control the Early Events of *Drosophila* Embryogenesis

Once we find a mutation that appears to affect development, we must have a way to detect the presence or absence of the related gene product in the developing embryo. A principal tool for doing this is autoradiography using $^{32}$P-labeled specific DNA probes. The embryo is labeled by *in situ* hybridization with the radioactive probe specific for the transcript of a gene. Both the location and intensity of the labeling give an indication of the level of expression of the gene. In some cases the protein product relating to a specific transcript can also be detected by use of a specific antibody-labeling technique. In this way the appearance of many regulatory genes involved in early development has been traced.

## Maternal-Effect Gene Products for Oocytes Are Frequently Made in Helper Cells

The first regulatory gene products to play a role in *Drosophila* development are active before fertilization. The bicoid (*bcd*) gene is a major determinant of the anterior–posterior pattern. Like many other maternally expressed genes, *bcd* is transcribed in the ovary in specialized cells that form a cluster around the future anterior pole of the developing oocyte. The *bcd* RNA synthesized in these so-called nurse cells passes into the oocyte through cytoplasmic canals and becomes localized at the anterior pole of the oocyte (fig. 31.28*a*). Similarly the regulatory gene product of *nos* becomes localized at the posterior pole of the egg (see fig. 31.28*a*). Females lacking a functional copy of the *bcd* gene produce eggs that develop into embryos with no head or thorax. Embryos produced by female mutants homozygous for *nos* develop normal head and thoracic segments but lack the entire abdomen.

**Figure 31.28**

Key events in the expression of developmental regulatory genes in the blastoderm embryo. Events take place in the order shown over a period of a few hours. Measurements are based on the use of probes for detecting the transcripts of various regulatory genes. Description of genes is given in table 31.2. (Source: Adapted from P. W. Ingham, The molecular genetics of embryonic pattern formation in Drosophila, *Nature* 335:25–34, 1988.)

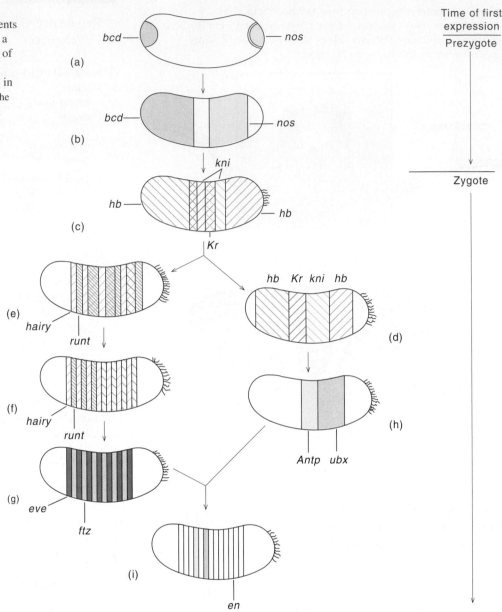

## Gap Genes Are the First Segmentation Genes to Become Active

The pattern-forming process continues on fertilization; as the *bcd* mRNA is translated, the protein diffuses from the anterior pole so that it becomes distributed over about half of the length of the egg (see fig. 31.28*b*). Simultaneously, *nos* gene product originally localized at the posterior pole begins to move forward.

By the beginning of the precellular blastoderm stage, gradients for maternally encoded products occur along the anterior–posterior axis of the embryo. This quantitative information is transformed into qualitative differences in the form of region-specific gene expression, by a process requiring interaction between the maternally derived products and the zygotic genome that resulted from fertilization.

The first segmentation genes to become active in the zygote belong to the *gap* class, so called because their mutants lack major regions of the body, thereby creating a gap in the anterior–posterior pattern. The three members of this class are hunchback (*hb*), Krüppel (*Kr*), and Knirps (*kni*). These genes are expressed in two division cycles prior to cellularization. Expression of *hb* is restricted to two regions: One extending from the anterior pole to 50% of the egg length, the other from the posterior pole to about 25% of the egg length (see fig. 31.28*c*). Initially the *Kr* gene is expressed in

## Figure 31.29

Influence of developmental regulatory genes on one another. An arrow with a plus sign at the arrowhead ⊕ indicates a positive effect on expression, whereas a minus sign ⊖ at the arrowhead indicates a negative effect on expression. The *gap* genes are essential for the establishment of the *hairy* and *runt* banded patterns, but the precise way in which they do this is unclear.

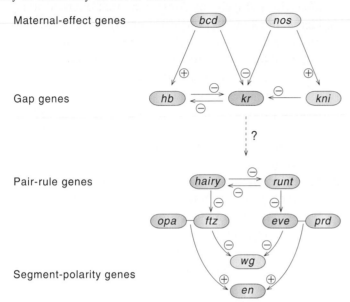

suggest that the *bcd* and *nos* group genes both act to repress transcription of *Kr* in the anterior and posterior regions, respectively, thereby restricting its region of expression to the central portion of the embryo. In contrast, *bcd* appears to act as a positive regulator of *hb,* and *nos* appears to be a positive regulator of *kni*. In normal embryos the transcriptional domains of *hb, Kr,* and *kni* narrow with time, giving rise to sharp boundaries of expression. This process is driven by negative effects between the gap genes (see fig. 31.28*d*), *hb* and *Kr,* mutually repressing each other, and *kni* acting as a negative regulator of *Kr*. Thus, the establishment of stable domains of *gap* gene expression is a two-step process: First, a differential response to graded levels of maternal determinants, and second, a mutual repression effect, leading to the generation of stable boundaries between adjacent domains.

The *gap* gene products regulate the position-specific expression of other genes, those belonging to the pair-rule class of segmentation genes.

By scrutinizing the location of different regulatory gene products as a function of time in both normal and mutant embryos, the patterns shown in figure 31.28 were established for all three classes of regulatory proteins that function in early development.

The positive and negative effects established between maternal-effect and pair-rule genes are summarized in figure 31.29. This diagram tells an incomplete story. The most serious omission in our knowledge is how the gradients of maternal-effect and *gap* genes help to deliver the signal for the establishment of the strict periodic patterns of expression seen for most of the segmentation genes (see fig. 31.28). It seems unlikely that the signals in the form of gradients could ever institute such a pattern without other factors being involved. Possibly some of the signals for expression of pair-rule genes are established in the oocyte before fertilization.

a single broad band in the middle of the embryo, whereas *kni* is expressed in two distinct domains, one anterior and one posterior to the *Kr* band. These transcriptional domains for *hb, Kr,* and *kni* are influenced by the maternally derived information encoded by the *bcd* and *nos* group genes. Thus, mutants that lack *bcd* activity do not express *hb,* whereas *Kr* is extended anteriorly in an unusually broad domain. The absence of the posterior maternal determinants results in the extension of the *Kr* domain posteriorly. These observations

## Summary

Some mechanisms of gene expression that are found in eukaryotes are rarely, if ever, seen in prokaryotes. Still, *trans*-acting regulatory proteins that bind to *cis* effector sites on the genome are present in eukaryotic systems as well as in *E. coli*. The best understood unicellular eukaryote is the budding yeast *Saccharomyces cerevisiae*. Gene regulation, particularly of development, can be quite complex in multicellular eukaryotes. Our discussion in this chapter focused on the following points.

1. In eukaryotes, individual genes encode transcripts for a single polypeptide chain. Although functionally related genes are clustered, each gene has its own promoter.

2. In yeast, genes of the *GAL* system are under the joint control of two *trans*-acting genes that encode regulatory proteins: *GAL4* and *GAL80*. GAL4 protein is an activator that binds at an upstream site, and GAL80 is a repressor that inhibits GAL4 action by binding to it. The ability of the GAL80 protein to bind to GAL4 is lost in the presence of galactose, which binds to the GAL80 protein, causing a reversible allosteric change in its structure.

3. Yeast has two haploid cells of opposite mating types and one nonmating diploid cell that results from the fusion of haploid cells of opposite mating type. The mating type is determined by the *MAT* locus. Information for the mating type is stored at other, silent loci and is expressed only if it is transposed to the *MAT* locus. Mating-type information is not expressed at the storage loci because of a complex repressor system of proteins interacting in *cis* fashion over a considerable distance at the control centers of the storage loci.

4. The greatest difference between the regulatory systems in yeast and *E. coli* is that yeast regulatory proteins can bind at a long distance from the RNA polymerase-binding site and still be effective.

5. Complex multicellular eukaryotes differentiate irreversibly so that different cell types express a different profile of genes. Genes that are expressed are usually associated with swollen chromatin. Proteins found in active regions of the genome show characteristic modifications.

6. Enhancers are elements of the genome that generally stimulate transcription. They resemble yeast UAS sequences in yeast in that they can function over long distances. In fact, they can function over even greater distances than UAS sequences, and they are effective in either orientation: Either upstream or downstream from the promoter. A possible mechanism for enhancer and UAS function over long distances is that the chromosome folds to bring the proteins bound at the enhancer site in close proximity to other regulatory proteins or RNA polymerase bound at the promoter.

7. A typical enhancer is composed of a cluster (usually two or three) of *cis* sites called enhansons; each enhanson contains binding sites for a unique combination of regulatory proteins. Enhansons must be clustered to be effective.

8. In chapter 30 we discussed DNA-binding proteins that regulate transcription in prokaryotes. In prokaryotes and eukaryotes specific recognition is dominated by H bond interactions that take place in the major groove of the DNA. In both cases the $\alpha$ helix is the most common element used for DNA recognition.

The most striking difference between DNA-binding proteins in prokaryotes and eukaryotes has to do with the symmetry of the interaction. In prokaryotes the binding proteins almost always interact in a symmetrical fashion with the DNA. In eukaryotes most of the cases that have been examined so far involve proteins that interact in an asymmetrical fashion with the DNA. In many cases the regulatory proteins interact in multisubunit complexes that contain nonidentical subunits. Four different types of structural motifs are discussed: The homeodomain, the zinc finger, the leucine zipper, and the helix-loop-helix.

9. Posttranscriptional regulation is an important mode of regulation in eukaryotes. Examples are given of three types of posttranscriptional regulation: Alternative modes of mRNA splicing, regulation at the translation level, and alternative modes of polypeptide processing.

10. Special combinations of regulatory factors give rise to developmental patterns of cell differentiation. Some organisms must amplify genes to keep pace with the high demand for certain gene products, especially in early development. Histones, ribosomal proteins, and ribosomal RNA genes are amplified.

11. The existence of many kinds of regulatory mutants has helped to advance our understanding of early development in the fruit fly, *Drosophila melanogaster*. Regulatory gene products are proteins that activate or repress other genes. Early development in *Drosophila* is a sequence of events in which different regulatory proteins gradually come into play in cascade fashion, controlling a wide range of enzymes and structural proteins and also influencing each other. Until the blastoderm stage the nuclei in a developing *Drosophila* embryo are not separated by cellular membranes. As a result, the regulatory proteins and other gene products may diffuse freely from their site of synthesis to other nuclei in the embryo. At the late blastoderm stage, the nuclei become cellularized. From this point on, the influence of regulatory proteins made in one cell must be exerted on another cell at the level of the cell membrane.

# Selected Readings

Achneider, R. J., and T. Shenk, Impact of virus infection on host cell protein synthesis. *Ann. Rev. Biochem.* 56:317–332, 1987.

Atchison, M. L., Enhancers: Mechanisms of action and cell specificity. *Ann. Rev. Cell. Biol.* 4:127–153, 1988.

Binétruy, B., T. Smeal, and M. Karin, Ha-Ras augments c-Jun activity and stimulates phosphorylation of its activation domain. *Nature* 351:122–127, 1991.

Bingham, P. M., T. Chou, I. Mims, and Z. Zachari, On/off regulation of gene expression at the level of splicing. *Trends Genet.* 4:134, 1988.

Brietbart, R. E., A. Andreadis, and B. Nadal-Ginard, Alternative splicing: A ubiquitous mechanism for the generation of multiple protein isoforms from single genes. *Ann. Rev. Biochem.* 56:467–495, 1987.

Brown, D. D., How a simple animal gene works. In *The Harvey Lectures,* Series 76, pp. 27–44. New York: Academic Press, 1982.

Chandler, V. L., B. A. Maler, and K. R. Yamamoto, DNA sequences bound specifically by glucocorticoid receptor *in vitro* render a heterologous promoter responsive *in vivo:* A steroid specific enhancer. *Cell* 33:489–499, 1983.

Chen, H-Z., T. Hoey, and G. Zubay, Purification and properties of the *Drosophila* zen protein. *Mol. Cell. Biochem.* 79:181–189, 1988. First evidence that a homeobox protein binds to DNA as a monomer.

Comai, L., N. Tanese, and R. Tjian, The TATA-binding protein and associated factors are integral components of the RNA polymerase I transcription factor, SL1. *Cell* 68:965–976, 1992.

Duncan, I., The bithorax complex. *Ann. Rev. Genet.* 21:285–320, 1987.

Evans, R. M., The steroid and thyroid hormone receptor superfamily. *Science* 240:889–895, 1988.

Ferre-D'Amare, A. R., G. C. Prendergast, E. B. Ziff, and S. K. Burley, Recognition by Max of its cognate DNA through a dimeric b/HLH/Z domain. *Nature* 363:38–44, 1993.

Forsburg, S. L., and L. Guarente, Communication between mitochondria and the nucleus in regulation of cytochrome genes in the yeast *Saccharomyces cerevisiae. Ann. Rev. Cell. Biol.* 5:153–180, 1989.

Funder, J. W., Mineralocorticoids, glucocorticoids, receptors and response elements. *Science* 259:1132–1133, 1993.

Gabrielsen, O. S., and A. Sentenac, RNA polymerase III (C) and its transcription factors. *Trends Biochem. Sci.* 16:412–416, 1991.

Gehring, U., Steroid hormone receptors: Biochemistry, genetics and molecular biology. *Trends Biochem. Sci.* 12:399–402, 1987.

Gehring, W., Homeo boxes in the study of development. *Science* 236:1245–1252, 1987.

Gehring, W. J., The molecular basis of development. *Sci. Am.* 253(4):152–162, 1985.

Gimeno, C. J., and G. R. Fink, The logic of cell division in the life cycle of yeast. *Science* 257:626, 1992.

Giniger, E., and M. Ptashne, Cooperative binding of the yeast transcriptional activator *GAL4. Proc. Natl. Acad. Sci.* 85:382–386, 1988.

Gruenberg, D. A., S. Natesan, C. Alexandre, M. Z. Gilman. Human and *Drosophila* homeo domain proteins that enhance the DNA-binding activity of serum response factor. *Science* 257:1089–1095, 1992.

Grunstein, M., Histones as regulators of genes. *Sci. Am.* 267:68–74, 1992.

Guarente, L., UASs and enhancers: Common mechanism of transcriptional activation in yeast and mammals. *Cell* 52:303–305, 1988.

Guarente, L. P., Regulatory proteins in yeast. *Ann. Rev. Genet.* 21:425–452, 1987.

Gurdon, J., Egg cytoplasm and gene control in development. The Croonian Lecture, 1976. *Proc. R. Lond. B.* 198:211–247, 1977.

Hahn, S., Structure and function of acidic transcription activators. *Cell* 72:481–483, 1993.

Hanes, S. D., and R. Brent, A genetic model for interaction of the homeodomain recognition helix with DNA. *Science* 251:426–430, 1991.

Harrison, S. C., A structural taxonomy of DNA-binding domains. *Nature* 353:715–719, 1991.

Herskowitz, I., Life cycle of the budding yeast *Saccharomyces cerevisiae. Microbiol. Rev.* 53:536–553, 1988.

Hinnebusch, A. G., Involvement of an initiation factor and protein phosphorylation in translational control of GCN4 mRNA. *Trends Biochem. Sci.* 15:148–152, 1990.

Horvitz, H. R., and P. W. Sternberg, Multiple intercellular signalling systems control development of the caenorhabditis elegans vulva. *Nature* 351:535–541, 1991.

Johnson, P. F., and S. L. McKnight, Eukaryotic transcriptional regulatory protein. *Ann. Rev. Biochem.* 58:799–839, 1989.

Karin, M., and T. Smeal, Control of transcription factors by signal transduction pathways: The beginning of the end. *Trends Biochem. Sci.* 17:418–422, 1992.

Karlsson, S., and A. W. Nienhius, Developmental regulation of human globin genes. *Ann. Rev. Biochem.* 54:1071–1108, 1985.

King, T., and R. Briggs, Serial transplantation of embryonic nuclei. *Cold Spring Harb. Symp. Quant. Biol.* 21:271–290, 1956.

Koleske, J., and R. A. Young, An RNA polymerase II holoenzyme responsive to activators. *Nature* 368:466–469, 1994.

Klevit, R. E., Recognition of DNA by Cys$_2$His$_2$ zinc fingers. *Science* 253:1367–1393, 1991.

Lai, E., and J. E. Darnell, Jr., Transcriptional control in hepatocytes: A window on development. *Trends Biochem. Sci.* 16:427–429, 1991.

Lamb, P., and S. L. McKnight, Diversity and specificity in transcriptional regulation: The benefits of heterotypic dimerization. *Trends Biochem. Sci.* 16:417–433, 1991.

Laybourn, P. J., and J. T. Kadonaga, Role of nucleosomal cores and histone H1 in regulation of transcription by RNA polymerase II. *Science* 254:238–245, 1991.

Lee, M. S., S. A. Kliewer, J. Provencal, P. E. Wright and R. M. Evans, Structure of the retinoid X receptor $\alpha$ DNA binding domain: A helix required for homodimeric DNA binding. *Science* 260:1117–1121, 1993.

Leuther, K. K., and S. A. Johnston, Nondissociation of GAL4 and GAL80 *in vivo* after Galactose Induction. *Science* 256:1333–1336, 1992.

Li, E., C. B. Beard, and R. Jaenisch, Role for DNA methylation in genomic imprinting. *Nature* 366:362–365, 1993.

Marmorstein, R., M. Carey, M. Ptashne, and S. C. Harrison, DNA recognition by GAL4: Structure of a protein-DNA complex. *Nature* 356:408–414, 1992.

Marzluff, W. F., and N. B. Pandey, Multiple regulatory steps control histone messenger-RNA concentrations. *Trends Biochem. Sci.* 12:49–51, 1988.

McClintock, B., Controlling elements and the gene. *Cold Spring Harb. Symp. Quant. Biol.* 21:197–216, 1956.

McKinney, J. D., and N. Heintz, Transcriptional regulation in the eukaryotic cell cycle. *Trends Biochem. Sci.* 16:430–434, 1991.

Melton, D. A., Pattern formation during animal development. *Science* 252:234–241, 1991.

Miner, J. N., and K. R. Yamamoto, Regulatory crosstalk at composite response elements. *Trends Biochem. Sci.* 16:423–426, 1991.

Mitchell, P. J., and R. Tjian, Transcriptional regulation in mammalian cells by sequence-specific DNA binding proteins. *Science* 245:371–378, 1989.

Nevins, J. R., Transcriptional activation by viral regulatory proteins. *Trends Biochem. Sci.* 16:435–439, 1991.

Parkhurst, S. M., D. Bopp, and D. Ish-Horowicz, X:A ratio, the primary sex-determining signal in *Drosophila,* is transduced by helix-loop-helix proteins. *Cell* 63:1179–1191, 1990.

Pearce, D., and K. R. Yamamoto, Mineralocorticoid and glucocorticoid receptor activities distinguished by nonreceptor factors at a composite response element. *Science* 259:1161–1164, 1993.

Prendergast, G. C., and E. B. Ziff, A new bind of Myc. *Trends Genet.* 8:92–96, 1992.

Ptashne, M., How gene activators work. *Sci. Am.* 260(1):41–47, 1989.

Raghow, R., Regulation of messenger RNA turnover in eukaryotes. *Trends Biochem. Sci.* 12:3358–3360, 1987.

Roeder, R. G., The complexities of eukaryotic transcription initiation: Regulation of preinitiation complex assembly. *Trends Biochem. Sci.* 16:402–407, 1991.

Ruvkun, G., and M. Finney, Regulation of transcription and cell identity by POU domain proteins. *Cell* 64:475–478, 1991.

Schler, A. F., and W. J. Gehring, Direct-homeo domain-DNA interaction in the autoregulation of the *fushi tarazu* gene. *Nature* 356:804–806, 1992.

Schwabe, J. W. R., and D. Rhodes, Beyond zinc fingers: Steroid hormone receptors have a novel structural motif for DNA recognition. *Trends Biochem. Sci.* 16:291–296, 1991.

Sherman, A., M. Shefer, S. Sagee, and Y. Kassir, Post-transcriptional regulation of IME1 determines initiation of meiosis in *Saccharomyces cerevisiae. Mol. Gen. Genet.* 237:375–384, 1993.

Singer, S. J., Intercellular communication and cell-cell adhesion. *Science* 255:1671–1677, 1992. Considers aspects of regulation we didn't have time to cover.

Stark, G., M. Debatisse, E. Giulotto, and G. M. Wahl, Recent progress in understanding mechanisms of mammalian gene amplification. *Cell* 57:901–908, 1989.

Struhl, K., Molecular mechanisms of transcriptional regulation in yeast. *Ann. Rev. Biochem.* 58:1051–1077, 1989.

Struhl, K., Helix-turn-helix, zinc-finger and leucine-zipper motifs for eukaryotic transcriptional regulatory proteins. *Trends Biochem. Sci.* 14:137–140, 1989.

Thompson, C. C., and S. I. McKnight, Anatomy of an enhancer. *Trends Genet.* 8:232–236, 1992.

Weinberg, R. A., Finding the anti-oncogene. *Sci. Am.* 259(3):44–51, 1988.

Weinzierl, R. O. J., B. D. Dynlacht, and R. Tjian, Largest subunit of *Drosophila* transcription factor IID directs assembly of a complex containing TBP and a coactivator. *Nature* 362:511–517, 1993.

## Problems

1. On the basis of what you know about genes in pro- karyotic and eukaryotic cells, define a gene. Make sure that your definition is brief and concise. Does your definition have any limitations or problems?

2. Why is attenuation control in eukaryotes unlikely?

3. The domain-swap experiment illustrated in figure 31.4 demonstrated that the *lexA* DNA-binding do- main from *E. coli* is functionally interchangeable with the equivalent domain from the GAL4 protein if and only if the *lexA* DNA recognition site replaces the GAL4 equivalent. Is the *GAL1* gene with a *lexA* binding site upstream transcribed in the presence of intact lexA protein (with no GAL4 activation do- main)? Why or why not?

4. A *GAL4* mutation (*GAL4$^c$*) leads to constitutive syn- thesis of the *GAL1* gene product in haploid yeast. Propose an explanation for the effect of this muta- tion.

5. How do you expect a deletion of *HMLE* to affect the expression of mating-type genes in yeast? Compare this effect with the deletion of the $\alpha_2$ gene from *MAT$_\alpha$*. Consider both homothallic and heterothallic backgrounds.

6. Color blindness is X-chromosome linked. Bearing in mind the phenomenon of X-chromosome inactiva- tion, suggest an explanation for the observation that females who are heterozygous for the defective gene show no signs of color blindness.

7. Briggs and King were able to grow a differentiating embryo from an egg of *Rana pipiens,* the chromo- somes of which were replaced with a single diploid nucleus from another embryo. What does this tell us about the state of the nucleus in the developing em- bryo?

8. There is no large difference in the frequency of his- tones in transcribed regions of the genome compared with untranscribed regions. Why don't nucleosomes interfere with transcription in eukaryotes?

9. High salt concentrations weaken the interaction of histones with DNA but have little effect on the bind-

ing of many regulatory proteins. Explain this obser- vation in terms of how these molecules interact with DNA.

10. List some characteristics that distinguish active from inactive chromatin.

11. The restriction endonuclease *Hpa*II cleaves the se- quence CCGG only if the second C is unmethylated. The enzyme *Msp*I cleaves the same sequence, whether or not it is methylated. How do the globin- specific sequences in erythroblast DNA (erythroblasts are red blood cell precursors) differ from those of other tissues in their susceptibility to these two en- zymes?

12. Explain how a DNA sequence (enhancer sequence) located 5,000 bp from a gene transcription start site can stimulate transcription even if its orientation is reversed.

13. Discuss the types of structural motifs found in eu- karyotic transcription factors.

14. What kind of changes have to be made in a typical eukaryotic structural gene for its protein product to be expressed in bacteria?

15. In the *Xenopus* oocyte a large number of ribosomes are made in a short time to handle the rapid demand for cell growth during cleavage stages. How is this large amount of rRNA made in such a short time?

16. Based only on the definition of maternal-effect genes, segmentation genes, and homeotic genes, which do you predict act earliest in development of the *Drosophila* embryo, and which act latest?

17. Given that specific subsets of homeotic genes are required for the development of specific segments of the *Drosophila* embryo, suggest a possible mecha- nism whereby the necessary spatially restricted pat- tern of homeotic gene expression might be achieved. Incorporate the observed effect of homeotic muta- tions on segment morphology into your model.

18. Given a cloned fragment of a *Drosophila* gene, how could you determine which chromosomal band(s) contain the gene?

# Principles of Physiology and Biochemistry: Immunobiology

When vertebrates are invaded by foreign agents, they can mobilize a versatile set of adaptive immune responses to form specifically reactive cells and proteins. These responses constitute the principal means of defense against pathogenic microorganisms and viruses and probably also against host cells that undergo transformation into cancer cells.

Throughout most of this century, the subject of immunology has attracted some of the keenest minds in biology. As a result, the intricacies of this fascinating subject have been largely unraveled, and its understanding is providing a strong bridge between the fields of biochemistry and physiology. In this chapter we give a brief introduction to some of the major findings in immunobiology and the experiments that have led to our current level of understanding.

## Overview of the Immune System

The immune system was first studied in humans, but mice became a popular subject for immune system studies when researchers began to appreciate how close the mouse and

**Figure S3.1**

B cells and T cells follow different pathways of development. In mammals, B lymphocytes mature in the bone marrow and then migrate to secondary lymphoid organs. On exposure to foreign substances known as antigens, they proliferate to produce immunoglobulins. T lymphocytes mature in the thymus gland. They also can be stimulated to proliferate by exposure to an appropriate antigen.

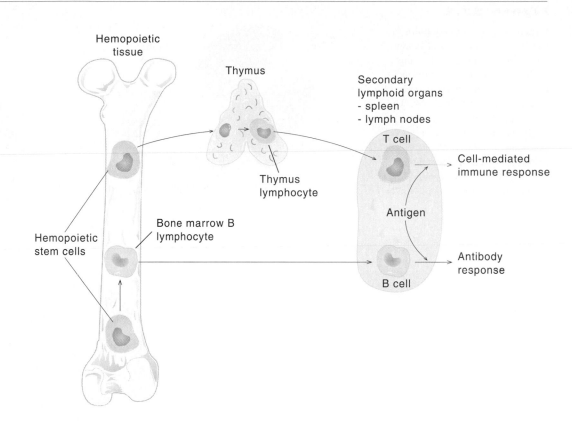

human systems were in their organization and action. Currently, both systems are under study in many different laboratories. Although we focus here on mice, we often refer to parallel observations on humans.

The immune system is an example of a developmental process that takes place in the mature organism. In the new cells that are constantly being generated, changes arise in the genes of the immune system. This variability results from DNA splicing and point mutations. It benefits the organism by providing the cells of the immune system with the widest possible range of specificities. As soon as the organism is invaded by foreign agents, usually viruses or bacteria, the immune system is activated. Those immune system cells that carry the specific immune receptors for interacting with the foreign agent are stimulated to proliferate. In a matter of days, clones of immune system cells with the appropriate specificity have been produced, and the organism fends off the invader with the specific immunologic tools provided by those cells. The clones of cells tend to persist for some time, usually months to years, which accounts for the fact that the second time the same foreign invader makes its presence known, it is usually rejected promptly and without crisis.

Two different classes of white cells, or lymphocytes are associated with the immune response: The *B cells* and the *T cells*. In mammals, B lymphocytes mature in the bone marrow and then migrate to secondary lymphoid organs

(fig. S3.1). On exposure to foreign substances known as antigens, they proliferate and produce immunoglobulin proteins known as antibodies, which they secrete into the bloodstream. T lymphocytes, by contrast, mature in the thymus gland before migrating to the secondary lymphoid organs. They also can be stimulated to proliferate by exposure to an appropriate antigen, but their specific effector molecules remain firmly bound to the cellular membrane, as opposed to being secreted. Three main types of T cells have been recognized; T killer cells, which specifically destroy target cells; T helper cells, which promote the maturation of antigen-stimulated B and T cells; and T suppressor cells, which block the effects of T helper cells.

## The Humoral Response: B Cells and T Cells Working Together

There are two basically different types of immune response. The first, called the humoral, or antibody, response, involves the concerted action of both B cells and T cells, and the active agents are the antibody, or immunoglobulin, proteins secreted by the B cells into the bloodstream. The combined B–T immune response is characteristic of most vertebrates. The second type of immune response, called the cell-mediated response, involves only T cells, and the active agent is the circulating T cell itself, which attacks the foreign agent. This type of immune re-

## Figure S3.2

Structure of immunoglobulin G (IgG). The light (L) and heavy (H) chains have repeating domains, each with about 110 amino acid residues and an approximately 60-member S—S bonded loop. The subscripts L and H refer to regions of the sequence that are relatively constant or quite variable, respectively, in different IgG species. (Illustration copyright by Irving Geis. Reprinted by permission.)

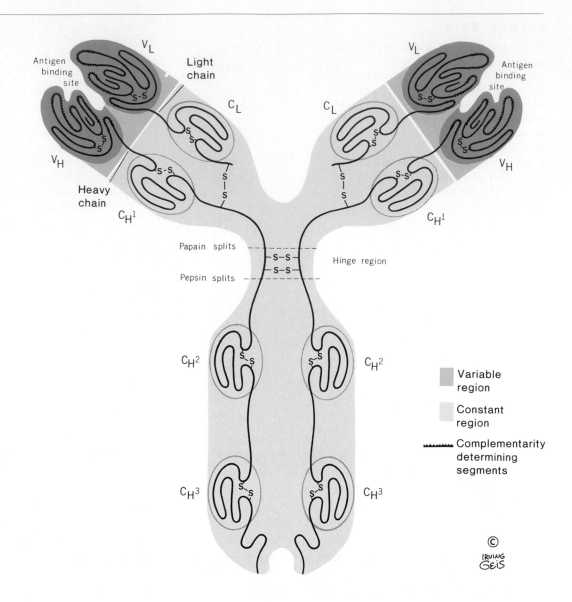

sponse is limited to certain groups of vertebrates, including mammals. We first discuss the B–T cell response mediated by antibodies.

## Immunoglobulins Are Extremely Varied in Their Specificities

A detailed investigation of immunoglobulin structure provided the first leads about how immunoglobulins of such a wide variety are synthesized. The most common type of immunoglobulin is the 7S molecule, known as immunoglobulin G (IgG). This immunoglobulin has a molecular weight of about 150,000 daltons. It is composed of four polypeptide chains: Two heavy (H) chains with molecular weights of about 50,000 and two light (L) chains with molecular weights of about 25,000 (fig. S3.2).

To obtain a detailed understanding of immunoglobulin structure, it was necessary to isolate pure antibody proteins for sequencing. Fortunately, it was discovered that certain plasma cell tumors known as myelomas produce enormous amounts of pure immunoglobulins, which have the same gross structure as the mixed immunoglobulins isolated from serum. Each myeloma appears to originate from a single cell turned cancerous. As a result, each myeloma serves as the source of a pure antibody protein, the sequence and other properties of which can be determined. In some cases myelomas synthesize both heavy and light chains, like normal antibody-forming cells, but in other cases they synthesize only one or the other type of chain. Comparison of the sequences from a number of different immunoglobulins derived from myelomas has shown that both the heavy and the light chains for immunoglobulins of the same class are di-

## Figure S3.3

Approximate locations of the hypervariable regions in the heavy and light chains of IgG. Each hypervariable segment is believed to make up part of the site that binds to antigen, known as the complementarity-determining region (CDR). The CDR regions are located in the loop regions of the variable domains where they can make close contact with antigen.

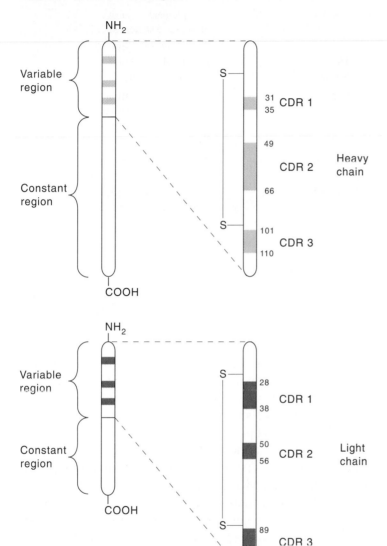

vided into regions of relatively constant sequence (the *C* segments) and regions of relatively variable sequence (the *V* segments). In the intact immunoglobulins the antigen-binding domain is composed exclusively of *V* segments originating from the H and L chains (see fig. S3.2), each tetramer containing two equivalent sites for the binding of a specific antigen.

As information on antibody sequences has accumulated, it has become increasingly clear that within the *V* regions of both the heavy and light chains, three segments account for most of the variability (fig. S3.3). These regions are called the hypervariable regions, and they are believed to be the parts of the antibody molecule in most direct contact with the antigen in the antigen–antibody complex.

Immunoglobulin G (IgG), the tetrameric species we have been discussing thus far, is not the only type of immunoglobulin found (table S3.1). Most of the other known antibodies—IgM, IgA, IgD, and IgE—also involve a closely related tetrameric structure, sometimes forming larger aggregates and always associated with different functions. For instance, IgM is a 19S antibody accounting for 5%–10% of the serum Ig. It is an aggregate of five tetramers that is formed early during the immune reaction, soon to be diminished in quantity and overshadowed by large amounts of IgG. IgA occurs in various polymeric forms. It normally accounts for about 15% of the total immunoglobulin found in serum, and in addition it is the principal immunoglobulin in exocrine secretions. Each of the different classes of immunoglobulin possesses distinct, heavy chains. Indeed, even within the human IgG class, antisera tests reveal four different types of IgG (IgG1 through IgG4), each with a distinctive H chain, comprising about 70, 19, 8, and 3%, respectively, of the total IgG proteins. The light chains, of which there are two types, are common to all classes of immunoglobulins.

Immunoglobulins that are produced in response to antigens are themselves highly immunogenic (that is, they stimulate antibody formation) when injected into genetically nonidentical organisms. The serological responses induced by using immunoglobulins as antigens or immunogens have been useful in classifying them according to their antigenic determinants.

So-called isotypic determinants are shared by all immunoglobulin molecules of a given class. For example, the human IgG molecules are classified into four isotypes because they are recognized by heterologous antisera produced in one species against the immunoglobulins of another species. Within a given isotype the different immunoglobulins can usually be separated into sets, called allotypes, that are distinguished by minor antigenic differences. Allotypic differences usually reflect alternative amino acid substitutions within otherwise quite similar amino acid sequences of the *C* region of the antibody. Allotypes show typical Mendelian inheritance patterns. Antisera useful for discriminating between allotypes are usually made by injecting an individual who lacks a specific allotype with immunoglobulins from an individual who carries the allotype. Finally, idiotype refers to the specific antigenic determinant of the antibody. Some idiotypic determinants are limited to a single immunoglobulin; others are shared by a small number of immunoglobulins. Whereas isotypic and allotypic differences usually result from differences in the

Different Isotypes Found in Humans

| Class | Heavy Chain | Light Chain | Molecular Formula | Molecular Weight (daltons) | Physiological Functions |
|-------|-------------|-------------|-------------------|----------------------------|-------------------------|
| IgG | $\gamma$ | $\kappa$ or $\lambda$ | $\gamma_2\kappa_2$ $\gamma_2\lambda_2$ | 150,000 | Complement fixation; placental transfer; stimulation of ingestion by macrophages |
| IgA | $\alpha$ | $\kappa$ or $\lambda$ | $\alpha_2\kappa_2$ $\alpha_2\lambda_2$ | 160,000 320,000 | Localized protection of external secretions |
| IgM | $\mu$ | $\kappa$ or $\lambda$ | $\mu_2\kappa_2$ $\mu_2\lambda_2$ | 900,000 | Complement fixation; early immune response; stimulation of ingestion by macrophages |
| IgD | $\delta$ | $\kappa$ or $\lambda$ | $\delta_2\kappa_2$ $\delta_2\lambda_2$ | 185,000 | Found on cell surfaces; function unknown |
| IgE | $\epsilon$ | $\kappa$ or $\lambda$ | $\epsilon_2\kappa_2$ $\epsilon_2\lambda_2$ | 200,000 | Stimulates mast cells to release histamines |

constant portion, or $C$ segments, of the immunoglobulin polypeptide chain, idiotypic differences usually result from differences in the variable portion, or $V$ segment. As we know already, this is the region that contains the binding site for the foreign agent or antigen.

## Antibody Diversity Is Augmented by Unique Genetic Mechanisms

Antibodies have been studied most extensively in the mouse, an ideal vertebrate for both genetic and biochemical manipulations. It is believed that mice can synthesize more than a million antibodies with different antigenic specificities. This enormous diversity of proteins is generated from a limited amount of genetic information with the help of two mechanisms: Somatic recombination and somatic mutation.

***DNA Splicing Brings Different Parts of the Antibody Gene Together.*** The involvement of somatic recombination was first demonstrated by examining the structure of a specific antibody gene in embryonic and adult immunoglobulin-forming tissue. The adult tissues favored for many studies are myelomas. These tumorous tissues, which we mentioned earlier, produce a homogeneous population of polypeptide chains. They have served as a convenient source of pure immunoglobulin polypeptide chains as well as their mRNAs.

Recall that the generalized structure of the predominant serum antibody consists of two identical heavy chains and two identical light chains (see fig. S3.2). In the mouse there are two classes of light chains, which differ appreciably in the constant regions of the polypeptide chains. These are referred to as the kappa ($\kappa$) and the lambda ($\lambda$) light chains. Using highly inbred, genetically identical (isogenic) strains of mice, S. Tonegawa and his co-workers isolated the DNA from embryonic cells and two different myeloma tumor cells: One that produces homogeneous $\lambda$ light chains (strain H2020) and one that produces $\kappa$ light chains (strain MOPC321). These DNAs were digested with the *Eco*R1 restriction enzyme and electrophoresed on an agarose gel. The gels contained an enormous variety of restriction fragments representing total nuclear DNA, and consequently no discrete pattern of bands could be seen with a stain for nucleic acid. However, when the electrophoresed gels containing the DNA were denatured and hybridized with [32]P-labeled DNA containing the sequences found in the RNA for the $\lambda$ chain, a specific pattern of bands showed up in autoradiographs (fig. S3.4). The R1 digest of DNA from the $\lambda$-containing myeloma (H2020) showed four bands; the DNAs of the embryo and the $\kappa$-containing myeloma (MOPC321) showed three bands in common with the first DNA but were missing the fourth band.

These results strongly suggested that at some point during development from the embryonic state, a rearrange-

## Figure S3.4

Analysis of DNA fragments containing $\lambda_1$ gene sequences from mouse embryo and myeloma cells. High-molecular-weight DNAs extracted from myeloma H2020, a $\lambda$-chain producer (A), from a 13-day-old BALB/c embryo (B), and from myeloma MOPC321, a $\kappa$-chain producer (C) were digested to completion with *Eco*R1, electrophoresed on agarose gel, and transferred to nitrocellulose membrane filters and hybridized with a nick-translated *Hha*I fragment of the plasmid B1 DNA. (Source: After C. Brack, M. Hirama, R. Lemhard-Schuller, and S. Tonegawa, A complete immunoglobulin gene is created by somatic recombination, *Cell* 15:1–14, 1978.)

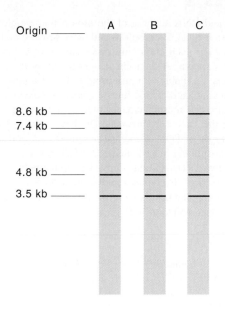

## Figure S3.5

Arrangement of mouse $\lambda_1$ gene sequences in embryos and $\lambda_1$ chain-producing plasma cells. The vertical arrows point to *Eco*R1 restriction sites. The blue dashed diagonal lines point to hypothesized splice points that explain the difference in structure of the region in the two cell types. The boxed regions represent coding regions, and $I_1$ and $I_2$ between boxed regions indicate first and second introns in the spliced gene.

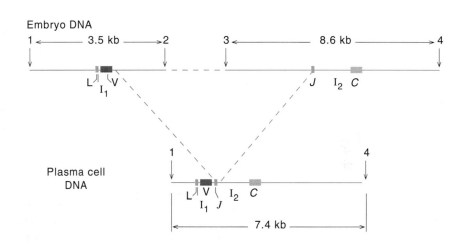

ment had taken place in the H2020 myeloma cells on one of the homologous pairs of chromosomes carrying the $\lambda$-chain gene. Further analyses with more specific radioactive probes, containing sequences from either $C_\lambda$ (the constant region of the $\lambda$ chain) or $V_\lambda$ (the variable region of the $\lambda$ chain), were done. Only the 7.4-kb fragment originating from the $\lambda$ myeloma cell hybridized to both probes. Thus, the *Eco*R1 fragment must contain both $V_\lambda$ and $C_\lambda$ DNA. The 8.6-kb fragment hybridized to the $C_\lambda$ but not the $V_\lambda$ probe. Therefore this fragment must contain $C_\lambda$ sequences but no $V_\lambda$ sequences. Both the 3.5-kb fragment and the 4.8-kb fragment hybridized exclusively to the $V_\lambda$ probe, so they must contain only $V_\lambda$ sequences. Still further analyses indicated

that the 7.4-kb fragment arose from a recombinational event between the 3.5-kb fragment and the 8.6-kb fragment (fig. S3.5). The 4.8-kb fragment originates from another $V_\lambda$ sequence located in the same chromosome. The mouse has only two $V_\lambda$ sequences in the embryo. As a rule, only one of the pairs of homologous chromosomes recombines to yield a productive antibody gene. This explains why the 3.5-kb and the 8.6-kb bands are still visible in the H2020 myeloma DNA preparation. The $\kappa$ myeloma cell producer shows the same pattern as the embryonic cell. Had it been probed with a DNA carrying the $V_\kappa$ antibody sequence, it would have shown differences from the embryonic pattern.

## Figure S3.6

Organization and DNA splicing of the mouse immunoglobulin genes. The organization is depicted before and after somatic cell rearrangement. For each class only one example of a rearranged chromosome is shown. A light chain is encoded by three distinct DNA elements, a variable ($V$), a joining ($J$), and a constant ($C$) element. Boxes indicate exons; lines connecting boxes indicate introns. Somatic cell rearrangement involves the joining of a $V$ and a $J$ segment on the same chromosome. The heavy chain is encoded by four distinct types of DNA elements: a $V$, a $D$, a $J$, and a $C$ element. In heavy-chain splicing a $V$ element, a $D$ element, and a $J$ element are joined in two DNA-splicing steps. The splice between $D$ and $J$ elements occurs first. The splices shown provide only one example of many that might occur in this system. (From G. Zubay, *Genetics,* Benjamin/Cummings, Menlo Park, Calif. 1987, p. 808.)

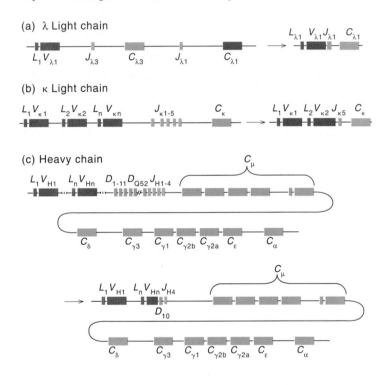

(a) λ Light chain

(b) κ Light chain

(c) Heavy chain

and a $J_L$ segment (encoding about 13 amino acids) are joined by DNA splicing. A complete light-chain gene transcript consists of three exons and two introns, arranged in the following order; $5'$L, $I_1$, $VJ$, $I_2$, $C$, $3'$. Here L refers to sequences in the leader, or signal, peptide, which is removed in the mature immunoglobulin, and $I_1$ and $I_2$ refer to the first and second introns, respectively, which are removed by RNA splicing. The heavy chain is encoded by four distinct DNA elements: a $V_H$, a $D$, a $J_H$, and a $C_H$ element. In heavy-chain variable region formation, a $V_H$ element (encoding about 99 amino acids), a $D$ element (encoding 1–15 amino acids), and a $J_H$ element (encoding about 15 amino acids) are joined in two DNA-splicing steps. The heavy-chain gene family has several closely linked $C_H$ genes, which determine the immunoglobulin class.

Class switching, involving an additional DNA-splicing operation, frequently occurs with heavy-chain genes. Thus, an immature B cell, which initially expresses a $\mu$ chain, results in IgM antibodies. Subsequently, the same cell can be induced to differentiate further so that the same $V_H$ region of the genome becomes relocated next to a $C_\gamma$ gene, causing the cell to produce an IgG antibody with the same $V_H$ region associated with a different $C_H$ region. In cases of class switching, the second DNA splicing involves removal of the intervening genes. Thus, in the case of $C_\mu$ to $C_\gamma$ switching, the region containing $C_\delta$ must be removed by the second splice reaction (fig. S3.7). In the case of $C_\mu$ to $C_\alpha$ class switching, all of the $C$ genes between $C_\mu$ and $C_\alpha$ must be removed by the second splice reaction.

Class switching of the heavy-chain genes illustrates a temporally regulated process associated with the DNA rearrangements that can occur in the antibody-forming genes. Because any given cell normally synthesizes only one type of antibody, a special type of regulatory mechanism must respond in a systematic way to antigenic stimulation. In B-cell development it is known that the $V_H$-$D$-$J_H$ joining occurs first, leading to the formation of $\mu$ heavy chains. Subsequently, $V_L$-$J_L$ joining occurs, with the production of functional light chains.

As a rule, the DNA-splicing reactions leading to the joining process occur in only one of the alleles for each of these chains. The mechanism for this phenomenon, known by the name allelic exclusion, is not understood. Another type of exclusion process results in the production of only $\kappa$ or $\lambda$ light chains in a given B cell. This process is called isotypic exclusion. Indirect evidence that the $\kappa$ gene family is expressed before the $\lambda$ gene family comes from the examination of human B-cell lines that are active $\kappa$- or $\lambda$-chain producers. In all B cells producing $\lambda$ chains, the $C_\kappa$ genes are either deleted or rearranged. By contrast, in all cells producing $\kappa$ chains, the $C_\lambda$ genes are in

The success achieved by Tonegawa and others in this type of sequence detection of the immunoglobulin genes led to a massive research effort and a general picture of antibody gene organization. All antibody polypeptide chains are derived from split genes. Antibody genes are encoded by three unlinked gene families located on different chromosomes. In the human, chromosomes 2, 22, and 14 encode those gene clusters associated with the kappa ($\kappa$) light chains, the lambda ($\lambda$) light chains, and the heavy chains, respectively.

The light chain is encoded by three distinct DNA elements—a variable ($V_L$), a joining ($J_L$), and a constant ($C_L$) element (fig. S3.6). During the differentiation of a B cell, a $V_L$ element (encoding the first 95 amino-terminal residues)

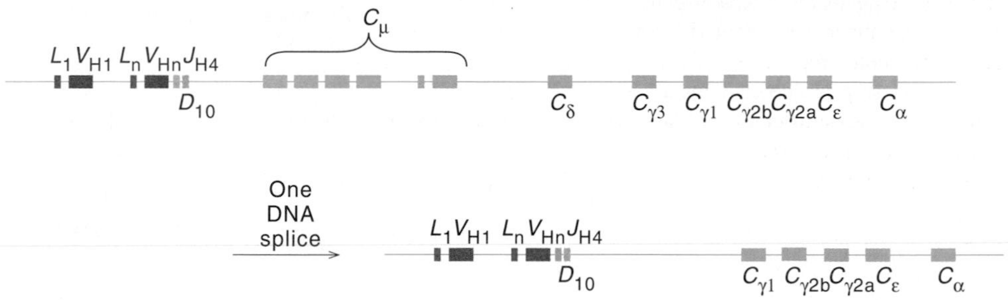

## Figure S3.7

Class switching of heavy-chain genes involves additional DNA splicing. Switching form $C_\mu$ to $C_{\gamma1}$ is illustrated. The additional splicing involves removal of the continuous segment carrying the $C_\mu$ through the $C_{\gamma3}$ genes. (From G. Zubay, *Genetics,* Benjamin/Cummings, Menlo Park, Calif. 1987, p. 809.)

the germ-line configuration. The obvious inference is that a B cell becomes a $\kappa$ producer first and then becomes a $\lambda$ producer. Other evidence exists for regulatory signals that may control isotypic exclusion. When a pre-B cell is fused with a myeloma cell that has been producing a functional heavy chain but no functional light chain, the hybrid cell is capable of expressing a new light chain as a result of a *V-J* joining of gene segments in the pre-B cell. If, however, the myeloma cell was already producing a functional light chain, no new light chains or DNA rearrangements result from the cell fusion process.

***Alternative Pathways Exist for RNA Splicing of Heavy-Chain mRNAs.*** The heavy-chain transcript contains numerous exons and introns. In addition to the introns between the $L_H$ and $V_H$ exons and between the $V_H$-$J_H$ and $C_H$ exons, each $C_H$ region contains several introns. The hinge regions of $\gamma$ chains (see fig. S3.2) and the intramembrane and cytoplasmic portions of all $C_H$ chains are encoded by independent exons. Most of the RNA-splicing operations are presumed to occur by standard mechanisms similar to those described in chapter 28. In the case of the heavy-chain genes, it has been shown that alternative RNA-splicing patterns involving the $C_H$ region can yield molecules that function as membrane-bound or secreted forms of the antibody. Thus, two membrane exons have been detected in the 3' flanking regions of both the $C_\mu$ and $C_\gamma$ genes. Alternative pathways of RNA splicing give rise to two mRNAs, which encode separately the membrane-bound and the secreted forms of the heavy chains. It seems likely that the alternative splicing pattern is temporally regulated, because the membrane-bound forms are favored before and the secreted forms are favored after antigenic stimulation. Amino acid as well as nucleotide sequencing of the $\mu_M$ chain ($\mu_M$ stands for the membrane-bound form of the $\mu$ chain) has revealed a 26-residue hydrophobic segment at the COOH terminus, which anchors in the lipid bilayer of the membrane.

***Somatic Mutation Contributes to Antibody Diversity.*** Various estimates, some exceeding a million, have been put

## Table S3.2

Estimated Number of Germ-Line Gene Segments for Different Mouse Antibody Genes

|   | Light Chains | | Heavy Chains |
|---|---|---|---|
|   | $\kappa$ | $\lambda$ | *H* |
| *C* | 1 | 4 | 8 |
| *V* | 90–300 | 2 | 100–200 |
| *J* | 5 | 4 | 4 |
| *D* | — | — | 12 |

From Geoffrey Zubay, *Genetics* (p. 810). Copyright © 1987 Benjamin/Cummings Publishing Company Inc., Menlo Park, Calif. Reprinted by permission of the author.

forward for the number of different antibodies an organism can make. A good deal of this diversity results from the different combinations of heavy and light chains that can be generated by making alternative splicings between the different gene segments that make up the light and heavy chains (table S3.2). Superimposed on the variation from this source, additional diversity is provided by somatic mutation within the coding regions. Several subsets of $\kappa$ and heavy chains and their germ-line *V* segments have been analyzed by cloning and sequencing. The results confirm that somatic mutations amplify the diversity encoded in the germ-line genome.

A $V_\kappa$ probe prepared for the $\kappa$ genes expressed in a particular myeloma, MOPC167, detected only one major band in the Southern gel blot analysis of total cellular DNA. This unusual situation permitted a relatively simple sequence comparison to be made between this segment, found in the germ-line, and its rearranged counterparts in two myeloma lines: MOPC167 and MOPC511. Four and five nucleotide differences were found, respectively, in the $V_\kappa$ regions of these two lines. The changes were completely

## Figure S3.8

DNA splicing of an antibody gene brings an enhancer element (E) close to the promoter region (P) of the gene. This step is depicted for a heavy-chain gene after the initial two splices (see fig. S3.6).

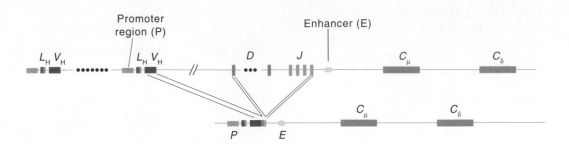

## Figure S3.9

Pluripotent stem cells in the marrow give rise to more stem cells and to various progenitors. The lymphocyte progenitors give rise to B-cell and T-cell lineages. The final differentiation step in both B-cell and T-cell development requires antigenic stimulation.

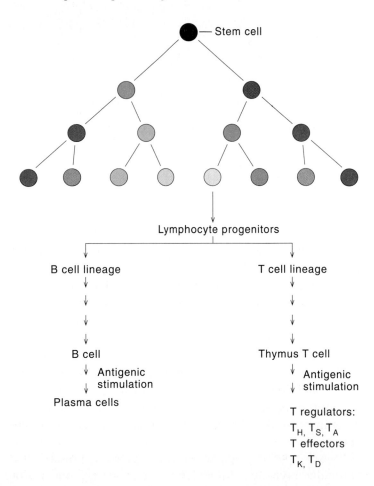

different in both lines. This investigation and other similarly directed investigations show that the regions in which mutations are introduced during somatic differentiation are highly restricted to the $V_1$-$J_1$ or $V_H$-$D$-$J_H$ segments. The density of mutations observed is about three times higher in the complementarity-determining regions (CDRs) of the vari-able segments than in the connecting framework regions (FRs) of the variable segments. (See fig. S3.3 for the approximate location of the CDR regions.) The CDRs are the regions in the immunoglobulin that play a critical role in the interaction with antigen. Mutations produced in these regions must occur either during or after DNA splicing, as they are not found in unrearranged (unspliced) DNA. Somatic mutation clearly increases the number of possible antibodies almost without limit. Most importantly, it leads to selectable changes in just those regions of the immunoglobulins that are responsible for antigen recognition. Several mechanisms for hypermutation have been proposed; none have been verified.

***DNA Splicing Brings an Enhancer Element Close to the Promoter.*** DNA sequences derived from the germ-line $J_H$-$C_\mu$ region are required for accurate and efficient transcription from a functionally rearranged $V_H$ promoter (fig. S3.8). Similar to viral transcriptional enhancer elements (see chapter 28 and supplement 4), these cellular sequences stimulate transcription from either the homologous $V_H$ gene segment promoter or a heterologous SV40 promoter. They are active when placed on either the 5′ or the 3′ side of the rearranged $V_H$ gene segment, and they function when their orientation is reversed. However, unlike viral enhancers, the immunoglobulin gene enhancer appears to act in a tissue-specific manner because it is active in mouse B cells but not in mouse fibroblasts.

The discovery of this enhancer suggests another benefit that results from splicing; it brings an enhancer in close proximity to the promoter just when that promoter activity becomes an important part of the cell's function.

## *Interaction of B Cells and T Cells Is Required for Antibody Formation*

Thus far, our discussion of antibody formation has centered on those reactions that occur at the gene level and on the structure of the antibody. However, since the immune response *in vivo* involves reactions between whole cells, we

## Figure S3.10

Kinetics of IgM and IgG appearance in the serum following a first and a second exposure to the same antigen. On first exposure a delay of several days occurs before any antibody appears in the serum. IgM appears before IgG. After many days the serum level falls. If the system is exposed to a second equivalent dose of antigen, the appearance of IgM and IgG is much faster, and the maximum level of IgG is much higher. (From G. Zubay, *Genetics,* Benjamin/Cummings, Menlo Park, Calif. 1987, p. 812.)

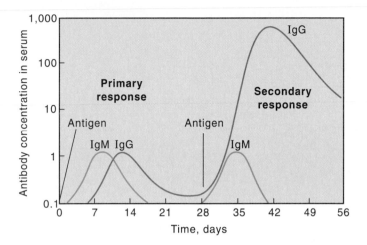

## Figure S3.11

An immature B cell (*a*) and a fully developed antibody-producing B cell (*b*). Most notable in (*b*) is the expanded cytoplasm with densely packed endoplasmic reticulum. (Electron micrographs courtesy of Dorthea Zucker-Franklin, New York University Medical Center.)

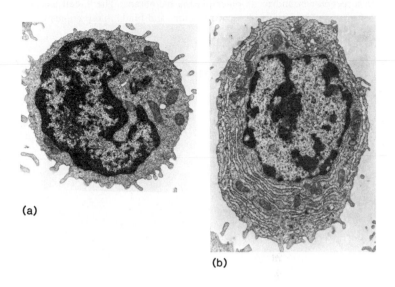

(a)

(b)

must also consider what happens at the level of cell-to-cell interactions. Some of the pluripotent stem cells of the bone marrow replicate in an undifferentiated state, maintaining their pluripotency, whereas others replicate and differentiate at the same time. The latter cells can differentiate along various pathways (fig. S3.9). Some of those that become committed to the lymphocyte pathway differentiate further into B-cell and T-cell lineages. T cells mature in the thymus, whereas B cells mature in the marrow in most vertebrates. Subsequently, both cell types relocate in secondary lymphatic organs, the spleen and the lymph nodes, and also more diffusely in all tissues. Here they remain dormant until they encounter specific antigens that trigger them to complete their differentiation into fully mature effector B cells and effector and regulator T cells. Each B cell contains immunoglobulins of a specific idiotype bound to its surface, usually of the IgM type and to a lesser extent of the IgD and IgG types.

***Antigens Stimulate the Formation of B-Cell Clones.*** The first time a foreign antigen is injected into the bloodstream, there is a long delay before specific antibody appears, and the response is fairly weak. The second time the same antigen is injected, the response is both more rapid and more intense (fig. S3.10). A great deal of research effort has gone into seeking an explanation for this difference between primary and secondary responses. Specific labeling studies show that secondary lymphatic organs contain a very heter-

ogeneous mixture of B cells, carrying membrane-bound antibodies of different idiotypes. When sufficient amounts of antigen bind to the surface antibody, the B cell becomes triggered to divide and make more cells of the same type. Some of these continue to divide, making clones of memory cells in the lymphatic organs, whereas others terminally differentiate into plasma cells richly laced with endoplasmic reticulum that is geared to the production of the antibody of the same idiotype as that which is membrane-bound (fig. S3.11). Whereas the idiotype and consequently the $V_L$ and $V_H$ parts of the antibody are the same as in the original, unstimulated B cell, the $C$ regions change. The first secreted antibody is of the secretory IgM type. This is gradually replaced by the secretory IgG type, which dominates in the secondary response. As we have noted (see fig. S3.7), the replacement of IgM by IgG is indicative of an additional DNA splicing.

***Helper T Cells Trigger B-Cell Division and Differentiation.*** The importance of T cells in mediating the B-cell response to most antigens is demonstrated by the fact that thymectomized animals, which are depleted in T cells, usually show a greatly attenuated reaction to foreign antigen. A normal response can be restored by transplantation of thymus tissue from a genetically identical animal. Activated T helper cells interact with B cells whose antibodies can recognize the same antigen that stimulated the helper T cell. The complex between the T and B cell results in the clustering of a large number of transmembrane proteins causing a

**Figure S3.12**

Stimulation of antibody-forming cells involves three types of cells. Typically, antigen is phagocytosed by a macrophage. The phagocyte partially digests the antigen and "presents" the processed antigen on its outer plasma membrane. A specific helper T cell binds the antigen to a receptor bound to its outer plasma membrane. The T cell usually has many of the same types of receptors and binds many copies of the antigen in a similar way. This stimulates T-cell proliferation. These T cells interact with B cells that display similar processed antigen. Several B-cell receptors involved in a similar way become focused in a local region of the outer plasma membrane, a process known as CAP formation. CAP formation triggers the proliferation and differentiation of the B cells involved.

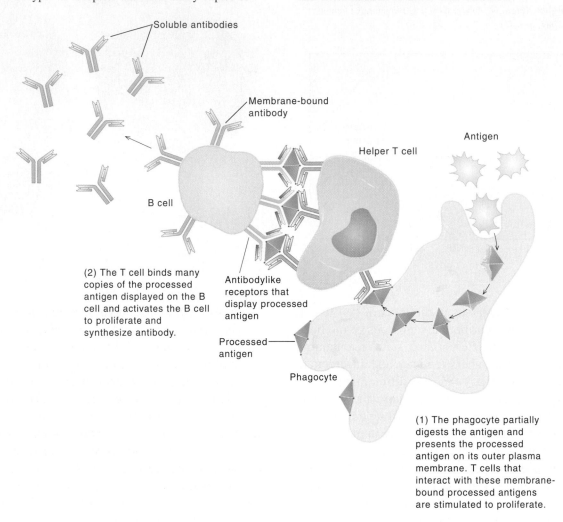

Soluble antibodies

Membrane-bound antibody

Helper T cell

Antigen

B cell

(2) The T cell binds many copies of the processed antigen displayed on the B cell and activates the B cell to proliferate and synthesize antibody.

Antibodylike receptors that display processed antigen

Processed antigen

Phagocyte

(1) The phagocyte partially digests the antigen and presents the processed antigen on its outer plasma membrane. T cells that interact with these membrane-bound processed antigens are stimulated to proliferate.

process known as CAP formation. CAP formation by some unknown means triggers the proliferation and differentiation of the B cells involved.

Support for the role of T helper cells in focusing the antigen comes from the observation that T-independent stimulation of B cells is frequently elicited by polymeric antigens. Such polymeric antigens constitute another way of presenting a large amount of antigen at one location on the B-cell membrane, thereby focusing the B-cell receptors.

Before they can help other lymphocytes respond to antigen, helper T cells must be activated themselves. This activation occurs when a helper T cell recognizes a foreign antigen bound on the surface of a specialized antigen-presenting cell. The latter cells are found in most tissues; they are derived from bone marrow and constitute a heterogeneous group, including dendritic cells in lymphoid organs and certain types of macrophages. Together with B cells, which can present antigen to helper T cells, these are the main cell types that react with helper T cells. The sequence of events involving antigen-processing cell, T helper cell, and antibody-forming B cell is depicted in figure S3.12.

Note that the mechanism proposed for helper T-cell function requires that T cells have surface receptors that recognize antigen-processing cells, antigen itself, and the appropriate B cell. We will shortly discuss the nature of T-cell receptors, as well as the surface structures they recognize, in a broader context.

**Figure S3.13**

The principal stages in complement activation. Complement activation occurs exclusively on the microbial cell membrane, where it is triggered by bound antibody or microbial envelope polysaccharides, both of which activate early complement components. Two sets of early components belong to two distinct pathways of complement activation. Activation of each complement system involves a cascade of proteolytic reactions. Each component of the complement system is a proenzyme that is activated by the preceding component of the chain by a limited proteolytic cleavage. The ultimate result of this chain reaction is the development of a complex that attacks the cell membrane.

Microbial polysaccharide pathway

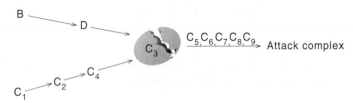

Antibody-binding pathway

## The Complement System Facilitates Removal of Microorganisms and Antigen–Antibody Complexes

Many antigen–antibody complexes are eliminated by phagocytosis. Others are attacked by complement, a group of serum proteins that aids in the defense against microorganisms. Individuals with a deficiency in their complement system are subject to repeated bacterial infections, just as are individuals deficient in antibodies themselves.

About 20 different proteins are included in the complement system: Proteins C1–C9, factors B and D, and a series of regulatory proteins. All these proteins are made in the liver, and they circulate freely in the blood and extracellular fluid. Activation of the complement system involves a cascade of proteolytic reactions. In addition to forming membrane attack complexes, the proteolytic fragments released during the activation process promote dilation of blood vessels and the accumulation of phagocytes at the site of infection.

The activation of complement occurs exclusively on the microbial cell membrane, where it is triggered either by bound antibody or microbial envelope polysaccharides, both of which activate early complement components. Two sets of early components belong to two distinct pathways of complement activation: C1, C2, and C4 belong to the pathway that is triggered by antibody binding; factors B and D belong to the alternative pathway that is triggered by micro-

bial polysaccharides (fig. S3.13). The early components of both pathways ultimately act on C3. The early components and C3 are proenzymes that are activated sequentially by limited proteolytic cleavage. As each proenzyme in the sequence is cleaved, it is activated to generate a serine protease, which cleaves the next proenzyme in the sequence. Many of these cleavages liberate small peptide fragments and expose a membrane-binding site on the larger fragment. The larger fragment binds tightly to the target cell membrane by its newly exposed membrane-binding site and helps to carry out the next reaction in the sequence. In this way complement activation is confined largely to the cell surface, where it began.

The components of the complement attack complex have a very short half-life if released from the complex. This limitation confines their destructive action to the point of assembly, the surface membrane of the microorganism.

## The Cell-Mediated Response: A Separate Response by T Cells

Most B cells are fixed in the secondary lymphatic organs: The spleen and the lymph nodes. They are stimulated by antigen only if the antigen is transported to the lymph nodes. In many cases, however, antigens are poorly transported. To rid the organism of antigens that are not readily transported, a second type of immune system exists that is independent of the B cells. This second type of immunity is not found in all vertebrates; it is restricted to mammals, birds, certain amphibians, and bony fishes. Long before the distinction between T and B cells was appreciated, it was known that certain types of immunity, which manifest a delayed-type hypersensitivity response, could be transferred from immunized animals to nonimmunized animals with leukocytes but not with antibody-containing serum. Subsequently, other forms of leukocyte-mediated immunity were discovered that involved increased phagocytic and bacteriocidal activity of macrophages, lysis of virus-infected cells and tumor cells, and the rejection of skin grafts from genetically dissimilar organisms.

The two types of cell-mediated immunity involve basically different types of T cells. In the delayed-type hypersensitivity response, the T cell, $T_D$, that reacts specifically with antigens secretes interleukins (see Box S3A) that attract and activate macrophages or other leukocytes, thereby causing a slowly developing inflammatory response. A second type of T cell, known as the T killer cell, $T_K$, reacts specifically with antigen that is bound to target cells, causing their lysis.

# S3A

# The Role of Interleukins

The interactions between immune and inflammatory cells are mediated in large part by certain proteins, called lymphokines, or interleukins (IL), that are able to promote cell growth, differentiation, and functional activation. Interleukins resemble hormones, which also function as intercellular messengers. In contrast to hormones, however, they are secreted by isolated cells rather than discrete glands. Several interleukins have been described; each has unique biological activities as well as some that overlap with the oth-ers. For example, macrophages produce IL-1 and IL-6, whereas T cells produce Il-2–IL-6, and bone marrow stromal cells produce IL-7. IL-1 and IL-6 not only play important roles in immune cell function, they also stimulate a spectrum of inflammatory cell types and induce fever. IL-2 is a potent proliferative signal for T cells. IL-1, Il-3, IL-4, and IL-7 enhance the development of a variety of hematopoietic precursors. IL-4–IL-6 also serve to enhance B-cell proliferation and anti-body production.

## Tolerance Prevents the Immune System from Attacking Self-Antigens

A thorough understanding of the immune system requires an explanation for how the immune system is able to discriminate between foreign antigens and its own antigens (self-antigens). The ability to recognize self-antigens and thus to avoid making an antagonistic response to them is called tolerance. Some serious diseases are believed to be due to a breakdown in the self-tolerance mechanism. Conversely, conditions occur under which a foreign antigen is able to establish itself so that it becomes tolerated, for either a short or a long time. Tolerance and rejection are so closely related that it seems likely that a full understanding of one entails a full understanding of the other.

Several conditions favor the establishment of tolerance: (1) Tolerance is much easier to establish in the fetus or the newborn than in the adult. Thus, if an organism is exposed as a fetus to a soluble foreign antigen or a foreign tissue graft, it may become indefinitely tolerant to the antigen. This situation is illustrated for two genetically nonidentical strains of mice in figure S3.14. Cells from the Y mouse are injected into a newborn X mouse. The same X mouse as an adult is the recipient of a skin graft (transplant) from a Y-type mouse. Normally such a graft would be rapidly rejected. However, because of the early exposure to the Y cells, the graft is tolerated indefinitely. This result indicates that the mouse that was exposed to the Y cells just after birth recognizes Y cells later in life as self-antigens.

(2) Very high levels of antigen frequently favor the development of tolerance, whereas intermediate levels of antigen favor the development of an immune response. (3) Frequently, the route of antigen administration is critical in the development of tolerance. An intravenously injected antigen is more likely to promote a tolerant condition than antigen injected subcutaneously.

Two cellular mechanisms may account for most forms of tolerance. In one of these the clones of T or B cells that could otherwise be activated by the antigen are destroyed or inactivated. In the other mechanism, the clones in question are still present but do not respond because they are blocked by $T_S$ cells. In general, B cells are more difficult to make tolerant than T cells. This difference is evident from observations that longer and higher doses of antigen are required to establish tolerance to the B-dependent system. Moreover, the period of tolerance is generally shorter when B cells are involved.

Adult mice may be made tolerant to foreign antigens by destroying their immune system with whole-body irradiation and then supplying them with transplants of bone marrow and thymus from a foreign donor, thus refurbishing their system with competent B and T cells, respectively. With the new immune system the animal is now tolerant to any tissue from a mouse that is isogenic with the mouse used to supply the marrow and the thymus.

The irradiated host organism also supplies a useful means of testing the state of B and T cells from a nonirradiated isogenic mouse. B cells may be effectively transferred

## Figure S3.14

Establishment of tolerance to tissue grafts. If a newborn X mouse is injected with Y cells when it is young, it is tolerant to Y tissue transplants when it becomes an adult. If the X mouse is not exposed to Y cells at an early age, it readily rejects the tissue graft.

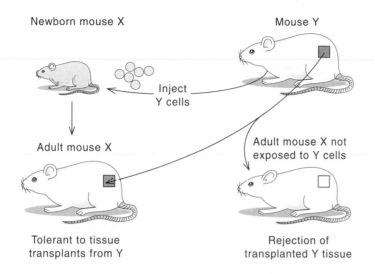

Newborn mouse X

Mouse Y

Inject Y cells

Adult mouse X

Adult mouse X not exposed to Y cells

Tolerant to tissue transplants from Y

Rejection of transplanted Y tissue

by bone marrow transplants, and T cells may be transferred by thymus tissue transplants. The donor cells may come from a normal, untreated isogenic strain. In this event the irradiated recipient shows normal immunological behavior. Alternatively, the donor cells may come from a donor that has been made either sensitive or tolerant to a particular antigen. By doing various combinations of experiments using different isogenic donors, it is possible to determine the relative importance of B-cell and T-cell involvement with different antigens.

## T Cells Recognize a Combination of Self and Nonself

All T cells, whether they fight infection directly or regulate the activity of other effector cells, recognize antigen only in combination with a cell membrane surface. Of greatest significance, the antigen must be associated with a cell from an organism that is genetically similar or identical to that of the T cell. This requirement was first demonstrated in an experiment where T cells were removed from a mouse that had been immunized with a virus. Immunization meant that the extracted T cells had been activated to direct themselves against cells infected with this particular virus. Thus, these T cells could kill fibroblasts infected with the virus in tissue culture. However, they could not kill fibroblasts infected with the virus if the fibroblasts came from a mouse with a different genetic background. Thus, sensitized killer T cells do not recognize foreign antigens alone; they recognize for-

eign antigens only in combination with determinants present on their host's own cells (fig. S3.15). These common determinants are a subset of the proteins encoded by the so-called major histocompatibility complex (MHC).

## MHC Molecules Account for Graft Rejection

MHC molecules were recognized long before their major biological function was appreciated. They were initially defined as the main target antigens in transplantation reactions, and because of this they became known as major histocompatibility antigens. Experiments on mice demonstrated that graft rejection is an immune response to the surface antigens of the grafted cells. Subsequently, it was shown that these reactions are mediated by T cells and that they are directed against a family of glycoproteins encoded by a group of genes called the major histocompatibility complex. MHC molecules are transmembrane proteins found on the cell membranes of all higher vertebrates.

A vertebrate does not normally need to be protected against invasion by foreign vertebrate cells, so the antagonistic reaction of its T cells to foreign MHC molecules was both an obstacle to transplant operations and an enigma to immunologists. The enigma was solved when it was discovered that MHC molecules contributed to the binding of T lymphocytes on those host cells that have foreign antigen on their surface, as in the case of the virus-infected cells we have just described (see fig. S3.15).

## There Are Two Major Types of MHC Proteins: Class I and Class II

There are two major classes of MHC proteins that have similar types of structures (fig. S3.16). Class I molecules contain three external domains, each about 90 residues in length; a transmembrane region; and a cytoplasmic domain. The third external domain is noncovalently associated with a small polypeptide known as the $\beta_2$ microglobulin. Class II molecules are composed of two noncovalently associated polypeptide chains: $\alpha$ and $\beta$, whose overall structure resembles that of the class I complex.

Several genes encode each type of MHC protein, and several alleles represent different MHC proteins for each of these genes. This gives rise to a tremendous variety of MHC proteins, with the consequence that it is extremely unlikely that any two members of the same species will possess the same assortment of MHC proteins. Thus, two humans are not likely to carry the same MHC proteins on their cell membranes unless they are identical twins. This is the reason why transplants between two individuals are almost

**Figure S3.15**

Activated T cells bind and kill only those cells that display a foreign antigen and a familiar cell surface antigen.

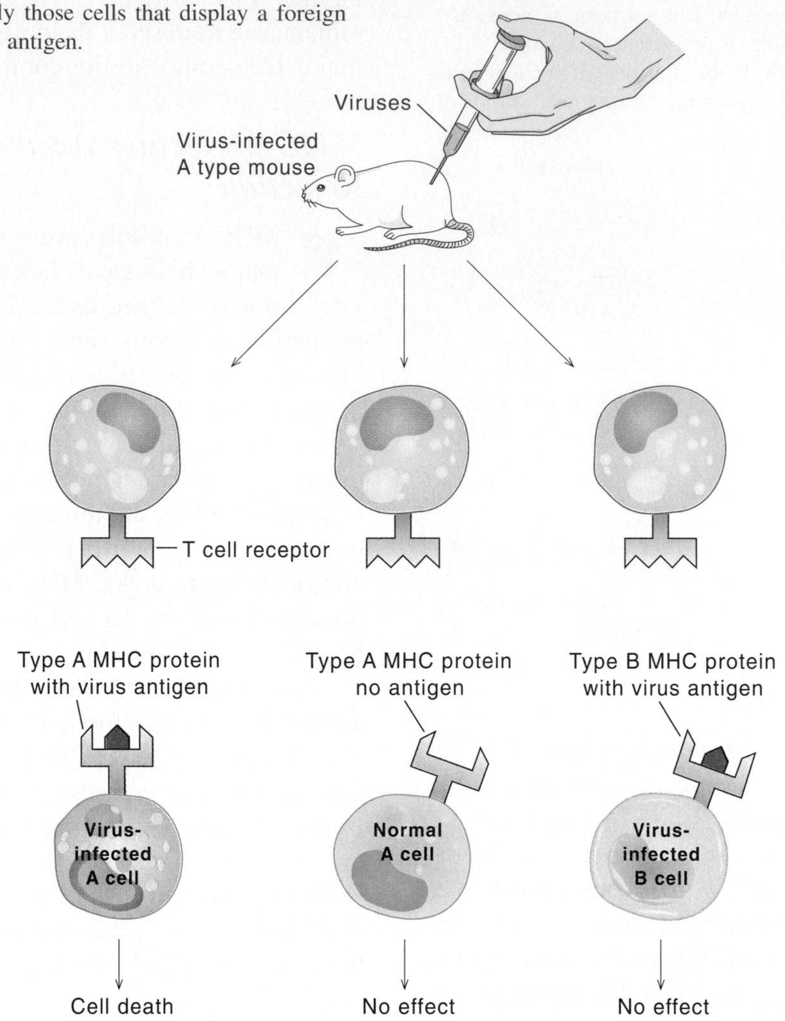

always rejected—it is usually just a matter of time. Highly inbred strains of mice, which are genetically identical, display the same MHC proteins and accordingly take grafts from each other without any rejection.

The two classes of MHC proteins are displayed on different cell types. Class I MHC proteins are found on almost all nucleated cells, including killer T cells. Class II MHC proteins are found mainly on cells involved in the immune response, including antigen-presenting cells, B cells, and T helper cells, but not T killer cells.

## T-Cell Receptors Resemble Membrane-Bound Antibodies

The specific receptors found on the membranes of T cells resemble the highly specific antibodies found on the membranes of B cells. Specific T-cell receptors (TCRs) are composed of two disulfide-linked peptide chains (called α and β). Each of these chains share with antibodies the distinc-

tive property of a variable amino-terminal region and a constant carboxyl-terminal region.

With one exception, all the mechanisms used by B cells to generate antibody diversity are also used by T cells to generate T-cell receptor diversity. The one mechanism that does not appear to operate in T-cell receptor diversification is somatic hypermutation. This is presumably because mutation would be likely to generate killer T cells that would wantonly attack self-molecules. This is much less of a problem for B cells, since most self-reactive B cells could not be activated without the aid of specific helper T cells.

## Additional Cell Adhesion Proteins Are Required to Mediate the Immune Response

In addition to membrane-bound antibodies, TCR proteins and MHC proteins, other transmembrane proteins are required to obtain the specific cell–cell interactions necessary to mediate the immune response. These additional proteins

**Figure S3.16**

Structures of class I and class II molecules as they would appear in the lipid bilayers of the cell membrane. Models show striking similarities between the two types of molecules. S—S indicates disulfide bridges in the class I and class II molecules and the $\beta_2$ microglobulin.

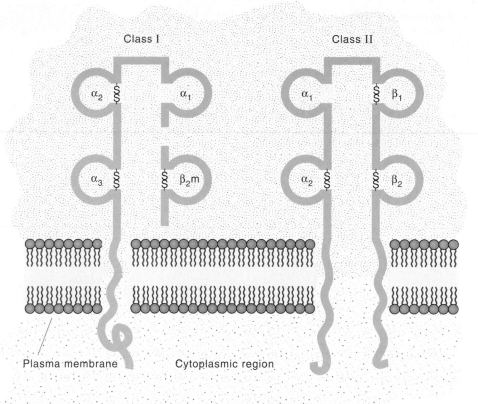

are also important in a number of nonimmune reactions in which cell–cell adhesion is important to the organism. We only consider a limited number of these additional proteins because of the enormous complexity of this subject. Three transmembrane proteins directly involved in the T-cell reaction with other cells are members of a family of cell–cell adhesion proteins (the CD family of proteins). The CD3 protein is found on both killer and helper T cells; it forms a complex with the membrane-bound TCR proteins. The CD8 and CD4 proteins interact with this complex, but unlike the CD3 proteins, CD8 is only found on killer T cells, whereas CD4 is only found on helper T cells. The presence of CD4 and CD8 proteins limits the range of binary complexes formed by T cells with other cells. This is because CD4 and CD8 have selective affinities for MHC class II and MHC class I proteins, respectively. As a result, killer T cells only form complexes with cells that display a processed foreign antigen in a complex with the MHC I protein; this includes most nucleated cells other than those that belong to the immune system. Similarly, helper T cells are limited to reactions with cells that display the appropriate processed antigen in a complex with the MHC II protein; this includes B cells or phagocytes (fig. S3.17).

From this discussion we can see that a productive interaction between T cells and other cells requires a highly specific reaction between a TCR protein and a processed antigen that is supplemented by less specific interactions between other transmembrane proteins on the two cell types.

Left out of this discussion is an explanation for the source of the processed antigen on the antigen-presenting B cell. Immature B cells display a membrane-bound antibody, which interacts with circulating complementary antigen. Once bound to the antibody, the antigen is internalized, processed, and ultimately displayed by the MHC II protein on the cell surface of the same B cell. It is the interaction between such a B cell and the appropriate helper T cell that triggers the B cell to divide and differentiate.

## Immune Recognition Molecules Are Evolutionarily Related

The striking structural similarity between immune recognition proteins almost certainly reflects their evolution from a common ancestor. All of them contain one or more immunoglobulinlike domains, each with about 90 amino acids stabilized by a conserved disulfide linkage. The simplest member of this family of genes is $\beta_2$-microglobulin. Possibly this or a similar protein was the ancestor for the remaining immune recognition proteins.

**Figure S3.17**

Antigen-specific reactions involving T cells. (*a*) Killer T cells interact and destroy target cells that display complexes of processed antigen with an MHC I protein. Effective interaction requires that the TCR protein interact specifically with the processed antigen. All of the remaining reactions between the two cells are general reactions that occur between other killer T cells and other target cells. (*b*) Helper T cells interact specifically with antigen-presenting B cells, provided a specific complex is made between the TCR protein and the processed antigen complexed to the MHC II protein and the processed antigen complexed to the MHC II protein on the B cell. The unlabeled membrane proteins in this figure represent other nonspecific protein–protein interactions that strengthen the binary cell reaction. Processed antigen is indicated in red.

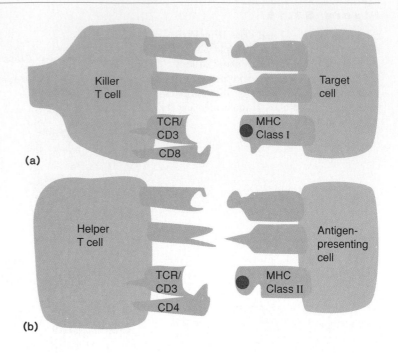

## Summary

In this supplement we considered the major findings that led to our understanding of the immune system in vertebrates, especially mice and humans. The following points are the highlights of our discussion.

1. Two different classes of white cells (lymphocytes) are associated with the immune response: The B cells and the T cells. B cells produce antibodies that are secreted and aggregated foreign substances known as antigens. T cells are of different types: Some of them assist B cells to become antibody-forming cells; others mount an attack on antigens of their own.

2. The most common type of antibody is a tetramer composed of two identical heavy (H) chains and two identical light (L) chains. Each chain is divided into a relatively constant sequence (the *C* segments) and a relatively variable sequence (the *V* segments). Within the variable regions there are subsegments known as the hypervariable regions. These subsegments are the regions that specifically bind different antigens.

3. According to some estimates, each organism is capable of making more than a million different antibodies from a limited amount of genetic information. This feat is accomplished with the help of somatic recombination and somatic mutation. Somatic recombination involves splicing at the DNA level. It brings specific *C* and *V* regions together to make a unit that can be transcribed. Somatic mutation is focused on those parts of the gene that encode the hypervariable regions.

4. T helper cells focus the antigen on an immature B cell. In response, some B cells turn into antibody-forming cells, and some B cells turn into memory cells for production of specific antibodies at a later date. The first time an organism is exposed to a specific antigen it takes longer to mount an immunological response. That is because of the lack of memory cells for making a specific antibody.

5. Complement is a group of serum proteins that aids in the defense against microorganisms and removal of antibody–antigen complexes.

6. T cells give rise to an immune response of their own. The two types of cell-mediated immunity involve basically different types of T cells. In the delayed-type response, the T cell reacts specifically with antigens and secretes lymphokines. These are substances that attract macrophages or other leukocytes, thus produc-

ing a slowly developing inflammatory response. A second type of T cell reacts specifically with antigen bound to target cells and causes their lysis.

7. Tolerance prevents the immune system from attacking self-antigens. To understand tolerance we must appreciate how T cells work. T cells recognize a combination of self and nonself. The cell surface antigens rec-

ognized by T cells are known as the major histocompatibility complex (MHC). If two organisms carry the same histocompatibility antigens, the tissues from the two organisms are completely compatible. The histocompatibility antigens resemble antibodies in structure.

## Selected Readings

Ada, G. L., and G. Nossal, The clonal selection theory. *Sci. Am.* 257(2):62–69, 1987.

Darnell, J., H. Lodish, and D. Baltimore, *Molecular Cell Biology.* New York: W.H. Freeman Co., Scientific American Books, 1990. For a more comprehensive textbook treatment of this subject, see chapter 25.

Golde, D. W., The stem cell. *Sci. Am.* 265(6)86–93 (December) 1991.

Honjo, T., and S. Habu, Origin of immune diversity: Genetic variation and selection. *Ann. Rev. Biochem.* 54:803–830, 1985.

Hood, L., M. Kronenberg, and T. Hunkapiller, T cell antigen receptors and the immunoglobulin super gene family. *Cell* 40:225–229, 1985.

Hunkapiller, T., and L. Hood, The growing immunoglobulin gene superfamily. *Nature* 323:15–17, 1986.

Kaappes, D., and J. L. Strominger, Human class II major histocompatibility genes and proteins. *Ann. Rev. Biochem.* 57: 991–1028, 1988.

Klein, J., *Immunology.* London: Blackwell Scientific Publications, 1990.

Leder, P., The genetics of antibody diversity. *Sci. Am.* 246(5):102–115, 1982.

Lieber, M. R., Site-specific recombination in the immune system. *FASEB J.* 5:2934–2944, 1991.

Milstein, C., Monoclonal antibodies. *Sci. Am.* 243(4):66–74, 1980. An exciting, tissue-culture technique for obtaining pure antibodies specific for an antigen of choice.

Mizel, S. B., The interleukins. *FASEB J.* 3:2379–2388, 1989.

Nowak, M. A., R. M. Anderson, A. R. McLean, T. F. W. Wolks, J. Goudsmit, and R. M. May, Antigenic diversity thresholds and the development of AIDS. *Science* 254:963–969, 1991.

Rini, J. M., U. Schulze-Gahmen, and I. A. Wilson, Structural evidence for induced fit as a mechanism for antibody–antigen recognition. *Science* 255:959–965, 1992.

Silver, M. L., H-C. Guo, J. L. Strominger, and D. C. Wiley, Atomic structure of a human MHC molecule presenting an influenza virus particle. *Nature* 360:367–372, 1992.

Singer, S. J., Intercellular communication and cell–cell adhesion. *Science* 255:1671–1677, 1992. A comprehensive review of the major cell–cell interactions important in immune reactions.

Springer, T. A., Adhesion receptors of the immune system. *Nature* 346:425–434, 1990. An excellent review.

Sutton, B. J., and H. J. Gould, The human IgE network. *Nature* 366:421–428, 1993. An understanding of the IgE system holds the key to understanding and intervening in the aetiology of allergic diseases.

Tonegawa, S., Somatic generation of antibody diversity. *Nature* 302: 801–803, 1983. A classic paper.

Tonegawa, S., The molecules of the immune system. *Sci. Am.* (December) 1985.

Weiss, R. A., How does HIV cause AIDS? *Science* 260:1273–1278, 1993. An exciting up-to-date account of the various routes whereby the HIV virus attacks and destroys the immune system.

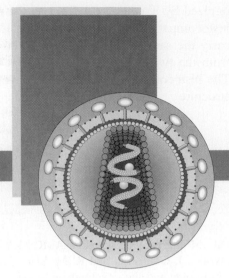

# Principles of Physiology and Biochemistry: Carcinogenesis and Oncogenes

In the complex developmental process that leads to a mature human being from a fertilized egg, each cell takes up a role in a specialized organ or tissue. Each group of cells proliferates to a certain point and then stops. In the case of a solid organ such as the liver, we see cells of several different types that grow until the organ takes on a fixed size and shape and then cease to grow. Normally, further growth occurs only at the very slow rate required for replacement of aging cells. However, if part of the liver is amputated, a rapid regenerative process ensues so that the liver regrows to approximately the same size it had before the amputation. Other organs or tissues may show more or less growth than the liver, but as a rule, new cells appear only as old and damaged cells are replaced. Furthermore, the cells of normal tissues do not as a rule mix.

Normally you do not find liver cells mixing with intestinal cells or any other types of cells, suggesting that cells

## Figure S4.1

A benign glandular adenoma (*a*) and a malignant glandular adenocarcinoma (*b*). In both cases the tumor cells differ morphologically from the surrounding tissue and show a pattern of disorganized growth.

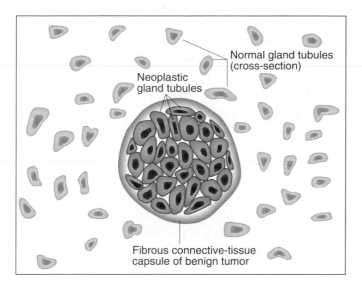

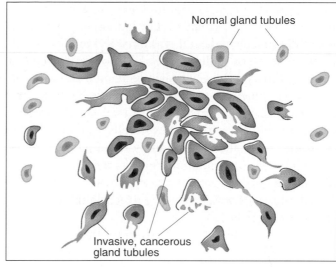

have surface properties that favor interaction of like cells. This was demonstrated in the 1940s by experiments in which embryonic tissues were dissociated into single cells and allowed to reassociate in the presence of other cell types; they usually reassociated with cells of a similar type. It is now known that adhesion molecules are located on cell surfaces that encourage such associations. This was alluded to in supplement S3 when we described the interaction of T cells with other cell types of the immune system. Cell adhesion molecules that favor the association of like cells are of two types: Some require $Ca^{2+}$ for adhesion, the cadherins, and others do not, the cell adhesion molecules (CAMs). When these molecules are not present, as in culture-transformed cell lines, like cells adhere poorly to each other or to other cultured cells. It seems likely that when tumor cells become invasive the character of their cell adhesion molecules has been altered. We mention this here because it is probably an important property of malignant cells, but we have little more to say about it in this supplement because this aspect of malignancy is not well characterized.

Cancer cells are relatively easy to recognize when they occur in groups in the organism (fig. S4.1). A cluster of cancer cells divides more rapidly, and it usually clashes with the architecture of normal cells in its immediate surroundings. From cytological inspection alone, we might speculate that cancer cells are breaking the rules for normal cell growth and associations. It appears that the processes that regulate cells are disrupted; cell growth is uncontrolled, and patches of morphologically distinct cells show by their invasive character that they do not recognize normal territorial boundaries between cells of different types. Cancer cells are deadly because ultimately they invade and disrupt a vital area. In this supplement we present a brief overview of some aspects of carcinogenesis and oncogenes.

## Cancers: Cells Out of Control

Cancer is not a stage in the orderly evolution of a complex organism; it is an organismic catastrophe. Perhaps it is only reasonable to expect that multicellular eukaryotes, with their many dividing cells, occasionally produce a mutant cell that gets out of control and destroys the system. The notion that cancer cells are out of control was proposed more than half a century ago. In the intervening time we have learned a great deal about the biochemistry of growth regulation. This knowledge has helped us to understand the molecular basis of cancer in many cases. Conversely, the study of cancer has added to our knowledge of normal growth regulation.

## Transformed Cells Are Closely Related to Cancer Cells

In the 1950s, systematic methods were developed for taking cells directly from the tissues of whole animals and inducing them to grow as single cells in liquid culture.

An outgrowth of this technology was the development of pure cell lines—cells that divide indefinitely. Primary cultures of dispersed single cells from tissue fragments usually die out after 50 or so duplications. However, during the growth of a primary culture, occasional altered cells appear that possess a somewhat different morphology; they frequently take over the culture, because they are capable of initiating new cultures with far fewer cells. A clone that is derived from one of these indefinitely dividing variants—which has in effect become immortal—is designated a cell line. The transition of primary culture cells to cells that can grow indefinitely is considered to be a step in the direction of becoming cancerous.

Typically, continuous cell lines grow in culture only on a solid support or "anchor" (the surface of a petri dish, for example) and in the presence of relatively high concentrations of nutrients. Even then, they divide only as long as the culture is sparse. When the cell density increases beyond a critical point, the growth rate decreases sharply; in fact, the cells of some continuous lines stop dividing altogether once they have formed a confluent monolayer.

At the density that normally halts growth, rare transformed cells continue to multiply and may reach cell densities up to 20 times higher than those of untransformed cells. By picking and subculturing such transformed cells, it is possible to establish clonal lines of transformants and to ask in what ways such cells differ from their untransformed progenitors.

The differences between transformed and untransformed cells can be classified in three ways: Changes in cell growth, changes in cell surface properties, and genetic alterations. These changes are too great in number and too diverse in quality to be caused directly by one or a few mutations. Clearly, then, most of the observed alterations must be secondary and tertiary events that have occurred as a consequence of some primary event. It is the aim of much current work with tumor-causing agents, especially certain viruses, to discover the molecular nature of this primary event.

Transformed cells sometimes, but not always, give rise to tumors when injected into an isogenic animal. There may be numerous reasons for the failure to produce tumors on all occasions. One of the best known reasons for lack of tumor production is immunological rejection. The transformed cell resulting from exposure to a tumor virus frequently displays viral antigens which are recognized as foreign to the animal host. This recognition excites a positive immune response that can lead to rejection of the transformed cells.

Just as transformed cells frequently do not develop into tumors when injected into whole animals, so is it that tumor cells do not always grow well when dispersed into tissue culture. Nevertheless, tumor cells usually adapt more readily to growth in tissue culture than normal cells taken from the whole animal.

## Environmental Factors Influence the Incidence of Cancers

It seems very unlikely that we will find cures for many types of cancer in the foreseeable future. Therefore it is a good idea to put considerable effort into cancer prevention, especially because there are reasons for believing that the vast majority of cancers are preventable. The frequency of different kinds of cancer varies enormously in different human populations (table S4.1). For example, the incidence of breast and colon cancer is much higher in the United States than in Japan, whereas that of stomach cancer is much higher in Japan. In these cases, diet is most likely the culprit. Liver cancer is highest in third world countries. Once again, diet seems the most likely cause. Lung cancer is more prevalent in countries where smoking is popular, and indeed, it is well documented that heavy smokers are much more likely to get lung cancer than nonsmokers. Overall, it is clear that much of the variation in cancer incidence is environmental rather than genetic, since these differences tend to disappear from one generation to the next in migrant populations. For example, the high incidence of stomach cancer in Japan is not matched by a similar figure among Japanese-Americans in the United States.

The conviction is growing that the carcinogenic agents in the environment are active as cancer-causing agents because they produce mutations. The hunt for carcinogens has been facilitated by Bruce Ames, who developed a test for carcinogens based on the mutagenic action of a compound on bacteria.

Ames tests have shown that many carcinogens originate from food and chemical pollutants. In addition to chemicals, ionizing radiation is carcinogenic. For example, the indications are strong that the ionizing radiation we get from the sun is the major cause of skin cancer. Thus, the incidence of skin cancer in the United States is much higher in the South than in the North. High dosages of radioactive materials or x-rays are also strongly correlated with the incidence of numerous forms of cancer.

In many cases, exposure to certain viruses is correlated with specific types of cancers. For example, liver cancer is 300 times more common in individuals who have previ-

**Table S4.1**

Variation in the Incidence of Cancers by Country

| Type of Cancer | Country of High Incidence | Country of Low Incidence | Ratio (High/Low) |
|---|---|---|---|
| Skin | Australia | India | 200 |
| Esophagus | Iran | Nigeria | 200 |
| Lung | England | Nigeria | 35 |
| Prostate | United States | Japan | 15 |
| Liver | Mozambique | England | 40 |
| Breast | Canada | Israel | 7 |
| Rectum | Denmark | Nigeria | 20 |
| Colon | United States | Nigeria | 10 |
| Ovary | Denmark | Japan | 6 |
| Stomach | Japan | Uganda | 25 |

ously been infected with hepatitis B virus. Although we can only speculate on the precise reasons for this, such numbers indicate that liver disease resulting from hepatitis B virus infection is almost certainly the major cause of human liver cancer.

A cancer of the B lymphocytes known as Burkitt's lymphoma is strongly correlated with infection by the Epstein-Barr virus. Once again, we do not know the precise chain of events that lead to cancer. Hepatitis B and Epstein-Barr are DNA viruses. Certain RNA viruses have also been shown to be correlated with cancer. The best known is the human immunodeficiency virus (HIV-1, the virus that causes AIDS.; fig. S4.2). The AIDS virus is believed to cause cancer indirectly by incapacitating the immune system. Various types of cancer follow infection by the AIDS virus, a fact testifying to the key role the immune system plays in protecting the body from cancer. The RNA virus known as human T-cell leukemia virus type I (HTLV-I) causes T-cell leukemia. It is believed that this virus carries a cancer-causing gene (an oncogene), the continuous expression of which upsets the normal metabolism, causing infected cells to become transformed to cancer cells.

## Cancerous Cells Are Genetically Abnormal

In the previous section we stated that a strong correlation exists between the mutagenic properties of a chemical and its potency as a carcinogen. This is consistent with other evidence that mutation often plays a causal role in carcinogenesis.

The most graphic illustrations of the correlation between genetic abnormality and cancer come from the frequent association of certain types of tumors with highly specific chromosomal translocations. Chromosomal translocations involve the movement of a segment of chromosomes from one location to another. Frequently, translocations occur between different chromosomes in a reciprocal fashion. This is the case for translocations between human chromosome 8 and the chromosomes that carry the antibody genes: 14, 2, and 22.

A reciprocal exchange of segments of chromosome from the ends of chromosomes 8 and 14 is frequently found in the genome of cancer cells of the Burkitt's lymphoma (fig. S4.3). Burkitt's lymphoma is a tumor of B lymphocytes; for unknown reasons, its occurrence is most common in Africa. The resulting tumor cells usually secrete antibodies. The fact that antibody genes, which are so active in B cells, map to the same chromosomal regions involved in the specific translocations in Burkitt's tumors led to the suggestion that a specific oncogene on chromosome 8 might be activated if it were placed near one of these antibody genes. This suggestion proved to be correct; the oncogene involved is *c-myc*. We have more to say about *c-myc* later.

The first chromosome abnormality found to be associated with a cancer was the Philadelphia chromosome, which arises as a result of a reciprocal translocation between chromosomes 9 and 22 (fig. S4.4). The tumor associated with this translocation is chronic myelogenous leukemia and results from the activation of the *c-abl* oncogene, which is normally located on chromosome 9.

Whereas it is a common belief that tumor formation involves mutation, certain properties associated with the etiology of teratocarcinoma do not fit with this concept. Tera-

## Figure S4.2

AIDS virion structure. Numbers associated with proteins are kilodalton masses (e.g., gp120 is a 120-kd protein). (© *Scientific American*, Inc., George V. Kelvin. Reprinted by permission.)

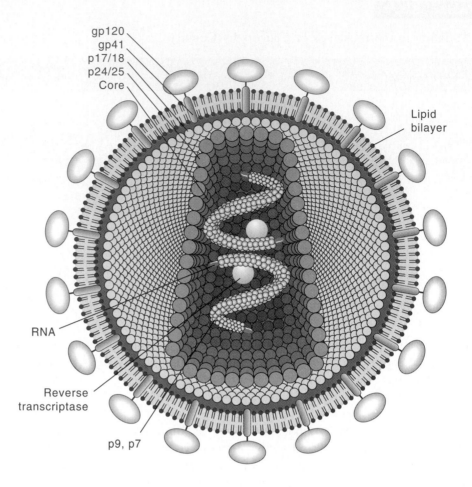

gp120
gp41
p17/18
p24/25
Core

Lipid bilayer

RNA

Reverse transcriptase

p9, p7

tocarcinoma is a unique type of tumor in which the tumor is associated with a cluster of cells of different types, including stem cells and differentiated cells of various types. Only the stem cells are oncogenic. These stem cells may be replicated *in vitro*. On subcutaneous injection these cells produce tumors, but on injection into a blastocyst embryo these stem cells revert to normal cells, which can be identified in the adult animal by the genetic markers they carry. If these stem cells bear a mutation that accounts for their oncogenicity, then it is hard to see how this mutation could revert to wild type by placing it in the cytoplasmic environment of the embryo. Possibly the teratoma is an exceptional situation in which the tumor is caused by abnormal development not necessitating a mutation.

## Some Tumors Arise from Genetically Recessive Mutations

The oncogenes *c-myc* and *c-abl* are dominant to their wild-type alleles. Many tumors arise from oncogenes that show recessive behavior when compared with their wild-type alleles. Such is the case with the genetic locus implicated in the malady known as bilateral retinoblastoma,

in which tumors develop in the retinas of both eyes. Humans with this condition usually inherit a genetically recessive mutation on chromosome 13. Sometimes the mutation is microscopically visible as a small chromosomal deletion. In general, recessive mutations are not apparent unless they are present in the homozygous state. However, in the course of the very extensive cell duplication that ensues in retinal development, there is a high probability that at least one cell in each retina will mutate in the wild-type allele of the genetic locus in question. This change results in a homozygous state and in the conversion of the normal cell to a tumorous cell that proliferates rapidly. As a result of this somatic mutation, both eyes usually develop tumors. It seems likely that genes that give rise to tumors when they are defective or lacking, as in retinoblastoma, are of the regulatory type. Some other tumors that result from recessive genetic lesions are listed in table S4.2).

## Many Tumors Arise by Mutational Events in Cellular Protooncogenes

A growing number of cellular genes have been implicated in a wide variety of cancers. These genes do not lead to cancers in their normal state but only after a mutational event

## Figure S4.3

Translocation associated with Burkitt's lymphoma. The protooncogene *c-myc,* on the end of chromosome 8q, is translocated to the end of chromosome 14q, adjacent to the immunoglobulin constant gene, *Ig-Cμ.* (From J. Yunis, The chromosomal basis of human neoplasia, *Science* 221:227–236, Jan. 1, 1983. Copyright 1983 by the AAAS. Reprinted by permission.)

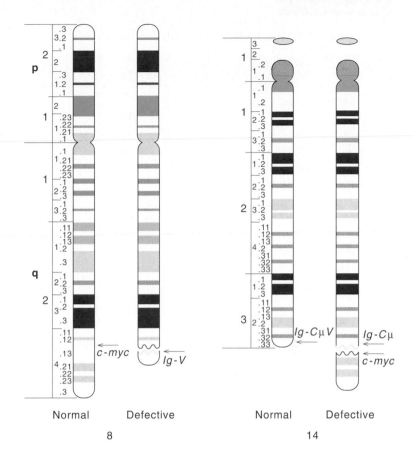

## Figure S4.4

The Philadelphia chromosome arises by a reciprocal translocation between chromosomes 9 and 22. Note that the *abl* protooncogene is moved in the translocation process.

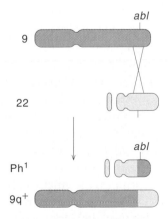

has somehow resulted in their altered expression. For that reason such genes are called <u>protooncogenes</u> in their normal state and <u>oncogenes</u> in their mutated, cancer-causing state. Most protooncogenes, including *c-abl* and *c-myc,* were not known until very similar genes were found as a result of studies of tumor-causing viruses.

## *Oncogenes Are Frequently Associated with Tumor-Causing Viruses*

The first virus linked to tumor production was discovered in 1908 by Ellerman and Bang. Their finding was followed in 1911 by the better known discovery by Peyton Rous of a virus that produces sarcomas in chickens (now known as Rous sarcoma virus, or RSV). Later (1932) Shope showed that a papilloma virus produced cutaneous tumors in rabbits, and Lucke (1934) showed that adenoviruses produced renal adenocarcinoma in the frog. These and other discoveries made in the 1940s and 1950s indicated that exposure to certain viruses could result in tumor production. However, viral oncogenes had not yet been identified, and skeptics

## Table S4.2

Recessive Genetic Lesions in Human Tumors

| Tumor | Chromosomal Locus |
|---|---|
| Retinoblastoma and osteosarcoma | 13 (q14) |
| Wilm's tumor (nephroblastoma) | 11 (p13) |
| Embryonal tumor of the kidneys, muscle, liver, and adrenal glands | 11p— |
| Bladder carcinoma | 11p |
| Acoustic neuroma or meningioma | 22 |

still classified cancer-causing viruses along with other carcinogens.

The development of tissue culture techniques in the 1950s permitted a systematic investigation of the causal link between tumor viruses and cellular transformation in culture. In many ways the properties of cells transformed in culture resembled those of tumor cells *in vivo,* and so the tissue culture system was considered to be a convenient way to explore the oncogenic properties of viruses. It was noted with some tumor viruses that the response of cells of different species was quite different. On cells that were normally permissive for virus replication, no tumors arose. This was hardly surprising, in view of the fact that the virus replication cycle, particularly for the DNA viruses, usually results in cell death. What was surprising was the finding that the same virus on another cell type, in which the virus could not replicate, sometimes resulted in cell transformation. Observations of this sort led Renato Dulbecco to hypothesize that a parallel existed between lysogeny of *Escherichia coli* by λ and transformation of animal cells by oncogenic viruses. In λ, the viral genome can exist either as an independent entity, leading to virus replication and cell death, or as an integrated passive state, leading to a lysogenic cell (see chapter 30). With λ the two states—lytic, or active, and lysogenic, or passive—are attainable in the same cell type. Tumor-causing viruses are capable of existing in two states also. However, in this case the two states, actively replicating and passively integrated, are not normally attainable in the same cell type. Usually two cell types of different species are

required. In one cell type the viral genome invariably exists in the actively replicating form, causing immediate cell death; in another cell type the viral genome can exist only in a passive integrated form. Dulbecco suggested that in the integrated state the viral genome expresses a limited number of gene products that lead to cellular transformation.

So far so good. Dulbecco's suggestion appeared to constitute a reasonable hypothesis for explaining the oncogenicity of certain DNA viruses. But it was also known that certain RNA viruses cause tumors. Indeed, one of the first known cancer-causing viruses, the Rous sarcoma virus, was an RNA virus. To explain the oncogenicity of RNA viruses, Dulbecco proposed that cancer-causing RNA viruses form a DNA copy of their genome that can be integrated into the host cellular genome. This bold proposal—that an RNA genome could be converted into a DNA copy—was confirmed by experiments in both David Baltimore's and Howard Temin's laboratories. Both groups found an enzyme, reverse transcriptase, uniquely associated with RNA tumor-causing viruses that could synthesize DNA from an RNA template.

The finding that many viruses cause cancer only after inserting their genome into the host genome has spurred new efforts in cancer research. More and more attention has been paid to viruses. With the help of genetic manipulations, the cancer-causing genes within the viruses have been identified and their gene products have been studied. The results of these studies are providing a broad platform of understanding about the kinds of biochemical processes that result in the conversion of normal cells to cancer cells.

## *The Role of DNA Viral Genes in Transformation Reflects Their Role in the Permissive Infectious Cycle*

Usually, tumor-causing DNA viruses possess oncogenes that bear no sequence relationships to any host genes. We can probably best understand the role of DNA viral oncogenes in cellular transformation by appreciating the role these genes play in the virus lytic cycle (table S4.3). First of all, the oncogenes we have found in the adenovirus, polyoma virus, and SV40 virus are all expressed early in virus infection. These genes also trigger a transient transformation process in most cells they infect, where a permissive virus replication cycle is not possible. Thus, it seems likely that these genes are primarily regulatory. In SV40 and polyoma the so-called big-T genes are clearly regulatory in productive infection because they are involved in regulating

**Table S4.3**

Properties of DNA Virus Oncogenes

| Viral Gene and/or Gene Product | Function in Lytic Infection | Function in Transformation | Other Properties |
| --- | --- | --- | --- |
| **Adenovirus** | | | |
| *E1A* 9 kd, 26 kd, 32 kd | Required for expression of other early genes | Immortalizes cells; partial transformation | General transcription activator |
| *E1B* 9 kd, 55 kd | Required for lytic infection | Required for full transformation and tumorigenic cells. Either 19 kd or 55 kd suffices | Associates with p53. 55-kd protein found primarily in nucleus; 19-kd protein in endoplasmic reticulum and in nucleus |
| **SV40** | | | |
| Big T 90 kd | Binds to specific sites; necessary for DNA synthesis; inhibition of early RNA synthesis and activation of late RNA synthesis | Immortalizes and transforms | Stimulates host DNA synthesis on microinjection; associates with p53 |
| Small t | None (?) | Required for full transformation on some cell lines | — |
| **Polyoma** | | | |
| Big T 100 kd | Binds to DNA; necessary for DNA synthesis and regulation of early and late RNA synthesis | Immortalizes cells; only amino terminal 40% fragment is required | Associates with p53; ATPase activity; gyrase activity |
| Middle T 56 kd | None (?) | Transforms cells and makes them tumorigenic | Associates with pp60$^{c-src}$ |
| Small t | None (?) | Transformed cells adhere less firmly when small t is present | Found in nucleus |

both viral DNA and RNA synthesis. In adenovirus there is no comparable gene, but the early genes *E1A* and *E1B* may play a similar, albeit indirect role. *E1A* functions as a transcription activator.

Another lead as to how a DNA virus transforms cells is to look at the cellular proteins they influence. There is a striking parallel here between the functions of adenovirus *E1B* and SV40 big T in this regard. These two genes are required for immortalization or transformation, and they both form a tight association with the cellular protein p53, which is considered to be a good candidate for a protooncogene-related protein.* Finally, the polyoma middle-T protein, which is required for transformation by polyoma virus,

forms a strong complex with the cellular protein encoded by the protooncogene *c-src*. It seems likely that these associations between viral oncogenes and host proteins result in changes in regulatory functions, and that if we knew what regulatory functions were being affected we would be on the threshold of understanding the biochemical mechanism(s) involved in transformation. We will have more to say about this after we have considered the retroviruses.

## p53 Is the Most Common Gene Associated with Human Cancers

The p53 protein acts as a negative regulator of cell growth; it was first detected through its association with the SV40 big-T oncoprotein in virus-transformed cells. Viral oncoproteins like SV40 T antigen and adenovirus *E1B* sequester

* In its normal protooncogenic form, p53 inhibits SV40 replication; in its mutated oncogenic state, it does not inhibit SV40 replication, nor does it bind to T antigen.

p53 protein in inactive complexes. Whereas this results in an increase in the steady-state concentration of p53, the p53 is no longer able to reach its normal site of action in the nucleus.

Many mutant *p53* alleles favor growth and transformation in cells that continue to carry intact, wild-type *p53* gene copies; accordingly, such mutant *p53* alleles act in a dominant fashion. The dominant behavior of mutant *p53* could be explained by the normal active form of the protein. In its active form p53 protein assembles into homotetramers and higher order homooligomeric structures. Defective subunits of such an oligomerizing protein (for example, mutant p53 molecules) may participate in forming a multisubunit complex together with wild-type monomers and, in so doing, poison the function of the complex as a whole.

Normal p53 protein binds DNA in a sequence-specific manner and thus most likely regulates gene transcription. Co-transfection experiments show that wild-type p53 activates the expression of genes adjacent to a p53 DNA-binding site. Cells bearing oncogenic forms of p53 have lost this activity.

The best demonstration that the loss of active normal p53 explains the oncogenic behavior of mutant p53 comes from studies in which a null mutation was introduced into the gene by homologous recombination in murine embryonic stem cells. Mice homozygous for the null allele appear to be normal but are prone to the development of a variety of neoplasms by 6 months of age. These observations suggest that a normal *p53* gene is dispensable for embryonic development but that its absence predisposes the animal to neoplastic disease.

Somatic mutation of *p53* has been implicated as a causal event in the formation of a large and ever-increasing number of common tumors, including those involving the hematopoietic organs, bladder, liver, brain, breast, lung, and colon. In view of the fact that many other oncogenes exist, we must ask why *p53* is so often implicated in commonly occurring tumors. This appears to stem from its genetic and biochemical traits. Point mutations create carcinogenic *p53*, and such simple genetic changes occur readily. Unlike *ras*, for example, in which point mutations productive for cancer are limited to specific changes in two or three codons, the cancer-favoring missense mutations of *p53* can occur in at least 30 distinct codons in its reading frame. In addition, these point mutations often create dominant alleles that produce shifts in cell phenotype even without a reduction to homozygosity.

# Retroviral-Associated Oncogenes that Are Involved in Growth Regulation

The first intensively studied oncogene associated with a retrovirus was the RSV *src* gene. It is clear that the *src* gene found in RSV, *v-src,* is structurally quite similar to a normal host gene found in the chicken genome called *c-src.* Subsequent work has shown that at least 30 other animal retroviruses exist that have acquired a cellular oncogene during their evolution. This uncanny association of retroviruses with cancer-causing genes has made retroviruses useful devices for scanning the cellular genome for the presence of protooncogenes (fig. S4.5).

A listing of some of the better characterized retrovirus-associated cellular oncogenes appears in table S4.4. In this table the name of the oncogene is indicated in the leftmost column. Proceeding to the right in the table are indicated: (1) The virus of origin, (2) the viral gene product, (3) the cellular homolog of the viral product; (4) the activity associated with the viral gene product, and (5) the subcellular location of the viral gene product, which in most cases is similar to the subcellular location of the cellular homolog.

## *The* src *Gene Product*

The protein encoded by the *src* gene has a molecular weight of 60 kd. Like many other oncogenic proteins, it is bound to the inner plasma membrane, and it possesses a tyrosine phosphokinase activity; that is, it catalyzes the addition of a phosphate group to the tyrosine hydroxyl groups of various proteins. It is considered highly likely that the kinase activity of the src protein is associated with its transforming activity because its loss in mutant *src* genes leads to loss of transforming activity. However, since many different proteins are phosphorylated by the src kinase, it is not clear which phosphorylations are crucial to the complex physiological response that triggers transformation.

## *The* sis *Gene Product*

Many oncogenes are associated with tumors produced from a restricted class of differentiated cells. Such is the case with the *sis* oncogene, found in the simian sarcoma virus. This oncogene is believed to be closely related to the gene for platelet-derived growth factor, PDGF. PDGF is a small protein, synthesized in platelets, that stimulates the growth and division of target cells that carry a membrane-bound

## Table S4.4

### Some Retroviral Oncogenes and Their Cellular Homologs

| Oncogene | Viral Origin[a] | Viral Gene Product | Cellular Homolog | Activity | Subcellular Location |
|---|---|---|---|---|---|
| *sis* | Simian sarcoma virus | p28$^{sis}$ | PDGF B-chain | PDGF agonist | Cytoplasm |
| *src* | Rous sarcoma virus | p60$^{v\text{-}src}$ | p60$^{c\text{-}src}$ | Tyrosine kinase | Plasma membrane |
| *fps*[b] | Fujinami sarcoma virus | p140$^{gag\text{-}fps}$ | | | |
| *abl* | Abelson murine virus | P120$^{gag\text{-}abl}$ | p150$^{c\text{-}abl}$ | Tyrosine kinase | Plasma membrane |
| *erbB* | Avian crythroblastosis virus | gp72$^{erbB}$ | Truncated EGF receptor | Tyrosine kinase | Plasma membrane |
| *myc* | Avian myelocytomatosis virus MC29 | P110$^{gag\text{-}myc}$ | p58$^{c\text{-}myc}$ | Binds DNA | Nucleus |
| *H-ras* | Harvey murine sarcoma virus | p21$^{v\text{-}Hras}$ | p21$^{c\text{-}H\text{-}ras}$ | Threonine kinase binds GDP or GTP | Plasma membrane |
| *K-ras* | Kirsten murine sarcoma virus | p21$^{v\text{-}Kras}$ | p21$^{v\text{-}Kras}$ | | |
| *N-ras* | See footnote *c*. | | p21$^{c\text{-}N\text{-}ras}$ | | |

[a] Only one example of a virus strain is given for each oncogene.
[b] *fps* and *fes* are homologous genes from chicken and cat, respectively.
[c] The transforming gene product, p21$^{N\text{-}ras}$, identified by transfection experiments.

## Figure S4-5

Hypothesis for how a retrovirus picks up a cellular protooncogene. The RNA–DNA cycle for retroviruses is shown in figure 26.25. The proviral DNA is believed to recombine the cellular DNA in such a way that it picks up a region of the cellular DNA containing a protooncogene. Red segments represent coding regions of C-*onc*.

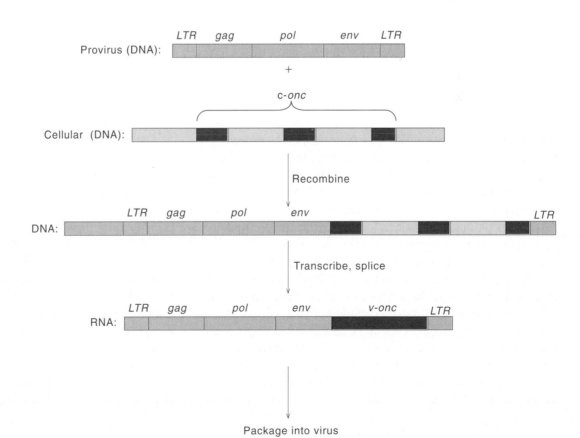

## Figure S4.6

Mechanism of mitogenesis in normal and transformed cells.
(*a*) Schematic representation of the growth-factor-related mitogenic pathway in normal cells. Here, (1) represents the growth factor, (2) the growth factor receptor, and (3) the intracellular messenger system that transmits the mitogenic signal from the receptor to the nucleus. (*b*) Schematic representation of a possible perturbation of the growth-factor-related mitogenic pathway in transformed cells. Here, (1) represents endogenous production of growth factor that may stimulate the cell. The endogenously produced factor may be secreted and interact with growth factor receptors at the cell surface (as shown) or, alternatively, activate the receptor in an intracellular compartment.

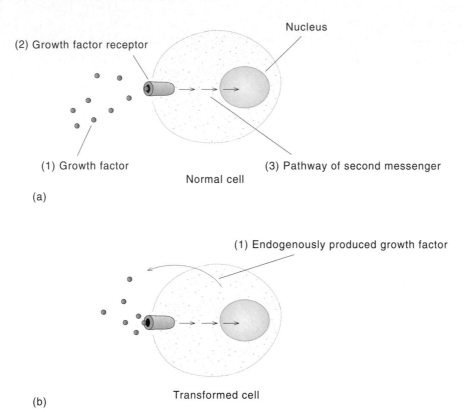

specific receptor for PDGF—in particular, the mesenchymal cells involved in wound healing. Tumors originate when the target cells carrying the PDGF receptor mutate to cells capable of synthesizing their own PDGF. This situation is believed to lead to uncontrolled growth because the factor continuously stimulates cell proliferation in the very cells in which it is synthesized (fig. S4.6).

The PDGF protein contains two polypeptide chains: A and B, with molecular weights of 28,000 and 32,000, respectively. The amino acid sequence of the PDGF protein is very similar to that of the predicted sequence of the transforming p23 sis protein of the SSV virus, indicating a common ancestral origin. Furthermore, it is believed that the SSV virus causes tumors by synthesizing large amounts of the viral p23 sis protein, which stimulates the unregulated proliferation of target cells that carry the PDGF receptor.

## The erbB *Gene Product*

Epidermal growth factor (EGF) is a small mitogenic protein that stimulates the proliferation of cells carrying specific membrane-associated EGF receptors. The EGF receptor has a strong amino acid sequence homology with gp65$^{erbB}$, the transforming protein of avian erythroblastosis virus (AEV). The EGF receptor has tyrosine kinase activity, like the src protein. In addition, the EGF receptor contains an extracellular domain that binds the EGF growth factor.

The *v-erbB* oncogene acts to expand a pool of highly mitotic, undifferentiated erythroid precursor cells, but these are poorly tumorigenic, because they differentiate at high rates into postmitotic, end-stage red cells. Another potential oncogene, *v-erbA,* blocks differentiation of erythroid precursors but creates no tumors because it is unable to provide the mitogenic impetus needed to expand the pool of stem cells. The two genes, *erbA* and *erbB,* carried into erythroid precursors together by avian erythroblastosis virus, act in concert to create an aggressive erythroleukemia; *v-erbB* drives expansion of the pool of undifferentiated precursor cells, whereas *v-erb4* blocks their conversion by the differentiation pathway. The *v-erbA* allele that participates in formation of chicken erythroleukemias is a mutant version of a transcriptional regulatory protein, the chicken thyroid hormone (triiodothyronine) receptor. Function of the wild-type receptor protein is blocked in the presence of *v-erbA*, because the latter occupies critical DNA-binding sites in a way that precludes association by the wild-type receptor protein and inhibits transcription.

## The **ras** *Gene Product*

One of the best understood protooncogenes is *ras*. The *ras* gene found in normal mammalian cells is closely related to the *ras* oncogenes of Harvey and Kirsten murine sarcoma viruses, *H-ras* and *K-ras,* respectively. *Ras* oncogenes have also been isolated with the help of DNA transfection techniques. Activated *ras* genes have been isolated from a wide variety of dissimilar neoplasms, including carcinomas, sarcomas, neuroblastomas, lymphomas, and leukemias. Some of these *ras* genes are associated with the murine sarcoma viruses (e.g., *N-ras* in table S4.4). All members of the *ras* gene family encode closely related proteins of approximately 21 kd, which have been designated p21. Unlike the oncogenes associated with many retroviruses, the level of p21 expression is similar in normal cells and in many different human tumor-cell lines. Nucleotide sequence analysis of the *H-ras* transforming gene isolated from a human bladder carcinoma cell line has shown that the transforming activity of this gene is the consequence of a point mutation altering amino acid 12 of wild-type p21 from glycine to valine (fig. S4.7).

Subcellular fractionation and immunofluorescence of normal cellular and viral ras proteins have indicated that p21 is localized on the inner face of the plasma membrane. Both the normal and the mutant ras proteins bind GTP specifically and strongly, but only the normal protein has GTPase activity.

We understand normal *ras* gene function better than that of many other protooncogenes because the yeast *Saccharomyces cerevisiae* carries *ras* genes, and many researchers have taken advantage of the powerful genetic tools available in yeast to study them. *S. cerevisiae* contains two closely related but distinct genes, *RAS1* and *RAS2*, that

**Figure S4.7**

Interaction between the *ras* protooncogene and GDP. The protein is shown as a green ribbon, interacting with GDP (dark purple, yellow, and orange). The dark purple rectangle represents the guanosine; the dark yellow pentagon, the ribose; and the orange circles, the phosphates of GDP. The domains of the protein interacting with each of these parts of GDP are represented by sleeves on the protein ribbon, color-coded to match the corresponding part of GDP. The P domain of the protein contains glycine 12; this amino acid is in a critical position next to the phosphates of the nucleotide. The N and C termini of the protein are labeled. (From L. Tong et al., Structure of the ras protein, *Science* 245:243, July 21, 1989. Copyright 1989 by the AAAS. Reprinted by permission.)

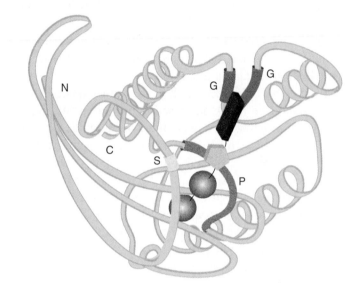

encode proteins that are highly homologous to the mammalian ras proteins. Although neither *RAS1* or *RAS2* by itself is an essential gene, *RAS* function is required for the continued growth and viability of yeast cells. Thus, *ras1⁻*, *ras2⁻* double mutants are nonviable unless they carry a suppressor mutation *bcyL*. The *bcyL* mutation suppresses lethality in adenylate-cyclase-deficient yeast. We will see the significance of this fact presently. Yeast strains that are *ras2⁻* are noticeably depressed in cAMP levels. In *ras1⁻*, *ras2⁻*, *bcyL* triple mutants, the levels of cAMP are virtually undetectable. But cells containing *RAS2 val19,* a *RAS2* allele with a missense mutation analogous to the one that activates the transforming potential of mammalian *ras* genes, have cAMP levels significantly higher than those observed in wild-type cells. Membranes from *ras1⁻*, *ras2⁻*, *bcyL* triple mutants lack the GTP-stimulated adenylate cyclase activity present in membranes from wild-type cells, and membranes from the *RAS2 val19* yeast strain have elevated levels of a GTP-independent adenylate cyclase activity. Mixing membranes from *ras1⁻*, *ras2⁻* yeast with membranes from an adenylate-cyclase-deficient yeast leads to the reconstitution of a GTP-dependent adenylate cyclase. It appears, then, that *RAS* encodes a protein or proteins that regulates activity of membrane-bound adenylate cyclase in a GTP-dependent manner. In the yeast mutant *RAS2 val19*, the regulatory properties of this protein appear to be disrupted so that the adenylate cyclase is overactive and no longer requires GTP for activation. Further experiments have been done to show that the *ras* genes from yeast cells can function in mammalian cells and *vice versa*. Thus, it has been shown that yeast carrying *ras1⁻*, *ras2⁻* mutations remain viable if they carry the mammalian *H-ras* gene. Conversely, a mutant yeast *RAS* gene has been isolated that can bring about tumorigenic transformation in mammalian cells.

From what has been said, you might expect that ras proteins would be involved in activating adenylate cyclase in mammalian cells. However, there is no indication that this is the case. In mammals, other heterotrimeric GTP-binding proteins are involved in the adenylate cyclase pathway (see table 24.5). Apparently, the *ras* system is a useful signaling module that has been adapted to various uses in different organisms; it could be likened to an electronic switching device, that serves different functions depending on the devices it is attached to. In this connection we note that a structural analysis indicates a similarity between p21$^{ras}$ structure and the G-domain of the protein synthesis elongation factor Tu(EF-Tu) from *E. coli*.

## The myc *Gene Product*

The *c-myc* gene, identified originally as the cellular homolog of the transforming determinant carried by avian myelocytomatosis virus, is altered in association with a broad spectrum of neoplasms. Consistent with the observation that altered *c-myc* is associated with tumors of diverse origin, it has been observed that normal *c-myc* is expressed in a variety of tissues. Thus, *c-myc* appears to encode a function associated with a ubiquitous metabolic pathway.

The *c-myc* gene encodes a 49-kd protein that is highly concentrated in the nucleus of the cell. The concentration of c-myc protein normally varies appreciably with the metabolic state of the cell, increasing by more than an order of magnitude in cells just prior to chromosome duplication and cell division.

In most human tumors associated with the *c-myc* gene, the concentration of c-myc protein is greatly amplified. Burkitt's lymphoma, which we discussed earlier, is a notable exception. In this case the c-myc transcript is marginally, and in some cases not at all, increased by comparison

with control lymphoblastoid cell lines. Recall that in Burkitt's lymphoma, reciprocal chromosomal translocations are found that involve a chromosome carrying *c-myc* proto-oncogene (chromosome 8) and one of the chromosome segments (usually chromosome 14) bearing immunoglobulin genes.

In at least one Burkitt's lymphoma cell line, the structure of the amino-acid-coding portion of the translocated *c-myc* gene, and hence its predicted protein product, has not been altered, a fact suggesting that activation of *c-myc* must be mediated via a regulatory disturbance. Normally, c-myc protein synthesis is strongly regulated with respect to the cell cycle and tightly repressed in quiescent cells. The translocated *c-myc* gene appears to be somewhat deregulated with respect to its expression during various phases of the cell cycle.

Members of the *myc* gene family have been implicated in the control of normal cell proliferation as well as in neoplasia. A more direct role for *myc* genes in transformation is indicated by their ability to transform primary rat embryo fibroblasts in association with the *c-H-ras* oncogene.

An important clue to *c-myc* function was the discovery in the conserved carboxy-terminal regions of three structural motifs, the leucine zipper (LZ), helix-loop-helix (HLH) and basic region (B). These motifs were originally defined in a number of other sequence-specific DNA-binding proteins but had not previously been found within a single protein.

The comparatively weak homooligomerization efficiency of the c-myc HLH/LZ suggested that a partner protein(s) might exist that heterooligomerizes with c-myc to form a specific DNA-binding complex. A protein termed max (for myc-associated "X" factor) was shown to interact with c-myc in a manner that required the integrity of the c-myc HLH/LZ motif. Max was also demonstrated to associate with N- and L-myc proteins, but not with the HLH/LZ proteins USF or AP-4, nor with several other HLH or leucine zipper proteins. In DNA-binding assays, c-myc-max specifically recognized a c-myc binding site (CACGTG) in a manner that required both an intact max basic region and the HLH/LZ motifs.

In yeast model systems it has been observed that myc is a transcription activator but only when present in a heterodimer with max. Max appears to be essential for DNA binding. Max dimer can bind to DNA on its own but it does not activate transcription on its own. Mammalian genes that are normally activated by myc are still not known.

It has been reported that the myc amino-terminal and central regions can specifically interact with the retinoblastoma (RB) tumor suppressor protein *in vitro*. This finding prompts the suggestion that RB may directly facilitate myc function, but it is not yet known whether the observed interaction reflects a physiologically relevant association *in vivo*.

## *The* jun *and* fos *Gene Products*

The *v-jun* oncogene was discovered as a 0.93-kb insert in the genome of a replication-defective retrovirus, avian sarcoma virus 17 (ASV17), isolated from chicken sarcoma. ASV17 causes fibrosarcomas in chickens and oncogenic transformation in cultured avian embyronic fibroblasts. The oncogenic potential of ASV17 is due to the presence of the *v-jun* gene, which is derived from the cellular *c-jun* gene. The gene is believed to be oncogenic in the virus because it is expressed there in higher amounts.

A great deal of excitement was generated by the discovery that the jun protein can substitute for the yeast transcription activator GCN4. Comparative analysis of the sequences of the two proteins showed that a strong homology occurs over about one-third of their length in the DNA-binding parts. Both of these proteins bind to the same consensus sequence, TGACTCA.

The oncogene of the FBJ murine osteosarcoma virus (*fos*) codes for a related nuclear protein that participates in transcriptional regulation. In human fibroblasts the fos protein is mostly associated with *c-jun.* The fos–jun complex binds specifically to DNA. Since fos alone does not show specific DNA binding, it is believed that jun is responsible for this affinity. Although jun can form homodimers that bind to DNA, the heterodimers formed between fos and jun show a greater affinity. The heterodimers are also more effective in transcription activation; therefore the heterodimer is probably the functionally relevant state of the jun and fos proteins.

Structural analysis indicates that the fos and jun proteins belong to a class of DNA-binding proteins that share the conserved structural motif known as the leucine zipper (see fig. 31.21). Thus, the dimerization of these two proteins is mediated by hydrophobic interaction between the leucine side chains of two leucine zipper domains.

The jun–fos protein complex interacts with regulatory regions of numerous genes. We have yet to find out which of these target genes are involved in aberrant cellular growth.

## The Transition from Protooncogene to Oncogene

All of the oncogenes thus far discovered are associated with a cellular homolog that is required for normal growth. The transition of the protooncogene to an oncogene is accompanied by abnormal expression of the gene products. In many cases, such as *src, jun,* or *sis,* excessive amounts of the gene product are synthesized. In some other cases, such as *H-ras* or *K-ras,* normal amounts of the oncogenic product are synthesized, but the protein encoded by the oncogene is altered so that it behaves differently. In still other cases such as *myc* the oncogene product is similar to the cellular homolog in amount as well as structure, but the time during the cell cycle when it is produced is altered.

Oncogene products assume specific locations within the cell. Usually they are associated with one of two locations: The plasma membrane or the nucleus. This specificity is consistent with the hypothesis that oncogene products are associated with normal cellular metabolism relating to regulation of cell proliferation; it seems likely that many of the elements regulating cell proliferation would be found at the membrane and nucleus of the cell.

It is not surprising to find oncogenes with varying specificity for producing tumors. Some components of the cell proliferation regulatory apparatus are probably quite general. An oncogene like *myc* is probably associated with one of these and is therefore associated with tumors of widely varying origins. By contrast, an oncogene like *sis* is specifically associated with cells that are designed to be triggered into proliferation by PDGF. Hence, tumors associated with this oncogene are limited to cell types possessing the PDGF membrane receptor.

## *Carcinogenesis Is a Multistep Process*

A wide range of observations indicate that tumorigenesis is a multistep process involving several mutations, each of which results in discrete changes in the cellular metabolism. If this is the case, then we might expect that any particular oncogene would have the capacity for affecting only one step in the overall process. We have seen that a multistep biochemical pathway ordinarily requires a separate enzyme for each step. Enzymes that function in the different steps of a pathway are sometimes said to complement one another. The question is, in carcinogenesis do different oncogenes show a similar complementation? For example, can purely cellular oncogenes such as *N-ras* complement one or the other of the polyoma oncogenes? In fact, middle-T antigen and *N-ras* oncogene produced no new phenotypes when they were cotransfected into rat embryo fibroblasts (REFs), but large-T antigen and *ras* together achieved dramatic results, producing rapidly expanding foci. This study shows that the conversion of a normal cell to a tumor cell can be achieved by the complementary action of two distinct oncogenes; in this case, one is cellular and one is viral.

Similar studies have shown that the *H-ras* and the *myc* oncogenes can cooperate to produce dense foci of morphologically transformed cells from REF cultures (fig. S4.8). Thus, two cellularly derived oncogenes have been shown to complement each other to produce a fully transformed phenotype. Parallel experiments have been done on whole mice to show that the combination of *ras* and *myc* is highly tumorigenic. Experiments of this sort suggest that at a mini-

## Figure S4.8

Rat embryo fibroblasts after several days of growth on plates. Normally these cells stop growing as they approach confluency (top frame). Overgrowth of the monolayer by rounded cells is indicative of cellular transformation. Two transformed foci are seen in cells pretreated with both *ras*- and *myc*-containing DNA (bottom frame). The transformed cells produced similar transformed cells when repleated and when injected into whole animals produce tumors with a much higher frequency than normal cells.

mum, two different oncogene functions are required to convert a normal cell into a tumorigenic one, but it is too early to say that only two are required in general. Nevertheless, these results are very exciting because they are the beginning of genetic complementation assays that should help us to classify oncogenes into different complementary functions.

One further comment on the multistep nature of carcinogenesis: In our discussion here, we focused on some of the early steps in the process that lead to uncontrolled growth but said nothing about those transitions that convert transformed cells into invasive cells. We also concentrated on oncogenes related to cancer-causing viruses because these are most likely to be the first understood in terms of their biochemical function.

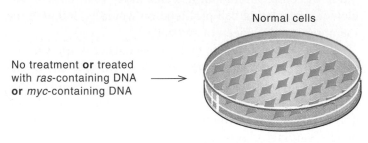

Normal cells

No treatment **or** treated with *ras*-containing DNA **or** *myc*-containing DNA

Flat and organized
Contact inhibited

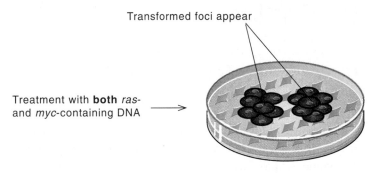

Transformed foci appear

Treatment with **both** *ras*- and *myc*-containing DNA

Cells in transformed foci are piled up and do not show contact inhibition

## Summary

From the time of their discovery, cancer cells have always appeared to be unruly. We have reached that stage in our understanding of cancer cells when we can point to specific aberrations in the genome. In this chapter we took the view that an understanding of cancer can be achieved by analyzing the regulatory pathways involved in growth control because most cancers appear to originate from mutations in specific genes involved in growth regulation. The following points are the highlights of our discussion.

1. Many properties of transformed cells grown in tissue culture resemble cancer cells. The factors that lead to uncontrolled growth *in vivo* can be studied *in vitro* by the effects they have on tissue culture cells.
2. To judge by the frequency of occurrence of cancers in different countries, environmental factors have more influence on the incidence of cancers than inherited genetic factors do. This conclusion is reinforced by studies of migrant populations.
3. Chromosomal translocations are frequently associated with specific types of cancer. This is direct evidence that genetic abnormalities can lead to cancer.
4. A number of tumors arise from recessive mutations in which the mutations appear to be in growth-control genes.
5. Growth-control genes that lead to cancers when they are altered in some way are referred to as proto-oncogenes. They become oncogenes, that is, cancer-causing genes, by mutation. Host protooncogenes are frequently very similar in structure to oncogenes carried by tumor causing viruses.
6. Dulbecco proposed that cancer-causing viruses insert their oncogenes into the host genome. It appears that

cancer-causing viruses are associated with a limited number of DNA viruses and RNA viruses known as retroviruses, which replicate through a DNA intermediate. It is this DNA intermediate that usually gets inserted into the host genome.

7. Abnormal expression accompanies the transition from protooncogene to oncogene. Three types of abnormalities occur: (1) Excessive production of the gene product; (2) altered behavior of the gene product, such as a change in its regulatory properties; and (3) expression at a time during the cell cycle when the gene is not normally expressed.

8. A fully developed cancer appears to arise in steps, each step showing a breakdown in normal regulation.

## Selected Readings

Aaronson, S. A., Growth factors and cancer. *Science* 254:1146–1152, 1991.

Amati, B., S. Dalton, M. W. Brooks, T. D. Littlewood, G. L. Evans, and H. Land, Transcriptional activation by the human c-myc oncoprotein in yeast requires interaction with max. *Nature* 359:423–426, 1992.

Cho, Y., Gorina, S., Jeffrey, P. D., and Pavletich, N. P., Crystal structure of a p53 tumor suppressor-DNA complex: Understanding tumorigenic mutations. *Science* 265:346–358, 1994.

Cobrinik, E., S. F. Dowdy, P. W. Hinds, S. Mittnacht, and T. A. Weinberg, The retinoblastoma protein and the regulation of cell cycling. *Trends Biochem. Sci.* 17:312–315, 1992.

Cooper, G. M., *Elements of Human Cancer.* Boston: Jones and Bartlett Publishers, 1991.

Donehower, L. A., M. Harvey, B. L. Slagle, M. J. McArthur, C. A. Montgomery, J. S. Butel, and A. Bradley, Mice deficient for p53 are developmentally normal but susceptible to spontaneous tumours. *Nature* 356:215–221, 1992. A remarkable new technique for obtaining null mutations demonstrates that embryogenesis is normal in the absence of p53. However, animals develop a variety of neoplasms in the first 6 months when p53 is lacking.

Downward, J., The ras superfamily of small GTP-binding proteins. *Trends Biochem. Sci.* 15:469–472, 1990.

Feig, L. A., The many roads that lead to ras. *Science* 260:757–758, 1993. Ras can be activated by a number of different transduction pathways.

Gallo, R. C., The AIDS virus. *Sci. Am.* 256(1):46–56, 1987.

Halauska, F. G., Y. Tsujimoto, and C. M. Croce, Oncogene activation by chromosome translocation in human malignancy. *Ann Rev. Genet.* 21:321–345, 1987.

Hausen, H., Viruses in human cancers. *Science* 254:1167–1173, 1991.

Henderson, B. E., R. K. Ross, and M. C. Pike, Toward the primary prevention of cancer. *Science* 254:1131–1144, 1991.

Jacks, T., A. Fazeli, E. M. Schmitt, R. T. Bronson, M. A. Goodell, and R. A. Weinberg, Effects of an Rb mutation in the mouse. *Nature* 359:295–300, 1992.

Lane, D. P., p53, guardian of the genome. *Nature* 358:15–16, 1992. Proposes that p53 monitors the integrity of the genome.

Levine, A. J., The p53 protein and its interaction with the oncogene products of the small DNA tumor viruses. *Virology* 177:419–426, 1990.

Linzer, D. I. H., The marriage of oncogenes and anti-oncogenes. *Trends Genet.* 4:245–247, 1988.

Liotta, L. A., Cancer cell invasion and metastasis. *Sci. Am.* 266:54–63, 1992. A most important aspect of carcinogenesis that we did not deal with in our short supplement.

Mack, D. H., J. Vartikar, J. M. Pipas, and L. A. Laimins, Specific repression of TATA-mediated but not initiator-mediated transcription by wild-type p53. *Nature* 363:281–283, 1993. The p53 protein may repress the activity of certain promoters by direct interaction with TATA box-dependent transcription machinery.

Makela, T. P., P. J. Koskinen, I. Vastrik, and K. Alitalo, Alternative forms of max as enhancers or suppressors of myc-ras cotransformation. *Science* 256:373–376, 1992.

Malkin, D., F. P. Li, L. C. Strong, J. F. Fraumeni, C. E. Nelson, D. H. Kim, J. Kassel, M. A. Gryka, F. Z. Bischoff, M. A. Tainsky, and S. H. Friend, Germ line p53 mutations in a familial syndrome of breast cancer, sarcomas, and other neoplasms. *Science* 250:1233–1238, 1990.

Milburn, M. V., L. Tong, A. M. deVos, A. Brunger, Z. Yamaizumi, S. Nishimura, and S. H. Kim, Molecular switch for signal transduction: Structural differences between active and inactive forms of protooncogenic ras proteins. *Science* 247:939–945, 1990.

Paparassiliou, A. G., M. Trier, C. Chavrier, and D. Bohmann, Targeted degradation of *c-fos,* but not *v-fos,* by a phosphorylation-dependent signal on *c-jun. Science* 258:1941–1949, 1992.

Solomon, E., J. Borrow, and A. D. Goddard, Chromosome aberrations and cancer. *Science* 254:1153–1160, 1991.

Varmus, H. E., Reverse transcription. *Sci. Am.* 257(3):56–64, 1987.

Weinberg, R. A., Finding the anti-oncogene. *Sci. Am.* 259(3):44–51, 1988.

Weinberg, R. A., The retinoblastoma gene and cell growth control. *Trends Biochem. Sci.* 15:199–202, 1990.

Weinberg, R. A., Tumor suppressor genes. *Science* 254:1138–1146, 1991. Provides an update on the mechanisms of action of a number of better known oncogenes.

Willinghofer, A., and E. F. Pai, The structure of ras protein: A model for a universal molecular switch. *Trends Biochem. Sci.* 16:382–387, 1991.

# Appendices

# Comparative Sizes of Biomolecules, Viruses and Cells

Frame 1

## Water, Amino Acids, DNA and Protein Structure

Starting at the far left, we see a water molecule, two common amino acids, alanine and tryptophan, a segment of a DNA double helix, a segment of a protein single helix, and the folded polypeptide chain of the enzyme copper, zinc superoxide dismutase or SOD. With respect to the relative sizes of some of these molecules and structures, the water molecule is roughly half a nanometer (nm) across, the DNA and protein helices are about 2 nm and 1 nm in diameter, respectively, and the SOD, a small, globular protein of about 150 amino acids, is about 6 nm in width. SOD catalyzes the breakdown of harmful, negatively charged oxygen radicals, thereby protecting people against neurodegenerative diseases such as Lou Gehrig's disease.

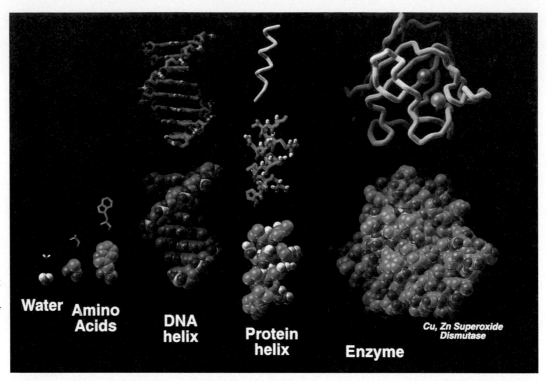

Frame 2

## Proteins and Viruses

At the far left, we can see the nucleic acid and protein structures shown in frame 1. In addition, we show a much larger protein, the immunoglobulin G antibody molecule. Four separate polypeptide chains join to make up an antibody molecule: two heavy chains (blue) of about 400 amino acids and two light chains (purple) of about 200 amino acids. The antibody is about 16 nm in width. Finally, at the far right, we show the core particle from a small plant virus, the reovirus. Only the icosahedral protein coat of the virus can be seen. The reovirus particle is about 60 nm across. The nucleic acids of the virus are sequestered inside the virus core. The reovirus family is unusual in that its nucleic acids are all double-stranded RNA molecules.

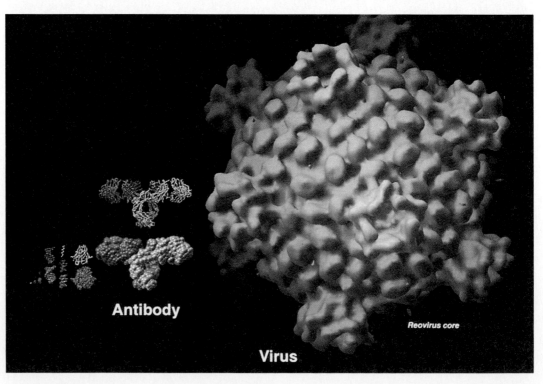

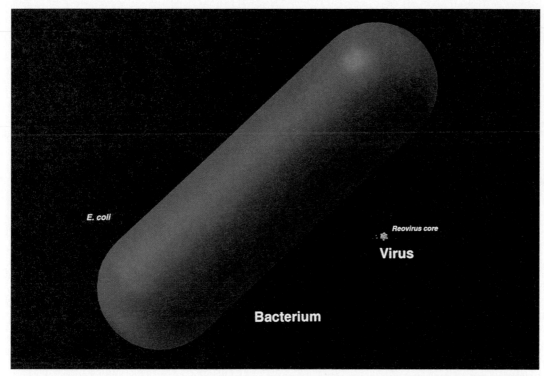

**A Bacterial Cell**

The common bacterium *E. coli,* shown here in a highly stylized form, is about 2 μm long. The vertebrate gut contains enormous numbers of *E. coli* cells which aid in digestive processes. To the right of the bacterial cell, the same reovirus core particle shown in frame 2 is displayed for size comparison.

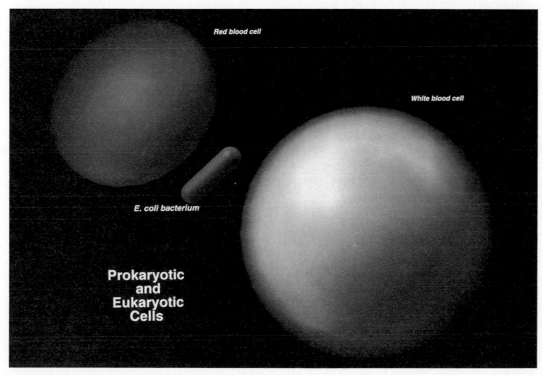

**Two Human Cells**

Two human blood cells, a white cell about 10 μm across and a red cell about 7 μm across, are shown here in highly stylized representations. The *E. coli* cell from frame 3 is shown between the two human blood cells for size comparison. (Frames 1–4: SOD enzyme by Parge, Hallewell, Getzoff, and Tainer. Antibody by Silverton, Navia, and Davies. Reovirus by Dryden and Yeager. Graphics by Michael Pique using AVS software. Concept by Arthur Olson. Visualization advice by David Goodsell. Images copyright/cW/1994 by The Scripps Research Institute.)

# Common Abbreviations in Biochemistry

| | | | |
|---|---|---|---|
| A | adenine | FADH$_2$ | flavin adenine dinucleotide (reduced form) |
| ACh | acetylcholine | FBP | fructose-1,6-bisphosphate |
| ACTH | adrenocorticotropic hormone | fMet | N-formylmethionine |
| ADP | adenosine-5′-diphosphate | FMN | flavin mononucleotide (oxidized form) |
| AIDS | acquired immune deficiency syndrome | FMNH$_2$ | flavin mononucleotide (reduced form) |
| Ala (A) | alanine | F1P | fructose-1-phosphate |
| AMP | adenosine-5′-monophosphate | F6P | fructose-6-phosphate |
| Asn (N) | asparagine | G | guanine |
| Asp (D) | aspartic acid | G protein | guanine-nucleotide binding protein |
| ATP | adenosine-5′-triphosphate | GDP | guanosine-5′-diphosphate |
| BChl | bacteriochlorophyll | Gly (G) | glycine |
| bp | base pair | GMP | guanosine-5′-monophosphate |
| BPG | D-2,3-bisphosphoglycerate | Gln (Q) | glutamine |
| C | cytosine | Glu (E) | glutamic acid |
| Cys (C) | cysteine | GSH | glutathione |
| CAP | catabolite gene activator protein | GSSG | glutathionine disulfide |
| CDP | cytidine-5′-diphosphate | GTP | guanosine-5′-triphosphate |
| CMP | cytidine-5′-monophosphate | Hb | hemoglobin |
| CTP | cytidine-5′-triphosphate | HDL | high-density lipoprotein |
| CoA or CoASH | coenzyme A | HETPP | hydroxyethylthiamine pyrophosphate |
| CoQ | coenzyme Q (ubiquinone) | HGPRT | hypoxanthine-guanosine phosphoribosyl transferase |
| cAMP | adenosine 3′,5′-cyclic monophosphate | | |
| cGMP | guanosine 3′,5′-cyclic monophosphate | His (H) | histidine |
| cyt | cytochrome | HIV | human immunodeficiency virus |
| d | 2′-deoxy- | HMG-CoA | β-hydroxy-β-methylglutaryl-CoA |
| DHAP | dihydroxyacetone phosphate | HPLC | high-performance liquid chromatography |
| DHF | dihydrofolate | IDL | intermediate-density lipoprotein |
| DHFR | dihydrofolate reductase | IF | initiation factor |
| DMS | dimethyl sulfate | IgG | immunoglobulin G |
| DNA | deoxyribonucleic acid | Ile (I) | isoleucine |
| cDNA | complementary DNA | IMP | inosine-5′-monophosphate |
| DNase | deoxyribonuclease | IP$_1$ | inositol-1-phosphate |
| DNP | 2,4-dinitrophenol | IP$_3$ | inositol-1,4,5-trisphosphate |
| ER | endoplasmic reticulum | IPTG | isopropylthiogalactoside |
| FAD | flavin adenine dinucleotide (oxidized form) | $K_m$ | Michaelis constant |

| | | | | |
|---|---|---|---|---|
| kb | kilobase pair | | RFLP | restriction-fragment length polymorphism |
| kDa | kilodaltons | | RNA | ribonucleic acid |
| LDL | low-density lipoprotein | | hnRNA | heterogeneous nuclear RNA |
| Leu (L) | leucine | | mRNA | messenger RNA |
| Lys (K) | lysine | | rRNA | ribosomal RNA |
| Man | mannose | | snRNA | small nuclear RNA |
| MHC | major histocompatibility complex | | tRNA | transfer RNA |
| Met (M) | methionine | | snRNP | small ribonucleoprotein |
| NAD$^+$ | nicotinamide-adenine dinucleotide (oxidized form) | | RNase | ribonuclease |
| | | | Ru1,5P | ribulose-1,5-bisphosphate |
| NADH | nicotinamide-adenine dinucleotide (reduced form) | | Ru5P | ribulose-5-phosphate |
| | | | R5P | ribose-5'-phosphate |
| NADP$^+$ | nicotinamide-adenine dinucleotide phosphate (oxidized form) | | RSV | Rous sarcoma virus |
| | | | s | Svedberg constant |
| NADPH | nicotinamide-adenine dinucleotide phosphate (reduced form) | | SAM | S-adenosylmethionine |
| | | | SDS | sodium dodecyl sulfate |
| NDP | nucleoside-5'-diphosphate | | Ser (S) | serine |
| NAM | N-acetylmuramic acid | | S7P | sedoheptulose-7-phosphate |
| NMR | nuclear magnetic resonance | | SRP | signal recognition particle |
| NTP | nucleoside-5'-triphosphate | | T | thymine |
| Phe (F) | phenylalanine | | THF | tetrahydrofolate |
| P$_i$ | inorganic orthophosphate | | Thr (T) | threonine |
| PEP | phosphoenolpyruvate | | TLC | thin-layer chromatography |
| PFK | phosphofructokinase | | TMV | tobacco mosaic virus |
| PG | prostaglandin | | TPP | thiamine pyrophosphate |
| 2PG | 2-phosphoglycerate | | Trp (W) | tryptophan |
| 3PG | 3-phosphoglycerate | | TTP | thymidine-5'-triphosphate |
| PIP$_2$ | phosphatidylinositol-4,5-bisphosphate | | Tyr (Y) | tyrosine |
| PK | pyruvate kinase | | U | uracil |
| PLP | pyridoxal-5-phosphate | | UDP | uridine-5'-diphosphate |
| PP$_i$ | inorganic pyrophosphate | | UDPG | UDP-glucose |
| Pro (P) | proline | | UMP | uridine-5'-monophosphate |
| PRPP | phosphoribosylpyrophosphate | | UQ | ubiquinone |
| PS | photosystem | | Val (V) | valine |
| Q | ubiquinone or plastoquinone | | VLDL | very-low-density lipoprotein |
| QH$_2$ | ubiquinol or plastoquinol | | XMP | xanthosine-5'-monophosphate |
| RER | rough endoplasmic reticulum | | Xu5P | xylulose-5'-phosphate |
| RF | release factor or replicative form | | | |

# Organic Chemistry and Its Relationship to Biochemistry

By definition organic chemistry deals with the chemistry of carbon regardless of its origin. Since biochemistry deals with carbon chemistry only insofar as it concerns living processes, it comprises a distinct subdivision of organic chemistry. In addition to this major difference between organic chemistry and biochemistry, there are two others. The range of conditions used in organic chemistry far exceeds that of biochemistry. Organic chemistry is the study of carbon chemistry in both organic and aqueous solvents, whereas biochemistry is the study of reactions that take place in aqueous solvents. Organic chemistry often involves the study of reaction conditions that are devised by the chemist in the laboratory; biochemistry is concerned exclusively with reactions that occur in living systems. Finally, in organic chemistry the reactions and the reaction conditions often serve no purpose except the ones intended by the chemist. In biochemistry the reactions are functionally related to the needs of the organism.

Regardless of these differences, organic chemistry and biochemistry overlap considerably. As a result, the principles that govern organic chemistry provide a strong foundation for exploring biochemical phenomena. Here we review some of the regions of overlap between the two disciplines. We start with the fundamental properties that affect the chemistry of atoms and molecules. We then consider some of the molecules, their functional groups, their reactions, and the catalysts that accelerate their reactions.

## Atoms Are Composed of Protons, Neutrons and Electrons

All atoms contain a centrally located nucleus composed of a mixture of protons and neutrons (except for the hydrogen nucleus, which contains only a single proton). Electrons surround the nucleus in a series of shells. Electrons in the innermost shell are the hardest ones to remove from the atom, and electrons in the outermost shell are the easiest to remove. The outermost shell is called the valence shell because, in most chemical reactions between atoms, electrons are either added to or removed from this shell.

Electrons rotate around the nucleus at high speeds, and determining their exact location at any given time is impossible. Nevertheless, for simplicity, electrons often are depicted as small spheres occupying discrete orbits (fig. 1). A more realistic depiction, the electron orbital model, indi-

**Figure A.1**

Bohr models of the six most common atoms found in living organisms.

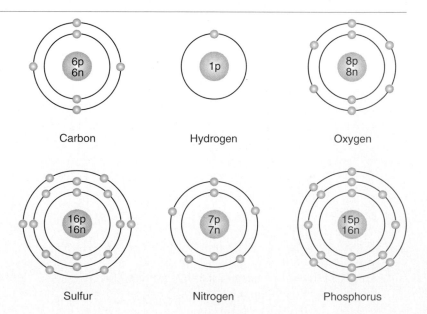

Carbon  Hydrogen  Oxygen

Sulfur  Nitrogen  Phosphorus

## Figure A.2

Electron orbitals. Model depicting the volume of space in which an electron is likely to be found 90% of the time. (*a*) The first energy level consists of one spherical orbital containing up to two electrons. The second energy level has four orbitals, each describing the distribution of up to two electrons. One of the orbitals of the second energy level is spherical; the other three are dumbbell-shaped and arranged perpendicular to one another, as the axis lines indicate (*b*). The nucleus is at the center, where the axes intersect.

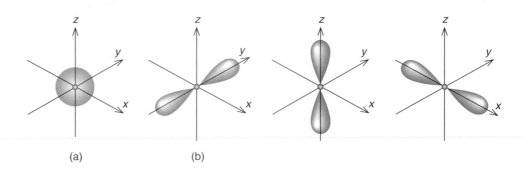

(a)                              (b)

cates the volume of space in which a particular electron is likely to be found most of the time (fig. 2). The shell closest to the nucleus consists of a single orbital containing one or two electrons. Both the second and third shells consist of four orbitals, each containing one or two electrons.

Subatomic particles from which all atoms are composed are distinguished by mass and electrical charge. A proton has a mass of one unit and a positive electrical charge of one unit. A neutron also has a mass of one unit, but it has no net charge. An electron has a mass only about 1/1800 that of a nuclear particle, but it has a negative charge equivalent in magnitude to the positive charge of the proton. The net charge and mass of an atom are the sums of the charges and masses of its constituent particles. In calculating mass we can usually ignore the weight of the electrons. Thus an atom's weight approximately equals the total number of protons and neutrons. Another term, atomic number, equals the number of protons in an atom's nucleus.

### Atoms Combine to Form Molecules

Except for the inert gases, atoms tend to interact with other atoms to form molecules. Hydrogen, oxygen, and nitrogen each readily form simple diatomic molecules. Invariably, molecules have properties that are quite different from those of the constituent elements. For example, a molecule of sodium chloride contains one atom of sodium (Na) and one atom of chlorine (Cl). Sodium is a highly reactive silvery metal, whereas chlorine is a corrosive yellow gas. When equal numbers of Na and Cl atoms interact, vigorous reaction occurs and white crystalline solid sodium chloride is formed.

Molecules are described by writing the symbols of the constituent elements and indicating the numbers of atoms of each element in the molecule as subscripts. For example, the sugar molecule glucose is represented as $C_6H_{12}O_6$, which indicates that it contains 6 atoms of carbon, 12 atoms of hydrogen, and 6 atoms of oxygen.

Atoms react with one another by gaining, losing, or sharing electrons to produce molecules. The type of chemi-cal bond that forms between atoms depends on the number of electrons the atoms have in their outermost valence shells.

### Ionic Bonds Are Formed Between Oppositely Charged Atoms (Ions)

Atoms of the elements prevalent in living things (carbon, nitrogen, oxygen, phosphorus, and sulfur) are most stable chemically when they have eight electrons in their valence shells. This chemical tendency to fill a valence shell is called the octet rule. One way that this is done is for an atom with one, two, or three electrons in this shell to lose them to an atom with, correspondingly, seven, six, or five electrons in the valence shell—a kind of chemical give-and-take. A sodium atom, for example, has a single electron in its outermost shell. As a result, sodium has a strong tendency to lose this single electron because then it will have 8 electrons in its outermost shell. By contrast chlorine has a strong tendency to gain a single electron to fill its valence shell, which contains only seven electrons.

The attraction between oppositely charged ions results in an ionic bond, such as the one that holds NaCl together. The oppositely charged ions $Na^+$ and $Cl^-$, attract each other in such an ordered manner that a crystal results (fig. 3).

### Covalent Bonds Are Formed Between Atoms That Share Electron Pairs

Atoms that have three, four, or five electrons in their valence shells are more likely to share electrons in a covalent bond than to swap them in the electronic give-and-take of an ionic bond. In both organic chemistry and biochemistry covalent bonds are much more common than ionic bonds. Carbon, with four electrons in its outer shell, can attain the stable eight-electron configuration in its outer shell by sharing electrons with four hydrogen atoms, each of which has one electron in its only shell. The resulting compound containing four single covalent bonds is methane, $CH_4$ (fig. 4).

Two or three electron pairs can also be shared in covalent bonds called double and triple bonds, respectively. A single atom of carbon forms a double bond with the oxygen

## Figure A.3

An ionically bonded molecule (NaCl). (*a*) A sodium atom (Na) can donate the one electron in its valence shell to a chlorine atom (Cl), which has seven electrons in its outermost shell. The resulting ions (Na$^+$ and Cl$^-$) bond to form the compound sodium chloride (NaCl). The octet rule has been satisfied. (*b*) The ions that constitute NaCl form a regular crystalline structure in the solid state.

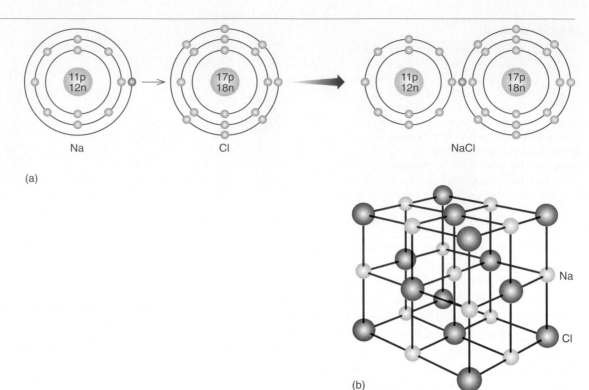

(a)

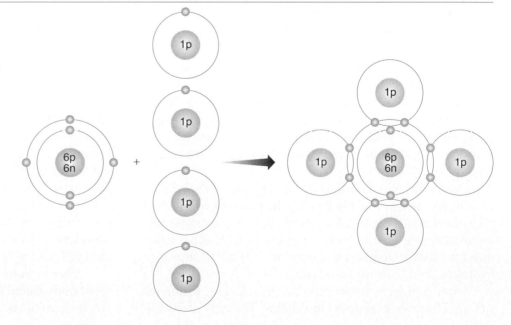

(b)

in formaldehyde and a triple bond with the nitrogen in hydrogen cyanide (fig. 5).

## Weak Forces Lead to Intramolecular or Intermolecular Interactions

The strong interactions that result in either ionic or covalent bonds determine the primary structure of molecules. Molecules often interact, not by forming more bonds of this type, but rather by forming weaker bonds by polar and apolar interactions. Polar interactions result in polar molecules when the electrons in a covalent bond are not equally shared. In a water molecule (H$_2$O), for example, the single oxygen atom has a greater affinity for the shared electrons than do the two hydrogen atoms. Thus most of the time the shared electrons are closer to the oxygen atom than to the hydrogen atoms. Because of this unequal sharing, the hy-

## Figure A.4

A covalently bonded molecule. By sharing electrons, one carbon and four hydrogen atoms complete their outermost shells to form methane. In carbon, the outermost shell contains four electrons whereas in hydrogen the outermost shell contains only one electron.

## Figure A.5

Covalent linkages formed by carbon with hydrogen, oxygen and nitrogen. Each shared electron pair is represented by two dots or a single straight line.

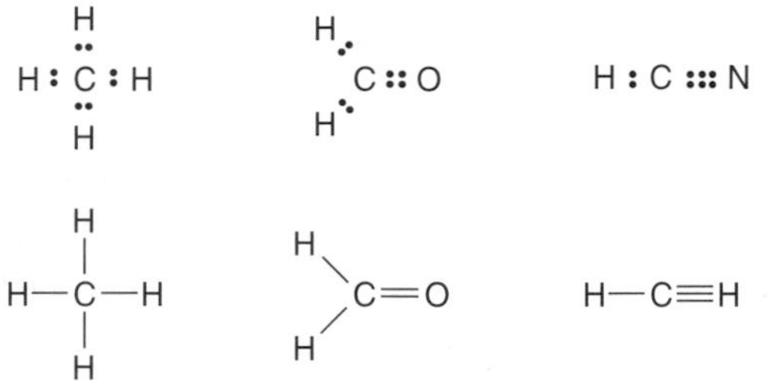

## Figure A.6

Two commonly occurring fatty acids (a) and comparable hydrocarbons (b).

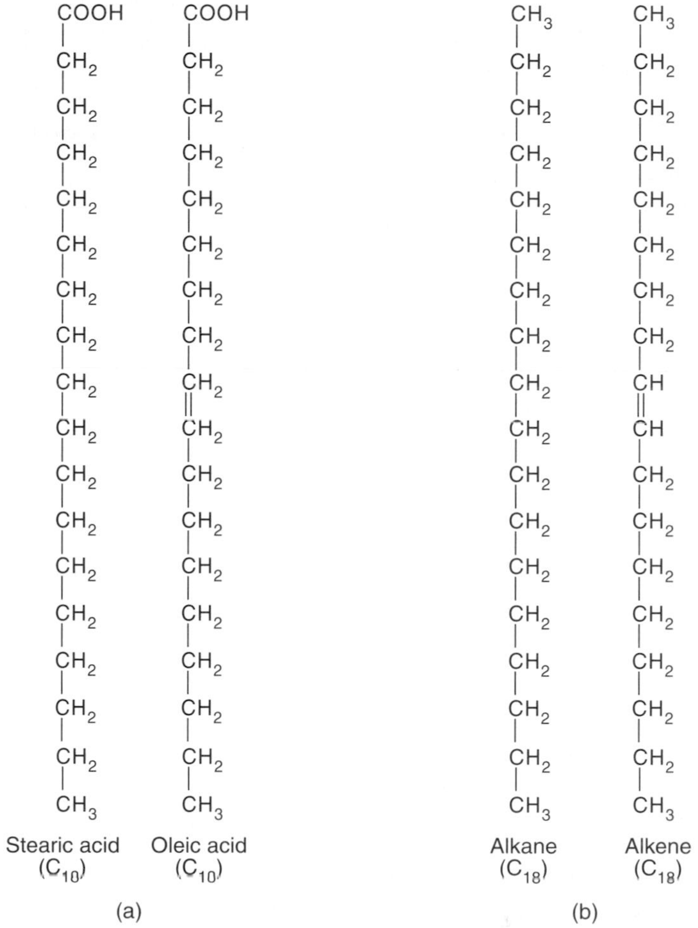

| COOH | COOH | | CH$_3$ | CH$_3$ |
|---|---|---|---|---|
| CH$_2$ | CH$_2$ | | CH$_2$ | CH$_2$ |
| CH$_2$ | CH$_2$ | | CH$_2$ | CH$_2$ |
| CH$_2$ | CH$_2$ | | CH$_2$ | CH$_2$ |
| CH$_2$ | CH$_2$ | | CH$_2$ | CH$_2$ |
| CH$_2$ | CH$_2$ | | CH$_2$ | CH$_2$ |
| CH$_2$ | CH$_2$ | | CH$_2$ | CH$_2$ |
| CH$_2$ | CH$_2$ | | CH$_2$ | CH$_2$ |
| CH$_2$ | CH$_2$ | | CH$_2$ | CH |
| CH$_2$ | CH$_2$ (double bond) | | CH$_2$ | CH (double bond) |
| CH$_2$ | CH$_2$ | | CH$_2$ | CH$_2$ |
| CH$_2$ | CH$_2$ | | CH$_2$ | CH$_2$ |
| CH$_2$ | CH$_2$ | | CH$_2$ | CH$_2$ |
| CH$_2$ | CH$_2$ | | CH$_2$ | CH$_2$ |
| CH$_2$ | CH$_2$ | | CH$_2$ | CH$_2$ |
| CH$_2$ | CH$_2$ | | CH$_2$ | CH$_2$ |
| CH$_2$ | CH$_2$ | | CH$_2$ | CH$_2$ |
| CH$_3$ | CH$_3$ | | CH$_3$ | CH$_3$ |
| Stearic acid (C$_{10}$) | Oleic acid (C$_{10}$) | | Alkane (C$_{18}$) | Alkene (C$_{18}$) |

(a)                         (b)

drogen atoms have a partial positive charge and the oxygen atoms have a partial negative charge. This polarity within individual water molecules results in electrostatic interactions between nonbonded H's and O's in liquid water. Apolar molecules are attracted to each other by so-called van der Waals forces. Whereas polar and apolar bonds are much weaker than ionic bonds and covalent bonds, they frequently outnumber the stronger bonds in the liquid state when molecules are nearly close packed. As a result, these weak forces exert a strong influence on the structure of water and the conformation of polymers in an aqueous solvent. Since water itself is polar, a polymer in water tends to fold in a way that exposes its polar groups to the aqueous solvent while burying its apolar groups so that they can interact with each other. In an apolar solvent the exact opposite occurs. The apolar groups on the polymer are exposed for interaction with the apolar organic solvent, whereas the polar groups are buried within the polymer structure so that they can interact with each other.

## Four Types of Organic Compounds Are Most Commonly Found in Biochemical Systems

Chemically, the simplest organic compounds found in biochemical systems are the fatty acids (fig. 6). For the most part fatty acids consist of long zig-zag chains with carbon backbones and hydrogen substituents. At one end they usually contain a carboxyl group. In organic chemistry the same compound without the carboxyl group is called an alkane. If two hydrogen atoms are removed from adjacent carbons in a fatty acid, the carbons form a double bond with one another. The comparable organic compound without the carboxyl group is called an alkene. Alkanes and alkenes are highly insoluble in aqueous solvents because of their uniformly apolar character. Fatty acids are partially soluble in aqueous media because of their polar carboxyl group. The carboxyl group also is an important functional group that participates in chemical reactions.

Sugars also possess a carbon backbone but they differ in major respects from fatty acids in that one of the hydrogens attached to each carbon usually is replaced by a hydroxyl group. This change has a dramatic effect on the solubility properties of sugars, making them highly soluble in aqueous solvents. Most sugars are much shorter than fatty acids because of the tendency of sugars to form five- or six-membered rings (fig. 7).

Amino acids are more varied in composition than either fatty acids or sugars (fig. 8). Most amino acids important to biochemistry contain a central carbon with four different substituents: a hydrogen, an amino group, a carboxyl group, and an R group that varies for different types of amino acids. These R groups may be positively charged, contain sulfur, be negatively charged, or be neutral. Neutral R groups are further divided into polar and apolar types. The water solubility of amino acids varies by R group.

At the highest level of complexity in biochemical systems are the nucleotides (fig. 9). These molecules contain three components: a five-carbon sugar to which are attached

## Figure A.7

Most sugars contain only 5 or 6 carbons. In aqueous solution, the circular hemiacetal form is strongly favored over the linear form.

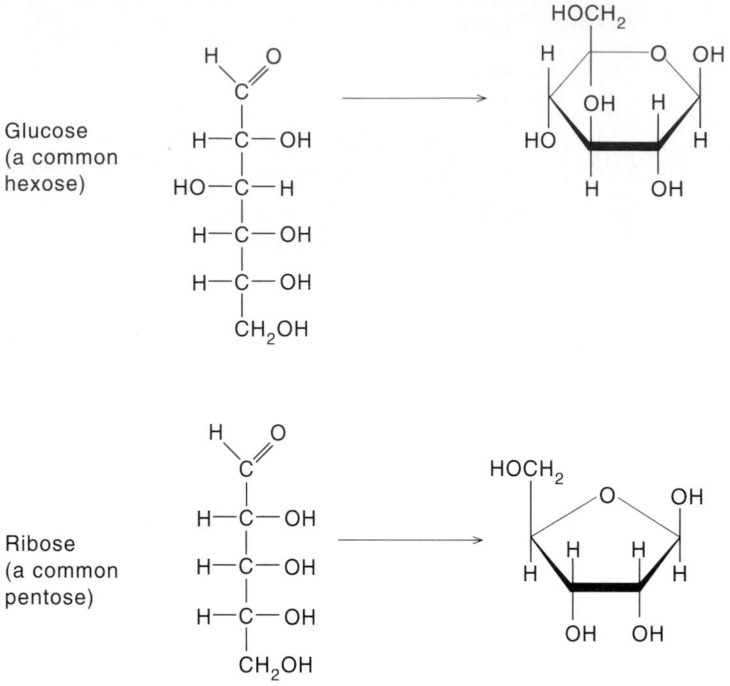

Glucose (a common hexose)

Ribose (a common pentose)

a phosphate group on one side and a heterocyclic base on the other side. Because of the phosphate group and the hydroxyl groups on the sugar, nucleotides tend to be highly soluble in water and poorly soluble in apolar organic solvents. Whereas twenty different types of amino acid are commonly found in biochemical systems, only five types of heterocyclic nitrogen bases are commonly found in nucleotides.

Each of the molecules described in this section represents a component of a considerably larger structure in living cells. Fatty acids are major components of lipids and membranes. Sugars are major components of polysaccharides and cell walls. Amino acids are the substance from which proteins are synthesized. Nucleotides are the building blocks of nucleic acids and chromosomes.

### Most Molecules Found in Cells Are Chiral

When you look into a mirror, you see a reflection that is very similar in appearance to yourself—the main difference being that your right and left sides are switched. No matter which way you turn, you cannot duplicate your mirror image. The same is true of most amino acids, which have four different substituents attached to their $\alpha$ carbon atoms. Amino acids exist in two chiral forms just like you and your mirror image. We can take this analogy a step further. Just as you know that you will never see anybody who looks just like your mirror image, so in biomolecules only one of two chiral forms is usually found. This characteristic is true of most of the biomolecules found in cells.

### Functional Groups Are Important in Both Organic Chemistry and Biochemistry

Biomolecules contain a limited number of functional groups that are the reactive centers of the molecules (table 1). Biochemical reactions involving these functional groups are closely related to reactions studied in organic chemistry. Here we describe some of the best known functional groups and the reactions in which they participate.

Alcohols, which contain a hydroxyl functional group, can undergo dehydration reactions with either carboxylic or phosphoric acid to form esters. The double arrows used in this and subsequent equations indicate that the reaction can go in either direction.

## Figure A.8

The structure of $\alpha$-amino acids found in proteins. A central carbon atom is linked to four different substituents (a). The R group substituents are of 20 different types.

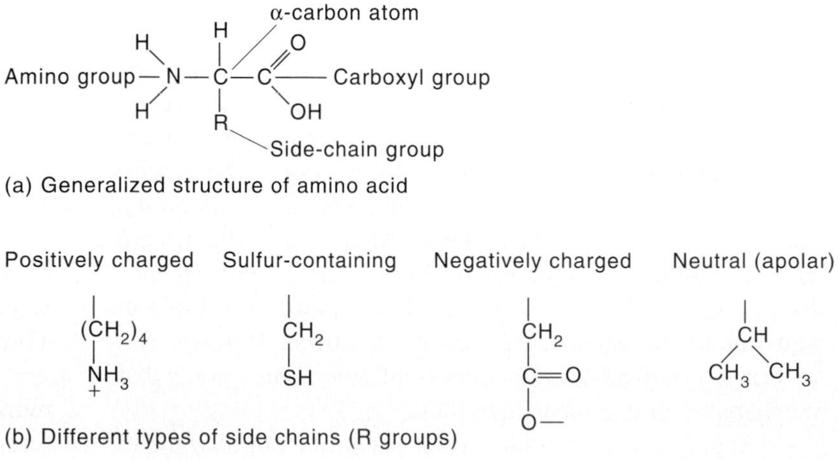

(a) Generalized structure of amino acid

(b) Different types of side chains (R groups)

## Figure A.9

The nucleotide is a three component structure (*a*). There are five different bases commonly found in nucleotides (*b*).

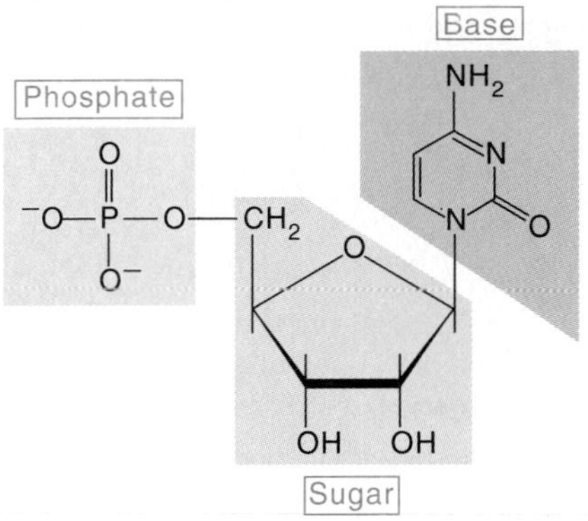

(a) Generalized structure of a nucleotide

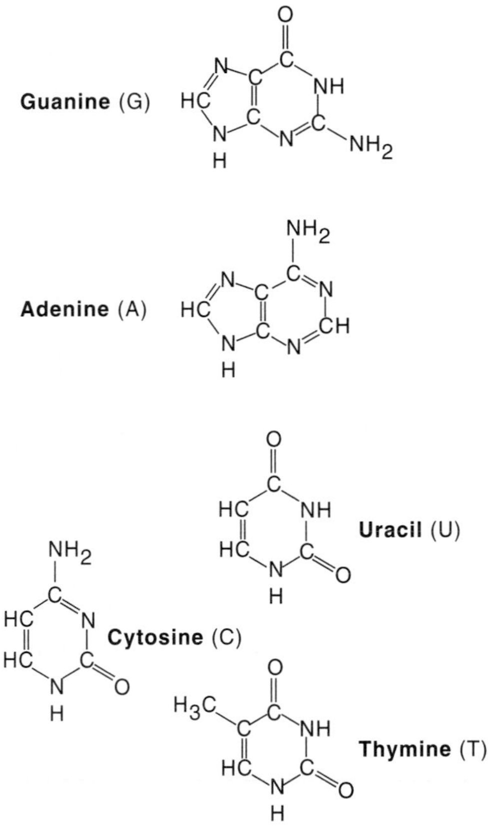

**Guanine** (G)

**Adenine** (A)

**Uracil** (U)

**Cytosine** (C)

**Thymine** (T)

(b) Different bases found in nucleotides

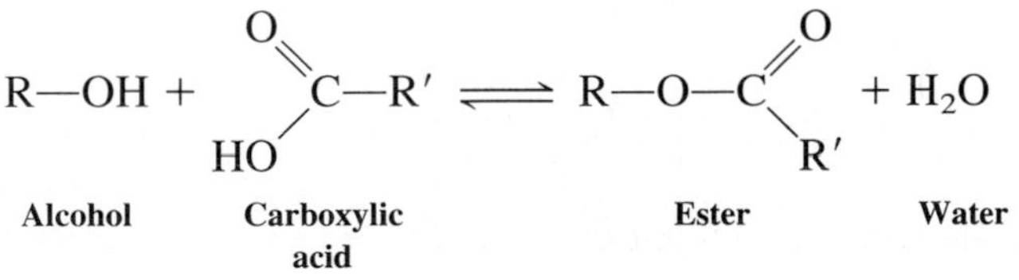

$$R\text{—OH} + \underset{HO}{\overset{O}{\underset{\|}{C}}}\text{—}R' \rightleftharpoons R\text{—O—}\overset{O}{\underset{\underset{R'}{\|}}{C}} + H_2O \quad (1)$$

Alcohol    Carboxylic    Ester    Water
         acid

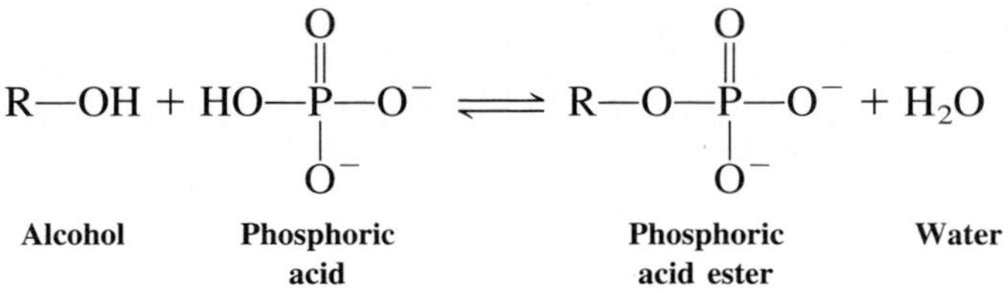

$$R\text{—OH} + HO\text{—}\overset{\overset{O}{\|}}{\underset{\underset{O^-}{|}}{P}}\text{—}O^- \rightleftharpoons R\text{—O—}\overset{\overset{O}{\|}}{\underset{\underset{O^-}{|}}{P}}\text{—}O^- + H_2O \quad (2)$$

Alcohol    Phosphoric    Phosphoric    Water
       acid    acid ester

Thiols contain sulfhydryl groups (—SH) and can substitute for alcohols in some reactions, leading to the formation of thiol esters.

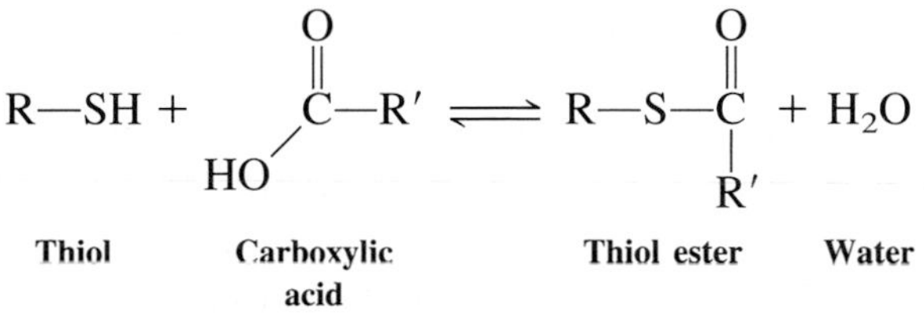

$$R\text{—SH} + \underset{HO}{\overset{O}{\underset{\|}{C}}}\text{—}R' \rightleftharpoons R\text{—S—}\overset{O}{\underset{\underset{R'}{\|}}{C}} + H_2O \quad (3)$$

Thiol    Carboxylic    Thiol ester    Water
    acid

Two alcohols can react with each other to form an ether.

$$R\text{—OH} + HO\text{—}R' \longrightarrow R\text{—O—}R' + H_2O \quad (4)$$

Alcohol$_1$    Alcohol$_2$    Ether

Alcohols may undergo a dehydrogenation reaction to form a carbonyl derivative (aldehyde or ketone).

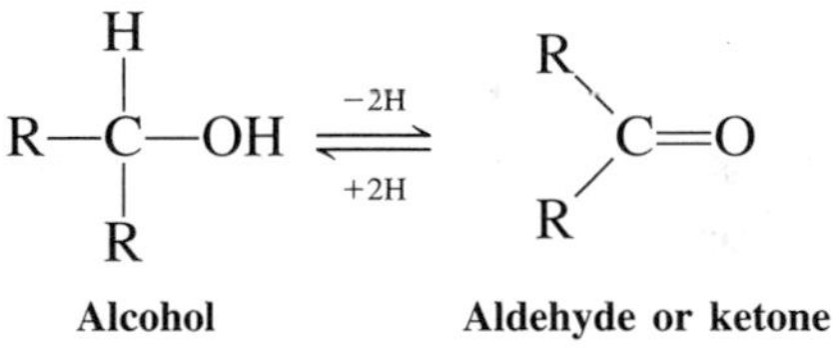

$$R\text{—}\overset{\overset{H}{|}}{\underset{\underset{R}{|}}{C}}\text{—OH} \underset{+2H}{\overset{-2H}{\rightleftharpoons}} \underset{R}{\overset{R}{>}}C\text{=}O \quad (5)$$

Alcohol    Aldehyde or ketone

Amines undergo reactions with carboxylic acids comparable to the formation of esters from alcohols. The product is known as an amide.

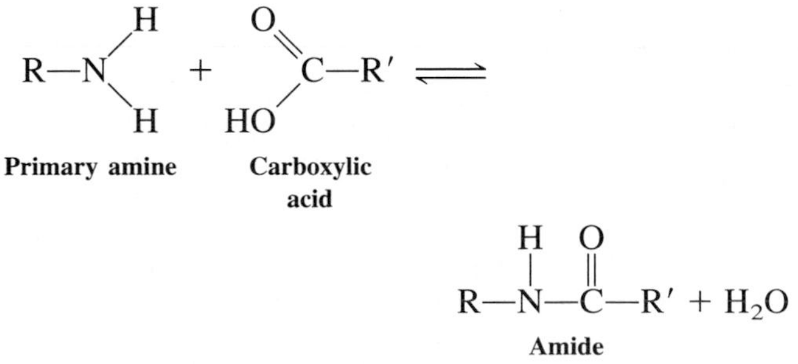

$$R\text{—}\overset{\overset{H}{|}}{\underset{\underset{H}{|}}{N}} + \underset{HO}{\overset{O}{\underset{\|}{C}}}\text{—}R' \rightleftharpoons$$

Primary amine    Carboxylic
       acid

$$R\text{—}\overset{\overset{H}{|}}{N}\text{—}\overset{\overset{O}{\|}}{C}\text{—}R' + H_2O \quad (6)$$

Amide

Amines may undergo dehydrogenation reactions leading to the formation of imines, which are frequently unstable in water and hydrolyze to ketones, or, in cases where one of the R groups is an H, to aldehydes.

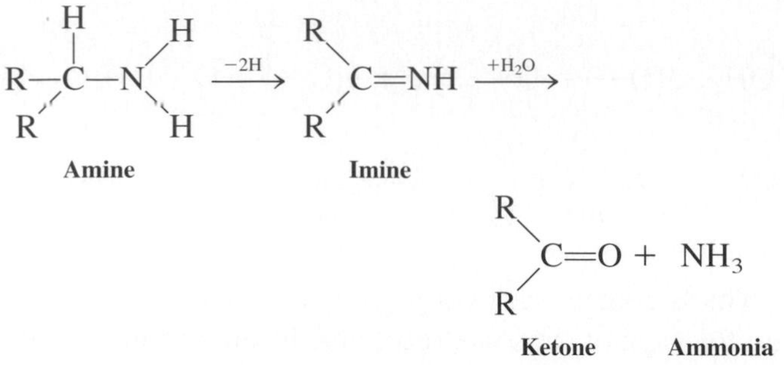

$$R-\underset{\underset{R}{|}}{\overset{\overset{H}{|}}{C}}-\underset{\underset{H}{|}}{\overset{\overset{H}{|}}{N}} \xrightarrow{-2H} \underset{R}{\overset{R}{>}}C=NH \xrightarrow{+H_2O}$$

**Amine**          **Imine**

$$\underset{R}{\overset{R}{>}}C=O + NH_3 \quad (7)$$

**Ketone    Ammonia**

Aldehydes and ketones both may be reduced to alcohols by hydrogenation (see the alcohol dehydrogenation reaction, equation 5). Aldehydes may react with either water or alcohol to form aldehyde hydrates or hemiacetals, respectively (also see figure 7 for intramolecular hemiacetals formed by sugars). Reaction of an aldehyde with two molecules of alcohol leads to acetal formation.

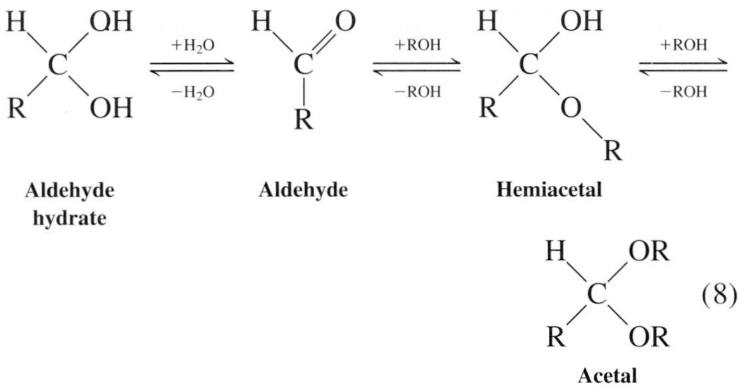

$$\underset{R}{\overset{H}{>}}C\underset{OH}{\overset{OH}{<}} \underset{-H_2O}{\overset{+H_2O}{\rightleftharpoons}} \underset{R}{\overset{H}{>}}C\overset{O}{\underset{}{\diagdown}} \underset{-ROH}{\overset{+ROH}{\rightleftharpoons}} \underset{R}{\overset{H}{>}}C\underset{O}{\overset{OH}{<}}_{R} \underset{-ROH}{\overset{+ROH}{\rightleftharpoons}}$$

**Aldehyde**      **Aldehyde**          **Hemiacetal**
**hydrate**

$$\underset{R}{\overset{H}{>}}C\underset{OR}{\overset{OR}{<}} \quad (8)$$

**Acetal**

Dehydrogenation of an aldehyde hydrate leads to carboxylic acid formation.

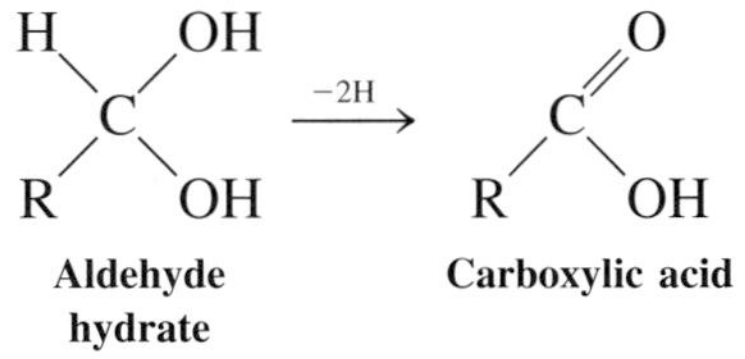

$$\underset{R}{\overset{H}{>}}C\underset{OH}{\overset{OH}{<}} \xrightarrow{-2H} \underset{R}{\overset{O}{>}}C\overset{O}{\underset{OH}{<}} \quad (9)$$

**Aldehyde**        **Carboxylic acid**
**hydrate**

$$R-C\overset{O}{\underset{OH}{<}} \underset{+H^+}{\overset{-H^+}{\rightleftharpoons}} \left[ R-C\overset{O}{\underset{O^-}{<}} \longleftrightarrow R-C\overset{O^-}{\underset{O}{<}} \right]$$

Aldehydes and ketones may also isomerize to the enol form as long as the adjacent carbon atom is bonded to at least one hydrogen atom. In the reaction a hydrogen migrates and the double bond shifts. At equilibrium the keto form is strongly preferred.

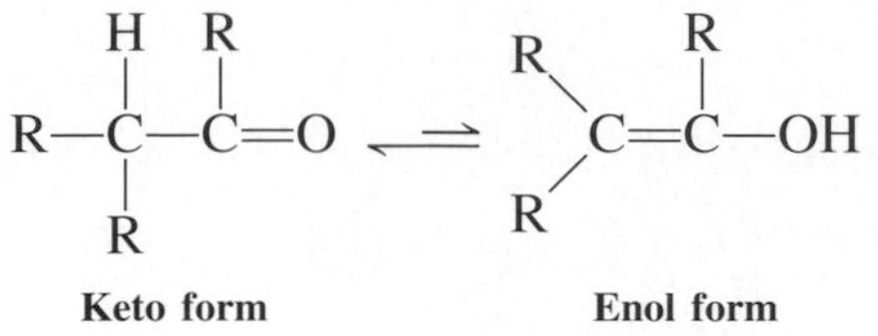

$$R-\underset{\underset{R}{|}}{\overset{\overset{H}{|}}{C}}-\overset{\overset{R}{|}}{C}=O \rightleftharpoons \underset{R}{\overset{R}{>}}C=\overset{\overset{R}{|}}{C}-OH \quad (10)$$

**Keto form**              **Enol form**

Pyrophosphates may hydrolyze to inorganic phosphoric acid (phosphate) and an organophosphoric acid.

$$R-O-\overset{\overset{O}{\|}}{\underset{\underset{O^-}{|}}{P}}-O-\overset{\overset{O}{\|}}{\underset{\underset{O^-}{|}}{P}}-O^- \underset{-H_2O}{\overset{+H_2O}{\rightleftharpoons}} R-O-\overset{\overset{O}{\|}}{\underset{\underset{O^-}{|}}{P}}-O^-$$

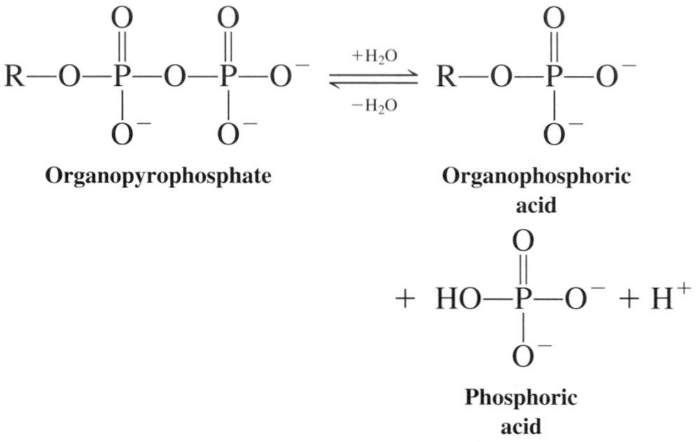

**Organopyrophosphate**      **Organophosphoric**
**acid**

$$+ HO-\overset{\overset{O}{\|}}{\underset{\underset{O^-}{|}}{P}}-O^- + H^+ \quad (11)$$

**Phosphoric**
**acid**

Such hydrolysis reactions yield considerable energy, which can be utilized in biosynthesis.

All the functional groups described are electrostatically neutral in organic solvents. However, in water many of these functional groups either lose or gain protons to become charged species. Such ionization reactions are extremely important in biochemical systems because they frequently influence solubility and reactivity.

Carboxylic and phosphoric acids lose one or more protons in water to become negatively charged. The ionized forms are stabilized by resonance as shown:

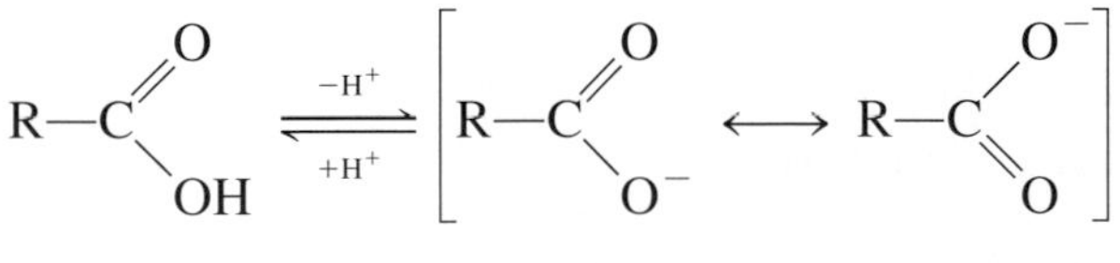

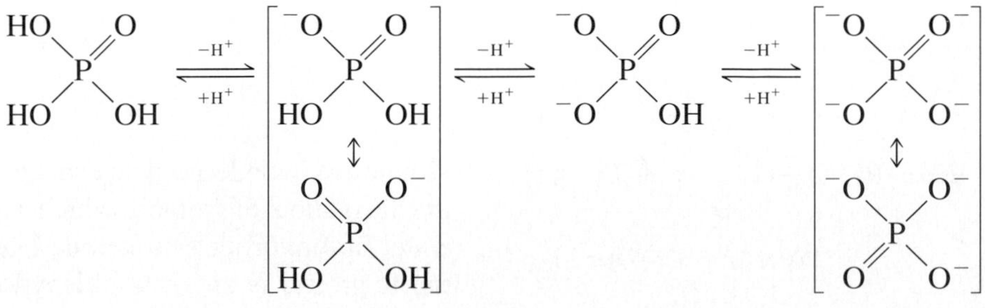

$$\underset{HO}{\overset{HO}{>}}\overset{\overset{O}{\|}}{P}\underset{OH}{\overset{O}{<}} \underset{+H^+}{\overset{-H^+}{\rightleftharpoons}} \left[\underset{HO}{\overset{^-O}{>}}\overset{\overset{O}{\|}}{P}\underset{OH}{\overset{O}{<}} \updownarrow \underset{HO}{\overset{O}{>}}\overset{P}{\underset{OH}{}}\overset{O^-}{}\right] \underset{+H^+}{\overset{-H^+}{\rightleftharpoons}} \underset{^-O}{\overset{^-O}{>}}\overset{\overset{O}{\|}}{P}\underset{OH}{\overset{O}{<}} \underset{+H^+}{\overset{-H^+}{\rightleftharpoons}} \left[\underset{^-O}{\overset{^-O}{>}}\overset{\overset{O^-}{\|}}{P}\underset{O^-}{\overset{O}{<}} \updownarrow \underset{^-O}{\overset{^-O}{>}}\overset{P}{\underset{O^-}{}}\overset{O^-}{}\right] \quad (12)$$

## Table 1

Organic Functional Groups of Biochemical Importance

| Type of Compound | General Structural Formula | Characteristic Functional Group | Name of Functional Group | Example Biomolecule with Functional Group | Chapter Reference |
|---|---|---|---|---|---|
| Alcohols | R—O—H | —OH | Hydroxyl group | Glycerol (polyhydroxyl)  H$_2$C—OH \| H—C—OH \| H$_2$C—OH | 17 |
| Ethers | R—O—R′ | —O— | Ether | Thromboxane A$_2$ (cyclic ether) | 19 |
| Thiols | R—S—H | —SH | Sulfhydryl | Coenzyme A  CoASH | 10 |
| Aldehydes | O=C—H \| R | O=C—H | Aldehyde | Glyceraldehyde-3-phosphate | 12 |

| Class | General structure | Functional group | Functional group name | Example | |
|---|---|---|---|---|---|
| Ketones | R—C(=O)—R′ | —C(=O)— | Keto | Fructose-6-phosphate | 12 |
| Carboxylic acids | R—C(=O)—OH | —C(=O)—OH | Carboxy | Stearic acid $CH_3(CH_2)_{15}CH_2$—C(=O)—OH | 17 |
| Mixed acid anhydride | R—C(=O)—O—P(=O)(OH)—OH | —C(=O)—O—P(=O)(OH)—OH | Phosphoric acid anhydride | Glycerate-1,3-bisphosphate | 12 |
| Esters | R—C(=O)—O—R′ | —C(=O)—O—R′ | Ester | Cholesterol ester | 20 |
| Hemiacetals | R—C(OH)(H)—O—R′ | —C(OH)(H)—O—R′ | Hemiacetal | Glucose | 12 |

## Table 1

Organic Functional Groups of Biochemical Importance (continued)

| Type of Compound | General Structural Formula | Characteristic Functional Group | Name of Functional Group | Example Biomolecule with Functional Group | Chapter Reference |
|---|---|---|---|---|---|
| Acetals | | | Acetal | Sucrose | 16 |
| Lactones | | | Lactone | 6-Phospho-D-gluconolactone | 12 |
| Amines | $R—NH_2$ | $—NH_2$ | Amino | Lysine | 3 |
| Amides | | | Amido | N-Acetyl-D-glucosamine | 16 |
| Alkenes | | | Alkenyl | Palmitoleic acid | 17 |

*Note:* R, R′, R″ are abbreviations for any alkyl or aryl group. R‴ is the abbreviation for any alkyl group or hydrogen.

Amines usually add a proton to become positively charged.

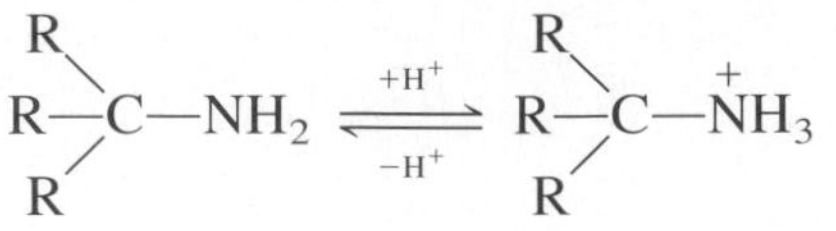

$$R-\underset{R}{\overset{R}{C}}-NH_2 \underset{-H^+}{\overset{+H^+}{\rightleftharpoons}} R-\underset{R}{\overset{R}{C}}-\overset{+}{N}H_3 \qquad (13)$$

Near neutrality ($10^{-7}$ M H$^+$), where most biochemical systems function, the carboxyl group exists mainly in the negatively charged form, phosphoric acid exists mainly in the diionized form, and amino groups exist mainly in the positively charged form. These conditions have interesting consequences for amino acids which contain one amino group and one carboxyl group. The amino acids are usually neutral overall, even though they contain two charged groups, one resulting from the deprotonation of the carboxyl group and the other resulting from the protonation of the amino group. Amino acids existing as dipolar ions are called zwitterions.

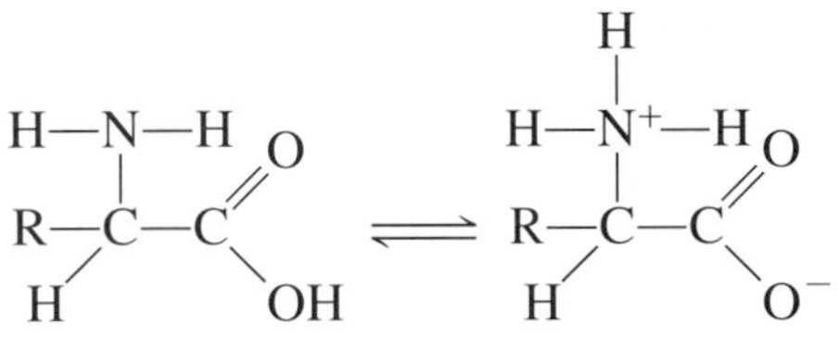

$$\underset{\textbf{Uncharged}}{\underset{H}{\overset{H-N-H}{R-\overset{|}{\underset{|}{C}}-\overset{O}{\underset{OH}{C}}}}} \rightleftharpoons \underset{\textbf{Zwitterion}}{\underset{H}{\overset{H-N^+-H}{R-\overset{|}{\underset{|}{C}}-\overset{O}{\underset{O^-}{C}}}}} \qquad (14)$$

The preceding are some of the more important reactions involving covalent bond breakage or formation in biochemistry. By now two things should be apparent about biochemical reactions: (1) the number of reactions in biochemistry is much more limited than in ordinary chemistry; and (2) as far as the reactants and products are concerned, biochemical reactions may be understood in the same terms as ordinary chemical reactions.

## Reaction Mechanisms, Arrow Formalism, and Catalysis

So far we have described reactions strictly in terms of their reactants and products. That doesn't tell us much about how or why a reaction proceeds from reactants to products. The reaction mechanism sometimes is quite a complex process, and for many well-known reactions we still don't understand the mechanism. We confine this discussion to consideration of the relatively simple mechanism of ester hydrolysis. Ester hydrolysis entails the reaction of a water molecule with an ester, leading to the production of an alcohol and a carboxylic acid. It is depicted in equation 1, going from right to left.

In attempting to find a mechanism for this or any other reaction we first must recall that all covalent bonds are formed by electrons; the rearrangement or breakage of such bonds starts with the migration of electrons. In the most general sense, reactive groups function either as electrophiles or nucleophiles. The former are electron-deficient substances that are attacked by electron-rich substances. The latter are electron-rich substances that attack electron-deficient substances.

In the case of ester hydrolysis the O of the water is an electron-rich substance and the carbonyl carbon of the ester is an electron-deficient substance. Ester hydrolysis occurs in three stages: (1) the initial stage in which electrons flow from the water molecule to the ester; (2) the intermediate stage in which the ester carbonyl forms a tetrahedral complex involving the hydroxyl group originating from the water molecule that releases a proton; and (3) the final stage in which carboxylic acid and alcohol are formed.

This reaction is kinetically favored because the oxygen of water is a nucleophile, whereas the carbonyl carbon is an electrophile. In the initial step the curved, colored arrows that represent the flow of electron pairs suggest three electron pair migrations that happen in quick succession (figure 10a). An electron pair migrates from the O—H of the water molecule to the O. An electron pair from the O attacks the carbonyl carbon, and an electron pair between the C and the O in the carbonyl migrates to the O of the carbonyl. The intermediate step involves the results of these and two additional electron pair migrations from the relatively unstable intermediate products that lead to the final products.

Normally, ester hydrolysis at ambient temperatures in water at neutral pH will occur very slowly—over a period of many days—unless something is done to accelerate the reaction. Usually, reactions are accelerated by raising the temperature because molecules react more vigorously as the temperature rises. However, a better way to accelerate a reaction is to add a catalyst, that is, a molecule that accelerates the reaction without undergoing any net consumption. Frequently, the main job of the catalyst is to make a potentially reactive center more attractive for reaction, i.e., to potentiate the electrophilic or nucleophilic character of the reactive centers of the reacting species.

In aqueous solution, protons (actually present as hydronium ions) or hydroxide ions are the catalysts most commonly used for nonenzymatic reactions. The way in which acid or base catalysts work in ester hydrolysis is illustrated in Figure 10b and c. As a result of the electronegativity of the oxygen atom in the ester $>C=O$ group, the oxygen has a fractional negative charge $\delta^-$ and the carbon has a fractional positive charge $\delta^+$. Either acid or base catalysts may be used to accelerate hydrolysis of the ester. In acid cataly-

**Figure A.10**

Reaction mechanisms for ester hydrolysis in the absence (*a*) and presence of catalysts (*b–e*).

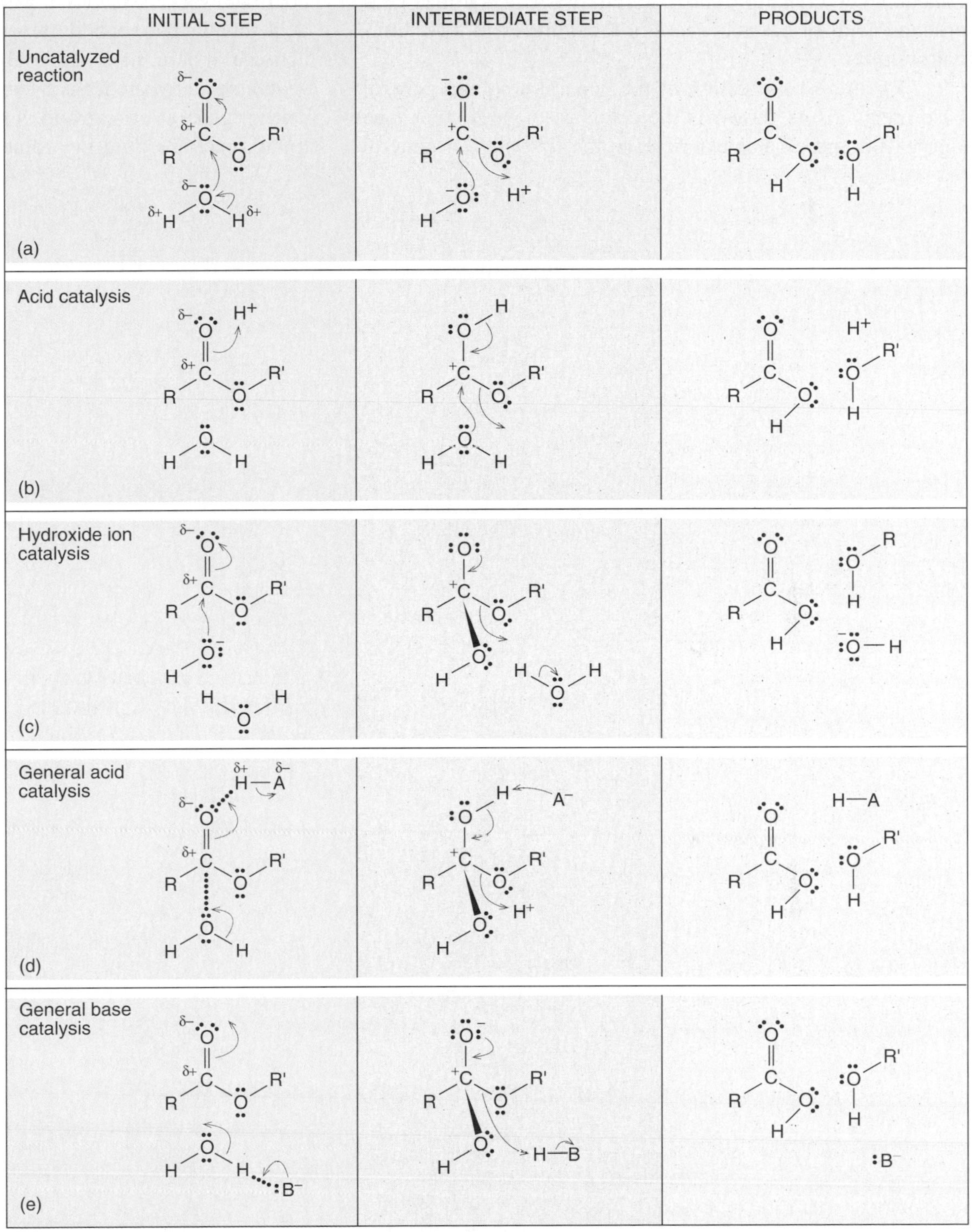

sis, a proton acting as an electrophile is attracted to the oxygen. That leads to an intermediate stage that accentuates the positive charge on the carbon atom, making it a more attractive electrophile to an attacking nucleophile—in this case, water. Water is a poor nucleophile and would attack the carbon slowly without such an inducement, so this step is the key to the catalysis. The remaining reactions leading to ester hydrolysis and regeneration of the catalyst occur

rapidly and spontaneously. In hydroxide ion catalysis of the ester, the carbon atom is attacked directly by a stronger nucleophile, OH⁻. Again after hydrolysis, the catalyst is regenerated.

To avoid the harshness of pH extremes, generalized acids or bases frequently are used to replace protons or hydroxide ions as catalysts. These compounds are capable of yielding protons or absorbing protons, respectively, during the course of a reaction. Enzymes almost always utilize generalized acids or bases because of the delicacy of biochemical systems.

An interesting feature of the general acid and general base mechanisms shown is the catalysis of both steps (see figure 10d and e). For example, in the first step of general acid catalysis, HA adds a proton and thus is acting as an acid, but in the second step $A^-$ removes a proton and is acting as a base. In the general base catalysis mechanism the sequence is reversed. Such sequential catalysis by a general acid or general base group is much more common in enzymatic reactions than in ordinary chemical reactions.

# Appendix A
## Some Landmark Discoveries in Biochemistry

In this appendix we list, in chronological order, some of the most important discoveries made in biochemistry during the past two centuries. It is impossible, for reasons of space, to give credit to every worker who has made a significant contribution, but it is possible to identify certain events as milestones, and thus to show how progress in this field has accelerated with the passage of time.

**1770–1774**
Priestly showed that oxygen is produced by plants and consumed by animals.

**1773**
Rouelle isolated urea from urine.

**1828**
Wohler synthesized the first organic compound, urea, from inorganic components.

**1838**
Schleiden and Schwann proposed that all living things are composed of cells.

**1854–1864**
Pasteur proved that fermentation is caused by microorganisms.

**1864**
Hoppe-Seyler crystallized hemoglobin.

**1866**
Mendel demonstrated the segregation and independent assortment of alleles in pea plants.

**1893**
Ostwald showed that enzymes are catalysts.

**1898**
Camillio Golgi described the Golgi apparatus.

**1905**
Knoop deduced the $\beta$ oxidation mechanisms for fatty acid degradation.

**1907**
Fletcher and Hopkins showed that lactic acid is formed quantitatively from glucose during anaerobic muscle contraction.

**1910**
Morgan discovered sex-limited inheritance in *Drosophila*.

**1912**
Warburg postulated a respiratory enzyme for the activation of oxygen.

**1913**
Michaelis and Menten developed a kinetic theory of enzyme action.

**1922**
McCollum showed that lack of vitamin D causes rickets.

**1926**
Sumner crystallized the first enzyme urease.

**1926**
Jansen and Donath isolated vitamin $B_1$ (thiamine) from rice polishings.

**1926–1930**
Svedberg invented the ultracentrifuge and used it to demonstrate the existence of macromolecules.

**1928**
Levene showed that nucleotides are the building blocks of nucleic acids.

**1928**
Szent-Gyorgyi isolated ascorbic acid (Vitamin C).

**1928–1933**
Warburg deduced the iron-prophyrin presence in the respiratory enzyme.

**1929**
Burr and Burr discovered that linoleic acid is an essential fatty acid for animals.

**1931**
Englehardt discovered that phosphorylation is coupled to respiration.

**1932**
Warburg and Christian discovered the "yellow enzyme," a flavoprotein.

**1933**
Krebs and Henseleit discovered the urea cycle.

**1933**

Embden and Meyerhof demonstrated the intermediates in the glycolytic pathway.

**1935**

Schoenheimer and Rittenberg first used isotopes as tracers in the study of intermediary metabolism.

**1935**

Stanley first crystallized a virus, tobacco mosaic virus.

**1937**

Krebs discovered the citric acid cycle.

**1937**

Warburg showed how ATP formation is coupled to the dehydrogenation of glyceraldehyde-3-phosphate.

**1938**

Hill found that cell-free suspensions of chloroplasts yield oxygen when illuminated in the presence of an electron acceptor.

**1939**

C. Cori and G. Cori demonstrated the reversible action of glycogen phosphorylase.

**1939**

Lipmann postulated the central role of ATP in the energy-transfer cycle.

**1939–1946**

Szent-Gyorgyi discovered actin and the actin-myosin complex.

**1940**

Beadle and Tatum deduced the one gene–one enzyme relationship.

**1942**

Bloch and Rittenberg discovered that acetate is the precursor of cholesterol.

**1943**

Chance applied spectrophotometric methods to the study of enzyme–substrate interactions.

**1943**

Martin and Synge developed partition chromatography.

**1944**

Avery, MacLeod, and McCarty demonstrated that bacterial transformation is caused by DNA.

**1947–1950**

Lipmann and Kaplan isolated and characterized coenzyme A.

**1948**

Leloir discovered the role of uridine nucleotides in carbohydrate metabolism.

**1948**

Hogeboom, Schneider, and Palade refined the differential centrifugation method for fractionation of cell parts.

**1948**

Kennedy and Lehninger discovered that the tricarboxylic acid cycle, fatty acid oxidation, and oxidative phosphorylation all take place in mitochondria.

**1949**

Christian deDuve discovered lysosomes.

**1950–1953**

Chargaff discovered the base equivalences in DNA.

**1951**

Pauling and Corey proposed the $\alpha$-helix structure for $\alpha$-keratins.

**1951**

Lynen postulated the role of coenzyme A in fatty acid oxidation.

**1952**

Palade, Porter, and Sjostrand perfected thin sectioning and fixation methods for electron microscopy of intracellular structures.

**1952–1954**

Zamecnik and his colleagues developed the first cell-free systems for the study of protein synthesis.

**1953**

Vincent du Vigneaud synthesized the first biologically active peptide hormone, ocytocin.

**1953**

Woodward and Bloch postulated a cyclization scheme for squalene, leading to cholesterol.

**1953**

Sanger and Thompson determined the complete amino acid sequence of insulin.

**1953**

Hokin and Hokin showed that acetylcholine induces the rapid biosynthesis of phosphatidylinositol in pigeon pancreas.

**1953**

Horecker, Dickens, and Racker elucidated the 6-phosphogluconate pathway of glucose catabolism.

**1953**

Watson and Crick and Wilkins determined the double-helix structure of DNA.

**1954**

Hugh Huxley proposed the sliding filament model for muscular contraction.

**1955**

Ochoa and Grunberg-Manago discovered polynucleotide phosphorylase.

**1955**

Kennedy and Weiss described the role of CTP in the biosynthesis of phosphatidylcholine.

**1956**

Kornberg discovered the first DNA polymerase.

**1956**

Umbarger reported that the end product isoleucine inhibits the first enzyme in its biosynthesis from threonine.

**1956**

Dorothy Crawfoot Hodgkin determined the structure of coenzyme B12.

**1956**

Ingram showed that normal and sickle-cell hemoglobin differ in a single amino acid residue.

**1956**

Anfinsen and White concluded that the three-dimensional conformation of proteins is specified by their amino acid sequence.

**1956**

Leloir determined the pathway to uridine diphosphate glucose (UDPG).

**1957**

Hoagland, Zamecnik, and Stephenson isolated tRNA and determined its function.

**1957**

Sutherland discovered cyclic AMP.

**1958**

Weiss, Hurwitz, and Stevens discovered DNA-directed RNA polymerase.

**1958**

Meselson and Stahl demonstrated that DNA is replicated by a semiconservative mechanism.

**1959**

Wakil and Ganguly reported that malonyl-CoA is a key intermediate in fatty acid biosynthesis.

**1960**

Kendrew reported the x-ray analysis of the structure of myoglobin.

**1961**

Jacob and Monod proposed the operon hypothesis.

**1961**

Jacob, Monod, and Changeux proposed a theory of the function and action of allosteric enzymes.

**1961**

Mitchell postulated the chemiosmotic hypothesis for the mechanism of oxidative phosphorylation.

**1961**

Nirenberg and Matthaei reported that polyuridylic acid codes for polyphenylalanine.

**1961**

Marmur and Doty discovered DNA renaturation.

**1962**

Racker isolated $F_1$ ATPase from mitochondria and reconstituted oxidative phosphorylation in submitochondrial vesicles.

**1966**

Maizel introduced the use of sodium dodecylsulfate (SDS) for high-resolution electrophoresis of protein mixtures.

**1966**

Crick proposed the wobble hypothesis.

**1966**

Gilbert and Muller-Hill isolated the lac repressor.

**1968**

Glomset proposed the theory of reverse cholesterol transport in which HDL is involved in the return of cholesterol to the liver.

**1968**

Meselson and Yuan discovered the first DNA restriction enzyme. Shortly thereafter Smith and Wilcox discovered the first restriction enzyme that cuts DNA at a specific sequence.

**1969**

Zubay and Lederman developed the first cell-free system for studying the regulation of gene expression.

**1970**

Howard Temin and David Baltimore discovered reverse transcriptase.

**1971**

Vane discovered that aspirin blocks the biosynthesis of prostaglandins.

**1972**

Jon Singer and Garth Nicolson proposed the fluid mosaic model for membrane structure.

**1973**

Cohen, Chang, Boyer, and Helling reported the first DNA cloning experiments.

**1975**

Brown and Goldstein described the low-density lipoprotein receptor pathway.

**1975**

Sanger and Barrell developed rapid DNA-sequencing methods.

**1976**

Michael Bishop and Harold Varmus discovered the c-src gene in uninfected cells, which is homologous to the v-src gene in the Rous sarcoma virus.

**1977**

Starlinger discovered the first DNA insertion element.

**1977**

McGarry, Mannaerts, and Foster discovered that malonyl-CoA is a potent inhibitor of $\beta$ oxidation.

**1977**

Splicing of RNA simultaneously discovered in Broker's and Sharp's laboratories.

**1977**

Nishizuka and coworkers reported the existence of protein kinase C.

**1978**

Shortles and Nathans did the first experiments in directed mutagenesis.

**1978**

Tonegawa demonstrated DNA splicing for an immunoglobulin gene.

**1981**

Cech discovered RNA self-splicing.

**1981**

Steitz determined the structure of CAP protein.

**1981–1982**

Palmiter and Brinster produced transgenic mice.

**1983**

Mullis amplified DNA by the polymerase chain reaction (PCR) method.

**1984**

Schwartz and Cantor developed pulsed field gel electrophoresis for the separation of very large DNA molecules.

**1984**

Michel, Deisenhofer, and Huber determined the structure of the photosynthetic reaction center.

**1984**

Blobel discovered the mechanism for protein translocation across the endoplasmic reticulum membrane—the signal hypothesis.

**1988**

Elion and Hitchings shared the Nobel Prize for design and synthesis of therapeutic purines and pyrimidines.

**1989**

Synder and colleagues purified and reconstituted the inositol-1,3,4-$P_3$ receptor.

# Appendix B
## A Guide to Career Paths in the Biological Sciences

### After You Receive Your Bachelor's Degree, What's Next?

When you receive your bachelor's degree in the biological sciences (e.g., biology, biochemistry, microbiology, etc.), you may choose to continue your education by attending graduate school, medical school, or some other professional school, or you may want to look for a job. If you are interested in going on to graduate school for a masters (MS) or doctorate (Ph.D.) degree in the biological sciences or to medical school for an MD or MD/Ph.D. degree, you can usually obtain information on specific schools and programs from your faculty advisor, on-campus career center, or library.

Job opportunities are available for bachelor's level biological sciences graduates if you are interested in finding employment directly after graduation. Because majors and job titles don't always match and colleges and universities don't typically have employment services, finding employment most often requires quite a bit of research on your part. Extensive job descriptions for virtually every type of position available are provided in *The Dictionary of Occupational Titles,* published by the U.S. Department of Labor. Also, research and reference firms such as Peterson's Guides, Inc., publish guides to job opportunities for science graduates.

### Job Opportunities for Bachelor's Level Biological Science Graduates

As indicated in *Peterson's Job Opportunities for Engineering, Science, and Computer Graduates 1993,\** job titles listed in employment advertisements can be categorized by functional occupational area. They are accounting/finance, administration, information systems/processing, marketing/sales, production/operations, research/development, and technical/professional services. All these functional areas are not directly applicable to people with degrees in the biological sciences, but thinking about jobs in terms of these areas can help broaden your career options.

The most applicable functional occupational areas for people with a degree in the biological sciences are mar-keting/sales, production/operations, research/development, and technical/professional services. A large number of companies and other organizations have entry-level positions available for bachelor's level biological sciences graduates, including the following.

### Marketing / Sales

American Cyanamid Company
1 Cyanamid Plaza
Wayne, NJ 07470
Contact: Office of Technical Recruiting

Eli Lilly and Company
Lilly Corporate Center
Indianapolis, IN 46285
Contact: Office of Corporate Recruiting

Johnson & Johnson
1 Johnson & Johnson Plaza
New Brunswick, NJ 08933
Contact: College Relations Coordinators in the Personnel Office of each Johnson & Johnson Division

Marion Merrell Dow, Inc.
10236 Marion Park Drive
Kansas City, MO 64137
Contact: Staffing Manager

Parke-Davis, Warner-Lambert Company
201 Tabor Road
Morris Plains, NJ 07950
Contact: Director, Human Resources, Sales & Marketing

### Production / Operations

Calgon Vestal Laboratories
Calgon Corporation
St. Louis, MO 63166
Contact: Personnel Department

Ciba-Geigy Corporation
444 Saw Mill River Road
Ardsley, NY 10502
Contact: Manager, College Relations and Staffing

*\* Peterson's Job Opportunities for Engineering, Science, and Computer Graduates 1993,* Copyright © 1992 Peterson's Guides, Inc., Princeton, New Jersey.

Genentech, Inc.
460 Point San Bruno Boulevard
South San Francisco, CA 94080
Contact: Human Resources Department

Merck & Co., Inc.
126 East Lincoln Avenue
Rahway, NJ 07065
Contact: Office of College Relations

Millipore Corporation
80 Ashby Road
Bedford, MA 01730
Contact: Manager of Employment

## Research / Development

Argonne National Laboratory, University of Chicago
9700 South Cass Avenue
Argonne, IL 60439
Contact: Recruiting Coordinator

Coca-Cola Foods
2000 St. James Place
Houston, TX 77056
Contact: Manager, Professional Staffing

General Mills, Inc.
One General Mills Boulevard
Minneapolis, MN 55446
Contact: Director of Recruitment and College Relations

Memorial Sloan-Kettering Cancer Center
1275 York Avenue
New York, NY 10021
Contact: Administrator, College Relations

Centers for Disease Control
Department of Health and Human Services
1600 Clifton Road, NE
Atlanta, GA 30333
Contact: Employment Office

National Cancer Institute, National Institutes of Health
9000 Rockville Pike
Bethesda, MD 20892
Contact: Personnel Staffing Specialist

## Technical / Professional

American Software USA, Inc.
470 East Paces Ferry Road, NE
Atlanta, GA 30305
Contact: Manager, Corporate Recruiting

Energy and Environment Analysis, Inc.
1655 North Fort Meyer Drive
Arlington, VA 22209
Contact: Recruiting Coordinator

General Electric Company
3135 Easton Turnpike
Fairfield, CT 06431
Contact: Recruiting Support Services

Patent and Trademark Office
U.S. Department of Commerce
2011 Crystal Drive, One Crystal Park
Arlington, VA 20231
Contact: Office of Personnel

Consumer Product Safety Commission
5401 Westband Avenue
Bethesda, MD 20816
Contact: Director, Division of Personnel Management

# Appendix C
## Answers to Selected Problems

**Chapter 2**

1. Chemists can often make a thermodynamically unfavorable reaction proceed to some extent by manipulating experimental reaction parameters such as pressure, temperature, or concentrations of reactants. Biochemists generally cannot alter such reaction parameters because organisms function within limited concentration ranges and at essentially constant pressure and temperature.

3. A state function describes the thermodynamic parameters of the system under consideration at a particular moment. Only the difference between initial and final states, not the path taken to achieve these states, is important in most thermodynamic considerations. Enthalpic contributions defining the thermodynamic state are considered only at the initial and final states. Enthalpy is independent of pathway and is therefore a state function.

5. In thermodynamics the total system is the universe, consisting of a particular system and its environment. If the particular system under study is a closed system, it can exchange internal energy as heat and work with its surroundings, but no exchange of matter can occur. An open system can also exchange matter with its surroundings. Therefore, although any energy lost by a particular system is gained by its environment (and vice versa), the energy of the universe remains constant. This is the first law of thermodynamics. Distinguishing between the system and the universe is important to differentiating between a situation in which energy can change and a situation in which energy is constant.

7. Entropy decreases because of hydration effects. The ionized species will order much of the water during hydration, decreasing the total number of free molecules.

9. The reaction will proceed toward oxaloacetate formation in the cell if low product concentration is maintained. Oxidation of NADH by the mitochondrial electron transport system and utilization of oxaloacetate in the formation of citrate shifts the malate-oxaloacetate reaction toward oxaloacetate production.

11. (a) For reaction (P1), $\Delta G^{\circ\prime} = -2.4$ kcal/mole;
    (b) $\Delta G^{\circ\prime}$ of ATP hydrolysis is $-7.9$ kcal/mole.

13. (a) $\Delta G^{\circ\prime} - 1.5$ kcal/mole; thermodynamically favorable as written.
    (b) $\Delta G^{\circ\prime} = +1.7$ kcal/mole; thermodynamically unfavorable as written.
    (c) $\Delta G^{\circ\prime} = -7.3$ kcal/mole; thermodynamically favorable as written.

15. In the first case, where the repressor protein is cut in half, the binding enthalpy for each part would be essentially half the enthalpy value for the intact repressor. The entropy would be less favorable (less positive) because of the chelation effect. As a result, the free energy would be less favorable (less negative).

    In the second case, where one of the binding sites on the DNA is eliminated, the binding enthalpy for the repressor would be approximately half that for repressor binding to the unmodified DNA sequence. The entropy would depend on the extent of hydration and the extent of mobility of the unbound portion of the repressor. Again, we would expect the free energy for binding to be less favorable.

**Chapter 3**

1. Berzelius's proposal of $C_{40}H_{62}N_{10}O_{12}$ has a molecular mass of 874 g/mole, of which 140 g/mole is nitrogen (or 16.0% nitrogen by mass).

3. $\text{pH} = 6.39 + \log \dfrac{0.0133 \text{ mole}/0.25 \text{ L}}{0.0060 \text{ mole}/0.25 \text{ L}} = 6.39 + 0.35$
$$= 6.74$$
$[H^+] = 10^{-6.74} = 1.82 \times 10^{-7} \text{ M}$

5. pI histidine $= [(\text{p}K_{\text{amino}}) + (\text{p}K_{\text{imidazole}})]/2$
$$\text{pI} = 7.69$$
pI aspartic acid $= [(\text{p}K_{\text{carboxyl}}) + (\text{p}K_R)]/2$
$$\text{pI} = 2.95$$
pI arginine $= (\text{p}K_{\text{amino}} + \text{p}K_R)/2$
$$\text{pI} = 10.74$$
The sum of positive and negative charge contributions is

| $\alpha$-amino group | $\alpha$-carboxyl | $\beta$-carboxyl | |
|---|---|---|---|
| (+1) | (−0.9) | (0.1) | = 0 |

These data demonstrate that the net charge on aspartic acid is zero at pH 2.95, which verifies 2.95 as the isoelectric pH.

7. Aspartic acid: pH 2, 4, 10. (The pH range 1 to 5 will be buffered by the $\alpha$-carboxyl and the $\beta$-carboxyl groups.)
   Histidine: pH 2, 9, 6 (imidazole side chain).
   Serine: pH 2, 9. (The p$K_a$ of the alcohol is outside the range of pH normally considered for buffers.)

9. Threonine and isoleucine

11. NH$_2$-Ala-Ala-Lys-Ala-Ala-Phe-Ala

## Chapter 4

1. Consider the entropic effect of decreasing water organization by moving the hydrophobic residue side chains from an aqueous to a nonaqueous environment.

3. The $\alpha$ helix is a rodlike element that cannot easily change direction. Loops, $\beta$ bends, and "random" structure break the helical structure and allow these directional changes.

5. An $\alpha$ helix broken at the Pro-Asn-Ala region with the hydrophobic residues on the exterior should insert into the membrane.

7. The right- or left-handedness of a helix is the same as a conventional screw or bolt. When turned clockwise, a right-handed screw advances. The same is true of a helix or a helical spring.

9. These substances can cause the loss of a protein's shape by disrupting hydrogen bonding and electrostatic interactions.

11. The detergent replaces the membrane, producing a soluble enzyme, and allows the enzymes to be purified free of the membrane.

## Chapter 5

1. The early chemists had no way of knowing that each hemoglobin contains four Fe$^{2+}$.

$$4.9 \text{ mg of Fe}_2\text{O}_3$$

3. (a) $CO_2 + H_2N\text{-hemoglobin} \longrightarrow$
   $$^-\text{O}\overset{\overset{\text{O}}{\|}}{\text{C}}\text{HN-hemoglobin} + H^+$$
   (b) $HOCO_2^- + H_2N\text{-hemoglobin} \longrightarrow$
   $$^-\text{O}\overset{\overset{\text{O}}{\|}}{\text{C}}\text{HN-hemoglobin} + H_2O$$

5. Red blood cells pass single file through the capillaries. In a sickle cell crisis the red blood cells "jam together," clogging the capillaries. The associated tissues become starved for oxygen, producing the pain of a sickle cell crisis.

7. Individuals with sickle cell anemia have a functional gene for the gamma chain. If the production of fetal hemoglobin could be "turned back on" the affected individuals could function normally except during pregnancy.

9. Unprotonated. The protonated form of histidyl F8 is positively charged and is less likely to interact favorably with the Fe$^{2+}$ of the heme.

11. 457 Å. Because the two 457-Å-long strands are twisted into a helix, the resulting TM is shorter than 457 Å.

13. Rigor is probably caused by the depletion of ATP and a considerable discharge of calcium from the sarcoplasmic reticulum.

## Chapter 6

1. It is often a rapid effective way to reduce the volume of crude extracts and at the same time eliminate a major portion of the total protein.

3. Difference in charge allows the separation of phosphorylase a from phosphorylase b with the use of DEAE-cellulose. Phosphorylase a and phosphorylase b should elute as a single peak upon gel filtration.

5. Heat treatment of protein solutions denatures and precipitates some of the proteins, while others remain both soluble and stable. Thermal lability is determined empirically for each enzyme or protein of interest.

7. The protein is positively charged at a pH more acidic than the pI. Therefore the protein will probably adhere to (bind to) the CM-cellulose if the pH is between 4 and 6.

9. The student's "pure" protein contains at least two components separable by the criterion of mass/charge but which share a common subunit molecular weight. The multiple protein bands appearing in the nondenaturing gel possibly arose through deamidation of glutamine or asparagine residue side chains.

11. (a) Proteins 1 and 3 should elute in the initial wash buffer, but proteins 2 and 4 are predicted to bind to the column. Based solely on isoelectric point, we might predict that protein 4, then protein 2, would be eluted in the salt gradient.
    (b) Proteins 2 and 4 would be eluted in the initial wash buffer from the column, whereas proteins 1 and 3 would be predicted to adhere to the column. Based solely on pI values, protein 3 would be predicted to elute prior to protein 1 in the KCl gradient.
    (c) Proteins with molecular weight greater than the limit (protein 62,000 $M_r$) are excluded from entry

into the gel and elute in the void volume ($V_o$). The other proteins in the solution will elute in the order protein 3, protein 1, and protein 4.

## Chapter 7

1. Reaction order is the power to which a reactant concentration is raised in defining the rate equation. The example is first order in A and B, second order overall, and second order in C.

3. Let $v = V_{max}/2$ and substitute into equation 25:

$$\frac{V_{max}}{2} = \frac{V_{max}[S]}{[S] + K_M}$$

Solving yields $K_M = [S]$.

5. No

7. Steady-state approximation is based on the concept that the formation of [ES] complex by binding of substrate to free enzyme and breakdown of [ES] to form product plus free enzyme occur at equal rates. A graphical representation of the relative concentrations of free enzyme, substrate, enzyme-substrate complex, and product is shown in figure 7.8 in the text. Derivation of the Michaelis-Menten expression is based on the steady-state assumption. Steady-state approximation may be assumed until the substrate concentration is depleted, with a concomitant decrease in the concentration of [ES].

9. $[S] = 1.0 \times 10^{-3}$ M, $[I] = 9.1 \times 10^{-4}$ M
   $[S] = 1.0 \times 10^{-5}$ M, $[I] = 1.8 \times 10^{-5}$ M
   $[S] = 1.0 \times 10^{-6}$ M, $[I] = 9.9 \times 10^{-6}$ M

11. An enzyme is a catalyst for a chemical reaction, so the rate of an enzyme catalyzed reaction increases with increased temperature. However, the catalyst, a protein, is structurally labile and is inactivated (denatured) at elevated temperatures. The precise temperature at which the enzyme is inactivated varies with the specific enzyme. There is no "temperature optimum" for a catalyst (enzyme).

13. $V_{max} = 3.3 \times 10^{-7}$ Ms$^{-1}$; $K_m = 2.4 \times 10^{-4}$ M; $k_{cat} = 3.3 \times 10^4$ s$^{-1}$; specificity constant = $1.4 \times 10^8$ M$^{-1}$s$^{-1}$

## Chapter 8

1. The substrate recognition site is a pocket on hexokinase into which glucose, then ATP, bind. Glucose and hexokinase bind together, changing the shape of both and forming a binding site for ATP.

3. Specificity of bond cleavage would most certainly change. The favorable electrostatic interaction between the Lys or Arg side chain of the peptide substrate and the aspartate in the binding pocket would be replaced by electrostatic repulsion upon substitution of the lysine residue. Thus selective binding of Arg or Lys to the substrate pocket would be precluded. The modified trypsin might therefore (a) cleave peptide bonds randomly, but exclude bonds on the carboxyl side of Lys or Arg residues, or (b) exhibit specificity for peptide cleavage on the carboxyl side of acidic amino acids (Asp, Glu). Side chains of these amino acids may fit well into the substrate binding pocket and have favorable electrostatic interaction with the lysine therein.

5. We expect iodoacetate and para-mercuribenzoate to inhibit plant proteases and diisopropylfluorophosphate to inhibit serine proteases.

7. Histidyl 12 accepts the 2′-OH hydrogen to start the reaction. The protonated histidyl 12 ultimately provides the hydrogen to the cyclic ester to produce the 3′-phosphate product.

9. Renaturation of denatured protein is dictated by the primary structure of the protein. The trypsin family of enzymes and carboxypeptidase A are synthesized as proenzymes that are proteolytically activated. The proteolyzed, active enzymes have primary structures different from the gene product and are not active upon renaturation. In addition, zinc is a cofactor required for carboxypeptidase A activity.

11. The structure of the transition-state analog is complementary to the structure of the active site. The analog thus binds tightly to the active site.

13. (a) Substitution of Asp for Lys 86 markedly decreases activity and several explanations are possible. The lysine is a critical residue either at or near the active site or is essential for maintaining a catalytically competent conformation of the enzyme. (b) Lysines 21 and 101 are probably outside the catalytic site and may not be evolutionarily conserved. Their replacement with aspartate yielded no great change in enzymatic activity. (c) Lysine 86 is essential for enzymatic activity and would be conserved.

## Chapter 9

1. The Ile in chymotrypsinogen is the 16th amino acid. During the conversion to chymotrypsin a 15 amino acid peptide is cleaved from chymotrypsinogen. The original amino acid number system was retained in chymotrypsin, making the N-terminus amino acid number 16.

3. Covalent modification requires an enzyme to control an enzyme. Allosterism requires only a binding site on an enzyme that interacts (via an equilibrium) with a particular small molecule.

A-28    APPENDIX C  Answers to Selected Problems

5. Phosphorylation of serine or threonine (and possibly tyrosine) on the target protein may be largely influenced by the amino acid sequence around these residues. These amino acid sequences may define a specific motif or recognition site for the protein kinase.

7. The substrate concentration required for half-maximal activity ($S_{0.5}$) of an allosterically regulated enzyme will depend on the cumulative effects of allosteric activators and/or inhibitors also present. Hence $S_{0.5}$ may be decreased with allosteric activators and may be increased with allosteric inhibitors.

9. Allosteric regulation of an enzyme having a binding site for a regulatory molecule and an active site on the same subunit is not uncommon. Regulatory molecules bind at sites separate from the active site and induce a conformational change that affects substrate binding to the active site on the same as well as adjacent subunits.

11. The methylene group prevents the elimination of a phosphate, which is required to convert the postulated intermediate into carbamoyl aspartate. The suggested oxygen analog might eliminate phosphate (i.e., be a substrate for aspartate carbamoyltransferase).

# Chapter 10

1. The following coenzymes contain the AMP moiety: $NAD^+$, NADH, $NADP^+$, NADPH, FAD, $FADH_2$, and CoASH.

3. (a) Refer to figure 10.5b in the text for the structures of each intermediate step.
   Steps 1. & 2. Form Schiff base
         3. Remove proton $\alpha$-C
         4. Protonate C-4'
         5. & 6. Hydrolyze Schiff base
              Release $\alpha$-keto acid and pyridoxamine phosphate
   Transfer of amino group from pyridoxamine phosphate to pyruvate, forming alanine occurs by reversal of these steps. Other transaminases use other $\alpha$-amino acids and $\alpha$-keto acids.
   (b) Figure 10.5b in the text gives the structures of intermediates.
      Steps   1. Decarboxylation of $\alpha$-$COO^-$
              2. Protonate $\alpha$-Carbon
              3. & 4. Hydrolyze Schiff base
   (c) $\beta$-Decarboxylation. Form Schiff base steps 1–3, solution 3a.

5. Refer to figure 10.2 in the text.
   (a) Steps 1. Deprotonation of TPP to ylid form, nucleophilic addition of ylid to $\alpha$-ketogroup

2. Decarboxylation
3. Resonance stabilization
4. Protonation, elimination of TPP

   (b) Steps 1. Deprotonation to ylid form, nucleophilic addition of ylid to fructose-6-phosphate
         2. Oxidation of C-3, release of erythrose-4-phosphate
         3. Resonance stabilization of intermediate
         4. Elimination of $-OH$ from C-2
         5. Transfer of acyl group to phosphate

7. (a) Thiamine pyrophosphate, lipoic acid, FAD, $NAD^+$, CoASH (an $\alpha$-ketoacid dehydrogenase); (b) biotin; (c) FAD.

9. (a) Reduced flavins in solution rapidly reduce $O_2$ to superoxide and $H_2O_2$, metabolites that are toxic to the cell. Enzyme-bound flavins are usually shielded from rapid oxidation by $O_2$. (b) Tightly bound NAD(P) is an advantage to enzymes catalyzing rapid $H:^-$ removal and readdition in a stereospecific fashion. Freely diffusing NADH is an advantage in transferring reducing equivalents among various enzyme-catalyzed reactions.

11. (a) Redox agents: FAD, FMN, $NAD^+$, $NADP^+$, and lipoyl; (b) acyl carriers: CoASH, lipoyl, and thiamine pyrophosphate; (c) both acyl carriers and redox agents: lipoyl.

13. The hydroxyl hydrogen on C-3 is the most acidic because the resulting anion is resonance stabilized.

# Chapter 11

1. The transition from left to right involves dehydration reactions; the transition from right to left involves hydrolysis reactions.

3. Catabolism involves pathways composed of enzymes and chemical intermediates that are involved primarily in the breakdown of large molecules into small molecules, often by oxidation processes. Anabolism is the collection of the enzymes and chemical intermediates involved in the biological synthesis of larger molecules from smaller molecules. Anabolism often involves reduction processes.

5. The advantage of subcellular compartments is that a specific pathway can be isolated from comparable pathways that might use similar or identical chemical intermediates. Compartmentalization simplifies regulation of the various processes.

7. (a) The concentration of most metabolites measured under steady-state conditions in the cell usually does not exceed the $K_m$ value. For an enzyme whose reaction can be described by simple Michaelis-Menten kinetics, the observed velocity, $v$, is $0.5V_{max}$ if substrate concentration equals the

$K_m$ value. The velocity of most enzymes is likely significantly less than $V_{max}$ *in vivo*.

(b) End-product inhibition usually occurs at the committed step in a metabolic pathway or at a branch point in the pathway.

(c) Catabolic pathways tend to be convergent rather than divergent. Metabolic convergence of precursors into common intermediates of a metabolic pathway provides an efficient route for the metabolism of a variety of metabolites by a limited number of enzymes.

(d) Enzymes that are regulated in metabolic pathways most frequently exhibit cooperative kinetic responses rather than hyperbolic responses with respect to substrate concentration and are frequently responsive to allosteric regulation by products, energy charge, or concentration ratio of $NAD(P)H/NAD(P)^+$. The rate of enzymatic activity over a narrow range of substrate concentration can be changed dramatically by allosteric activators or inhibitors. Such dramatic changes in velocity in response to small changes in substrate concentration is not observed with enzymes exhibiting Michaelis-Menten kinetics.

(e) Low energy charge signals the cell that a need for ATP formation exists, and pathways (glycolysis and Krebs cycle) leading to ATP formation are activated. Anabolic pathways that demand high ATP concentrations are inhibited at low energy charge. In the latter case, the ATP required to drive biosynthesis is in low supply. Conversely, increased energy charge inhibits pathways leading to ATP formation and activates anabolic pathways.

(f) Formation of multienzyme complexes is a strategy frequently used for efficient catalysis and control of metabolic pathways. Substrates diffuse shorter distances between active sites in multienzyme complexes than if the enzymes were not organized. Frequently, intermediates covalently bound to cofactors (e.g., lipoamide, biocytin) are moved among active sites within the complex, effectively trapping intermediates in the complex and increasing the concentration of substrates at the enzyme active site.

(g) Separation of catabolic and anabolic pathways diminishes the likelihood of futile cycling of metabolites. Enzymes catalyzing the $\beta$-oxidation of fatty acids are located in the mitochondrial matrix, whereas enzymes catalyzing the synthesis of palmitate are located in the cytosol.

9. The "committed step" in a reaction sequence steers the metabolite to a sequence of reactions whose in-termediates have no other function in the cell. Control of the committed step prevents wasteful accumulation of these single-purpose intermediates and obviates the necessity of controlling each enzyme in a pathway.

**Chapter 12**

1. The number of sugars is $2^n$ where $n$ is the number of chiral centers but does not include the chiral carbon involved in producing the D series of sugars. Because C-2 of the ketoses is a carbonyl group, only the geometry of C-3 and C-4 remain to produce the four D-ketohexoses ($2^2 = 4$). The carbonyl group on C-1 of the aldoses allow C-2, C-3, and C-4 to produce eight D-aldohexoses ($2^3 = 8$).

3. (a) Trehalose is two $\alpha$-D-glucoses linked through C-1 of each sugar. However, the correct answer to this question would also include (b) two $\beta$-D-glucoses linked through C-1 and (c) one $\alpha$-D-glucose linked through C-1 to C-1 of a $\beta$-D-glucose.

5. Maltose, lactose, and cellobiose are reducing sugars; sucrose and trehalose (problem 3) are nonreducing sugars.

7. Glucose C-3 and C-4 are lost as $CO_2$ during ethanol production.

9. By assessing the rate of $^{14}CO_2$ production from $^{14}C$-1-glucose versus $^{14}C$-6-glucose, we can determine the relative significance of the different pathways in a tissue.

11. Glycerate-2,3-bisphosphate was encountered as an allosteric effector of hemoglobin.

13. In the first example the actual chemistry is identical: an aldose-ketose interconversion. Note that the geometry of the alcohol on C-2 is identical. The second example has essentially the same chemistry. Logically, to do the same chemistry, nature could have stumbled upon the same mechanism twice. More likely, however, the mechanism evolved once and, following gene duplication, the segments of the gene controlling the substrate specificity changed on one copy of the gene.

15. The carboxyl group of pyruvate is lost as $CO_2$ during the pyruvate decarboxylase–catalyzed formation of acetaldehyde. The pyruvate carboxyl group is formed by the oxidation of glyceraldehyde-3-phosphate. The aldehyde (C-1) carbon is derived directly from the aldolase-dependent cleavage of the fructose-1,6-bisphosphate. In this cleavage, C-4 of glucose becomes C-1 of glyceraldehyde-3-$P_i$ and C-3 of glucose becomes C-3 of dihydroxyacetone phosphate. DHAP is isomerized to Ga3P$_i$. In this isomerization, C-3 of DHAP (originally C-3 of glucose) becomes C-1 of Ga3P. Thus, labeling either C-3 or C-4 of

glucose will ensure that label is released as $CO_2$ upon fermentation to ethanol.

17. (a) Triose phosphate isomerase deficiency would inhibit conversion of DHAP to Ga3P and would cause accumulation of DHAP, preventing half of the glucose molecule (C1-C3) from being metabolized through the remainder of the glycolytic pathway. There would be a recovery of only 2 of the possible 4 moles of ATP from glucose, resulting in no net formation of ATP. In addition, DHAP, a product of the aldolase reaction, would likely reverse the aldolase (reaction) and eventually inhibit glycolysis. Either result would be lethal to a cell whose only energy source was glycolysis.

(b) The small amount of TPI activity would likely allow glycolysis to proceed slowly, but low energy (ATP) level will limit the growth rate under anaerobiosis. However, the yield of ATP is significantly greater when the pyruvate, formed during glycolysis, is oxidized to $CO_2$ and $H_2O$. Hence, the growth rate of the mutant should be correspondingly greater under aerobic growth conditions but not as great as the wild type.

(c) Cells expressing DHAP phosphatase would likely not grow anaerobically if glycolysis of glucose to lactate were the only pathway for ATP formation. The combined activities of TPI and DHAP phosphatase would be predicted to deplete the pool of triosephosphate and the yield of ATP per glucose would likely be less than 1.

19. (a) $K'_{eq} = 30$

(b) Hexoses brought into the glycolytic pathway must be phosphorylated to provide the appropriate substrate for the glycolytic enzymes and to trap the sugar within the cell. Phosphorylation at the expense of ATP or group translocation at the expense of PEP are common methods to activate the sugar molecules. Sucrose phosphorylase uses the exergonic lysis of the glycosidic bond between the hemiacetyl OH group of glucose and the hemiketal OH group of fructose to drive the endergonic phosphorylation of the hemiacetal C-1 OH group of glucose. Transfer of the phosphate to C-6, catalyzed by phosphoglucomutase, provides substrate for entry into the glycolytic pathway without addition of ATP. The net ATP yield will therefore be 3 rather than 2 moles of ATP per mole of glucose derived from sucrose. The fructose can be phosphorylated by ATP and used in the glycolytic pathway.

21. The free energy available from a reaction depends on the energies of the products compared with the substrates. Dehydration of 2-phosphoglycerate "traps" phospho-(enol)pyruvate in the enolate form. The hydrolysis products of PEP are phosphate and the enol form of pyruvate, but (enol) pyruvate is significantly less stable than (keto) pyruvate and rapidly tautomerizes to the more stable keto form. The tautomerization drives the reaction strongly toward products, resulting in a larger free energy difference between substrate and product.

23. Given the ratio of ATP/ADP and the $K'_{eq}$ of $10^6$, the equilibrium ratio of [Pyry]/[PEP] would be about $10^5$. This calculation supports the metabolic irreversibility of the pyruvate kinase reaction.

25. 6-phosphogluconate + $NADP^+$ $\longrightarrow$
$NADPH + H^+ + CO_2 + $ Ribulose-5-phosphate

27. (a) Unregulated hepatic PK theoretically could become part of a futile cycle. Net: GTP $\longrightarrow$ GDP + $P_i$ plus formation of cytosolic NADH at the expense of mitochondrial NADH.

(b) Activation by fructose-1,6-bisphosphate decreases $S_{0.5}$ for PEP and increases PK activity at a given PEP concentration. During gluconeogenesis, the fructose-1,6-bisphosphate concentration should diminish, owing to the hydrolytic activity of the FBPase-1. The low FBP concentration, coupled with the elevated ATP levels, could inhibit the hepatic pyruvate kinase.

## Chapter 13

1. The oxidation occurs first, creating a $\beta$-keto-carboxylate, which readily loses $CO_2$ via a resonance-stabilized carbanion.

3. Citrate is a tertiary alcohol, which does not oxidize readily.

5. The following sequence will do the task: $\alpha$-ketoglutarate, the tricarboxylic acid cycle to oxaloacetate, to phosphoenolpyruvate, to pyruvate, to acetyl-CoA, into the tricarboxylic acid cycle.

7. Both of the oxidative-decarboxylation steps in the tricarboxylic acid cycle are bypassed by the glyoxylate cycle.

9. (a) Increased concentrations of acetyl-CoA slow the activity of pyruvate dehydrogenase.

(b) The removal of succinyl-CoA for heme synthesis removes any control it has over lowering the activity of $\alpha$-ketoglutarate dehydrogenase and citrate synthase.

11. (a) False. Lipoamide transacetylase catalyzes reduction of the disulfide on lipoamide concomitantly with oxidation and transfer of the hydroxyethyl group from thiamine pyrophosphate. Dihydrolipoamide dehydrogenase catalyzes oxidation of dihydrolipoamide and reduction of $NAD^+$. (b) False. Hydrolysis of

acetyl-CoA thioester should yield as much free energy as succinyl-CoA hydrolysis, a process coupled to ADP phosphorylation (succinate thiokinase). (c) False. The methyl group of acetyl-CoA could be derived from pyruvate, from $\beta$ oxidation of long-chain fatty acids, or from amino acid metabolism. (d) False. If aconitase failed to discriminate between the ($-CH_2-COO^-$) groups, half the $CO_2$ would arise from oxaloacetate and half from the acetate carboxylate. The two $CO_2$ molecules released by oxidative decarboxylation of isocitrate and $\alpha$-ketoglutarate are derived from the carboxyl groups of oxaloacetate with which the acetyl-CoA was condensed. (e) False. Malate can easily be dehydrated to fumarate by reversal of the fumarase reaction.

13. (a) Without malate synthase, the yeast would be unable to grow on 2-carbon precursors as the sole carbon source because TCA cycle intermediates could not be synthesized.

(b) TCA cycle activity would markedly diminish if the cycle intermediates were being used in biosynthetic pathways. In addition, the biosynthetic pathways (lipids, amino acids, and carbohydrates) dependent on TCA cycle intermediates would also be inhibited by the lack of metabolites.

(c) If the PDH were inhibited more strongly than usual by acetyl-CoA, we might suspect that acetyl-CoA concentration in the mitochondrial matrix would markedly decrease, in turn limiting activity of citrate synthase and diminishing TCA cycle activity. Hence growth of the organism may be limited by lowered energy production and by diminished concentrations of biosynthetic precursors supplied by the TCA cycle.

15. (a) Fumarate + NADH + $H^+$ $\longrightarrow$ succinate + $NAD^+$.

(b) Phosphoenolpyruvate + $CO_2$ $\longrightarrow$ oxaloacetate + $P_i$
(PEP carboxylase)
Oxaloacetate + NADH + $H^+$ $\longrightarrow$ L-malate + $NAD^+$
(malate dehydrogenase)
L-Malate $\longrightarrow$ fumarate + HOH
(fumarase)
Fumarate + NADH + $H^+$ $\longrightarrow$ succinate + $NAD^+$
(fumarate reductase)

(c) In the reactions shown in part (b), four reducing equivalents (two hydride groups) are transferred to carbon acceptors. Reduction of oxaloacetate to malate by malate dehydrogenase (MDH) and reduction of fumarate to succinate by fumarate reductase each requires hydride (or equivalent) transfer from NADH to the organic substrate. In the reduction of pyruvate to lactate via LDH

only one hydride is used. Thus, two equivalents of $NAD^+$ are resupplied to glycolysis by the activities of MDH and fumarate reductase, whereas only one equivalent of $NAD^+$ is regenerated by LDH. Fumarate is one of the terminal electron acceptors used during anaerobic respiration in *E. coli*. However, each mole of PEP carboxylated is at the expense of 1 mole equivalent of ATP that could have been formed as a product of pyruvate kinase.

17. The committed step in a metabolic pathway is usually under metabolic control. Inhibition of the committed step in a metabolic sequence or pathway prevents the accumulation of unneeded intermediates and effectively precludes activity of the enzymes using those intermediates as substrates. The decarboxylation of pyruvate and the oxidative transfer of the hydroxyethyl group by pyruvate dehydrogenase constitutes the committed step in the pyruvate dehydrogenase catalytic sequence and is a logical control point.

## Chapter 14

1. In hemoglobin and myoglobin the heme serves as a carrier with oxygen–heme iron ($Fe^{2+}$) interacting in a ligand-metal coordination relationship. In the cytochromes heme plays a redox role, with the iron interconverting between $Fe^{2+}$ and $Fe^{3+}$.

3. (a) 12.5
(b) 16
(c) 10

5. Both iron and copper have two common stable ions; the transition between $Fe^{2+}$ and $Fe^{3+}$ or $Cu^{1+}$ and $Cu^{2+}$ are both redox reactions.

7. (a) Heme of the mitochondrial *b*-type cytochrome interacts hydrophobically with adjacent hydrophobic residues from the membrane-spanning $\alpha$ helices. The heme iron is fully coordinated through two imidazole groups from histidines in the protein. Heme in the *c*-type cytochromes is covalently bound to the protein through thioethers formed by the addition of cysteinyl sulfhydryl groups to the vinyl substituents on the heme ring. The iron is also fully liganded to a nitrogen from the imidazole group of histidine and a sulfur from the thioether linkage of methionine providing the fifth and sixth ligands. Heme *a* differs from the protoheme (heme) of the *b*- and *c*-type cytochromes by substitution of a formyl group at ring position 8 and a 17-carbon isoprenoid chain at position 2.

(b) Neither CO nor $CN^-$ at low concentration interact with the cytochrome $b$ or $c$ heme iron because no open ligand position is available to the iron in these heme proteins.

9. (a) In biological systems, the iron-sulfur centers are obligatorily single electron donors/acceptors, regardless of the number of Fe atoms in the center or their initial oxidation state.

(b) The 4Fe-4S cluster is found in iron-sulfur proteins that transfer electrons at low and at high potential. The reduction potential is a measure of the ease of addition of an electron to the couple, compared to the standard hydrogen electrode. Thus the protein component of the iron-sulfur protein affects the reduction potential.

11. $E°$ (pH 6) = +170 mV; $E°$ (pH 8) = 50 mV.

13. (a) The large amount of UQ is necessary to ensure efficient transfer of electrons from the mitochondrial dehydrogenases to complex III.

(b) (i) If UQ (ubiquinone) were limiting, the rate of oxidation of NADH by the mitochondrial electron-transfer system would also be limited. NADH concentration would increase, the $NAD^+$ supply would decrease, and the $NAD^+$-dependent dehydrogenases would be inhibited. Subsequently, the rate of oxidation of NADH-producing substrates would decrease.

(ii) Electrons from succinate oxidation are also transferred to ubiquinone from the succinate dehydrogenase, so a deficiency of the quinone would limit succinate oxidation.

(iii) Ascorbate plus a redox mediator reduces cytochrome $c$ but does not transfer electrons to ubiquinone. Limiting amounts of the quinone should not affect ascorbate-dependent reduction of cytochrome $c$.

(c) The deficiency of UQ will decrease the rate of extramitochondrial NADH oxidation resulting in an increase in the amount of pyruvate reduced to lactate. The lactate content would be expected to increase rapidly upon mild exercise.

15. (a) The uncoupler 2,4-dinitrophenol circumvents respiratory control in the mitochondria by short-circuiting the proton gradient. The lipophilic weak acid transports $H^+$ across the membrane, bypassing the $F_1$-$F_0$ complex. Substrates will be oxidized independently of ADP or ATP concentrations, and $O_2$ reduction will be more rapid than in state 3 respiration.

(b) Uncoupled mitochondria oxidize NADH and succinate but fail to phosphorylate ADP. Cellular processes will continue to utilize ATP, causing an accumulation of ADP and AMP. Increased levels of $NAD^+$ and ADP or AMP activate both the TCA cycle and glycolysis. The rate of oxidation of carbohydrates and fatty acids would markedly increase.

(c) The uncoupled mitochondria use little, if any, energy to phosphorylate ADP. The energy is dissipated as heat, leading to elevated body temperature (hyperthermia) and profuse perspiration in an effort to decrease body temperature.

(d) 2,4-Dinitrophenol is a lipid soluble weakly acidic compound thought to allow equilibration of protons across the inner mitochondrial membrane. Although protons are translocated from the matrix across the inner membrane during electron transfer, the proton gradient would immediately be depleted without passing through the $F_0$-$F_1$-dependent ADP phosphorylation system. Respiratory (ADP) control would be lost.

17. The transport process is electrogenic if the export of one molecule coupled with the import of another molecule yields a net charge difference across the membrane. In general terms, transfer of $A^{3-}$ from the matrix and $A^{3-}$ into the matrix yields a net negative charge on the cytoplasmic side of the membrane. Electrogenic processes are driven by the membrane potential ($\Delta\Psi$).

Neutral transport processes exchange molecules of net identical charge (sign and magnitude) in the opposite vectorial direction or oppositely charged molecules in the same vectorial direction. Processes coupled to the export of $OH^-$ are equivalent to $H^+$ import, that is, to an energetically favorable decrease in the proton gradient.

## Chapter 15

1. The word fixed refers to converting a gas into a liquid or solid.

3. Chlorophyll molecules are flat, completely conjugated ring compounds that obviously have a resonance form. Chlorophylls have a ring containing $9\pi$ bonds or $18\pi$ electrons, which fits the $n = 4$ situation in the Huckel ($4n + 2$) rule. The (cis, trans, trans)$_3$ double bond pattern of the chlorophylls also fits that of [18]annulene, which is aromatic.

5. Upon illumination, the chromatophore P870 is activated by absorption of a photon of light. The absorbance at 870 nm decreases because the $\pi$-cation radical of the oxidized chromatophore has a lower absorbance at that wavelength. Thus the trace monitoring 870 nm decreases upon illumination of the

chromatophore. The *c*-type cytochrome is added initially in the reduced form (cyt $c^{2+}$) and absorbs at 550 nm. Oxidized cytochrome $c$(cyt $c^{3+}$) has only a small absorbance at 550 nm. Electron transfer from reduced cytochrome $c$ to the $\pi$-cation radical (P870$^+$) regenerates the ground state P870 and forms oxidized cytochrome $c$. The absorbance of the P870 increases to the initial level and the absorbance at 550 nm of the cytochrome $c$ pool decreases.

7. (a) The absorbance at 275 nm decreases because the concentration of ubiquinone is diminished and because the semiquinone radical does not absorb 275 nm light. The absorbance increase at 450 nm is consistent with the formation of semiquinone radical that absorbs 450 nm light. The second flash activates transfer of a second electron from the photocenter to reduce the bound semiquinone to dihydroquinone. The dihydroquinone absorbs at neither 275 nm nor 450 nm. Reduction of the semiquinone form abolishes the absorbance at 450 nm. The decrease in absorbance at 275 nm is consistent with the decrease in oxidized (ubiquinone) concentration.

   (b) In this experiment, reduced cytochrome $c$ is the electron donor to P870$^+$. Were the reductant omitted, the P870$^+$ might oxidize the reduced quinone, or the activated reaction center P870* might return to ground state by emission of fluorescence.

9. Singlet-state oxygen causes cumulative oxidative damage to the chloroplast. Carotenoids compete with the oxygen for the triplet-state chlorophyll and inhibit singlet oxygen production. Plants deficient in carotenoids risk photooxidative damage because of an increased flux of singlet oxygen.

11. The mechanisms are different, but the energetic outcome is the same.

13. The enolate intermediate (figure 15.26) reacts with oxygen to produce a cycloperoxide intermediate that decomposes into glycerate-3-phosphate and glycolate-2-phosphate (figure 15.27).

15. Some of the many differences include the involvement of erythrose-4-phosphate and dihydroxyacetone phosphate in the production of sedoheptulose-1,7-bisphosphate, phosphorylation of ribulose-5-phosphate to ribulose-1,5-bisphosphate, the addition of $CO_2$ to ribulose-1,5-bisphosphate, and the production of two glyceraldehyde-3-phosphates.

## Chapter 16

1. The sequence, bond geometry, and linkage type of monosaccharides in an oligosaccharide or polysaccharide is determined by the specificity of the glycosyltransferases involved.

3. In general, amino groups are added to biological molecules in locations once occupied by keto groups. Because ketoses have C-2 as the carbonyl group the majority of the amino sugars in nature are 2-amino sugars.

5. Sugars have a large number of hydroxyl groups that form glycosidic bonds with the anomeric carbons of other sugars; these anomeric hydroxyl groups can be either $\alpha$ or $\beta$ conformation. More than eighty different glycosidic linkages have been identified.

7. Patients with I-cell disease do not phosphorylate the mannose residues on the glycoproteins that are lysosome-bound. These "lysosomal hydrolases" are therefore secreted.

9. Gluconic acid cannot form a cyclic hemiacetal and cannot form the glucosidic bonds required for participation in oligo- or polysaccharides.

11. The undecaprenol phosphate functions as a carrier and is acted upon by the series of reactions.

13. The major problem in the synthesis of complex carbohydrates outside the cell is the lack of an external energy source such as ATP. The biosynthesis of a bacterial cell wall such as the peptidoglycan occurs mainly inside the cell, the final step being the cross-linking of the peptidoglycan strands outside the cell. This reaction is a transpeptidation, which does not require any energy source.

## Chapter 17

1. The lecithin serves as an emulsifying agent that allows the aqueous and lipid phases to be dispersed in each other and increases the time required for phase separation.

3. The fatty acyl groups are placed in these two different positions by different enzymes.

5. The hydrophobic amino acid side chains on the exterior of the integral membrane protein interact with the hydrophobic lipid of the membrane exterior and are stable in the nonaqueous environment. These residues pack in the interior, hydrophobic environment of globular proteins.

7. Peripheral proteins are bound to the inner or outer aspects of the membrane through weak ionic interactions that include association with phospholipid head groups, by electrostatic or ionic interaction with a hydrophilic region of an integral membrane protein or through divalent metal ion bridging to the membrane surface. Peripherally bound proteins may be released without disrupting the membrane. Thus in-

creased salt concentration shields ionic charges and weakens the charge–charge interactions between the peripheral protein and the membrane components.

Integral proteins are dissolved into the lipid bilayer of the membrane through interactions of the hydrophobic amino acid side chains and fatty acyl groups of phospholipids. In order to remove integral membrane proteins, the membrane must be disrupted by addition of detergents or other chaotropic reagents to solubilize the protein and to prevent aggregation and precipitation of the hydrophobic proteins upon their removal from the membrane.

9. In principle, placing a hydrophilic residue in a nonaqueous environment is energetically unfavorable. In integral proteins with multiple $\alpha$ helices that span the membrane, hydrophilic side chains from different helical segments may interact and in some cases form a channel through which ions may diffuse. Portions of the helical segments exposed to the lipid will contain primarily hydrophobic amino acid residues.

11. Visualizing a protein actually revolving in a controlled manner within the membrane is difficult.

13. (a) $\Delta\Psi = +18$ mV; (b) from side 1 to side 2; (c) Side 1: 64.3 mM $K^+$, 50 mM $Na^+$, 144.3 mM $Cl^-$. Side 2: 85.7 mM $K^+$, 85.7 mM $Cl^-$. $\Delta\Psi = 7.5$ mV.

15. Sucrose uptake should be inhibited by a proton ionophore if uptake is by a proton symport. If a proteinbinding system was operational, membrane vesicles or cells subjected to osmotic shock would be defective in uptake. If a $Na^+$ symport was involved, uptake would be dependent on extracellular $Na^+$. If a PTS was operational, sucrose phosphorylation would be dependent on PEP and not ATP in a crude cell extract.

## Chapter 18

1. Arachidic acid produces 134 moles of ATP/mole, whereas arachidonic acid gives 126 moles of ATP/mole of fatty acid. This difference in ATP quantities is rather minor. Therefore any biological difference caused by dietary saturated versus unsaturated fats being due to their ATP yields is unlikely.

3. (a) 29.5 or 30 moles ATP
   (b) 64 moles ATP
   (c) 33% and 32%, respectively
   (d) 2.25 versus 2.13

5. (a) Oxidation of 1 mole of glucose yields 32 moles of ATP. 350 kcal of energy are stored as ATP or 1.9 kcal/g. (b) Oxidation of 1 mole of palmitate yields 106 moles of ATP. 1200 kcal of energy are stored as ATP, or 4.7 kcal/g. (c) Lipids are more highly reduced than are carbohydrates and supply more reducing equivalents to the electron-transport system than do carbohydrates. (d) Lipids have approximately 2.5 times greater energy storage capacity per gram than do carbohydrates and are stored as compact, hydrophobic globules. Storage of an equivalent energy as carbohydrate would require at least 2.5 times the mass, not considering the water of hydration that would accompany the carbohydrate.

7. The availability of citrate has no relationship to the flow of metabolites through the tricarboxylic acid cycle. Increased citrate concentrations result in increased cytoplasmic acetyl-CoA concentrations, which in turn increases fatty acid biosynthesis.

9. From a reaction sequence viewpoint it is a cyclic process. From a substrate viewpoint it is a decreasing spiral because the substrate's length decreases with each turn around the spiral.

11. Many mechanisms are conceivable. One possibility is the approach of a phosphate on succinyl-CoA to produce succinyl phosphate (a mixed anhydride).

13. (a) Ketone body formation in liver supplies an easily transported, water-soluble, energy-rich metabolite that can be used in lieu of glucose in many nonhepatic tissues. (b) $\beta$-hydroxybutyrate supplies an additional hydride (2 reducing equivalents) compared with acetoacetate. (c) Consider the reactions catalyzed by $\beta$-hydroxybutyrate dehydrogenase, 3-ketoacyl-CoA transferase, and thiolase.

15. (a) The $^{14}C$-labeled methyl group of acetyl-CoA will be C-16 of palmitate. (b) Only one deuterium atom from each labeled malonyl-CoA will remain in the reduced lipid chain. Carbons 2, 4, 6, 8, 10, 12, and 14 of palmitic acid will each have one deuterium label. (c) The $^{14}C$ label will be lost by decarboxylation and no label will remain in the palmitate.

17. Glucose-6-phosphate dehydrogenase, 6-phosphogluconate dehydrogenase, and the $NADP^+$-malic enzyme are sources of the 14 moles of NADPH required for biosynthesis of palmitate.

19. The thioesterase activity of the fatty acid synthase prefers the palmitoyl acyl carrier protein thioester as substrate.

## Chapter 19

1. Phosphatidylserine decarboxylase is a pyridoxal phosphate enzyme.

3. The synthesis of phosphatidylcholine is thermodynamically feasible because of the ATP used to phosphorylate choline and the CTP used to form CDP-choline.

5. One possibility is to remove the fatty acyl groups that have been damaged by oxidation.

7. Carriers of a defective gene for the hexosaminidase A enzyme still have a functional copy of the gene.

9. The activated component is different. In the first case (figure 19.4) the minor component is activated, whereas in the production of phosphatidylinositol (figure 19.6) the diacylglycerol is activated.

11. Arachidonic acid is stored in membranes as phospholipids with $C_{20}$ polyunsaturated fatty acids in the SN-2 position. Phospholipase $A_2$ releases arachidonic acid, which is then used to synthesize prostaglandins, which induce inflammation.

## Chapter 20

1. No

3. These frequencies are consistent if one assumes that the gene has only two alleles: $A + B = 1$ and $A^2 + 2AB + B^2 = 1$ describes the population where $A^2$ are "normal," $2AB$ are heterozygous and $B^2$ are homozygous for familial hypercholesterolemia. $2AB = 1/500$, $AB = 0.001$, $A = 0.999$ and $B = 0.001$. Therefore, $B^2 = (0.001)^2 = 0.000001$, which is the one in a million indicated in the text. These frequencies would be inconsistent if the gene has more than two alleles.

5. If a normal person is given an inhibitor for HMG-CoA reductase, cholesterol synthesis is inhibited in the liver. Lower levels of cholesterol then signal the synthesis of increased levels of LDL receptors. This increases the uptake of LDL into the liver and reduces serum LDL. In a patient with FH, this has little effect because there are no LDL receptors. The only effect is that the liver does not make as much cholesterol and does not contribute as much to serum LDL levels. The new liver will make normal amounts of LDL receptors and have normal uptake of LDL from the blood. This result will dramatically lower serum LDL levels and prevent the new heart from developing coronary artery disease. If the liver transplant had not been done, the heart transplant would have been to no avail.

7. The $^{14}C$ in HMG-CoA derived from 2-[$^{14}$-C]-acetate is marked in the structure.

$$\overset{-}{O}-\overset{O}{\underset{}{\overset{\|}{C}}}-\underset{CH_2}{\overset{\overset{*}{CH_3}}{\underset{|}{\overset{|}{\underset{|}{C}}}}}-\overset{OH}{\underset{CH_2}{\overset{|}{\underset{|}{C}}}}-\overset{O}{\underset{}{\overset{\|}{C}}}-S-CoA$$

$* = {}^{14}C$

9. The evolution of membranes, organelles, and organs combined with the evolutionary development of new metabolic sequences from previously developed enzymatic mechanisms probably best explains this phenomenon.

## Chapter 21

1. The oxide ion is a poor leaving group.

3. The two sequences differ by an ATP.

5. Presumably the availability of valine inhibited an enzyme, such as acetohydroxy acid synthase, in the early stages of valine, isoleucine and leucine biosynthesis.

7. Only when the environmental tryptophan is depleted will material flow through the pathway, or in this case, because of the auxotroph, only part way through the pathway.

9. In this two-step pathway the acetyl group is added in the first reaction and an acetate leaves, and an $HS^-$ is added in the second step. The acetyl group facilitates the removal of the serine oxygen. The OH group is a poor leaving group; acetate is better.

11. Glutamine, the product of glutamine synthase, is the source of nitrogen required in the synthesis of a number of diverse, structurally unrelated compounds synthesized by different pathways. Total inhibition of the glutamine synthase by a single product would in turn inhibit the synthesis of all compounds requiring glutamine.

13. Pyruvate, from glycolysis of glucose, is carboxylated to oxaloacetate or oxidized to acetyl-CoA. These metabolites enter the Krebs cycle, are metabolized to $\alpha$-ketoglutarate and oxaloacetate, then transaminated to aspartate or glutamate. Asn, Gln, and Pro are synthesized from Asp or Glu. The cycle replenishes intermediates via the anaplerotic reactions (e.g., carboxylation of pyruvate to form oxaloacetate).

15. Hydroxyproline is formed by a posttranslational modification of proline residues in the protein. The $^{14}C$-labeled hydroxyproline is not incorporated directly into the collagen because there is no genetic codon to specify the incorporation of hydroxyproline.

## Chapter 22

1. The *N*-acetyl groups in the *de novo* pathway prevent the spontaneous cyclization of the semialdehyde intermediate.

3. One of the nitrogens enters the urea cycle via carbamoyl phosphate that comes from ammonia. The second urea nitrogen enters the urea cycle as part of aspartic acid that can come from transamination of oxaloacetate. Glutamate is the source of the amino group in the transamination.

5. Because of the importance of the urea cycle, the capacity to convert ornithine into arginine is obvious. Complete loss of the ability to produce ornithine (a catalyst or carrier in the urea cycle) would limit the organism's control over production of its nitrogen waste product.

7. Based on figure 22.11 Thr, Ala, Ser, Gly, Cys, Asn, Asp, Gln, Glu, His, Arg, Pro, Val, and Met are glucogenic; Lys, Trp, and Leu are ketogenic; and Phe, Tyr, and Ile are both ketogenic and glucogenic.

9. Because of its critical role in ATP production, a homozygotic defect in a gene for a protein involved in glycolysis, the citric acid cycle, or the electron transport chain probably leads to the death of the cell(s) soon after fertilization.

11. Pyridoxal phosphate forms a Schiff base (imine) with the glycine. A carbon-bound hydrogen is labile, and the resulting carbanion stabilized by resonance back into the pyridoxal phosphate. The carbanion approaches the carbonyl carbon of the succinyl-CoA. Following the elimination of the CoASH, the intermediate shown in figure 22.13 is formed. The intermediate then loses a $CO_2$, forming a carbanion that is resonance stabilized back into the pyridoxal phosphate.

13. We would predict an increased arginase activity in the liver of the untreated diabetic animal. The untreated diabetic animal synthesizes glucose primarily by hepatic gluconeogenesis utilizing amino acids derived from protein catabolism as the carbon source. The urea cycle activity must increase to accommodate the increased flux of amino groups removed from the amino acids. Arginase catalyzes the hydrolysis of arginine yielding urea plus ornithine and is the rate-limiting step in the urea cycle.

15. (a) L-glutathione is synthesized in successive steps, catalyzed by γ-glutamyl cysteine synthase and glutathione synthase. Glutathione synthesis is directed by the substrate specificity of these enzymes. (b) Decreased glutathione synthesis would increase the probability of oxidative damage to the cell.

## Chapter 23

1. Carbamoyl phosphate synthase contributes to two processes: (a) the initial enzyme in the biosynthesis of pyrimidines and (b) a component in the synthesis of arginine biosynthesis or the urea cycle. In bacteria both of these processes occur within the same compartment. In human beings the carbamoyl phosphate synthase involved in the urea cycle is contained in the mitochondria; it is isolated from the cytosol counterpart that is involved in the biosynthesis of pyrmidines. Because the two carbamoyl phosphate synthases are in separate cellular compartments in human beings, control of the cytosol carbamoyl phosphate synthase by pyrimidine pathway products has no impact on the urea cycle.

3. Typically, the mechanism for amine addition involves a nucleophilic approach by a nitrogen, requiring a lone pair of electrons on the nitrogen. Ammonium ions are protonated at physiological pH and do not have a lone pair. The amide of glutamine is not protonated and carries a lone pair of electrons.

5. The phosphorylation step utilizing ATP converts the oxygen into a much better leaving group.

7. This oxygen becomes incorporated into an $H_2O$ molecule.

9. Because Lesch-Nyhan patients lack the enzyme hypoxanthine-guanine phosphoribosyltransferase, they accumulate high levels of PRPP, which stimulates purine biosynthesis to high levels, leading to production of large amounts of uric acid. The brain may not have high levels of *de novo* purine biosynthesis and probably relies on the salvage pathway enzymes for its purine nucleotides.

11. The product would be 3-amino-2-methylpropanoic acid.

## Chapter 24

1. In a liver cell, fatty acid synthesis takes place in the cytosol. It uses acetyl-CoA carboxylase and a large, multifunctional polypeptide fatty acid synthase. The first committed step is acetyl-CoA carboxylase, which is highly regulated by hormonal control; this results in the phosphorylation of the enzyme. Citrate is transported from the mitochondria and is used to generate acetyl-CoA and reducing power in the form of NADPH (NADPH is used in biosynthetic reactions instead of NADH). Citrate activates the carboxylase, but the end product, palmitate, inhibits the reaction. Fatty acid degradation occurs in the matrix of the mitochondria. A key point of regulation is on the uptake of the fatty acid into the matrix of the mitochondria. Malonyl-CoA, the product of the acetyl-CoA carboxylase, inhibits uptake and prevents the newly made palmitic acid from being degraded.

Gluconeogenesis utilizes many of the glycolytic enzymes, yet three of these enzymes in glycolysis have large negative free energy changes in the direction of pyruvate formation. These reactions must be replaced in gluconeogenesis to make glucose formation thermodynamically favorable. Replacement allows glycolysis and gluconeogenesis to be thermodynamically favorable and at the same time permits the pathways to be independently regulated to avoid a futile cycle.

3. A hormone receptor must do two things if it is to function properly:

(1) Distinguish the hormone from all other surrounding chemical signals and bind it with a very high affinity ($K_d$ ranges from $1 \times 10^{-7}$ M to $1 \times 10^{-12}$ M).

(2) Upon binding the hormone, undergo a conformational change into an active form that can then interact with other molecules that initiate the molecular events leading to the hormone's elicited response.

Proteins are the only macromolecules that can exhibit this kind of behavior (specific binding and conformation change).

5. (1) A polypeptide signal sequence must be present if the protein is to be transported into the endoplasmic reticulum and subsequently secreted.

(2) Additional polypeptide sequences are necessary for proper peptide chain folding (e.g., C peptide of insulin).

(3) Cleavage allows control of hormones from inactive to active form (e.g., thyroxine).

(4) Production of a number of different hormones from the same precursor allows coordinate production of several hormones. Specific cleavage by the cell allows control of which peptides are produced (e.g., cleavage of prepro-opiocortin to corticotropin, $\beta$-lipotropin, $\gamma$-lipotropin, $\alpha$-MSH, $\beta$-MSH, $\gamma$-MSH, endorphin, and enkephalin).

(5) A large precursor of the hormone can serve as a storage form (e.g., thyroglobulin).

7. The same amino acid, named as 5-oxoproline, is an intermediate in the $\gamma$-glutamyl cycle. The name pyroglutamate suggests formation involving dehydration via heat (fire) from glutamate.

9. Vitamin D can be considered both a hormone and a vitamin. Its mode of action is like that of other steroid hormones, and it is synthesized in the body. It can be given in the diet (e.g., in supplemented milk) and would then be called a vitamin.

## Chapter 25

1. Avery, working with two different strains of pneumococcus, was able to show that a fraction isolated from the pathogenic S strain that transformed the nonpathogenic R strain was DNA. This transforming activity was not affected by RNase, proteases, or enzymes that degrade capsular polysaccharides but was destroyed by treatment with DNase. Purified DNA from S cells was able to transform R cells into S cells *in vitro*.

3. dpCpGpTpA or, in abbreviated form, CGTA.

5. (a) In the major groove of B-form DNA:

Adenine   $N^7$; $N^6$
Cytosine   $N^4$
Guanine   $N^7$, $O^6$
Thymine   $O^4$

In the minor groove of B-form DNA:

Adenine   $N^3$
Cytosine   $O^2$
Guanine   $N^3$; $N^2$
Thymine   $O^2$

(b) Although the numbers of electronegative atoms capable of forming hydrogen bonds in the major and minor groove are similar, access to those in the minor groove is hindered by ribose moiety.

7. The melting temperature ($T_m$) of DNA is affected by the base composition, with the G-C-rich DNA having a higher $T_m$ than the A-T-rich DNA. The DNA samples could be heated in a spectrophotometer and the increase in absorbance of ultraviolet light (hyperchromism) could be plotted against temperature. The A-T-rich DNA would have a lower $T_m$ than the G-C-rich DNA.

9. A simple approach would be to take small aliquots of the sample and treat them with the enzymes ribonuclease (RNase) or deoxyribonuclease (DNase). Digestion of the sample by one of these enzymes would indicate whether the sample is RNA or DNA. Another method is to treat a small sample with alkali, which degrades RNA to mononucleotides, but only denatures DNA to the single-stranded form. One way to detect whether these treatments had any effect on the nucleic acid is to subject it to electrophoresis on an agarose gel. Free nucleotides, or even small oligonucleotides, are not visible on agarose gels, whereas the original viral nucleic acid should yield one (or more) discrete high molecular weight bands.

To determine whether the nucleic acid is single-stranded or double-stranded, it could be heated. A sharp increase in absorbance at 260 nm would indicate a double-stranded RNA or DNA; a broader melting curve would suggest a single-stranded nucleic acid. We could also analyze the nucleotide composition of the nucleic acid. Equivalence between A and T and between G and C would strongly suggest that the nucleic acid is double-stranded.

11. The 2'-OH group found on the ribose in RNA sterically prevents the B duplex from forming.

13. The differences in Ethidium (Et) binding capacities between linear duplex DNA and covalently closed circular DNA (cccDNA) can be understood in terms of the differences in topological constraints imposed on the two molecules. By intercalating between two adjacent base pairs, Et unwinds the double helix, which results in an increase in the length of the helix (pitch). For cccDNA, the conformational stress introduced by unwinding is compensated for by a change in tertiary structure of the molecule (i.e., supercoiling).

   At some point, the torsional stress caused by the positive supercoils will become energetically unfavorable and the tendency of the molecule will be toward winding, thus preventing further binding of Et. Linear duplex DNA does not undergo this torsional stress because it is not covalently closed and would be expected to have a greater binding capacity for Et.

15. The ratio of the histones in chromatin supports the model proposed for nucleosome structure. The core of the nucleosome is made up of an octamer of two molecules each of H2A, H2B, H3, and H4. The H1 histone seals off the nucleosome (i.e., only one H1 per nucleosome).

# Chapter 26

1. If DNA replication were dispersive, each strand of each daughter molecule would have had an intermediate density in the experiment. The fact that after one generation, one strand of the DNA still had the same density as the parental ($^{15}$N) DNA served to further corroborate the conclusion that DNA replication is semiconservative.

3. As has been seen in many other biochemical reactions in the cell, the generation of pyrophosphate coupled with its hydrolysis by pyrophosphatase is the major driving force for DNA synthesis. The $Mg^{2+}$ can bind to the transition state intermediate (trigonal bipyramidal intermediate) and stabilize it, lowering the energy of activation. The metal ion can also promote the reaction by charge shielding; the $Mg^{2+}NTP$ complex is the actual substrate, with the metal reducing the negative charge on the phosphate groups so

as not to repel the electron pair of the attacking nucleophile.

5. The use of a primer increases fidelity in DNA synthesis by providing a more extensive stacked double helix to which the first few bases of DNA are added, allowing the proofreading exonuclease activity of the polymerase to evaluate more accurately the stability of the newly synthesized DNA. The reason that the primer is made of RNA and not DNA may be because RNA, even in a double-stranded nucleic acid, is readily recognized as being different from DNA (RNA–DNA duplexes adopt a different conformation because of the presence of the 2'-OH on the ribonucleotides.) A proofreading DNA polymerase recognizes the differences in the RNA primer and replaces it with DNA.

7. About 33 min would be required to replicate the chromosome because you have a bidirectional mode of replication with two replication forks. If you had multifork replication (one round of replication starting before the other finished), a 20-min division time would be possible.

9. $2.9 \times 10^9$ bp $\times$ 1 sec/60 bp $\times$ 1 h/3600 sec = 13,400 h
   The human genome would need at least 13,400 replication origins to be completely replicated in one hour.

11. DNA in eukaryotic chromosomes is complexed with histone proteins in complexes called nucleosomes. These DNA-protein complexes are disassembled directly in front of the replication fork. The nucleosome disassembly may be rate-limiting for the migration of the replication forks, as the rate of migration is slower in eukaryotes than prokaryotes. The length of Okazaki fragments is also similar to the size of the DNA between nucleosomes (about 200 bp). One model that would allow the synthesis of new eukaryotic DNA and nucleosome formation would be the disassembly of the histones in front of the replication fork and then the reassembly of the histones on the two duplex strands. Histone synthesis is closely coupled to DNA replication.

13. The SOS response is reversed when the protease activity of recA can no longer be activated because most or all of the damaged DNA has been repaired or eliminated and intact lexA protein levels begin to rise. Then lexA can act as a repressor, binding to the gene control regions of all of the genes it regulates, including its own and that of recA.

15. The recA protein is very important in DNA repair. An insult to the DNA leads to the activation of the

protease function of recA, which then cleaves lexA protein, turning on the genes in the SOS response. Once the DNA is repaired, the recA protease is inactivated and new lexA protein is made, repressing the DNA repair genes again. RecA mutations are totally deficient in homologous recombination, demonstrating the important role of the recA protein in this process. The purified recA protein will catalyze the exchange between duplex and single-stranded DNAs with the hydrolysis of ATP. Also, recA protein will form a complex between two circular helices if one helix is gapped on one strand. These functions of recA protein would place it at the hub of activities in recombination. RecA mutants would be very useful in genetic research because mutants generated would not be repaired, and in recombinant DNA cloning recombinational events between vector recombinant DNA and host genomic DNA would not occur.

## Chapter 27

1. Spaces are introduced every 10 nucleotides for clarity: 5'-CAAAAAACGG ACCGGGTGTA CAACTTTTAC TATGGCGTGA CACCTAAATT ATAGGCAGAA ATAAGTACAT GACTATTGGG AGGAGCAGGA ACAAGTAGG-3'.

3. The frequency of occurrence of the sequence -CTGCAG- (a PstI site) in DNA that is 80% G + C is $0.4 \times 0.1 \times 0.4 \times 0.4 \times 0.1 \times 0.4$, or $(0.4)^4(0.1)^2$ = once every 3906 bp. An AAGCTT (HindIII) site should occur $(0.4)^2(0.1)^4$ = once every 62,500 bp.

   The *E. coli* genome composition is approximately 26% G, 26% C, 24% A, and 24% T. PstI would cleave this genome $(0.26)^4(0.24)^2$ = once every 3,800 bp, and HindIII, $(0.26)^2(0.24)^4$ = once every 4,500 bp.

5. There is no difference between the single-stranded ends generated by BamHI and by MboI. These ends are complementary, and they can anneal and be ligated together by DNA ligase. The resulting sequence contains an MboI site and can be cleaved by that enzyme. It has only a 25% probability of restoring a BamHI site, however, depending on the nucleotide located adjacent to the MboI site.

7. In order to isolate a 15-kb gene from a genome containing $3 \times 10^9$ bp, it would be necessary to isolate $3 \times 10^9/15,000$ or 200,000 fragments per genome, or 200,000 clones. We recommend that a library 3 to 10 times the minimum size should be prepared, to ensure a high probability that a given fragment will be represented at least once. In the case of the gene

specified, the library should therefore contain between $6 \times 10^5$ and $2 \times 10^6$ clones.

9. (a) 2, 4
   (b) 1, 5
   (c) 1, 3

11. To reduce the hybridization stringency, researchers choose conditions that stabilize double-stranded DNA, allowing sequences that share only partial complementarity to form base pairs. Examples of such conditions include reducing the hybridization temperature and increasing the salt concentration in solution.

13. Chromosome jumping is a useful procedure to traverse long distances and skip troublesome regions of the genome (repetitious sequences, etc.). One approach with this technique is to digest the genome with rare cutting restriction enzymes (*Not*I recognizes an 8-base sequence and generates an average fragment size of 500 kb of DNA). The fragments are circularized with a small marker DNA between the ends. The marker DNA contains sequences necessary for cloning in λ. These circular DNA fragments are then cleaved with another restriction enzyme that produces fragments small enough to clone in λ. With this procedure only clones that contain the marker DNA (i.e., the ends of the original fragment) are isolated. This method would permit only one jump; thus another method used with this technique is called a linking library. The same sample of DNA is digested with a restriction enzyme that gives smaller fragments. These smaller fragments are then circularized with the same marker DNA used in the jumping library. The circular DNA is then digested with *Not*I to linearize the fragments for insertion into the vector. The linking library carries sequences on both sides of the *Not*I restriction sites while the jumping library carries sequences from one side of two adjacent *Not*I sites.

## Chapter 28

1. RNA and DNA polymerases catalyze the same reaction mechanistically, involving hydrolysis of a nucleotide triphosphate to release pyrophosphate and form a phosphodiester bond. In both cases, the order of nucleotide addition is specified by the template, and synthesis of the growing nucleic acid chain is in a 5' to 3' direction (the enzymes move in a 3' to 5' direction along the template strand). In addition to the obvious difference in substrates (RNA polymerase utilizes ribonucleotides, whereas DNA polymerase utilizes deoxyribonucleotides), these two enzymes differ in their requirements for initiating synthesis:

RNA polymerase initiates *de novo,* and does not require a primer, unlike DNA polymerase. Finally, RNA polymerase does not perform any proofreading activity, unlike DNA polymerases, which generally have a 3′ to 5′ exonuclease activity to remove misincorporated nucleotides.

3. mRNA has a short half-life (a couple of minutes); tRNA and rRNA are very stable (half-life measured in hours) and accumulate to make up most of the RNA in the cell (95%).

5. The bases within the base-paired region of each arm of the tRNA cloverleaf stack in a manner similar to the base stacking described in Chapter 25 for DNA. In addition to the base stacking within base-paired regions, there is also stacking of one helix on top of another in the tRNA molecule. In particular, the acceptor stem stacks with the TΨC stem and loop to form one nearly continuous stacked double helix. The anticodon stem and the D stem also stack on top of one another.

7. The D and T loops in tRNA interact with each other to form the tertiary structure, leaving only the anticodon with a single-stranded loop able to be cleaved by RNase.

9. The RNA polymerase binds to the same side of the duplex at the -10 and -35 regions with about two turns of the duplex helix between these boxes. This binding site can be determined by a variety of ''footprint'' experiments. DNA that is 5′-labeled can be mixed with the polymerase, and regions that are protected from digestion with an enzyme like DNase I can be determined on a sequencing gel. Sequences with tight contact with the RNA polymerase are observed as a series of blank spots in the sequencing ladder that look like footprints.

11. The 5′ terminus is capped ($G^m$pppX—), generating a guanosine nucleoside with 2′ and 3′-OH groups and a 5′-5′ pyrophosphate linkage. The other end of the mRNA has poly(A) added, so the 3′ end is adenosine. Therefore most of the mRNA in the cell has a guanosine and an adenosine 2′ and 3′-OH and no typical 5′-triphosphate ending (pppN—).

13. By flattening and widening the DNA minor groove, TATA binding protein may assist other factors in forming an open complexlike structure with separated strands. The bending induced by TATA binding protein would bring the DNA upstream and downstream of the TATA box closer together, promoting interactions between proteins bound to upstream elements and the start site of transcription.

15. If an intron or part of an intron containing a stop signal for translation was not removed, this mRNA would be longer but would yield a shorter polypeptide. Alternative splicing would remove the intron or use an alternative splice site, generating a shorter mRNA and a longer polypeptide.

## Chapter 29

1. The difference in the mechanism of translation initiation in prokaryotes compared to eukaryotes has profound consequences for the strategy used to coordinate the expression of a set of genes in the two systems. In prokaryotes this coordination is achieved by organizing genes into transcription units that are transcribed to give polycistronic mRNAs. Each cistron within a polycistronic mRNA begins with a Shine-Dalgarno sequence and initiation AUG codon. In contrast, in eukaryotes, in which translation begins almost invariably (and exclusively) at the 5′-most AUG codon, monocistronic mRNAs are necessarily the order of the day. Genes that must be coordinately expressed are consequently not organized into transcriptional units, and coordination must be achieved in some other way.

3. This serine tRNA has inosine at the 5′ position of its anticodon, which can pair with U, C, or A. Thus the anticodon of this tRNA is most likely IGA (given in the correct 5′ to 3′ direction).

5. A number of variations have been found in the universal genetic code in genes located in mitochondria and chloroplast. These variations in the meaning of some of the code words represent divergences from the standard genetic code and not an independent origin of another genetic code. These divergences probably arose in these organelles because of the limited number of genes coded and requirements for the synthesis of ribosomes and tRNA. Clearly, something was unusual when only 24 types of tRNAs were found in mitochondria. Mitochondria do not use all 61 codons.

7. Met  Val  Glu  Ile  Arg  Asp  Thr  His  Leu  Lys  Lys  Gln  Ile  Ala  Phe  Ter  Ter

9. (5′)AAY TGG GCN CAR TGY AAY CC(3′). R is the abbreviation for a mixture of A or G (R = puRine), Y is the abbreviation for a mixture of C and T (Y = pYrimidine), and N represents a mixture of all four Nucleotides.

11. The x-ray crystal structure of EF-Tu bound to GDP and GTP is known. Based on this structure, many of the amino acids shared among IF-2, EF-Tu, EF-G, and RF-3 apparently are somehow involved in binding of GTP and GDP. Thus these proteins (and many others, including the proto-oncogene, Ras) share a similar GTP-binding domain.

13. Import into the endoplasmic reticulum requires an N-terminal signal sequence that contains a long stretch of hydrophobic amino acids. The mitochondrial transit peptide is a hydrophilic sequence rich in serine and threonine, with regularly spaced basic amino acids. Import into the ER requires the signal recognition particle and its receptor, but mitochondrial import does not require the SRP and presumably uses a different receptor. Import into mitochondria requires a membrane potential, but import into the ER does not.

15. GAA encodes a glutamic acid residue. If this glu is essential for the catalytic activity of the protein, any mutation (except GAA to GAG) will be harmful to the activity of the protein. If the glu residue is not essential for catalysis or proper folding, the possible mutations, in order of increasing potential severity, are:

| GAA changes to | Glu changes to |
| --- | --- |
| GAG | Glu |
| GAT or GAC | Asp |
| CAA | Gln |
| GCA | Ala |
| AAA | Lys |
| GTA | Val |
| GGA | Gly |
| TAA | Terminator |

## Chapter 30

1. The existence of a repressor in *lac* operon regulation was first suggested by the results of studies of merodiploids. In merodiploids of the type $i^+z^-/Fi^-z^+$, Jacob and Monod were able to demonstrate that the $i^+$ (inducible) allele is dominant to the $i^-$ (constitutive) allele when on the same chromosome (*cis*) or on a different chromosome (*trans*) with respect to the $z^+$ allele.

3. (a) Cells with the genotype $i^s o^+ z^+$ have a "superrepressor" $i^s$ mutation, which causes the repressor to be insensitive to an inducer. Thus even though the operator and $\beta$-galactosidase loci are wild-type, no $\beta$-galactosidase will be produced in this mutant. On media containing X-gal, with or without IPTG, colonies will be white.

   (b) Cells with the genotype $i^s o^c z^+$ still have the superrepressor, but the $\beta$-galactosidase gene is under the control of a constitutive operator, $o^c$. This mutation interferes with the ability of the repressor to bind and repress transcription. Therefore, $\beta$-galactosidase probably would be produced continuously. On X-gal, with or without IPTG, colonies will be blue.

   (c) Merodiploids with the genotype $i^s o^c z^-/i^+ o^+ z^+$ would behave like the mutant described in part (a).

   (d) Merodiploid cells with the genotype $i^+ o^c z^-/i^- o^+ z^+$ would exhibit $\beta$-galactosidase regulation essentially identical to that of wild-type cells; on X-gal alone, colonies would be white. In the presence of an inducer, the $\beta$-galactosidase gene would be induced; on X-gal and IPTG, colonies would be blue.

5.

| | Uninduced | | Induced | |
| --- | --- | --- | --- | --- |
| Strains | Enz A | Enz B | Enz A | Enz B |
| **Haploid** | | | | |
| (1) $R^+O^+A^+B^+$ | 1 | 1 | 100 | 100 |
| (2) $R^+O^cA^+B^+$ | 1–100 | 1–100 | 100 | 100 |
| (3) $R^-O^+A^+B^+$ | 100 | 100 | 100 | 100 |
| **Diploid** | | | | |
| (4) $R^+O^+A^+B^+/$ $R^+O^+A^+B^+$ | 2 | 2 | 200 | 200 |
| (5) $R^+O^cA^+B^+/$ $R^+O^+A^+B^+$ | 2–101 | 2–101 | 200 | 200 |
| (6) $R^+O^+A^-B^+/$ $R^+O^+A^+B^+$ | 1 | 2 | 100 | 200 |
| (7) $R^-O^+A^+B^+/$ $R^+O^+A^+B^+$ | 2 | 2 | 200 | 200 |

7. The explanation for the less than perfect match of most promoters to the consensus sequence is to be found in the need to regulate transcription. Transcriptional regulation is achieved in many instances by the selective improvement of the affinity of specific promoters for RNA polymerase. Such selective improvement is well illustrated in the case of regulation of the *lac* operon.

9. One possibility is the binding of the repressor nonspecifically to the DNA and then searching in one dimension (binding to the DNA and then sliding along until the promoter is reached). Also, a long section of DNA would not be randomly distributed but would form a loose ball of DNA that would define a domain much smaller than the solution in the test tube. When the repressor was released from the DNA it could more quickly find another strand of DNA to bind (effectively giving a much higher concentration of DNA).

11. The synthesis of ribosomal proteins is regulated in *E. coli* by translational regulation (i.e., free ribosomal proteins inhibit the translation of their own

mRNA). As long as rRNA is being made, these proteins bind to the rRNA and the translation of the ribosomal proteins continues. The genes for the ribosomal proteins are clustered in a number of operons that produce polycistronic mRNAs. One of the simplest operons is $P_{L11}$, which codes for proteins L1 and L11. L1 is the regulatory protein and can bind to the 23S rRNA or to the 5′ end of its own polycistronic mRNA. If the levels of L1 increase, it binds to its own mRNA and inhibits translation of both L1 and L11 proteins. This mechanism keeps the levels of L1 and L11 in register with the amount of rRNA. The other ribosomal proteins are regulated in a similar manner.

13. Translation is an amplification process, in that each molecule of ribosomal protein mRNA can yield many copies of the corresponding protein.

15. The genes encoded by bacteriophage lambda DNA are organized so that gene products required for the same function or process are clustered. Clustering facilitates regulation because all the gene products required at the same time can be induced simultaneously. Clustering also leads to more efficient organization and tighter packing of the bacteriophage genome.

17. The final proof of the two-site model came with the cocrystallization of the regulatory protein and its DNA-binding site. The protein binds on one side of the DNA in two adjacent major grooves. The hydrogen-bonding groups exposed in the major groove present many possibilities for interactions with the amino acid side chains of the protein.

## Chapter 31

1. One definition is: *the DNA (or RNA) sequences necessary to produce a peptide (or RNA).* Some viruses have RNA as their genetic material and some genes do not code for a protein but make a functional RNA such as tRNA or rRNA.

3. No, this construct would not be activated by binding of *lexA* because *lexA* does not have a eukaryotic activation domain.

5. A deletion of *HMLE* would remove the element repressing transcription of $HML_\alpha$ at the "storage" location, and result in the constitutive expression of $HML_\alpha$ genes from this locus. The consequences of such constitutivity would depend on the mating type of the mutant. If $HML_\alpha$ were at the *MAT* locus ($MAT_\alpha$), the arrangement characteristic of the α mating type, the cell exhibits the behavior expected of the α mating type. However, if $HMR_a$ were at the

*MAT* locus ($MAT_a$), the arrangement giving rise to the a mating type, the cell would resemble the diploid, expressing both a and α genes simultaneously, and thus be sterile.

Mutants that began as α-type would become sterile at a rate conditioned by the frequency of transposition of the $HMR_a$ to the *MAT* locus. In homothallic strains this frequency is very high, approximately once per cell division, while in heterothallic strains it is considerably lower. Deletion of the gene encoding α2 from $MAT_\alpha$ would result in a failure to inhibit expression of the a-specific genes. Haploid mutants that contained such a deletion would therefore resemble the diploid, and be sterile. If the α2 gene contained by the $HML_\alpha$ locus were similarly mutated, mutants of this type would be unable to switch to the α-type. However, transposition of $HMR_a$ to the *MAT* locus would allow for expression of the a mating type. Again the frequency of transposition would condition the rate of switching between sterile and a-type, the frequency being far greater in homothallic strains than in heterothallic strains.

7. The results of the nuclear transplantation experiments of Briggs and King indicated that nuclei, to a certain stage in embryonic development, remain totipotent, in the sense that they can support the normal development of enucleated differentiating embryos. They found that blastula nuclei, although already "committed" to differentiated pathways, were capable of being "reprogrammed" by a proper environment to allow for some degree of dedifferentiation. Gastrula nuclei, in contrast, were found to support normal development only at a much reduced efficiency.

9. Histones bind to DNA by electrostatic interaction of basic amino acids (arginine and lysine) with the negative charge on the phosphate. These electrostatic (negative-positive charge) interactions are weakened by high salt, which will allow the histone proteins to disassociate from the DNA. Many regulatory proteins may initially interact in a nonspecific fashion with DNA until they locate their high-affinity binding sites. Once these proteins bind at their specific recognition sites, the major interaction is through hydrogen bonding and hydrophobic interactions. These hydrophobic interactions are stabilized by high salt.

11. MspI would cleave at every CCGG sequence in or near the globin genes, regardless of the cell type from which the DNA had been isolated because this enzyme is insensitive to methylation. Globin genes isolated from erythroblast cells would not differ from those of other tissues in their cleavage pattern with MspI.

HpaII does not recognize CCGG sequences when the second C is methylated. In tissues in which genes are *not* transcribed, CpG sequences tend to be methylated more often than in tissues where the same genes are transcribed. Thus HpaII is expected to recognize and cleave more sites in globin genes in erythroblast DNA than in DNA from other cell types (where the globin genes are not transcribed and the corresponding DNA is more highly methylated).

13. Most transcription factors have two-domain structures, one for DNA binding and another for protein binding. The DNA-binding domains involve various structural motifs that interact with the DNA in the major groove. The type of structure previously discussed in prokaryotes is the helix-turn-helix motif, which is also found in some eukaryotic transcription factors. A more common motif in eukaryotes is the zinc finger (zinc forms a tetrahedral complex with histidines and cysteines). Another type of structure found is the leucine zipper (a highly conserved stretch of amino acids with net basic charge followed by a region of four leucine residues at intervals of seven amino acids). All of these motifs found in these transcription factors involve interactions in the major groove of the DNA with either an $\alpha$-helix segment (more common) or a two-stranded antiparallel $\beta$ sheet.

15. One strategy employed to generate sufficient amounts of a product required at a specific stage of development is gene amplification. An alternative strategy used to provide large amounts of required products early in development is the accumulation of mRNAs or proteins prior to need.

17. Although the precise regulatory relationships between homeotic genes remain unclear, it is apparent that a hierarchy of interactions exists among gap, pair-rule, and segment polarity genes. Each gene class is influenced by the action of earlier acting genes that control larger units of pattern, and by some members of the same class. Thus gap genes that are expressed early influence the pattern of expression of pair-rule, (e.g., fushi terazu, *ftz*), segment polarity (e.g., engrailed, *en*) and homeotic genes (e.g., ultrabithorax, *ubx*), which are expressed later in a hierarchical fashion. Although segment polarity genes do not affect the expression of gap or pair-rule genes that occupy a higher position in the hierarchy, some gap genes, and presumably genes at other levels in the hierarchy, have been found to be mutually negative regulators of each other.

# Glossary

## A

**A form.** A duplex DNA structure with right-handed twisting in which the planes of the base pairs are tilted about 70° with respect to the helix axis.

**Acetal.** The product formed by the successive condensation of two alcohols with a single aldehyde. It contains two ether-linked oxygens attached to a central carbon atom.

**Acetyl-CoA.** Acetyl-coenzyme A, a high-energy ester of acetic acid that is important both in the tricarboxylic acid cycle and in fatty acid biosynthesis.

**Actin.** A protein found in combination with myosin in muscle and also found as filaments constituting an important part of the cytoskeleton in many eukaryotic cells.

**Actinomycin D.** An antibiotic that binds to DNA and inhibits RNA chain elongation.

**Activated complex.** The highest free energy state of a complex in going from reactants to products.

**Active site.** The region of an enzyme molecule that contains the substrate-binding site and the catalytic site for converting the substrate(s) into product(s).

**Active transport.** The energy-dependent transport of a substance across a membrane.

**Adenine.** A purine base found in DNA or RNA.

**Adenosine.** A purine nucleoside found in DNA, RNA, and many cofactors.

**Adenosine diphosphate (ADP).** The nucleotide formed by adding a pyrophosphate group to the 5′-OH group of adenosine.

**Adenosine triphosphate (ATP).** The nucleotide formed by adding yet another phosphate group to the pyrophosphate group on ADP.

**Adenylate cyclase.** The enzyme that catalyzes the formation of cyclic 3′,5′-adenosine monophosphate (cAMP) from ATP.

**Adipocyte.** A specialized cell that functions as a storage depot for lipid.

**Aerobe.** An organism that utilizes oxygen for growth.

**Affinity chromatography.** A column chromatographic technique that employs attached functional groups that have a specific affinity for sites on particular proteins.

**Alcohol.** A molecule with a hydroxyl group attached to a carbon atom.

**Aldehyde.** A molecule containing a doubly bonded oxygen and a hydrogen attached to the same carbon atom.

**Alleles.** Alternative forms of a gene.

**Allosteric enzyme.** An enzyme the active site of which can be altered by the binding of a small molecule at a non-overlapping site.

**Allosteric site.** Location on an allosteric enzyme where the allosteric effector binds.

**Aminotransferase.** An enzyme that catalyzes the transfer of an amino group from an α-amino acid to an α-keto acid.

**Amphibolic pathway.** A metabolic pathway that functions in both catabolism and anabolism.

**Anabolism.** Metabolism that involves biosynthesis.

**Anaerobe.** An organism that does not require oxygen for maintenance or growth.

**Anaplerotic reaction.** An enzyme-catalyzed reaction that replenishes the intermediates in a cyclic pathway.

**Angstrom (Å).** A unit of length equal to $10^{-8}$ cm.

**Anomers.** The sugar isomers that differ in configuration about the carbonyl carbon atom. This carbon atom is called the anomeric carbon atom of the sugar.

**Antibiotic.** A natural product that inhibits bacterial growth (is bacteriostatic) and sometimes results in bacterial death (is bacteriocidal).

**Antibody.** A specific protein that interacts with a foreign substance (antigen) in a specific way.

**Anticodon.** A sequence of three bases on the transfer RNA that pair with the bases in the corresponding codon on the messenger RNA.

**Antigen.** A foreign substance that triggers antibody formation and is bound by the corresponding antibody.

**Antiparallel $\beta$-pleated sheet ($\beta$ sheet).** A hydrogen-bonded secondary structure formed between two or more extended polypeptide chains.

**Antiport.** A protein system that can transport different molecules in opposite directions.

**Apoactivator.** A regulatory protein that stimulates transcription from one or more genes in the presence of a coactivator molecule.

**Asexual reproduction.** Growth and cell duplication that does not involve the union of nuclei from cells of opposite mating types.

**Asymmetrical carbon.** A carbon that is covalently bonded to four different groups.

**Attenuator.** A provisional transcription stop signal.

**Autoradiography.** The technique of exposing film in the presence of disintegrating radioactive particles. Used to obtain information on the distribution of radioactivity in a gel or a thin cell section.

**Autoregulation.** The process by which a gene regulates its own expression.

**Autotroph.** An organism that can form its organic constituents from $CO_2$.

**Auxin.** A plant growth hormone usually concentrated in the apical bud.

**Auxotroph.** A mutant that cannot grow on the minimal medium on which a wild-type member of the same species can grow.

**Avogadro's number.** The number of molecules in a gram molecular weight of any compound ($6.022 \times 10^{23}$).

## B

**$\beta$ bend.** A characteristic way of turning an extended polypeptide chain in a different direction, involving the minimum number of residues.

**$\beta$ oxidation.** Oxidative degradation of fatty acids that occurs by the successive oxidation of the $\beta$-carbon atom.

**$\beta$ sheet.** A sheetlike structure formed by the interaction between two or more extended polypeptide chains.

**B cell.** One of the major types of cells in the immune system. B cells can differentiate to form memory cells or antibody-forming cells.

**B form.** The most common form of duplex DNA, containing a right-handed helix and about 10 (10.5 exactly) bp per turn of the helix axis.

**Base analog.** A compound, usually a purine or a pyrimidine, that differs somewhat from a normal nucleic acid base.

**Base stacking.** The close packing of the planes of base pairs, commonly found in DNA and RNA structures.

**Bidirectional replication.** Replication in both directions away from the origin, as opposed to replication in one direction only (unidirectional replication).

**Bilayer.** A double layer of lipid molecules with the hydrophilic ends oriented outward, in contact with water, and the hydrophobic parts oriented inward toward each other.

**Bile salts.** Derivatives of cholesterol with detergent properties that aid in the solubilization of lipid molecules in the digestive tract.

**Biochemical pathway.** A series of enzyme-catalyzed reactions that results in the conversion of a precursor molecule into a product molecule.

**Bioluminescence.** The production of light by a biochemical system.

**Blastoderm.** The stage in embryogenesis when a unicellular layer at the surface surrounds the yolk mass.

**Bond energy.** The energy required to break a bond.

**Branchpoint.** An intermediate in a biochemical pathway that can follow more than one route in subsequent steps.

**Buffer.** A conjugate acid–base pair that is capable of resisting changes in pH when acid or base is added to the system. This tendency is maximal when the conjugate forms are present in equal amounts.

## C

**cAMP.** $3',5'$ cyclic adenosine monophosphate. The cAMP molecule plays a key role in metabolic regulation.

**CAP.** The catabolite gene activator protein, sometimes incorrectly referred to as the CRP protein. The latter term, in small letters (*crp*), should be used to refer to the gene but not to the protein.

**Capping.** Covalent modification involving the addition of a modified guanine group in a $5'-5''$ linkage. It occurs only in eukaryotes, primarily on mRNA molecules.

**Carbanion.** A negatively charged carbon atom.

**Carbohydrate.** A polyhydroxy aldehyde or ketone.

**Carboxylic acid.** A molecule containing a carbon atom attached to a hydroxyl group and to an oxygen atom by a double bond.

**Carcinogen.** A chemical that can cause cancer.

**Carotenoids.** Lipid-soluble pigments that are made from isoprene units.

**Catabolism.** That part of metabolism concerned with degradation reactions.

**Catabolite repression.** The general repression of transcription of genes associated with catabolism that is seen in the presence of glucose.

**Catecholamines.** Hormones that are amino derivatives of catechol, for example, epinephrine or norepinephrine.

**Catalyst.** A compound that lowers the activation energy of a reaction without itself being consumed.

**Catalytic site.** The site of the enzyme involved in the catalytic process.

**Catenane.** An interlocked pair of circular structures, such as covalently closed DNA molecules.

**Catenation.** The linking of molecules without any direct covalent bonding between them, as when two circular DNA molecules interlock like the links in a chain.

**cDNA.** Complementary DNA, made *in vitro* from the mRNA by the enzyme, reverse transcriptase, and deoxyribonucleotide triphosphates.

**Cell commitment.** That stage in a cell's life when it becomes committed to a certain line of development.

**Cell cycle.** All of those stages that a cell passes through from one cell generation to the next.

**Cell line.** An established clone originally derived from a whole organism through a long process of cultivation.

**Cell lineage.** The pedigree of cells resulting from binary fission.

**Cell wall.** A tough outer coating found in many plant, fungal, and bacterial cells that accounts for their ability to withstand mechanical stress or abrupt changes in osmotic pressure. Cell walls always contain a carbohydrate component and frequently also a peptide and a lipid component.

**Channeling.** The direct transfer of a reaction intermediate from one enzyme to the next.

**Chelate.** A molecule that contains more than one binding site and frequently binds to another molecule through more than one binding site at the same time.

**Chemiosmotic coupling.** The coupling of ATP synthesis to an electrochemical potential gradient across a membrane.

**Chimeric DNA.** Recombinant DNA the components of which originate from two or more different sources.

**Chiral compound.** A compound that can exist in two forms that are nonsuperimposable images of one another.

**Chlorophyll.** A green photosynthetic pigment that is made of a magnesium dihydroporphyrin complex.

**Chloroplast.** A chlorophyll-containing photosynthetic organelle, found in eukaryotic cells, that can harness light energy.

**Chromatin.** The nucleoprotein fibers of eukaryotic chromosomes.

**Chromatography.** A procedure for separating chemically similar molecules. Segregation is usually carried out on paper or in glass or metal columns with the help of different solvents. The paper or glass columns contain porous solids with functional groups that have limited affinities for the molecules being separated.

**Chromosome.** A threadlike structure, visible in the cell nucleus during metaphase, that carries the hereditary information.

**Chromosome puff.** A swollen region of a giant chromosome; the swelling reflects a high degree of transcription activity.

**Cis dominance.** Property of a sequence or a gene that exerts a dominant effect on a gene to which it is linked.

**Cistron.** A genetic unit that encodes a single polypeptide chain.

**Citric acid cycle.** *See* tricarboxylic acid (TCA) cycle.

**Clone.** One of a group of genetically identical cells or organisms derived from a common ancestor.

**Cloning vector.** A self-replicating entity to which foreign DNA can be covalently attached for purposes of amplification in host cells.

**Closed system.** A system that exchanges neither matter nor energy with its surroundings.

**Coactivator.** A molecule that functions in conjunction with a protein apoactivator. For example, cAMP is a coactivator of the CAP protein.

**Codon.** In a messenger RNA molecule, a sequence of three bases that represents a particular amino acid.

**Coenzyme.** An organic molecule that associates with enzymes and affects their activity.

**Cofactor.** A small molecule required for enzyme activity. It could be organic, like a coenzyme, or inorganic, like a metallic cation.

**Competitive inhibition.** An inhibitor that competes with the substrate for binding to the enzyme.

**Complementary base sequence.** For a given sequence of nucleic acids, the nucleic acids that are related to them by the rules of base pairing.

**Configuration.** The spatial arrangement of atoms in a molecule that can only be changed by breaking and reforming covalent bonds.

**Conformation.** The spatial arrangement of groups in a molecule that can be changed without breaking covalent bonds. Often molecules with the same configuration can have more than one conformation.

**Consensus sequence.** In nucleic acids, the "average" sequence that signals a certain type of action by a specific protein. The sequences actually observed usually vary around this average.

**Constitutive enzymes.** Enzymes synthesized in fixed amounts, regardless of growth conditions.

**Cooperative binding.** A situation in which the binding of one substituent to a macromolecule favors the binding of another. For example, DNA cooperatively binds histone molecules, and hemoglobin cooperatively binds oxygen molecules.

**Coordinate induction.** The simultaneous expression of two or more genes.

**Cosmid.** A DNA molecule with *cos* ends from λ bacteriophage that can be packaged *in vitro* into a virus for infection purposes.

**Cot curve.** A curve that indicates the rate of DNA–DNA annealing as a function of DNA concentration and time.

**Covalent bond.** A chemical bond that involves sharing electron pairs.

**Cytidine.** A pyrimidine nucleoside found in DNA and RNA.

**Cytochromes.** Heme-containing proteins that function as electron carriers in oxidative phosphorylation and photosynthesis.

**Cytokinin.** A plant hormone produced in root tissue.

**Cytoplasm.** The contents enclosed by the plasma membrane, excluding the nucleus.

**Cytosine.** A pyrimidine base found in DNA and RNA.

**Cytoskeleton.** The filamentous skeleton, formed in the cytoplasm, that is largely responsible for controlling cell shape.

**Cytosol.** The liquid portion of the cytoplasm, including the macromolecules but not including the larger structures, such as subcellular organelles or cytoskeleton.

# D

**D loop.** An extended loop of single-stranded DNA displaced from a duplex structure by an oligonucleotide.

**Dalton.** A unit of mass equivalent to the mass of a hydrogen atom ($1.66 \times 10^{-24}$ g).

**Dark reactions.** Reactions that can occur in the dark, in a process that is usually associated with light, such as the dark reactions of photosynthesis.

*De novo* **pathway.** A biochemical pathway that starts from elementary substrates and ends in the synthesis of a biochemical.

**Deamination.** The enzymatic removal of an amine group, as in the deamination of an amino acid to an $\alpha$-keto acid.

**Dehydrogenase.** An enzyme that catalyzes the removal of a pair of electrons (and usually one or two protons) from a substrate molecule.

**Deletion mutation.** A mutation in which one or more nucleotides is removed from a region of the gene.

**Denaturation.** The disruption of the native folded structure of a nucleic acid or protein molecule; may be due to heat, chemical treatment, or change in pH.

**Density-gradient centrifugation.** The separation, by centrifugation, of molecules according to their density, in a gradient varying in solute concentration.

**Dialysis.** Removal of small molecules from a macromolecule preparation by allowing them to pass across a semipermeable membrane.

**Diauxic growth.** Growth on a mixture of two carbon sources in which one carbon source is used up before the other one is mobilized. For example, in the presence of glucose and lactose, *E. coli* utilizes the glucose before the lactose.

**Difference spectra.** Display comparing the absorption spectra of a molecule or an assembly of molecules in different states, for example, those of mitochondria under oxidizing or reducing conditions.

**Differential centrifugation.** Separation of molecules or organelles by sedimentation rate.

**Differentiation.** A change in the form and pattern of a cell and the genes it expresses as a result of growth and replication, usually during development of a multicellular organism. Also occurs in microorganisms (e.g., in sporulation).

**Diploid cell.** A cell that contains two chromosomes ($2N$) of each type.

**Dipole.** A separation of charge within a single molecule.

**Directed mutagenesis.** In a DNA sequence, an intentional alteration that can be genetically inherited.

**Dissociation constant.** An equilibrium constant for the dissociation of a molecule into two parts (e.g., dissociation of acetic acid into acetate anion and proton).

**Disulfide bridge.** A covalent linkage formed by oxidation between two SH groups either in the same polypeptide chain or in different polypeptide chains.

**DNA.** Deoxyribonucleic acid. A polydeoxyribonucleotide in which the sugar is deoxyribose; the main repository of genetic information in all cells and most viruses.

**DNA cloning.** The propagation of individual segments of DNA as clones.

**DNA library.** A mixture of clones, each containing a cloning vector and a segment of DNA from a source of interest.

**DNA polymerase.** An enzyme that catalyzes the formation of 3'-5' phosphodiester bonds from deoxyribonucleotide triphosphates.

**DNA supercoiling.** The coiling of double helix DNA upon itself.

**Domain.** A segment of a folded protein structure showing conformational integrity. A domain can comprise the entire protein or just a fraction of the protein. Some proteins, such as antibodies, contain many structural domains.

**Dominant.** Describing an allele the phenotype of which is expressed regardless of whether the organism is homozygous or heterozygous for that allele.

**Double helix.** A structure in which two helically twisted polynucleotide strands are held together by hydrogen bonding and base stacking.

**Duplex.** Synonymous with double helix.

**Dyad symmetry.** Property of a structure that can be rotated by 180° to produce the same structure.

# E

**Ecdysone.** A hormone that stimulates the molting process in insects.

**Edman degradation.** A systematic method of sequencing proteins, proceeding by stepwise removal of single amino acids from the amino terminal of a polypeptide chain.

**Eicosanoid.** Any fatty acid with 20 carbons.

**Electron carrier.** A protein or coenzyme that can reversibly gain and lose electrons and that serves the function of carrying electrons from one site to another.

**Electron donor.** A substance that donates electrons in an oxidation–reduction reaction.

**Electrophoresis.** The movement of particles in an electrical field. A commonly used technique for analysis of mixtures of molecules in solution according to their electrophoretic mobilities.

**Elongation factors.** Protein factors uniquely required during the elongation phase of protein synthesis. Elongation factor G (EF-G) brings about the movement of the peptidyl-tRNA from the A site to the P site of the ribosome.

**Eluate.** The effluent from a chromatographic column.

**Embryo.** Plant or animal at an early stage of development.

**Enantiomers.** Isomers that are mirror images of each other.

**Endergonic reaction.** A reaction with a positive free energy change.

**Endocrine glands.** Specialized tissues the function of which is to synthesize and secrete hormones.

**Endonuclease.** An enzyme that breaks a phosphodiester linkage at some point within a polynucleotide chain.

**Endopeptidase.** An enzyme that breaks a polypeptide chain at an internal peptide linkage.

**Endoplasmic reticulum.** A system of double membranes in the cytoplasm that is involved in the synthesis of transported proteins. The rough endoplasmic reticulum has ribosomes associated with it. The smooth endoplasmic reticulum does not.

**End-product (feedback) inhibition.** The inhibition of the first enzyme in a pathway by the end product of that pathway.

**Energy charge.** The fractional degree to which the AMP–ADP–ATP system is filled with high-energy phosphates (phosphoryl groups).

**Enhancer.** A DNA sequence that can bind protein factors that stimulate transcription at an appreciable distance from the site where it is located. It acts in either orientation and either upstream or downstream from the promoter.

**Entropy.** A thermodynamic measure of the randomness of a system.

**Enzyme.** A protein that contains a catalytic site for a biochemical reaction.

**Epimers.** Two stereoisomers with more than one chiral center that differ in configuration at one of their chiral centers.

**Equilibrium.** In chemistry the point at which the concentrations of two compounds are such that the interconversion of one compound into the other compound does not result in any change in free energy.

*Escherichia coli (E. coli).* A gram negative bacterium commonly found in the vertebrate intestine. It is the bacterium most frequently used in the study of biochemistry and genetics.

**Established cell line.** A group of cultured cells derived from a single origin and capable of stable growth for many generations.

**Ether.** A molecule containing two carbons linked by an oxygen atom.

**Eukaryote.** A cell or organism that has a membrane-bound nucleus.

**Excision repair.** DNA repair in which a damaged region is replaced.

**Excited state.** An energy-rich state of an atom or a molecule, produced by the absorption of radiant energy.

**Exergonic reaction.** A chemical reaction that takes place with a negative change in free energy.

**Exon.** A segment within a gene that carries part of the coding information for a protein.

**Exonuclease.** An enzyme that breaks a phosphodiester linkage at one or the other end of a polynucleotide chain so as to release a single nucleotides or small oligonucleotides.

## F

**F factor.** A large bacterial plasmid, known as the sex-factor plasmid because it permits mating between $F^+$ and $F^-$ bacteria.

**Facilitated diffusion.** Diffusion of a substance across a membrane through a protein transporter.

**Facultative aerobe.** An organism that can use molecular oxygen in its metabolism but that also can live anaerobically.

**Fatty acid.** A long-chain hydrocarbon containing a carboxyl group at one end. Saturated fatty acids have completely saturated hydrocarbon chains. Unsaturated fatty acids have one or more carbon–carbon double bonds in their hydrocarbon chains.

**Feedback inhibition.** *See* end-product inhibition.

**Fermentation.** The energy-generating breakdown of glucose or related molecules by a process that does not require molecular oxygen.

**Fingerprinting.** The characteristic two-dimensional paper chromatogram obtained from the partial hydrolysis of a protein or a nucleic acid.

**Fluorescence.** The emission of light by an excited molecule in the process of making the transition from the excited state to the ground state.

**Footprinting.** A technique that results in a DNA sequence ladder in which part of the ladder is missing due to the binding of protein to the DNA before processing.

**Frameshift mutations.** Insertions or deletions of genetic material that lead to a shift in the translation of the reading frame. The mutation usually leads to nonfunctional proteins.

**Free energy.** That part of the energy of a system that is available to do useful work.

**Furanose.** A sugar that contains a five-member ring as a result of intramolecular hemiacetal formation.

**Futile cycle.** *See* pseudocycle.

## G

**$G_1$ phase.** That period of the cell cycle in which preparations are being made for chromosome duplication, which takes place in the S phase.

**$G_2$ phase.** That period of the cell cycle between S phase and mitosis (M phase).

**Gametes.** The ova and the sperm, haploid cells that unite during fertilization to generate a diploid zygote.

**Gel exclusion chromatography.** A technique that makes use of certain polymers that can form porous beads with varying pore sizes. In columns made from such beads, it is possible to separate molecules, which cannot penetrate beads of a given pore size, from smaller molecules that can.

**Gene.** A segment of the genome that codes for a functional product.

**Gene amplification.** The duplication of a particular gene within a chromosome two or more times.

**Gene splicing.** The cutting and rejoining of DNA sequences.

**General recombination.** Recombination that occurs between homologous chromosomes at homologous sites.

**Generation time.** The time it takes for a cell to double its mass under specified conditions.

**Genetic map.** The arrangement of genes or other identifiable sequences on a chromosome.

**Genome.** The total genetic content of a cell or a virus.

**Genotype.** The genetic characteristics of an organism (distinguished from its observable characteristics, or phenotype).

**Globular protein.** A folded protein that adopts an approximately globular shape.

**Gluconeogenesis.** The production of sugars from nonsugar precursors, such as lactate or amino acids. Applies more specifically to the production of free glucose by vertebrate livers.

**Glycogen.** A polymer of glucose residues in 1,4 linkage and 1,6 linkage at branchpoints.

**Glycogenic.** Describing amino acids the metabolism of which may lead to gluconeogenesis.

**Glycolipid.** A lipid containing a carbohydrate group.

**Glycolysis.** The catabolic conversion of glucose to pyruvate with the production of ATP.

**Glycoprotein.** A protein linked to an oligosaccharide or a polysaccharide.

**Glycosaminoglycans.** Long, unbranched polysaccharide chains composed of repeating disaccharide subunits in which one of the two sugars is either N-acetylglucosamine or N-acetylgalactosamine.

**Glycosidic bond.** The bond between a sugar and an alcohol. Also the bond that links two sugars in disaccharides, oligosaccharides, and polysaccharides.

**Glyoxylate cycle.** A pathway that uses acetyl-CoA and two auxiliary enzymes to convert acetate into succinate and carbohydrates.

**Glyoxysome.** An organelle containing key enzymes of the glyoxylate cycle.

**Goldman equation.** An equation expressing the quantitative relationship between the concentrations of charged species on either side of a membrane and the resting transmembrane potential.

**Golgi apparatus.** A complex series of double-membrane structures that interact with the endoplasmic reticulum and that serve as a transfer point for proteins destined for other organelles, the plasma membrane, or extracellular transport.

**Gram molecular weight.** For a given compound, the weight in grams that is numerically equal to its molecular weight.

**Ground state.** The lowest electronic energy state of an atom or a molecule.

**Growth factor.** A substance that must be present in the growth medium to permit cell proliferation.

**Growth fork.** The region on a DNA duplex molecule where synthesis is taking place. It resembles a fork in shape because it consists of a region of duplex DNA connected to a region of unwound single strands.

**Guanine.** A purine base found in DNA or RNA.

**Guanosine.** A purine nucleoside found in DNA and RNA.

# H

**Hairpin loop.** A single-stranded complementary region that folds back on itself and base-pairs into a double helix.

**Half-life.** The time required for the disappearance of one half of a substance.

**Haploid cell.** A cell containing only one chromosome of each type.

**Heavy isotopes.** Forms of atoms that contain greater numbers of neutrons (e.g., $^{15}N$, $^{13}C$).

**Helix.** A spiral structure with a repeating pattern.

**Heme.** An iron–porphyrin complex found in hemoglobin and cytochromes.

**Hemiacetal.** The product formed by the condensation of an aldehyde with an alcohol; it contains one oxygen linked to a central carbon in a hydroxyl fashion and one oxygen linked to the same central carbon by an ether linkage.

**Henderson-Hasselbach equation.** An equation that relates the $pK_a$ to the pH and the ratio of the proton acceptor $(A^-)$ and the proton donor (HA) species of a conjugate acid–base pair.

**Heterochromatin.** Highly condensed regions of chromosomes that are not usually transcriptionally active.

**Heteroduplex.** An annealed duplex structure between two DNA strands that do not show perfect complementarity. Can arise by mutation, recombination, or the annealing of complementary single-stranded DNAs.

**Heteropolymer.** A polymer containing more than one type of monomeric unit.

**Heterotroph.** An organism that requires preformed organic compounds for growth.

**Heterozygous.** Describing an organism (a heterozygote) that carries two different alleles for a given gene.

**Hexose.** A sugar with a six-carbon backbone.

**High-energy compound.** A compound that undergoes hydrolysis with a high negative standard free energy change.

**Histones.** The family of basic proteins that is normally associated with DNA in most cells of eukaryotic organisms.

**Holoenzyme.** An intact enzyme containing all of its subunits with full enzymatic activity.

**Homeobox.** A conserved sequence of 180 bp encoding a protein domain found in many eukaryotic regulatory proteins.

**Homologous chromosomes.** Chromosomes that carry the same pattern of genes but not necessarily the same alleles.

**Homopolymer.** A polymer composed of only one type of monomeric building block.

**Homozygous.** Describing an organism (a homozygote) that carries two identical alleles for a given gene.

**Hormone.** A chemical substance made in one cell and secreted so as to influence the metabolic activity of a select group of cells located elsewhere in the organism.

**Hormone receptor.** A protein that is located on the cell membrane or inside the responsive cell and that interacts specifically with the hormone.

**Host cell.** A cell used for growth and reproduction of a virus.

**Hybrid (or chimeric) plasmid.** A plasmid that contains DNA from two different organisms.

**Hydrogen bond.** A weak attractive force between one electronegative atom and a hydrogen atom that is covalently linked to a second electronegative atom.

**Hydrolysis.** The cleavage of a molecule by the addition of water.

**Hydrophilic.** Preferring to be in contact with water.

**Hydrophobic.** Preferring not to be in contact with water, as is the case with the hydrocarbon portion of a fatty acid or phospholipid chain.

## I

**Ion-exchange resin.** A polymeric resinous substance, usually in bead form, that contains fixed groups with positive or negative charge. A cation exchange resin has negatively charged groups and is therefore useful in exchanging the cationic groups in a test sample. The resin is usually used in the form of a column, as in other column chromatographic systems.

**Isoelectric pH.** The pH at which a protein has no net charge.

**Isomerase.** An enzyme that catalyzes an intramolecular rearrangement.

**Isomerization.** Rearrangement of atomic groups within the same molecule without any loss or gain of atoms.

**Isoprene.** The hydrocarbon 2-methyl-1,3-butadiene, which in some form serves as the precursor for many lipid molecules.

**Isozymes.** Multiple forms of an enzyme that differ from one another in one or more properties.

## K

$K_m$. *See* Michaelis constant.

**Ketogenic.** Describing amino acids that are metabolized to acetoacetate and acetate.

**Ketone.** A functional group of an organic compound in which a carbon atom is double-bonded to an oxygen. Neither of the other substituents attached to the carbon is a hydrogen. Otherwise the group would be called an aldehyde.

**Ketone bodies.** Refers to acetoacetate, acetone, and $\beta$-hydroxybutyrate made from acetyl-CoA in the liver and used for energy in nonhepatic tissues.

**Ketosis.** A condition in which the concentration of ketone bodies in the blood or urine is unusually high.

**Kilobase.** One thousand bases in a DNA molecule.

**Kinase.** An enzyme catalyzing phosphorylation of an acceptor molecule, usually with ATP serving as the phosphate (phosphoryl) donor.

**Kinetochore.** A structure that attaches laterally to the centromere of a chromosome; it is the site of chromosome tubule attachment.

**Krebs cycle.** *See* tricarboxylic acid (TCA) cycle.

## L

**Lampbrush chromosome.** Giant diplotene chromosome found in the oocyte nucleus. The loops that are observed are the sites of extensive gene expression.

**Law of mass action.** The finding that the rate of a chemical reaction is a function of the product of the concentrations of the reacting species.

**Leader region.** The region of an mRNA between the 5′ end and the initiation codon for translation of the first polypeptide chain.

**Lectins.** Agglutinating proteins usually extracted from plants.

**Ligase.** An enzyme that catalyzes the joining of two molecules together. In DNA it joins 3′-OH to 5′ phosphates.

**Linkers.** Short oligonucleotides that can be ligated to larger DNA fragments, then cleaved to yield overlapping cohesive ends, suitable for ligation to other DNAs that contain comparable cohesive ends.

**Linking number.** The net number of times one polynucleotide chain crosses over another polynucleotide chain. By convention, right-handed crossovers are given a plus designation.

**Lipid.** A biological molecule that is soluble in organic solvents. Lipids include steroids, fatty acids, prostaglandins, terpenes, and waxes.

**Lipid bilayer (*see* Bilayer).** Model for the structure of the cell membrane based on the hydrophobic interaction between phospholipids.

**Lipopolysaccharide.** Usually refers to a unique glycolipid found in Gram negative bacteria.

**Lyase.** An enzyme that catalyzes the removal of a group to form a double bond, or the reverse reaction.

**Lysogenic virus.** A virus that can adopt an inactive (lysogenic) state, in which it maintains its genome within a cell instead of entering the lytic cycle. The circumstances that determine whether a lysogenic (temperate) virus adopts an inactive state or an active lytic state are often subtle and depend on the physiological state of the infected cell.

**Lysosome.** An organelle that contains hydrolytic enzymes designed to break down proteins that are targeted to that organelle.

**M**

**M phase.** That period of the cell cycle when mitosis takes place.

**Meiosis.** Process in which diploid cells undergo division to form haploid sex cells.

**Membrane transport.** The facilitated transport of a molecule across a membrane.

**Merodiploid.** An organism that is diploid for some but not all of its genes.

**Mesosome.** An invagination of the bacterial cell membrane.

**Messenger RNA (mRNA).** The template RNA carrying the message for protein synthesis.

**Metabolic turnover.** A measure of the rate at which already existing molecules of the given species are replaced by newly synthesized molecules of the same type. Usually isotopic labeling is required to measure turnover.

**Metabolism.** The sum total of the enzyme-catalyzed reactions that occur in a living organism.

**Metamorphosis.** A change of form, especially the conversion of a larval form to an adult form.

**Metaphase.** That stage in mitosis or meiosis when all of the chromosomes are lined up on the equator (i.e., an imaginary line that bisects the cell).

**Micelle.** An aggregate of lipids in which the polar head groups face outward and the hydrophobic tails face inward; no solvent is trapped in the center.

**Michaelis constant ($K_m$).** The substrate concentration at which an enzyme-catalyzed reaction proceeds at one-half maximum velocity.

**Michaelis-Menten equation** (also known as the Henri-Michaelis-Menten equation). An equation relating the reaction velocity to the substrate concentration of an enzyme.

**Microtubules.** Thin tubules, made from globular proteins, that serve multiple purposes in eukaryotic cells.

**Mismatch repair.** The replacement of a base in a heteroduplex structure by one that forms a Watson-Crick base pair.

**Missense mutation.** A change in which a codon for one amino acid is replaced by a codon for another amino acid.

**Mitochondrion.** An organelle, found in eukaryotic cells, in which oxidative phosphorylation takes place. It contains its own genome and unique ribosomes to carry out protein synthesis of only a fraction of the proteins located in this organelle.

**Mitosis.** The process whereby replicated chromosomes segregate equally toward opposite poles prior to cell division.

**Mixed-function oxidases.** Enzymes that use molecular oxygen to oxidize two different molecules simultaneously, usually a substrate and a coenzyme.

**Mobile genetic element.** A segment of the genome that can move as a unit from one location on the genome to another, without any requirement for sequence homology.

**Molecularity of a reaction.** The number of molecules involved in a specific reaction step.

**Monolayer.** A single layer of oriented lipid molecules.

**Mutagen.** An agent that can bring about a heritable change (mutation) in an organism.

**Mutagenesis.** A process that leads to a change in the genetic material that is inherited in subsequent generations.

**Mutant.** An organism that carries an altered gene or change in its genome.

**Mutarotation.** The change in optical rotation of a sugar that is observed immediately after it is dissolved in aqueous solution, as the result of the slow approach of equilibrium of a pyranose or a furanose in its $\alpha$ and $\beta$ forms.

**Mutation.** The genetically inheritable alteration of a gene or group of genes.

**Myofibril.** A unit of thick and thin filaments in a muscle fiber.

**Myosin.** The main protein of the thick filaments in a muscle myofibril. It is composed of two coiled subunits ($M_r$ about 220,000) that can aggregate to form a thick filament that is globular at each end.

**N**

**Nascent RNA.** The initial transcripts of RNA, before any modification or processing.

**Negative control.** Regulation of the activity by an inhibitory mechanism.

**Negative feedback.** Regulation of a reaction or a pathway by the end product.

**Nernst equation.** An equation that relates the redox potential to the standard redox potential and the concentrations of the oxidized and reduced form of the couple.

**Nitrogen cycle.** The passage of nitrogen through various valence states, as the result of reactions carried out by a wide variety of different organisms.

**Nitrogen fixation.** Conversion of atmospheric nitrogen into a form that can be converted by biochemical reactions to an organic form. This reaction is carried out by a very limited number of microorganisms.

**Nitrogenous base.** An aromatic nitrogen-containing molecule with basic properties. Such bases include purines and pyrimidines.

**Noncompetitive inhibitor.** An inhibitor of enzyme activity the effect of which is not reversed by increasing the concentration of substrate molecule.

**Nonsense mutation.** A change in the base sequence that converts a sense codon (one that specifies an amino acid) to one that specifies a stop (a nonsense codon). There are three nonsense codons.

**Northern blotting.** *See* Southern blotting.

**Nuclease.** An enzyme that cleaves phosphodiester bonds of nucleic acids.

**Nucleic acids.** Polymers of the ribonucleotides or deoxyribonucleotides.

**Nucleohistone.** A complex of DNA and histone.

**Nucleolus.** A spherical structure visible in the nucleus during interphase. The nucleolus is associated with a site on the chromosome that is involved in ribosomal RNA synthesis.

**Nucleophilic group.** An electron-rich group that tends to attack an electron-deficient nucleus.

**Nucleosome.** A complex of DNA and an octamer of histone proteins in which a small stretch of the duplex is wrapped around a molecular bead of histone.

**Nucleotide.** An organic molecule containing a purine or pyrimidine base, a five-carbon sugar (ribose or deoxyribose), and one or more phosphate groups.

**Nucleus.** In eukaryotic cells, the centrally located organelle that encloses most of the chromosomes. Minor amounts of chromosomal substance are found in some other organelles, most notably the mitochondria and the chloroplasts.

## O

**Okazaki fragment.** A short segment of single-stranded DNA that is an intermediate in DNA synthesis. In bacteria, Okazaki fragments are 1,000–2,000 bases in length; in eukaryotes, 100–200 bases in length.

**Oligonucleotide.** A polynucleotide containing a small number of nucleotides. The linkages are the same as in a polynucleotide; the only distinguishing feature is the small size.

**Oligosaccharide.** A molecule containing a small number of sugar residues joined in a linear or a branched structure by glycosidic bonds.

**Oncogene.** A gene of cellular or viral origin that is responsible for rapid, unruly growth of animal cells.

**Operon.** A group of contiguous genes that are coordinately regulated by two cis-acting elements: A promoter and an operator. Found only in prokaryotic cells.

**Optical activity.** The property of a molecule that leads to rotation of the plane of polarization of plane-polarized light when the latter is transmitted through the substance. Chirality is a necessary and sufficient property for optical activity.

**Organelle.** A subcellular membrane-bound body with a well-defined function.

**Osmotic pressure.** The pressure generated by the mass flow of water to that side of a membrane-bound structure that contains the higher concentration of solute molecules. A stable osmotic pressure is seen in systems in which the membrane is not permeable to some of the solute molecules.

**Oxidation.** The loss of electrons from a compound.

**Oxidative phosphorylation.** The formation of ATP as the result of the transfer of electrons to oxygen.

**Oxido-reductase.** An enzyme that catalyzes oxidation–reduction reactions.

## P

**Palindrome.** A sequence of bases that reads the same in both directions on opposite strands of the DNA duplex (e.g., GAATTC).

**Pentose.** A sugar with five carbon atoms.

**Pentose phosphate pathway.** The pathway involving the oxidation of glucose-6-phosphate to pentose phosphates and further reactions of pentose phosphates.

**Peptide.** An organic molecule in which a covalent amide bond is formed between the $\alpha$-amino group of one amino acid and the $\alpha$-carboxyl group of another amino acid, with the elimination of a water molecule.

**Peptide mapping.** Same as fingerprinting.

**Peptidoglycan.** The main component of the bacterial cell wall, consisting of a two-dimensional network of heteropolysaccharides running in one direction, cross-linked with polypeptides running in the perpendicular direction.

**Periplasm.** The space between the inner and outer membranes of a bacterium.

**Permease.** A protein that catalyzes the transport of a specific small molecule across a membrane.

**Peroxisomes.** Subcellular organelles that contain flavin-requiring oxidases and that regenerate oxidized flavin by reaction with oxygen.

**Phenotype.** The observable trait(s) that result from the genotype in cooperation with the environment.

**Phenylketonuria.** A human disease caused by a genetic deficiency in the enzyme that converts phenylalanine to tyrosine. The immediate cause of the disease is an excess of phenylalanine. The condition can be alleviated by a diet low in phenylalanine.

**Pheromone.** A hormonelike substance associated with insects that acts as an attractant.

**Phosphodiester.** A molecule containing two alcohols esterified to a single molecule of phosphate. For example, the backbone of nucleic acids is connected by 5′-3′ phosphodiester linkages between the adjacent individual nucleotide residues.

**Phosphogluconate pathway.** Another name for the pentose phosphate pathway. This name derives from the fact that 6-phosphogluconate is an intermediate in the formation of pentoses from glucose.

**Phospholipid.** A lipid containing charged hydrophilic phosphate groups; a component of cell membranes.

**Phosphorolysis.** Phosphate-induced cleavage of a molecule. In the process the phosphate becomes covalently linked to one of the degradation products.

**Phosphorylation.** The formation of a phosphate derivative of a biomolecule.

**Photoreactivation.** DNA repair in which the damaged region is repaired with the help of light and an enzyme. The lesion is repaired without excision from the DNA.

**Photosynthesis.** The biosynthesis that directly harnesses the chemical energy resulting from the absorption of light. Frequently used to refer to the formation of carbohydrates from $CO_2$ that occurs in the chloroplasts of plants or the plastids of photosynthetic microorganisms.

**Pitch length (or pitch).** The number of base pairs per turn of a duplex helix.

**pK.** The negative logarithm of the equilibrium constant.

**Plaque.** A circular clearing on a lawn of bacterial or cultured cells, resulting from cell lysis and production of phage or animal virus progeny.

**Plasma membrane.** The membrane that surrounds the cytoplasm.

**Plasmid.** A circular DNA duplex that replicates autonomously in bacteria. Plasmids that integrate into the host genome are called episomes. Plasmids differ from viruses in that they never form infectious nucleoprotein particles.

**Polar group.** A hydrophilic (water-loving) group.

**Polar mutation.** A mutation in one gene that reduces the expression of a gene or genes distal to the promoter in the same operon.

**Polarimeter.** An instrument for determining the rotation of polarization of light as the light passes through a solution containing an optically active substance.

**Polyamine.** A hydrocarbon containing more than two amino groups.

**Polycistronic messenger RNA.** In prokaryotes, an RNA that contains two or more cistrons; note that only in prokaryotic mRNAs can more than one cistron be utilized by the translation system to generate individual proteins.

**Polymerase.** An enzyme that catalyzes the synthesis of a polymer from monomers.

**Polynucleotide.** A chain structure containing nucleotides linked together by phosphodiester (5′-3′) bonds. The polynucleotide chain has a directional sense, with a 5′ and a 3′ end.

**Polynucleotide phosphorylase.** An enzyme that polymerizes ribonucleotide diphosphates. No template is required.

**Polypeptide.** A linear polymer of amino acids held together by peptide linkages. The polypeptide has a directional sense, with an amino- and a carboxyl-terminal end.

**Polyribosome (polysome).** A complex of an mRNA and two or more ribosomes actively engaged in protein synthesis.

**Polysaccharide.** A linear or branched-chain structure containing many sugar molecules linked by glycosidic bonds.

**Porphyrin.** A complex planar structure containing four substituted pyrroles covalently joined in a ring and frequently containing a central metal atom. For example, heme is a porphyrin with a central iron atom.

**Positive control.** A system that is turned on by the presence of a regulatory protein.

**Posttranslational modification.** The covalent bond changes that occur in a polypeptide chain after it leaves the ribosome and before it becomes a mature protein.

**Primary structure.** In a polymer, the sequence of monomers and the covalent bonds.

**Primer.** A structure that serves as a growing point for polymerization.

**Primosome.** A multiprotein complex that catalyzes synthesis of RNA primer at various points along the DNA template.

**Prochiral molecule.** A nonchiral molecule that may react with an enzyme so that two groups that have a mirror image relationship to each other are treated differently.

**Prokaryote.** A unicellular organism that contains a single chromosome, no nucleus, no membrane-bound organelles, and has characteristic ribosomes and biochemistry.

**Promoter.** The region of the gene that signals RNA polymerase binding and the initiation of transcription.

**Prophage.** The silent phage genome. Some prophages integrate into the host genome; others replicate autonomously. The prophage state is maintained by a phage-encoded repressor.

**Prophase.** The stage in meiosis or mitosis when chromosomes condense and become visible as refractile bodies.

**Proprotein.** A protein that is made in an inactive form, so that it requires processing to become functional.

**Prostaglandin.** An oxygenated eicosanoid that has a hormonal function. Prostaglandins are unusual hormones in that they usually have effects only in that region of the organism where they are synthesized.

**Prosthetic group.** Synonymous with coenzyme except that a prosthetic group is usually more firmly attached to the enzyme it serves.

**Protamines.** Highly basic, arginine-rich proteins found complexed to DNA in the sperm of many invertebrates and fish.

**Protein subunit.** One of the components of a complex multicomponent protein.

**Protein targeting.** The process whereby proteins following synthesis are directed to specific locations.

**Proteoglycan.** A protein-linked heteropolysaccharide in which the heteropolysaccharide is usually the major component.

**Protist.** A relatively undifferentiated organism that can survive as a single cell.

**Proton acceptor.** A functional group capable of accepting a proton from a proton donor molecule.

**Proton motive force (Δ$p$).** The thermodynamic driving force for proton translocation. Expressed quantitatively as $\Delta G_{H^+}/F$ in units of volts.

**Protooncogene.** A cellular gene that can undergo modification to a cancer-causing gene (oncogene).

**Pseudocycle.** A sequence of reactions that can be arranged in a cycle but that usually do not function simultaneously in both directions. Also called a futile cycle because the net result of simultaneous functioning in both directions would be the expenditure of energy without accomplishing any useful work.

**Pulse-chase.** An experiment in which a short labeling period is followed by the addition of an excess of the same, unlabeled compound to dilute out the labeled material.

**Purine.** A heterocyclic ring structure with varying functional groups. The purines adenine and guanine are found in both DNA and RNA.

**Puromycin.** An antibiotic that inhibits polypeptide synthesis by competing with aminoacyl-tRNA for the ribosomal binding site A.

**Pyranose.** A simple sugar containing the six-member pyran ring.

**Pyrimidine.** A heterocyclic six-member ring structure. Cytosine and uracil are the main pyrimidines found in RNA, and cytosine and thymine are the main pyrimidines found in DNA.

**Pyrophosphate.** A molecule formed by two phosphates in anhydride linkage.

# Q

**Quaternary structure.** In a protein, the way in which the different folded subunits interact to form the multisubunit protein.

# R

**R group.** The distinctive side chain of an amino acid.

**R loop.** A triple-stranded structure in which RNA displaces a DNA strand by DNA–RNA hybrid formation in a region of the DNA.

**Rapid-start complex.** The complex that RNA polymerase forms at the promoter site just before initiation.

**Recombination.** The transfer to offspring of genes not found together in either of the parents.

**Redox couple.** An electron donor and its corresponding oxidized form.

**Redox potential ($E$).** The relative tendency of a pair of molecules to release or accept an electron. The standard redox potential ($E°$) is the redox potential of a solution containing the oxidant and reductant of the couple at standard concentrations.

**Regulatory enzyme.** An enzyme in which the active site is subject to regulation by factors other than the enzyme substrate. The enzyme frequently contains a nonoverlapping site for binding the regulatory factor that affects the activity of the active site.

**Regulatory gene.** A gene the principal product of which is a protein designed to regulate the synthesis of other genes.

**Renaturation.** The process of returning a denatured structure to its original native structure, as when two single strands of DNA are reunited to form a regular duplex, or the process by which an unfolded polypeptide chain is returned to its normal folded three-dimensional structure.

**Repair synthesis.** DNA synthesis following excision of damaged DNA.

**Repetitive DNA.** A DNA sequence that is present in many copies per genome.

**Replica plating.** A technique in which an impression of a culture is taken from a master plate and transferred to a fresh plate. The impression can be of bacterial clones or phage plaques.

**Replication fork.** The Y-shaped region of DNA at the site of DNA synthesis; also called a growth fork.

**Replicon.** A genetic element that behaves as an autonomous replicating unit. It can be a plasmid, phage, or bacterial chromosome.

**Repressor.** A regulatory protein that inhibits transcription from one or more genes. It can combine with an inducer (resulting in specific enzyme induction) or with an operator element (resulting in repression).

**Resonance hybrid.** A molecular structure that is a hybrid of two structures that differ in the locations of some of the electrons. For example, the benzene ring can be drawn in two ways, with double bonds in different positions. The actual structure of benzene is a blending of these two equivalent structures.

**Restriction-modification system.** A pair of enzymes found in most bacteria (but not eukaryotic cells). The restriction enzyme recognizes a certain sequence in duplex DNA and makes one cut in each unmodified DNA strand at or near the recognition sequence. The modification enzyme methylates (or modifies) the recognition sequence, thus protecting it from the action of the restriction enzyme.

**Reverse transcriptase.** An enzyme that synthesizes DNA from an RNA template, using deoxyribonucleotide triphosphates.

**Rho factor.** A protein involved in the termination of transcription of some messenger RNAs.

**Ribose.** The five-carbon sugar found in RNA.

**Ribosomal RNA (rRNA).** The RNA parts of the ribosome.

**Ribosomes.** Small cellular particles made up of ribosomal RNA and protein. They are the site, together with mRNA, of protein synthesis.

**RNA (ribonucleic acid).** A polynucleotide in which the sugar is ribose.

**RNA polymerase.** An enzyme that catalyzes the formation of RNA from ribonucleotide triphosphates, using DNA as a template.

**RNA splicing.** The excision of a segment of RNA, followed by a rejoining of the remaining fragments.

**Rolling-circle replication.** A mechanism for the replication of circular DNA. A nick in one strand allows the 3′ end to be extended, displacing the strand with the 5′ end, which is also replicated, to generate a double-stranded tail that can become larger than the unit size of the circular DNA.

## S

**S phase.** The period during the cell cycle when the chromosome is replicated.

**Salting in.** The increase in solubility that is displayed by typical globular proteins on the addition of small amounts of certain salts such as ammonium sulfate.

**Salting out.** The decrease in protein solubility that occurs when salts such as ammonium sulfate are present at high concentrations.

**Salvage pathway.** A family of reactions that permits nucleosides or purine and pyrimidine bases resulting from the partial breakdown of nucleic acids, to be reutilized in nucleic acid synthesis.

**Satellite DNA.** A DNA fraction the base composition of which differs from that of the main component of DNA, as revealed by the fact that it bands at a different density in a CsCl gradient. Usually repetitive DNA or organelle DNA.

**Scissile.** Capable of being cut smoothly or split easily.

**Second messenger.** A diffusible small molecule, such as cAMP, that is formed at the inner surface of the plasma membrane in response to a hormonal signal.

**Secondary structure.** In a protein or a nucleic acid, any repetitive folded pattern that results from the interaction of the corresponding polymeric chains.

**Semiconservative replication.** Duplication of DNA in which the daughter duplex carries one old strand and one new strand.

**Sigma factor.** A subunit of bacterial RNA polymerase that recognizes specific sites on DNA for initiation of RNA synthesis.

**Signal sequence.** A sequence in a protein that serves as a signal to guide the protein to a specific location.

**Signal transduction.** The process by which an extracellular signal is converted into a cellular response.

**Single-copy DNA.** A region of the genome the sequence of which is present only once per haploid complement.

**Somatic cell.** Any cell of an organism that cannot contribute its genes to a subsequent generation.

**SOS system.** A set of DNA repair enzymes and regulatory proteins that regulate their synthesis so that maximum synthesis occurs when the DNA is damaged.

**Southern blotting.** A method for detecting a specific DNA restriction fragment, developed by Edward Southern. DNA from a gel electrophoresis pattern is blotted onto nitrocellulose paper; then the DNA is denatured and fixed on the paper. Subsequently, the pattern of specific sequences in the Southern blot can be determined by hybridization to a suitable probe and autoradiography. A northern blot is similar, except that RNA is blotted instead onto the nitrocellulose paper.

**Splicing.** *See* RNA splicing.

**Sporulation.** Formation from vegetative cells of metabolically inactive cells that can resist extreme environmental conditions.

**Stacking energy.** The energy of interaction that favors the face-to-face packing of purine and pyrimidine base pairs.

**Steady-state.** In enzyme kinetic analysis, the time interval when the rate of reaction is approximately constant with time. The term is also used to describe the state of a living cell in which the concentrations of many molecules are approximately constant because of a balancing between their rates of synthesis and breakdown.

**Stem cell.** A cell from which other cells stem or arise by differentiation.

**Stereoisomers.** Isomers that are nonsuperimposable mirror images of each other.

**Steroids.** Compounds that are derivatives of a tetracyclic structure composed of a cyclopentane ring fused to a substituted phenanthrene nucleus.

**Structural domain.** An element of protein tertiary structure that recurs in many structures.

**Structural gene.** A gene encoding the amino acid sequence of a polypeptide chain.

**Structural protein.** A protein that serves a structural function.

**Subunit.** Individual polypeptide chains in a protein.

**Supercoiled DNA.** Supertwisted, covalently closed duplex DNA.

**Suppressor gene.** A gene that can reverse the phenotype of a mutation in another gene.

**Suppressor mutation.** A mutation that restores a function lost by an initial mutation and that is located at a site different from the initial mutation.

**Svedberg unit (S).** The unit used to express the sedimentation constant $s$: $1 \text{ S} = 10^{-13}$ s. The sedimentation constant $s$ is proportional to the rate of sedimentation of a molecule in a given centrifugal field and is related to the size and shape of the molecule.

**Synapse.** The chemical connection for communication between two nerve cells or between a nerve cell and a target cell such as a muscle cell.

**Synapsis.** The pairing of homologous chromosomes, seen during the first meiotic prophase.

## T

**Tandem duplication.** A duplication in which the repeated regions are immediately adjacent to one another.

**TCA cycle.** *See* tricarboxylic acid cycle.

**Template.** A polynucleotide chain that serves as a surface for the absorption of monomers of a growing polymer and thereby dictates the sequence of the monomers in the growing chain.

**Termination factors.** Proteins that are exclusively involved in the termination reactions of protein synthesis on the ribosome.

**Terpenes.** A diverse group of lipids made from isoprene precursors.

**Tertiary structure.** In a protein or nucleic acid, the final folded form of the polymer chain.

**Tetramer.** Structure resulting from the association of four subunits.

**Thioester.** An ester of a carboxylic acid with a thiol or mercaptan.

**Thymidine.** One of the four nucleosides found in DNA.

**Thymine.** A pyrimidine base found in DNA.

**Topoisomerase.** An enzyme that changes the extent of supercoiling of a DNA duplex.

**Transamination.** Enzymatic transfer of an amino group from an $\alpha$-amino acid to an $\alpha$-keto acid.

**Transcription.** RNA synthesis that occurs on a DNA template.

**Transduction.** Genetic exchange in bacteria that is mediated via phage.

**Transfection.** An artificial process of infecting cells with naked viral DNA.

**Transfer RNA (tRNA).** Any of a family of low-molecular-weight RNAs that transfer amino acids from the cytoplasm to the template for protein synthesis on the ribosome.

**Transferase.** An enzyme that catalyzes the transfer of a molecular group from one molecule to another.

**Transformation.** Genetic exchange in bacteria that is mediated via purified DNA. In somatic cell genetics the term is also used to indicate the conversion of a normal cell to one that grows like a cancer cell.

**Transgenic.** Describing an organism that contains transfected DNA in the germ line.

**Transition state.** The activated state in which a molecule is best suited to undergoing a chemical reaction.

**Translation.** The process of reading a messenger RNA sequence for the specified amino acid sequence it contains.

**Transport protein.** A protein the primary function of which is to transport a substance from one part of the cell to another, from one cell to another, or from one tissue to another.

**Tricarboxylic acid (TCA) cycle.** The cyclical process whereby acetate is completely oxidized to carbon dioxide and water, and electrons are transferred to $NAD^+$ and FAD. The TCA cycle is localized to the mitochondria in eukaryotic cells and to the plasma membrane in prokaryotic cells. Also called the Krebs cycle.

**Trypsin.** A proteolytic enzyme that cleaves peptide chains next to the basic amino acids arginine and lysine.

**Tryptic peptide mapping.** The technique of generating a chromatographic profile characteristic of the fragments resulting from trypsin enzyme cleavage of the protein.

**Tumorigenesis.** The mechanism of tumor formation.

**Turnover number.** The maximum number of molecules of substrate that can be converted to product per active site per unit time.

## U

**Ultracentrifuge.** A high-speed centrifuge that can attain speeds up to 60,000 rpm and centrifugal fields of 500,000 times gravity. Useful for characterizing and separating macromolecules.

**Uncoupler.** A substance that uncouples phosphorylation of ADP from electron transfer; for example, 2,4-dinitrophenol.

**Unidirectional replication.** *See* bidirectional replication.

**Unwinding proteins.** Proteins that help to unwind double-stranded DNA during DNA replication.

**Urea cycle.** A metabolic pathway in the liver that leads to the synthesis of urea from amino groups and $CO_2$. The function of the pathway is to convert the ammonia resulting from catabolism to a nontoxic form, which is subsequently secreted.

**UV irradiation.** Electromagnetic radiation with a wavelength shorter than that of visible light (200–390 nm). Causes damage to DNA (mainly pyrimidine dimers).

## V

**van der Waals forces.** Refers to two types of interactions, one attractive and one repulsive. The attractive forces are due to favorable interactions among the induced instantaneous dipole moments that arise from fluctuations in the electron charge densities of neighboring nonbonded atoms. Repulsive forces arise when noncovalently bonded atoms come too close together.

**Viroids.** Pathogenic agents, mostly of plants, that consist of short (usually circular) RNA molecules.

**Virus.** A nucleic acid–protein complex that can infect and replicate inside a specific host cell to make more virus particles.

**Vitamin.** A trace organic substance required in the diet of some species. Many vitamins are precursors of coenzymes.

## W

**Watson-Crick base pairs.** The type of hydrogen-bonded base pairs found in DNA, or comparable base pairs found in RNA. The base pairs are A-T, G-C, and A-U.

**Wild-type gene.** The form of a gene (allele) normally found in nature.

**Wobble.** A proposed explanation for base-pairing that is not of the Watson-Crick type and that often occurs between the 3′ base in the codon and the 5′ base in the anticodon.

# X

**X-ray crystallography.** A technique for determining the structure of molecules from the x-ray diffraction patterns that are produced by crystalline arrays of the molecules.

# Y

**Ylid.** A compound in which adjacent, covalently bonded atoms, both having an electronic octet, have opposite charges.

# Z

**Z form.** A duplex DNA structure in which the usual type of hydrogen bonding occurs between the base pairs but in which the helix formed by the two polynucleotide chains is left-handed rather than right-handed.

**Zwitterion.** A dipolar ion with spatially separated positive and negative charges. For example, most amino acids are zwitterions, having a positive charge on the $\alpha$-amino group and a negative charge on the $\alpha$-carboxyl group but no net charge on the overall molecule.

**Zygote.** A cell that results from the union of haploid male and female sex cells. Zygotes are diploid.

**Zymogen.** An inactive precursor of an enzyme. For example, trypsin exists in the inactive form trypsinogen before it is converted to its active form, trypsin.

# Credits

Portions of this text have been adapted from Geoffrey Zubay, *Genetics.* To be published.

**Molecular Modeling**

**Molecular Graphics** The molecular graphics photos listed below were developed by Michael Pique at the SCRIPPS Research Institute using software by Yng Chen, Michael Connolly, Michael Carson, Alex Shah, and AVS, Inc. Images copyright 1994 by the SCRIPPS Research Institute. Visualization advice was provided by Holly Miller, Wake Forest University Medical Center.

**Atomic Coordinates** All atomic coordinates for the molecular graphics images were obtained from the Protein Data Bank at Brookhaven National Laboratory, Upton, N.Y.

**4.21**: Phosphoglycerate kinase: entry 3PGK of July 1992. T. N. Bryant, et al. (To be published). **7.7**: Thermolysin: entry 1TLP of June 1987. D. E. Tronrud, et al., *European Journal of Biochemistry,* 157:261, 1986. **8.6a**: Trypsin: entry 3PTB of Sept. 1982. M. Marquart, et al., *ACTA Crystallography,* Section B, 39:480, 1983: Chymotrypsin: entry 4CHA of Nov. 1984. H. Tsukada & D. M. Blow, *Journal of Molecular Biology,* 184:703, 1985; Elastase: entry 2EST of Mar. 1986. D. L. Hughes, et al., *Journal of Molecular Biology,* 162:645, 1982. **8.6b**: Trypsin: entry 3PTB of Sept. 1982. M. Marquart, et al., *ACTA Crystallography,* Section B, 39:480, 1983; Subtilisin: entry 1SBC of May 1988. D. J. Neidhart & G. A. Petsko, (To be published). **8.7a-b**: Trypsin: entry 3PTB of Sept. 1982. M. Marquart, et al., *ACTA Crystallography,* Section B, 39:480, 1983. **8.19a-b**: Ribonuclease A: entry 6RSA of Feb. 1986. B. Borah, et al., *Biochemistry,* 24:2058, 1985. **8.20a-b**: Triose phosphate isomerase: entry 1TIM of Sept. 1976. D. W. Banner, et al., *Biochemical & Biophysical Research Communications,* 72:146, 1976. **9.9**: Phosphofructokinase: entry 4PFK of Jan. 1988. P. R. Evans, et al., *Philosophical Transactions of the Royal Society of London,* 293:53, 1981. **21.4a-b**: Glutamine synthetase: entry 2GLS of May 1989. M. M. Yamashita, et al., *Journal of Biological Chemistry,* 264:17681, 1989.

## Chapter 1
**1.1**: From Sylvia S. Mader, *Inquiry into Life,* 7th edition. Copyright © 1994 Wm. C. Brown Communications, Inc., Dubuque, Iowa. All Rights Reserved. Reprinted by permission.
**1.10**: From R. E. Dickerson and I. Geis, *The Structure and Action of Proteins,* Benjamin/Cummings, Menlo Park, Calif., 1969. Illustration copyright by Irving Geis. Reprinted by permission.
**1.15**: From R. E. Dickerson and I. Geis, *The Structure and Action of Proteins,* Benjamin/Cummings, Menlo Park, Calif., 1969. Coordinates courtesy of D. C. Phillips, Oxford. Illustration copyright by Irving Geis. Reprinted by permission.

## Chapter 3
**3.6**: From R. H. Haschenmeyer and A. E. V. Haschenmeyer, *A Guide to Study by Physical and Chemical Methods,* John Wiley & Sons, New York, 1973.

## Chapter 4
**4.2**: From R. E. Dickerson and I. Geis, *The Structure and Action of Proteins,* Benjamin/Cummings, Menlo Park, Calif., 1969. Illustration copyright by Irving Geis. Reprinted by permission.
**4.5**: From R. E. Dickerson and I. Geis, *The Structure and Action of Proteins,* Benjamin/Cummings, Menlo Park, Calif., 1969. Illustration copyright by Irving Geis. Reprinted by permission.
**4.24a**: From Lansing M. Prescott, John P. Harley, and Donald A. Klein, *Microbiology,* 2d edition. Copyright © 1993 Wm. C. Brown Communications, Inc., Dubuque, Iowa. All Rights Reserved. Reprinted by permission.
**4.25**: From Lansing M. Prescott, John P. Harley, and Donald A. Klein, *Microbiology,* 2d edition. Copyright © 1993 Wm. C. Brown Communications, Inc., Dubuque, Iowa. All Rights Reserved. Reprinted by permission.
**4a**: From C. H. Bamford et al., *Synthetic Polypeptides,* Academic Press, Orlando, Fla., 1956.

## Chapter 5
**5.15**: From John W. Hole, Jr., *Human Anatomy and Physiology,* 5th edition. Copyright © 1990 Wm. C. Brown Communications, Inc., Dubuque, Iowa. All Rights Reserved. Reprinted by permission.

## Chapter 6
**6.2**: From E. J. Cohn and J. T. Edsall, *Proteins, Amino Acids, and Peptides as Ions and Dipolar Ions,* Reinhold, New York, 1942.

## Chapter 7
**7.5**: From *Enzyme Structure and Mechanism,* 2d edition, by Alan Ferscht. Copyright © 1985 by W. H. Freeman and Company. Reprinted with permission.
**7.3**: From *Enzyme Structure and Mechanism* by Alan Ferscht. Copyright © 1985 by W. H. Freeman and Company. Reprinted with permission.

## Chapter 9
**9.20**: From N. B. Madsen in *The Enzymes,* 3d edition, vol. XVII, ed. by P. D. Boyer and E. G. Krebs, Academic Press, New York, 1986.

## Chapter 16
**16.5**: Reproduced with permission from B. Alberts, D. Bray, J. Lewis, M. Raff, K. Roberts, and J. D. Watson, *Molecular Biology of the Cell* (New York: Garland Publishing, 1989).

## Chapter 17
**17.3**: Adapted from M. K. Jain and R. C. Wagner, *Introduction to Biological Membranes,* John Wiley & Sons, New York, 1980.
**17.18**: From J. T. Segrest and L. D. Kohn, Protein-lipid interactions of the membrane penetrating MN-glycoprotein from the human erythrocyte, *Protides of the Biological Fluids,* 21st colloquium, ed. by J. Peeters, Pergamon Press, New York, 1973.
**17.4**: Adapted from M. K. Jain and R. C. Wagner, *Introduction to Biological Membranes,* John Wiley & Sons, New York, 1980.
**17.23**: From B. Alberts et al., *Molecular Biology of the Cell,* 2d edition, Garland Publishing, New York, 1989.

## Supplement 1
**S1.3**: Adapted from S. W. Kuffler and J. G. Nicholls, *From Neuron to Brain,* Sinauer Associates, Sunderland, Mass., 1976.

## Chapter 25
**25.16**: From Geoffrey Zubay, *Genetics,* Benjamin/Cummings, Menlo Park, Calif., 1987. Reprinted by permission of the author.
**25.18b**: From Robert F. Weaver and Philip W. Hedrick, *Basic Genetics.* Copyright © 1991 Wm. C. Brown Communications, Inc., Dubuque, Iowa. All Rights Reserved. Reprinted by permission.

## Chapter 26
**26.16**: From Robert F. Weaver and Philip W. Hedrick, *Basic Genetics.* Copyright © 1991 Wm. C. Brown Communications, Inc., Dubuque, Iowa. All Rights Reserved. Reprinted by permission.
**26.21**: From Geoffrey Zubay, *Genetics,* Benjamin/Cummings, Menlo Park. Calif., 1987. Reprinted by permission of the author.
**26.22**: From Geoffrey Zubay, *Genetics,* Benjamin/Cummings, Menlo Park, Calif., 1987. Reprinted by permission of the author.

## Chapter 27
**27.4**: From Robert F. Weaver and Philip W. Hedrick, *Basic Genetics.* Copyright © 1991 Wm. C. Brown Communications, Inc., Dubuque, Iowa. All Rights Reserved. Reprinted by permission.

## Chapter 28
**28.2**: From Geoffrey Zubay, *Genetics,* Benjamin/Cummings, Menlo Park, Calif., 1987. Reprinted by permission of the author.

# Index

Note: Page numbers followed by F indicate illustrations; page numbers followed by T indicate tables.